Lecture Notes in Artificial Intelligence   3518

Edited by J. G. Carbonell and J. Siekmann

Subseries of Lecture Notes in Computer Science

Tu Bao Ho   David Cheung   Huan Liu (Eds.)

Advances in Knowledge Discovery and Data Mining

9th Pacific-Asia Conference, PAKDD 2005
Hanoi, Vietnam, May 18-20, 2005
Proceedings

 Springer

Series Editors

Jaime G. Carbonell, Carnegie Mellon University, Pittsburgh, PA, USA
Jörg Siekmann, University of Saarland, Saarbrücken, Germany

Volume Editors

Tu Bao Ho
Japan Advanced Insitute of Science and Technology
1-1 Asahidai Tatsunokuchi, Ishikawa 923-1292, Japan
E-mail: bao@jaist.ac.jp

David Cheung
University of Hong Kong
Pokfulam Road, Hong Kong, China
E-mail: dcheung@csis.hku.hk

Huan Liu
Arizona State University
Tempe, AZ 85287-8809, USA
E-mail: hliu@asu.edu

Library of Congress Control Number: Applied for

CR Subject Classification (1998): I.2, H.2.8, H.3, H.5.1, G.3, J.1, K.4

ISSN   0302-9743
ISBN-10   3-540-26076-5 Springer Berlin Heidelberg New York
ISBN-13   978-3-540-26076-9 Springer Berlin Heidelberg New York

This work is subject to copyright. All rights are reserved, whether the whole or part of the material is concerned, specifically the rights of translation, reprinting, re-use of illustrations, recitation, broadcasting, reproduction on microfilms or in any other way, and storage in data banks. Duplication of this publication or parts thereof is permitted only under the provisions of the German Copyright Law of September 9, 1965, in its current version, and permission for use must always be obtained from Springer. Violations are liable to prosecution under the German Copyright Law.

Springer is a part of Springer Science+Business Media

springeronline.com

© Springer-Verlag Berlin Heidelberg 2005
Printed in Germany

Typesetting: Camera-ready by author, data conversion by Scientific Publishing Services, Chennai, India
Printed on acid-free paper   SPIN: 11430919   06/3142   5 4 3 2 1 0

Preface

The Pacific-Asia Conference on Knowledge Discovery and Data Mining (PAKDD) is a leading international conference in the area of data mining and knowledge discovery. It provides an international forum for researchers and industry practitioners to share their new ideas, original research results and practical development experiences from all KDD-related areas including data mining, data warehousing, machine learning, databases, statistics, knowledge acquisition and automatic scientific discovery, data visualization, causality induction, and knowledge-based systems. This year's conference (PAKDD 2005) was the ninth of the PAKDD series, and carried the tradition in providing high-quality technical programs to facilitate research in knowledge discovery and data mining. It was held in Hanoi, Vietnam at the Melia Hotel, 18–20 May 2005.

We are pleased to provide some statistics about PAKDD 2005. This year we received 327 submissions (a 37% increase over PAKDD 2004), which is the highest number of submissions since the first PAKDD in 1997) from 28 countries/regions: Australia (33), Austria (1), Belgium (2), Canada (11), China (91), Switzerland (2), France (9), Finland (1), Germany (5), Hong Kong (11), Indonesia (1), India (2), Italy (2), Japan (21), Korea (51), Malaysia (1), Macau (1), New Zealand (3), Poland (4), Pakistan (1), Portugal (3), Singapore (12), Taiwan (19), Thailand (7), Tunisia (2), UK (5), USA (31), and Vietnam (9). The submitted papers went through a rigorous reviewing process. Each submission was reviewed by at least two reviewers, and most of them by three or four reviewers. The Program Committee members were deeply involved in a highly engaging selection process with discussions among reviewers, and, when necessary, additional expert reviews were sought. As a result, the PAKDD 2005 Program Committee accepted for publication and oral presentation 48 regular papers and 49 short papers, representing 14.6% and 14.9% acceptance rates, respectively. The PAKDD 2005 program also included two workshops ("Knowledge Discovery and Data Management in Biomedical Science" and "Rough Set Techniques in Knowledge Discovery"), and four tutorials ("Graph Mining Techniques and Their Applications," "Rough Set Approach to KDD," "Web Delta Mining: Opportunities and Solutions," and "Advanced Techniques for Information and Image Classification for Knowledge Management and Decision Making").

PAKDD 2005 would not have been possible without the help of many people and organizations. First and foremost, we would like to thank the members of the Steering Committee, the Program Committee and external reviewers for their invaluable contributions. We wish to express our gratitude to:

- Honorary conference chairs: Dang Vu Minh (President of the Vietnamese Academy of Science and Technology, Vietnam) and Hoang Van Phong (Minister of Science and Technology, Vietnam);
- Conference chairs: Hiroshi Motoda (Osaka University, Japan) and Phan Dinh Dieu (Vietnam National University, Hanoi, Vietnam);

- Keynote and invited speakers: Tom Mitchell (Carnegie Mellon University, USA), Nada Lavrac (J. Stefan Institute, Slovenia) and Unna Huh (Information and Communications University, Korea);
- Local organizing committee chairs: Luong Chi Mai (Institute of Information Technology, Hanoi, Vietnam) and Nguyen Ngoc Binh (Hanoi University of Technology, Vietnam);
- Workshop chair: Kyuseok Shim (National Korean University, Korea);
- Tutorial chair: Takashi Washio (Osaka University, Japan);
- Industrial chair: Wee Keong Ng (Nanyang Technological University, Singapore);
- Publicity chair: Tran Tuan Nam (Japan Advanced Institute of Science and Technology, Japan);
- Publication chair: Saori Kawasaki (Japan Advanced Institute of Science and Technology, Japan);
- Registration chair: Nguyen Trong Dung, Institute of Information Technology, Hanoi, Vietnam);
- Award selection committee: David Cheung (University of Hong Kong, China), Huan Liu (Arizona State University, USA) and Graham Williams (ATO, Australia);
- Chani Johnson for his tireless effort in supporting Microsoft's Conference Management Tool;
- Workshop organizers: Kenji Satou and Tu Bao Ho (Japan Advanced Institute of Science and Technology, Japan), Marcin S. Szczuka and Nguyen Hung Son (Warsaw University, Poland). Tutorialists: Sharma Chakravarthy (University of Texas at Arlington, USA), Sanjay Madria (University of Missouri-Rolla, USA), Nguyen Hung Son and Marcin S. Szczuka (Warsaw University, Poland) and Parag Kulkarni (Capsilon India, India).
- External reviewers.

We greatly appreciate the financial support from various sponsors: Japan Advanced Institute of Science and Technology (JAIST), Vietnamese Academy of Science and Technology (VAST), Ministry of Science and Technology of Vietnam (MoST), Hanoi University of Technology (HUT), AFOSR/AOARD, IBM and Oracle Vietnam.

Last but not least, we would like to thank all authors, and all conference attendees for their contribution and participation. Without them, we would not have had this conference. We hope all attendees took time to exchange ideas with each other and enjoyed PAKDD 2005.

May 2005   Tu Bao Ho, David Cheung, Huan Liu

Organization

PAKDD 2005 Conference Committee

Honorary Chairs

Dang Vu Minh	President of Vietnamese Academy of Science and Technology
Hoang Van Phong	Minister of Science and Technology, Vietnam

Conference Chairs

Phan Dinh Dieu	Vietnam National University, Hanoi, Vietnam
Hiroshi Motoda	Osaka University, Japan

Program Committee Chairs

Ho Tu Bao	Japan Advanced Institute of Science and Technology, Japan
David Cheung	University of Hong Kong, China
Huan Liu	Arizona State University, USA

Local Organizing Committee Chairs

Luong Chi Mai	Institute of Information Technology, VAST, Vietnam
Nguyen Ngoc Binh	Hanoi University of Technology, Vietnam

Workshop Chair

Kyuseok Shim	National Korean University, Korea

Tutorial Chair

Takashi Washio	Osaka University, Japan

Industrial Chair

Wee Keong Ng	Nanyang Technological University, Singapore

Publicity Chair

Tran Tuan Nam	Japan Advanced Institute of Science and Technology, Japan

Publication Chair

Saori Kawasaki	Japan Advanced Institute of Science and Technology, Japan

Registration Chair

Nguyen Trong Dung	Institute of Information Technology, VAST, Vietnam

PAKDD 2005 Steering Committee

Hiroshi Motoda (Chair)	Osaka University, Japan
David Cheung (Co-chair)	University of Hong Kong, China
Hongjun Lu (Treasurer)	Hong Kong University of Science & Technology, China
Arbee L.P. Chen	National Chengchi University, Taiwan
Ming-Syan Chen	National Taiwan University, Taiwan
Jongwoo Jeon	Seoul National University, Korea
Masaru Kitsuregawa	Tokyo University, Japan
Rao Kotagiri	University of Melbourne, Australia
Takao Terano	University of Tsukuba, Japan
Kyu-Young Whang	Korea Advanced Institute of Science and Technology, Korea
Graham Williams	ATO, Australia
Ning Zhong	Maebashi Institute of Technology, Japan
Chengqi Zhang	University of Technology Sydney, Australia

PAKDD 2005 Program Committee

Hiroki Arimura	Hokkaido University, Japan
Ho Tu Bao	Japan Advanced Institute of Science and Technology, Japan
Nguyen Ngoc Binh	Hanoi University Technology, Vietnam
Pavel Brazdil	University of Porto, Portugal
Tru Hoang Cao	Ho Chi Minh City University of Technology, Vietnam
Nicholas Cercone	Dalhousie University, Canada
Arbee L.P. Chen	National Chengchi University, Taiwan
Ming-Syan Chen	National Taiwan University, Taiwan
David Cheung	University of Hong Kong, China
Vic Ciesielski	RMIT University, Australia
Vincent Corruble	University of Paris 6, France
Jirapun Daengdej	Assumption University, Thailand
Honghua Dai	Deakin University, Australia
Manoranjan Dash	Nanyang Technological University, Singapore
AnHai Doan	University Illinois Urbana, USA
Guozhu Dong	Wright State University, USA
Nguyen Trong Dung	Institute of Information Technology, VAST, Vietnam
Peter A. Flach	University of Bristol, UK
Eibe Frank	University of Waikato, New Zealand
Joao Gama	University of Porto, Portugal
Minos Garofalakis	Bell Laboratories, USA
Sudipto Guha	University of Pennsylvania, USA
Dimitrios Gunopulos	University of California, Riverside, USA
Shyam Kumar Gupta	Indian Institute of Technology, Delhi, India
Peter Haddawy	Asian Institute of Technology, Thailand

Jiawei Han	University of Illinois, Urbana-Champaign, USA
Doan B. Hoang	University of Technology, Sydney, Australia
Thu Hoang	University of Paris 5, France
Achim Hoffmann	University of New South Wales, Australia
Se June Hong	IBM T.J. Watson Research Center, USA
Wynne Hsu	National University of Singapore, Singapore
Joshua Z. Huang	University of Hong Kong, China
Siu Cheung Hui	Nanyang Technological University, Singapore
San-Yih Hwang	National Sun Yat-Sen University, Taiwan
Jongwoo Jeon	Seoul National University, Korea
Rong Jin	Michigan State University, USA
Hiroyuki Kawano	Nanzan University, Japan
Gabriele Kern-Isberner	University of Dortmund, Germany
Hoang Kiem	Vietnam National University HCM, Vietnam
Boonserm Kijsirikul	Chulalongkorn University, Thailand
Myoung Ho Kim	Korea Advanced Institute of Science and Technology, Korea
Yasuhiko Kitamura	Kwansei Gakuin University, Japan
Masaru Kitsuregawa	University of Tokyo, Japan
Rao Kotagiri	University of Melbourne, Australia
Marzena Kryszkiewicz	Warsaw University of Technology, Poland
Vipin Kumar	University of Minnesota, USA
Jonathan Lawry	University of Bristol, UK
Aleksandar Lazarevic	University of Minnesota, USA
Doheon Lee	Korea Advanced Institute of Science and Technology, Korea
Geuk Lee	Hannam University, Korea
Kwang Hyung Lee	Korea Advanced Institute of Science and Technology, Korea
Sang Ho Lee	Soongsil University, Korea
Yoon-Joon Lee	Korea Advanced Institute of Science and Technology, Korea
Jinyan Li	Institute for Infocomm Research, Singapore
Tsau Young Lin	San Jose State University, USA
Bing Liu	University of Illinois at Chicago, USA
Huan Liu	Arizona State University, USA
Hongjun Lu	Hong Kong University of Science and Technology, China
Luong Chi Mai	Vietnamese Academy of Science and Technology, Vietnam
Yuji Matsumoto	Nara Institute of Science and Technology, Japan
Hiroshi Motoda	Osaka University, Japan
Tetsuya Murai	Hokkaido University, Japan
Yoshiteru Nakamori	Japan Advanced Institute of Science and Technology, Japan
Huynh Van Nam	Japan Advanced Institute of Science and Technology, Japan

Douglas Newlands	Deakin University, Australia
Wee Keong Ng	Nanyang Technological University, Singapore
Zaiqing Nie	Microsoft Research Asia, China
Monique Noirhomme	University of Notre Dame de la Paix, Belgium
Masayuki Numao	Osaka University, Japan
Takashi Okada	Kwansei Gakuin University, Japan
Dino Pedreschi	Universitá di Pisa, Italy
T.V. Prabhakar	Indian Institute of Technology Kanpur, India
Joel Quinqueton	University of Montpellier 2, France
Rajeev Rastogi	Bell Laboratories, USA
Kenji Satou	Japan Advanced Institute of Science and Technology, Japan
Michele Sebag	University of Paris, Orsay, France
Rudy Setiono	National University of Singapore, Singapore
Kyuseok Shim	Seoul National University, Korea
Akira Shimazu	Japan Advanced Institute of Science and Technology, Japan
Masashi Shimbo	Nara Institute of Science and Technology, Japan
Simeon J. Simoff	University of Technology, Sydney, Australia
Andrzej Skowron	Warsaw University, Poland
Nguyen Hung Son	Warsaw University, Poland
Takao Terano	Tsukuba University, Japan
Nguyen Thanh Thuy	Hanoi University Technology, Vietnam
Hiroshi Tsukimoto	Tokyo Denki University, Japan
Shusaku Tsumoto	Shimane Medical University, Japan
Anh Vo	University of Melbourne, Australia
Zhi-Hai Wang	Beijing Jiaotong University, China
Takashi Washio	Osaka University, Japan
Kyu-Young Whang	Korea Advanced Institute of Science and Technology, Korea
Graham Williams	ATO, Australia
Xindong Wu	University of Vermont, USA
Takehisa Yairi	University of Tokyo, Japan
Seiji Yamada	National Institute of Informatics, Japan
Takahira Yamaguchi	Keio University, Japan
Yiyu Yao	University of Regina, Canada
Tetsuya Yoshida	Hokkaido University, Japan
Philip S. Yu	IBM T.J. Watson Research Center, USA
Mohammed J. Zaki	Rensselaer Polytechnic Institute, USA
Chengqi Zhang	University of Technology, Sydney, Australia
Bo Zhang	Tsinghua University, China
Ning Zhong	Maebashi Institute of Technology, Japan
Zhi-Hua Zhou	Nanjing University, China
Djamel A. Zighed	University of Lyon 2, France

PAKDD 2005 External Reviewers

Alexandre Termier
Arkadiusz Wojna
Asad Satti
Aysel Ozgur
Benjarath Bphhphakdee
Bi-Ru Dai
Carlos Pinto
Chen Chen
Chiara Renso
Chung-Wen Cho
Daan He
Dae-Won Kim
Dayang Iskandar
Ding-Ying Chiu
Eamonn Keogh
Ellery Chen
Feng Gao
Francesco Bonchi
Gang Li
Gaurav Pandey
Georges Koepfler
Graham Cormode

Gulisong Nansierding
Gyorgy Simon
Han Liu
Hidenao Abe
Ho Wai Shing
Hui Xiong
Hui Zhang
Hung-Chen Chen
I-Jen Chiang
Iko Pramudiono
J. Gama
Jean-Daniel Zucker
Jeng-Kuen Chiu
Jia Hu
Jiri Navratil
Jiye Li
Juliana Hsieh
Justin Zobel

Karlton Sequeira
Kouzou Ohara
Kun-Ta Chuang
Kwoh Chee Keong
Lance Parsons
Le Anh Cuong
Le Minh Hoang
Le Si Quang
Lei Tang
Lei Yu
Levent Ertoz
Li Lin
Ling Zhuang
Lizhuang Zhang
Lizhuang Zhao
Mark Hsieh
Masahi Toyoda
Masaki Kuremtasu
Maurizio Atzori
Michael Steinbach
Miho Ohsaki
Mikihiko Mori

Milton Silva
Min-Ling Zhang
Mintz Hsieh
Mirco Nanni
Miriam Baglioni
Miyuki Nakano
Mohammed Zaki
Muneaki Ohshima
Nagender Parimi
Naoki Fukuta
Narendra S. Chaudhari
Nenad Stankovic
Ng Wil Lie
Nguyen Canh Hao
Nguyen Duc Dung
Nguyen Le Minh
Nguyen Minh
Nguyen Thi Minh Hai

Noboru Nakajima
Noriko Imafuji
P. Rodrigues
Pabitra Mitra
Phu Chien Nguyen
Pusheng Zhang
Remco Bouckaert
Rohit Gupta
Ronaldo Prati
Salvatore Ruggieri
Sangkyum Kim
Shichao Zhang
Shuigeng Zhou
Shyam Boriah
Steffi Soo
Surendra Singhi
Tadashi Ohmori
Takeshi Sagara
Varun Chandola
Vic Ciesielski
Vincent S.M. Tseng
Vineet Chaoji
Vivekanand
 Gopalkrishnan
Vlado Keselj
Vu Van Thinh
Xiaolei Li
Xifeng Yan
Xingquan Zhu
Yasufumi Takama
Yen-Kuang Lu
Yi Liu
Yi Xia
Yifan Li
Ying Lu
Yitong Wang
Yongdai Kim
Yun Chi
Yunxiao Ma
Zheng Shao
Zheng Zhao

In Loving Memory of
Professor Hongjun Lu
1945–2005

Professor Lu was one of the founders of the PAKDD conference series. He played a key leadership role in nurturing and establishing the PAKDD conferences to become a world-recognized forum. He served as the Steering Committee Co-chair (1998–2001), and as Chair (2001–2003) of PAKDD. He was the Program Co-chair of the inaugural PAKDD (1997). He was honored with the inaugural PAKDD Distinguished Contribution Award (2005) for his significant and ongoing contributions in research and services to the advancement of the PAKDD community and series of conferences.

Professor Lu also served in many important and influential positions in the research community. He was elected as a Trustee of the VLDB Endowment in 2000. He was a member of the Advisory Board of ACM SIGMOD (1998–2002). He was an Editor for IEEE Transactions on Knowledge and Data Engineering (TKDE) (1996–2000) and for Knowledge and Information Systems: An International Journal (1998–2001). He has served on the program committees of numerous international conferences in databases.

Professor Lu passed away on March 3 from complications arising from his treatment for cancer. His research has made an impact in many areas, especially in the many important issues related to query processing and optimization, data warehousing and data mining. His long-term contributions through over 200 research publications in scientific journals, conferences and workshop proceedings have provided the foundations for many other researchers, and will be an ongoing contribution to our scientific endeavors for many years to come.

He will always be remembered as a great scholar, researcher, teacher and leader, and as a caring, considerate and compassionate friend to very many.

Table of Contents

Keynote Speech and Invited Talks

Machine Learning for Analyzing Human Brain Function
Tom Mitchell 1

Subgroup Discovery Techniques and Applications
Nada Lavrač 2

IT Development in the 21st Century and Its Implications
Unna Huh 15

Theoretic Foundations

Data Mining of Gene Expression Microarray via Weighted Prefix Trees
Tran Trang, Nguyen Cam Chi, Hoang Ngoc Minh 21

Automatic Extraction of Low Frequency Bilingual Word Paris from Parallel Corpora with Various Languages
Hiroshi Echizen-ya, Kenji Araki, Yoshio Mornouchi 32

A Kennel Function Method in Clustering
Ling Zhang, Tao Wu, Yanping Zhang 38

Performance Measurements for Privacy Preserving Data Mining
Nan Zhang, Wei Zhao, Jianer Chen 43

Extraction of Frequent Few-Overlapped Monotone DNF Formulas with Depth-First Pruning
Yoshikazu Shima, Kouichi Hirata, Masateru Harao 50

Association Rules

Rule Extraction from Trained Support Vector Machines
Ying Zhang, HongYe Su, Tao Jia, Jian Chu 61

Pruning Derivative Partial Rules During Impact Rule Discovery
Shiying Huang, Geoffrey I. Webb 71

IGB: A New Informative Generic Base of Association Rules
Gh. Gasmi, S. Ben Yahia, E. Mephu Nguifo, Y. Slimani 81

A Divide and Conquer Approach for Deriving Partially Ordered
Sub-structures
Sadok Ben Yahia, Yahya Slimani, Jihen Rezgui 91

Finding Sporadic Rules Using Apriori-Inverse
Yun Sing Koh, Nathan Rountree 97

Automatic View Selection: An Application to Image Mining
Manoranjan Dash, Deepak Kolippakkam 107

Pushing Tougher Constraints in Frequent Pattern Mining
Francesco Bonchi, Claudio Lucchese 114

An Efficient Compression Technique for Frequent Itemset Generation
in Association Rule Mining
Mafruz Zaman Ashrafi, David Taniar, Kate Smith 125

Mining Time-Profiled Associations: An Extended Abstact
Jin Soung Yoo, Pusheng Zhang, Shashi Shekhar 136

Online Algorithms for Mining Inter-stream Associations from Large
Sensor Neworks
K. K. Loo, Ivy Tong, Ben Kao 143

Mining Frequent Ordered Patterns
Zhi-Hong Deng, Cong-Rui Ji, Ming Zhang, and Shi-Wei Tang 150

Biomedical Domains

Conditional Random Fields for Transmembrane Helix Prediction
Lior Lukov, Sanjay Chawla, W. Bret Church 155

A DNA Index Structure Using Frequency and Position Information
of Genetic Alphabet
*Woo-Cheol Kim, Sanghyun Park, Jung-Im Won, Sang-Wook Kim,
Jee-Hee Yoon* 162

An Automatic Unsupervised Querying Algorithm for Efficient Information
Extraction in Biomedical Domain
Min Song, Il-Yeol Song, Xiaohua Hu, Robert B. Allen 173

Voting Fuzzy k-NN to Predict Protein Subcellular Localization from
Normalized Amino Acid Pair Compositions
Thai Quang Tung, Doheon Lee, Dae-Won Kim, Jong-Tae Lim 180

Comparison of Tree Based Methods on Mammography Data
Richard De Veaux, Thu Hoàng 186

Bayesian Sequence Learning for Predicting Protein Cleavage Points
Michael Mayo 192

A Novel Indexing Method for Efficient Sequence Matching in Large DNA Database Environment
Jung-Im Won, Jee-Hee Yoon, Sanghyun Park, Sang-Wook Kim 203

Classification and Ranking

Threshold Tuning for Improved Classification Association Rule Mining
Frans Coenen, Paul Leng, Lu Zhang 216

Using Rough Set in Feature Selection and Reduction in Face Recognition Problem
Le Hoai Bac, Nguyen Anh Tuan 226

Analysis of Company Growth Data Using Genetic Algorithms on Binary Trees
Gerrit K. Janssens, Kenneth Sörensen, Arthur Limère, Koen Vanhoof 234

Considering Re-occurring Features in Associative Classifiers
Rafal Rak, Wojciech Stach, Osmar R. Zaïane, Maria-Luiza Antonie 240

A New Evolutionary Neural Network Classifier
Arit Thammano, Asavin Meengen 249

A Privacy-Preserving Classification Mining Algorithm
Weiping Ge, Wei Wang, Xiaorong Li, Baile Shi 256

Combining Classifiers with Multi-representation of Context in Word Sense Disambiguation
Cuong Anh Le, Va-Nam Huynh, Akira Shimazu 262

Automatic Occupation Coding with Combination of Machine Learning and Hand-Crafted Rules
Kazuko Takahashi, Hiroya Takamura, Manabu Okumura 269

Retrieval Based on Language Model with Relative Entropy and Feedback
Hua Huo, Boqin Feng 280

Text Classification for DAG-Structured Categories
Cao D. Nguyen, Tran A. Dung, Tru H. Cao 290

Sentiment Classification Using Word Sub-sequences and Dependency Sub-trees
Shotaro Matsumoto, Hiroya Takamura, Manabu Okumura 301

Improving Rough Classifiers Using Concept Ontology
Nguyen Sinh Hoa, Nguyen Hung Son 312

QED: An Efficient Framework for Temporal Region Query Processing
Yi-Hong Chu, Kun-Ta Chuang, Ming-Syan Chen 323

Clustering

A MPAA-Based Iterative Clustering Algorithm Augmented by Nearest Neighbors Search for Time-Series Data Streams
Jessica Lin, Michai Vlachos, Eamonn Keogh, Dimitrios Gunopulos, Jianwei Liu, Shoujian Yu, Jiajin Le 333

Locating Motifs in Time-Series Data
Zheng Liu, Jeffrey Xu Yu, Xuemin Lin, Hongjun Lu, Wei Wang 343

Stochastic Local Clustering for Massive Graphs
Satu Elisa Schaeffer 354

A Neighborhood-Based Clustering Algorithm
Shuigeng Zhou, Yue Zhao, Jihong Guan, Joshua Huang 361

Improved Self-splitting Competitive Learning Algorithm
Jun Liu, Kotagiri Ramamohanarao 372

Speeding-Up Hierarchical Agglomerative Clustering in Presence of Expensive Metrics
Mirco Nanni 378

Dynamic Cluster Formation Using Level Set Methods
Andy M. Yip, Chris Ding, Tony F. Chan 388

A Vector Field Visualization Technique for Self-organizing Maps
Georg Pölzlbauer, Andreas Rauber, Michael Dittenbach 399

Visualization of Cluster Changes by Comparing Self-organizing Maps
Denny McG. Squrie, David McG. Squire 410

An Incremental Data Stream Clustering Algorithm Based on Dense Units Detection
Jing Gao, Jianzhong Li, Zhaogong Zhang, Pang-Ning Tan 420

Visual Interactive Evolutionary Algorithm for High Dimensional Data Clustering and Outlier Detection
Lydia Boudjeloud, François Poulet 426

Approximated Clustering of Distributed High-Dimensional Data
Hans-Peter Kriegel, Peter Kunath, Martin Pfeifle, Matthias Renz 432

Dynamic Data Mining

Improvements of IncSpan: Incremental Mining of Sequential Patterns
in Large Database
Son N. Nguyen, Xingzhi Sun, Maria E. Orlowska 442

Efficient Sampling: Application to Image Data
Surong Wang, Manoranjan Dash, Liang-Tien Chia 452

Cluster-Based Rough Set Construction
Qiang Li, Bo Zhang 464

Graphic Model Discovery

Learning Bayesian Networks Structures from Incomplete Data:
An Efficient Approach Based on Extended Evolutionary Programming
Xiaolin Li, Xiangdong He, Senmiao Yuan 474

Dynamic Fuzzy Clustering for Recommender Systems
Sung-Hwan Min, Ingoo Han 480

Improving Mining Quality by Exploiting Data Dependency
Fang Chu, Yizhou Wang, Carlo Zaniolo, D. Stott Parker 486

High Dimensional Data

Feature Selection for High Dimensional Face Image Using
Self-organizing Maps
Xiaoyang Tan, Songcan Chen, Zhi-Hua Zhou, Fuyan Zhang 500

Progressive Sampling for Association Rules Based on
Sampling Error Estimation
Kun-Ta Chuang, Ming-Syan Chen, Wen-Chieh Yang 505

CLeVer: A Feature Subset Selection Technique for Multivariate Time Series
Kiyoung Yang, Hyunjin Yoon, Cyrus Shahabi 516

Covariance and PCA for Categorical Variables
Hirotaka Niitsuma, Takashi Okada 523

Integration of Data Warehousing

ADenTS: An Adaptive Density-Based Tree Structure for Approximating
Aggregate Queries over Real Attributes
Tianyi Wu, Jian Xu, Chen Wang, Wei Wang, Baile Shi 529

Frequent Itemset Mining with Parallel RDBMS
Xuequn Shang, Kai-Uwe Sattler 539

Knowledge Management

Using Consensus Susceptibility and Consistency Measures
for Inconsistent Knowledge Management
Ngoc Thanh Nguyen, Michal Malowiecki 545

WLPMiner: Weighted Frequent Pattern Mining with Length-Decreasing
Support Constraints
Unil Yun, John J. Leggett 555

Machine Learning Methods

A Framework for Incorporating Class Priors into Discriminative
Classification
Rong Jin, Yi Liu 568

Increasing Classification Accuracy by Combining Adaptive Sampling
and Convex Pseudo-Data
Chia Huey Ooi, Madhu Chetty 578

Kernels over Relational Algebra Structures
Adam Woźnica, Alexandros Kalousis, Melanie Hilario 588

Adaptive Nonlinear Auto-Associative Modeling Through Manifold Learning
Junping Zhang, Stan Z. Li 599

Maximizing Tree Diversity by Building Complete-Random Decision Trees
Fei Tony Liu, Kai Ming Ting, Wei Fan 605

SETRED: Self-training with Editing
Ming Li, Zhi-Hua Zhou 611

Adjusting Mixture Weights of Gaussian Mixture Model via
Regularized Probabilistic Latent Semantic Analysis
Luo Si, Rong Jin 622

Training Support Vector Machines Using Greedy Stagewise Algorithm
Liefeng Bo, Ling Wang, Licheng Jiao 632

Cl-GBI: A Novel Approach for Extracting Typical Patterns
from Graph-Structured Data
Phu Chien Nguyen, Kouzou Ohara, Hiroshi Motoda, Takashi Washio 639

Improved Bayesian Spam Filtering Based on Co-weighted
Multi-area Information
Raju Shrestha, Yaping Lin 650

Novel Algorithms

An Efficient Framework for Mining Flexible Constraints
Arnaud Soulet, Bruno Crémilleux 661

Support Oriented Discovery of Generalized Disjunction-Free
Representation of Frequent Patterns with Negation
Marzena Kryszkiewicz, Katarzyna Cichoń 672

Feature Selection Algorithm for Data with Both Nominal
and Continuous Features
Wenyin Tang, Kezhi Mao 683

A Two-Phase Algorithm for Fast Discovery of High Utility Itemsets
Ying Liu, Wei-keng Liao, Alok Choudhary 689

On Multiple Query Optimization in Data Mining
Marek Wojciechowski, Maciej Zakrzewicz 696

USAID: Unifying Signature-Based and Anomaly-Based Intrusion Detection
Zhuowei Li, Amitabha Das, Jianying Zhou 702

Spatial Data

Mining Mobile Group Patterns: A Trajectory-Based Approach
San-Yih Hwang, Ying-Han Liu, Jeng-Kuen Chiu, Ee-Peng Lim 713

Can We Apply Projection Based Frequent Pattern Mining Paradigm
to Spatial Co-location Mining?
Yan Huang, Liqin Zhang, Ping Yu 719

PatZip: Pattern-Preserved Spatial Data Compression
Yu Qian, Kang Zhang, D. T. Huynh 726

Temporal Data

A Likelihood Ratio Distance Measure for the Similarity Between
the Fourier Transform of Time Series
Gareth J. Janacek, Anthony J. Bagnall, Michael Powell 737

The TIMERS II Algorithm for the Discovery of Causality
Howard J. Hamilton, Kamran Karimi 744

A Recent-Based Dimension Reduction Technique for Time Series Data
Yanchang Zhao, Chengqi Zhang, Shichao Zhang 751

Graph Partition Model for Robust Temporal Data Segmentation
Jinhui Yuan, Bo Zhang, Fuzong Lin 758

Accurate Symbolization of Time Series
Xinqiang Zuo, Xiaoming Jin 764

A Novel Bit Level Time Series Representation with Implication
of Similarity Search and Clustering
*Chotirat Ratanamahatana, Eamonn Keogh, Anthony J. Bagnall,
Stefano Lonardi* 771

Finding Temporal Features of Event-Oriented Patterns
Xingzhi Sun, Maria E. Orlowska, Xue Li 778

An Anomaly Detection Method for Spacecraft Using Relevance
Vector Learning
Ryohei Fujimaki, Takehisa Yairi, Kazuo Machida 785

Cyclic Pattern Kernels Revisited
Tamás Horváth 791

Text and Web Data Mining

Subspace Clustering of Text Documents with Feature Weighting
K-Means Algorithm
Liping Jing, Michael K. Ng, Jun Xu, Joshua Zhexue Huang 802

Using Term Clustering and Supervised Term Affinity Construction
to Boost Text Classification
Chong Wang, Wenyuan Wang 813

Technology Trends Analysis from the Internet Resources
*Shin-ichi Kobayashi, Yasuyuki Shirai, Kazuo Hiyane, Fumihiro Kumeno,
Hiroshi Inujima, Noriyoshi Yamauchi* 820

Dynamic Mining Hierarchical Topic from Web News Stream Data
Using Divisive-Agglomerative Clustering Method
Jian-Wei Liu, Shou-Jian Yu, Jia-Jin Le 826

Collecting Topic-Related Web Pages for Link Structure Analysis
by Using a Potential Hub and Authority First Approach
Leuo-Hong Wang, Tong-Wen Lee 832

A Top-Down Algorithm for Mining Web Access Patterns from Web Logs
*Jian-Kui Guo, Bei-jun Ruan, Zun-ping Cheng, Fang-zhong Su,
Ya-qin Wang, Xu-bin Deng, Shang Ning, Yang-Yong Zhu* 838

Kernel Principal Component Analysis for Content Based Image Retrieval
Guang-Ho Cha 844

Mining Frequent Trees with Node-Inclusion Constraints
Atsuyoshi Nakamura, Mineichi Kudo 850

Author Index 861

Machine Learning for Analyzing Human Brain Function

Tom Mitchell

Center for Automated Learning and Discovery,
Carnegie Mellon University, USA

Abstract. A major opportunity for knowledge discovery and data mining over the coming decade is to accelerate scientific discovery by providing new computer tools to analyze experimental data. Scientific fields from astronomy to cell biology to neuroscience now collect experimental data sets that are huge when compared to the data sets available just a decade ago. New data mining tools are needed to interpret these new data sets.

This talk presents our own research in one such scientific subfield: studying the operation of the human brain using functional Magnetic Resonance Imaging (fMRI). A typical fMRI experiment captures three-dimensional images of human brain activity, once per second, at a spatial resolution of a few millimeters, providing a 3D movie of brain activity. We present our recent research exploring the question of how best to analyze fMRI data to study human cognitive processes. We will first describe our recent successes training machine learning classifiers to distinguish cognitive subprocesses based on observed fMRI images. For example, we have been able to train classifiers to discriminate whether a person is reading words about tools, or words about buildings, based on their observed fMRI brain activation. We will then introduce an algorithm for learning a new class of probabilistic time series models called Hidden Process Models, and discuss their use for tracking multiple hidden cognitive processes from observed fMRI brain image data.

Subgroup Discovery Techniques and Applications

Nada Lavrač[1,2]

[1] Jožef Stefan Institute, Jamova 39, 1000 Ljubljana, Slovenia
[2] Nova Gorica Polytechnic, Vipavska 13, 5000 Nova Gorica, Slovenia

Abstract. This paper presents the advances in subgroup discovery and the ways to use subgroup discovery to generate actionable knowledge for decision support. Actionable knowledge is explicit symbolic knowledge, typically presented in the form of rules, that allow the decision maker to recognize some important relations and to perform an appropriate action, such as planning a population screening campaign aimed at detecting individuals with high disease risk. Two case studies from medicine and functional genomics are used to present the lessons learned in solving problems requiring actionable knowledge generation for decision support.

1 Introduction

Rule learning is an important form of *predictive* machine learning, aimed at inducing classification and prediction rules from examples [2]. Developments in *descriptive induction* have recently also gained much attention of researchers interested in rule learning. These include mining of association rules [1], subgroup discovery [11, 4, 6] and other approaches to non-classificatory induction.

This paper discusses actionable knowledge generation by means of subgroup discovery. The term *actionability* is described in [10] as follows: "a pattern is interesting to the user if the user can *do something with it* to his or her advantage." As such, actionability is a subjective measure of interestingness.

The lessons in actionable knowledge generation, described in this paper, were learned from two applications that motivated our research in actionable knowledge generation for decision support. In an ideal case, the induced knowledge should enable the decision maker to perform an action to his or her advantage, for instance, by appropriately selecting individuals for population screening concerning high risk for coronary heart disease (CHD). Consider one rule from this application:

$$\text{CHD} \leftarrow \text{body mass index} > 25 \; kgm^{-2} \; \& \; \text{age} > 63 \text{ years}$$

This rule is actionable as the general practitioner can select from his patients the overweight patients older than 63 years.

This paper provides arguments in favor of actionable knowledge generation through recently developed subgroup discovery approaches, where a subgroup discovery task is informally defined as follows [11, 4, 6]: Given a population of individuals and a specific property of individuals that we are interested in, find population subgroups that are statistically 'most interesting', e.g., are as large as possible and have the most unusual distributional characteristics with respect to the property of interest.

We restrict the subgroup discovery task to learning from class-labeled data, and induce individual rules (describing individual subgroups) from labeled training examples (labeled positive if the property of interest holds, and negative otherwise), thus targeting the process of subgroup discovery to uncovering properties of a selected target population of individuals with the given property of interest. Despite the fact that this form of rules suggests that standard supervised classification rule learning could be used for solving the task, the goal of subgroup discovery is to uncover individual rules/patterns, as opposed to the goal of standard supervised learning, aimed at discovering rulesets/models to be used as accurate classifiers of yet unlabeled instances [4].

In subgroup discovery, the induced patterns must be represented in explicit symbolic form and must be relatively simple in order to be recognized as actionable for guiding a decision maker in directing some targeted campaign. We provide arguments in favour of actionable knowledge generation through recently developed subgroup discovery algorithms, uncovering properties of individuals for actions like population screening and functional genomics data analysis. For such tasks, actionable rules are characterized by the experts' choice of the 'actionable' attributes to appear in induced subgroup descriptions, as well as by high coverage (support), high sensitivity and specificity[1], even if this can be achieved only at a price of lower classification accuracy, which is the quality to be optimized in classification and prediction tasks.

This paper is structured as follows. Two applications that have motivated our research in actionable knowledge generation are described in Section 2. Section 3 introduces the ROC and the TP/FP space needed for better understanding of the task and results of subgroup discovery. Section 6 introduces the functional genomics domain in more detail, where the task is to distinguish between different cancer types.

2 Two Case Studies

The motivation for this work comes from practical data mining problems in a medical and a functional genomics domain.

---

[1] *Sensitivity* measures the fraction of positive cases that are classified as positive, whereas *specificity* measures the fraction of negative cases classified as negative. If TP denotes true positives, TN true negatives, FP false positives, FN false negatives, Pos all positives, and Neg all negatives, then $Sensitivity = TPr = \frac{TP}{TP+FN} = \frac{TP}{Pos}$, and $Specificity = \frac{TN}{TN+FP} = \frac{TN}{Neg}$, and $FalseAlarm = FPr = 1 - Specificity = \frac{FP}{TN+FP} = \frac{FP}{Neg}$. Quality measures in association rule learning are *support* and *confidence*: $Support = \frac{TP}{Pos+Neg}$ and $Confidence = \frac{TP}{TP+FP}$.

The medical problem domain is first outlined: the problem of the detection and description of Coronary Heart Disease (CHD) risk groups [4]. Typical data collected in general screening include anamnestic information and physical examination results, laboratory tests, and ECG at-rest test results. In many cases with significantly pathological test values (especially, for example, left ventricular hypertrophy, increased LDL cholesterol, decreased HDL cholesterol, hypertension, and intolerance glucose) the decision is not difficult. However, the hard problem in CHD prevention is to find endangered individuals with slightly abnormal values of risk factors and in cases when combinations of different risk factors occur. The results in the form of risk group models should help general practitioners to recognize CHD and/or to detect the illness even before the first symptoms actually occur. Expert-guided subgroup discovery discovery is aimed at easier detection of important risk factors and risk groups in the population.

In functional genomics, gene expression monitoring by DNA microarrays (gene chips) provides an important source of information that can help in understanding many biological processes. The database we analyze consists of a set of gene expression measurements (examples), each corresponding to a large number of measured expression values of a predefined family of genes (attributes). Each measurement in the database was extracted from a tissue of a patient with a specific disease; this disease is the class for the given example. The domain, described in [9, 5] and used in our experiments, is a typical scientific discovery domain characterised by a large number of attributes compared to the number of available examples. As such, this domain is especially prone to overfitting, as it is a domain with 14 different cancer classes and only 144 training examples in total, where the examples are described by 16063 attributes presenting gene expression values. While the standard goal of machine learning is to start from the labeled examples and construct models/classifiers that can successfully classify new, previously unseen examples, our main goal is to uncover interesting patterns/rules that can help to better understand the dependencies between classes (diseases) and attributes (gene expressions values).

3 Background: The ROC and the TP/FP Space

A point in the ROC space (ROC: Receiver Operating Characteristic) [8] shows classifier performance in terms of false alarm or *false positive rate* $FPr = \frac{|FP|}{|TN|+|FP|} = \frac{|FP|}{|N|}$ (plotted on the X-axis), and sensitivity or *true positive rate* $TPr = \frac{|TP|}{|TP|+|FN|} = \frac{|TP|}{|P|}$ (plotted on the Y-axis).

A point (FPr, TPr) depicting rule R in the ROC space is determined by the covering properties of the rule. The ROC space is appropriate for measuring the success of subgroup discovery, since rules/subgroups whose TPr/FPr tradeoff is close to the diagonal can be discarded as insignificant; the reason is that the rules with TPr/FPr on the diagonal have the same distribution of covered positives and negatives as the distribution in the training set. Con-

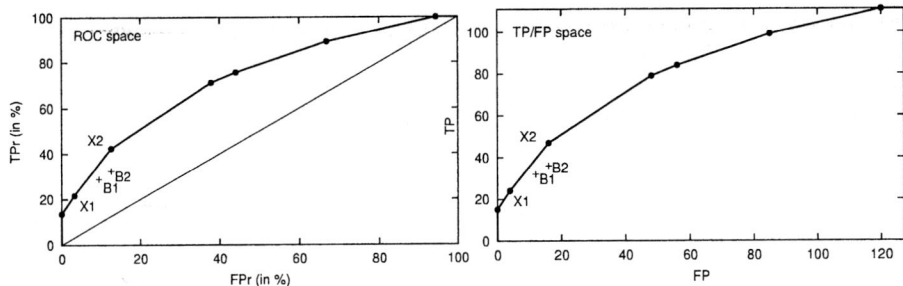

Fig. 1. The left-hand side figure shows the ROC space with a convex hull formed of seven rules that are optimal under varying TPr/FPr tradeoffs, and two suboptimal rules B1 and B2. The right-hand side presents the positions of the same rules in the corresponding TP/FP space

versely, significant rules/subgroups are those sufficiently distant from the diagonal. Subgroups that are optimal under varying TPr/FPr tradeoffs form a convex hull called the ROC curve. Figure 1 presents seven rules on the convex hull (marked by circles), including $X1$ and $X2$, while two rules $B1$ and $B2$ below the convex hull (marked by +) are of lower quality in terms of their TPr/FPr tradeoff.

It was shown in [6] that for rule R, the vertical distance from the (FPr, TPr) point to the ROC diagonal is proportional to the significance of the rule. Hence, the goal of a subgroup discovery algorithm is to find subgroups in the upper-left corner area of the ROC space, where the most significant rule would lie in point $(0, 1)$ representing a rule covering only positive and none of the negative examples ($FPr = 0$ and $TPr = 1$).

An alternative to the ROC space is the so-called TP/FP space (see the right-hand side of Figure 1), where FPr on the X-axis is replaced by $|FP|$ and TPr on the Y-axis by $|TP|$.[2] The TP/FP space is equivalent to the ROC space when comparing the quality of subgroups induced in a single domain. The reminder of this paper considers only this simpler TP/FP space representation.

4 Constraint-Based Subgroup Discovery

Subgroup discovery is a form of supervised inductive learning of subgroup descriptions of the target class. As in all inductive rule learning tasks, the language bias is determined by the syntactic restrictions of the pattern language and the vocabulary of terms in the language. In this work the hypothesis language is restricted to simple if-then rules of the form $Class \leftarrow Cond$, where $Class$ is the target class and $Cond$ is a conjunction of features. Features are logical condi-

---

[2] The TP/FP space can be turned into the ROC space by simply normalizing the TP and FP axes to the [0,1]x[0,1] scale.

tions that have values *true* or *false*, depending on the values of attributes which describe the examples in the problem domain: subgroup discovery rule learning is a form of two-class propositional inductive rule learning, where multi-class problems are solved through a series of two-class learning problems, so that each class is once selected as the target class while examples of all other classes are treated as non-target class examples.

This section briefly outlines a recently developed approach to subgroup discovery that can be applied to actionable knowledge generation.

4.1 Constraint-Based Subgroup Discovery with the SD Algorithm

In this paper, subgroup discovery is performed by SD, an iterative beam search rule learning algorithm [4]. The input to SD consists of a set of examples E and a set of features F that can be constructed for the given example set. The output of the SD algorithm is a set of rules with optimal covering properties on the given example set. The SD algorithm is implemented in the on-line Data Mining Server (DMS), publicly available at http://dms.irb.hr.[3]

In a constraint-based data mining framework [3], a formal definition of subgroup discovery involves a set of constraints that induced subgroup descriptions have to satisfy. The following constraints are used to formalize the SD constraint-based subgroup discovery task.

Language Constraints

- Individual subgroup descriptions have the form of rules $Class \leftarrow Cond$, where $Class$ is the property of interest (the target class), and $Cond$ is a conjunction of features (conditions based on attribute value pairs) defined by the language describing the training examples.
- For discrete (categorical) attributes, features have the form $Attribute = value$ or $Attribute \neq value$, for continuous (numerical) attributes they have the form $Attribute > value$ or $Attribute \leq value$. Note that features can have values *true* and *false* only, that every feature has its logical complement (for feature f_1 being $A_1 = v_1$ its logical complement $\overline{f_1}$ is $A_1 \neq v_1$, for $A_2 > v_2$ its logical complement is $A_2 \leq v_2$), and that features are different from binary valued attributes because for every attribute at least two different features are constructed.
- To simplify rule interpretation and increase rule actionability, subgroup discovery is aimed at finding short rules. This is formalized by a language constraint that every induced rule R has to satisfy: rule size (i.e., the number of features in $Cond$) has to be below a user-defined threshold: $size(R) \leq MaxRuleLength$.

---

[3] The publicly available Data Mining Server and its constituent subgroup discovery algorithm SD can be tested on user submitted domains with up to 250 examples and 50 attributes. The variant of the SD algorithm used in gene expression data analysis was not limited by these restrictions.

Evaluation/Optimization Constraints

- To ensure that induced subgroups are sufficiently large, each induced rule R must have high support, i.e., $sup(R) \geq MinSup$, where $MinSup$ is a user-defined threshold, and $sup(R)$ is the relative frequency of correctly covered examples of the target class in examples set E:

$$sup(R) = p(Class \cdot Cond) = \frac{n(Class \cdot Cond)}{|E|} = \frac{|TP|}{|E|}$$

- Other evaluation/optimization constraints have to ensure that the induced subgroups are highly significant (ensuring that the class distribution of examples covered by the subgroup description will be statistically significantly different from the distribution in the training set). This could be achieved in a straight-forward way by imposing a significance constraint on rules, e.g., by requiring that rule significance is above a user-defined threshold. Instead, in the SD subgroup discovery algorithm [4] the following rule quality measure assuring rule significance, implemented as a heuristic in rule construction, is used:

$$q_g(R) = \frac{|TP|}{|FP| + g} \quad (1)$$

In Equation 1, TP are true positives (target class examples covered by rule R), FP are false positives (non-target class examples covered by rule R), and g is a user defined generalization parameter. High quality rules will cover many target class examples and a low number of non-target examples. The number of tolerated non-target class cases, relative to the number of covered target class cases, is determined by parameter g. It was shown in [4] that by using this optimization constraint (choose the rule with best $q_g(R)$ value in beam search of best rule conditions), rules with a significantly different distribution of covered positives and negatives, compared to the prior distribution in the training set, are induced.

5 Experiments in Patient CHD Risk Group Detection

Early detection of artherosclerotic coronary heart disease (CHD) is an important and difficult medical problem. CHD risk factors include artherosclerotic attributes, living habits, hemostatic factors, blood pressure, and metabolic factors. Their screening is performed in general practice by data collection in three different stages.

A Collecting anamnestic information and physical examination results, including risk factors like age, positive family history, weight, height, cigarette smoking, alcohol consumption, blood pressure, and previous heart and vascular diseases.

B Collecting results of laboratory tests, including information about risk factors like lipid profile, glucose tolerance, and thrombogenic factors.

C Collecting ECG at rest test results, including measurements of heart rate, left ventricular hypertrophy, ST segment depression, cardiac arrhythmias and conduction disturbances.

In this application, the goal was to construct at least one relevant and interesting CHD risk group for each of the stages A, B, and C, respectively.

A database with 238 patients representing typical medical practice in CHD diagnosis, collected at the Institute for Cardiovascular Prevention and Rehabilitation, Zagreb, Croatia, was used for subgroup discovery [4]. The database is in no respect a good epidemiological CHD database reflecting actual CHD occurrence in a general population, since about 50% of gathered patient records represent CHD patients. Nevertheless, the database is very valuable since it includes records of different types of the disease. Moreover, the included negative cases (patients who do not have CHD) are not randomly selected persons but individuals considered by general practitioners as potential CHD patients, and hence sent for further investigations to the Institute. This biased dataset is appropriate for CHD risk group discovery, but it is inappropriate for measuring the success of CHD risk detection and for subgroup performance estimation in general medical practice.

5.1 Results of Subgroup Discovery

The process of expert-guided subgroup discovery was performed as follows. For every data stage A, B and C, the SD algorithm was run for values g in the range 0.5 to 100 (values 0.5, 1, 2, 4, 6, ...), and a fixed number of selected output rules equal to 3. The rules induced in this iterative process were shown to the expert for selection and interpretation. The inspection of 15–20 rules for each data stage triggered further experiments, following the suggestions of the medical expert to limit the number of features in the rule body and avoid the generation of rules whose features would involve expensive and/or unreliable laboratory tests.

In the iterative process of rule generation and selection, the expert has selected five most interesting CHD risk groups. Table 1 shows the induced subgroup descriptions. The features appearing in the conditions of rules describing the subgroups are called the *principal factors*. The described iterative process was successful for data at stages B and C, but it turned out that medical history data on its own (stage A data) is not informative enough for inducing subgroups, i.e., it failed to fulfil the expert's subjective criteria of interestingness. Only after engineering the domain, by separating male and female patients, interesting subgroups $A1$ and $A2$ have actually been discovered.

Separately for each data stage A, B and C, we have investigated which of the induced rules are the best in terms of the TP/FP tradeoff, i.e., which of them are used to define the convex hull in the TP/FP space. At stage B, for instance, seven rules (marked by +) are on the convex hull of the TP/FP space shown in Figure 1. Notice that the expert-selected subgroups B1 and B2 are significant, but are not among those lying on the convex hull in Figure 1. The reason for selecting exactly those two rules at stage B are their simplicity (con-

Table 1. Induced subgroups in the form of rules. Rule conditions are conjunctions of principal factors. Subgroup A1 is for male patients, subgroup A2 for female patients, while subgroups B1, B2, and C1 are for both male and female patients. The subgroups are induced from different attribute subsets (A, B and C, respectively) with different g parameter values (14, 8, 10, 12 and 10, respectively)

	Expert Selected Subgroups
A1	CHD ← positive family history & age over 46 year
A2	CHD ← body mass index over 25 kgm^{-2} & age over 63 years
B1	CHD ← total cholesterol over 6.1 $mmolL^{-1}$ & age over 53 years & body mass index below 30 kgm^{-2}
B2	CHD ← total cholesterol over 5.6 $mmolL^{-1}$ & fibrinogen over 3.7 gL^{-1} & body mass index below 30 kgm^{-2}
C1	CHD ← left ventricular hypertrophy

sisting of three features only), their generality (covering relatively many positive cases) and the fact that the used features are, from the medical point of view, inexpensive laboratory tests. Additionally, rules B1 and B2 are interesting because of the feature *body mass index below 30 kgm^{-2}*, which is intuitively in contradiction with the expert knowledge that both increased body weight as well as increased total cholesterol values are CHD risk factors. It is known that increased body weight typically results in increased total cholesterol values while subgroups B1 and B2 actually point out the importance of increased total cholesterol when it is not caused by obesity as a relevant disease risk factor.

5.2 Statistical Characterization of Subgroups

The next step in the proposed subgroup discovery process starts from the discovered subgroups. In this step, statistical differences in distributions are computed for two populations, the target and the reference population. The target population consists of true positive cases (CHD patients included into the analyzed subgroup), whereas the reference population are all available non-target class examples (all the healthy subjects). Statistical differences in distributions for all the descriptors (attributes) between these two populations are tested using the χ^2 test with 95% confidence level ($p = 0.05$).

To enable testing of statistical significance, numerical attributes have been partitioned in up to 30 intervals so that in every interval there are at least 5 instances. Among the attributes with significantly different value distributions there are always those that form the features describing the subgroups (the principal factors), but usually there are also other attributes with statistically significantly different value distributions. These attributes are called *supporting*

Table 2. Statistical characterizations of induced subgroup descriptions (supporting factors)

	Supporting Factors
A1	psychosocial stress, cigarette smoking, hypertension, overweight
A2	positive family history, hypertension, slightly increased LDL cholesterol, normal but decreased HDL cholesterol
B1	increased triglycerides value
B2	positive family history
C1	positive family history, hypertension, diabetes mellitus

attributes, and the features formed of their values that are characteristic for the discovered subgroups are called *supporting factors*.

Supporting factors are very important for subgroup descriptions to become more complete and acceptable for medical practice. Medical experts dislike long conjunctive rules which are difficult to interpret. On the other hand, they also dislike short rules providing insufficient supportive evidence. In this work, we found an appropriate tradeoff between rule simplicity and the amount of supportive evidence by enabling the expert to inspect all the statistically significant supporting factors, whereas the decision whether they indeed increase the user's confidence in the subgroup description is left to the expert. In the CHD application the expert has decided whether the proposed supporting factors are meaningful, interesting and actionable, how reliable they are and how easily they can be measured in practice. Table 2 lists the expert selected supporting factors.

6 Experiments in Functional Genomics

The gene expression domain, described in [9,5] is a domain with 14 different cancer classes and 144 training examples in total. Eleven classes have 8 examples each, two classes have 16 examples and only one has 24 examples. The examples are described by 16063 attributes presenting gene expression values. In all the experiments we have used gene presence call values (A, P, and M) to describe the training examples. The domain can be downloaded from http://www-genome.wi.mit.edu/cgi-bin/cancer/datasets.cgi. There is also an independent test set with 54 examples. The standard goal of machine learning is to start from such labeled examples and construct classifiers that can successfully classify new, previously unseen examples. Such classifiers are important because they can be used for diagnostic purposes in medicine and because they can help to understand the dependencies between classes (diseases) and attributes (gene expressions values).

6.1 Choice of the Description Language of Features

Gene expression scanners measure signal intensity as continuous values which form an appropriate input for data analysis. The problem is that for continuous valued attributes there can be potentially many boundary values separating the classes, resulting in many different features for a single attribute. There is also a possibility to use presence call (signal specificity) values computed from measured signal intensity values by the Affymetrix GENECHIP software. The presence call has discrete values A (absent), P (present), and M (marginal). Subgroup discovery as well as filtering based on feature and rule relevancy are applicable both for signal intensity and/or the presence call attribute values. Typically, signal intensity values are used [7] because they impose less restrictions on the classifier construction process and because the results do not depend on the GENECHIP software presence call computation. For subgroup discovery we prefer the later approach based on presence call values. The reason is that features presented by conditions like $Gene = P$ is $true$ (meaning that $Gene$ is present, i.e., expressed) or $Gene = A$ is $true$ (meaning that $Gene$ is absent, i.e., not expressed) are very natural for human interpretation and that the approach can help in avoiding overfitting, as the feature space is very strongly restricted, especially if the marginal value M is encoded as value unknown.

In our approach, the M value is handled as an unknown value, as we do not want to increase the relevance of features generated from attributes with M values. The M values are therefore handled as unknown values as follows: unknown values in positive examples are replaced by value $false$, while unknown values in negative examples are replaced by value $true$. As for the other two values, A and P, it holds that two features for gene X, $X = A$ and $X \neq P$, are identical. Consequently, for every gene X there are only two distinct features $X = A$ and $X = P$.

6.2 The Experiments

The experiments were performed separately for each cancer class so that a two-class learning problem was formulated where the selected cancer class was the target class and the examples of all other classes formed non-target class examples. In this way the domain was transformed into 14 inductive learning problems, each with the total of 144 training examples and between 8 and 24 target class examples. For each of these tasks a complete procedure consisting of feature construction, elimination of irrelevant features, and induction of subgroup descriptions in the form of rules was repeated. Finally, using the SD subgroup discovery algorithm [4], for each class a single rule with maximal q_g value was selected, for $q_g = \frac{|TP|}{|FP|+g}$ being the heuristic of the SD algorithm and $g = 5$ the generalization parameter default value. The rules for all 14 tasks consisted of 2–4 features. The procedure was repeated for all 14 tasks with the same default parameter values. The induced rules were tested on the independent example set.

Table 3. Covering properties on the training and on the independent test set for rules induced for three classes with 16 and 24 examples. Sensitivity is $\frac{|TP|}{|P|}$, specificity is $\frac{|TN|}{|N|}$, while precision is defined as $\frac{|TP|}{|TP|+|FP|}$

Cancer	Training set			Test set		
	Sens.	Spec.	Prec.	Sens.	Spec.	Prec.
lymphoma	16/16	128/128	100%	5/6	48/48	100%
leukemia	23/24	120/120	100%	4/6	47/48	80%
CNS	16/16	128/128	100%	3/4	50/50	100%

There are very large differences among the results on the test sets for various classes (diseases) and the precision higher than 50% was obtained for only 5 out of 14 classes. There are only three classes (lymphoma, leukemia, and CNS) with more than 8 training cases and all of them are among those with high precision on the test set, while for only two out of eleven classes with 8 training cases (colorectal and mesothelioma) high precision was achieved. The classification properties of rules induced for classes with 16 and 24 target class examples (lymphoma, leukemia and CNS) are comparable to those reported in [9] (see Table 3), while the results on eight small example sets with 8 target examples were poor. An obvious conclusion is that the use of the subgroup discovery algorithm is not appropriate for problems with a very small number of examples because overfitting can not be avoided in spite of the heuristics used in the SD algorithm and the additional domain-specific techniques used to restrict the hypothesis search space. But for larger training sets the subgroup discovery methodology enabled effective construction of relevant rules.

6.3 Examples of Induced Rules

For three classes (lymphoma, leukemia, and CNS) with more than 8 training cases the following rules were induced by the constraint-based subgroup discovery approach involving relevancy filtering and handling of unknown values described in this chapter.

Lymphoma class:
(CD20_receptor EXPRESSED) AND
(phosphatidylinositol_3_kinase_regulatory_alpha_subunit NOT EXPRESSED)
Leukemia class:
(KIAA0128_gene EXPRESSED) AND
(prostaglandin_d2_synthase_gene NOT EXPRESSED)
CNS class:
(fetus_brain_mRNA_for_membrane_glycoprotein_M6 EXPRESSED) AND
(CRMP1_collapsin_response_mediator_protein_1 EXPRESSED)

The expert interpretation of the results yields several biological observations: two rules (for the lymphoma and leukemia classes) are judged as reassuring and one (the CNS class) has a plausible, albeit partially speculative explanation. Namely, the best-scoring rule for the lymphoma class in the multi-class cancer recognition problem contains a feature corresponding to a gene routinely used as a marker in diagnosis of lymphomas (CD20), while the other part of the conjunction (phosphatidylinositol, the PI3K gene) seems to be a plausible biological co-factor. The best-scoring rule for the leukemia class contains a gene whose relation to the disease is directly explicable (KIAA0128, Septin 6). Both M6 and CRMP1 appear to have multifunctional roles in shaping neuronal networks, and their function as survival (M6) and proliferation (CRMP1) signals may be relevant to growth promotion and CNS malignancy.

Both good prediction results on an independent test set as well as expert interpretation of induced rules prove the effectiveness of described methods for avoiding overfitting in scientific discovery tasks.

Acknowledgments

The paper describes joint work with Dragan Gamberger from Rudjer Bošković Institute, Zagreb, Croatia, supported by the Slovenian Ministry of Higher Education, Science and Technology.

References

1. R. Agrawal, H. Mannila, R. Srikant, H. Toivonen, and A.I. Verkamo. Fast discovery of association rules. In U.M. Fayyad, G. Piatetski-Shapiro, P. Smyth and R. Uthurusamy, editors, *Advances in Knowledge Discovery and Data Mining*, 307–328. AAAI Press, 1996.
2. P. Clark and T. Niblett. The CN2 induction algorithm. *Machine Learning*, 3(4):261–283, 1989.
3. R.J. Bayardo, editor. *Constraints in Data Mining. Special issue of* SIGKDD Explorations, 4(1), 2002.
4. D. Gamberger and N. Lavrač. Expert-guided subgroup discovery: Methodology and application. *Journal of Artificial Intelligence Research* 17: 501–527, 2002.
5. D. Gamberger, N. Lavrač, F. Železný, and J. Tolar. Induction of comprehensible models for gene expression datasets by the subgroup discovery methodology. *Journal of Biomedical Informatics* 37:269–284, 2004.
6. N. Lavrač, B. Kavšek, P. Flach and L. Todorovski. Subgroup discovery with CN2-SD. *Journal of Machine Learning Research*, 5: 153–188, 2004.
7. J. Li and L. Wong. Geography of differences between two classes of data. In *Proc. of 6th European Conference on Principles of Data Mining and Knowledge Discovery (PKDD2002)*, Springer, 325–337, 2002.
8. F. Provost and T. Fawcett. Robust classification for imprecise environments. *Machine Learning*, 42(3): 203–231, 2001.
9. S. Ramaswamy et al. Multiclass cancer diagnosis using tumor gene expression signitures. In *Proc. Natl. Acad. Sci. USA*, 98(26): 15149–15154, 2001.

10. A. Silberschatz and A. Tuzhilin. On subjective measures of interestingness in knowledge discovery. In *Proc. First International Conference on Knowledge Discovery and Data Mining (KDD)*, 275–281, 1995.
11. S. Wrobel. An algorithm for multi-relational discovery of subgroups. In *Proceedings of the 1st European Symposium on Principles of Data Mining and Knowledge Discovery*, Springer, 78–87, 1997.

IT Development in the 21st Century and Its Implications

Unna Huh

President, Information and Communications University, Korea

Abstract. This talk discusses general IT development in the 21st century and its positive and negative effects. It also talks about Korea's IT development and makes suggestions for Korea's further advances in IT. The three keywords to describe the IT development in the 21st century are digitalization, convergence, and ubiquitous revolution. This IT development has presented new opportunities for our society, corporations and individuals while posing a threat to us. That is, IT revolution may dramatically improve the quality of human life and allow amazing degree of comfort in our lives, but like a double-edged sword, IT may also be misused and have disastrous impacts on our daily lives, such as invasion of privacy, leakage and abuse of personal information, and hacking. In dealing with these problems, technological advances alone may not be sufficient. There is a need for innovative education regionally and worldwide to cultivate wisdom in U-citizens so that they can use the modern convenience of IT with strong ethics. We also need to establish new laws and societal systems appropriate for the ubiquitous era.

Ladies and Gentlemen!
It is my great pleasure and honor to be here. I am Unna Huh, President of Information and Communications University (ICU), Korea. First of all, I would like to express my sincere gratitude to the Pacific Asia conference on Knowledge Discovery & Data Mining (PAKDD) for giving me this opportunity to share my views on the 21st century's IT development.

1 Introduction

I have been asked to speak about general IT development. In the next few minutes, I will discuss the trends of IT development in the 21st century and analyze both the positive and the negative effects of the development. Then I will focus on the IT development in Korea and suggest some strategies for future IT development.

In my opinion, the three keywords to describe the IT development in the 21st century are digitalization, convergence, and ubiquitous revolution. This IT development has presented new opportunities for our society, corporations and individuals while posing a threat to us. In order to maximize the positive effects of IT development and to minimize its negative effects, I believe society must develop general coping strategies.

2 The Trends of IT Development in the 21st Century

The rapid IT development in the 20th century brought about drastic changes in the operational styles of organizations, society's infrastructure and culture, and individuals' life styles. There are several rules of thumb that we have gathered from the past IT development and that would help us predict the future trends of IT development in terms of quantity, and these are as follows:

- Moore's Law: Every 18 months computer chips fall in price half, but the performance doubles.
- Glider's Law: In the near future, the overall performance (bandwidth) of the IT system will triple every 12 months.
- Metcalfe's Law: The value of IT network increases non-linearly as the number of users increases; therefore, the future value of IT network will make an explosive leap.

Let me review now the three key elements of the IT development in the 21st century.

2.1 Digitalization

First of all, I expect that there will be rapid digitalization of IT equipment. We will soon see digital TVs, digital camera/phone/DVDs, and digital household electronic appliances. There will also be digitalization of IT networks, such as digital broadcasting stations, digital cable TVs, digital mobile communication (2.5G/3G), IP-connected wired telephone network, and digital satellites. Furthermore, the digitalization of information or contents will be accelerated; thus, there will be widespread use of MP3 music files, digital photos and animation, and e-books.

2.2 Convergence

Secondly, we have come to witness the phenomenon of convergence as the computer and communication are converged. Especially thanks to the far-reaching digitalization, there will be active device convergence, and it will be difficult to determine the specific characteristics of a product. For example, we will be using smart phones in which mobile phones and PDA are integrated and Internet TVs, a combination of the Internet and TVs. Moreover, the different infrastructures of various communication networks that have existed separately will become broadband, converged, and composite as in the cases of the convergence of communication and broadcasting and the wireless and wired integration. As digital contents with identical standards have become available, they are used in different media and different types of equipment. We now can take photos with a digital camera, edit them in the computer, and send them via mobile phones. Presently beyond the boundary of the existing IT industries, IT and banking are being converged, and we also see super convergence taking place between IT, NT, and BT.

2.3 Ubiquitous Revolution

In the core of the 21st century's information and communication technologies lie ubiquitous services combining the strengths of the Internet and wireless and wired IT. In other words, we should expect to live in an era of ubiquitous revolution soon, where we can be connected to anyone anywhere any time. This will be made possible by the combination of wired and wireless communication technologies and computing technologies. RFID in particular is getting attention as one of the most important technological, industrial elements in the next-generation ubiquitous environment and is expected to create new markets in various areas.

3 The Implications of IT Development

I have so far described the three major phenomena of IT development in the 21st century. This is an overview of both the constructive aspects and the threats and dangers against our society, corporations, and individuals that the 21st century's IT development has brought to us.

3.1 The Impact of Digitalization

First, digitalization will dramatically increase the efficiency and performance of IT equipment and network. As the content compatibility and the extent of inter-operation among different types of equipment improve, we are likely to enjoy more the benefits of having networks. We will get exposed to more and better-quality information that is generated and disseminated, and it will be much easier for the users to have interactions with one another. We will also benefit from affordable e-Life, i.e., distance e-health services made available by information and communication technologies, and thus enjoy better-quality life.

On the other hand, it will be critical to determine when to raise investment funds and how to measure the outcome of future investment as we expect a considerable amount of investment in digital upgrade. Some of the traditional businesses, such as conventional photo studios, video rentals, film manufacturers, and analog watch vendors, will go out of business thanks to the emergence of digital technologies. The number of the video rentals in Korea in 1996, which exceeded 32,000, considerably decreased to 8000 rentals by February 2003. In addition, there will be a problem of digital divide between those who have access to and make use of digital technologies and those who do not. In the digital era, where intellectual property is one's major asset, we will be faced with frequent problems of hacking and plagiarism, made easier by the characteristics of digital technologies.

3.2 The Impact of Convergence

Secondly, it is possible to create a new service or product using digital technologies as value transfer has already begun in all types of industries and fields. Conventional manufacturing industries are given new opportunities to overcome the existing boundaries and to explore a new aspect of growth: Telematics is an integration of auto and IT technologies; cyber apartments are an outcome of combining construction and Internet technologies; the home appliance industry is leading home networking.

However, some of the traditional companies have had to deal with competition that they can't possibly win because of the collapse of the boundaries among the industries. Credit companies, for example, have suffered a serious blow because of the advent of a payment system via portable phones. The value chain of the existing industries has been destroyed, and multiple value chains are being converged and combined into a new value network, which has put traditional companies in a difficult position to compete against new breeds of businesses. We have seen the severe competition to occupy the wireless portal market between mobile companies and traditional Internet portals.

3.3 The Impact of Ubiquitous Revolution

Finally, with the advent of ubiquitous services that overcome spatial, time constraints, the efficiency of companies and society in general will greatly improve. In a ubiquitous environment, there will be a large number of devices and sensors surrounding the user. As the user moves around, the user device will be able to access the network that can provide the best quality of service at the most economic rate in a seamless manner. Ubiquitous computing also establishes USN (ubiquitous sensor network), where invisible computers automatically detect time-varying contexts and makes it possible to execute information sharing, behavior proposals, and action measures necessary for human life. In sum, ubiquitous services will contribute to the improvement of the quality of individuals' everyday lives.

Unfortunately, omnipresent computing technologies have also had negative effects, such as the broadening of information gap and digital waste created by the digitalization and networking of equipment. The right of being unconnected is not guaranteed, which may give rise to serious violations of individual privacy. Furthermore, IT development may aggravate people's feelings of isolation, the phenomenon called digital isolation. Ubiquitous computing is defined as "connected to any device anytime and anywhere." Because of these all-pervading characteristics, it is likely that new kinds of problems in information protection will arise; the seriousness of the existing issues will be amplified. RFID in specific has innate problems of information leakage, distortion, and supervision. If we do not take proper measures to deal with these problems, they will act as stumbling blocks in the expansion of RFID and in our entry into ubiquitous society.

4 Korea's Current State of IT Technologies and Issues

To this point I have discussed the general IT development, and the rest of my talk will concentrate on the IT development in Korea and its related issues.

4.1 Major Achievements

The most noteworthy IT achievement that Korea has made in recent years is the fact that Korea has achieved the commercialization and localization of major digital technologies, such as CDMA, DRAM, and TDX, for the first time in the world.

Furthermore, mobile communication terminals have grown to be a strategically important item of exports, taking up approximately 8% of the total national exports, and become high-quality, high-end products in the world's market.

In addition, the number of domestic users of mobile communication services has reached 34 million; the majority of them are subscribers of digital communications. The number is close to 73% of the overall distribution rate, and more than the half of them are wireless Internet subscribers. One out of every two Korean citizens (59%) is using the Internet, and 73% of the households are subscribers of VHS Internet. Korea has taken the fifth place in Internet use and risen as the world's digital power.

Also, Korea boasts of the world's best information infrastructure: the Korean government has created a world-class service infrastructure to be an electronic government and established a VHS information communication network connecting 144 major cities across the nation in 2000. This was the world's first achievement of its kind.

4.2 Some Issues of Note

While Korea has had some remarkable achievements in IT, there are also some limitations that should be overcome for further speedy advance. Most of all, there is a lack of core technologies to carry out digital convergence. In the case of semi-conductor technology, which is considered a core technology, the domestic semi-conductor industry focuses on manufacturing of memory chips but heavily depends on the imports of ASIC (SOC) while the proportion of non-memory and memory chips in the world market is 80 vs. 20.

Insufficient manpower and technological prowess in the area of software, the axis of digital convergence, are other major problems. To set off a rapid growth of software industry, we are in urgent need of more software experts, especially high-quality manpower ready with both knowledge of software engineering technologies and project management skills. In the next four years, the domestic software industry will suffer from a shortage of human resources, up to 56,000 people.

Finally, although digital contents are at the heart of IT industry as digitalization, convergence, and ubiquitous revolution are in progress, Korea does not hold strong competitive power over emotional contents, where culture, technology and economy meet.

4.3 Suggestions for Solutions and Coping Strategies

Above all, I must say that it is necessary to establish academic departments of composite disciplines in higher education and to train manpower of complex knowledge and skills in order to carry out the digitalization and convergence. Digitalization and convergence are not passing fads. Society needs experts who can draw the overall picture for the general public. That is to say what leads to the success of information and communication technologies, including digitalization and convergence, is not the technology itself but the humans. We need no other choice but to make a bold investment in IT education. Korea in large part has maintained the traditional education system, and only a few institutions produce human resources equipped with knowledge of composite technologies. We will need to encourage multifaceted learning and produce manpower of complex knowledge beyond the boundary of the existing traditional education system. The case of Information and Communications University (ICU), where disciplinary education in the true sense is provided between the disciplines of IT engineering and IT management, is worth notice. ICU is Korea's most prominent IT institution and strives to become world's foremost research-centered university in the field of IT. Its mission is to educate world's future IT leaders.

Moreover, an emphasis has to be placed on the industry of digital contents. It is recommended that Korea learn to train/produce rule makers, e.g., J. K. Rowling, author of Harry Potter series, and Steven Spielberg, that is, composite talents who can lead world's trends and take hold of cultural power in the wave of digitalization, convergence, and ubiquitous revolution. Also, it is necessary to discover novel, unique ideas, materialize them into digital contents, and service and utilize them via digital media.

Furthermore, IT industry needs to shift gear from manufacturing-oriented to marketing-oriented. It needs to learn and find out people's digital needs. When one makes a large amount of investment in a technology without an appropriate viewpoint of marketing, it may help facilitate the widespread use of the technology, but it is not certain whether or not such development is useful or will eventually improve the quality of life. Therefore, it is important for the industry to find out what the consumers ultimately want from digitalization and convergence.

Lastly, there should be deterrence measures to discourage indiscriminate digitalization, convergence, and ubiquitous revolution. There is no guarantee that the digitalization or convergence of particular items will lead to success in the market. Several years ago, a digital camera equipped with a MP3 player was on the market, but it turned out a disastrous failure. It is not necessary to do convergence on all the items. Some may need convergence and others, divergence. This requires a balanced approach on the basis of an appropriate strategy.

In this talk, I have discussed the IT development as a worldwide phenomenon and the Korean example. I have also illustrated the positive and negative aspects of the advances in digital technologies, convergence, and ubiquitous computing. IT revolution may dramatically improve the quality of human life and allow amazing degree of comfort in our lives, but like a double-edged sword, IT may also be misused and have disastrous impacts on our daily lives, such as invasion of privacy, leakage and abuse of personal information, and hacking. In dealing with these problems, technological advances alone may not be sufficient. There is a need for innovative education regionally and worldwide to cultivate wisdom in U-citizens so that they can use the modern convenience of IT with strong ethics. We also need to establish new laws and societal systems appropriate for the ubiquitous era. As we are heading for a U-society, where one can be connected to any one any time and any where, the task for all of us who are here today is to establish collaboration among different regions, nations, and continents to facilitate the constructive, beneficial aspect of IT advances and to discuss how to do so effectively from our points of view. I hope you will have a wonderful, meaningful stay during the conference. Thank you very much for listening.

Data Mining of Gene Expression Microarray via Weighted Prefix Trees

Tran Trang, Nguyen Cam Chi, and Hoang Ngoc Minh

Centre Intégré de BioInformatique,
Centre d'Etude et de Recherche en Informatique Médicale,
Université de Lille 2, 1 Place Verdun, 59045 Lille Cedex, France
{ttran, cnguyen, hoang}@univ-lille2.fr

Abstract. We used discrete combinatoric methods and non numerical algorithms [9], based on *weighted* prefix trees, to examine the data mining of DNA microarray data, in order to capture biological or medical informations and extract new knowledge from these data. We describe hierarchical cluster analysis of DNA microarray data using structure of weighted trees in two manners : classifying the degree of overlap between different microarrays and classifying the degree of expression levels between different genes. These are most efficiently done by finding the characteristic genes and microarrays with the maximum degree of overlap and determining the group of *candidate genes* suggestive of a pathology.

Keywords: combinatoric of words, weighted trees, data mining, cluster analysis, DNA microarrays.

1 Introduction

DNA microarray is technology used to measure simultaneously the expression levels of thousands of genes under various conditions and then provide genome-wide insight. Microarray is a microscope slide to which the thousands of DNA fragments are attached. The DNA microarrays are hybridized with fluorescently labelled cDNA prepared from total mRNA of studied cells. The cDNA of the first cell sample is labelled with a green-fluorescent dye and the second with a red-fluorescent dye. After hybridization, the DNA microarrays are placed in a scanner to create a digital image of the arrays. The intensity of fluorescent light varies with the strength of the hybridization. The measure of expression level of a gene is determined by the logarithm of the ratio of the luminous intensity of the red fluorescence I_R to the luminous intensity of the green fluorescence I_G, $E = \log_2(\frac{I_R}{I_G})$. In this work, a change of differential expression of a gene between two cell samples by a factor of greater than 2 was considered significant. Thus, if $E \geq 1$, the gene is said *up-regulated*; if $-1 < E < 1$, it is said *no-regulated*; if $E \leq -1$, it is said *down-regulated* in the second cell sample [3].

The important application of microarray techology is an organization of microarray profiles and gene expression profiles into different clusters according to degree of expression levels. Since some microarray experiments can contain up to 30 000 target spots, the data generated from a single array mounts up quickly. The interpretation of the experiment results requires the mathematic methods and software programs to capture the biological or medical information and to extract new knowledge from these data. There exist already several statistical methods and software programs to analyze the microarrays. Examples include 1) Unsupervised learning methods : hierarchical clustering and k-means clustering algorithms based on distance similarity metrics [3, 4] are used to search for genes with maximum similarity. 2) Supervised learning methods : support vector machine approaches based on kernel function [5], the bayesian naive approach based on maximum likelihood method [6, 7] are used to classify the gene expression into a training set. These methods are based essentially on *numerical* algorithms and do not really require structured data.

In this paper, we used *combinatoric* methods and *non-numerical* algorithms, based on structure of *weighted* prefix trees, to analyze microarray data, *i.e* :

1. defining the distances to compare different genes and different microarrays,
2. organizing the microarrays of a pathology into the different clusters,
3. classifying the genes of a pathology into the different clusters,
4. researching the characteristic genes and/or characteristic microarrays,
5. determining the group of candidate genes suggestive of a pathology.

The key to understanding our approach is the information on gene expression in a microarray is represented by a *symbolic* sequence, and a collection of microarrays is viewed as a *language*, which is implementated by weighted prefix trees [8, 9]. In fact, the expression levels of a gene will be modeled as an alphabet whose size is the number of expression levels. Thus, a microarray profile or a gene expression profile is viewed as a word over this alphabet and a collection of microarrays forms a language that is called a *DNA microarray language*. In this study, we chose three symbols to represent the three expression levels of a gene according to : a spot representing a gene *up-regulated* is encoded by $+$; a gene *no-regulated* is encoded by $\circ$; a gene *down-regulated* is encoded by $-$. From this modeling, we obtain an alphabet $\mathfrak{X} = \{+, \circ, -\}$ and a microarray is then represented by a *ternary* word over $\mathfrak{X}$. And a collection of microarrays is represented as the set of ternary words. The encoding by symbol sequences permits the analysis of the microarrays by using *ternary* weighted trees. These trees provide a visual of a set of data and also are tools to classify enormous masses of words according to the overlap degree of their prefixes. Thus, these structures open a new way to examine the data mining step in the process of knowledge discovery in databases to understand general characteristics of DNA microarray data. In other words, they permit the extraction of hidden biological and medical informations from the mass of this data. They also address the automatic learning and pattern recognition problems. This paper describes two manners of hierarchical cluster analysis of DNA microarray data, using weighted prefix trees, includes clustering the profile of microarrays and clustering genes expression profiles. The next section recalls elements of words and languages. Section 3 presents weighted prefix trees. Section 4 describes the experimental results on a DNA microarray data of *breast cancer* cells.

2 Elements of Words and Languages

2.1 Words and Precoding of Words

Let $\mathcal{X} = \{x_1, \ldots, x_m\}$ be an alphabet of the size m. A w is a sequence. The lenght of the *word* w over $\mathcal{X}$ is $|w|$. In particular, the empty word is denoted by ε. The set $\mathcal{X}^*$ is the set of words over $\mathcal{X}$. For any $h \geq 0$, we denote $\mathcal{X}^h$ the set $\{w \in \mathcal{X}^*, \text{s.t.} |w| = h\}$. The *concatenation* of word $x_{i_1} \ldots x_{i_k}$ and $x_{j_1} \ldots x_{j_l}$ is the word $x_{i_1} \ldots x_{i_k} x_{j_1} \ldots x_{j_l}$. Equipped with the concatenation product, $\mathcal{X}^*$ is a monoid whose element neuter is the empty word ε. A word u (resp. v) is called *prefix*, or *left factor* (resp. *suffix*, or *right factor*) of w if $w = uv$. For any $u, v \in \mathcal{X}^*$, let a be the *longest left common factor* of u and v, i.e $u = au'$ and $v = av'$. For $x_i \in \mathcal{X}$, let $\text{precod}(x_i) = i$ be the precoding of x_i, for $i = 1, \ldots, m$. The *precoding* $\text{precod}(w)$ of w in base $m = \text{Card}\,\mathcal{X}$ is defined as $\text{precod}(\varepsilon) = 0$ and $\text{precod}(w) = m\,\text{precod}(u) + \text{precod}(x)$, if $w = ux$, for $u \in \mathcal{X}^*$ and $x \in \mathcal{X}$.

2.2 Languages

Let $\mathcal{L}$ be a language containing N words over $\mathcal{X}^*$. For $u \in \mathcal{X}^*$, let us consider $N_u = \text{Card}\{w \in \mathcal{L} | \exists v \in \mathcal{X}^*, w = uv\}$, in particular $N_\varepsilon = N$. Thus, for $u \in \mathcal{L}$,

$$N_u = \sum_{x \in \mathcal{X}} N_{ux} \text{ and for any } h \geq 0, N = \sum_{u \in \mathcal{X}^h} N_u. \qquad (1)$$

Let $\mu : \mathcal{L} \to \mathbb{N}$ be the *mass* function defined as $\mu(u) = N_{x_{i_1}} + \cdots + N_{x_{i_h}}$, for all $u = x_{i_1} \ldots x_{i_h} \in \mathcal{L}$. For $u \in \mathcal{X}^*$, let us consider also the ratios $P_u = N_u/N$, in particular $P_\varepsilon = 1$. For $u \in \mathcal{X}^*, x_i \in \mathcal{X}$, to simplify the notation, let

$$p = \text{precod}(u), \quad q_i = \text{precod}(ux_i) = mp + \text{precod}(x_i). \qquad (2)$$

and we consider the ratios

$$P_{p,q_i} = N_{ux_i}/N_u, \text{ for } i = 1, \ldots, m. \qquad (3)$$

By the formula (1), since $\sum_{x_i \in \mathcal{X}} N_{ux_i} = N_u$ then the ratios P_{p,q_i} define the *discrete probability* over $\mathcal{X}^*$: $0 \leq P_{p,q_i} \leq 1$ and $\sum_{u \in \mathcal{X}^*, x_i \in \mathcal{X}} P_{p,q_i} = 1$. For $u = x_{i_1} \ldots x_{i_h} \in \mathcal{X}^h$, the *appearance probability* of u is computed by

$$P_u = P_{q_{i_0}, q_{i_1}} \ldots P_{q_{i_{h-1}}, q_{i_h}}. \qquad (4)$$

Note that $\sum_{w \in \mathcal{L}} P_w = 1$ and for any $h \geq 0, \sum_{u \in \mathcal{X}^h} P_u = 1$.

2.3 Rearrangement of Language

Let $\mathcal{L} = \{w_1, \ldots, w_N\}$ be the language such that $|w_i| = L$, for $i = 1, \ldots, N$. Let $\mathfrak{S}_L$ denote the set of permutations over $[1, \ldots, L]$. Let $\sigma \in \mathfrak{S}_L$ and let $w = x_{i_1} \ldots x_{i_L}$. Then $\sigma w = x_{\sigma(i_1)} \ldots x_{\sigma(i_L)}$. We extend this definition over $\mathcal{L}$ as $\sigma \mathcal{L} = \{\sigma w\}_{w \in \mathcal{L}}$. There exists a permutation $\sigma \in \mathfrak{S}_L$ such that $\sigma w_1 = av_1, \ldots, \sigma w_N = av_N$, where

$v_1, \ldots, v_N \in \mathcal{X}^*$ and a is the *left longest factor* of $\mathcal{L}$. The σ is not unique. The Rearrangement($\mathcal{L}$) algorithm is proposed in [9] as : for $h \leq L$ and for $x \in \mathcal{X}$, let $n_h(x)$ be the number of letters x in the position h of the words in $\mathcal{L}$ and let $n_h = \max_{x \in \mathcal{X}} n_h(x)$ and $\sum_{x \in \mathcal{X}} n_h(x) = N$. One rearranges then $n_1, \ldots, n_L$ appearing in the order $n_{\sigma(1)} \geq \cdots \geq n_{\sigma(L)}$ by *sorting* algorithm [2].

Example 1. Let $\mathcal{L} = \{++-\circ, \circ++\circ, -++\circ\}$. One has $n_2(+) = n_4(\circ) = 3 > n_3(+) = 2 > n_1(+) = n_1(\circ) = n_1(-) = 1$, so one permutes the position 1 and 4. One obtains $\sigma = \begin{pmatrix} 1 & 2 & 3 & 4 \\ 4 & 2 & 3 & 1 \end{pmatrix}$ and $\sigma\mathcal{L}$ has $a = \circ +$ as a longest common prefix.

3 Weighted Prefix Trees [9]

3.1 Counting Prefix Trees

Let $\mathcal{L} \subseteq \mathcal{X}^*$ be a language. The *prefix tree* $\mathcal{A}(\mathcal{L})$ associated to $\mathcal{L}$ is usually used to optimize the storage of $\mathcal{L}$ and defined as follows

- the *root* is initial node which contains the empty word ε,
- the set of the nodes corresponds to the prefixes of $\mathcal{L}$,
- the set of the terminal nodes represents $\mathcal{L}$,
- the transitions are of form $\mathrm{precod}(u) \xrightarrow{x} \mathrm{precod}(ux)$, for $x \in \mathcal{X}, u \in \mathcal{X}^*$.

Equipped with the number N_{ux} as defined in (1), the prefix tree $\mathcal{A}(\mathcal{L})$ becomes a *counting tree*. To enumerate the nodes of a tree, we use the precoding of words defined in (2). The *internal* nodes $p = \mathrm{precod}(u)$ associated to prefix u are nodes such that $N_u \geq 2$ and the *simple internal* nodes q are nodes such that $N_u = 1$. Thus, an internal node $p = \mathrm{precod}(u)$ corresponds to N_u words starting with the same prefix u stored in sub-tree p. The counting tree of the language $\mathcal{L}$ is constructed by Insert($\mathcal{L}, \mathcal{A}$) algorithm in [8,9]. By this construction, the transitions between the nodes on a counting tree have the form : for $x \in \mathcal{X}$ and $u \in \mathcal{X}^*, \mathrm{precod}(u) \xrightarrow{x, N_{ux}} \mathrm{precod}(ux)$. Counting prefix trees permits the comparison of all words of $\mathcal{L}$ according to the mass of their prefixes as defined in section 2.2. We proposed the Characteristic-Words($\mathcal{A}(\mathcal{L})$) algorithm in [8,9], which returns the words having the maximum number of occurences, to extract *characteristic words*.

3.2 Probabilistic Prefix Trees

To compute the appearance probability of an output word over an alphabet $\mathcal{X}$, we introduce the *probabilistic tree*. Augmented with the probability $P_{p,q}$ defined in (3), the counting tree $\mathcal{A}(\mathcal{L})$ becomes a probabilistic tree. The labelled probability is estimated by the maximum likelihood method (see (3)). It is the conditional probability that the word w accepts the common prefix ux knowing the common prefix u. The transitions between the nodes on a probabilistic tree have the form : for $x \in \mathcal{X}$ and $u \in \mathcal{X}^*, p = \mathrm{precod}(u) \xrightarrow{x, P_{p,q}} q = \mathrm{precod}(ux)$. The appearance probability of a word is computed by (4).

Example 2. Let $\mathcal{L} = \{+-+\circ, ++\circ\circ, ++\circ\circ, ++\circ\circ, +-+\circ, +-+\circ, ++\circ\circ, ++-\circ\}$ be the language over $\mathfrak{X} = \{+, \circ, -\}$, we have the weighted trees as below.

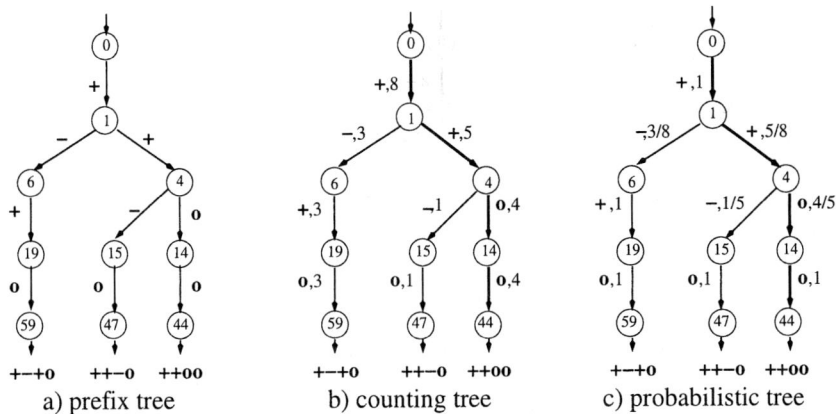

Fig. 1. Weighted trees associated to $\mathcal{L}$. The sequence of nodes $0 \to 1 \to 4 \to 14 \to 44$ represents the characteristic word $++\circ\circ$ with the maximum mass

3.3 Trees Having Longest Prefix

Consider the schema as follows

$$\mathcal{A}(\mathcal{L}) \longleftrightarrow \mathcal{L} \xrightarrow{\sigma} \sigma\mathcal{L} \longleftrightarrow \mathcal{A}(\sigma\mathcal{L}),$$

where $\mathcal{A}(\mathcal{L})$ (resp. $\mathcal{A}(\sigma\mathcal{L})$) is the tree associated with $\mathcal{L}$ (resp. $\sigma\mathcal{L}$). Where the tree $\mathcal{A}(\sigma\mathcal{L})$ represents the longest prefix of $\sigma\mathcal{L}$ (see section 2.3).

Example 3. Let $\mathcal{L}$ given in Example 2 and let $\sigma = \begin{pmatrix} 1 & 2 & 3 & 4 \\ 1 & 4 & 2 & 3 \end{pmatrix}$. One presents these trees in the Figure 2.

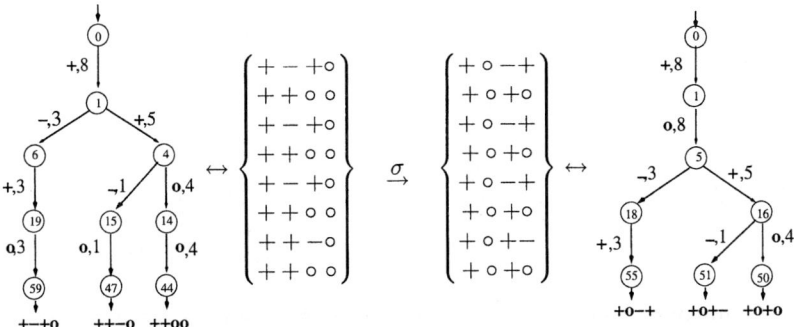

Fig. 2. Schema obtaining the tree $\mathcal{A}(\sigma\mathcal{L})$

We have proposed the Longest-Prefix-Tree($\mathcal{L} \cup w$) algorithm to insert a word into a longest prefix tree and to give a new longest prefix tree [9].

4 Experimental Results

4.1 Descriptions

We employ the weighted trees, to organize microarrays (resp. genes) expression profiles into different clusters such that each cluster contains all microarrays (resp. genes) which represents a highly similar expression in degree of overlap. Here, we describe the cluster according to two manners including 1) the cluster analysis of DNA microarrays for searching common profile of microarray data; 2) the cluster analysis of gene expression profiles to search common expression profile of genes. Consider a collection of L genes across in N different measure experiments. The gene expression profiles or the gene expression patterns is the matrix $E = \left(\log_2 I_{R_{ij}} / I_{G_{ij}} \right)_{1 \leq i \leq L, 1 \leq j \leq N}$, where $I_{R_{ij}}$ (resp. $I_{G_{ij}}$) is the luminous intensity of the red (resp. green) fluorescence dye of spot i in experiment j. By symbolic represention, the gene expression profiles is represented as a language of L words of length N over alphabet $\{+, \circ, -\}$. In the same way, the profile of microarray data is the transposition of the matrix E, and the profile of microarray data is representated by a language of N words of lenght L. These two languages are implemented by use of *longest prefix tree* permitting automatically to classify profile of microarrays (resp. genes) according to common expression levels as Figure 3, where the longest common expression indicates the similarity between microarrays (resp. genes). In the case of gene expression profiles (Section 4.3), the prefixe indicate also the *co-regulated* genes. This method returns then the hierarchical clustering using weighted trees which is an unsupervised learning technique and thus it does not requires *a priori* knowledge of cluster number before clustering. This criterion is important in DNA microarray data analysis since the characteristics of the data are often unknown.

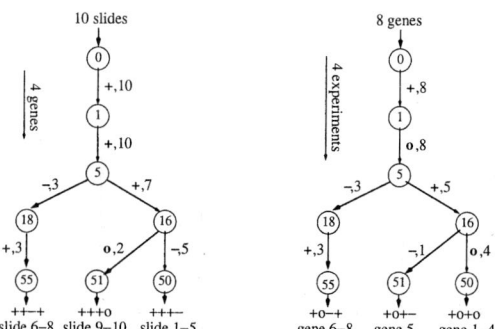

Fig. 3. Counting trees of microarray data. Two sub-trees associated to node 5 define two clusters corresponding. The process of nodes $0 \to 1 \to 5 \to 16 \to 50$ represents 5 microarrays (resp. 4 genes) having the same expression levels $+ + +-$ (resp. $+ \circ + \circ$) with maximum degree of overlap

4.2 Clustering of DNA Microarrays

As an example, we analyze the data of 77 microarrays of 9216 genes of *breast cancer* cells coming from 77 patients available at the website http://genome-www.stanford.edu/breast-cancer/. The cluster analysis and the characteristic microarrays were represented in the Figures 4, 5, 6 and Table 1.

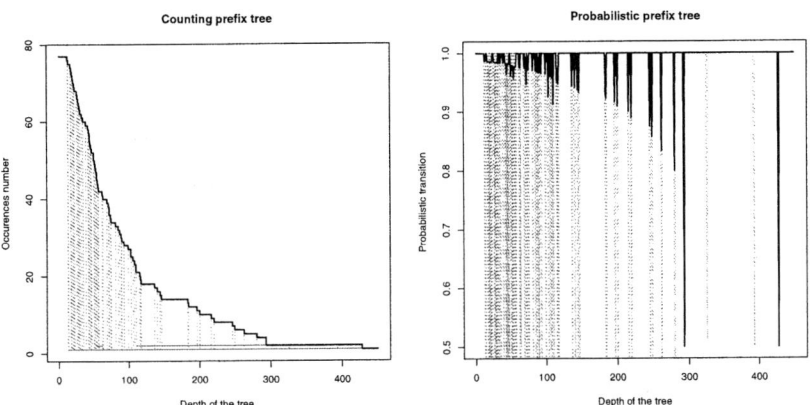

Fig. 4. Counting and probabilistic tree. Each depth of tree gives the number (probability) of microarrays having the common expression. The microarrays having a maximal common expression are represented by a dark line

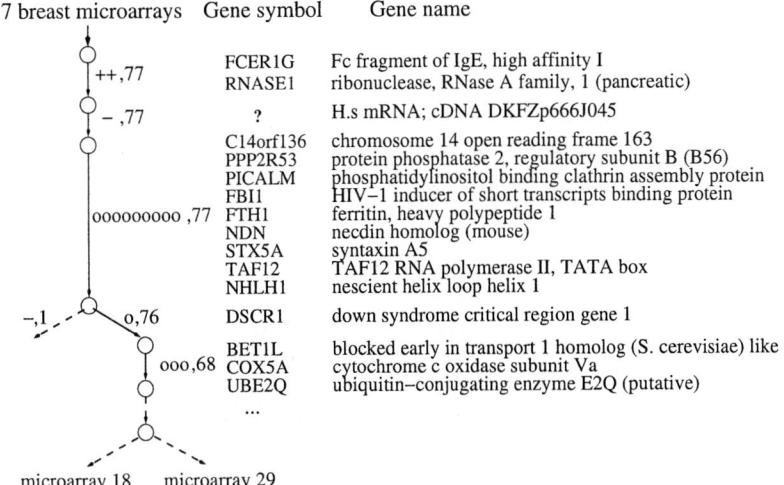

Fig. 5. Visual of microarrays. The 12 first depth of counting tree includes two genes FCER1G, RNASE1 are *up-regulated*, one gene is *down-regulated* and 9 genes are *no-regulated* on 77 experiments, etc... These three first genes can be viewed as the group of candidate genes of breast cancer microarray

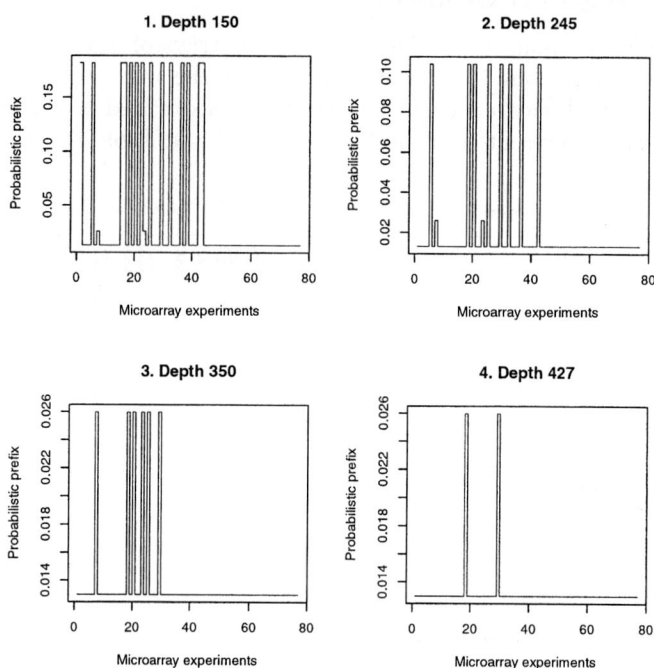

Fig. 6. Appearance probability of prefixes. At the depth **427** there are two microarrays, 18 and 29, with coincident probability of **0.026**. It can be considered as the *characteristic microarrays*

Table 1. Group of similar microarrays. At depth 245 there are 4 groups of microarray with highly degree of overlap. Microarray 18 and 29 are most similarity : there are 427 first genes having the same expression levels in which 35 genes are *up-regulated*, 377 genes are *no-regulated* and 15 genes are *down-regulated*

cluster	slideID	overlap	degree $(+, \circ, -)$
1	18	427	(35,377,15)
	29	427	(35,377,15)
2	20	392	(30,350,12)
	25	392	(30,350,12)
3	7	324	(29,284,11)
	23	324	(29,284,11)
4	5	261	(25,266,10)
	42	261	(25,266,10)
	32	245	(25,210,10)
	36	248	(25,213,10)

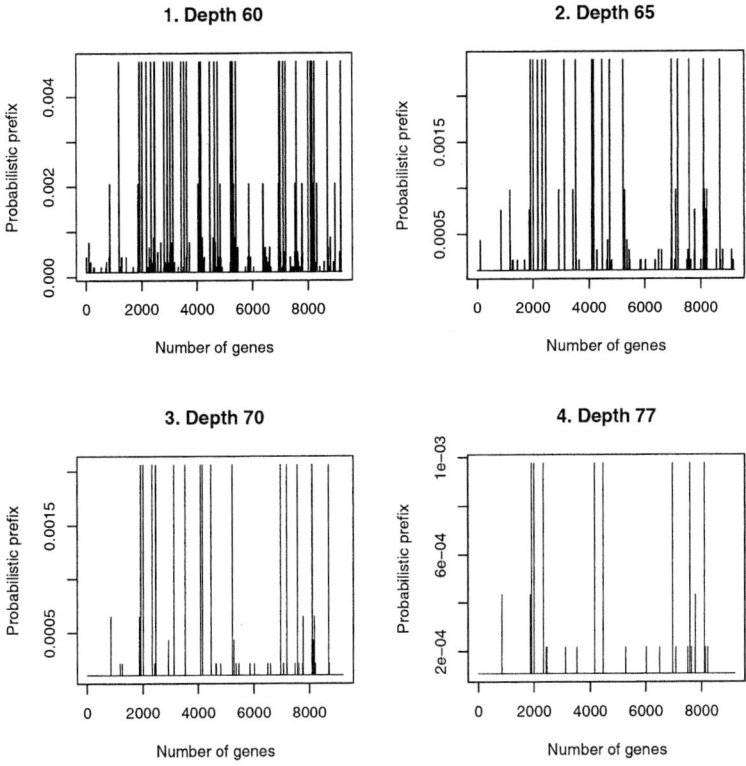

Fig. 7. Cluster of gene expression profiles. The graphs represent the appearance probability of expression prefixes of genes. The weighted tree gives the genes with the maximum degree of overlap of *co-upregulated* and *co-downregulated* expression levels which are represented in Table 2

4.3 Clustering Gene Expression Profiles

The Figure 7 and Table 2 give the cluster analysis of gene expression profiles.

4.4 Observations and Notes

There are two groups of *co-regulated* genes. In Table 2, the maximum *co-upregulated* (resp. *co-downregulated*) genes are represented on the left (resp. right) table. Each cluster represents the maximum co-regulated genes with the corresponding microarray identity : the cluster A (resp. a) of Table 2 represents 2 up-regulated (resp. 1 down-regulated) genes over 77 microarrays corresponding to three first depths of Figure 5. These maximum co-regulated genes could be then considered as characteristic genes of breast microarrays. These results permit also to isolate the groups of *no-regulated* genes : from the depth 4 to the depth 12,

Table 2. Groups of the co-regulated genes. Left table represents 5 clusters of 21 co-upregulated genes. Right table represents 4 clusters of 21 co-downregulated genes

cluster	gene symbol	overlap	(+, ○, −)	cluster	gene symbol	overlap	(+, ○, −)
A	FCER1G	77	(77,0,0)	a	?	77	(0,0,63)
	RNASE1	77	(77,0,0)		DLG7	63	(0,0,63)
B	C1orf29	58	(58,0,0)		AHSG	63	(0,0,63)
	G1P2	58	(58,0,0)		RPS6KA3	63	(0,0,63)
	MGP	58	(58,0,0)		GASP	63	(0,0,63)
	ISLR	58	(58,0,0)	b	MPO	49	(0,0,49)
	C1QG	58	(58,0,0)		HBZ	49	(0,0,49)
	SPARCL1	58	(58,0,0)		SERPINE2	49	(0,0,49)
C	HLA-DQA2	50	(50,0,0)		DHFR	49	(0,1,48)
	IGHG3	50	(50,0,0)		ESDN	49	(0,1,48)
	?	50	(50,0,0)	c	FGA	45	(0,1,44)
	CD34	50	(50,0,0)		LGALS4	45	(0,0,45)
	C1QB	50	(49,1,0)		SLCA5	45	(0,0,45)
D	SFRP4	41	(41,0,0)	d	APOB	29	(0,0,29)
	FCGR3A	41	(41,0,0)		GPA33	29	(0,0,29)
E	MS4A4A	32	(32,0,0)		MAL	29	(0,0,29)
	FBLN1	32	(32,0,0)		ARD1	29	(0,1,28)
	FLJ27099	32	(32,0,0)		ETV4	29	(0,1,28)
	FCGR2A	32	(32,0,0)		ASGR	29	(0,1,28)
	CTSK	32	(32,0,0)		APOH	29	(0,1,28)
	FGL2	32	(32,0,0)		AFP	29	(0,1,28)

there are 9 no-regulated genes over 77 microarrays and at the depth 5500 of the counting tree there are 4348 no-regulated genes (48%) in least 50 microarrays (65%).

5 Conclusions

We used the weighted prefix trees to examine the data mining in the process of knowledge discovery in DNA microarray data. The hierarchical clustering using weighted trees gives a tool to cluster gene expression microarray data. The longest prefix tree is used to establish the characteristic genes and/or characteristic microarrays that have the longest common expression and the maximum degree of overlap. It permits also to determine the groups of candidate (and no-regulated) genes of pathologic condition. We anticipate that with further refinement these methods may be extremely valuable in analysing the mass of DNA microarray data, with possible significant clinical applications. In addition to application on microarrays, weighted prefix tree could be also used to explore other kinds of genomic data and they are pontentially usefull in other classification problems.

Acknowledgements. Many thanks to J. Soula for help in the data conversion.

References

1. M.Crochemore and al. : Algorithmique du texte, Vuibert Informatique, 2001
2. P.Flajolet and R.Sedgewick : An introduction to the analysis of algorithms, Addison-Wesley, 1996
3. Jonathan R.Pollack and al. : *Microarray analysis reveals a major direct role of DNA copy number alteration in the transcriptional program of human breast tumors*, Proc. Natl. Acad. Sci. USA, 2002
4. M.Eisen, David Botstein and al. : *Cluster analysis and display genome-wide expression patterns*, Stanford, 1998
5. M.Brown, W.Grundy and al. : *Knowledge-based analysis of microarray gene expression data by using support vector machines*, University of California, 1999
6. Inaki Inza and al. : *Filter versus wrapper gene selection approaches in DNA microarray domains*, Artificial Intelligence in Medecine, Elsevier, 2004
7. P.Walker, A. Famili and al. : *Data mining of gene expression changes in Alzheimer brain*, Artificial Intelligence in Medecine, Elsevier, 2004
8. Tran Trang, Nguyen Cam Chi, Hoang Ngoc Minh, al. : *The management and the analysis of DNA microarray data by using weighted trees*, Hermes Publishing, 2004
9. Tran Trang, Nguyen Cam Chi, Hoang Ngoc Minh : *Management and analysis of DNA microarray data by using weighted trees*, to appear in Journal of Global Optimization : Modeling, Computation and Optimization in Systems Engineering

Automatic Extraction of Low Frequency Bilingual Word Pairs from Parallel Corpora with Various Languages

Hiroshi Echizen-ya[1], Kenji Araki[2], and Yoshio Momouchi[1]

[1] Dept. of Electronics and Information, Hokkai-Gakuen University, S26-Jo, W11-Chome, Chuo-ku, Sapporo, 064-0926 Japan
{echi, momouch}@eli.hokkai-s-u.ac.jp
[2] Graduate School of Information Science and Technology, Hokkaido University, N14-Jo, W9-Chome, Kita-ku, Sapporo, 060-0814 Japan
araki@media.eng.hokudai.ac.jp

Abstract. In this paper, we propose a new learning method for extraction of low-frequency bilingual word pairs from parallel corpora with various languages. It is important to extract low-frequency bilingual word pairs because the frequencies of many bilingual word pairs are very low when large-scale parallel corpora are unobtainable. We use the following inference to extract low frequency bilingual word pairs: the word equivalents that adjoin the source language words of bilingual word pairs also adjoin the target language words of bilingual word pairs in local parts of bilingual sentence pairs. Evaluation experiments indicated that the extraction rate of our system was more than 8.0 percentage points higher than the extraction rate of the system based on the Dice coefficient. Moreover, the extraction rates of bilingual word pairs for which the frequencies are one and two respectively improved 11.0 and 6.6 percentage points using AIL.

1 Introduction

Use of parallel corpora with various languages is effective to build dictionaries of bilingual word pairs because bilingual sentence pairs that are pairs of source language (SL) sentences and target language (TL) sentences include natural equivalents and novel equivalents. Moreover, it is important to extract low-frequency bilingual word pairs because the frequencies of many bilingual word pairs are extremely low when large-scale parallel corpora are unobtainable. Consequently, systems based on similarity measures [1,2] fall into the sparse data problem because bilingual word pair candidates with close similarity value increase when many low-frequency bilingual word pairs exist.

From the perspective of learning [3], we propose a new method for extraction of low-frequency bilingual word pairs from parallel corpora. We call this new learning method **A**djacent **I**nformation **L**earning (AIL). The AIL is based on the inference that the equivalents of the words that are adjacent the SL words of bilingual word pairs also adjoin the TL words of bilingual word pairs in local parts of bilingual sentence pairs. Our method easily acquires such adjacent information solely from parallel corpora. Moreover, our system can extract not only high-frequency bilingual word pairs, but also low-frequency bilingual word pairs, which typically have the sparse data problem. Thereby, our system can limit the search scope for the decision of equivalents in bilingual sentence pairs.

Evaluation experiments using five kinds of parallel corpora indicated that the extraction rate of our system using AIL was more than 8.0 percentage points higher than the extraction rate of a system based on the Dice coefficient. Moreover, the extraction rate of bilingual word pairs for which the frequencies are one and two respectively improved 11.0 and 6.6 percentage points using AIL. We thereby confirmed that our method is effective to extract low-frequency bilingual word pairs efficiently.

2 Outline

Our system consists of four processes: a method based on templates, a method based on two bilingual sentence pairs, a decision process of bilingual word pairs, and a method based on similarity measures.

First, the user inputs SL words of a bilingual word pair. In the method based on templates, the system extracts bilingual word pairs using the bilingual sentence pairs, the templates, and the SL words. In this paper, templates are defined as rules to extract new bilingual word pairs. Similarity between SL words and TL words is determined in all extracted bilingual word pairs. In the method based on two bilingual sentence pairs, the system obtains bilingual word pairs and new templates using two bilingual sentence pairs and the SL words. Similarity is determined in all templates. Moreover, during the decision process of bilingual word pairs, the system chooses the most suitable bilingual word pairs using their similarity values from among all extracted bilingual word pairs. The system then compares similarity values of chosen bilingual word pairs with a threshold value. Consequently, the system registers the chosen bilingual word pairs to the dictionary for bilingual word pairs when their similarity values are greater than the threshold value.

The system extracts bilingual word pairs using the Dice coefficient with bilingual sentence pairs and the SL words in the method based on similarity measures. It does so when their similarity values are not over the threshold or when no bilingual word pairs are extracted.

3 Extraction Process of Bilingual Word Pairs

3.1 Method Based on Two Bilingual Sentence Pairs

In the method based on two bilingual sentence pairs, the system obtains bilingual word pairs and templates using two bilingual sentence pairs. Details of the method based on two bilingual sentence pairs are the following:

(1) The system selects bilingual sentence pairs for which the SL words exist.
(2) The system compares the bilingual sentence pairs selected by process (1) with other bilingual sentence pairs in the parallel corpus. The system selects those bilingual sentence pairs that have the same word strings as those adjoining the SL words, *i.e.*, the common parts, and those that have parts in common with TL sentences.
(3) The system extracts the TL words that correspond to the SL words using the common parts from the bilingual sentence pairs selected through process (1). When the system uses the common parts that exist near words at the beginning of a sentence, it extracts, from the TL sentence, those parts between words at the beginning of a sentence and words that adjoin the left sides of the common parts. When the system uses the common parts that exist near words at the end of a sentence, it extracts, from the TL sentence, those parts between words that adjoin the right sides of common parts and words at the end of a sentence. When the system uses several common parts, it extracts, from the TL sentence, those parts between the two common parts.
(4) The system only selects parts that are nouns, verbs, adjectives, adverbs, or conjunctions.
(5) The system calculates the similarity values between the SL words and the parts selected by process (4) using the Dice coefficient [1].
(6) The system replaces the extracted bilingual word pairs with variables in bilingual sentence pairs.
(7) The system acquires templates by combining common parts and variables.
(8) The system calculates the similarity values between SL words and TL words in the acquired templates using the Dice coefficient; it registers the templates to the template dictionary.

Figure 1 shows an example of acquisition of template: (by @; @ *de*) is acquired as the template. The system replaces the SL word "air mail" and the TL word "*koukubin*" with the variable "@" in bilingual sentence pair by process (6). In this case, "by" and "*de*" are common parts between two bilingual sentence pairs. Consequently, the system obtains (by @; @ *de*) as a template by combining "by @" and "@ *de*." In this paper, the parts extracted from SL sentences are called SL parts; the parts extracted from TL sentences are called TL parts.

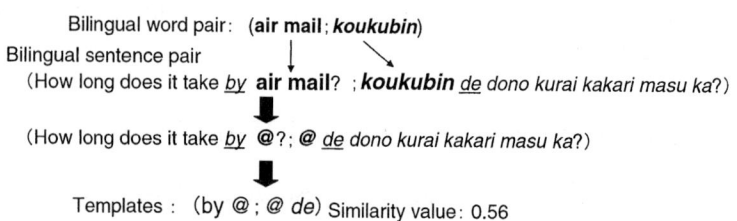

Fig. 1. An example of template acquisition

3.2 Method Based on Templates

In the method based on templates, the system extracts bilingual word pairs using the acquired templates. Details of the extraction process of bilingual word pairs using templates are the following:

(1) The system selects bilingual sentence pairs for which the SL words exist.
(2) The system compares the bilingual sentences selected by process (1) with the templates in the dictionary. Subsequently, the system selects the templates for which SL parts have the same parts as those adjoining the SL words, and for which TL parts have the same parts as those in TL sentences.
(3) The system extracts TL words that correspond to the SL words. The system extracts words that adjoin the left sides of common parts from TL sentences when variables exist on the left sides in TL parts of templates. The system extracts words that adjoin the right sides of common parts from TL sentences when variables exist on the right sides in TL parts of templates.
(4) The system calculates similarity values between the SL words and the parts extracted from TL sentences using the Dice coefficient.

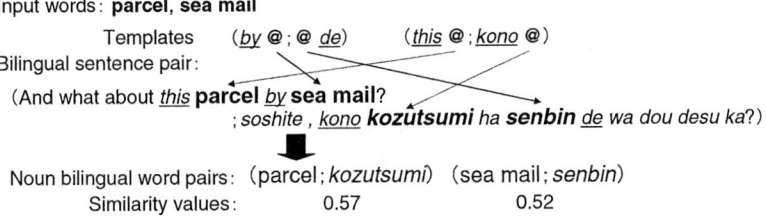

Fig. 2. Examples of extraction of bilingual word pairs

Figure 2 shows examples of extraction of bilingual word pairs from English – Japanese bilingual sentence pairs. In Fig. 2, (parcel;*kozutsumi*) and (sea mail;*senbin*) are extracted respectively as the noun bilingual word pairs using the templates (by @;@ *de*) and (this @;*kono* @). The template (by @;@ *de*) has information that equivalents of words, which adjoin the right side of "by", exist on the left side "*de*" in TL sentences. This fact indicates that the acquired templates have bilingual knowledge that can be used to process the differing word orders of SL and TL. Moreover, our system using AIL can extract bilingual word pairs efficiently without depending on the frequencies of bilingual word pairs using the templates.

3.3 Decision Process of Bilingual Word Pairs and the Method Based on Similarity Measures

In the decision process of bilingual word pairs, the most-suitable bilingual word pairs are selected according to similarity values when several bilingual word pairs are extracted. The extracted bilingual word pairs are sorted so that bilingual word pairs with the largest similarity values are ranked highest.

Moreover, in the method based on similarity measures, the system extracts bilingual word pairs using only the Dice coefficient without AIL when the similarity values are not greater than a threshold value or when no bilingual word pairs are extracted.

4 Performance Evaluation and Conclusion

Five kinds of parallel corpora were used in this paper as experimental data. These parallel corpora are for English – Japanese, French – Japanese, German – Japanese, Shanghai-Chinese – Japanese and Ainu – Japanese. They were taken from textbooks containing conversational sentences. The number of bilingual sentence pairs was 1,794. We inputted all SL words of nouns, verbs, adjectives, adverbs, and conjunctions to our system using AIL and the system based on the Dice coefficient, respectively. The initial conditions of all dictionaries are empty. Moreover, our system using AIL uses 0.5^1 as its best threshold value. We repeated the experiments for each parallel corpus using each system. We evaluated whether correct bilingual word pairs are obtained or not, and calculated the extraction rate for all SL words.

Experimental results indicated that the extraction rate of our system using AIL was more than 8.0 percentage points (from 52.1% to 60.1%) higher than that of the system based on the Dice coefficient. Moreover, in each parallel corpus, the extraction rates improved using AIL. Therefore, our method is effective when using parallel corpora of various languages.

Tables 1 and 2 show extraction rate details in our system using AIL and the system based on the Dice coefficient. In Tables 1 and 2, the extraction rates of the bilingual word pairs for which the frequencies are one and two respectively improved 11.0 and 6.6 percentage points using AIL. This result verified that our system using AIL can extract low-frequency bilingual word pairs efficiently.

In related works, K-vec [4] is applied only to bilingual word pairs for which the frequencies are greater than three. Therefore, it is insufficient in terms of extraction of low-frequency bilingual word pairs. In one study [5] that acquired low-frequency bilingual terms, a bilingual dictionary and MT systems were used for measuring similarity. Therefore, it is difficult to deal with various languages because of the use of large-scale translation knowledge. On the other hand, one study [6] that uses the co-occurrence of words depends on the number of co-occurrence words and their frequency. Therefore, such a method is insufficient in terms of efficient extraction of bilingual word pairs. In contrast, AIL merely requires a one-word string as the co-occurrence word, e.g., only "by" and "this", as shown in Fig. 2. Moreover, AIL can extract bilingual word pairs even when the frequencies of the pairs of the co-occurrence words and the bilingual word pairs are only one. In Fig. 2, the respective frequencies of "by sea mail" and "this

---

[1] This value was obtained through preliminary experiments.

Table 1. Details of extraction rates in our system using AIL

Frequency	English	French	German	Sh.-Chinese	Ainu	Total	Number of bilingual word pairs
1	46.4%	49.4%	51.3%	49.1%	56.9%	**50.4%**	681
2	71.4%	80.0%	71.4%	90.7%	74.4%	**78.6%**	168
others	89.7%	73.5%	79.2%	82.1%	61.5%	75.4%	232
Total	58.0%	56.7%	61.0%	62.9%	61.5%	60.1%	1,081

Table 2. Details of extraction rates in the system based on the Dice coefficient

Frequency	English	French	German	Sh.-Chinese	Ainu	Total	Number of bilingual word pairs
1	35.7%	37.5%	39.5%	40.0%	45.0%	**39.4%**	681
2	64.3%	80.0%	67.9%	74.4%	71.8%	**72.0%**	168
others	89.7%	73.5%	79.2%	83.9%	58.5%	75.0%	232
Total	49.7%	47.9%	53.3%	54.9%	54.0%	52.1%	1,081

parcel", which are pairs formed by the co-occurrence of words and the SL words of bilingual word pairs, are only one. The method [7] that acquires templates requires many similar bilingual sentence pairs to extract effective templates.

Future studies will apply this method to a multilingual machine translation system.

References

1. Manning, C. D. and Schütze, H. 1999. Foundations of Statistical Natural Language Processing. The MIT Press.
2. Smadja, F., K. R. McKeown and V. Hatzivassiloglou. 1996. Translating Collocations for Bilingual Lexicons: A Statistical Approach. *Computational Linguistics*, vol.22, no.1, pp.1–38.
3. Echizen-ya, H., K. Araki, Y. Momouchi, and K. Tochinai. 2002. Study of Practical Effectiveness for Machine Translation Using Recursive Chain-link-type Learning. In *Proceedings of COLING '02*, pp.246–252.
4. Pedersen, T. and N. Varma. 2003. K-vec++: Approach For Finding Word Correspondences. Available at . http://www.d.umn.edu/ tpederse/Code/Readme.K-vec++.v02.txt
5. Utsuro, T., K. Hino, and M. Kida. 2004 Integrating Cross-Lingually Relevant News Articles and Monolingual Web Documents in Bilingual Lexicon Acquisition. In *Proceedings of COLING'04*, pp.1036–1042.
6. Tanaka, K and H. Iwasaki 1996. Extraction of Lexical Translation from Non-Aligned Corpora. In *Proceedings of COLING'96*, pp.580–585.
7. McTait, K. and A. Trujillo. 1999. A Language-Neutral Sparse-Data Algorithm for Extracting Translation Patterns. In *Proceedings of TMI'99*, pp.98–108.

A Kernel Function Method in Clustering[1]

Ling Zhang, Tao Wu, and Yanping Zhang

Artificial Intelligence Institute Anhui University, Anhui, China
zling@ahu.edu.cn

Abstract. Cluster analysis is one of main methods used in data mining. So far there have existed many cluster analysis approaches such as partitioning method, density-based, k-means, k-nearest neighborhood, etc. Recently, some researchers have explored a few kernel-based clustering methods, e.g., kernel-based K-means clustering. The new algorithms have demonstrated some advantages. So it's needed to explore the basic principle underlain the algorithms such as whether the kernel function transformation can increase the separability of the input data in clustering and how to use the principle to construct new clustering methods. In this paper, we will discuss the problems.

Keywords: Data mining, clustering, supper-sphere, kernel function.

1 Introduction

Cluster analysis is one of main methods used in data mining. So far there have existed many cluster analysis approaches. For example, partitioning method [1][2], density-based [3][4], k-means [5], k-nearest neighborhood [4], neural networks [6], fuzzy clustering [7] etc. For each kind of clustering, the key is to define a specific metric to measure the similarity (or dissimilarity) among objects. So far various metrics have been adopted such as Euclidean distance, Manhattan distances, inner product, fuzzy membership function, etc. No matter what kind of measurement is used, in principle, there are basically two kinds: one for measuring the similarity between two objects (two data), the other for measuring the similarity between an object and a cluster (a set of data). It's known that the possible regions partitioned by a clustering are limited. For example, K-means algorithm [5] can only partition the data into elliptical regions. So it's hard to use these kinds of clustering to complex clustering.

Kernel-based methods have wisely been used in machine learning [8][9]. So far they were used to supervised learning (classification) mainly [10]. In kernel-based classification algorithms, the input data set is mapped into a high dimensional space by a kernel function. The basic principle underlain the algorithms is to increase the separability of the input data by the non-linear transformation. For example, the SVM is one of the well-known supervised learning algorithms [10]. In the algorithm, by

---

[1] Supported by National Nature Science Foundation of China (Grant No. 60135010), Chinese National Key Foundation Research Plan (2004CB318108) and Innovative Research Team of 211 Project in Anhui University.

using the kernel function transformation the input data become linearly separable on the new space whereas the same data set is non-linearly separable on the original space. Therefore, the SVM algorithms have demonstrated more efficiently. Recently some researchers have explored a few kernel-based unsupervised learning algorithms, e.g., kernel-based K-means clustering [8][11]. So it's needed to explore the basic principle underlain the algorithms such as whether the kernel function transformation can increase the separability of the input data in clustering and how to use the principle to construct new clustering methods. In this paper, we will discuss the problems.

2 Two Propositions

Definition 1: Given a space X, a set $D \subset X$ and a criteria G, the clustering problem can be stated as follows: find a partition $P = \{C_1, C_2, ..., C_m\}$ of X such that the points within C_i belong to the same class and the partition is optimal in some sense under the criteria G. Set C_i is called a cluster (class) of D and P is the clustering of D.
Then we have the following propositions.

Proposition 1: Given a metric space X. Assume that cluster C consists of n super-spheres $S_1, S_2, ..., S_n$ with r as their radius and $x_1, x_2, ..., x_n$ as their centers, respectively. There exists a kernel function K and its corresponding map ϕ. Space X is mapped into a feature space Z by ϕ such that $\phi(C)$ can be approached by the intersection of a super-sphere B and $\phi(X)$ (see fig. 1,2).

Proof: We first prove that the trajectory of $f(x) = e^{-1}$ and the boundary of C are approximate.
 Construct a function $f_i(x) = \exp(-(x-x_i)^2/r^2), i = 1, 2, ..., n$.
 Let $f(x) = f_1(x) + f_2(x) + ,..., + f_n(x)$.
 Obviously, the solution of $f_i(x) = e^{-1}$ and that of $(x - x_i)^2 = r^2$ are equivalent, i.e., a super "spherical surface". Letting $f(x) = e^{-1}$, x falls into the boundary of C. Assume that x falls into the boundary $(x - x_i)^2 = r^2$ of S_i but does not fall into the inside of S_j. Omitting the effect of $f_j(x), j \neq i$, then $f(x) \approx e^{-1}$. Therefore, the super spherical surface $f(x) = e^{-1}$ can be approached by the boundary of set C that consists of n super-spheres. That is, C can represent set $\{x | f(x) \geq e^{-1}\}$ approximately.
 Secondly, we prove that $\phi(C)$ is the intersection of some super-sphere and $\phi(X)$ in space Z.
 Given a kernel function $K(x, y) = \exp(-(x-y)^2/r^2)$ and ϕ is its corresponding map. Then $\exp(-(x-y)^2/r^2) = <\phi(x), \phi_i(x)>$.

$$f(x) = f_1(x) + f_2(x) + ,..., + f_n(x) = \sum_i \exp(-(x-x_i)^2/r^2)$$
$$= \sum <\phi(x), \phi_i> = <\phi(x), \sum \phi_i(x)> \geq e^{-1}$$

Letting the point $\sum_i \phi(x_i)$ in the feature space Z be ϕ_0, then
$<\phi(x), \phi_0> \geq e^{-1}$ is a super-plane in Z with ϕ_0 as its normal in the feature space Z.
$<\phi(x), \phi_0> \geq e^{-1}$ is a half space S of Z. Then $\phi(C)$ falls into S.

Since $K(x,x) = 1$, $\phi(X)$ falls into a unit super-sphere of Z. The intersection of the unit super-sphere and the half space of Z is a super-sphere neighborhood. Therefore, $\phi(C)$ falls into the intersection of the super-sphere neighborhood and $\phi(X)$.

Proposition 2: X is a bounded Euclidean space and $D \subset X$. $P = \{C_1, C_2, ..., C_m\}$ is a clustering of D. There exists a kernel function K. X is mapped into a feature space Z by its corresponding map ϕ such that each $\phi(C_i)$ can be approached by the intersection of $\phi(X)$ and a super spherical surface with d as its radius in Z.

Proof: Since X is bounded, C_i is bounded as well. Let C_i' be the closure of C_i. C_i' is a bounded close and compact set in the Euclidean space. For C_i', construct a ε-cover $B = \{B(x, \varepsilon), x \in C_i'\}$. According to the bounded covering theorem in compact set, we may choose a limit number $\{B_1, B_2, ..., B_k\}$ of covers from B such that they cover C_i'. That is, C_i' can be represented by the union of a limit number of super-spheres approximately. From proposition 1, we prove proposition 2.

3 Kernel-Based Clustering

From the above propositions, we know that in clustering the separability of input data is also increased by using the kernel function transformation, since a simple sphere-like region can represent any cluster approximately in the new transformed space. Therefore, the principle can be used to improve the clustering algorithms.

As we known, some simple clustering algorithms such as K-means [5] can only partition the data into elliptical regions. It's difficult to use these kinds of algorithms to complex clustering problems. Kernel-based clustering algorithms can overcome the drawback, since a sphere-like region can represent any cluster in the transformed space in despite of the complex clustering problem.

In neural networks, we presented a constructive learning algorithm based on the same principle as presented in Section 2. We transform the training data into a high dimensional space. In the new space the data can be covered (partitioned) by a set of simple sphere neighborhoods easily. Therefore the algorithm is quite efficient [12][13].

4 Conclusion

By using kernel-based clustering, the input data are transformed into a new high dimensional space. In the new space, we can always use a simple supper-sphere to describe a cluster approximately no matter how complex the original cluster is. In data mining, each cluster generally represents a rule among a data set, this means that we can use less and simpler rules to describe the same data by using kernel-based clustering.

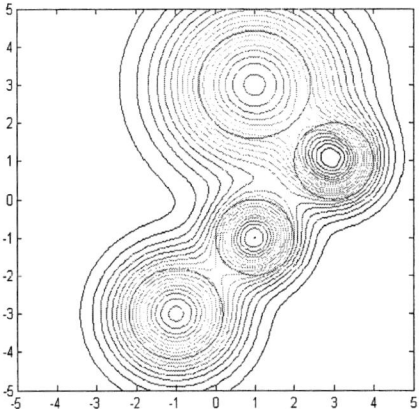

Fig 1. The contour lines

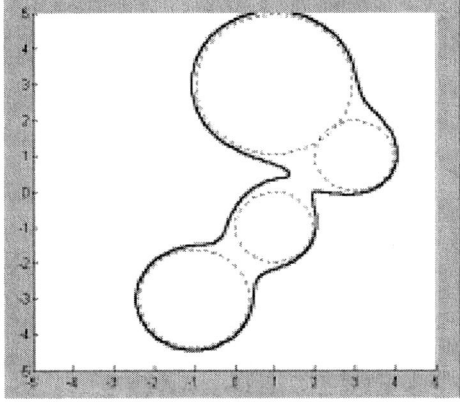

Fig. 2. A cluster

References

[1] MacQueen J, Some methods for classification and analysis of multivariate observations. Proc. 5th Berkley Symp. Math. Statist. Prob., 1967,1:281-297.
[2] C. J. Veenman et al, A maximum variance cluster algorithm, IEEE Trans. on PAMI, vol24, no9, Sept. 2002, 1273-1280.

[3] Jain A. K, Dubes R. C, Algorithms for clustering, Englewood, Cliffs N. J, Prentice Hall.1988.
[4] Jain A. K, Murry M. N, Flynn P. J, Data clustering: A survey, ACM Comput. Surve.,1999, 31: 264-323.
[5] Ester M, Kriegel HP, Sander J, Xu X.X, Density based algorithm for discovering clusters in large spatial databases with noise. In: Simoudis E, Han JW, Fayyad UM, eds., Proceedings of the 2nd International Conference on Knowledge Discovery and Data Mining, Portland : AAAI Press, 1996:226-231.
[6] J-H Wang, et al, Two-stage clustering via neural networks, IEEE Trans. on Neural Networks, vol14, no3, May 2003, 606-615.
[7] Frank Hoppner, Frank Klawonn, Rudolf Kruse and Thomas Runkler, *Fuzzy Cluster Analysis*, Wiley 1999.
[8] K-R Muller, et al, An introduction to kernel-based learning algorithms, *IEEE Trans. on Neural Networks*, vol.12, no. 2. March 2001, pp.181-201.
[9] B. Scholkopf, et al, Input space versus feature space in kernel-based methods, IEEE Trans. on Neural Networks, vol. 10, no. 5, Sept. 1999, pp.1000-1016.
[10] V. N. Vapnik, *Statistical Learning Theory*, John Wiley & Sons, Inc. 1998.
[11] M. Girolami, Mercer kernel-based clustering in feature space, IEEE Trans. on Neural Networks, vol. 13, no. 3, May 2002, pp.780-784.
[12] Ling Zhang and Bo Zhang, A geometrical representation of McCulloch-Pitts neural model and its applications, *IEEE Trans. on Neural Networks,* 10(4), pp.925-929, 1999.
[13] L. Zhang and B. Zhang, Neural network based classifiers for a vast amount of data, in *Lecture Notes in Artificial Intelligence* 1574, pp. 238-246, Ning Zhong (Eds.), Springer 1999.

Performance Measurements for Privacy Preserving Data Mining

Nan Zhang, Wei Zhao, and Jianer Chen

Department of Computer Science, Texas A&M University,
College Station, TX 77843, USA
{nzhang, zhao, chen}@cs.tamu.edu

Abstract. This paper establishes the foundation for the performance measurements of privacy preserving data mining techniques. The performance is measured in terms of the accuracy of data mining results and the privacy protection of sensitive data. On the accuracy side, we address the problem of previous measures and propose a new measure, named "effective sample size", to solve this problem. We show that our new measure can be bounded without any knowledge of the data being mined, and discuss when the bound can be met. On the privacy protection side, we identify a tacit assumption made by previous measures and show that the assumption is unrealistic in many situations. To solve the problem, we introduce a game theoretic framework for the measurement of privacy.

1 Introduction

In this paper, we address issues related to the performance measurements of privacy preserving data mining techniques. The objective of privacy preserving data mining is to enable data mining without violating the privacy of data being mined.

We consider a distributed environment where the system consists of a data miner and numerous data providers. Each data provider holds one private data point. The data miner performs data mining tasks on the (possibly perturbed) data provided by the data providers. A typical example of this kind of system is online survey, as the survey analyzer (data miner) collects data from thousands of survey respondents (data providers). Most existing privacy preserving algorithms in such system use an randomization approach which randomizes the original data to protect the privacy of data providers [1, 2, 3, 4, 5, 6, 8].

In this paper, we establish the foundation for analyzing the tradeoff between the accuracy of data mining results and the privacy protection of sensitive data. Our contribution can be summarized as follows.

- On accuracy side, we address the problem of previous measures and propose a new accuracy measure, named "effective sample size", to solve this problem. We show that our new measure can be upper bounded without any knowledge of the data being mined and discuss when the bound can be met.
- On privacy protection side, we show that a tacit assumption made by previous measures is that all adversaries use the same intrusion technique to invade privacy. We address the problems of this assumption and propose a game theoretic formulation which takes the adversary behavior into consideration.

2 System Model

Let there be n data providers $C_1, \ldots, C_n$ and one data miner S in the system. Each data provider C_i has a private data point (e.g., transaction, data tuple, etc.) x_i. We consider the original data values $x_1, \ldots, x_n$ as n independent and identically distributed (i.i.d.) variables that have the same distribution as a random variable X. Let the domain of X (i.e., the set of all possible values of X) be V_X, and the distribution of X be p_X. As such, each data point x_i is i.i.d. on V_X with distribution p_X.

Due to the privacy concern of data providers, we classify the data miners into two categories. One category is honest data miners. These data miners always act honestly in that they only perform regular data mining tasks and have no intention to invade privacy. The other category is malicious data miners. These data miners would purposely compromise the privacy of data providers.

3 Related Work

To protect the privacy of data providers, countermeasures must be implemented in the data mining system. Randomization is a commonly used approach. It is based on an assumption that accurate data mining results can be obtained from a robust estimation of the data distribution [2]. Thus, the basic idea of randomization approach is to distort individual data values but keep an accurate estimation of the data distribution.

Based on the randomization approach, the privacy preserving data mining process can be considered as a two-step process. In the first step, each data provider C_i perturbs its data x_i by applying a randomization operator $R(\cdot)$ on x_i, and then transfers the randomized data $R(x_i)$ to the data miner. We note that $R(\cdot)$ is known by both the data providers and the data miner. Let the domain of $R(x_i)$ be V_Y. The randomization operator $R(\cdot)$ is a function from V_X to V_Y with transition probability $p[x \rightarrow y]$. Existing randomization operators include random perturbation operator [2], random response operator [4], MASK distortion operator [8], and "select-a-size" operator [6].

In the second step, a honest data miner first employs a distribution reconstruction algorithm on the aggregate data, which intends to reconstruct the original data distribution from the randomized data. Then, the honest data miner performs the data mining task on the reconstructed distribution. Various distribution reconstruction algorithms have been proposed [1,2,6,4,8]. Also in the second step, a malicious data miner may invade privacy by using a private data recovery algorithm. This algorithm is used to recover individual data values from the randomized data supplied by the data providers.

Clearly, any privacy preserving data mining technique should be measured by its capability of both constructing the accurate data mining results and protecting individual data values from being compromised by the malicious data miners.

4 Quantification of Accuracy

In previous studies, several accuracy measures have been proposed. We classify these measures into two categories. One category is application-specified accuracy measures

[8]. Measures in this category are similar to those in systems without privacy concern and are specific to a data mining task (e.g., classification, association rule mining, etc.). The other category is general accuracy measures. Measures in this category can be applied to any privacy preserving data mining systems based on the randomization approach. An existing measure in this category is information loss measure [1], which is in proportion to the expected error of the reconstructed distribution.

We remark that the ultimate goal of the performance measurements is to help the system designers to choose the optimal randomization operator. As we can see from the privacy preserving data mining process, the randomization operator has to be determined before any data is transferred from the data providers to the data miner. Thus, in order to reach its goal, a performance measure must be estimated or bounded without any knowledge of the data being mined. As we can see, the application-specified accuracy measures depend on both the reconstructed data distribution and the performance of data mining algorithm. The information loss measure depends on both the original distribution and the reconstructed distribution. Neither measure can be estimated or bounded when the original data distribution is not known. Thus, previous measures cannot be used by the system designers to choose the optimal randomization operator.

We propose effective sample size as our new accuracy measure. Roughly speaking, given the number of randomized data points, the effective sample size is in proportion to the minimum number of original data points needed to make an estimate of the data distribution as accurate as the distribution reconstructed from the randomized data points. The formal definition is stated as follows.

Definition 1. *Given randomization operator $R : V_X \to V_Y$, let $\tilde{p}$ be the maximum likelihood estimate of the distribution of x_i reconstructed from $R(x_1), \ldots, R(x_n)$. Let $\tilde{p}_0(k)$ be the maximum likelihood estimate of the distribution based on k variables randomly generated from the distribution p_X. We define the effective sample size r as the minimum value of k/n such that*

$$D_{\mathrm{Kol}}(\tilde{p}_0(k), p_X) \leq D_{\mathrm{Kol}}(\tilde{p}, p_X) \qquad (1)$$

where D_{Kol} is the Kolmogorov distance [7], which measures the distance between an estimated distribution and the theoretical distribution [1].

As we can see, effective sample size is a general accuracy measure which measures the accuracy of the reconstructed distribution. We now show that the effective sample size can be strictly bounded without any knowledge of p_X.

Theorem 1. *Recall that $p[x \to y]$ is the probability transition function of $R : V_X \to V_Y$. An upper bound on the effective sample size r is given as follows.*

$$r \leq 1 - \sum_{y \in V_Y} \min_{x \in V_X} p[x \to y]. \qquad (2)$$

Due to space limit, please refer to [9] for the proof of this theorem.

---

[1] Other measures of such distance (e.g., Kuiper distance, Anderson-Darling distance, etc) can also be used to define the effective sample size. The use of other measures does not influence the results in this paper.

5 Quantification of Privacy Protection

In previous studies, two kinds of privacy measures have been proposed. One is information theoretic measure [1], which measures privacy disclosure by the mutual information between the original data x_i and the randomized data $R(x_i)$ (i.e., $I(x_i; R(x_i))$). This measure was challenged in [5], where it is shown that certain kinds of privacy disclosure cannot be captured by the information theoretic measure. The other kind of privacy measure can be used to solve this problem [5, 10, 2]. In particular, the privacy breach measure [5] defines the level of privacy disclosure as $\max_{x, x' \in V_X} p[x \to y]/p[x' \to y]$ for any given $y \in V_Y$. This measure captures the *worst case* privacy disclosure but is (almost) independent of the average amount of privacy disclosure.

Note that the data miner has the freedom to choose different intrusion techniques in different circumstances. As such, the privacy protection measure should depend on two important factors: a) the privacy protection mechanism of the data providers, and b) the unauthorized intrusion technique of the data miner. However, previous measures do not follow this principle. Instead, they make a tacit assumption that all data miners will use the same intrusion technique. This assumption seems to be reasonable as a (rational) data miner will always choose the intrusion technique which compromises the most private information. However, as we will show in the following example, the optimal intrusion technique *varies* in different circumstances. Thereby, the absence of consideration of intrusion techniques results in problems of privacy measurement.

Example 1. Let there be $V_X = \{0, 1\}$. The original data x_i is uniformly distributed on V_X. The system designer needs to determine which of the following two randomization operators, R_1 and R_2, discloses less private information.

$$R_1(x) = \begin{cases} x, \text{ with probability } 0.70, \\ \bar{x}, \text{ with probability } 0.30. \end{cases} \quad R_2(x) = \begin{cases} 0, \text{ if } x = 0, \\ 1, \text{ if } x = 1, \text{ with probability } 0.01, \\ 0, \text{ if } x = 1, \text{ with probability } 0.99. \end{cases}$$

In the example, we have $I(x; R_1(x)) \gg I(x; R_2(x))$. Due to the information theoretic measure, R_2 discloses less privacy. However, R_2 discloses more privacy due to the privacy breach measure. The reason is that if the data miner receives $R_2(x_i) = 1$, then it can always infer that $x_i = 1$ with probability of 1. We now show that whether R_1 or R_2 discloses more private information actually *depends* on the system setting. In particular, we consider the following two system settings.

1. The system is an online survey system. The value of x_i indicates whether a survey respondent is interested in buying certain merchandise. A malicious data miner intends to make unauthorized advertisement to data providers with such interest.
2. The system consists of n companies as the data providers and a management consulting firm as the data miner. The original data x_i contains the expected profit of the company which has not been published yet. A malicious data miner may use x_i to make investment on a high-risk stock market. The profit from a successful investment is tremendous. However, an unsuccessful investment results in a loss five times greater than the profit the data miner may obtain from a successful investment.

In the first case, an advertisement to a wrong person costs the data miner little. As we can see, R_1 discloses the original data value with probability of 0.7, which is greater than that of R_2 (0.501). Thus, R_2 is better than R_1 in the privacy protection perspective.

In the second case, the data miner will not perform the intrusion when R_1 is used by the data providers. The reason is that the loss from an incorrect estimate of x_i is too high to risk. As we can see, when R_1 is used, the expected net benefit from an unauthorized intrusion is less than 0. However, the data miner will perform the intrusion when R_2 is used. The reason is that when $R_2(x_i) = 1$, the data miner has a fairly high probability (99%) to make a successful investment. If a randomized data $R_2(x_i) = 0$ is received, the data miner will simply ignore it. As such, in this case, R_1 is better than R_2 in the privacy protection perspective.

As we can see from the example, the data miner will choose different privacy intrusion techniques in different system settings (in the above example, there is an intrude-or-not selection), which will result in different performance of randomization operators. Thus, the system setting has to be considered in the measurement of privacy disclosure.

In order to introduce the system setting and the privacy intrusion technique to our privacy measure, we propose a game theoretic framework to analyze the strategies of the data miner (i.e., privacy intrusion technique). Since we are studying the privacy protection performance of the randomization operator, we consider the randomization operator as the strategy of the data providers.

We model the privacy preserving data mining process as a non-cooperative game between the data providers and the data miner. There are two players in the game. One is the data providers. The other is the data miner. Since we only consider the privacy measure, the game is zero-sum in that the data miner can only benefit from the violation of privacy of the data providers. Let S_c be the set of randomization operators that the data providers can choose from. Let S_s be the set of the intrusion techniques that the data miner can choose from. Let u_c and u_s be the utility functions (i.e., expected benefits) of the data providers and the data miner, respectively. Since the game is zero-sum, we have $u_c + u_s = 0$. We remark that the utility functions depend on both the strategies of the players and the system setting.

We assume that both the data providers and the data miner are rational. As such, given a certain randomization operator, the data miner always choose the privacy intrusion technique which maximizes u_s. Given a certain privacy intrusion technique, the data providers always choose the randomization operator which maximizes u_c. We now define our privacy measure based on the game theoretic formulation.

Definition 2. *Given a privacy preserving data mining system $G \langle S_s, S_c, u_s, u_c \rangle$, we define the privacy measure l_p of a randomization operator R as*

$$l_p(R) = u_c(R, L_0), \tag{3}$$

where L_0 is the optimal privacy intrusion technique for the data miner when R is used by the data providers, u_c is the utility function of the data providers when R and L_0 are used.

As we can see, the smaller $l_p(R)$ is, the more benefit is obtained by the data miner from the unauthorized intrusion. Let σ be the ratio between the benefit obtained by a malicious data miner from a correct estimate and the loss of it from an incorrect estimate. A useful theorem is provided as follows.

Theorem 2. *Let there be* $\max_{x_0 \in V_X} \Pr\{x_i = x_0\} = p_m$ *in the original data distribution. We have* $l_p(R) = 0$ *if the randomization operator* $R : V_X \to V_Y$ *satisfies*

$$\max_{y \in V_Y} \frac{\max_{x \in V_X} p[x \to y]}{\min_{x \in V_X} p[x \to y]} \leq \frac{1 - p_m}{\sigma p_m}. \qquad (4)$$

Please refer to [9] for the proof of this theorem.

6 Conclusion

In this paper, we establish the foundation for the measurements of accuracy and privacy protection in privacy preserving data mining. On accuracy side, we address the problem of previous accuracy measures and solve the problem by introducing an effective sample size measure. On privacy protection side, we present a game theoretic formulation of the system and propose a privacy protection measure based on the formulation. Our work is preliminary, and there are many possible extensions. We are currently investigating using our performance measurements to derive the optimal trade-off between accuracy and privacy which can be achieved by the randomization approach.

References

1. D. Agrawal and C. C. Aggarwal. On the design and quantification of privacy preserving data mining algorithms. In *Proceedings of the twentieth ACM SIGMOD-SIGACT-SIGART symposium on Principles of database systems*, pages 247–255. ACM Press, 2001.
2. R. Agrawal and R. Srikant. Privacy-preserving data mining. In *Proceedings of the 2000 ACM SIGMOD international conference on Management of data*, pages 439–450. ACM Press, 2000.
3. W. Du and M. Atallah. Privacy-preserving cooperative statistical analysis. In *Proceedings of the 17th Annual Computer Security Applications Conference*, page 102, Washington, DC, USA, 2001. IEEE Computer Society.
4. W. Du and Z. Zhan. Using randomized response techniques for privacy-preserving data mining. In *Proceedings of the ninth ACM SIGKDD international conference on Knowledge discovery and data mining*, pages 505–510, New York, NY, USA, 2003. ACM Press.
5. A. Evfimievski, J. Gehrke, and R. Srikant. Limiting privacy breaches in privacy preserving data mining. In *Proceedings of the twenty-second ACM SIGMOD-SIGACT-SIGART symposium on Principles of database systems*, pages 211–222, New York, NY, USA, 2003. ACM Press.
6. A. Evfimievski, R. Srikant, R. Agarwal, and J. Gehrke. Privacy preserving mining of association rules. *Inf. Syst.*, 29(4):343–364, 2004.
7. F. Massey. The kolmogorov-smirnov test for goodness of fit. *Journal of the American Statistical Association*, 46(253).
8. S. Rizvi and J. Haritsa. Maintaining data privacy in association rule mining, 2002.

9. N. Zhang, W. Zhao, and J. Chen. On the performance measurement for privacy preserving data mining. technical report, 2004.
10. Y. Zhu and L. Liu. Optimal randomization for privacy preserving data mining. In *Proceedings of the 2004 ACM SIGKDD international conference on Knowledge discovery and data mining*, pages 761–766, New York, NY, USA, 2004. ACM Press.

Extraction of Frequent Few-Overlapped Monotone DNF Formulas with Depth-First Pruning*

Yoshikazu Shima[1], Kouichi Hirata[2], and Masateru Harao[2]

[1] Graduate School of Computer Science and Systems Engineering
[2] Department of Artificial Intelligence, Kyushu Institute of Technology,
Kawazu 680-4, Iizuka 820-8502, Japan
shima@dumbo.ai.kyutech.ac.jp
{hirata, harao}@ai.kyutech.ac.jp

Abstract. In this paper, first we introduce *frequent few-overlapped monotone DNF formulas* under the *minimum support* σ, the *minimum term support* τ and the *maximum overlap* λ. We say that a monotone DNF formula is *frequent* if the support of it is greater than σ and the support of each term (or itemset) in it is greater than τ, and *few-overlapped* if the overlap of it is less than λ and $\lambda < \tau$. Then, we design the algorithm *ffo_dnf* to extract them. The algorithm *ffo_dnf* first enumerates all of the *maximal* frequent itemsets under τ, and secondly connects the extracted itemsets by a disjunction $\vee$ until satisfying σ and λ. The first step of *ffo_dnf*, called a *depth-first pruning*, follows from the property that every pair of itemsets in a few-overlapped monotone DNF formula is incomparable under a subset relation. Furthermore, we show that the extracted formulas by *ffo_dnf* are *representative*. Finally, we apply the algorithm *ffo_dnf* to bacterial culture data.

1 Introduction

The purpose of *data mining* is to extract hypotheses that explain a database. An *association rule* is one of the most famous forms of hypotheses in data mining or association rule mining [1, 2, 6]. In order to extract association rules from a transaction database, the algorithm APRIORI, introduced by Agrawal *et al.* [1, 2], first enumerates *frequent itemsets* as sets of items satisfying the *minimum support*. Then, by dividing items in each frequent itemset into a *premise* and a *consequence* of an association rule, we can extract the association rules satisfying both the minimum support and the *minimum confidence*.

However, the frequent itemset is inappropriate when we extract hypotheses that explain a database *nearly overall*, because it just reflects the items with

---

* This work is partially supported by Grand-in-Aid for Scientific Research 15700137 and 16016275 from the Ministry of Education, Culture, Sports, Science and Technology, Japan.

very high frequency, which are not interesting in general. Furthermore, when we deal with a reconstructed transaction database by paying our attention to the specified attribute value v, it is natural to extract an association rule $X \to v$ with the consequence v rather than a standard association rule $X - Y \to Y$ ($Y \subseteq X$) from a frequent itemset X of the database.

In our previous works [7, 12], we have given an appropriate form of hypotheses in the above case, by regarding an itemset as a *monotone term* and by extending it to a *monotone DNF formula* as a disjunction of monotone terms (or itemsets). We say that a monotone DNF formula $f = X_1 \vee \cdots \vee X_m$ is *frequent* if each X_i is a frequent itemset under τ, that is, $supp(X_i) \geq \tau$, and $supp(f) \geq \sigma$, where $supp(f)$ denotes the *support* of f. We call such σ and τ the *minimum support* and the *minimum term support*, respectively.

In order to reduce a search space to extract frequent monotone DNF formulas, we have introduced the *overlap* $ol(f)$ of a monotone DNF formula f, and dealt with a frequent monotone DNF formula f satisfying that $ol(f) \leq \lambda$. We call such a λ the *maximum overlap*. By using σ, τ and λ, we have designed the algorithms *dnf_cover* [7] and *cdnf_cover* [12] to extract frequent and frequent closed monotone DNF formulas under σ, τ and λ from a transaction database.

It is known *disjunction-free* [4, 5, 8] and *generalized disjunction-free* [9, 10] itemsets as the researches to introduce a disjunction into itemsets. The difference between their works and this paper (and our previous works [7, 12]) is that their disjunction has been introduced into the *conclusion* in association rules, whereas our disjunction is into the *premise*. Furthermore, the main purpose of their works is to formulate the *concise* or *condensed representations*.

In the algorithms *dnf_cover* [7] and *cdnf_cover* [12], we have adopted a *breadth-first pruning*. The algorithm *dnf_cover* (*resp.*, *cdnf_cover*) first not only extracts itemsets satisfying σ by APRIORI [1, 2] (*resp.*, CHARM [16]), but also stores itemsets not satisfying σ but satisfying τ (*resp.*, and that are closed) to a *seed*. Next, it constructs monotone DNF formulas by connecting each element of a seed to a disjunction $\vee$ until satisfying σ and λ. Note that *dnf_cover* and *cdnf_cover* do not store all itemsets not satisfying σ but satisfying τ to a seed. They store just itemsets satisfying τ when not satisfying σ by APRIORI and CHARM.

In this paper, we extract monotone DNF formulas under σ, τ and λ from a transaction database with another pruning. We pay our attention to a natural assumption that $\lambda < \tau$. If not, then it is possible to extract a monotone DNF formula f containing a redundant itemset X such that $\tau \leq supp(X) \leq \lambda$. We call a monotone DNF formula satisfying that $\lambda < \tau$ and $ol(f) \leq \lambda$ *few-overlapped*.

Then, we design the algorithm *ffo_dnf* to extract frequent few-overlapped monotone DNF formulas under σ, τ and λ. The algorithm *ffo_dnf* first enumerates all of the *maximal* frequent itemsets [3, 11] under τ, and secondly connects the extracted itemsets by a disjunction $\vee$ until satisfying σ and λ.

Under the assumption that $\lambda < \tau$, every pair of itemsets in a few-overlapped monotone DNF formula f is always incomparable under a subset relation, that is, neither $X \subseteq Y$ nor $Y \subseteq X$ for each $X, Y \in f$ such that $X \neq Y$. Hence, *ffo_dnf* first enumerates all maximal frequent itemsets but not all frequent itemsets. We call this pruning a *depth-first pruning*. Furthermore, we show that the extracted formulas by *ffo_dnf* are *representative*.

Finally, we implement the algorithm *ffo_dnf* and apply it to bacterial culture data, which are full version of data in [13, 14, 15] and have given the empirical results in [7, 12]. We use 5 kinds of data, which are reconstructed data of which detected bacterium are MRSA (methicillin-resistant Staphylococcus aureus), Bacteroides, Fusobacterium, Prevotella and Streptococcus, respectively. The last 4 data are spices of Anaerobes. The number of records in them is 118, 498, 154, 157 and 155, respectively. In particular, we also use *all data* consisting of 118 MRSA data and 4886 MSSA (methicillin-suspectible Staphylococcus aureus) and *initial data* from a patient consisting of 35 MRSA data and 1613 MSSA data. Then, we verify the extracted formulas from MRSA data to MSSA data.

This paper is organized as follows. In Section 2, we show the several properties of overlaps and introduce frequent few-overlapped monotone DNF formulas. In Section 3, we design the algorithm *ffo_dnf* to extract frequent few-overlapped monotone DNF formulas under σ, τ and λ with depth-first pruning, and show that the extracted formulas are representative. In Section 4, we give some empirical results by applying the algorithm *ffo_dnf* to bacterial culture data.

2   Frequent Few-Overlapped Monotone DNF Formulas

Let $\mathcal{X}$ and $\mathcal{I}$ be finite sets. We call an element of $\mathcal{X}$ an *item* and $\mathcal{I}$ a *transaction id* (*tid*, for short). Also we call $X \subseteq \mathcal{X}$ an *itemset* and $I \subseteq \mathcal{I}$ a *tidset*. We call $\{x \in \mathcal{X} \mid (i,x) \in \mathcal{D}\} \subseteq \mathcal{X}$ a *transaction* of a tid i. Then, $\mathcal{D} \subseteq \mathcal{I} \times \mathcal{X}$ is a *transaction database*. For a tidset $I \subseteq \mathcal{I}$ and an itemset $X \subseteq \mathcal{X}$, we define the following function $tid : 2^{\mathcal{X}} \to 2^{\mathcal{I}}$.

$$tid_{\mathcal{D}}(X) = \{i \in \mathcal{I} \mid \forall x \in X, (i,x) \in \mathcal{D}\}.$$

Then, the *frequency* and the *support* of X in $\mathcal{D}$ are defined as $freq_{\mathcal{D}}(X) = |tid_{\mathcal{D}}(X)|$ and $supp_{\mathcal{D}}(X) = freq_{\mathcal{D}}(X)/|\mathcal{D}|$, respectively. In the remainder of this paper, we omit the phrases 'of X' and 'in $\mathcal{D}$,' and the subscript $\mathcal{D}$.

Definition 1 (Agrawal *et al.* [1, 2]). We say that an itemset X is *frequent* if $supp(X) \geq \sigma$. Here, we call the threshold σ ($0 \leq \sigma \leq 1$) the *minimum support*.

We denote the set of all frequent itemsets under σ by $Freq_{\sigma}$.

As a special frequent itemset, we introduce a *maximal frequent* itemset.

Definition 2 (Pasquier *et al.* [3, 11]). We say that a frequent itemset $X \in Freq_{\sigma}$ is *maximal* if there exists no frequent itemset $Y \in Freq_{\sigma}$ such that $Y \supset X$.

We denote the set of all maximal frequent itemsets under σ by $MaxFreq_{\sigma}$.

In this paper, we regard an item $x \in \mathcal{X}$ as a *variable* and an itemset $X \subseteq \mathcal{X}$ as a *monotone term* over $\mathcal{X}$, that is, a conjunction of variables. Then, we extend a monotone term over $\mathcal{X}$ to a *monotone DNF formula* $X_1 \vee \cdots \vee X_m$ (or $\{X_1, \ldots, X_m\}$) over $\mathcal{X}$ as a disjunction of monotone terms $X_1, \ldots, X_m$.

Let f be a monotone DNF formula $X_1 \vee \cdots \vee X_m$. Then, we extend the function tid as $tid(f) = tid(X_1) \cup \cdots \cup tid(X_m)$. The *frequency* of f in $\mathcal{D}$ and the *support* of f in $\mathcal{D}$ are defined as $freq(f) = |tid(f)|$ and $supp(f) = freq(f)/|\mathcal{D}|$.

Definition 3. We say that a monotone DNF formula $f = X_1 \vee \cdots \vee X_m$ is *frequent* if $supp(f) \geq \sigma$ and $supp(X_i) \geq \tau$ for each i ($1 \leq i \leq m$). Here, we call the minimum support τ ($0 \leq \tau \leq 1$) for each X_i the *minimum term support*.

Here, the condition $\tau > \sigma$ is meaningless, so we implicitly assume that $\tau \leq \sigma$.

Next, we introduce another measure *overlap* [7]. The *overlap set* $ols_\mathcal{D}(X,Y)$ of itemsets X and Y in $\mathcal{D}$ and the *overlap set* $ol_\mathcal{D}(f)$ of a monotone DNF formula $f = X_1 \vee \cdots \vee X_m$ in $\mathcal{D}$ are defined in the following way.

$$ols_\mathcal{D}(X,Y) = tid_\mathcal{D}(X) \cap tid_\mathcal{D}(Y),$$
$$ols_\mathcal{D}(f) = \bigcup_{1 \leq i < j \leq m} ols_\mathcal{D}(X_i, X_j).$$

The *overlap* of f in $\mathcal{D}$ is defined as $ol_\mathcal{D}(f) = |ols_\mathcal{D}(f)|/|\mathcal{D}|$. As similar as tid, we omit the subscript $\mathcal{D}$.

Theorem 1 (Hirata et al. [7]). *The overlap is monotonic, that is, it holds that $ol(f) \leq ol(f \vee g)$ for monotone DNF formulas f and g.*

Theorem 2 (Hirata et al. [7]). *For a monotone DNF formula f and an itemset X, it holds that $ols(f \vee X) = ols(f) \cup \left(\bigcup_{Y \in f} ols(Y, X) \right)$.*

Theorem 3. *For a monotone DNF formula f and an itemset X, it holds that $tid(f \vee X) = tid(f) \cup tid(X)$ and $ols(f \vee X) = ols(f) \cup (tid(f) \cap tid(X))$.*

Proof. The first statement is obvious by the definition of tid. By Theorem 2, the following equation holds for a monotone DNF formula $f = X_1 \vee \cdots \vee X_m$.

$$\begin{aligned}
& ols(f \vee X) \\
&= \left(\bigcup_{1 \leq i < j \leq m}(tid(X_i) \cap tid(X_j)) \right) \cup \left(\bigcup_{1 \leq i \leq m}(tid(X_i) \cap tid(X)) \right) \\
&= ols(f) \cup \left(\left(\bigcup_{1 \leq i \leq m} tid(X_i) \right) \cap tid(X) \right) \\
&= ols(f) \cup (tid(f) \cap tid(X)).
\end{aligned}$$

Hence, the second statement holds. □

Theorem 4. *Let $X_1, \ldots, X_m$ and $Y_1, \ldots, Y_m$ be sequences of itemsets such that $X_i \cap X_j \subseteq Y_i \cap Y_j$ for each i and j ($1 \leq i, j \leq m$) and f a monotone DNF formula. Then, it holds that $supp(f \vee X_1 \vee \cdots \vee X_m) \geq supp(f \vee Y_1 \vee \cdots \vee Y_m)$ and $ol(f \vee X_1 \vee \cdots \vee X_m) \geq ol(f \vee Y_1 \vee \cdots \vee Y_m)$.*

Proof. Since $X_i \subseteq Y_i$ (just the case that $j = i$), it holds that $tid(X_i) \supseteq tid(Y_i)$, so $tid(X_1 \vee \cdots \vee X_m) \supseteq tid(Y_1 \vee \cdots \vee Y_m)$. Hence, the first statement holds.

We show the second statement by induction on m. If $m = 1$, then it holds that $X_1 \subseteq Y_1$, so $tid(X_1) \supseteq tid(Y_1)$. Theorem 2 implies that $ols(f \vee X_1) = ols(f) \cup \left(\bigcup_{Z \in f} ols(Z, X_1)\right) \supseteq ols(f) \cup \left(\bigcup_{Z \in f} ols(Z, Y_1)\right) = ols(f \vee Y_1)$.

Let g and h be monotone DNF formulas $f \vee X_1 \vee \cdots \vee X_{m-1}$ and $f \vee Y_1 \vee \cdots \vee Y_{m-1}$, respectively. Suppose that $ols(g) \supseteq ols(h)$. By Theorem 2, the following equations hold.

$$ols(g \vee X_m) = ols(g) \cup \left(\bigcup_{Z \in f} ols(Z, X_m)\right) \cup \left(\bigcup_{1 \le i \le m-1} ols(X_i, X_m)\right),$$
$$ols(h \vee Y_m) = ols(h) \cup \left(\bigcup_{Z \in f} ols(Z, Y_m)\right) \cup \left(\bigcup_{1 \le i \le m-1} ols(Y_i, Y_m)\right).$$

Since $X_i \subseteq Y_i$, it holds that $ols(Z, X_i) \supseteq ols(Z, Y_i)$ for each $Z \in f$ and $1 \le i \le m-1$. Furthermore, since $X_i \cap X_m \subseteq Y_i \cap Y_m$, it holds that $ols(X_i, X_m) \supseteq ols(X_i, Y_m)$ for each $1 \le i \le m-1$. By induction hypothesis, it holds that $ols(g) \supseteq ols(h)$. Hence, it holds that $ols(g \vee X_m) \supseteq ols(h \vee Y_m)$. □

Definition 4. We say that a monotone DNF formula $f = X_1 \vee \cdots \vee X_m$ satisfying τ is *few-overlapped* under λ if $\lambda < \tau$ and $ol(f) \le \lambda$. Here, we call the threshold λ ($0 \le \lambda \le 1$) the *maximum overlap*.

In Definition 4, we adopt a natural assumption that $\lambda < \tau$. If not, then it is possible to extract a monotone DNF formula f containing a redundant itemset X such that $\tau \le supp(X) \le \lambda$. In the remainder of this paper, we deal with a *frequent few-overlapped monotone DNF formula* under σ, τ and λ, that is, a monotone DNF formula $f = X_1 \vee \cdots \vee X_m$ such that $supp(f) \ge \sigma$, $supp(X_i) \ge \tau$ for each i ($1 \le i \le m$), $ol(f) \le \lambda$ and $\lambda < \tau$.

3 Extraction Algorithm with Depth-First Pruning

In this section, we design the algorithm *ffo_dnf* to extract frequent few-overlapped monotone DNF formulas under σ, τ and λ described as Fig. 1. In the algorithm *ffo_dnf*, the set *FFO* of frequent few-overlapped monotone DNF formulas is constructed by a simple depth-first search on the overlap, which follows from the monotonicity of the overlap (Theorem 1). Here, for an itemset X, we set $ols(X) = \emptyset$ and $ol(X) = 0$, and, for a set $M = \{X_1, \ldots, X_m\}$, we set $M[i] = \{X_i, \ldots, X_m\}$ ($1 \le i \le m$) and $M[m+1] = \emptyset$.

In the algorithm *ffo_dnf*, we implicitly store the elements of $MaxFreq_\tau$ as a set of ordered pairs $(X, tid(X))$ for a maximal itemset X, so we enumerate $MaxFreq_\tau$ by the improvement of the algorithm CHARM [16], because it deals with directly such an ordered pair. Here, we add the check whether or not an itemset of each leaf in the search tree under τ is maximal to CHARM. Then, we maintain an ordered triple $(f, tid(f), ols(f))$ to construct a monotone DNF formula f in *ffo_search*. By Theorem 3, we can obtain $(f \vee X, tid(f \vee X), ols(f \vee X))$ from $(f, tid(f), ols(f))$ and $(X, tid(X))$.

We call a pruning to first enumerate $MaxFreq_\tau$ but not $Freq_\tau$ in *ffo_dnf* a *depth-first pruning*. On the other hand, a *breadth-first pruning* in the algorithm

```
procedure ffo_dnf(σ, τ, λ)
M ← MaxFreq_τ; ffo_search(∅, M, σ, λ);
procedure ffo_search(f, M, σ, λ);
if M = ∅ then halt;
for i = 1 to m do begin /* M = {X_1, ..., X_m}, M[i] = {X_i, ..., X_m} */
    if ol(f ∨ X_i) ≤ λ then
        if supp(f ∨ X_i) ≥ σ then FFO ← FFO ∪ {f ∨ X_i};
        ffo_search(f ∨ X_i, M[i + 1], σ, λ);
    else ffo_search(f, M[i + 1], σ, λ);
end /* for */
return FFO;
```

Fig. 1. The algorithm *ffo_dnf*

dnf_cover [7] checks whether or not $X \cup \{x\} \in Freq_\tau$ for just an itemset $X \cup \{x\}$ such that $X \in Freq_\sigma$ but $X \cup \{x\} \notin Freq_\sigma$, instead of searching for $Freq_\tau$. Also a breadth-first pruning in the algorithm *cdnf_cover* [12] checks whether or not $X \cup \{x\} \in FreqClosed_\tau$ for just an itemset $X \cup \{x\}$ such that $X \in FreqClosed_\sigma$ but $X \cup \{x\} \notin FreqClosed_\sigma$, instead of searching for $FreqClosed_\tau$. Here, $FreqClosed_\sigma$ is the set of all frequent closed itemsets under σ [11, 16].

Hence, the algorithm *ffo_dnf* with depth-first pruning searches for itemsets satisfying τ, while the algorithms *dnf_cover* and *cdnf_cover* with breadth-first pruning searches for just itemsets satisfying σ.

The depth-first pruning in *ffo_dnf* is based on the following theorem.

Theorem 5. *Let X and Y be itemsets such that $supp(X) \geq \tau$ and $supp(Y) \geq \tau$ and f a monotone DNF formula such that $X \in f$. If either $X \subseteq Y$ or $Y \subseteq X$, then it holds that $ol(f \vee Y) > \lambda$.*

Proof. If $X \subseteq Y$, then $tid(Y) \subseteq tid(X)$, so it holds that $ol(f \vee Y) \geq ol(X \vee Y) = |tid(X) \cap tid(Y)| = |tid(Y)| \geq \tau > \lambda$. If $Y \subseteq X$, then $tid(X) \subseteq tid(Y)$, so it holds that $ol(f \vee Y) \geq ol(X \vee Y) = |tid(X) \cap tid(Y)| = |tid(X)| \geq \tau > \lambda$. □

Theorem 5 claims that every pair of itemsets in a few-overlapped monotone DNF formula f is always incomparable under a subset relation, that is, neither $X \subseteq Y$ nor $Y \subseteq X$ for each $X, Y \in f$ ($X \neq Y$). Hence, one of the reasons why we adopt a depth-first pruning is that every maximal frequent itemset is incomparable.

Furthermore, another reason follows from the following theorem.

Theorem 6. *For every few-overlapped monotone DNF formula f, there exists a few-overlapped monotone DNF formula g such that for every $X \in f$, there exists an itemset $Y \in g$ such that $X \subseteq Y$ and Y is maximal under τ.*

Proof. Let f be a few-overlapped monotone DNF formula $X_1 \vee \cdots \vee X_m$. For $l \leq m$, let p be a mapping $\{1, \ldots, m\} \to \{1, \ldots, l\}$ such that $X_i \subseteq Y_{p(i)}$ and $Y_{p(i)}$ is maximal, and g a monotone DNF formula $Y_1 \vee \cdots \vee Y_l$. Since $tid(X_i) \supseteq tid(Y_{p(i)})$, it holds that $tid(X_i) \cap tid(X_j) \supseteq tid(Y_{p(i)}) \cap tid(Y_{p(j)})$. Then, it holds that

$ols(f) = \bigcup_{1 \leq i < j \leq m}(tid(X_i) \cap tid(X_j)) \supseteq \bigcup_{1 \leq i < j \leq m}(tid(Y_{p(i)}) \cap tid(Y_{p(j)})) = \bigcup_{1 \leq i < j \leq l}(tid(Y_i) \cap tid(Y_j)) = ols(g)$, so it holds that $ol(f) \geq ol(g)$. Since $ol(f) \leq \lambda$, it holds that $ol(g) \leq \lambda$. Hence, g is few-overlapped. □

Theorem 6 claims that every few-overlapped monotone DNF formula f has a corresponding few-overlapped monotone DNF formula g such that each itemset in g is maximal under τ. Hence, the extracted formulas by *ffo_dnf* are *representative*.

By Theorem 4, 5 and 6, the algorithm *ffo_dnf* extracts frequent few-overlap monotone DNF formulas under σ, τ and λ that are representative.

4 Empirical Results from Bacterial Culture Data

In this section, we give the empirical results obtained by applying the algorithm *ffo_dnf* to bacterial culture data, which are full version in [15] and have given the empirical results in [7,12]. The computer environment is that CPU and RAM are Pentium 4 2.8 GHz and 2 GB, respectively.

We use 5 kinds of data, MRSA (methicillin-resistant Staphylococcus aureus) Bacteroides (Bact), Fusobacterium (Fuso), Prevotella (Prev) and Streptococcus (Stre) data. The number of records in them is 118, 498, 154, 157 and 155, respectively. The last 4 data are a part of Anaerobes data corresponding to 4 species of Anaerobes.

In particular, concerned with MRSA data, we use *all data* consisting of 118 MRSA data and 4886 MSSA (methicillin-suspectible Staphylococcus aureus) and *initial data* from a patient consisting of 35 MRSA data and 1613 MSSA data. Here, we use the MSSA data to verify the extracted formulas from MRSA data. We refer all data and initial data for MRSA (*resp.*, MSSA) to a_MRSA (*resp.*, a_MSSA) and i_MRSA (*resp.*, i_MSSA).

All of them consist of data between 4 years (from 1995 to 1998) with 93 attributes, containing 17 antibiotics for benzilpenicillin (PcB), synthetic penicillins (PcS), augmentin (Aug), anti-pseudomonas penicillin (PcAP), 1st generation cephems (Cep1), 2nd generation cephems (Cep2), 3rd generation cephems (Cep3), 4th generation cephems (Cep4), anti-pseudomonas cephems (CepAP), aminoglycosides (AG), macrolides (ML), tetracyclines (TC), lincomycins (LCM), chloramphenicols (CP), carbapenems (CBP), vancomycin (VCM) and RFP/FOM (RFPFOM). Here, the above antibiotics have the value of *resistant* (R), *intermediate* (I) or *suspectible* (S).

Fig. 2 describes the number of frequent few-overlapped monotone DNF formulas extracted by *ffo_dnf*. Here, τ is fixed to 25%. # Max and # DNF denote the number of maximal itemsets and frequent few-overlapped monotone DNF formulas. Note that # DNF is not always increasing when # Max is increasing.

Fig. 3 describes all occurrences of the items for the sensitivity of antibiotics in the frequent few-overlapped monotone DNF formulas extracted by *ffo_dnf* under $(\sigma, \tau, \lambda) = (70, 25, 20)$. Note that Fig. 3 contains more information of the resistant for Anaerobes than the result in our previous work [12], and such information extracted by just *ffo_dnf* is pruned by *cdnf_cover* [12].

data	# Max	σ	λ	# DNF	time (sec.)
a_MRSA	73	80	10	0	0.08
			15	0	0.10
			20	0	0.21
		75	10	2	0.07
			15	2	0.09
			20	10	0.21
		70	10	5	0.08
			15	27	0.08
			20	39	0.21
i_MRSA	21	80	10	1	0.00
			15	2	0.00
			20	11	0.02
		75	10	7	0.01
			15	11	0.01
			20	48	0.02
		70	10	9	0.01
			15	17	0.01
			20	84	0.01
Bact	132	80	10	4	0.36
			15	4	1.04
			20	18	5.71
		75	10	29	0.38
			15	55	1.05
			20	137	5.74
		70	10	159	0.34
			15	780	1.09
			20	2007	5.43

data	# Max	σ	λ	# DNF	time (sec.)
Fuso	39	80	10	0	0.04
			15	0	0.04
			20	0	0.06
		75	10	0	0.05
			15	0	0.04
			20	0	0.05
		70	10	4	0.04
			15	6	0.04
			20	6	0.05
Prev	61	80	10	1	0.09
			15	2	0.11
			20	22	0.18
		75	10	20	0.10
			15	23	0.12
			20	68	0.17
		70	10	107	0.11
			15	159	0.14
			20	228	0.21
Stre	37	80	10	10	0.01
			15	10	0.02
			20	24	0.03
		75	10	10	0.01
			15	13	0.03
			20	37	0.01
		70	10	15	0.01
			15	75	0.04
			20	203	0.04

Fig. 2. The number of frequent few-overlapped monotone DNF formulas extracted by *ffo_dnf* under the minimum term support τ is fixed to 25%

data	PcB	PcS	PcAP	Cep1	Cep2	Cep3	AG	ML	TC	LCM	CP	CBP	VCM	RFPFOM
a_MRSA	R	R		R			R	R	R	R		S	S	RS
i_MRSA	R	R		R			R	R	R	R		S	S	RS
Bact	R		RS	R	S	S		RS	S	RS	S	S		
Fuso	S		S	S	S	S		RS	S	S	S	S		
Prev	RS		S	RS	S	S		S	S	S	S	S		
Stre	S	S	S	S	S	S		S	S	S		S		S

Fig. 3. The sensitivity of antibiotics appearing in the extracted formulas under $(\sigma, \tau, \lambda) = (75, 25, 20)$

The frequent few-overlapped monotone DNF formulas extracted by *ffo_dnf* under $(\sigma, \tau, \lambda) = (70, 25, 20)$ except from Stre have the following characterization, which is the similar characterization by *cdnf_cover* [12].

1. The extracted formulas always contains the items with information in 17 antibiotics. For MRSA, note that if Staphylococcus aureus are resistant for PcB, PcS, Cep1 and AG, then it is determined to MRSA.
2. The extracted formulas are always *non-redundant*, that is, they contain no formulas such as year = 95 ∨ year = 96 ∨ year = 97 ∨ year = 98 or male ∨ female. Because all of the above itemsets are not maximal.

a_MRSA	a_MSSA	ol	
70.34	9.82	11.86	$(CBP = S \wedge Cep1 = R \wedge LCM = R \wedge PcB = R \wedge PcS = R \wedge VCM = S$ $\wedge ML = R \wedge RFPFOM = S \wedge TC = R \wedge \beta = 0 \wedge male)_{30.51}$ $\vee (Cep1 = R \wedge LCM = R \wedge PcB = R \wedge PcS = R \wedge TC = R \wedge VCM = S$ $\wedge dis = 33)_{25.42} \quad \vee (RFPFOM = R \wedge \beta = 0)_{26.27}$
70.34	12.27	12.71	$(CBP = S \wedge Cep1 = R \wedge LCM = R \wedge PcB = R \wedge PcS = R \wedge VCM = S$ $\wedge ML = R \wedge RFPFOM = S \wedge TC = R \wedge \beta = 0 \wedge male)_{30.51}$ $\vee (Cep1 = R \wedge \beta = 0 \wedge dis = 33)_{26.27} \quad \vee (RFPFOM = R \wedge \beta = 0)_{26.27}$
70.34	12.45	11.86	$(CBP = S \wedge Cep1 = R \wedge LCM = R \wedge PcB = R \wedge PcS = R \wedge VCM = S$ $\wedge ML = R \wedge RFPFOM = S \wedge TC = R \wedge \beta = 0 \wedge male)_{30.51}$ $\vee (AG = R \wedge Cep1 = R \wedge LCM = R \wedge PcB = R \wedge dis = 33)_{25.42}$ $\vee (RFPFOM = R \wedge \beta = 0)_{26.27}$
73.73	13.01	4.24	$(CBP = S \wedge Cep1 = R \wedge LCM = R \wedge PcB = R \wedge PcS = R \wedge VCM = S$ $\wedge ML = R \wedge RFPFOM = S \wedge TC = R \wedge \beta = 0 \wedge spl = 5)_{26.27}$ $\vee (Cep1 = R \wedge LCM = R \wedge PcB = R \wedge PcS = R \wedge VCM = S$ $\wedge spl = 1)_{25.42} \quad \vee (RFPFOM = R \wedge \beta = 0)_{26.27}$
i_MRSA	i_MSSA	ol	
72.73	7.51	15.15	$(CBP = S \wedge Cep1 = R \wedge LCM = R \wedge ML = R \wedge PcB = R \wedge PcS = R$ $\wedge VCM = S \wedge dis = 17)_{27.27}$ $\vee (AG = R \wedge Cep1 = R \wedge TC = R \wedge LCM = R \wedge PcB = R \wedge PcS = R$ $\wedge VCM = S \wedge dis = 7)_{27.27} \quad \vee (RFPFOM = R \wedge \beta = 0)_{33.33}$
72.73	7.56	15.15	$(AG = R \wedge Cep1 = R \wedge TC = R \wedge LCM = R \wedge PcB = R \wedge PcS = R$ $\wedge VCM = S \wedge dis = 7)_{27.27}$ $\vee (Cep1 = R \wedge LCM = R \wedge PcS = R \wedge PcB = R \wedge \beta = 0$ $\wedge year = 95)_{33.33} \quad \vee (RFPFOM = R \wedge \beta = 0)_{33.33}$
72.73	7.83	18.18	$(AG = R \wedge CBP = S \wedge TC = R \wedge Cep1 = R \wedge LCM = R \wedge PcB = R$ $\wedge PcS = R \wedge VCM = S \wedge male \wedge year = 95)_{30.30}$ $\vee (ML = R \wedge dis = 7)_{30.30} \quad \vee (RFPFOM = R \wedge \beta = 0)_{33.33}$
72.73	8.21	12.12	$(Cep1 = R \wedge LCM = R \wedge PcS = R \wedge PcB = R \wedge \beta = 0 \wedge dis = 7)_{27.27}$ $\vee (CBP = S \wedge Cep1 = R \wedge LCM = R \wedge ML = R \wedge PcB = R \wedge PcS = R$ $\wedge VCM = S \wedge age = 70s)_{27.27} \quad \vee (RFPFOM = R \wedge \beta = 0)_{33.33}$

Fig. 4. The extracted formulas under $(\sigma, \tau, \lambda) = (75, 25, 20)$ from a_MRSA (*resp.*, i_MRSA) of which support in a_MSSA (*resp.*, i_MSSA) is smaller than others

Also we can give the following characterizations of the extracted formulas under $(\sigma, \tau, \lambda) = (70, 25, 20)$, not found in our previous works [7, 12].

1. From a_MRSA and i_MRSA, items for samples that spl=1 (catheter) and spl=5 (respiratory) are extracted from a_MRSA, while just an item spl=1 is extracted from i_MRSA. In particular, for a_MRSA, the formula containing spl=1 also contains spl=5 in another itemset. Here, a_MRSA and i_MRSA contain 6 kinds of the items for samples.
2. From Bact, items that a β-lactamese is either 1 or 3 are extracted. Note that just an item that a β-lactamese is 0 is extracted from other data.
3. From Fuso, for 6 extracted formulas, 3 formulas contain the resistant for macrolides (Fig. 3).
4. From Prev, the formula containing PcB=S also contains PcB=R in another itemset. Also the occurrence of Cep1=S is independent from one of Cep1=R.
5. From Stre, there are formulas not containing the items with the information for antibiotics, for example, $(\beta = 0 \wedge ctr = 1) \vee (\beta = 0 \wedge wcl = 4) \vee (\beta = 0 \wedge male \wedge spl = 1)$.

Fig. 4 describes the extracted formulas from a_MRSA (*resp.*, i_MSSA) of which support in a_MSSA (*resp.*, i_MSSA) is smaller than others. Here, items dis=7, dis=17 and dis=33 denote that the disease is a tumor, respiratory and postoperative, respectively. Note that the items for samples appear in a_MRSA, while the items for ages and years appear in i_MRSA.

All of the formulas in Fig. 4 contain a term $X = $ (RFPFOM $=$ R $\land$ $\beta = 0$). Here, $supp_{\text{a\_MSSA}}(X) = 5.65\%$, $supp_{\text{i\_MSSA}}(X) = 4.97\%$, $supp_{\text{a\_MRSA}}(X) = 26.27\%$ and $supp_{\text{i\_MRSA}}(X) = 33.33\%$. On the other hand, for a term $Y = $ (ML $=$ R $\land$ dis $= 7$) appearing in i_MRSA but not in a_MRSA, $supp_{\text{i\_MRSA}}(Y) = 30.30\%$, $supp_{\text{i\_MSSA}}(Y) = 2.81\%$, $supp_{\text{a\_MRSA}}(Y) = 24.58\%$ and $supp_{\text{a\_MSSA}}(Y) = 3.32\%$. Hence, the reason why X appears in both a_MRSA and i_MRSA but Y appears in just i_MRSA is to extract few-overlapped formulas.

5 Conclusion

In this paper, we have introduced *frequent few-overlapped monotone DNF formulas* and designed the algorithm *ffo_dnf* to extract the formulas that are *representative*. We have adopted a *depth-first pruning* in the algorithm *ffo_dnf*, different from a *breadth-first pruning* adopted in *dnf_cover* [7] and *cdnf_cover* [12]. Finally, we have applied it to bacterial culture data, the MRSA data and the 4 species of Anaerobes data, and evaluated the extracted formulas.

We have adopted the depth-first pruning based on Theorem 5 in the algorithm *ffo_dnf*. Let f be a monotone DNF formula $X_1 \lor \cdots \lor X_m$ such that $ol(f) \leq \lambda$ and $X_i \in \textit{MaxFreq}_\sigma$. Such an f can be extracted as *FO* by adding the statement "$FO \leftarrow FO \cup \{f \lor X_i\};$" between "**if** $ol(f \lor X_i) \leq \lambda$ **then**" and "**if** $supp(f \lor X_i) \geq \sigma$ **then**" in *ffo_dnf*. By Theorem 6, we can enumerate all of the frequent few-overlapped monotone DNF formulas as $f_X = (f - \{Y \in f \mid X \subset Y\}) \lor X$ for every f and X. It is a future work to *efficiently* enumerate all of the frequent few-overlapped monotone DNF formulas *without pruning*.

References

1. R. Agrawal, H. Mannila, R. Srikant, H. Toivonen, A. I. Verkamo: *Fast discovery of association rules*, in [6], 307–328.
2. R. Agrawal, R. Srikant: *Fast algorithms for mining association rules in large databases*, Proc. 20th VLDB, 487–499, 1994.
3. D. Burdick, M. Calimlim, J. Gehrke: *MAFIA: A maximal frequent itemset algorithm for transaction databases*, Proc. ICDE2001, 443–452, 2001.
4. A. Bykowski, C. Rigotti: *A condensed representation to find frequent patterns*, Proc. PODS2001, 267–273, 2001.
5. A. Bykowski, C. Rigotti: *DBC: A condensed representation of frequent patterns for efficient mining*, Information Systems **28**, 949-977, 2003.
6. U. M. Fayyed, G. Piatetsky-Shapiro, P. Smyth, R. Uthurusamy (eds.): *Advances in knowledge discovery and data mining*, AAAI/MIT Press, 1996.
7. K. Hirata, R. Nagazumi, M. Harao: *Extraction of coverings as monotone DNF formulas*, Proc. DS2003, 165–178, 2003.
8. M. Kryszkiewicz: *Concise representation of frequent patterns based on disjunction-free generators*, Proc. ICDM2001, 305–312, 2001.

9. M. Kryszkiewicz, M. Gajek: *Concise representation of frequent patterns based on generalized disjunction-free generators*, Proc. PAKDD-02, 159–171, 2002.
10. M. Kryszkiewicz, M. Gajek: *Why to apply generalized disjunction-free generators representation of frequent patterns?*, Proc. ISMIS2002, 383–392, 2002.
11. N. Pasquier, Y. Bastide, R. Taouil, L. Lakhal: *Discovering frequent closed itemsets for association rules*, Proc. ICDT'99, 398–416, 1999.
12. Y. Shima, S. Mitsuishi, K. Hirata, M. Harao: *Extracting minimal and closed monotone DNF formulas*, Proc. DS2004, 298–305, 2004.
13. E. Suzuki: *Mining bacterial test data with scheduled discovery of exception rules*, in [14], 34–40.
14. E. Suzuki (ed.): Proc. KDD Challenge 2000.
15. S. Tsumoto: *Guide to the bacteriological examination data set*, in [14], 8–12.
16. M. J. Zaki, C.-J. Hsiao: *CHARM: An efficient algorithm for closed itemset mining*, Proc. SDM2002, 457–478, 2002.

Rule Extraction from Trained Support Vector Machines

Ying Zhang, HongYe Su, Tao Jia, and Jian Chu

National Laboratory of Industrial Control Technology, Institute of Advanced Process Control,
Zhejiang University Yuquan Campus, Hangzhou 310027, P.R. China
{zhangying, hysu, Chujian}@iipc.zju.edu.cn

Abstract. Support vector machine (SVM) is applied to many research fields because of its good generalization ability and solid theoretical foundation. However, as the model generated by SVM is like a black box, it is difficult for user to interpret and understand how the model makes its decision. In this paper, a hyperrectangle rules extraction (HRE) algorithm is proposed to extract rules from trained SVM. Support vector clustering (SVC) algorithm is used to find the prototypes of each class, then hyperrectangles are constructed according to the prototypes and the support vectors (SVs) under some heuristic conditions. When the hyperrectangles are projected onto coordinate axes, the if-then rules are obtained. Experimental results indicate that HRE algorithm can extract rules efficiently from trained SVM and the number and support of obtained rules can be easily controlled according to a user-defined minimal support threshold.

1 Introduction

Support vector machine (SVM) is a new class of machine learning algorithms, motivated by results of statistical learning theory [1], which is originally developed for pattern recognition. Because of its good generalization and solid theoretical foundation, SVM is applied to many research fields. However, a problem that SVM must face is that, after training, it is usually difficult to understand the concept representations and give a reasonable explanation. Like the neural networks, SVM generates a black box model. Usually, a concept representation learned by SVM is difficult to understand because the representation is encoded by a large number of real-valued parameters. However, it is important to be able to understand a learned concept definition. For example, for medical diagnosis, the users must understand how the system makes its decisions in order to be confident in its predictions [2].

In order to overcome the limitations of SVM, the hypothesis generated by SVM could be transferred into a more comprehensible representation. These conversion methods are known as rule extraction. In the last few years, some methods of rule extraction from the trained neural networks have been proposed [3, 4]. However, considering the difference between SVM and neural networks, most of them can not be directly applied to SVM. Nunez [5] introduced an approach for rule extraction from SVM, in which the K-means clustering is used to determine prototypes for each class, and then prototypes are combined with support vectors to define an ellipsoid which are then mapped to if-then rules in the input space. But this approach does not scale well and the generated ellipsoid rules seriously overlap each other. Furthermore, as the solution quality of k-means excessively depends on initial values for centers, it is hard to control the number and quality of obtained rules.

In this paper, a hyperrectangle rules extraction (HRE) algorithm is proposed to extract rules from trained SVM. Support vector clustering (SVC) algorithm [9] is adopted to find the prototypes of each class, then hyperrectangles are constructed on the base of prototypes and support vectors (SV) under some heuristic limitations. Even when training set contains outliers, HRE can also generate high quality rules because of merits of the SVC clustering algorithm. Experimental results show that it is easy for HRE to control the number and the support of the generated rules.

This paper is organized as follows: Section 2 briefly reviews the basic theory of SVM. In section 3, The HRE algorithm based on SVC is presented. Section 4 describes experimental results of HRE algorithm on some benchmark data sets. Finally, some conclusions are given in section 5.

2 Support Vector Machines

Assume that a training data set is given as $\chi = \{(\mathbf{x}_1, y_1), ...(\mathbf{x}_i, y_i), ..., (\mathbf{x}_r, y_r)\}$, $i = 1,...,r$, where $\mathbf{x}_i \in \mathbf{R}^n$ and $y_i \in \{-1,1\}$ is class label. The goal of SVM is to find an optimal classification hyperplane. For binary classification case, to find the optimal hyperplane is equal to solve the quadratic programming (QP) problem as follows [6]:

$$\min \ \Phi = \frac{1}{2}\|\mathbf{w}\|^2 + G \cdot \sum_{i=1}^{r} \xi_i$$
$$\text{s.t.} \ y_i(\mathbf{w} \bullet \mathbf{x}_i) + b - 1 > \xi_i \qquad (1)$$
$$\xi_i \geq 0, i = 1,...,r.$$

where ξ_i are slack variables, $\|\mathbf{w}\|^2$ is structure risk which characterizes the complexity of model and $\sum_{i=1}^{r} \xi_i$ is experience risk which represents the training error. Parameter G is slack factor which determines the trade off between structure and experience risk. Minimizing (1) captures the main insight of statistical learning theory: in order to obtain a small risk function, one needs to control both training error and model complexity. Introducing Lagrange multipliers and making some substitutions, we can obtain the Wolf dual of the optimization problem (1) as follows:

$$\max \ Q(\alpha) = \sum_{i=1}^{r} \alpha_i - \frac{1}{2}\sum_{i,j=1}^{r} \alpha_i \alpha_j y_i y_j \mathbf{x}_i \bullet \mathbf{x}_j$$
$$\text{s.t.} \ \sum_{i=1}^{r} y_i \alpha_i = 0 \qquad (2)$$
$$0 \leq \alpha_i \leq G, i = 1,...,r.$$

In nonlinear case, data points in input space are mapped into a high dimension feature space via mapping $\varphi: \mathbf{R}^n \to \mathbf{H}$. Then in the space $\mathbf{H}$, the optimal classification hyperplane is constructed. According to Mercer condition [5], there exist a mapping φ and a kernel function $K(\cdot,\cdot)$ which satisfies $K(\mathbf{x}_i, \mathbf{x}_j) = \varphi(\mathbf{x}_i) \cdot \varphi(\mathbf{x}_j)$. The decision function can be written as

$$f(\mathbf{x}) = \operatorname{sign}(\sum_{i=1}^{r} \alpha_i y_i K(\mathbf{x}_i, \mathbf{x}) + b) \qquad (3)$$

3 The HRE Algorithm

3.1 Hyperrectangle Rule

Salaberg describes a family of learning algorithms based on nested generalized exemplars (NGE) [7]. In NGE, an exemplar is a single training example, and a generalized exemplar is an axis-parallel hyperrectangle that may cover several training examples. The NGE algorithm grows the hyperrectangles incrementally as training examples are processed. Once the generalized exemplars are learned, a test example can be classified by computing the Euclidean distance between the example and each of the generalized exemplars. If an example is contained inside a generalized exemplar, the distance to that generalized exemplar is zero. The class of the nearest generalized exemplar is output as the predicted class of the test example. Each hyperrectangle $\mathbf{H}^{j,L_j}$ which is labeled with class label L_j is represented by its lower left cornet $\mathbf{H}^{j,L_j}_{lower}$ and upper right cornet $\mathbf{H}^{j,L_j}_{upper}$. The distance between $\mathbf{H}^{j,L_j}$ and an example $\mathbf{x} = (x_1, \ldots, x_F)^T$ is defined as follows [8]:

$$D(\mathbf{x}, \mathbf{H}^{j,L_j}) = \sqrt{\sum_{i=1}^{F} (w_{f_i} \times (d_i(\mathbf{x}, \mathbf{H}^{j,L_j}))^2)} \qquad (4)$$

where w_{f_i} is weight of ith feature and

$$d_i(\mathbf{x}, \mathbf{H}^{j,L_j}) = \begin{cases} x_i - H^{j,L_j}_{upper,i} & \text{if } x_i > H^{j,L_j}_{upper,i} \\ H^{j,L_j}_{lower,i} - x_i & \text{if } x_i < H^{j,L_j}_{lower,i} \\ 0 & \text{otherwise} \end{cases}, \qquad (5)$$

where $H^{j,L_j}_{lower,i}$ is the ith element of $\mathbf{H}^{j,L_j}_{lower}$. In order to obtain the hyperrectangle rules from trained SVM, a support vector clustering (SVC) algorithm is used to determine the clusters and prototypes of each class. And then the hyperrectangle rules are constructed according to the obtained prototypes and SVs.

3.2 SVC Algorithm

The SVC clustering algorithm is presented by Ben-Hui [9]. Training set $T=\{x_1,...,x_i,...,x_l\}$ in input space R^n is mapped into a high dimension feature space F via mapping $\Phi: R^n \to F$. Then in the space F, the smallest enclosing hypersphere with center a and radius R is determined by solving the quadratic programming problem as follows [10]:

$$\max\ Q(\beta) = \sum_{i=1}^{l} K(x_i,x_i)\beta_i - \sum_{i,j=1}^{l} \beta_i\beta_j K(x_i,x_j) \qquad (6)$$
$$\text{s.t.}\ \ 0 \leq \beta_i \leq C, i=1,2,...,l$$

where C is slack factor which determines the trade off between the volume of the hypersphere and the number of target objects rejected. After solving the quadratic programming, the distance between a data sample and the center of feature space hypersphere is computed by

$$D(x_i) = \sqrt{\sum_{i,j=1}^{l} \beta_i\beta_j K(x_i,x_j) + K(x_i,x_i) - 2\sum_{j=1}^{l} K(x_j,x_i)\beta_j} \qquad (7)$$

The radius of smallest enclosing hypersphere in feature space is determined by $R = D(x_i) | \forall 0 < \beta_i < C$, and the contours that enclose the points in input space are defined by set $\{x | D(x) = R\}$. The adjacency matrix between pairs of points x_i and x_j whose images lie in or on the hypersphere in feature space is defined as follows:

$$A_{i,j} = \begin{cases} 1 & \text{for all } z \text{ on the line segment connecting } x_i \text{ and } x_j, \text{ if } D(z) \leq R \\ 0 & \text{otherwise} \end{cases} \qquad (8)$$

Clusters are now defined as the connected components of the graph induced by $A_{i,j}$. Checking the line segment is implemented by sampling a number of points. For each two points in the sample set, we take 10 points that lie on the line connecting these points and check whether they are within the hypersphere. If all the points on that line are within the hypersphere, the two sample points are assumed to belong to the same cluster. The most usually used kernel in SVC is Gaussian kernel $K(x_i,x_j) = e^{-p\|x_i-x_j\|^2}$. As discussed below, the parameters p and C control the number and support of rules.

3.3 Confidence and Support

Each discovered rule should have a measure of certainty associated with it which assesses the validity of the rule. There are two objective measures based on the structure of discovered rules and the statistics underlying them. One of them is the confidence. Confidence is referred to as reliability or accuracy which represents the strength or quality of a rule. Another is support which represents the percentage of data samples that the given rule satisfies. In order to control the number and validity of the hyperrectangles, we define the support and confidence of $\mathbf{H}^{j,L_j}$ as follows:

$$\text{conf.}(\mathbf{H}^{j,L_j}) = \frac{\text{samples with class label } y_j \text{ within } \mathbf{H}^{j,L_j}}{\text{all sample within } \mathbf{H}^{j,L_j}} \quad (9)$$

$$\text{supp.}(\mathbf{H}^{j,L_j}) = \frac{\text{samples within } \mathbf{H}^{j,L_j}}{\text{all of the samples with class label } L_j}$$

Rules that satisfy both a user-specified minimum confidence threshold (MCT) and minimum support threshold (MST) are referred to as strong association rules, and are considered interesting. On the contrary, rules with low support likely represent noises, or exceptional cases.

3.4 Construction of Hyperrectangle

For the training samples, its classification hyperplane can be obtained by SVM algorithm. Then for each class of samples with different class label, we use SVC algorithm to obtain its clusters and prototypes. Given a MCT and MST, the HRE algorithm follows an incremental procedure to generate $\mathbf{H}^{j,L_j}$. Beginning with a small hyperrectangel around the prototype, a partition test is applied on it. For every sample belong to the cluster, if the distance from it to the hyperrectangle is not equal zero, the hyperrectangle is extended to include the sample. Otherwise, another point in the cluster is selected to test using the same way. Repeat this procedure until one of the following conditions is satisfied,

(a) All of the samples of the cluster are covered by $\mathbf{H}^{j,L_j}$ and the confidence of $\mathbf{H}^{j,L_j}$ is smaller than user-specified MCT (as $\mathbf{H}^{2,-1}$ in Fig.1.).
(b) One of support vectors of the cluster is covered by $\mathbf{H}^{j,L_j}$ (as $\mathbf{H}^{1,-1}$ in Fig.1.).
(c) Some samples with opposite class label are covered by $\mathbf{H}^{j,L_j}$ and confidence of $\mathbf{H}^{j,L_j}$ is smaller than user-specified MCT (as $\mathbf{H}^{3,-1}$ in Fig.1.).
(d) A prototype of another cluster is covered by $\mathbf{H}^{j,L_j}$.

Under condition (a), $\mathbf{H}^{j,L_j}$ will cover all of the samples and have larger support, but perhaps it will cover some opposite samples. In condition (a) and (d), a MCT is specified to ensure $\mathbf{H}^{j,L_j}$ has a higher confidence. If a cluster does not contain any support vectors, in that case, condition (d) is applied to limit the size of $\mathbf{H}^{j,L_j}$ and it can avoid the hyperrectangle largely overlapping with other hyperrectangle.

Fig.1 shows the rules extracted from a trained SVM. Samples with negative class are separated into three clusters which are covered by hyperrectangles. Although some conditions are adopted to avoid the hyperrectangles overlapping each other, there still exist some overlaps and some samples lie in several hyperrectangles with different class labels. In this case, the distances between the samples and those hyperrectangles are calculated and the samples are assigned the same class label as the hyperrectangle which has the smallest distances to it. For example, sample $\mathbf{x}_1$ and $\mathbf{x}_2$ belong to both hyperrectangle $\mathbf{H}^{1,1}$ and $\mathbf{H}^{3,-1}$, because of $D(\mathbf{x}_1, H^{1,1}) < D(\mathbf{x}_1, H^{3,-1})$ and $D(\mathbf{x}_2, H^{1,1}) < D(\mathbf{x}_2, H^{3,-1})$, $\mathbf{x}_1$ and $\mathbf{x}_2$ belong to $\mathbf{H}^{1,1}$.

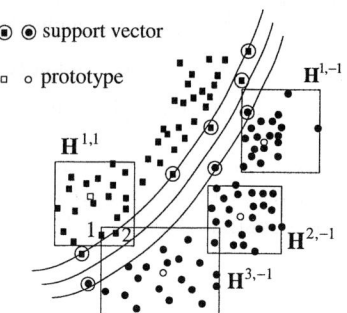

Fig. 1. Hyperrectangle rules exaction from the trained SVM. Samples with negative class are separated into three clusters

3.5 Parameters in HRE

There are two important parameters in HRE algorithm. One is the scale parameter p of Gaussian kernel and another is the penalty factor C in SVC. As discussed in [9], parameter p determines the number of the clusters. When p is small, there is only one cluster and the generated hyperrectangle rule has a larger support value (as shown in Fig.2 (a)), but it cannot cover all of the samples. With the increase of p the clustering boundary fit the data more tightly and splits into more clusters (as shown in Fig.2 (b)). Although the support value of each generated hyperrectangle rules is smaller, most of samples are covered by those hyperrectangle rules.

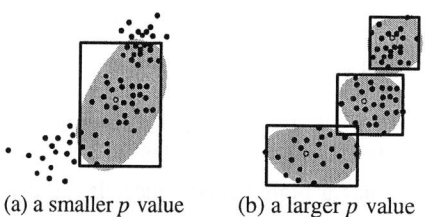

(a) a smaller p value   (b) a larger p value

Fig. 2. Influence of the scale parameter p of Gaussian kernel. With the increase of p the clustering boundary fit the data more tightly and splits into more clusters

As discussed in [10], the penalty factor C can be determined by setting a priori maximal allowed rejection rate of the error on the clusters. When parameter C is larger than 1, all of the samples will belong to the generated cluster including some noises and outliers. Furthermore, Scholkopf [11] proofs that the parameter $1/_{lC}$ is an upper bound for the fraction of samples outside the clusters. As shown in Fig.3 (b), when parameter C equals 0.5, there are two outliers are excluded from the cluster, and the hyperrectangle fit more tightly to the samples.

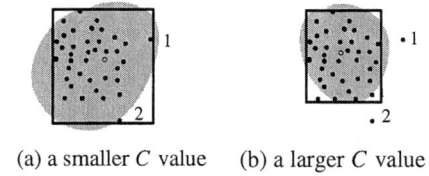

(a) a smaller C value   (b) a larger C value

Fig. 3. Influence of the scale parameter C. When parameter C equals 0.5, there are two outliers are excluded from the cluster, and the hyperrectangle fit more tightly to the samples

4 Experimental Results

In this section, we furthermore evaluate the performance of HRE algorithm in experiments. HRE is tested on UCI machine learning benchmark data sets [12] and Statlog collection [13]. Although HRE is originally designed for rule extraction rather than for classifying, in order to evaluate its performance, we use the hyperrectangle rules generated by HRE to predict unlabeled samples comparing with some common classification algorithms, such as CART, C4.5, k-NN, BP-neural network, naive bayes and radial basis function method.

Table 1. Domain characteristics of training set

Training Set	Training Set Size	Test Set Size	Number of Features
Iris	105	45	4
waveform	300	100	21
Adult	500	200	14
DNA	2000	1186	180
Australian	490	200	14
diabetes	668	100	8
heart disease	170	100	13
Letter	15000	5000	16
segment	1810	500	19
satimage	4435	2000	36
shuttle	43500	14500	9
vehicle	646	200	18

We present experiments on data sets: iris, waveform, adult, DNA, Australian, diabetes, heart disease, letter recognition, satimage, segment, shuttle and vehicle from UCI [12]. The training sets and their data features are reported in Table 1.

For each trial, the examples were divided into a training set and a test set. 70% of the examples were randomly chosen for each trial to be in the training set. Four different trials were run, and the final results were an average of those trials [7]. After training, the percentage of correct classification on the test set was measured. Table 2 gives the prediction error of those algorithms and HRE. Usually, the Gauss RBF kernel is more widely used than other kernel [9]. So we choose it as the kernel function in HRE algorithm. The parameters of HRE are obtained by ten-fold cross-validation. From Table 2, we can see the comparing performance of HRE with other algorithms. In some training sets such as iris, DNA, diabetes and satimage, the HRE algorithm has highest accuracy.

The prediction error of HRE on iris data set is 0.033, which is smaller than that of ellipse (0.040) and interval (0.040) rules reported in [5]. And the rules generated by HRE have less overlap. It shows that the quality of rules generated by HRE is better than that of ellipse and interval rules.

Note that the HRE algorithm isn't originally designed for a learning method i.e. a classification target. As discussed in [6,9], SVC clustering algorithm used in HRE spends most of training time on calculating kernel functions matrix which has a dimension equal to the number of training examples. So the training process of HRE is very time-consuming. But if the most usually used kernel functions are cached to memory in the process of training SVM, the training time of HRE algorithm will be significantly reduced as it needn't calculate kernel functions again.

Table 2. Predicted error on different training sets

Training Set	CART	C4.5	k-NN	Naive Bayes	BP-neural network	Radial Basis Function	HRE
Iris	0.070	0.081	0.064	0.067	**0.033**	0.041	**0.033**
Waveform	**0.068**	0.102	0.137	0.183	0.163	0.147	0.158
Adult	**0.083**	0.089	0.091	0.104	0.133	0.197	0.094
DNA	0.075	0.076	0.146	0.068	0.088	0.041	**0.040**
Australian	0.145	0.099	--	0.136	**0.087**	0.107	0.108
Diabetes	0.227	0.131	0.324	0.239	0.198	0.218	**0.113**
Heart disease	0.045	0.078	0.048	**0.037**	0.057	0.078	0.053
Letter	--	0.132	**0.068**	0.529	0.327	0.233	0.244
Segment	**0.040**	**0.040**	0.077	0.265	0.054	0.069	0.084
Satimage	0.138	0.150	0.094	0.287	0.139	0.121	**0.086**
Shuttle	**0.080**	0.100	0.440	0.450	0.430	0.140	0.143
Vehicle	0.235	0.266	0.275	0.558	**0.207**	0.307	0.376

Some hyperrectangle rules for iris data set with different parameters C and p are shown in Table 3, 4 and 5. The MCT and MST are set as 0.9 and 0.1 respectively. We can see that when $p=10$, there are only three rules which have larger support. With the increase of p, more rules are obtained and each of them has smaller support.

Table 3. Hyperrectangle rules for Iris data set with $C = 0.5$ and $p = 10$

Hyperrectangle Rules	[supp., conf.]
1. if $x_1 \in [4.4,5.7]$ $x_2 \in [2.9,4.4]$ $x_3 \in [1.2,1.9]$ $x_4 \in [0.1,0.5]$ then class 1	[0.91, 0.97]
2. if $x_1 \in [5.2,6.9]$ $x_2 \in [2.3,3.4]$ $x_3 \in [3.5.5.0]$ $x_4 \in [1.0,1.8]$ then class 2	[0.76, 0.94]
3. if $x_1 \in [5.6,7.4]$ $x_2 \in [2.5,3.4]$ $x_3 \in [4.8,6.3]$ $x_4 \in [1.5,2.5]$ then class 3	[0.78, 0.91]

Table 4. Hyperrectangle rules for Iris data set with $C = 0.5$ and $p = 60$

Hyperrectangle Rules	[supp., conf.]
1. if $x_1 \in [4.6,5.0]$ $x_2 \in [3.0,3.4]$ $x_3 \in [1.2,1.6]$ $x_4 \in [0.1,0.3]$ then class 1	[0.31, 1.00]
2. if $x_1 \in [5.0,5.4]$ $x_2 \in [3.3,3.8]$ $x_3 \in [1.4,1.7]$ $x_4 \in [0.2,0.5]$ then class 1	[0.34, 0.99]
3. if $x_1 \in [6.2,7.0]$ $x_2 \in [2.8,3.3]$ $x_3 \in [4.3,5.0]$ $x_4 \in [1.3,1.7]$ then class 2	[0.25, 1.00]
4. if $x_1 \in [5.4,5.9]$ $x_2 \in [2.3,3.0]$ $x_3 \in [3.7,4.5]$ $x_4 \in [1.0,1.5]$ then class 2	[0.35, 1.00]
5. if $x_1 \in [6.2,7.2]$ $x_2 \in [2.7,3.4]$ $x_3 \in [5.1,6.0]$ $x_4 \in [1.8,2.5]$ then class 3	[0.43, 0.99]
6. if $x_1 \in [6.0,6.3]$ $x_2 \in [2.5,3.0]$ $x_3 \in [4.8,5.1]$ $x_4 \in [1.5,1.9]$ then class 3	[0.12, 0.99]

As discussed in section 3, parameter $1/lC$ is an upper bound for the fraction of sample outside the clusters, so the smaller the parameter C is, the less support of a rule has. From Table 5 we can see that with $C = 0.1$, support values of the three rules in Table 3 are reduced to 0.72, 0.63 and 0.67.

Table 5. Hyperrectangle rules for Iris data set with $C = 0.1$ and $p = 10$

Hyperrectangle Rules	[supp., conf.]
1. if $x_1 \in [4.6,5.5]$ $x_2 \in [3.0,4.2]$ $x_3 \in [1.3,1.7]$ $x_4 \in [0.1,0.5]$ then class 1	[0.72, 0.99]
2. if $x_1 \in [5.4,6.9]$ $x_2 \in [2.4,3.4]$ $x_3 \in [3.5,4.9]$ $x_4 \in [1.0,1.6]$ then class 2	[0.63, 1.00]
3. if $x_1 \in [5.6,6.9]$ $x_2 \in [2.5,3.4]$ $x_3 \in [4.8,6.0]$ $x_4 \in [1.5,2.5]$ then class 3	[0.67, 1.00]

5 Conclusions

In this paper, a hyperrectangle rules extraction algorithm is proposed to extract rules from the trained SVM. A support vector clustering algorithm is used to find the prototypes of each class samples, and then hyperrectangle rules are constructed according the prototypes and SVs. Experimental results indicate that the HRE algorithm can extract rules efficiently and it is easy to control the number and support of the obtained rules.

Acknowledgements

This work is supported by the New Century 151 Talent Project of Zhejiang Province and the National Outstanding Youth Science Foundation of China (NOYSFC: 60025308).

References

1. Vapnik V. *The Nature of Statistical Learning Theory*. Springer, NewYork, 1995.
2. Craven M. W. and Shavlik J. W. Using Sampling and Queries to Extract Rules From Trained Neural Networks. In *Proceedings of the 11th International Conference on Machine Learning*, New Brunswick, NJ, pages 37-45, 1994.
3. R. Andrews, J. Diederich and A. Tickle. A Survey and Critique of Techniques for Extracting Rules from Trained Artificial Neural Networks, *Knowledge-Based Systems*, 8(6), pages 373-389, 1995.
4. Fu L. Rule learning by searching on adapted nets. *Proceedings of the 9th National Conference on Artificial Intelligence*, Anaheim, CA, pages 590-595, 1991.
5. H. Núñez, C. Angulo, A. Catala, Rule-extraction from Support Vector Machines, *The European Symposiumon Artificial Neural Networks*, Burges, pages 107-112, 2002.
6. C.J.C. Burges. A Tutorial on Support Vector Machines for Pattern Recognition. *Data Mining and Knowledge Discovery*, 2(2):1-47, 1998.
7. S. Salzberg, A Nearest Hyperrectangle Learning Method, *Machine Learning*, Kluwer Academic Publisher, Boston, pages 251-276, 1991.
8. Wettschereck D, Dietterich T. An Experimental Comparison of The Nearest-Neighbor and Nearest-hyperrectangle Algorithms. *Machine Learning,* 19(1):5–27, 1995.
9. A.Ben-Hui D, Horn H.T.Sidgelmann et.al. Support vector clustering. *Journel of machine learning research*, pages 125-137, 2001.
10. D.M.J. Tax and R.P.W. Duin. Data Domain Description by Support Vectors. In Proceedings ESANN, pages 251-256, 1999.
11. Bernhard Schölkopf and Alexander J. Smola. Learning with Kernels: Support Vector Machines, *Regularization, Optimization, and Beyond*. MIT Press, Cambridge, MA, 2002.
12. C. J. Merz and P. M. Murphy, UCI Repository for Machine Learning Data-Bases. Irvine, CA: University of California, Department of Information and Computer Science, http://www.ics.uci.edu/~mlearn/MLRepository.html, 1998.
13. D. Michie, D. J. Spiegelhalter, and C. C. Taylor. Machine Learning, *Neural and Statistical Classification*. Prentice Hall, Englewood Cliffs, N.J., 1994. Data available at http://www.ncc.up.pt/liacc/ML/statlog/datasets.html.

Pruning Derivative Partial Rules During Impact Rule Discovery

Shiying Huang and Geoffrey I. Webb

Faculty of Information Technology, Monash University,
Melbourne VIC 3800, Austrailia
{Shiying.Huang, Geoff.Webb}@infotech.monash.edu.au

Abstract. Because exploratory rule discovery works with data that is only a sample of the phenomena to be investigated, some resulting rules may appear interesting only by chance. Techniques are developed for automatically discarding statistically insignificant exploratory rules that cannot survive a hypothesis with regard to its ancestors. We call such *insignificant* rules *derivative extended rules*. In this paper, we argue that there is another type of derivative exploratory rules, which is *derivative* with regard to their children. We also argue that considerable amount of such derivative partial rules can not be successfully removed using existing rule pruning techniques. We propose a new technique to address this problem. Experiments are done in impact rule discovery to evaluate the effect of this derivative partial rule filter. Results show that the inherent problem of too many resulting rules in exploratory rule discovery is alleviated.

Keywords: Exploratory rule discovery, impact rules, rule significance, derivative rules.

1 Introduction

Exploratory rule discovery seeks to retrieve all implicit patterns and regularities that satisfy some user-defined set of constraints in a population, with respect to a set of available sample data. The best known such approach is *association rule discovery* [1]. Most approaches seeks rules $A \rightarrow C$ for which there is a correlation between the antecedent A and the consequent C. However, whenever one such rule is found, there is a risk that many derivative and potentially uninteresting rules $A' \rightarrow C'$ will also be found. These *derivative rules* are those for which there is a correlation between the antecedent and the consequent only by virtue of there being a correlation between A and C. For example, if A and C are correlated then for any term B that is unrelated to either A or C, AB will also turn out to be correlated with C.

Considerable research has been devoted to automatically identify and discard such derivative rules. The closed itemset techniques [12, 3, 16] can identify rules for which some elements can be removed without changing the support of the rule. Minimum improvement techniques [6] can to identify rules for which some elements can be removed without decreasing rule confidence. However, since exploratory rule discovery seeks to discover rules characterizing the features in a population, with respect to a given sample, rules may happen to be interesting simply due to sampling fluctuation. Statistical tests are also applied to assess whether there is evidence that no elements can be removed without significantly altering the status of the rule with respect to the population from which the sample data is drawn [11, 5, 9]. However, all these techniques relate only to identifying rules that are derivative due to the addition of irrelevant or unproductive elements.

There exists, however, another type of derivative rules that may also result in many rules that are likely to be of little interest to the user. For any rule $AB \rightarrow C$ which is not derivative from another rule and for which there is a correlation between the antecedent and the consequent, both A and B may each be correlated with C solely due to correlation between AB and C. In this case $A \rightarrow C$ and $B \rightarrow C$ will both be potentially uninteresting derivative rules that may be discovered by an exploratory rule discovery system.

The following example illustrates an occasion where such a potentially uninteresting rule may be generated.

Example 1. Suppose a retailer is trying to identify the groups of customers who is likely to buy some new products. After applying the impact rule discovery with the rule filters proposed by Huang and Webb [9, 10], two rules are identified as solutions:

$$District = A \rightarrow profit(coverage = 200, mean = 100)$$
$$District = A \ \& \ age < 50 \rightarrow profit(coverage = 100, mean = 200)$$

Although these two rules are both "significant" as is identified by the rule filter proposed by Huang and Webb [9], the first rule, which is an ancestor of the second one is misleading. Actually, no profit is produced by customers who belong to district A and are older than 50 years! The retailer's attention should more sensibly focus on the group of customers who are under age 50 in district A, instead of on all those in district A. Keeping the first rule in the resulting solutions may confuse the decision makers.

Impact rule discovery is a type of exploratory rule discovery that seeks rules for which the consequent is an undiscretized quantitative variable, referred to as the *target* and is described using its distribution. This paper investigates the identification of the second type of derivative rules in the context of impact rule discovery [9, 14].

The rest of this paper is organized like this: a brief introduction to exploratory rule discovery related concepts is presented in section 2. The definitions and notations of impact rule discovery is charaterized in section 3. Derivative impact rules are defined and relationship between different rules are clarified in section 4, together with the implementation of the derivative rule filter in section 3. Experimental results are evaluated in section 5, which is followed by our conclusions in section 6.

2   Exploratory Rule Discovery

Many machine learning systems discover a single model from the available data that is expected to maximize some objective function of interestingness on unknown future data. Predictions or classifications are done on the basis of this single model [15]. However, alternative models may exist that perform equally well. Thus, it is not always sensible to choose only one of the "best" models. Moreover the criteria for deciding whether a model is best or not also varies with the context of application. Exploratory rule discovery techniques overcome this problem by searching for multiple models which satisfy certain user-defined set of constraints and present all these models to the users to provide them with alternative choices. Greater flexibility is achieved in this way.

Exploratory rule discovery techniques [9] are classified into propositional rule discovery which seeks rules with qualitative attributes only and distributional-consequent rule discovery which seeks rules with undiscretized quantitative variables as consequent. Propositional rules are composed of Boolean conditions only. While the status or performance of the undiscretized quantitative attributes in distributional-consequent rules are described with their distributions. *Association rule discovery* [1], *contrast sets discovery* [5] and *correlation rule discovery* [8] are examples of propositional exploratory rule discovery, while *impact rule* [14] or *quantitative association rule discovery* [2], as is variously known, belongs to the class of distributional-consequent rule discovery. It is argued that distributional-consequent rules are able to provide better descriptions of the interrelationship between quantitative variables and qualitative attributes.

Considering the differences between propositional rule discovery and distributional-consequent rule discovery, there are inherent differences between the techniques for propositional and distributional-consequent rule pruning and optimizations. Researchers have devoted extensive efforts to develop rule pruning and optimization techniques. Reviews of such work can be found in many related works [9].

We define some key notions of exploratory rule discovery as follows:

1. For propositional rule discovery, a *record* is an element to which we apply Boolean predicates called conditions, while for distributional-consequent rule discovery, a record is a *pair* $<c, v>$, where c is the nonempty set of Boolean conditions, and v is a set of values for the quantitative variables in whose distribution the users are interested.
2. Rule r_1 is a parent of r_2 if the body of r_1 is a subset of the body of r_2. If the cardinality of the body of r_1 is smaller than that of r_2 by 1, then the second rule is referred to as a *direct parent* of the first rule, otherwise, it is a non-direct ancestor of the first rule.
3. We use the notion $coverset(A)$, where A is a conjunction of conditions, to represent the set of records that satisfy A. If a record x is in $coverset(A)$, we say that x is *covered* by A. If A is an $\emptyset$, $coverset(A)$ includes all the records in the database. $Coverage(A)$ is the number of records covered by A. $coverage(A) = |coverset(A)|$.

3   Impact Rule Discovery

We construct our impact rule discovery algorithm on the basis of OPUS [13] search algorithm, which enables successful discovery of the top k impact rules that satisfy a certain set of user-specified constraints.

We characterized the terminology of k-optimal impact rule discovery to be used in this paper as follows:

1. An impact rule takes the form of $A \rightarrow target$, where the target is describe by the following measures: *coverage, mean, variance, maximum, minimum, sum* and *impact*. This is an example of impact rules discovered by our algorithm:

$Address = Brighton$ & $profession = programmer \rightarrow income$
$(coverage : 23\%, mean : 60000, variance : 4000, max : 75000,$
$min : 44000, sum : 1380000, impact : 3903.98)$

2. *Impact* is an interestingness measure suggested by Webb [14][1]: $impact(A \rightarrow target) = (mean(A \rightarrow target) - \overline{targ}) \times coverage(A))$.
3. An k-optimal impact rule discovery task is a 6-tuple: $KOIRD(D, \mathcal{C}, \mathcal{T}, \mathcal{M}, \lambda, k)$.

   D: is a nonempty set of records, which is called the database. A record is a pair $< c, v >, c \subseteq \mathcal{C}$ and v is a set of values for $\mathcal{T}$. D is an available sample from the global population $\mathcal{D}$.

   $\mathcal{C}$: is a nonempty set of Boolean conditions, which are the set of available conditions for impact rule antecedents, which is generated from the given data in D.

   $\mathcal{T}$: is a nonempty set of the variables in whose distribution we are interested.

   $\mathcal{M}$: is a set of constraints. A constraint is a criteria which the resulting rules must satisfy.

   λ: $\{X \rightarrow Y\} \times \{D\} \rightarrow \mathcal{R}$ is a function from rules and databases to values and defines an interestingness metric such that the greater the value of $\lambda(X \rightarrow Y, D)$ the greater the interestingness of this rule given the database.

   k: is a user specified integer number denoting the number of rules in the ultimate set of solutions for this rule discovery task.

Pseudo code of the original algorithm for impact rule discovery is described in table 1. In this table, *current* is the set of conditions, whose supersets (children) are currently being explored. *Available* is the set of conditions that may be added to *current*. By adding the conditions in *available* to *current* one by one, the antecedent of the *current rule*: $New \rightarrow target$, is produced. *Rule_list* is an ordered list of the top-k interesting rules we have encountered by now.

The search space of this algorithm is illustrated in figure 1. Each node in this search space is connected with a potential impact rule, whose antecedent

---

[1] In this formula, $mean(A \rightarrow target)$ denotes the mean of the *targets* covered by A, and $coverage(A)$ is the number of the records covered by A.

Table 1. OPUS_IR

```
Algorithm: OPUS_IR(Current, Available, M)

  1. SoFar := ∅
  2. FOR EACH P in Available
     2.1 New := Current ∪ P
     2.2 IF current rule New → target does not satisfy any of the prunable constraints in
         M
             THEN go to step 2.
     2.4 ELSE IF current rule New → target satisfies all the nonprunable constraints in M
             Record New → target in the rule_list;
     2.5 OPUS_IR(New, SoFar, (M));
     2.6 SoFar := SoFar ∪ P
     2.7 END IF
  3. END FOR
```

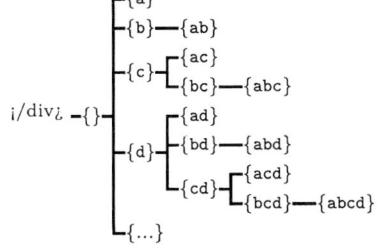

Fig. 1. Fixed search space for OPUS_IR

is composed of the conditions between the braces. By performing a depth-first search through such a search space, the algorithm is guarantee to access every nodes and generate all potential impact rules. Based on the OPUS structure, powerful search space pruning is facilitated [13], making it suitable for discovering impact rules in vary large, dense databases. The completeness of OPUS based algorithms is proved by Webb [13].

4 Derivative Partial Impact Rules

Techniques for automatically discarding potentially uninteresting rules are extensively explored, examples are the constraint-based techniques, the non-redundant techniques and the techniques regarding the rule improvement and statistically significance. The first classes of techniques seek to identify whether a rule r fails to satisfy the constraints in $\mathcal{M}$. The second class of techniques assess whether the resulting rules are redundant or not by reference to the sample data. Example of non-redundant rule discovery techniques are the closed set related techniques and the trivial rule filter. Each assessment of whether r is desirable is not always free from the risk that the rule is not correct with respect to $\mathcal{D}$ due to the sampling fluctuation [15].

The third class was proposed to reduce the influence of sampling on resulting rules. Statistical tests have been utilized for discarding potentially uninteresting rules generated due to sampling fluctuation, both in context of propositional and distributional-consequent rule discovery. For propositional rule discovery, Brin et al. [8] proposed a pruning technique for removing insignificant correlation rules using a chi-square test; Liu et al. [11] also made use of the chi-square test to identify the significance of association rules with fixed consequents. Bay and Pazzani [5] applied a significance test to remove the insignificant contrast sets in STUCCO. Webb [15] sought to control the number of potentially uninteresting association rules which happen to be interesting due to the sampling by applying a Fisher exact test. For distributional consequent rule discovery, Aumann and Lindell [2] applied a standard z test to quantitative association rule discovery and Huang and Webb [9] also developed an insignificance filter in impact rule discovery whose efficiency is considerably improved by introducing several efficiency improving techniques for rule discovery in very large, dense databases.

However, the techniques mentioned above can only successfully remove a subset of derivative rules.

4.1 Relationship Among Rules

As is argued in the introduction, there are derivative rules other than the derivative extended rules that the existing techniques cannot successfully remove. Even after both rules, $A \rightarrow target$ and $A \& B \rightarrow target$, have been identified as nonderivative extended rules, there is still a risk that either or both of them are potentially uninteresting. For example, if the target mean of $coverset(A\&\neg B)$ is not significantly higher than the target mean of $coverset(\neg A)$, it can be asserted that the notably high target mean for $coverset(A)$ derives solely from that of $coverset(A\&B)$, which is only a subset of $coverset(A)$. Such rules are defined as *derivative partial rules*, which are insignificant compared to fundamental rules which are their children.

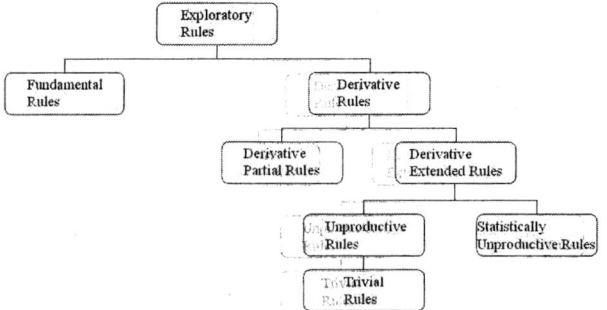

Fig. 2. Relationship of different rules

The inter-relationships among different rules are explained in figure 2. In this figure, fundamental rules can also be regarded as non-derivative rules. Derivative extended rules are those referred to as insignificant rules in previous research. Unproductive rules are those exhibit no improvement in target mean comparing with their parent rules. Trivial rules are rules whose antecedents cover exactly the same records as one of their parent rules. As was proved by Huang and Webb [9], trivial rules are a subset of unproductive rules. while those that are productive, with respect to the sample, but fail the significance test are all classified as statistically unproductive.

4.2 Derivative Partial Rules and Implementation

We first define derivative partial impact rules:

Definition 1. *A non-derivative extended impact rule, $A \rightarrow target$ is an derivative partial rule, iff there exists a condition x, not included in A, where the target mean for $coverset(A) - coverset(A\&x)$ is not higher than the target mean for $coverset(\neg A)$ at a user specified level of significance.*

$$InsigPartial(A \rightarrow target) = \exists x \in (\mathcal{C} - \{A\}),$$
$$TarMean(coverset(A\&\neg x)) \succ TarMean(coverset(\neg A))$$

Statistical Test. Since by performing the exploratory rule discovery, we are aiming at discovering rules that characterize the features of the population with reference to sample data, hypothesis tests must be done to identify whether an impact rule is derivative or not. A t test is applied to assess whether a partial rule is derivative with regard to its children.

Table 2. Derivative rule Filter

```
Algorithm: OPUS_IR_Filter(Current, Available, M)

1. SoFar := ∅
2. FOR EACH P in Available
   2.1 New := Current ∪ P
   2.2 IF New satisfies all the prunable constraints in M except the nontrivial
       constraint THEN
       2.2.1 current_rule = New → target
       2.2.2 IF the mean of current_rule is significantly higher than all its direct parents
             THEN
             2.2.2.1 add the parent rules to the parent_rule_list
             2.2.2.2 IF the rule satisfies all the other non-prunable constraints in M.
                     THEN record Rule to the ordered rule_list
             2.2.2.3 END IF
             2.2.2.4 FOR EACH parent_rule in parent_rule_list
                     IF parent_rule is a derivative partial rule with regard to
                     current_rule
                     THEN delete parent_rule from rule_list.
                     END IF
             2.2.2.5 END FOR
       2.2.3 OPUS_IR(New, SoFar, M)
       2.2.4 SoFar := SoFar ∪ P
       2.2.5 END IF
   2.3 END IF
3. END FOR
```

Implementation. The new algorithm with derivative partial rule filter is provided in table 2. In this algorithm all the parent rules of the *current_rule* are stored in the *parent_rule_list* while checking whether *current_rule* is a derivative extended rule or not. After *current_rule* is identified as perspectively *fundamental* the derivative partial rule filter is then applied to check whether the parents are derivative with regard to *current_rule*. Derivative parent rules are deleted from the *rule_list*. Since all the parent rules of *current_rule* has already been explored before *current_rule* (please refer to the search space of OPUS_IR), every derivative rule is guaranteed to be removed.

5 Experimental Evaluations

We study the effectiveness of the algorithm in table 2 for the derivative partial rule filter by applying it to 10 large databases chosen from KDD archives [4] and UCI machine learning repository [7], in which many attributes are quantitative. Great differences exist among these databases with the smallest database in size having less than 300 records and the greatest having 2000 times as many records as that of the smallest. Number of attributes vary from only 9 to almost 90. Since complex interrelationships exist among the data, there is a strong necessity for rule pruning. We choose a target attribute from among the quantitative attributes in each database, and discretize the rest using a 3-bin equal-frequency discretization. After discretization the numbers of available conditions turn out to be over 1500 for some of the databases. The significance level for the derivative rule filters is 0.05.

We did the experiments using following protocol. First, the program in table 1 is run using the insignificance filter proposed by Huang and Webb [9] to find the top 1000 significance rules from each database, with maximum number of conditions on rule antecedent set to 3, 4 and 5. Then, the algorithm with derivative partial rule filter in table 2 is executed to remove derivative partial rules from the resulting solutions. Results are organized in table 4. The numbers of fundamental rules found after both filters are applied are those before the slashes. Integers after the slashes are those found using the insignificance filter only. Decreases in resulting rules are also presented in percentage.

Here is an example of an impact rule which is discarded as derivative partial

$Sex = M \rightarrow Shucked\_weight(coverage : 1528, mean : 0.432946,$
$variance : 0.049729, min : 0.0065, max : 1.351, sum : 661.542, impact : 112.428)$

It is derivative regarding its parent rule:
$Sex = M \& 1.0295 <= Whole\_weight \rightarrow Shucked\_weight(coverage : 687,$
$mean : 0.619229, variance : 0.0284951, min : 0.315, max : 1.351,$
$sum : 425.411, impact : 178.525)$

In this example, if an abalone is male but have a whole weight less than 1.0295 cannot have a very high shucked weight. The first rule is thus misleading!

From the experimental results in table 4, we make the following observation: When the number of maximum conditions on rule antecedent increases, gen-

Table 3. Basic information of the databases

database	records	attributes	conditions	Target
Abalone	4117	9	24	Shuckedweight
Heart	270	13	40	Max heart rate
Housing	506	14	49	MEDV
German credit	1000	20	77	Credit amount
Ipums.la.97	70187	61	1693	Total income
Ipums.la.98	74954	61	1610	Total income
Ipums.la.99	88443	61	1889	Total income
Ticdata2000	5822	86	771	Ave. income
Census income	199523	42	522	Wage per hour
Covtype	581012	55	131	Elevation

Table 4. Experimental results

database	MNC=3		MNC=4		MNC=5	
Abalone	82/86	4.65%	127/138	7.97%	149/173	13.87%
Heart	43/57	24.56%	63/80	21.25%	81/100	19.0%
Housing	131/171	23.39%	168/255	34.12%	192/288	33.33%
German credit	152/197	22.84%	213/273	21.98%	222/295	24.75%
Ipums.la.97	949/1000	5.1%	867/1000	13.3%	809/1000	19.1%
Ipums.la.98	944/1000	5.6%	890/1000	11.0%	761/1000	23.9%
Ipums.la.99	959/1000	4.1%	930/1000	7.0%	896/1000	10.4%
Ticdata2000	803/1000	19.7%	739/1000	26.1%	674/1000	32.6%
Census income	894/1000	10.6%	776/1000	22.4%	744/1000	25.6%
Covtype	918/1000	8.2%	829/1000	17.1%	733/1000	26.7%

erally, more derivative partial rules are produced by the impact rule discovery system. The greatest change for the numbers of resulting rules after the derivative partial rule filter is applied is as much as 34%. Even the database with a slightest change saw a decrease of over 4%. This justify the argument that there are considerable amount of derivative partial rules still exist in the resulting rules even after the derivative extended rule filter (insignificance filter) is applied. The derivative partial rules can be pruned using our proposed algorithm in reasonable period of time.

6 Conclusions

Exploratory rule discovery searches for multiple models within a set of given data to represent the underlying patterns or regularities. However, it often results in large numbers of rules. Sometimes, the resulting rules are too numerous for human to analysis. Research has investigated techniques for automatically discarding potentially uninteresting rules, thus reducing the number of rules and removing those that are unlikely to be of fundamental interest. One class of these techniques is to apply statistical tests to the resulting models, so as to alleviate the risk of accepting rules which appear to be interesting by reference to the given data which is only a sample, instead of the real world population. In this paper, we argued that there is a type of potentially uninteresting rules which existing techniques fail to remove. We call these rules derivative rules. A derivative rule filter is developed in a impact rule discovery system. Experiments showed a considerable decrease in the number of resulting rules.

References

1. R. Agrawal, T. Imielinski, and A. N. Swami. Mining association rules between sets of items in large databases. In *Proceedings of the 1993 ACM SIGMOD International Conference on Management of Data*.
2. Y. Aumann and Y. Lindell. A statistical theory for quantitative association rules. In *Knowledge Discovery and Data Mining*, pages 261–270, 1999.
3. Yves Bastide, Nicolas Pasquier, Rafik Taouil, Gerd Stumme, and Lotfi Lakhal. Mining minimal non-redundant association rules using frequent closed itemsets. pages 972–986, 2000.
4. S. D. Bay. The uci kdd archive [http://kdd.ics.uci.edu], 1999.
5. S.D. Bay and M.J. Pazzani. Detecting group differences: Mining contrast sets. In *Data Mining and Knowledge Discovery*, pages 213–246, 2001.
6. Roberto J. Bayardo, Jr., Rakesh Agrawal, and Dimitrios Gunopulos. Constraint-based rule mining in large, dense databases. *Data Min. Knowl. Discov.*, 4(2-3):217–240, 2000.
7. C.L. Blake and C.J. Merz. UCI repository of machine learning databases, 1998.
8. S. Brin, R. Motwani, and C. Silverstein. Beyond market baskets: Generalizing association rules to correlations. In Joan Peckham, editor, *SIGMOD 1997, Proceedings ACM SIGMOD International Conference on Management of Data, May 13-15, 1997, Tucson, Arizona, USA*, pages 265–276. ACM Press, 1997.
9. Shiying Huang and Geoffrey I. Webb. Discarding insignificant rules during impact rule discovery in large database, 2004. Accepted for publication in SIAM Data Mining Conference 2005.
10. Shiying Huang and Geoffrey I. Webb. Efficiently identification of exploratory rules' significance, 2004.
11. B. Liu, W. Hsu, and Y. Ma. Pruning and summarizing the discovered associations. In *Knowledge Discovery and Data Mining*, pages 125–134, 1999.
12. Nicolas Pasquier, Yves Bastide, Rafik Taouil, and Lotfi Lakhal. Closed set based discovery of small covers for association rules. In *Proc. 15emes Journees Bases de Donnees Avancees, BDA*, pages 361–381, 1999.
13. G. I. Webb. OPUS: An efficient admissible algorithm for unordered search. *Journal of Artificial Intelligence Research*, 3:431–465, 1995.
14. G. I. Webb. Discovering associations with numeric variables. In *Proceedings of the seventh ACM SIGKDD international conference on Knowledge discovery and data mining*, pages 383–388. ACM Press, 2001.
15. G.I. Webb. Statistically sound exploratory rule discovery, 2004.
16. M. J. Zaki. Generating non-redundant association rules. In *Knowledge Discovery and Data Mining*, pages 34–43, 2000.

$\mathcal{IGB}$: A New Informative Generic Base of Association Rules

Gh. Gasmi[1], S. Ben Yahia[1,2], E. Mephu Nguifo[2], and Y. Slimani[1]

[1] Départment des Sciences de l'Informatique,
Faculté des Sciences de Tunis,
Campus Universitaire, 1060 Tunis, Tunisie
{sadok.benyahia, yahya.slimani}@fst.rnu.tn
[2] Centre de Recherche en Informatique de Lens-IUT de Lens,
Rue de l'Université SP 16, 62307 Lens Cedex
mephu@cril.univ-artois.fr

Abstract. The problem of the relevance and the usefulness of extracted association rules is becoming paramount, since an overwhelming number of association rules may be derived from even reasonably sized real-life databases. A possible solution consists in using results of Formal Concept Analysis to generate a generic base of association rules. This set, of reduced size, makes it possible to derive all the association rules via an adequate axiomatic system. In this paper, we introduce a novel generic and informative base of association rules, conveying two types of knowledge: "factual" and "implicative". We present also a valid and complete axiomatic system allowing to derive the set of all association rules. Results of the experiments carried out on real-life databases showed important profits in terms of compactness of the introduced generic base.

Keywords: Association rules, Generic base, Galois connection, Axiomatic system.

1 Introduction

The problem of the relevance and usefulness of extracted association rules is of primary importance. Indeed, in most real life databases, thousands and even millions of high-confidence rules are generated, among which many are redundant. This problem encouraged the development of tools for rule classification, according to their properties, for rule selection according to user-defined criteria, and for rule visualization. The *Selection without loss of information* is mainly based on the extraction of a generic subset of all association rules, called *generic base*, from which the remaining (redundant) association rules are generated.

In this paper, we introduce a new generic base of association rules called $\mathcal{IGB}$. Through $\mathcal{IGB}$, we introduce a novel characterization of generic association rules instead of the classical one, *i.e.*, exact and approximative. In fact, we shall distinguish between the *"factual"* and the *"implicative"* generic association rule. Indeed, a factual generic association rule, fulfilling the premise part emptiness, permits to highlight item correlation without any conditionality. However, for an implicative generic association rule, where the premise part is not empty, item correlation is conditioned by the existence of premise items. The introduced generic base $\mathcal{IGB}$ fulfills the "informativeness property", *i.e.*, the support and the confidence of the derived association rules can be exactly retrieved. In order to derive valid association rules from the $\mathcal{IGB}$ base, we introduce an axiomatic system, which it is shown to be valid and complete.

We conducted several experiments on typical benchmarking datasets to assess the $\mathcal{IGB}$ compactness. The introduced generic rule characterization permitted to explain the "atypical" behavior of the variation of the reported generic association rules number versus the variation of the *minconf* value, *i.e.*, the number of the reported rules does not necessarily decrease with the augmentation of the *minconf* value.

The remainder of the paper is organized as follows. Section 2 presents the basic mathematical foundations for the derivation of generic bases of association rules. We devote section 3 to a review of the literature relating to the extraction of the generic bases. Section 4 introduces a novel informative base of generic association rules and the associated axiomatic system. Results of the experiments carried out on real-life databases are reported in section 5. The conclusion and future work are presented in section 6.

2 Mathematical Background

Due to lack of available space, interested reader for key results from the Galois lattice-based paradigm in FCA is referred to [1].

Frequent Closed Itemset: An itemset $I \subseteq \mathcal{I}$ is said to be *closed* if $I = \omega(I)$, and is said to be *frequent* with respect to the *minsup* threshold if support(I)= $\frac{|\psi(I)|}{|\mathcal{O}|} \geq minsup$ [2].

Minimal Generator: An itemset $g \subseteq \mathcal{I}$ is said to be *minimal generator* of a closed itemset I, if and only if $\omega(g) = I$ and does not exist g' $\subseteq$ g such that $\omega(g') = I$ [3, 4].

Iceberg Galois Lattice: When only frequent closed itemsets are considered with set inclusion, the resulting structure $(\hat{\mathcal{L}}, \subseteq)$ only preserves the *join* operator [1]. This is called a join semi-lattice or upper semi-lattice. In the remaining of the paper, such structure is referred to as "*Iceberg Galois Lattice*" [5]. Therefore, given an Iceberg Galois lattice in which each closed itemset is "decorated" with its associated list of minimal generators, generic bases of association rules can be derived in a straightforward manner. Indeed, generic approximative rules represent "inter-node" implications, assorted with the confidence measure, between two comparable equivalence relation classes, *i.e.*, from a sub-closed-itemset to a super-closed-itemset when starting from a given node in the partially ordered structure [6].

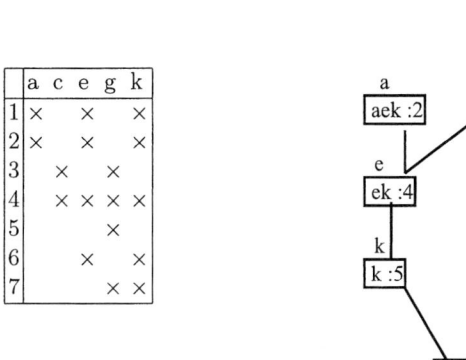
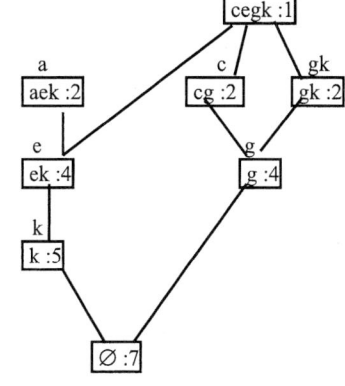

Fig. 1. Left: The formal context $\mathcal{K}$ **Right:** Associated Iceberg Galois lattice for $minsup = 1$

Example 1. Let us consider the formal context $\mathcal{K}$ given by Figure 1 (Left). The associated Iceberg Galois lattice, for minsup=1, is depicted by Figure 1 (Right)[1]. Each node in the Iceberg is represented as a couple (closed itemset, support) and is decorated with its associated minimal generator list.

3 Extraction of Generic Bases of Association Rules

The problem of the relevance and the utility of the association rules is of primary importance. This is due to the high number of the association rules extracted from the real-life databases and the presence of the high percentage of redundant rules conveying the same information. In the literature, we can witness the presence of some techniques to prune such set of rules, mainly based on statistical metrics [7]. In what follows, we put the focus on the results issued from the FCA, to retrieve a reduced set of rules without information loss. Indeed, this reduced set, called *base*, is composed of generic rules and should fulfill the following requirements:

- *"Informativeness":* The generic base of association rules, allows to retrieve exactly the support and confidence of the derived (redundant) rules.
- *"Derivability"*: An inference mechanism should be provided (*e.g.*, an axiomatic system). The axiomatic system has to be valid (*i.e.*, should forbid derivation of non valid rules) and complete (*i.e.*, should enable derivation of all redundant valid rules).

A critical review of the dedicated literature permitted to mainly report the following previous works:

1. **Representative Rules:** In [8], Kryszkiewicz introduced a new syntactic derivation operator, called the "*Cover*", defined as follows: Cover(X ⇒Y) = $\{X \cup Z \Rightarrow V \mid Z, V \subseteq Y \wedge Z \cap V = \emptyset \wedge V \neq \emptyset\}$.

---

[1] We use a separator-free form for sets, e.g., *cg* stands for $\{c, g\}$.

Based on the Cover operator, the author defined a minimal base of rules called *representative rules (RR)*, such that: $\mathcal{RR}=\{R \in \mathcal{AR} \mid \nexists\ R' \in \mathcal{AR}, R \neq R'$ and $R \in Cover(R')\}$ where $\mathcal{AR}$ is the set of all valid association rules. $\mathcal{RR}$ was redefined in [9] under the name of *"representative basis (RB)"*. However, the premise and the conclusion parts of the generic rules of $\mathcal{RB}$ are not necessarily disjoint. To derive redundant rules from $\mathcal{RB}$, the author proposed an axiomatic system composed of Left augmentation and Decomposition axioms

As pointed out in [10], $\mathcal{RR}$ is the smallest rule set that covers all valid association rules by means of the Cover operator. However, it is not informative (*i.e.,* it does not allow to retrieve exactly the support and the confidence of derived rules). In [8,9], the authors were the first to consider generic association rules whose premise part can be empty. However, no attention was paid to a semantic interpretation of this type of knowledge.

2. **Generic basis for exact association rules and Informative basis for approximative association rules:** Bastide *et al.* characterized what they called *"Generic basis for exact association rules"* (adapting the global implication base of *Duquenne and Guigues* [11]) which is defined as follows:
Let $\mathcal{FCI}$ be the set of frequent closed itemsets extracted from the context and, for each frequent closed itemset I, let us denote $\mathcal{G}_I$ the set of minimal generators of I.

$$\mathcal{GBE} = \{R : g \Rightarrow (I - g) \mid I \in \mathcal{FCI} \wedge g \in \mathcal{G}_I \wedge g \neq I\}.$$

The authors also characterized what they called *"Informative basis for approximative association rules"* (adapting the partial implications of *Luxenberger* [12]) which is defined as follows: $\mathcal{GBA} = \{R : X \stackrel{c}{\Rightarrow} (Y\text{-}X), Y \in \mathcal{FCI} \wedge \omega(X) \subset Y \wedge c = \text{confidence}(R) \geq \text{minconf}\}$.

As pointed out in [10], by using the Cover operator as axiomatic system, the couple, proposed by Bastide *et al.* [3], ($\mathcal{GBE},\mathcal{GBA}$) forms a subset of associative rules which is information lossless. A couple of valid and complete axiomatic systems for $\mathcal{GBE}$ and $\mathcal{GBA}$, respectively, was given in [13]. However, the couple ($\mathcal{GBE},\mathcal{GBA}$) suffers from a huge number of generic rules, in particular, when we consider dense databases.

4 A New Generic Base

The couple of generic bases proposed by Bastide *et al.* is informative. However, it suffers from a huge number of generic association rules. The extraction of generic base $\mathcal{RR}$ ($\mathcal{RB}$) is the smallest rule set covering all valid association rules by means of the cover operator. However, it is not information lossless. Hence, our contribution consists in introducing a new approach permitting to extract informative generic base for association rules which is more compact than ($\mathcal{GBE},\mathcal{GBA}$). Thus, we begin by redefining association rule-redundancy as follows:

Definition 1. *Let $\mathcal{AR}$ be the set of all association rules that can be derived from an extraction context $\mathcal{K}$. $R: X \Rightarrow Y \in \mathcal{AR}$ is redundant with respect to R_1: $X_1 \Rightarrow Y_1$ if it fulfills the following conditions:*

1. $\omega(XY) = X_1 Y_1$
2. $X_1 \subseteq X \wedge Y \subset Y_1$

Based on the former definition, we introduce the following definition of the new generic base called $\mathcal{IGB}$:

Definition 2. *Let $\mathcal{FCI}$ be the set of frequent closed itemsets and $\mathcal{G}_I$ the set of minimal generators of a frequent closed itemset I.*
$\mathcal{IGB} = \{ R: g_s \Rightarrow (I\text{-}g_s) \mid I \in \mathcal{FCI} \wedge I \neq \emptyset \wedge g_s \in \mathcal{G}_{I'}, I' \in \mathcal{FCI} \wedge I' \subseteq I \wedge confidence(R) \geq minconf \wedge \nexists\, g' / g' \subset g_s \wedge confidence(g' \Rightarrow I\text{-}g') \geq minconf \}$.

Proposition 1. *The generic base $\mathcal{IGB}$ is informative, i.e., the support and the confidence of all derived rules can be exactly retrieved from $\mathcal{IGB}$.*

Proof. Our approach consists in finding for each non empty frequent closed itemset I, the smallest minimal generator g_s of a frequent closed itemset I' subsumed by I and fulfilling the *minconf* constraint. Thus, generic association rules of $\mathcal{IGB}$ have the following form: $g_s \Rightarrow I\text{-}g_s$. Therefore, we are able to reconstitute all frequent closed itemset by concatenation of the premise and the conclusion parts of a generic rule. Since the support of a frequent itemset is equal to the support of the smallest frequent closed itemset containing it, then the support and the confidence of all derived rules can be exactly retrieved.

4.1 The $\mathcal{IGB}$ Generic Base Construction

The $\mathcal{IGB}$ construction algorithm takes as input the set of all frequent closed itemsets $\mathcal{FCI}$ extracted by using one of the dedicated algorithms[2].

Proposition 2. *Let I be a non empty frequent closed itemset, if $support(I) \geq minconf$, then the generic association rule $R: \emptyset \Rightarrow I \in \mathcal{IGB}$.*

Proof. Proposition 2 derives straightforwardly from Definition 2. Since confidence $(R:\emptyset \Rightarrow I) = support(I)$, then the generic rule $\emptyset \Rightarrow I$ is valid. Hence, R presents the largest conclusion that can be drawn from the frequent closed itemset I since there is no another rule $R':X' \Rightarrow Y'$ such that $X' \subset \emptyset$ and $I \subseteq Y'$.

The $\mathcal{IGB}$ construction algorithm is based on Proposition 2. So, it considers the set of frequent closed itemsets $\mathcal{FCI}$. For each non empty closed itemset I, it checks whether its support is greater than or equal to *minconf*. If it is the case, then we generate the generic rule $R:\emptyset \Rightarrow I$. Otherwise, it has to look for the smallest minimal generator g_s, associated to a frequent closed itemset subsumed by I, and then generates the generic rule $R:g_s \Rightarrow I\text{-}g_s$ if the *minconf* threshold is reached.

---

[2] A critical survey of these algorithms can be found in [6].

Example 2. Let us consider the extraction context given by Figure 1 (Left). Table 1 shows the running process of the $\mathcal{IGB}$ construction algorithm for minsup $=\frac{1}{7}$ and minconf $=\frac{1}{5}$.

Table 1. Running process of the $\mathcal{IGB}$ construction algorithm

Iteration	non empty I $\in \mathcal{FCI}$	support(I)> minconf	$I_1 \subseteq I$	$L_{smallest-gen}$	$\mathcal{IGB}$
1	cegk	No	k, ek, cg, g, gk, cegk	k, e, c, g	$k \overset{\frac{1}{5}}{\Rightarrow} ceg$ $e \overset{\frac{1}{4}}{\Rightarrow} cgk$ $c \overset{\frac{1}{2}}{\Rightarrow} egk$ $g \overset{\frac{1}{4}}{\Rightarrow} cek$
2	aek	yes			$\emptyset \overset{\frac{2}{5}}{\Rightarrow} aek$
3	ek	yes			$\emptyset \overset{\frac{4}{7}}{\Rightarrow} ek$
4	k	yes			$\emptyset \overset{\frac{5}{7}}{\Rightarrow} k$
5	cg	yes			$\emptyset \overset{\frac{2}{5}}{\Rightarrow} cg$
6	gk	yes			$\emptyset \overset{\frac{2}{5}}{\Rightarrow} gk$
7	g	yes			$\emptyset \overset{\frac{4}{7}}{\Rightarrow} g$

4.2 Generic Association Rule Semantics

In the following, we have to discuss semantics attached to an association rule R:X$\overset{c}{\Rightarrow}$Y-X. Usually, R expresses that the probability of finding Y with a value c depends on the presence of X. Thus, X constitutes a constraint for the correlation Y items. In the $\mathcal{IGB}$ base, we can find generic association rules whose premise part can be empty. Such rules were considered in [9, 10], but very little attention was paid to a semantic interpretation attached to this type of knowledge.

Let us consider the extraction context given by Figure 1 (Left). For *minconf*=$\frac{1}{7}$ and applying Bastide *et al.* approach, we obtain among the possibly extracted generic association rules, c $\Rightarrow$ egk, e $\Rightarrow$ cgk, g $\Rightarrow$ cek, k $\Rightarrow$ ceg. However, does the correlation probability of the items c, e, g and k depend on the presence of c, e, g or k? Actually, the probability of the correlation of c, e, g and k with a value greater than or equal to *minconf* does not depend on any condition. Thus, we propose to represent such type of correlation by only one generic association rule, *i.e.*, R: $\emptyset \Rightarrow$ cegk. The generic base $\mathcal{IGB}$ contains then two types of knowledge: (*i*) "Implicative knowledge" represented by a generic association rule whose the premise part is not empty. (*ii*) "Factual knowledge" represented by a generic association rule whose the premise part is empty.

4.3 Redundant Association Rule Derivation

In order to derive the set of all valid redundant association rules, we propose in what follows an axiomatic system and we prove that it is valid (*i.e.*, should forbid derivation of non valid rules) and that it is complete (*i.e.*, should enable derivation of all the valid rules).

Proposition 3. *Let us consider the generic base denoted by $\mathcal{IGB}$ and the set of all valid association rules extracted from $\mathcal{K}$, denoted by $\mathcal{AR}$. The following axiomatic system is valid.*

A0. Conditional reflexivity: *If $X \stackrel{c}{\Rightarrow} Y \in \mathcal{IGB} \wedge X \neq \emptyset$ then $X \stackrel{c}{\Rightarrow} Y \in \mathcal{AR}$*

A1. Augmentation: *If $X \stackrel{c}{\Rightarrow} Y \in \mathcal{IGB}$ then $X \cup Z \stackrel{c'}{\Rightarrow} Y\text{-}\{Z\} \in \mathcal{AR}, Z \subset Y$.*

A2. Decomposition: *If $X \stackrel{c}{\Rightarrow} Y \in \mathcal{AR}$ then $X \stackrel{c}{\Rightarrow} Z \in \mathcal{AR}, Z \subset Y \wedge \omega(XZ) = XY$.*

Proof. **A0. Conditional reflexivity:** follows from the proper definition of the $\mathcal{IGB}$.

A1. Augmentation: Since $X \stackrel{c}{\Rightarrow} Y \in \mathcal{IGB}$ then confidence($X \stackrel{c}{\Rightarrow} Y$)=c $\Leftrightarrow$ $\frac{support(Y)}{support(X)}$ = c $\geq$ minconf. Since $X \subset XZ$, then support(X)> support(XZ) and minconf < $\frac{support(XY)}{support(X)}$ < $\frac{support(XY)}{support(XZ)}$. Thus, $X \cup Z \stackrel{c'}{\Rightarrow} Y\text{-}\{Z\}$ is a valid association rule having a confidence value equal to c'= $\frac{support(XY)}{support(XZ)}$.

A2. Decomposition: Since, $X \stackrel{c}{\Rightarrow} Y \in \mathcal{AR}$ then confidence($X \stackrel{c}{\Rightarrow} Y$)=c $\geq$ minconf, c=$\frac{support(XY)}{support(X)}$ then support(XY)=c × support(X). Also, we have $\omega(XZ) = \omega(XY)$, then support(XZ)=support(XY) consequently, support(XZ) = c × support(X). Thus, $X \stackrel{c}{\Rightarrow} Z$ is a valid associative rule.

Remark 1. The constraint of non emptiness of the the premise was introduced in respect to an implicit "habit" stipulating that the premise part of an association rule is usually non empty.

Proposition 4. *The proposed axiomatic system is complete: the set of all associative rules extracted from $\mathcal{K}$ are derivable from $\mathcal{IGB}$ by using the proposed axiomatic system.*

Proof. Let $\mathcal{IGB}$ be the generic base extracted from the extraction context $\mathcal{K}$ for given minsup et minconf. $\mathcal{AR}$ denotes the set of all association rules extracted from $\mathcal{K}$ and $\mathcal{FCI}$ the set of frequent closed itemsets.

Let R:X$\Rightarrow$Y-X $\in \mathcal{AR}$. In the following, we have to show that R can be derived from a generic association rule of $\mathcal{IGB}$ by the application of the proposed axiomatic system.

- If Y ∈ $\mathcal{FCI}$ then two cases are possible:
  1. Then, there is no rule R':X'⇒Y-X' ∈ $\mathcal{AR}$ such that X'⊂X, then
     - if support(Y)< minconf then R:X⇒Y-X ∈ $\mathcal{IGB}$. R:X⇒Y-X ∈ $\mathcal{AR}$ by application of the conditional reflexivity axiom.
     - Else it exists a rule R":∅⇒Y∈ $\mathcal{IGB}$. By application of the augmentation axiom to R", we obtain the rule R:X⇒Y-X.
  2. It exists a rule R":X"⇒Y-X"∈ $\mathcal{IGB}$ such that X"⊂X' ∧ X"⊂X. By application of the augmentation axiom to R":X"⇒Y-X", we obtain R:X⇒Y-X.
- Otherwise, it exists a rule R':X⇒Y'-X∈ $\mathcal{AR}$ such that Y'∈ $\mathcal{FCI}$ ∧ ω(Y') = ω(Y). Then, it exists a rule R":X"⇒Y'-X"∈ $\mathcal{IGB}$ such that X"⊆X. We apply firstly, the augmentation axiom to R":X"⇒Y'-X"(if X"⊂X) in order to obtain R':X⇒Y'-X. Next, we apply the decomposition axiom to R' to find R:X⇒Y-X.

5 Experimental results

We carried out experimentations on benchmarking datasets, in order to evaluate the number of generic association rules. We implemented algorithms in the C language under Linux Fedora Core 2. Physical characteristics of the machine are: a PC pentium 4 with a CPU clock rate of 3.06 Ghz and a main memory of 512 Mo.

In the following, we put the focus on the variation of the reported generic rule number of the different generic bases versus the minconf value variation.

- $\mathcal{IGB}$: For *minconf=minsup*, $\mathcal{IGB}$ contains only factual generic association rules. The number of factual generic association rules is equal to the number of frequent closed itemsets. This can be explained by the fact that all frequent closed itemset supports are equal to or greater than *minconf*.
- By increasing the *minconf* value, the number of factual generic rules decreases until reaching 0 when *minconf*=1. Indeed, by varying *minconf*, each factual generic rule is substituted by a number of implicative generic rules equal to the cardinality of $L_{smallest-gen}$. Thus, the more this cardinality is important, the more increases the number of generic association rules of $\mathcal{IGB}$. A singularity for the Mushrooms dataset is noteworthy. In fact, the number of factual generic rules is equal to 1 even for *minconf*=1 (usually, it is equal to 0). This can be explained by the fact that the item coded by '85' appears in all dataset transactions. Thus, for any value of *minconf*, the factual generic rule ∅ →85 is always valid.
- ($\mathcal{GBE}$,$\mathcal{GBA}$): We note that the number of exact generic rules of $\mathcal{GBE}$ is insensitive to the variation of *minconf* value. However, the more *minconf* increases, the more decreases the number of approximate generic rules of $\mathcal{GBA}$. Indeed, by increasing *minconf*, the number of minimal generators satisfying the *minconf* constraint decreases.
- $\mathcal{RR}$: When *minconf=minsup*, the set of generic rules of $\mathcal{RR}$ is composed only of factual generic association rules and their number is equal to that

of maximal frequent closed itemset. Following the variation of *minconf*, the number of implicative generic rules, substituting factual generic rules, determines the number of generic rules of $\mathcal{RR}$.

In what follows, we discuss the compactness degree of generic bases. We note that the gap between $\mathcal{IGB}$ and the set of all valid rules extracted using the Apriori algorithm is more important in dense datasets. Indeed, when *minconf* is less than 100%, the compactness degree of $\mathcal{IGB}$ ranges from 0.4% to 80%. Conversely to sparse datasets, compactness degree is limited to a value varying between 41% and 100%. This gap widens by lowing support values. The compactness degree of $\mathcal{RR}$ base ranges from 0.06% to 80% for dense datasets, while it ranges from 26% to 100% for sparse datasets. For sparse datasets, $(\mathcal{GBE},\mathcal{GBA})$ contains the set of all valid rules. This is can be explained by the fact that for sparse datasets, the set of frequent itemsets is equal to the set of frequent closed itemsets and the set of minimal generators. However, for dense datasets, the compactness degree of $(\mathcal{GBE},\mathcal{GBA})$ varies between 8% and 80%.

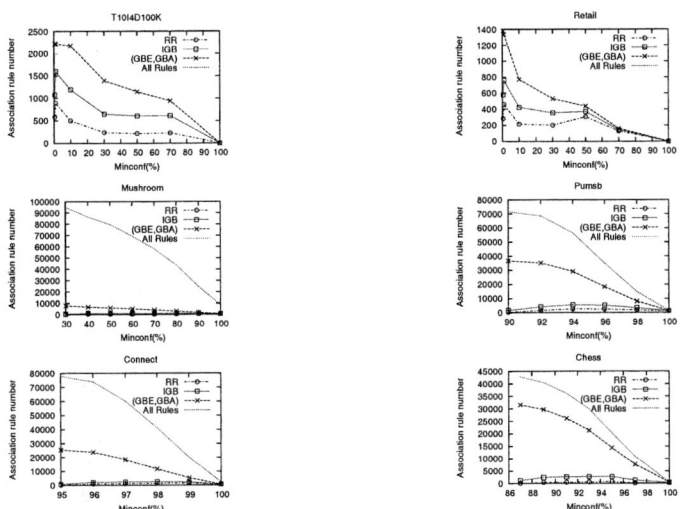

Fig. 2. The generic association rules variation versus minconf variation

6 Conclusion

In this paper, we introduced a novel informative generic base, for association rules which is more compact than $(\mathcal{GBE},\mathcal{GBA})$. We also provided a valid and complete axiomatic system, composed of the conditional reflexivity, augmentation and decomposition axioms, permitting to derive the set of all valid rules. We distinguished between two types of generic rules; "*factual*" and "*implicative*" ones. We also implemented algorithms of $\mathcal{IGB}$, $(\mathcal{GBE},\mathcal{GBA})$ and $\mathcal{RR}$ construction. Experimental results carried out on benchmarking datasets showed important profits in terms of compactness of the introduced generic base. In the near future, we plan to examine the potential benefits from integrating the new base in a query expansion system and in a generic bases visualization environment [14].

References

1. Ganter, B., Wille, R.: Formal Concept Analysis. Springer-Verlag (1999)
2. Pasquier, N., Bastide, Y., Touil, R., Lakhal, L.: Pruning closed itemset lattices for association rules. In Bouzeghoub, M., ed.: Proceedings of 14th Intl. Conference Bases de Données Avancées, Hammamet, Tunisia. (1998) 177–196
3. Bastide, Y., Pasquier, N., Taouil, R., Lakhal, L., Stumme, G.: Mining minimal non-redundant association rules using frequent closed itemsets. In: Proceedings of the Intl. Conference DOOD'2000, LNCS, Springer-verlag. (2000) 972–986
4. Godin, R., Mineau, G.W., R.Missaoui, Mili, H.: Méthodes de Classification Conceptuelle Basées sur le Treillis de Galois et Applications. Revue d'intelligence Artificielle **9** (1995) 105–137
5. Liquière, M., Nguifo, E.M.: Legal (learning with galois lattice): Un système d'apprentisage de concepts à partir d'exemples. In: Proceedings of the Intl. 5th Journées Francaises de l'apprentissage, Lannion, France. (1990) 93–114
6. BenYahia, S., Nguifo, E.M.: Approches d'extraction de règles d'association basées sur la correspondance de galois. Ingénierie des Systèmes d'Information (ISI), Hermès-Lavoisier **3–4** (2004) 23–55
7. Tan, P., Kumar, V., Srivastava, J.: Selecting the right interestingness measure for association patterns. In: Proceedings of the Eight ACM SIGKDD International Conference on Knowledge Discovery and Data Mining (ICDM'02), ACM Press. (2002) 32–41
8. Kryszkiewicz, M.: Representative association rules. In: Research and Development in Knowledge Discovery and Data Mining. Proc. of Second Pacific-Asia Conference (PAKDD). Melbourne, Australia. (1998) 198209
9. Luong, V.P.: Raisonnement sur les règles d'association. In: Proceedings 17ème Journées Bases de Données Avancées BDA'2001, Agadir (Maroc), Cépaduès Edition. (2001) 299–310
10. Kryszkiewicz, M.: Concise representations of association rules. In Hand, D.J., Adams, N., Bolton, R., eds.: Proceedings of Pattern Detection and Discovery, ESF Exploratory Workshop, London, UK. Volume 2447 of Lecture Notes in Computer Science., Springer (2002) 92–109
11. Guigues, J., Duquenne, V.: Familles minimales d'implications informatives résultant d'un tableau de données binaires. Mathématiques et Sciences Humaines (1986) 5–18
12. Luxenburger, M.: Implication partielles dans un contexte. Mathématiques et Sciences Humaines **29** (1991) 35–55
13. BenYahia, S., Nguifo, E.M.: Revisiting generic bases of association rules. In: Proceedings of 6th International Conference on Data Warehousing and Knowledge Discovery (DaWaK 2004),LNCS 3181, Springer-Verlag, Zaragoza, Spain. (2004) 58–67
14. BenYahia, S., Nguifo, E.M.: Emulating a cooperative behavior in a generic association rule visualization tool. In: Proceedings of 16th IEEE International Conference on Tools with Artificial Intelligence (ICTAI'04), Boca Raton, Florida. (2004) 148–155

A Divide and Conquer Approach for Deriving Partially Ordered Sub-structures

S. Ben Yahia, Y. Slimani, and J. Rezgui

Département des Sciences de l'Informatique,
Faculté des Sciences de Tunis,
Campus Universitaire, 1060 Tunis, Tunisie
{sadok.benyahia, yahya.slimani}@fst.rnu.tn

Abstract. The steady growth in the size of data has encouraged the emergence of advanced main memory trie-based data structures. Concurrently, more acute knowledge extraction techniques are devised for the discovery of compact and lossless knowledge formally expressed by generic bases. In this paper, we present an approach for deriving generic bases of association rules. Using this approach, we construct small partially ordered sub-structures. Then, these ordered sub-structures are parsed to derive, in a straightforward manner, local generic association bases. Finally, local bases are merged to generate the global one. Extensive experiments carried out essentially showed that the proposed data structure allows to generate a more compact representation of an extraction context comparatively to existing approaches in literature.

1 Introduction

Classical approaches for extraction of such implicit knowledge suffer from the huge number of potentially interesting correlations (specially association rules) that can be drawn from a dataset. In order to limit the number of the reported rules, while conserving the "informativeness" property, a battery of results, provided by the mathematical foundations of the Formal Concept Analysis, yielded a compact and lossless subset of association rules, called *generic bases of association rules* [1]. In order to derive generic bases of association rules, the extraction of knowledge base problem can be reformulated as follows:(*1*)Discover two distinct "closure systems", *i.e.*, sets of sets which are closed under the intersection operator,: the set of closed itemsets and the set of associated minimal generators. Also, the *upper covers* (Cov^u) of each closed itemset should be available. (*2*)From all the information discovered in the first step, *i.e.*, two closure systems and the upper covers sets, derive generic bases of association rules (from which all the remaining rules can be derived). It is noteworthy that the recently proposed approaches advocate the use of advanced data structure, essentially based on tries structures, to store compactly in main memory input dataset [2, 3] or to store partial outputs (*e.g.*,[4]).

In this paper, we propose a new trie-based data structure called ITEMSET-TRIE. The ITEMSET-TRIE extends the idea claimed by the authors of FP-TREE [2] and CATS [3] structures, aiming to improve storage compression and to allow (closed) frequent pattern mining without "explicit" candidate itemset generation step. Next, we propose an algorithm, falling in the characterization "Divide and Conquer" to extract frequent closed itemsets **with their associated minimal generators**. Hence, the derivation of approximative generic association rules is based on the exploration of such closed itemsets organized upon their natural partial order (also called *precedence relation*). That's why we construct on the fly, concurrently with the closed itemsets discovery process, the local "iceberg lattice" [5]. Such local ordered sub-structures can be drawn quite naturally in a parallel manner. Then, these ordered sub-structures are parsed to derive, in a straightforward manner, local generic bases of association rules. Finally, local bases are merged to generate the global one. Such process can be recapitulated as follows: (*i*)Construct the ITEMSET-TRIE, (*ii*)Construct the local ordered structures,(*iii*)Merge the local generic association rules to derive a global one.

The remainder of the paper is organized as follows : In Section 2, we present the ITEMSET-TRIE. Section 3 introduces the construction of the partially ordered structures topic[1]. Section 4 discusses preliminary results on the practical performances of the presented algorithms. Section 5 concludes the paper and points out future directions to follow.

2 Itemset-Trie Data Structure

In the context of mining frequent (closed) patterns in transaction databases or many other kinds of databases, an important number of studies rely on Apriori-like "test-and-generate" approach[2]. However, this approach suffers from a very expensive candidate set generation step, especially with long patterns or under low user-requirements. This drawback is reinforced with tediously repeated disk-stored database scans. To avoid the approach bottleneck, recent studies (e.g, the pioneering work of Han *et al.* and its FP-TREE structure [2]) proposed to adopt an advanced data structure, where the database is compressed in order to achieve pattern mining. The idea behind the compact data structure FP-TREE is that when multiple transactions share an identical frequent itemset, they can be merged into one with a registered number of occurrences. Beside a costly sorting step, the proposed FP-TREE structure is unfortunately not suited for an interactive mining process, in which a user may be interested in varying the support value. In this case, the FP-TREE should be rebuilt since its construction is support dependent. Although the work presented in [3] tackles this insufficiency, the proposed structure, called CATS, in which a single item is represented in a node. That's why we introduce a, support independent, more compact structure called ITEMSET-TRIE, in which each node is composed by an itemset. To illustrate this compactness, let us consider the extraction context given by Figure 1(Up). Figure 1(a) depicts the associated FP-TREE, while Figure 1(b) represents the associated ITEMSET-TRIE. Indeed, we remark that the

---

[1] Please note that algorithm pseudo-codes are omitted due to lack of available space.
[2] For a critical overview of these approaches, please refer to [6].

associated ITEMSET-TRIE is more compact than the corresponding FP-TREE, since it contains only 7 nodes and 3 levels while the FP-TREE contains 12 nodes and 6 levels.

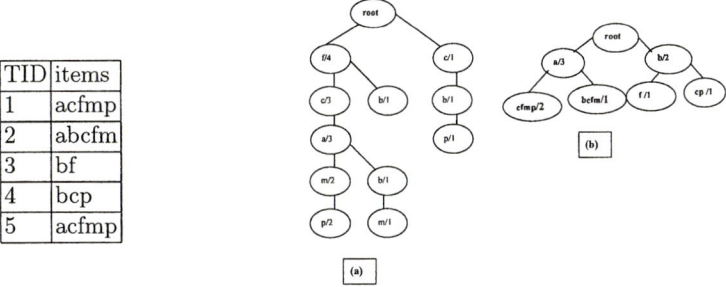

Fig. 1. (**Left**) The extraction context $\mathcal{K}$ (**Right**) FP-TREE and the ITEMSET-TRIE associated to the extraction context $\mathcal{K}$

3 Construction of the Partially Ordered Structures

As output of the first step, we constructed the ITEMSET-TRIE. In order to perform a generic association rule base extraction (specially the approximative one), we need to construct partially ordered structures based on the precedence relation. As we work only with closed itemsets, the order construction needs to retrieve the precedence relation from the family of closed itemsets. The main objective (and contribution also) of our approach is to discover the closed itemsets and to order them on the fly. This is performed in a gradual process, *i.e.*, by linking one closed itemset at a time to a structure which is only partially finished. We do not aim at constructing only one ordered structure from the input relation (which turns to construct the *Hasse* diagram), but instead, we look for constructing several ordered structures. Of course, some redundancy will appear, *i.e.*, a given closed itemset can appear in more than one ordered structure, but we avoid the expensive cost of *Hasse* diagram construction [7].

Once the ITEMSET-TRIE tree is built, it can be used to mine closed itemsets and their associated minimal generators repeatedly for different support thresholds settings without the need to rebuild the tree. Like FP-GROWTH [2] and FELINE [3], the proposed algorithm falls in the association rules mining algorithms characterization "Divide-and-conquer". The initial ITEMSET-TRIE is fragmented into conditional sub-tries. Indeed, given a pattern called **p**, a **p**'s conditional ITEMSET-TRIE tree is built, representing faithfully all transactions that contain the pattern p. For example, given the extraction context given by Figure 1(Up), the set of 1-itemsets, with their associated supports, is as follows: $< a/3; b/3; c/4; f/4; m/3; p/3 >$. Hence, we have to derive the **a**'s, **b**'s and so on conditional itemset-tries.

It is noteworthy that unlike FP-GROWTH [2] and CLOSET [8] algorithms, we consider only the lexicographic order and we consider that in a given conditional trie all the remaining 1-itemset should be included. For example, in the above mentioned algorithms (*i.e.*, FP-GROWTH and CLOSET), the conditional **b**'s trie will not include the 1-itemset {**a**} and that of **c** will exclude both {**a**} and {**b**}. The authors, aiming to discover only frequent closed itemsets, argue that there is no need to include the 1-itemset {a} in the b's one, since all frequent closed itemsets containing {a} have been already extracted for the a's conditional trie. In our approach, we aim to extract closed itemsets and their associated minimal generators to construct their associated ordered structure (*i.e.*, Hasse diagram). Since we plan to lead the mining process in a parallel manner, by assigning to each processor a subset of the conditional tries set, each sub-trie should contain an exhaustive description to ensure frequent closed itemsets discovery correctness and to minimize the inter-processors communication cost to check itemsets inclusions.

Example 1. Let us consider the extraction context given by Figure 1(Up). Below, we describe the ordered structures construction for minsup=1. The set of 1-itemsets, with their associated supports, is defined as follows: <a/3 ;b/3 ;c/4 ;f/4 ;m/3 ;p/3>. Then starting with the a's conditional ITEMSET-TRIE, we can find the associated itemset L_a list :$< b/1; c/3; f/3; m/3; p/2 >$. From such list we remark that the 1-itemsets c,f and m are as frequent as the 1-itemset a. Hence, they constitute a closed itemset {$acfm$} with a support equal to 2 and with the 1-itemset {a} as its minimal generator. The 1-itemsets c,f and m are removed from L_a. Since it is not empty, we have to go recursively further in depth and to construct the sub-tries, respectively for the 2-itemsets {ab} and {ap}. From L_{ab}, we discover the closed itemset {$abcfm$} with a support equal to 1 and with the 2-itemset {ab} as its minimal generator. While from L_{ap}, we discover the closed itemset {$acfmp$} with support equal to 1 and with the 2-itemset {ap} as its minimal generator. The treatment of L_a ends since there are no more elements to handle. As output, the local Hasse diagram (associated with the a's conditional ITEMSET-TRIE) can be drawn incrementally. Indeed, the in-depth of L_a list enables to connect, first, the closed itemsets {$acfm$} and {$abcfm$}, and second to connect {$acfm$} and {$acfmp$}. The algorithm has to deal next with the L_b list :$< a/1; c/2; f/2; m/1; p/1 >$, extracted from the conditional trie. We can easily check that no 1-itemset is so frequent as **b** and then {b} is a closed itemset. Since the remaining list to develop is not empty, we go further in depth and we start with the 2-itemset {ab}. L_{ab} is defined as follows: $< c/1; f/1; m/1 >$ from that list we discover the closed itemset {$abcfm$} with a support equal to 1 and with the 2-itemset {ab} as its minimal generator. There is no more exploration of this list since it is empty. The closed itemset {b} is connected to the closed itemset {$abcfm$}. Next, we have to tackle L_{bc} which is equal to $< a/1; f/1; m/1; p/1 >$. Any 1-itemset in this list is so frequent as {bc} and then we can conclude that {bc} is a closed itemset with a support equal to 2 and having {bc} as its minimal generator. The list with which to go further in depth remains unchanged. We have respectively to handle L_{abc},

L_{bcf} and L_{bcm} lists, all yielding the closed itemset $\{abcfm\}$. The closed itemset $\{bc\}$ is connected to that of $\{abcfm\}$. Next, we have to connect $\{b\}$ to $\{bc\}$. This is performed after systematically checking whether they share a common immediate successor, which is the case in this example. In fact, $\{bc\}$ and $\{b\}$ are connected respectively to their immediate successor which is $\{abcfm\}$. That's why we have to delete the link between $\{b\}$ and $\{abcfm\}$. The processing of the L_{bc} list ends by launching the L_{bcp} list, which gives the closed itemset $\{bcp\}$ with a support equal to 1 and with $\{bcp\}$ as its minimal generator.

4 Experimental Results

This section presents some experimental results of the proposed algorithms. ITEMSET-TRIE and SUB-TRIE algorithms were written in C, and were running on a parallel machine IBM SP2 with 32 processors. All experiments have been conducted on both sparse and dense datasets[3].

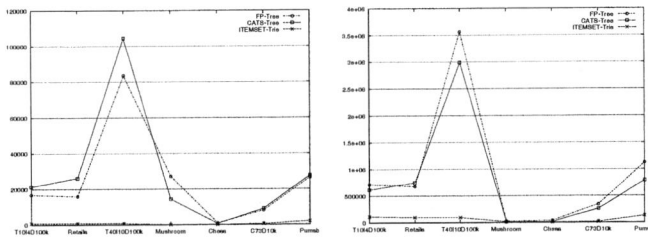

Fig. 2. (**Left**)Number of nodes of ITEMSET-TRIE vs those of FP-TREE and CATS, respectively. (**Right**)Size of ITEMSET-TRIE vs those of FP-TREE and CATS, respectively

From reported statistics illustrated by Figure 2, we can remark that the proposed data structure is by far more compact than FP-TREE and CATS, by reducing both the number of nodes and the required main memory for storing input datasets[4]. In fact, sparse datasets in average are represented at most with 24.4% from their actual size. While, dense datasets in average are represented at most with 18.8% from their actual size. Figure2 assessing ITEMSET-TRIE construction algorithm performances, shows that the latter is correlated to the number of levels. Moreover, it is highly dependent of the actual size of the dataset. We also remark that the number of levels for dense datasets is higher than that of sparse datasets. In fact, the more dense the dataset is, the higher the number of levels (*c.f.*, the CHESS base reaches 25 levels). In the sequel, we are interested in the study of another aspect putting the focus on the construction of ordered substructures algorithm performances. To provide a better idea about trends that lay behind the construction algorithm, we recorded statistics for sparse datasets

---
[3] Freely downloadable at: http://fimi.cs.helsinki.fi/data.
[4] All structures were constructed for minsup=1.

of variable size, by adding increments of fixed size (*e.g.*, 1000 transactions for the T10I4D100K base). We remarked that performances are linearly dependent of the number of transactions. However, this augmentation tends to stagnate reaching a given threshold number of transactions. For dense datasets, we were constrained, due to a lack of available main memory, to evaluate performances by varying the number of items. We remarked that the execution time is linearly dependent of the number of items. Also, execution times obtained for the C73D10K base are less important than those obtained for the CHESS base. This difference can be simply explained by the density, highlighted by the corresponding number of levels, of the considered bases (15 levels for the C73D10K base vs 25 levels for the CHESS base).

5 Conclusion

We presented in this paper a new data structure to extract frequent closed itemsets in order to generate generic bases of association rules. Then, we proposed an algorithm to construct local ordered structures from which it is possible to derive generic bases of association rules. We examined benefits from implementing the proposed approach on an *MIMD* machine (IBM SP2). Indeed, the construction method leads to a natural parallelization, in the sense where each processor of a parallel architecture can construct locally its ordered structure. Once the local structures are constructed, a master processor can merge them to derive a global generic base of association rules.

References

1. Pasquier, N., Bastide, Y., Taouil, R., Lakhal, L.: Efficient Mining of Association Rules Using Closed Itemset Lattices. Information Systems Journal **24** (1999) 25–46
2. Han, J., Pei, J., Yin, Y.: Mining frequent patterns without candidate generation. In: Proceedings of the ACM-SIGMOD Intl. Conference on Management of Data (SIGMOD'00), Dallas, Texas. (2000) 1–12
3. Cheung, W., Zaiane, O.: Incremental mining of frequent patterns without candidate generation or support constraint. In: Proceedings of the Seventh International Database Engineering and Applications Symposium (IDEAS 2003), Hong Kong, China. (2003)
4. Grahne, G., Zhu, J.: Efficiently using prefix-trees in mining frequent itemsets. In Goethals, B., Zaki, M.J., eds.: Proceedings of Workshop on Frequent Itemset Mining Implementations (FIMI '03), Florida, USA, IEEE (2003)
5. Stumme, G., Taouil, R., Bastide, Y., Pasquier, N., Lakhal, L.: Computing iceberg concept lattices with TITANIC. J. on Knowledge and Data Engineering (KDE) **2** (2002) 189–222
6. BenYahia, S., Nguifo, E.M.: Approches d'extraction de règles d'association basées sur la correspondance de galois. Ingénierie des Systèmes d'Information (ISI), Hermès-Lavoisier **3–4** (2004) 23–55
7. Ganter, B., Wille, R.: Formal Concept Analysis. Springer-Verlag (1999)
8. Pei, J., Han, J., Mao, R., Nishio, S., Tang, S., Yang, D.: Closet: An efficient algorithm for mining frequent closed itemsets. In: Proceedings of the ACM SIGMOD DMKD'00, Dallas,TX. (2002) 21–30

Finding Sporadic Rules Using Apriori-Inverse

Yun Sing Koh and Nathan Rountree

Department of Computer Science, University of Otago, New Zealand
{ykoh, rountree}@cs.otago.ac.nz

Abstract. We define sporadic rules as those with low support but high confidence: for example, a rare association of two symptoms indicating a rare disease. To find such rules using the well-known Apriori algorithm, minimum support has to be set very low, producing a large number of trivial frequent itemsets. We propose "Apriori-Inverse", a method of discovering sporadic rules by ignoring all candidate itemsets above a maximum support threshold. We define two classes of sporadic rule: perfectly sporadic rules (those that consist only of items falling below maximum support) and imperfectly sporadic rules (those that may contain items over the maximum support threshold). We show that Apriori-Inverse finds all perfectly sporadic rules much more quickly than Apriori. We also propose extensions to Apriori-Inverse to allow us to find some (but not necessarily all) imperfectly sporadic rules.

1 Introduction

Association rule mining has become one of the most popular data exploration techniques, allowing users to generate unexpected rules from "market basket" data. Proposed by Agrawal et al. [1, 2], association rule mining discovers all rules in the data that satisfy a user-specified minimum support (minsup) and minimum confidence (minconf). Minsup represents the minimum amount of evidence (that is, number of transactions) we require to consider a rule valid, and minconf specifies how strong the implication of a rule must be to be considered valuable.

The following is a formal statement of association rule mining for transactional databases. Let $I = \{i_1, i_2, \ldots, i_m\}$ be a set of items and D be a set of transactions, where each transaction T is a set of items such that $T \subseteq I$. An association rule is an implication of the form $X \to Y$, where $X \subset I$, $Y \subset I$, and $X \cap Y = \emptyset$. X is referred to as the *antecedent* of the rule, and Y as the *consequent*. The rule $X \to Y$ holds in the transaction set D with *confidence* $c\%$ if $c\%$ of transactions in D that contain X also contain Y. The rule $X \to Y$ has *support* of $s\%$ in the transaction set D, if $s\%$ of transactions in D contain $X \cup Y$ [2]. One measure of the predictive strength of a rule $X \to Y$ is its *lift* value, calculated as confidence($X \to Y$) / support(Y). Lift indicates the degree to which Y is more likely to be present when X is present; if lift is less than 1.0, Y is less likely to be present with X than Y's baseline frequency in D. The task of generating association rules is that of generating all rules that meet minimum support and minimum confidence, and perhaps meet further requirements such as having lift greater than 1.0.

The Apriori algorithm and its variations are used widely as association rule mining methods. However, several authors have pointed out that the Apriori algorithm, by definition, hinders us from finding rules with low support and high confidence [3, 4, 5]. Apriori generates *frequent* itemsets (i.e. those that will produce rules with support higher than minsup) by joining the frequent itemsets of the previous pass and pruning those subsets that have a support lower than minsup [2]. Hence, to generate rules that have low support, minsup must be set very low, drastically increasing the running time of the algorithm. This is known as the *rare item problem*. It means that, using the Apriori algorithm, we are unlikely to generate rules that may indicate events of potentially dramatic consequence. For example, we might miss out rules that indicate the symptoms of a rare but fatal disease due to the frequency of incidences not reaching the minsup threshold. Some previous solutions to this problem are reviewed in Section 2.

The aim of our research is to develop a technique to mine low support but high confidence rules effectively. We call such rules "sporadic" because they represent rare cases that are scattered sporadically through the database but with high confidence of occurring together. In order to find sporadic rules with Apriori, we have to set a very low minsup threshold, drastically increasing the algorithm's running time. In this paper, we adopt an Apriori-Inverse approach: we propose an algorithm to capture rules using a *maximum* support threshold. First, we define the notion of a *perfectly* sporadic rule, where the itemset forming the rule consists only of items that are all below the maximum support threshold. To enable us to find *imperfectly* sporadic rules, we allow maximum support to be increased slightly to include itemsets with items above maximum support. Finally, we demonstrate that Apriori-Inverse lets us find sporadic rules more quickly than using the Apriori algorithm.

2 Related Work

The most well-known method for generating association rules is the Apriori algorithm [2]. It consists of two phases: the first finds itemsets that satisfy a user-specified minimum support threshold, and the second generates association rules that satisfy a user-specified minimum confidence threshold from these "frequent" itemsets. The algorithm generates all rules that satisfy the two thresholds and avoids generating itemsets that do not meet minimum support, even though there may be rules with low support that have high confidence. Thus, unless minimum support is set very low, sporadic rules will never be generated. There are several proposals for solving this problem. We shall discuss the MSApriori (Multiple Supports Apriori), RSAA (Relative Support Apriori Algorithm) and Min-Hashing approaches.

Liu et al. [4] note that some individual items can have such low support that they cannot contribute to rules generated by Apriori, even though they may participate in rules that have very high confidence. They overcome this problem with a technique whereby each item in the database can have a minimum item support (MIS) given by the user. By providing a different MIS for different items, a higher minimum support is tolerated for rules that involve frequent items and lower minimum support for rules that involve less frequent items. The MIS for each data item i is generated by first specifying LS (the lowest allowable minimum support), and a value $\beta, 0 \leq \beta \leq 1.0$. $MIS(i)$ is then set according to the following formula:

$$M(i) = \beta \times f \quad (O \leq \beta \leq 1)$$
$$MIS = M(i), \text{if}(M(i) > LS)$$
$$= LS, \text{otherwise}$$

The advantage of the MSApriori algorithm is that it has the capability of finding some rare-itemset rules. However, the actual criterion of discovery is determined by the user's value of β rather than the frequency of each data item. Thus Yun et al. [5] proposed the RSAA algorithm to generate rules in which significant rare itemsets take part, without any "magic numbers" specified by the user. This technique uses *relative support*: for any dataset, and with the support of item i represented as $sup(i)$, relative support ($RSup$) is defined as:

$$RSup\{i_1, i_2, \ldots, i_k\} = max(\ sup(i_1, i_2, \ldots, i_k)/sup(i_1),$$
$$sup(i_1, i_2, \ldots, i_k)/sup(i_2),$$
$$\ldots,$$
$$sup(i_1, i_2, \ldots, i_k)/sup(i_k))$$

Thus, this algorithm increases the support threshold for items that have low frequency and decreases the support threshold for items that have high frequency. Like Apriori and MSApriori, RSAA is exhaustive in its generation of rules, so it spends time looking for rules which are not sporadic (i.e. rules with high support and high confidence). If the minimum-allowable relative support value is set close to zero, RSAA takes a similar amount of time to that taken by Apriori to generate low-support rules in amongst the high-support rules.

Variations on Min-Hashing techniques were introduced by Cohen [3] to mine significant rules without any constraint on support. Transactions are stored as a 0/1 matrix with as many columns as there are unique items. Rather than searching for pairs of columns that would have high support or high confidence, Cohen et al. search for columns that have high *similarity*, where similarity is defined as the fraction of rows that have a 1 in both columns when they have a 1 in either column. Although this is easy to do by brute-force when the matrix fits into main memory, it is time-consuming when the matrix is disk-resident. Their solution is to compute a hashing signature for each column of the matrix in such a way that the probability that two columns have the same signature is proportional to their similarity. After signatures are calculated, candidate pairs are generated, and then finally checked against the original matrix to ensure that they do indeed have strong similarity.

It should be noted that, like MSApriori and RSAA above, the hashing solution will produce many rules that have high support and high confidence, since only a *minimum* acceptable similarity is specified. It is not clear that the method will extend to rules that contain more than two or three items, since ^mC_r checks for similarity must be done where m is the number of unique items in the set of transactions, and r is the number of items that might appear in any one rule. Removing the support requirement entirely is an elegant solution, but it comes at a high cost of space: for n transactions containing an average of r items over m possible items, the matrix will require $n \times m$ bits, whereas the primary data structure for Apriori-based algorithms will require $n \times \log_2 m \times r$ bits. For a typical application of $n = 10^9$, $m = 10^6$ and $r = 10^2$, this is 10^{15} bits versus approximately 2×10^{12} bits.

For our application, we are interested in generating *only* sporadic rules, without having to wade through a lot of rules that have high support (and are therefore not sporadic), without having to generate any data structure that would not normally be generated in an algorithm like Apriori, and without generating a large number of *trivial* rules (e.g. those rules of the form A → B where the support of B is very high and the support of A rather low). In the next section, we propose a framework for finding certain types of sporadic rules.

3 Proposal of Apriori-Inverse

In the previous section, the techniques discussed generate all rules that have high confidence and support. Using them to find sporadic rules would require setting a low minsup. As a result, the number of rules generated can be enormous, with only a small number being significant sporadic rules. In addition, not all rules generated with these constraints are interesting. Some of the rules may correspond to prior knowledge or expectation, refer to uninteresting attributes, or present redundant information [6].

3.1 Types of Sporadic Rule

We refer to all rules that fall *below* a user-defined *maximum* support level (maxsup) but *above* a user-defined minimum confidence level (minconf) as *sporadic rules*. We further split sporadic rules into those that are *perfectly sporadic* (have no subsets above maxsup) and those that are *imperfectly sporadic*. We then demonstrate an algorithm, which we call Apriori-Inverse, that finds all perfectly sporadic rules.

Definition:

$A \rightarrow B$ is *perfectly sporadic* for maxsup s and minconf c iff

$$\text{confidence}(A \rightarrow B) \geq c, \text{ and}$$
$$\forall x : x \in (A \cup B), \text{ support}(x) < s$$

That is, support must be *under* maxsup and confidence *at least* minconf, and no member of the set of $A \cup B$ may have support above maxsup. Perfectly sporadic rules thus consist of antecedents and consequents that occur rarely (that is, less often than maxsup) but, when they do occur, tend to occur together (with at least minconf confidence).

While this is a useful definition of a particularly interesting type of rule, it certainly does not cover all cases of rules that have support lower than maxsup. For instance, suppose we had an itemset $A \cup B$ with support$(A) = 12\%$, support$(B) = 16\%$, and support$(A \cup B) = 12\%$, with maxsup $= 12\%$ and minconf $= 75\%$. Both $A \rightarrow B$ (confidence $= 100\%$) and $B \rightarrow A$ (confidence $= 75\%$) are sporadic in that they have low support and high confidence, but neither are *perfectly* sporadic, due to B's support being too high. Thus, we define *imperfectly* sporadic rules as the following:

Definition:

$A \rightarrow B$ is *imperfectly sporadic* for maxsup s and minconf c iff

$$\text{confidence}(A \rightarrow B) \geq c, \quad \text{and}$$
$$\text{support}(A \cup B) < s, \quad \text{and}$$
$$\exists x : x \in (A \cup B), \quad \text{support}(x) \geq s$$

That is, a rule is imperfectly sporadic if it meets the requirements of maxsup and minconf but has a subset of its constituent itemsets that has support above maxsup. Clearly, some imperfectly sporadic rules could be completely trivial or uninteresting: for instance, when the antecedent is rare but the consequent has support of 100%. What we should like is a technique that finds all perfectly sporadic rules and some of the imperfectly sporadic rules that are *nearly* perfect.

3.2 The Apriori-Inverse Algorithm

In this section, we introduce the Apriori-Inverse algorithm. Like Apriori, this algorithm is based on a level-wise search. On the first pass through the database, an inverted index is built using the unique items as keys and the transaction IDs as data. At this point, the support of each unique item (the 1-itemsets) in the database is available as the length of each data chain. To generate k-itemsets under maxsup, the $(k-1)$-itemsets are extended in precisely the same manner as Apriori to generate candidate k-itemsets. That is, a $(k-1)$-itemset i_1 is turned into a k-itemset by finding another $(k-1)$-itemset i_2 that has a matching prefix of size $(k-2)$, and attaching the last item of i_2 to i_1. For example, the 3-itemsets $\{1,3,4\}$ and $\{1,3,6\}$ can be extended to form the 4-itemset $\{1,3,4,6\}$, but $\{1,3,4\}$ and $\{1,2,5\}$ will not produce a 4-itemset due to their prefixes not matching right up until the last item.

These candidates are then checked against the inverted index to ensure they at least meet a *minimum absolute support* requirement (say, at least 5 instances) and are pruned if they do not (the length of the *intersection* of a data chain in the inverted index provides support for a k-itemset with k larger than 1). The process continues until no candidate itemsets can be generated, and then association rules are formed in the usual way.

It should be clear that Apriori-Inverse finds all perfectly sporadic rules, since we have simply inverted the downward-closure principle of the Apriori algorithm;

rather than all subsets of rules being over minsup, all subsets are under maxsup. Since making a candidate itemset longer cannot increase its support, all extensions are viable *except* those that fall under our minimum absolute support requirement. Those exceptions are pruned out, and are not used to extend itemsets in the next round.

```
Algorithm Apriori-Inverse
Input: Transaction Database D, maxsup value
Output: Sporadic Itemsets

(1) Generate inverted index I of (item, [TID-list]) from D.
(2) Generate sporadic itemsets of size 1:
       S₁ = ∅
       for each item i ∈ I do begin
           if count(I,i)/|D| < maximum support and
              count(I,i) > minimum absolute support
           then S₁ = S₁ ∪ i
       end
(3) Find Sₖ, the set of sporadic k-itemsets where k ≥ 2:
       for (k = 2; Sₖ₋₁ ≠ ∅; k++) do begin
           Sₖ = ∅
           for each i ∈ {itemsets that are extns of Sₖ₋₁} do begin
               if all subsets of i of size k − 1 ∈ Sₖ₋₁
                  and count(I,i) > minimum absolute support
               then Sₖ = Sₖ ∪ i
           end
       end
       return ⋃ₖ Sₖ
```

Apriori-Inverse does not find any imperfectly sporadic rules, because it never considers itemsets that have support above maxsup; therefore, no subset of any itemset that it generates can have support above maxsup. However, it can be extended easily to find imperfectly sporadic rules that are nearly perfect: for instance, by setting $maxsup_i$ to maxsup/minconf where $maxsup_i$ is maximum support for imperfectly sporadic rules and maxsup is maximum support for reported sporadic rules.

3.3 The "Less Rare Itemset" Problem

It is, of course, true that rare itemsets may be formed by the combination of less rare itemsets. For instance, itemset A may have support 11%, itemset B support 11%, but itemset $A \cup B$ only 9%, making $A \cup B$ sporadic for a maxsup of 10% and $A \rightarrow B$ a valid imperfectly sporadic rule for minconf of 80%. However, Apriori-Inverse will not generate this rule if maxsup is set to 10%, for $A \rightarrow B$ is an imperfectly sporadic rule rather than a perfectly sporadic one.

It would be nice to be able to generate imperfectly sporadic rules as well. We note, however, that not all imperfectly sporadic rules are necessarily interesting:

in fact, many are not. One definition of rules that are trivial is proposed by Webb and Zhang [7], and a similar definition for those that are redundant given by Liu, Shu, and Ma [8]. Consider an association rule $A \rightarrow C$ with support 10% and confidence 90%. It is possible that we may also generate $A \cup B \rightarrow C$, with support 9% and confidence 91%: however, adding B to the left hand side has not bought us very much, and the shorter rule would be preferred. Another situation in which trivial rules may be produced is where a very common item is added to the consequent; it is possible that $A \rightarrow C$ has high confidence because the support of C is close to 100% (although in this case, it would be noticeable due to having a lift value close to 1.0). Therefore, we do not necessarily wish to generate *all* imperfectly sporadic rules.

We propose three different modifications of Apriori-Inverse, all of which produce rules that are not-too-far from being perfect. We refer to the modifications as "Fixed Threshold", "Adaptive Threshold", and "Hill Climbing". In general, we adjust the maxsup threshold to enable us to find at least some imperfectly sporadic rules: specifically, those that contain subsets that have support just a little higher than maxsup.

Fixed Threshold: In this modification, we propose adjusting the maximum support threshold before running Apriori-Inverse to enable us to find more rare itemsets. The maximum support threshold is adjusted by taking the proportion of the maximum support threshold and the minconf threshold. For example, given a minsup threshold of 0.20 and a minconf of 0.80, the new minsup threshold would be set to $0.2/0.8 = 0.25$. However, during the generation of rules, we only consider itemsets that satisfy the original maximum support threshold. Rules that have supports which are higher than the *original* maxsup are not generated.

Adaptive Threshold: In this modification, we propose changing the maximum support by a small increment η (typically 0.001) at each value of k during the generation of sporadic k-itemsets. The threshold is increased until the number of itemsets in the current generation does not change when compared to the previous generation. In general, we search for a plateau where the number of itemsets found does not change.

Hill Climbing: Hill Climbing is an extension of Adaptive Threshold; it adjusts the maximum support threshold by adding an increment that is the product of a rate-variable η (like the learning constant for a gradient descent algorithm; but typically 0.01) and the gradient of the graph of the number of itemsets generated so far. Like the previous method we modify the threshold until the number of itemsets reaches a plateau. Using this method in a large dataset the plateau is likely to be found sooner, since the increment used becomes greater when the gradient is steep and smaller when the gradient becomes less steep.

4 Results and Discussion

In this section, we compare the performance of the standard Apriori algorithm program with the proposed Apriori-Inverse. We also discuss the results of three the different variation of Apriori-Inverse. Testing of the algorithms was carried out on six different datasets from the UCI Machine Learning Repository [9].

Table 1 displays results from implementations Apriori-Inverse and the Apriori algorithms. Each row of the table represents an attempt to find perfectly sporadic rules—with maxsup 0.25, minconf 0.75, and lift greater than 1.0—from the database named in the left-most column. For Apriori-Inverse, this just involves setting maxsup and minconf values. For the Apriori algorithm, this involves setting minsup to zero (conceptually; in reality, the algorithm has been adjusted to use a minimum absolute support of 5), generating all rules, then pruning out those that fall above maxsup. In each case, this final pruning step is not counted in the total time taken. In the first three cases, Apriori was able to generate all frequent itemsets with maxsup greater than 0.0, but for the final three it was not clear that it would finish in reasonable time. To give an indication of the amount of work Apriori is doing to find low-support rules, we lowered its minsup threshold until it began to take longer than 10,000 seconds to process each data set.

Table 1. Comparison of results of Apriori-Inverse and Apriori

Dataset	Apriori-Inverse (maxsup=0.25, minconf=0.75)				Apriori (minconf=0.75)					
	Rules	Passes	Average Sporadic Itemsets	Time (sec)	Min Sup	Rules with Min Sup < 0.25	Rules	Passes	Average Frequent Itemsets	Time (sec)
TeachingEval.	11	3	12	0.01	0	281	294	4	68	0.32
Bridges	9	3	8	0.01	0	24086	24436	9	405	6.44
Zoo	79	4	11	0.03	0	40776255	42535557	17	34504	8380.64
Flag	2456	7	128	1.32	0.11	16427058	16944174	14	57765	11560.77
Mushroom	1142015	13	3279	225.20	0.15	28709481	31894347	16	21654	11489.32
Soybean-Large	37859	10	307	6.51	0.43	0	101264259	17	46310	11550.22

Using Apriori, we were able to find all rules below a support of 0.25 for the Teaching Assistant Evaluation dataset, Bridges dataset, and Zoo dataset. However, using Apriori on the Flag dataset and Mushroom dataset, we could only push the minimum support down to 0.11 and 0.15 respectively, before hitting the time constraint of 10 thousand seconds. Compare this to Apriori-Inverse, finding all perfectly sporadic rules in just a few minutes for the Mushroom database. For the Soybean-Large dataset, no rules below a support of 43% could be produced in under 10 thousand seconds.

We conclude that, while Apriori is fine for discovering sporadic rules in small databases such as the first three in Table 1, a method such as Apriori-Inverse is required if sporadic rules under a certain maximum support are to be found in larger or higher-dimensional datasets. We also note that Apriori is finding a much larger number of rules under maxsup than Apriori-Inverse; this is, of course, due to Apriori finding all of the imperfectly sporadic rules as well as the perfectly sporadic rules. To take the Teaching Evaluation Dataset as an example, Apriori finds

$$\{\text{course}=11\} \rightarrow \{\text{instructor}=7\}$$
$$\{\text{course}=11, \text{nativeenglish}=2\} \rightarrow \{\text{instructor}=7\}$$
$$\{\text{course}=11\} \rightarrow \{\text{instructor}=7, \text{nativeenglish}=2\}$$

whereas, from this particular grouping, Apriori-Inverse only finds

$$\{\text{course}=11\} \rightarrow \{\text{instructor}=7\}$$

However, since the second and third rules found by Apriori have the same support, lift, and confidence values as the first, they both count as trivial according to the definitions given in [8] and [7]. Apriori-Inverse has ignored them (indeed, has never spent any time trying to generate them) because they are imperfect.

Table 2 shows a comparison of the methods used to allow Apriori-Inverse to find some imperfectly sporadic rules. The Fixed Threshold method finds the largest number of sporadic rules, because it is "overshooting" the maxsup thresholds determined by the two adaptive techniques, and therefore letting more itemsets into the candidate group each time. As a result, it requires fewer passes of the inverted index, but each pass takes a bit longer, resulting in longer running times. However, the times for the Fixed Threshold version seem so reasonable that we are not inclined to say that the adaptive techniques give any significant advantage. Determining a principled way to generate imperfectly sporadic rules—and determining a good place to *stop* generating then—remains an open research question. Nevertheless, we note that the time taken to generate all of the imperfectly sporadic rules by all three methods remains very much smaller than the time taken to find them by techniques that require a *minimum* support constraint.

Table 2. Comparison of results of extensions to Apriori-Inverse

Dataset	Fixed Threshold				Adaptive Threshold ($\eta = 0.001$)				Hill Climbing ($\eta = 0.01$)			
	Rules	Passes	Avg Spdc Sets	Time (sec)	Rules	Passes	Avg Spdc Sets	Time (sec)	Rules	Passes	Avg Spdc Sets	Time (sec)
TeachingEval.	46	4	22	0.01	11	6	12	0.03	11	6	12	0.04
Bridges	104	5	14	0.03	30	11	8	0.04	30	11	8	0.04
Zoo	203	5	15	0.03	203	19	13	0.14	203	19	13	0.12
Flag	12979	9	268	4.86	5722	31	165	9.49	13021	42	228	19.81
Mushroom	1368821	13	7156	791.82	1142015	26	3279	445.95	1142015	26	3279	474.88
Soybean-Large	1341135	11	801	31.47	95375	52	425	63.09	56286	30	352	27.38

5 Conclusion and Future Work

Existing association mining algorithms produce all rules with support *greater* than a given threshold. But, to discover rare itemsets and sporadic rules, we should be more concerned with *infrequent* items. This paper proposed a more efficient algorithm, Apriori-Inverse, which enables us to find perfectly sporadic rules without generating all the unnecessarily frequent items. We also defined the notion of imperfectly sporadic rules, and proposed three methods of finding them using Apriori-Inverse: Fixed Threshold, Adaptive Threshold, and Hill Climbing.

With respect to finding imperfectly sporadic rules, our proposed extensions to Apriori-Inverse are—at best—heuristic. More importantly, there are some types of imperfectly sporadic rule that our methods will not find at all. Our future work will involve ways of discovering rules such as $A \cup B \rightarrow C$ where neither A nor B is rare, but their association is, and C appears with $A \cup B$ with high confidence. This is the case of a rare association of common events (A and B) giving rise to a rare event (C). It is a particularly interesting form of imperfectly sporadic rule, especially in the fields of medicine (rare diseases) and of process control (disaster identification and avoidance).

References

1. Agrawal, R., Imielinski, T., Swami, A.: Mining association rules between sets of items in large databases. In Buneman, P., Jajodia, S., eds.: Proceedings of the 1993 ACM SIGMOD International Conference on Management of Data. (1993) 207–216
2. Agrawal, R., Srikant, R.: Fast algorithms for mining association rules. In Bocca, J.B., Jarke, M., Zaniolo, C., eds.: Proceedings of the 20th International Conference on Very Large Data Bases, VLDB'94. (1994) 487–499
3. Cohen, E., Datar, M., Fujiwara, S., Gionis, A., Indyk, P., Motwani, R., Ullman, J.D., Yang, C.: Finding interesting association rules without support pruning. IEEE Transactions on Knowledge and Data Engineering **13** (2001) 64–78
4. Liu, B., Hsu, W., Ma, Y.: Mining association rules with multiple minimum supports. In: Proceedings of the 5th ACM SIGKDD International Conference on Knowledge Disconvery and Data Mining. (1999) 337–341
5. Yun, H., Ha, D., Hwang, B., Ryu, K.H.: Mining association rules on significant rare data using relative support. The Journal of Systems and Software **67** (2003) 181–191
6. Toivonen, H., Klemettinen, M., Ronkainen, P., Hatonen, K., Mannila, H.: Pruning and grouping of discovered association rules. In: ECML-95 Workshop on Statistics, Machine Learning, and Knowledge Discovery in Databases. (1995) 47–52
7. Webb, G.I., Zhang, S.: Removing trivial associations in association rule discovery. In: Abstracts Published in the Proceedings of the First International NAISO Congress on Autonomous Intelligent Systems (ICAIS 2002). (2002)
8. Liu, B., Hsu, W., Ma, Y.: Pruning and summarizing the discovered associations. In: Proceedings of the 5th ACM SIGKDD International Conference on Knowledge Discovery and Data Mining. (1999) 125–134
9. Blake, C., Merz, C.: UCI repository of machine learning databases. http://www.ics.uci.edu/~mlearn/MLRepository.html, University of California, Irvine, Department of Information and Computer Sciences (1998)

Automatic View Selection: An Application to Image Mining

Manoranjan Dash[1] and Deepak Kolippakkam[2]

[1] School of Computer Engineering,
Nanyang Technological University, Singapore 639798
asmdash@ntu.edu.sg
[2] Department of Computer Science and Engg.,
Arizona State University, Tempe, AZ 85387-5406
n.kolippakkam@asu.edu

Abstract. In this paper we discuss an image mining application of Egeria detection. Egeria is a type of weed found in various lands and water regions over San Joaquin and Sacramento deltas. The challenge is to find a view to accurately detect the weeds in new images. Our solution contributes two new aspects to image mining. (1) Application of view selection to image mining: View selection is appropriate when a specific learning task is to be learned. For example, to look for an object in a set of images, it is useful to select the appropriate views (a view is a set of features and their assigned values). (2) Automatic view selection for accurate detection: Usually classification problems rely on user-defined views. But in this work we use association rule mining to automatically select the best view. Results show that the selected view outperforms other views including the full view.

Keywords: View, feature selection, image classification.

1 Introduction

With the advent of the Internet and rapid advances made in storage devices, non-standard data such as images and videos have grown significantly. The process of discovering valuable information from image data is called image mining, which finds its sources in data mining, content-based image retrieval (CBIR), image understanding and computer vision. The tasks of image mining are mostly concerned with *classification* problems such as "labeling" regions of an image based on presence or absence of some characteristic patterns, and with *image retrieval* problems where "similar" images are identified. In image mining, "training" images are used for learning, and results (knowledge obtained from training) are applied to a large number of new images to fulfill the required tasks.

In order to efficiently mine from images, we need to convert the image into data that can be processed by image mining systems. The conventional data format used is the *feature-value* format (i.e., the tabular format). The attributes (or columns or features) are some characteristics representing an image object (or instance or rows) having corresponding to those attributes. Features describe the image pixel data. There could be hundreds of different features for an image. These features may include: *color* (in various channels), *texture*, etc. Domain experts can usually manually identify a set of relevant features in an ad hoc manner. In general, no single feature can describe an image in its entirety.

With many possible features for an image, often only a small number of features are useful in an application. Using all available features may negatively affect a mining algorithm's performance (e.g., time and accuracy). In addition, it takes time to extract these features, though many features may be irrelevant or redundant. Furthermore, a large number of features would require a huge number of instances in order to learn properly from data. This is known as the curse of dimensionality [4]. With a limited number of instances but a large number of features, many mining algorithms will suffer from data over-fitting. Hence, it is imperative to choose a set of relevant features (i.e., feature selection [1]).

In this paper we discuss how to automatically identify a particular type of weed called Egeria densa [5] in color infrared (CIR) images. These are available as aerial photographs of various land and water regions over San Joaquin and Sacramento deltas. The weed grows in water bodies and is present in various parts of the image: clusters, patches, or sometimes as a 'mesh'. When we apply feature selection to this problem using some efficient algorithm [1, 2], we obtain a number of features. One applies a learning algorithm to determine the exact feature-value combinations for determining any particular task. An alternative solution that we propose in this paper is to select a set of feature-value combinations *directly*. This combines the effect of feature selection that selects the best features and classification that takes decision based on feature values. This is termed as *view selection* [6] which is an instantiated feature set. It is appropriate when there is a specific learning task to be learned. For example, in a set of images it is suitable to select the appropriate views.

In image mining, domain experts often play an important role in deciding relevant features in a time-consuming process. Automatic view selection aims to find relevant feature-values by learning from training images. Therefore, domain experts are not required to specify relevant features, but only to identify or label instances of interest in an image. The latter is a much easier task than the former for experts, and its results are also more reliable because it directly deals with image classification. However, for k features with binary values, the number of possible views for a given class is 2^k. Because k can be as large as hundreds, enumeration of all views is clearly an insurmountable task. Hence, we need to find an efficient search technique for view selection. In this work we use association rule mining to automatically select the best view. The solution is a combination of top association rules that gives the maximum accuracy. We test our method on Egeria application. Results show that the selected view outperforms other views including the full view that contains all features.

2 View Validation via Learning

Here we define a view, and discuss how to learn good views. We give related work on view selection in http://www.ntu.edu.sg/home/asmdash/pakdd_expanded.pdf.

Views. Given a set of k features $f_1, f_2, f_3, ..., f_k$ (feature vector $\mathbf{fv}^k$) and t possible class labels $C_1, C_2, ..., C_t$ for an image of t types of objects, the problem of classifying an instance to one of the t classes can be defined as assigning a class with maximum posterior probability, $c = \arg\max_i P(C_i|\mathbf{fv}^k)$ where $P(C_i|\mathbf{fv})$ is the posterior probability of class label C_i, given the feature vector $\mathbf{fv}^k$.
    A view is an *instantiation* of a subset of features for a class. In other words, a *view* V for class C_i is: $V_{C_i} = \{f_1 = v_1, f_2 = v_2, ..., f_m = v_m\}$. where $m \leq k$, v_j is a value from the domain of feature f_j, and each f_j corresponds to a feature describing the objects to be detected in an image I. A *full view* contains all k feature values, i.e., $m = k$.

View Selection. In the search for good views, we need to find a suitable performance measure that can differentiate good views from the others. The *goal* of view selection is to find a *sufficient and necessary* subset of features with suitable values that make good views. The necessity requirement indicates that the subset of features should be as small as possible; and the sufficiency requirement suggests that a good view can guarantee the attainment of specified accuracy in image classification/detection. In image mining, positive means the presence of an object of interest, negative otherwise. If an object of interest is to be detected in an image, one of the four possibilities can arise: we may *correctly* classify/detect the object's presence (True Positive - TP); we may *falsely* detect the object's presence (False Positive - FP); we may *miss* the detection of the object which should be *present* (False Negative - FN); we may *correctly* detect that the object is absent (True Negative - TN).
    In image classification, we are concerned about the presence or absence of a particular object in a *two-class* classification problem. In building a good classification system, we wish to maximize TP and TN, and minimize FP and FN (errors). Using TP, TN, FP, and FN, we define accuracy $A = \frac{TP+TN}{TP+FP+TN+FN}$.

Learning Good Views from Image Data. With the performance measure defined, we can systematically generate views and estimate their accuracy based on training data. We would prefer a view with the smallest number of features. Therefore, we can start evaluating views of 1 feature, increasing the number of features one at a time until all features are considered. However, exhaustive search of all views is impractical when the number of features is moderately large (e.g., $k > 20$) as the search space grows exponentially. There are many search procedure in the feature selection literature [2] that are not exhaustive yet produce reasonably good feature subsets. In this work we use association rule mining (ARM) for selecting the best view. Association rules are mined according to two measures: *Support* which measures generality of a rule and *Confidence* which measures precision of the rule. For a given class C and feature-values $f'_1 = v_1, f'_2 = v_2, ..., f'_m = v_m$, where $m \in \{1, ..., k\}$ and $f'_1, ..., f'_m$ are a subset of k features; let fv'_j be $f'_j = v_j$. Support and confidence are defined as $P(fv'_1, fv'_2, ..., fv'_m, C)$ and $P(C|fv'_1, fv'_2, ..., fv'_m)$, respectively. An association rule with high support and confidence is both general and precise. Thus, a good association rule in the pre-specified form of $fv'_1, fv'_2, ..., fv'_m \Rightarrow C$ can define a good view. In addition, association rules avoid exhaustive search and can be mined with efficient algorithms [3]. Thus, we have successfully mapped the problem of view selection to a problem of ARM in the specific form of $fv'_1, fv'_2, ..., fv'_m \Rightarrow C$ without resorting to exhaustive search. We can apply any existing ARM algorithm to achieve automatic view selection. We evaluate our method to detect Egeria weeds in aerial images.

3 Experimental Evaluation and Conclusion

In the empirical study, we investigate (1) how to learn good views, and (2) what is the accuracy over testing data. We first describe the application domain used in the experiments, next explain the conventional ways of identifying features and detecting objects of interest, then present results of learning and evaluation.

Application - Detecting Egeria. In this application, the problem is to automatically identify the Egeria weeds in aerial images of land and water regions over San-Joaquin and Sacremento delta. A set of *cover* images corresponding to 51 images are provided in which the weeds are manually detected by experts for evaluation purposes. One image is used for training to obtain *good views* via learning and the remaining 50 are used for testing in experiments.

Images are available in TIF format with varying sizes: 1 of size 1644×1574, 30 of size 300×300, the remaining 20 of varying sizes $528 - 848 \times 312 - 444$. The largest image of size 1644×1574 was selected by the domain expert to be used as the training image for learning. This large image covers multiple regions of the delta, and is the most comprehensive image that contains a considerable amount of information for defining features to extract Egeria from the image.

See http://www.ntu.edu.sg/home/asmdash/pakdd_expanded.pdf for discussion on feature identification and extraction.

Evaluation Results of Learning. In order to search for the best view, we apply an ARM algorithm to the training data. We find association rules that describe the relationships between feature-values with class label 1 (presence of Egeria), i.e., in the association rule, the antecedent consists of the feature-values and the consequent consists of the class label 1. As we discussed in the previous section, the best view is the rule with the highest support and confidence. Support is the first criterion, i.e., if a rule has higher support than all other rules then it is chosen first. If two rules have equal support then the rule with higher confidence is chosen. The *best view* found by ARM is: color_1 = 1 $\wedge$ color_2 = 1 $\wedge$ color_3 = 1 $\wedge$ texture_1b = 0 $\Rightarrow$ 1. This view (R1) has only 4 features: 3 color features (all 3 channels) should be *present*, and the first channel of the second texture feature should be *absent*. This view has support=0.751, confidence=0.95.

It has the highest support among all views, indicating that three color features and one texture feature (absence of water body) are sufficient to detect the weed in the training data. The single texture feature *has* to be 0 in the combination with the color features in order to predict the occurrence of the weed accurately. This is quite different from the scenario where only the three color features are considered in the view, or if the texture feature has a value 1. This view has effectively filtered the possible inclusion of water body in our coverage of the weed. Though 14 features are defined based on the domain, view selection suggests that these 4 features with their proper values should be sufficient for detecting weeds in this image.

The training image was a normal image (in terms of brightness), and hence R1 works well. However, as we explained in the previous section, some of the images are dark. R1 is not satisfactory for extracting features from dark images. We observed that the next best view (second best association rule) contains the edge feature (along with other feature-values) which is useful for dark images. *Next best view* found by ARM is: $color\_1 = 1 \wedge color\_2 = 1 \wedge color\_3 = 1 \wedge edge = 1 \wedge texture\_2a = 0 \wedge texture\_1a = 0 \Rightarrow 1$.

This rule (R2) contains the three color features combined with the edge feature and two texture features. It has a support of 0.61 and a confidence of 0.93. The regions which were misclassified as FP in the image (dark) when we used the first view, are now labelled as TN using this view. We combine the two rules with logical AND ($\wedge$), i.e., $R1 \wedge R2$ and then we choose that view which has higher accuracy, i.e., $A_V = \max(A_{R1}; A_{R1 \wedge R2})$.

Note that depending on applications a user can adapt the above rule. For example, there may be cases where R1 fails to detect correctly (i.e., FN) but R2 detects correctly (i.e., TP). In such cases instead of AND we use OR ($\vee$) to get the next best view. So, one may have to choose that view which gives the highest accuracy in the training image among the three: $R1$, $R1 \wedge R2$, and $R1 \vee R2$, i.e., $\max(A_{R1}; A_{R1 \wedge R2}; A_{R1 \vee R2})$.

An issue here is to determine how many rules to select. The following algorithm is used to select the top rules in the Egeria application.

1. Estimate the base accuracy A_0 using a full view.
2. Continue selecting top rules until accuracy of the new view A is larger than base accuracy A_0 by more than a user defined δ, i.e., $A - A_0 > \delta$.

A proper δ value is dependent on the application. But usually it is small. We design experiments to verify in test images if this view indeed outperforms other views including the full view and if it does so not *by chance*. So, we conducted three comparative experiments: comparing the above best view (1) with the full view, (2) with random views of four features, and (3) with the best random view. We select random values (0 or 1) for the four randomly selected features. Each image has a manually determined cover image (ground truth) that indicates where Egeria blocks are. For the purpose of the experiments, this cover image is defined as the perfect result achievable for Egeria detection. We use *accuracy gain* to evaluate the difference between any two views. If the accuracy measures of two views are A and A_0, $Gain$ is defined as: $Gain = \frac{A - A_0}{A_0}$. Results show

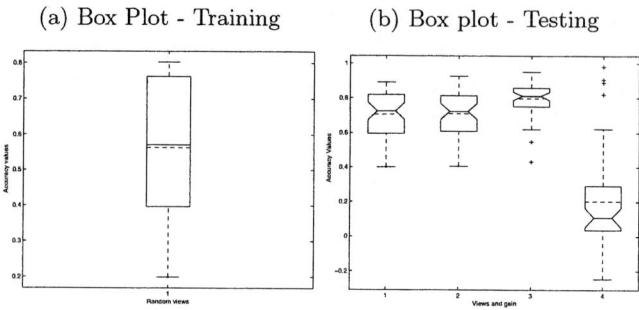

Fig. 1. Training: Mean - dotted line, Median - Solid line; Testing: 1 - Accuracy of Full View, 2 - Accuracy of R1, 3 - Accuracy of R2, 4 - Gain of best-among R1, R2 over full

accuracy is – for best view (using our method): 0.92, for full view: 0.81, (gain for our method = 14%); average of 30 random views: 0.56 (gain = 19%); and for best random view: 0.8 (gain = 14%). For the group of 30 randomly selected views, box plots of accuracy values are shown in Figure 1 (a). These two comparative experiments show that the best view cannot be found by chance.

Evaluation Results for New Images. The best view is obtained from the training image. We now evaluate its accuracy for the 50 testing images. The performance is evaluated by comparing the accuracy gains of the full View and the selected best view. Though there are some cases where we have negative gains, on average the gain is about 22%. We compare the cover which we obtained (using block processing) with the ground truth cover (which is in pixels). Hence there would always be some approximation in our covers - especially in images which have Egeria in thinner patches. These are more or less the images which have negative gains too. The box-plot of the results for the 50 images is shown in Figure 1 (b). Details of the experimental work can be found in http://www.ntu.edu.sg/home/asmdash/pakdd_expanded.pdf.

Conclusion. We define concepts of view and good view, identify the need for automatic view selection, show that a full view can make image mining unnecessarily more complicated and less efficient and effective, then propose and verify a learning approach to automatic view selection. Our system can learn good views from a training image data set and then the views can be applied to unseen images for mass processing of image mining tasks. With this approach, domain experts do not need to provide accurate information about what relevant features are, but only need to focus on the tasks they are best at (in this case, identifying if a block is Egeria or not in a training image). The presented approach is independent of type of feature and hence can be applied to many applications to determine good views. A future work is to consider the large number of rules that are usually generated as part of association rule mining. We acknowledge the contributions of Prof Huan Liu and Prof Trishi Foschi.

References

1. A.L. Blum and P. Langley. Selection of relevant features and examples in machine learning. *Artificial Intelligence*, 97:245–271, 1997.
2. M. Dash and H. Liu. Feature selection methods for classifications. *Intelligent Data Analysis: An International Journal*, 1(3):131–156, 1997.
3. J. Han, J. Pei, and Y. Yin. Mining frequrent patterns without candidate generation. In *SIGMOD'00*, pages 1–12, 2000.
4. T. Hastie, R. Tibshirani, and J. Friedman. *The Elements of Statistical Learning*. Springer, 2001.
5. N. Kolippakkam and A. Mandvikar. Egeria densa mining project webpage, 2002. http://www.eas.asu.edu/ egeria.
6. I. Musleaa, S. Minton, and C.A. Knoblock. Adaptive view validation: A first step towards automatic view detection. In *Proceedings of ICML*, pages 443–450, 2002.

Pushing Tougher Constraints in Frequent Pattern Mining

Francesco Bonchi[1] and Claudio Lucchese[2]

[1] Pisa KDD Laboratory, ISTI - C.N.R., Area della Ricerca di Pisa, Italy
[2] Department of Computer Science, University Ca' Foscari, Venezia, Italy

Abstract. In this paper we extend the state-of-art of the constraints that can be pushed in a frequent pattern computation. We introduce a new class of tough constraints, namely *Loose Anti-monotone* constraints, and we deeply characterize them by showing that they are a superclass of convertible anti-monotone constraints (e.g. constraints on *average* or *median*) and that they model tougher constraints (e.g. constraints on *variance* or *standard deviation*). Then we show how these constraints can be exploited in a level-wise Apriori-like computation by means of a new data-reduction technique: the resulting algorithm outperforms previous proposals for convertible constraints, and it is to treat much tougher constraints with the same effectiveness of easier ones.

1 Introduction

Frequent itemsets play an essential role in many data mining tasks that try to find interesting patterns from databases, such as association rules, correlations, sequences, episodes, classifiers, clusters and many more. Although the collection of all frequent itemsets is typically very large, the subset that is really interesting for the user usually contains only a small number of itemsets. This situation is harmful for two reasons. First, performance degrades: mining generally becomes inefficient or, sometimes, simply unfeasible. Second, the identification of the fragments of interesting knowledge, blurred within a huge quantity of mostly useless patterns, is difficult. Therefore, the paradigm of constraint-based mining was introduced. Constraints provide focus on the interesting knowledge, thus reducing the number of patterns extracted to those of potential interest. Additionally, they can be pushed deep inside the pattern discovery algorithm in order to achieve better performance [9, 10, 14, 15, 16, 17, 18].

Constrained frequent pattern mining is defined as follows. Let $\mathcal{I} = \{x_1, ..., x_n\}$ be a set of distinct literals, usually called *items*, where an item is an object with some predefined attributes (e.g., price, type, etc.). An *itemset* X is a non-empty subset of $\mathcal{I}$. If $|X| = k$ then X is called a *k-itemset*. A constraint on itemsets is a function $\mathcal{C} : 2^{\mathcal{I}} \rightarrow \{true, false\}$. We say that an itemset I satisfies a constraint if and only if $\mathcal{C}(I) = true$. We define the *theory* of a constraint as the set of itemsets which satisfy the constraint: $Th(\mathcal{C}) = \{X \in 2^{\mathcal{I}} \mid \mathcal{C}(X)\}$. A *transaction database* $\mathcal{D}$ is a bag of itemsets $t \in 2^{\mathcal{I}}$, usually called *transactions*.

The *support* of an itemset X in database $\mathcal{D}$, denoted $supp_{\mathcal{D}}(X)$, is the number of transactions which are superset of X. Given a user-defined *minimum support* σ, an itemset X is called *frequent* in $\mathcal{D}$ if $supp_{\mathcal{D}}(X) \geq \sigma$. This defines the minimum frequency constraint: $\mathcal{C}_{freq[\mathcal{D},\sigma]}(X) \Leftrightarrow supp_{\mathcal{D}}(X) \geq \sigma$. When the dataset and the minimum support threshold are clear from the context, we indicate the frequency constraint simply $\mathcal{C}_{freq}$. Thus with this notation, the *frequent itemsets mining problem* requires to compute the set of all frequent itemsets $Th(\mathcal{C}_{freq})$. In general, given a conjunction of constraints $\mathcal{C}$ the *constrained frequent itemsets mining problem* requires to compute $Th(\mathcal{C}_{freq}) \cap Th(\mathcal{C})$.

Related Work and Constraints Classification. A first work defining classes of constraints which exhibit nice properties is [15]. In that paper is introduced an Apriori-like algorithm, named CAP, which exploits two properties of constraints, namely *anti-monotonicity* and *succinctness*, in order to reduce the frequent itemsets computation. Given an itemset X, a constraint $\mathcal{C}_{AM}$ is **anti-monotone** if $\forall Y \subseteq X : \mathcal{C}_{AM}(X) \Rightarrow \mathcal{C}_{AM}(Y)$. The frequency constraint is the most known example of a $\mathcal{C}_{AM}$ constraint. This property, *the anti-monotonicity of frequency*, is used by the Apriori [1] algorithm with the following heuristic: if an itemset X does not satisfy $\mathcal{C}_{freq}$, then no superset of X can satisfy $\mathcal{C}_{freq}$, and hence they can be pruned. Other $\mathcal{C}_{AM}$ constraints can easily be pushed deeply down into the frequent itemsets mining computation since they behave exactly as $\mathcal{C}_{freq}$: if they are not satisfiable at an early level (small itemsets), they have no hope of becoming satisfiable later (larger itemsets).

A **succinct** constraint $\mathcal{C}_S$ is such that, whether an itemset X satisfies it or not, can be determined based on the singleton items which are in X. A $\mathcal{C}_S$ constraint is *pre-counting pushable*, i.e. it can be satisfied at candidate-generation time: these constraints are pushed in the level-wise computation by substituting the usual *generate_apriori* procedure, with the proper (w.r.t. $\mathcal{C}_S$) candidate generation procedure. Constraints that are both anti-monotone and succinct can be pushed completely in the level-wise computation before it starts (at pre-processing time). For instance, consider the constraint $min(S.price) \geq v$: if we start with the first set of candidates formed by all singleton items having price greater than v, during the computation we will generate only itemsets satisfying the given constraint. Constraints that are neither succinct nor anti-monotone are pushed in the CAP [15] computation by inducing weaker constraints which are either anti-monotone and/or succinct.

Monotone constraints work the opposite way of anti-monotone constraints. Given an itemset X, a constraint $\mathcal{C}_M$ is monotone if: $\forall Y \supseteq X : \mathcal{C}_M(X) \Rightarrow \mathcal{C}_M(Y)$. Since the frequent itemset computation is geared on $\mathcal{C}_{freq}$, which is anti-monotone, $\mathcal{C}_M$ constraints have been considered more hard to be pushed in the computation and less effective in pruning the search space [2, 8, 7, 12]: while anti-monotone constraints can be used to effectively prune the search space to a small downward closed collection, the upward closed collection of the search space satisfying the monotone constraints cannot be pruned at the same time. Recently, it has has been shown that a real synergy of these two opposite types of constraints exists and can be exploited by reasoning on both the itemset search space and the input database *together*, using the ExAnte data-reduction technique [4]. Using data reduction techniques, anti-monotone and monotone pruning strengthen each other recursively [3, 5].

In [16, 17] the class of **convertible** constraints is introduced, and an FP-growth based methodology to push such constraints is proposed. A constraint $\mathcal{C}_{CAM}$ is convertible anti-monotone provided there is an order $\mathcal{R}$ on items such that whenever an itemset X satisfies $\mathcal{C}_{CAM}$, so does any prefix of X. A constraint $\mathcal{C}_{CM}$ is convertible monotone provided there is an order $\mathcal{R}$ on items such that whenever an itemset X violates $\mathcal{C}_{CM}$, so does any prefix of X. In [16, 17], two FP-growth based algorithms are introduced: $\mathcal{FIC}^A$ to mine $Th(\mathcal{C}_{freq}) \cap Th(\mathcal{C}_{CAM})$, and $\mathcal{FIC}^M$ to mine $Th(\mathcal{C}_{freq}) \cap Th(\mathcal{C}_{CM})$. A major limitation of any FP-growth based algorithm is that the initial database (internally compressed in the prefix-tree structure) and all intermediate projected databases must fit into main memory. If this requirement cannot be met, these approaches can simply not be applied anymore. This problem is even harder with $\mathcal{FIC}^A$ and $\mathcal{FIC}^M$: in fact, using an order on items different from the frequency-based one, makes the prefix-tree lose its compressing power. Thus we have to manage much greater data structures, requiring a lot more main memory which might not be available. This fact is confirmed by our experimental analysis reported in Section 4: sometimes $\mathcal{FIC}^A$ is slower than FP-growth, meaning that having constraints brings no benefit to the computation. Another important drawback of this approach is that it is not possible to take full advantage of a conjunction of different constraints, since each constraint in the conjunction could require a different ordering of items.

The first (and to our knowledge unique) work, trying to address the problem of how to push constraints which are **not convertible**, is [13]. The framework proposed in that paper is based on the concept of finding a *witness*, i.e. an itemset such that, by testing whether it satisfies the constraint we can deduce information about properties of other itemsets, that can be exploited to prune the search space. This idea is embedded in a depth-first visit of the itemsets search space. The main drawback of the proposal is the following: it may require quadratic time in the number of frequent singleton items to find a witness. The cost can be amortized if items are reordered, but this leads to the same problems discussed for FP-growth based algorithms. Moreover, even if a nearly linear time search is performed, this is done without any certainty of finding a witness which will help to prune the search space. In fact, if the witness found satisfies the given constraint, no pruning will be possible and the search time will be wasted. Our approach is completely orthogonal: while they try to explore the exponentially large search space in some smart way, we massively reduce the dataset as soon as possible, reducing at the same time the search space and obtaining a progressively easier mining problem.

Paper Contribution. The contribution of this paper is threefold. First, we extend the actual state-of-art classification of constraints that can be pushed in a frequent pattern computation, by showing how to push tough constraints as those ones based on *variance* or *standard deviation*. Second, we show that it is possible to push convertible constraints in a level-wise Apriori-like computation, outperforming previously proposed FP-growth based algorithms [16, 17]. Third, we propose a general Apriori-like algorithm, based on data-reduction techniques, which is able to push all possible kinds of constraint studied so far.

2 Loose Anti-monotone Constraints

In this Section we introduce a new class of tougher constraints, which is a proper superclass of convertible anti-monotone. The following example shows that exist interesting constraints which are not convertible, and thus cannot be exploited within a prefix pattern framework.

Example 1 (var constraint is not convertible). Calculating the variance is an important task of many statistical analysis: it is a measure of how spread out a distribution is. The variance of a set of number X is defined as:

$$var(X) = \frac{\sum_{i \in X}(i - avg(X))^2}{|X|}$$

A constraint based on var is not convertible. Otherwise there is an order $\mathcal{R}$ of items such that $var(X)$ is a prefix increasing (or decreasing) function. Consider a small dataset with only four items $\mathcal{I} = \{A, B, C, D\}$ with associated prices $P = \{10, 11, 19, 20\}$. The lexicographic order $\mathcal{R}_1 = \{ABCD\}$ is such that $var(A) \leq var(AB) \leq var(ABC) \leq var(ABCD)$, and it is easy to see that we have only other three orders with the same property: $\mathcal{R}_2 = \{BACD\}, \mathcal{R}_3 = \{DCBA\}, \mathcal{R}_4 = \{CDBA\}$. But, for $\mathcal{R}_1$, we have that $var(BC) \not\leq var(BCD)$, which means that var is not a prefix increasing function w.r.t. $\mathcal{R}_1$. Moreover, since the same holds for $\mathcal{R}_2, \mathcal{R}_3, \mathcal{R}_4$, we can assert that there is no order $\mathcal{R}$ such that var is prefix increasing. An analogous reasoning can be used to show that it neither exists an order which makes var a prefix decreasing function.

Following a similar reasoning we can show that other interesting constraints, such as for instance those ones based on *standard deviation (std)* or *unbiased variance estimator (var_{N-1})* or *mean deviation (md)*, are not convertible as well. Luckily, as we show in the following, all these constraints share a nice property that we name *"Loose Anti-monotonicity"*. Recall that an anti-monotone constraint is such that, if satisfied by an itemset then it is satisfied by *all* its subsets. We define a loose anti-monotone constraint as such that, if it is satisfied by an itemset of cardinality k then it is satisfied by *at least one* of its subsets of cardinality $k - 1$. Since some of these interesting constraints make sense only on sets of cardinality at least 2, in order to get rid of such details, we shift the definition of loose anti-monotone constraint to avoid considering singleton items.

Definition 1 (Loose Anti-monotone constraint). *Given an itemset X with $|X| > 2$, a constraint is loose anti-monotone (denoted $\mathcal{C}_{LAM}$) if: $\mathcal{C}_{LAM}(X) \Rightarrow \exists i \in X : \mathcal{C}_{LAM}(X \setminus \{i\})$*

The next proposition and the subsequent example state that the class of $\mathcal{C}_{LAM}$ constraints is a proper superclass of $\mathcal{C}_{CAM}$ (convertible anti-monotone constraints).

Table 1. Classification of commonly used constraints.

Constraint	Anti-monotone	Monotone	Succinct	Convertible	$\mathcal{C}_{LAM}$
$min(S.A) \geq v$	yes	no	yes	strongly	yes
$min(S.A) \leq v$	no	yes	yes	strongly	yes
$max(S.A) \geq v$	no	yes	yes	strongly	yes
$max(S.A) \leq v$	yes	no	yes	strongly	yes
$count(S) \leq v$	yes	no	weakly	$\mathcal{A}$	yes
$count(S) \geq v$	no	yes	weakly	$\mathcal{M}$	no
$sum(S.A) \leq v \; (\forall i \in S, i.A \geq 0)$	yes	no	no	$\mathcal{A}$	yes
$sum(S.A) \geq v \; (\forall i \in S, i.A \geq 0)$	no	yes	no	$\mathcal{M}$	no
$sum(S.A) \leq v \; (v \geq 0, \forall i \in S, i.A\theta 0)$	no	no	no	$\mathcal{A}$	yes
$sum(S.A) \geq v \; (v \geq 0, \forall i \in S, i.A\theta 0)$	no	no	no	$\mathcal{M}$	no
$sum(S.A) \leq v \; (v \leq 0, \forall i \in S, i.A\theta 0)$	no	no	no	$\mathcal{M}$	no
$sum(S.A) \geq v \; (v \leq 0, \forall i \in S, i.A\theta 0)$	no	no	no	$\mathcal{A}$	yes
$range(S.A) \leq v$	yes	no	no	strongly	yes
$range(S.A) \geq v$	no	yes	no	strongly	yes
$avg(S.A)\theta v$	no	no	no	strongly	yes
$median(S.A)\theta v$	no	no	no	strongly	yes
$var(S.A) \geq v$	no	no	no	no	yes
$var(S.A) \leq v$	no	no	no	no	yes
$std(S.A) \geq v$	no	no	no	no	yes
$std(S.A) \leq v$	no	no	no	no	yes
$var_{N-1}(S.A)\theta v$	no	no	no	no	yes
$md(S.A) \geq v$	no	no	no	no	yes
$md(S.A) \leq v$	no	no	no	no	yes

Proposition 1. *Any convertible anti-monotone constraint is trivially loose anti-monotone: if a k-itemset satisfies the constraint so does its $(k-1)$-prefix itemset.*

Example 2. We show that the constraint $var(X.A) \leq v$ is a $\mathcal{C}_{LAM}$ constraint. Given an itemset X, if it satisfies the constraint so trivially does $X \setminus \{i\}$, where i is the element of X which has associated a value of A which is the most far away from $avg(X.A)$. In fact, we have that $var(\{X \setminus \{i\}\}.A) \leq var(X.A) \leq v$, until $|X| > 2$. Taking the element of X which has associated a value of A which is the closest to $avg(X.A)$ we can show that also $var(X.A) \geq v$ is a $\mathcal{C}_{LAM}$ constraint. Since the standard deviation std is the square root of the variance, it is straightforward to see that $std(X.A) \leq v$ and $std(X.A) \geq v$ are $\mathcal{C}_{LAM}$. The mean deviation is defined as: $md(X) = (\sum_{i \in X} |i - avg(X)|) / |X|$. Once again, we have that $md(X.A) \leq v$ and $md(X.A) \geq v$ are loose anti-monotone. It is easy to prove that also constraints defined on the unbiased variance estimator, $var_{N-1} = (\sum_{i \in X}(i - avg(X))^2) / (|X| - 1)$ are loose anti-monotone.

In Table 1 we update the state-of-art classification of commonly used constraints. The next key Theorem indicates how a $\mathcal{C}_{LAM}$ constraint can be exploited in a level-wise Apriori-like computation by means of data-reduction. It states that if at any iteration $k \geq 2$ a transaction is not superset of at least one frequent k-itemset which satisfy the $\mathcal{C}_{LAM}$ constraint (a solution), then the transaction can be deleted from the database.

Theorem 1. *Given a transaction database $\mathcal{D}$, a minimum support threshold σ, and a $\mathcal{C}_{LAM}$ constraint, at the iteration $k \geq 2$ of the level-wise computation, a transaction $t \in \mathcal{D}$ such that: $\nexists X \subseteq t, |X| = k, X \in Th(\mathcal{C}_{freq[\mathcal{D},\sigma]}) \cap Th(\mathcal{C}_{LAM})$ can* be pruned away from $\mathcal{D}$, since it will never be superset of any solution itemsets of cardinality $> k$.

Proof. Suppose that exists $Y \subseteq t, |Y| = k+j, Y \in Th(\mathcal{C}_{freq[\mathcal{D},\sigma]}) \cap Th(\mathcal{C}_{LAM})$. For loose anti-monotonicity this implies that exists $Z \subseteq Y, |Z| = k+j-1$ such that $\mathcal{C}_{LAM}(Z)$. Moreover, for anti-monotonicity of frequency we have that $\mathcal{C}_{freq[\mathcal{D},\sigma]}(Z)$. The reasoning can be repeated iteratively downward to obtain that must exist $X \subseteq t, |X| = k, X \in Th(\mathcal{C}_{freq[\mathcal{D},\sigma]}) \cap Th(\mathcal{C}_{LAM})$.

Note that a conjunction of loose anti-monotone constraint is not a loose anti-monotone constraint anymore, and therefore each constraint in a conjunction must be treated separately. However, a transaction can be pruned whenever Theorem 1 does not hold for even only one constraint in the conjunction (this is implemented by line 14 of the pseudo-code in Figure 1).

In the next Section we exploit such property of $\mathcal{C}_{LAM}$ constraints in a level-wise Apriori-like computation by means of data-reduction.

3 The $ExAMiner^{\mathcal{LAM}}$ Algorithm

The recently introduced algorithm ExAMiner [3], aimed at solving the problem $Th(\mathcal{C}_{freq}) \cap Th(\mathcal{C}_M)$ (conjunction of anti-monotonicity and monotonicity), generalizes the ExAnte idea to reduce the problem dimensions at all levels of a level-wise Apriori-like computation. This is obtained by coupling the set of data reduction techniques in Table 2 (see [3] for the proof of correctness), which are based on the anti-monotonicity of $\mathcal{C}_{freq}$, with the data reduction based on the $\mathcal{C}_M$ constraint. Here, in order to cope with the mining problem $Th(\mathcal{C}_{freq}) \cap Th(\mathcal{C}_{LAM})$, we couple the same set of $\mathcal{C}_{freq}$-based data reduction techniques with the $\mathcal{C}_{LAM}$-based data reduction technique described in Theorem 1. The resulting algorithm is named $ExAMiner^{\mathcal{LAM}}$.

Essentially $ExAMiner^{\mathcal{LAM}}$ is an Apriori-like algorithm, which at each iteration $k-1$ produces a reduced dataset $\mathcal{D}_k$ to be used at the subsequent iteration k. Each transaction in $\mathcal{D}_k$, before participating to the support count of candidate itemsets, is reduced as much as possible by means of $\mathcal{C}_{freq}$-based data reduction, and only if it survives to this phase, it is effectively used in the counting phase.

Table 2. Data-reduction techniques based on the anti-monotonicity of $\mathcal{C}_{freq}$

$\mathcal{G}_k(i)$	an item which is not subset of at least k frequent k-itemsets can be pruned away from all transactions in $\mathcal{D}$.
$\mathcal{T}_k(t)$	a transaction which is not superset of at least $k+1$ frequent k-itemsets can be removed from $\mathcal{D}$.
$\mathcal{L}_k(i)$	given an item i and a transaction t, if the number of frequent k-itemsets which are superset of i and subset of t is less than k, then i can be pruned away from transaction t.

Procedure count&reduce$^{\mathcal{LAM}}$

Input: $\mathcal{D}_k, \sigma, \mathcal{C}_{LAM}, \mathcal{C}_M, C_k, V_{k-1}$

1. forall $i \in \mathcal{I}$ do $V_k[i] \leftarrow 0$
2. forall tuples t in $\mathcal{D}_k$ do
3.   forall $C \in \mathcal{C}_{LAM}$ do $t.lam[C] \leftarrow false$
4.   forall $i \in t$ do if $V_{k-1}[i] < k-1$
5.   then $t \leftarrow t \setminus i$
6.   else $i.count \leftarrow 0$
7.   if $|t| \geq k$ and $\mathcal{C}_M(t)$ then forall $X \in C_k, X \subseteq t$ do
8. $X.count$++; $t.count$++
9.   forall $C \in \mathcal{C}_{LAM}$ do
10.   if $\neg t.lam[C]$ and $C(X)$ then $t.lam[C] \leftarrow true$
11.   forall $i \in X$ do $i.count$++
12.   if $X.count = \sigma$ then
13. $L_k \leftarrow L_k \cup \{X\}$; forall $i \in X$ do $V_k[i]$++
14.   if $\forall C \in \mathcal{C}_{LAM} : t.lam[C]$ then
15.   if $|t| \geq k+1$ and $t.count \geq k+1$ then
16.   forall $i \in t$ if $i.count < k$ then $t \leftarrow t \setminus i$
17.   if $|t| \geq k+1$ and $\mathcal{C}_M(t)$ then write t in $\mathcal{D}_{k+1}$

Fig. 1. Pseudo-code of procedure count&reduce$^{\mathcal{LAM}}$

Each transaction which arrives to the counting phase, is then tested against the $\mathcal{C}_{LAM}$ property of Theorem 1, and reduced again as much as possible, and only if it survives to this second set of reductions, it is written to the transaction database for the next iteration $\mathcal{D}_{k+1}$. The procedure we have just described, is named count&reduce$^{\mathcal{LAM}}$, and substitutes the usual support counting procedure of the Apriori algorithm from the second iteration on ($k \geq 2$). Therefore to illustrate the $ExAMiner^{\mathcal{LAM}}$ algorithm we just provide the pseudo-code of the count&reduce$^{\mathcal{LAM}}$ procedure (Figure 1), avoiding to provide the well-known Apriori algorithm pseudo-code [1]. We just highlight the we adopt the usual notation of the Apriori pseudo-code: C_k: to denote the set of *candidate* itemsets, and L_k to denote the set of *frequent* (or large) itemsets at iteration k.

In the pseudo-code in Figure 1, the count&reduce$^{\mathcal{LAM}}$ procedure, at iteration k takes in input the actual database $\mathcal{D}_k$, the minimum support threshold σ, a user-defined conjunction of loose anti-monotone constraints $\mathcal{C}_{LAM}$, a user-defined conjunction of monotone constraints $\mathcal{C}_M$, the actual set of candidate itemsets C_k, and an array of integers V_{k-1} of the size of $\mathcal{I}$. Such array is used in order to implement the data-reduction $\mathcal{G}_k(i)$. The array V_k records, for each singleton item, the number of frequent k-itemsets in which it appears. This information is then exploited during the subsequent iteration $k+1$ for the global pruning of items from all transaction in $\mathcal{D}_{k+1}$ (lines 4 and 5 of the pseudo-code). On the contrary, data reductions $\mathcal{T}_k(t)$ and $\mathcal{L}_k(i)$ are put into effect during the same iteration in which the information is collected. Unfortunately, they require information (the frequent itemsets of cardinality k) that is available only at the end of the actual counting (when all transactions have been used). However, since the set of frequent k-itemsets is a subset of the set of candidates C_k, we can use such data reductions in a relaxed version: we just check the number of candidate itemsets X which are subset of t ($t.count$ in the pseudo-code, lines 8 and 15) and which are superset of i ($i.count$ in the pseudo-code, lines 6, 11 and 16). Analogously, the data reduction based on loose anti-monotonicity described in Theorem 1, is exploited in the same relaxed version with candidates instead of frequent itemsets. In the pseudo-code, for each constraint C in the given conjunction of loose anti-monotone constraints C_{LAM}, we have a flag $t.lam[C]$ which is set to $true$ as soon as an itemset $X \in C_k$, such that $X \subseteq t, X \in Th(C)$, is found (line 10). A transaction which has even only one of the $t.lam[C]$ flags set to false after the counting phase, will not enter in the database for the next iteration $\mathcal{D}_{k+1}$ (line 14 of the pseudo-code). In fact, such a transaction has not covered any candidate itemset which satisfies the constraint C, for some C in the conjunction C_{LAM}, therefore it will not support any itemset satisfying such constraint, and thus any solution itemset.

4 Experimental Analysis

In this Section we describe in details the experiments we have conducted in order to assess loose anti-monotonicity effectiveness on both convertible constraints (e.g. $avg(X.A) \geq m$) and tougher constraints (e.g. $var(X.A) \leq m$). The results are reported in Figure 2. All the tests were conducted on a Windows XP PC equipped with a 2.8GHz Pentium IV and 512MB of RAM memory, within the $cygwin$ environment. The datasets used in our tests are those ones of the FIMI repository[1], and the constraints were applied on attribute values generated randomly with a gaussian distribution within the range $[0, 150000]$.

In Figure 2(a) and (b) are reported the tests with the C_{LAM} constraint $var(X.A) \leq m$. We compare $ExAMiner^{\mathcal{LAM}}$ against two unconstrained computation: FP-Growth and ExAMiner without constraint (i.e. it only exploits C_{freq}-based data reduction). Such tests highlight the effectiveness of loose anti-monotonicity: we have a speed up of much more than one order of magnitude, and a data reduction rate up to four order of magnitude.

This behavior is reflected in run-time performances: $ExAMiner^{\mathcal{LAM}}$ is one order of magnitude faster than ExAMiner as reported in Figure 2(c). Conversely, $\mathcal{FIC}^A$ is not able to bring such improvements. In Figure 2(d) we report the speed-up of $ExAMiner^{\mathcal{LAM}}$ w.r.t. ExAMiner and $\mathcal{FIC}^A$ w.r.t. FP-growth. The tests conducted on various datasets show that exploiting loose anti-monotonicity property brings a higher speed up than exploiting convertibility. In fact, $ExAMiner^{\mathcal{LAM}}$ exhibits in average a speed up of factor 100 against its own unconstrained computation, while $\mathcal{FIC}^A$ always provides a speed up w.r.t. FP-growth of a factor lower than 10, and sometimes it is even slower than its unconstrained version. In other words, FP-Growth with a filtering of the output

---

[1] http://fimi.cs.helsinki.fi/data/

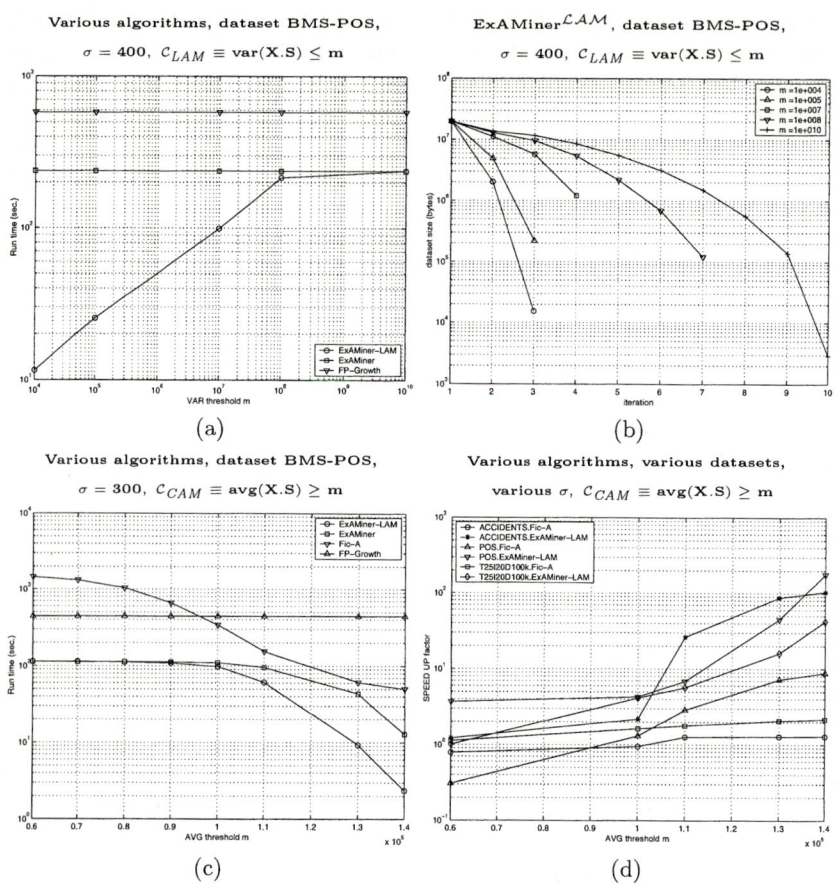

Fig. 2. Loose anti-monotonicity: experimental analysis results

in some cases is better that its variant $\mathcal{FIC}^\mathcal{A}$, which is explicitly geared on constrained mining. As discussed before, this is due to the items ordering based on attribute values and not on frequency.

5 Pushing Multiple Constraints

As already stated, one of the most important advantage of our methodology is that, pushing constraints by means of data-reduction in a level-wise framework, we can exploit different properties of constraints all together, and the total benefit is always greater than the sum of the individual benefits. In other words, by means of data-reduction we exploit a real synergy of all constraints that the user defines for the pattern extraction: each constraint does not only play its part in reducing the data, but this reduction in turns strengthens the pruning power of the other constraints. Moreover data-reduction induces a pruning of the search space, and the pruning of the search space in turn strengthens future data reductions.

Note that in the pseudo-code in Figure 1 we pass to the procedure both a set of $\mathcal{C}_{LAM}$ and a set of $\mathcal{C}_M$ constraints: obviously if the set of $\mathcal{C}_{LAM}$ constraints is empty we obtain the standard ExAMiner *count&reduce* [3] (no $\mathcal{C}_{LAM}$ data reduction); while if we have an empty set of $\mathcal{C}_M$ constraints, the $\mathcal{C}_M$ testing (lines 7 and 17 of the pseudo code) always succeed and thus the μ-reduction is never applied. Whenever we have both $\mathcal{C}_M$ and $\mathcal{C}_{LAM}$ constraints (i.e. a query corresponding to the mining problem $Th(\mathcal{C}_{freq}) \cap Th(\mathcal{C}_M) \cap Th(\mathcal{C}_{LAM}))$ we can benefit of all the data-reduction techniques together, obtaining a stronger synergy.

Example 3. The constraint $range(S.A) \geq v \equiv max(S.A) - min(S.A) \geq v$, is both monotone and loose anti-monotone. Thus, when we mine frequent itemsets which satisfy such constraint we can exploit the benefit of having together, in the same *count&reduce*$^{\mathcal{LAM}}$ procedure, the $\mathcal{C}_{freq}$-based data reductions of Table 2, the μ-reduction for monotone constraints, and the reduction based on $\mathcal{C}_{LAM}$.

Being a level-wise Apriori-like computation, our framework can exploit all different properties of constraints all together. In other words, our contribution can be easily integrated with previous works (e.g. [15, 3]), in a unique Apriori-like computational framework able to take full advantage by any conjunction of possible constraints. In particular, anti-monotone ($\mathcal{C}_{AM}$) constraints are exploited to prune the level-wise exploration of the search space together with the frequency constraint ($\mathcal{C}_{freq}$); succinct ($\mathcal{C}_S$) constraints are exploited at candidate generation time as done in [15]; monotone ($\mathcal{C}_M$) constraints are exploited by means of data reduction as done in [3]; convertible anti-monotone ($\mathcal{C}_{CAM}$) and Loose anti-monotone ($\mathcal{C}_{LAM}$) constraints are exploited by means of data reduction as described in this paper.

At Pisa KDD Laboratory we are currently developing such unified computational framework (within the P^3D project[2]) which will be soon made available to the community.

References

1. R. Agrawal and R. Srikant. Fast Algorithms for Mining Association Rules in Large Databases. In *Proceedings of VLDB'94*.
2. F. Bonchi, F. Giannotti, A. Mazzanti, and D. Pedreschi. Adaptive Constraint Pushing in frequent pattern mining. In *Proceedings of PKDD'03*.
3. F. Bonchi, F. Giannotti, A. Mazzanti, and D. Pedreschi. ExAMiner: Optimized level-wise frequent pattern mining with monotone constraints. In *Proceedings of ICDM'03*.
4. F. Bonchi, F. Giannotti, A. Mazzanti, and D. Pedreschi. ExAnte: Anticipated data reduction in constrained pattern mining. In *Proceedings of PKDD'03*.
5. F. Bonchi and B. Goethals. FP-Bonsai: the art of growing and pruning small fp-trees. In *Proceedings PAKDD'04*.

---

[2] http://www-kdd.isti.cnr.it/p3d/index.html

6. F. Bonchi and C. Lucchese. On closed constrained frequent pattern mining. In *Proceedings of ICDM'04*.
7. C. Bucila, J. Gehrke, D. Kifer, and W. White. DualMiner: A dual-pruning algorithm for itemsets with constraints. In *Proceedings of ACM SIGKDD'02*.
8. L. DeRaedt and S. Kramer. The levelwise version space algorithm and its application to molecular fragment finding. In *Proceedings of IJCAI'01*.
9. G. Grahne, L. Lakshmanan, and X. Wang. Efficient mining of constrained correlated sets. In *16th International Conference on Data Engineering (ICDE' 00)*, pages 512–524. IEEE, 2000.
10. J. Han, L. V. S. Lakshmanan, and R. T. Ng. Constraint-based, multidimensional data mining. *Computer*, 32(8):46–50, 1999.
11. J. Han, J. Pei, and Y. Yin. Mining frequent patterns without candidate generation. In *Proceedings of ACM SIGMOD'00*.
12. B. Jeudy and J.-F. Boulicaut. Optimization of association rule mining queries. *Intelligent Data Analysis Journal*, 6(4):341–357, 2002.
13. D. Kifer, J. Gehrke, C. Bucila, and W. White. How to quicklyfind a witness. In *Proceedings of PODS'03*.
14. L. V. S. Lakshmanan, R. T. Ng, J. Han, and A. Pang. Optimization of constrained frequent set queries with 2-variable constraints. *SIGMOD Record*, 28(2), 1999.
15. R. T. Ng, L. V. S. Lakshmanan, J. Han, and A. Pang. Exploratory mining and pruning optimizations of constrained associations rules. In *Proceedings of the ACM SIGMOD'98*.
16. J. Pei and J. Han. Can we push more constraints into frequent pattern mining? In *Proceedings of ACM SIGKDD'00*.
17. J. Pei, J. Han, and L. V. S. Lakshmanan. Mining frequent item sets with convertible constraints. In *(Proceedings of ICDE'01)*.
18. R. Srikant, Q. Vu, and R. Agrawal. Mining association rules with item constraints. In *Proceedings of ACM SIGKDD'97*.

An Efficient Compression Technique for Frequent Itemset Generation in Association Rule Mining

Mafruz Zaman Ashrafi, David Taniar, and Kate Smith

School of Business Systems, Monash University, Clayton,
VIC 3800, Australia
{Mafruz.Ashrafi, David.Taniar,
Kate.Smith}@infotech.monash.edu.au

Abstract. Association Rule mining is one of the widely used data mining techniques. To achieve a better performance, many efficient algorithms have been proposed. Despite these efforts, we are often unable to complete a mining task because these algorithms require a large amount of main memory to enumerate all frequent itemsets, especially when dataset is large or the user-specified support is low. Thus, it becomes apparent that we need to have an efficient main memory handling technique, which allows association rule mining algorithms to handle larger datasets in main memory. To achieve this goal, in this paper we propose an algorithm for vertical association rule mining that compresses a vertical dataset in an efficient manner, using bit vectors. Our performance evaluations show that the compression ratio attained by our proposed technique is better than those of the other well known techniques.

1 Introduction

One of the widely used data mining techniques is association rule mining [1]. Association rule mining algorithms iterate dataset many times to enumerate frequent itemsets that exist in the transactions of a given dataset. However, dataset scan is considered as an I/O exhaustive process [1]. Therefore, the performance degrades if mining algorithm requires multiple dataset scans.

Since main memory plays a significant role in the association rule mining performance, in recent year several novel techniques have been proposed [4, 6, 9, 10] in order to efficiently use main memory. These techniques generally cut down the dataset size, so that the mining algorithms will be able to finish the mining task on bigger datasets or with a low support. We can categorize these existing techniques into three groups: (*i*) vertical compression [8, 9], and (*ii*) horizontal compression [7] and (*iii*) vertical tid compression [4, 10].

Vertical Compression: It uses vertical bit vector to represent the presence and absence of an item, and adopts a lossless compression. However, several researches show that these compression techniques heavily depend on user specified support or dataset characteristics [6, 7]. When the support is low or the dataset is sparse, these compression techniques may cause expansion, not compression [4].

Horizontal Compression: It uses a tree based data structure [7] to compress horizontal dataset. As a result, they are able to condense a dense dataset when the user specified support is high. On the other hand, it also causes an expansion rather than compression, if the dataset is sparse or the user support is low [10].

Vertical Tid Compression: It stores the vertical tid dataset in an alternative way, for example, diffset [4] stores tids from transactions of a dataset where a particular item is absent. And subsequently, it gains a good compression ratio if the user specified is high. However, when the support is low, the diffset technique is unable to compress the dataset, because the number of diffset is larger than the actual tids (i.e. number of times each of the items occurs in the transaction).

Since all of the abovementioned compression techniques heavily depend on user specified support and/or dataset characteristics, these techniques often do not allow us to enhance main memory utilization. However, from the above discussion it is apparent that without main memory enhancement techniques, it becomes quite difficult to achieve any performance improvement.

We are motivated by the abovementioned fact that main memory is an important resource and to improve performance we need to use it in an efficient way without exceeding its capacity. To enhance main memory capacity, in this paper we propose an algorithm that uses a bit-oriented approach to compress vertical tid dataset. The proposed technique keeps track of difference between two tids and converts the differences into a bit format, and finally, stores these bits into a bit vector in an efficient way, so the resultant bit vector has only a few unused bits.

The important outcome of this method is that the proposed technique is not bias, which means that it does not depend on a particular dataset characteristics (i.e. dense or sparse) or the user specified support. Rather, it has an ability to compress the original dataset regardless of dataset size, type or user specified support. Our performance evaluation also shows that it achieves good compression ratio in all scenarios. Therefore, it is able to keep large datasets and allows the mining algorithms to perform mining tasks on such datasets.

The rest of paper is organized as follows. We describe the reason why we need efficient main memory in section 2. Next, we present our proposed efficient main memory compression algorithm. Then, we presented the performance evaluation and comparison followed by the conclusion.

2 Rationale of Our Work and Analysis of Existing Work

Performance improvement of association rule mining can normally be attained in two ways including the use of efficient mining techniques and the reduction in using main memory. However, during the mining task, if the algorithm exceeds the main memory limit, the mining process will take a long time regardless of how efficient the mining technique is. In other words, efficient mining techniques will only be efficient if and only if there is an abundant space in the main memory, so that the mining process will not exceed the main memory limit.

Before we detail of our proposed memory enhancement technique, let us first analyze some of the existing algorithms and corresponding main memory management techniques. In the following few paragraphs we will discuss a number of well known association rule mining algorithms and the amount of main memory needed by these algorithms to perform the mining task.

The Apriori algorithm [2] uses a horizontal dataset layout and a breadth-first bottom-up search technique to enumerate all frequent itemsets that meet the user specified support. It achieves good performance with a high support threshold when the total number frequent itemsets and the length of frequent itemsets are small. However, it will exceed main memory when the number of frequent items is large.

Another novel algorithm that uses a horizontal dataset layout but employs a depth-first search technique is FP growth [7]. It aims to compress the horizontal dataset into FP tree after the first dataset scan. After that, it generates all frequent itemsets using the compressed dataset and it consequently solves the multiple dataset scan problem. However, the main problem of this approach is the size of compressed dataset. It is often unable to reduce the size if the number of items after the first iteration is large.

A few algorithms [4, 5, 8, 9] use a vertical dataset layout to generate frequent itemset. For example, Eclat [4] uses a depth-first search technique. However, the main problem of such approach is that when the dataset is dense, it has too many *tids*, and holding the intermediate results of these *tids* often exceeds the main memory capacity. To overcome such a memory constraint, Zaki et al [5] present another algorithm known as dEclat [5]. This algorithm stores diffset of two itemsets and generates frequent itemsets from diffset, not from the entire *tid*. However, this algorithm is still not able to reduce the size of the difsets dataset compared to *tid* dataset, especially when the dataset is sparse and/or the support is low.

VIPER [9] algorithm uses a vertical bit vector to represent items occurrence in the transactions of a dataset. For example, if item "A" appears in the "n^{th}" transaction, then it sets "*1*" to the nth position of A. This approach also requires a huge memory, when the number of transactions and the number of attributes in a dataset is large. VIPER algorithm then uses a compression technique known as *skinning* to reduce the size of the bit vector. However, in some cases this compression technique is not able to compress the bit vector [5], but rather its size increases.

From the above discussions, it is clear that main memory is one of the key resources to improve performance of association rule mining. However, due to a limited main memory capacity, the mining algorithms experience performance bottleneck particularly when performing the mining task on a large dataset. To enhance main memory capacity, in this paper we propose an algorithm that uses a bit-oriented approach to compress vertical tid dataset. The proposed technique keeps track of the difference between two tids and converts the differences into a bit format, and finally, stores these bits into a bit vector in an efficient way, so the resultant bit vector has only a few unused bits.

3 The Proposed Technique

To increase the performance of association rule mining, it is necessary to exploit main memory efficiently without exceeding its capacity. In order to achieve this goal in this paper we propose a main memory optimization technique to be used to perform the first phase of any association rule mining task, that is generating the frequent itemset.

We use a vertical dataset layout to compress the dataset because it allows us to perform intersection between one or more items in order to find their co-occurrence in the dataset. We can represent the vertical dataset into two different ways: vertical bitmap and vertical tids. Let us first analyze each of the vertical dataset representations and the amount of memory needed by them in order to find the possibilities where we can optimize the memory usage.

Since vertical bitmaps vector needs to register both the absence and presence of an item in a transaction [9], it will reduce the dataset size when dataset is dense or the user specified support is high. In other words, if all items of the dataset occur many times, that is the bitmap vector has more '1's than '0's, subsequently the size of vertical bitmaps is smaller than the dataset size. On contrary, it has more '0's than '1's when dataset is sparse or user specified support is low, vertical bitmaps size will go beyond the original dataset size. The total number of bits we require to hold the whole dataset in main memory can be calculated using the following formulae:

$$I_{BV} = \sum x_1 + \sum x_0 \qquad (1)$$

$$T_{BV} = \sum_{I=1}^{N} I_{BV} \qquad (2)$$

where I_{BV} is total number of bits require to hold all x_1 and x_0 (i.e. the number of transaction where item I is present and absent) and T_{BV} total number of bits we need to hold N number of items in the main memory.

The vertical tids is an alternative representation of vertical bit vectors. Each item of this representation has a list, consisting of all transaction ids where that item appears. Since each item only represents the transaction id where it appears, we need less memory compared to vertical bit vectors when dataset is sparse and/or when the user specified support is low. However, this representation becomes more expensive compare to bit vector in terms of space if user specified support is more than 3%. Because each entry is a word, and to hold this in main memory we need 32 bits [8]. We can calculate the total number of bits we need to hold vertical tids in main memory in the following formulae:

$$I_{TID} = m * 32 \qquad (3)$$

$$T_{TID} = \sum_{I=1}^{N} I_{TID} \qquad (4)$$

where m is number of times item I occur in the dataset, I_{TID} is the total number of bits required to hold item I and T_{TID} is total number of bit we need to hold all N number of items in the main memory.

3.1 Observation

The vertical tid representation always needs a minimum 32 bits to hold each occurrence of an item (i.e. word) appeared in the transaction regardless of the dataset size. However, when dataset is small, one can reduce a significant amount of space, if bits are used instead of integer to represent a word. The rationale behind this is that in a small dataset, the total number of transactions is also small, and hence to accommodate these tids into main memory, we need less than 32 bits, if we convert integer value of all tids to corresponding bits. Converting any integer to bits or vice versa is quite straightforward; one can do it on the fly without any performance degradation. For simplicity, we name this representation as *tid-bits*. Nevertheless, this representation requires less than 32 bits to hold a word, and therefore it becomes less expensive in terms of space compared to the bit vectors, if the user specified support is less than $(100/n)\%$, where n is number of bits when converting the last transaction id to bit. To illustrate the above rationale more clearly, let us consider the following example:

Suppose a dataset has 1,000 items and 1 million transactions with an average transaction size of 20. Now, we will calculate number of bits required to hold this dataset in the main memory using the abovementioned three different techniques: *vertical bitmaps*, *vertical tids* and *tid-bits* format. Firstly, we employ formula (2) and calculate the total number of bits required in vertical bitmaps which is equal to 1,000,000,000 bits. Secondly, using formula (4) we found that vertical tids needs 640,000,000 bits. Finally to find the total number of bits required in tid-bits format, we first convert the last tid of that dataset into bits; in this case we convert 1,000,000 to bit. Since the last transaction id is 20 bit long, therefore we can accommodate a word (i.e. any tid of this dataset) within 20 bits. And to hold entire dataset we only need 400,000,000 bits.

From the above example, it is clear that tid-bits approach requires fewer bits than the other two vertical representations. However, we have not yet used any user specified support, and hence one may think that this calculation may not be appropriate in the presence of user specified support. In this regard, we would like to mention here that tid-bits approach requires less space than vertical tids no matter what the user specified support is. In contrast, with bitmap vector, it requires less space as long as the user specified support is less than $100/n\%$ (where n is number of bits when converting the last transaction id to bit). Because indeed the bitmap vector representations will have more '0' than '1' when user specified support is less than $100/n$ %.

3.2 Algorithm

Before we move to the proposed algorithm in details, it is important to explain why we need a specific number of bits for all tids (i.e. equal to the bit representation of last tid of a dataset) in the tid-bits approach. For example, when a dataset has 100,000 transactions, we need 17 bits for each entry, and it increases up to 20 bits when dataset has 1,000,000 transactions, although we can accommodate the first 100,000 transactions within 17 bits. However we can not put different number of bits (i.e. size) for different tids, because when we convert each entry of bit vector to find its corresponding tid, we need to read a specific number of bits for every entry; otherwise it is difficult to convert these bits to its corresponding tids.

For example, an item that occurs in the 1st and 100th transaction of a dataset is converted its tids into a bit format (i.e. *1* and *1100100*) and put these bits directly to a vector. However, doing such an insertion causes ambiguity when we try to convert entries of that vector to its corresponding tids, as we will not know how many tids there are or how many bits each entry has. Thus, we are unable to obtain the tid values of each entry of that vector. To overcome from such ambiguity, we can only use a specific number of bits for all entries regardless the exact bit size of each individual entry. And if we come across a bit representation of a tid that is less than that number then we pad it with '*0*'.

Since the length of bits (i.e. each entry) increases as tids increases, we can only achieve scalability when we are able to find an alternative approach that does not increase bits length as tids increases in the dataset. On the other hand, the above goal is only achievable when we are able to compress bit vector. In other words, we need to condense the bits representation of all tids, in such a way that its size remains constant as the number of tids increases.

3.2.1 Dif-Tid

To achieve the above goal, we propose an enhancement technique that compresses tids in an efficient way. The proposed technique does not convert any tid into a bit format on the fly as it does with the tid-bits. Rather, during the dataset reading it finds the difference between current (i.e. transaction number of current reading position) and previous (i.e. the transaction number where this item appears last time) tid of every item and converts that value into a bit representation. For simplicity we called it *dif-bit*. Finally, it places these dif-bit into a bit vector. The analogy behind this can be described in the following property:

Property 1: The difference between two tids is always smaller than the largest tid of those two.

Rationale: Suppose T_1 and T_2 are two tid of an item '*A*' and '*D*', the difference between those two tids is $D = T_2 - T_1$. Since D is the subtraction of two tids, it is always smaller than T_2.

In addition, when we keep the difference of two tids, it will be simple to find the exact tids from those values, that is by adding *n* number of differences. For a better understanding, let us consider the following example:

Item	TID
A	1 3 5 6 7
B	1 2 5 7 8 10
C	2 4 8 10
D	1 2 4 7 8
E	8 10 11
F	1 2 3 9
G	11
H	1 3 6 9
I	3 4 6 9 10 12
J	11

(a)

Item	TID-Difference
A	1 2 2 1 1
B	1 1 3 1 2 2
C	2 2 4 2
D	1 1 2 3 1
E	8 2 1
F	1 1 1 3
G	11
H	1 2 3 3
I	3 1 2 3 1 2
J	11

(b)

Item	Diffrence Bits
A	1 1 0 1 0 1 1
B	1 1 1 1 1 1 0 1 0
C	1 0 1 1 0 1 0 0 0 1 0
D	1 1 1 1 0 1 1 1
E	1 0 0 0 1 0 1
F	1 1 1 1 1 1 0
G	1 0 1 1
H	1 1 0 1 1 0 1 1 1 0 1
I	1 0 1 1 1 0 1 0 1 1 1 1 0
J	1 0 1 1

(c)

Fig. 1. (a)Dataset; (b)Difference between two TIDs and (c)Corresponding Bit Vector

Suppose, we have a dataset, as shown in figure 1(a), where item "*A*" appears in the {1st, 3rd, 5th, 6th and 7th} transactions. To store these tids directly in the main memory, we need at least 32 × 5 = 160 bits. However, if we only keep the difference of each two tids and convert these into bits (i.e dif-bit), then we can accommodate all appearance of item "A" within 7 bits as shown in figure 1(b and c).

The above example clearly depicts that if we put dif-bit rather than the original tid value in the bit vector, we will be able to reduce bit vector size significantly. In addition, it is worth to mention that the proposed method only needs 32 bits when the difference between the two tids is more than 2.3 billions. But, it is quite uncommon that an item occurs in the dataset after such a large interval. Nevertheless, if any item occurs in the dataset after such a long interval, the corresponding item support is also low. For example, if an item always appears in the dataset after 2.3 billions transactions, its support must be less than $4.4 \times 10\text{-}9\%$. Indeed, it is apparent that association rule mining algorithms rarely or never use such small support threshold for generating frequent itemsets, thus the proposed technique rarely requires 32 bits to store dif-bits no matter how big the dataset is.

Inserting difference bits into a bit vector reduces its size. On the other hand, it is difficult to convert bit vector to its original tids format because from that vector we do not know the exact number of bits we need to construct the original differences of two tids. To elevate this, we have modified our proposed technique and put n number of bits in the bit vector for every entry before inserting the original difference in a bit form, where n specify the number of bits the bit representation of difference value has. Therefore, during the conversion we know the exact size of the difference if we read that n number of bits in advance for every entry.

Since inserting n bits in advance for every entry incurs some overhead, one may raise a question about the efficiency of our proposed technique. In this regard we argue that the size of n is very small. For example if n is equal 5, then we can put dif-bit that has 32 bits. However from previous discussion we say that we rarely need 32 bits because using 32 bits we can represent a difference of two tids that occur after 2.3 billions transactions.

Nevertheless, we can further reduce the dif-bit size, if we remove the left most bit. Since the left most bit of dif-bit always has "*1*", hence we remove that from the dif-bit. And when we convert dif-bit to its corresponding tid, we simply add "*1*" in the same position (i.e. left most bit) of the dif-bit, so that we obtain the original dif-bit value. For example, if the difference between two tids is 9 (i.e. *1001*) and n is 3 (i.e. *11*), then we put "*11001*" to the bit vector rather than "*111001*" and subsequently reduce "*1*" bit from each tid (i.e. dif-bit). Finally, when we convert "*11001*" to its corresponding tid, we place "*1*" after the n bits, and in this case after "*11*" we obtain the original dif-bit value i.e. "*1001*".

The pseudocode of the proposed algorithm is shown in figure 2. Initially, it reads the dataset and finds the difference of each item (i.e. subtract current tid with the previous tid where that item appears last time) during reading. Then, it converts those tid difference values into bit format and finds the corresponding range bits. Finally, those bits and range bits are inserted into the bit vector of the corresponding item. During the conversion, it first reads the range bits, and from that range bits it finds the bit size of next difference. Subsequently, it reads that specific number of bits and converts it to the number format. Finally, it finds the exact tid by adding that difference value with the previous tid.

```
procedure add()                              procedure read()
input : Dataset(D),Items(I), Range(r) and    input : A bit Vectors(V), Range(r);
        Minimum Support(s);                  output :Set of Tids(T);
output : Set of Bit Vectors F;               i_p = 0;
for i ∈ I occuring in D                      while (V ≠ {}) do
    //i_c current tid; i_p previous tid          n = read_range_bit(r);
    Compute D_i = i_c − i_p ; i_p = i_c ;        s = convert_to_no(n);
    b = convert_to_bits( D_i );                  b = read_bits (s);
    n = find_range_size( b );                    no = convert_to_no(b);
    add_bit_vector( n,b );                       i_p = i_p + no;
End for;                                     End do;
while (F ≠ {}) do
    for all I ∈ F
        sup = support(i);
        if sup π s
            delete(I);
    End for;
End do;
```

Fig. 2. The Proposed Algorithm

4 Performance Evaluation

We have done an extensive performance study on our proposed technique to confirm our analysis of its effectiveness. Four datasets are chosen for this evaluation study. Table 1 shows the characteristics of each datasets that are used in our evaluation. It describes the number of items, the average size of each transaction, and the number of transactions of each dataset has.

Table 1. Dataset Characteristics

Name	Transaction Size avg.	Number of Distinct Items	Number of Records
Cover Type	55	120	581012
Connect-4	43	130	67557
T40I10D100K	40	1000	100000
Kosarak	8	41000	990000

When we put dif-bit into a bit vector we put 'n' number of bits prior to dif-bit insertion in order to specify the size of that dif-bit. However, for choosing the value of 'n' we have two alternatives: exact size and range. The former specifies the exact size of a dif-bit, hence the size of 'n' never goes beyond 5, since using 5 bits we can express the exact bit size of any number that has 32 or less bits. The latter approach is specified a dif-bit size, and therefore the size of 'n' becomes smaller. For example, 32 bits can be divided into 4 groups (i.e. 1-8, 9-16, etc.) and to represent each group we need 2 bits only. And during the insertion, a particular range value that is suitable for a dif-bit size is found and placed before the dif-bit. Since the latter approach needs fewer bits, in the performance evaluation we adopt the range approach.

In the first experimentation, we evaluate the efficiency of our proposed method by comparing it with the tid-based approaches. It is worth to mention that this tid based approach uses a well known vertical dataset layout and have been used in many different association mining algorithms [4, 5, 10]. These algorithms discover frequent itemsets in two phases: at first it stores tids of all items separately into the main memory, then intersects tids of one item with other item's and find the exact support.

Because the aims of this experiment is to find out the total memory that each of the approaches consumes, therefore we are only keen to know how much memory space those tid-based algorithms [4] need in order to complete the first phase. In figure 3, we plot a detail comparison between dif-bit (our proposed method) and tid. It shows how much memory each of the approaches takes at different support thresholds.

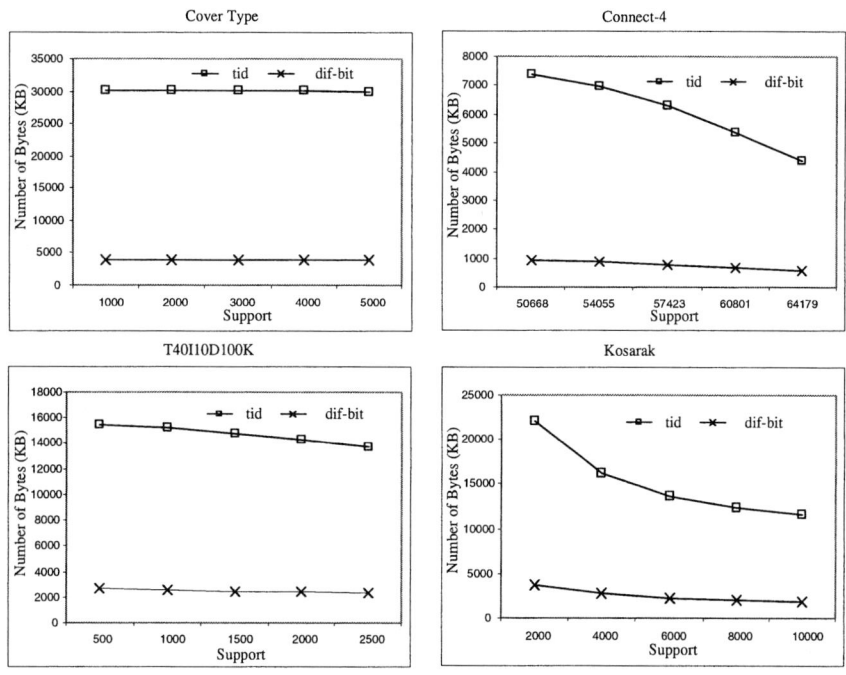

Fig. 3. Comparison dif-bit vs. tid

From the graph shown in figure 3, it is clear that the dif-bit approach always requires less memory than those of the tid approach. The size of dif-bit approach is 5-8 times smaller than the corresponding tid approach, as the dif-bit approach finds the difference of two tids, converts that difference in a bit format and finally puts those bits into a bit vector. Therefore, it often requires only fewer bits to represent a difference value. In contrast, the tid approach requires 4 bytes (32 bits) to hold each of the word (i.e each appearance), hence requires more memory. In addition, the proposed dif-bit approach increases the compression ratio as the support increases because in a higher support the average difference value reduces and subsequently requires fewer bits to represent a difference. For example, when we consider T40I10D100K dataset at 500 (500/100,000 = 0.005%) support the proposed dif-bit size is 5.85 times smaller than the tid approach, whereas its size is 6 times smaller than the corresponding tid approach when support is 2,500 (2,500/100,000 = 0.025%).

5 Conclusion

Main memory is one of the important resources that can be used to improve the performance of association rule mining. However, due to small main memory capacity, mining algorithms often experience performance bottleneck when performing a mining task on a large dataset. Since association mining algorithms inheritably dependable on amount of main memory, hence when that amount is not sufficient then these algorithms will be unable to finish the mining task. Therefore, we critically need efficient memory enhancement techniques to make any mining algorithm complete the mining task on large datasets. To achieve this goal in this paper we present a technique, known as *"Diff-tid"*, which gain good compression ratio regardless of dataset characteristics or user specified support. The performance evaluation confirms that the proposed algorithm needs only a small amount of memory compared to other compression techniques that had been used by many association rule mining.

References

1. M. J. Zaki, "Parallel and Distributed Association Mining: A Survey". IEEE Concurrency, October-December 1999.
2. R. Agrawal and R. Srikant, "Fast Algorithms for Mining Association Rules in Large Database". In Proc. of the 2oth VLDB, pp. 407-419, Santiago, Chile, 1994.
3. R. Agrawal, T. Imielinski and A. N. Swami, "Mining Association Rules Between Sets of Items in Large Databases". In Proc. of the ACM SIGMOD, pp 207-216, 1993.
4. M. J. Zaki, "Scalable Algorithms for Association Mining". IEEE Transactions on Knowledge and Data Engineering, Vol. 12 No.2 pp. 372-390, 2000.
5. M. J. Zaki and K. Gouda, "Fast Vertical Mining Using Diffsets". In Proc. of the 9th ACM SIGKDD, ACM Press 2003.
6. M. El-Hajj and O. R. Zaiane, "Inverted matrix efficient discovery of frequent items in large datasets in the context of interactive mining". In Proc. of the 9th ACM SIGKDD, 2003.
7. J. Han, J. Pei, and Y. Yin, "Mining Frequent Patterns without Candidate Generation". In Proc. of ACM SIGMOD, pp 1-12, 2000.
8. B. Doug, C. Manuel and G. Johannes "MAFIA: a maximal frequent itemset algorithm for transactional databases" In Proc. ICDE pp. 443-452, 2001.
9. P. Shenoy, J.R. Haritsa, S. Sudarshan, G. Bhalotia, M. Bawa, and D. Shah, "Turbo-charging vertical mining of large databases". In Proc. ACM SIGMOD, 2000.
10. B Geothals, "Memory Efficient Association Mining", In Proc. ACM SAC, 2003.
11. P. S. Jong, C. Ming-Syan and Y. S. Philip "An Effective Hash Based Algorithm for Mining Association Rules", In Proc. ACM SIGMOD, pp. 175-186, San Jose, California, 1995.

12. S. Ashoka, O. Edward and N. B. Shamkant, "An Efficient Algorithm for Mining Association Rules in Large Databases", In Proc. of the VLDB, pp. 432-444, Zurich, Swizerland, 1995.
13. C. L. Blake and C. J. Merz. "UCI Repository of Machine Learning Databases", University of California, Irvine, Department. of Information and Computer Science, www.ics.uci.edu/~mlearn/MLRepository.html, 1998.
14. B.Goethals and M. J. Zaki, "FIMI Repository", http://fimi.cs.helsinki.fi/fimi03/, 2003.
15. B.Goethals "Survey on Frequent Pattern Mining", University of Helsinki, Finland.

Mining Time-Profiled Associations: An Extended Abstract*

Jin Soung Yoo, Pusheng Zhang, and Shashi Shekhar

Computer Science & Engineering Department, University of Minnesota,
200 Union Street SE, Minneapolis, MN 55455, U.S.A
{jyoo, pusheng, shekhar}@cs.umn.edu

Abstract. A time-profiled association is an association pattern consistent with a query sequence over time, e.g., identifying the interacting relationship of droughts and wild fires in Australia with the El Nino phenomenon in the past 50 years. Traditional association rule mining approaches reveal the generic dependency among variables in association patterns but do not capture the evolution of these patterns over time. Incorporating the temporal evolution of association patterns and identifying the co-occurring patterns consistent over time can be done by time-profiled association mining. Mining time-profiled associations is computationally challenging due to the large size of the itemset space and the long time points in practice. In this paper, we propose a novel one-step algorithm to unify the generation of statistical parameter sequences and sequence retrieval. The proposed algorithm substantially reduces the itemset search space by pruning candidate itemsets based on the monotone property of the lower bounding measure of the sequence of statistical parameters. Experimental results show that our algorithm outperforms a naive approach.

1 Introduction

A *time-profiled association* is an association pattern [2] consistent with a query sequence over time. One example is the frequent co-occurrences of climate features with the El Nino phenomenon over the last 50 years [10]. El Nino, an abnormal warming in the eastern tropical Pacific Ocean[1], has been linked to climate phenomena such as droughts and wild fires in Australia and heavy rainfall along the eastern coast of South America in the past 50 years. Transaction data are implicitly associated with time, i.e., any transaction is associated with a certain time slot. Thus the association patterns might change over time. For example, a sales association between diaper and beer is high only in the evening

---

* This work was partially supported by NSF grant 0431141 and Army High Performance Computing Research Center contract number DAAD19-01-2-0014. The content of this work does not necessarily reflect the position or policy of the government and no official endorsement should be inferred. AHPCRC and Minnesota Supercomputer Institute provided access to computing facilities. Readers may refer to the technical report [9] for more details.

but not in other time slots. Association patterns found might have different popularity levels over time. These variations in time are not captured under traditional association rule mining[2]. Hence time-profiled association mining can be used to discover interacting relationships consistent with a query prevalence sequence over time. Mining time-profiled associations is crucial to many applications which analyze temporal trends of interactions among variables, including Earth science, climatology, public health, and commerce.

Mining time-profiled associations is computationally challenging since the sizes of itemset space and temporal space are extremely large in practice. In the example of the El Nino investigation, there are millions of spatial units with climate features (e.g., temperature and precipitation), each having 50 years worth of daily observations, i.e.,50*12*365=21,900 time points. An observation at one time point in a specific location can be treated as one transaction, so there are more than millions of transactions globally at one time snapshot. Therefore, exploring a pair of climate features will involve about a trillion itemset space and long time series, and exploring all relationships among features would be even more exorbitant.

To our knowledge, there is no prior work directly tackling the problem of mining time-profiled associations. Some relevant work has attempted to capture the temporal dynamics of association patterns, including active data mining [4], cyclic association rule mining [8], and calendar-based association rule mining [7]. However, these approaches do not appear to be directly applicable for identifying consistent associations over time with a query sequence.

A naive approach to mining time-profiled associations can be characterized using a two-phase paradigm. The first phase updates the history of the statistical parameters (e.g., support) for rules at different time points using a traditional *Apriori* [2] approach, and generates a sequence of statistical parameters. The second phase matches the sequences of statistical parameters to find time-profiled associations with the query sequence. However, exponentially increasing computational costs of generating all combinatorial candidate itemsets become prohibitively expensive. We propose a novel one-step algorithm to unify the generation of statistical parameter sequences and sequence searching. The proposed algorithm prunes the candidate itemsets by using the monotone property of the lower bounding measure of the sequence of statistical parameters. It substantially reduces the search space of itemsets, and is efficient in terms of the number of candidate itemset generations. Experimental results show that our algorithm outperforms the naive approach.

2 Problem Statement

A time-profile association is an association pattern consistent with a specific time sequence over time. The problem of mining time-profiled association patterns is to find all itemsets whose time sequences of prevalence measures are similar to a user specified query sequence under a given similarity threshold. The detailed problem definition is described as follows. We assume that a query time sequence Q is in the same scale as the prevalence measures or can be transformed to the same scale

Given:

1) A set of items $E = \{e_1, \ldots, e_m\}$.
2) A time-stamped transaction database D. Each transaction $d \in D$ is a tuple $< time\text{-}stamp, itemset >$ where $time\text{-}stamp$ is a time that the transaction d is executed and $itemset$ is a set of items which is a subsets of E.
3) A time unit t. The ith time slot t_i, $0 \leq i < n$, corresponds to the time interval $[i \cdot t, (i+1) \cdot t)$. The set of transactions executed in t_i is denoted by D_i.
4) A query time sequence $\boldsymbol{Q} =< q_0, \ldots, q_{n-1} >$ over time slots $t_0, \ldots, t_{n-1}$.
5) A threshold of similarity value θ.

Find: A complete and correct set of itemsets $I \subseteq E$ where $f_{similar}(\boldsymbol{P_I}, \boldsymbol{Q}) \leq \theta$, where $\boldsymbol{P_I} =< p_0^I, \ldots, p_{n-1}^I >$ is the time sequence of prevalence values of an itemset I over time slots $t_0, \ldots, t_{n-1}$ and $f_{similar}(\boldsymbol{P_I}, \boldsymbol{Q})$ is a similarity function between two sequences $\boldsymbol{P_I}$ and $\boldsymbol{Q}$.

Objective: Minimize computational cost.

3 Properties of Time-Profiled Associations

3.1 Basic Concepts

Support Time Sequence : We use support as the prevalence measure of an itemset since it represents how statistically significant a pattern is, and it has an anti-monotone property [2].

Definition 1. *Given a time-stamped transaction database* $D = D_0 \cup \ldots \cup D_{n-1}$, *the support time sequence* $\boldsymbol{P_I}$ *of an itemset* I *is the sequence of support values of an itemset* I *over* $D_0, \ldots, D_{n-1}$ *such that*

$$\boldsymbol{P_I} =< support_{D_0}(I), \ldots, support_{D_{n-1}}(I) >$$

where $support_{D_i}(I) = |\{d \in D_i | I \subseteq d\}|/|D_i|$.

Choice of Similarity Measure: Several similarity measures have been proposed in the time series literature [6]. We propose using Euclidean distance as the similarity measure between two sequences since it is a typical similarity measure and is useful in many applications [3,5]. For two time sequences $\boldsymbol{X} =< x_0, \ldots, x_{n-1} >$ and $\boldsymbol{Y} =< y_0, \ldots, y_{n-1} >$, the Euclidean similarity measure is defined as $f_{similar}(\boldsymbol{X}, \boldsymbol{Y}) = D(\boldsymbol{X}, \boldsymbol{Y}) = \sqrt{\sum_{i=0}^{n-1}(x_i - y_i)^2}$. If this distance is below a user-defined threshold θ, we say that the two sequences are similar.

3.2 Upper Bound Time Sequence and Lower Bounding Measure

Lemma 1. *Let* I_{k+1} *be a size* $k+1$ *itemset* $\subseteq E$ *and* $\{I_k^1, \ldots, I_k^{k+1}\}$ *be a set of all size* k *sub itemsets of* I_{k+1}, *where* $I_k \subset I_{k+1}$. *Let* $\boldsymbol{P_{I_{k+1}}} = < p_0^{I_{k+1}}, \ldots, p_{n-1}^{I_{k+1}} >$

be the support time sequence of I_{k+1} and $\boldsymbol{P_{I_k}} = <p_0^{I_k},\ldots,p_{n-1}^{I_k}>$ be the support time sequence of I_k. The upper bound sequence of $\boldsymbol{P_{I_{k+1}}}$, $\boldsymbol{U_{I_{k+1}}} = <u_0^{I_{k+1}},\ldots,u_{n-1}^{I_{k+1}}>$ is $<min\{p_0^{I_k^1},\ldots,p_0^{I_k^{k+1}}\},\ldots,min\{p_{n-1}^{I_k^1},\ldots,p_{n-1}^{I_k^{k+1}}\}>$.

Definition 2. *Given a query time sequence $\boldsymbol{Q}$, the lower bounding measure between $\boldsymbol{Q}$ and the support time sequence $\boldsymbol{P_I}$ of an itemset I is defined as*

$$D_{lb}(\boldsymbol{Q},\boldsymbol{P_I}) = \sqrt{\sum_{i=0}^{n-1}(q_i-u_i)^2},\ q_i \geq u_i,$$

where i is a time slot, $q_i \in \boldsymbol{Q} =<q_0,\ldots,q_{n-1}>$ and $u_i \in \boldsymbol{U_I} =<u_0,\ldots,u_{n-1}>$, the upper bound time sequence of $\boldsymbol{P_I}$.

Lemma 2. *For the true similarity measure $D(\boldsymbol{Q},\boldsymbol{P_I})$ and the lower bounding measure $D_{lb}(\boldsymbol{Q},\boldsymbol{P_I})$ of a query time sequence $\boldsymbol{Q}$ and the support time sequence $\boldsymbol{P_I}$ of an itemset I, the following inequality holds:*

$$D_{lb}(\boldsymbol{Q},\boldsymbol{P_I}) \leq D(\boldsymbol{Q},\boldsymbol{P_I})$$

3.3 Monotone Property of the Lower Bounding Measure

Lemma 3. *Let $\boldsymbol{P_{I_k}}$ be the support time sequence of a size k itemset I_k and $\boldsymbol{P_{I_{k+1}}}$ be the support time sequence of a size $k+1$ itemset I_{k+1}, where $I_{k+1} = I_k \cup I_1$ and $I_1 \notin I_k$. The following inequality holds:*

$$D_{lb}(\boldsymbol{Q},\boldsymbol{P_{I_k}}) \leq D_{lb}(\boldsymbol{Q},\boldsymbol{P_{I_{k+1}}})$$

It is clear by Lemma 1 and Definition 2. The upper bound of the support time sequence of an itemset decreases with increasing itemset size. As a result, the lower bounding measure does not decrease with increasing size of itemset. For a similarity threshold θ, if $D_{lb}(\boldsymbol{Q},\boldsymbol{P_{I_k}}) > \theta$, then $D_{lb}(\boldsymbol{Q},\boldsymbol{P_{I_{k+1}}}) > \theta$. Lemma 3 ensures that the lower bounding measure can be used to effectively prune the search space and efficiently find interesting itemsets.

4 Time-Profiled Association Mining Algorithm

We propose a one-step algorithm to combine the generation of support time sequences and the time sequence search. Our algorithm prunes the candidate itemsets by using the monotone property of the lower bounding measure of support time sequences without scanning the transaction database and even without computing their true similarity measure. The following is the simple description of the algorithm.

Generation of Support Time Sequences of Single Items: In the first scan of a time-stamped database, the supports of all single items ($k = 1$) are counted per each time slot and their support time sequences are generated. If the lower bounding measure between a query sequence and the support time sequence is greater than a given similarity threshold value, the single item is pruned from the candidate set. If the true similarity value between them satisfies the threshold, the item is added to a result set.

Generation of Candidate Itemsets: All size $k+1$ candidate itemsets are generated using size k candidate itemsets.

Generation of Upper Bound Sequences: The upper bound time sequences of size $k+1$ candidate itemsets are generated using the support sequences of their size k subsets.

Pruning of Candidate Itemsets Using the Lower Bounding Measure: Calculate the lower bounding measure between the upper bound sequence of the candidate itemset and the query time sequence. If the lower bounding measure is greater than the similarity threshold, the candidate itemset is eliminated from the set of candidate itemsets.

Scanning the Database and Finding Itemsets Showing Similar Support Time Sequences: The supports of candidate itemsets after pruning are counted from the database and their support time sequences are calculated. If the similarity value between the support sequences and the query sequence is less than the threshold value, the itemset is included in the result set. The size of examined itemsets is increased to $k = k + 1$ and the above procedures are repeated until no candidate itemset remains in the previous pass.

5 Experimental Evaluation

Our experiments were performed to examine the effect of different threshold values and the effect of itemset pruning by the lower bounding measure. The results were compared with the naive method. The dataset was generated using the transaction generator designed by the IBM Quest project used in [2]. We added a time slot parameter for generating time-stamped transactions. All experiments were performed on a workstation with 4 processors, each an Intel Xeon 2.8 GHz with 3 Gbytes of memory running the Linux operating system.

Effect of Similarity Threshold: The effect of similarity measure was examined with different similarity thresholds using a synthetic dataset in which the total number of transactions was 100,000, the number of items was 20, the average size of transaction was 10 and the number of time slots was 10. The query sequence was chosen near the median spport value of single items at each time slot. In Fig. 1 (a), our method showed dramatically less execution time compared with the naive approach. With the increase in the similarity threshold, the execution time increased. Otherwise, the naive approach showed stable execution time because the approach calculated all time sequences of all combination itemsets independent of the threshold value

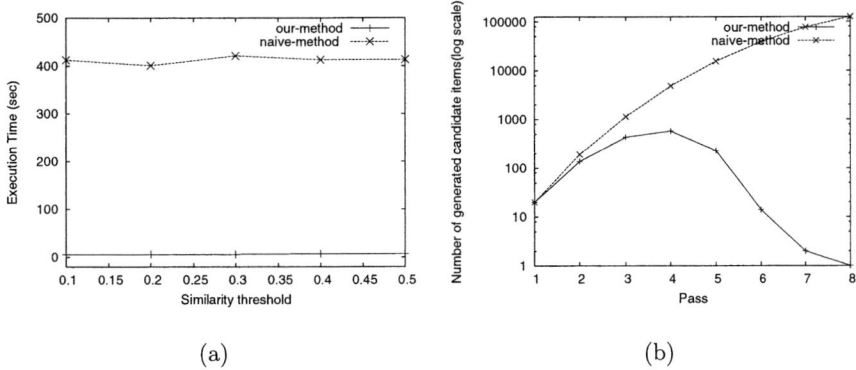

Fig. 1. Experiment Results: (a) Effect of threshold (b) Effect of pruning

Effect of Lower Bounding Pruning: Fig. 1 (b) shows the number of generated candidate itemsets per each pass in the experiment using the same dataset. Note that the y value is in log scale. Our method generated much fewer candidate itemsets compared with the naive method.

6 Conclusions

We introduced the problem of mining time-profiled association patterns and proposed a one-phase algorithm to efficiently discover time-profiled associations. The proposed algorithm substantially reduced the search space by pruning candidate itemsets based on the monotone property of the lower bounding measure of the sequence of statistical parameters. Experimental results showed that our algorithm outperformed the naive approach.

References

1. NOAA El Nino Page. http://www.elnino.noaa.gov/.
2. R. Agarwal and R. Srikant. Fast algorithms for Mining association rules. In *Proc. of the 20th VLDB Conference*, 1994.
3. R. Agrawal, C. Faloutsos, and A.Swami. Efficient Similarity Search in Sequence Databases . In *Proc. Int. Conference on Foundations of Data Organization*, 1993.
4. R. Agrawal and G. Psaila. Active Data Mining. In *Proc. The First International Conference on Knowledge Discovery and Data Mining*, 1995.
5. C. Faloutsos, M. Ranganathan, and Y.Manolopoulos. Fast subsequence matching in time-series database. In *Proc. ACM SIGMOD Conference*, 1993.
6. D. Gunopulos and G. Das. Time Series Similarity Measures and Time Series Indexing. *SIGMOD Record*, 30(2), 2001.
7. Y. Li, P. Ning, X. Wang, and S. Jajodia. Discovering Calendar-Based Temporal Assocation Rules. In *Proc. Symp. Temporal Representation and Reasoning*, 2001.
8. B. Ozden, S. Ramaswamy, and A. Silberschatz. Cyclic Association Rules. In *Proc. of IEEE Int. Conference on Data Engineering*, 1998.

9. J.S. Yoo, P. Zhang, and S. Shekhar. Mining Time-Profiled Associations: A Preliminary Study. In *University of Minnesota*, 2005.
10. P. Zhang, M. Steinbach, V. Kumar, S. Shekhar, P. Tan, S. Klooster, and C. Potter. Discovery of Patterns of Earth Science Data Using Data Mining. In M. Kantardzic and J. Zurada, editors, *in Next Generation of Data Mining Applications*. IEEE Press, 2005.

Online Algorithms for Mining Inter-stream Associations from Large Sensor Networks*

K. K. Loo, Ivy Tong, and Ben Kao

Department of Computer Science,
The University of Hong Kong, Hong Kong
{kkloo, smtong, kao}@cs.hku.hk

Abstract. We study the problem of mining frequent value sets from a large sensor network. We discuss how sensor stream data could be represented that facilitates efficient online mining and propose the interval-list representation. Based on Lossy Counting, we propose ILB, an interval-list-based online mining algorithm for discovering frequent sensor value sets. Through extensive experiments, we compare the performance of ILB against an application of Lossy Counting (LC) using a weighted transformation method. Results show that ILB outperforms LC significantly for large sensor networks.

1 Introduction

Data mining is an area of active database research because of its applicability in various areas. One of the important tasks of data mining is to extract frequently occurring *patterns* or *associations* hidden in very large datasets. In recent years, stream processing and in particular sensor networks has attracted much research interest [1, 3]. Stream processing poses challenging research problems due to large volumes of data involved and, in many cases, on-line processing requirements.

Any device that detects and reports the state of a monitored attribute can be regarded as a sensor. In our model, we assume that a sensor only takes on a finite number of discrete states. Also, we assume that a sensor only reports state changes. A sensor stream can thus be considered as a sequence of updates such that each update is associated with a time at which the state change occurs. Figure 1 shows a system of six sensors $(S_1, ..., S_6)$, each could be in one of the two possible states "low" (L) and "high" (H). Our goal is to discover associations among sensor values that co-exist during a significant portion of time.

If one considers a sensor value, such as "$S_1 = H$", as an item, mining frequent value sets is similar to mining frequent itemsets. One possible approach is to transform the stream data into a dataset of transactions, and then apply a traditional mining algorithm like Apriori to the resulting dataset. A straightforward data transformation would generate transactions by taking snapshots of sensor

---

* This research is supported by Hong Kong Research Grants Council grant HKU 7040/02E.

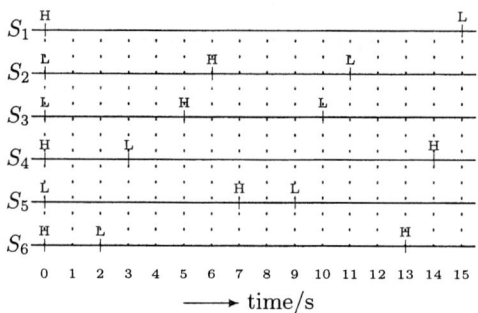

Fig. 1. Sensor data

Table 1. A simple transformation

Transaction ID	S_1	S_2	S_3	S_4	S_5	S_6
1	H	L	L	H	L	H
2	H	L	L	H	L	H
3	H	L	L	H	L	L
4	H	L	L	L	L	L
5	H	L	L	L	L	L
6	H	L	H	L	L	L

Table 2. A weighted transformation

Transaction ID	S_1	S_2	S_3	S_4	S_5	S_6	weight
1	H	L	L	H	L	H	2
2	H	L	L	H	L	L	1
3	H	L	L	L	L	L	2
4	H	L	H	L	L	L	1

states at regular intervals. Alternatively, one can derive a transaction only when there is an update (from any sensor). With this approach, different snapshots of the sensor states could have different life-spans, which are taken as the *weights* of respective transactions. Tables 1 and 2 demonstrate these transformations.

A major problem with the two transformations is that the derived data contain a lot of redundancy. One can observe from Tables 1 and 2 that successive transactions only differ by at most one sensor value. The redundancy causes traditional mining algorithms to perform badly.

A third transformation that could avoid the redundancy problem is to represent a stream by an *interval list*. Given a sensor S and a value v, the interval list $IL(S = v)$ is a list of (*start-time, end-time*) pairs. Each pair specifies the start time and the end time of a time interval during which S assumes the value v.

Intuitively, the interval list representation can potentially support a more efficient mining algorithm in a large sensor network environment over the traditional Apriori-based approaches. This is because the representation avoids data redundancy which leads to a much smaller dataset. Moreover, determining the support of a sensor value set is achieved by list intersection. This avoids the large number of redundant subset testing performed by Apriori-based algorithms.

Due to the reactive nature of monitoring systems and the large volume of data generated by a massive sensor network, data analysis algorithms should be online and one-pass. In [4], Manku et al. proposed the Lossy Counting algorithm, which is an online, one-pass procedure for finding frequent itemsets from a data stream of transactions. In this paper, we study how the Lossy Counting framework can be used to derive online one-pass algorithms for mining large sensor streams under the two data representations (weighted transactions and interval list).

The rest of the paper is structured as follows. We formally define the problem of finding frequently co-existing sensor value sets in Section 2. In Section 3, we review the Lossy Counting algorithm. In Section 4, we define interval list and propose an interval-list-based algorithm for solving the problem. Section 5 reports the experimental results. Finally, Section 6 concludes the paper.

2 Problem Definition

We define a sensor as a device for monitoring and reporting the states of some physical attribute. The set of all possible states of a sensor S is called the *domain* of S. We assume that every sensor has a finite domain. (If the domain of a sensor is continuous, we assume that it can be appropriately quantized and mapped to a finite one.) We assume that the state of a sensor changes at discrete time instants called *updates*. The state of a sensor stays the same between updates.

A sensor value is a state reported by a sensor. We denote a sensor value by $S = v$ where S is a sensor and v is the state reported. If the state of sensor S is v at a certain time t, we say that the sensor value $S = v$ is *valid* at time t. Given a time interval I, if $S = v$ is valid at every instant of I, we say that $S = v$ is valid in I. A sensor network consists of a number of sensors. A set of sensor values $\mathcal{V}$ is valid in an interval I if all sensor values in $\mathcal{V}$ are valid in I.

We assume that all sensors in a sensor network start reporting values at time 0. At any time instant $T\ (> 0)$, the *support duration* of a value set $\mathcal{V}$, denoted by $SD(\mathcal{V})$, is the total length of all non-overlapping intervals within $[0, T]$ in which $\mathcal{V}$ is valid. We define the *support* of $\mathcal{V}$, denoted by $sup(\mathcal{V})$, as $SD(\mathcal{V})/T$. A value set $\mathcal{V}$ is *frequent* if $sup(\mathcal{V}) \geq \rho_s$, a user specified support threshold.

Under the stream environment, finding the *exact* set of frequent value sets would require keeping track of all value sets that have ever occurred. The high memory and processing requirements render this approach infeasible. As an alternative, we adopt the Lossy Counting framework proposed in [4] and report all value sets that are frequent plus some value sets whose supports are guaranteed to be not less than $\rho_s - \epsilon$ for some user-specified error bound ϵ.

3 Lossy Counting

In [4], Manku and Motwani propose Lossy Counting, a simple but effective algorithm for counting approximately the set of frequent itemsets from a stream of transactions. Since our algorithms use the framework of Lossy Counting, we briefly describe the algorithm in this section.

With Lossy Counting, a user specifies a support threshold ρ_s and an error bound ϵ. Itemsets' support counts are stored in a data structure D. We can consider D as a table of entries of the form (e, f, Δ), where e is an itemset, f is an *approximate* support count of e, and Δ is an error bound of the count. The structure D is maintained such that if N is the total number of transactions the system has processed, the structure D satisfies the following properties:

P1: If the entry (e, f, Δ) is in D, then $f \leq f_e \leq f + \Delta$, where f_e is the exact support count of e in the N transactions.

P2: If the entry (e, f, Δ) is not in D, then f_e must be less than ϵN.

The data structure D is initially empty. To update D, transactions are divided into batches. The size of a batch is limited by the amount of memory available. Figure 2(a) illustrates the update procedure of D. Let B be a batch

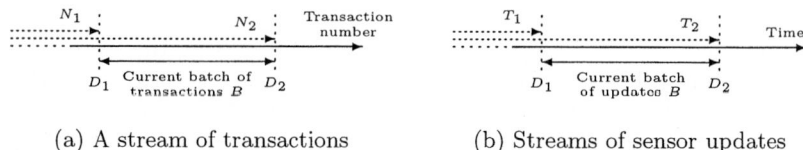

Fig. 2. Lossy Counting

of transactions. Let N_1 denote the number of transactions before B and let D_1 denote the data structure D before B is processed. Lossy Counting enumerates itemsets that are present in B and counts their supports in the batch. Let e be an itemset that appears in B whose support count w.r.t. B is f_B. D is then updated by the following simple rules (D_2 denotes the updated D in the figure):

Insert: If D_1 does not contain an entry for e, the entry $(e, f_B, \epsilon N_1)$ is created in D unless $f_B + \epsilon N_1 \leq \epsilon N_2$, where N_2 is the total number of transactions processed including those in B.

Update: Otherwise, the frequency f of e in D_1 is incremented by f_B.

Delete: After all updates, an entry (e, f, Δ) in D is deleted if $f + \Delta \leq \epsilon N_2$.

For an efficient implementation of Lossy Counting, certain optimization is done. Details of which can be found in [4]. Besides, to apply Lossy Counting to our frequent value set mining problem, a few modifications have to be made. Due to space limitation, readers are referred to [2] for details.

4 Interval List

In this section we formally define interval lists and discuss how they could be used to mine frequent value sets under the Lossy Counting framework.

An *interval* is a continuous period of time. We denote an interval I by $(t, \bar{t})$, where t and $\bar{t}$ are the start time and the end time of the interval, respectively. The duration of I is given by $\delta(I) = \bar{t} - t$. Given two intervals $I_1 = (t_1, \bar{t}_1)$ and $I_2 = (t_2, \bar{t}_2)$ such that $t_1 \leq t_2$, they are *overlapping* if $t_2 < \bar{t}_1$ and their *intersection* is given by $I_1 \cap I_2 = (t_2, \min(\bar{t}_1, \bar{t}_2))$.

An *interval list* is a sequence of non-overlapping intervals. The intervals in an interval list are ordered by their start time. The duration of an interval list IL is given by $\delta(IL) = \sum \delta(I) \mid I \in IL$. Given two interval lists IL_1 and IL_2, their intersection is defined as: $IL_1 \cap IL_2 = \bigcup \{I_1 \cap I_2 \mid I_1 \in IL_1 \wedge I_2 \in IL_2\}$.

Given a set of sensor value $\mathcal{V}$, we use the notation $IL(\mathcal{V})$ to denote the interval list that contains all and only those intervals in which the value set $\mathcal{V}$ is valid. We call such an interval list *the interval list of $\mathcal{V}$*. Given two sensor value sets, $\mathcal{V}_1$ and $\mathcal{V}_2$, it can be easily verified that the interval list of $\mathcal{V}_1 \cup \mathcal{V}_2$ can be obtained by intersecting the interval lists of $\mathcal{V}_1$ and $\mathcal{V}_2$. That is, $IL(\mathcal{V}_1 \cup \mathcal{V}_2) = IL(\mathcal{V}_1) \cap IL(\mathcal{V}_2)$.

```
1  C_1 ← set of all size-1 value sets;
2  B ← {IL(V) | V ∈ C_1};
3  i ← 1;
4  While C_i ≠ ∅ do
5     Foreach V ∈ C_i do
6        IL(V) = ⋂{IL(v) | v ∈ V};
7        SD = δ(IL(V));
8        Update(D, V, SD, T_1, T_2, ε);
9     end-for
10    D_i ← {(V, f, Δ) | (V, f, Δ) ∈ D ∧ |V| = i};
11    C_{i+1} ← ApGen(D_i, i + 1);
12    i ← i + 1;
13 end-while
```

Fig. 3. Procedure for updating D using the interval list representation

```
1  function Update (D, V, SD, T_1, T_2, ε)
2     if (∃(V, f, Δ) ∈ D) do
3        f ← f + SD;
4        if (f + Δ < εT_2) do
5           remove all entries (X, ., .) from D
            where X ⊇ V;
6        end-if
7     else if (SD ≥ ε(T_2 − T_1)) do
8        D = D ∪ (V, SD, εT_1);
9     end-if
```

Fig. 4. Function Update()

The interval list representation can be used to mine frequent value sets under the Lossy Counting framework in the following way. First of all, time is partitioned into a number of intervals, each corresponds to a batch of sensor updates (see Figure 2(b)). Instead of representing a batch of updates as a set of weighted transactions, the updates are represented by the interval lists of the sensor values. Similar to the case of Lossy Counting, the size of a batch is limited by the amount of buffer memory available. Also, a data structure D is again used that keeps track of certain sensor value sets' support durations. The function and properties of D is the same as those described in Section 3.

Figure 3 outlines our procedure for updating D. The number of sensor values in a value set V is its *size*. The procedure starts by collecting all size-1 value sets into a set of candidates, C_1. The batch B is represented by a set of interval lists, one for each sensor value. The procedure then executes a while loop. During each iteration, a set of candidate value sets, C_i, is considered. Essentially, each value set V in C_i is of size i and that V's support duration up to time T_2 has the potential of exceeding ϵT_2. The procedure then verifies whether V should be included in D by finding its support duration in batch B. This is achieved by computing $IL(V)$ in B through intersecting the interval lists of relevant sensor values, followed by determining the total length of all the intervals in $IL(V)$. D is then updated by function Update(), described in Figure 4, which essentially follows the three update rules listed in Section 3.

After all candidates in C_i are processed, all the entries in D for size-i value sets are properly updated. These entries are collected in D_i. The set D_i is used to generate the candidate set C_{i+1} for the next iteration. More specifically, a size-$(i+1)$ value set V is put into C_{i+1} unless there is a size-i subset V' of V that is not in D_i. This is because by Property P2 of D (see Section 3), if the entry (V', f, Δ) is not in D_i, we know that the support duration of V' w.r.t. time T_2 must be smaller than ϵT_2. Since the support duration of V cannot be larger than the support duration of its subset V', the support duration of V is smaller than ϵT_2 as well. That is, V should be left out of D and needs not be considered.

There are a number of optimizations that can be applied to speed up our procedure of processing a batch. Interested readers are referred to [2] for details.

5 Results

We performed extensive experiments comparing the performance of the mining algorithms using different data representations. Due to space limitation, we highlight some of the findings here. A detailed analysis can be found in [2].

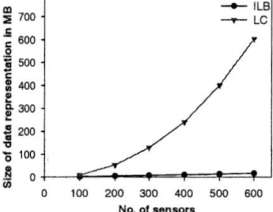 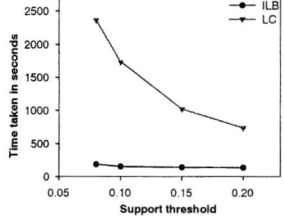 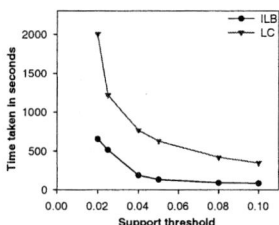

Fig. 5. Dataset size under two representations

Fig. 6. Running time vs. ρ_s (600-sensor network)

Fig. 7. Running time vs. ρ_s (400-sensor network)

As we have alluded to earlier, one major advantage of the interval list representation is that it is more space-efficient than the weighted representation. Figure 5 shows the size of the dataset generated for a stream history of 92000 time units when the number of sensors in the network varies from 100 to 600. The dataset size grows linearly w.r.t. the number of sensors under the Interval-List-Based Lossy Counting algorithm (ILB) described in Section 4. The weighted transformation representation (LC), however, does not scale well.

The dataset size has a significant impact on the algorithms' performance. As an example, Figure 6 shows that ILB is much more efficient than LC for a 600-sensor network, especially when ρ_s is small. It is because, as shown in Figure 5, the dataset size for LC is much (31 times in this case) larger than that for ILB. Hence, for LC, a batch contains 31 times fewer updates than that of ILB. The small batch size is undesirable for the Lossy Counting framework because of the *false alarm* effect, as discussed in [2].

Figure 7 compares the performance of ILB and LC for a 400-sensor network. ILB still outperforms LC although the margin is less drastic than the 600-sensor case. It is because, for a 400-sensor network, the dataset size is much smaller for LC. This allows a larger batch and thus the effect of false alarm is ameliorated.

6 Conclusion

We study the problem of mining frequent sensor value sets from a massive sensor network. We discuss methods for representing sensor stream data. We derive online mining algorithms, namely, LC and ILB and evaluate the algorithms' performance through experiments. The results show that ILB could outperform LC by a significant margin, particularly for large sensor networks. The interval list representation is thus a viable option in representing a massive sensor network.

References

1. Donald Carney et al. Monitoring streams - a new class of data management applications. In *VLDB*, pages 215–226, 2002.
2. K. K. Loo, Ivy Tong, Ben Kao, and David Cheung. Online algorithms for mining inter-stream associations from large sensor networks. Technical Report TR-2005-02, Dept. of Computer Science, The University of Hong Kong, 2005.
3. Samuel Madden and Michael J. Franklin. Fjording the stream: An architecture for queries over streaming sensor data. In *ICDE*, pages 555–566, 2002.
4. Gurmeet Singh Manku and Rajeev Motwani. Approximate frequency counts over data streams. In *VLDB*, pages 346–357, 2002.

Mining Frequent Ordered Patterns

Zhi-Hong Deng[1,2], Cong-Rui Ji[1], Ming Zhang[1], and Shi-Wei Tang[1,2]

[1] School of Electronics Engineering and Computer Science,
Peking University, Beijing 100871, China
[2] National Laboratory of Machine Perception, Peking University,
Beijing 100871, China
zhdeng@cis.pku.edu.cn, {crji, mzhang}@db.pku.edu.cn,
tsw@pku.edu.cn

Abstract. Mining frequent patterns has been studied popularly in data mining research. All of previous studies assume that items in a pattern are unordered. However, the order existing between items must be considered in some applications. In this paper, we first give the formal model of ordered patterns and discuss the problem of mining frequent ordered patterns. Base on our analyses, we present two efficient algorithms for mining frequent ordered patterns. We also present results of applying these algorithms to a synthetic data set, which show the effectiveness of our algorithms.

1 Introduction

Mining frequent patterns from a large database is important and interesting to the fundamental research in the mining of association rule. It is also play an essential role in many other important data mining tasks [1-3]. Since the first introduction of mining of frequent itemsets in [4], various algorithms [5-7] have been proposed to discover frequent itemsets efficiently.

However, very little work has been done on the problem of mining in the case that some order existing between items of a transaction. Agrawal [8] have studied the problem of mining sequential patterns. But, the order in sequential patterns mining is the sequence of transactions according to increasing transaction-times. In sequential patterns mining, items of a transaction are out-of-order. In some real applications, such as Web user navigation modeling, items of a set (or transaction) are ordered and the order existing between items must be considered in mining useful patterns.

In this paper, we first bring forward the problem of mining frequent ordered patterns and design efficient algorithms for this problem. As far as we know, this paper is the first one that systemically presents the study of mining frequent ordered patterns. There is no related study has been reported.

The organization of the rest of the paper is as follows. In section 2, we give a formal statement of the problem. In section 3, we present our algorithms for mining frequent ordered patterns. In section 4, we present some experimental results. In section 5, we conclude with a summary and point out some future research issues.

2 Basic Definitions

Before embarking on describing the definition of frequent ordered patterns, we will briefly state some terminology. Let $I = \{a_1, a_2, \ldots, a_m\}$ be a set of items.

Definition 1 (position order): Let A ($\subseteq I$) be a set of items. According to the positions of items in A, we define a binary relation $<$ on A as follows: For any two items $a_i, a_j \in A$, if a_i occurs in A before a_j then $a_i < a_j$, or else $a_j < a_i$. The set A with $<$ is called an **ordered pattern**. The number of items in A is called its length. An ordered pattern with length l is also denoted l-pattern. For the convenience of discussion, we use expression $a_i <_A a_j$ to denote that there exists $a_i < a_j$ in A.

Definition 2: Given two ordered patterns A and B. B is a subset of A if and only if: (1) $B \subseteq A$; (2) $\forall a_i, a_j \in B$, $a_i <_B a_j \Leftrightarrow a_i <_A a_j$.

The first condition requires that B should be a subset of A from the point of view of the set theory. The second condition requires that the position relations of any two items should be consistent in A and B. Let $A = <a_3, a_1, a_2>$ and $B = <a_3, a_2>$ be two ordered patterns. It is obvious that B is a subset of A. But, $C = <a_2, a_3>$ is not a subset of A because $a_2 <_C a_3$ and $a_3 <_A a_2$ violate the second condition of definition 2. If B is a subset of A, We also call A contains B and denote $B \{ A$.

Lemma 1: Given three ordered patterns A, B, and C, $B \{ A \wedge C \{ B \quad C \{ A$.

Proof. $B \{ A$ implies that $B \subseteq A$. $C \{ B$ implies that $B \subseteq A$. As a result, we have $C \subseteq A$. For $a_i, a_j \in C$, we have $a_i <_C a_j \Leftrightarrow a_i <_B a_j$ because of $C \{ B$. In addition, we have $a_i <_B a_j \Leftrightarrow a_i <_A a_j$ because of $B \{ A$. Therefore, we have $a_i <_C a_j \Leftrightarrow a_i <_A a_j$.

Lemma 1 shows that subset relation between ordered patterns is transitive.

Definition 3 (ordered transaction): a transaction T, which contains a set of items in I, with the relation $<$ is called an ordered transaction. A database consisting of ordered transactions is called an ordered transaction database.

Based on above definitions, we have the definition of frequent ordered patterns as follows.

Definition 4: Given an ordered transaction database ODB and an ordered pattern A, the **support** of A is the number of ordered transactions containing A in ODB. A is a **frequent ordered pattern** if A's support is no less than a predefined *minimum support threshold* ξ.

Given an ordered transaction database ODB and a minimum support threshold ξ, the problem of finding the complete set of frequent ordered patterns is called the frequent ordered pattern mining problem.

3 Discovering Frequent Ordered Patterns

3.1 Some Properties

Before embarking on the algorithm description, we will briefly introduction some properties of frequent ordered patterns. Let ξ be the predefined threshold, and a database $ODB = \{T_1, T_2, \ldots, T_n\}$, where T_i ($i \in [1..n]$) is an ordered transaction. We have the following property.

Property 1(anti-monotone): if A is infrequent, any ordered patterns containing A, which are also called supersets of A, must be infrequent.

Rational. Let B is a superset of A and we assume B is frequent. For any transaction T_i containing B, T_i must contain A by Lemma 1. That is, The support of A must be great than that of B. We know B is frequent. So, we have A must be frequent. It conflict with the fact that A is infrequent. Therefore, our assumption is wrong and B should be infrequent.

Property 1 shows that the Apriori property, which is the basic of mining of frequent patterns, is still reserved in ordered patterns. We can exploit this property to mining all ordered frequent patterns from short patterns to long patterns.

3.2 ABMFOP Algorithm

For mining frequent ordered patterns, we employ an iterative approach known as a *level-wise* search, where i-patterns are used to explore $(i+1)$-patterns. First, the set of frequent ordered 1-pattern is found. This set is denoted OL_1. OL_1 is used to find OL_2, which is the set of frequent ordered 2-patterns, and then OL_2 is used to find OL_3, and so on, until no more frequent ordered i-patterns can be found. The finding of each OL_i requires one scan of the database. To improve the efficiency of the level-wise generation of frequent ordered patterns, Property 1 are used to narrow the range of search.

Based on above discussion, we design an algorithm called ABMFOP. ABMFOP is abbreviation for Apriori-Based Mining of Frequent Ordered Patterns. The main idea of ABMFOP is almost the same as Apriori algorithm [5] used to mine Frequent Patterns. But, there are some differences betweem ABMFOP and Apriori algorithm because the order between items must be regarded in ABMFOP. First, the procedure of generating candidate pattern is different. Second, the procedure of judging the subset relation of two patterns is different. **Gen_Candidate** and **Is_Subset()** have more details.

Algorithm: ABMFOP
Input: Database, ODB, of ordered transactions; minimum support threshold ξ.
Output: OL, frequent ordered patterns in ODB.
Method:
    OL_1 = find_frequent_ordered_1-pattern(ODB);
    for ($i = 2$; $OL_{i-1} \neq \varnothing$; i++)
        SC_i = **Gen_Candidate**(OL_{i-1});
        for each ordered transaction $T \in ODB$
            for each candidate pattern $C \in SC_i$
                if (**Is_Subset**(C, T) then C.support++;
        $OL_i = \{C \in SC_i \mid C.\text{support} \geq \xi\}$
    return $OL = \cup_i OL_i$;

Procedure Gen_Candidate(OL_{i-1}) // generate candidate ordered patterns
    for each frequent ordered pattern $A_1(=<x_1, x_2, \ldots x_{i-2}, x_{i-1}>) \in OL_{i-1}$
        for each frequent ordered pattern $A_2(=<y_1, y_2, \ldots y_{i-2}, y_{i-1}>) \in OL_{i-1}$
            if $((x_1 = y_1) \wedge (x_2 = y_2) \wedge \ldots \wedge (x_{i-2} = y_{i-2}) \wedge (x_{i-1} \neq y_{i-1}))$ then
                $X = <x_1, x_2, \ldots x_{i-2}, x_{i-1}, y_{i-1}>$;
                If $\neg$ (**Has_Infrequent_Subset**(X, OL_{i-1})) then $SC_i = SC_i \cup \{X\}$;
                $Y = <x_1, x_2, \ldots x_{i-2}, y_{i-1}, x_{i-1}>$;
                If $\neg$ (**Has_Infrequent_Subset**(Y, OL_{i-1})) then
                      $SC_i = SC_i \cup \{Y\}$;
    return SC_i;

Procedure Has_Infrequent_Subset(A, OL_{i-1}).
    Let $A==<a_1, a_2, \ldots a_{i-1}, a_i>$;
    for ($k = 1; k \leq i; k$++) {
        $A_k = <a_1, a_2, \ldots, a_{k-1}, a_{k+1}, \ldots, a_i>$;
        If $A_k \notin OL_{i-1}$ then reture TRUE;
    reture FALSE;

Procedure Is_Subset(Y, X)
    Let $Y = <y_1, y_2, \ldots x_{m-1}, y_m>$ and $X = <x_1, x_2, \ldots x_{n-1}, x_n>$;
    if $n < m$ then
        reture FALSE;
    $A = X$;
    for ($k = 1; k \leq m; k$++)
        if $y_k \notin A$ then reture FALSE;
        else   // there must exist $x_j \in A$ such that $x_j = y_k$;
            Let x_j be the item in A that is equal to y_k;
            if ($j > (n - m + k)$) then reture FALSE;
            else $A = <x_{j+1}, x_{j+2}, \ldots x_{n-1}, x_n>$;
    reture TRUE;

3.3 ABMFOP_F Algorithm

In experiment, we find that finding frequent ordered 2-patterns is the bottleneck of ABMFOP algorithm. Too many candidate 2-patterns make it time-consumed to find frequent ordered 2-pattern from them. Instead of generating candidate 2-patterns from frequent ordered 1-patterns, we design a more efficient algorithm called ABMFOP_F, which generates candidate 2-patterns directly from ordered transactions. This strategy greatly decreases the time for finding frequent ordered 2-patterns. ABMFOP_F is almost the same as ABMFOP except the process of 2-patterns. With the limitation of space, we do not describe ABMFOP_F in detail in this paper.

4 Experiments

We experimented with the mining algorithms using a synthetic data set. All the experiments are performed on a Pentium 4 PC running MS XP. All the programs are written in Microsoft/Visual C++ 6.0. The synthetic data set, T50.I10.L100k.D10K, is generated using the procedure described in [5]. We randomly change the orders of items in each transaction in order to mimic reality more effectively. ABMFOP and ABMFOP_F showed good scalability as the minimum support threshold decreased from 500 to 300. ABMFOP_F performs about 1.5 times faster than ABMFOP on average. The reason is that ABMFOP_F adopts a better way, which is called subset-generating-and-counting, for finding all frequent ordered 2-patterns.

5 Conclusions

In this paper, we study the problem of mining frequent ordered patterns, which has not ever been proposed as far as we know. Based on the characteristic of ordered patterns, we have designed two algorithms for deal with the problem. Our performance study shows the effectiveness of our algorithms. In the future, we will take efforts towards more efficient algorithms by adopting useful ideas from many proposed algorithms of mining frequent patterns.

References

1. S. Brin, R. Motwani, and C. Silverstein. Beyond market basket: Generatalizing association rules to correlations. In SIGMOD'97, pp. 265-276.
2. R. Agrawal and R. Srikant. Mining sequential patterns. In ICDE'95, pp. 3-14.
3. J. Han, G. Dong, and Y. Yin. Efficient mining of partical periodic patterns in time series database. In ICDE'99, pp. 106-115.
4. R. Agrawal, T. Imielinski, and A. Swami. Mining Association Rules between Set of Items in Large Databases. In SIGMOD'93, pp. 207-216.
5. R. Agrawal and R.Srikant. Fast algorithm for mining Association rules. In VLDB'94, pp. 487-499.
6. J. Han, J. Pei, and Y. Yin. Mining frequent patterns without candidate generation. In SIGMOD'00, pp. 1-12.
7. M. J. Zaki. Scalable algorithms for association mining. IEEE TKDE, 12(3): 372-390, 2000.
8. R. Agrawal and R.Srikant. Mining sequential patterns. In ICDE'95, pp. 3-14.

Conditional Random Fields for Transmembrane Helix Prediction

Lior Lukov[1], Sanjay Chawla[1], and W. Bret Church[2]

[1] University Of Sydney[†]
{lior, chawla}@it.usyd.edu.au
[2] University Of New South Wales[‡]
b.church@unsw.edu.au

[†] School of Information Technologies, Sydney University, NSW 2006, Australia
[‡] Department of Physiology and Pharmacology, University of New South Wales, NSW 2052, Australia

Abstract. It is estimated that 20% of genes in the human genome encode for integral membrane proteins (IMPs) and some estimates are much higher. IMPs control a broad range of events essential to the proper functioning of cells, tissues and organisms and are the most common target of clinically useful drugs [1]. However there is a dearth of high-resolution 3D structural information on the IMPs. Therefore good prediction methods of IMPs structures are to be highly valued. In this paper we apply Conditional Random Fields (CRFs) to build a probabilistic model to solve the membrane protein helix prediction problem. The advantage of CRFs is that it allows seamless and principled integration of biological domain knowledge into the model. Our results show that the CRF model outperforms other well known helix prediction approaches on several important measures.

1 Introduction

A number of high throughput projects have been positioned to assist in the interpretation of the human genome sequence data. Structural determination of integral membrane proteins can be problematic due to difficulties in obtaining sufficient amounts of sample. Protein sequence analysis methods extended by our knowledge of protein structure may be suited to contribute significantly to these aspects of protein structure and function.

In this paper we cast the protein helix prediction task as a binary sequential classification problem and use Conditional Random fields (CRFs) to solve it [2]. Given a set of membrane proteins sequences, each single record in the set contains pair of sequences: The observation sequence, represented by x and the label sequence, represented by y. The protein observation sequence is a sequence of amino acids, represented by 20 different letters. The label sequence is a transmembrane helical/non-helical structure sequence represented by binary labels 0/1 respectively. This data, called the training data, is represented by

$T = (x^{(k)}, y^{(k)})_{k=1}^{N}$, where N is the total number of proteins. Our goal is to predict the helical structure of a target set, which has observation sequences only.

2 The Sequential Classification Problem

The sequential classification problem is well known in many different fields such as computational linguistics, part of speech tagging, computational biology and many more. Given set of observation sequences, goal here is to find corresponding label sequences to these observations. A very common approach is using generative models, such as Hidden Markov Models (HMMs), finding the joint probability distribution $p(X, Y)$ where X and Y are random variables describing the observation and the labelled sequences respectively. This approach suffers from a major drawback that in order to find the joint distribution, a generative model has to calculate all possible observation sequences, which may be not practical [3]. In contrast, the conditional models specify the probability of a label given an observation sequence $p(Y|X)$. Thus, no effort is spent on modelling all possible observation sequences, but only on selecting the labels which maximize the conditional probability [2].

3 Conditional Random Fields (CRFs)

Conditional Random Fields (CRFs) is a probabilistic framework for labelling sequential data. CRFs is a form of undirected graphical state model that defines a log-linear distribution for each state over the label sequence based on the observation sequence [3]. CRFs main advantage over other non-generative finite-state models based on directed graphical models, such as Maximum Entropy Markov Models (MEMMs), is by avoiding a weakness called the label bias problem. The Markovian assumptions in MEMMs and similar state-conditional models separate the decision making at one step from future dependent decisions of consecutive steps, and may be biased towards states with fewer outgoing transitions. In contrast, CRFs have a single exponential model for the joint probability of the entire sequence of labels given the observation sequence [2].

Formally, we define $G = (V, E)$ to be an undirected graph when $v \in V$ corresponding to each of the random variables representing a label sequence Y_v from Y and $e \in E$ corresponding to the transition between a given label to the next one. Even though in theory the structure of graph G may be arbitrary, in our application the graph is a simple chain, where each node corresponds to a label [3].

3.1 Definition

Let $G = (V, E)$ be a graph that $Y = (Y_v) v \in V$. If each random variable Y_v in the graph G obeys the Markov property, then (Y, X) is a conditional random field F in which $p(Y_v|X, Y_w, w \neq v) = p(Y_v|X, Y_w, w \sim v)$, where $w \sim v$ are neighbors in G. A clique c in the graph G is defined as a subset of vertices which are completely connected. In a chain graph the cliques are either from first order (single vertex) or second order neighbors (two neighbor vertices).

From the definition of Gibbs Random Fields (GRFs), a set of random variables f is said to be a Gibbs random field if and only if its configuration obey a Gibbs distribution of the form:

$$P(f) = Z^{-1} \times e^{-\frac{1}{T}U(f)} \tag{1}$$

where Z is a normalizing factor: $Z = \sum_{f \in F} e^{-\frac{1}{T}U(f)}$, T is a constant called the temperature which equals to 1 in the most simple case and $U(f)$ is the energy function. By the The Hammersley-clifford theorem if f obeys the Markov property (and positivity) then the physical topology (chain) coincides with the logical topology and the energy function can be expressed as a sum of the cliques's neighbors order:

$$U(f) = \sum_{\{v\} \in C_1} V_1(f_v) + \sum_{\{v,w\} \in C_2} V_2(f_v, f_w) \tag{2}$$

[4]. Since conditional random fields also hold the conditions of Markov random field, then according to Hammersley-clifford theorem, they have a Gibbs distribution, leading us to the fundamental theorem of random fields:

$$p_\theta(y|x) \propto exp\left(\sum_j \lambda_j f_j(y_{i-1}, y_i, x, i) + \sum_k \mu_k g_k(y_i, x, i)\right) \tag{3}$$

where $f_j(y_{i-1}, y_i, x, i)$ is a transition feature function of the entire observation sequence and the labels at positions i and $i-1$, $g_k(y_i, x, i)$ is a state feature function of the entire observation sequence and the label at position i. λ_j and μ_k are estimated from the training data. We assume that the feature functions f_k and g_k are given and fixed [3].

3.2 Feature Functions and Model Estimation

Each potential function actually represents a constraint on subset of random variables on which it operates. Thus, by satisfying a constraint we actually increase the likelihood of the global configuration. In what follows, we look at the transition function as a general case of the state function by writing $g(y_i, x, i) = g(y_{i-1}, y_i, x, i)$. We also define the sum of a feature over the sequence by $F_j(y, x) = \sum_{i=1}^n f_j(y_{i-1}, y_i, x, i)$ where $f_j(y_{i-1}, y_i, x, i)$ refers to either transition or state function [3]. Therefore, the probability of a label sequence y given the observation sequence x is in the form

$$p(y|x, \lambda) = \frac{1}{Z(x)} exp(\sum_j \lambda_j F_j(y, x)) \tag{4}$$

where $Z(x) = \sum_y exp(\sum_j \lambda_j F_j(y, x))$. The parameters ($\lambda_j$) are computed by maximizing the log-likelihood with the training data using either iterative scaling or conjugate gradient methods [5, 3]. The most likely label sequence $\hat{y}$ for input sequence x is

$$\hat{y} = arg\max_y p(y|x, \lambda) = arg\max_y \sum_j \lambda_j \cdot F_j(y, x)$$

3.3 Feature Integration with the Model

The most important aspect of specifying the model is selecting the set of features that capture the important relationships among the observation and the label sequences, in our case the protein sequence and the helical structure respectively [6]. In our work we have selected a basic set of features capturing the model's constraints and divided them into several groups:

Start, End and Edge Features. By using these features we capture the probability of starting/ending a sequence with assigning a given label or the transition probability for moving from one state to the consecutive state. For instance, the start unigram feature has the form:

$$u_{start}(x, i) = \begin{cases} 1 \text{ if the Amino Acid at position i is the first in the sequence} \\ 0 \text{ otherwise} \end{cases}$$

The relationship between the observation and a potential helix membrane structure is described in the feature:

$$f_{start_H}(y_i, x, i) = \begin{cases} u_{start}(x, i) \text{ if } y_i = \text{Helix membrane} \\ 0 \qquad\qquad \text{otherwise} \end{cases}$$

Similarly, we define another set of features for the relationship with a non-helix membrane structure.

The Edge feature in contrast, is a bigram feature which depends on two consecutive labels:

$$f_{edge_{H-H}}(y_{i-1}, y_i, x, i) = \begin{cases} u_{edge}(x, i) \text{ if } y_{i-1} = \text{Helix membrane and } y_i = \text{Helix membrane} \\ 0 \qquad\qquad \text{otherwise} \end{cases}$$

Basic Amino Acid Feature. Amino acids have different tendencies to populate one membrane helical structure in preference to another. Since our language contains 20 possible amino acids, we have 20 different unigram features from this type. The unigram feature of amino acid n in position i is:

$$u_n(x, i) = \begin{cases} 1 \text{ if the Amino Acid in sequence } x \text{ at position i is from type } n \\ 0 \text{ otherwise} \end{cases}$$

Amino Acid Property Feature. Amino acids differ one from another in their chemical structure expressed by their side chains, providing them different properties. The fact that amino acids from the same classification group tend to appear in similar locations, motivated us to create special property features. We have adopted the properties classification taken from Sternberg [7] classifying the amino acids into nine groups[1], each group described by a unigram feature. Note that some amino acids may appear in more than one group simultaneously.

The hydrophobicity property for instance, is described in the feature:

$$u_{Hydrophobic}(x,i) = \begin{cases} 1 \text{ if the Amino Acid in } x \text{ at position } i \in (M,I,L,V,A,G,F,W,Y,H,K,C) \\ 0 \text{ otherwise} \end{cases}$$

4 Experiments, Results and Analysis

We now report on our experiments to test the effectiveness of features proposed in Section 3.3, embedded in a CRF model, to predict the location of membrane helical regions in protein sequences.

4.1 Data Set

The data set consists of a set of 148 transmembrane protein sequences with experimentally confirmed transmembrane regions, which are significantly non-similar, based on pairwise similarity clustering compiled by Möller et al [8]. The data set can be accessed via ftp://ftp.ebi.ac.uk/databases/testsets/trans membrane. We randomly picked 24 sequences out of the 148 and grouped them as a test set, using the remaining 124 sequences as the training set. We repeated this procedure ten times, having a cross validation test of ten independent experiments and calculated the average values of these measurements.

4.2 Results and Analysis

In our experiment we have evaluated the prediction accuracy of the test set with the experimentally confirmed results based on two two main approaches: *per-residue accuracy* and *per-segment accuracy* as described in Chen, Kernytsky and Rost (henceforth referred as CKR) [9]. In per-residue accuracy the predicted label and actual label are compared by residue. In per-segment accuracy we determine how accurately a method correctly predicts the location of a transmembrane helix (referred as TMH) region. We have used two popular methods to score per-segment accuracy. The first method requires a minimal overlap of 3 residues between the two corresponding segments and does not allow the same helix to be counted twice, as used in the paper of Chen et al. [9]. This method we refer as $3R$. The second method requires minimal overlap of 9 residues but does allow counting the same helix twice, indicated by $9R$. For our comparison we will closely follow the CKR paper as it has collated results of several methods for transmembrane helix prediction on a common benchmark data set displayed in the following table:

---

[1] Aromatic (F,W,Y,H), Hydrophobic (M,I,L,V,A,G,F,W,Y,H,K,C), Positive (H,K,R), Polar (W,Y,C,H,K,R,E,D,S,Q,N,T), Charged (H,K,R,E,D), Negative (E,D), Aliphatic (I,L,V), Small (V,A,G,C,P,S,D,T,N), Tiny (A,G,S).

Per-Residue Accuracy					Per-Segment Accuracy				
Q_2	$Q_{2T}^{\%obs}$	$Q_{2T}^{\%prd}$	$Q_{2N}^{\%obs}$	$Q_{2N}^{\%prd}$	$Q_{ok}^{(3R)}$	$Q_{tmh}^{\%obs(3R)}$	$Q_{ok}^{(9R)}$	$Q_{tmh}^{\%obs(9R)}$	$Q_{tmh}^{\%prd}$
83	67	74	92	88	28	43	44	72	99

In order to compare our results with other available methods, we consider the work of Chen et al. [9] and methods contained within as a reference. In the "Per-Residue Accuracy" results we have achieved high prediction accuracy for both transmembrane and non-transmembrane residues, lower accuracy of transmembrane residues only, and higher accuracy of non-transmembrane residues. In the "Per-Segment Accuracy" results we can see a considerable difference between the $3R$ test and the $9R$ test. The figures in $Q_{tmh}^{\%prd}$ indicate high precision of true prediction among those helices who were detected by the model. When comparing our prediction results with the other methods, our model performed well with high percentage of accuracy on the per-residue test. **The CRFs model achieved the highest score among all 28 other methods in the overall percentage of residues predicted correctly in both transmembrane and non-transmembrane helices (Q_2) with 83% of true prediction.** On the per-segment test, our model achieved high precision but low prediction score compared to the other models. Notice that some methods may have involved use of proteins from the data set as training so their results may be overestimates.

5 Conclusions

In this paper we introduced the Conditional Random Fields (CRFs) technique which has found good application in the solution of sequential mining problems. We used CRFs to segment and label sequence data to solve the membrane protein helix prediction problem. Our results look promising compared to currently available methods, and as such will motivate the future use of CRFs to solve sequential labelling data problems. For more information on this paper please check our website on http://www.it.usyd.edu.au/~chawla/publications/crf1.pdf.

References

1. Chen, C.P., Rost, B.: State-of-the-art in membrane protein prediction. Applied Bioinformatics **1** (2002) 21–35
2. Lafferty, J., McCallum, A., Pereira, F.: Conditional random fields: Probabilistic models for segmenting and labeling sequence data. In: Proc. 18th International Conf. on Machine Learning, Morgan Kaufmann, San Francisco, CA (2001) 282–289
3. Wallach, H.M.: Conditional random fields: An introduction. Technical Report MS-CIS-04-21, University of Pennsylvania (2004)
4. Li, S.: Markov random field modeling in computer vision. Springer-Verlag New York (1995)
5. Berger, A.: The improved iterative scaling algorithm: A gentle introduction. Technical report, Carnegie Mellon University (1997)

6. Buehler, E.C., Ungar, L.H.: Maximum entropy methods for biological sequence modeling. In: BIOKDD. (2001) 60–64
7. Sternberg, M.J.: Protein Structure Prediction: A Practical Approach. Oxford University Press (1996)
8. Moller, S., Kriventseva, E.V., Apweiler, R.: A collection of well characterized integral membrane proteins. Bioinformatics **16** (2000) 1159–1160
9. Chen, C.P., Kernytsky, A., Rost, B.: Transmembrane helixpredictions revisited. Protein Science **11** (2002) 2774–2791

A DNA Index Structure Using Frequency and Position Information of Genetic Alphabet[*]

Woo-Cheol Kim[1], Sanghyun Park[1], Jung-Im Won[1],
Sang-Wook Kim[2], and Jee-Hee Yoon[3]

[1] Department of Computer Science,
Yonsei University, Korea
{twelvepp, sanghyun, jiwon}@cs.yonsei.ac.kr
[2] College of Information and Communications,
Hanyang University, Korea
wook@hanyang.ac.kr
[3] Division of Information Engineering and Telecommunications,
Hallym University, Korea
jhyoon@hallym.ac.kr

Abstract. Exact match queries, wildcard match queries, and k-mismatch queries are widely used in lots of molecular biology applications including the searching of ESTs (Expressed Sequence Tag) and DNA transcription factors. In this paper, we suggest an efficient indexing and processing mechanism for such queries. Our indexing method places a sliding window at every possible location of a DNA sequence and extracts its signature by considering the occurrence frequency of each nucleotide. It then stores a set of signatures using a multi-dimensional index, such as the R*-tree. Also, by assigning a weight to each position of a window, it prevents signatures from being concentrated around a few spots in indexing space. Our query processing method converts a query sequence into a multi-dimensional rectangle and searches the index for the signatures overlapped with the rectangle.

Keywords: DNA database, indexing, exact match, wildcard match, k-mismatch.

1 Introduction

DNA sequences hold the code that determines the characteristics of living organisms, and can be represented as a long list over the four-letter alphabet of A, C, G, and T known as nucleotides. *DNA sequence searching* is an operation that finds, from a DNA database, DNA (sub-)sequences whose nucleotide

---

[*] This work was supported by the Korea Research Foundation Grant (KRF-2004-003-D00302), the Basic Research Program Grant (Grant R04-2003-000-10048-0), and the IT Research Center via Cheju National University.

arrangements are similar to a given query sequence. To cater for the evolutionary mutations and noises in DNA sequences, approximate match queries are preferred to exact match queries for DNA sequence searching.

The most fundamental way for processing approximate match queries is to use the *Smith-Waterman alignment algorithm* [12], a dynamic programming approach for finding an optimal local alignment between two sequences. This algorithm, however, takes a long processing time of $O(mn)$, where m and n are the lengths of the two sequences to be aligned, respectively. A natural idea to resolve this kind of drawbacks is to employ the *filtering and refinement approach*. *BLAST* [4,5] is a typical example that follows this approach. Due to performance reasons, it uses a heuristic algorithm based on a similarity model that is slightly different from the one adopted in the Smith-Waterman alignment algorithm. Recently, Kaheci et al. [10] proposed the *MR-Index* for efficient processing of k-difference queries. A *k-difference query* is to find data subsequences that can be matched to a given query sequence by performing at most k replacing, inserting, and deleting operations.

In this paper, we proposes an approach for efficient processing of DNA sequence searching, especially exact match queries, wildcard match queries, and k-mismatch queries. *Exact match queries* search a DNA database for the subsequences that are exactly matched to a query sequence. *Wildcard match queries* contain wildcard characters marked as '*' in a query sequence, and find the subsequences that are matched to a query sequence. Note that a wildcard matches with any single nucleotide. *K-mismatch queries* retrieve the data subsequences that have at most k nucleotides mismatched to those of a given query sequence. These queries are widely used in various molecular biology applications such as retrieval of expressed sequence tags and DNA transcription factors [8].

2 Definitions

The *alphabet* $\sum$ of nucleotides consists of 15 characters that can occur in DNA sequences (See Table1). Four characters, A, C, G, and T, are used to express the regions of a DNA sequence whose characteristics are discovered completely. We call these four characters as *principal nucleotides*.

A *DNA sequence* $T = \langle t_1, t_2, \cdots, t_n \rangle$ is an ordered list of characters in the alphabet $\sum$. $|T|$ denotes the length of T. We use T' to denote a contiguous subsequence of T. A *window* is defined as a subsequence of a fixed length taken from a DNA sequence. W and $|W|$ denote a window and its length, respectively. The window beginning at the ith position of a DNA sequence is denoted as W_i.

Any two characters s and q are said to be *matched* if the intersection of the set of characters represented by s and the set of characters represented by q is not empty. Given a DNA data sequence T and a query sequence Q, the *DNA sequence searching problem* is to find all subsequences T' of T that satisfy both of the following conditions: (1) $|Q| = |T'|$, and (2) for each i between 1 and $|Q|$, the ith character of Q matches the ith character of T'.

Table 1. Characters included in the alphabet of nucleotides

Code	Bases	Code	Bases	Code	Bases
A	A	Y	C or T	B	C or G or T
C	C	S	G or C	D	A or G or T
G	G	W	A or T	H	A or C or T
T	T	K	G or T	V	A or C or G
R	A or G	M	A or C	N	any base

3   Related Work

The *Boyer-Moore algorithm* [7] and the *Knuth-Morris-Pratt(KMP) algorithm* [11] have been devised for exact match queries. Their worst-case time complexity proved to be linear to the length of data sequence. These algorithms, however, should access the entire data sequences from disk because they are based on the sequential scan.

The method combining the *Aho-Corasick algorithm* [3] and the *scan vector* has been proposed for processing wildcard match queries [8]. By eliminating all the wildcards from a query sequence, this method first obtains a set of subpatterns and their starting positions within a query sequence. Next, by using an one-dimensional array called a scan vector, it finds the data subsequences, each of which contains all those subpatterns in order. This method, however, has a large storage overhead since it maintains the scan vector as large as the data sequence. Also, it requires much processing time because it accesses the whole data sequences from disk.

For processing k-mismatch queries, the *suffix-tree-based method* [13] constructs a suffix tree on data and query sequences. Next, it finds from the suffix tree the lowest one among the common ancestor nodes of both sequences. It then traverses down the subtree of that node until it encounters k mismatches. This method can be applied to the processing of exact match and wildcard match queries in a similar way. However, it suffers from a large storage overhead and high cost for maintaining and traversing a huge suffix tree.

4   Basic Signature Index

This section proposes a new indexing method called BSI (Basic Signature Index) and also suggests a query processing method based on the proposed index structure.

To construct an index, we first locate a sliding window of size $|W|$ on every possible position of data sequence T. We then extract a *basic signature* from each window, considering the minimum and maximum frequencies of each principal nucleotide.

Definition 1. *Basic Signature: BS*
Let $BS(W_i)$ be a *basic signature* of window W_i. $BS(W_i)$ is expressed as follows:

$$BS(W_i) = (([min_A, max_A], [min_C, max_C], [min_G, max_G], [min_T, max_T]), i)$$

Here, min_A and max_A denote the minimum and maximum numbers of occurrences of character A, respectively, in W_i. The meanings of min_C, max_C, min_G, max_G, min_T, and max_T are analogous.

$BS(W_i)$ is regarded as a 4-dimensional rectangle of $([min_A, max_A], [min_C, max_C], [min_G, max_G], [min_T, max_T])$ along with the identifier i and thus can be stored in a multi-dimensional index such as the R*-tree [9] and the X-tree [6]. The total number of windows taken from a data sequence T is $|T| - |W| + 1$. Since $|T| \gg |W|$ in most cases, the index for T could be much larger than T itself.

To reduce this storage space, we only store the MBRs (Minimum Bounding Rectangles) which cover the signatures for consecutive c data windows extracted from a data sequence. Note that the signatures for consecutive two data windows are not that different from each other and thus are located closely in the 4-dimensional indexing space. Therefore, we expect that the MBR covering consecutive c signatures will not be enlarged much. By using this approach, we are able to reduce storage space for indexing to $1/c$. We call c the *index compression coefficient*.

The first step for query processing is to construct a query rectangle from a query sequence Q. A query rectangle is formed in a different way according to the types of a query submitted. Let us first suppose that $|Q| = |W|$.

o **Exact match query:** We construct a 4-dimensional query rectangle, $([min_A, max_A], [min_C, max_C], [min_G, max_G], [min_T, max_T])$, from the query sequence.
o **Wildcard match query:** We first construct a 4-dimensional query rectangle by using the procedure for exact match queries. We then increase max_A, max_C, max_G, and max_T by the number of occurrences of the wildcard on the query sequence.
o **K-mismatch query:** We construct a 4-dimensional query rectangle by using the procedure for wildcard match queries. We then increase max_A, max_C, max_G, and max_T by the value of k, and also decrease min_A, min_C, min_G, and min_T by the value of k. This implies that each principal nucleotide in a data window is allowed to occur k times more or less than that in a query signature by k mismatches. If an adjusted minimum value becomes less than 0, we set it to 0.

After constructing a query rectangle from a query sequence, we search the index for the data rectangles overlapping with the query rectangle. We call them *candidate rectangles*. Then, we perform a post-processing step to discard false alarms, those candidates that are not real answers. Using the identifier of each candidate rectangle, this step reads its corresponding data window from the database, and then verifies whether the data window actually matches with the query sequence. Only the candidate rectangles which pass this verification are returned as final answers.

The identifier of each candidate rectangle is the beginning position of its consecutive c data windows. Therefore, by using the identifier of each candidate rectangle, we actually retrieve and verify the corresponding c data windows together in the post-processing step.

Until now, we assumed $|Q| = |W|$. When $|Q| < |W|$, we generate a new query sequence Q' of length $|W|$ by appending $|W| - |Q|$ wildcard characters '*' to the end of Q and then apply the above query processing procedure to Q'. When $|Q| > |W|$, we first partition a query sequence Q into p sub-query sequences, $Q_1, Q_2, \cdots, and\ Q_p$, such that $p = \lceil |Q|/|W| \rceil$ and $|Q_i| = |W|$ for every i between 1 and p. Here, the last sub-query sequence Q_p can be overlapped with Q_{p-1} to make the constraint $|Q_i| = |W|$ satisfied. Next, we apply the above query processing procedure to every sub-query sequence, and then obtain the final answers by merging all the results.

5 Weighted Signature Index

Let us first mention a couple of drawbacks of BSI. First, in BSI, the signature of a window is decided only by the number of occurrences of each principal nucleotide. Therefore, there may be a great number of windows that are different from one another but are represented as the same signature. It causes a large number of false alarms, resulting in high index-searching and post-processing costs. Second, in most DNA sequences, the occurrence ratios of the four principal nucleotides, A, C, G, and T, are roughly 30%, 20%, 20%, and 30%, respectively. The windows taken from such sequences also show similar occurrence ratios regardless of their beginning positions. Therefore, it is likely that lots of windows are represented by the signatures close to the center $(0.3 \times |W|, 0.2 \times |W|, 0.2 \times |W|, 0.3 \times |W|)$.

To overcome the above limitations, we need to increase the number of distinct signatures and spread them evenly on the indexing space.

5.1 Basic Strategy

The simplest way to overcome the limitations of BSI is to extract more features from windows. However, this increases the dimensionality of the underlying index, and thus leads to the well-known dimension curse. To represent windows more discriminatively without increasing the dimensionality, we propose a simple but effective method that assigns a weight to each position within a window. This makes it possible to express both occurrence frequencies and occurrence positions of nucleotides with a signature of the same dimensionality. To incorporate this method into our indexing approach, we first define a weight function $w(j)$ $(1 \leq j \leq |W|)$ which assigns a weight to each position i within a window. We then extract a weighted signature from each window.

Definition 2. *Weighted Signature: WS*
Let $WS(W_i)$ be a weighted signature of window W_i. $WS(W_i)$ is expressed as follows:

$$WS(W_i) = (([wmin_A, wmax_A], [wmin_C, wmax_C], [wmin_G, wmax_G], \\ [wmin_T, wmax_T]), i)$$

Here, $wmin_A$ is the sum of the weights of the positions at which character A **must** occur in window W_i, and $wmax_A$ is the sum of the weights of the positions at which character A **may** occur in W_i. The meanings of $wmin_C$, $wmax_C$, $wmin_G$, $wmax_G$, $wmin_T$, and $wmax_T$ are analogous.

By taking the above weighting scheme, disparate windows that were represented by the same basic signature may now be expressed by different weighted signatures. We incorporate this weighing scheme into the proposed index structure, thus producing a very effective index structure called WSI (Weighted Signature Index). WSI solves the problems of BSI by scattering the disparate windows, which were represented by the same basic signature, over the indexing space.

The query processing algorithm for WSI is not that different from that for BSI. However, when we construct a query rectangle for answering a k-mismatch query, we need to consider the positions at which mismatches may occur. The procedure to build a query rectangle for a k-mismatch query is skipped due to space limitation.

5.2 Weight Function

Since the weight function determines the distribution of signatures in indexing space, it has to be carefully designed. Consider a set of data windows which have the same *basic* signature. Their weighted signatures get scattered over the indexing space by the weight function. Let us consider an MBR that covers all such weighted signatures. Larger MBR implies that the weighted signatures are scattered over larger space. However, if the weighted signatures are scattered too much, the corresponding MBR may overlap with its neighboring MBRs, producing new false alarms. Therefore, we have to choose a weight function which enlarges MBRs as much as possible without making them overlap with their neighboring MBRs.

Let us give a formal discussion on this issue. For each principal character X, let $R_{min}(X, s)$ denote the minimum of all $wmin_X$ values obtained from a set of all windows in which X occurs s times. That is, $R_{min}(X, s) = \sum_{j=1}^{s} sw(j)$ where $sw(j)$ denotes the jth smallest weight in a window. Similarly, let $R_{max}(X, s)$ denote the maximum of all $wmax_X$ values obtained from a set of all windows in which X occurs s times. That is, $R_{max}(X, s) = \sum_{j=|W|-s+1}^{|W|} sw(j)$.

To prevent neighboring MBRs from being overlapped, $R_{max}(X, s) < R_{min}(X, s+1)$ should be satisfied for every s between 0 and $|W| - 1$. Supposing $w(j) = j + C$, let us solve the inequality. Note that $sw(j)$ is identical to $w(j)$ in this case.

$$R_{max}(X,s) < R_{min}(X, s+1)$$
$$\Leftrightarrow \sum_{j=|W|-s+1}^{|W|} sw(j) - \sum_{j=1}^{s+1} sw(j) < 0$$
$$\Leftrightarrow \sum_{j=|W|-s+1}^{|W|} w(j) - \sum_{j=1}^{s+1} w(j) < 0$$
$$\Leftrightarrow -C - (s^2 + (1-|W|)s + 1) < 0$$

Since the above inequality should be satisfied for every s between 0 and $|W|-1$, we obtain $C > \frac{(|W|-1)^2}{4} - 1$. Among the values of C which satisfy the inequality, we choose $|W|^2$ for the sake of simplicity. That is, we use $w(j) = j + |W|^2$ for a weight function.

6 Performance Evaluation

In our experiments, as a data sequence T, we used six sets of DNA sequences downloaded from NCBI [1]: human chromosome 3 (2.5Mbp), 17 (5Mbp), 1 (7.5Mbp), 2 (10Mbp), 10 (20Mbp), and 5 (40Mbp). As a query sequence, we used 1,000 DNA sequences of length 256 to 2,048. A half of them were randomly selected from T, and the other half were obtained from DNA sequences [2] frequently used by biologists at laboratories.

We evaluated performances of four approaches: BSI, WSI, SeqScan, and Suffix. SeqScan is the sequential scan based method, and Suffix is the method that uses the suffix tree as an index structure.

6.1 Parameter Settings

It is desirable to set the window size slightly smaller than a typical size of a query sequence. For determining a window size, we analyzed the lengths of 35,685 query sequences downloaded from [2]. From the results, we observed that 62% of them have the lengths of 256 to 2,048. Thus, we set the basic window size to 256 for further experiments.

In order to find a good value for the index compression coefficient, while changing the index compression coefficient, we evaluated the k-mismatch query processing time of BSI and WSI using human chromosome 2 of 10Mbp as a data sequence and 1% of the length of a query sequence as the value of k. As shown in Fig. 1, as the compression coefficient increases up to 80, the query processing time of both BSI and WSI decreases. From that point, however, their query processing time increases as the compression coefficient gets larger. Therefore, we set the base value for the compression coefficient to 80.

6.2 Results and Analyses

Experiment 1: Query Processing Time with Various Query Size
In this experiment, we compared query processing times of different approaches while changing the length of query sequences. We used human chromosome 2 of 10Mbp as a data sequence. Also, we set both k for k-mismatch queries and the number of wildcard characters for wildcard match queries to 10, which is 1% of

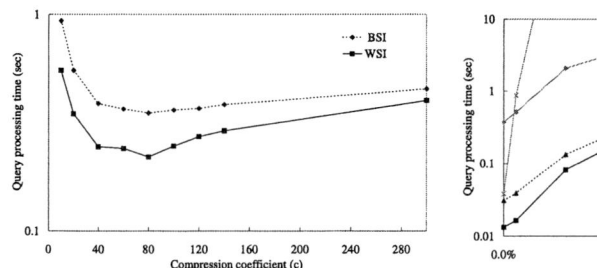 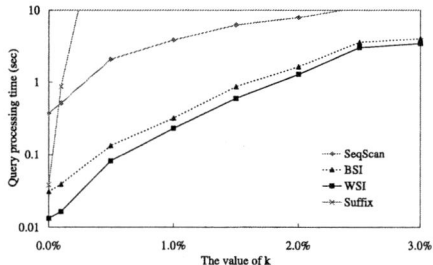

Fig. 1. Query processing time with various values for compression coefficient

Fig. 2. Processing time of k-mismatch with various k values

the average length of query sequences. Fig. 3 depicts query processing times of all the approaches for exact match, wildcard match, and k-mismatch queries.

In exact match queries, SeqScan and Suffix show nearly constant performance regardless of the length of query sequences. In BSI and WSI, we observe that the query processing time shrinks until the length of a query sequence reaches a point (i.e., 512), and then grows gradually after that point.

In wildcard match queries, every approach spends more query processing time compared with that of exact match queries. In BSI and WSI, wildcard characters in a query enlarge the corresponding query rectangle and increase the number of candidates, which leads to a large query processing time. As a query sequence gets longer, however, the number of candidates decreases remarkably. Thus, the performance improves significantly.

K-mismatch queries take a processing time much bigger than exact match and wildcard match queries. In particular, Suffix shows performance worse than even SeqScan since the part of the index to be traversed increases explosively. In BSI and WSI, however, their performance is shown to be nearly constant, and is not that affected by the changes of the length of query sequences.

In exact match queries, the results show that WSI outperforms SeqScan, Suffix, and BSI 19 to 44 times, 2.9 to 6.1 times, and 2.2 to 2.7 times, respectively. In wildcard match queries, WSI performs better than SeqScan, Suffix, and BSI 4 to 21 times, 1.4 to 4.5 times, and 1.5 to 1.8 times, respectively. Also, in k-mismatch queries, BSI performs faster than SeqScan, Suffix, and BSI 7 to 28 times, several thousand times, and 1.3 to 1.6 times, respectively.

Experiment 2: Processing Time of k-mismatch with Various k Value
In this experiment, we compared the processing times of k-mismatch queries of different approaches with various k values. We used human chromosome 2 of 10Mbp as a data sequence. Fig. 2 shows an average query processing time of each approach while setting k as 0%, 1%, 2%, and 3% of the length of a query sequence. We observe that the query processing time of WSI, BSI, Suffix, and SeqScan gets higher as k grows. In WSI and BSI, a higher k value makes the part of an index to be traversed increased, and thus increases the query processing time gradually. In Suffix, however, as k grows, the part of an index to be

traversed becomes explosively larger, and thus, the query processing time grows abruptly. The results reveal that WSI shows the best performance, and performs better than SeqScan, Suffix, and BSI 3.6 to to 31 times, 3 to several thousand times, and 1.1 to 2.3 times, respectively.

Experiment 3: Query Processing Time with Various Lengths of Data Sequences

In this experiment, we measured the query processing times of different approaches with various data sizes. We excluded Suffix in this experiment since its performance degradation in performing k-mismatch queries on a large database is too serious to conduct experiments. Here, we set both k for k-mismatch queries

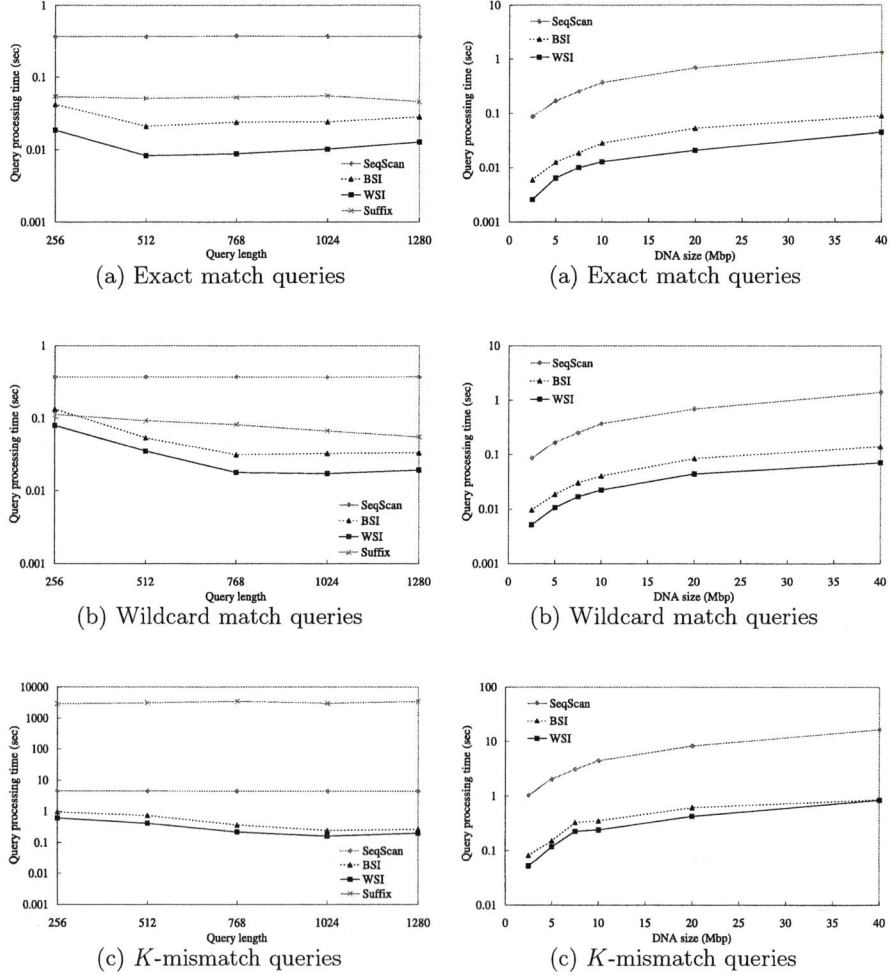

Fig. 3. Query processing time with various lengths of query sequences

Fig. 4. Query processing time with various data sizes

and the number of wildcard characters for wildcard match queries to 10, which is 1% of the average length of query sequences. Fig. 4 shows an average processing time of each approach for exact match, wildcard match, and k-mismatch queries.

The processing time of BSI and WSI for three kinds of queries increases almost linearly as the data size grows. WSI performs better than the other approaches in processing all kinds of queries. In exact match queries, WSI runs faster than SeqScan and BSI 25 to 33 times and 1.8 to 2.5 times, respectively. In wildcard match queries, WSI outperforms SeqScan and BSI 15 to 19 times and 1.7 to 1.9 times, respectively. Also, in k-mismatch queries, WSI performs better than SeqScan and BSI 13 to 20 times and 1.0 to 1.5 times, respectively.

7 Conclusion

Exact match queries, wildcard match queries, and k-mismatch queries are widely used in lots of molecular biology applications including the searching of ESTs (Expressed Sequence Tag) and DNA transcription factors.

In this paper, we proposed an efficient indexing and processing technique for processing such queries on large DNA databases. The proposed indexing method places a sliding window at every possible location of a data sequence, and extracts its signature by considering the occurrence frequency of each nucleotide character. It then stores and manages a set of signatures using a multi-dimensional index, such as R*-tree. Especially, by assigning a weight to each position of a window, it scatters the signatures over the index space and thus reduces false alarms. The experiments with real biological data sets revealed that the proposed method is at least 2.9 times, 1.4 times, and several orders of magnitude faster than the suffix-tree-based method in performing exact match, wildcard match, and k-mismatch queries, respectively.

References

1. http://www.ncbi.nlm.nih.gov
2. ftp://ftp.ensembl.org
3. A. Aho, M. Corasick, "Efficient string matching: an aid to bibliographic search", Communications of the ACM, Vol. 18, pp. 333-40, 1975.
4. S.F. Altschul, T.L. Madden, A.A. Schaffer, J. Zhang, Z. Zhang, W. Miller, and D.J. Lipman, "Gapped BLAST and PSI-BLAST: A new generation of protein database search programs", Nucleic Acids Research, 25(17), 1997.
5. S. Altschul, W. Gish, W. Miller, E. Myers, and D. Lipman, "Basic local alignment search tool", Journal of Molecular Biology, Vol. 215, pp. 403-410, 1990.
6. S. Berchtold, D.A. Keim, and Hans-Peter Kriegel, "The X-tree: An index structure for high-dimensional data", VLDB, pp 28-39, 1996.
7. R.S. Boyer, J.S. Moore, "A fast string searching algorithm", Communications of the ACM, Vol. 20, pp. 762-772, 1977.
8. D. Gusfield, Algorithms on Strings, Trees, and Sequences: Computer Science and Computational Biology, Cambridge University Press, 1997.

9. A. Guttman, "R*-Trees, A dynamic index structure for spatial searching", ACM SIGMOD, pp. 47-57, 1984.
10. T. Kaheci, A. K. Singh, "An efficient index structure for string databases", VLDB, 2001.
11. D. E. Knuth, J. H. Morris, V. B. Pratt, "Fast pattern matching in strings", SIAM J. Comput., Vol. 6, pp. 323-350, 1977.
12. T. Smith and M. Waterman, "Identification of common molecular subsequences", Journal of Molecular Biology, Vol. 147, pp. 195-197, 1981.
13. G. A. Stephen, String Searching Algorithm, World Scientific Publishing, 1994.

An Automatic Unsupervised Querying Algorithm for Efficient Information Extraction in Biomedical Domain

Min Song, Il-Yeol Song, Xiaohua Hu, and Robert B. Allen

College of Information Science and Technology, Drexel University,
Philadelphia, PA 19104
{min.song, song, xiaohua.hu, rba}@drexel.edu

Abstract. In the domain of bioinformatics, extracting a relation such as protein-protein interations from a large database of text documents is a challenging task. One major issue with biomedical information extraction is how to efficiently digest the sheer size of unstructured biomedical data corpus. Often, among these huge biomedical data, only a small fraction of the documents contain information that is relevant to the extraction task. We propose a novel query expansion algorithm to automatically discover the characteristics of documents that are useful for extraction of a target relation. Our technique introduces a hybrid query re-weighting algorithm combining the modified Robertson Sparck-Jones query ranking algorithm with a keyphrase extraction algorithm. Our technique also adopts a novel query translation technique that incorporates POS categories to query translation. We conduct a series of experiments and report the experimental results. The results show that our technique is able to retrieve more documents that contain protein-protein pairs from MEDLINE as iteration increases. Our technique is also compared with SLIPPER, a supervised rule-based query expansion technique. The results show that our technique outperforms SLIPPER from 17.90% to 29.98 better in four iterations.

1 Introduction

Rich information is embedded in unstructured text collections, and this information is often discovered in a structured or relational form. With more than 12 million abstracts in MEDLINE, processing time becomes a bottleneck in exploiting IE to leverage extracted information with relational databases. Current IE approaches, however, are not flexible for huge online biomedical text databases, which cover so many sub-domains [1]. We introduce an automatic querying technique, called DocSpotter, to identify the promising documents for the extraction of a relation from text. DocSpotter only requires an initial query to the text database provided by a user. Our technique is an unsupervised querying technique for retrieving useful documents for information extraction from large biomedical databases.

There are several key advantages of using key phrases for queries in an iterative search process. First, it is relatively robust in that average performance of queries tends to be improved using this type of expansion. Second, it is a novel automatic query expansion technique that combines global analysis with local analysis. Global analysis refers to a technique of extracting key phrases from the whole collection and then maps it out to synonymously as well as hierarchically related concepts. Local analysis means a technique to weigh noun phrases and converted noun phrases from verbs found in the same passages that contains the matched query. Our technique also translates candidate keyphrases to Disjunctive Normal Form (DNF) with Part-Of-Speech (POS) categories. Third, our approach is based on the technique of iterative and exhaustive keyphrase extraction. This approach fits well in the target application that requires the greedy and comprehensive retrieval results.

The rest of the paper is organized as follows: Section 2 describes the overall architecture of DocSpotter. Section 3 describes the query expansion procedures. Section 4 reports on the experiments. Section 5 concludes the paper.

2 The System Architecture

DocSpotter is a novel querying technique to iteratively retrieve promising documents for information extraction. Figure 1 illustrates how a novel query expansion algorithm works in DocSpotter.

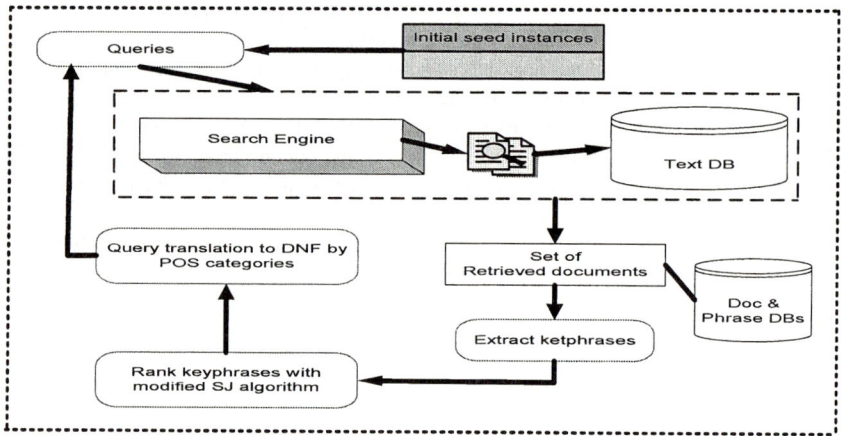

Fig. 1. The System architecture of DocSpotter

The outline of the approach described in Figure 1 is as follows:

- Step 1: Starting with a set of user-provided seed instances our system retrieves a sample of documents from the databases.
- Step 2: On the retrieved document set, we parse each document into sentences and apply the keyphrase extraction technique proposed in [6] to extract keyphrases from the input documents.
- Step 3: Applying a hybrid querying expansion algorithm that combines the modified Robertson and Spark-Jones ranking algorithm with Information Gain-based keyphrase ranking to derive queries targeted to match—and retrieve— additional documents similar to the positive examples.

- Step 4: Running the information extraction over the documents, it produces a set of extracted patterns from the documents and these patterns are kept in the pattern base.
- Step 5: Run the new queries from Step 4 to retrieve a set of promising documents form the databases, go to Step 2. The whole procedure repeats until the no new additional documents are retrieved.

3 Query Expansion Procedure

In this section, we discuss what techniques and procedures are used for query expansion. The following three subsections gave detailed descriptions of the techniques used for keyphrase extracting, query re-weighting, and query translating in DocSpotter.

3.1 Keyphrase Extraction Procedures

As illustrated in Figure 2, keyphrase extraction in DocSpotter comprises the following two stages: 1) building extraction model and 2) extracting keyphrases. In Figure 2, the dotted line represents the processing logic for "building extraction model" whereas the solid line indicates the processing logic for "extracting keyphrases." The detail descriptions are provided in the following subsections. These two stages are fully automated.

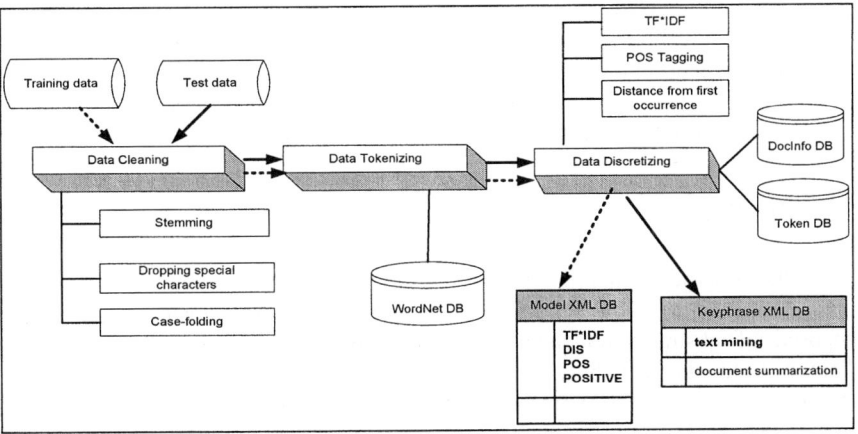

Fig. 2. Keyphrase extraction procedure adopted in DocSpotter

3.2 Keyphrase Ranking

Automatic query expansion requires a term selection stage where the system selects the terms based on some criterion. The ranked order of terms is of primary importance in that the terms that are most likely to be useful are close to the top of the list. We exploit a state-of-art term weighting scheme from IR to re-weight candidate keyphrases. While there are many promising alternatives to this weighting scheme in the IR literature [5], we chose Robertson-Spark Jones algorithm as our base because it has been demonstrated to perform well and is naturally well suited to our task. In addition, incorporating other term weighting schemes into DocSpotter is easy and does not require changes to our model. F4point-5 formula is first proposed by Robertson and Spark Jones [5]. It has been widely used in IR systems with some modifications (Okapi). We modified F4point-5 formula for information extraction tasks.

3.3 Query Translation into DNF

Query reformulation specifically in Boolean IR systems has been the subject of study for over 25 years. The algorithm for automating Boolean query formulation was proposed in 1970. This method employs a term weighting function first described in [4] to decide the "importance" of terms which have been identified. The terms are then aggregated into "sub-requests" and combined into a Boolean expression in disjunctive normal form (DNF). The algorithms proposed to translate a query to DNF include classification-based [4], decision-tree [2], and thesaurus-based [7] algorithms. Our POS category-based translation technique is differentiated from others in that ours is unsupervised which does not require training and easily integrated into other domains. In our technique, there are four different phrase categories defined; 1) MESH term category, 2) Non-MESH noun category, 3) Non-MESH proper noun category, and 4) Verb category.

For the top N ranked keyphrases, DocSpotter looks up MESH to determine whether any corresponding heading to each keyphrase exists. If there is a corresponding heading, the keyphrase is categorized as MESH term category. Non-MESH keyphrases are then classified into three categories depending on the POS class of keyphrases. Keyphrases within the category are translated into DNF and categories are then translated into Conjunctive Normal Form (CNF).

4 Evaluation

We conducted experiments to evaluate the performance of DocSpotter on the task of protein-protein interaction extraction. The protein-protein interaction data sets are composed of abstracts collected from the MEDLINE.]The protein names are collected from the Database of Interacting Proteins (DIP) and Protein-Protein Interaction Database (PPID) databases. We also compare DocSpotter with SLIPPER, a supervised rule-based query expansion technique [3]. SLIPPER is one of the well-accepted query expansion techniques.

4.1 SLIPPER

We chose SLIPPER to compare the performance of DocSpotter in generating queries. SLIPPER is an efficient rule-learning system, which is based on confidence-ruled boosting, a variant of AdaBoost [3]. SLIPPER learns concise rules such as *"protein AND interacts" --> Useful*, which shows that if a document contains both term protein and term interacts, it is declared to be useful. These classification rules generated by SLIPPER are then translated into conjunctive queries in the search engine syntax. For instance, the above rule is translated into a query "protein AND interacts."

4.2 Experimental Results

The experimental results are shown in Table 1. The first column, iteration, indicates the order of query expansion. There are four iterations in terms of query expansion. The second column is the number of retrieved documents from MEDLINE for iteration. The third column means the number of documents containing protein-protein pairs in the retrieved documents by DocSpotter.

Table 1. Experimental results for four consecutive iteration of querying

Iteration	No of retrieved documents	No of documents containing protein-protein pairs
First	30	18
Second	609	289
Third	832	352
Fourth	1549	578

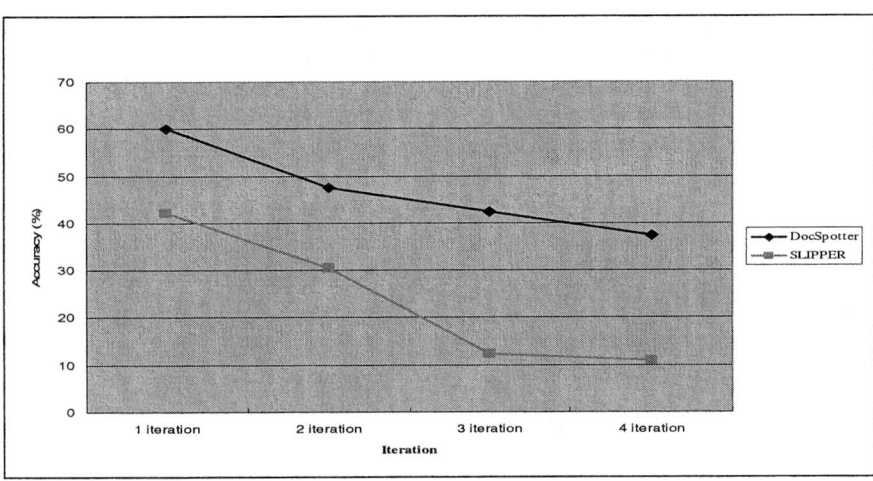

Fig. 3. The Performance Comparison between DocSpotter and SLIPPER

As shown in Table 1, our approach is able to retrieve more documents that contain protein-protein pairs as the number of iterations increases. In first iteration, the initial query retrieved 30 documents. Within the 30 documents retrieved, 18 documents contain protein-protein pairs. In fourth iteration, out of 1549 documents retrieved, 578 documents contain protein-protein pairs. Although accuracy seems decrease from first to fourth iteration, our interest is in ability of retrieving more documents containing protein-protein pairs. The experimental results seem encouraging.

We also conduct another set of experiments to compare DocSpotter with SLIPPER. SLIPPER is a supervised rule-based query expansion technique. Fig 3 shows the results of the performance comparison between DocSpotter and SLIPPER. The y axis represents the accuracy calculated by dividing the total number of relevant documents by the total number of retrieved documents. DocSpotter outperforms SLIPPER from 17.90% to 29.98% better in all four cases. The accuracy in Fig 3 is calculated by dividing the documents containing protein-protein pairs by the total number of retrieved documents. It appears that accuracy for SLIPPER drops significantly in third iteration whereas accuracy for DocSpotter looks steady with marginal decreases.

5 Conclusion

We proposed novel effective query technique for information extraction. Our approach automatically discovers the characteristics of documents that are useful for extraction of a target relation. Our technique starts with only a handful of user-provided examples of instances of the relation to extract. Using these seed instances, our system retrieves a sample of documents from the database. Then we apply machine learning and information retrieval techniques to learn queries that will tend to match additional useful documents. Our technique is different from other query expansion techniques in the following aspects. First, it proposes a hybrid query expansion algorithm combining the Robertson Sparck-Jones query re-ranking and the keyphrase extraction algorithm by Information Gain. Second, our technique adopts a query translation technique that converts candidate keyphrases to Disjunctive Normal Form (DNF) with POS categories. Third, our approach is based on the technique of iterative and exhaustive keyphrase extraction. This approach fits well in the target application that requires the greedy and comprehensive retrieval results.

We conducted a series of experiments to examine how accurately our technique retrieves a set of documents containing protein-protein pairs. The results show that DocSpotter is able to retrieve more documents that contain protein-protein pairs from MEDLINE as iteration increases. In another set of experiments, DocSpotter is compared with SLIPPER, a supervised rule-based query expansion technique. The results show that DocSpotter outperforms SLIPPER from 17.90% to 29.98 better in terms of accuracy in all four iterations.

References

[1] Agichtein, E. and Gravano, L. (2003). Querying Text Databases for Efficient Information Extraction, in: *Proceedings of the 19th IEEE International Conference on Data Engineering (ICDE)*, 113-124.

[2] Chang, K.C., Garcia-Molina, H., and Paepcke, A. (1996). Boolean Query Mapping Across Heterogeneous Information Sources, *IEEE Transactions on Knowledge and Data Engineering*, 8(4): 515-521.

[3] Cohen, W.W. and Singer, Y. (1999). Simple, Fast, and Effective Rule Learner, In: Proceedings of the Sixteenth National Conference on Artificial Intelligence and Eleventh Conference on Innovative Applications of Artificial Intelligence, July 18-22, 335-342.
[4] French, J.C., Brown , D.E., and Kim, N.H. (1997). A Classification Approach to Boolean Query Reformulation, *Journal of the American Society for Information Science*, 48(8): 694-706.
[5] Salton, G. , Buckley, C., and Fox, E.A. (1983). Automatic query formulations in information retrieval. *Journal of the American Society for Information Science*, 34(4):262-280, July 1983.
[6] Song, M, Song, I.Y., and Hu, T. (2003) KPSpotter: A Flexible Information Gain-based Keyphrase Extraction System, *Fifth International Workshop on Web Information and Data Management (WIDM'03)*, 50-53.
[7] Van Der Pol, R. (2003). Dipe-D: a Tool For Knowledge-based Query Formulation, *Information Retrieval*, 6, 21-47.

Voting Fuzzy k-NN to Predict Protein Subcellular Localization from Normalized Amino Acid Pair Compositions*

Thai Quang Tung[1], Doheon Lee[1], Dae-Won Kim[1], and Jong-Tae Lim[2]

[1] Department of BioSystems,
KAIST 373-1 Guseong-dong Yuseong-gu Daejeon 305-701, Korea
[2] Department of Computer Engineering, Kongju National University,
182 Shinkwan-dong, Kongju-si, Chungcheonam-do 314-701, Korea
{tqtung, dhlee}@biosoft.kaist.ac.kr

Abstract. There are a huge number of protein sequences in databanks whose functions are not known. Since the biological functions of these proteins are closely correlated with their subcellular localization, it is important to develop a system to automatically predict subcellular localization from sequences for large-scale genome analysis. In this paper, we first propose a new formula to estimate the composition of amino acid pairs for feature extraction, and then we present a voting scheme that combines a set of fuzzy k-nearest-neighbor (k-NN) classifiers to predict subcellular locations. In order to detect sequence-order features, individual classifier is constructed using different types of features, including amino acid and amino acid pair compositions. We apply our method to several datasets and significant improvements are achieved.

1 Introduction

Proteins are essential polymers involved in almost all biological functions. Directly finding out the function of a protein is not easy, but knowing the localization of a protein in a cell can give us hints to understand the protein functions. Experimental determination of subcellular location is time-consuming and costly. With the number of protein sequences entering databanks rapidly increasing, the importance of developing a powerful tool to automatically identify protein subcellular location from protein sequences has become self-evident.

So far there have been many methods and systems being developed to predict protein subcellular locations. Most of them are grounded on global sequence properties, in particular, the compositions of amino acids under investigation.

---

* This work was supported by the Korean Systems Biology Research Grant (M1-0309-02-0002) from the Ministry of Science and Technology. We would like to thank CHUNG Moon Soul Center for BioInformation and BioElectronics and the IBM SUR program for providing research and computing facilities.

The amino acid consists of 20 components, each representing the occurrence frequency of one of the 20 native amino acids in a given protein. This approach was first suggested by Nakashima and Nishikwa [8]. They found that the intracellular and extracellular proteins could be accurately discriminated with only by amino acid composition. Several machine learning methods were then employed to improve the prediction accuracy. Cedano [2] adopted a statistical method with Mahalanobis distance for prediction; Reinhardt [9] used neural networks while Hua [5] constructed a prediction system using SVMs.

It is known that when a protein sequence is encoded in terms of amino acid composition, its sequence-order properties will be lost. Hence, it is expected that a higher accuracy should be gained with new sequence encoding schemes that can capture the sequence order features. Park [6] introduced the concept of amino acid pair compositions and applied SVMs classifier for prediction. Recently, Ying [10] employed a fuzzy k-NN algorithm based on amino acid dipeptide composition and his result was superior compared to the other methods.

In his work, Ying defined the composition of amino acid dipeptide as number of dipeptide occurrences. In fact, the length of protein sequences is so varied that we need to consider it when estimating the dipeptide composition. Hereby, we propose a new representation of amino acid pair (including dipeptide) compositions using normalization by sequence length. Furthermore, in order to detect more sequence-order features, we use different types of compositions to construct multiple classifiers, and then a combination method is employed expecting that prediction accuracy can be improved.

2 Feature Extraction

Proteins are sequences that consist of a chain of units called amino acids. An amino acid is labeled as a letter in {A,C,D,E,F,G,H,I,K,L,M,N,P,Q,R,S,T,V,W,Y} thus a protein is presented as a string of letters. The length of proteins is varied from 50 to thousands so that it is not efficient to pass the whole sequences to the prediction system. Thus, to encode protein sequences, the compositions of a single amino acid and amino acid pair were proposed in previous works (see Introduction).

Ying [10] simply defined the composition of amino acid dipeptide (0-gapped amino acid pair) as number of dipeptide occurrences. If two proteins, one is long and the other is short, are in the same class, the dipeptide occurrences are so different that distance-based classifiers, such as fuzzy k-NN, could misclassify them. Therefore, we should consider the sequence length when defining the composition. Statistically, there must be a relation between dipeptide occurrences and the sequence length. To simulate this relation we devise a positive strength parameter α which determine how heavily the sequence length is weighted when calculating the composition. The composition of n-gapped amino acid pair is then defined as follow:

composition of n-gapped pair (a_i, a_j) =

$$\frac{\text{occurrences of } (a_i, a_j)}{(\text{total number of possible } n\text{-gapped amino acid pairs in protein})^{\alpha}}$$

In our research, 4 types of amino acid pair compositions (0-3 gapped amino acid pair compositions) are proposed in order to extract more sequence-order features. In a sequence, an n-gapped amino acid pair is a couple of amino acids between which there are n other amino acids located. There are 20 natural amino acids, so a sequence is represented by a 400-dimensonal vector of amino acid pair compositions.

3   Classification Algorithm

3.1   Fuzzy k-Nearest Neighbor Algorithm

The k-nearest neighbor (k-NN) rule [4] is one of the oldest and simplest methods for performing nonparametric classification. The main idea of k-NN can be stated as following: given a test pattern x with unknown label, its label is assigned according to the labels of its k nearest neighbors in the training set. The k-NN is widely used in machine learning and has many variations. Among them, fuzzy k-NN [7] usually gives better classification performance, especially in biological and medical data classification problems [10].

Let $\{x_1, x_2, .., x_N\}$ be the set of N already labeled pattern (training data), $\{c_1, c_2, .., c_c\}$ be the result classification space and x be the pattern to be classified. At the beginning, the fuzzy k-NN assigns membership values for each pattern to different categories rather than a particular class as in k-NN rule. The membership value of a pattern x_j to class c_i, denoted as $v_i(x_j)$, can be estimated in several ways. In crisp initialization, $v_i(x_j)$ is assigned to 1 if x_j belongs to class c_i, otherwise it is assign to 0. After initializing membership values for all patterns in the training data, membership value of x to class c_i is calculated as following equation:

$$\mu_i(x) = \frac{\sum_{j=1}^{k} v_i(x_j) d(x, x^{(j)})^{2/(1-m)}}{\sum_{j=1}^{k} d(x, x^{(j)})^{2/(1-m)}} \quad (1)$$

where k is the number of nearest neighbors; m is a fuzzy-strength parameter which determines how heavily the distance is weighted when calculating each nearest neighbor's contribution to the membership value; $d(x, x^{(j)})$ is the distance between x and its jth nearest neighbor $x^{(j)}$. Finally, the pattern x is classified to class to which the membership value of x is maximum.

3.2   Voting Fuzzy kNN Classifiers

As stated above, we use amino acid composition and 0-3 gapped amino acid pair compositions as features to build prediction system. For each type of compositions we construct a fuzzy k-NN classifier then a voting scheme is applied to combine the classifiers. By incorporating many classifiers, the model is expected to capture more sequence-order features and be more stable.

For combining multiple classifiers, instead of using majority voting scheme, we assign membership values for input pattern to every classes. Given unlabeled pattern x, let $\mu_i^{(j)}(x)$ be the membership value that jth classifier assign for x to class i, the membership value assigned by combined classifier is estimated as Eq. 2, and x is predicted as the class to which the total membership value is maximum.

$$\mu_i(x) = \sum_{j=1}^{5} \mu_i^{(j)}(x) \qquad (2)$$

4 Result and Discussion

Three datasets that were used in the previous works are investigated, including FuzzyLoc dataset [10], PLoc dataset [6] and Reinhardt's dataset [9]. In FuzzyLoc dataset, there are 7203 eukaryotic proteins classified in 11 subcellular locations. The PLoc dataset contains 7579 eukaryotic proteins in 12 locations and Reinhardt's dataset has 2427 protein located in 4 locations. Jackknife test (leave-one-out) is employed to evaluate the algorithm performance.

4.1 Searching for Optimal α Value

To search for the value of α that can best adapt to fuzzy k-NN classifier we applied that classification algorithm to the three datasets with various α values.

Euclidian measure is used for distance measuring. At first we use amino acid dipeptide compositions as features for classifying. While calculating dipeptide compositions, α value is varied from 0 to 1 with step 0.1. Classification algorithm performance is showed in Fig. 1. Ignoring sequence length factor ($\alpha=0$), the accuracies are 85.2 for Reinhardt's dataset, 80.1% for PLoc dataset and 80.1% for FuzzyLoc dataset, almost the same as Ying's result. On the contrary, when the composition is estimated as fraction of dipeptide ($\alpha=1$), the accuracies are dramatically decreased. Maximum accuracies are reached when α is equal to 0.5. We have investigated this test with various values of fuzzy strength parameter m and number of nearest neighbor parameter k. In all the cases 0.5 is still the best value. Therefore, when estimating the composition of dipeptides, we select α as 0.5.

4.2 Prediction Accuracy

Our voting procedure is devised in order to best utilize the potentials of 5 different representations of sequences. To show the effectiveness of our voting scheme, here, we apply it to FuzzyLoc dataset then analyze the result. The accuracies of individual classifiers and the voted one are shown in Table 1 (column 2 to 7). As shown in the table, there is a definite trend of improved accuracies in every protein classes. Apparently, the incorporation of different types of compositions

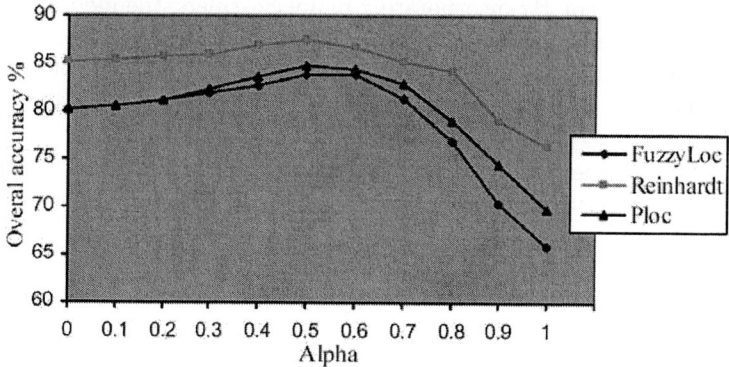

Fig. 1. Overall accuracies of fuzzy k-NN using amino acid dipeptide composition with different α values for feature extraction

Table 1. The performace of individual classifiers and voted classifier. A comparison bettween our result with Ying's work

Locations (No. of entries)	Individual fuzzy k-NN classifier					Proposed (Vote)	Ying's Method
	Amino acid dipeptide	1-gap	2-gap	3-gap			
Extracelluar(2134)	79.0	93.6	93.4	92.5	92.8	**94.3**	93.3
Nuclear(2149)	87.3	87.5	86.6	84.6	86.2	**89.9**	81.9
Mitochondrial(692)	51.5	66.9	66.5	63.7	63.7	**69.1**	59.0
Cytoplasmic(1251)	65.3	75.9	74.3	74.4	77.0	**78.8**	70.2
ER(82)	42.7	58.5	63.4	61.0	59.8	**62.2**	57.3
Chloroplast(645)	71.2	87.4	86.2	87.0	87.0	**89.5**	84.7
Cytoskeleton(10)	50.0	50.0	50.0	50.0	50.0	**50.0**	40.0
Peroxisomal(81)	35.8	60.5	60.5	61.7	63.0	**64.2**	56.8
Golgi apparatus(31)	9.7	12.9	12.9	6.5	12.9	**12.9**	16.1
Lysosomal(83)	67.5	73.5	77.1	69.9	71.1	**78.3**	67.5
Vacuolar(41)	12.2	34.5	31.4	39.0	36.6	**36.6**	34.1
Overall accuracy	74.0	83.9	83.2	82.0	83.1	**85.8**	80.1

can better capture sequence-order effects in protein sequences. The last column in Table 1 shows Ying's results. He used fuzzy k-NN based on amino acid dipeptide composition, but the equation of the composition is different from ours. By proposing a new formula and adding more sequence-order features, we can make 5.7% improvement compared to Ying's result.

Beside the FuzzyLoc dataset, we also perform experiment on PLoc dataset and Reinhardt's dataset then compare the result with other previous works. Park and his colleague used SVMs with the same features as ours, but they did not explain the formula for calculating the amino acid pair compositions. The overall accuracy they got was 78.2% on PLoc, their own dataset. Our new formula and voting fuzzy k-NN method can achieve 8.8% improvement compared to Park's result. On Reinhardt's dataset, there were many methods being applied including neural networks, SVMs, Markov chain, fuzzy k-NN. Among them, our method can give the best performance (result is not shown here). Although this dataset may be old and cover only 4 protein classes, the result can demonstrate applicability of our relative simple method.

5 Conclusion

In this paper, for classifying protein subcellular locations we first proposed an equation for estimating the composition of amino acid pairs from protein sequences. Then, we employed a voting procedure for combining fuzzy k-NN classifiers which were built from multiple amino acid pair compositions. This method could take the advantage of sequence-order and sequence-length effects in sequences. We have applied our method to several datasets and high predictive accuracies were achieved using a jackknife test. Our method is simple and practical because it just needs raw sequence data. In the future we will use this method to annotate protein database.

References

1. Boeckmann,B., et al.: The SWISS-PROT protein knowledge base and its supplement TrEMBL in 2003. In: Nucleic Acids Res., Vol. 31. (2003) 365–370.
2. Cadeno J., Alloy P., P'erez-Pons J.A., Querol E.: Relation between amino acid composition and cellular location of proteins. In: J Mol Biol. Vol. 266. (1997) 594–600.
3. Chou, K. C. and Elrod, D. W.: Using discriminant function for prediction of subcellular location of prokaryotic proteins. In: Biochem. Biophys. Res. Commun. Vol. 252. (1998) 63–68.
4. Duda,R.O., Hart,P.E. and Stork,D.G.: Pattern Classification, 2nd edn. Wiley, New York, (2000).
5. Hua S., Sun Z.: Support vector machines approach for protein subcellular localization prediction. In: Bioinformatics, Vol. 17. (2001) 721–728.
6. Keun-Joon Park and Minoru Kanehisa: Prediction of protein subcellular locations by support vector machines using compositions of amino acids and amino acid pairs. In: Bioinformatics, Vol. 19. (2003) 1656–1663.
7. Keller J.M., Gray M.R. and Givens J.A.: A fuzzy k-nearest neighbor algorithm. In: IEEE Trans. Syst. Man Cybern., Vol. 15. (1985) 580–585.
8. Nakashima, H., Nikishawa, K.: Discrimination of intracellular and extracellular proteins using amino acid composition and residue pair frequencies. In:J. Mol. Biol. Vol. 238. (1994) 54–61.
9. Reinhardt A., Hubbard T.: Using neural networks for prediction of the subcellular location of proteins. In: Nucleic Acids Res. Vol. 26. (1998) 2230–2236.
10. Ying Huang: Prediction of protein subcellular locations using fuzzy k-NN method. In: Bioinformatics. Vol. 20. (2004) 21–28.

Comparison of Tree Based Methods on Mammography Data

Richard De Veaux[1] and Thu Hoàng[2]

[1] Williams College, Williamstown, MA 01267, USA
deveaux@williams.edu
[2] Laboratoire MAP5, UFR Biomédicale,
CNRS & Université René Descartes, 75006 Paris, France
hoang@biomedicale.univ-paris5.fr

Abstract. X-ray film mammography and physical examination of the breast are the mainstays for early detection of breast cancer. Unfortunately, error rates for mammograms read by radiologists are high. We examine a particularly difficult to read series of 1618 mammograms where in order to achieve a false positive rate lower than 50%, the false negative rate of radiologists is nearly 25%. We examine a variety of automatic data mining tools in an attempt to improve the accuracy of the diagnosis. Our results suggest that roughly the same or higher accuracy rate than the radiologists can be attained at a much reduced cost. This potential cost savings could have a major financial impact for health care in developing nations.

1 Introduction

X-ray film mammography and physical examination of the breast are the mainstays for early detection of breast cancer. They contribute to the improvement of survival in breast carcinoma by allowing early treatment of occult lesions. One of the most important tasks in mammography is the assessment of calcifications and 10% to 40% lesions thus identified turn out to be malignant in subsequent biopsies.

A mammogram requires minimal cost (compared with newer instruments) and intervention. Other than the instrument itself, the main cost of the mammogram is the radiologist, who decides whether a suspicious area exists. A positive reading often leads to more expensive and invasive biopsies.

Unfortunately the error rates for mammograms read by radiologists are not low. The data we examined are the part of a series of 8421 mammograms [1] that entailed both false positive and false negative rates near 25%. In western countries it has been reported that missed cancers (false negatives) are particularly common in pre-menopausal women due to dense and highly glandular structure of their breasts [2] [3]. They are also common among post-menopausal women on estrogen replacement therapies [4]. In day-to-day practice in the USA, mammograms can miss more than a quarter (28%) of all tumors [5]. Because early detection of breast cancer is so crucial to its successful treatment, a false negative has serious health and cost consequences. On the other hand, false positives, (mistakenly diagnosed cancers from healthy tissue) are particularly common in the same two groups [2] [4]. A false positive can lead to needless anxiety, more X-ray exposure and unnecessary biopsies. As many as three-fourths of all post-mammogram biopsy results turn out to be non-cancerous lesions [6]. It has been estimated that the cumulative risk of false positives increases to "as high as 100 percent" over a decade of screening tests [7]. In Asian women, in whom smaller breasts result in higher density, the sensitivity of mammography is correspondingly lowered [8] [9]. For applicability to Asian populations of any breast cancer diagnostic system based on mammography it is therefore of interest to see whether data mining techniques may help improving interpretation accuracy for a mammogram series with high error rates of reading by experienced radiologists. Providing accurate screening of breast cancer from mammograms could have a significant impact on the health and well being of women throughout the developing world.

In this paper we will look at the relationship between radiologic parameters and the presence of malignant breast diseases. We compare several data mining techniques including classification trees (section 3), bagged and boosted trees (section 4), and neural networks (section 5). We summarize and suggest some directions for future research in section 6.

2 Data

The data consist of records of 1618 mammograms showing clustered microcalcifications on which follow up excisions were performed. The data were provided to us by Dr. Bernard Asselain, Head of the Biostatistics Department of the Institut Curie, and by Dr. Michèle Le Gal, senior radiologist now retired from the same institution, who introduced in 1976 a five way classification of microcalcifications based on the probability of malignancy [10] (see Fig. 1 where type 1 to 5 denote increasing probability of malignancy). When assessing the Le Gal's classification against the BI-RADS (breast imaging reporting and data system) introduced by the American College of Radiology (ACR) in 1992, Gülsün et al. (2003, [11]) found the former to entail higher positive predictive value for types 4-5 lesions and better interradiologist agreement than the latter system. About 30% of the mammograms were obtained in the Institut Curie with a Diagnost UM dedicated unit (Philips Medical Systems), while 70% were obtained elsewhere with various units of equivalent quality. The film used at the Institut Curie was Ortho MA (Eastman Kodak, Rochester, NY) with Min R screens (Eastman Kodak). Craniocaudal, lateral and oblique mediolateral views were obtained. The response variable is whether the tumor was benign or malignant as resulting from posterior histological examination. In addition, 11 other predictor variables are available:

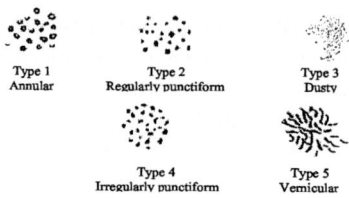

Fig. 1. Morphologic type of clusters according to Le Gal's classification

- two clinical variables pertaining to the patient's age and the indication about the left or right side of the breast being examined,
- the main usual radiographic features, i.e., appearance of tissue density, size, number, shape and location of both the clusters and the microcalcifications,
- the clusters' morphology according to Le Gal's classification. ([12]; Fig. 1)

3 Classification Trees

A tree model splits the observations into bins, defined by combinations of values on the predictor variables. It then estimates the probability of malignancy simply by calculating the proportion of malignant cases in each bin. The depth of the tree can be preselected but is usually chosen via cross-validation, by trying trees of various depths and then choosing the depth that has lowest average error across a series of test subsets.(see [15] for details) We found that using a minimum subset size of 50 was optimal for our data set. Using all the predictor variables, with a minimum subset size of 50, resulted in the tree shown in Fig. 2.

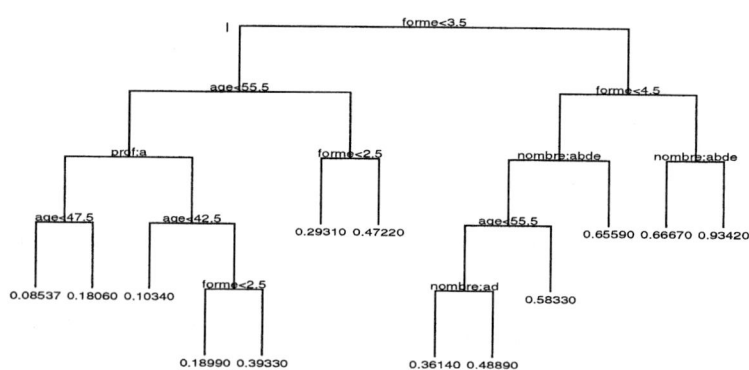

Fig. 2. A classification tree for the microcalcification data using 999 training cases

Of course, this tree was fit to just one random split of the data into training and test sets. To compare the performance of a tree to other methods, we

Table 1. Error Rates for Neural Networks and Trees

	False Positive Rate	False Negative Rate
Simple tree	0.322	0.337
Bagged trees	0.193	0.288
Boosted trees	0.249	0.325
Neural net 1	0.255	0.317
Radiologists	0.224	0.358

repeated the random splitting 1000 times, selecting 999 cases for training and 600 cases for testing each time. We report the average error rate for all methods, including the radiologists' assessment on the same 1000 test sets in Table 1.

4 Bagging and Boosting

Ensemble averaging methods have become widely used. Rather than training one tree in order to predict on a test set, we trained an ensemble of trees, selecting a bootstrapped (sampled with replacement) training set from the original training set, each time predicting on the same test set. To estimate a case in the test set, we fit each of the generated trees and take the model average prediction. This is a case of bagging or bootstrap aggregation (see [14] or [13]). We also generated each tree with a random set of predictor variables as well. In this case there are three parameters to vary: the number of trees generated, the number of predictors in each tree and the size of the tree. After some exploration, we settled on choosing 10 of the 11 predictors at random, a minimum final node size of 50 as before and 1000 trees. For the prediction, we found that rather than using the majority vote, we had lower total error using a cutoff of 30%. That is we took as predicting malignancy if 35% or more of the trees predicted malignancy. (Actually using 40% lowered the total error slightly, but raised the false negative rate significantly). The 35% rule gave error rates shown in Table 1, substantially lower than either the single tree or the radiologists.

Boosting is another strategy for building an ensemble of models. Unlike bagging, the data are reweighted after the first tree is fit, with the observations that are misclassified given *higher* weights than the ones correctly classified. The model is then refit and this process is repeated many time. At the end, we have a sequence of predictions from the reweighted fits. These models are then weighted by their misclassification rates and a weighted vote is produced for each new case from the test set. (see [16], [15] and [13]).

In the case of trees, we used the commercial implementation of boosted trees by Salford Systems, Treenet©. The best fit was found after 120 trees were fit and the classification errors on the test set are shown in table 1.

5 Neural Networks

As a contrast to tree methods, we also considered fitting multilayer feedforward neural networks. For both the hidden layer and the response, we used a sigmoidal activation function. To avoid overfitting we used weight decay, which adds a penalty term to the residual sum of squares, before minimization, choosing the weight decay parameter via cross-validation.

We used the implementaion of neural networks available in JMP©. We adjusted the output so that a case was predicted to be malignant if the predicted probability of malignancy was at least 0.40, similar to the value we took for the trees. The error rates are shown in Table 1. One of the advantages of this neural network implementation is the ability to look at the generated function via a contour profiler. For this model, this is shown in Fig. 3. From the profiler, we can see the positive correlation of malignant probability on the variables *age* and *forme* quite clearly. Notice the non linear increase in risk for older women (here selected for *forme* = 4).

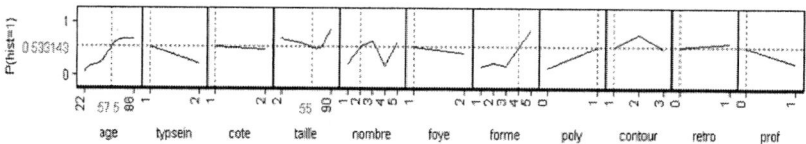

Fig. 3. A profiler for the neural network

6 Discussion

We investigated several data mining strategies in the hopes of providing more automatic, and hence less costly diagnostic capability from mammograms. Our results indicate that bagged trees may provide a diagnostic accuracy matching or eventually better than that of radiologists. We are encouraged from our findings that such automatic diagnostic capability might be realized. This has the potential of providing breast cancer diagnostic/screening to developing nations at a long far less than is currently available. Further research could be conducted to assess the potential for combining these models into a committee of experts, comprised perhaps from bagged trees, boosted trees and neural networks, using the model diagnostic as a guide.

Acknowledgments. We thank Dr. B. Asselain and Dr. Le Gal for providing us with the data.

References

1. Langer-Cherbit A, Le Gal M, Asselain B, Neuenschwander S. Breast cancer: zones of increased density mammographic features, correlated to clinical TNM and prognosis. *Eur J Radiol.* **24**(1):48-53, 1997 Jan.

2. Vogel, V. G. Screening younger women at risk for breast cancer. *J. Natl. Cancer Inst. Monogr.* **16**: 55- 60, 1994.
3. Baines, C. J., and Dayan, R. A tangled web: Factors likely to affect the efficacy of screening mammography. *J. Natl. Cancer Inst.* **91**(10): 833- 838, 1999.
4. Laya, M. B. Effect of estrogen replacement therapy on the specificity and sensitivity of screening mammography. *J. Natl. Cancer Inst.* **88**(10), 643- 649, 1996.
5. Poplack S et al., "Mammography in 53,803 Women from the New Hampshire Mammography Network," *Radiology*, **217**, 832-840, 2000 Dec.
6. Institute of Medicine/National Resource Council, *Mammography and Beyond*, National Academy Press: Washington, DC., p. 39, 2001.
7. Christiansen, C. L., et al. Predicting the cumulative risk of false-positive mammograms. *J. Natl. Cancer Inst.* **92**(20), 1657- 1666, 2000.
8. Alagaratnam TT, Wong J. Limitations of mammography in Chinese females. *Clin Radiol* **36**,1757, 1985.
9. Yah-Yuen Tan, Siew-Bock Wee, Mona P.C. Tan and Bee-Kiang Chong Positive Predictive Value of BI-RADS Categorization in an Asian Population *Asian Journal of Surgery* **27**(3) 186-191, 2004.
10. Le Gal M, Durand JC, Laurent M, Pellier D. Management following mammography revealing grouped microcalcifications without palpable tumor *Nouv Presse Med.* 5(26):1623-7, 1976 Jun 26.
11. Gülsün M. Demirkazik F.B. Ariyurek M. Evaluation of breast microcalcifications according to breast imaging reporting data system criteria and Le Gal's classification *European Journal of Radiology*, **47** (3) 227 - 231, 2003.
12. Le Gal M, Chavanne G, Pellier D. Diagnostic value of clustered microcalcifications discovered by mammography (apropos of 227 cases with histological verification and without a palpable breast tumor)] *Bull Cancer*;**71**(1),57-64, 1984.
13. De Veaux, R.D., "Bagging and Boosting", Encyclopedia of Biostatistics Second Edition, John Wiley and Sons, New York, 2004.
14. Breiman, L. Bagging Predictors. *Machine Learning*, **24**(2), 123-140, 1996.
15. Hastie, T., Tibshirani, R. and J. Friedman, *The Elements of Statistical Learning*, Springer-Verlag, New York, 2001.
16. Shapire, R.E., The strength of weak learnability, *Machine Learning* **5**(2), 197-227, 1990.

Bayesian Sequence Learning for Predicting Protein Cleavage Points

Michael Mayo

Dept. of Computer Science, University of Waikato, New Zealand
mmayo@cs.waikato.ac.nz

Abstract. A challenging problem in data mining is the application of efficient techniques to automatically annotate the vast databases of biological sequence data. This paper describes one such application in this area, to the prediction of the position of signal peptide cleavage points along protein sequences. It is shown that the method, based on Bayesian statistics, is comparable in terms of accuracy to the existing state-of-the-art neural network techniques while providing explanatory information for its predictions.

1 Introduction

The amount of sequence data generated by experimental biologists and made available via Internet databases is growing at an increasing rate. For example, SWISS-PROT [2], the leading protein sequence database, consists of 170140 entries, with an additional 1.6 million sequences in a supplementary database [2] awaiting addition. One of the significant issues with data of this nature is how to annotate sequences with properties that can occur anywhere along the length of the sequence. Manual experimental annotation in a biologist's laboratory is reliable but time consuming and expensive. Automatic annotation is fast and cheap.

The case study presented in this paper is the problem of determining signal peptides. Given a database of protein sequences with the signal peptides annotated, can a machine learning system discover the rules underlying the form and nature of a signal peptide?

Signal peptides are important because they direct proteins to their correct destination within the cell. Proteins need to have this "address" because they serve a multitude of functions, such as being reaction catalysts and transport molecules [12]. They are also the basic building blocks of the cell itself, and signal peptide failures can lead to diseases such as cystic fibrosis [3]. Knowledge of how signal peptides work is also useful when designing new drugs, which are often created in the form of proteins and therefore must have the correct signal attached to them [3].

Once a protein reaches its destination, its signal peptide is no longer needed. By a careful process of alignment, the signal peptide is cleaved off, severing it from the rest of the protein. An important point is that the signal peptide is always cleaved at exactly the same point along the protein sequence. The question posed here is: is it possible to predict this unique cleavage point for a newly sequenced protein?

The basic process described in this paper involves firstly extracting features from the training sequences. The frequencies of the features are determined and converted into probabilities, and then Bayes' Theorem is applied to predict the posterior probability of a cleavage point given each feature. When a test sequence is presented, the posterior probability of a cleavage point at each position along the sequence can be calculated and the position with the highest posterior is taken to be the predicted cleavage site.

This relatively simple Bayesian method is comparable to state-of-the-art neural network methods. Furthermore, this method can provide rudimentary explanations (in terms of ranked features) for its predictions. Such explanations are important for biologists trying to understand the nature of signal peptides.

In the next section, the biological and machine learning background to this paper is reviewed. Section 3 describes my proposed Bayesian method, and Section 4 reports on some results using a signal peptide dataset. Section 5 is the conclusion and mentions some issues for future research to address.

2 Background

2.1 Biological Background

All protein molecules are made up of a linear sequence of smaller molecules called amino acid residues. There are twenty amino acid residues in total. Each residue by convention has two abbreviations: a three-letter abbreviation and a one-letter abbreviation. For example, the abbreviations of Alanine are *Ala* and *A*. All twenty residues and their standard abbreviations are listed in Table 1.

Computationally speaking, a protein sequence can be viewed as a string of symbols (the residues) drawn from an alphabet of size twenty. Although it is also possible to augment each residue with a set of its properties, in this paper I consider only the basic sequence itself.

Signal peptides have a known structure that can aid in predicting the cleavage point, but within that structure there is considerable variability that makes the task difficult. In this study's datasets, the length of the signal peptide varies from five residues up to 90 residues. The average length is approximately 25 residues. In contrast, the total length of a protein can be thousands of residues. Signal peptides always occur at the beginning (the *N-terminal*) of the protein.

According to von Heijne [13] and Neilson & Krogh [7], a signal peptide consists of three main regions. Firstly, there is the *n-region* near the N-terminal, which comprises positively charged residues and is the greatest contributor

Table 1. Amino acid residue abbreviations

Residue	Abbreviations	
Alanine	*Ala*	*A*
Arginine	*Arg*	*R*
Asparagine	*Asn*	*N*
Aspartic acid	*Asp*	*D*
Cysteine	*Cys*	*C*
Glutamic acid	*Glu*	*E*
Glutamine	*Gln*	*Q*
Glycine	*Gly*	*G*
Histidine	*His*	*H*
Isoleucine	*Ile*	*I*
Leucine	*Leu*	*L*
Lysine	*Lys*	*K*
Methionine	*Met*	*M*
Phenylalanine	*Phe*	*F*
Proline	*Pro*	*P*
Serine	*Ser*	*S*
Threonine	*Thr*	*T*
Tryptophan	*Trp*	*W*
Tyrosine	*Tyr*	*Y*
Valine	*Val*	*V*

to the variability in the length of a signal peptide [3]. This is followed by a so-called *h-region*, which is a longer stretch of eight to fifteen hydrophobic residues. Finally, near the cleavage point, there is typically a *c-region*, consisting of around five mostly uncharged amino acids. This structure is depicted in Figure 1, using the sequence for human growth hormone as an example.

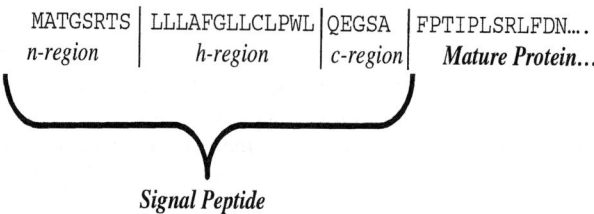

Fig. 1. Structure of a signal peptide for human growth hormone

The most important part of the signal peptide is the h-region: it serves the dual purpose of both encoding the protein's destination, and it is also used to align the signal peptide for cleavage when it finally arrives [12].

It should be noted that each of these regions are not necessarily a contiguous run of like residues. The hydrophobic h-region, for example, can be interrupted more than once by sequences of non-hydrophobic residues. This contributes to the difficulty of making predictions.

2.2 Signal Peptide Prediction Background

The earliest signal peptide prediction method was known as "the (-3,-1) rule" [12, 8]. This basically followed from the observation that positions −3 and −1 upstream (i.e. to the left) of the cleavage point were often "small and neutral". Using this simple rule seemed sufficient when the number of known signal peptides was small, but it has proved inadequate as the amount of data has increased.

Chou [3] extended the (-3,-1) rule when he introduced the subsite coupling approach. Basically, he formulated an algorithm which takes into account additional positions such as +1, as well as the expected lengths of each of the regions. The algorithm outputs the position on the sequence most likely to be the actual cleavage point. Although Chou reports that the results are encouraging, this method was trained on different data than the other methods were trained on and so it is difficult to make comparisons. An important point is that both of these approaches operate directly on variable-length sequences.

In contrast, more recent machine learning approaches do not operate directly on the variable length sequences but instead preprocess the sequences into fixed length records and transform the problem into one of classification rather than sequence annotation. For example, if the fixed record size is ten positions and the original sequence length is, say, 34, then 24 fixed length records would be produced from this single original sequence. Such preprocessing fits well with existing machine learning tools because they demand fixed-length data, but it does have a number of drawbacks.

The main one is that since each original sequence only has a single cleavage point, there is going to be a high abundance of negative examples (in a single sequence, only one fixed length record ends in the cleavage point and is therefore labelled as positive; the rest are labelled negative). Many machine learning algorithms given this biased data may simply predict every sequence as negative in order to obtain a high level of testing accuracy. To eliminate this problem and balance the classes more evenly, a considerable number of negative examples have to be discarded – a situation that could result in important information being lost.

The currently best-known and most widely used machine learning solution is the SignalP suite [1, 7, 8]. SignalP version 1 was a solely neural network approach. The neural network had a feedforward architecture and was trained on fixed length, sparsely encoded records derived from a "moving window" [8]. Hidden Markov models were added as a second predictor in SignalP version 2 [7], which increased accuracy slightly but also had the added benefit of being able to discriminate with high accuracy between signal sequences and non-signal sequences. SignalP version 3 [1] is a refinement of both the neural network and hidden Markov model approaches, with a claimed significant increase in prediction accuracy. Table 2 summarises the prediction accuracy results as reported by Bendtsen et al. [1]. Different neural network architectures have failed to provide a significant improvement over Signal P (see, e.g., [4, 9]).

Table 2. Best recorded accuracies of the SignalP suite of predictors

	Eukaryotyes	Gram-	Gram+
SignalP1	70.2	79.3	67.9
SignalP2	72.4	83.4	67.4
SignalP3	79.0	92.5	85.0

There are number of points worth mentioning about these results. Firstly, separate predictors were trained from data from three different sources: Eukaryotes (being all organisms except viruses, bacteria, and blue-green algae), and two types of Prokaryotes (bacteria): Gram-positive and Gram-negative. Other approaches do not subdivide the data at all and therefore the results are not directly comparable.

One significant weakness of the SignalP evaluations was that they performed only five-fold cross validation. In most cases, 10-fold cross validation is the minimum required for statistical significance [14].

Support Vector Machines (SVMs) have also been applied to this problem. Vert [11] developed a new SVM kernal for strings and applied his method to cleavage point prediction. His dataset was the same as that used to train SignalP1, but he did not subdivide the data. He reports 68% accuracy in predicting the cleavage point.

Some authors have attempted to incorporate residue properties into their systems to improve prediction accuracy. Recently, Smith [10] used a naïve Bayes-based text mining approach and reported accuracy comparable to Vert's SVM approach described above. Maetschke et al. [6] compared a number of different encoding of Blomaps using the WEKA machine learning workbench [14] and came to the conclusion that a particular encoding called BLOSUM62 combined with naïve Bayes produced the best results.

One difficulty when comparing these approaches is the lack of a standard benchmark dataset. It should be noted that Vert [11], Smith[10], and SignalP version 1 [8] all use the same dataset, namely that developed for SignalP version 1. Other authors have generated their own datasets from the SWISS-PROT database, and therefore it is quite possible that differences in accuracy are largely due to differences in data. To date, the SignalP2 dataset is publicly available but the SignalP3 dataset is not available.

3 Fast Bayesian Cleavage Point Prediction

3.1 Dataset Description

Before describing the method, it is necessary to briefly describe the sequence data used.

The dataset in this study is the same dataset used to train SignalP version 2. Each record consists of three lines: firstly, a biological description of the sequence in English which ties the sequence to its original record in SWISS-PROT; second, the residues sequence from the N-terminal all the way to position 29 downstream of the cleavage site; and thirdly, an annotation showing which residues are part of the signal peptide, which are part of the mature protein, and which is the cleavage site (defined as the first residue of the mature protein). Figure 2 below depicts two sample records taken from the dataset.

```
51 11SB_CUCMA     21 11S GLOBULIN BETA SUBUNIT PRECURSOR.
MARSSLFTFLCLAVFINGCLSQIEQQSPWEFQGSEVWQQHRYQSPRACRLE
SSSSSSSSSSSSSSSSSSSSCMMMMMMMMMMMMMMMMMMMMMMMMMMMMMM

54 41BB_MOUSE     24 T CELL ANTIGEN 4-1BB PRECURSOR.
MGNNCYNVVVIVLLLVGCEKVGAVQNSCDNCQPGTFCRKYNPVCKSCPPSTFSS
SSSSSSSSSSSSSSSSSSSSSSSCMMMMMMMMMMMMMMMMMMMMMMMMMMMMMM
```

Fig. 2. Two records taken from the dataset

The method by which this dataset was derived is worth briefly mentioning. SWISS-PROT contains protein sequences both with experimentally verified cleavage points, and without them. The database is constantly being updated as new protein sequences are added, and errors in annotations of existing sequences are corrected. Only sequences with experimentally-verified cleavage points were included in the dataset. Furthermore, some sequences were removed if they met certain criteria, for example having an origin in a virus gene [7].

The next step in the dataset creation was homology reduction. Many protein sequences occurring in nature are homologous, that is, they share long common subsequences which may include the signal peptide. This means that simple string alignment could result in a very high accuracy when predicting cleavage points on test sequences homologous to the training sequences. To eliminate this potential source of bias, for all pairs of homologous sequences in the dataset, Neilson & Krogh [7] discarded one of the sequences. By this method, more than 50% of the sequences in the dataset were discarded.

The final dataset contains 1666 protein sequences, of which 1137 are Eukaryote sequences, 697 are Gram negative Prokaryote sequences, and 280 are Gram positive Prokaryote sequences.

3.2 Training Method and Model

I will now describe the Bayesian method used for building a model based on the training data, and applying it to the prediction of signal peptide cleavage points. This approach is relatively simple, fast to train, and as shall been seen in the next section, has accuracy comparable to existing systems.

The basic idea is to define a set of features that protein sequences can have, extract from the training set the frequencies of those features, and convert those frequencies into posterior probabilities. This set of features and their posteriors will be referred to as the model. The model is then used to predict the final posterior probability of a cleavage point at each position along a test sequence given all the features on the test sequence.

What are the features? I define two types of feature: a pattern of residues that may occur anywhere along a sequence, and a pattern of residues at a fixed position relative to some other position. Table 3 gives some examples of features extracted from the human growth hormone sequence depicted in Figure 1. I have used an "@" symbol to denote patterns with a position specified.

Table 3. Examples of features extracted from training dataset

Feature	Description
A	The residue Alanine.
C_L	Cysteine, followed some other residue, followed by Leucine.
L@-10	Leucine at position –10 relative to some position c.
C_L@-3	Cysteine at position –3 and Leucine at position –1, both relative to some position c.

The following features were extracted from the training set because they resulted in the best accuracies during informal testing: all of the features comprising single residues, without any limits on the distance of the residue from the cleavage point (e.g. see the first and third rows of Table 3); and all the diresidue sequences separated by exactly one position (e.g. see second and fourth rows of Table 3). However, only the position-specific diresidue sequences (i.e. those with an "@" symbol) starting at – 3 were extracted. The reasoning for this is that such an approach makes the standard simplifying naïve Bayes assumption (i.e. that the occurrence of a residue at a particular position relative to the cleave point is independent of the residues at other positions given the cleave point). However, this does not hold for (-3,-1), which are considered non-independent. By having a specific feature for the diresidue pattern at (-3,-1), the system can therefore effectively model the (-3,-1) rule mentioned earlier.

Now, for every feature, a probability is calculated. Suppose f is a single residue or pattern of residues without a specific position, and $f@p$ is the same pattern with a specific relative position. The probability *P(f)* is defined as the prior probability of *f@p*, and is determined by calculating the total fraction of occurrences of *f* in the training set, in both signal and non-signal portions of the sequences. For example, if *f=A*, then *P(f)* is simply the total fraction of residues in the training set that is Alanine.

For each feature, a conditional probability is also calculated. Let *cleave(c)* denote the proposition that position *c* on the sequence is the cleavage point. *P(f@p | cleave(c))* is defined as the fraction of occurrences in the training set of the feature at a particular fixed position relative to the known cleavage point.

For example, from the dataset, the prior probability of the single-residue feature *L*, *P(L)*, is 0.127, but *P(L@-1|cleave(0))* = 0.019 and *P(L@-15|cleave(0))* is 0.285. While the priors capture the general abundance of residues in the training data, the conditionals capture the distribution of residues across positions relative to the cleavage point. I also compute conditional probabilities for the patterns occurring at positions (-3,-1), as mentioned above.

It is now time to explain how the posterior probabilities used for prediction are computed. Essentially, this is an application of Bayes' Theorem. Equation (1) shows how the priors and conditionals are combined to compute the overall probability of a cleave at some position *c*. *F* is defined as the set of all features on a particular sequence with positions relative to some position *c*. The training model consists of a posterior probability for every feature present in the training data.

$$P(cleave(c) | F) \propto \prod_{f@p \in F} \frac{P(f @ p | cleave(c))}{P(f)} \qquad (1)$$

We now come to the prediction algorithm. Given a test sequence with an unknown cleavage point, the system predicts a score for every position *c* on the test sequence. The score is the posterior probability as defined in Equation (1) above. When every position is scored, the posteriors are normalised and the position with the highest posterior probability is the predicted cleavage point. Figure 3 depicts the output of the system when tested on the sequence for human growth hormone depicted in Figure 1 after training on the entire dataset minus the human growth hormone sequence. The predicted probability of a cleave at the actual cleavage site is 0.87.

```
MATGSRTSLLLAFGLLCLPWLQEGSAFPTIPLSRLFDNAMLRAHRLHQLAFDTYQE
SSSSSSSSSSSSSSSSSSSSSSSSSSSSCMMMMMMMMMMMMMMMMMMMMMMMMMMM
...
W   S        0.00277371
L   S        0.000196609
Q   S        0.00609159
E   S        0.00524503
G   S        0.0381914
S   S        0.0577272
A   S        0.0125238
F   C        0.874029           ************
P   M        0.000244162
T   M        0.0015616
...
```

Fig. 3. Normalised predictions for human growth hormone. Only residues with a non-negligible probability of being the cleavage point are shown

4 Results

I evaluated the method described in the previous section using Leaving One Out Cross Validation (LOOCV) on the SignalP version 2 dataset (the dataset for SignalP3 is different and currently unavailable). LOOCV was applied to the entire dataset, as well as the same three subsets that SignalP was trained on, namely the Eukaryote, Gram positive Prokaryotes, and Gram negative Prokaryotes subsets.

4.1 Accuracy

Compared to computationally more expensive methods such as neural networks, this approach results in comparable testing accuracy. Table 4 compares the accuracies achieved by SignalP version 2 and this method, both of which were trained on the same dataset. A comparison with other versions of SignalP is not as useful because of the different datasets being used.

Table 4. Comparison of SignalP2 and the Bayesian method described in this paper

	Eukaryotyes	Gram-	Gram+
SignalP2	72.4	83.4	67.4
Bayesian	69.2	81.5	66.5

As can be observed, the Bayesian method is consistently 1-2% less accurate than SignalP2. However, such a slight difference is likely to be a reflection of the statistical variation arising from Neilson & Krogh's [7] use of the less-rigorous five-fold cross validation for testing. In contrast, the Bayesian method utilised the more reliable LOOCV method. The difference may also reflect the independence assumption made about all positions except (-3,-1): it is possible that including additional diresidue features could further increase accuracy. (Interestingly, treating positions (-3, -1) as non-independent contributes to a large proportion of the accuracy. If this feature is not extracted, and instead only two independent features for positions −3 and −1 are used, then the accuracy is reduced by about 25%.)

I also tested the predictive performance of the Bayesian approach when trained on the entire SignalP2 dataset without subdivision. Again, LOOCV was applied. The accuracy for this experiment was 71.2%, which compares favourably with Vert's SVM approach [11] that achieved 68% accuracy, albeit on the (mostly similar) SignalP1 dataset.

Aside from raw accuracy, one can also consider how close erroneous predictions are from the actual predictions. In Figure 4, the distribution of predicted cleavage sites against proximity to the real cleavage site are depicted following LOOCV on the entire SignalP2 dataset. The diagram clearly shows that the majority of predictions (91.4%) lie within −5 and +5 of the actual cleavage site even though the raw accuracy is 71.2%. It is quite possible that many of these predictions are correct, but have been misclassified by the experimental biologist, as suggested by Hiller et al. [4].

Finally, it is necessary to comment on the relationship between the value of the posterior probability and the confidence of the prediction. In other words, is the posterior probability calculated a good indicator of the reliability of the prediction? I performed an analysis on the results of the LOOCV experiment applied to the entire dataset, and found a positive correlation between posterior probability and true positive rate. The result of this analysis is depicted graphically in Figure 5.

Clearly, predictions with a high posterior probability are to be considered more confident than predictions with a low posterior probability. For example, where the best predicted cleavage point has a probability of only 0.5 or above, the true positive rate was only 75%. However, for predictions with a posterior of 0.95 and above, the true positive rate is between 85% and 90% - quite a significant increase.

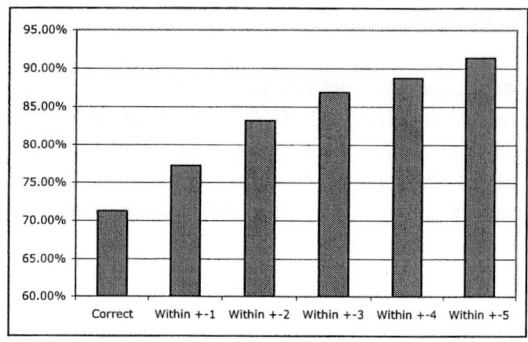

Fig. 4. Accuracy of prediction vs. percentage frequency after LOOCV on entire dataset

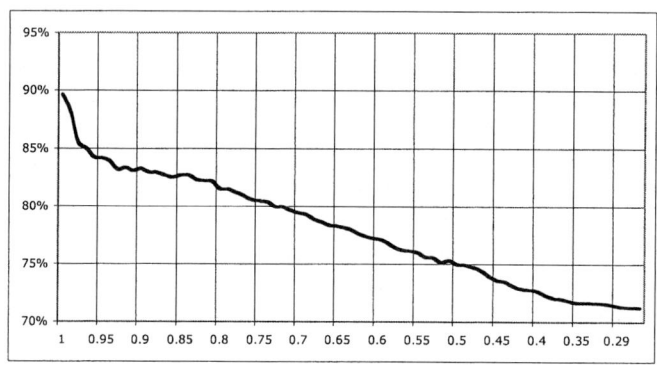

Fig. 5. Posterior probability of predicted cleavage site vs. true positive percentage

4.2 Explanations

The Bayesian method has one significant advantage over neural network approaches: namely, the ability to extract the reason for the system making a particular prediction. Since the overall posterior probability of a cleave is simply the product of the individual posteriors of a cleave given a single feature, it is possible to rank the features in a test sequence by how much they contribute to the final prediction. In Figure 6, the features contributing to the human growth hormone prediction shown in Figure 3 are listed in decreasing order of individual posterior.

It can be seen that the biggest contributor to the prediction is the presence if Ala at position −1. The pattern of Gly and Ala at positions (−3, −1) is the second largest predictor, and this is followed by the occurrence of Leu at multiple positions from –6 to –19, which is where the hydrophobic region is expected to be. Pro at positions 1 and 4 also has a high posterior.

```
A@-1 (5.3), GA@-3,-1 (3.7), L@-12 (2.8), P@1 (2.8), L@-11 (2.6),
L@-9 (2.6), L@-16 (2.1), L@-17 (1.6), L@-6 (1.6), P@4 (1.5),
F@-14 (1.4), Q@-5 (1.4), T@2 (1.4), L@-18 (1.4), W@-7 (1.3),
A@-15 (1.3), S@-2 (1.2), S@-19 (1.1), R@7 (0.9), S@6 (0.9),
C@-10 (0.9), T@-20 (0.9), G@-3 (0.8), F@9 (0.8),I@3 (0.7),
G@-13 (0.6), F@0 (0.6), E@-4 (0.5), L@8 (0.5), L@5 (0.4), P@-8 (0.2)
```

Fig. 6. Features and their posteriors ranked from most significant to least significant, for the human growth hormone prediction

5 Conclusion

To conclude, an efficient and effective method of predicting signal peptide cleavage points along protein sequences has been presented. I have shown that computationally more expensive approaches are not necessarily better in terms of accuracy than simpler Bayesian approaches, and the Bayesian approach described here can offer some degree of explanations for its predictions. Some of the issues involved in applying data mining techniques to biological datasets (such as dealing with variable length sequences) have also been explored.

References

1. Bendtsen J., Neilson H., von Heijne G., Brunak S.: Improved Prediction of Signal Peptides – SignalP 3.0. Journal of Molecular Biology **340** (2004), 783-795.
2. Boekmann B. et al.: The SWISS-PROT Protein Knowledgebase and its supplement TrEMBL in 2003. Nucleic Acids Research **31**(1) (2003) 365-370.
3. Chou K.: Prediction of Protein Signal Sequences. Current Protein and Peptide Science **3** (2003), 615-622.
4. Hiller K., Grote A., Scheer M., Munch R., Jahn D.: PrediSi: Prediction of Signal Peptides and their Cleavage Positions. Nucleic Acids Res. **1** (2004) W375-W379.
5. Hua S., Sun Z.: Support Vector Machine Approach for Protein Subcellular Localization Prediction. Bioinformatics **17**(8) (2001), 721-728.
6. Maetschke S., Towsey M., Boden M: BLOMAP: An Encoding of Amino Acids which Improves Signal Peptide Cleavage Site Prediction. In Chen Y., Wong L: Proc. 3rd Asia-Pacific Bioinformatics Conference, Imperial College Press (2005).
7. Neilson H., Krogh A.: Prediction of Signal Peptides and Signal Anchors by a Hidden Markov Model. In: Glasgow J et al.: Proc Sixth Int. Conf. on Intelligent Systems for Molecular Biology. AAAI Press (1998), 122-130.
8. Nielson H., Englebrecht J., Brunak S., von Heijne G.: Identification of Prokaryotic and Eukaryotic Signal Peptides and Prediction of their Cleavage Sites. Protein Engineering **10**(1) (1997), 1-6.

9. Reczko M., Fiziev P., Staub E., Hatzigeorgiou A: Finding Signal Peptides in Human Protein Sequences using Recurrent Neural Networks. In Guigo R., Gusfield D.: Algorithms in Bioinformatics, Proceedings of the 2nd Int. Workshop WABI 2002, Rome, Italy, Lecture Notes in Computer Science, Springer, **2452** (2002), 60-67.
10. Smith T: A Text-Classification Approach to the Prediction and Characterization of Signal Peptides. In: Informatica 2004: World Congress on Bioinformatics, Havana, Cuba (2004).
11. Vert J.: Support Vector Machine Prediction of Signal Peptide Cleavage Sites Using a New Class of Kernals for Strings. In: Proc. Pacific Sym. on Biocomputing (2002), 649-660.
12. von Heijne G.: Life and Death of a Signal Peptide. Nature **396** (1998) 111-113.
13. von Heijne, G: A New Method for Predicting Signal Sequence Cleavage Sites. Nucleic Acids Research **14**(11) (1986), 4683-4690.
14. Witten I., Frank E.: Data Mining: Practical Machine Learning Tools and Techniques with Java Implementations. Morgan Kauffman (1999).

A Novel Indexing Method for Efficient Sequence Matching in Large DNA Database Environment

Jung-Im Won[1], Jee-Hee Yoon[2], Sanghyun Park[1], and Sang-Wook Kim[3]

[1] Department of Computer Science,
Yonsei University, Korea
{jiwon, sanghyun}@cs.yonsei.ac.kr
[2] Division of Information Engineering and Telecommunications,
Hallym University, Korea
jhyoon@hallym.ac.kr
[3] College of Information and Communications,
Hanyang University, Korea
wook@hanyang.ac.kr

Abstract. In molecular biology, DNA sequence matching is one of the most crucial operations. Since DNA databases contain a huge volume of sequences, fast indexes are essential for efficient processing of DNA sequence matching. In this paper, we first point out the problems of the suffix tree, an index structure widely-used for DNA sequence matching, in the respects of the storage overhead, search performance, and difficulty in seamless integration with DBMS. Then, we propose a new index structure that resolves such problems. The proposed index structure consists of the two parts: the primary part realizes the *trie* as binary bit-string representation without any pointers, and the secondary part helps fast accesses of leaf nodes of the trie that need to be accessed for post-processing. We also suggest efficient algorithms based on that index for DNA sequence matching. To verify the superiority of the proposed approach, we conduct performance evaluation via a series of experiments. The results reveal that the proposed approach, which requires smaller storage space, can be a few orders of magnitude faster than the suffix tree.

Keywords: DNA databases, DNA sequence matching, indexing.

1 Introduction

DNA sequences hold the code that determines life characteristics of every living organism. A DNA sequence is represented as a string of a four-character alphabet of A, C, G, and T known as the nucleotide bases. The DNA database contains a huge volume of DNA sequences. Historically, the database has roughly doubled in size every 14 months, and the increasing rate is growing gradually [3]. Since the size of DNA databases increases considerably as such, fast indexing is crucial for an efficient information retrieval from those databases. DNA subsequence matching is an operation that is most frequently performed on a DNA database [7][20]. Given a database S, a query sequence Q, and a tolerance T, it finds subsequences S' of S whose dissimilarity with some subsequences Q' of Q is less than T.

BLAST [1] is a de-facto standard tool widely used by molecular biologists to perform DNA subsequence matching. BLAST provides high performance by using a heuristic algorithm, however, does not guarantee accuracy; i.e, it may loose some true answers. The most popular algorithm that guarantees accuracy is the Smith-Waterman algorithm [16]. The Smith-Waterman algorithm uses a dynamic programming approach for finding an optimal local alignment between S and Q of the two sequences. However, it suffers from a long processing time of $O(|Q| \times |S|)$.

The suffix tree has been known to be a good index structure for efficient DNA subsequence matching [5][11]. The suffix tree is a compressed digital trie whose set of keywords comprises the suffixes of given sequences. The suffix tree shows reasonable performance in finding all the matched subsequences. Moreover, it is ready to be applied to applications that necessitate DNA subsequence matching since approximate matching algorithms for it have already been proposed [18][8]. The elapsed time of subsequence matching by using such algorithms, however, increases dramatically as the length of a query sequence and a tolerance increase. To alleviate this problem, reference [13] proposed a hybrid indexing method that divides a query sequence into multiple smaller pieces, performs their subsequence matchings with a smaller tolerance, and then integrates the results thus obtained. Also, reference [12] suggested a method that applies the best-first(A*) search method [9] in traversing a suffix tree. It shows the performance of subsequence matching comparable to that of BLAST in case of short query sequences. Moreover, it guarantees accuracy as in the Smith-Waterman algorithm.

The suffix tree still has the following drawbacks due to its structural characteristics: (1) **Storage space:** The suffix tree requires a large storage space; It is often several ten times larger than a database [10][13][6]. Hunt et al. [8] reported that a suffix tree required 19G bytes when they built it on DNA sequences of 286M bases. (2) **Search performance:** The large storage space required by a suffix tree inversely affects the search performance. In addition, the poor locality of the suffix tree causes a significant loss of efficiency in respect of disk accesses [6]. Thus, overall search performance deteriorates in DNA databases. (3) **Integration with DBMS:** DBMS uses a page as a unit for storing all kinds of data on disk. In contrast, the suffix tree has a difficulty in employing a page as a storage unit due to its structural characteristics [17][19]. Thus, the suffix tree has a problem in integrating itself with DBMS seamlessly.

In this paper, we propose a novel index structure that supports DNA subsequence matching efficiently as well as resolves the above drawbacks of the suffix tree. The proposed index adopts a trie [17] as its conceptual structure and realizes the trie by binary bit-string representation without pointers. In addition, it employs a multi-dimensional index as a secondary structure for fast accesses of the target leaf nodes when traversing the trie. With these characteristics, the proposed index successfully solves all the problems in the respects of the storage space, search performance, and integration with DBMS. We also propose algorithms that effectively process both exact and approximate DNA subsequence matching by using the proposed index. Through extensive experiments, we quantitatively verify the effectiveness of our approach in comparison with the previous ones. The results reveal that, compared with the previous ones, our approach requires smaller storage space and achieves several times to several ten times improvement in DNA subsequence matching performance.

2 Indexing Method

2.1 Binary Suffix Trie

A trie is defined as a $|\sum|$-ary tree in which each edge has a symbol from the alphabet $\sum$ and symbols in each root-to-leaf path form a key. Here, $|\sum|$ is the alphabet size. A selection of subtries at level i is determined only by the ith symbol of the search key, not the whole key. The most straightforward implementation of $|\sum|$-ary tries is to store $|\sum|$ pointers in each node. This method enables to select a child node in constant time. However, it is not space-efficient because trie nodes may contain lots of NULL pointers when $|\sum|$ is large. An alternative is to use dynamic data structures such as linked lists. In the linked list representation, each trie node stores two pointers, one to its leftmost right sibling and one to its leftmost child. This implementation reduces a lot of NULL pointers and therefore requires lesser storage space especially when $|\sum|$ is large. However, it cannot select a child node in constant time. In the worst case, all the child nodes have to be examined.

Shang et al. [15] suggested pointerless binary tries which attained competitive search speed with a minimal storage requirement. Pointerless binary tries require the alphabet $\sum$ to have only two symbols, 0 and 1. Therefore, every node has at most two outgoing edges. In the pointerless binary bit-string representation, the symbols on the edges do not have to be stored explicitly by enforcing the following rules: (1) the outgoing edge labeled with 0 connects to the left child node, and (2) the outgoing edge labeled with 1 connects to the right child node. More specifically, the trie node storing the two-bit data '10' has only one child which is on its left, and the node storing the two-bit data '01' has only one child which is on its right. Similarly, the trie node with '11' has both left child and right child, and the node with '00' has no child.

In this paper, we propose an index structure for efficient DNA sequence matching, exploiting the basic concepts of pointerless binary tries. Our aim is to efficiently find the subsequences matched exactly or approximately to a query sequence. Therefore, we extract all the *suffixes* from the DNA sequences and insert each one of them into the trie. Since the suffixes are the inputs to the trie construction algorithm, the resultant tree has the properties of suffix tries [17]. Suffix tries compress the input data set substantially when the input sequences have lots of common prefixes. A DNA sequence can be considered as a string from the alphabet $\sum = \{A, C, G, T\}$. Since the alphabet size is small (which

Symbol	Binary Code
$	000
A	001
C	010
G	011
N	100
T	101
S	110
Y	111

Suffix	Binary Representation
S1: ACGT$	001010011101000
CGT$	010011101000
GT$	011101000
T$	101000
S2: ACT$	001010101000
CT$	010101000
T$	101000

Fig. 1. Binary code of each symbol in the alphabet

Fig. 2. Binary representations of the suffixes from $S_1 =$ 'ACGT' and $S_2 =$ 'ACT'

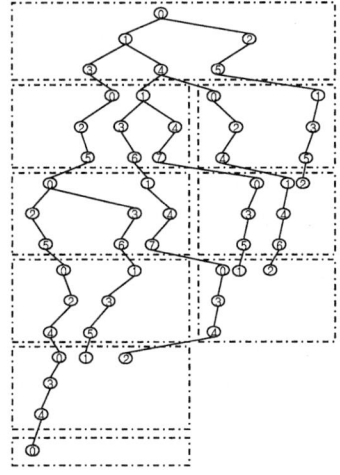

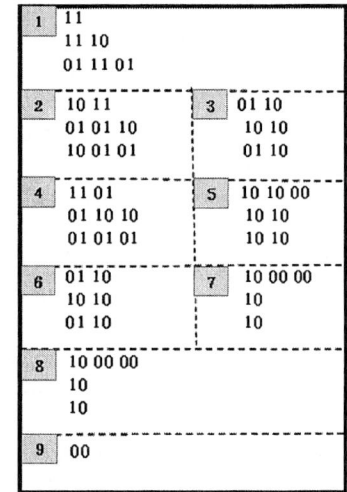

Fig. 3. Binary suffix trie constructed from the suffixes of Figure 2

Fig. 4. Internal representations of the binary suffix trie in Figure 3

is 4), it is highly possible that there exist a considerable number of common prefixes in the suffixes of the input data set.

In this research, we use the minimum number of bits to represent each symbol rather than using a character of 8 bits, to obtain higher compression ratio. Note that DNA sequences may contain wild-card characters as well as the four typical symbols of A, C, G, and T. For example, the wild-card N denotes one from A, C, G, and T, and B denotes one from C, G, and T. Although wild-card characters do not occur frequently, we need to uniquely encode each wild-card character in addition to the typical four characters. For instance, when the number of disparate symbols occurring in the DNA sequences to be indexed is at most seven, we can use 3 bits to encode each symbol uniquely. If we construct the suffix trie from DNA sequences encoded binary, we can expect a higher compression ratio due to the increased number of common prefixes.

Let us examine the steps to build a binary suffix trie using an example. Figure 1 shows a binary code of each symbol in our alphabet. Here, '$' is a special character used as an end marker of every suffix. Given two sequences $S_1 = $ 'ACGT' and $S_2 = $ 'ACT', we first convert all of their suffixes into the corresponding binary bit-string representations as shown in Figure 2. We then construct the trie through successive insertions of binary suffixes according to their lexicographic order. Insertions based on the lexicographic order make the trie grow only one direction and thus facilitate the disk-based trie construction. Figure 3 shows the binary suffix trie constructed from the suffixes of Figure 2, and Figure 4 shows its internal representation.

For the trie construction, we use a disk-based algorithm [4]. Therefore, whenever the main memory space of a predetermined size (i.e. page size) becomes occupied by a sub-trie, it is written onto a secondary storage (i.e. disk). To prevent a sub-trie larger than a page from being written onto a disk page, we precalcuate the maximum number of trie levels and the maximum number of trie nodes that can be stored within a single page. In each page, the child nodes of each level are either entirely on or entirely off that page. In other words, edges can only cross the horizontal boundaries of pages, not the vertical boundaries. This restriction is to reduce the number of disk pages to be read during query processing. Since the trie is partitioned into a set of pages, it is necessary to maintain the *page table* [15] to figure out the page connections. Each entry of the page table corresponds to a page and stores information related to that page, and each entry is filled right after the corresponding page has been written on the disk.

2.2 Storing Leaf Nodes

Each suffix is identified by the pair of the sequence identifier and the starting offset. When a suffix is inserted into the trie, its identifier is stored in the corresponding leaf node. However, every trie node is represented by a two-bit number in our indexing scheme. Therefore, suffix identifiers have to be kept separately from the trie, using, i.e., a leaf node table.

When a query sequence is given, we traverse down the trie to find a node beyond which more comparisons are meaningless. When the matching is successful, a series of labels on the path between the root node and the node visited last becomes the subsequence we are looking for in the database. To find the locations at which the subsequences matched to a query sequence start, we need to retrieve all the leaf nodes under the node visited last and get the suffix identifiers stored in these leaf nodes. When the index is large and the traversal ends at a position not deep, a large portion of the trie has to be visited.

In this work, we propose to use a multi-dimensional index to speed up the operation that retrieves all the leaf nodes under a given internal node. By regarding a binary bit-string representation of a suffix as a multi-dimensional key, we build a multi-dimensional index from a set of suffixes. Notice that suffixes do not have the same length. Therefore, we need the following scheme to convert a suffix of variable length into a set of predetermined k integers: (1) **When the binary bit-string representation of a suffix is shorter than k-integer length**, we append multiple 0s to the end of a binary bit-string to make it be of k-integer length. (2) **When the binary bit-string representation of a**

suffix is longer than k-integer length, we cut out the rightmost bits so that the resultant binary bit-string becomes of k-integer length.

3 Query Processing Method

3.1 Exact Subsequence Matching

Since each trie node is represented by a two-bit number in the proposed index, the pointers from parents to children are not stored explicitly. The information on the trie levels is not stored explicitly, either. Therefore, while traversing down the index to find the subsequences matched to a query sequence, the algorithm has to fetch the corresponding page and then extract those *implicit* information using the data in the page.

Algorithm 1. Query processing algorithm Search-Trie

Input : binary suffix trie T, query sequence Q, page table P
Output: set of answers

1 initialize C_0, $N_{0,c}$, S_0, and $N_{0,f}$;
2 **for** $j := 0;\ j < p\_Height;\ j{++}$ **do**
3   **if** $j > 0$ **then**
4   page_change(P);
5   reset C_0, $N_{0,c}$, S_0, and $N_{0,f}$;
6   **for** $i := 0;\ i < n\_Height;\ i{++}$ **do**
7   **while** isBefore($N_{i,c}$) **do**
8   increase C_i;
9   update S_i;
10   **if** !(match(node($N_{i,c}$), Q_i)) **then**
           return {};
11   **if** isLast(Q_i) **then**
           return find_answers();
12   get(Q_{i+1}); increase C_i; update S_i;
13   **while** isBefore($N_{i,f}$) **do**
           update S_i;
14   **if** $i < (n\_Height - 1)$ **then**
           reset C_{i+1}, $N_{i+1,c}$, S_{i+1}, and $N_{i+1,f}$;

The algorithm Search-Trie which traverses the binary suffix trie T to retrieve the subsequences matched to a query sequence is shown in Algorithm 1. We assume that the query sequence Q has been already converted to its binary form. Remember that the information related to the page partitioning is maintained in the page table P. Let L_i denote the ith trie level in the page that is being examined. The algorithm uses the following four variables to figure out the internal structure of the page. The variable S_i stores the total number of nodes located at L_i. If a node at L_i has the value '11', it will increase S_{i+1} by one. On the contrary, if a node at L_i has the value '00', it will decrease S_{i+1} by one. The variable $N_{i,f}$ denotes the position of the rightmost node at L_i. $N_{i+1,f}$ is simply computed by summing $N_{i,f}$ and S_{i+1}. The variable $N_{i,c}$ indicates the position of the node at L_i that should be compared with the ith query bit. The variable C_i stores the total number of 1 bits counted from the leftmost node at L_i to the node positioned at $N_{i,c}$. $N_{i+1,c}$ is obtained by summing $N_{i,f}$ and C_i.

The algorithm Search-Trie operates as follows. We assume that the index has p_Height page levels and each page level has n_Height node levels. First, we initialize all the variables according to the fact that the first node of the first page in the index is the root (line 1). The lines 3-5 in the external for loop (lines 2-14) replace the current page level with the next page level. The function page_change(P) in line 4 computes the location of the next page using the information in the page table P, and reads in the next page. Next, all the variables are updated before entering into the stage of traversing the nodes in the new page. The internal for loop (lines 6-14) is for handling a node level, and it consists of the following four steps. Increasing C_i and updating S_i, the first step (lines 7-9) sequentially reads the nodes positioned before $N_{i,c}$. The second step (lines 10-12) checks whether the node $N_{i,c}$ matches the ith query bit Q_i or not. If not matched, the statement in line 10 is executed. If matched, the algorithm checks if there are more query bits to be examined. If there is no more query bit left, the function find_answers() is called in line 11. The function find_answers() retrieves the suffix identifiers from the leaf nodes under $N_{i,c}$. If there are more query bits to be examined, the statement in line 12 is executed where the next query bit is read and the variables S_i and C_i are updated and increased respectively. While updating the variable S_i, the third step in line 13 sequentially reads the nodes positioned before $N_{i,f}$. The final step in line 14 resets all the variables if there remain more node levels in the current page.

3.2 Direct Access of Leaf Nodes

The algorithm Search-Trie has the step to retrieve all the leaf nodes under the node $N_{i,c}$ at which the last query bit is matched successfully. This operation is mainly performed in the function find_answers(). The multi-dimensional index introduced in Section 2.2 enables direct retrieval of the leaf nodes under $N_{i,c}$. When the path p from the root to $N_{i,c}$ matches the query sequence, we take one of the following three options according to the length of p. (1) **When p has the length shorter than k-integers**: Let p_0 denote the binary bit-string of k-integer length obtained by appending multiple 0s to the end of p. And let p_1 denote the binary bit-string of k-integer length obtained by appending multiple 1s to the end of p. From the multi-dimensional index, we retrieve all the leaf nodes having the values between p_0 and p_1. (2) **When p has the length of k-integers**: From the multi-dimensional index, we retrieve all the leaf nodes having the value p. (3) **When p has the length longer than k-integers**:

Let p_k be the prefix of p with k-integer length. From the multi-dimensional index, we retrieve all the leaf nodes having the value p_k. Then, we perform the post-processing to detect and discard false matches.

3.3 Approximate Subsequence Matching

The basic method for approximate subsequence matching in DNA databases is the dynamic programming (DP) technique. Given two sequences Q and S, the DP technique finds their optimal distance by building a two-dimensional DP table of $|Q|+1$ rows and $|S|+1$ columns. The recurrence relations corresponding to the similarity measure of a target application are used to fill in each cell of the DP table. The edit distance function [8][17] is a popular similarity measure for approximate subsequence matching.

There have been several approaches [18][13][8] which employ the suffix tree as an index to speed up approximate subsequence matching. They traverse the suffix tree in the depth-first order and build-up the DP table between a query sequence and a path from the root node of the suffix tree. The proposed binary suffix trie also can be used as an index structure for approximate subsequence matching. However, since every node is represented by a two-bit number in the binary suffix trie, we need to access more than one node to append a new column to the DP table.

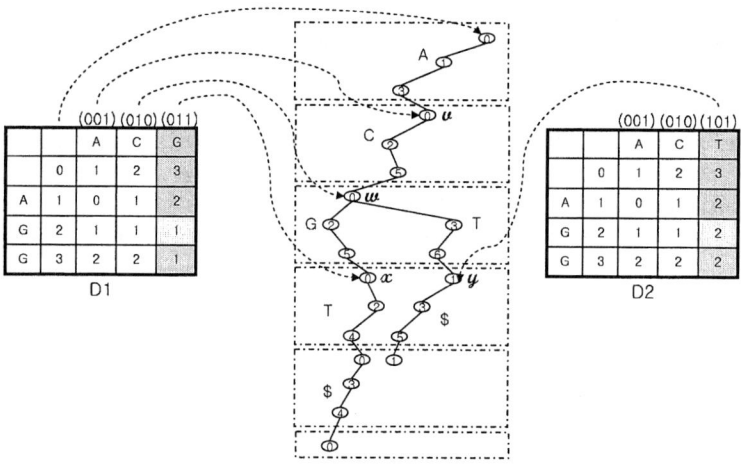

Fig. 5. DP tables constructed from the binary suffix trie of Figure 3

Let us use an example to explain the proposed approximate subsequence matching algorithm. Suppose that we want to find the subsequences whose edit distances to the query sequence 'AGG' are not larger than 1. Figure 5 shows how the DP tables are constructed during the traversal of the binary suffix trie shown in Section 2. Since every symbol is encoded by three bits, the algorithm accesses three successive nodes to append a new column to the existing DP table. That is, the columns for the symbols 'A(001)', 'C(010)', and 'G(011)' are appended individually to the DP table when the algorithm reaches the nodes v, w, and x, respectively. D_1 in Figure 5 is the resultant DP table. Whenever a new column is added to the DP table, we check whether or not the cell at the last row of the newly added column has a value not larger than a distance threshold. If so, all the leaf nodes under the node being visited satisfy the query. We use the multi-dimensional index to directly retrieve such leaf nodes. In D_1 of Figure 5, the column for the symbol 'G(011)' is the newly added column. Since the value of the cell at its last row is 1, all the leaf nodes under the node x satisfy the query. The DP table D_2 is obtained when the node y is visited. Since all the cells in the last column have values larger than 1, the traversal stops at the node y and comes back to its parent. Note that the first two columns of D_1 and D_2 tables are identical. These two columns are shared by the two tables to save space and time.

4 Performance Evaluation

In our experiments, we have used DNA sequences of human chromosomes 18, 19 and 21 downloaded from GenBank [14]. From those data sequences, we have randomly extracted some subsequences of arbitrary lengths as query sequences. The DNA sequences used in our experiments consist of four frequent characters A, C, G, and T, and also contain some infrequent wild-card characters such as N, S, and Y. In addition, we have used a special character $ for representing the end of a sequence. Thus, 8 different characters may appear within the DNA sequences in our experiments. The hardware platform is the Pentium IV 2GHz PC equipped with 1 Gbyte main-memory and 40 Gbyte hard disk. The software platform is the Windows 2000 Server.

In experiment 1, we have compared three approaches Trie-Rtree, Trie-Naive, and Suffix in the respect of the index size. Trie-Rtree represents our approach that employs the trie using pointerless binary bit-string representation in conjunction with a multi-dimensional index. As a multi-dimensional index, we have used the R*-tree [2], a most-widely used in the literature. Trie-Naive also represents our approach that uses just the trie using pointerless binary bit-string representation without employing a multi-dimensional index. Finally, Suffix is the previous approach based on the suffix tree. We have applied an incremental disk-based algorithm [4] for suffix tree construction, and also have allocated 32 byte memory chunk for each node in the suffix tree.

Figure 6 shows the change of the index sizes in the three approaches with different data sizes. We have set the page size for each index to 4K bytes. The suffix tree in Suffix consists of internal nodes and leaf nodes. The index in Trie-Naive consists of a binary suffix trie, a page table, and a leaf node table. The index in Trie-Rtree contains those used in Trie-Naive, and also maintains an additional R*-tree for fast accesses of leaf nodes of the trie. In the figure, we observe that the index size increases almost linearly in proportion to the data size in all the

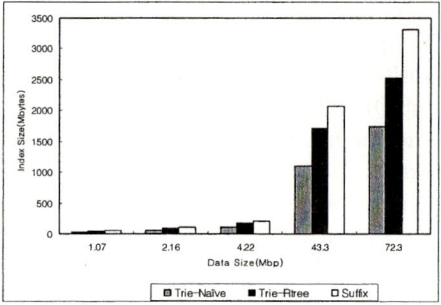 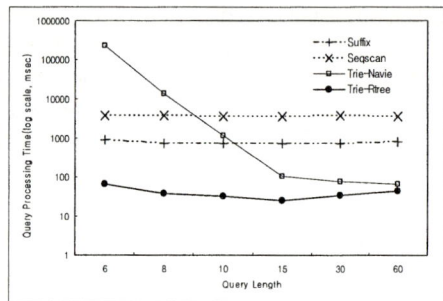

Fig. 6. Index sizes with different data sizes

Fig. 7. Elapsed times of exact subsequence matching with different query sequence lengths

approaches. The results show that our Trie-Naive and Trie-Rtree achieve around 48% and 24% savings in storage space, respectively, in comparisons with Suffix.

In experiment 2, we have compared the three approaches along with Seqscan in terms of the elapsed time for exact subsequence matching. Seqscan is the simplest baseline method for DNA sequence matching, which is based on sequential scan. For this experiment, we have used human chromosome 21 of 43.3 Mbp as a data sequence. The total elapsed time is the time spent in finding the offsets in the DNA sequence from which subsequences exactly matched to a query sequence start.

Figure 7 shows the result. Seqscan performs poorly regardless of query sequence lengths. Trie-Navie performs well with long query sequences, but performs poorly with short query sequences due to its high overhead for post-processing. On the other hand, Trie-Rtree shows good performance regardless of query sequence lengths, and achieves 13 to 29 times speedup compared with Suffix, 54 to 145 times speedup compared with Seqscan.

In experiment 3, we have compared two approaches Trie-Rtree and Suffix in terms of the elapsed time for approximate subsequence matching. We have employed two different approaches: one is to find all the subsequences whose edit distances to a query sequence are not larger than k, which has been commonly used in many DNA subsequence matching, and the other is to find the similar subsequences using the best-first(A*) search algorithm. The data sequence used in the experiment is human chromosome 21 of 43.3Mbp.

Figure 8 shows the elapsed times of approximate subsequence matching by Suffix and Trie-Rtree for finding all the subsequences whose edit distances to a query sequence are not larger than 1. In the current experiment, we follow the behavior of reference [13], considering only short query sequences with a small tolerance. The elapsed time here is the total time required for obtaining pairs <sequence number, offset> of all the similar subsequences. The values within parentheses represent the post-processing time spent in finding leaf nodes. The result shows that Suffix has a large elapsed time for short query sequences due to a big post-processing time. On the other hand, Trie-Rtree shows better

Query Length	Total hits	Query Processing Time(msec)	
		Trie-Rtree	Suffix
6	388,321	817.4(623)	14,248.5(7446)
8	33,422	854(412.7)	3,120.2(674.9)
10	3,857	1,157.5(365)	3,055.6(109)
15	22	1,216.9(3.1)	3,456.8(0.4)

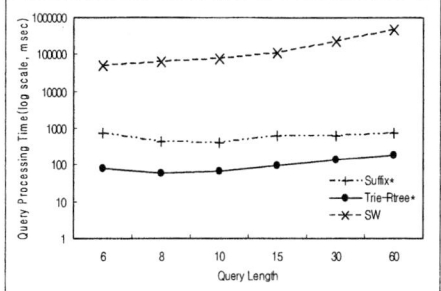

Fig. 8. Elapsed times spent in finding all the subsequences whose edit distances to a query sequence are not larger than 1

Fig. 9. Elapsed times spent in finding the subsequence most similar to a query sequence

performance due to direct accesses of leaf nodes by using the R*-tree. For long query sequences, however, a large number of bit operations increase the time for traversing the suffix trie, and subsequently enlarge the entire elapsed time.

Figure 9 depicts the result of comparing the elapsed times of Suffix*, Trie-Rtree* and SW. Here, the elapsed time is the total time required to find a set of subsequences, each of which is most similar to a query sequence in each data sequence, from a DNA database. Trie-Rtree* and Suffix* represent the elapsed time of approximate subsequence matching by Trie-Rtree and Suffix, respectively, that employ the best-first(A*) search algorithm [12]. Also, SW represents elapsed time of approximate subsequence matching by the Smith-Waterman algorithm. The result shows that Trie-Rtree* performs better than Suffix*. This is because the way for storing nodes in the suffix trie harmonizes with the level-first traversal fashion of the best-first(A*) search algorithm. That is, as mentioned in Section 2.1, all the child nodes of each level of a page are either entirely on or entirely off that page. This is quite effective in such environment where all the sibling nodes are accessed together as in the best-first(A*) search. The result shows that, compared with Suffix* and SW, Trie-Rtree* performs about 4 to 9 times and about 592 to 2,505 times better, respectively.

5 Conclusions

In this paper, we first have pointed out the problems occurring in the suffix tree for DNA sequence matching: (1) high storage overhead, (2) low search performance, (3) difficulty in seamless integration with DBMS. Then, we have proposed a novel index structure that resolves them. Our index employs a trie as its primary structure and implements it by using binary bit-string representation without pointers. Major advantages of this implementation are to reduce the storage overhead considerably and to build its structure easily in page units. Also, our index employs a multi-dimensional index as a secondary structure for fast accesses of the target leaf nodes after traversing the trie. With the proposed index, we can successfully alleviate the three problems of the suffix tree. We also have proposed the algorithms that process DNA sequence matching effectively based on the proposed index. To verify the effectiveness of our approach, we have performed a series of experiments. The results reveal that the proposed approach, which requires smaller storage space, can be a few orders of magnitude faster than the suffix tree. In case of exact matching, **Trie-Rtree**, our enhanced approach, runs 13 to 29 times faster than the Suffix. In case of approximate matching, it achieves 4 to 9 times speedup over Suffix.

Acknowledgments. This work was supported by the Basic Research Program(Grant R04-2003-000-10048-0) of KOSEF, the ITRC support program (MSRC) of IITA, the Hanyang University(HY-2003-T), and the Korea Research Foundation Grant (KRF-2004-003-D00302).

References

1. S. Altschul, W. Gish, W. Miller, E. Myers, and D. Lipman, "Basic local alignment search tool", *Journal of Molecular Biology*, 215, pp. 403-410, 1990.
2. N. Beckmann, H. Kriegel, R. Schneider, and B. Seeger, "The R*-tree: An efficient and robust access method for points and rectangles", *Proc. ACM SIGMOD International Conference on Management of Data*, pp. 322-331, 1990.
3. D. A. Benson, M. S. Boguski, D. J. Lipman, J. Ostell, and B. F. Quellette, "Genbank", *Nucleic Acids Research*, Vol. 26, No. 1, pp. 1-7, 1998.
4. P. Bieganski, J. Riedl, and J. V. Carlis, "Generalized suffix trees for biological sequence data: applications and implementation", *Proc. Hawaii International Conference on System Sciences*, 1994.
5. A. L. Delcher, S. Kasif, R. D. Fleischmann, J. Peterson, O. White, and S. L. Salzberg, "Alignment of whole genomes", *Nucleic Acids Research*, 27, pp. 2369-2376, 1999.
6. R. Giegerich, S. Kurtz, and J. Stoye, "Efficient Implementation of Lazy Suffix Trees", *Softw. Pract. Exp.*, Vol 33, pp. 1035-1049, 2003.
7. R. S. C. Goble, P. Baker, and Brass, "A Classification of tasks in bioinformatics", *Bioinformatics*, Vol. 17, No. 2, pp. 180-188, 2001.
8. E. Hunt, M. P. Atkinson and R. W. Irving, "Database indexing for large DNA and protein sequence collections", *VLDB Journal*, Vol. 11, No. 3, pp. 256-271, 2002.
9. K. Kelly and P. Labute, "The A* Search and Applications to Sequence Alignment", http://www.chemcomp.com/article/astar.htm, 1996.
10. S. Kurtz and C. Schleiermacher, "REPuter: fast computation of maximal repeats in complete genomes", *Bioinformatics*, Vol. 15, No. 5, pp.426-427, 1999.
11. S. Kurtz, J. Choudhuri, E. Ohlebusch, C. Schleiermacher, J. Stoye, and R. Giegerich, "REPuter: the manifold applications of repeat analysis on a genome scale", *Nucleic Acids Research*, Vol. 29, No. 22, pp. 4633-4642, 2001.
12. C. Meek, J. M. Patel, and S. Kasetty, "OASIS: An Online and Accurate Technique for Local-Alignment Searches on Biological sequences", *Proc. VLDB Conference*, pp. 920-921, 2003.
13. G. Navarro and R. Baeza-Yates, "A Hybrid Indexing Method for Approximate String Matching", *Journal of Discrete ALgorithms*, Vol. 1, No. 1, pp.205-239, 2000.

14. http://www.ncbi.nlm.nih.gov
15. H. Shang and T. H. Merrett, "Tries for approximate string matching", *IEEE Trans. on Knowledge and Data Engineering*, Vol. 8, No. 4, pp. 540-547, 1996.
16. T. Smith and M. Waterman, "Identification of Common Molecular Subsequences", *Journal of Molecular Biology* 147, pp. 195-197, 1981.
17. G. A. Stephen, *String Searching Algorithms*, World Scientific Publishing, 1994.
18. E. Ukkonen, "Approximate string matching over suffix trees", *Proc. Combinatorial Pattern Matching*, pp. 228-242, 1993.
19. H. Wang et al., "BLAST++: A Tool for BLASTing Queries in Batches", *Proc. Asia-Pacific Bioinformatics Conference*, pp. 71-79, 2003.
20. H. E. Williams and J. Zobel, "Indexing and Retrieval for Genomic Databases", *IEEE TKDE*, Vol. 14, No. 1. pp. 63-78, 2002.

Threshold Tuning for Improved Classification Association Rule Mining

Frans Coenen[1], Paul Leng[1], and Lu Zhang[2]

[1] Department of Computer Science, The University of Liverpool,
Liverpool L69 3BX, UK
{frans, phl}@csc.liv.ac.uk

[2] Department of Computer Science and Technology,
Peking University Beijing 100871, P.R. China
zhanglu@sei.pku.edu.cn

Abstract. One application of Association Rule Mining (ARM) is to identify Classification Association Rules (CARs) that can be used to classify future instances from the same population as the data being mined. Most CARM methods first mine the data for candidate rules, then prune these using coverage analysis of the training data. In this paper we describe a CARM algorithm that avoids the need for coverage analysis, and a technique for tuning its threshold parameters to obtain more accurate classification. We present results to show this approach can achieve better accuracy than comparable alternatives at lower cost.

1 Introduction

An Association Rule (AR) is a way of describing a relationship that can be observed between database attributes [1], of the form "if the set of attribute-values A is found together in a database record, then it is likely that the set B will be present also". A rule of this form, $A \to B$, is of interest only if it meets at least two threshold requirements: *support* and *confidence*. The support for the rule defines the number of database records within which the association can be observed. The confidence in the rule is the ratio of its support to that of its antecedent. Association Rule Mining (ARM) aims to uncover all such relationships that are present in a database, for specified thresholds of support and confidence.

One application of ARM is to define rules that will *classify* database records. A Classification Association Rule (CAR) is a rule of the form $X \to c$, where X is a set of attribute-values, and c is a class to which database records (instances) can be assigned. Mining of CARs usually proceeds in two steps. First, a training set of database records is mined to find all ARs for which one of the target classes is the consequent, and which satisfy specified thresholds of support and confidence. This stage is essentially similar to ARM in the more general case, with the classes c treated as attribute-values, and the restriction that the only rules we need consider are those for which the consequent is one of these. A second stage then sorts and reduces the set of rules found, with the aim of producing a consistent set that will enable efficient and reliable classification of future instances.

The CBA algorithm described in [6] exemplifies the approach. First, a version of the well-known Apriori algorithm [2] is used to generate a set of *ruleitems* that satisfy a required support threshold, where a ruleitem is a set of items (attribute-values) associated with a class label, which thus defines a potential CAR. The rules thus generated are pruned, using the calculated confidence to eliminate those that fail to meet a required confidence threshold or which conflict with higher-confidence rules. Finally, a classifier is built by selecting an ordered subset of the remaining CARs. This process involves *coverage analysis* in which each candidate CAR is examined in turn, to find a set of CARs that cover the dataset fully. The CMAR algorithm of [7] has a similar general form, using a version of the FP-growth algorithm [5] to generate the candidate CARs.

Results presented in [6] and[7] show that classification using CARs seems to offer greater accuracy, in many cases, than other methods such as C4.5 [8]. The problem with both CBA and CMAR, however, is that the cost of the coverage analysis is essentially a product of the size of the dataset and the number of candidate CARs being considered. The CPAR algorithm [10] has a different approach to generating rules, using a procedure derived from the FOIL algorithm [9] rather than a classical ARM method. Although this improves performance by generating a smaller set of rules, the performance is still $O(nmr)$, where n is the number of records, m the number of items, and r the number of candidate rules. The RIPPER algorithm [4] incorporates a pruning strategy that can be applied in a manner independent of the rule generation strategy used. All these methods follow an *overfit and prune* strategy that will be costly if used to construct classifiers from the very large and wide datasets that are characteristic of classical ARM.

Thus, we identify three general problems of CAR mining. First, the ARM task is inherently costly because of the exponential complexity of the search space. Second, it is very likely that this first stage will generate a very large number of candidate rules, and so the selection of a suitable subset for classification may also be computationally expensive. Finally, the reliability of the resulting classifier depends in some degree on the rather arbitrary choice of support and confidence thresholds used in the mining process.

In this paper we describe a new approach to the generation of CARs, that significantly reduces the cost of mining the training data by using both support and confidence thresholds in the first stage of mining, to produce a small set of rules without the need for coverage analysis. We present results to show that this approach achieves a classification accuracy that is comparable with other methods, but in many cases at much lower cost. We also describe a strategy for tuning support and confidence thresholds to obtain a best accuracy.

2 Generating Classification Association Rules Using TFPC

We begin with the observation that, if the support and confidence thresholds have been selected correctly, then the existence of a rule $X \rightarrow c1$ should make it unnecessary to consider any other rules whose antecedent is a superset of

X. In practice, however, we may still find a rule $Y \to c2$, say, where Y is a superset of X, which has higher confidence and to which we would wish to give higher precedence. It remains possible, also, that there will be a further rule $Z \to c1$, where Z is a superset of Y, with still higher confidence, and so on. This reasoning leads other methods to a process in which all possible rules are first generated and then evaluated. In this paper we adopt an alternative heuristic: *If we can identify a rule $X \to c$ which meets the required support and confidence thresholds, then it is not necessary to look for other rules whose antecedent is a superset of X and whose consequent is c.* It will still of course be necessary to continue to look for rules that select other classes. This heuristic both reduces the number of candidate rules to be considered, and the risk of overfitting.

We use a method derived from our TFP (Total From Partial) algorithm to generate a set of CARS. This method, described in [3], first builds a set-enumeration tree structure, the *P-tree*, that contains an incomplete summation of support-counts for relevant sets. Using the P-tree, the algorithm uses an Apriori-like procedure to build a second set enumeration tree, the *T-tree*, that finally contains all the frequent sets (i.e. those that meet the required threshold of support), with their support-counts. The T-tree is built level by level, the first level comprising all the single items (attribute-values) under consideration. In the first pass, the support of these items is counted, and any that fail to meet the required support threshold are removed from the tree. Candidate-pairs are then generated from remaining items, and appended as child nodes. The process continues, as with Apriori, until no more candidate sets can be generated.

Figure 1 illustrates the form of a T-tree, for the set of items $\{A, B, C, x, y\}$, where x and y are class identifiers. This tree is complete, i.e it includes all possible itemsets, except for those including both x and y which we will assume cannot occur. In practice, an actual T-tree would include only those nodes representing the frequent sets. For example, if the set $\{A, C\}$ fails to reach the required support threshold, then the node labelled AC would be pruned from the tree, and the nodes ABC, ACx, ACy, $ABCx$ and $ABCy$ would not be created. All the candidate itemsets that include the class-identifier x can be found in the subtree rooted at x, and thus all the rules that classify to x can be derived from this subtree (and likewise for y).

The algorithm used to build the T-tree in Figure 1 is a modification of the original TFP approach. As each pass is concluded, we first remove from the tree all those nodes representing sets that fail to meet the support threshold. The remaining (frequent) sets that are included within the class-identifier subtrees (x and y in this example) define possible classification rules: for example, the set Bx corresponds to a rule $B \to x$. We now calculate the confidence of all such rules, i.e. in the case of Bx, the ratio $(support of Bx)/(support of B)$. If this rule exceeds the required confidence threshold, we add the rule to our target set, and remove the corresponding node (Bx) from the tree. The effect of this is that when the next level is generated, supersets of Bx (ie ABx and BCx) will not be added to the tree.

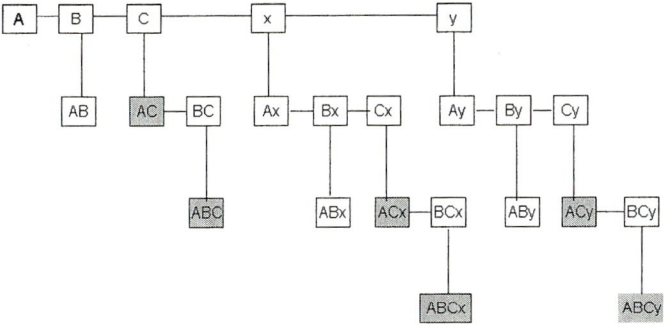

Fig. 1. Form of a T-tree for $\{A, B, C, x, y\}$

This algorithm, which we call TFPC (Total From Partial Classification), generates far fewer frequent sets than would be produced from a generic ARM algorithm such as Apriori or TFP. Moreover, the sets left on the tree can immediately be used to define classification rules which meet our threshold requirements for support and confidence. To classify using these rules, the rule set is sorted by confidence. Between two rules of equal confidence, the more specific takes precedence: i.e. if $B \rightarrow x$, say, and $BD \rightarrow y$, then the latter will be selected.

3 Experimental Results

To investigate the performance of TFPC, we carried out experiments using a number of data sets taken from the UCI Machine Learning Repository. The implementation of TFPC was as a Java program, and for comparison purposes we have used our own (Java) implementations of the published algorithms for CMAR [7] and CPAR [10]. In the first set of experiments, as in [6] and [7], we have assumed a support threshold of 1% and a confidence threshold of 50%. For the implementation of CPAR, we used the same parameters used in [10], i.e. minimum gain threshold = 0.7, total weight threshold = 0.05, decay factor = 2/3, and similarity ratio 1 : 0.99. We tried to use the same data sets as those used to analyse CMAR and CPAR. However in many cases the data sets that were used in [7] and [10] appear to be no longer available, and in others were found not have identical parameters to those reported.

The data sets chosen therefore comprise a subset of those used in [7] and [10], augmented by a further selection from the UCI repository. The choice of additional data sets concentrated on larger/denser data sets (2000+ records) because the majority of the data sets used in the reported analysis of CMAR and CPAR were relatively small (less than 1000 records). The sets chosen were discretized using the LUCS-KDD DN software [1], where appropriate continuous attributes

---

[1] Available at http://www.csc.liv.ac.uk/~frans/KDD/Software/LUCS-KDD-DN/lucs-kdd_DN.html

were ranged using five sub-ranges. The programs were run on a 1.2 GHz Intel Celeron CPU with 512 Mbyte of RAM running under Red Hat Linux 7.3.

The results from these experiments are shown in Table 1. The row labels describe the key characteristics of the data, in the form into which it was discretized. For example, the label **anneal.D106.N798.C6** denotes the 'anneal' data set, which includes 798 records in 6 classes, with attributes that for our experiments have been discretized into 106 binary categories.

The last three columns in Table 1 tabulate the accuracy of the classification obtained by the three methods. In all cases, the figure shown is the average obtained from a *10-fold cross-validation* using the full data set. The accuracy obtained using the TFPC method is in most cases comparable with that obtained from the other methods investigated. Although the average accuracy of the method was slightly lower than the others, in 5 cases TFPC gave an accuracy as high as or higher than both other methods, and only in two cases (ionosphere and wine) was it markedly worse than both.

The significance of these results is that this level of accuracy was obtained from a very efficient rule-generation process. The first three columns in Table 1 show the execution times for the three methods. In each case, the time shown is that obtained for the full experimental evaluation using 10-fold cross-validation. As can be seen, the performance of TFPC (in comparable implementations) is

Table 1. Results for support= 1%, confidence= 50%

Data Set	Time			Number of rules				Accuracy		
	CMAR	CPAR	TFPC	CMAR	CPAR	TFP	TFPC	CMAR	CPAR	TFPC
adult.D131.N48842.C2	2088.6	809	80	3063	183	20530	82	71.2	76.7	76.1
anneal.D106.N798.C6	150.5	1.8	12.1	319	34	20525	61	83.5	90.2	84.6
breast.D48.N699.C2	7.5	0.7	1	191	16	4008	43	85.2	94.8	95.9
connect4.D129.N67557.C3	1449.2	24047	206.4	821	816	3600	106	66.9	54.3	65.9
heart.D53.N303.C5	36.2	1	11.4	305	53	13429	193	56.8	51.1	57.1
hepatitus.D58.N155.C2	160	0.3	15.2	99	14	31126	58	76.3	76.5	77.3
horseColic.D94.D368.C2	40.3	0.6	6.3	387	18	17219	99	74.5	82.3	78.7
ionosphere.D104.N351.C2	28.4	1.1	8.4	214	26	8581	380	96	92.9	83.8
iris.D23.N150.C3	0.2	0.2	0.1	69	11	200	20	94.7	94.7	94.7
led7.D24.N3200.C10	1.6	5.7	0.7	229	31	464	29	73.7	71.2	69
mushroom.D127.N8124.C2	1045.5	15.4	31.6	171	31	25253	116	99.1	98.8	96.1
nursery.D32.N12960.C5	22.2	51.7	5.2	487	84	1709	30	91.4	78.5	77.9
pageBlocks.D55.N5473.C5	25.3	15.5	2.4	243	56	7882	19	89.8	76.2	89.8
penDigits.D90.N10992.C10	222.7	101.9	56.8	2052	167	7667	3015	79.1	83	79.7
pimaIndians.D42.N768.C2	3.2	1	1.3	540	23	3693	34	77.4	75.6	74.9
waveform.D108.N5000.C3	1149	38.1	86.9	1186	114	22876	662	71.2	75.4	71.7
wine.D68.N178.C3	1018.7	0.3	8.7	110	18	31961	164	97.1	92.5	86.3
zoo.D43.N101.C7	780.5	0.2	6	31	19	25050	250	92	96	93
Average	457.2	1394.0	30.0	584.3	95	15361	298	82.0	81.2	80.7

markedly superior to that of CMAR (in all cases) and CPAR (in many). The improvement over CMAR results from the smaller number of rules that tend to be produced using TFPC and because this approach eliminates the expensive coverage analysis carried out in CMAR. The columns headed 'Number of rules' show the average number of rules included in the final classifiers generated, for the three methods. For comparison, the column headed TFP shows the total number of classification rules generated by the TFP algorithm, which defines the total number of candidate classification rules produced by a straightforward ARM algorithm using the given support and confidence thresholds. A comparison of this column with that for TFPC shows the advantage gained during rule generation from the heuristics used in the latter. In most (although not all) cases, TFPC also produces fewer rules than CMAR.

CPAR, conversely, usually produces fewer rules than TFPC, and this is reflected in the faster execution times it achieves in some cases. The advantage of TFPC, however, is that by dispensing with coverage analysis the performance of the method scales better for larger data sets. In all the cases where the data included more than 10000 records (adult, connect4, nursery, pendigits), TFPC is much faster than CPAR, sometimes by an order of magnitude or more. CPAR also performed relatively poorly in the cases (pageblocks, led) in which there were both a moderately large number of records and a relatively large number of classes. In all these cases, TFPC achieves a classification accuracy close to or superior to CPAR at much lower cost.

4 Finding Best Support and Confidence

The results presented above show that it is possible to obtain a set of rules that will in most cases provide acceptable classification accuracy, using CARM techniques, without further coverage analysis. However, because the tree generation heuristic used in TFPC stops looking for more specific rules once a general rule with satisfactory confidence has been found, it may sometimes fail to find high-confidence rules that could be significant in special cases. The method may therefore be more sensitive to the choice of support and confidence thresholds used than is the case for methods that rely on coverage analysis to derive the final ruleset. To investigate this further, we performed a series of experiments to identify the combination of support and confidence threshold that would lead to the highest classification accuracy in each of the data sets studied. Figure 2 shows the results obtained, in the form of 3-D plots, for a number of example data sets selected to demonstrate the variety of the results obtained. For each plot the X and Y axes represent support and confidence thresholds ranging from 100 to 0%, and the Z axis the classification accuracy obtained.

From Figure 2 it can be seen how the classification accuracy produced using TFPC may vary significantly depending on the choice of support and confidence thresholds. The extent of the variability depends on the characteristics of the data. For example, the 'adult' data set (Figure 2(a)) shows a substantial plateau of support-confidence values within which accuracy is constant (although there is

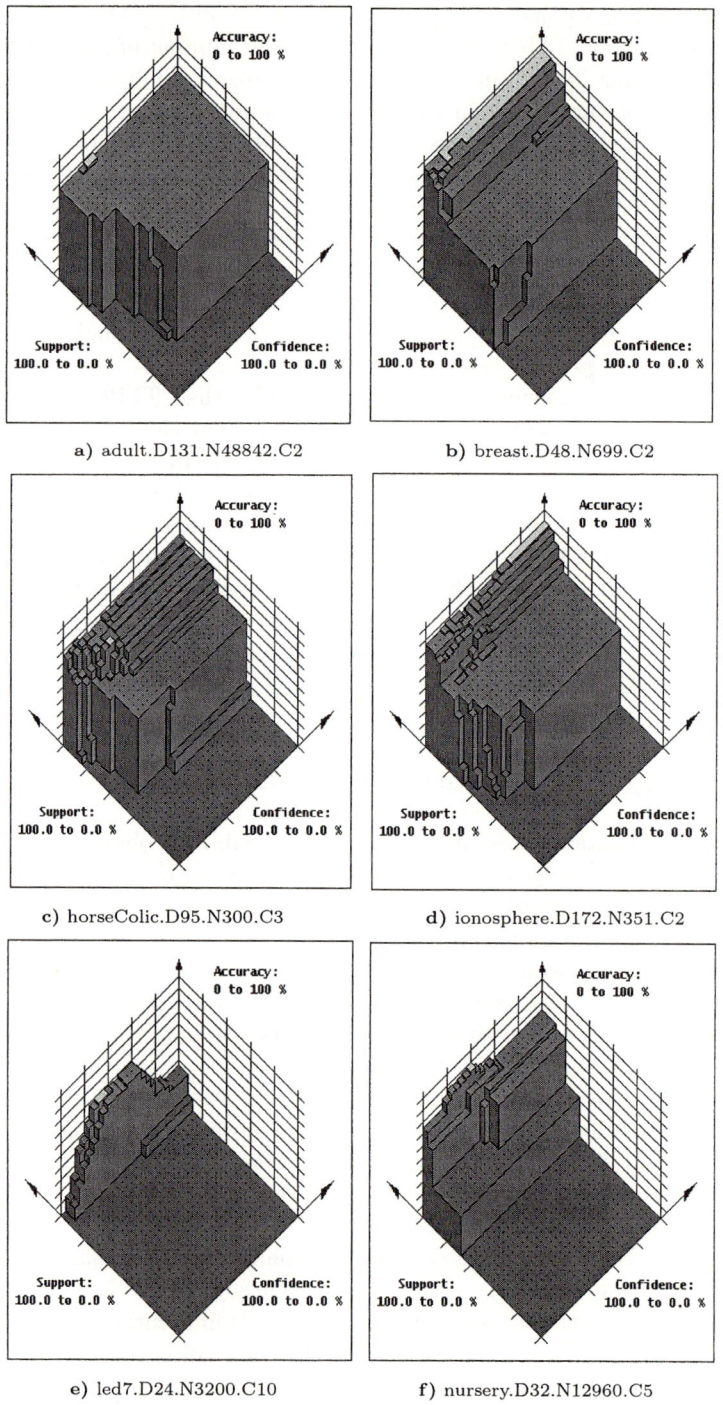

Fig. 2. 3-D plots of classification accuracy v. support/confidence thresholds

a small peak, highlighted in white, at which a higher accuracy can be obtained). Conversely, the 'led7' data set (e) illustrates a case in which the method is highly sensitive to the choice of support and confidence threshold. In this case, the values 1% and 50% used for the experiments of Table 1 were, fortuitously, close to the peak, so reasonably good results were obtained. This was also the case for the 'breast' set (b) and 'horsecolic (c). In the cases of 'ionosphere' (d) and nursery (f) conversely, the relatively poor accuracy obtained using TFPC is a consequence of a poor choice of support and confidence thresholds.

To obtain a best classification the TFPC algorithm was applied in an iterative manner in conjunction with a "hill-climbing" procedure. The hill climbing technique makes use of a 3-D playing area measuring $100x100x100$, as visualised in the plots of Figure 2. The axes of the playing area represent percentage values for support thresholds, confidence thresholds and accuracies. The technique commences with initial support and confidence thresholds, with an associated accuracy, that describes a current location in the playing area. The procedure then traverses the playing area with the aim of maximising the accuracy value. To do this it continuously generates data for a "grid" of eight test locations, obtained by applying to the current location positive and negative increments to the support and confidence thresholds. The current and test locations thus form a $3x3$ location grid centred on the current location. The threshold increments are reduced progressively as the procedure converges.

Table 2 shows the results obtained using TFPC-HC (columns headed THC) for all the data sets included in Table 1. For comparison, the results obtained from TFPC (without hill climbing), and from CMAR and CPAR, with 1% support and 50% confidence, are included.

The results confirm the theory demonstrated by the plots presented in Figure 2. In those cases (e.g adult, breast, led7, horsecolic) where, by chance, the (1%, 50%) pair gives a near-best result, little improvement is obtained from the hill-climbing procedure. In the cases of ionosphere and nursery, conversely, and also anneal, hepatitis and pendigits, a very substantial increase in accuracy is obtained. It is noteworthy, also, that both overall and in many particular cases, TFPC-HC achieves a higher classification accuracy than either CMAR or CPAR.

Although these experiments show the gain that can be obtained by tuning the selection of support and confidence thresholds, the hill-climbing procedure is, of course, more time-consuming than CMAR, CPAR or TFPC without hill climbing. To reduce the cost of this the algorithm was further adapted, TFPC-HC+, so that hill-climbing was only carried out using a single division of the data set into a (90%) training set and (10%) test set. The hill-climbing procedure was used to find the support and confidence thresholds that led to the most accurate classification of this test set. These thresholds were then used to apply the TFPC method to the full data set, with ten cross-validation. The results are included in the columns of Table 2 headed HC+.

As can be seen, this more restricted hill-climbing (essentially, on a sample of the data) is still relatively effective in obtaining a near-optimal support and confidence pairing for use in the method. Overall, the results from this show

Table 2. Comparison of methods of classification

Data Set	Accuracy					Execution times				
	CMAR	CPAR	TFPC	THC	HC+	CMAR	CPAR	TFPC	THC	HC+
adult.D131.N48842.C2	71.2	76.7	76.1	76.9	76.8	2089	809	80	1187	347
Anneal.D106.N798.C6	83.5	90.2	84.6	91.3	90.9	150	2	12	374	52
breast.D48.N699.C2	85.2	94.8	95.9	96.4	96.1	8	1	1	9	2
Connect4.D129.N67557.C3	66.9	54.3	65.9	65.9	65.9	1449	24047	206	11816	906
heart.D53.N303.C5	56.8	51.1	57.1	61.0	57.4	36	1	11	285	43
hepatitus.D58.N155.C2	76.3	76.5	77.3	86,0	77.3	160	1	15	249	24
horseColic.D94.D368.C2	74.5	82.3	78.7	80.9	79.5	40	1	6	186	41
ionosphere.D104.N351.C2	96.0	92.9	83.8	91.5	88.9	28	1	8	674	135
Iris.D23.N150.C3	94.7	94.7	94.7	94.7	94.7	1	1	1	6	1
Led7.D24.N3200.C10	73.7	71.2	69.0	71.3	68.8	2	6	1	7	3
mushroom.D127.N8124.C2	99.1	98.8	96.1	99.3	96.1	1046	15	32	1313	127
Nursery.D32.N12960.C5	91.4	78.5	77.9	92.2	90.5	22	52	5	263	22
pageBlocks.D55.N5473.C5	89.8	76.2	89.8	89.8	89.8	25	16	2	6	3
penDigits.D90.N10992.C10	79.1	83.0	79.7	89.0	88.1	223	102	57	2546	303
pimaIndians.D42.N768.C2	77.4	75.6	74.9	78.9	74.9	3	1	1	28	2
waveform.D108.N5000.C3	71.2	75.4	71.7	75.1	75.6	1149	38	87	5670	611
wine.D68.N178.C3	97.1	92.5	86.3	95.1	92.1	1019	1	9	55	6
Zoo.D43.N101.C7	92.0	96.0	93.0	96.0	93.0	781	1	6	148	14
Average	82.0	81.2	80.7	85.1	83.1	457	1394	30	1379	147

greater accuracy than either CMAR or CPAR, as well as improvements on TFPC with the (1%, 50%) thresholds. The times obtained show that TFPC-HC+ is relatively efficient, especially with respect to large datasets such as the adult and connect4 sets.

5 Conclusions

Previous research has shown that ARM can be an effective route to accurate classification. A drawback of existing methods is that they involve detailed coverage analysis of candidate rules to obtain a final set of classification rules. This procedure is expensive: prohibitively so when very large training sets are involved. We have here introduced an algorithm, TFPC, which obtains a set of classification rules directly from an ARM procedure, without further coverage analysis. Our results show that the classification accuracy obtained is comparable with, or close to, that obtained from established methods. The TFPC method, however, is very much faster than alternatives that involve coverage analysis, especially when dealing with large data sets. We believe this method offers a realistic approach to deriving classifiers from extremely large data sets, for which existing methods would be inapplicable.

The accuracy of classification obtained using TFPC is, however, relatively sensitive to the choice of support and confidence thresholds used when mining the classification rules. We have described a 'hill-climbing' procedure we have used to select the best values for these thresholds. Our results show that, using these optimal values, the accuracy of classification of TFPC is improved very substantially, in most cases improving on the best current methods. The significance of these results is that they demonstrate how performance can be improved by careful selection of these threshold values. Although the hill-climbing process is, in general, too time-consuming to be used routinely for this purpose, we have shown that a more limited version of this (TFPC-HC+) can give results that are almost as good with a speed that, for large data sets, improves on methods requiring coverage analysis.

It may also be the case that the tuning of support and confidence thresholds with respect to accuracy, may improve the performance of other methods. There is scope for further research in this direction, and also to investigate other ways to perform this optimisation efficiently.

References

1. Agrawal, R., Imielinski, T. and Swami, A.: Mining Association Rules between Sets of Items in Large Databases. Proc. ACM SIGMOD Conference on Management of Data, 207-216, 1993.
2. Agrawal, R. and Srikant, R.: Fast Algorithms for Mining Association Rules. Proc. of the 20th VLDB Conference, Santiago, Chile, 487-499, 1994.
3. Coenen, F., Goulbourne, G., and Leng, P.: Computing Association Rules using Partial Totals. PKDD 2001, pages 54-66, 2001.
4. Cohen, W.W.: Fast Effective Rule Induction. Proc. of the 12th Int. Conf. on Machine Learning, pages115-123, 1995.
5. Han, J., Pei, J. and Yin, Y.: Mining Frequent Patterns without Candidate Generation. In Proc. of the ACM SIGMOD Conference on Management of Data, Dallas, pages 1-12, 2000.
6. Liu, B., Hsu, W. and Ma, Y.: Integrating Classification and Association Rule Mining. Proc KDD 1998, 80-86
7. Li, W., Han, J. and Pei, J.: CMAR: Accurate and Efficient Classification Based on Multiple Class-Association Rules. Proc ICDM'01, 2001, 369-376
8. Quinlan, J.R. C4.5: Programs for Machine Learning. Morgan Kaufmann, 1993.
9. Quinlan, J.R. and Cameron-Jones, R.M.: FOIL: A midterm report. Proc ECML 1993, 3-20.
10. Yin, X. and Han, J.: CPAR: Classification Based on Predictive Association Rules. Proc SIAM Int Conf on Data Mining (SDM'03), 2003, 331-335

Using Rough Set in Feature Selection and Reduction in Face Recognition Problem

Le Hoai Bac and Nguyen Anh Tuan

Faculty of Information Technology, University of Natural Sciences, HCM City, Vietnam
{lhbac, natuan}@fit.hcmuns.edu.vn

Abstract. Feature selection and reduction are fundamental steps in pattern recognition problems. The idea of reducts in rough set theory has encouraged many researchers in studying the effectiveness of rough set theory in the problem mentioned above. Through results of experiments in this article, we will show that rough set theory, accompanied by appropriate heuristics, can increase significantly the system's recognition accuracy.

Keywords: rough set, feature selection, feature reduction, face recognition.

1 Introduction

Many researchers have proposed a lot of methods for feature selection. Rough set theory, with its idea of reducts, becomes an attractive and potential approach for this problem. In this article, we construct a recognition model in which rough set based criteria are used to reduce the set of features which are extracted by Principle Component Analysis. The simple nearest – neighbor classifier is used in recognition phase. Finally, we end with some remarks about the recognition results that emphasize the advantages of heuristic factor combined with rough set in the recognition accuracy and rough set based algorithms in choosing significant features.

2 Definitions in Rough Set Theory

In this section, we review some definitions in rough set theory ([1]), which are the basis of the presented reduction algorithms in the next sections.

Information system: An information system is a pair $\mathcal{A} = (U, A)$, in which U, called *universe*, is the non-empty finite set of *objects* concerned, and A is the non-empty finite set of all *attributes* (or *features*) used to evaluate objects such that $a : U \rightarrow V_a, \forall a \in A$ where V_a is the value set of attribute a. An information system is called a *decision system* if $A = C \cup D$, where C is the set of all *conditional attributes* and D is the set of all *decision attributes*.

In the next sections, we let $\mathcal{A} = (U, A)$ denote the information system being concerned.

Discernibility: With any $B \subseteq A$, there is an equivalence relation $IND(B)$ called $B-indiscernibility\ relation$ which is defined as followed:

$$IND(B) = \{(x,x') \in U^2 \mid \forall a \in B, a(x) = a(x')\} \quad (1)$$

Set approximation: Let $B \subseteq A$, $X \subseteq U$. We can approximate X using only the information contained in B by constructing the $B-lower$ and $B-upper$ approximation of X, denoted $\underline{B}X$ and $\overline{B}X$ respectively, and are defined as followed:

$$\underline{B}(X) = \{x \mid [x]_B \subseteq X\} \quad (2)$$

$$\overline{B}(X) = \{x \mid [x]_B \cap X \neq \emptyset\} \quad (3)$$

Positive region: Let $C, D \subseteq A$. The $C-positive$ region of D is defined as followed:

$$POS_C(D) = \bigcup_{X \in U|D} C(X) \quad (4)$$

Relative core: The attribute a is called $Q-dispensible$ in P if $POS_P(Q) = POS_{P-\{a\}}(Q)$, otherwise it is called $Q-indispensible$. The set of all $Q-indispensible$ objects in P is called $Q-$relative core or $Q-$core of P và is denoted $CORE_Q(P)$.

3 Recognition Model Using Rough Set Theory

In our testing model, we use the Principle Components Analysis (PCA) ([5]) to get feature vector for each face image. The features of these vectors will compose the conditional attribute set C, and the feature used to determine to which person a particular face image belongs will compose the decision attribute set D. The set of feature vectors, together with their personal identification feature, then constitutes a decision system $(U, A = C \cup D)$ in which U is the set of all feature vectors of face images, C and D are respectively the conditional attribute set and decision attribute set. In the discretization stage, we use a simple algorithm to divide each feature's value set into several equal ranges, so all real values in the same range will get the same discrete value after the discretization process. After that, reduction algorithms based on rough set theory will be applied to the information system to get the reduced conditional attribute set R ($R \subseteq C$). In the training phase, we use Learning Vector Quantization ([6]) to find reference vectors describing the distributions of feature vectors of face images of each person in the reduced decision system

$(U, A' = R \cup D)$. These reference vectors, associated with the nearest neighbors classifier, will be used to test the performance of the system in the recognition phase. All of the two phases are discribed in Figure 1 and 2.

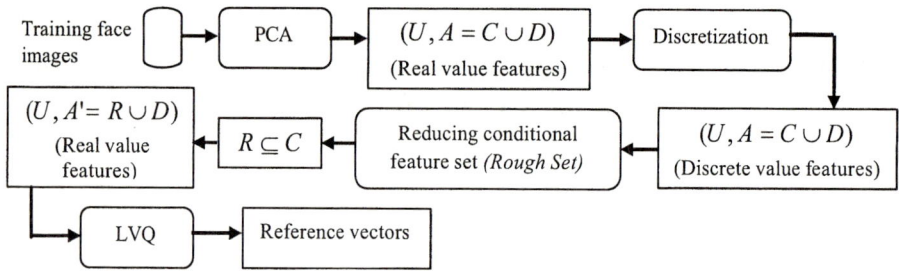

Fig. 1. In the training phase, rough set theory is used to select conditional features. The Learning Vector Quantization will be applied to the reduced decision system to get reference vectors

In the next sections, we introduce three feature set reduction algorithms used in our model : Johnson algorithm ([4]) based on greedy strategy, random algorithm ([4]) and the algorithm that combines rough set theory and heuristic for generating rules ([3]).

3.1 Johnson Algorithm

Let $(U, C \cup D)$ be an information system, in which U is the set of all discrete value feature vectors of face images, C and D respectively are sets of conditional and decision features. Our goal is to find a reduced set $R \subseteq C$ such that: with any given two arbitrary vectors $v_i, v_j \in U$ of two different persons, if they differ from each other in at least one feature in C, so do they in R (i.e., there's at least one feature in R in which v_i and v_j differ from each other).

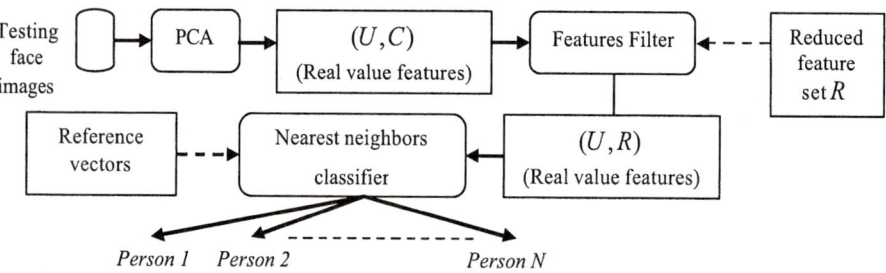

Fig. 2. In the testing phase, Principle Components Analysis will be applied to the testing face images to get information system which is reduced by just remain features in reduced feature set. The nearest neighbors classifier, associated with reference vectors found earlier, is used to classify image into particular person

The Johnson algorithm is based on the greedy strategy in the sense that at each step, we will choose the feature in which there are the most feature vectors differing from each other.

Let $W(B)$ be the number of pairs of feature vectors u, v belonging to two different persons and differing from each other in any features in B. At step i, assume that R_i is the current reduced feature set. The attribute $a_{i+1} \in C \setminus R_i$ will be selected at the next step $i+1$ if:

$$a_{i+1} = \arg\max\{W(R_i \cup \{a\}) \mid a \in C \setminus R_i\} \tag{5}$$

Johnson Algorithm
   Input: Information system $(U, C \cup D)$
   Output: Reduced feature set $R \subseteq C$
      Step 1: $R = \emptyset$
      Step 2: Repeat
         $a_{max} = \arg\max\{W(R \cup \{a\}) \mid a \in C \setminus R\}$
         $R = R \cup \{a_{max}\}$
         <u>Until</u> $(R = C)$ or (For any $u, v \in U$ belonging to two different persons, u and v differ
            from each other in at least one feature in R)

3.2 Random Algorithm

This algorithm use the value $W(B)$ above to evaluate the probability to choose feature at each step. At step $i+1$, the probability to choose the feature $a \in C \setminus R_i$ is:

$$P(a) = \frac{W(R_i \cup \{a\})}{\sum\limits_{a' \in C \setminus R_i} W(R_i \cup \{a'\})}, \quad \forall a \in C \setminus R_i \tag{6}$$

Random algorithm
   Input: Information system $(U, C \cup D)$
   Output: Reduced feature set $R \subseteq C$
      Step 1: $R = \emptyset$
      Step 2: Repeat
         Use $P(a)$, $\forall a \in C \setminus R$ in (6) to choose feature a_{max}.
         $R = R \cup \{a_{max}\}$
         <u>Until</u> $(R = C)$ or (For any $u, v \in U$ belonging to two different persons, u and v differ from each other in at least one feature in R)

3.3 Combining Rough Set Theory and Heuristic for Generating Rules

Note that the ability of conditional attribute set C to classify feature vectors into classes of persons is measured by the cardinal of the set $POS_C(D)$. On the other hand, the relative core $CORE_D(C)$ is the set of features which will change the positive region $POS_C(D)$ if being eliminated. So, the initial reduced feature set must be $CORE_D(C)$: $R = CORE_D(C)$.

Next, we will introduce the heuristic for adding the other features into R gradually until it becomes a reduct of C, or the following condition must satisfy: $POS_R(D) = POS_C(D)$. This standard is a variant of one used in rule discovery system GDT-RS. The attribute $a_0 \in C \setminus R$ will be chosen based on two comments:

1. In order to get as small feature set as possible, we prefer to choose feature a_0 so that after adding a_0 into R, the number of consistent objects grows fastest, expressed in the coefficient:

$$v_a = card(POS_{R \cup \{a\}}(D)), a \in C \setminus R \qquad (7)$$

2. For any $a_0 \in C \setminus R$, the partition of all consistent objects (or feature vectors) created by equivalence relation $IND(R \cup \{a_0\} \cup D)$, or the set $POS_{R \cup \{a_0\}}(D) \mid IND(R \cup \{a_0\} \cup D)$, will change, and so does the set of generated rules. Among equivalence classes in this new partition, let M be the class that has the most feature vectors and r be the rule generated from M. We see that, the bigger the size of M is, the higher the coverage of r is, or in particular, the more feature vectors satisfy the rule r. Thus, we can use the size of M as the second standard for choosing features:

$$m_a = \max\_size(POS_{R \cup \{a\}}(D) \mid IND(R \cup \{a\} \cup D)), a \in C \setminus R \qquad (8)$$

These two coefficients are contrary to each other, so we consider $v_a \times m_a$ to be the final standard to choose features in $C \setminus R$.

Algorithm
  Input:
  - Information system $(U, C \cup D)$.
  - Positive value *threshold*: the percentage of consistent objects.

  Output: Reduced feature set $R \subseteq C$.
    Step 1: $R = CORE_D(C)$, $P = C \setminus R$
    Step 2: Remove consistent objects : $U = U \setminus POS_R(D)$

Step 3: $k = \dfrac{card(POS_R(D))}{card(U)}$

If $(k \geq threshold)$ or $(POS_R(D) = POS_C(D))$
Then : Stop
End if

Step 4: For each $a \in P$
$$v_a = card(POS_{R \cup \{a\}}(D))$$

$$m_a = \max\_size(POS_{R \cup \{a\}}(D) \mid IND(R \cup \{a\} \cup D))$$
End for each

Step 5: $a_0 = \arg\max_{a \in P}(v_a x m_a)$

Step 6: $R = R \cup \{a_0\}$, $P = P \setminus \{a_0\}$

Step 7: Go to step 2.

4 Testing Results

We use the face image database ORL (available at *http://www.uk.research.att.com/facedatabase.html*) for testing. Because the number of face image of each person in this database is quite small, we change some images before proceeding. In table 1 and 2 below, the second column *(size of initial feature vectors)* corresponds to the number of best eigen vectors remained in the phase of extracting feature vectors using Principle Components Analyisis. The third one is the recognition accuracy when we use these original feature vectors. Finally, the next columns are the size of reduced feature vectors and the recognition accuracy of three selection algorithms described above using these reduced vectors.

Some good results in the above tables are in bold type. We have following remarks.

Remark 1. Using the same size of feature vectors, the recognition accuracy obtained from reduced feature vectors is higher than from the original feature vectors (e.g. six – feature vectors: 94.33 % vs 87% in Table 1, ten – feature vectors: 95.5 % vs 85.25 % in Table 2).

Remark 2. Reducing the size of feature vectors increase the recognition accuracy (e.g. from 12 to 7 features: 92% vs 97.67% in Table 1; from 15 to 10 features: 86.5 % vs 95.5 % in Table 2).

Remark 3. For the reduced feature vectors, the recognition accuracy from ones generated by combining rough set theory and heuristic is higher than ones generated by Johnson and Random algorithms.

Table 1. The recognition accuracy of original and reduced feature vectors. The number of persons : 15, face images per person used in training phase : 24, face images per person used in recognition phase : 20, reference vectors (LVQ) : 10, the number of discrete ranges used in discretization phase : 3, training epoches : 8000

No.	Size of initial feature vectors	Recognition accuracy (%) (without reducing feature set)	Using Rough set to reduce feature set					
			Johnson		Rough set + Heuristic		Random	
			Size of reduced feature vectors	Recognition accuracy (%)	Size of reduced feature vectors	Recognition accuracy (%)	Size of reduced feature vectors	Recognition accuracy (%)
4	9	90.67	9	90.67	8	90.33	9	90.67
5	10	91.33	8	89.67	6	**94.33**	8	**97.67**
6	11	91.33	8	89.67	6	**94.33**	9	90.67
7	12	92	8	89.67	7	**97.67**	8	90.33
8	13	92.33	8	89.67	7	**97.67**	9	90.67
9	14	92.33	9	91.67	7	**97.67**	10	88.67
10	15	92.33	9	91.67	7	**97.67**	8	89.67

Table 2. The recognition accuracy of original and reduced feature vectors. The number of persons: 20, face images per person used in training phase : 24, face images per person used in recognition phase: 20, reference vectors (LVQ): 10, the number of discrete ranges used in discretization phase: 3, training epoches : 10000

No.	Size of initial feature vectors	Recognition accuracy (%) (without reducing feature set)	Using Rough set to reduce feature set					
			Johnson		Rough set + Heuristic		Random	
			Size of reduced feature vectors	Recognition accuracy (%)	Size of reduced feature vectors	Recognition accuracy (%)	Size of reduced feature vectors	Recognition accuracy (%)
4	13	92.33	10	85.5	8	89.25	10	90.67
5	14	87.25	11	81.25	9	94.5	11	88.67
6	15	86.5	11	81.25	10	95.5	13	89.67
7	16	86.5	11	81.25	10	95.5	12	82
8	17	86.5	10	81.75	10	95.5	11	89.5
9	18	87	10	81.75	11	90.25	13	91.5
10	19	87.25	11	81.75	11	90.25	12	84.5
11	20	87.25	11	81.75	11	90.25	12	79.5

An important note is that the feature corresponding to the largest eigen value in the phase of feature extracting using PCA, i.e. the feature best describes the distribution of feature vectors in their space, always belongs to the reduced feature set. This is actually the interesting characteristic proving the potential success of rough set theory in features selection and reduction for recognition problem.

5 Conclusion

In this article, we see that rough set theory, accompanied by appropriate heuristic, is a potential approach in feature selection for recognition systems. An interesting result is that the reduced feature sets created by all rough set based approachs contain the most significant features which corresponds to the largest eigen values in the phase of feature extracting using PCA. However, we actually need to study more investigation into the problem of using rough set in feature selection in recognition systems.

References

1. Jan Komorowski, Lech Polkowski, Andrzej Skowron: Rough Sets: A Tutorial.
2. Roman W. Swiniarski : Rough Set Methods in Feature Reduction and Classification – Int. Appl. Math. Comput. Sci., 2001, Vol 11, No.3, 565 – 582.
3. Ning Zhong, Juzhen Dong, Setsuo Ohsuga: Using Rough Sets with Heuristic for Feature Selection. Journal of Intelligent Information System, 16, 199 – 214 (2001).
4. Nguyen Sinh Hoa, Nguyen Hung Son – Institute of Computer Sciences, Wasaw University, Poland: Some Efficient Algorithms for Rough Set Methods.
5. Matthew Turk, Alex Pentland: Eigenfaces for Recognition (1991).
6. LVQ_Pak, Neural Network Research Centre – Laboratory of Computer and Information Science, Helsinki University of Technology (*http://www.cis.ut.fi/research/lqv_pak*).

Analysis of Company Growth Data Using Genetic Algorithms on Binary Trees

Gerrit K. Janssens[1], Kenneth Sörensen[2], Arthur Limère[3], and Koen Vanhoof[1]

[1] Limburg University Centre, Faculty of Applied Economics,
Data Analysis and Modelling Research Group (DAM),
B–3590 Diepenbeek, Belgium
[2] University of Antwerp, Faculty of Applied Economics,
B–2000 Antwerp, Belgium
[3] Limburg University Centre, Faculty of Applied Economics,
Financial Management Research Group (FIM),
B–3590 Diepenbeek, Belgium

Abstract. This paper investigates why some companies grow faster than others, by data mining a survey of a large number of companies in Flanders (the northern part of Belgium). Faster or slower average growth over a time period is explained by building a classification tree containing several categorical variables (both quantitative and qualitative). The technique used – called genAID – splits the population at different levels. It is inspired by the Automatic Interaction Detector (AID) technique to find trees that explain the variability in average growth but uses a genetic algorithm to overcome some of the drawbacks of AID.

Classical AID or other tree-growing techniques usually generate a single tree for interpretation. This approach has been criticized because, due to the artifacts of data, spurious interactions may occur. genAID offers the user-analyst a set of trees, which are the best ones found over a number of generations of the genetic algorithm. The user-analyst is then offered the choice of choosing a tree by trading off explanatory power against either the ease of understanding or the conformity with an existing theory.

1 Introduction

Data mining is concerned with the development of tools and techniques to automate the extraction of information from data [1]. In recent years, a multitude of data mining techniques have been developed. For a detailed overview we refer to [2].

The genAID technique [14] used in this paper is based on the *Automatic Interaction Detector* (AID) technique, originally developed by [8, 13]. AID was developed to solve some of the shortcomings of statistical regression by building a *classification tree*, of which the leaves represent classes in which the observations are classified by repeated binary splits. Classification tree methods have been used in machine learning (e.g. ID3 and C4.5 [10, 11]).

2 Data Mining with the Automatic Interaction Detector (AID) - Technique

The AID technique mimics the steps taken by an experienced data analyst to determine the relationship between a set of *binary independent (predictor) variables* and a *continuous dependent variable*. Given a set of observations, the variance of the dependent variable is explained by classifying the observations according to a *binary classification tree*. Such trees classify a set of observations into a finite number of classes, represented by the leafs (or terminal nodes) of the tree. Non-terminal nodes are labeled with a predictor variable, and the edges emanating from a given node represent the possible values for this predictor value. Observations are classified by following a path from the root to a leaf, taking the path determined by the value of the predictor variables that correspond to the observation.

In an AID tree, the leafs represent the classes and the nodes represent predictor variables that have a good explanatory power. A predictor variable splits a (sub)group in two parts, labeled + and − respectively. AID works by a recursive exhaustive search of all predictor variables. In a first step, every possible predictor variable x is tested to see which one has the strongest predictive power. This is measured by eq. 1.

$$P(x) = \frac{n_1(\bar{x}_1 - \bar{x})^2 + n_2(\bar{x}_2 - \bar{x})^2}{ns^2}, \qquad (1)$$

where $\bar{x}_1$ and $\bar{x}_2$ are the averages in subgroups 1 and 2, n_1 and n_2 are the number of subjects in subgroups 1 and 2, $\bar{x}$ is the population average, n is the total number of subjects in the population, s^2 is the population variance. The population is then split into two classes according to this predictor variable. Both subsets of observations that result are then split again according to the best predictor variable for that subset. This process is repeated until some stopping criterion is satisfied.

Later versions of the AID technique (CHAID and THAID, e.g.) improved on some of its deficiencies [3, 4] by combining it with statistical hypothesis testing methods [5]. Notwithstanding this fact, the AID technique only ever reports a single tree.

3 genAID: An AID Based on Genetic Algorithms

To solve some of the shortcomings of AID, genAID was developed by [14]. This technique uses a genetic algorithm (GA) to develop a diverse population of AID classification trees. It uses a set of specifically constructed tree-based genetic operators, borrowed from genetic programming [6] and the work of [15]. The fitness of a tree in the genAID population is determined by a formula that borrows from one-way ANOVA:

$$f(A) = \frac{\sum_{i=1}^{K} n_i(\bar{x}_i - \bar{x})^2}{\sum_{i=1}^{K} \sum_{j=1}^{n_i} (x_{ij} - \bar{x})^2}, \qquad (2)$$

where A is the tree, n_i is the number of observations in class i, K is the number of classes, x_{ij} is the j-th observation in class i, $\bar{x}_i$ is the class i sample mean and $\bar{x}$ is the overall sample mean. Calculating the fitness of a classification tree is a computationally difficult process because each subject in the population has to be classified into one of the resulting classes.

genAID first builds a population of random trees and uses specifically tailored genetic operators to combine parent trees into new (and hopefully better) offspring trees. GenAID uses two types of operators: *macro-operators* exchange entire subtrees in and between trees, *micro-operators* do not operate on subtrees, but on labels only. For a complete description of genAID, we refer to [14].

4 The Analysis of Company Growth Data

The analysis uses data obtained both from quantitative financial statement data and from qualitative data from a large-scale survey executed in April 2001. This data set includes about 18% of all incorporated Flemish companies (1997) and proved to be representative for the whole population of Flemish companies in terms of size, economic activities and location.

In the analysis, the dependent variable expresses the company's growth, which can be seen as a very important measure of firm performance. In this paper, growth is measured by the average growth rate of total assets over the period 1993-1999. The use of growth rate of total assets as a measure for firm growth has already been used in previous research on Belgian companies [7, 9]. An average growth over a number of years is preferred to a growth percentage of a single year because growth percentages may fluctuate strongly from one year to another.

As an example, the following list contains the names of some of the independent variables, as well as their meaning and their set of potential values in the analysis. The total number of variables in the study was 17.

1. vindwn: relates to the problem of finding suitable employees, associated to growth (values: 1 = no problems up to 5 = a lot of problems);
2. karasom: relates to the characterization of the environment (values: 3 to 15; the higher the value, the more benign the environment);
3. hfdstrat: relates to the main strategy regarding the company's most important product line (values: 1= innovative and risk seeking up to 4= conservative and risk avoiding);
4. finken2: relates to the use of type of financial ratios in the company (values: 1 up to 4, depending whether at least one ratio of a certain type is used. Four types are identified: profitability, solvency, liquidity and added value);

5. ondind1: relates to the use of performance indicators in marketing, distribution, sales and manufacturing (values: 1 up to 8, depending how many indicators are used out of a list of eight).

4.1 Encoding of Predictor Variables

Most of the data analyzed is of an ordinal nature. This is a result of the fact that the data were collected from a survey, asking the respondents to answer on a scale of varying size (e.g. totally agree – agree – indifferent – disagree – strongly disagree). AID classifies a set of observations of a continuous dependent variable based on a number of binary independent (predictor) variables. This implies that all independent variables have to be binary encoded. Ordinal variables are encoded using the so-called *thermometer coding* [12]. For a variable x that can take n values, $n-1$ binary variables are created. A binary variable x_i is set to 1 if $x > i + 1$. E.g. if x can take 5 values, 4 variables are created. This is illustrated in Fig. 1.

x	x_1	x_2	x_3	x_4
1	0	0	0	0
2	1	0	0	0
3	1	1	0	0
4	1	1	1	0
5	1	1	1	1

Fig. 1. Thermometer encoding for an ordinal variable taking 5 possible values

With this encoding, the variables are interpretable. The binary variable x_2, e.g. splits the observations in a *low* and a *high* group: those having $x < 3$ and those having $x \geq 3$.

4.2 Results and Discussion

In an experiment, genAID is run for 1000 generations with a constant population size equal to 50. genAID creates perfect binary trees of height three. This means that seven (not necessarily different) variables can be used to define splits in order to partition the population. During the process of subsequent generations in the genetic algorithm the best 50 solutions in terms of predictive power are kept in memory. The trees are presented to the user-analyst to choose a tree which suits him best. The user-analyst may use secondary objective criteria to make his choice, or he can use subjective knowledge to prefer one tree above another. For each case, we work out an example.

Variables which appear in various branches of the splits of the tree represent a main effect. If the same variables appear only in one branch they are considered to represent an interaction effect. The user-analyst might, for example, be interested in finding as many main effects as possible and not too many interactions, of which some might be spurious and, at least, maybe difficult to interpret. A tree of height three, as we are using in the experiment, can involve at maximum seven variables, while the tree in Fig. 2 involves only five. In this reasoning minimizing the number of different variables in a tree could serve as a secondary objective criterion.

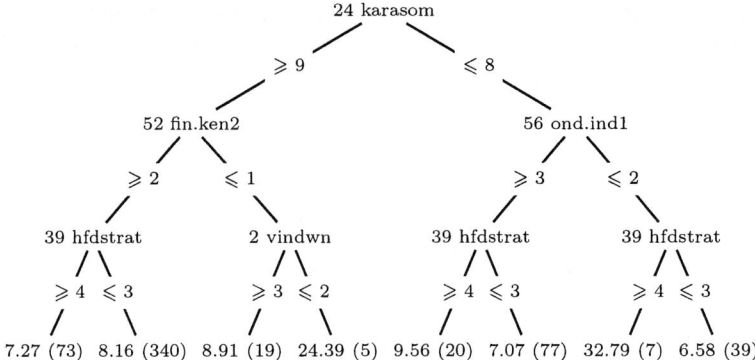

Fig. 2. A resulting tree

A second approach involves a user-analyst with some theories in mind and exploring structures of trees in order to retrieve those who represent a confirmation of some theories with a high level of predictive power. Figure 2 shows the tree preferred by such an analyst. This tree provides an interesting interpretation of enablers and disablers of the growth process in companies and sheds some new light on possible growth strategies.

Not surprisingly, the variable KARASOM is selected as the root variable, because it has been found to be an important factor in previous research [7]. It reflects the general perception of the CEO about the environment being either benign or hostile. After this first split in the tree, companies with strong growth can still be found in both branches, but depending on different conditions (variables).

If the environment is rather friendly, the strong growing companies apply a more defensive, conservative strategy as opposed to being a prospector, a reactor or an analyst (variable HFDSTRAT). The highest growth is found in companies which have (or need) a rather limited follow-up of the company-policies that are been carried out. In terms of performance they indicate in the survey a limited use of quantitative performance indices (variable OND.IND1). On the other hand, if the environment is hostile, strong growing companies have to watch their financial performance intensively by using various financial ratios (variable FIN.KEN2) and by applying more aggressive strategies. This latter strategy allows for more than half of the sample (340) to realize a higher than average growth figure. The best five growing companies in a hostile environment however operate in a good labour market, where they can find suitable labour forces (variable VINDWN). These companies spend less effort in monitoring financial ratios.

5 Conclusions

In this paper, the genAID technique is applied to a set of company growth data to be explained by subjective assessments from the presidents of companies through a large survey. While the genAID technique uses binary splits on variables and produces a binary classification tree, it is shown, through the right encoding technique, how this data mining technique can be used if variables are categorical but not binary.

Better growth patterns have been detected in companies that operate in a benign environment and they can afford to apply a defensive strategy. The best performing companies (7) carry out their policies by following up a limited number of indices in their balanced scorecard. If the environment is hostile, offering less business opportunities, at least the labour market and labour climate should be friendly. If not so, growth can only be realized through a more aggressive strategy and by a good follow-up of financial performances.

References

1. P. Adriaans and D. Zantinge. *Data Mining*. Addison-Wesley, Harlow, 1996.
2. M-S. Chen, Han J., and Yu P.S. Data mining: an overview from a database perspective. *IEEE Transactions on Knowledge and Data Engineering*, 8:866–883, 1996.
3. H.J. Einhorn. Alchemy in the behavioral sciences. *Public Opinion Quarterly*, 36:367–378, 1972.
4. G.V. Kass. Significance testing in automatic interaction detection (AID). *Applied Statistics*, 24:178–189, 1975.
5. G.V. Kass. An exploratory technique for investigating large quantities of categorical data. *Applied Statistics*, 29:119–127, 1980.
6. J.R. Koza. *Genetic Programming*. MIT Press, Cambridge, MA, 1992.
7. E. Laveren, A. Limère, K. Cleeren, and E. Van Bilsen. Growth factors of flemish enterprises: an exploratory study over the periode 1993-1997. *Brussels Economic Journal-Cahiers Economiques de Bruxelles*, 46(1):5–38, 2003.
8. J.N. Morgan and J.A. Sonquist. Problems in the analysis of survey data, and a proposal. *Journal of the American Statistical Association*, 58:415–435, 1963.
9. H. Ooghe, E. Verbaere, and M. Croucke. Ondernemingsdimensie en financiële structuur. *Maandblad voor Accountancy en Bedrijfseconomie*, 3:62–77, 1988. in Dutch.
10. J.R. Quinlan. Induction of decision trees. *Machine Learning*, 1:81–106, 1986.
11. J.R. Quinlan. *C4.5: Programs for machine learning*. Morgan Kaufmann, San Francisco, 1993.
12. M. Smith. *Neural networks for statistical modeling*. Thomson, Boston, 1996.
13. J.A. Sonquist, E. Baker, and J. Morgan. Searching for structure. Technical report, Institute for Social Research, University of Michigan, Ann Arbor, 1973.
14. K. Söorensen and G.K. Janssens. Data mining with genetic algorithms on binary trees. *European Journal of Operational Research*, 151:253–264, 2003.
15. H. Van Hove and A. Verschoren. Genetic algorithms and trees: part 1: recognition trees (the fixed width case). *Computers and Artificial Intelligence*, 13:453–476, 1994.

Considering Re-occurring Features in Associative Classifiers

Rafal Rak, Wojciech Stach, Osmar R. Zaïane, and Maria-Luiza Antonie

University of Alberta, Edmonton, Canada
{rrak, wstach}@ece.ualberta.ca
{zaiane, luiza}@cs.ualberta.ca

Abstract. There are numerous different classification methods; among the many we can cite associative classifiers. This newly suggested model uses association rule mining to generate classification rules associating observed features with class labels. Given the binary nature of association rules, these classification models do not take into account repetition of features when categorizing. In this paper, we enhance the idea of associative classifiers with associations with re-occurring items and show that this mixture produces a good model for classification when repetition of observed features is relevant in the data mining application at hand.

1 Introduction

Classification is one of the most common tasks in data mining and machine learning. By and large, it consists of extracting relevant features from labelled training data to build a model that discriminates between classes for unlabelled observed objects. Myriad techniques have been proposed and while there are, in general, better approaches than others, there is no clear winner in terms of correctness and usability given a particular problem application.

Associative classification is a relatively new method. The main objective is to discover strong patterns that are associated with the class labels in the training set. The training set is modeled into transactions with items being the observed features. As a final classification model, one obtains a set of association rules associating features with class labels. In the literature, there are few known classifiers based on the above-mentioned idea, i.e. CBA [4], CMAR [3], and ARC-AC/ARC-BC [9].

One considerable limitation of all these algorithms is that they do not handle the observations with repeated features. In other words, if a data object is described with repeated features, only the presence of the feature is considered, but not its repetition. However, in many applications such as medical image categorization or other multimedia classificaiton problems, the repetition of the feature may carry more information than the existence of the feature itself [10]. Also in text mining and information retrieval, it is widely recognized that the repetition of words is significant and symptomatic, hence the common use of TF/IDF (i.e. the frequency of a term in a document relative to the frequency of the term in a collection).

Associative classifiers use association rule mining to build a classification model. However, association rule mining typically considers binary transactions; transactions that indicate presence or absence of items. Binary transactions simply do not model repetitions. A few approaches to mining association rules with re-occurring items have been proposed, such as MaxOccur [10], FP'-tree [5] and WAR [8]. The main goal of our research is to devise a classifier that combines the idea of associative classification and association rules with reoccurring items. Our contributions presented in this paper exploit, combine, and extend the ideas mentioned above, especially ARC-BC and MaxOccur algorithms. We also suggest new strategies to select rules for classification from the set of discovered association rules. Our hypothesis is that associative classifiers with recurrent items have more discriminatory power since they maintain and exploit more information about both objects and rules.

A delicate issue with associative classifiers is the use of a subtle parameter: support. Support is a difficult threshold to set, inherited from association rule mining. It indicates the proportion of the database transactions that support the presence of an item (or object). It is known in the association rule mining field that the support threshold is not obvious to tune in practice. In the associative classification literature it has been commonly and arbitrarily set to 0.1%. However, the accuracy of the classifier can be very sensitive to this parameter. In the case of re-occurring items, there are two ways of calculating support: transaction-based support and object-based support [10] (i.e. either the proportion of transactions or the proportion of objects that support the existence of an object in the database). Our experiments show that an associative classifier that considers re-occurrence of features is considerably less sensitive to the variation of support. This leads to more practical applications and eventually the possibility to automatically determine and tune this parameter.

The remainder of the paper is organized as follows: Section 2 presents the problem statement: the model of an associative classifier and the consideration in the model of recurrent items. Related work on associative classification and mining association rules with repetitions is also presented in Section 2. We present our new approach ACRI in Section 3. The experiments showing the performance of our approach are presented in Section 4. Section 5 offers some conclusions.

2 Problem Statement and Related Work

The first known classifier using association rules was introduced in [4]. The main idea was to modify the form of transactions known from the traditional approach to the form of $< \{i_1, i_2, ...i_n\}, c >$, where i_k is an item in a transaction and c is a class label. In other words, objects in a training set are represented by sets of features appended with the observed class label. All the rules generated from frequent itemsets are restricted to those with a class label as a consequent.

Our task is to combine the associative classification with the problem of recurrent items. More formally, it can be stated that our goal is to modify the original approach using transactions to the form of $< \{o_1i_1, o_2i_2, ...o_ni_n\}, c >$, where o_k is the number of the occurrences of the item i_k in the transaction.

Association rules have been recognized as a useful tool for finding interesting hidden patterns in transactional databases. Several different techniques have been introduced. However less research has been done considering transactions with reoccurrence of items. In [8], the authors assign weights to items in transactions and introduce the WAR algorithm to mine the rules. This method is two fold: in the first step frequent itemsets are generated without considering weights and then weighted association rules (WARs) are derived from each of these itemsets. MaxOccur algorithm [10] is an efficient Apriori-based method for discovering association rules with recurrent items. It reduces the search space by effective usage of joining and pruning techniques. The FP'-tree approach presented in [5] extends the FP-tree design [2] with a combination from the MaxOccur idea. For every distinct number of occurrences of given item, the separated node is created. In case when a new transaction is inserted into the tree, it might increase support count for the different path(s) of the tree as well. This is based on the intersection between these two itemsets. Given the complete tree, the enumeration process to find frequent patterns is similar to that from the FP-tree approach. None of the existing associative classifier uses reoccurrence.

3 The Proposed Approach

Our approach, ACRI (Associative Classifier with Reoccurring Items), consists of two modules: Rule generator and classifier. We decided to base our algorithm for mining associations with reoccurring items on Apriori-based MaxOccur. The building of the classification model follows our previous ARC-BC approach. The rational is based on the efficiency of this method in the case of non-evenly distributed class labels. Indeed other associative classification methods are biased towards dominant classes in the case when rare classes exist. Rare classes are classes with very few representatives in the training set. MaxOccur run on transactions from each known class separately makes the core of our rule generator module. It mines the set of rules with reoccurring items from the training set. These rules associate a condition set with a class label such that the condition set may contain items preceded by a repetition counter. The classification process might be considered as plain matching of the rules in the model to the features of an object to classify. Different classification rules may match, thus the classifier module applies diverse strategies to select the appropriate rules to use. In addition, simple matching is sometimes not possible because there is no rule that has the antecedent contained in the feature set extracted from the object to classify. With other associative classifiers, a default rule is applied, either the rule with the highest confidence in the model or simply assigning the label of the dominant class. Our ACRI approach has a different strategy allowing partial matching or closest matching by modeling antecedents of rules and new objects in a vector space. The following elaborates on both modules.

Rule Generator: This module is designed for finding all frequent rules in the form of $< \{o_1i_1, o_2i_2, \ldots, o_ni_n\}, c >$ from a given set of transactions. The modules's general framework is based on ARC-BC [9]: transactions are divided into N subsets - each for one given class (N is equal to the number of classes); Once rules are generated for each individual class, the rules are merged to form a classification model. The rule generator for each class C_x is an Apriori-based algorithm for mining frequent itemsets that extends the original method by taking into account reoccurrences of items in a single transaction à la MaxOccur [10]. In order to deal with this problem, the support count was redefined. Typically, a support count is the number of transactions that contain an item. In our approach, the main difference is that single transactions may increase the support of a given itemset by more than one. The formal definition of this approach is as follows. A transaction $T =< \{o_1i_1, o_2i_2, \ldots, o_ni_n\}, c >$ supports itemset $I = \{l_1i_1, l_2i_2, \ldots, l_ni_n\}$ if and only if $\forall i = 1..n \, l_1 \leq o_1 \wedge l_2 \leq o_2 \wedge \ldots \wedge l_n \leq o_n$. The number t by which T supports I is calculated according to the formula: $t = min[\frac{o_i}{l_i}] \forall i = 1..n, l_i \neq 0 \wedge o_i \neq 0$.

The Classifier: This module labels new objects based on the set of mined rules obtained from the rule generator. An associative classifier is a rule-based classification system, which means that an object is labelled on the basis of a matched rule (or set of rules in case of multi-class classification). This task is simple if there is an exact match between a rule and an object. The model, however, often does not include any rule that matches a given object exactly. In such a case, in order to make the classification, all rules are ranked according to a given scenario and the best one (or several) is matched to a given object. Rule ranking might be performed following different strategies, which associate each rule to a number that reflects its similarity to a given object. These strategies may be used either separately or in different combinations. We have tested *cosine measure, coverage, dominant matching class, support* and *confidence*. Let us consider the rule $< \{o_1i_1, o_2i_2, \ldots, o_ni_n\}, c >$ and the object to be classified $< l_1i_1, l_2i_2, \ldots, l_ni_n >$. The corresponding n-dimensional vectors can be denoted as $\vec{0} = [o_1, o_2, \ldots, o_n]$ and $\vec{l} = [l_1, l_2, \ldots, l_n]$. The **Cosine measure (CM)** assigns a value that is equal to the angle between these two vectors, i.e. The smaller the CM value is, the smaller the angle, and the closer these vectors are in the n-dimensional space. **Coverage (CV)** assigns a value that is equal to the ratio of the number of common items in the object and rule to the number of items in the rule (ignoring reoccurrences). In this case, the larger the CV ratio is, the more items are common for the rule and the object. CV=1 means that the rule is entirely contained in the object. With **Dominant matching class**, the class label is assigned to the object by choosing the one being the most frequent from the set of rules matching the new object. Notice that dominance can be counted by simply enumerating the matching rules per class or a weighted count using the respective confidences of the matching rules. The **support** and **confidence** are used to rank rules. They refer to the rule property only and do not depend on the classified object. Thus, they have to be used with other measures that prune the rule set.

4 Experiments

We tested ACRI on different datasets to evaluate the best rule selection strategy as well as compare ACRI with an associative classifier like ARC-BC. As an example, we report here an experiment with the mushroom dataset from the UCI repository [7]. It appears that the rule selection strategies have roughly similar performance in terms of accuracy. However, this accuracy varies with the support threshold. The lower the support, the more rules are discovered allowing a better result using selection based on cosine measure for example. Using the dominant matching class was also doing well, confirming the benefit of the dominance factor introduced in [9]. The selection based on best rule support was not satisfactory in general and is not reported here. We also observed that coverage (CV) gave better results when set to 1. Thus all results reported herein have CV set to 1. The other measures are comparable in performance and trend, except for best confidence. When the support threshold is high, fewer rules are discovered and confidence tends to provide better results while the cosine measure returns matches that have big angles separating them from the object to classify, hence the lower accuracy. Figure 1 on the left shows the superiority of the rule selection strategy dominant matching class up to a support threshold of 25%, beyond which best confidence becomes a winning strategy. Figure 1 on the right shows how the more rules are discovered the more effective in terms of accuracy the strategies dominant matching class and cosine measure becomes in comparison to best confidence approach. The number of rules is correlated with support.

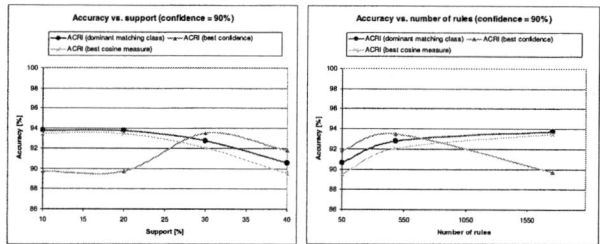

Fig. 1. Accuracy of rule selection strategies vis-à-vis support and number of rules

We compare ACRI with ARC-BC using the Reuters-21578 text collection as in the paper presenting ARC-BC for text categorization using the top 10 topics [6]. The total of 9980 documents is split into two sets: 7193 and 2787 for a training and test set respectively. At first, we tested both approaches ACRI and ARC-BC using relatively high support. We produced several different sets of rules to be used in the classifier. For ARC-BC we chose the support threshold range from 10 to 30% with the step of 5%; and 15 to 65% with the same step for our approach. The difference between the support thresholds lies in the definition of support for mining rules with recurrent items. A single document can support a set of words more than once. Therefore, if we consider support as the ratio of support count to the total number of transactions, as it was introduced in [5], we may encounter support more than 100% for some itemsets. On the other hand, if we choose the definition presented in [10], i.e., the ratio of support count to the number of distinct items (words), the support will never reach 100%. Actually, in practice, the latter support definition requires for setting very small thresholds to obtain reasonable results. Hence, we decided to use the first one as it is more similar to the "classical" definition of support. It is important to notice that no matter which definition we choose, it eventually leads to setting the same support count with ARC-BC. For each support threshold we set three different confidence thresholds: 0, 35, and 70%. The latter threshold was used in [9] as minimum reasonable threshold for producing rules; the first one (no threshold) was introduced to observe the reaction of the classifier for dealing with a large number of rules; and the threshold of 35% is simply the middle value between the two others. For each single experiment we tried to keep the level of more then 98% of classified objects, which resulted in manipulating the coverage CV from 0.3 to 1. We discarded cases for which it was not possible to set CV to satisfy the minimum number of classified objects. More than 90% of the remaining results had CV = 1. We also performed experiments without specifying CV (using different methods of choosing applicable rules); however, they eventually produced lower accuracy than those with specified CV ¿ 0.3. We used different classification techniques for choosing the most applicable rule matching the object. Best confidence and dominant matching class matching methods were utilized for both ARC-BC and ACRI approaches. Additionally, ACRI was tested with the cosine measure technique. So for all experiments herein reported the coverage (CV) is set to 1. In other words, for a rule to be selected for classification, all features expressed in the antecedent of the rule have to be observed in the new object to classify. We also performed tests with combination of matching techniques with different tolerance factors for each test. An example scenario in Figure 2 A, combines cosine measure, dominant matching class and best confidence: (1) choose top 20% of rules with the best cosine measure, then (2) choose 50% of the remaining rules with the highest confidence, and then (3) choose the rule based on the dominant class technique. We also did a battery of tests using relatively low supports. This significantly increases the number of classification rules. We varied the support between 0 and 0.1% and compared the harmonic average of precision and recall (F1 measure) for the same cases as before: Best confidence and dominant matching class for both ARC-BC and ACRI approaches, and the cosine measure technique for ACRI (Figures 2 E-F).

Categorizing documents from the Reuters dataset was best performed when the confidence level of the rules was at the 35% threshold for both the ACRI and ARC-BC approaches. For ARC-BC classifier, the best strategy was to use dominant factor, whereas in case of ACRI combination of cosine measure and confidence factors worked best. Figure 2 A shows the relationship between support and accuracy for these approaches. Comparing the best-found results, ARC-BC slightly outperforms the ACRI using the dominant matching class strategy at the 20% support level. However, ARC-BC seems to be more sensitive to changes of the support threshold. The accuracy of ACRI virtually does not depend on the

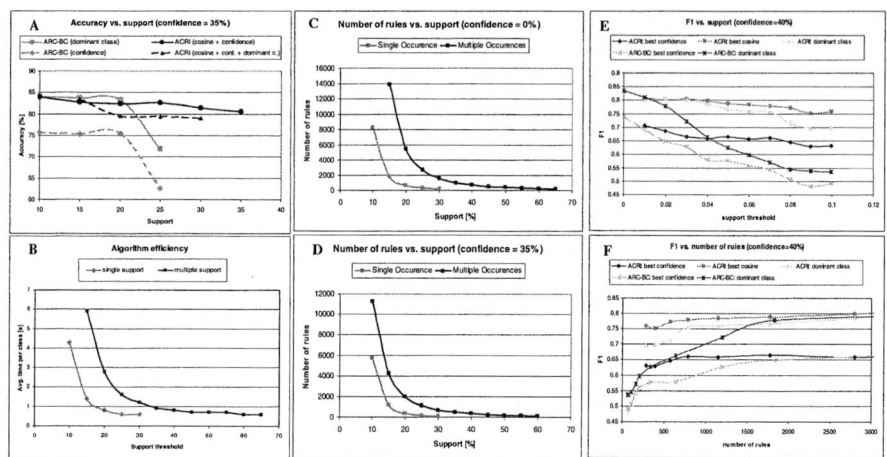

Fig. 2. A: Accuracy of ACRI and ARC-BC with high support; B: Algorithms CPU time efficiency; C: Number of rules with confidence > 0%; D: Number of rules with confidence > 35%; E: Effectiveness at low support; F: Effectiveness versus size of model

support threshold and is stable. In the case of ARC-BC the accuracy decreases significantly when this support is greater than 20%.

Figures 2 C and D show the number of generated rules with and without recurrent items. Considering recurrences results in having more rules, this has its origin in different support definition. The other interesting relationship is that by increasing the confidence threshold from 0% to 35%, the difference between number of rules decreases more rapidly for ACRI.

Experiments using low support thresholds confirm the stability of ACRI with regard to support. When varying the support from 0% to 0.1% ARC-BC loses in precision and recall while ACRI remains relatively consistent or looses effectiveness on a slower pace. Figure 2 E also shows that ACRI outperforms ARC-BC at these lower support thresholds. Using the cosine measure for selecting rules appears to be the best strategy. The cosine measure is also the best rule selection strategy when considering the number of rules discovered. In addition, the more rules are available the more effective the cosine measure becomes at selecting the right discriminant rules.

Figure 2 B shows the relationship between running time for rule generator with and without considering recurrent items. The algorithm with recurrences is slower, since it has to search a larger space, yet the differences become smaller when increasing the support threshold.

The best results for ACR-BC were found in [9] for confidence threshold greater than 70%. However, our experiments show that effectiveness is better on lower confidence for both ARC-BC and ACRI approaches. In other words, some classification rules with low confidence have more discriminant power and are selected by our rule selection strategies. This discrepancy with previous results may be explained by the use of the different method of counting support and confidence or/and by the fact that our classifier ACRI with re-occurring items and without re-occurrence consideration to simulate ARC-BC is using a different setup for rule selections.

5 Conclusion and Future Work

In this paper we introduced the idea of combining associative classification and mining frequent itemsets with recurrent items. We combined these two and presented ACRI, a new approach of associative classification with recurrent items. We also suggest new strategies to select classification rules during the classification phase. In particular, using the cosine measure to estimate the similarity between objects to classify and available rules is found very effective for associative classifiers that consider re-occurrence. When comparing our ACRI approach with other associative classifiers represented by ARC-BC we found that considering repetitions of observed features is beneficial. In particular in the case of text categorization, repetition of words has discriminant power and taking these repetitions in consideration can generated good classification rules. Our experiments also show that ACRI becomes more effective as the number of rules increases in particular with our cosine measure for rule selection. Moreover, ACRI seems to be less sensitive, with respect to accuracy, to the support threshold, while other associative classifiers are typically very sensitive to the support threshold which is very difficult to determine effectively in practice. This research is still preliminary. We intend to investigate the possibility to eliminate the need for the support threshold by automatically selecting an optimal support based on available data. This is in part possible because ACRI is not substantially sensitive to the variation of the support. We are also investigating other rule selection strategies since selecting the right rules has a paramount effect on the precision of a classifier. Moreover, pruning the large set of classification rules can improve the accuracy and speed of the classifier.

References

1. Antonie, M.-L., Zaïane, O. R. Text document categorization by term association. *IEEE International Conference on Data Mining*, (2002) 19–26
2. Han, J., Pei J., Y., Yin: Mining Frequent Patterns Without Candidate Generation, *ACM Intl' Conf. on Management of Data*, (2000)
3. Li, W., Han, J., Pei J.: CMAR: Accurate and efficient classification based on multiple class-association rules. *IEEE International Conference on Data Mining*, (2001)
4. Liu, B., Hsu, H., Ma, Y.: Integrating classification and association rule mining. In *4th Intl. Conf. on Knowledge Discovery and Data Mining*, (1998) 80–86.
5. Ong K.-L., Ng, W.-K., Lim, E.-P., Mining Multi-Level Rules with Recurrent Items Using FP'-Tree, ICICS (2001)
6. Reuters-21578 Top 10 Topics Collection, http://www.jihe.net/datasets.htm
7. UCI repository, http://www.ics.uci.edu/~mlearn/MLRepository.html

8. Wang, W., Yang, J., Yu, P., WAR: Weighted Association Rules for Item Intensities, Knowledge and Information Systems, vol. 6, (2204) 203-229
9. Zaïane, O. R., and Antonie, M.-L. Classifying text documents by associating terms with text categories. In *Proc. of the Thirteenth Australasian Database Conference (ADC'02)*, (2002) 215–222.
10. Zaïane, O. R., Han, J., Zhu, H. Mining recurrent items in multimedia with progressive resolution refinement. In *Int. Conf. on Data Engineering*, (2000) 461–470

A New Evolutionary Neural Network Classifier

Arit Thammano* and Asavin Meengen**

Faculty of Information Technology
King Mongkut's Institute of Technology Ladkrabang,
Bangkok, 10520 Thailand
`arit@it.kmitl.ac.th`*, `asv_kmitl@hotmail.com`**

Abstract. This paper proposes two new concepts: (1) the new evolutionary algorithm and (2) the new approach to deal with the classification problems by applying the concepts of the fuzzy c-means algorithm and the evolutionary algorithm to the artificial neural network. During training, the fuzzy c-means algorithm is initially used to form the clusters in the cluster layer; then the evolutionary algorithm is employed to optimize those clusters and their parameters. During testing, the class whose cluster node returns the maximum output value is the result of the prediction. This proposed model has been benchmarked against the standard backpropagation neural network, the fuzzy ARTMAP, C4.5, and CART. The results on six benchmark problems are very encouraging.

1 Introduction

In the past decades, data are being collected and accumulated at a dramatic pace. Therefore, there is an urgent need for a new generation of computational techniques and tools to assist humans in extracting useful information (knowledge) from the rapidly growing volumes of data [1]. This arouses many researchers to study into the area of data mining. One of the data mining functionalities, which plays an important role in business decision-making tasks, is classification. Classification is the process of finding a set of models that describe and distinguish data classes or concepts, for the purpose of being able to use the model to predict the class of objects whose class label is unknown [2]. A variety of techniques have been applied to deal with the classification problems, such as neural networks, decision trees, and statistical methods. However, many previous research works show that neural network classifiers have a better performance, lower classification error rate, and more robust to noise than the other two methods mentioned above. The proposed evolutionary neural network classifier described in this paper employs the concept of the fuzzy c-mean clustering and the evolutionary algorithm to find and optimize the center and the standard deviation of each cluster. The performance of the proposed network is evaluated against the fuzzy ARTMAP, the backpropagation neural network, C4.5, and CART.

This paper is organized as follows. Following this introduction, section 2 presents the architecture of the evolutionary neural network classifier. The learning algorithm is described in section 3. In section 4, the experimental results are demonstrated and discussed. Finally, section 5 is the conclusions.

2 The Proposed Model

The architecture of the evolutionary neural network classifier is a three-layer feedforward neural network as shown in Figure 1. The first layer is the input layer, which consists of N nodes. Each node represents a feature component of the input data. The second layer is the cluster layer. The nodes in this second layer are constructed during the training phase; each node represents a cluster that belongs to one of the classes. The third layer is the output layer. Each node in the output layer represents a class. In this paper, the input vector is denoted by $X_i = (x_{i1}, ..., x_{iN})$, where i is the ith input pattern, and N is the number of features in X. The nodes in the cluster layer are fully connected to the nodes in the input layer. Therefore, once the model receives the input and its associated target output (X_i, Y_i), the input vector X_i is directly transmitted to the cluster layer via these connections. Each node in the cluster layer then calculates the membership degree to which the input vector X_i belongs to its cluster j.

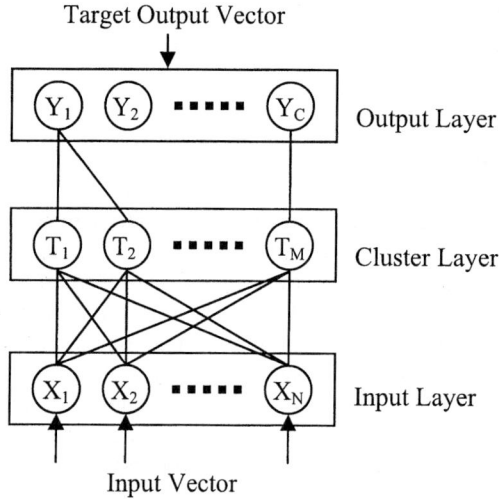

Fig. 1. Architecture of the evolutionary neural network classifier

$$T_j = \mu_j(X_i) = \text{Gaussian}(X_i, c_j, \sigma_j) = \exp\left[\frac{-d_{ji}^2}{2\sigma_j^2}\right]. \qquad (1)$$

where j = 1, 2, ..., M. It denotes the jth node in the cluster layer.

M is the total number of nodes in the cluster layer.

$d_{ji} = \|X_i - c_j\|$. It is the distance between the input vector X_i and the center of the jth cluster.

c_j is the center of the jth cluster.

σ_j is the width of the jth cluster.

The cluster node with the highest degree of membership is selected to be the winning node.

$$T_J = \max\{T_j : j = 1, 2, ..., M\}. \tag{2}$$

Then, the class of the input vector X_i is predicted as the one whose cluster node exhibits the highest degree of membership (the winning node).

3 The Learning Algorithm

During the training period of the evolutionary neural network classifier, two phases of learning are involved: Phase 1 (Initialize the cluster node) and Phase 2 (Optimize the cluster node).

3.1 Phase 1: Initialize the Cluster Node

The fuzzy c-means algorithm is used in this phase to partition the input space into M clusters, which are then used to be the nodes in the cluster layer. The goal of the fuzzy c-means algorithm is to minimize the objective function J_m, which is

$$J_m = \sum_{j=1}^{M}\sum_{i=1}^{P}\mu_{ji}^{m}d_{ji}^{2}. \tag{3}$$

where P is the total number of patterns in the training data set.

m is a weight exponent in the fuzzy membership. It can be any real number which is greater than 1.

μ_{ji} is the membership value of the ith data in the jth cluster.

d_{ji} is the distance between the ith data and the center of the jth cluster.

The detailed procedure of the fuzzy c-means algorithm is as follows:

A. Initialize the center values of all clusters.
B. Calculate μ_{ji} as follows:

$$\mu_{ji} = \left[\sum_{k=1}^{M}(d_{ji}/d_{ki})^{2/(m-1)}\right]^{-1}. \tag{4}$$

C. Update the center value c_j by using the following formula:

$$c_j = \sum_{i=1}^{P} \mu_{ji}^m x_i \Big/ \sum_{i=1}^{P} \mu_{ji}^m .\qquad(5)$$

where x_i is the ith data in the series.

D. Calculate J_m using Equation 3.
E. If a convergence criterion is reached, stop the loop. Otherwise return to step B.

After the centers of the clusters are obtained, the width of the jth cluster is calculated as follows:

$$\sigma_j = \min_{k=1}^{M} (\| c_j - c_k \|), \text{ k is not in the same class as j}.\qquad(6)$$

3.2 Phase 2: Optimize the Cluster Node

After the centers and the widths of the cluster nodes are initially determined in Phase 1, this phase employs the evolutionary algorithm to optimize those clusters and their parameters (c_j and σ_j) so that the prediction error of the system is minimal. A description of the process used in this phase is given below.

A. Define the size (S) of the initial population. Encode the initial parameters of all clusters in the cluster layer into a starting chromosome. A gene in the chromosome represents the center and the width of each cluster. For each cluster node j, create a list of input patterns whose membership value μ_{ji} are in the top S-1. Then repeat the following steps S-1 times:
  A.1. Randomly select an input pattern from each list, and assign it to be the center of the cluster. Then calculate the width of the cluster by using equation 6.
  A.2. Encode the centers and the widths of all clusters obtained from step A.1 into a new chromosome.
B. Combine the newly generated S-1 chromosomes with the starting chromosome to form the initial population.
C. Evaluate the fitness of each chromosome s in the population. The fitness function used in this research is as follows:

$$f(s) = \left[\sum_{q=1}^{S} (w_q + e_q) \right] - (w_s + e_s).\qquad(7)$$

$$w_s = \sum_{i=1}^{P} A_i .\qquad(8)$$

$$A_i = \begin{cases} 1, & \text{if } J \notin Y_i \\ 0, & \text{otherwise} \end{cases}\qquad(9)$$

$$e_s = \sum_{i=1}^{P} \frac{T_b(X_i)}{T_g(X_i)} \Big/ \sum_{q=1}^{S}\sum_{i=1}^{P} \frac{T_b(X_i)}{T_g(X_i)} .\qquad(10)$$

$$T_g(X_i) = \max_{\forall j, j \in Y_i} (T_j(X_i)). \tag{11}$$

$$T_b(X_i) = \max_{\forall j, j \notin Y_i} (T_j(X_i)). \tag{12}$$

where S is the total number of chromosomes in the population.

J is the winning cluster node.
Y_i is the target output of the ith input pattern.

If the highest fitness value of the current population is equal to that of the last generation, increase the size (S) of the current population by 2 chromosomes. As long as there is no improvement in the fitness value of the current population, the size of the current population is continuously increased by 2 chromosomes. On the contrary, when the improvement is significant enough, the size of the current population will be reduced back to the initial stage.
D. Create a new population by repeating the following steps S/2 times:
   D.1. Select two parent chromosomes from a current population using the roulette wheel technique.
   D.2. With a crossover probability, cross over the parent chromosomes to form the new offspring.
   D.3. Randomly pick a number from a set of {1, 2, 3}. If the picked number is 1, perform the mutation on the new offspring as follows:
      - Randomly select mutation positions.
      - With a mutation probability, mutate the selected positions by adding or subtracting a small random number to/from the original value of the selected positions.

      If the picked number is 2, add a new gene to the end of the new offspring. However, if the picked number is 3, delete a randomly-selected gene from the new offspring.
   D.4. Place the new offspring into a new population.
E. Combine a new population with the current population. Then select S chromosomes from the combined list according to their fitness to form the next generation.
F. If a predetermined number of iteration is reached or the end condition is satisfied, stop the loop and return the best chromosome in the current population. The genes of the best chromosome are then used as the parameters of the nodes in the cluster layer. If not, go to step C.

4 Experimental Results

To test the performance of the proposed approach, the experiments have been conducted on 6 benchmark data sets: the iris data [3], the vowel recognition problem [4], the Pima Indians diabetes database [5], the ionosphere data [5], the BUPA liver disorders data [5], and the heart disease problem [6].

Results of the experiments are shown in Table 1. For the iris data (data set 1), all three neural network models -- the fuzzy ARTMAP, the backpropagation neural network (BPNN), and the proposed model – give a perfect accuracy (100%). However, it is the opposite way round for the data set 2, where both decision trees outpace all three neural network models. For the data set 3, the proposed model comes out to be the best among the compared methods, while the performance of the rest is about the same. For the data set 4, there is no significant difference in the performance of the three neural network methods. For the data set 5, the best prediction performance (71.68%) is obtained from the fuzzy ARTMAP neural network using the vigilance parameter (ρ_a) = 0.5 and the learning rate (β) = 0.7. However, the proposed model is not far behind the performance of the fuzzy ARTMAP (50 versus 49 misclassifications). For the data set 6, the proposed model, which yields an accuracy of 85.56%, outperforms other methods by a wide margin.

Table 1. Experimental results

Data Set	% Correct				
	C4.5	CART	Fuzzy ARTMAP	BPNN	Proposed Model
1	95.0 [7]	93.0 [8]	100	100	100
2	80.0 [7]	78.2 [9]	56.71	55.19	66.23
3	73.0 [9]	74.5 [9]	75.52	76.04	80.73
4	94.9 [9]	88.9 [9]	98.68	99.34	100
5	65.0 [7]	-	71.68	67.05	71.10
6	77.5 [10]	-	77.78	78.88	85.56

5 Conclusions

In this paper, the new evolutionary neural network, which applies the concepts of the fuzzy c-means algorithm and the evolutionary algorithm to the artificial neural network, is proposed and its performances are compared with two of the best decision trees and two of the best neural networks. The results of the proposed evolutionary neural network are the best among those of the compared methods.

References

1. Fayyad, U., Piatetsky-Shapiro, G., Smyth, P.: Knowledge Discovery and Data Mining: Towards a Unifying Framework. Proceedings of the Second International Conference on Knowledge Discovery and Data Mining. (1996)
2. Han, J., Kamber, M.: Data Mining: Concepts and Techniques. Academic Press. (2001)
3. Fisher, R. A.: The Use of Multiple Measurements in Taxonomic Problems. Annual Eugenics, Vol. 7, Part II. (1936) 179-188
4. Deterding, D. H.: Speaker Normalization for Automatic Speech Recognition. Ph.D. Dissertation. (1989)
5. Blake, C. L., Merz, C. J.: UCI Repository of Machine Learning Databases [http://www.ics.uci.edu/~mlearn/MLRepository.html]. University of California, Department of Information and Computer Science. (1998)

6. Statlog Project Datasets: Retrieved March 28, 2003. From http://www.liacc.up.pt/ML/statlog/datasets/heart/
7. Ventura, D., Martinez, T. R.: An Empirical Comparison of Discretization Methods. Proceedings of the Tenth International Symposium on Computer and Information Sciences. (1995) 443-450
8. Tschichold-Gürman, N.: Fuzzy RuleNet: An Artificial Neural Network Model for Fuzzy Classification. Proceedings of the 1994 ACM Symposium on Applied Computing. (1994) 145-149
9. Datasets used for Classification: Comparison of Results: Retrieved March 5, 2005. From http://www.phys.uni.torun.pl/kmk/projects/datasets.html
10. Last, M., Maimon, O.: A Compact and Accurate Model for Classification. IEEE Transactions on Knowledge and Data Engineering, Vol. 16, No. 2. (2004) 203-215

A Privacy-Preserving Classification Mining Algorithm[1]

Weiping Ge, Wei Wang, Xiaorong Li, and Baile Shi

Department of Computing and Information Technology, Fudan University,
No.220 Handan Road, Shanghai, 200433, China
{geweiping, weiwang1, 021021093, bshi}@fudan.edu.cn

Abstract. Privacy-preserving classification mining is one of the fast-growing sub-areas of data mining. How to perturb original data and then build a decision tree based on perturbed data is the key research challenge. By applying transition probability matrix this paper proposes a novel privacy-preserving classification mining algorithm which suits all data types, arbitrary probability distribution of original data, and perturbing all attributes (including label attribute). Experimental results demonstrate that decision tree built using this algorithm on perturbed data has comparable classifying accuracy to decision tree built using un-privacy-preserving algorithm on original data.

1 Introduction

Privacy and security has become the focus of many data mining researches [1~4]. It is well known that data mining requires aggregated data other than individual data. So, this research will focus on how to make individual information private and secure while maintaining a high level of accessibility for data miners.

A novel privacy-preserving classification mining algorithm is proposed in this paper. The main idea of this algorithm consists of two parts. The first part focuses on how to perturb the original data to preserve information privacy. Firstly, "single attribute transition probability matrix" is proposed. Secondly, "multiple split attributes joint transition probability matrix" is proposed to express multiple attributes' joint perturbing probability. Thirdly, a data perturbing method is described to perturb original data by applying "single attribute transition probability matrix". The second part focuses on how to recover the original support count of attributes value from perturbed data to build a decision tree. Firstly, a formula is derived to recover the original support count of attributes value from perturbed data. Secondly, another formula is derived to calculate Gain by the original support count of attributes value to choose the best split attribute and split point. Thirdly, a narrative privacy-preserving decision tree classification algorithm-PPCART is given.

---

[1] This paper was supported by the National Natural Science Foundation of China (No.69933010, 60303008) and China National 863 High-Tech Projects (No.2002AA4Z3430).

2 Related Works

R. Agrawal proposed a privacy-preserving classification mining algorithm in which firstly sensitive values of a user's record was distorted by a randomizing function, and then the probability distribution of the original data was reconstructed by Bayesian rule [1]. However this algorithm has the following shortcomings: (1)It can't perturb label attribute, so it can't avoid privacy breaches. (2)It iteratively integrates probability density functions and differentiates distribution functions to reconstruct the probability distribution of the original data, so the reconstruction accuracy depends on the distribution of the original data. (3)It doesn't fit boolean and categorical type attributes well. Y. Lindell proposed a kind of privacy-preserving classification algorithm about building decision tree from two horizontally partitioned data sets by using security multi-party computation [2]. D. Agrawal recovered original data probability distribution by EM [3], and said the algorithm outperformed R. Agrawal's in reconstruction accuracy, but this algorithm didn't rectify the deficiencies of R. Agrawal's. W.L. Du proposed a kind of privacy-preserving classification mining algorithm based on randomized response techniques [4], but this algorithm only fitted boolean type attributes.

As is mentioned in above section, our algorithm is superior to all the others in that it rectifies their deficiencies and combines their merits.

3 Definition

3.1 Support Count of Attributes Value

Definition 1. Support count of attributes value
Let $\{A_1, A_2, \ldots, A_k\}$ denotes full attributes set X, $Y \subseteq X$, $Y \neq \emptyset$, and y_i denotes one of values of Y. Then we define the support count of y_i as the number of records which values in Y are equal to y_i.

3.2 Attribute Transition Probability Matrix

Definition 2. Single attribute transition probability matrix
Assume A is one of attributes and A has n different values a_j $(1 \leq j \leq n)$, then we define A's attribute transition probability matrix P_A as $\begin{pmatrix} \lambda_{11} & \cdots & \lambda_{1n} \\ \vdots & \ddots & \vdots \\ \lambda_{n1} & \cdots & \lambda_{nn} \end{pmatrix}$, where λ_{kl} $(1 \leq k \leq n, 1 \leq l \leq n)$ denotes the probability of attribute value a_k becoming a_l, $\lambda_{kl} \geq 0$; $\sum_{l=1}^{n} \lambda_{kl} = 1$ for any k; and P_A's inverse matrix P_A^{-1} exists.

Definition 3. Single split attribute transition probability matrix

Assume a_j is the first split point, and attribute A is split into two intervals [2] $-(a_1,a_j)$ and (a_{j+1},a_n), then we define and calculate A's split attribute transition probability matrix $P_A = \begin{pmatrix} \lambda'_{11} & \lambda'_{12} \\ \lambda'_{21} & \lambda'_{22} \end{pmatrix}$, where

$$\lambda'_{11} = (\sum_{k=1}^{j} r_k * (\sum_{i=1}^{j} \lambda_{ki}))/\sum_{k=1}^{j} r_k, \quad \lambda'_{12} = (\sum_{k=1}^{j} r_k * (\sum_{i=j+1}^{n} \lambda_{ki}))/\sum_{k=1}^{j} r_k,$$

$$\lambda'_{21} = (\sum_{k=j+1}^{n} r_k * (\sum_{i=1}^{j} \lambda_{ki}))/\sum_{k=j+1}^{n} r_k, \quad \lambda'_{22} = (\sum_{k=j+1}^{n} r_k * (\sum_{i=j+1}^{n} \lambda_{ki}))/\sum_{k=j+1}^{n} r_k,$$

r_k ($1 \leq k \leq n$) is the distribution probability of attribute value a_k in original data.

When attribute A splits again, we could continuously calculate A's split attribute transition probability matrix in the same way.

Definition 4. Double split attributes joint transition probability matrix

Assume attribute A has n split intervals a_k ($1 \leq k \leq n$), λ_{kl} ($1 \leq k \leq n, 1 \leq l \leq n$) is the element of row k and column l in P_A; attribute B has m split intervals b_i ($1 \leq i \leq m$), β_{ij} ($1 \leq i \leq m, 1 \leq j \leq m$) is the element of row k and column l in P_B. Then we define "double split attributes joint transition probability matrix" of A and B as $P_{(A,B)}$ which is a square matrix of $m*n$ dimensions, and its element of row ($(i-1)*n + k$) and column ($(j-1)*n + l$) is equal to $\beta_{ij} * \lambda_{kl}$ which denotes the probability of attribute B's value b_i becoming b_j and attribute A's value a_k becoming a_l.

Definition 5. Multiple split attributes joint transition probability matrix

Similarly, we could define and calculate "multiple split attributes joint transition probability matrix" - $P_{(A1,A2,...,Ak)}$.

3.3 Data Perturbing Method

Let original data set $T=\{T_1,T_2,...,T_n\}$, full attributes set $X=\{A_1,A_2,...,A_k\}$. Assume attribute A_j ($1 \leq j \leq k$) has n different values a_k ($k=1,2,...,n$), and A_j's transition probability matrix $P_{(Aj)} = \begin{pmatrix} \lambda_{11} & \cdots & \lambda_{1n} \\ \vdots & \ddots & \vdots \\ \lambda_{n1} & \cdots & \lambda_{nn} \end{pmatrix}$.

To each record $T_i = (T_{i(A1)}, T_{i(A2)},...,T_{i(Ak)})$ ($1 \leq i \leq n$), we independently and randomly make each attribute value $T_{i(Aj)}$ become $D_{i(Aj)}$ by the probability defined in $P_{(Aj)}$ where $1 \leq j \leq k$. Thus we get $D_i = (D_{i(A1)}, D_{i(A2)},...,D_{i(Ak)})$. Eventually we get the whole perturbed data set $D=\{D_1,D_2,...,D_n\}$.

---

[2] If A is categorical type, then A is split into two non-intersected and non-sequential subsets.

3.4 Measure of Privacy-Preserving Level

Definition 6. Measure of privacy-preserving level
Privacy-preserving level of an attribute = "square root of the number of non-zero elements of an attribute transition probability matrix"/the domain size of the attribute in original data.

4 Privacy-Preserving Classification Mining Algorithm

4.1 Method to Recover the Original Support Count of Attributes Value from Perturbed Data

Theorem 1: $P_{(A1,A2,...,Ak)}^{-1} = P_{(A1}^{-1}{}_{,A2}^{-1}{}_{,...,Ak}^{-1})$. (1)

Theorem 2: $T_{(A1,A2,...,Ak)} = D_{(A1,A2,...,Ak)} * P_{(A1,A2,...,Ak)}^{-1}$. (2)

Where $k \geq 1$, $T_{(A1,A2,...,Ak)} / D_{(A1,A2,...,Ak)}$ denotes the row matrix of $\prod_{i=1}^{k} |A_i|$ elements, and each element stands for the support count of each joint value of attributes $A_1, A_2, ..., A_k$ in original / perturbed data (If A_i is a split attribute, then $|A_i|$ is equal to the number of split intervals, otherwise $|A_i|$ is equal to the number of different attribute values).

4.2 How to Choose Split Attribute and Split Point by the Original Support Count of Attributes Value (Using CART as Prototype)

(1) Calculating gini
Assume S is a data set which has s samples. Label attribute C has m different values- $\{c_1, c_2, ..., c_m\}$, related to m different classes-$C_i(i=1,...,m)$. Then

$$\text{gini}(S) = 1 - \sum_{i=1}^{m} p_i^2 \text{, where } p_i = s_i/s \text{ (s_i is the support count of c_i).}$$

(2) Choose split attribute and split point by calculating $\text{gini}_{\text{split}}$ and Gain
Assume attribute A has v different values -$\{a_1, a_2, ..., a_v\}$ which divide S into v subsets - $\{R_1, R_2, ..., R_v\}$. s_{ij} is the number of samples which label attribute value is equal to c_i in R_j, and is actually equal to support count of $a_j c_i$, which is the joint value of attribute A and C. If there is a split point a_k which divide S into two subsets- $\{S_1, S_2\}$, where $S_1 = \bigcup_{j=1}^{k} R_j$ and $S_2 = \bigcup_{j=k+1}^{v} R_j$. Then

$$\text{gini}(S_1) = 1 - \sum_{i=1}^{m} p_i^2 \text{, where } p_i = (\sum_{j=1}^{k} s_{ij}) / (\sum_{i=1}^{m} \sum_{j=1}^{k} s_{ij}).$$

$$\text{gini}(S_2) = 1 - \sum_{i=1}^{m} p_i^2 \text{, where } p_i = (\sum_{j=k+1}^{v} s_{ij}) / (\sum_{i=1}^{m} \sum_{j=k+1}^{v} s_{ij}).$$

$$\text{gini}_{\text{split}}(S) = (\sum_{i=1}^{m}\sum_{j=1}^{k} s_{ij} / \sum_{i=1}^{m}\sum_{j=1}^{v} s_{ij})*\text{gini}(S_1) + (\sum_{i=1}^{m}\sum_{j=k+1}^{v} s_{ij} / \sum_{i=1}^{m}\sum_{j=1}^{v} s_{ij})*\text{gini}(S_2).$$

$$\text{Gain}(S, A) = \text{gini}(S) - \text{gini}_{\text{split}}(S). \quad (3)$$

The largest Gain corresponds to the best split attribute and split point. Sequentially, a decision tree is created in the same way.

4.3 Privacy-Preserving Decision Tree Classification Algorithm-PPCART (Using CART as Prototype)

Partition(S, split_attr_list&value&flag)

Input: Perturbed samples set S, split attributes, split point values and flag;

Output: a decision tree;

(1) Create node N;
(2) If (One of the support count of label attribute values is obviously large) Then (Most samples in S are the same class C), so Return (Node N as a leaf node labeled as class C);
(3) For each attribute A do
    a) Scan S and count the support counts of attributes values (here, attributes includes all split attributes up to now, attribute A and label attribute);
    b) Calculate "multiple split attributes joint transition probability matrix" [3];
    c) Recover the original support count of attributes values from perturbed data by formula (2);
    d) Calculate Gain, and find out the best split attribute and split point by above support counts by formula (3);
(4) Label node N as the best attribute & split point, and append the best attribute & split point into parameters - split_attr_list&value;
(5) Partition(S, split_attr_list&value&0) and Partition(S, split_attr_list&value&1);

5 Experimental Results

All attributes values (including label attribute values) are perturbed to preserve privacy in our experiments. Boolean, categorical, and numeric type attributes are adopted to test the adaptability of PPCART algorithm in this experiment (Please refer to document [5] for detailed attribute descriptions). In order to test the classification performance in different conditions, we adopt five classification functions introduced in document [5] to assign values to the label attribute "Group".

---

[3] In order to calculate the matrix, we need firstly recover each single attribute' r_k (described in Definition 3) by formula (2).

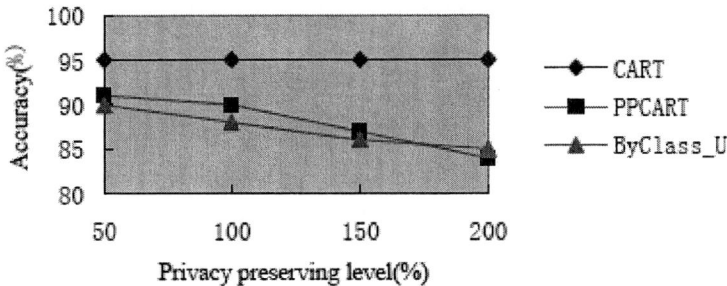

Fig. 1. Algorithm Accuracy. The figure shows the average classification accuracy of five functions to different privacy-preserving levels among PPCART, CART, and ByClass [2] on the basis of 100,000 uniformly distributed training sample records and 5,000 test sample records. Experimental results show classification accuracy of PPCART is better than ByClass. Furthermore PPCART perturbs all attributes including the label attribute -"group" to preserve privacy, and boolean, categorical and numeric type attributes are adopted in this experiment. So PPCART is superior to ByClass. Although the average classification accuracy of PPCART is 5% behind CART for privacy-preserving level of 100%, PPCART still keep 90% average classification accuracy. It shows that PPCART is reliable and practical

Other experimental results show that classification accuracy of PPCART has little bearing on the samples' distribution.

6 Conclusions and Future Work

The main topic of this paper is about privacy-preserving classification mining research. Applying transition probability matrix to privacy-preserving association rules mining and OLAP is our future work.

References

1. R. Agrawal, R. Srikant: Privacy-Preserving Data Mining. In Proc. of the ACM SIGMOD Conference on Management of Data, Dallas, Texas, May 2000. 439-450.
2. Y. Lindell and B. Pinkas: Privacy preserving data mining. In Advances in Cryptology – Crypto2000. 36-54.
3. D. Agrawal and C. Aggarwal: On the design and quantification of privacy preserving data mining algorithms. In Proceedings of the 20th Symposium on Principles of Database Systems, Santa Barbara, California, USA, May 2001.
4. W.L. Du and Z.J. Zhan: Using Randomized Response Techniques for Privacy-Preserving Data Mining. In Proceedings of the 9th ACM SIGKDD Conference on Knowledge Discovery in Databases and Data Mining, Washington, DC, USA, August 24–27 2003.
5. R. Agrawal, S. Ghost, T. Imielinski, B. Iyer, and A. Swami: An interval Classifier for database mining applications. In Proc. of the VLDB Conference, Vancouver, British Columbia, Canada, August 1992. 560-573.

Combining Classifiers with Multi-representation of Context in Word Sense Disambiguation

Cuong Anh Le[1,3], Van-Nam Huynh[2], and Akira Shimazu[1]

[1] School of Information Science
[2] School of Knowledge Science,
Japan Advanced Institute of Science and Technology,
Tatsunokuchi, Ishikawa, 923-1292, Japan
{cuonganh, huynh, shimazu}@jaist.ac.jp
[3] College of Technology, Vietnam National University, Hanoi,
144 Xuan Thuy, Cau Giay, Hanoi, Vietnam

Abstract. In this paper, we first argue that various ways of using context in WSD can be considered as distinct representations of a polysemous word under consideration, then all these representations are used jointly to identify the meaning of the target word. Under such a consideration, we can then straightforwardly apply the general framework for combining classifiers developed in Kittler et al. [5] to WSD problem. This results in many commonly used decision rules for WSD. The experimental result shows that the multi-representation based combination strategy of classifiers outperform individual ones as well as known techniques of classifier combination in WSD.

1 Introduction

Word sense disambiguation involves the association of a given word in a text or discourse with a particular sense among numerous potential senses of that word. As mentioned in [4], this is an "intermediate task" necessarily to accomplish most natural language processing tasks. Since its inception, many methods involving WSD have been developed in the literature (see, e.g., [4] for a survey). During the last decade, many supervised machine learning algorithms have been used for this task. As observed in studies of machine learning systems, although one could choose one of learning systems available to achieve the best performance for a given pattern recognition problem, the set of patterns misclassified by the different classification systems would not necessarily overlap. This means that different classifiers may potentially offer complementary information about patterns to be classified. This observation highly motivated the interest in combining classifiers during the recent years. Especially, classifier combination for WSD has been unsurprisingly received much attention recently from the community as well, e.g., [3, 10, 6, 2, 11].

As is well-known, there are basically two classifier combination scenarios. In the first scenario, all classifiers use the same representation of the input pattern. In the context of WSD, the work by Klein et al. [6], and Florian and

Yarowsky [2] could be grouped into this first scenario. In the second scenario, each classifier uses its own representation of the input pattern. An important application of combining classifiers in this scenario is the possibility to integrate physically different types of features. In this sense, the work by Pedersen [10], Wang and Matsumoto [11] can be considered as belonging to this scenario. In this paper, we focus on classifier combination for WSD in the second scenario. Particularly, we consider various ways of using context in WSD as distinct representations of a polysemous word. This allows us to immediately use the common theoretical framework for combining classifiers developed in Kittler et al. [5] to WSD problem. The experimental result shows that combining classifiers with multi-representation of context significantly improves the accuracy of WSD.

2 Classifier Combination for WSD

2.1 WSD with Multi-representation of Context

Given an ambiguous word w, which may have m possible senses (classes): c_1, $c_2, \ldots, c_m$, in a context C, the task is to determine the most appropriate sense of w. For a target word w, we may have different representations of context C corresponding to different views of context. Assume that we have such R representations: $\mathbf{f}_1, \ldots, \mathbf{f}_R$, serving for the aim of identifying the right sense of w. The set of features $\mathbf{f}_i$ is used by the i-th classifier. Due to the interpretation of $\mathbf{f}_i$'s and the role of context in WSD, quite naturally, we shall assume that the classification models are mutually exclusive, i.e. that only one model can be associated with each target w.

Under such a mutually exclusive assumption, given representations $\mathbf{f}_i$ ($i = 1, \ldots, R$), the Bayesian theory suggests that the word w should be assigned to class c_j provided the a posteriori probability of that class is maximum, namely

$$j = \arg\max_k P(c_k|\mathbf{f}_1, \ldots, \mathbf{f}_R) \tag{1}$$

Then the following decision rule is derived due to Bayes theorem:

$$j = \arg\max_k P(\mathbf{f}_1, \ldots, \mathbf{f}_R|c_k)P(c_k) \tag{2}$$

Further, we also easily see that the theoretical framework for combining classifiers developed in [5] can be applied for WSD problem as in the following.

2.2 Basic Combination Schemes

Product Rule. As we see, $P(\mathbf{f}_1, \ldots, \mathbf{f}_r|c_k)$ represents the joint probability distribution of the representations extracted by the classifiers. Assume that the representations used are conditional independent, so that the decision rule (2) can be rewritten as follows:

$$j = \arg\max_k P(c_k) \prod_{i=1}^{R} P(\mathbf{f}_i|c_k) \tag{3}$$

Using Bayes rule, we obtain the decision rule (4) that quantifies the likelihood of a hypothesis by combining a posteriori probabilities generated by the individual classifiers by means of a product rule:

$$j = \arg\max_k P(c_k) \prod_{i=1}^{R} \frac{P(c_k|\mathbf{f}_i)P(\mathbf{f}_i)}{P(c_k)} = \arg\max_k [P(c_k)]^{-(R-1)} \prod_{i=1}^{R} P(c_k|\mathbf{f}_i) \quad (4)$$

Sum Rule. Let us return to the decision rule (4), in some application it may be appropriate further to assume that a posteriori probabilities computed by the respective classifiers will not deviate dramatically from the prior probabilities because of high levels of noise of information used for making decisions [5]. In such a situation it can be assumed that the a posteriori probabilities can be expressed as:

$$P(c_k|\mathbf{f}_i) = P(c_k)(1 + \delta_{ki}) \quad (5)$$

where $\delta_{ki} \ll 1$. If we expand the product and neglect any terms of second and higher order, we can obtain the sum rule as follows:

$$j = \arg\max_k \left[(1-R)P(c_k) + \sum_{i=1}^{R} P(c_k|\mathbf{f}_i) \right] \quad (6)$$

2.3 Derived Combination Strategies

Mathematically, it is easy to see the following relation holds

$$\prod_{i=1}^{R} P(c_k|\mathbf{f}_i) \leq \min_{i=1}^{R} P(c_k|\mathbf{f}_i) \leq \frac{1}{R} \sum_{i=1}^{R} P(c_k|\mathbf{f}_i) \leq \max_{i=1}^{R} P(c_k|\mathbf{f}_i) \quad (7)$$

This relationship has suggested in [5] that the product and sum decision rules can be approximated by the upper or lower bounds appropriately, and under the assumption of equal priors, we can derive the following decision rules:

Max Rule $j = \arg\max_k \left[\max_i P(c_k|\mathbf{f}_i) \right]$

Min Rule $j = \arg\max_k \left[\min_i P(c_k|\mathbf{f}_i) \right]$

Median Rule $j = \arg\max_k \left[\frac{1}{R} \sum_{i=1}^{R} P(c_k|\mathbf{f}_i) \right]$

Majority Vote Rule $j = \arg\max_k \sum_i \Delta_{ki}$

where functions Δ_{ki} is defined by

$$\Delta_{ki} = \begin{cases} 1, & \text{if } P(c_k|\mathbf{f}_i) = \max_j P(c_j|\mathbf{f}_i) \\ 0, & \text{otherwise} \end{cases}$$

3 Representations of Context for WSD

In [10], Pedersen used the topic context with different sizes of context windows, grouping into three groups: small (with sizes of 0, 1, 2), medium (with sizes of 3, 4, 5), and large (with sizes of 10, 25, 50), for creating different representations of a polysemous word. For the purpose of experimental comparison, we simply consider the maximum window size in each group and generate nine representations by combining different sizes of left and right windows. We call this Pedersen's multi-representation of context.

On the other hand, we observe that two of the most important information sources for determining the sense of a polysemous word are the topic of context and relational information representing the structural relations between the target word and the surrounding words in a local context. Under such an observation, we have experimentally designed five kinds of representation defined as follows: $\mathbf{f}_1$ is a set of unordered words in the large context; $\mathbf{f}_2$ is a set of words assigned with their positions in the local context; $\mathbf{f}_3$ is a set of part-of-speech tags assigned with their positions in the local context; $\mathbf{f}_4$ is a set of collocations of words; $\mathbf{f}_5$ is a set of collocations of part-of-speech tags. Symbolically, we have

- $\mathbf{f}_1 = \{w_{-n_1}, \ldots, w_{-2}, w_{-1}, w_1, w_2, \ldots, w_{n_1}\}$
- $\mathbf{f}_2 = \{(w_{-n_2}, -n_2), \ldots, (w_{-1}, -1), (w_1, 1), \ldots, (w_{n_2}, n_2)\}$
- $\mathbf{f}_3 = \{(p_{-n_3}, -n_3), \ldots, (p_{-2}, -2), (p_{-1}, -1), (p_1, 1), (p_2, 2), \ldots, (p_{n_3}, n_3)\}$
- $\mathbf{f}_4 = \{w_{-l} \cdots w_{-1} w w_1 \cdots w_r | \ l + r \leq n_4\}$
- $\mathbf{f}_5 = \{p_{-l} \cdots p_{-1} w p_1 \cdots p_r | \ l + r \leq n_5\}$

where w_i is the word at position i in the context of the ambiguous word w and p_i be the part-of-speech tag of w_i, with the convention that the target word w appears precisely at position 0 and i will be negative (positive) if w_i appears on the left (right) of w. In the experiment, we design the window size of topic context (for both left and right windows) as 50 for the representation $\mathbf{f}_1$, i.e. $n_1 = 50$, while the window size n_i of local context as 3 for remaining representations.

4 Experimental Results

We tested on the datasets of four words, namely *interest*, *line*, *serve*, and *hard*, which are used in numerous comparative studies of word sense disambiguation methodologies such as Pedersen [10], Ng and Lee [9], Bruce & Wiebe [1], and Leacock and Chodorow [7]. In the experiments, we use a 10-folds cross validation and the experimental results are given in Tables 1 and 2.

From these results, we see that while the best combination strategies for Pedersen's multi-representation of context correspond to Sum rule and Median rule, that for our multi-representation of context in most cases corresponds to Product rule, with the exception of Median rule for *hard*. Turning back to Pedersen's method, we see that different representations have some overlaps between them, so that the conditional independence assumption imposed on individual classifiers may not be suitable for this multi-representation and, consequently,

Table 1. Results using Pedersen's multi-representation

	Best individual classifier (%)	Product (%)	Sum (%)	Max (%)	Min (%)	Median (%)	Majority Voting (%)
interest	86.3	90.4	**90.6**	87.9	87.6	88.8	87.8
line	82.7	87.3	**87.4**	84.1	83.7	84.6	81.6
hard	89.4	88.8	86.0	89.2	89.1	**90.4**	90.1
serve	83.7	86.2	86.5	85.6	85.0	**87.6**	85.8

Table 2. Results using our multi-representation

	Best individual classifier (%)	Product (%)	Sum (%)	Max (%)	Min (%)	Median (%)	Majority Voting (%)
interest	86.8	**91.4**	89.2	90.0	89.9	90.2	88.7
line	82.8	**89.4**	81.4	86.6	87.0	83.9	79.8
hard	90.2	89.5	85.2	89.8	89.2	**91.0**	90.4
serve	84.4	**89.6**	86.9	87.5	87.9	88.6	85.4

Table 3. The comparison with previous studies

(%)	BW	M	NL	LC	P	Best combined classifiers	
						Pedersen's representation	Our representation
interest	78	–	87	–	89	90.6	**91.4**
line	–	72	–	84	88	87.4	**89.4**
hard	–	–	–	83	–	90.4	**91.0**
serve	–	–	–	83	–	87.6	**89.6**

the Product rule does not yield the best result. On the contrary, in our multi-representation of context, each individual classifier corresponds to a distinct type of features so that the conditional independence assumption seems to be realistic.

Table 3[1] shows the comparison of results from the best classifier combination with Pedersen's method and our method of multi-representation with previous WSD studies tested on the same datasets. It is shown that the best classifier combination according to our method gives the highest accuracy in all cases.

---

[1] In Table 3, BW, M, NL, LC, and P respectively abbreviate for Bruce & Wiebe [1], Mooney [8], Ng & Lee [9], Leacock & Chodorow [7], and Pedersen [10].

5   Conclusion

In this paper we have argued that various ways of using context in WSD can be considered as distinct representations of a polysemous word for jointly using to identify its meaning. This consideration allowed us to apply a common theoretical framework for combining classifiers developed in [5] to develop numerous strategies of classifier combination for WSD. In parallel with the experiment conducted on Pedersen's multi-representation of context, we have experimentally designed a set of individual classifiers corresponding to distinct representation types of context considered in the WSD literature. It has been shown that this multi-representation of context significantly improves the accuracy of WSD by combining classifiers.

Acknowledgement

This research is partly conducted as a program for the "Fostering Talent in Emergent Research Fields" supported by the Japanese Ministry of Education, Culture, Sports, Science and Technology.

References

1. Bruce, R. and Wiebe, J. 1994. Word-Sense Disambiguation using Decomposable Models. *Proceedings of the 32nd Annual Meeting of the Association for Computational Linguistics (ACL)*, pp. 139–145.
2. Florian, R., and D. Yarowsky, Modeling consensus: Classifier combination for Word Sense Disambiguation, *Proceedings of EMNLP 2002*, pp. 25–32.
3. Hoste, V., I. Hendrickx, W. Daelemans, and A. van den Bosch, Parameter optimization for machine-learning of word sense disambiguation, *Natural Language Engineering* **8** (3) (2002) 311–325.
4. Ide, N., J. Véronis, Introduction to the Special Issue on Word Sense Disambiguation: The State of the Art, *Computational Linguistics* **24** (1998) 1–40.
5. Kittler, J., M. Hatef, R. P. W. Duin, and J. Matas, On combining classifiers, *IEEE Transactions on Pattern Analysis and Machine Intelligence* **20** (3) (1998) 226–239.
6. Klein, D., K. Toutanova, H. Tolga Ilhan, S. D. Kamvar, and C. D. Manning, Combining heterogeneous classifiers for Word-Sense Disambiguation, *ACL WSD Workshop*, 2002, pp. 74–80.
7. Leacock, C., M. Chodorow, and G. Miller, Using corpus statistics and WordNet relations for Sense Identification, *Computational Linguistics* (1998) 147–165.
8. Mooney, R. J., Comparative experiments on Disambiguating Word Senses: An illustration of the role of bias in machine learning, *Proceedings of the Conference on Empirical Methods in Natural Language Processing (EMNLP)*, 1996, pp. 82–91.
9. Ng, H. T., and H. B. Lee, Integrating multiple knowledge sources to Disambiguate Word Sense: An exemplar-based approach, *Proceedings of the 34th Annual Meeting of the Society for Computational Linguistics (ACL)*, 1996, pp. 40–47.

10. Pedersen, T., A simple approach to building ensembles of Naive Bayesian classifiers for Word Sense Disambiguation, *Proceedings of the North American Chapter of the Association for Computational Linguistics (NAACL)*, 2000, pp. 63–69.
11. Wang, X. J., and Y. Matsumoto, Trajectory based word sense disambiguation, *Proceedings of the 20th International Conference on Computational Linguistics*, Geneva, August 2004, pp. 903–909.

Automatic Occupation Coding with Combination of Machine Learning and Hand-Crafted Rules

Kazuko Takahashi[1], Hiroya Takamura[2], and Manabu Okumura[2]

[1] Keiai University, Faculty of International Studies,
1-9 sanno, Sakura, Japan
takak@u-keiai.ac.jp
[2] Tokyo Institute of Technology, Precision and Intelligence Laboratory,
4259 nagatsuta-cho midori-ku, Yokohama, Japan
{takamura, oku}@pi.titech.ac.jp

Abstract. We apply a machine learning method to the occupation coding, which is a task to categorize the answers to open-ended questions regarding the respondent's occupation. Specifically, we use Support Vector Machines (SVMs) and their combination with hand-crafted rules. Conducting the occupation coding manually is expensive and sometimes leads to inconsistent coding results when the coders are not experts of the occupation coding. For this reason, a rule-based automatic method has been developed and used. However, its categorization performance is not satisfiable. Therefore, we adopt SVMs, which show high performance in various fields, and compare it with the rule-based method. We also investigate effective combination methods of SVMs and the rule-based method. In our methods, the output of the rule-based method is used as features for SVMs. We empirically show that SVMs outperform the rule-based method in the occupation coding and that the combination of the two methods yields even better accuracy.

1 Introduction

Occupation is a very important attribute in sociology. In social surveys, data samples on occupation are mainly collected as responses to open-ended questions. Researchers then assign one of nearly 200 occupation codes to each sample [3]. The reason why respondents are not supposed to choose an occupation code in a questionnaire is that they often misunderstand their own occupation codes. This classification task by researchers is called *occupation coding*, which must be conducted immediately and accurately to statistically process occupations data as well as other variables [5]. The manual occupation coding has two problems. First, for coders, the task is time-consuming and complicated especially in large-scale surveys. Second, the results are not always consistent when coders are not experts of the occupation coding.

To solve these problems, a rule-based system has been developed, which has a rule set derived from the definitions of the occupations and heuristic knowledge of domain experts [12]. In this system, most of the rules are expressed as the form of case frames (i.e., verb-object structures). The system tries to transform responses to open-ended questions into case frames. The system has been applied to 6 surveys including JGSS surveys (Japan General Social Surveys) [1]. Although the accuracy of the system is 65~70%, it saves coders' labor and produces consistent coding results [13, 14, 15, 16]. However, the system still has three problems. First, creating accurate rules is quite difficult, because formalization of all the knowledge about respondents' occupations used in the occupation coding is almost beyond our power. Second, the system can hardly deal with responses which are not transformed into the form of case frame. [2] Third, the system requires constant efforts to maintain the rule set and the thesaurus, because both terms and expressions which respondents use in describing their occupations change with the times. Therefore, we apply a machine learning method to the occupation coding. We select SVMs, which show high performance in document classification [7, 11], since this task can be regarded as classification of very short documents. For example, the average length of occupation data in JGSS is approximately 15 characters, while that of a newspaper article in the Mainichi Shinbun Newspaper in Japanese published in 2000 [20] is approximately 550 characters. However, the rule-based system has many effective rules containing domain experts' knowledge. We apply various combination methods of SVMs and the rule-based method to the occupation coding, making use of information provided by this system.

The following are our contributions. First, we develop a new automatic occupation coding system based on SVMs, which is superior to the rule-based method in accuracy. Second, we show that SVMs are also effective in the classification task of very short "text" such as the occupation coding. Third, we show that the combination of SVMs and the rule-based method works well in the classification task. When we apply these methods to the occupation coding, we have to prepare a large training data set with correct codes. Although we expect that the more amount of a training data set, the higher an accuracy, it is desirable that the amount of a training data set is smaller for coders. Therefore, we investigate the relationship between the amount of the training data set and an accuracy.

2 Related Work

Giorgetti and Sebastiani proposed a method for automatically assigning a proper code to the response to an open-ended question in a survey [2]. As survey coding is a difficult task, they formulated the problem of automated survey coding as a text categorization task using supervised machine learning techniques. Compar-

---

[1] JGSS-2000, JGSS-2001 and JGSS-2002 are available at
http://jgss.daishodai.ac.jp .
They do not contain raw text but only coded data.

[2] In 1995SSM (Social Stratification and social Mobility survey in 1995), nearly 20% responses were not expressed in the form of a case frame [12].

ing the supervised machine learning approaches, which were Naive Bayesian classifiers and SVMs, with the dictionary-based approaches through an experiment using a corpus of social surveys (namely, General Social Surveys (GSS)), they have shown that supervised machine learning approaches significantly outperformed dictionary-based approaches. In dictionary-based approaches, response were classified according to the similarity between the feature vectors of the response and the category vectors, whose elements were the manually collected words. Park and Zhang [10] applied a combination method to Korean chunking. They forwarded the samples that could not be handled well by their rule-based method to a machine-learning method. They showed that the combination method was better in F-score than the rules or various machine learning methods alone. Isozaki and Hirao [6] applied a combination method to Zero Pronoun Resolution in Japanese. They first sorted antecedent candidates by using rules, which ranked candidates in the order of priority for applying. Second, they applied SVMs to the candidates in the order. Third, if SVMs judged that a candidate was a positive example, they selected the candidate and stopped there. They showed that this combination gave better performance than either of the two previous approaches.

3 The Occupation Coding

The occupation coding is a task in which researchers assign proper occupation codes to occupation data collected in social surveys. One of the most common occupation code sets in Japan is the SSM code set [3], which is based on a national census. It is quite hard for coders to completely learn the definitions of all the occupation codes, because the number of occupation codes is nearly 200 in both SSM and JGSS.[3] A data sample on occupation consists of responses to various questions regarding the respondents' occupations [4]:

- "job task" (open-ended),
- "industry" (open-ended),
- "employment status" (close-ended),
- "job title" (close-ended),
- "firm size" (close-ended).

We show an example of an occupation data sample and occupation code. If an occupation data is as follows:

- "job task" is "to arrange the delivery vehicles",
- "industry" is "load and unload of luggage",
- "employment status" is "2: Regular employee",
- "job title" is "1: No managerial post",
- "firm size" is "8: From 500 to 999",

then the occupation code is determined to "563" (a transportation clerk).

---
[3] The numbers of occupations are 188 in SSM and 194 in JGSS.

Although coders judge mainly from responses to "job task", they also have to consider other responses. Coders cannot readily get skill in the occupation coding, because the task is complicated. In large-scale surveys such as JGSS or SSM, coding work is conducted to the same data more than twice in order to make the results more reliable. Furthermore, in addition to the current occupations of respondents, their first occupations, current occupations of their spouses and the occupations of their parents are often asked in such surveys. Therefore, coders' labor is huge. Consequently, since the task is conducted by many coders over a long period of time, the results of the occupation coding are sometimes inconsistent with each other.

4 Rule-Based Method

In this paper, we call the previous system using rules of the form of case frame a rule-based method. At first the system was experimentally developed using a part of occupation data of SSM in 1995 [12] and has manually been improved every time a survey was conducted. In the system, it is supposed that a number of responses are represented by a verb and the noun sub-categorized by the verb. For example, "teach a tea-ceremony" corresponds to "539" (a tea master), while "teach at a primary school" corresponds to "521" (a primary school teacher).

The process of the system is as follows. First, the system extracts a triplet of a verb, the sub-categorized noun and the case of the noun from the response to "job task". The case is a shallow case such as "wo" or "de", which are postpositional particles in Japanese. Second, the system generalizes the verb to a verb class using a thesaurus [1]. Then the system searches for a rule that matches the generalized triplet. We call this type of rule $rule$-α[4] :

<verb-class, case, noun> ⇒ <occupation code>.

If a rule-α is found, the occupation code is assigned to the sample, otherwise the noun is also generalized using a thesaurus. If no rule matches the generalized triplet, "undetermined" is assigned. The system also use information of responses to "industry", if necessary. Finally, for some occupation codes, the system checks other occupation variables such as "employment status", "job title" and "firm size".[5] We call this type of rule $rule$-β[6] :

<occupation code, employment status, job title, firm size > ⇒ <occupation code>.

We illustrate the rule-based method with the above-mentioned sample. First, the system extracts a triplet that verb is "arrange", case is "wo (accusative)" and

---
[4] There are 3524 rules in the rule set.
[5] For example, if "firm size" is small, the respondent is not regarded as "manager", even if the respondent answers "management work" in "job task" .
[6] There are 27 rules.

noun is "haisha (delivery vehicle)". Second, the verb "arrange" is generalized to a verb-class "class-arrange" by a thesaurus. Then the system finds a rule-α: <class-arrange, wo, haisha(delivery vehicles)> $\Rightarrow$ <563> and assigns the occupation code "563" to the response. In this case, the system does not need to generalize the noun because a rule matching the triplet is found before the noun is generalized. For this occupation code, the system does not need to change the first determined occupation code by the rule-β. The final output of the rule-based method is "563".

In the rule-based method, it is important to enrich both the rule set and the thesaurus. We have to make a constant effort to maintain them, because new words or expressions with which respondents describe their jobs are frequently created. Moreover, this system has another problem that it can deal with only the responses transformed into the form of case frames.

The rule-based system has been applied to 6 surveys including JGSS surveys. Table 1 shows the performance of the rule-based method applied to JGSS [15, 16, 17]. Both total accuracy and accuracy for the label-assigned samples of the other 3 surveys are similar to those of JGSS [13]. The accuracy in this paper is defined as the number of correctly-classified samples divided by the number of all samples. The accuracies in Table 1 are not so high, because this system does not produce any codes for some samples. If the accuracy is measured for only the samples to which the system assigns a code, its value reaches nearly 80%.

Table 1. Performance of the rule-based method (the label-assigned samples mean samples assigned some label by rules.)

	JGSS-2000	JGSS-2001	JGSS-2002
Total accuracy	67.3	65.8	66.1
Accuracy for the label-assigned samples	80.9	79.7	79.8
Total number of samples	6,848	6,448	6,770

5   Machine Learning Method

Compared with the rule-based method, machine learning methods have the advantage that human is not required to create rules. Consequently, we do not have to make effort to maintain rules. Furthermore, machine learning methods are applicable to many domains. Although machine learning requires a large amount of training data for learning, there are both data samples on occupation and correct codes available, because in social surveys, the occupation codes have been manually checked after an automatic coding and we can use the labeled data of the previous years as training data.

There are various methods in machine learning such as decision tree or neural network. Among them, we use SVMs in the occupation coding, because SVMs [18] are superior to the other methods in accuracy in many tasks including the document classification and the dependency structure analysis [9].

5.1 Application of SVMs

We apply SVMs to the occupation coding in the following way. First, we create *basic* features from responses:

- words in responses to "job task"
- words in responses to "industry"
- responses to "employment status" and "job title"

Next, we train SVMs, which then determine the occupation codes of test samples.

We have to extend SVMs to a multi-classifier for the occupation coding, because SVMs are a binary-classifier. We use the one-versus-rest method [8].

5.2 The Combinations of SVMs and Rule-Based Method

Although the accuracy of the rule-based method is not satisfiable, its precision for "code-assigned samples" (i.e., the samples to which the rule-based method assigned a unique label) is quite high (see Table 1). We therefore propose the following four combination methods of the rule-based method and SVMs. In the first three methods, the output of the rule-based method is used as features for SVMs. In the last method, the rule-based method and SVMs are used sequentially. The proposed methods are as follows :

- *add-code* : the occupation codes provided by the rule-based method are added to the feature set of SVMs,
- *add-rule* : the rules used to determine occupation codes in the rule-based method are added to the feature set of SVMs,
- *add-code-rule* : both the occupation codes and the used rules are added to the feature set of SVMs,
- *seq* : SVMs are applied only when the rule-based method cannot determine a unique occupation.[7]

We show examples using above-mentioned sample. First, *add-code* adds a new feature "563", which is determined as occupation code by the rule-based system, to basic features of SVMs. Second, *add-rule* adds the ID of rule-α that is used in decision by the rule-based system as a new feature to basic features of SVMs. Although only one rule is used in this case, two rules are added if rule-β is used also. Third, *add-code-rule* adds two new features, which are "563" and the ID of rule-α, to basic features of SVMs. Finally, *seq* outputs "563" as the final occupation code, without proceeding to SVM classification, because the rule-based system can determine one occupation code in this case. If the rule-based system outputs two different codes or if it outputs the undetermined occupation code "999", seq uses SVMs.

We consider that seq is a sort of ensemble learning, which uses sequentially results of more than two methods, although it is not a combination of machine learning methods [11]. Similarly, we consider that add-code is a sort of stacking [19].

---

[7] There are two cases. One is the case that the rule-based method outputs undetermined code. The other is the case that it outputs more than two occupation codes.

6 Experiments

6.1 Experimental Settings and Preliminary Experiments

We conduct two kinds of experiments. In the first experiment (Experiment 1), we compare SVMs[8] with the rule-based system in terms of accuracy. We also investigate effective combinations of SVMs and the rule-based method. In the second experiment (Experiment 2), we investigate the relationship between the size of a training data set and categorization accuracy.

The dataset we used here consists of three datasets: JGSS-2000, JGSS-2001 and JGSS-2002. Each dataset has approximately 6000 to 7000 samples[9], and the total number of the samples is 20066. In Experiment 1, we use JGSS-2000 and JGSS-2001 as the training data set and JGSS-2002 as the test data set.

The purpose of the Experiment 2 is to estimate the required amount of training data in order to conduct coding work smoothly. We examine two cases; one is the case where the coded samples of the previous surveys are available (Case 1), and the other is where coders have to conduct coding work from the scratch (Case 2). In Case 1, the training dataset is JGSS-2000, JGSS-2001 and a part of JGSS-2002. The test dataset is the rest of JGSS-2002. In Case 2, the training dataset is a part of JGSS-2002, and the test dataset is the rest of JGSS-2002. Thus, both cases are the simulation of coding JGSS-2002 with or without JGSS-2000 and JGSS-2001. In both cases, we conduct n-fold cross validations (n=2, · · ·, 10), splitting JGSS-2002 in two parts in each fold. The two parts are later exchanged for experiments with small training datasets. We conduct Experiment 2 with four methods: add-code, add-rule, add-code-rule and SVMs.

We use the linear kernel in SVMs and change the soft margin parameter C within the range from 0.1 to 1.0 in Experiment 1, while we set C as the best value tuned to each method in Experiment 2. A soft margin parameter C represents a degree of allowance for exceptional samples. The smaller the value, the lighter the weight for exceptional samples.

Before Experiment 1, we conducted the following preliminary experiments using JGSS-2001.

- Experiment 2-gram/3-gram+basic features :
  We add 2-gram/3-gram in responses to "job task" and "industry" to basic features.

- Experiment kana-basic features :
  We use kana (Japanese cursive) instead of Chinese character, which are used in kana-basic features.

- Experiment 2-gram/3-gram+kana-basic features :
  We add 2-gram/3-gram in responses to "job task" and "industry" to kana-basic features.

---

[8] http://chasen.org/~taku/software/TinySVM/
[9] We excluded samples in which respondents are unemployed or students.

In these experiments, no significant improvements in results have not been observed. Feature selection by Information Gain [11] did not increase accuracy, either. Therefore, we use basic features in the following experiments.

6.2 Experiment 1: Results and Discussion

At $C = 1.0$, the accuracy of SVMs is 71.9%, which is 5.8 % higher than that of the rule-based method. One of the reasons is that according to the strategy of the rule-based method, the rule-based method does not assign any occupation code to difficult samples, while SVMs automatically assigns a code to all the samples. The rule-based method assigns a code to only 74.6% of all the samples, and 79.8% of them are correct (see Table 1). For those code-assigned samples, SVMs yield the accuracy of 78.5%, which is slightly less than that of the rule-based method.

Figure 1 displays that the accuracy values of the four combination methods as well as SVMs. These methods are, in terms of accuracy at $C = 1.0$, ranked in the following order: add-code-rule > add-code > seq ≈ add-rule > SVM > (the rule-based method). We call SVMs that are not combined with the rule-based method as SVM.

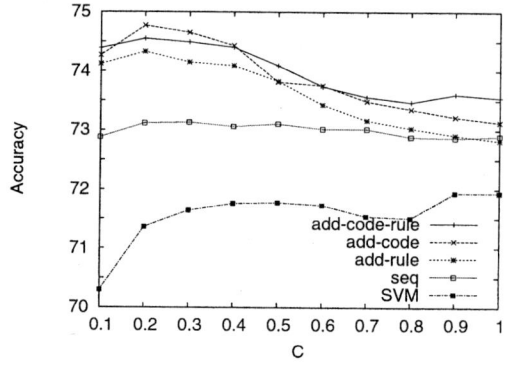

Fig. 1. Accuracy of each method with different values of C

The combination methods of SVMs and the rule-based method are effective, because the combination methods are superior to SVMs at each point of C. The accuracy of add-code is superior to that of add-rule at each point of C. A possible reason is that the added features of add-rule are more widely distributed and the reliable learning is not successful due to the lack of enough training data.[10] The accuracy of add-code-rule is mostly the same as that of add-code. This fact shows that little additional information is provided by the features of add-rule in case there are add-code features.

---

[10] Both 459 classes of rule-α rules and 18 classes of rule-β are used in add-rule, while less than 200 classes of occupation codes are used in add-code.

Since add-code, add-rule, add-code-rule and SVM are sensitive to the value of C, we need to select the best value of C beforehand. Therefore, we conducted additional experiments for tuning the value of C, splitting the training dataset into two, where we use JGSS-2000 as a temporary training dataset and JGSS-2001 as a temporary test dataset. Table 2 shows the accuracy at the predicted best value of C in each method. The accuracy of each method at these values of C corresponds to nearly the maximum accuracy.

Table 2. The predicted best values and the actual best values of C and the corresponding values of accuracy. The values of C are predicted using JGSS-2000 and JGSS-2001. The values of accuracy are calculated on JGSS-2002

		add-code	add-rule	add-code-rule	SVM
predicted best	C	0.4	0.3	0.2	0.6
	Accuracy	74.4	74.2	74.5	71.7
actual best	C	0.2	0.2	0.2	1.0
	Accuracy	74.8	74.3	74.5	71.9

In seq, 26% of all the samples are not code-assigned by the rule-based method and henceforth forwarded to SVMs. Approximately, 30% of these forwarded samples have been assigned to multiple codes and 70% to "undetermined". When seq uses SVMs, two types of samples can be used as a training data set. One consists of all the samples, and the other consists of only undetermined samples. Table 3 shows that accuracy of the former is higher than that of the latter.

Table 3. Accuracy of seq in two types of samples used as a training data set

	all the samples	only undetermined samples
$C = 1.0$	72.9	71.1
max	73.1	71.9
min	72.9	70.5

6.3 Experiment 2: Results and Discussion

Figure 2 shows the relationship between the size of a training data set and the accuracy of add-code, add-rule, add-code-rule and SVM in Case 1 and Case 2 (C is set to the best value in each method).

The accuracy increases as a size of a training data set becomes larger. The difference between accuracies of Case 1 and those of Case 2 decreases as the size of the training data set becomes larger. In both cases, add-code and add-code-rule are constantly better than add-rule and SVM. That difference is clearer in Case 2 than in Case 1. Therefore, it is strongly recommended to use add-code or add-code-rule especially when no previously coded samples are available. In each

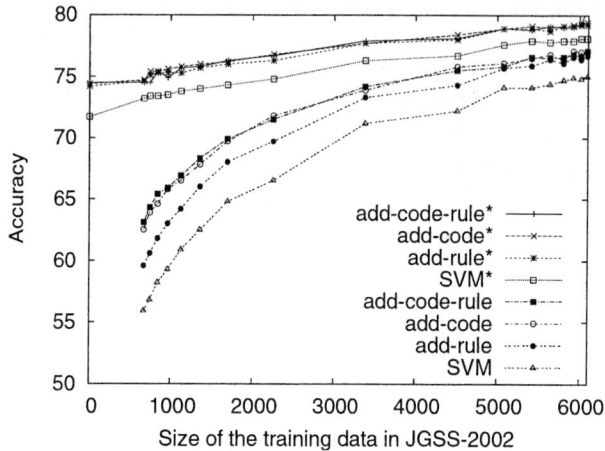

Fig. 2. Relationship between the size of training data and accuracy (the methods marked with * belong to Case 1)

method, the accuracy in Case 2 with the half of the newly-added training data approximately equals to the accuracy in Case 1 without any newly-added training samples (left-most of Figure 2). Therefore, if we cannot use coded samples, conducting the occupation coding by a half size of all the samples is effective.

7 Conclusion

We have applied SVMs to the occupation coding and shown that SVMs are superior to the rule-based method in terms of categorization accuracy also when a document is very short. We have also applied the combinations of SVMs and the rule-based method to the occupation coding and shown that each of the combination methods is superior to SVMs. Furthermore, we have conducted experiments to investigate effects of feedback and shown that a feedback is effective. In future work, we would like to find a method for measuring confidence for each output of these automatic methods, in order to support manual check of the results. We will also adopt active learning in the feedback process.

Acknowledgements

The Japanese General Social Surveys JGSS are designed and carried out at the Institute of Regional Studies at Osaka University of Commerce in collaboration with the Institute of Social Science at Tokyo University under the direction of I. TANIOKA, M. NITTA, H. SATO and N. IWAI. The project is financially assisted by Gakujutsu Frontier Grant from the Japanese Ministry of Education, Culture, Sports, Science, and Technology for 1999-2003 academic years.

References

1. The National Institute for Japanese Language Publications, editor. *Word List by Semantic Principles*. Shuei Press, 1964.
2. D. Giorgetti and F. Sebastiani. Multiclass text categorization for automated survey coding. In *Proceedings of the 18th ACM Symposium on Applied Computing (SAC-03)*, pages 798–802, 2003.
3. 1995SSM Survey Research Group, *SSM Industry and Occupation Classification (the 1995 edition)*. 1995SSM Survey Research Group, 1995.
4. 1995SSM Survey Research Group, *Codebook for 1995SSM Survey*. 1995SSM Survey Research Group, 1996.
5. J. Hara and M. Umino. *Social Surveys Seminar*. University of Tokyo Press, 1984.
6. H. Isozaki and T. Hirao. Japanese zero pronoun resolution based on ranking rules and machine learning. In *Proceedings of the 8th Conference on Empirical Methods in Natural Language Processing (EMNLP-03)*, pages 184–191, 2003.
7. T. Joachims. Text categorization with support vector machines: Learning with many relevant features. In *Proceedings of the European Conference on Machine Learning (ECML-98)*, pages 137–142, 1998.
8. U. Kressel. Pairwise classification and support vector machines. In B. Schölkopf, C. J. C. Burgesa, and A. J. Smola, editors, *Advances in Kernel Methods -Support Vector Learning*, pages 255–268. The MIT Press, 1999.
9. T. Kudo and Y. Matsumoto. Chunking with support vector machines. *Journal of Natural language Processing*, 9(5):3–22, 2002.
10. S.-B. Park and B.-T. Zhang. Text chunking by combining hand-crafted rules and memory-based learning. In *Proceedings of the 41th Annual Meeting of the Association for Computational Linguistics (ACL-03)*, pages 497–504, 2003.
11. F. Sebastiani. Machine learning automated text categorization. *ACM Computing Surveys*, 34(1):1–47, 2002.
12. K. Takahashi. A supporting system for coding of the answers from an open-ended question : An automatic coding system for SSM occupation data by case frame. *Sociological Theory and Methods*, 15(1):149–164, 2000.
13. K. Takahashi. Automatic coding system for open-ended answers : Occupation data coding in the health and stratification survey. *Keiai University International Studies*, 8(1):31–52, 2001.
14. K. Takahashi. Applying automatic occupation/industry coding system. In *Proceedings of the 8th Annual Meeting of the Association for Natural Language Processing*, pages 491–494, 2002.
15. K. Takahashi. Applying the automatic occupational/industrial coding system to JGSS-2000. *Japanese Values and Behavioral Pattern Seen in the Japanese General Social Surveys in 2000*, pages 171–184, 2002.
16. K. Takahashi. Applying the automatic occupational/industrial coding system to JGSS-2001. *Japanese Values and Behavioral Pattern Seen in the Japanese General Social Surveys in 2001[2]*, pages 179–192, 2003.
17. K. Takahashi. A combination of ROCCO-system and support vector machines in occupation coding. *Japanese Values and Behavioral Pattern Seen in the Japanese General Social Surveys in 2002[3]*, pages 163–174, 2004.
18. V. Vapnik. *Statistical Learning Theory*. John Wiley, New York, 1998.
19. D. Wolpert. Stacked generalization. *Neural Networks*, 5:241–259, 1992.
20. Mainichi, *CD Mainichi Shinbun 2000*. Nichigai Associates Co., 2001.

Retrieval Based on Language Model with Relative Entropy and Feedback*

Hua Huo[1,2] and Boqin Feng[1]

[1] Department of Computer Science,
Xi'an Jiaotong University, Xi'an, P.R.China
hhuo@mail.xjtu.edu.cn
[2] Institute of Electronics and Information,
Henan University of Science and Technology, Luo yang, P.R.China

Abstract. A new method for information retrieval which is on the basis of language model with relative entropy and feedback is presented in this paper. The method builds a query language model and document language models respectively for the query and the documents. We rank the documents according to the relative entropies of the estimated document language models with respect to the estimated query language model. The feedback documents are used to estimate a query model by the approach that we assume that the feedback documents are generated by a combined model in which one component is the feedback document language model and the other is the collection language model. Experimental results show that the method is effective for feedback documents and performs better than the basic language modeling approach. The results also indicate that the performance of the method is sensitive to both the smoothing parameters and the interpolation coefficients used to estimate the values of the language models.

1 Introduction

The language modeling approach to information retrieval has recently been proposed as a new alternative to traditional vector space models and other probabilistic models owing to its sound theoretical basis and good empirical success. Although the language modeling approach has performed well empirically, a significant amount of performance increase is often due to feedback [5]. Several recent papers[3][4] have presented techniques for improving language modeling techniques using relevance or pseudo relevance feedback. In the most of existing work, feedback has so far only been deal with heuristically within the language modeling approach, and it has been incorporated in an unnatural way: by expanding a query with a set of terms. Such an expansion-based feedback strategy

---

* This research is supported by the Natural Science Foundation Program of the Henan Provincial Educational Department in China(200410464004) and the Science Research Foundation Program of Henan University of Science and Technology in China(2004ZY041).

is generally not very compatible with the essence of the language modeling approach, which is model estimation. As a result, the expanded query usually has to be interpreted differently than the original query. This is in contrast to the natural way of performing feedback in the classical relevance-based probabilistic model, such as the binary independence model [9].

We propose a new retrieval method based on language model with relative entropy and feedback. Different from the traditional language method in which the query is assumed being generated from the document language model, in our method we assume that the query and the document are respectively generated from the query language model and the document language model. We rank the documents according to the relative entropies of the estimated document language models with respect to the estimated query language model. We think that our assumption that the query is generated from the query model is more reasonable than the assumption that the query is generated from the document language model, because the query and the document are not completely same. In order to better capture the important processes behind relevance feedback and query expansion, we believe it is important to view the query as a sample from a model of the information need. So, when estimating a query model, we develop a natural approach to perform feedback, in which we assume the feedback documents are generated by a combined model in which one component is the feedback document language model and the other is the collection language model.

2   Language Model for Retrieval

The general idea of using language model for information retrieval is to build a language model M_d for each document d, and rank the documents according to how likely the query q can be generated from each of these document models, i.e. $p(q|M_d)$. In different models, the probability is calculated in different ways. There are two typical methods for doing it. For example, Ponte and Croft [6] treat the query as a set of unique terms, and use the product of two probabilities – the probability of producing the query terms and the probability of not producing other terms – to approximate $p(q|M_d)$. The formula of this method is

$$p(q|M_d) = \prod_{w \in q} p(w|M_d) \prod_{w \, not \in q} (1.0 - p(w|M_d)) \qquad (1)$$

Song and Croft [6] treat the query as a sequence of independent terms, taking into account possibly multiple occurrences of the same term. Thus the query probability can be obtained by multiplying the individual term probabilities, and the formula can be written as

$$p(q|M_d) = \prod_{i=1}^{n} p(w_i|M_d) \qquad (2)$$

where w_i is the ith term in the query.

In our approach, we also use language modeling for retrieval, but it is different from the above approaches in which both the query and the document are assumed to be generated from the document language model. We assume that the query is generated from the query language model while the document is generated from the document language model.

3 Relative Entropy

The relative entropy is a measure of the distance between two distributions. In statistics, it arises as an expected logarithm of the likelihood ratio. The relative entropy $D(p||q)$ is a measure of the inefficiency of assuming that the distribution is q when the true distribution is p.

Definition 1. *The relative entropy between two probability mass functions $p(x)$ and $q(x)$ is defined as $D(p||q) = \sum_x p(x) \log \frac{p(x)}{q(x)}$*

In the above definition, we use the convention (based on continuity arguments) that $0 \log \frac{0}{0} = 0$ and $p \log \frac{p}{0} = \infty$.

The relative entropy is always non-negative and is zero if and only if $p = q$. However, it is not a true distance between distributions since it is not symmetric and does not satisfy the triangle inequality. Nonetheless, it is often useful to think of relative entropy as a "distance" between distributions [1].

4 Retrieval Based on Relative Entropy

4.1 Ranking Model with Relative Entropy

Let's suppose that a query q is generated by a generative model $p(q|M_q)$ with M_q denoting the parameters of the query unigram language model. Similarly, assume that a document d is generated by a generative model $p(d|M_d)$ with M_d denoting the parameters of the document unigram language model. Let $\hat{M}_q$ and $\hat{M}_d$ be the estimated query language model and document language model respectively, so, the relevance value of d with respect to q can be measured by the following function:

$$R(\hat{M}_q||\hat{M}_d) = -D(M_q||M_d) \qquad (3)$$

$$= -\sum_w p(w|M_q) \log \frac{p(w|M_q)}{p(w|M_d)} \qquad (4)$$

$$= \sum_w p(w|M_q) \log p(w|M_d) - \sum_w p(w|M_q) \log p(w|M_q) \qquad (5)$$

The second term of the formula (5) is a query-dependent constant, or more specifically, the entropy of the query model M_q. It can be ignored for the purpose of ranking documents, so we have a ranking formula such as

$$R(\hat{M}_q||\hat{M}_d) \propto \sum_w p(w|M_q) \log p(w|M_d) \tag{6}$$

From the formula (6), we can observe that how to estimate the values of $p(w|M_d)$ and $p(w|M_q)$ will influences the retrieval accuracy of the ranking model.

4.2 Estimating the Document Model

In the language model, the simplest method to estimate $p(w|M_d)$, which is a unigram language model based on the given document d, is the maximum likelihood estimator, simply given by relative counts

$$p(w|M_d) = P_{ml}(w|M_d) \tag{7}$$

$$= \frac{tf(w,d)}{\sum_{w'} tf(w',d)} \tag{8}$$

where $tf(w,d)$ is the number of times the word w occurs in the document d, $\sum_{w'} tf(w',d)$ is the total number of times all words occur in the document d, it is essentially the length of the document d.

However, one obstacle in applying the maximum likelihood estimator to estimate $p(w|M_d)$ is the problem of zero probability [9]. From the formula (8), we can see that if a word is unseen in the document d, we will get a zero probability according the maximum likelihood estimator. But a zero value of $p(w|M_d)$ is not permitted to appear in the formula (6). To the problem, we use smoothing methods in the estimating of $p(w|M_d)$.

Some smoothing methods, such as Good-Turing method, Jelinek-Mercer method, and Absolute discounting etc., have been proposed, mostly in the context of speech recognition tasks. In general, all smoothing methods are trying to discount the probabilities of the words seen in the document, and then to assign the extra probability mass to the unseen words according to some "fallback" model [7]. For information retrieval, it makes much sense, and is very common, to exploit the collection language model as the fallback model. Because a retrieval task typically requires efficient computations over a large collection of documents, our study is constrained by the efficiency of the smoothing method. In this paper, we select the Jelinek-Mercer method and Absolute discounting method which are popular and relatively efficient to implement.

The *Jelinek − Mercer method* involves a linear interpolation of maximum likelihood model with the collection model, using a coefficient λ to control the influence of each model [8].The method is given by

$$p(w|M_d) = (1-\lambda) p_{ml}(w|M_d) + \lambda p(w|C) \tag{9}$$

$$p(w|C) = \frac{tf(w,C)}{\sum_{w'} tf(w',C)} \tag{10}$$

The idea of the *Absolute discounting method* is to lower the probability of seen words by subtracting a constant from their counts. It is similar to the Jelinek-Mercer method, but differs in that it discounts the seen word probability by subtracting a constant instead of multiplying it by $1-\lambda$. The model is given by

$$p(w|M_d) = \frac{max(tf(w,d) - \delta, 0)}{\sum_{w'} tf(w',d)} + \sigma p(w|C) \qquad (11)$$

where $\delta \in [0,1]$ is a discount constant and $\sigma = \frac{\delta |d|_u}{|d|}$, so that all probabilities sum to one. Here, $|d|_u$ is the number of unique terms in the document d, and $|d| = \sum_{w'} tf(w',d)$ is the total count of words in the document [11].

4.3 Estimating the Query Model

The simplest way to estimate $p(w|M_q)$ is also the maximum likelihood estimator, which gives us

$$p(w|M_q) = P_{ml}(w|M_q) \qquad (12)$$

$$= \frac{tf(w,q)}{\sum_{w'} tf(w',q)} \qquad (13)$$

But for using feedback documents to improve retrieval performance, we explore a new method to exploit feedback documents when estimating the query language model, which is different from the methods in [3] and [4]. In the new method, we assume the feedback documents are generated by a combined model in which one component is the feedback document language model and the other is the collection language model.

Let q_0 be the original query, and $p(w|M_{q_0})$ be the original query language model, q be the updated query, and $p(w|M_q)$ be the updated query language model. We assume that $F = (f_1, f_2, ..., f_n)$ is the set of feedback documents which are judged to be relevant by a user, or which are the top documents from an initial retrieval, and $p(w|M_F)$ is the language model of the set F. We employ a linear interpolation strategy for combining the language model of the feedback documents set with the language model of the original query. Then, the updated query model $p(w|M_q)$ is

$$p(w|M_q) = (1-\alpha)p_{ml}(w|M_{q_0}) + \alpha p(w|M_F) \qquad (14)$$

where α controls the influence of the feedback documents set model to $p(w|M_q)$. We will describe how to estimate $p(w|M_F)$ as follows.

For estimating $p(w|M_F)$, we assume that the feedback documents are generated by a probabilistic model $p(F|M_F)$. Specifically, assume that each word in F is generated independently according to M_F by a generative model which is a unigram language model. That is,

$$p(F|M_F) = \prod_i \prod_w p(w|M_F)^{tf(w,f_i)} \qquad (15)$$

where $tf(w, f_i)$ is the number of times word w occurs in the document f_i.

However, not all information contained in the feedback documents is relevant to the query. Some information may be "background noise". The "background noise" is not considered in the above model, so, it is not very reasonable. A more reasonable model would be a combined model that generates a feedback document by combining the feedback document model $p(w|M_F)$ with a collection language model $p(w|C)$. For most of the information contained in the collection is irrelevant, it is reasonable to use the collection language model as the model of the "background noise" in a feedback document. Under the simple combined model, $p(F|M_F)$ is described as

$$p(F|M_F) = \prod_i \prod_w ((1-\beta)p(w|M_F) + \beta p(w|C))^{tf(w,f_i)} \tag{16}$$

where β is a parameter that indicates the amount of "background noise" in the feedback documents, and that needs to be set empirically.

Now, for the given β, the feedback documents set F, and the collection language model $p(w|C)$, we can use EM(Expectation Maximum) algorithm to compute the maximum likelihood estimate of M_F. The estimated M_F is as

$$\hat{M_F} = argmax_{M_F} \log p(F|M_F) \tag{17}$$

The EM updating formulas for $p_\beta(w|M_F)$ are as follows [9].

$$h^{(n)}(w) = \frac{(1-\beta)p_\beta^{(n)}(w|M_F)}{(1-\beta)p_\beta^{(n)}(w|M_F) + \beta p(w|C)} \tag{18}$$

$$p_\beta^{(n+1)}(w|M_F) = \frac{\sum_{j=1}^{n} tf(w,f_j)h^{(n)}(w)}{\sum_i \sum_{j=1}^{n} tf(w_i,f_j)h^{(n)}(w_i)} \tag{19}$$

We use the result of $p_\beta(w|M_F)$ computed iteratively by the above EM updating formulas as a substitute for $p(w|M_F)$ of the formula (14). Then we can obtain the value of $p(w|M_q)$ by using the formula (14).

5 Experiments and Results

Our goal is to study the performance of our method presented in this paper by comparing it with other retrieval methods. For the convenience of describing the experiments, we call our method using the Jelinek-Mercer smoothing technique the REJM method, and call our method using the Absolute discounting smoothing technique the READ method, while we call the method (with feedback) in [4] the BLM method. We use precision-recall plot and average precision as two performance measures to evaluate the above three methods.

5.1 Data Sets

We experiment over three data sets taken from TREC. They are the Associate Press Newswire (AP) 1988-90 with queries 51-150, the Financial Times (FT) 1991-94 with queries 301-400, and the Federal Register (FR) 1988-89 with queries 51-100. The first two data sets are news corpora and they are homogenous. In contrast, FR is a heterogeneous collection consisting of long documents than can span different subject areas. Queries are taken from the title field of TREC topics. Relevance judgments are taken from the judged pool of top retrieved documents by various participating retrieval system from previous [2]. Detail information of the data sets is given in Table 1.

Table 1. Information of data sets

Data set	Contents	Number of Docs	Size	Queries
AP	Associate Press Newswire 1988-90	242918	0.73GBytes	TREC topics 51-150
FT	Financial Times 1991-94	210158	0.56GBytes	TREC topics 301-400
FR	Federal Register 1988-89	45820	0.47GBytes	TREC topics 51-100

5.2 Experimental Setup

Two sets of experiments are performed in this paper. The first set of experiments is to compare the performances of the READ method and the REMJ method with the performance of the BLM method. We get different results with different parameter setting in the READ method and the REJM method, but we use the best results of them to compare their performances. The second set of experiments investigates whether the performances of the READ method and the REMJ method are sensitive to the parameters which are λ, δ, α, and β.

For convenience of showing how the average precision changes according to different values of the two interpolation coefficients α and β, in the experiment, we assign fixed values (which make the methods have best retrieval performance empirically) to a coefficient while we change the values of the other coefficient.

5.3 Experimental Results

Results of the first set of experiments are shown in Fig.1 and Table 2. From the precision-recall plots in Fig.1, we can observe that the performances of the REJM method and the READ method are better than that of the BLM method on the three data sets. In Table 2, it is shown that the average precision of the READ and the REJM on the three data sets are all better than the BLM method. We think that the improvements are mainly attributed to that the methods of performing feedback in REJM and READ are more compatible with the essence of the language modeling approach than that in BLM . And, we also note that

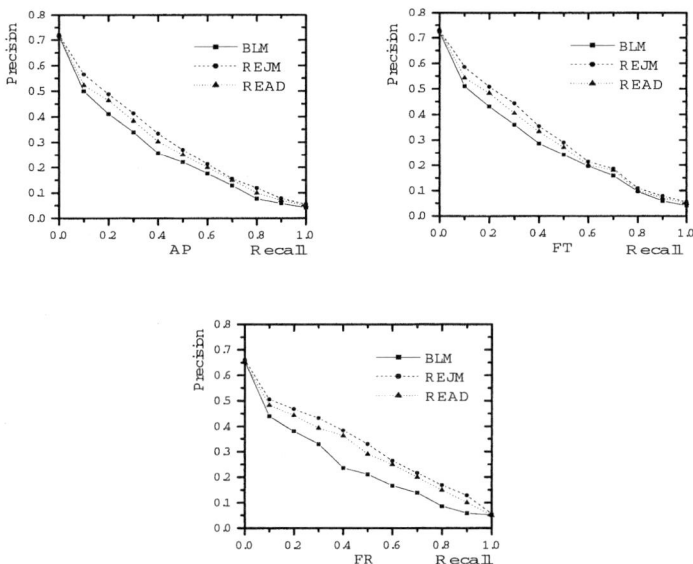

Fig. 1. Precision-Recall plots of BLM, REJM, and READ on AP, FT and FR

Table 2. The average precision of BLM, READ, and REJM on AP, FT and FR

Data set	BLM	READ	Chg1	REJM	Chg2
AP	0.265	0.291	+9%	0.310	+17%
FT	0.282	0.305	+8%	0.322	+14%
FR	0.250	0.306	+22%	0.328	+31%

the REJM method performs better in performance than the READ method. We think it is a result of that the Jelinek-Mercer smoothing method performs better than the Absolute discounting smoothing method.

Results of the second set of experiments are presented in Fig.2. The plots of part (a) in Fig.2 show the average precision of the REJM method on the three data sets according to different values of λ, while $\alpha=0.7$ and $\beta=0.6$. We can observe that the average precision of the REJM method is quite sensitive to the setting of λ, and that the average precision is better when the value of λ approximates to 0.6.

The plots of part (b) in Fig.2 show the average precision of the READ method on the three data sets for different settings of the parameter δ, while the parameters α and β are set to be 0.7 and 0.6 respectively. Similarly, we can also observe that the average precision of the READ method is sensitive to the setting of δ.

The plots of part (c) in Fig.2 show the average precision of the REJM method and the READ method for different values of α, while $\lambda=0.6$, $\delta=0.7$, and $\beta=0.55$. We find that the average precision of the REJM method and the READ method is significantly sensitive to the setting of α on both AP data set and FT data

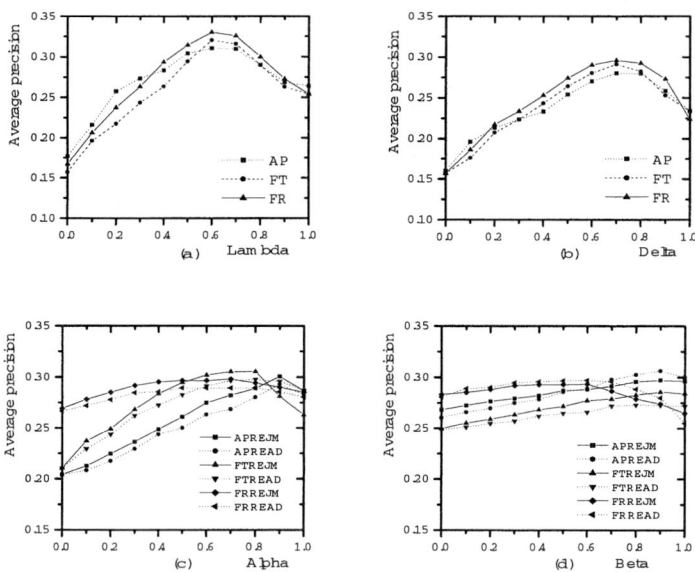

Fig. 2. Sensitivity of average precision to the parameters on AP, FT, and FR

set, but it is relatively sensitive on FT data set. We think that the main reason may be the AP data set and the FT data set are homogenous, but the FR data set is heterogeneous.

The plots of part (d) in Fig.2 show the average precision of the REJM method and the READ method on the three data sets for different values of β, while $\lambda=0.6$, $\delta=0.7$, and $\alpha=0.7$. It is obvious that the average precision of the two methods is relatively sensitive to the setting of β on three data sets, and the average precision of the two methods decreases while the value of β is larger than 0.55 on the FR data set. The phenomenon may be also relevant to that the FR data set is heterogeneous.

6 Conclusions

We have presented a new method for information retrieval based on language model with relative entropy and feedback. Experimental results show that the method performs better than the method in [4]. Analysis of the results indicates that the performance is sensitive to the smoothing parameters used to estimate the value of the document language model. And the analysis also indicates that the performance is not always very sensitive to the interpolation coefficients which are used to estimate the value of the query language model. It sometimes is only relatively sensitive on different data sets.

References

1. Cover, T.M., Thomas, J.A.: Elements of Information Theory. Beijing, Tsinghua University Press, (2003)15-18.
2. Liu,X., Croft, W.B.: Cluster-based retrieval using language models. In proceedings of ACM SIGIR'04 conference, (2004)276-284.
3. Miller, D.H., Leek,T. & Schwartz, R.: A hidden Markov model information retrieval system. In proceedings of ACM SIGIR'99 , (1999)214-221.
4. Ponte, J.: Language Models for Relevance Feedback. In W.B.Croft (Ed.), Advances in Information Retrieval: Recent Research from the CIIR. Kluwer Academic Publishers, chapter3, (2000)73-95.
5. Ponte,J., Croft,W.B.: A language modeling approach to information retrieval. In proceedings of ACM SIGIR'98, (1998)275-281.
6. Song, F., Croft,W.B.: A general language model for information retrieval. In proceedings of the 22nd annual international ACM-SIGIR'99,(1999)279-280.
7. Zaragoza,H., Hiemstra,D. & Tipping,M.: Bayesian extension to the language model for ad hoc information retrieval. In proceedings of ACM SIGIR'03.(2003)325-327.
8. Zhai,C., Lafferty,J.: A study of smoothing methods for language models applied to ad hoc information retrieval. In proceding of SIGIR'01, (2001)334-342.
9. Zhai,C., Lafferty,J.: Model-based feedback in the language modeling approach to information retrieval. In proceding of SIGIR'01, (2001)403-410.

Text Classification for DAG-Structured Categories

Cao D. Nguyen, Tran A. Dung, and Tru H. Cao

Faculty of Information Technology,
Ho Chi Minh City University of Technology, Vietnam
tru@dit.hcmut.edu.vn

Abstract. Hierarchical text classification concerning the relationship among categories has become an interesting problem recently. Most research has focused on tree-structured categories, but in reality directed acyclic graph (DAG) - structured categories, where a child category may have more than one parent category, appear more often. In this paper, we introduce three approaches, namely, flat, tree-based, and DAG-based, for solving the multi-label text classification problem in which categories are organized as a DAG, and documents are classified into both leaf and internal categories. We also present experimental results of the methods using SVMs as classifiers on the Reuters-21578 collection and our data set of research papers in Artificial Intelligence.

Keywords: text classification, hierarchies, multi-labels, SVM.

1 Introduction

The problem of text classification has been examined for a long time but most of studies have focused on *flat classification* ([24]), in which the predefined categories were treated in isolation, with no consideration about the relationship among them. For this problem, one-against-the-rest and pairwise classification methods ([11], [14]) have been widely applied. Based on the classification result of each binary classifier, those methods provided different ways to determine the categories to which a document should be assigned.

Meanwhile, categories are usually organized in a hierarchical structure, usually like a tree or a directed acyclic graph (DAG). Therefore, researchers have turned into a new classification approach, namely *hierarchical classification*, in which the subsumption relationship among categories was taken into account for classification. Using the top-down level-based approach, a hierarchical classification method constructed some classifiers at each level of the category hierarchy. Beginning from the root category, the classifiers at one level determined if the documents presented to them should be passed to the classifiers at the next lower level for further classification. The categories of a document were those assigned to it during such a classification process.

Some methods have been developed for solving the classification problem on hierarchical categories ([2], [3], [8], [13]). In [13], the classification method assigned a document to only one category, which was a leaf category, assuming that the document also belonged to all the ancestors of that leaf category. For each internal category, a Naïve Bayes classifier was built to predict the probabilities of its child categories. Then the child category with the highest probability was selected for further classification at the next level. That classification process continued until a leaf category was reached. In order to classify a document to more than one leaf category, [8] improved [13] by defining a threshold for each level of the category tree, and considering all the child categories at the next level that exceeded the threshold for further classification, where SVMs were used as binary classifiers.

That a document was assigned to only leaf categories was not satisfactory when, in practice, a document might not belong to a leaf category but an ancestor of that leaf category instead. In [3], classification was viewed as a mapping from a document into the leaf categories of a subtree of a category tree. As such, a document could be classified into both leaf and internal categories of the original tree. However, that did not allow a document to have both a category and some of its ancestors as its labels. Meanwhile, in [2] a document could be labeled by any subset of categories in a category tree of discourse.

However, all the above-mentioned methods were for tree-structured categories only, where a category had at most one parent category. Ontologically, a category may have multiple parent categories, as often encountered in the real world. In this paper we introduce three approaches that can solve the multi-label text classification problem for DAG-based categories and analyze their performances. The first approach is called *DAG-based*, which manipulates the category DAG directly. The second approach is called *tree-based*, which transforms the category DAG into an equivalent tree and adapts the approach in [2] for it. The third approach is called *flat*, which converts the problem into a flat classification one.

The paper is organized as follows. Firstly, Section 2 reviews the hierarchical method in [2] for category trees. Section 3 presents and analyzes our three methods for category DAGs. Section 4 shows experimental results using SVMs as binary classifiers. Finally, Section 5 concludes the paper and suggests future work.

2 A Classification Approach for Category Trees

The approach proposed in [2] was hierarchical classification for category trees where documents were assigned to both leaf and internal categories. All involved classifiers were binary ones. The subsumption relationship among categories was used during the training and classification phases of those binary classifiers.

Building Classifiers

- For each category C_i in the category tree of discourse, the coverage of C_i, denoted by $Coverage(C_i)$, is the set of those categories in the subtree rooted at C_i, including C_i. Function $Parent(C_i)$ returns the parent category of C_i.
- For each internal category C_i, a binary classifier called *subtree-classifier* is built to determine whether a document should be assigned to any category in $Coverage(C_i)$.
- For each category C_i, another binary classifier called *local-classifier* is built to determine whether a document should be assigned to C_i.

Training Phase
Appropriate positive and negative training documents, respectively denoted by *+ve* and *–ve*, are selected for each kind of the above-mentioned classifiers.
- Subtree classifier of an internal category C_i:
  - *+ve*: all documents d_j such that d_j is labeled by a category in *Coverage(C_i)*.
  - *–ve*: all document d_j such that d_j is not labeled by any category in *Coverage(C_i)* but by a category in *Coverage(Parent(C_i))*.
- Local classifier of an internal category C_i:
  - *+ve*: all documents d_j such that d_j is labeled by C_i.
  - *–ve*: all documents d_j such that d_j is not labeled by C_i but by a category in *Coverage(C_i)*.
- Local classifier of a leaf category C_i:
  - *+ve*: all documents d_j such that d_j is labeled by C_i.
  - *–ve*: all documents d_j such that d_j is not labeled by C_i but by a category in *Coverage(Parent(C_i))*.

Classification Phase
Information about the tree structure of categories is used during the classification phase. It is a top-down level-based classification process in which a document is presented to the classifiers of a category. If the classification process at that category cannot go further down, the classification along that branch will stop without consideration of its next level classifiers. Starting from the root category, for each document d_j presented to a category C_i, do the followings:

- If C_i is an internal category:
  The subtree classifier of C_i is used to classify d_j. If d_j is negatively classified, then return. Otherwise:
  - Use the local classifier of C_i to classify d_j. If d_j is positively classified, then add C_i into the set of categories for d_j.
  - Continue classifying d_j by the classifiers of the child categories of C_i.
- If C_i is a leaf category:
  Use the local classifier of C_i to classify d_j. If d_j is positively classified, then add C_i into the set of categories for d_j.

The approach was realized using SVM binary classifiers. The experiment results showed that it performed well for the Reuters-21578 collection [18] if given enough training documents. However, the method cannot be used directly for categories organized as a DAG, in which a category may have more than one parent. We have adapted it for DAG-structured categories as presented in the next section.

3 Classification Approaches for Category DAGs

3.1 Tree-Based Approach

The approach proposed in [2] can be adapted for a category DAG by transforming the graph into an equivalent tree. Each category having multiple parents is copied into different nodes whose number is equal to the number of branches from the root leading to that category. The tree can be created by traversing the graph in the depth-first order. Each time a category is visited, a copy of its is made and indexed by the visiting time. Figure 3.1 illustrates such a transformation. Then, although the copies of a category are physically separated, they are logically treated as the same label in the training and classification phases.

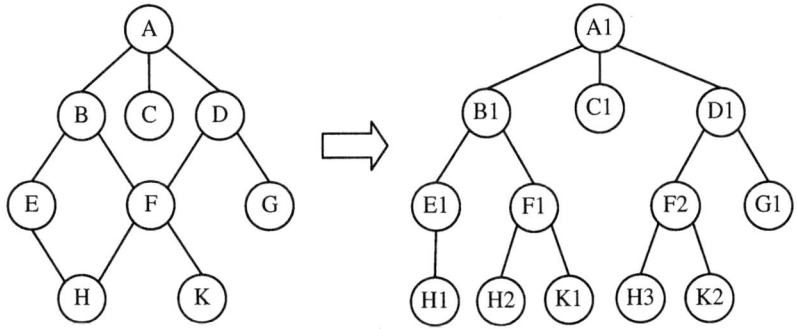

Fig. 3.1. DAG-to-Tree Transformation

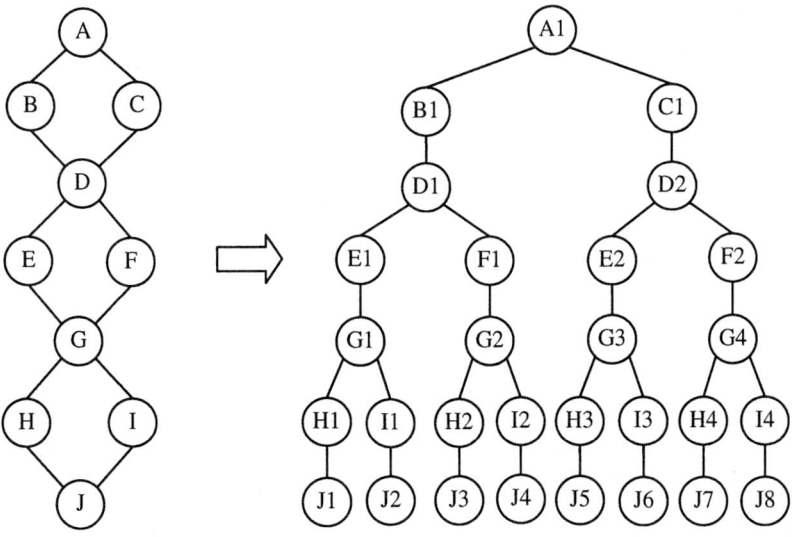

Fig. 3.2. A Large Category Tree Transformed from a Category DAG

However, this adapted classification approach has two main drawbacks. Firstly, the resulted tree may be very large if the original DAG contains cascaded nodes with multiple parents as exemplified in Figure 3.2. Secondly, such a category tree has completely similar classifiers. For example in Figure 3.1, the classifiers of F1, F2 are similar because they are trained with the same training set. That wastes computation and time because, when a document is classified into the subtree rooted at F1, it does not need to be considered by any classifier in the subtree rooted at F2, and vice versa.

3.2 DAG-Based Approach

We propose another approach also adapted from [2] but manipulates directly on a category DAG, without transformation into a tree. The main idea is that each internal category (or leaf category) with n parents is equipped with n subtree classifiers (or local classifiers, respectively). So, not as in the tree-based approach, there can be more than one subtree classifier at an internal category (and more than one local classifier at a leaf category). A document is also classified into both leaf and internal categories.

Building Classifiers

- For each category C_i in a given category DAG, the coverage of C_i, denoted by $Coverage(C_i)$, is the set of the categories in the subgraph rooted at C_i, including C_i.
- For each internal category C_i and a parent category C_k of C_i, a binary classifier called *subtree-classifier$_{ik}$* is built to determine whether a document passed from C_k should be classified into any category in $Coverage(C_i)$.
- For each internal category C_i, another binary classifier called *local-classifier* is built to determine whether a document should be classified into C_i.
- For each leaf category C_i and a parent category C_k of C_i, a binary *local-classifier$_{ik}$* is built to determine whether a document passed from C_k should be classified into C_i.

Training Phase

Appropriate positive and negative training documents, respectively denoted by +ve and –ve, are selected for each kind of the above-mentioned classifiers.

- Subtree-classifier$_{ik}$ of an internal category C_i and its parent category C_k:
  - +ve: all documents d_j such that d_j is labeled by a category in $Coverage(C_i)$.
  - –ve: all documents d_j such that d_j is not labeled by any category in $Coverage(C_i)$ but by a category in $Coverage(C_k)$.
- Local classifier of an internal category C_i:
  - +ve: all documents d_j such that d_j is labeled by C_i.
  - –ve: all documents d_j such that d_j is not labeled by C_i but by a category in $Coverage(C_i)$.
- Local-classifier$_{ik}$ of a leaf category C_i and its parent category C_k:
  - +ve: all documents d_j such that d_j is labeled by C_i.
  - –ve: all documents d_j such that d_j is not labeled by C_i but by a category in $Coverage(C_k)$.

Classification Phase

Information about the DAG structure of categories is used during the classification phase. Not as in the tree-based approach, the classification along the branch leading to a category will stop in two cases: (1) it cannot go further down; or (2) that category has already been in the list of categories assigned to the document in the previous steps. For each document d_j presented to a category C_i from its parent category C_k, do the followings:

- If C_i is an internal category:
  Subtree-classifier$_{ik}$ of C_i is used to classify d_j. If d_j is negatively classified, then return. Otherwise:
  - Use the local classifier of C_i to classify d_j. If d_j is positively classified, then add C_i into the set of categories for d_j.
  - Continue classifying d_j by the classifiers at the child categories of C_i.
- If C_i is a leaf category:
  Use local-classifier$_{ik}$ of C_i to classify d_j. If d_j is positively classified, then add C_i into the set of categories for d_j.

3.3 Flat Approach

Usually, a category is not fully covered by its child categories. That is, given a document of a category, it is not necessarily classifiable into any of its child categories. For each internal category, one can add in a dummy child category to represent the remaining unspecified subcategory of that category. Then, all the child categories of a category collectively cover it, and all the leaf categories, including the dummy ones, form an exhaustive set of categories for all documents in a domain of discourse.

For example, in Figure 3.3 the nodes in dashed ovals represent supplemented dummy child categories. As such, the hierarchical text classification problem can be reduced to the flat one on the set of all leaf categories supplemented with dummy ones. If a document is classified as "Expert System" and "Data Mining^c", for instance, then that means it is about Expert System and a subject of Data Mining other than Information Extraction.

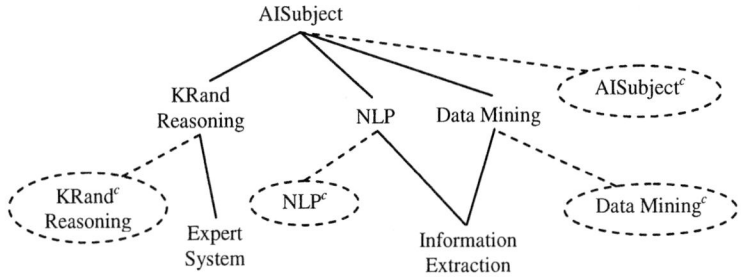

Fig. 3.3. A Category DAG with Dummy Categories

The advantage of this approach is its simplicity. However, when a category DAG is flattened and only leaf categories are used for classification, each classifier of a category has a very large number of negative training examples in comparison with positive ones, making a great bias in learning. Also, when the number of categories considered at the same level is large, it is hard to distinguish them among each other, reducing classification accuracy.

4 Experiments

There have been various learning algorithms for text classification, among which SVMs ([21], [22]) has been shown to be one of the fastest and most effective algorithms ([7], [10]). Therefore, we have realized the three approaches mentioned above using SVM classifiers. We have implemented the SVM classifiers using the library LIBSVM provided by [5]. We have also employed the automatic model selection proposed in [12] for choosing C and kernel parameters of the SVM classifiers.

Two data sets have been used to test and compare the performance of the three implemented methods, namely tree-based SVM, DAG-based SVM, and flat SVM. One data set is the commonly used Reuters-21578 ([18]) with 1327 documents. The other data set comprises 1,000 research papers in Artificial Intelligence (AI) based on the category DAG designed in [4]. Those AI papers have been retrieved from CiteSeer ([6]), MIT library ([17]), ACM Portal ([1]) and some technical reports of the Computer Science Department, Stanford University ([20]). We have extracted the abstracts of those papers and manually labeled them as our training set. The preprocessing phase (removing stop words and stemming) has been implemented using the BOW library ([15]). Based on the experiment of [9], the Document Frequency feature selection method has been used to build the feature vectors for our training set.

In order to evaluate the performance of the presented methods, we have used the precision, recall and F_β measures proposed in [19]. Let us recall the definitions of those parameters as follows:

$$\Pr{}_c = \frac{\sum_{i}^{|C|} TP_i}{\sum_{i}^{|C|} (TP_i + FP_i)} \qquad \operatorname{Re}_c = \frac{\sum_{i}^{|C|} TP_i}{\sum_{i}^{|C|} (TP_i + FN_i)} \qquad F_\beta = \frac{(\beta^2 + 1) \cdot \Pr_c \cdot \operatorname{Re}_c}{\beta^2 \cdot \Pr_c + \operatorname{Re}_c}, \quad \beta \in [0, \infty)$$

where:
  $|C|$ is the number of categories
  TP_i is the number of documents positively classified into category C_i
  FP_i is the number of documents negatively classified into category C_i
  FN_i is the number of documents belonging to category C_i but negatively classified
  β is the user-defined importance of precision and recall.

In our experiments, for the same importance of precision and recall, we set $\beta = 1$. Then F_1 is given below:

$$F_1 = \frac{2 \cdot \Pr_c \cdot \operatorname{Re}_c}{\Pr_c + \operatorname{Re}_c}$$

For the Reuters-21578, as in other research works, we have chosen 6 categories that have the highest numbers of documents. The tree structure of these categories is depicted in Figure 4.1. Since the tree-based SVM is a special case of the DAG-based SVM, which perform the same on a tree, we compare only the DAG-based SVM and the flat SVM as reported in Table 4.1. The result shows that the DAG-based SVM performs well, and a bit better than the flat SVM.

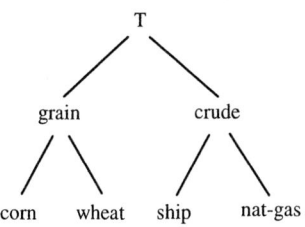

Fig. 4.1. A Category Tree for Reuters-21578

Table 4.1. Flat SVM and DAG-Based SVM Performances on Reuters-21578

Category	Flat SVM		DAG-based SVM	
	Pr	Re	Pr	Re
grain	1.0	0.872	1.0	0.879
corn	0.925	0.661	0.921	0.625
wheat	0.906	0.676	0.879	0.718
crude	0.911	0.809	0.812	0.968
ship	0.868	0.888	0.854	0.854
nat-gas	0.913	0.833	0.92	0.767
overall	*0.9*	*0.8*	*0.889*	*0.854*
F_1	0.8512		0.872	

For the AI paper data set, we have tested and compared the three methods on three of its subsets with different selected category labels. The subsets comprise 1000 documents on 57 categories, 610 documents on 9 categories (Figure 4.2), and 502 documents on 6 categories (Figure 4.3). Since the number of documents is quite small, we have used cross validation to evaluate the methods' performance. Following [23], the number of folds we have chosen is 10 and, in order to get accurate results, we have run the cross validation for several times (1, 10, 50, 100). The performances of the three methods on the three data subsets are illustrated in Figures 4.4, 4.5, and 4.6, respectively.

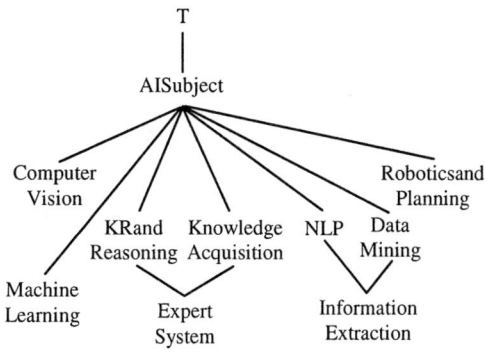

Fig. 4.2. A 9-Category DAG

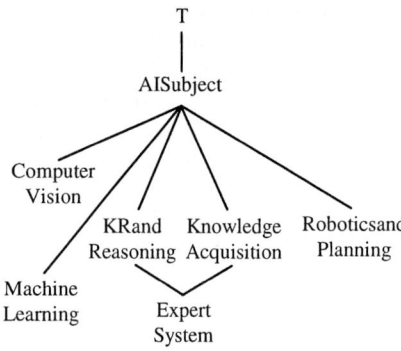

Fig. 4.3. A 6-Category DAG

Figure 4.4 shows that the flat SVM has the poorest performance, with F_1 being about 0.396, and the tree-based and DAG-based performances are nearly the same, with F_1 being about 0.434. However, all the three methods have quite low F_1 values, due to a large number of categories involved. As shown in Figures 4.5 and 4.6, when the number of categories is reduced, the performances of the methods are improved and the gap between the flat SVM and structure-based SVM (i.e., tree-based or DAG-based) is also reduced.

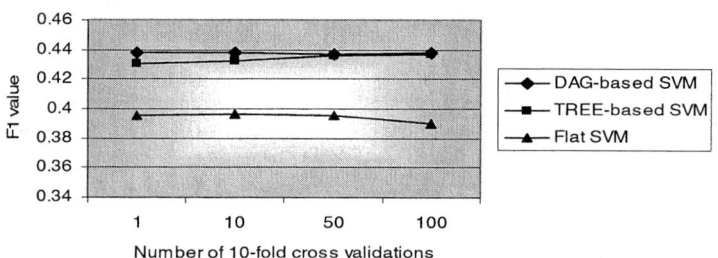

Fig. 4.4. Performances on the 57-Category DAG

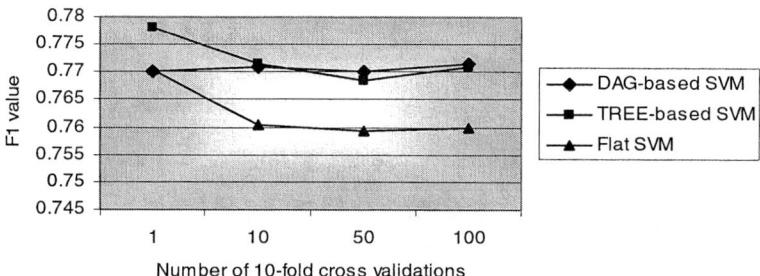

Fig. 4.5. Performances on the 9-Category DAG

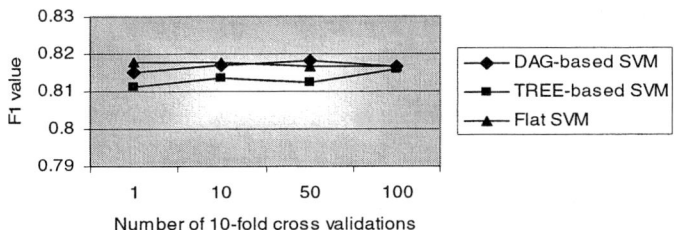

Fig. 4.6. Performances on the 6-Category DAG

5 Conclusion

We have presented three approaches to tackle the text classification problem where categories are organized as a DAG. The tree-based approach adapts the method for tree-structured categories in [2] to work on the equivalent tree transformed from an original DAG. The DAG-based approach modifies the method in [2] to manipulate directly a category DAG. The flat approach adds in dummy leaf categories and reduces the problem to flat classification.

We have conducted experiments on the Reuters-21578 data set and our constructed AI paper data sets, using SVMs as classifiers. The results show that the flat approach, which is simple, has a comparable performance to the hierarchical approaches when the number of categories involved is small. The tree-based and DAG-based approaches have nearly the same classification accuracy, but the former tends to produce large trees.

The performance of a hierarchical approach depends on the performance of internal category classifiers. The closer a category is to the root, the more important its classifier is; if it makes a wrong decision, that will effect all the classifiers at the lower levels. We are researching a way to minimize the error at each internal category classifier. Also, we need to do experiments on larger data sets with more complicated hierarchical structures such as Yahoo or Google web pages. Those are among the topics that are being investigated.

References

1. ACM Portal: http://portal.acm.org/portal.cfm.
2. Aixin, S., Ee-Peng, L.: Hierarchical Text Classification and Evaluation. In Proceedings of the 2001 IEEE International Conference on Data Mining (2001) 521-528.
3. Blockeel, H., Bruynooghe, M., Dzeroski, S., Ramon, J., Struyf: Hierarchical Multi-Classification. In Proceedings of the 1st SIGKDD Workshop on Multi-Relational Data Mining (2002) 21-35.
4. Cao, T.H., Ta, H.D.N., Tran, C.T.Q: Searching the Web: a Semantics-Based Approach. In Proceedings of the 2003 International Conference on High Performance Scientific Computing, Springer-Verlag (2004) 57-68.
5. Chang, C.C., Lin, C.J.: LIBSVM - A Library for Support Vector Machines. http://www.csie.ntu.edu.tw/~cjlin.

6. Citeseer: http://citeseer.ist.psu.edu/.
7. Cristianini, N., Taylor, J.S: An Introduction to Support Vector Machines and Other Kernel-Based Learning Methods. Cambridge University Press (2000).
8. Dumais, S., Chen, H.: Hierarchical classification of Web content. In Proceedings of the 23rd ACM Internation Conference on Research and Development in Information Retrieval (2000) 256-263.
9. Huynh, T.N., Vu, D.Q., Cao, T.H.: Automatic Topics Extraction from Artificial Intelligence Research Papers. In Proceedings of the 2004 International School on Computational Sciences and Engineering (2004) 133-139.
10. Joachims, T.: Learning to Classify Text Using Support Vector Machines. Kluwer Academic Publishers (2001).
11. Joachims, T.: Text Categorization with Support Vector Machines: Learning with Many Relevent Features. European Conference on Machine Learning (1998).
12. Keerthi, S.S., Lin, C.J: Asymptotic Behaviors of Support Vector Machines with Gaussian Kernel. In Proceedings of the 15th International Conference on Neural Computation (2003) 1667-1689.
13. Koller, D., Sahami, M.: Hierarchically Classifying Documents Using a Very Few Words. In Proceedings of the 14th International Conference on Machine Learning (1997) 170-178.
14. Krebel, U.: Pairwise Classification and Support Vector Machines. In Scholkopf, B., Burges, C.J.C., Smola, A.J. (eds): Advances in Kernel Methods – Support Vector Learnings, MIT Press (1999) 255-268.
15. McCallum, A.: Bow: A Toolkit for Statistical Language Modeling, Text Retrieval, Classification and Clustering. http://www.cs.cmu.edu/~mccallum/bow.
16. McCallum, A., Rosenfeld, R., Mitchell, T., Andrew, Y.N.: Improving Text Classification by Shrinkage in a Hierarchy of Classes. In Proceedings of the 15th International Conference on Machine Learning (1998) 359-367.
17. MIT AI Library: http://www.ai.mit.edu/.
18. Reuters-21578. http://www.daviddlewis.com/resources/testcollections/reuters221578/.
19. Rijsbergen, C.J.V.: Information Retrievel (2nd ed.). Butterworths (1979).
20. Stanford CS Technical Reports. http://www-db.standford.edu/TR/.
21. Vapnik, V.: The Nature of Statistical Learning Theory. Springer Verlag (1995).
22. Vapnik, V.: Statistical Learning Theory. John Wiley (1998).
23. Witten, I. H., Frank, E.: DataMining: Practical Machine Learning Tools and Techniques with Java Implementations. Morgan Kaufmann Publishers (1999).
24. Yang, Y.: An Evaluation of Statistical Approaches to Text Classification. Journal of Information Retrieval, Vol. 1, No. 1/2 (1999) 67-88.

Sentiment Classification Using Word Sub-sequences and Dependency Sub-trees

Shotaro Matsumoto, Hiroya Takamura, and Manabu Okumura

Tokyo Institute of Technology, Precision and Intelligence Laboratory, 4259
Nagatsuta-cho Midori-ku Yokohama, Japan
shotaro@lr.pi.titech.ac.jp
{takamura, oku}@pi.titech.ac.jp

Abstract. Document sentiment classification is a task to classify a document according to the positive or negative polarity of its opinion (favorable or unfavorable). We propose using syntactic relations between words in sentences for document sentiment classification. Specifically, we use text mining techniques to extract frequent word sub-sequences and dependency sub-trees from sentences in a document dataset and use them as features of support vector machines. In experiments on movie review datasets, our classifiers obtained the best results yet published using these data.

1 Introduction

There is a great demand for information retrieval systems which are able to handle reputations behind documents such as customer reviews of products on the web. Since sentiment analysis technologies which identify sentimental aspects of a text are necessary for such a system, the number of researches for them has been increasing. As one of the problems of sentiment analysis, there is a document sentiment classification task to label a document according to the positive or negative polarity of its opinion (favorable or unfavorable). A system using document sentiment classification technology can provide quantitive reputation information about a product as the number of positive or negative opinions on the web.

In the latest studies on document sentiment classification, classifiers based on machine learning (e.g., [1], [6], [10]), which have been successful in other document classification tasks, showed higher performance than rule-based classifiers. Pang et al. [2] reported 87% accuracy rate of document sentiment classification of the movie reviews by their classifier using word unigram as feature for support vector machines (SVMs).

For these classifiers, a document is represented as a bag-of-words, where a text is regarded as a set of words. Therefore, the document representation ignores word order and syntactic relations between words appearing in a sentence included in the original document. However, not only a bag-of-words but also word order and syntactic relations between words in a sentence are intuitively

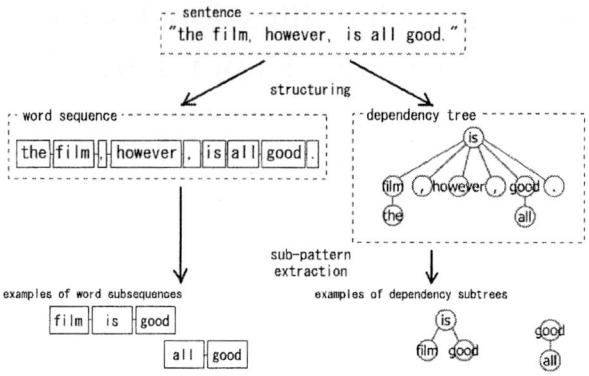

Fig. 1. A word sequence and a dependency tree representation of a sentence and examples of sub-patterns of the sentence

important and useful for sentiment classification. Thus, there appears to remain considerable room for improvement by incorporating such information.

To incorporate such information into document sentiment classification, we propose to use a word sequence and a dependency tree as structured representations of a sentence (we call simply "a sentence" below) and mining frequent sub-patterns from the sentences in a document dataset as features for document sentiment classification. We believe that the extracted set of frequent sub-patterns includes subjective expressions and idioms in the domain.

As shown in Figure 1, we regard a sentence as a word sequence and a dependency tree. We then extract frequent sub-patterns from these structured representations of sentences.

The rest of the paper is organized as follows. In the next section, we show related works on sentiment classification. In Section 3, we describe our approach to handle word order and syntactic relations between words in a sentence included in a document. In Section 4, we report and discuss the experimental results of sentiment classification. Finally, Section 5 gives conclusion.

2 Related Work

Sentiment classification is a task of classifying a target unit in a document to positive (favorable) or negative (unfavorable) class. Past researches mainly treated three kinds of target units: a word, a sentence and an overall document. to positive or negative.

Word Sentiment Classification. Hatzivassiloglou et al. [11] used conjunctive expressions such as "smart and beautiful" or "fast but inaccurate" to extract sentiment polarities of words.

Turney [7] determined the similarity between two words by counting the number of results returned by web searches. The relationship between a polarity-unknown word and a set of manually-selected seeds was used to classify the polarity-unknown word into a positive or negative class.

Sentence Sentiment Classification. Kudo et al. [9] used subtrees of word dependency trees as features for sentence-wise sentiment polarity classification. They used boosting algorithm with the subtree-based decision stamps as weak learners.

Document Sentiment Classification. Pang et al. [1] attempted to classify movie reviews. They applied to document sentiment classification a supervised machine learning method which had succeeded in other document classification tasks (e.g., on the task classifying articles of Reuters to 10 categories, Dumais et al [8] achieved F-measure of 0.92 with SVMs.). They used a *word N-gram* in the dataset as bag-of-words features for their classifier. A word N-gram is a set of N continuous words extracted from a sentence. The best results came from word unigram-based model run through SVMs, with 82.9% accuracy.

Pang [2] also attempted to improve their classifier by using only subjective sentences in the review. But accuracy of their method is less than that of the classifier using full reviews, which was introduced in their former study [1].

Dave et al. [6] used machine learning methods to classify reviews on several kinds of products. Unlike Pang's research, they obtained the best accuracy rate with word bigram-based classifier on their dataset. This result indicates that the unigram-based model does not always perform the best and that the best settings of the classifier is dependent on the data.

To use the prior knowledge besides a document, Mullen and Collier [10] attempted to use the semantic orientation of words defined by Turney [7] and several kinds of information from Internet and thesaurus. They evaluated on the same dataset used in Pang et al.'s study [1] and achieved 84.6% accuracy with the lemmatized word unigram and the semantic orientation of words.

To our knowledge, word order and syntactic relations between words in a sentence have not been used for the document sentiment classification.

3 Our Approach

We propose to use word order and syntactic relations between words in a sentence for a machine learning based document sentiment classifier. We give such information as frequent sub-patterns of sentences in a document dataset: word subsequences and dependency subtrees.

3.1 Word Subsequence

As shown in Figure 2, a word sequence is a structured representation of a sentence. From the word sequence, we can obtain ordered words in the sentence.

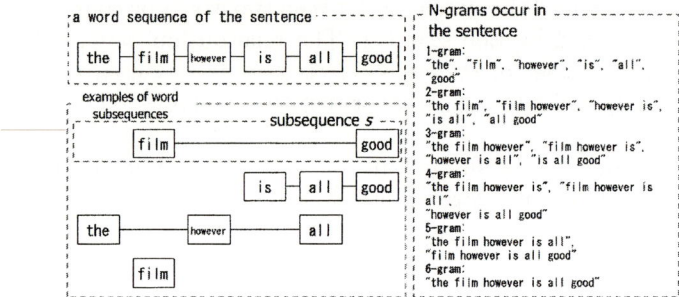

Fig. 2. A word sequence of a sentence "*The film however is all good*" and examples of subsequences

We define a *word subsequence* of a word sequence as a sequence obtained by removing zero or more words from the original sequence. In word subsequences, the word order of the original sentence is preserved.

While word N-grams cover only cooccurrences of N *continuous* words in a sentence, word subsequences cover cooccurrences of an arbitrary number of *non-continuous* words as well as continuous words. Therefore incorporating the occurrences of subsequences into the classification appears to be effective.

For example, N-grams do not cover cooccurrence of "*film*" and "*good*", when another word appears between the two words as in Figure 2. On the contrary, subsequences cover the pattern "*film-good*", denoted by s in the figure.

3.2 Dependency Subtree

As shown in Figure 3, a dependency tree is a structured representation of a sentence. The dependency tree expresses dependency between words in the sentence by child-parent relationships of nodes. We define a *dependency subtree* of a dependency tree as a tree obtained by removing zero or more nodes and branches from the original tree. The dependency subtree preserves partial dependency between the words in the original sentence. Since each node corresponding to a word is connected by a branch, a dependency subtree would give richer syntactic information than a word N-gram and a word subsequence.

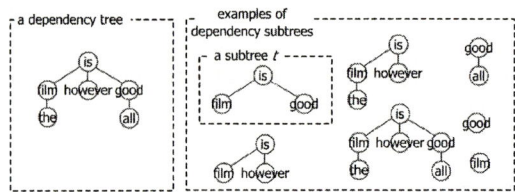

Fig. 3. A dependency tree of a sentence "*The film however is all good*" and examples of subtrees

For example, in Figure 3, to express the relation between the words *"good"* and *"film"*, a dependency subtree t (denoted as *(is(film)(good))*) does not only show the cooccurrence of *"good"* and *"film"*, but also guarantees that *"good"* and *"film"* are syntactically connected by the word *"is"*.

3.3 Frequent Pattern Mining

The number of all sub-patterns of sentences in a document dataset tends to be very large. Thus we consider not all sub-patterns but only all *frequent* sub-patterns in the dataset. A sentence *contains* a pattern if and only if the pattern is a subsequence or a subtree of the sentence. We then define the *support* of a sub-pattern as the number of sentences containing the sub-pattern. If a support of a sub-pattern is a given *support threshold* or more, the sub-pattern is *frequent*.

We mine all frequent sub-patterns from the dataset by the following mining algorithms.

Frequent Subsequence Mining: Prefixspan [4]

Prefixspan introduced by Pei et al. [4] is an efficient algorithm for mining all the frequent subsequences from a dataset consisting of sentences. First, the algorithm starts with a set of frequent subsequences consisting of single items (in this paper, corresponding to words). Then the algorithm expands each already-obtained frequent subsequence of size k by attaching a new item to obtain frequent sequence of size $k + 1$. By repeating the latter step recursively, the algorithm obtains all frequent subsequences.

However, expanding a subsequence by attaching a new item to an arbitrary position leads to duplicated enumeration of the same candidate subsequence. To avoid such enumeration, the algorithm restricts the position to attach a new item to the end of newly-obtained subsequence in left-to-right order.

Frequent Subtree Mining: FREQT [5]

FREQT introduced by Abe et al. [5] is an efficient algorithm to mine all frequent subtrees from a dataset consisting of trees. First, the algorithm starts with a set of frequent subtrees consisting of single nodes (in this paper, corresponding to words). Then the algorithm expands each already-obtained frequent subtree of size k by attaching a new node to obtain frequent tree of size $k + 1$. By repeating the latter step recursively, the algorithm obtains all frequent subtrees.

However, expanding a subtree by attaching a new node to an arbitrary position of the subtree leads to duplicated enumeration of the same candidate subtree. To avoid such enumeration, the algorithm restricts the position to attach a new node to the end of newly-obtained subtree in depth-first order.

4 Experiment

4.1 Movie Review Dataset

We prepared two movie review datasets.

The first dataset, used by Pang et al. [1] and Mullen et al. [10], consists of 690 positive and 690 negative movie reviews. Following the experimental settings presented in Pang et al. [1] and Mullen et al. [10], we used 3-fold cross validation with this dataset for the evaluation.

The second dataset, used by Pang et al. [2], consists of 1000 positive and 1000 negative movie reviews. Following the experimental settings presented in Pang et al. [2], we used 10-fold cross validation with this dataset for the evaluation.

4.2 Features

We extract word unigram, bigram, word subsequence and dependency subtree patterns from the sentences in the dataset for features of our classifiers. Each type of features is defined as follows.

- **word unigram**: *uni*
  Unigram patterns which appear in at least 2 distinct sentences in the dataset.
- **word bigram**: *bi*
  Bigram patterns which appear in at least 2 distinct sentences in the dataset.
- **frequent word subsequence**: *seq*
  Frequent word subsequence patterns whose sizes are 2 or more. These pat-

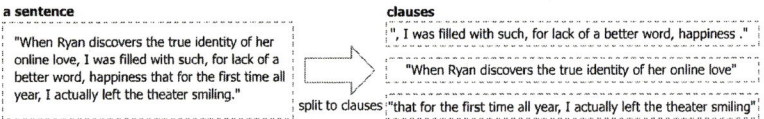

Fig. 4. An example of a sentence and clauses obtained by splitting the sentence

terns are extracted from the dataset in the fashion mentioned in Section 3.3[1]. We set the support threshold to 10. Since the number of frequent subsequences usually grows exponentially with lengths of sequences for mining, we do not use sentences but short *clauses* as the sequences. For example in Figure 4, we regard each clause of this sentence as a word sequence. We split a sentence to clauses with occurrences of the nodes labeled *'SBAR'*, which indicates a root of a subordinate clause, in a phrase structure tree of the sentence [2]. To further shorten sequences, we removed punctuations and words tagged with a part-of-speech in Table 1. Words of these parts-of-speech are presumably not a constituent of a subjective expression.

---

[1] We use Prefixspan, which is available at http://www.chasen.org/~taku/software/.
[2] In this paper, a phrase structure tree and part-of-speech tags of a sentence is given with Charniak parser [3].

Table 1. A list of Part-Of-Speech tags removed from word sequences of clauses

POS tag	example	POS tag	example
AUX	do done have is	NNPS	Americans Amharas
CC	and both but either	PDT	all both half many
CD	one-tenth ten million 0.5	POS	' 's
DT	all an the them these this	PRP	hers herself him himself
EX	there	PRP$	her his mine my
FW	gemeinschaft hund ich jeux	RP	aboard about across along
IN	astride among uppon whether out	SYM	% & ' "
LS	SP-44005 SP-44007 Second Third	TO	to
NNP	Motown Venneboerger Ranzer	WDT	that what whatever which

- **frequent dependency tree:** *dep*
  Frequent dependency subtree patterns whose sizes are 2 or more. These patterns are extracted from the dataset in the fashion mentioned in Section 3.3[3]. We set the support threshold to 10. To avoid mining noisy patterns, we remove punctuations from the dataset.

We also extract another feature set whose elements are features consisting of lemmatized words. As in the extraction of the features *uni, bi, seq, dep* described above, we extract these lemmatized features (uni_l, bi_l, seq_l, dep_l).

To convert features of a document to an input of the machine-learning classifier, we define a feature vector representation of the document. Each dimension corresponds to a feature. The ith dimension's value d_i is set to 1 if the ith feature appears in this document, otherwise 0.

4.3 Classifiers and Tests

We used support vector machines (SVMs) with the linear kernel as a classifier and the feature vector representation of each document normalized by 2-norm. The linear kernel has a learning parameter C (called a soft margin parameter), which needs adjustment. Since the results of the preliminary experiments indicated the performance of the classifier are dependent on this parameter, we carried out three kinds of cross-validation tests:

- test 1: Following past researches ([1], [2], [10]), we fix C as 1 in all learning steps in each fold of the dataset. The result is used for comparison to the past researches.
- test 2: Assuming that we can select the best value for C out of $\{e^{-2.0}, e^{-1.5}, \ldots, 1, \ldots, e^{2.0}\}$, we report the average of the best accuracy rates of their folds as final result. The result shows the potential performance of the classifier with the features.

---
[3] We use FREQT, which is available at http://www.chasen.org/~taku/software/.

- test 3: In each fold, out of $\{e^{-2.0}, e^{-1.5}, ..., 1, ..., e^{2.0}\}$, we predict a proper value of C which gives the best accuracy rate of 5-fold cross-validation of the training data. We then use the predicted C for learning the training data. We report an average of these accuracy rates of their folds as final result. Since proper C cannot be obtained before experiment, the result gives the practical performance of the classifier with the features.

Since it takes too much time to evaluate all combinations of features, we first select the best combination of bag-of-words features (we denote these features as bow) according to the accuracy rate of test 2. We then evaluate the classifier using combinations of the bow and word subsequence and/or dependency subtree pattern features. We finally discuss the improvement of performance by adding the sub-pattern features with the accuracy rate of test 1 and test 3.

4.4 Result and Discussion

The results of the experiment on the dataset 1 are shown in Table 2. The results of the experiment on the dataset 2 are shown in Table 3.

Obviously, our approach is successful in both of the datasets.

In the test 1, our best classifier obtains 87.3% accuracy on the dataset 1 and 92.9% accuracy on the dataset 2. The comparison between these results and the results of past researches ([1] [10] [2]) indicates that our method is more effective for document sentiment classification than the past researches.

Table 2. Results for dataset 1

Features	Acurracy(%)		
	test1	test2	test3
Pang et al. [1]	82.9	N/A	N/A
Mullen et al. [10]	84.6	N/A	N/A
word unigram ($= uni$)	83.0	83.7	83.0
lemma unigram ($= uni_l$)	82.8	83.8	83.2
word bigram ($= bi$)	79.6	80.4	80.1
lemma bigram ($= bi_l$)	80.4	80.9	80.7
$uni + bi$	83.8	84.6	84.0
$uni + bi_l$	83.6	84.2	83.5
$uni_l + bi$	84.4	84.8	84.6
$uni_l + bi_l$ ($= bow$)	84.0	84.9	84.2
$bow + seq$	84.1	85.3	84.9
$bow + seq_l$	84.4	85.7	84.9
$bow + dep$	86.6	87.6	87.5
$bow + dep_l$	87.3	88.3	88.0
$bow + seq + dep$	86.2	87.2	87.2
$bow + seq + dep_l$	87.0	87.5	87.5
$bow + seq_l + dep$	86.5	87.5	87.0
$bow + seq_l + dep_l$	87.0	87.6	87.0

Table 3. Results for dataset 2

Features	Acurracy(%)		
	test1	test2	test3
Pang et al. [2]	87.1	N/A	N/A
word unigram ($= uni$)	87.1	88.1	87.0
lemma unigram ($= uni_l$)	86.4	86.9	85.9
word bigram ($= bi$)	84.2	85.3	85.1
lemma bigram ($= bi_l$)	84.3	85.2	84.7
$uni + bi$ ($= bow$)	88.1	88.8	88.0
$uni + bi_l$	87.8	88.6	87.8
$uni_l + bi$	87.3	88.2	87.3
$uni_l + bi_l$	87.7	88.3	87.9
$bow + seq$	88.2	89.4	88.3
$bow + seq_l$	88.5	89.8	88.5
$bow + dep$	92.4	93.7	92.7
$bow + dep_l$	92.8	93.7	92.9
$bow + seq + dep$	92.6	93.5	92.8
$bow + seq + dep_l$	92.9	93.7	93.2
$bow + seq_l + dep$	92.6	93.2	93.0
$bow + seq_l + dep_l$	92.9	93.3	93.1

In the test 3, our best classifier obtains 88.0% accuracy on the dataset 1 and 93.2% accuracy on the dataset 2. The comparison between these results and the results obtained by the best bag-of-words classifiers in the test 3 indicates that our classifiers are more effective for document sentiment classification than the bag-of-words based classifiers. The contribution of dependency subtree feature is large. Only using this feature with the best bag-of-words feature, we obtained obviously better performance than the preceding bag-of-words based classifiers.

Adding the word subsequence features slightly improves the baseline classifier with bag-of-words features. Since our word pruning strategies is naive, there may exist a more sophisticated strategy which gives higher performance.

The classifier using both word subsequences and dependency subtrees with the best bag-of-words features yields almost the same performance as the classifier using dependency trees and the best bag-of-words features. It suggests that there exists large overlap between these two types of pattern features.

Opposite to Pang et al. [1], using word bigrams yields good influence to the classification performance. We consider that the main reason of the difference is the setting of support threshold used to extract bigram patterns. While our method used all bigram patterns which occur at least twice in the dataset, Pang et al [1] used only patterns which occur at least 7 times in the dataset.

Lemmatized features are not always more effective for classification than the original ones. If the dataset is large, lemmatizing words may be harmful because it ignores information in the conjugated forms. If the dataset is small, sub-patterns consisting of unlemmatized words tend to be infrequent. Thus there is a risk of missing sub-patterns which are useful for classification.

4.5 Weighted Patterns

A classifier, obtained by SVMs with the linear kernel, labels either of two distinct classes to examples based on a weighted voting, where each voter corresponds to a feature of SVMs. The absolute value of each weight indicates how large the contribution of the feature is.

We observed pattern features with large weights in the SVM classifier with the following features: *uni*, *bi*, *seq* and dep_l[4]. We used all reviews in the dataset 2 as training data. The value of C was set to 1. In Table 4, we show some patterns along with their weights. We could find several heavily-weighted patterns which appear to be effective to detect sentiment polarity (e.g., *"stern"*, *"pull off"*, *"little-life"*, *"(without(doubt))"* in Table 4). We also found an overlap of patterns used for the classification. For instance, unigram pattern *"bad"* and *"movie"*, bigram pattern *"bad movie"*, word subsequence pattern *"bad-movie"* and dependency subtree pattern *"(movie(bad))"* overlapped each other.

---

[4] This combination of features follows our best classifier's settings on the dataset 2.

Table 4. Examples of patterns with their weights in the weight vector

the type of the pattern	weight	pattern
Word unigram (*uni*)	1.0618	*hilarious*
	0.6221	*masterpiece*
	-1.4265	*stern*
	-0.3749	*movie*
	-1.9937	*bad*
Word bigram (*bi*)	8.0561	*pull off*
	0.8565	*one of*
	-0.1150	*little life*
	-0.3330	*bad movie*
Word subsequence (*seq*)	-7.0200	*little-life*
	0.2340	*not-only-also*
	0.3497	*good-film*
	0.3243	*film-good*
	-0.5828	*bad-movie*
	-0.0053	*movie-bad*
Dependency subtree (dep_l)	5.9784	*(without(doubt))*
	0.0236	*(film(good))*
	-1.1406	*(should(have))*
	-0.8306	*(movie(bad))*

5 Conclusion

In this paper, we have shown the methods for incorporating word order and syntactic relations between words in a sentence into the classification. We have obtained sub-pattern features as information of word order and syntactic relations between words in a document by mining frequent sub-patterns from word sequences and dependency trees in the dataset. In the experiments on the movie review domain, our classifier with a bag-of-words feature and sub-pattern features showed better performance than past classifiers. In future work, we would like to incorporate discourse structures in a document into the classifier.

References

1. Bo Pang, Lillian Lee, and Shivakumar Vaithyanathan. Thumbs up? Sentiment Classification using Machine Learning Techniques. *Proc. of 7th EMNLP*, pp.79–86, 2002.
2. Bo Pang, Lillian Lee. A Sentimental Education: Sentiment Analysis Using Subjectivity Summarization Based on Minimum Cuts. *Proc. of 42nd ACL*, pp. 271–278, 2004.
3. Eugene Charniak. A Maximum-Entropy-Inspired Parser. *Proc. of 1st NAACL*, pp.132–139, 2000.

4. Jian Pei, Jiawei Han, Behzad Mortazavi-Asl, Helen Pinto, Qiming Chen, Umeshwar Dayal, and Mei-Chun Hsu. Prefixspan: Mining Sequential Patterns Efficiently by Prefix-Projected Pattern Growth. *Proc. of 17th ICDE*, pp.215–224, 2001.
5. Kenji Abe, Shinji Kawasoe, Tatsuya Asai, and Hiroki Arimura, Setsuo Arikawa. Optimized substructure discovery for semi-structured data. *Proc. of 6th PKDD*, pp.1–14, 2002.
6. Kushal Dave, Steve Lawrence, and David Pennock. Mining the peanut gallery: opinion extraction and semantic classification of product reviews. *Proc. of 12th WWWC*, pp.519–528, 2003.
7. Peter Turney. Thumbs up or thumbs down? semantic orientation applied to unsupervised classification of reviews. *Proc. of the 40th ACL*, pp.417–424, 2002.
8. Susan Dumais, John Platt, David Heckerman, and Mehran Sahami. Inductive Learning Algorithms and Representations for Text Categorization. *Proc. of 7th CIKM*, pp. 148–155, 1998.
9. Taku Kudo and Yuji Matsumoto. A Boosting Algorithm for Classification of Semi-Structured Text. *Proc. of 9th EMNLP*, pp.301–308, 2004.
10. Tony Mullen and Nigel Collier. Sentiment Analysis using Support Vector Machines with Diverse Information Sources. *Proc. of 9th EMNLP*, pp.412–418, 2004.
11. Vasileios Hatzivassiloglou and Kathleen McKeown. Predicting the Semantic Orientation of Adjectives. *Proc. of 35th ACL and 8th EACL*, pp. 174–181, 1997.

Improving Rough Classifiers Using Concept Ontology

Nguyen Sinh Hoa[1] and Nguyen Hung Son[2]

[1] Polish-Japanese Institute of Information Technology,
Koszykowa 86, 02-008, Warsaw, Poland
[2] Institute of Mathematics, Warsaw University,
Banacha 2, 02-097 Warsaw, Poland
{hoa, son}@mimuw.edu.pl

Abstract. We present a method of classifier synthesis based on rough set theory and hierarchical learning idea. The improvement of the generated classifiers is achieved by using concept ontology as a domain knowledge. We examine the effectiveness of the proposed approach by comparing it with standard learning approaches with respect to different criteria. Our experiments are performed on benchmark data set as well as on artificial data sets generated by a road traffic simulator.

1 Introduction

Rough set theory has been introduced by [9] as a tool for concept approximation from uncertainty. Till now, one can find many efficient applications of rough sets in machine learning and data mining, since many problems like classification, clustering or regression can be formulated as concept approximation problem [4]. In a typical process of concept approximation we assume that there is given information consisting of values of conditional and decision attributes on objects from a finite subset (training set) of the universe and using this information one should induce approximations of the concept over the whole universe.

In some learning tasks, e.g., identification of dangerous situations on the road by unmanned vehicle aircraft (UAV), the target concept is too complex and it can not be approximated directly from feature value vectors. The difficulty is based either on the unlearnability of the hypothesis space or on the high complexity of the the learning algorithm. In such cases, there is a need of using a domain knowledge to improve the learning process. In this paper, we assume that domain knowledge is given as a concept ontology, which can be understood as a treelike structure with the target concept located at the root, with attributes (variables, features) located at leaves, and with some additional concepts located in internal nodes. With this assumption, the layered learning [15] seen as a generalization of standard approach to concept approximation.

Given the concept ontology, the main idea is to gradually synthesize a target concept from simpler ones. The importance of hierarchical concept synthesis is now well recognized by researchers (see, e.g., [8] [11]). An idea of hierarchical concept synthesis, in the rough mereological and granular computing frameworks has been developed (see, e.g., [11] [14]) and problems connected with compound concept approximation are discussed, e.g., in [1] [8] [13].

In this paper we concentrate on concepts that are specified by decision classes in decision systems [9]. The crucial for inducing concept approximations is to create the description of concepts in such a way that makes it possible to maintain the acceptable level of imprecision along all the way from basic attributes to final decision. We discuss some strategies for concept composing based on rough set theory. The effectiveness of layered learning approach and the comparison with standard rule-based learning approach are performed with respect to generality of concept approximation, preciseness of concept approximation, computation time required for concept induction and concept description lengths.

2 Basic Notions

The problem of concept approximation can be treated as a problem of searching for description (expressible in a given language) of an unknown concept.

Formally, given an universe $\mathcal{X}$ of objects and a concept C which can be interpreted as a subset of $\mathcal{X}$, the problem is to find a description of C which can be expressed in a predefined descriptive language $\mathcal{L}$. We assume that $\mathcal{L}$ consists of such formulas that are interpretable as subsets of $\mathcal{X}$. The approximation is required to be as *close* to the original concept as possible.

In this paper, we assume that objects from $\mathcal{X}$ are described by finite set of attributes (features) $A = \{a_1, .., a_k\}$. Each attribute $a \in A$ corresponds to the function $a : \mathcal{X} \to V_a$ where V_a is called the *domain* of a. For any non-empty set of attributes $B \subseteq A$ and any object $x \in \mathcal{X}$, we define the *B-information vector* of x by: $inf_B(x) = \{(a, a(x)) : a \in B\}$. The set $INF_B(\mathbb{S}) = \{inf_B(x) : x \in U\}$ is called the *B-information set*. The language $\mathcal{L}$, which is used to describe approximations of the given concept, consists of Boolean expressions over descriptors of the form $(attribute = value)$ or $(attribute \in set\_of\_values)$.

Usually, the concept approximation problem is formulated as an *inductive learning problem*, i.e., the problem of searching for a (approximated) description of a concept C based on a *finite set of examples* $U \subset \mathcal{X}$, called the training set. The closeness of the approximation to the original concept can be measured by different criteria like accuracy, description length, etc., which can be also estimated by *test examples*.

The input data for concept approximation problem is given by *decision table* which is a tuple $\mathbb{S} = (U, A, dec)$, where U is a non-empty, finite set of *training objects*, A is a non-empty, finite set, of *attributes* and $dec \notin A$ is a distinguished attribute called *decision*. If $C \subset \mathcal{X}$ is a concept to be approximated, then the decision attribute dec is a characteristic function of concept C, i.e., if $x \in C$ we have $dec(x) = yes$, otherwise $dec(x) = no$. In general, the decision attribute dec can describe several disjoint concepts. Therefore, without loss of generality, we assume that the domain of the decision dec is finite and equal to $V_{dec} = \{1, \ldots, d\}$. For any $k \in V_{dec}$, the set $CLASS_k = \{x \in U : dec(x) = k\}$ is called the kth *decision class of* $\mathbb{S}$. The decision *dec* determines a partition of U into decision classes, i.e., $U = CLASS_1 \cup \ldots \cup CLASS_d$.

The approximated description of a concept can be induced by any learning algorithm from inductive learning area. In the next Section we concentrate on methods based on layered learning and rough set theory.

3 Rough Sets and Concept Approximation Problem

Let $C \subseteq \mathcal{X}$ be a concept and let $\mathbb{S} = (U, A, dec)$ be a decision table describing the training set $U \subseteq \mathcal{X}$. Any pair $\mathbb{P} = (\mathbf{L}, \mathbf{U})$ is called *rough approximation of* C (see [1] [9]) if it satisfies the following conditions:

1. $\mathbf{L} \subseteq \mathbf{U} \subseteq \mathcal{X}$;
2. $\mathbf{L}, \mathbf{U}$ are expressible in the language $\mathcal{L}$;
3. $\mathbf{L} \cap U \subseteq C \cap U \subseteq \mathbf{U} \cap U$;
4. $\mathbf{L}$ is maximal and $\mathbf{U}$ is minimal among those $\mathcal{L}$-definable sets satisfying 3.

The sets $\mathbf{L}$ and $\mathbf{U}$ are called the *lower approximation* and the *upper approximation* of the concept C, respectively. The set $\mathbf{BN} = \mathbf{U} - \mathbf{L}$ is called the *boundary region of approximation* of C. For objects $x \in \mathbf{U}$, we say that "probably, x is in C". The concept C is called *rough* with respect to its approximations $(\mathbf{L}, \mathbf{U})$ if $\mathbf{L} \neq \mathbf{U}$, otherwise C is called *crisp* in $\mathcal{X}$.

The condition (4) in the above list can be substituted by inclusion to a degree to make it possible to induce approximations of higher quality of the concept on the whole universe $\mathcal{X}$. In practical applications the last condition in the above definition can be hard to satisfy. Hence, by using some heuristics we construct sub-optimal instead of maximal or minimal sets.

3.1 Rough Classifier

The rough approximation of a concept can be also defined by means of a rough membership function. A function $\mu_C : \mathcal{X} \to [0, 1]$ is called a rough membership function of the concept $C \subseteq \mathcal{X}$ if, and only if $(\mathbf{L}_{\mu_C}, \mathbf{U}_{\mu_C})$ is a rough approximation of C, where $\mathbf{L}_{\mu_C} = \{x \in \mathcal{X} : \mu_C(x) = 1\}$ and $\mathbf{U}_{\mu_C} = \{x \in \mathcal{X} : \mu_C(x) > 0\}$ (see [1]). The rough membership function can be treated as a fuzzyfication of rough approximation. It makes the translation from rough approximation into membership function. The main feature that stands out rough membership functions is related to the fact that it is derived from data. Any algorithm that computes the value of a rough membership function $\mu_C(x)$ having information vector $inf(x)$ of an object $x \in \mathcal{X}$ as an input, is called *the rough classifier*.

Rough classifiers are constructed from training decision table. Many methods of construction of rough classifiers have been proposed, e.g., the classical method based on reducts [9][10], the method based on k-NN classifiers [1], or the method based on decision rules [1]. Let us remind the Rough Set based algorithm, called *RS algorithm*, that constructs rough classifiers from decision rules. This method will be improved in the next section.

Let $\mathbb{S} = (U, A, dec)$ be a given decision table. The first step of RS algorithm is construction of some decision rules, i.e., implications of a form

$$\mathbf{r} \equiv_{df} (a_{i_1} = v_1) \wedge ... \wedge (a_{i_m} = v_m) \Rightarrow (dec = k) \quad (1)$$

where $a_{i_j} \in A$, $v_j \in V_{a_{i_j}}$ and $k \in V_{dec}$. Searching for *short*, *strong* decision rules *with high confidence* from a given decision table is a big challenge for data mining. Some methods based on rough set theory have been presented in [3] [5] [10] [12]. Let **RULES**($\mathbb{S}$) be a set of decision rules induced from $\mathbb{S}$ by one of the mentioned rule extraction methods. One can define the rough membership function $\mu_k : \mathcal{X} \to [0,1]$ for the concept determined by $CLASS_k$ as follows:

1. For any object $x \in \mathcal{X}$, let $MatchRules(\mathbb{S}, x)$ be the set of rules which are supported by x. Let $\mathbf{R}_{yes}$ be the set of all decision rules from $MatchRules(\mathbb{S}, x)$ for k^{th} class and let $\mathbf{R}_{no}$ be the remainder of $\mathbf{R}_{yes}$.
2. We define two real values w_{yes}, w_{no} by

$$w_{yes} = \sum_{\mathbf{r} \in \mathbf{R}_{yes}} strength(\mathbf{r}) \text{ and } w_{no} = \sum_{\mathbf{r} \in \mathbf{R}_{no}} strength(\mathbf{r})$$

where $strength(\mathbf{r})$ is a normalized function depending on *length*, *support*, *confidence* of $\mathbf{r}$ and some global information about the decision table $\mathbb{S}$ like table size, class distribution (see [12][1]).
3. The value of $\mu_k(x)$ is defined by:

$$\mu_k(x) = \begin{cases} \text{undetermined if } \max(w_{yes}, w_{no}) < \omega \\ 0 \quad \text{if } w_{no} \geq \max\{w_{yes} + \theta, \omega\} \\ 1 \quad \text{if } w_{yes} \geq \max\{w_{no} + \theta, \omega\} \\ \frac{\theta + (w_{yes} - w_{no})}{2\theta} \text{ in other cases} \end{cases}$$

Parameters ω, θ should be tuned by the user to control of the size of boundary region. They are very important in layered learning approach based on rough set theory.

3.2 Construction of Complex Rough Classifier from Concept Ontology

In this section we describe a strategy that learns to approximate the concept established on the higher level of a given ontology by composing approximations of concepts located at the lower level. We will discuss the method that gives us the ability to control the level of the approximation quality along all the way from attributes (basic concepts) to the target concept.

Let us assume that a concept hierarchy (or a ontology of concepts) is given. The concept hierarchy should contain either inference diagram or dependence diagram that connects the target concept with input attribute through intermediate concepts. Formally, any concept hierarchy can be treated as a treelike structure $\mathbb{H} = (\mathcal{C}, \mathcal{R})$, where $\mathcal{C}$ is a set of all concepts in the hierarchy including basic concepts (input attributes), intermediated concepts and target concept and $\mathcal{R} \subset \mathcal{C} \times \mathcal{C}$ is a dependency relation between concepts from $\mathcal{C}$. Usually, concept hierarchy is a rooted tree including target concept at root and input attributes at leaves. We also assume that concepts are divided into levels in such a way that every concept is connected with concepts in the lower levels only. Some examples of concept hierarchy are presented in Fig. 2 and Fig. 4.

In Section 3.1, we presented a classical approach (the RS algorithm) to concept approximation problem. This algorithm works for *flat* hierarchy of concepts (i.e., the target concept (decision attribute) is connected directly to input attributes). The specification of RS algorithm is as follows:

Input: Given decision table $\mathbb{S}_C = (U, A_C, dec_C)$ for a flat concept hierarchy (containing C on the top and attributes from A_C on the bottom);
Parameters: ω_C, θ_C;
Output: Approximation of C, i.e., such a set of hypothetical classifiers h_C that indicates the membership of any object x (x not necessary belongs to U) to the concept C. Let us suppose that $h_C(x) = \{\mu_C(x), \mu_{\overline{C}}(x)\}$, where $\overline{C}$ is a complement of the concept C.

For more complicated concept hierarchies, we can use the RS algorithm as a building block to develop a layered learning algorithm. The idea is to apply the RS algorithm to approximate the successive concepts through the hierarchy (from leaves to target concepts). Let $prev(C) = \{C_1, ..., C_m\}$ be the set of concepts in the lower layers, which are connected with C in the hierarchy. The rough approximation of the concept C can be determined by two steps:

1. Construct a decision table $\mathbb{S}_C = (U, A_C, dec_C)$ appropriated the concept C;
2. Apply RS algorithm to extract an approximation of C from $\mathbb{S}_C$;

The main trouble in layered learning algorithms is based on the construction of an adequate decision table. In this paper, we assume that the set of training objects U is common for the whole hierarchy. The set of attributes A_C is strictly related to concepts $C_1, ..., C_m$ in the set $prev(C)$, i.e., $A_C = h_{C_1} \cup h_{C_2} \cup ... \cup h_{C_m}$, where h_{C_i} denotes the set of hypothetical attributes related to the concept C_i. If C_i is an input attribute $a \in A$ then $h_{C_i}(x) = \{a(x)\}$, otherwise $h_C(x) = \{\mu_C(x), \mu_{\overline{C}}(x)\}$. The idea is illustrated in Fig. 1.

The problem which often occurs in layered learning algorithm is related to the lack of decision attributes for intermediate concepts (see Section 4.1). In such situations, we use a supervised clustering algorithm (using decision attribute of the target concept as a class attribute) to create a synthetic decision attribute.

A training set for layered learning is represented by decision table $\mathbb{S}_\mathbb{H} = (U, A, D)$, where D is a set of decision attributes corresponding to all intermediate concepts and to the target concept. Decision values indicate if an object belong to the given concept in the ontology. The most advanced feature of the

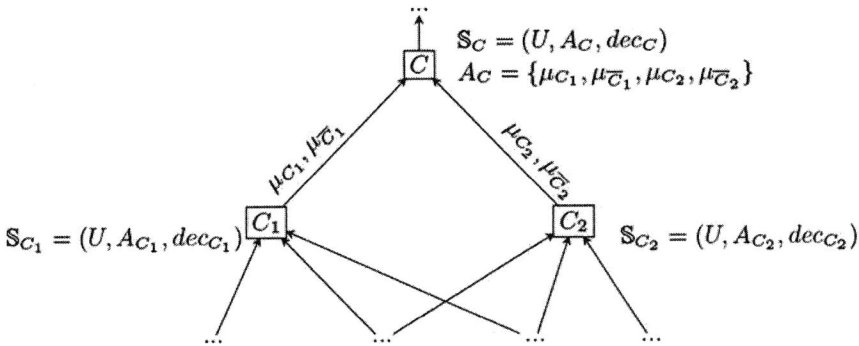

Fig. 1. The construction of decision table for a higher level concept using rough approximation of concepts from lower level

proposed method is the possibility of tuning the quality of concept approximation process via the parameters ω_C, θ_C. More details about this problem will be discussed in our next contribution. The layered learning algorithm based on rough set theory is presented in Algorithm 1.

Algorithm 1. Layered learning algorithm

Input: Decision system $\mathbb{S} = (U, A, D)$, concept hierarchy $\mathbb{H}$;
Output: Hypothetical attributes of all concepts in the hierarchy
1: **for** $l := 0$ to max_level **do**
2:   **for** (any concept C_k at the level l in H) **do**
3:     **if** $(C_k = a \in A)$ **then**
4:       $h_k := \{a\}$ // C_k is an input attribute
5:     **else**
6:       $U_k := U; \quad A_k := \bigcup_{C \in prev(C_k)} h_C;$
7:       Apply the RS algorithm to decision table $\mathbb{S}_{C_k} = (U_k, A_k, dec_{C_k})$ to generate the rough approximation μ_{C_k} of the concept C_k and $\mu_{\overline{C}_k}$ of its complement;
8:       set $h_k(x) := \mu_{C_k}, \mu_{\overline{C}_k}$ for all objects $x \in U$;
9:     **end if**
10:   **end for**
11: **end for**

---

4 Experimental Results

We have implemented the proposed solution on the basis of RSES system [2]. To verify a quality of hierarchical classifiers we performed the following experiments.

4.1 Nursery Data Set

This is a real-world model developed to rank applications for nursery schools [7]. The concept ontology is presented in Figure 2. The data set consists of

```
NURSERY              : not_recom, recommend, very_recom, priority, spec_prior
|→ EMPLOY            : Undefined (employment of parents and child's nursery)
|   |→ parents       : usual, pretentious, great_pret
|   |→ has_nurs      : proper, less_proper, improper, critical, very_crit
| STRUCT_FINAN       : Undefined (family structure and financial standings)
|   |→ STRUCTURE     : Undefined (family structure)
|   |   |→ form      : complete, completed, incomplete, foster
|   |   |→ children  : 1, 2, 3, more
|   |→ housing       : convenient, less_conv, critical
|   |→ finance       : convenient, inconv
|→ SOC_HEALTH        : Undefined (social and health picture of the family)
    |→ social        : non-prob, slightly_prob, problematic
    |→ health        : recommended, priority, not_recom
```

Fig. 2. The ontology of concepts in NURSERY data set

Table 1. Comparison results for Nursery data set achieved on 50% cases for training

	rule-based classifier using original attributes only	Layered learning using intermediate concepts
Classification Accuracy	83.4	99.9%
Coverage	85.3%	100%
Nr of rules	634	42 (for the target concept) 92 (for intermediate concepts)

12960 objects and 8 input attributes which are printed in lowercase. Besides the target concept (NURSERY) the model includes four *undefine intermediate concepts*: EMPLOY, STRUCT_FINAN, STRUCTURE, SOC_HEALTH. To approximate intermediate concepts we have applied a supervised clustering algorithm, in which the similarity between two vectors is determined by a distance between their class distributions. Next, we use rule based algorithm to approximate the target concept. The comparison results are presented in Table 1.

4.2 Road Simulator

Learning to recognize and predict traffic situations on the road is the main issue in many unmanned vehicle aircraft (UVA) projects. It is a good example of hierarchical concept approximation problem. Some exemplary concepts and a dependency diagram between those concepts are shown in Fig. 4. Definitions of concepts are given in a form of a question which one can answer YES, NO or NULL (does not concern). We demonstrate the proposed layered learning approach on the simulation system called *road simulator*. The detail description of road simulator has been presented in [6].

Road simulator is a computer tool generating data sets consisting of recording vehicle movements on the roads and at the crossroads. Such data sets are next used to learn and test complex concept classifiers working on information coming

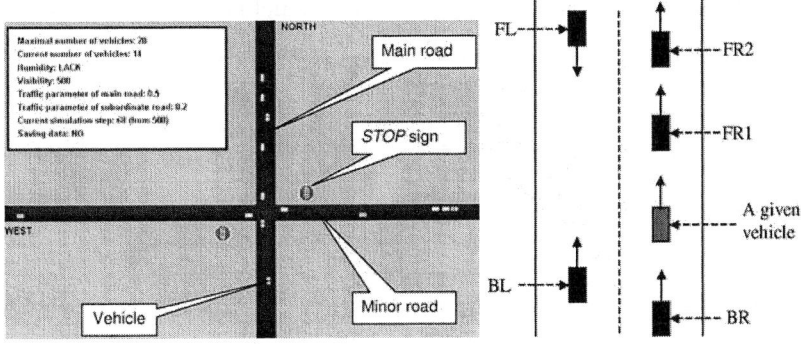

Fig. 3. Left: the board of simulation; Right: given vehicle and five vehicles around him

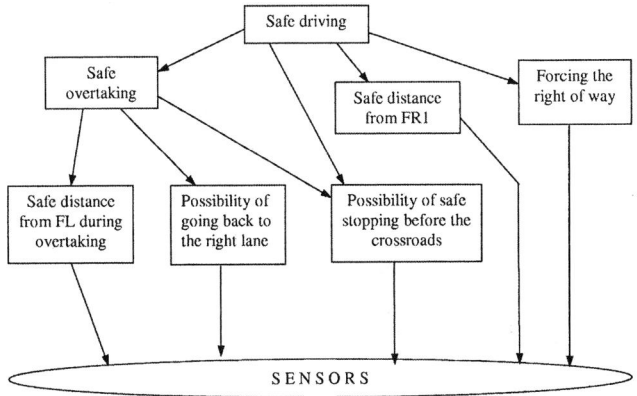

Fig. 4. The relationship diagram for presented concepts

from different devices (sensors) monitoring the situation on the road. During the simulation data may be generated and stored in a text file in a form of a rectangular table (information system). Each line of the table depicts the situation of a single vehicle and contains the sensors' and concepts' values for the vehicle and its neighboring vehicles, see Fig. 3.

Experiment Setup: We have generated 6 training data sets: *c10_s100, c10_s200, c10_s300, c10_s400, c10_s500, c20_s500* and 6 corresponding testing data sets named by *c10_s100N, c10_s200N, c10_s300N, c10_s400N, c10_s500N, c20_s500N*. All data sets consists of 100 attributes. The smallest data set consists of above 700 situations (100 simulation units) and the largest data set consists of above 8000 situations (500 simulation units).

We compare the accuracy of two classifiers, i.e., **RS:** the standard classifier induced by the rule set method, and **RS-L:** the hierarchical classifier induced by the RS-layered learning method. The comparison results are performed with respect to the accuracy of classification, covering rate of new cases, and computing time necessary for classifier synthesis.

Table 2. Classification accuracy of a standard and hierarchical classifiers

	Accuracy				Coverage			
	Total		Class NO		Total		Class NO	
	RS	RS-L	RS	RS-L	RS	RS-L	RS	RS-L
c10_s100N	0.94	0.97	0	0	0.44	0.72	0.50	0.38
c10_s200N	0.99	0.96	0.75	0.60	0.72	0.73	0.50	0.63
c10_s300N	0.99	0.98	0	0.78	0.47	0.68	0.10	0.44
c10_s400N	0.96	0.77	0.57	0.64	0.74	0.90	0.23	0.35
c10_s500N	0.96	0.89	0.30	0.80	0.72	0.86	0.40	0.69
c20_s500N	0.99	0.89	0.44	0.93	0.62	0.89	0.17	0.86
Average	0.97	0.91	**0.34**	**0.63**	**0.62**	**0.79**	**0.32**	**0.55**

Table 3. Time for standard and hierarchical classifier generation (all experiments were performed on computer with processor AMD Athlon 1.4GHz., 256MB RAM)

Tables	RS	RS-L	Speed up ratio
c10_s100	94 s	2.3 s	40
c10_s200	714 s	6.7 s	106
c10_s300	1450 s	10.6 s	136
c10_s400	2103 s	34.4 s	60
c10_s500	3586 s	38.9 s	92
c20_s500	10209 s	98s	104
Average			90

Classification Accuracy: Similarly to real life situations, we are interested on the accuracy and the coverage of classifiers on the decision class "safe driving = NO", i.e., dangerous situations. This is a issue for this problem, since datasets are is unbalanced (the concept states only 4% - 9% of training sets).

Table 2 presents the classification accuracy of RS and RS-L classifiers. One can see that the hierarchical classifier showed to be much better than the standard classifier for this class. Accuracy of "NO" class of the hierarchical classifier is quite high when training sets reach a sufficient size. The generality of classifiers usually is evaluated by the recognition ability for unseen objects. One can observe the similar scenarios to the accuracy degree. The recognition rate of dangerous situations is very poor in the case of the standard classifier. One can see in Table 2 the improvement on coverage of the hierarchical classifier.

Computing Speed: The layered learning approach also shows a tremendous advantage with respect to the computation time. One can see in Table 3 that speed up ratio of the layered learning approach to the standard one reaches from 40 to 130 times.

5 Conclusion

We presented a new method for improving rough classifiers using concept ontology. It is based on the layered learning approach. Unlike traditional approach, in the layered learning approach the concept approximations are induced not only from accessed data sets but also from expert's domain knowledge, which is necessary to created an concept ontology. In the paper, we assume that knowledge is represented by concept dependency hierarchy. The layered learning approach showed to be promising for the complex concept synthesis. The advantages of this new approach in comparison to the standard approach have been illustrated by experiments with the road traffic simulator.

Acknowledgements. The research has been partially supported by the grant 3T11C00226 from Ministry of Scientific Research and Information Technology of the Republic of Poland. The authors are deeply grateful to Dr. J. Bazan for his road simulator system and Prof. A. Skowron for valuable discussions on the layered learning approach.

References

1. J. Bazan, H. S. Nguyen, A. Skowron, and M. Szczuka. A view on rough set concept approximation. In Wang G., Liu Q., Yao Y., and Skowron A. (eds), RSFD-GrC'2003, Chongqing, China, volume 2639 of *LNAI*, Springer-Verlag Heidelberg 2003, pp. 181–188
2. J. G. Bazan and M. Szczuka. RSES and RSESlib - a collection of tools for rough set computations. Ziarko W. and Yao Y., editors, RSCTC'02, volume 2005 of *LNAI*, pages 106–113, Banff, Canada, October 16-19 2000. Springer-Verlag.
3. Grzymała-Busse J., A New Version of the Rule Induction System LERS Fundamenta Informaticae, Vol. 31(1), 1997, pp. 27–39
4. W. Kloesgen and J. Żytkow, editors. *Handbook of Knowledge Discovery and Data Mining*. Oxford University Press, Oxford, 2002.
5. J. Komorowski, Z. Pawlak, L. Polkowski, A. Skowron, Rough sets: A tutorial. In: Pal S.K., Skowron A. (eds.), Rough - fuzzy hybridization: A new trend in decision making, Springer-Verlag, Singapore, 1999, pp. 3–98.
6. S.H. Nguyen, J. Bazan, A. Skowron, and H.S. Nguyen. Layered learning for concept synthesis. In Peters J., Skowron A., Grzymala-Busse J., Kostek B., Swiniarski R., Szczuka M. (eds), *Transactions on Rough Sets I*, volume LNCS 3100 of *Lecture Notes on Computer Science*, Springer, 2004, pp. 187–208.
7. M. Olave, V. Rajkovic and M. Bohanec. An application for admission in public school systems. In I. T. M. Snellen, W. B. H. J. van de Donk and J.-P. Baquiast (eds), Expert Systems in Public Administration, Elsevier Science Publishers (North Holland), 1989, pp. 145–160.
8. S. K. Pal, L. Polkowski, and A. Skowron, editors. *Rough-Neural Computing: Techniques for Computing with Words*. Cognitive Technologies. Springer-Verlag, Heidelberg, Germany, 2003.
9. Z. Pawlak. *Rough Sets: Theoretical Aspects of Reasoning about Data*, volume 9 of *System Theory, Knowledge Engineering and Problem Solving*. Kluwer Academic Publishers, Dordrecht, The Netherlands, 1991.

10. Z. Pawlak and A. Skowron. A rough set approach for decision rules generation. In *Proc. of IJCAI'93*, pages 114–119, Chambéry, France, 1993. Morgan Kaufmann.
11. L. Polkowski and A. Skowron. Rough mereology: A new paradigm for approximate reasoning. *International Journal of Approximate Reasoning*, 15(4):333–365, 1996.
12. L. Polkowski and A. Skowron, editors. *Rough Sets in Knowledge Discovery 1: Methodology and Applications*. Physica-Verlag, Heidelberg, Germany, 1998.
13. A. Skowron. Approximation spaces in rough neurocomputing. In M. Inuiguchi, S. Tsumoto, and S. Hirano, editors, *Rough Set Theory and Granular Computing*, pages 13–22. Springer-Verlag, Heidelberg, Germany, 2003.
14. A. Skowron and J. Stepaniuk. Information granules and rough-neural computing. In Pal et al. [8], pages 43–84.
15. P. Stone. *Layered Learning in Multi-Agent Systems: A Winning Approach to Robotic Soccer*. The MIT Press, Cambridge, MA, 2000.
16. Wróblewski J., Covering with reducts - a fast algorithm for rule generation. In: Polkowski L., Skowron A.(eds.): Proc. of RSCTC'98, Warsaw, Poland. Springer-Verlag, Berlin, 1998, pp. 402–407.

QED: An Efficient Framework for Temporal Region Query Processing

Yi-Hong Chu, Kun-Ta Chuang, and Ming-Syan Chen

Department of Electrical Engineering,
National Taiwan University,
Taipei, Taiwan, ROC
{yihong, doug}@arbor.ee.ntu.edu.tw, mschen@cc.ee.ntu.edu.tw

Abstract. In this paper, we explore a new problem of *"temporal dense region query"* to discover the dense regions in the constrainted time intervals which can be separated or not. A *Querying tEmporal Dense Region framework* (abbreviated as *QED*) proposed to deal with this problem consists of two phases: (1) an offline maintaining phase, to maintain the statistics of data by constructing a number of summarized structures, RF-trees; (2) an online query processing phase, to provide an efficient algorithm to execute queries on the RF-trees. The QED framework has the advantage that by using the summarized structures, RF-trees, the queries can be executed efficiently without accessing the raw data. In addition, a number of RF-trees can be merged with one another efficiently such that the queries will be executed efficiently on the combined RF-tree. As validated by our empirical studies, the QED framework performs very efficiently while producing the results of high quality.

Keywords: Temporal dense region query, dense region query.

1 Introduction

Region-oriented queries have been recognized as important operations in many applications for the purpose of locating the regions satisfying certain conditions, e.g., those on density, total area, etc. [1][4][5][6][7]. In this paper, we focus on the query to find all dense regions whose densities are larger than their surrouding regions. In the literature, such queries are usually called as *"dense region queries"*, and the dense region can be identified by examining whether its density exceeds a density threshold. The previous work STING [6] conducted a study on speeding up the process of answering region-oriented queries by constructing an index structure to capture the statistical information of the data. Thus the queries can be executed efficiently by using the index structure without directly accessing the raw data.

However, previous research tends to ignore the time feature of the data. They treat all data as one large segment, and execute queries over the entire database. However, in practice, the characteristic of the data may change over time. It is noted that some dense regions may only exist in certain time intervals but will not be discovered if taking all data records into account since their average densities may not exceed the density threshold. Therefore, discovering dense regions over different time intervals is crucial for users to get the interesting patterns hidden in data.

To discover such temporal patterns, we explore in this paper a novel problem, named "*temporal dense region query*" to address the dense region discovery in constrainted time intervals. The time constraint will be represented by a continuous time interval such as (6AM~8AM), or a set of *separated* time intervals, such as (6AM~8AM) and (9AM~10AM). The problem of temporal dense region query is to find the dense regions in the queried time intervals. However, it is infeasible for the previous work on dense region queries to efficiently execute the temporal dense region queries. Because the queried time intervals are unknown in advance, the direct extension of their methods would be to delay the construction of index structures until the user queries the dataset, which is, however, inefficient for an interactive query environment. To the best of knowledge, despite of its importance, the problem of temporal dense region query had not been explicitly studied before.

Consequently, we devise in this paper a "*Querying tEmporal Dense region*" framework (abbreviated as *QED*), to execute temporal dense region queries. The QED framework consists of two phases: (1) an offline maintaining phase, to maintain the summarized structures of the evolving data; (2) an online query processing phase, to provide an efficient algorithm to execute queries on the summarized structures. Note that since the query processing is only applied to the summarized structures rather than to the original data points, the QED framework proposed is very efficient in practice. Furthermore, in order to support the diverse kinds of time intervals for dense region discovery, a number of base time slots are provided in advance for users to specify the time periods of interest, where the time slots are derived by segmenting the data into a number of partitions. Thus users will specify the time intervals by a number of time slots which can be separated or not. However the queried time periods are unknown when the data are summarized in the offline maintaining phase. It is challenging to summarize adequate statistics for queries with different time slots to be executed efficiently in the online phase. Therefore a novel summarized structure, referred to as *RF-tree* (standing for *Region Feature tree*), is proposed, and it has the property that a number of RF-trees can be merged with one another efficiently. Explicitly, in the offline phase a RF-tree will be constructed for each time slot, and in the online phase the queries will be executed efficiently on the RF-tree derived by merging the RF-trees with respect to the queried time slots. As validated by our experiments, the QED framework performs very efficiently while producing query results of very high quality.

The rest of this paper is organized as follows. In Section 2, the problem of temporal dense region query is explored. The offline maintaining phase of the QED framework is presented in Section 3, and the online discovering phase is described in Section 4. In Section 5 the empirical studies are conducted to evaluate the performance of QED. This paper concludes with Section 6.

2 Temporal Dense Region Query

2.1 Problem Description

There are two concepts in the temporal dense region query: one is the dense region; the other is the set of time intervals to be queried for dense regions.

Assume that the data records contained in the d−dimensional dataset are viewed as points in the d-dimensional data space constructed by the d attributes. In this paper, we use the grid-based approach to discover the dense regions. Initially the data space is partitioned into a number of small and non-overlapping cells which are obtained by partitioning each of the d dimensions into δ equal-length intervals. A cell is called a "*dense cell*" if it contains points exceeding a predetermined density threshold ρ. Thus based on these dense cells, the dense regions will be formed by a set of connected dense cells.

In addition, a number of "*time slots*" are provided in advance for users to specify a variety of time periods of interest. These *time slots* are obtained by segmenting the data points with a time granularity, e.g. week, month, year, etc. Two time slots are called *separated* if there are one or more slots between them. Thus the time period specified in the query will be represented by one or some separated time slots.

Problem Definition: (Temporal Dense Region Query)
Given a set of time slots, and the density threshold ρ, find the dense regions in those data points contained in the queried time slots, where each of the dense regions is formed by a set of connected dense cells with the number of data points in each of them exceeding ρ.

2.2 Overview of the QED Framework

A "*Querying tEmporal Dense region*" framework (abbreviated as *QED*), is proposed in this paper to deal with the temporal dense region query. The QED framework consists of two phases: (1) an offline maintaining phase, to maintain the statistics of the data; (2) an online query processing phase, to execute temporal dense region queries on the summarized statistics.

Phase 1: Offline maintaining phase
The main task of the offline maintaining phase is to maintain the statistics of the data such that queries can be executed on the summarized information instead of the original data points, thereby enabling the dense region discovery to be very efficient. It is a two-step approach described as follows:

Step1: Partition the data set: The data set is partitioned into *time slots*.

Step2: Construct the RF-tree: For each time slot, a summarized structure, RF-tree is constructed to maintain the statistics of the data points. The RF-tree has the advantage that a number of RF-trees can be merged with one another efficiently. Therefore in the online phase, the overall statistics in the queried time slots can be obtained by combining the corresponding RF-trees, and then queries can be executed on the combined RF-tree.

Phase 2: Online query processing phase
When the user issues the queries, it is the main task of the online processing phase to process these queries. The online query processing phase is also a two-step approach:

Step1: Combine the RF-trees: Those RF-trees in the queried time slots are combined into a RF-tree by an efficient algorithm to describe in Section 4.1.

Step2: Execute the query: The query is executed on the combined RF-tree to discover the dense regions with respect to the density threshold δ specified in the query.

3 Offline Maintaining Phase

In the offline maintaining phase, an RF-tree will be constructed for each time slot to maintain the statistics of the data. In Section 3.1, we will give the definition of the uniform region which is discovered in the RF-tree for summarizing the cells in the data space. The algorithm for constructing the RF-tree will be described in Section 3.2.

3.1 The Definition of the Uniform Region

In the RF-tree, the entire data space is represented by a number of non-overlapped *regions* which are defined as follows:

Definition 1 *(region)*: A **region** in the d-dimensional data space is defined by the intersection of one interval from each of the d attributes, and can be represented by a Disjunctive Normal Form expression $(l_1 \leq A_1 < h_1) \wedge \cdots \wedge (l_d \leq A_d < h_d)$.

Definition 2 *(region feature)*: A **Region Feature** of a region R in the d-dimensional data space is defined as a binary: **RF** $=(N_R, N_c)$ where N_R is the number of data points contained in region R, and N_c is the corresponding number of cells.

It is noted that by setting δ to be a larger value the cells in the data space will become smaller. Therefore, if the cells in a region have nearly the same number of data contained in them, the total data points in this region can be viewed as approximately uniformly distributed in it. Such a region is called as *"uniform region"*. In a uniform region, the number of data points contained in each of the cells will be very close to the value N_R/N_c, which is the average number of data points in a cell and is calculated from the RF of this uniform region. Thus, in the RF-tree the uniform region is used to summarize the cells contained in it by taking the average value, N_R/N_c, to approximate the number of data points in each cell.

To identify the uniform regions in the data space, we use an entropy-based technique. The entropy is in essence a measure of the uncertainty of a random variable[2]. When the random variable follows the uniform distribution, we are most uncertain about the outcome and the corresponding entropy is the highest. In light of the entropy principle, a uniform region will typically has a higher entropy than a non-uniform one. In the following, we first define the entropy of a region, and then use it to identify a uniform region. Let the RF of a region be denoted as (N_R, N_c). Also, $n(c_i)$ denotes the number of data points contained in the cell c_i. Therefore the entropy of the region R, $H(R)$, can be defined as follows:

Definition 3: *(entropy of a region)*

$$H(R) = -\sum_{i=1}^{N_c} \frac{n(c_i)}{N_R} \times \log \frac{n(c_i)}{N_R}, \quad if \ n(c_i) \neq 0.$$

Consequently, we can judge whether a region R is a uniform region or not by first calculating the maximal entropy of this region, $H_{\max}(R)$, and then comparing $H(R)$ with $H_{\max}(R)$. Note that the maximum entropy of a random variable is proved to be the value, $-\log \frac{1}{|\chi|}$, where $|\chi|$ denotes the number of possible outcomes of this random variable [2]. Analogously, the maximum entropy of a region R is defined as follows:

Definition 4 *(maximum entropy of a region):* As defined above, N_c is the number of cells contained in the region R. The maximum entropy of the region R is defined as

$$H_{\max}(R) = -\log \frac{1}{N_c}.$$

Then, with a given similarity threshold θ, a uniform region is defined as follows:

Definition 5 *(uniform region):* A region R is a uniform region if

$$\frac{H(R)}{H_{\max}(R)} \geq \theta.$$

3.2 Algorithm for Constructing the RF-Tree

In this section we will introduce the RF-tree constructed for each time slot in the offline maintaining phase. The RF-tree is a hierarchical structure constructed to discover the uniform regions by summarizing the cells in the data space.

The process of constructing the RF-tree is a top-down and recursive approach. Initially, the root of the RF-tree is set as the entire data space. Let the root be at level 1 and its children at level 2, etc. Each node in the RF-tree represents a region in the data space, and the node in level i corresponds to the union of regions of its children at level $i + 1$. Each node in the RF-tree will be examined whether it is a uniform region by applying the Definition 5 with the similarity threshold θ. Thus, if it is examined as a uniform region, it will become a leaf node; otherwise, it will be a nonleaf node and its children will be derived by partitioning the region into a set of subregions, which are obtained by segmenting the interval of each of the dimension of this region into two intervals.

4 Online Query Processing Phase

The main task of the online query processing phase is to execute the query by using the RF-trees. Note that the overall statistics of the data in the queried time slots for dense region discovery can be derived by combining the RF-trees in the queried time slots. Section 4.1 introduces an efficient algorithm for combining the RF-trees in the queried time slots. Then, the dense region discovery will be executed on the combined RF-tree, which is described in Section 4.2.

4.1 Algorithm for Combining the RF-Trees

In the following, algorithm **COMB** (combine RF-trees) outlined below is to combine two RF-trees, and it can be easily extended to deal with more than two RF-trees. Algorithm **COMB** is a top-down approach. Let the roots of the two RF-trees be denoted as r_1, and r_2, and the root of the combined RF-tree be denoted as r_b. Initially in Step 3, r_b is set as the entire data space. In Step 4 and Step 5, the RF (Region Feature) of r_b is set up according to the RF of r_1 and r_2. Then in Step 6, procedure **SC** (Set up Children) is called to set up the children of r_b by taking r_1, r_2, and r_b as its inputs n_1, n_1, and n_b.

Procedure **SC** is to set up the children of input node n_b by taking into consideration of the three cases of n_1, and n_2: (1) n_1 and n_2 are both uniform (Step 3 to Step 5); (2) n_1 and n_2 are both non-uniform (Step 6 to Step 25); (3) only one of n_1 and n_2 is uniform (Step 26 to Step 38).

<Case 1>: n_1 and n_2 are both uniform. That is, the data points are uniformly distributed in n_1 and n_2. It will be also uniform if all data points are taken into consideration such that n_b will be a uniform region. Thus, procedure **SC** terminates.

<Case 2>: n_1 and n_2 are both non-uniform. Thus, n_1 and n_2 will both have at most 2^d children. In Step 7, the 2^d children of n_b are first generated by partitioning the region of n_b into 2^d subregions. For each child C, in Step 10 and Step 11, the node C_1 and C_2 which are child nodes of n_1 and n_2 with respect to the same region of C are identified. Then, in Step 12 to Step 24, the RF of C will be set up according to the RF of C_1 and C_2. Specifically, in Step 14, only C_1 is identified such that the children of C will be just set up with respect to the corresponding children of C_1 by calling procedure **AC** (Assign Children), and the descendants of C are set up recursively. Otherwise, in Step 24, the children of C will be set up by calling procedure **SC** to recursively combine the identified C_1 and C_2.

<Case 3>: only one of n_1 and n_2 is uniform. Without loss of generality, assume that n_1 is non-uniform and n_2 is uniform such that n_1 will have at most 2^d children. In Step 28, the 2^d children of n_b are first generated, and then for each child C, the RF of C will be set up in Step 30 to Step 37. Specifically, in Step 31, for each child C, $C.p$ denotes the average number of data derived from the uniform node n_2. Note that in Step 37 by calling procedure **AC**, the children of C will be just set up with respect to the corresponding children of C_1, and $C.p$

will be further averaged to the children of C if C will have children set up in procedure **AC**.

Note that procedure **SC** is applied recursively to combine the nodes in two RF-trees. It will terminate if Case 1 is encountered as shown in Step 3, or n_1 and n_2 are cells as shown in Step 1.

Algorithm COMB: Combine RF-trees
Input: RF-tree1, RF-tree2
Output: the combined RF-tree
1.   // r_1 is the root of the RF-tree1, r_2 is the root of the RF-tree2
2.   // r_b is the root of the combined RF-tree
3. r_b = entire data space
4. $r_b.N_R = r_1.N_R + r_2.N_R$
5. $r_b.N_c = r_1.N_c$
6. $SC(r_1, r_2, r_b)$

Procedure SC: Set up Children
Input: (node n_1, node n_2, node n_b)
1.   if (n_1 is a cell & n_2 is a cell)
2.   return
3.   if (n_1 is uniform & n_2 is uniform) { //Case 1
4.   return
5.   }
6.   else if (n_1 is non-uniform & n_2 is non-uniform) { //Case 2
7.   Generate 2^d children of node n_b
8.   For each child node C
9. $C.N_c = (n_b.N_c)/2^d$
10. C_1 = the child of n_1 with respect to the same region of C
11. C_2 = the child of n_2 with respect to the same region of C
12.   if (C_1 == null && C_2 == null)
13.   remove the chid node C from n_b
14.   else if (C_1!= null && C_2 == null)
15. $C.N_R = C_1.N_R$
16. $C.p = 0$
17. $AC(C_1, C)$
18.   else if (C_1 == null && C_2!= null)
19. $C.N_R = C_2.N_R$
20. $C.p = 0$
21. $AC(C_2, C)$
22.   else
23. $C.N_R = C_1.N_R + C_2.N_R$
24. $SC(C_1, C_2, C)$
25.   }
26.   else { // Case 3
27.   // Suppose n_1 is non-uniform, and n_2 is uniform
28.   Generate 2^d children of node n_b
29.   For each child node C

30. $C.N_c = (n_b.N_c)/2^d$
31. $C.p = (n_2.N_R)/2^d$
32. $C_1 = $ find a child of n_1 with respect to the same region of C
33.   if ($C_1 == $ null)
34. $C.N_R = C.p$
35.   else if ($C_1 != $ null)
36. $C.N_R = C.p + C_1.N_R$
37.   AC(C_1, C)
38   }

Procedure AC: Assign Children
Input: (node n_1, node n_b)
1.   if (n_1 has no children)
2.   return
3.   else
4.   For each child node C_1 of n_1
5.   Generate a child C with respect to the same region of C_1
6.   if ($n_b.p != 0$)
7. $C.p = (n_b.p)/2^d$
8.   else
9. $C.p = 0$
10. $C.N_R = C_1.N_R + C.p$
11. $C.N_c = C_1.N_c$
12.   AC(C_1, C)

4.2 Execute the Temporal Dense Region Query

After the combing process, the query will be executed on the combined RF-tree. Initially all leaf nodes in the combined RF-tree are examined to discover the dense cells in the data space, and then the leaf nodes containing dense cells will be put into a queue for further dense region discovery. Note that the leaf nodes will be of two cases : (1) a cell, and (2) a uniform region. For the case (1), if the number of data points it contains exceeds the density threshold ρ, it is identified as a dense cell and is put into the queue. For the case (2), the cells contained in this uniform region will have the same average number of data points, i.e. N_R/N_c, calculated from its RF. Thus these cells will be identified as dense ones if the value of N_R/N_c exceeds the density threshold ρ, and then this uniform region will be put into the queue.

After all leaf nodes are examined, the dense regions can be discovered by grouping the connected ones in the queue. This can be executed by a breadth-first search. Each time we take out a leaf node n_i from the queue, and examine the rest ones whether they are connected to n_i. If no one is identified, output the node n_i to be a dense region. Otherwise, the identified nodes will be first taken out from the queue, and then the rest nodes are recursively examined on whether they are connected to the previous identified nodes. Finally, the leaf nodes connected to n_i will become a dense region.

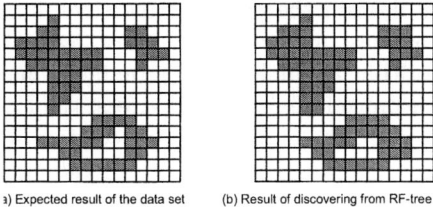

Fig. 1. Illustration of the quality of the RF-tree

5 Experiments

5.1 Quality of the RF-Tree

To evaluate the quality of the RF-tree, we generate a two dimensional data set of 5000 data points, and δ is set to 16. The density threshold ρ is set to 1.1 times the expectation of data points in each cell under uniform distribution, i.e. $1.1 \times (5000/16^2)$. The expected result is shown in Figure 1(a). Figure 1(b) is the result of executing the query on the RF-tree constructed by setting θ to 0.9. From Figure 1, it is seen that the RF-tree is able to successfully discover the dense regions and the query result is very close to the expected one.

5.2 Performance of Combining RF-Trees

In this experiment the scalability of the QED on queries with different number of time slots is studied. One real data set is used, which is the census-income data of 25,0000 data points from the UCI KDD Archive [3] with three attributes, i.e. age, income, weight. The total data points are partitioned into five time slots $W_1 \ldots W_5$ such that there are 50000 data points in each one. We test the performance with queries with the number of queried time slots varying from 1 to 5. There are five queries with one time slot, i.e. W_1, W_2, W_3, W_4, and W_5. There are ten ones with two time slots, i.e., $W_1W_2, W_1W_3, W_1W_4,\ldots$,etc. For each query, the density threshold ρ is set to 1.1 times of the expectation of data points in each cell under uniform distribution.

Figure 2(a) shows the relative execution time of STING and RF-tree on varied number of time slots. The execution time for queries with the same number of time slots is averaged. In this experiment, STING is extended to deal with the temporal dense region queries with two steps: (1) constructing their proposed

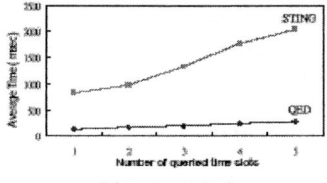

Num of time slots	recall	precision	F-score
1	1	0.9719	0.9845
2	1	0.9655	0.9828
3	1	0.9665	0.9829
4	1	0.9661	0.9827
5	1	0.9746	0.9872

(a) Time complexity (b) Quality of discovered dense regions

Fig. 2. The performance of querying on the combined RF-tree

index hierarchy on all data in the queried time slots, and (2) using the index hierarchy to answer queries. As shown in Figure 2(a), QED outperforms the extended STING algorithm about by very prominent margin. Note that with the efficient combining procedure for the RF-trees, the execution time of QED only slightly increases when the number of queried time slots increases.

In addition, in Figure 2(b) we use the metric of *recall* and *precision* to evaluate the qualities of query result of the QED framework. *Recall* is defined as the percentage of the expected dense cells identified by QED. *Precision* is defined as the percentage of the dense cells identified in QED truly dense expectedly. *F-score* is defined as the value of (2×recall×precision)/(recall+precision). As shown in Figure 2(b), all of the expected dense cells are discovered by QED because the recall are all ones. Very few cells are falsely identified as dense by QED as expected, and the precisions are very close to 1. Moreover, QED is still robust when the number of queried time slots increases since recall and precison remain very close to 1 as shown in the figure. Therefore, as validated in these experiments QED is effective and efficient for temporal dense region queries.

6   Conclusion

In this paper, the problem of temporal dense region query is explored to discover dense regions in the queried time slots. We also propose the QED framework to execute temporal dense region queries. QED is advantageous in that various queries with different density thresholds and time slots can be efficiently supported by using the concept of time slot and proposed RF-tree. With the merit of the efficiency of combining RF-trees, the QED framework scales well with respect to varied number of time slots. As evaluated by the synthetic and real data, QED is powerful in discovering the dense regions and outperforms prior methods significantly.

Acknowledgements. The work was supported in part by the National Science Council of Taiwan, R.O.C., under Contracts NSC93-2752-E-002-006-PAE.

References

1. D.-S. Cho, B.-H. Hong, and J. Max. Efficient Region Query Processing by Optimal Page Ordering. *In Proc. of ADBIS-DASFAA*, 2000.
2. T. M. Cover and J. A. Thomas. Elements of Information Theory. *Wiley*, 1991.
3. S. Hettich and S. Bay. The UCI KDD archive. *[http://kdd.ics.uci.edu]*, 1999.
4. G. G. Lai, D. Fussel, and D. F. Wong. HV/VH Trees: A New Spatial Data Structure for Fast Region Queries. *In Proc. of Design Automation Conference*, 1993.
5. B. S. Lee, R. R. Snapp, R. Musick, and T. Critchlow. Metadata Models for Ad Hoc Queries on Terabyte-Scale Scientific Simulations. *J. Braz. Comp. Soc.*, 2002.
6. W. Wang, J. Yang, and R. Muntz. STING: A Statistical Information Grid Approach to Spatial Data Mining. *In Proc. of VLDB*, 1997.
7. A.M. Yip, E.H. Wu, M.K. Ng, and T.F. Chan. An Efficient Algorithm for Dense Regions Discovery from Large-Scale Data Streams. *In Proc. of PAKDD*, 2004.

A MPAA-Based Iterative Clustering Algorithm Augmented by Nearest Neighbors Search for Time-Series Data Streams

Jessica Lin[1], Michai Vlachos[1], Eamonn Keogh[1], Dimitrios Gunopulos[1],
Jianwei Liu[2], Shoujian Yu[2], and Jiajin Le[2]

[1] Department of Computer Science and Engineering University of California,
Riverside Riverside, CA 92521
{jessica, mvlachos, eamonn, dg}@cs.ucr.edu
[2] College of Computer Science & Technology,
Donghua University
liujw@mail.dhu.edu.cn

Abstract. In streaming time series the Clustering problem is more complex, since the dynamic nature of streaming data makes previous clustering methods inappropriate. In this paper, we propose firstly a new method to evaluate Clustering in streaming time series databases. First, we introduce a novel multi-resolution PAA (MPAA) transform to achieve our iterative clustering algorithm. The method is based on the use of a multi-resolution piecewise aggregate approximation representation, which is used to extract features of time series. Then, we propose our iterative clustering approach for streaming time series. We take advantage of the multiresolution property of MPPA and equip a stopping criteria based on Hoeffding bound in order to achieve fast response time. Our streaming time-series clustering algorithm also works by leveraging off the nearest neighbors of the incoming streaming time series datasets and fulfill incremental clustering approach. The comprehensive experiments based on several publicly available real data sets shows that significant performance improvement is achieved and produce high-quality clusters in comparison to the previous methods.

1 Introduction

Numerous clustering algorithms of time series have been proposed, the majority of them work in relatively static model, while many current and emerging applications require support for on-line analysis of rapidly changing streaming time series. In this paper, we present a new approach for cluster streaming time series datasets.

Our work is motivated by the recent work by Jessica Lin and Eamonn Keogh on iterative incremental clustering of time series [1]. While we speed up clustering process by examining the time series at increasingly finer levels of approximation using multi-solution piecewise aggregate approximation (MPAA). We argue that MPAA has all the pruning power of Wavelet transform dimensionality reduction, but is also able to handle arbitrary length queries, is much faster to compute and can support a more general distance measures. Although there has been a lot of work on more flexible distance measures using Wavelet [2, 3], none of these techniques are indexable. While time series databases are often extremely large, any dimensionality reduction technique should support index method. For the task of indexing MPAA has all the advantages of Wavelet with none of the drawbacks.

Our work addresses four major challenges in applying their ideas for clustering time series in a streaming environment. Specifically, our work has fourfold main contribution:

Clustering Time Series in Streaming Environment: Streaming time-series are common in many recent applications, e.g., stock quotes, e-commerce data, system logs, network traffic management, etc [4]. Compared with traditional datasets, streaming time-series pose new challenges for query processing due to the streaming nature of data which constantly changes over time. Clustering is perhaps the most frequently used data mining algorithm. Surprisingly, clustering streaming time-series still have not explored thoroughly, to the best of our knowledge, no previous work has addressed this problem.

MPAA-based Iterative Time Series Clustering: PAA (Piecewise Aggregate Approximation) [5] transformation produces a piecewise constant approximation of the original sequence. In this paper, we introduce a novel multi-resolution PAA (MPAA) transform to achieve our iterative clustering algorithm.

Proposed stopping criteria for multi-level iterative clustering: We solve the difficult problem of deciding exactly how many levels are necessary at each node in iterative clustering algorithm by using a statistical result known as the Hoeffding bound [6].

Time Series Clustering augmented Nearest Neighbor: Our proposed inline clustering algorithm exploits characteristic of a neighborhood and significantly reduce clustering construction time and improve clustering quality.

The rest of the paper is organized as follows. In section 2, we develop enhanced iterative clustering and streaming clustering algorithm. Section 4 presents the experimental evaluation of our proposed algorithms both in offline and online form. We conclude in Section 5 with some summary remarks and future research directions.

2 Streaming Iterative Clustering Method

2.1 MPAA -Based Dimensionality Reduction

Our MPAA-based time series representation work is derived from the recent work by Eamonn Keogh [5] and Yi and Faloutsos [7] on segmenting time series representations of dimensionality reduction.

The basic idea on which their work develops is as follows. Suppose, we denote the set of time series which constitute the database as $X = \{X_1, \cdots, X_n\}$. A time series

X_i of length n is represented in N space by a vector $\overline{X}_i = \overline{x}_{i1}, \cdots, \overline{x}_{iN}$. The ith element of $\overline{X}_i$ is calculated by the following equation:

$$\overline{X}_i = \frac{N}{n} \sum_{j=\frac{n}{N}(i-1)+1}^{\frac{n}{N}i} x_j \qquad (1)$$

Our MPPA method divides time series X_i of length n into a series of lower-dimensional signal with different resolution N. where $N \in \{1, \cdots, n\}$. Simply stated, in first level, the data is divided into N "frames", whose sizes need not be contiguous and equal. The mean value of the data falling within a frame is calculated and a vector of these values becomes the data reduced representation. Then Recursively applying the above pairwise averaging process on the lower-resolution array containing the averages, we get a multi-resolution representation of time series.

We give a simple example to illustrate the MPAA decomposition procedure in Table 1. Suppose we are given a time series containing the following eight values A= [3, 5, 2, 6, 4, 8, 7, 1] and we initiate divide it into 4 segments. The MPAA transform of A can be computed as follows. We first average the values together pairwise to get a new "lower-resolution" representation of the data with the following average values [4, 4, 6, 4]. In other words, the average of the first two values (that is, 3 and 5) is 4 and that of the next two values (that is, 6 and 4) is 5, and so on. Recursively applying the above pairwise averaging process on the lower-resolution array containing the averages, we get the following full decomposition:

Table 1. A simple example to illustrate the MPAA decomposition procedure

Resolution	MPAA Values
8	3,5,2,6,4,8,7,1
4	4,4,6,4
2	4,5
1	4.5

The MPAA approximation scheme has some desirable properties that allow incremental computation of the solution. These properties are necessary in order for the algorithm to be able to operate efficiently on large datasets and streaming environment.

2.2 Enhanced Iterative Clustering Methods

Our iterative clustering method is similar to [1]. The algorithm works by leveraging off the multiresolution property of MPPA.

Note that an open problem that arise with this sort of iterative models is the definition of a minimum number of observations, i.e., devising an objective functions that determine the quality of clustering results from the previous stages to eliminate the need to re-compute all the distances.

Due to us perform the k-Means clustering algorithm, starting at the second level and gradually progress to finer levels, in order to find the stopping resolutions as low as possible to complete a good k-means clustering, it may be sufficient to consider only a small subset of the multi-level clustering examples that pass through the level of decomposition tree. We solve the difficult problem of deciding exactly how many levels are necessary at each node by using a statistical result known as the Hoeffding bound or additive Chernoff bound [6], which have in fact be successfully used in online decision trees [8][9]. After n independent observations of a real-valued random variable r with range R, the Hoeffding bound ensures that, with confidence $1-\delta$, the true mean of r is at least $\bar{r} - \varepsilon$, where $\bar{r}$ is the observed mean of the samples and

$$\varepsilon = \sqrt{\frac{R^2 In(1/\delta)}{2n}} \quad (2)$$

This is true irrespective of the probability distribution that generated the observations.

Table 2. The enhanced iterative clustering algorithms

	Algorithm SI-kMeans
1	Decide on a value for k.
2	Perform MPAA decomposition on raw data
3	Initialize the k cluster centers (randomly, if necessary).
4	Compute the hoeffding bound(ε)
5	Run the k-Means algorithm on the level i of MPAA representation of the data
6	Use final centers from level i as initial centers for level i+1. This is achieved by projecting the k centers returned by k-Means algorithm for the 2^i space in the 2^{i+1} space.
7	Compute the distance D_{center} between initial centers of level i and initial centers for level i+1
8	Compute respectively maximum values of the sum of squared intra-cluster errors in j^{th} iterative clustering and $(j+1)^{th}$ iterative clustering, i.e. $E_{max(i)}$ and $E_{max(i+1)}$
9	If $E_{max(i+1)} - E_{max(i)} > \varepsilon$, exit.
10	If $D_{center} > \varepsilon$, goto 3.

We call the new iterative clustering algorithm supporting stopping criteria SI-kMeans, where S stands for "stopping criteria.", and I stands for "interactive." Table 2 gives a skeleton of this idea.

2.3 Proposed Streaming Clustering Algorithm

A key challenging issue with streaming time series clustering algorithm is the high rate of input sequences insertion.

To illustrate our application, consider the following issue. Most streaming time series are related to previously arrived time series or future ones, hence, this strong temporal dependency between the streaming time series should not be ignored when clustering streaming data collection. This issue can be addressed by considering the nearest neighbor. A simple distance metric between two new arriving time series and the clustering center will show how much they are related to each other. Hence, the nearest neighbor analysis allows us to automatically identify related cluster.

Below, we give a more formally definition in order to depict our Streaming Clustering algorithm.

Definition 1. Similarity measure: To measure closeness between two sequences, we use correlation between time series as a similarity measure. Supposed that time-series T_i and T_j in a sliding window which length is w is represented respectively by $\{<u_{i1},t_{i1}>,\cdots,<u_{in},t_{in}>\}$ and $\{<v_{j1},t_{j1}>,\cdots,<v_{jn},t_{jn}>\}$ The Similarity between two time series T_i and T_j is defined by

$$similarity(T_i, T_j) = \frac{\sum_{k=1}^{w} u_{ik} \cdot v_{jk} - w\bar{u}\bar{v}}{\sqrt{\sum_{k=1}^{w} u^2_{ik} - w\bar{u}^2} \cdot \sqrt{\sum_{k=1}^{w} v^2_{ik} - w\bar{v}^2}} \qquad (3)$$

Definition 2. Similar: If $similarity(T_i, T_j) \geq \varsigma$, then a time series T_i is referred to as similar to a time series T_j.

Based on the definition of similar in **Definition 1**, we can define the ς-neighborhood $N_\varsigma(T_i)$ as follows:

Definition 3. ς-neighborhood $N_\varsigma(T_i)$: ς-neighborhood for a time series T_i is defined as a set of sequences $\{X_j : similarity(X_j, T_i) \geq \varsigma\}$.

Our proposed clustering algorithm exploits characteristic of a neighborhood. It is based on the observation that a property of a time-series would be influenced by its neighbors. Examples of such properties are the properties of the neighbors, or the percentage of neighbors that fulfill a certain constraint. The above idea can be translated into clustering perspective as follows: a cluster label of a time-series depends on the cluster labels of its neighbors.

The intuition behind this algorithm originates from the observation that the cluster of time series sequences can often be approximately captured by performing nearest neighbor search. In what follows, our idea is explained in detail.

Initially, we assume that only time series in now window is available. Thus, we implement SI-kMeans clustering on these sequences itself and form k clusters. Adding new sequences to existing cluster structure proceeds in three phases: neighborhood search, identification of an appropriate cluster for a new sequences, and re-clustering based on local information. The proposed incremental clustering algorithm STSI-kMeans (streaming time series iterative K-means clustering algorithm) can be discribed as follow:

Step 1. Initialization. Get next new sequences $\{T_c-w+1,\cdots,T_c\}$ in now window.

Step 2. Neighborhood search. Given a new incoming sequences $\{T_c-w+1,\cdots,T_c\}$ and let C_{K_j} be the set of clusters containing any time series belonging to $N_\zeta(T_j)$, obtain $\{N_\zeta(T_c-w+1), N_\zeta(T_c-w+2),\cdots,N_\zeta(T_c)\}$ by performing a neighborhood search on $\{T_c-w+1,\cdots,T_c\}$, and find the candidate cluster C_{K_j} which can host a new sequence $T_j \in \{T_c-w+1,\cdots,T_c\}$, that mean to identify $C_{K_j} \supset N_\zeta(T_j)$.

Step 3. Identifying an appropriate cluster. Cluster If there exists a cluster C_K that can host a sequence T_j, and then add T_j to the cluster C_K. Otherwise, create a new cluster C_{new} for T_j.

To identify a cluster C_K which can absorb the new time-series T_j from the set of candidate clusters C_{K_j}, we employ a simple but effective approach, which measures the Euclidean distance between the center of candidate clusters and the new time-series T_j, the cluster which returns the minimum distance is selected as a cluster C_K which can absorb the new time-series T_j.

Step 4. Re-clustering over affected cluster. If T_j is assigned to C_K or create a new cluster C_{new} for T_j, then a merge operation needs to be triggered. This is based on a locality assumption [10]. Instead of re-clustering the whole dataset, we only need to focus on the clusters that are affected by the new time-series. That is, a new time-series is placed in the cluster, and a sequence of cluster re-structuring processes is performed only in regions that have been affected by the new time-series, i.e., clusters that contain any time-series belonging to the neighborhood of a new time-series need to be considered.

Note that based on SI-kMeans re-clustering, the number of clusters, k' value Decide by the number of affected clusters k'' by absorbing the new time-series. Where $k' = k''$.

Step 5. Repetition. Repeat Step 2-4 whenever new sequences available in the next window.

3 Experimental Evaluation

In this section, we implemented our algorithms SI-kMeans and STSI-kMeans, and conducted a series of experiments to evaluate their efficiency. We also implemented the I-kMeans algorithm, to compare against our techniques. When not explicitly mentioned, the results reported are averages over 100 tests.

3.1 Datasets

The data using in our experiment is similar to [1]. We tested on two publicly available, real datasets: JPL datasets and heterogeneous datasets [11]. The dataset cardinalities range from 1,000 to 8,000. The length of each time series has been set to 512 on one dataset, and 1024 on the other. Each time series is z-normalized to have mean value of 0 and standard deviation of 1.

3.2 Offline Clustering Comparison

To show that our SI-kMeans approach is superior to the I-kMeans algorithm for clustering time series in offline form, in the first set of experiments, we performed a series of experiments on publicly available real datasets. After each execution, we compute the error and the execution time on the clustering results.

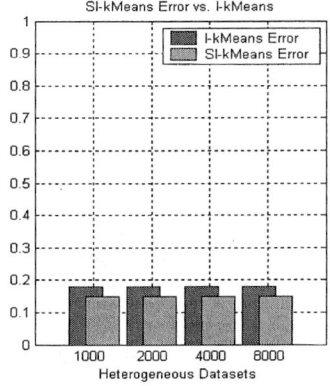 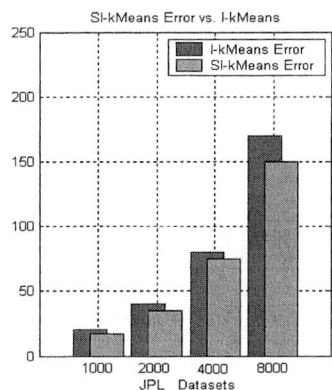

Fig. 1. Comparison of the clustering approximation error between SI-kMeans and I-kMeans. (a) Error of SI-kMeans algorithm on the Heterogeneous dataset, presented as fraction of the error from the I-kMeans algorithm. (b) Objective functions of SI-kMeans algorithm on the JPL dataset, presented as fraction of error from the I-kMeans algorithm

Figure 1 illustrates the results of clustering approximation error. As it can be seen, our algorithm achieves better clustering accuracy.

Figure 2 shows Speedup of SI-kMeans against I-kMeans. the SI-kMeans algorithm finds the best result in relatively early stage and does not need to run through all levels.

3.3 Online Clustering Comparison

In the next set of experiments, we compare the inline performance of STSI-kMeans to I-kMeans, which is essentially a comparison between an online and the corresponding offline algorithm. Since original I-kMeans algorithm is not suitable for online clustering streaming time series, we revise it and adapt it to online clustering.

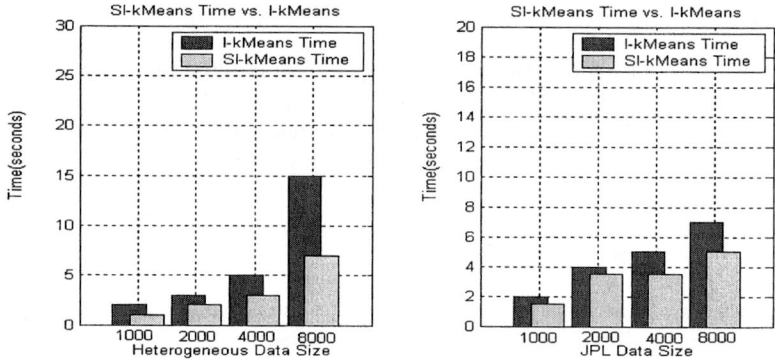

Fig. 2. Speedup of SI-kMeans against I-kMeans. (a) SI-kMeans vs. I-kMeans algorithms in terms of clustering error and running time for in the Heterogeneous dataset. (b) SI-kMeans vs. I-kMeans algorithms in terms of objective function and running time for JPL dataset

We quantify firstly the differences in the performance of the two algorithms. We report the cumulative relative error over count-based or sequence-based windows, which measure the relative increase in the cumulative error when using STSI-kMeans and I-kMeans.

$$CRE = \left| \frac{\sum_{j=1}^{q} Error_{\text{STSI-kMeans}}(w_j) - \sum_{j=1}^{q} Error_{\text{I-kMeans}}(w_j)}{\sum_{j=1}^{q} Error_{\text{I-kMeans}}(w_j)} \right| \times 100 \qquad (4)$$

Where, q is the number of elapsed windows. In Figure 3, we depict CRE as a function of q and k. In the experiment of Figure 5, the length of streaming time series 1000,2000,4000,8000 points, through, for increasing q we observe a very slow build-up of the relative error. Our algorithm performs better as the number of q increases.

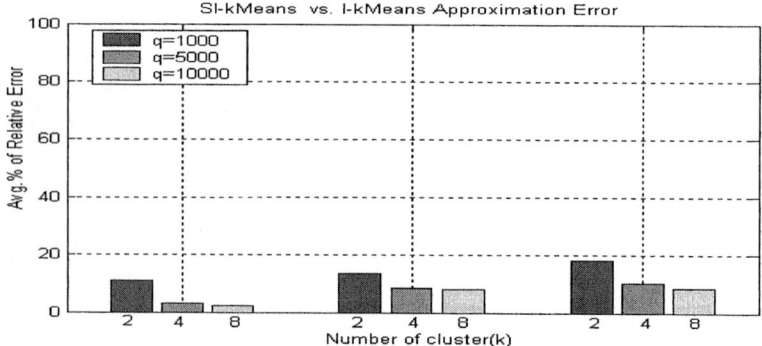

Fig. 3. Comparison of the clustering approximation error between STSI-kMeans and I-kMeans

The second measure of interest is the speedup, which measures how many times faster STSI-kMeans is when compared to I-kMeans.

$$speedup = \frac{\sum_{j=1}^{q} time_{\text{STSI-kMeans}}(w_j)}{\sum_{j=1}^{q} time_{\text{I-kMeans}}(w_j)} \qquad (5)$$

Figure 4 shows the speedup that our algorithm achieves, which translates to one or two orders of magnitude faster execution than the offline I-kMeans algorithm (for the experiments we ran). The STSI-kMeans algorithm is 10-30 times faster than I-kMeans. We observe that the speedup increases significantly for decreasing k. This is because the amount of work that STSI-kMeans does remains almost constant, while I-kMeans requires lots of extra effort for smaller values of k. As expected, the speedup gets larger when we increase q.

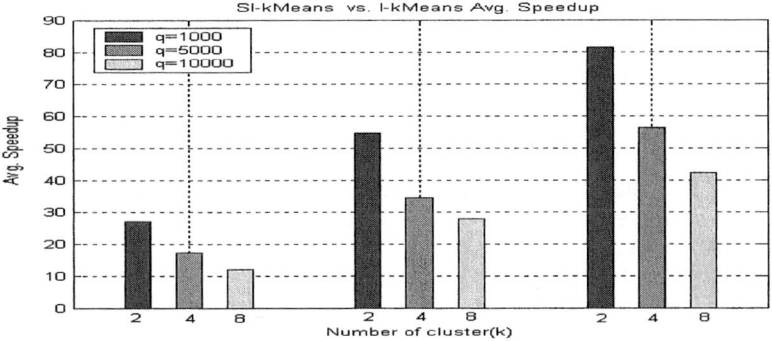

Fig. 4. Speedup of STSI-kMeans against I-kMeans

4 Conclusions

In this paper, we have presented firstly an approach to perform incremental clustering of time-series at various resolutions using the multi-resolution piecewise aggregate transform. The algorithm equipping a stopping criteria based on Hoeffding bound stabilizes at very early stages, eliminating the needs to operate on high resolutions. This approach resolves the dilemma associated with the choices of initial centers for k-Means and at which stage terminate the program for I-kMeans. This allows our algorithm to terminate the program at early stage with quality guarantee, thus eliminate the need to re-compute all the distances and significantly improves the execution time and clustering quality. We also expend our method to streaming time series environment. Our streaming time-series clustering algorithm works by leveraging off the nearest neighbors of the incoming streaming time series datasets and fulfill incremental clustering approach. Our experimental results based on several publicly available real data sets shows that significant performance improvement is achieved and produce high-quality clusters in comparison to the previous methods.

References

1. Lin, J., Vlachos, M., Keogh, E., & Gunopulos, D.: Iterative Incremental Clustering of Time Series. In proceedings of the IX Conference on Extending Database Technology (EDBT 2004). Crete, Greece. (2004) 14-18
2. Huhtala, Y., Kärkkäinen, J., & Toivonen. H.: Mining for Similarities in Aligned Time Series Using Wavelets. In Data Mining and Knowledge Discovery: Theory, Tools, and Technology. SPIE Proceedings Series Vol. 3695. Orlando, Florida. (1999)150 - 160
3. Struzik, Z. , Siebes, A.: The Haar Wavelet Transform in The Time Series Similarity Paradigm. In Proc 3rd European Conference on Principles and Practice of Knowledge Discovery in Databases. (1999)12-22
4. D. Carney, U. Cetinternel, M. Cherniack, C. Convey, S. Lee, G. Seidman, M. Stonebraker, N. Tatbul, S. Zdonik.: Monitoring streams: A New Class of Data Management Applications. In Proc. 28th Int. Conf. on Very Large Data Bases, (2002) 215-226.
5. Keogh, E., Chakrabarti, K. Pazzani, M , Mehrotra, S.: Dimensionality Reduction for Fast Similarity Search in Large Time Series Databases. Journal of Knowledge and Information Systems. Vol. 3, No. 3. (2001) 263-286
6. Hoeffding, W.: Probability inequalities for sums of bounded random variables. Journal of the American Statistical Association (1963) 13-30
7. Yi, B. , Faloutsos, C.: Fast Time Sequence Indexing for Arbitrary Lp Norms. In proceedings of the 26th Int'l Conference on Very Large Databases. Cairo, Egypt, Sept 10-14. pp 385-394.l Database Management. Berlin, Germany, Jul 26-28. (2000)55-68.
8. Domingos, P., Hulten, G.: Mining High-Speed Data Streams. In: Proceedings of the Sixth International Conference on Knowledge Discovery and Data Mining, Boston, MA, ACM Press (2000) 71-80
9. Gama, J., Medas, P., Rodrigues, P.: Concept Drift in Decision-Tree Learning for Data Streams. In: Proceedings of the Fourth European Symposium on Intelligent Technologies and their implementation on Smart Adaptive Systems, Aachen, Germany, Verlag Mainz (2004) 218-225
10. L. Ralaivola, F. dAlche-Buc.: Incremental Support Vector Machine Learning: A Local Approach. In Proceedings of the Annual Conference of the European Neural Network Society. (2001) 322-329
11. Bay, S. D.: The UCI KDD Archive [http://kdd.ics.uci.edu]. Irvine, CA: University of California, Department of Information and Computer Science. (1999)

Locating Motifs in Time-Series Data

Zheng Liu[1], Jeffrey Xu Yu[2], Xuemin Lin[1], Hongjun Lu[3], and Wei Wang[1]

[1] The University of News South Wales, Sydney, Australia
{zliu, lxue, weiw}@cse.unsw.edu.au
[2] The Chinese University of Hong Kong, Hong Kong, China
yu@se.cuhk.edu.hk
[3] The Hong Kong University of Science and Technology, Hong Kong, China
luhj@cs.ust.hk

Abstract. Finding motifs in time-series is proposed to make clustering of time-series subsequences meaningful, because most existing algorithms of clustering time-series subsequences are reported meaningless in recent studies. The existing motif finding algorithms emphasize the efficiency at the expense of quality, in terms of the number of time-series subsequences in a motif and the total number of motifs found. In this paper, we formalize the problem as a continuous top-k motif balls problem in an m-dimensional space, and propose heuristic approaches that can significantly improve the quality of motifs with reasonable overhead, as shown in our experimental studies.

1 Introduction

Data clustering is one of the primary data mining tasks [4]. In [6], Keogh et al. made a surprising claim that clustering of time-series subsequences is meaningless. Their claim is based on the fact that a data point at a certain time in a time-series appears in m adjoining sliding windows where m is the window size. The mean of all such time-series subsequences will be an approximately constant vector, which makes any time-series subsequence clustering approaches meaningless. Finding motifs is proposed as an effective solution for this problem [2,7,8]. For finding motifs, a time-series subsequence of length m (or a time-series subsequence in a sliding window of size m) is treated as a data point in an m-dimensional space. Two time-series subsequences of length m are similar, if the two corresponding m-dimensional data points are similar. The similarity is controlled by their distances, and the similar time-series subsequences are grouped as m-dimensional data points in a ball of radius r. A motif in a time-series dataset is then a dense ball with most data points after removing trivial matches in an m-dimensional space. Here, removing trivial matches is a process of removing those meaningless time-series subsequences that should not contribute to the density, and therefore is a solution to the meaningless time-series subsequence clustering problem [6]. For example, let a time-series subsequence of length m from a position l denoted as $t[l]$ and assume that it is mapped into an m-dimension data point s_l. The data point s_{i+1} is s_i's trial match, if s_i and s_{i+1}

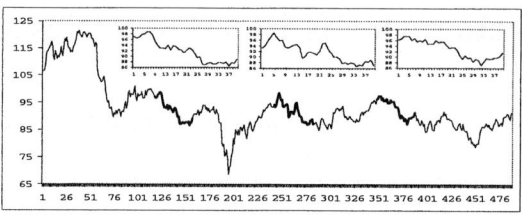

Fig. 1. Motifs in a real time-series data

are similar (within distance r). Therefore, s_{i+1} should be removed, because the two corresponding time-series subsequences $t[i+1]$ and $t[i]$ are almost the same, and counting such trivial matches makes clustering of time-series subsequences meaningless. Fig. 1 shows an example of a motif, which appears 3 times in a time-series after trivial matches have been removed.

One problem of the existing motif finding approaches is that they may miss motifs (dense balls) or report the density of balls that is less than it should be, because they only consider existing data points as potential centers to find their neighbors in radius of r [2,7,8]. In real time-series data, it is highly possible that there exist most dense balls where no existing data points can be their centers. In this paper, we formalize this problem as a problem of finding top-k motif balls in a dataset $\mathcal{S}$, in an m-dimensional space, R^m, where a motif ball is a ball, with a user given radius r, that contains most data points and is not contained in other balls. We do not limit the number of potential centers of balls to $|\mathcal{S}|$. In other words, the potential centers are countless, because they can be anywhere in the region covered by the dataset $\mathcal{S}$. That makes it possible for us to find better and/or more motif balls and is challenging.

Contributions: The main contributions include the following.

- We formalized the problem of finding motifs as finding continuous top-k motif balls by allowing non-existing data points in a dataset $\mathcal{S}$ as centers of balls of radius r.
- We proposed effective heuristic approaches to find continuous top-k motif balls. Our approach first identifies a set of candidate sets where each candidate set is enclosed by a ball whose maximal radius is $\sqrt{\frac{1}{\frac{1}{m}+1}}\sqrt{2}r$ for m-dimensions. The set of candidate sets identified are minimized in order to reduce redundant computation. Second, we approach two simple but effective heuristics to find motif balls of radius r within each candidate set.
- We conducted extensive experimental studies and our results showed that our approaches can significantly improve the quality of motifs found in time-series datasets.

Organization: Section 2 gives the problem definition. The algorithms for finding k-BALL are given in Section 3. Experimental results are given in Section 4. Finally, Section 5 concludes the paper.

2 Problem Definition

Formally, given a time-series $T = t_1, t_2, \cdots$, let the length of T be $|T|$. A subsequence of length m of T, $t_i, \cdots, t_{i+m-1}$, is denoted by $t[i]$, for $1 \le i \le |T|-m+1$. We say $t[i] \preceq t[j]$ if $i < j$. Consider a subsequence $t[i]$ as an m-dimensional data point s_i. In other words, there exists a one-to-one mapping τ from $t[i]$ to s_i such as $\tau(t[i]) = s_i$ and $\tau^{-1}(s_i) = t[i]$. We call $s_i \preceq s_j$ if $t[i] \preceq t[j]$. Let $\mathcal{S} = \{s_1, s_2, \cdots s_n\} \subset R^m$ where $s_i = \tau(t[i])$ and $n = |T| - m + 1$.

A *ball* is a set of data points around a center $c \in R^m$, denoted as B, such as $\{s_j \mid d(c, s_j) \le r \wedge s_j \in \mathcal{S}\}$ where r is a user given radius for measuring similarity between two data points, and $d()$ is a distance function (for example, Euclidean distance). Note: as a unique feature of this problem, a ball B is defined with a center c which does not necessarily belong to $\mathcal{S}$ but R^m. In the following, we call a ball whose center is in $\mathcal{S}$ a *discrete ball*, and a ball whose center is not in $\mathcal{S}$ a *continuous ball*. A *discrete ball* is contained by some *continuous ball* when their radii are the same. Below, otherwise stated, a ball is a continuous ball.

Let B be a ball, we use $cent(B)$, $rd(B)$ and $sp(B)$ to denote its center c, its radius and the set of points on the boundary of the ball, respectively. The density of a ball is the total number of data points in B, denoted as $|B|$. In addition, given a set of data points D, we use $ball(D)$ to denote the smallest enclosing ball that contains all data points in D. A ball function can be implemented using a move-to-front strategy as the *miniball* function given in [9, 3]. Let $ball_r(D)$ be a boolean function which returns **true** if the radius of $ball(D)$ is less than or equal to r ($rd(ball(D)) \le r$) and otherwise **false**. We simply say a set of data points, D, is a r-ball if $ball_r(D)$ is **true**. A motif ball is a r-ball, $\mathcal{B}$, after removing trivial matches from r-ball. Here, a trivial match is defined as follows. Let $s_i = \tau(t[i])$, $s_j = \tau(t[j])$, and $s_k = \tau(t[k])$. s_j is s_i's trivial match if s_i and s_j are in a r-ball and there does not exist s_k outside the r-ball such as $t[i] \preceq t[k] \preceq t[j]$.

The problem of locating *top-k* motif balls, denoted by k-BALL, is to find a set of motif balls in a dataset $\mathcal{S}$, $\{\mathcal{B}_1, \cdots, \mathcal{B}_k\}$. Here, the top-1 motif ball $\mathcal{B}_1$ has the highest density. $\mathcal{B}_i$ is the top-1 motif ball after the corresponding r-balls for $\mathcal{B}_1, \cdots \mathcal{B}_{i-1}$ are removed from $\mathcal{S}$.

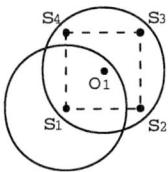

Fig. 2. Finding $\mathcal{B}$'s for $cent(\mathcal{B}) \notin \mathcal{S}$

Consider finding continuous k-BALL (top-k continuous motif balls), in comparison with finding discrete k-BALL. Figure 2 shows an example where $\mathcal{S} = \{s_1, s_2, s_3, s_4\}$. Assume that the distances, $d(s_1, s_2)$, $d(s_2, s_3)$, $d(s_3, s_4)$ and

$d(s_4, s_1)$, are all equal to $r + \epsilon$, for a very small $\epsilon > 0$. There does not exist a discrete ball $\mathcal{B}_i$, where $s_i = cent(\mathcal{B}_i)$ and $rd(\mathcal{B}_i) \leq r$, that contains more than one data point. In other words, all the discrete motif balls contain only a single data point. But, in fact, the four data points belong to a continuous motif ball whose center is o_1 – the center of the four data points in $\mathcal{S}$. Note $o_1 \notin \mathcal{S}$ and $d(o_1, s_i) < r$ for $s_i \in \mathcal{S}$. Such a ball $\mathcal{B}$ at center o_1 is the motif ball to be found.

Let $\mathcal{S}$ be a dataset in an m-dimensional space, and $n = |\mathcal{S}|$. In terms of the number of times of calling the distance function $d()$, the time complexity of discrete motif balls is $O(n^2)$, because it basically needs to find those data points within a distance r for every data point in a dataset. Several efficient methods were proposed in [2,7]. The time complexity of finding continuous motif balls is $O(2^n)$, in terms of the number of times of calling the function $ball()$, because it needs to check if any subset of the dataset can be a motif ball. It is important to note that the cost for continuous motif balls to check whether a set of data points is a r-ball is the complexity of the $ball$ function which is $O((m+1)!(m+1)n)$ in an m-dimensional space [3]. When the dimensions are less than 30, the $ball$ function [3] is efficient. When the dimensions are large, the $ball$ function becomes noticeably slow. Hence, effective heuristics are on demand to find better and more motif balls with reasonable overhead, which is the focus of this paper.

3 Finding k-BALL

Our k-BALL algorithm is given in Algorithm 1, which takes three parameters: a data set $\mathcal{S}$, a radius r (> 0) and a positive number $k > 0$. First, it generates a set of candidates (line 2). Second, it calls allBall() to find potential motif balls which will be stored in a tree structure (lines 3-5). Third, it calls top-k-BALL() to report k-BALLs (line 6). Below, we discuss i) generating candidate sets (the function gen), ii) finding motif balls, and iii) identifying k-BALLs.

Algorithm 1. k-BALL $(\mathcal{S}, r, k)$

Input: a set of data points, $\mathcal{S}$, a radius r, and k.
Output: a set of motif balls.
1: Let $\mathcal{T}$ be a tree maintaining motif balls, which is set $\emptyset$;
2: $\mathcal{C} \leftarrow gen(\mathcal{S})$;
3: **for** each $\mathcal{C}_i \in \mathcal{C}$ **do**
4:     $\mathcal{B} \leftarrow allBall(\mathcal{C}_i, r)$;
5:     insert $\mathcal{B}$ into $\mathcal{T}$;
6: **return** top-k-BALL$(\mathcal{T})$;

3.1 Candidate Set Generation

We take a divide-and-conquer approach to divide the entire set of $\mathcal{S}$ into subsets, $\mathcal{C}_1, \cdots, \mathcal{C}_{|\mathcal{S}|}$, and identify motif balls in each $\mathcal{C}_i$ such as the set of motif balls is the union of motif balls found in every $\mathcal{C}_i$. In doing so, for a given dataset $\mathcal{S}$, the complexity of the continuous motif balls changes from 2^n to $n \cdot 2^l$, with the hope that $l \ll n$, where $n = |\mathcal{S}|$ and l is the maximum size of a subset C_i. The similar approach was also taken in [2, 7]. Here, we call C_i a *candidate set* and define it as $C_i = \{s_j \mid d(s_i, s_j) \leq 2r\}$. In other words, a candidate set C_i is enclosed in a smallest enclosing ball, $ball(C_i)$, whose center is $s_i \in \mathcal{S}$, and its radius, $rd(ball(C_i))$, is less than or equal to $2r$.

Lemma 1. *Any motif ball, if it exists in $\mathcal{S}$, then it exists in a candidate set C_i.*

It is worth of noting that the number of data points in a candidate set C_i, in a ball of radius $2r$, can be large, and the number of motif balls of radius r in a radius-$2r$ ball can be considerably large. In this paper, we further divide C_i into a set of subsets, C_{i_j}, such as the distance of *every* two data points in C_{i_j} is less than or equal to $2r$. We call such a C_{i_j} as a *min candidate set* in a sense that it may miss a motif ball if we further remove a data point from C_{i_j}. There are many such min candidate sets in C_i, we call a min candidate set, C_{i_j} ($\subseteq C_i$), as a *maxmin candidate set* if there does not exist C_{i_k} ($\subseteq C_i$) such as $C_{i_j} \subseteq C_{i_k}$. Identifying a maxmin candidate set can significantly reduce computational cost, because all motif balls found in C_{i_j} can be identified in C_{i_k}.

Lemma 2. *Any motif ball B, if it exists in $\mathcal{S}$, exists at least in a maxmin candidate set C_{i_j}.*

Following Lemma 2, the entire set of $\mathcal{S}$ is divided into a set of maxmin candidate sets. The set of motif balls to be found is the union of all motif balls found in all maxmin candidate sets.

Lemma 3. *In an m-dimensional space, the maximum radius of the smallest enclosing ball for a maxmin candidate set is $\sqrt{\frac{1}{\frac{1}{m}+1}}\sqrt{2}r$ ($> r$). When m approaches $+\infty$, the maximum radius is $\sqrt{2}r$.*[1]

Given two candidate sets C_i and C_j. Each will generate a set of maxmin candidate sets: $C_i = \{C_{i_1}, C_{i_2}, \cdots C_{i_l}\}$ and $C_j = \{C_{j_1}, C_{j_2}, \cdots C_{j_k}\}$. Here, each C_{l_k} is a maxmin candidate set in a candidate set C_l. It is possible that the same maxmin candidate set may appear in different candidate sets, such as $C_{i_l} = C_{j_k}$ for $i \neq j$, or a C_{i_l} in a candidate set C_i is a subset of C_{j_k}, because we use a divide-and-conquer approach which first generates a set of candidate sets and then generates a set of maxmin candidate sets from each candidate set. The issue here is how to reduce such redundant computation. A naive solution is to generate a set of all maxmin candidate sets first and then remove redundant and obtain maxmin candidate sets. However, it requires a large memory space and long CPU time. We propose a simple but effective pruning strategy to ensure that a C_{i_l} is only computed once if there $C_{i_l} = C_{j_k}$, and C_{i_l} will not be computed if $C_{i_l} \subset C_{j_k}$. The strategy is given below. Recall C_i is a candidate set with a point s_i as its center. Therefore, any maxmin candidate set C_{i_l} of C_i must contain the data point s_i.

---

[1] Due to the space limitation, the details of the proof are omitted and the interested readers may email the authors to get a full version of this paper.

Remark 1. (**Filtering maxmin candidate sets**) Given C_{i_l}, and let s_k be the first data point in C_{i_l} following the order $\preceq$. The strategy is: C_{i_l} does not need to be computed if $s_k \neq s_i$ where s_i is the center of C_i. It does not miss any motif balls because there must exist a maxmin candidate set C_{k_j} in C_k such as $C_{j_k} = C_{i_l}$ and the first data point in C_{k_j} is s_k.

Based on Remark 1, we only need to process all C_{i_l} whose first data point is s_i following $\preceq$. Suppose we filter maxmin candidate sets using Remark 1, we further claim that there is no such $C_{i_l} \subset C_{j_k}$, where C_{i_l} is a maxmin candidate set in C_i whose center is s_i, and C_{j_k} is a maxmin candidate set in C_j whose center is s_j. We consider two cases, i) $s_i \preceq s_j$ and ii) $s_j \preceq s_i$. There is no case i), because if $C_{i_l} \subset C_{j_k}$ then $s_i \in C_{i_l}$ must belong to C_{j_k}. But s_j is the smallest element which indicates that s_i cannot be in C_{j_k}, because of the filtering. There is no case ii), because if $C_{i_l} \subset C_{j_k}$, then the distance between s_j and any data point in C_{j_k} is $< 2r$. So, C_{i_l} is supposed to include s_j. But, C_{i_l} cannot contain s_j because, if so, there must exists $C'_{i_l} \supset C_{i_l}$ so C_{i_l} is removed, and C'_{i_l} is removed because its first data point is s_j but not s_i. It implies that C_{i_l} does not exist.

Lemma 4. *Based on filtering maxmin candidate sets (Remark 1), the union of remaining maximin candidate sets $\{C_{i_1}, C_{i_2}, \cdots\}$ from all candidate sets C_i are sufficient to answer continuous k-BALL and the entire maxmin candidate sets left are minimal such as there does not exist $C_{i_j} \subseteq C_{k_l}$.*

The maxmin candidate sets can be found from a candidate set C_i, using the A-priori property. The problem becomes to find all min candidate subsets that are not included in any other candidate subsets using existing maximal frequent itemsets mining approaches. Alternatively, the maxmin candidate sets can be identified as finding cliques in an undirected graph, $G(V, E)$, for every candidate set C_i. Here, $V = C_i$ (for a candidate set) and $E \subseteq V \times V$ such that $(v_i, v_j) \in E$ if $d(v_i, v_j) \leq 2r$. A clique in a graph is a maximal complete subgraph and a maxmin candidate is such a clique.

The main concern is the cost for reducing the radius of a candidate set from $2r$ to $\sqrt{\frac{1}{m+1}}\sqrt{2}r$. However, in handling our problem in a high dimensional space, it can be processed efficiently for the following reasons. First, in a high dimensional space, due to the curse of dimensionality [5], the data is sparse. Second the radius r for measuring similarity is considerably small. Therefore, the number of data points in a candidate set (within radius of $2r$) is limited. Note: the complexity of clique is linear with the number of nodes in a graph and is exponential with k which is the maximal complete subgraph. The number of nodes is small in our problem. The clique can be efficiently identified because possible maximum k is small accordingly. We adapt the clique algorithm [1] which uses branch-and-bound technique to cut off branches that cannot lead to a clique, and shows efficiency for identifying maxmin candidate sets in our experimental studies.

3.2 Finding All Balls

Given a dataset $\mathcal{S}$, we can obtain a set of maxmin candidate sets, $\mathcal{C} = \{\mathcal{C}_1, \mathcal{C}_2, \cdots\}$. In this section we present two heuristic algorithms, as an alternative of the implementation of allBall() in Algorithm 1, to find motif balls in a maxmin candidate set $\mathcal{C}_l$. The all motif balls found in $\mathcal{S}$ are the union of all motif balls found in $\mathcal{C}_l$.

- **approx-Ball-1**: Given a maxmin candidate set $\mathcal{C}_l$, let P be the data points on the boundary of the smallest enclosing ball of $\mathcal{C}_l$. It assumes that the estimated center of the smallest enclosing ball of $\mathcal{C}_l$ is the mean vector of the data points in $\mathcal{C}_l$, and removes a data point in P from $\mathcal{C}_l$ that is the furthest to the mean vector until a r-ball is identified. The algorithm is given in Algorithm 2.
- **approx-Ball-mean**: Given a maxmin candidate set $\mathcal{C}_l$, it also assumes that the estimated center of the smallest enclosing ball of $\mathcal{C}_l$ is the mean vector of the data points in $\mathcal{C}_l$. In each iteration, it computes the mean vector of data points and treats it as the center followed by removing those data points that are not within the distance r of the center and adding those within the distance r. The algorithm is given in Algorithm 3.

Algorithm 2. approx-Ball-1$(\mathcal{C}_l, r)$

1: $\mathcal{B} \leftarrow ball(\mathcal{C}_l)$;
2: $r' \leftarrow rd(\mathcal{B})$;
3: $P \leftarrow sp(\mathcal{B})$;
4: **while** $r' > r$ **do**
5:    let c be the mean vector of $\mathcal{C}_l$;
6:    remove $p \in P$, which is furthest from c, from $\mathcal{C}_l$;
7:    $\mathcal{B} \leftarrow ball(\mathcal{C}_l)$;
8:    $r' \leftarrow rd(\mathcal{B})$;
9:    $P \leftarrow sp(\mathcal{B})$;
10: **return** $\mathcal{C}_l$;

Algorithm 3. approx-Ball-mean$(\mathcal{C}_l, r)$

1: $B' \leftarrow \mathcal{C}_l$;
2: Let c be the mean vector of B';
3: $B \leftarrow \{s_k \mid d(s_k, c) \leq r, s_k \in \mathcal{C}_l\}$;
4: **while** $B' \neq B$ **do**
5:    $B' \leftarrow B$;
6:    Let c be the mean vector of B;
7:    $B = \{s_k \mid d(s_k, c) \leq r, s_k \in \mathcal{C}_l\}$;
8: **return** B;

3.3 CB-TREE

In this section, we show how we maintain a set of motif balls for the final process of identifying k-BALLS. A CB-TREE is a tree that maintains all motif balls that are not contained by any others. In a CB-TREE, a node represents a data point, a path from the root to a leaf represents a motif ball. The data structure for a CB-TREE consists of a header table and a tree structure. A node in the tree contains pointers to its children and a pointer to its parent. For each data point, there is an entry in the header table that contains a list of pointers pointing to the occurrences of the data points in different paths in the tree. The center and radius of a ball are kept in the leaf node of the corresponding path. Let $\mathcal{T}$ be a CB-TREE, and $\mathcal{B}$ be a motif ball (a dataset). Several primitive operations are defined on $\mathcal{T}$: search($\mathcal{T}, \mathcal{B}$) to search whether the superset of $\mathcal{B}$ (including itself) exists in $\mathcal{T}$ as a path, insert($\mathcal{T}, \mathcal{B}$) to insert $\mathcal{B}$ into $\mathcal{T}$, subsets($\mathcal{T}, \mathcal{B}$) to identify a set of motif balls (paths) in $\mathcal{T}$ that are contained in $\mathcal{B}$, delete($\mathcal{T}, \mathcal{B}$) to delete $\mathcal{B}$ from $\mathcal{T}$, and condense($\mathcal{T}$) to condense $\mathcal{T}$. A tree $\mathcal{T}$ is not condensed if there are some edges that can be removed to represent the same balls. For example, consider two balls, $\{s_1, s_2, s_3\}$ and $\{s_1, s_3, s_5\}$. The tree can be represented as either two paths $s_1.s_2.s_3$ and $s_1.s_3.s_5$ where both paths only have s_1 as their common prefix, or two paths $s_1.s_3.s_2$ and $s_1.s_3.s_5$ where both paths have $s_1.s_3$ as their common prefix. The latter is condensed. The insertion of a motif ball into $\mathcal{T}$ in Algorithm 1 (line 5) is implemented as follows. First, it checks if there is a superset of $\mathcal{B}$ in $\mathcal{T}$ already by calling search($\mathcal{T}, \mathcal{B}$). If there is such a path representing a superset of $\mathcal{B}$, then there is no need to do insertion. Otherwise, it inserts $\mathcal{B}$ into $\mathcal{T}$ and at the same time removes all motif balls (paths) that are contained in subsets($\mathcal{T}, \mathcal{B}$). Finally, it calls condense($\mathcal{T}$) to make the tree condensed. We maintain a list of CB-TREEs where each CB-TREE maintains a set of motif balls for a candidate set C_i with a header and a tree.

3.4 Identifying Top k-BALLs

Identifying top k-BALLs can be implemented as operations on the list of CB-TREEs. The longest path found in the list of CB-TREEs is the top-1 motif ball. We remove edges from the leaf node of the longest paths if they are not shared by others, and keep its center. Suppose that we have already identified l motif balls where $l < k$. We can easily identify the next motif ball if it has longest path after removing the nodes that are already in the former l motif balls. Trivial matches will be removed in this step.

4 Performance Study

We implemented our continuous k-BALL algorithm (Algorithm 1) with two heuristic algorithms for allBll(), namely, approx-Ball-1 (Algorithm 2) and approx-Ball-mean (Algorithm 3). Below we denote the two continuous k-BALL algorithms as Ball-1 and Ball-m, respectively. Also, based on [2,7], we implemented a discrete k-BALL algorithm, denoted Ball-d, which first identifies a set of motif balls for $\mathcal{S}$ where a motif ball $\mathcal{B}_i = \{s_j \mid d(s_i, s_j) \leq r\}$ is at the center $s_i \in \mathcal{S}$, second, maintains the motif balls found in a CB-TREE, and third identifies k-BALL using the same routine of top-k-BALL() in Algorithm 1. All algorithms use Euclidean distance function and were implemented using C++ including some optimizing techniques in [2,7]. All experiments were performed on a Pentium IV 2.53GHz PC with 512MB memory, running Microsoft Windows 2000. We report our findings for the three algorithms Ball-1, Ball-m and Ball-d.

We conducted extensive testing using a two year New York Stock Exchange tick-by-tick real dataset (year 2001 and year 2002). We report our finding using two representative stocks, A and B. The lengths of A and B are 12,802, as shown in Fig. 3. Based on a time-series, T, of size $|T|$, we generate an m-dimensional dataset $\mathcal{S}$ of size $n = |T| - m + 1$, where m is the sliding window size. Due to space limit, we only report the results for $m = 128$. We normalized time-series subsequences in every sliding window into the range [-1,1]. We show our testing results using a radius r in the range of 0.8 and 1.15, because the similarity between every two adjoining time-series subsequences in the two time-series, A and B, forms a normal distribution where the peak value is 1.4 and majority values are in the range of 0.7 and 2.3. And $r = 0.8$ is the smallest radius that the density of motif balls is greater than 1.

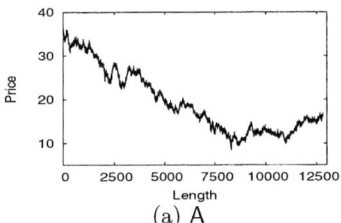
(a) A

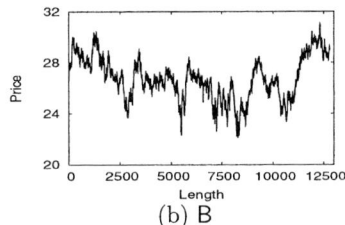
(b) B

Fig. 3. Two Representative Stocks

In addition to the total number of motif balls, the density of motif balls, and the CPU time, we measure the quality of a motif ball using a *centric* value. The centric value takes two factors into consideration: the distance between the mean of all data points in a motif ball and the center of the motif ball, and the density of a motif ball. Here, all data points in a motif ball mean all remaining data points after removing the trivial matches and the density of a motif ball is the density of the remaining data points. Suppose $\mathcal{B} = \{s_1, s_2, \cdots, s_n\}$ is a motif ball, with n m-dimensional data points. Let the mean of $\mathcal{B}$ be $mean(\mathcal{B}) = \frac{1}{n}\sum_{i=1}^{n} s_i$. The centric of $\mathcal{B}$ is given below. The smaller the centric value is, the better quality of $\mathcal{B}$ is $centric(\mathcal{B}) = \frac{d(mean(\mathcal{B}),cent(\mathcal{B}))}{|\mathcal{B}|}$.

4.1 The Total Number of Motif Balls and The CPU Time

Fig. 4 (a) and (b) show the total number of motif balls found for time-series A and B. Both *Ball-1* and *Ball-m* significantly outperform *Ball-d*. The curves of *Ball-1* and *Ball-m* decrease sometimes when $r > 1.0$ in Fig. 4 (a) and (b), because when the radius r becomes larger, a motif ball can contain more data points. The CPU time for *Ball-1*, *Ball-m* and *Ball-d* are shown in Fig. 4 (c) and (d). The CPU time includes the time for generating candidate sets, finding motif balls and inserting motif balls into CB-TREEs. The CPU time does not include the time to report top-k balls from CB-TREEs, because it penalizes the algorithms that find more and better motif balls. The CPU time for *Ball-1* is higher than that

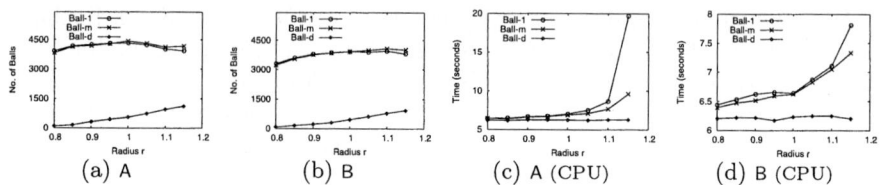

Fig. 4. The Total Number of Motif Balls and The CPU Time

for *Ball-m*, because *Ball-1* needs to call *ball* function repeatedly, in particular when the radius of the smallest enclosing ball for a candidate set is large. The time of computing cliques takes such little time that could be ignored, because in a high-dimensional dataset, there are not many data points in a candidate set C_i when the radius r is small, due to the curse of dimensionality [5].

4.2 Top-k Motif Balls

Fig. 5 (a) and (b) show the density of the top-1 motif ball for A and B separately, while varying the radius r, where $m = 128$. In all cases, *Ball-1* and *Ball-m* outperform *Ball-d*, because both can report motif balls with more data points. Fig. 5 (c) and (d) show the centric values. Also, *Ball-1* and *Ball-m* outperform *Ball-d*. Fig. 5 (c) and (d) suggest that the data points in the motif balls found by *Ball-1* and *Ball-m* are close to the center of the motif balls, whereas the data points in the motif balls found by *Ball-1* can be rather various. The centric value of *Ball-m* in Fig. 5 (c) when $r = 0.8$ is close to one of *Ball-d* is because there is only one remaining point in the top-1 motif ball after removing trivial matches as we can refer to Fig. 5 (a) and the mean of all data points is itself.

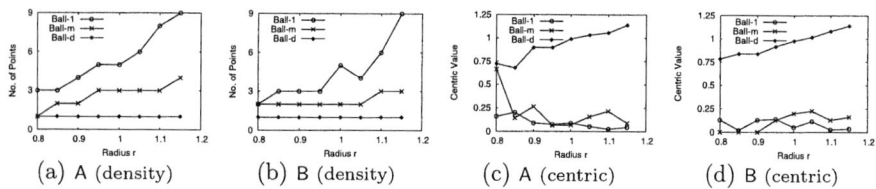

Fig. 5. Top-1 Motif Balls While Varying r

The results for top-k motif balls are shown in Fig. 6, where k is 15. Fig. 6 (a) and (b) show the density, and Fig. 6 (c) and (d) show the centric values for the top-15 motif balls for $r = 1.1$. The top-k motif balls found by *Ball-d* is inferior to both *Ball-1* and *Ball-m*, in terms of both the density and the centric values.

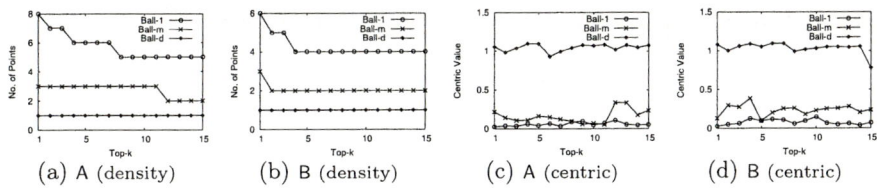

(a) A (density) (b) B (density) (c) A (centric) (d) B (centric)

Fig. 6. Top-15 Motif Balls When $r = 1.1$

5 Conclusion

In this paper, we formalized the problem of locating motifs as finding continuous top-k motif balls by allowing non-existing data points in a dataset S as centers of balls of radius r, and proposed effective heuristic approaches to find continuous top-k motif balls. First, our approach identifies a minimal set of maxmin candidate sets where each maxmin candidate set is enclosed by a ball whose max radius is $\sqrt{\frac{1}{\frac{1}{m}+1}}\sqrt{2}r$ for m-dimensions. Second, in order to compromise efficiency, we propose two heuristics to find motifs within each maxmin candidate set. Our experimental results showed that our approaches can significantly improve the quality of motifs found in time-series datasets.

Acknowledgment. The work described in this paper was supported by grants from the Research Grants Council of the Hong Kong SAR (CUHK4229/01E) and ARC Discovery Grant (DP0346004).

References

1. C. Bron and J. Kerbosch. Algorithm 457: finding all cliques of an undirected graph. *Communications of the ACM*, 16(9):575–577, 1973.
2. B. Chiu, E. Keogh, and S. Lonardi. Probabilistic discovery of time series motifs. In *Proc. of KDD'03*, 2003.
3. B. Gärtner. Fast and robust smallest enclosing balls. In *Proc. of ESA'99*, 1999.
4. J. Han and M. Kamber. *Data Mining: Concepts and Techniques*. 2001.
5. A. Hinnerburg and D. A. Keim. Optimal grid-clustering: Towards breaking the curse of dimensionality in high-dimensional clustering. In *Proc. of VLDB'99*, 1999.
6. E. Keogh, J. Lin, and W. Truppel. Clustering of time series subsequences is meaningless: Implications for past and future research. In *Proc. of ICDM'03*, 2003.
7. J. Lin, E. Keogh, S. Lonardi, and P. Patel. Finding motifs in time series. In *Proc. of the 2nd Workshop on Temporal Data Mining, at ACM SIGKDD'02*, 2002.
8. P. Patel, E. Keogh, J. Lin, and S. Lonardi. Mining motifs in massive time series database. In *Proc. of ICDM'02*, 2002.
9. E. Welzl. Smallest enclosing disks (balls and ellipsoids). In *New Results and New Trends in Computer Science*. Springer-Verlag, 1991.

Stochastic Local Clustering for Massive Graphs

Satu Elisa Schaeffer

Helsinki University of Technology, Espoo, Finland
elisa.schaeffer@tkk.fi
http://www.tcs.hut.fi/~satu/

Abstract. Most graph-theoretical clustering algorithms require the complete adjacency relation of the graph representing the examined data. This is infeasible for very large graphs currently emerging in many application areas. We propose a local approach that computes clusters in graphs, one at a time, relying only on the neighborhoods of the vertices included in the current cluster candidate. This enables implementing a local and parameter-free algorithm. Approximate clusters may be identified quickly by heuristic methods. We report experimental results on clustering graphs using simulated annealing.

1 Introduction

Many practical applications of information processing involve massive amounts of data, much of which tends to be noise, and only a small fraction contains semantically interesting information. In general, *clustering* is the process of organizing such data into meaningful groups in order to interpret properties of the data. Some general clustering methods operate online [2, 9], but practically all existing methods require some parameters in addition to a distance measure, such as the number of clusters to produce. In graph clustering, the data consists of a set of n *vertices* V connected by a set of m *edges* E. A *cluster* in a graph $G = (V, E)$ is considered to be a set of vertices that have relatively many connections among themselves with respect to the graph structure on a global scale. The existing methods are mostly global and rely on the full adjacency relation of the graph or a derived measure, such as the graph's eigenvalue spectrum [3].

2 Local Clustering

For many applications, only a small subset of vertices needs to be clustered instead of the whole graph. These include locating documents or genes closely related to a given "seed" data set. This motivates the use of a *local* approach for finding a good cluster containing a specified vertex or a set of vertices by examining only a limited number of vertices at a time, proceeding in the "vicinity" of the seed vertex. The scalability problem is avoided, as the graph as a whole never needs to be processed and clusters for different seeds may be obtained by parallel computation.

We adopt the following notation: in a graph $G = (V, E)$, a cluster candidate is a set of vertices $\mathcal{C} \subseteq V$. The *order* of the cluster is the number of vertices included in the cluster, denoted by $|\mathcal{C}|$. Two vertices u and v are said to be *neighbors* if $(u, v) \in E$. Following common criteria [3, 5, 8], we want the clusters to be vertex sets that are connected in G by many internal connections and only few connections outside. We define the *internal degree* of a cluster $\mathcal{C}$ to be the number of edges connecting vertices in $\mathcal{C}$ to each other:

$$\deg_{\text{int}}(\mathcal{C}) = |\{(u, v) \in E \mid u, v \in \mathcal{C}\}|. \tag{1}$$

The *local density*[1] of a cluster $\mathcal{C}$ is

$$\delta_\ell(\mathcal{C}) = \frac{\deg_{\text{int}}(\mathcal{C})}{\binom{|\mathcal{C}|}{2}} = \frac{2 \deg_{\text{int}}}{|\mathcal{C}|(|\mathcal{C}| - 1)}. \tag{2}$$

Clearly, optimizing $\delta_\ell \in [0, 1]$ alone makes small cliques superior to larger but slightly sparser subgraphs, which is often impractical. For clusters to have only a few connections to the rest of the graph, one may optimize the *relative density* $\delta_r(\mathcal{C})$ (see [6] and the references therein); it is defined in terms of the internal degree $\deg_{\text{int}}$ (Equation 1) and *external degree* of a cluster $\mathcal{C}$,

$$\deg_{\text{ext}}(\mathcal{C}) = |\{(u, v) \in E \mid u \in \mathcal{C}, v \in V \setminus \mathcal{C}\}|, \tag{3}$$

as the ratio of the internal degree to the number of edges incident to the cluster,

$$\delta_r(\mathcal{C}) = \frac{\deg_{\text{int}}(\mathcal{C})}{\deg_{\text{int}}(\mathcal{C}) + \deg_{\text{ext}}(\mathcal{C})}, \tag{4}$$

which favors subgraphs with few connections to other parts of the graph. Possible combinations of the above measures are numerous; in this paper we use the product as a cluster quality measure:

$$f(\mathcal{C}) = \delta_\ell(\mathcal{C}) \cdot \delta_r(\mathcal{C}) = \frac{2 \deg_{\text{int}}(\mathcal{C})^2}{|\mathcal{C}|(|\mathcal{C}| - 1)(\deg_{\text{int}}(\mathcal{C}) + \deg_{\text{ext}}(\mathcal{C}))}. \tag{5}$$

The complexity of optimizing Equation 5 can be studied through the decision problem of whether a given graph G has a k-vertex subgraph $\mathcal{C}$ for which $f(\mathcal{C}) \geq \gamma$ for some fixed $k \in \mathbf{N}$ and $\gamma \in [0, 1]$. Especially, we are interested to know whether there is such a subgraph that contains a given vertex v. Both $\delta_\ell(\mathcal{C})$ and $\delta_r(\mathcal{C})$ alone correspond to NP-complete decision problems; the complexities of these and other cluster fitness measures are discussed in a separate paper [11].

2.1 Computation by Local Search

Calculation of the proposed fitness measure only requires the adjacency lists of the included vertices. Therefore, a good approximation of the optimal cluster

---

[1] For $|\mathcal{C}| \in \{0, 1\}$, we set $\delta_\ell(\mathcal{C}) = 0$.

for a given vertex can be obtained by *local search*. To locate a cluster containing a given vertex $v \in V$ from a graph $G = (V, E)$, we stochastically examine subsets of V containing v, and choose the candidate with maximal f as $\mathcal{C}(v)$. The initial cluster $\mathcal{C}'(v)$ of a vertex v contains v itself and all vertices adjacent to v. Each search step may either add a new vertex that is adjacent to an already included vertex, or remove an included vertex. Upon the removal of $u \in \mathcal{C}'(v)$, $u \neq v$, the connected component containing v becomes the next cluster candidate. Redefining $\deg_{\text{ext}} = |\{\langle u, v \rangle \in E \mid u \in \mathcal{C}, v \in V \setminus \mathcal{C}\}|$ allows clustering directed graphs. Our clustering of a 32,148-vertex directed graph representing the Chilean inter-domain link structure is discussed in [12].

The method is well-suited for memory-efficient implementation: if the graph is stored as adjacency lists, $\langle v : w_1, w_2, \ldots, w_{\deg(v)} \rangle$, only one such entry at a time needs to be retrieved from memory. For n vertices, the entries can be organized into a search tree with $\mathcal{O}(\log n)$ access time. The search needs to maintain only the following information:

(a) the list of currently included vertices $\mathcal{C}$,
(b) the current internal degree $\deg_{\text{int}}(\mathcal{C})$ (Equation 1), and
(c) the current external degree $\deg_{\text{ext}}(\mathcal{C})$ (Equation 3).

When a vertex v is considered for addition into the current cluster candidate $\mathcal{C}$, its adjacency list is retrieved and the degree counts for the new candidate $\mathcal{C}' = \mathcal{C} \cup \{v\}$ are calculated as follows:

$$\deg_{\text{int}}(\mathcal{C}') = \deg_{\text{int}}(\mathcal{C}) + k, \qquad \deg_{\text{ext}}(\mathcal{C}') = \deg_{\text{ext}}(\mathcal{C}) - k + \ell, \qquad (6)$$

where $k = |\mathcal{C} \cap \Gamma(v)|$ and $\ell = \deg(v) - k$, $\Gamma(v)$ denoting the set of neighbors of vertex v. The removal of vertices from a cluster candidate is done analogously, subtracting from the internal degree the lost connections and adding them to the external degree. The memory consumption is determined by the local structure of the graph. The order of the initial cluster is limited from above by the maximum degree of the graph Δ plus one; in natural graphs, usually $\Delta \ll n$ and $|\mathcal{C}| \ll n$. Hence examining the adjacency lists of the vertices included in the final cluster candidate takes $\mathcal{O}(\Delta \cdot |\mathcal{C}|)$ operations. The extent to which the graph is traversed depends on the local search method applied.

3 Experiments

We have conducted experiments on natural and generated nonuniform random graphs. As natural data, in addition to the web graph discussed in Section 2, we used *collaboration graphs* [7]. We guide the local search with *simulated annealing* [4]. For generalized caveman graphs [13] consisting of a set of interconnected dense subgraphs of varying order, the method correctly identifies any dense "cave" regardless of the starting point; an example is shown in Figure 1. For illustrations of collaboration graph clusterings, see [13]. In other work [14], we also discuss the applicability of the clustering method in mobile ad hoc networks for improved routing.

 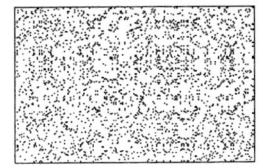 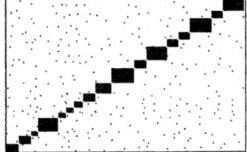

Fig. 1. On the left, an example caveman graph with 55 vertices and 217 edges; each cave (encompassed by a dotted line) is correctly identified as a cluster by the local method. On the right, the adjacency matrix of a caveman graph with 210 vertices and 1,505 edges; the left one uses random vertex order, reflecting very little structure, whereas the one on the right is sorted by the clusters found and reveals the graph structure

As local search procedures are not prohibited from traversing further away in the graph or revisiting parts of the graph, it is interesting to examine whether the extent to which the search traverses the graph has a significant effect on the clusters that the algorithm chooses. We clustered the largest connected component of a scientific collaboration graph with 108,624 vertices and 333,546 edges. We varied the number of independent restarts $R \in \{20, 40, \ldots, 100\}$ per search vertex and the number of cluster modification steps S (from 200 to 1,000 in increments of 100) taken after each restart for simulated annealing. The Figure 2 shows the ratio of the number of vertices visited during the search to the final cluster order, averaged over 100 vertices selected uniformly at random; the final orders are plotted for reference. Figure 2 plots the ratio of the number of vertices visited and the final order of the selected cluster using R restarts of S steps averaged over 50 randomly selected vertices. The extent to which the graph is traversed grows much slower than the number of modification steps taken, implying high locality of the search. As the iteration count is increased, the relative difference gets smaller, which indicates that the number of vertices visited practically stops growing if the increasing possibility for random fluctuations is ignored. The distributions of the cluster orders over three R/S-pairs of the same graph are shown on the right in Figure 2; the distribution changes very little as the parameters are varied, indicating high stability of the method.

We compared the clusterings obtained with the local method to the clusterings of *GMC (Geometric Minimum Spanning Tree Clustering)* with additional linear-time post-processing and *ICC (Iterative Conductance Cutting)* [1] for caveman graphs of different orders. For each graph, we compared the clusters of each vertex obtained with the three methods by calculating what fraction (shown in percentages) of the vertices of a cluster $\mathcal{A}$ determined by one method are also included in the cluster $\mathcal{B}$ determined by another method. Table 1 shows the results for a caveman graph with 1,533 vertices and 50,597 edges; the results

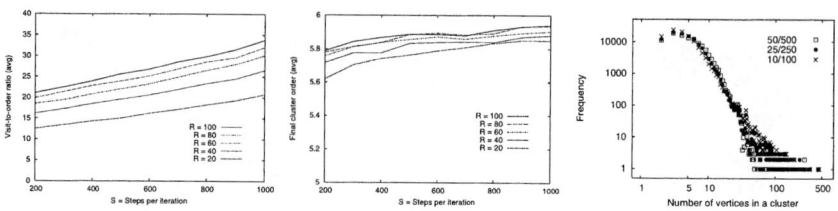

Fig. 2. On the left, the ratio of the number of vertex visited (i.e., the visit count for an (R, S)-pair) to that of the number of vertices selected in the final cluster (i.e., the cluster order) averaged over 100 vertices selected uniformly at random and repeated 50 times per vertex. In the middle, the average final cluster orders of the same experiment set. On the right, the distribution of the number of vertices per cluster for a large collaboration graph (with 108,624 vertices and 333,546 edges) for three different R/S-pairs, where $R \in \{10, 25, 50\}$ and $S = 10R$.

Table 1. Denote by $\mathcal{A}$ the cluster chosen by one method for vertex v, and by $\mathcal{B}$ the cluster chosen for v by another method. If the two methods agree, the *overlaps* $a = |\mathcal{A} \cap \mathcal{B}|/|\mathcal{B}|$ and $b = |\mathcal{A} \cap \mathcal{B}|/|\mathcal{A}|$ are high. For three clusterings of a caveman graph, the percentages p of vertices for which the values a and b fall into a certain range are shown. The values are to be interpreted as follows: if $a = a_1$ and $b = b_1$, then a_1 percent of cluster $\mathcal{B}$ (the method of the right column) is included in $\mathcal{A}$ (the method of the left column) and b_1 percent of cluster $\mathcal{A}$ is included in $\mathcal{B}$. The figure on the right shows a single cave in a 649-vertex graph; the small circles are neighbors in other caves. The shape of the vertex indicates its cluster for the post-processed GMC (with three clusters overlapping the cave) and the color indicates the clustering of ICC (seven clusters overlap); the local method selects the entire cave for any start vertex

Local			Local	ICC		GMC	ICC	
a	b	p	a	b	p	a	b	p
all	all	74	all	(11, 14)	45	all	(5, 14)	71
all	(74, 95)	14	all	(22, 27)	12	all	(22, 34)	10
all	(2, 24)	4	all	(5, 7)	36	all	(40, 55)	2
(86, 97)	all	5	all	(46, 54)	6	(80, 91)	(7, 31)	7
(3, 57)	(5, 87)	3				[50, 67)	(5, 20]	4
						(71, 89)	(45, 55]	3
						(9, 46)	(2, 100]	3

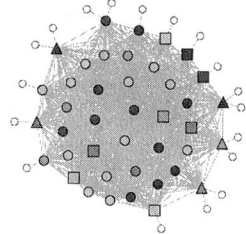

for the smaller graphs allow the same conclusions. ICC splits the caves into small clusters, which is a sign that it fails to recognize the cave boundary on which the density jump takes place. GMC and the local method agree in a majority of cases exactly in cluster selection, and even when they differ, one is usually a subset of the other. GMC and ICC agree poorly with each other.

4 Conclusions

In this paper we have defined a local method for clustering. The method requires no parameters to be determined by the user, nor does it estimate parameters from the dataset; only the local search method being employed may require parameters. No knowledge of the structure of the graph as a whole is required, and the implementation of the method can be done efficiently. The experiments show that approximate clustering with our measure produces intuitively reasonable clusters without extensive traversing of the graph. Employing a local method in the presented way is likely to produce an approximation of some global clustering method; we hope to determine as further work how our local method relates to other methods, such as spectral clustering discussed in [3]. For another local method, see our recent paper [10].

Acknowledgments

This research was supported by the Academy of Finland under grant 206235 and the Nokia Foundation. We thank Marco Gaertler for providing the GMC and ICC clusterings, Kosti Rytkönen for the assisted use of his graph visualization software, and Pekka Orponen and Christos Faloutsos for their valuable comments on this work.

References

1. U. Brandes, M. Gaertler, and D. Wagner. Experiments on graph clustering algorithms. In G. Di Battista and U. Zwick, eds, *Proc. of ESA*, vol. 2832 of *LNCS*, pp. 568–579, Heidelberg, Germany, 2003. Springer-Verlag.
2. S. Guha, et al. Clustering data streams. In *Proc. of FOCS*, pp. 359–366, Los Alamitos, USA, 2000. IEEE Comp. Soc. Press.
3. R. Kannan, S. Vempala, and A. Vetta. On clusterings — good, bad and spectral. In *Proc. of FOCS*, pp. 367–377, Los Alamitos, USA, 2000. IEEE Comp. Soc. Press.
4. S. Kirkpatrick, C. D. Gelatt Jr., and M. P. Vecchi. Optimization by simulated annealing. *Science*, 220(4598):671–680, 1983.
5. J. Kleinberg and S. Lawrence. The structure of the web. *Science*, 294(5548):1849–1850, 2001.
6. M. Mihail, et al. On the semantics of Internet topologies. Tech. report GIT-CC-02-07, Atlanta, USA, 2002.
7. M. E. J. Newman. The structure of scientific collaboration networks. *Proc Natl Acad Sci USA*, 98(2):404–409, 2001.
8. M. E. J. Newman. Fast algorithm for detecting community structure in networks. *Phys. Rev. E*, 69:066133, 2004.
9. L. O'Callaghan, et al. Streaming-data algorithms for high-quality clustering. In *Proc. of ICDE*, pp. 685–694, Los Alamitos, USA, 2002. IEEE Comp. Soc. Press.
10. P. Orponen and S. E. Schaeffer. Local clustering of large graphs by approximate Fiedler vectors. In S. Nikoletseas, ed., *Proc. of WEA*, *LNCS*, Heidelberg, Germany, to appear. Springer-Verlag.

11. J. Šíma and S. E. Schaeffer. On the NP-completeness of some cluster fitness measures, 2005. In preparation.
12. S. E. Virtanen. Clustering the Chilean web. In *Proc. of LA-Web*, pp. 229–231, Los Alamitos, USA, 2003. IEEE Comp. Soc. Press.
13. S. E. Virtanen. Properties of nonuniform random graph models. Research Report A77, TCS, TKK, Espoo, Finland, 2003.
14. S. E. Virtanen and P. Nikander. Local clustering for hierarchical ad hoc networks. In *Proc. of WiOpt'04*, pp. 404–405, 2004.

A Neighborhood-Based Clustering Algorithm*

Shuigeng Zhou[1], Yue Zhao[1], Jihong Guan[2], and Joshua Huang[3]

[1] Dept. of Computer Sci. and Eng., Fudan University, Shanghai 200433, China
{sgzhou, zhaoyue}@fudan.edu.cn
[2] Dept. of Computer Sci. and Eng., Tongji University, Shanghai 200092, China
jhguan@tongji.edu.cn
[3] E-Business Technology Institute, The University of Hong Kong, Hong Kong, China
jhuang@eti.hku.hk

Abstract. In this paper, we present a new clustering algorithm, NBC, *i.e.*, Neighborhood Based Clustering, which discovers clusters based on the neighborhood characteristics of data. The NBC algorithm has the following advantages: (1) NBC is effective in discovering clusters of arbitrary shape and different densities; (2) NBC needs fewer input parameters than the existing clustering algorithms; (3) NBC can cluster both large and high-dimensional databases efficiently.

1 Introduction

As one of the most important methods for knowledge discovery in databases (KDD), clustering is very useful in many data analysis scenarios, including data mining, document retrieval, image segmentation, and pattern classification [1]. Roughly speaking, the goal of a clustering algorithm is to group the objects of a database into a set of meaningful clusters each of which contains objects as similar as possible according to a certain criterion. Currently, mainly four types of clustering algorithms have been developed, including hierarchical, partitioning, density-based and grid-based algorithms.

With the fast development of data collection and data management technologies, the amount of data stored in various databases increases rapidly. Furthermore, more and more new types of data come into existence, such as image, CAD data, geographic data, and molecular biology data. The hugeness of data size and the variety of data types arise new and challenging requirements for clustering algorithms. Generally, a good clustering algorithm should be *Effective*(e.g. be able to discover clusters of arbitrary shape and different distributions), *Efficient*(e.g. be able to handle either very large databases or high-dimensional data-bases, and *Easy to use*(e.g. need no or few input parameters).

However, there are few current clustering algorithms can meet fully the 3-E criteria above-mentioned. In this paper, we present a new clustering algorithm:

---

* This work is supported by the Natural Science Foundation of China under grant No. 60373019 and 60496325, and partially supported by IBM-HKU Visiting Scholars Program.

Neighborhood-Based Clustering algorithm (NBC in abbr.). The NBC algorithm uses the neighborhood relationship among data objects to build a neighborhood based clustering model to discover clusters. The core concept of NBC is the *Neighborhood Density Factor* (NDF in abbr.). NDF is a measurement of *relative local density*, which is quite different the *absolute global density* used in DB-SCAN [2]. In this sense, NBC can still be classified into density based clustering algorithms. However, comparing with DBSCAN(the pioneer and representative of the existing density based clustering algorithms), NBC boasts of the following advantages:

- NBC can automatically discover clusters of arbitrary distribution, it can also recognize clusters of different local-densities and multi-granularities in one dataset, while DBSCAN uses global parameters, it can not distinguish small, close and dense clusters from large and sparse clusters. In this sense, NBC is closer to the *Effective* criterion than DBSCAN.

  To support this point, let us see a dataset sample shown in Fig. 4(a). In this dataset, there are totally five clusters, in which three are dense and close to each other (near the center of the figure) and the other two are much sparse and locate far away (locating near the upper-right angle and the upper-left angle of the figure respectively). Distance between any two of the three dense clusters is not larger than the distance between any two points in the two sparse clusters. With such a dataset, no matter what density threshold is taken, DBSCAN can not detect all the five clusters. In fact, when the density threshold is selected low, DBSCAN can find the two sparse clusters, but the three dense clusters are merged into one cluster; In contrast, when the threshold is set high, DBSCAN can find the three dense clusters, but all data points in the two sparse clusters are labelled as noise. However, NBC can easily find all the five clusters. We will give the clustering results in the performance evaluation section.

- NBC needs only one input parameter(the k value), while DBSCAN requires three input parameters(the k value, the radius of the neighborhood, and the density threshold). That is, NBC needs fewer input parameters than DBSCAN, so NBC is advantageous over DBSCAN in view of the *Easy to Use* criterion.

- NBC uses cell-based structure and VA file [3] to organize the targeted data, which makes it be efficient and scalable even for very large and high dimensional databases.

With all these advantages, we do not intend to replace the existing algorithms with NBC. Instead, we argue that NBC can be a good complement to the existing clustering methods.

2 A Novel Algorithm for Data Clustering

2.1 Basic Concepts

The key idea of neighborhood-based clustering is that: for each object p in a cluster, the number of objects whose k-nearest-neighborhood contains p should not less than the number of objects contained in p's k-nearest-neighborhood. In what follows, we give the formal definition of neighborhood-based cluster and its related concepts.

Given a dataset $D=\{d_1, d_2, \ldots, d_n\}$, p and q are two arbitrary objects in D. We use Euclidean distance to evaluate the distance between p and q, denoted as $dist(p, q)$. We will first give the definitions of k-nearest neighbors set and reverse k-nearest neighbors set. Although similar definitions were given in the literature, we put them here to facilitate the readers to understand our new algorithm.

Definition 1 (k-Nearest Neighbors Set, or simply kNN). The k-nearest neighbors set of p is the set of k ($k > 0$) nearest neighbors of p, denoted by kNN(p). In other words, kNN(p) is a set of objects in D such that
(a) $|k\text{NN}(p)| = k$;
(b) $p \notin k\text{NN}(p)$;
(c) Let o and o' be the k-th and the ($k+1$)-th nearest neighbors of p respectively, then $dist(p, o') \geq dist(p, o)$ holds.

Definition 2 (Reverse k-Nearest Neighbors Set, or simply R-kNN). The reverse k-nearest-neighbors set of p is the set of objects whose kNN contains p, denoted by R-kNN(p), which can be formally represented as

$$R\text{-}kNN(p) = \{q \in D | p \in kNN(q) \text{ and } p \neq q\}. \tag{1}$$

Note that in the literature, reverse kNN is usually abbreviated as RNN. Here we use R-kNN rather than RNN because every RNN set is evaluated based on a certain k value.

kNN(p) and R-kNN(p) expose the relationship between object p and its neighbors in a two-way fashion. On one hand, kNN(p) describes who makes up of its own neighbors; On the other hand, R-kNN(p) indicates whose neighborhood p belongs to. This two-way description of the relationship between an arbitrary object and its neighborhood gives a clearer and more precise picture of its position in the dataset both locally and globally, which depends on the value of k, than simply using only kNN. In what follows, we will give the definitions of an object's neighborhood.

Definition 3 (r-Neighborhood, or simply rNB). Given a positive real number r, the neighborhood of p with regard to (abbreviated as w.r.t. in the rest part of this paper) r, denoted by rNB(p), is the set of objects that lie within the circle region with p as the center and r as the radius. That is,

$$rNB(p) = \{q \in D | dist(q, p) \leq r \text{ and } q \neq p\}. \tag{2}$$

Definition 4 (k-Neighborhood, or simply kNB). For each object p in dataset D, $\exists o \in k\text{NN}(p)$, $r' = dist(p, o)$ such that $\forall o' \in k\text{NN}(p)$, $dist(p, o') \leq r'$. The k-neighborhood of p, written as kNB(p), is defined as r'NB(p), i.e., kNB(p)= r'NB(p). We call kNB(p) as p's k-neighborhood w.r.t. kNN(p).

Definition 3 and Definition 4 define two different forms of neighborhood for a given object from two different angles: rNB(p) is defined by using an explicit radius; In contrast, $kNB(p)$ is defined by using an implicit radius, which corresponds to the circle region covered by $kNN(p)$. It is evident that $|kNB(p)| \geq k$ because there could be more than one object locating on the edge of the neighborhood (the circle). Accordingly, we define the reverse k-Neighborhood as follows.

Definition 5 (Reverse k-Neighborhood, or simply R-kNB). The reverse k-neighborhood of p is the set of objects whose kNB contains p, denoted by R-$kNB(p)$, which can be formally written as

$$R\text{-}kNB(p) = \{q \in D | p \in kNB(q) \text{ and } p \neq q\}. \quad (3)$$

Similarly, we have $|\text{R-}kNB(p)| \geq |\text{R-}kNN(p)|$.

In a local sense, data points in a database can be exclusively classified into three types: *dense point, sparse point* and *even (distribution) point*. Intuitively, points within a cluster should be dense or even points; and points on the boundary area of a cluster are mostly sparse points. Outliers and noise are also sparse points. Currently, most density-based clustering algorithms (*e.g.* DBSCAN) use an intuitive and straightforward way to measure density, *i.e.*, a data object's density is the number of data objects contained in its neighborhood of a given radius. Obviously, this is a kind of absolute and global density. Such a density measurement makes DBSCAN unable to detect small, close and dense clusters from large and sparse clusters. In this paper, we propose a new measurement of density: *Neighborhood based Density Factor* (or simply NDF), which lays the foundation of our new clustering algorithm NBC.

Definition 6 (Neighborhood-based Density Factor, or simply NDF). The NDF of point p is evaluated as follows:

$$NDF(p) = \frac{|R\text{-}kNB(p)|}{|kNB(p)|}. \quad (4)$$

Then what is the implication of NDF? Let us check it. $|kNB(p)|$ is the number of objects contained in p's k-nearest neighborhood. For most data objects, this value is around k (According to Definition 4, it maybe a little greater, but not less than k). $|\text{R-}kNB(p)|$ is the number of objects contained in p's reverse k-nearest neighborhood, *i.e.*, the number of objects taking p as a member of their k-nearest neighborhoods. This value is quite discrepant for different data points. Intuitively, the larger $|\text{R-}kNB(p)|$ is, which implies that the more other objects take p as a member of their k-nearest neighborhoods, that is, the denser p's neighborhood is, or the larger $NDF(p)$ is. In such a situation, $NDF(p) > 1$. For uniformly distributed points, if q is in $kNB(p)$, then p is most possibly in $kNB(q)$, therefore, $kNB(p) \approx \text{R-}kNB(p)$, that is $NDF(p) \approx 1$. Thus, NDF is actually a measurement of the density of any data object's neighborhood, or data object's local density in *relative(not absolute)* sense. Furthermore, such a measurement is intuitive(easy understanding), simple(easy implementation) and effective(being able to find some cluster structures that DBSCAN can not detect).

To demonstrate the capability of NDF as a measurement of local density, we give an example in Fig. 1. Fig.1(a) is a dataset that contains two clusters C_1,

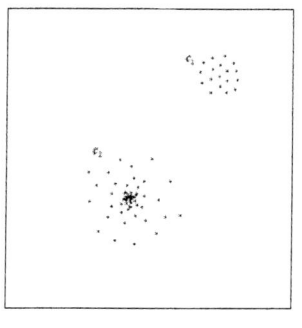
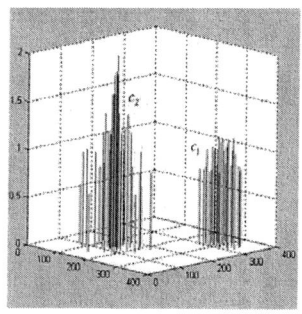

(a) Dataset (b) NDF values of data points

Fig. 1. An illustration of NDF

C_2. Data in C_1 is uniformly distributed; and data in C_2 conforms to Gaussian distribution. Fig. 1(b) shows the NDF values of all data points in the dataset. As we can see, data points locate within cluster C_1 have NDF values approximately equal to 1, while data points locating on boundary of C_1 have smaller NDF values. For cluster C_2, the densest point is near the centroid of C_2, which has the largest NDF value, while other objects have smaller NDF values, and the further the points locate from the centroid, the smaller their NDF values are.

With NDF, in what follows, we give the definitions of three types of data points in local sense: *local event points*, *local dense points* and *local sparse points*.

Definition 7 (Local Dense Point, simply DP). Object p is a *local dense point* if its NDF(p) is greater than 1, also we call p a dense point w.r.t. kNB(p), denoted by DP w.r.t. kNB(p). The larger NDP(p) is, the denser p's k-neighborhood is.

Definition 8 (Local Sparse Point, simply SP). Object p is a *local sparse point* if its NDF(p) is less than 1, and we call p a sparse point w.r.t. kNB(p), denoted by SP w.r.t. kNB(p). The smaller NDP(p) is, the sparser p's k-neighborhood is.

Definition 9 (Local Even Point, simply EP). Object p is a local even point if its NDF(p) is equal (or approximately equal) to 1. We call p an even point w.r.t. kNB(p), denoted by EP w.r.t. kNB(p).

With the concepts defined above, in what follows, we introduce the concepts of neighborhood-based cluster. Our definition follows the way of DBSCAN.

Definition 10 (Directly neighborhood-based density reachable). Given two objects p and q in dataset D, p is *directly neighborhood-based density reachable* (*directly ND-reachable* in abbr.) from q w.r.t. k, if

(a) q is a DP or EP, and
(b) $p \in k$NB(q).

Definition 11 (Neighborhood-based density reachable). Given two objects p and q in dataset D, p is *neighborhood-based density reachable* (*ND-reachable* in abbr.) from q w.r.t. k, if there is a chain of objects $p_1, \cdots, p_n, p_1 = p, p_n = q$ such that p_i is directly ND-reachable from p_{i+1} w.r.t. k.

According to Definition 11, if object p is directly ND-reachable from object q, p is surely ND-reachable from q.

Definition 12 (Neighborhood-based density connected). Given two objects p and q in a dataset D, p and q are *neighborhood-based density connected (ND-connected* in abbr.) w.r.t. k, if p is ND-reachable from q w.r.t. k or q is ND-reachable from p w.r.t. k or there is a third object o such that p and q are both ND-reachable from o w.r.t. k.

With the concepts above, now we are able to define the *neighborhood-based cluster* as follows.

Definition 13 (Neighborhood-based cluster). Given a dataset D, a cluster C w.r.t. k is a non-empty subset of D such that

(a) for two objects p and q in C, p and q are ND-connected w.r.t. k, and
(b) if $p \in C$ and q is ND-connected from p w.r.t. k, then $q \in C$.

The definition above guarantees that a cluster is the maximal set of ND-connected objects w.r.t. k.

2.2 The NBC Algorithm

The NBC algorithm consists of two major phases:

- *Evaluating NDF values.* We search kNB and R-kNB for each object in the target dataset, and then calculate its NDF.
- *Clustering the dataset.* Fetch an object p randomly, if p is a DP or EP (NDF(p)$\geq$1), then create a new cluster, denoted as p's cluster, and continue to find all other objects that are ND-reachable from p w.r.t. k, which involves all objects belonging to p's cluster. Otherwise, if p is a SP, then just put it aside temporarily, and continue to retrieve the next point to process. This work is recursively done until all clusters are discovered. More concretely, given a DP or EP p from the database, first finding the objects that are directly ND-reachable from p w.r.t. k. Objects in kNB are the first batch of such objects, which will be moved into p's cluster. Then finding the other objects directly ND-reachable from each DP or EP in p's cluster until there is no more object can be added into p's cluster. Second, from the rest of the dataset, fetching another DP or EP to build another cluster. When there is no more DP or EP to fetch to create clusters, the algorithm comes to an end. Points belonging to none cluster are noise or outliers. Fig.2 outlines the NBC algorithm in C pseudo-code.

Here, *Dataset* indicates the dataset clustered, k is the only input parameter used in NBC to evaluate kNB and R-kNB. The value of k can be set by experts on the database at the very beginning or by experiments. The determination of parameter k will be discussed in next subsection. *DPset* keeps the DPs or

EPs of the currently processed cluster. The objects in *DPset* are used to expand the corresponding cluster. Once a DP or EP's kNB is moved into the current cluster, it is removed from *DPset*. A cluster is completely detected when there is no object in *DPset*. When the NBC algorithm comes to stop, the unclassified objects whose *clst_no* property is NULL are regarded as noises or outliers.

The NBC algorithm starts with the *CalcNDF* function to calculate kNB, R-kNB and NDF of each object in *Dataset*. Among the traditional index structures, R^*-Tree and X-tree are usually used to improve the efficiency of kNB query processing over relatively low dimensional datasets. However, there is few index structure works efficiently over high-dimensional datasets. To tackle this problem, we employ a cell-based approach to support for kNB query processing. The data space is cut into high-dimensional cells, and VA file [3] is used to organize the cells. Due to space limitation, we neglect the detail here.

```
NBC(Dataset, k) {
   for each object p in Dataset
      p.clst_no=NULL; // initialize cluster number for each object

   CalcNDF(Dataset, k); // calculate NDF
   NoiseSet.empty(); // initialize the set for storing noise
   Cluster_count = 0; // set the first cluster number to 0
   for each object p in Dataset{ // scan dataset
      if(p.clst_no!=NULL or p.ndf < 1) continue;
      p.clst_no = cluster_count; // label a new cluster
      DPSet.empty(); // initialize DPSet

      for each object q in kNB(p){
        q.clst_no = cluster_count;
        if(q.ndf>=1) DPset.add(q)}

      while (DPset is not empty){ // expanding the cluster
        p = DPset.getFirstObject();
        for each object q in kNB(p){
           if(q.clst_no!=NULL)continue;
           q.clst_no = cluster_count;
           if(q.ndf>=1) DPset.add(q);}
        DPset.remove(p);
      }
      cluster_count++;
   }

   for each object p in Dataset{ // label noise
      if(p.clst_no=NULL) NoiseSet.add(p);}
}
```

Fig. 2. The NBC algorithm in C pseudo-code

2.3 Algorithm Analysis

The Determination of k Value. The parameter k roughly determines the size of the minimal cluster in a database. According to the neighborhood-based notion of cluster and the process of the NBC algorithm, to find a cluster, we must first find at least one DP or EP whose R-kNB is larger than or equal to its kNB(*i.e.*, the value of NDF not less than 1). Suppose C is the minimal cluster w.r.t. k in database D, and p is the first DP or EP found to expand cluster C. All objects in kNB(p) are naturally assigned to C. Considering p itself, therefore the minimal size of C is $k+1$. So we can use the parameter k to limit the size of the minimal cluster to be found.

A cluster is a set of data objects that show some similar and unique pattern. If the size of a cluster is too small, its pattern is not easy to demonstrate. In such a case, the data behaves more like outliers. In experiments, we usually set k to 10, with which we can find most meaningful clusters in the databases.

Complexity. The procedure of the NBC algorithm can be separated into two independent parts: calculating NDF and discovering clusters. The most time-consuming work of calculating NDF is to evaluate kNB queries. Let N be the size of the d-dimension dataset D. Mapping objects into appropriate cells takes $O(N)$ time. For a properly settled value of the cell length l, in average, cells of 3 layers are needed to search and each cell contains k objects. Therefore, the time complex of evaluating kNB query is $O(mN)$ where $m = k * 5^d$. For large datasets, $m \ll N$, it turns to $O(N)$. However, considering that $m \gg 1$, so time complexity of CalcNDF is $O(mN)$. The recursive procedure of discovering cluster takes $O(N)$. Therefore, the time complexity of the NBC algorithm is $O(mN)$.

3 Performance Evaluation

In this section, we evaluate the performance of the NBC algorithm, and compare it with DBSCAN. In the experiments, we take k=10. Considering that k value mainly affect the minimal cluster to find, we do not give the clustering results for different k values.

To test NBC's capability of discovering clusters of arbitrary shape, we use a synthetic dataset that is roughly similar to the database 3 in [2], but more complicated. In our dataset, there are five clusters and the noise percentage is 5%. The original dataset and the clustering result of NBC are shown in Fig.3. As is shown, NBC discovered all clusters and recognized the noise points.

To demonstrate NBC's outstanding capability of discovering all clusters of different densities in one dataset, we use the synthetic dataset sample shown in Fig. 4(a). The clustering results by DBSCAN and NBC are shown in Fig. 4(b)(corresponding to a relatively low density threshold) and Fig. 4(c) respectively. We can see that NBC discovered all the five clusters. As for DBSCAN, no matter what density threshold we take, it can not detect all the five clusters. When the density threshold is selected low, DBSCAN can find the two sparse

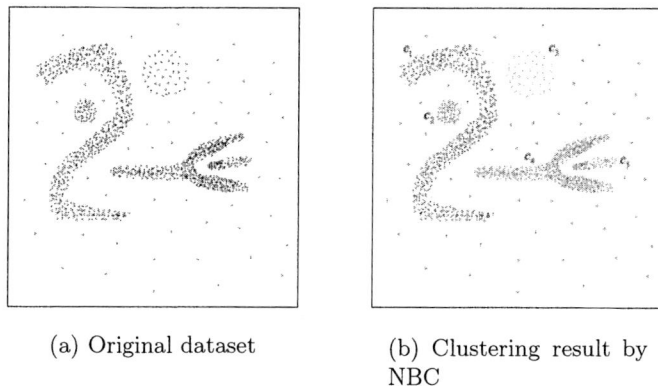

Fig. 3. Discoverying clusters of arbitrary shape

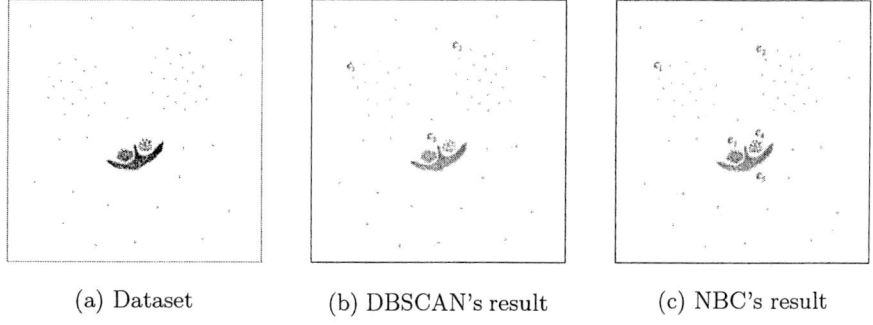

Fig. 4. Discoverying clusters of different densities(NBC vs. DBCSAN)

clusters(C_1 and C_2), but the three dense clusters are merged into one cluster C_3(See Fig. 4(b)); While the threshold is set high, DBSCAN can find the three dense clusters(C_3, C_4 and C_5), but all data points in the two sparse clusters are labelled as noise(We do not illustrate the result here due to space limit).

To test the efficiency of NBC, we use the SEQUOIA 2000 benchmark databases [4] to compare NBC with DBSCAN. The time costs of NBC and DBSCAN over these datasets are shown in Fig. 5(a). As the size of dataset grows, time-cost of NBC increases slowly, while the time-cost of DBSCAN climbs quite fast. In Fig.5(b), we show the time cost of NBC on larger datasets (number of data points varies form 10,000 to 60,000). Here, we give only NBC's results because with the size of dataset increases, the discrepancy of time cost between DBSCAN and NBC is so large that their time costs cannot be shown properly in the drawing. In stead, we show the *speedup* of NBC over DBSCAN in Fig. 5(c). The time cost of NBC is linearly proportional to the size of dataset. And with the size of dataset increase, NBC has larger and larger speedup over DBSCAN.

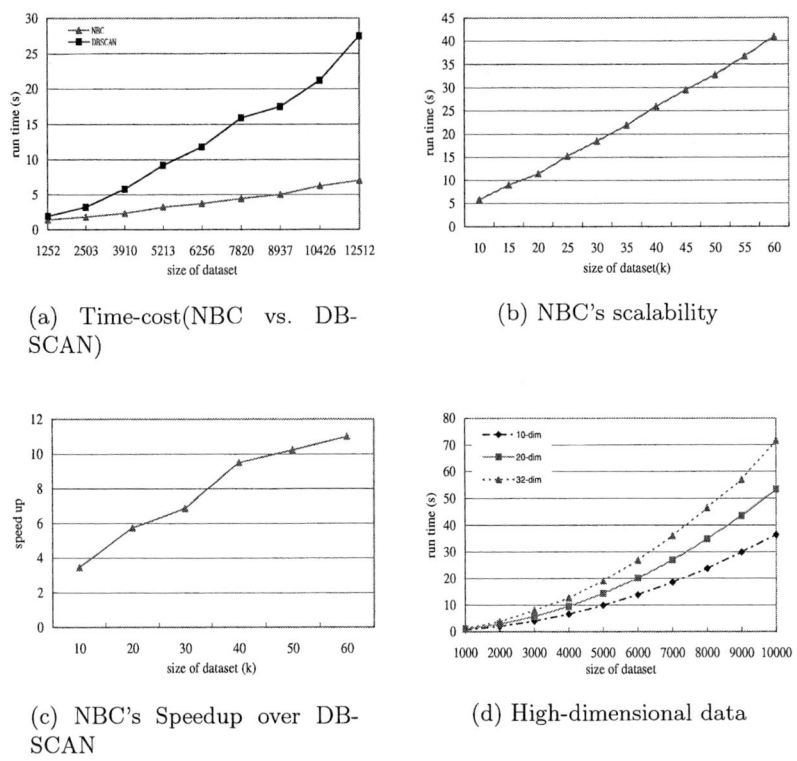

Fig. 5. NBC clustering efficiency and scalability

To test the efficiency and scalability of NBC over high-dimensional datasets, we use the UCI Machine Learning Databases [5]. The results are in Fig.5(d), which show that the run-time of NBC on high-dimensional datasets is approximately linear with the size of database for middle high-dimensional data (d=10-20). But as the number of dimensions increases, the curve turns steeper. The reason is that higher dimensional data space will be divided into more cells, which means more neighbor cells will be searched for evaluating kNB queries.

4 Conclusion

We present a new clustering algorithm, NBC, *i.e.*, Neighborhood Based Clustering, which discovers clusters based on the neighborhood relationship among data. It can discover clusters of arbitrary shape and different densities. Experiments show that NBC outperforms DBSCAN in both clustering effectiveness and efficiency. More importantly, NBC needs fewer input parameter from the users than the existing methods.

References

1. J. Han and M. Kamber. Data mining: concepts and techniques. Morgan Kaufmann Publishers, 2000.
2. Ester M., Kriegel H., Sander J., and Xu X. A density-based algorithm for discovering clusters in large spatial databases with noise. In Proc. KDD'96, pages 226-231, Portland, Oregon, 1996.
3. R. Weber, H.-J. Schek and S. Blott. A quantitative analysis and performance study for similarity-search methods in high-dimensional spaces. In Proc. of VLDB'98, pages 194-205, New York City, NY, August 1998.
4. M. Stonebraker, J. Frew, K. Gardels, and J. Meredith. The SEQUOIA 2000 Storage Benchmark. In Proc. of SIGMOD'93, pages 2-11, Washington D.C., 1993.
5. C. Merz, P. Murphy, and D. Aha. UCI Repository of Machine Learning Databases. At http://www.ics.uci.edu/ mlearn/MLRepository.html.

Improved Self-splitting Competitive Learning Algorithm

Jun Liu and Kotagiri Ramamohanarao

Department of Computer Science and Software Engineering,
The University of Melbourne, Victoria, 3010, Australia
{junliu, rao}@cs.mu.oz.au

Abstract. The Self-Splitting Competitive Learning (SSCL) is a powerful algorithm that solves the difficult problems of determining the number of clusters and the sensitivity to prototype initialization in clustering. The SSCL algorithm iteratively partitions the data space into natural clusters without *a prior* information on the number of clusters. However, SSCL suffers from two major disadvantages: it does not have a proven convergence and the speed of learning process is slow. We propose solutions for these two problems. Firstly, we introduce a new update scheme and lead a proven convergence of Asymptotic Property Vector. Secondly, we modify the split-validity to accelerate the learning process. Experiments show these techniques make the algorithm faster than the original one.

1 Introduction

Clustering is the unsupervised classification of patterns (observations, data items, or feature vectors) into subgroups (clusters). It has important applications in many problem domains, such as data mining, document retrieval, image segmentation and pattern classification. One of the well-known methods is the k-means algorithm [3], which iteratively reassigns each data point to the cluster whose center is closest to the data point and then recomputes the cluster centers.

Several algorithms have been proposed previously to determine cluster number (called k) automatically. Bischof *et al.* [2] use a Minimum Description Length (MDL) framework, where the description length is a measure of how well the data are fit by the model optimized by the k-means algorithm. Pelleg and Moore [4] proposed a regularization framework for learning k, which is called X-means. The algorithm searches over many values of k and scores each clustering model. X-means chooses the model with the best score on the data.

Recently, Zhang and Liu presented the SSCL algorithm [6] based on the One Prototype Takes One Cluster (OPTOC) learning paradigm. The OPTOC-based learning strategy has the following two main advantages: 1) it can find natural clusters, and 2) the final partition of the data set is not sensitive to initialization.

Although promising results have been obtained in some applications [6], we show in this study that the SSCL does not have a proven convergence and the learning speed is slow. This paper will improve the SSCL by giving a new update scheme to make sure the convergence and increase the learning speed.

The remainder of this paper is organized as follows. In Section 2, the original and the improved SSCL algorithms are introduced. Their performance in identifying Gaussian clusters is compared in Section 3. Finally, Section 4 presents the conclusions.

2 SSCL Algorithm and Its Improvement

2.1 Original SSCL

Clustering is an unsupervised learning process [1]. Given a data set of N dimensions, the goal is to identify groups of data points that aggregate together in some manner in an N-dimensional space. We call these groups "natural clusters." In the Euclidean space, these groups form dense clouds, delineated by regions with sparse data points.

The OPTOC idea proposed in [6] allows one prototype to characterize only one natural cluster in data set, regardless of the number of clusters in the data. This is achieved by constructing a dynamic neighborhood using an online learning vector $\boldsymbol{A_i}$, called the Asymptotic Property Vector (APV), for the prototype $\boldsymbol{P_i}$, such that patterns inside the neighborhood of $\boldsymbol{P_i}$ contribute more to its learning than those outside. Let $|\boldsymbol{XY}|$ denote the Euclidean distance from $\boldsymbol{X}$ to $\boldsymbol{Y}$, and assume that $\boldsymbol{P_i}$ is the winning prototype for the input pattern $\boldsymbol{X}$ based on the minimum-distance criterion. The APV $\boldsymbol{A_i}$ is updated by

$$\boldsymbol{A_i^*} = \boldsymbol{A_i} + \frac{1}{n_{\boldsymbol{A_i}}} \cdot \delta_i \cdot (\boldsymbol{X} - \boldsymbol{A_i}) \cdot \Theta(\boldsymbol{P_i}, \boldsymbol{A_i}, \boldsymbol{X}) \tag{1}$$

where Θ is a general function given by

$$\Theta(\boldsymbol{\mu}, \boldsymbol{\nu}, \boldsymbol{\omega}) = \begin{cases} 1 & \text{if } |\boldsymbol{\mu\nu}| \geq |\boldsymbol{\mu\omega}|, \\ 0 & \text{otherwise}, \end{cases} \tag{2}$$

and δ_i, within the range $0 < \delta_i \leq 1$, is defined as

$$\delta_i = \left(\frac{|\boldsymbol{P_i A_i}|}{|\boldsymbol{P_i X}| + |\boldsymbol{P_i A_i}|} \right)^2. \tag{3}$$

$n_{\boldsymbol{A_i}}$ is the winning counter which is initialized to zero and is updated as follow:

$$n_{\boldsymbol{A_i}} = n_{\boldsymbol{A_i}} + \delta_i \cdot \Theta(\boldsymbol{P_i}, \boldsymbol{A_i}, \boldsymbol{X}). \tag{4}$$

The winning prototype $\boldsymbol{P_i}$ is then updated by

$$\boldsymbol{P_i^*} = \boldsymbol{P_i} + \alpha_i (\boldsymbol{X} - \boldsymbol{P_i}), \tag{5}$$

where,

$$\alpha_i = \left(\frac{|\boldsymbol{P_i A_i^*}|}{|\boldsymbol{P_i X}| + |\boldsymbol{P_i A_i^*}|} \right)^2. \tag{6}$$

If the input pattern X is well outside the dynamic neighborhood of P_i, i.e., $|P_iX| \gg |P_iA_i|$, it would have very little influence on the learning of P_i since $\alpha_i \to 0$. On the other hand, if $|P_iX| \ll |P_iA_i|$, i.e., X is well inside the dynamic neighborhood of P_i, both A_i and P_i would shift toward X according to Equations (1) and (5), and P_i would have a large learning rate α_i according to Equation (5). During learning, the neighborhood $|P_iA_i|$ will decrease monotonically. When $|P_iA_i|$ is less than a small quantity ε, P_i would eventually settle at the center of a natural cluster in the input pattern space.

When cluster splitting occurs, the new prototype is initialized at the position specified by a Distant Property Vector (DPV) R_i associated with the mother prototype P_i. The idea is to initialize the new prototype far away from its mother prototype to avoid unnecessary competition between the two. Initially, the DPV is set to be equal to the prototype to which it is associated with. Then each time a new pattern X is presented, the R_i of the winning prototype P_i is updated as follows:

$$R_i^* = R_i + \frac{1}{n_{R_i}} \cdot \rho_i \cdot (X - R_i) \cdot \Theta(P_i, X, R_i), \tag{7}$$

where

$$\rho_i = \left(\frac{|P_iX|}{|P_iX| + |P_iR_i|} \right)^2, \tag{8}$$

and n_{R_i} is the number of patterns associated with the prototype P_i. Note that unlike A_i, R_i always try to move away from P_i. After a successful split, the property vectors (A_i, R_i) of every prototype P_i are reset and the OPTOC learning loop is restarted.

2.2 Improved SSCL

According to Equation (1) and (5), when P_i is changing, the statement that the neighborhood $|P_iA_i|$ will decrease monotonically [6] during learning is not true. Suppose one input pattern X is outside the dynamic neighborhood of P_i, A_i keeps unmoved and P_i may move away from A_i. This may also happen, even when the input pattern is inside the dynamic neighborhood. Sometimes, this will result in oscillation.

The oscillation can be eliminated by only updating prototype P_i in the dynamic neighbor area of P_i with radius $|P_iA_i|$. This will enhance the influence of factor, α_i in equation (5). Therefore, the new update scheme of prototype P_i is as follow,

$$P_i^* = P_i + \alpha_i(X - P_i) \cdot \Theta(P_i, A_i, X), \tag{9}$$

and A_i is updated by

$$A_i^* = A_i + \alpha_i(X - A_i) \cdot \Theta(P_i, A_i, X). \tag{10}$$

With this learning scheme, we can see that A_i always shifts towards the patterns located in the neighborhood of its associated prototype P_i and gives up those data points out of this area. In other words, A_i tries to move closer to P_i

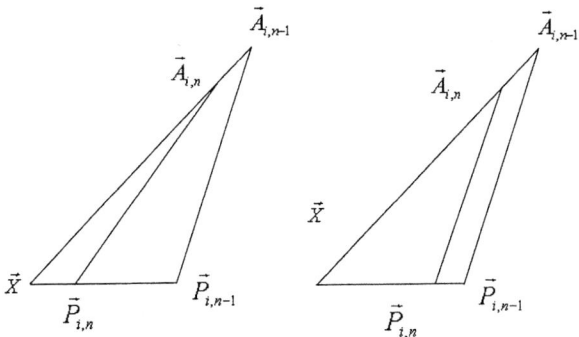

Fig. 1. Update scheme of A_i and P_i. Left is the old update scheme and right is the new one

by recognizing those "inside" patterns that may help A_i to achieve its goal while ignoring those "outside" patterns that are of little benefit. Suppose that $P_i^1, P_i^2, \ldots, P_i^n$ and $A_i^1, A_i^2, \ldots, A_i^n$ are the sequential locations of P_i and A_i respectively during the learning period. According to Equations (1) and (9), in the learning progress, if P_i^{n-1} and A_i^{n-1} are updated, vector $P_i^n A_i^n$ is always parallel $P_i^{n-1} A_i^{n-1}$, which means $|P_i^n A_i^n| < |P_i^{n-1} A_i^{n-1}|$. From the learning rule described above, we have,

$$|P_i^n A_i^n| \leq |P_i^{n-1} A_i^{n-1}| \leq \cdots \leq |P_i^1 A_i^1|.$$

It can be observed that in the finite input space, the APV A_i always tries to move towards P_i, i.e., A_i has the *asymptotic* property with respect to its associated prototype P_i. The geometry description of two different update algorithms are illustrated in Fig. 1.

There are several possible strategies to incorporate the new update procedure to the SSCL algorithm. The first idea is to replace the old update scheme with the new one. However, only updating prototype when the presented pattern is inside the dynamic neighbor area will bring a severe bias. The worst scene is that when there are no patterns are inside the neighbor area of prototype P_i and $|P_i A_i|$ is quite larger than convergence threshold, the prototype P_i will become a dead node. To avoid this problem, the second idea is to transfer to new update procedure when $|P_i A_i|$ is less than a small value γ, which is set $3 \cdot \varepsilon$. This algorithm is called M2 of improved SSCL (ISSCL-M2). The macro view of the asymptotic trajectory of A_i in Fig. 2 shows the convergence of ISSCL-M2 compared with SSCL. The third idea is applying the new update procedure when the distance $|P_i A_i|$ is increasing in some runs and then switch back to normal update scheme in SSCL. This algorithm is referred as M12 of improved SSCL (ISSCL-M12). This will accelerate the speed of convergence and eliminate the possible oscillation of prototypes.

Recall that after a successful split, the property vectors A_i of every prototype P_i are reset randomly far from P_i. We only need reset the A_{i+1} far from new

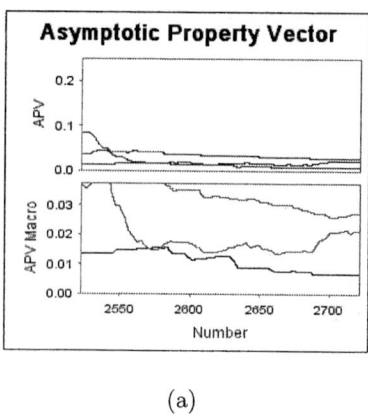 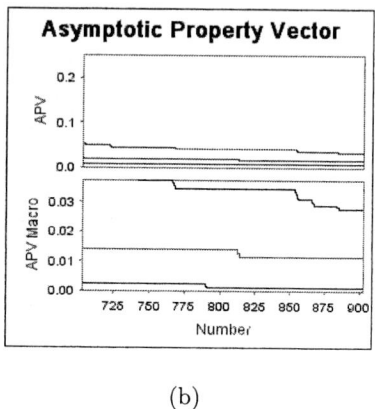

(a) (b)

Fig. 2. The asymptotic trajectory of A_i: SSCL vs ISSCL-M2. (a) There are oscillation phenomena of $|P_iA_i|$ during the SSCL learning process; (b) The $|P_iA_i|$ decreases monotonically at the last stage of ISSCL-M2 learning

prototype P_i and reset other converged A_i a proper range outside of P_i. Here this range λ is set as double splitting threshold, i.e. $\lambda = 2\xi$.

3 Experimental Results

We have conducted experiments on randomly-generated data, as described in [5]. The synthetic experiments were conducted in the following manner. First, a data-set was generated using 72 randomly-selected points (class centers). For each data-point, a class was first selected at random. Then, the point coordinates were chosen independently under a Gaussian distribution with mean at the class center.

As far as speed is concerned, ISSCL scales much better than SSCL. One data set generated as described above with 2 dimensions contained different number of points, from 2000 to 100000, respectively drawn this way. The deviation δ equals to 0.1 and each dimension data rage is (-10, 10). The SSCL, ISSCL-M2 and ISSCL-M12 are run on this data-set and measured for speed. The experiment is repeated 30 times and averages are taken. Fig. 3 shows the run-times of ISSCL and SSCL with convergence threshold ε set 0.025 and 0.05 respectively. It takes a longer time for both ISSCL and SSCL to reach the smaller convergence threshold. From the Fig. 3(a), SSCL runs faster with the number of samples increasing, because SSCL converges fast on high density data set. But when the number of samples reaches to a certain number, it will take SSCL more time to converge with the number of samples increasing. Compared with SSCL, ISSCL-M12 and ISSCL-M2 performs better especially when the convergence threshold is very tiny.

Since new update scheme is used only in last stage of ISSCL-M2, ISSCL-M12 performs a bit better than ISSCL-M2, which is shown in Fig. 3.

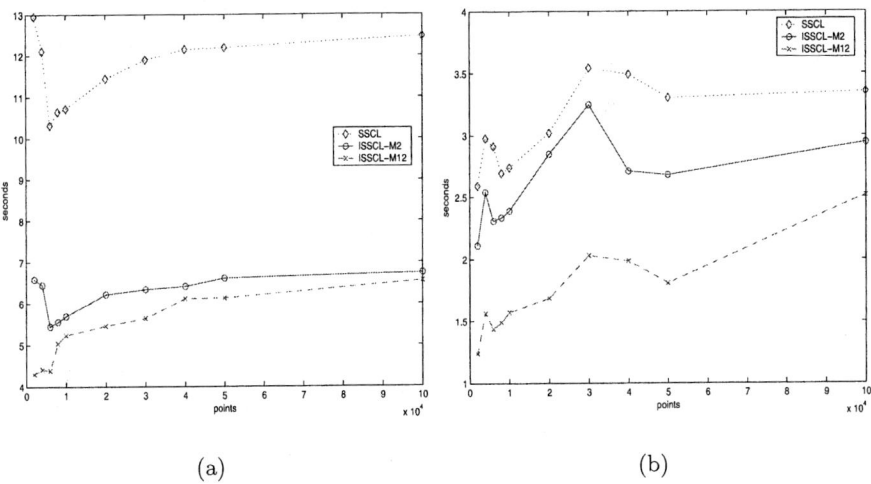

Fig. 3. Average run-times are shown for 2 dimensions and 20 classes with convergence threshold (a) $\varepsilon = 0.025$; (b) $\varepsilon = 0.050$

4 Conclusion

In this paper, we show that the original SSCL has oscillation phenomena during the learning process. We have presented an improved SSCL algorithm by incorporating new update scheme to eliminate the oscillation in the SSCL and achieved a stable convergence. We also modify the split-validity to accelerate the learning process. Our experimental results on random generated data show that both ISSCL-M2 and ISSCL-M12 perform faster than SSCL. ISSCL algorithm performs even better when the convergence threshold is set to a very small value.

References

[1] Barlow, H.B.: Unsupervised learning. Neural Computation **1** (1989) 295–311
[2] Bischof, H., Leonardiš, A., Selb, A.: MDL principle for robust vector quantization. Pattern Analysis and Applications **2** (1999) 59–72
[3] MacQueen, J.: Some methods for classification and analysis of multivariate observations. In: Proc. 5th Berkeley Symposium on Mathematics, Statistics and Probability, Berkeley, CA: Univ.California Press (1967) 282–297
[4] Pelleg, D., Moore, A.: X-means: Extending K-means with efficient estimation of the number of clusters. In: Proceedings of the 17th International conference on Machine Learning, Morgan Kaufmann, San Francisco, CA (2000) 727–734
[5] Pelleg, D., Moore, A.: Accelerating exact k-means with geometric reasoning. Technical report, Carnegie Mellon University, Pittsburgh, PA. (2000)
[6] Zhang, Y.J., Liu, Z.Q.: Self-splitting competitive learning: A new on-line clusteringparadigm. IEEE Trans. on Neural Networks **13** (2002) 369–380

Speeding-Up Hierarchical Agglomerative Clustering in Presence of Expensive Metrics

Mirco Nanni

ISTI-CNR, Pisa, Italy
mirco.nanni@isti.cnr.it

Abstract. In several contexts and domains, hierarchical agglomerative clustering (HAC) offers best-quality results, but at the price of a high complexity which reduces the size of datasets which can be handled. In some contexts, in particular, computing distances between objects is the most expensive task. In this paper we propose a pruning heuristics aimed at improving performances in these cases, which is well integrated in all the phases of the HAC process and can be applied to two HAC variants: single-linkage and complete-linkage. After describing the method, we provide some theoretical evidence of its pruning power, followed by an empirical study of its effectiveness over different data domains, with a special focus on dimensionality issues.

1 Introduction

In several domains, hierarchical agglomerative clustering algorithms are able to yield best-quality results. However, this class of algorithms is characterized by a high complexity which reduces the size of datasets which can be handled. In the most standard cases, such complexity is $O(dN^2 + N^2 \log N)$, N being the number of objects in the dataset and d the cost of computing the distance between two objects, which is the result of $O(N^2)$ distance computations followed by $O(N^2)$ selection steps, each having cost $O(\log N)$. In typical settings, d is either a constant or very small w.r.t. $\log N$, so that the algorithm complexity is usually simplified to $O(N^2 \log N)$.

In some contexts, however, computing distances can be a very expensive task, such as in the case of high-dimensional data or complex comparison functions, e.g., the edit distance between long strings. In these cases, the computation of all object-to-object distances dominates the overall cost of the clustering process, and so any attempt to improve performances should aim at saving a significant portion of distance computations. To the best of our knowledge, this aspect has not been explicitly studied in literature, yet, despite the fact that it has been marginally mentioned in several works (e.g., most of those described in Section 2.2).

In this work, we will consider two popular instances of the general hierarchical agglomerative algorithms family, namely the single- and complete-linkage versions, and propose a simple pruning strategy that improves their performances by reducing the number of object-to-object distances to compute without affecting the results. A formal proof of its effectiveness under some assumptions will also be given, together with an extensive experimental session to test it on different contexts and conditions.

2 Background and Related Work

In this section we will provide a short description of the general hierarchical agglomerative clustering schema, instantiating it to the two specific cases discussed in this paper. Finally, a brief summary of related work will follow.

2.1 Hierarchical Agglomerative Clustering (HAC)

The objective of hierarchical clustering algorithms is to extract a multi-level partitioning of data, i.e., a partitioning which groups data into a set of clusters and then, recursively, partitions them into smaller sub-clusters, until some stop criteria are satisfied [3]. Hierarchical algorithms can be divided in two main categories: agglomerative and divisive. Agglomerative algorithms start with several clusters containing only one object, and iteratively two clusters are chosen and merged to form one larger cluster. The process is repeated until only one large cluster is left, that contains all objects. Divisive algorithms work in the symmetrical way. In this paper we will focus on the former class of algorithms.

Algorithm: Hierarchical Agglomerative Clustering
Input: a dataset D
Output: a tree structure T

1. $C := \{\{o\}|\ o \in D\}$ and $T = \emptyset$;
2. **while** $|C| > 1$ **do**
3.   Select best couple (a, b) s.t. $a, b \in C$;
4.   Create a new cluster $c = a \cup b$, and let a and b be children of c in T;
5. $C := C \cup \{c\} \setminus \{a, b\}$;
6.   **foreach** $x \in C$ **do**
7.   Compute the distance between x and c;
8. **return** T;

Fig. 1. General schema for hierarchical agglomerative clustering algorithms

The general structure of an agglomerative clustering algorithm can be summarized as in Figure 1. As we can notice, there are two key operations in the general schema which still need to be instantiated: the choice of the *best* couple of clusters, and the computation of the distances between the new cluster and the existing ones. Each different instantiation of these two operations results into a different agglomerative clustering algorithm. In this paper, the cluster selection in step 3 is performed by selecting the closest pair of clusters, while the distance computation in step 7 is performed in two alternative ways: by extracting the distance between the closest pair of objects (excluding couples belonging to the same cluster), which yields a so called *Single-linkage* algorithm; and by extracting the distance between the farthest pair of objects, which yields a *Complete-linkage* algorithm. In particular, the complete-linkage algorithm in general produces tightly bound or compact clusters, while the single-link algorithm, on the contrary, suffers from a *chaining effect*, i.e., it has a tendency to produce clusters that are straggly or elongated [3].

2.2 Related Work

Efficiency is a strong issue in hierarchical clustering, and it has been treated in literature in many different ways. In the following we summarize some of the main approaches to the problem.

Some approaches seek slight computational complexity improvements for the HAC problem. For example, [2] introduces a data structure for dynamic closest pair retrieval, which is directly applicable to hierarchical clustering, and which is shown to reach a $O(n^2)$ complexity for simple aggregation operators (e.g., maximum, minimum and average). For specific contexts, even faster solutions have been proposed, such as a sub-quadratic single-linkage method for low-dimensional data [4], and a $O(n \log n)$ complete-linkage solution for $\mathcal{R}^d$ spaces ($d \geq 1$) with L_1 and L_∞ metrics [5]. We remark that these approaches do not take into account the (pragmatic) possibility of having very expensive distance computations, which is exactly the context we will focus on in this paper. When some degree of approximation in the hierarchical clustering structure can be tolerated, several approximation approaches can be followed, which mainly try to reduce the size of data: from data simple sampling methods to data aggregation solutions, such as (i) *grid-based* clustering solutions for vectorial datasets [3], and (ii) the *data bubbles* approach [1], which extends the grid-based approach to non-vectorial data.

3 HAC with Enhanced Distance Management

The basic assumption of our method is that our distance function is a metric. Then, the key idea is that from the exact distances of a limited number of couples it is possible to derive useful approximated values for all object-to-object distances. Such approximations can be easily updated at each iteration of the HAC algorithm, and can be used to effectively limit the number of exact distance computations needed along the whole process.

3.1 Distance Approximations

As basic means for estimating unknown distances, we propose to use the triangular inequality, a property satisfied by all metrics: $\forall a, b, p \in D : d(a,b) \leq$

$d(a,p) + d(p,c)$, where d is a metric defined over a domain D. With some simple math and exploiting the symmetry property of metrics, we can rewrite it as

$$\forall a,b,p \in D: \;|d(p,a) - d(p,c)| \leq d(a,b) \leq d(p,a) + d(p,c) \qquad (1)$$

Now, assuming to know all $|D|$ distances $d(p,a)$ for some fixed element $p \in D$, which we will call *pivot*, the above formula can be directly used to provide a bounding interval for the distance between any couple (a,b) of objects. Henceforth, we will refer to such bounding intervals as *approximated distances* or simply *approximations*. In particular, we notice that if some object a is very close to the pivot, the $d(p,a)$ values in (1) will be very small, and therefore the approximation of any distance $d(a,b)$ from a will be very tight.

In our approach, the computation of all object-to-object distances, usually performed at the beginning of HAC algorithms, is replaced by (i) the computation of the $|D|$ exact distances relative to a randomly chosen pivot, and (ii) the approximation of all other distances by following the method outlined above.

3.2 Enhanced Distance Management

The method shown in the previous section can be used to provide an initial set of approximations aimed at replacing as much as possible the full matrix of distances. In the following we will describe: (i) how such approximations can be used to save exact distance computations in the couple selection phase (step 3 in Figure 1); (ii) how they can be composed to derive approximations for a newly created cluster (steps 6–7); and (iii) how to exploit them also in the *on demand* computation of exact distances between compound clusters, when they are required in the couple selection phase.

Enhanced Couple Selection. Both the single- and complete-linkage algorithms, at each iteration find the couple of clusters with minimal distance, and merge them. A simple method for searching such couple exploiting the approximated distances, is the following:

1. Select the couple (a,b) which has the lowest-bounded approximation;
2. **if** the approximation is perfect
3.   **then** return (a,b);
4.   **else** compute the exact $d(a,b)$ and return to step 1;

Essentially, a two-steps selection is performed: a first selection of the most promising candidate couple is performed by means of the known approximations; if the *best* approximation is perfect, then all other couples certainly have an equal or greater distance, and therefore we can safely choose the selected couple for the merging phase; otherwise, another step is necessary, i.e., the exact distance of the couple needs to be computed and checked to be still the best candidate. The last test is implicitly performed by immediately repeating the selection step.

Deriving New Approximations. When two clusters are merged, all distances from the resulting new cluster have to be computed, exact or approximated, so that it can be considered in the next iterations of the selection-merging process. Analogously to the case of exact distances, the approximations for the new cluster can be derived by aggregating the already known approximations of the two clusters it originated from. In particular, we exploit a simple property of the max and min aggregation operators, that are used in the single- and complete-linkage HAC algorithms[1]:

Proposition 1. *Let* $x, y, l_1, u_1, l_2, u_2 \in \mathcal{R}$, $x \in [l_1, u_1]$ *and* $y \in [l_2, u_2]$. *Then:*

$$\min\{x, y\} \in [\min\{l_1, l_2\}, \min\{u_1, u_2\}] \qquad (2)$$
$$\max\{x, y\} \in [\max\{l_1, l_2\}, \max\{u_1, u_2\}] \qquad (3)$$

In the single-linkage algorithm, the distance between two clusters c and c' is computed as the minimum of the object-to-object distances between elements of the two clusters, i.e., $d(c, c') = \min_{a \in c, b \in c'} d(a, b)$. If c is obtained by merging clusters c_1 and c_2, then we can write $d(c, c') = \min_{a \in c_1 \cup c_2, b \in c'} d(a, b)$, and therefore $d(c, c') = \min\{d(c_1, c'), d(c_2, c')\}$. This property, together with (2), provides a straightforward means for approximating all distances $d(c, c')$ from c, given that we know an approximation for both its components c_1 and c_2. A completely symmetrical reasoning can be repeated for the complete-linkage algorithm, which makes use of inequality (3).

Enhanced Distance Computation. In the (enhanced) selection step it is often necessary to compute the exact distance between two clusters. That happens whenever the best candidate couple found is associated with only an approximated distance. The trivial way to do it, consists in computing all distances between each object in the first cluster and each object in the second one and aggregating them with the proper operator (min or max). An obvious drawback of this solution is that it easily leads to compute all $\frac{|D| \cdot (|D|-1)}{2}$ object-to-object distances, which is exactly what we wanted to avoid. A surprisingly effective enhancement can be obtained by exploiting the following simple fact:

Proposition 2. *Let* c_1, c_2, c' *be three distinct clusters and* $c = c_1 \cup c_2$, $d(c_1, c') \in [l_1, u_1]$, $d(c_2, c') \in [l_2, u_2]$. *If* $u_1 \leq l_2$, *then: (i) in the single-linkage algorithm it holds that* $d(c, c') = d(c_1, c')$, *and (ii) in the complete-linkage* $d(c, c') = d(c_2, c')$.

The basic idea is to compute the distance between compound clusters by recursively analyzing their components (i.e., the two sub-clusters they originated from), until we reach simple objects. At each step of the recursion, the above property allows to prune unnecessary distance computations. The process for single-linkage HAC can be summarized as in Figure 2. If the clusters to compare

---
[1] Due to space limits, all the proofs are omitted here, and can be found in [8].

Algorithm: EDC
Input: two clusters a and b
Output: the exact distance $d(a,b)$

1. **if** a and b contain only one object **then** Stop and **return** $d(a,b)$;
2. **if** a contains only one object **then** Swap a and b;
3. Let a_1, a_2 be the clusters which compose a, i.e., $a = a_1 \cup a_2$;
4. Let $d(a_1, b) \in [l_1, u_1]$ and $d(a_2, b) \in [l_2, u_2]$;
5. **if** $l_1 > l_2$ **then** Swap a_1 and a_2;
6. $d_1 := EDC(a_1, b)$;
7. **if** $d_1 < l_2$ **then** Stop and **return** d_1;
8. $d_2 := EDC(a_2, b)$;
9. **return** $\min\{d_1, d_2\}$;

Fig. 2. Enhanced Distance Computation (EDC) for single-linkage HAC

contain single objects, then the algorithm simply computes their distance (step 1), otherwise it breaks down one of the compound clusters into its components (steps 2–4), and recursively analyzes them. In the analysis of sub-components, priority is given to the *most promising* one, i.e., that with the smaller lower bound distance (step 5), to the purpose of maximizing the pruning opportunities offered by Proposition 2. Step 7 implements that by avoiding to compute the distance for the *less promising* component when it is not strictly necessary.

The complete-linkage version of the algorithm is essentially the same, and can be obtained by just modifying the conditions of step 5 and 7 with, respectively, $(u_1 < u_2)$ and $(d_1 > u_2)$, and by replacing min with max in step 9.

3.3 Selecting Pivots

As we noticed in Section 3.1, the approximations computed before the clustering process can have variable tightness. In particular, the approximations for objects close to the *pivot* will be tight, while the others will be looser. A natural extension of the method consists in choosing more than one pivot, so that a larger number of objects will have a pivot near to them, and therefore a larger quantity of approximated distances will result tight. The expected consequence is that the pruning strategies described in the previous sections will be more effective.

Choosing several pivots, we obtain several approximations for the same distance – one for each pivot – so they need to be composed together in some way. The approximation computed by means of each pivot represents a constraint that the real distance must satisfy. Therefore, the composition of approximations corresponds to the conjunction of the constraints they represent, which is simply implemented by intersecting of the available approximations.

A more difficult problem is the choice of the pivots. While a simple, repeated random choice would be a possible solution, it provides no guarantee on the results. On the contrary, assuming that a dataset is really composed of a number

Algorithm: Pivots Selection
Input: a dataset D and an integer n
Output: a set P of n pivots

1. Randomly select an object $p_0 \in D$;
2. $P := \{p_0\}$;
3. **while** $|P| < n$ **do**
4. $\quad p = \arg\max_{o \in D}\{\min_{p' \in P} d(p', o)\}$;
5. $\quad P := P \cup \{p\}$;
6. **return** P;

Fig. 3. Algorithm for selecting the initial pivots

of clusters, an optimal choice for pivots would assign at least one pivot to each cluster. The key idea of our pivot selection heuristics is the following: assuming to have very well defined clusters in our data, each point is expected to be far from the objects of other clusters, at least if compared with the distance from other objects in the same cluster. Therefore, given a set of pivots, we can reasonably search a new good pivot, i.e., a pivot which belongs to an *uncovered* cluster, among those objects which are far from all existing pivots. These are essentially the same ideas applied in [6], where a similar approach has been used for approximated clustering. Figure 3 shows our pivot selection method.

The very first pivot is chosen randomly (steps 1–2), while the following ones are chosen as mentioned above. In particular, the *furthest* object from the existing set of pivots is selected, i.e., the object which maximizes the distance from the closest pivot (step 4). This simple algorithm seems to capture reasonably well the cluster structure of data, at least for clean-cut clusters, as indicated by the property proven below.

Definition 1 (δ-separateness). *Given a set of objects D and a distance $d()$, D is called δ-separated if it can be split into at least two clusters, such that the following holds: $\forall a, b, a', b' \in D$: if a and b belong to the same cluster while a' and b' do not, then $d(a', b') > \delta \cdot d(a, b)$.*

Essentially, δ-separateness requires that the minimum distance between clusters is at least δ times larger than the maximum diameter of clusters.

Proposition 3. *Let D be a 1-separated dataset composed of n clusters, and $k \geq n$. Then, PivotsSelection(D,k) returns at least one object from each cluster.*

4 Performance Evaluation

In this section we provide some experimental and theoretical evaluations of the performances of the HAC algorithms with enhanced distance management described in this work.

4.1 Theoretical Evaluation

While any realistic context usually shows some kind of irregularity, such as noise (i.e., objects that do not clearly belong to any cluster) and dispersion (i.e., largely dispersed clusters, possibly without clear boundaries), it is useful to have some theoretical estimation of performances also on ideal datasets: on one hand, it provides at least a comparison reference for empirical studies; on the other hand, it helps to understand where are the weak and strong points of the algorithm analyzed. In this section, we provide one of such theoretical hints.

First of all, we introduce a slight variant of HAC algorithms:

Definition 2 (k-HAC). *Given a HAC algorithm and a parameter k, we define the corresponding k-HAC algorithm as its variant which stops the aggregation process when k clusters are obtained. That corresponds to replace step 2 in the general HAC algorithm (Figure 1) with the following:* **while** $|C| > k$ **do**.

In practice, such generalization is quite reasonable, since usually it is easy to provide some a priori lower bound on the number of clusters we are interested in – obviously at least 2, but often it is much larger.

Proposition 4. *Given a 3-separated dataset D with n clusters, and a parameter $k \geq n$, the execution of an optimized k-HAC algorithm over D with k initial pivots requires $O(N_1^2 + \cdots + N_k^2)$ object-to-object distance computations, where $(N_i)_{i=1,\ldots,k}$ are the sizes of the k top level clusters returned by the algorithm.*

In summary, when clusters are very compact our pruning strategy allows to limit the distance computations just to couples within the same cluster. That results in a considerable reduction factor, as stated by the following:

Corollary 1. *Under the assumptions of Proposition 4, if $k = n$ and the clusters in D have balanced sizes (i.e., $\forall i : N_i \sim N/k$), then the k-HAC algorithm with enhanced distance computation requires a fraction $O(1/k)$ of the distances required by the simple HAC algorithm.*

We notice that the above analysis does not take in consideration the pruning capabilities of the Enhanced Distance Computation algorithm. As the next section will show, in some cases this second component allows to obtain much larger reduction factors.

4.2 Experimental Evaluation

In order to study the effectiveness of our pruning heuristics, and to understand which factors can affect it, we performed several experiments over datasets of different nature with corresponding distance functions:

- 2D points: the dataset contains points in the $\mathcal{R}^2$ space, and the standard Euclidean distance is applied. Data were randomly generated into 10 spherical, normally-distributed clusters with a 5% of random noise. Although Euclidean metrics are not expensive, this kind of metric space provides a good example of low-dimensional data, so it is useful to evaluate the pruning power of our heuristics on it and to compare the results with the other data types.

- Trajectories: each element describes the movement of an object in a bi-dimensional space, and is represented as a sequence of points in space-time. The distance between two objects is defined as the average Euclidean distance between them. Data were synthesized by means of a random generator, which created 10 clusters of trajectories (see [7] for the details).

For each data domain, datasets of different sizes were generated (from 400 to 3200 objects) and the single- and complete-linkage versions of a 10-HAC algorithm were applied, with a variable number of pivots (from 4 to 48). Figure 4 depicts the results of our experiments for single-linkage, which are evaluated by means of the ratio between the total number of distance computations required by the basic HAC algorithms and the number of distances computed by their enhanced version. We will refer to such ratio as *gain factor*, and each value is averaged over 16 runs. Due to space limitations, the results for the complete-linkage algorithm are not reported here, since they are quite similar to the single-linkage case. The interested reader can find them in [8], together with tests on other datasets. We can summarize the results as it follows:

- For 2D data (Figure 4 left), a very high gain factor is obtained for all settings of the parameters. In particular, the gain factor grows very quickly with the size of the database, and the best results are obtained with the minimal number of pivots. The latter fact essentially means that the pruning power of the EDC procedure (Figure 2) is so high in this context, that only a very small number of exact distances are needed to capture the structure of data, and so the $k|D|$ distances computed in the initialization phase ($|D|$ for each pivot) become a limitation to the performances.
- For trajectory data (Figure 4 right), the gain factor is moderately high, and the enhanced HAC algorithms reduce the number of computed distances of around an order of magnitude. In this contexts, we notice that the best results are obtained with a number of pivots around 10–20, and both smaller and higher values yield a decrease in performances.

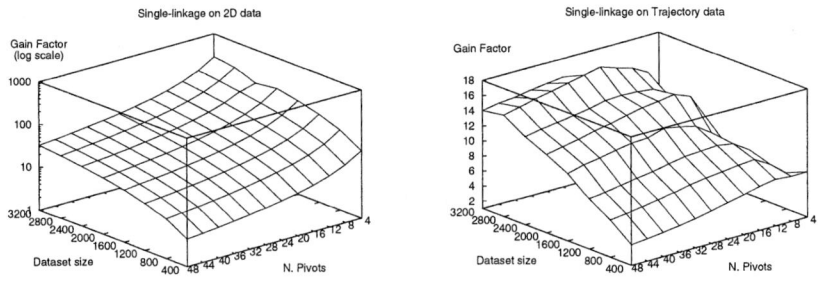

Fig. 4. Gain Factor of Single-linkage HAC on 2D and trajectory data

Due to space limitations, no analysis of execution times is provided here. We summarize our results as follows: with 2D data the gain in running times is slightly negative, because the Euclidean metric is extremely cheap, and then, even though the overhead of our heuristics results to be small, the distances saved cannot compensate it; with other data, instead, the gain in running times is almost identical to the gain factor, since the distances are more complex, and saving even a few of them is enough to balance all the overhead.

5 Conclusions

In this paper we introduced an optimization technique for two popular hierarchical clustering algorithms, and studied its potentialities and its limitations by means of both theoretical and empirical means. Our optimization technique tries to save as many distance computations as possible, which is particularly important for contexts where distances are time-consuming, and we showed that on reasonably dense datasets it is able to achieve good performances.

As future work, we plan (i) to perform a systematic study aimed at understanding more precisely which statistical properties of data influence the performances of our pruning heuristics (as suggested in the previous section and confirmed by additional tests in [8], dimensionality is one of them); (ii) to empirically evaluate the pruning power of the heuristics over several real world datasets, having different characteristics; and, finally, (iii) to extend the heuristics to other variants of HAC and, if possible, to other clustering approaches.

References

1. M. M. Breunig, H.-P. Kriegel, P. Krüger, and J. Sander. Data bubbles: quality preserving performance boosting for hierarchical clustering. In *SIGMOD '01: Proc. of the 2001 ACM SIGMOD Int' Conf. on Management of data*, pages 79–90, 2001.
2. David Eppstein. Fast hiearchical clustering and other applications of dynamic closet pairs. In *SODA '98: Proceedings of the ninth annual ACM-SIAM symposium on Discrete algorithms*, pages 619–628, 1998.
3. A. K. Jain, M. N. Murty, and P. J. Flynn. Data clustering: a review. *ACM Comput. Surv.*, 31(3):264–323, 1999.
4. D. Krznaric and C. Levcopoulos. The first subquadratic algorithm for complete linkage clustering. In *ISAAC '95: Proceedings of the 6th International Symposium on Algorithms and Computation*, pages 392–401. Springer-Verlag, 1995.
5. D. Krznaric and C. Levcopoulos. Optimal algorithms for complete linkage clustering in d dimensions. *Theor. Comput. Sci.*, 286(1):139–149, 2002.
6. R. R. Mettu and C. G. Plaxton. Optimal time bounds for approximate clustering. *Machine Learning*, 56(1–3):35–60, 2004.
7. M. Nanni. *Clustering methods for spatio-temporal data*. PhD thesis, Dipartimento di Informatica, Università di Pisa, 2002.
8. M. Nanni. Hierarchical clustering in presence of expensive metrics. Technical report, ISTI-CNR, 2005. http://ercolino.isti.cnr.it/mirco/papers.html.

Dynamic Cluster Formation Using Level Set Methods*

Andy M. Yip[1,2], Chris Ding[1], and Tony F. Chan[2]

[1] Computational Research Division, Lawrence Berkeley National Laboratory,
Berkeley, CA 94720
chqding@lbl.gov
[2] Department of Mathematics, University of California,
Los Angeles, CA 90095-1555
{mhyip, chan}@math.ucla.edu

Abstract. Density-based clustering has the advantages for (i) allowing arbitrary shape of cluster and (ii) not requiring the number of clusters as input. However, when clusters touch each other, both the cluster centers and cluster boundaries (as the peaks and valleys of the density distribution) become fuzzy and difficult to determine. In higher dimension, the boundaries become wiggly and over-fitting often occurs. We introduce the notion of *cluster intensity function* (CIF) which captures the important characteristics of clusters. When clusters are well-separated, CIFs are similar to density functions. But as clusters touch each other, CIFs still clearly reveal cluster centers, cluster boundaries, and, degree of membership of each data point to the cluster that it belongs. Clustering through bump hunting and valley seeking based on these functions are more robust than that based on kernel density functions which are often oscillatory or over-smoothed. These problems of kernel density estimation are resolved using *Level Set Methods* and related techniques. Comparisons with two existing density-based methods, valley seeking and DBSCAN, are presented to illustrate the advantages of our approach.

1 Introduction

Recent computer, internet and hardware advances produce massive data which are accumulated rapidly. Applications include genomics, remote sensing, network security and web analysis. Undoubtedly, knowledge acquisition and discovery from such data become an important issue. One common technique to analyze data is clustering which aims at grouping entities with similar characteristics together so that main trends or unusual patterns may be discovered.

---

* This work has been partially supported by grants from DOE under contract DE-AC03-76SF00098, NSF under contracts DMS-9973341, ACI-0072112 and INT-0072863, ONR under contract N00014-03-1-0888, NIH under contract P20 MH65166, and the NIH Roadmap Initiative for Bioinformatics and Computational Biology U54 RR021813 funded by the NCRR, NCBC, and NIGMS.

Among various classes of clustering algorithms, density-based methods are of special interest for their connections to statistical models which are very useful in many applications. Density-based clustering has the advantages for (i) allowing arbitrary shape of cluster and (ii) not requiring the number of clusters as input, which is usually difficult to determine. Examples of density-based algorithms can be found in [1, 2, 3].

There are several basic approaches for density-based clustering. (A1) A common approach is so-called bump-hunting: first find the density peaks or "hot spots" and then expand the cluster boundaries outward until they meet somewhere, presumably in the valley regions (local minimums) of density contours. The CLIQUE algorithm [3] adopted this methodology. (A2) Another direction is to start from valley regions and gradually work uphill to connect data points in low-density regions to clusters defined by density peaks. This approach has been used in Valley Seeking [4] and DENCLUE [2]. (A3) A recent approach, DBSCAN [1], is to compute reachability from some seed data and then connect those "reachable" points to their corresponding seed. Here, a point p is reachable from a point q with respect to *MinPts* and *Eps* and there is a chain of points $p_1 = q, p_2, \ldots, p_n = p$ such that, for each i, the *Eps*-neighborhood of p_i contains at least *MinPts* points and contains p_{i+1}.

When clusters are well-separated, density-based methods work well because the peak and valley regions are well-defined and easy to detect. When clusters touch each other, which is often the case in real situations, both the cluster centers and cluster boundaries (as the peaks and valleys of the density distribution) become fuzzy and difficult to determine. In higher dimension, the boundaries become wiggly and over-fitting often occurs.

In this paper, we apply the framework of bump-hunting but with several new ingredients adopted to overcome problems that many density-based algorithms share. The major steps of our method are as follows: (i) obtain a density function by *Kernel Density Estimation* (KDE); (ii) identify peak regions of the density function using a surface evolution equation implemented by the *Level Set Methods* (LSM); (iii) construct a distance-based function called *Cluster Intensity Function* (CIF) based on which valley seeking is applied. In the followings, we describe each of the above three notions. An efficient graph-based implementation of the valley seeking algorithm can be found in [4].

Kernel Density Estimation (KDE). In density-based approaches, a general philosophy is that clusters are high density regions separated by low density regions. We particularly consider the use of KDE, a non-parametric technique to estimate the underlying probability density from samples. More precisely, given a set of data $\{\mathbf{x}_i\}_{i=1}^{N} \subset \mathbb{R}^p$, the KDE is defined to be $f(\mathbf{x}) := 1/(Nh_N^p) \sum_{i=1}^{N} K((\mathbf{x} - \mathbf{x}_i)/h_N)$ where K is a positive kernel and h_N is a scale parameter. Clusters may then be obtained according to the partition defined by the valleys of f.

There are a number of important advantages of kernel density approach. Identifying high density regions is independent of the shape of the regions. Smoothing

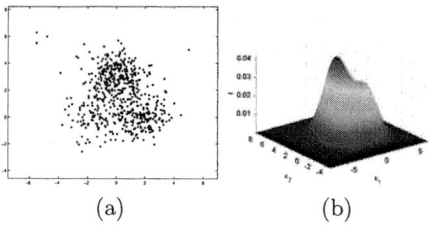

Fig. 1. (a) A mixture of three Gaussian distributions. (b) The KDE f using Gaussian kernel with window size $h = 1$. In (b), peaks and valleys corresponding to the two smaller large clusters are very vague that the performance of applying bump-hunting and/or valley seeking algorithm based on the KDE is expected to be poor. Clusters obtained from our method are shown in Fig. 2

effects of kernels make density estimations robust to noise. Kernels are localized in space so that outliers do not affect the majority of the data. The number of clusters is automatically determined from the estimated density function, but one needs to adjust the scale parameter h_N to obtain a good estimate.

Despite the numerous advantages of kernel density methods, there are some fundamental drawbacks which deteriorate the quality of the resulting clusterings. KDEs are very often oscillatory (uneven) since they are constructed by adding many kernels together. Such oscillatory nature may lead to the problem of overfitting, for instance, when clusters touch each other, a smooth cluster boundary between the clusters are usually preferred than an oscillatory one. Last but not least, valleys and peaks of KDEs are often very vague especially when clusters touch each other.

In Fig. 1, we show a dataset drawn from a mixture of three Gaussian components and the KDE f. We observe that the valleys and peaks correspond to the two smaller large clusters of the KDE are very vague or may even not exist. Thus, the performance of KDE-based bump-hunting and/or valley seeking could be poor when the clusters are overlapped.

Level Set Methods (LSM). We recognize that the key issue in density-based approach is how to advance the boundary either from peak regions outward towards valley regions, or the other way around. In this paper, we employ LSM, which are effective tools for computing boundaries in motion, to resolve the boundary advancing problem. LSM have well-established mathematical foundations and have been successfully applied to solve a variety of problems in image processing, computer vision, computational fluid dynamics and optimal design. LSM use implicit functions to represent complicated boundaries conveniently. While implicit representation of static surfaces has been widely used in computer graphics, LSM move one step further allowing the surfaces to dynamically evolve in an elegant and highly customizable way, see [5] for details.

Advantages of LSM include: (i) the boundaries in motion can be made smooth conveniently and smoothness can be easily controlled by a parameter that characterizes surface tension; (ii) merging and splitting of boundaries can be easily done in a systematical way. Property (ii) is very important in data clustering as clusters can be merged or split in an automatic fashion. Furthermore, the advancing of boundaries is achieved naturally within the framework of partial differential equation (PDE) which governs the dynamics of the boundaries.

Cluster Intensity Functions (CIF). We may use LSM strictly as an effective mechanism for advancing boundaries. For example, in the above approach (A1), once the density peaks are detected, we may advance cluster boundaries towards low-density regions using LSM. However, it turns out that utilizing LSM we can further develop a new and useful concept of *cluster intensity function*. A suitably modified version of LSM becomes an effective mechanism to formulate CIFs in a dynamic fashion. Therefore our approach goes beyond the approaches (A1)–(A3) described earlier.

CIFs are effective to capture important characteristics of clusters. When clusters are well-separated, CIFs become similar to density functions. But as clusters touch each other, CIFs still clearly describe the cluster structure whereas density functions and hence cluster structure become blurred. In this sense, CIFs are a better representation of clusters than density functions.

CIFs resolve the problems of KDEs while advantages of KDEs are inherited. Although CIFs are also built on the top of KDEs, they are cluster-oriented so that only information contained in KDEs that is useful for clustering is kept while other irrelevant information is filtered out. We have shown that such a filtering process is very important in clustering especially when the clusters touch each other. On the other hand, it is well-known that when the clusters are well-separated, then valley seeking on KDEs results in very good clusterings. Since the valleys of CIFs and KDEs are very similar, if not identical, when the clusters are well-separated, clustering based on CIFs is as good as that based on KDEs. However, advantages of CIFs over KDEs become very significant when the clusters touch each other.

In our method, once the CIF is obtained, cluster labels can be easily assigned by applying the valley seeking algorithm [4] but with the density function replaced the distance-based CIF.

2 Cluster Formation

In this section, we describe our methodology to construct clusters using LSM. We start by introducing some terminologies. A *cluster core contour* (CCC) is a closed surface surrounding the core part/density peak of a cluster at which density is relatively high. A *cluster boundary* refers to the interface between two clusters, i.e., a surface separating two clusters. A CCC is usually located near a density peak while a cluster boundary is located at the valley regions of a density distribution. Here, a point $\mathbf{x}$ is said to belong to a valley region of f if there exists a direction along which f is a local minimum. The gradient and the Laplacian of a function g are denoted by ∇g and Δg respectively.

Our method consists of the following main steps which will be elaborated in details in the next subsections: (i) initialize CCCs to surround high density regions; (ii) advance the CCCs using LSM to find density peaks; (iii) apply valley seeking algorithm on the CIF constructed from the final CCCs to obtain clusters.

2.1 Initialization of Cluster Core Contours (CCC)

We now describe how to construct an initial cluster core contours Γ_0 effectively. The basic idea is to locate the contours at which f has a relatively large (norm of) gradient. In this way, regions inside Γ_0 would contain most of the data points — we refer these regions as *cluster regions*. Similarly, regions outside Γ_0 would contain no data point at all and we refer them as *non-cluster regions*. Such an interface Γ_0 is constructed as follows.

Definition 1. *An initial set of CCCs Γ_0 is the set of zero crossings of Δf, the Laplacian of f. Here, a point $\mathbf{x}$ is a zero crossing if $\Delta f(\mathbf{x}) = 0$ and within any arbitrarily small neighborhood of $\mathbf{x}$, there exist $\mathbf{x}^+$ and $\mathbf{x}^-$ such that $\Delta f(\mathbf{x}^+) > 0$ and $\Delta f(\mathbf{x}^-) < 0$.*

We note that Γ_0 often contains several closed surfaces, denoted by $\{\Gamma_{0,i}\}$. The idea of using zero crossings of Δf is that it outlines the shape of datasets very well and that for many commonly used kernels (e.g. Gaussian and cubic B-spline) the sign of $\Delta f(\mathbf{x})$ indicates whether $\mathbf{x}$ is inside or outside Γ_0.

Complete reasons for using zero crossings of Δf to outline the shape of datasets are several folds: (a) the solution is a set of surfaces at which $\|\nabla f\|$ is relatively large; (b) the resulting Γ_0 is a set of closed surfaces; (c) Γ_0 well captures the shape of clusters; (d) the Laplacian operator is an isotropic operator which does not bias towards certain directions; (e) the equation is simple and easy to solve; (f) it coincides with the definition of edge in the case of image processing. In fact, a particular application of zero crossings of Laplacian in image

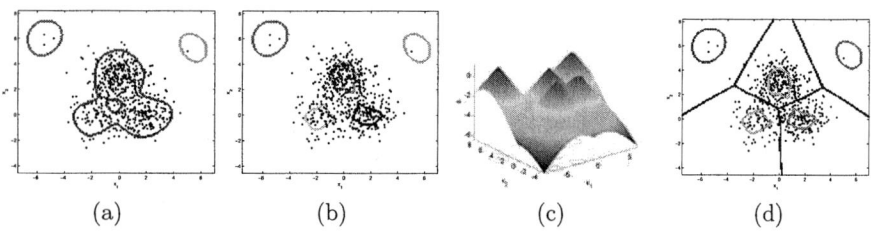

(a) (b) (c) (d)

Fig. 2. Evolution of cluster core contours (CCC) using the bump hunting PDEs of Eq. (2). The dataset is the one used in Fig. 1. (a) Initial CCC. (b) Final CCCs. (c) CIF constructed from the contours in (b). Peaks corresponding to the three large clusters are clearly seen. (d) Valleys of the CIF. We observe that the initial CCCs capture the shape of the dataset and that the resulting boundaries capture the hot spots of the dataset very well. In (d), the three cluster cores are well-discovered

processing is to detect edges to outline objects;(g) the sign of $\Delta f(\mathbf{x})$ indicate whether $\mathbf{x}$ is inside (negative) or outside (positive) of a cluster region.

In Fig. 2(a), we show the initial CCCs juxtaposed with the dataset shown in Fig. 1(a). We observe that the CCCs capture the shape of the dataset very well.

2.2 Advancing Cluster Core Contours

Next, we discuss how to advance the initial CCCs to obtain peak regions through hill climbing in a smooth way. We found that this is a key issue in density-based approaches and is also how ideas from LSM come into play. More precisely, we employ PDE techniques to advance contours in an elegant way.

Since each initial CCC $\Gamma_{0,i}$ in Γ_0 changes its shape as evolution goes on, we parameterize such a family of CCCs by a time variable t, i.e., the i-th CCC at time t is denoted by $\Gamma_i(t)$. Moreover, $\Gamma(0) \equiv \Gamma_0$.

In LSM, $\Gamma(t)$ is represented by the zero level set of a Lipschitz function $\phi = \phi(\mathbf{x}, t)$, i.e., $\Gamma(t) = \{\mathbf{x} : \phi(\mathbf{x}, t) = 0\}$. The value of ϕ at non-zero level sets can be arbitrary, but a common practice is to choose ϕ to be the *signed distance function* $\psi_{\Gamma(t)}(\mathbf{x})$ for numerical accuracy reasons [5]. In general, the signed distance function with respect to a set of surfaces Γ is defined by

$$\psi_\Gamma(\mathbf{x}) = \begin{cases} \min_{\mathbf{y} \in \Gamma} \|\mathbf{x} - \mathbf{y}\| & \text{if } \mathbf{x} \text{ lies inside } \Gamma \\ -\min_{\mathbf{y} \in \Gamma} \|\mathbf{x} - \mathbf{y}\| & \text{if } \mathbf{x} \text{ lies outside } \Gamma \end{cases}, \quad (1)$$

To evolve $\Gamma(t)$ (where $\Gamma(0)$ is the initial data) with speed $\beta = \beta(\mathbf{x}, t)$, the equation is given by $\partial \phi / \partial t = \beta \|\nabla \phi\|$ which is known as the level set equation [5]. Our PDE also takes this form.

Using a level set representation, the *mean curvature* $\kappa = \kappa(\mathbf{x}, t)$ (see [5]) of $\Gamma(t)$ at $\mathbf{x}$ is given by $\kappa(\mathbf{x}, t) = \nabla \cdot (\nabla \phi(\mathbf{x}, t) / \|\nabla \phi(\mathbf{x}, t)\|)$.

Given the initial CCCs $\Gamma(0)$ represented by the zero level set of $\phi(\mathbf{x}, 0)$ (which is chosen to be $\psi_{\Gamma(0)}(\mathbf{x})$), the time dependent PDE that we employ for hill climbing on density functions is given by

$$\frac{\partial \phi}{\partial t} = \left(\frac{1}{1 + \|\nabla f\|} + \alpha \kappa\right) \|\nabla \phi\|, \quad \phi(\mathbf{x}, 0) = \psi_{\Gamma(0)}(\mathbf{x}). \quad (2)$$

This equation is solved independently for each cluster region defined according to $\Gamma(t)$. During evolution, each contour and hence each cluster region may split or merge. Evolution is stopped when no further splitting occurs.

The aim of the factor $1/(1 + \|\nabla f\|)$ is to perform hill climbing to look for density peaks. Moreover, the factor also adjusts the speed of each point on the CCCs in such a way that the speed is lower if $\|\nabla f\|$ is larger. Thus the CCCs stay in steep regions of f where peak regions are defined better. In the limiting case where f has a sharp jump ($\|\nabla f\| \to \infty$), the CCCs actually stop moving at the jump. We remark that in traditional steepest descent methods for solving minimization problems, the speed (step size) is usually higher if $\|\nabla f\|$ if larger, which is opposite to what we do. This is because our goal is to locate steep regions of f rather than local minimums.

The curvature κ exerts surface tension to smooth out the CCCs. In contrast, without exerting surface tension, the CCCs could become wiggly which may lead to the common problem of over-fitting of KDEs. Therefore, we employ the term κ to resolve such a problem. In fact, if ϕ is kept to be a signed distance function for all t, i.e., $\|\nabla \phi\| \equiv 1$, then $\kappa = \Delta \phi$ so that ϕ is smoothed out by Gaussian filtering. In the variational point of view, the curvature term exactly corresponds to minimization of the length (surface area in general) of the CCCs.

The scalar $\alpha \geq 0$ controls the amount of tension added to the surface and will be adjusted dynamically during the course of evolution. At the beginning of evolution of each $\Gamma_i(0)$, we set $\alpha = 0$ in order to prevent smoothing out of important features. After a CCC is split into pieces, tension is added and is gradually decreased to 0. In this way, spurious oscillations can be removed without destroying other useful features. Such a mechanism is similar to cooling in simulated annealing.

In summary, the PDE simply (i) moves the initial CCCs uphill in order to locate peak regions; (ii) adjusts the speed according to the slope of the KDE; (iii) removes small oscillations of the CCCs by adding tension so that hill climbing is more robust to the unevenness of the KDE (c.f. Examples 1 and 2 in §3). In addition to these, the use of LSM allows the CCCs to be split and merged easily.

In Fig. 2(a)–(b), we show the CCCs during the course of evolution governed by Eq. (2). The two CCCs correspond to outliers are freezed quickly. We observe that the contours are attracted to density peaks. When a contour is split into several contours, the pieces are not very smooth near the splitting points. Since tension is added in such cases, the contours are straighten out quickly.

2.3 Cluster Intensity Functions

In non-parametric modelling, one may obtain clusters by employing valley seeking on KDEs. However, as mentioned above, such methods perform well only when the clusters are well-separated and of approximately the same density in which case peaks and valleys of the KDE are clearly defined. On the other hand, even though we use the density peaks identified by our PDE Eq. (2) as a starting point, if we expand the CCCs outward according to the KDE, we still have to face the problems of the KDE; we may still get stuck in local optimum due to its oscillatory nature.

In this subsection, we further explore cluster intensity functions which are a better representation of clusters than that by KDEs. Due to the advantages of CIFs, we propose to perform valley seeking on CIFs to construct clusters, rather than on KDEs. Here, CIFs are constructed based on the final CCCs obtained by solving the PDE Eq. (2).

CIFs capture the essential features of clusters and inherit advantages of KDEs while information irrelevant to clustering contained in KDEs is filtered out. Moreover, peaks and valleys of CIFs stand out clearly which is not the case for KDEs. The principle behind is that clustering should not be done solely based on density,' rather, it is better done based on density and distance. For example, it is well-known that the density-based algorithm DBSCAN [1] cannot separate clusters that are closed together even though their densities are different.

CIFs, however, are constructed by calculating *signed distance* from CCCs (which are constructed based on density). Thus, CIFs combine both density and distance information about the dataset. This is a form of regularization to avoid over-specification of density peaks.

The definition of a CIF is as follows. Given a set of closed hypersurfaces Γ (zero crossings of Δf or its refined version), the CIF ϕ with respect to Γ is defined to be the signed distance function (1), $\phi = \psi_\Gamma$.

The value of a CIF at $\mathbf{x}$ is simply the distance between $\mathbf{x}$ and Γ with its sign being positive if $\mathbf{x}$ lies inside Γ and negative if $\mathbf{x}$ lies outside Γ. Roughly speaking, a large positive (respectively negative) value indicates that the point is deep inside (respectively outside) Γ while a small absolute value indicates that the point lies close to the interface Γ.

In Fig. 2(c), the CIF constructed from the CCCs in Fig. 2(b) is shown. The peaks correspond to the three large clusters can be clearly seen which shows that our PDE is able to find cluster cores effectively. Based on the CIF, valley seeking (c.f. §1) can be easily done in a very robust way. In Fig. 2(d), we show the valleys of the CIF juxtaposed with the dataset and the final CCCs.

We remark that the CCCs play a similar role as cluster centers in the k-means algorithm. Thus, our method generalizes the k-means algorithm in the sense that a "cluster center" may be of arbitrary shape instead of just a point.

Under LSM framework, valleys and peaks are easily obtained. The valleys are just the singularities of the level set function (i.e. CIF) having negative values. On the other hand, the singularities of the level set function having positive values are the peaks or ridges of the CIF (also known as skeleton).

3 Experiments

In addition to the examples shown in Figs. 1–2, we give more examples to further illustrate the usefulness of the concepts introduced. Comparisons with valley seeking [4] and DBSCAN [1] algorithms (c.f. §1) are given in Examples 1 and 2. Clustering results of two real datasets are also presented. For visualization of CIFs which is one dimension higher than the datasets, two dimensional datasets are used while the theories presented above apply to any number of dimensions.

When applying DBSCAN, the parameter $MinPts$ is fixed at 4 as suggested by the authors in [1].

Example 1. We illustrate how the problem of over-fitting (or under-fitting) of KDEs is resolved using our method. In Fig. 3, we compare the clustering results of valley seeking using the scale parameter $h = 0.6, 0.7$ and the DBSCAN algorithm using $Eps = 0.28, 0.29$. The best result is observed in Fig. 3(a) but it still contains several small clusters due to the spurious oscillations of the KDE. For other cases, a mixture many small clusters and some over-sized clusters are

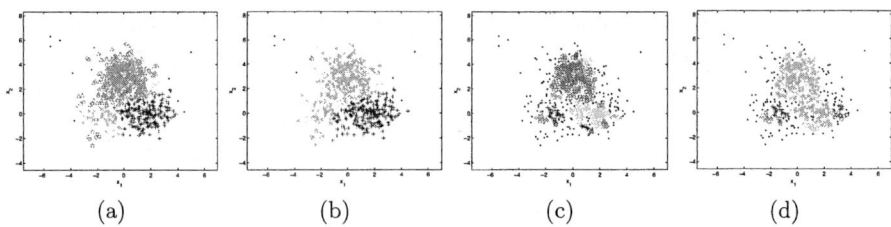

(a) (b) (c) (d)

Fig. 3. Clusters obtained from applying valley seeking and DBSCAN to the dataset in Fig. 1. (a) Valley seeking with $h = 0.6$. (b) Valley seeking with $h = 0.7$. (c) DBSCAN with $Eps = 0.28$. (d) DBSCAN with $Eps = 0.29$. In (a) and (c), many small clusters are present due to unevenness of the density functions. These results are not as good as the results using our method as shown in Fig. 2

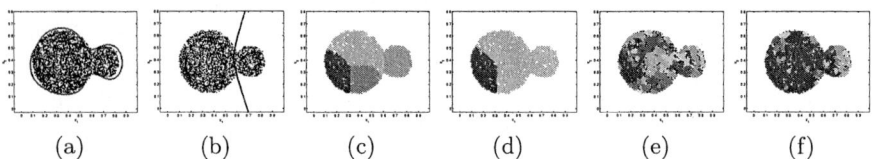

(a) (b) (c) (d) (e) (f)

Fig. 4. Comparisons of our method with valley seeking and DBSCAN. (a) Dataset with the zero crossing of Δf superimposed. (b) Final CCCs (the closed curves in red and blue) and the valleys of the CIF (the line in black). (c) Valley seeking with $h = 0.05$. (d) Valley seeking with $h = 0.06$. (e) DBSCAN with $Eps = 0.010$. (f) DBSCAN with $Eps = 0.011$. The CCCs are able to capture the cluster shape while valley seeking and DBSCAN seem to suffer from over-fitting and result in many small spurious clusters

present. In contrast, our method (shown in Fig. 2) resolves these problems by (i) outlining the shape of the dataset well by keeping the CCCs smooth; (ii) using curvature motion to smooth out oscillations due to unevenness of KDEs.

Example 2. A dataset with 4000 uniformly distributed points lying in two touching circles is considered. The dataset together with the zero crossings of Δf are shown in Fig. 4(a). The result of our method is in Fig. 4(b). We observe that the final CCCs adapt to the size of the clusters suitably. The results of valley seeking on KDEs ($h = 0.05, 0.06$) are shown in Fig. 4(c) and (d) where the unevenness of the KDEs result in either 2 or 4 large clusters. The results of DBSCAN with $Eps = 0.010, 0.011$ are in Fig. 4(e) and (f) which contain many small clusters. In addition, this example also makes it clear that density functions must be regularized which is done implicitly by adding surface tension in our method.

Example 3. This example uses a dataset constructed from the co-expression patterns of the genes in yeast during cell cycle. Clusters of the data are expected to reflect functional modules. The results are shown in Fig. 5. We observe that the valleys of the CIF are right on the low density regions and thus a reasonable clustering is obtained.

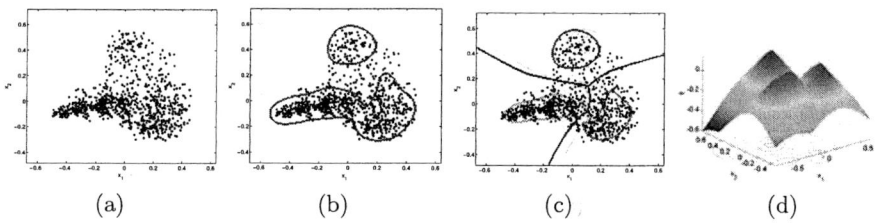

Fig. 5. (a) DNA gene expression dataset. (b) Zero crossings of Δf. (c) The final CCCs and the valleys of the CIF. (d) The CIF. The cluster cores are well-retrived and the valleys successfully separate the data into clusters of relatively high density

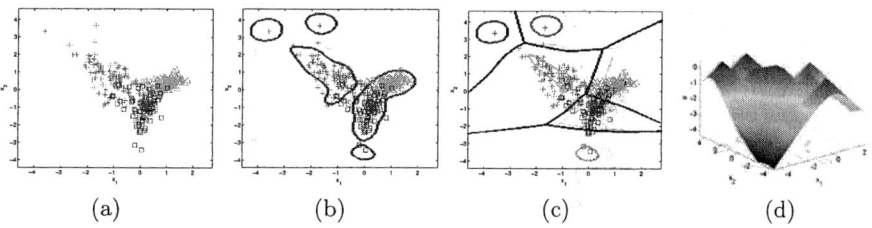

Fig. 6. (a) A dataset of 3 internet newsgroups with 100 news items in each group (news items in the same group are displayed with the same symbol). (b) The zero crossings of Δf. (c) Clustering results (lines are valleys of the final CIF and closed curves are the cluster cores). (d) The CIF

Example 4. Our next example uses a real dataset from text documents in three newsgroups. For the ease of visualization, the dataset is first projected to a 2-D space using principle component analysis. The results in Fig. 6 show that the clustering results agree with the true clustering very well.

4 Concluding Remarks

In the paper, we introduced level set methods to identify density peaks and valleys in density landscape for data clustering. The method relies on advancing contours to form cluster cores. One key point is that during contour advancement, smoothness is enforced via LSM. Another point is that important features of clusters are captured by cluster intensity functions which serve as a form of regularization. The usual problem of roughness of density functions is overcome. The method is shown to be much more robust and reliable than traditional methods that perform bump hunting or valley seeking on density functions.

Our method can also identify outliers effectively. After the initial cluster core contours are constructed, outliers are clearly revealed and can be easily identified. In this method, different contours evolve independently. Thus outliers do not affect normal cluster formation via contour advancing; This nice property does not hold for clustering algorithms such as the k-means where several outliers could skew the clustering.

Our method for contour advancement Eq. (2) is based on the dynamics of interface propagation in LSM. A more elegant approach is to recast the cluster core formation as a minimization problem where the boundary advancement can be derived from first principles which will be presented in a later paper.

References

1. Ester, M., Kriegel, H., Sander, J., Xu, X.: A density-based algorithm for discovering clusters in large spatial databases with noise. In: Int. Conf. Knowledge Discovery and Data Mining, Portland, OR, AAAI Press (1996) 226–231
2. Hinneburg, A., Keim, D.A.: An efficient approach to clustering in large multimedia databases with noise. In: Int. Conf. Knowledge Discovery and Data Mining, New York City, NY, AAAI Press (1998) 58–65
3. Agrawal, R., Gehrke, J., Gunopulos, D., Raghavan, P.: Automatic subspace clustering of high dimensional data for data mining applications. In: Proc. ACM-SIGMOD Int. Conf. Management of Data, Seattle, WA, ACM Press (1998) 94–105
4. Fukunaga, K.: Introduction to Statistical Pattern Recognition. 2nd edn. Boston Academic Press (1990)
5. Osher, S., Fedkiw, R.: Level Set Methods and Dynamic Implicit Surfaces. Spring Verlag, New York (2003)

A Vector Field Visualization Technique for Self-organizing Maps

Georg Pölzlbauer[1], Andreas Rauber[1], and Michael Dittenbach[2]

[1] Department of Software Technology,
Vienna University of Technology,
Favoritenstr. 11-13, Vienna, Austria
{poelzlbauer, rauber}@ifs.tuwien.ac.at
[2] eCommerce Competence Center – ec3,
Donau-City-Str. 1, Vienna, Austria
michael.dittenbach@ec3.at

Abstract. The Self-Organizing Map is one of most prominent tools for the analysis and visualization of high-dimensional data. We propose a novel visualization technique for Self-Organizing Maps which can be displayed either as a vector field where arrows point to cluster centers, or as a plot that stresses cluster borders. A parameter is provided that allows for visualization of the cluster structure at different levels of detail. Furthermore, we present a number of experimental results using standard data mining benchmark data.

1 Introduction

The Self-Organizing Map (SOM) [1] is a valuable tool in data analysis. It is a popular unsupervised neural network algorithm that has been used in a wide range of scientific and industrial applications [3], like Text Mining [6], natural language processing and monitoring of the condition of industrial plants and processes. It provides several beneficial properties, such as vector quantization and topology preserving mapping from a high-dimensional input space to a two-dimensional output space. This projection can be visualized in numerous ways in order to reveal the characteristics of the input data or to analyze the quality of the obtained mapping.

Our method is based on the SOM codebook and the neighborhood kernel, which induces a concept of proximity on the map. For each map unit, we compute a vector pointing to the direction of the most similar region in output space. We propose two methods of visualizing the results, a vector field plot, which can be seen analogous to flow visualization and gradient visualization, and a plot that emphasizes on the cluster structure of the map. The SOMs used for demonstration and experiments are trained on Fisher's well-known Iris data.

The rest of this paper is organized as follows. Section 2 describes several visualization techniques for SOMs and related work. Section 3 gives an overview of neighborhood kernel functions and their parametrization. In Section 4, our visualization method is introduced, along with a description of its properties and interpretations. Section 5 presents experimental results, where the the influence of choices of neighborhood kernel, neighborhood radius and map size are investigated. Finally, Section 6 gives a short summary of the findings presented in this paper.

2 Related Work

In this section, we briefly describe visualization concepts for SOMs related to our novel method. The most common ones are component planes and the U-Matrix. Both take only the prototype vectors and not the data vectors into account. Component planes show projections of singled out dimensions of the prototype vectors. If performed for each individual component, they are the most precise and complete representation available. However, cluster borders cannot be easily perceived, and high input space dimensions result in lots of plots, a problem that many visualization methods in multivariate statistics, like scatterplots, suffer from. The U-Matrix technique is a single plot that shows cluster borders according to dissimilarities between neighboring units. The distance between each map unit and its neighbors is computed and visualized on the map lattice, usually through color coding. Recently, an extension to the U-Matrix has been proposed, the U*-Matrix [8], that relies on yet another visualization method, the P-Matrix [7]. Other than the original, it is computed by taking both the prototype vectors and the data vectors into account and is based on density of data around the model vectors. Interestingly, both the U*-Matrix and our novel method, among other goals, aim at smoothing the fine-structured clusters that make the U-Matrix visualization for these large SOMs less comprehensible, although the techniques are conceptually totally different. Other visualization techniques include hit histograms and Smoothed Data Histograms [4], which both take the distribution of data into account, and projections of the SOM codebook with concepts like PCA or Sammon's Mapping, and concepts that perform labeling of the SOM lattice [6]. For an in-depth discussion, see [9].

In Figure 1, the hit histogram and U-Matrix visualizations are depicted for SOMs trained on the Iris data set with 30×40 and 6×11 map units, respectively. The feature dimensions have been normalized to unit variance. The U-Matrix reveals that the upper third of the map is clearly separated from the rest of the map. The hit histogram shows the projection of the data samples onto the map lattice. It can be seen that this SOM is very sparsely populated, because the number of map units is higher than the number of data samples. When the two methods are compared, it can be observed that the fine cluster structures in the U-Matrix occur exactly between the map units that are occupied by data points. It is one of the goals of this work to create a representation that allows a more global perspective on these kinds of maps and visualize it such that the intended level of detail can be configured.

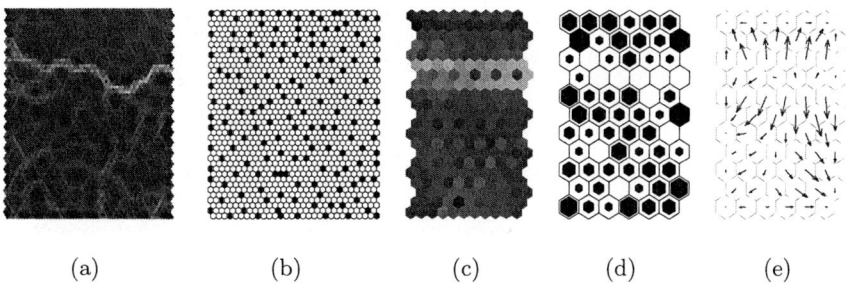

Fig. 1. 30 × 40 SOM: (a) U-Matrix, (b) Hit histogram, 6 × 11 SOM: (c) U-Matrix, (d) Hit histogram, (e) Vector Field with Gaussian kernel ($\sigma = 2$), see Section 5

To our best knowledge, the neighborhood kernel that is described in the next section has not been used for visualization purposes. Apart from the SOM training algorithm the neighborhood function is applied in the SOM Distortion Measure [2], which is the energy function of the SOM with fixed radius, where the neighborhood kernel is aggregated and serves as a weighting factor comparable to the one we use in this paper.

3 SOM Neighborhood Kernels

A particularly important component of the Self-Organizing Map is the concept of adjacency in output space, i.e. the topology of the map lattice, and its definition of neighborhood that affects the training process. Our visualization technique heavily depends on this neighborhood kernel as a weighting factor. The neighborhood kernel is a parameterized function that takes the distance between two map units on the lattice as input and returns a scaling factor that determines by which amount the map unit is updated for each iteration. The parameter the kernel depends on is the neighborhood radius $\sigma(t)$, which is itself a monotonically decreasing function over time t. σ controls the width of the kernel function, such that high values lead to kernels that are stretched out and low values result in sharply peaked kernels. In this work, we will not consider the radius as a function of time as the training process does, but rather as a parameter that has to be specified before the visualization can be applied.

The kernel function $h_\sigma(d_{input})$ has the property of decreasing monotonically with increasing distance d_{input}. This distance will be formally defined in the next section, but can be roughly envisioned as the number of units that lie between two map units. Examples of neighborhood kernels are the Gaussian kernel, the bubble function, and the inverse function. The Gaussian kernel is the most frequently used kernel for the SOM. It has the well-known form of the Gaussian Bell-Shaped Curve, formally

$$h_\sigma^G(d_{input}) = \exp\left(-\frac{d_{input}^2}{2\sigma}\right) \tag{1}$$

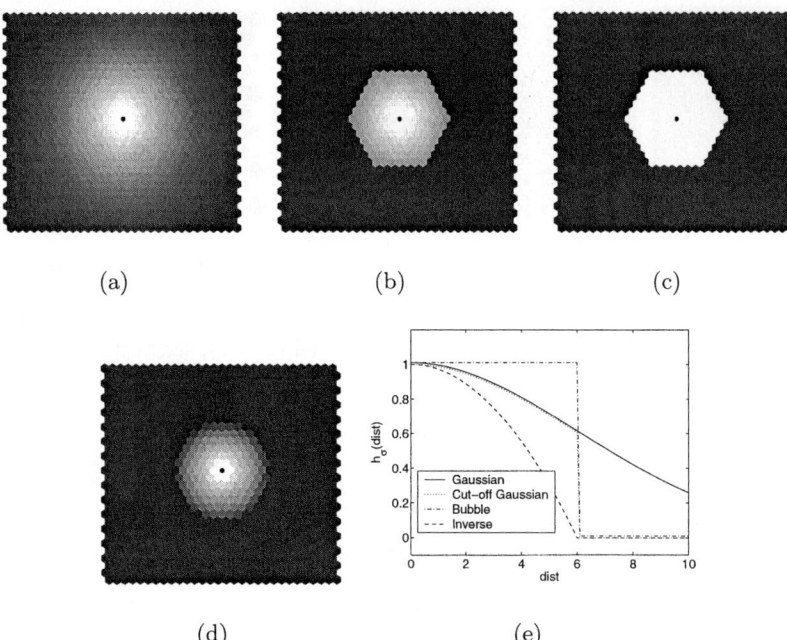

Fig. 2. Overview of different kernel functions: (a) Gaussian kernel, (b) cut-off Gaussian kernel, (c) bubble function, (d) inverse function, (e) comparison of neighborhood functions

Since the returned value is exponentially decreasing for higher values of d_{input}, the effects on the training process are neglegible. Thus, the kernel is frequently modified to cut off the function at input values greater than σ:

$$h_\sigma^{c/G}(d_{input}) = \begin{cases} h_\sigma^G(d_{input}) & \text{if } d_{input} \leq \sigma \\ 0 & \text{otherwise} \end{cases} \quad (2)$$

Another kernel is the bubble function, which exclusively relies on this principle of cutting off at radius σ. It is a simple step function, formally

$$h_\sigma^B(d_{input}) = \begin{cases} 1 & \text{if } d_{input} \leq \sigma \\ 0 & \text{otherwise} \end{cases} \quad (3)$$

Another option is the inverse proportional function:

$$h_\sigma^I(d_{input}) = \begin{cases} 1 - \frac{d_{input}^2}{\sigma^2} & \text{if } d_{input} \leq \sigma \\ 0 & \text{otherwise} \end{cases} \quad (4)$$

which shows a sharper decrease than the Gaussian kernel.

The different kernel functions are depicted in Figure 2, which shows the values of the kernel for the map unit located in the center, indicated by a black dot.

Figure 2(e) shows a plot of the kernels as function of the distance between the units and fixed neighborhood radius. All the graphics use the same value of 6 for parameter σ.

4 A Vector Field Based Method for Visualization

In this section, we introduce our visualization technique for the SOM. Similar to the U-Matrix, only the prototype vectors and their pairwise similarities are investigated. In the U-Matrix, only the differences between direct neighbors are considered. We aim to extend this concept to include the region around the units according to the neighborhood kernel. Furthermore, we wish to obtain the direction for each unit where the most similar units are located. The resulting visualization is analogous to gradient vector fields where units are repelled from or attracted to each other.

First, we have to make some formal definitions. The type of SOM that we will consider has a two-dimensional lattice, consisting of a number M of map units p_i, where i is between 1 and M. Each of the map units is linked to a model vector m_i of input dimension N. Each of the m_i is linked to the output space by its position on the map. To distinguish between feature space and map lattice, we explicitly write p_i for the position vector of map unit that represents prototype vector m_i; the index i connects input and output space representation. We denote the horizontal and vertical coordinates of the map unit as p_i^u and p_i^v, respectively. Thus, the distance between two prototype vectors m_i and m_j, or p_i and p_j, can be determined both in input and output space:

$$d_{input}(m_i, m_j) = ||m_i - m_j||_{input} \tag{5}$$

where $||.||_{input}$ is a suitable distance metric and

$$d_{output}(p_i, p_j) = \sqrt{(p_i^u - p_j^u)^2 + (p_i^v - p_j^v)^2} \tag{6}$$

which is the Euclidean Distance.

The neighborhood kernel requires the distance between the model vectors' positions on the map lattice $d_{output}(p_i, p_j)$ as its input. This kernel function computes how much the prototype vectors influence each other during the training process. We will use it as a weighting function that allows us to compute the similarity (in terms of input space distance) of map units that are close to each other on the map.

Our technique plots arrows for each map unit like in gradient field visualizations. A unit's arrow points to the region where the most similar prototype vectors are located on the map. The length of this arrow shows the degree of how much the area it is pointing to is more similar to it than the opposite direction.

Each arrow is computed for unit p_i as a two-dimensional vector a_i. It can be decomposed in u and v coordinates, denoted as a_i^u and a_i^v. For each of the two axes, we compute the amount of dissimilarity along positive and negative directions. Our method determines these vectors in a two-step process: First, the computations for each map unit are performed separately for the positive and negative directions of axes u and v, and finally, these components are aggregated by a weighting scheme to calculate the coordinates of a_i.

The angle α that identifies the direction of p_j seen from p_i on the map lattice is defined in basic trigonometry as

$$\alpha(p_i, p_j) = \arctan(\frac{p_j^v - p_i^v}{p_j^u - p_i^u}) \qquad (7)$$

The influence of the neighborhood kernel projected onto the u and v axes is computed as

$$w^u(p_i, p_j) = \cos(\alpha(p_i, p_j)) \cdot h_\sigma(d_{output}(p_i, p_j)) \qquad (8)$$

$$w^v(p_i, p_j) = \sin(\alpha(p_i, p_j)) \cdot h_\sigma(d_{output}(p_i, p_j)) \qquad (9)$$

Here, the influence of the neighborhood kernel is distributed among the two axes according to the position of p_i and p_j on the map and serves as a weighting factor in the following steps. The neighborhood kernel relies on the width parameter σ, which determines the influence of far-away map units.

Then, we decompose the amount of dissimilarity in its positive and negative direction for both axes for each pair of map units p_i, p_j:

$$con_+^u(p_i, p_j) = \begin{cases} d_{input}(m_i, m_j) \cdot w^u(p_i, p_j) & \text{if } w^u(p_i, p_j) > 0 \\ 0 & \text{otherwise} \end{cases} \qquad (10)$$

$$con_-^u(p_i, p_j) = \begin{cases} -d_{input}(m_i, m_j) \cdot w^u(p_i, p_j) & \text{if } w^u(p_i, p_j) < 0 \\ 0 & \text{otherwise} \end{cases} \qquad (11)$$

where con_+^u denotes the contribution of map unit p_j's dissimilarity in positive direction along u, and con_-^u in negative direction. The definition of con_+^v and con_-^v follows analogously. For example, a map unit p_j that lies to the lower right of p_i results in $con_-^u(p_i, p_j) = con_+^v(p_i, p_j) = 0$, and some positive values for $con_+^u(p_i, p_j)$ and $con_-^v(p_i, p_j)$ according to the distance in output space, which is weighted through the neighborhood kernel, and also its distance in input space, which is directly measured by the factor d_{input}.

Next, the sum of contributions in both directions is computed for each p_i

$$diss_+^u(p_i) = \sum_{j=1...M, j \neq i} con_+^u(p_i, p_j) \qquad (12)$$

$$diss_-^u(p_i) = \sum_{j=1...M, j \neq i} con_-^u(p_i, p_j) \qquad (13)$$

Again, $diss_+^v$ and $diss_-^v$ are defined analogously. The variable $diss_+^u(p_i)$ indicates how much m_i is dissimilar from its neighbors on the side in the positive u direction. In a gradient field analogy, this value shows how much it is repelled from the area on the right-hand side.

Next, we aggregate both negative and positive components into the resulting vector a_i. Normalization has to be performed, because units at the borders of the map lattice would have components pointing outside of the map equal to zero, which is not intended. The sums of the neighborhood kernel weights w_i pointing in positive and negative directions are

$$w_+^u(p_i) = \sum_{j=1...M, j \neq i} \begin{cases} w^u(p_i, p_j) & \text{if } w^u(p_i, p_j) > 0 \\ 0 & \text{otherwise} \end{cases} \quad (14)$$

$$w_-^u(p_i) = \sum_{j=1...M, j \neq i} \begin{cases} -w^u(p_i, p_j) & \text{if } w^u(p_i, p_j) < 0 \\ 0 & \text{otherwise} \end{cases} \quad (15)$$

Finally, the u component of the gradient vector a is computed as

$$a_i^u = \frac{diss_-^u(p_i) \cdot w_+^u(p_i) - diss_+^u(p_i) \cdot w_-^u(p_i)}{diss_+^u(p_i) + diss_-^u(p_i)} \quad (16)$$

and likewise for the v direction. The weighting factor w_+^u is multiplied with the component in the other direction to negate the effects of units close to the border in which case the sum of the neighborhood kernel is greater on one side. If this normalization would be omitted, the vector a would be biased towards pointing to the side where units are missing. For map units in the center of the map's u-axis, where w_+^u and w_-^u are approximately equal, Equation (16) can be approximated by this simpler formula

$$a_i^u \approx \mu \cdot \frac{diss_-^u(p_i) - diss_+^u(p_i)}{diss_+^u(p_i) + diss_-^u(p_i)} \quad (17)$$

where μ is a constant factor equal to $\frac{w_+^u + w_-^u}{2}$ and is approximately the same for all units in the middle of an axis.

The results obtained for different ratios and proportions of $diss_+$ and $diss_-$ are briefly described:

- If negative and positive dissimilarities are roughly equal, the resulting component of a will be close to zero.
- If the positive direction is higher than the negative one, a will point into the negative direction, and vice versa. The reason for this is that the prototype vectors on the negative side of the axis are more similar to the current map unit than on the positive side.
- If one side dominates, but the second side still has a high absolute value, the normalization performed in the denominator of Equation (16) decreases the length of the vector.

In Figure 3(b), our visualization technique is shown for the 30 × 40 SOM trained on the Iris data set with a Gaussian kernel with $\sigma = 5$. If compared to the U-Matrix in Figure 1(a), it can be seen that the longest arrows are observed near the cluster borders, pointing to the interior of their cluster and away from

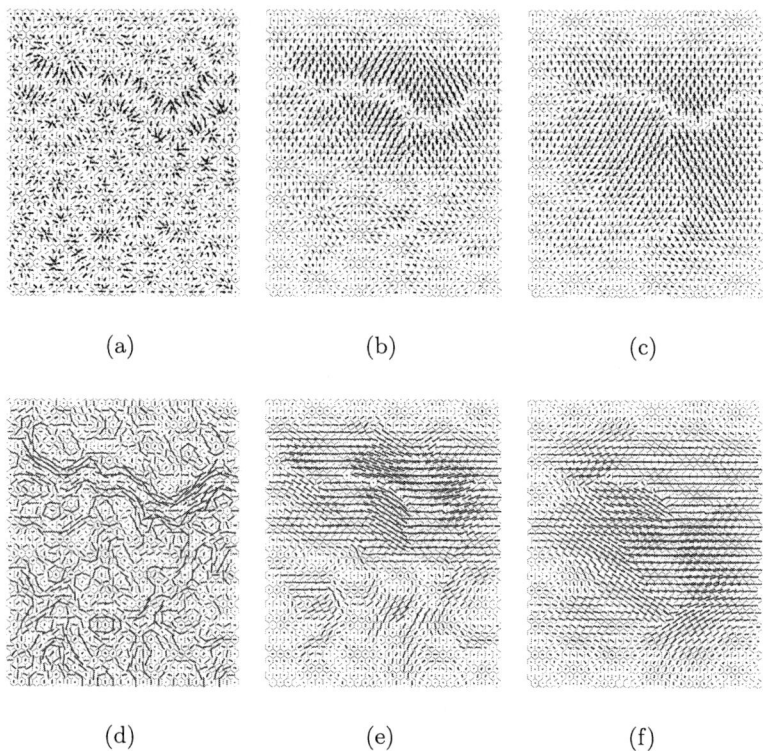

Fig. 3. 30 × 40 SOM trained on Iris data (a)-(c) Vector field representation with $\sigma = 1, 5, 15$, (d)-(f) Border representation with $\sigma = 1, 5, 15$

these borders. Adjacent units, for which the arrow points in different directions, are clearly along a cluster border. The length of the arrows indicates how sharp the border is. In the middle of these transitions, arrows are sometimes drawn with almost no distinguishable length or direction. The corresponding prototype vectors are likely to be very far away from either cluster, and are referred to as interpolating units, since they do not represent any data vectors in a vector quantization sense, but are only a link connecting two distant data clouds. Cluster centers also have small dot-like arrows pointing in no distinguishable direction, but the difference is that the surrounding arrows are pointing in their direction, and not away from them. Another property of this visualization is that the units on the edges of the map never point outside of it, which is desired and stems from the normalization performed in (16).

One interesting extension to our visualization method is that the results can also be depicted to show the cluster borders themselves with a slight modification in the representation by depicting not the direction of the gradient, but rather the hyperplane obtained by rotation of 90 degrees in either direction. In our case of a two-dimensional map lattice, the hyperplane is a one-dimensional line. We choose to depict this line with length proportional to the original arrow. The result is visualized in Figure 3(e). The emphasis of this dual representation is stressing cluster borders, while information on directions is omitted.

We have found that our method is most useful when applied in combination with other visualization techniques, such as hit histograms and component planes. What can be learned from comparing the positions of the different Iris species to our method is that the class membership of the data samples correlates with the cluster structure in case of the Setosa species, while Versicolor and Virginica do not show a distinguishable separation. This is of course a well-known fact about the Iris data set, and application of our technique to more complex data is subject to further research and is addressed in [5].

5 Experiments

In this section, we will investigate the empirical results of our method applied to SOMs of different sizes, as well as how the choice of parameter σ influences the visualization, and the effects of different kernel functions.

First, we examine the effect of the map size, i.e. the number of prototype vectors. The data vectors remain the same for both maps. The smaller version of the SOM consists of 6×11 units, and the larger one of 30×40 units. In the latter case, the number of data vectors (150) is much lower than the number of map units (1200). The visualization for the smaller version is depicted in Figure 1(e). U-Matrix and vector field plots for the larger map are shown in Figures 1(a) and 3, respectively. In the smaller SOM the gap between the upper third part representing the well-separated Setosa species and the lower two-thirds of the map can clearly be distinguished, as in the larger SOM. However, the larger version of the SOM gives more insight into the structure of the data. Transitions and gradual changes in directions and length can be distinguished more easily at this higher granularity.

In the next experiment, we investigate the influence of the width parameter σ. In Figure 3, the large Iris SOM is visualized with three different values of σ. Figures 3(a), (d) show the two methods for $\sigma = 1$. The visualization with this width is the one most closely related to the U-Matrix technique, since only distances between direct neighbors are regarded, while the influence of slightly more distant units is neglected. Of all the visualizations shown here, these two are chiseled the most and are least smooth. The frequent changes in direction of neighboring arrows is due to the very local nature of this kernel. In Figures 3(b), (e) the visualization is shown for $\sigma = 5$, where the increased neighborhood radius produces a smoothing effect over the vector field. Here, changes in direction between close arrows can be better distinguished and result in a visually more comprehensible picture. The set of arrows is perceived as a whole and as less chaotic. It gives the impression of visualizing a somewhat more global structure. Finally, the visualization for $\sigma = 15$ is depicted in Figures 3(c), (f), where only big clusters can be perceived. The effect of σ can be summarized as follows: For a value of 1, the cluster representation is very similar to the U-Matrix, which is the method relying mostly on local differences. With higher values of σ, the kinds of perceived cluster structures gradually shift from local to global. The choice of σ has a deep impact on this visualization method and is dependant on the map size. Further experiments have shown that good choices are close to one tenth of the number of map units in the axis of the map lattice with fewer map units, but it also depends on the desired level of granularity.

Finally, we investigate the influence of the type of neighborhood function on the visualization. The examples in this paper so far are all performed with Gaussian kernels. Surprisingly, the differences to the inverse function and cut-off Gaussian kernel are so minimal that they are hardly distinguishable. The only exception is the bubble function, which is actually a very unusual choice for a neighborhood kernel during training. Since all the map units are treated equally within the sphere of this radius, and nodes on the borders of this circle are not weighted less than near the center, the visualization is harder to interpret than the other kernels. During training, cluster structures are introduced that are not present in the data set. We find that the bubble function is not appropriate for this kind of visualization, and conclude that the neighborhood kernel should be a continuous function.

6 Conclusion

In this paper, we have introduced a novel method of displaying the cluster structure of Self-Organizing Maps. Our method is distantly related to the U-Matrix. It is based on the neighborhood kernel function and on aggregation of distances in the proximity of each codebook vector. It requires a parameter σ that determines the smoothness and the level of detail of the visualization. It can be displayed either as a vector field as used in flow visualizations or as a plot that highlights the cluster borders of the map. In the former case, the direction of the most similar region is pointed to by an arrow. Our experiments have shown that this method is especially useful for maps with high numbers of units and that the choice of the neighborhood kernel is not important (as long as it is continuous), while the neighborhood radius σ has a major impact on the outcome.

Acknowledgements

Part of this work was supported by the European Union in the IST 6. Framework Program, MUSCLE NoE on Multimedia Understanding through Semantics, Computation and Learning, contract 507752.

References

1. T. Kohonen. *Self-Organizing Maps, 3rd edition.* Springer, 2001.
2. J. Lampinen and E. Oja. Clustering properties of hierarchical self-organizing maps. *Journal of Mathematical Imaging and Vision*, 2(2–3):261–272, 1992.

3. M. Oja, S. Kaski, and T. Kohonen. Bibliography of self-organizing map (SOM) papers: 1998-2001 addendum. *Neural Computing Surveys*, 3:1–156, 2001.
4. E. Pampalk, A. Rauber, and D. Merkl. Using smoothed data histograms for cluster visualization in self-organizing maps. In *Proc. Intl. Conf. on Artifical Neural Networks (ICANN'02)*, Madrid, Spain, 2002. Springer.
5. G. Pölzlbauer, A. Rauber, and M. Dittenbach. A visualization technique for self-organizing maps with vector fields to obtain the cluster structure at desired levels of detail. In *International Joint Conference on Neural Networks (IJCNN2005)*, Montral, Canada, 2005.
6. A. Rauber and D. Merkl. Automatic labeling of self-organizing maps: Making a treasure-map reveal its secrets. In *Pacific-Asia Conference on Knowledge Discovery and Data Mining (PAKDD'99)*, Bejing, China, 1999. Springer.
7. A. Ultsch. Maps for the visualization of high-dimensional data spaces. In *Proc. Workshop on Self organizing Maps*, Kyushu, Japan, 2003.
8. A. Ultsch. U*-matrix: a tool to visualize clusters in high dimensional data. Technical report, Departement of Mathematics and Computer Science, Philipps-University Marburg, 2003.
9. J. Vesanto. *Data Exploration Process Based on the Self-Organizing Map*. PhD thesis, Helsinki University of Technology, 2002.

Visualization of Cluster Changes by Comparing Self-organizing Maps

Denny[1,2] and David McG. Squire[2]

[1] Faculty of Computer Science, University of Indonesia, Indonesia
denny@cs.ui.ac.id
[2] School of Computer Science and Software Engineering,
Faculty of Information Technology, Monash University, Australia
David.Squire@csse.monash.edu.au

Abstract. In this paper we introduce Self-Organizing Map-based techniques that can reveal structural cluster changes in two related data sets from different time periods in a way that can explain the new result in relation to the previous one. These techniques are demonstrated using a real-world data set from the World Development Indicators database maintained by the World Bank. The results verify that the methods are capable of revealing changes in cluster strucure and membership, corresponding to known changes in economic fortunes of countries.

1 Introduction

In today's fast-moving world, organizations need knowledge of change so that they can quickly adapt their strategies. If an organization has devised marketing strategies based on a clustering of the past year's customer data, it is important to know if the current year's clustering differs from the last, in order to review, and perhaps revise, those strategies. Knowing what has changed, particularly if this has not been discovered by competitors, would be a major advantage [1].

Simply clustering a new data set does not, of itself, solve this problem:. The user must be able to relate the new clustering result to the previous one. It is difficult simply to compare cluster centroids obtained by a using a k-means technique—particularly if the number of clusters has changed—or to compare dendrograms obtained by hierarchical clustering algorithms. This is particularly problematic if the organization has already implemented a strategy based on an earlier clustering result. If users cannot relate new clustering results to older ones, it is difficult to revise existing strategies. Therefore, methods that can relate and contrast new clustering results with earlier ones are needed.

Here we consider various types of cluster changes, such as migration of individuals between clusters, the introduction of new clusters, and the disappearance of clusters. We introduce Self-Organizing Map (SOM) based techniques that can reveal structural cluster changes in two related data sets from different time periods in a way that can explain the new result in relation to the previous one.

2 Change Detection

Change mining, in general, can be categorized as change point detection, or change mining via the comparison models obtained via data mining. In change point detection, finding the time points at which something changed is more important than discovering the causes of the changes. Approaches to change point detection include fitting segmented models [2], where change points are defined as the points between consecutive segments. This has been applied to traffic data. Another approach is to compare values estimated using a learned model to actual observations [3], where large moving-average deviation is interpreted as evidence of change. This has been applied successfully to stock market data.

Change detection via the comparison of learned models is more closely related to the approach we presented. Since a data mining model is designed to capture specific characteristics of the data set, changes in the underlying data sets can be detected by comparing the models learned from the data sets [4]. In the field of association rule discovery (ARD), "emerging patterns" have been defined as rules the supports of which increase significantly from one data set to another [5]. Others have compared decision trees learned with two related data sets. Several C4.5-based variants of this technique have been introduced [6].

Techniques to detect two kinds of change using SOMs have been introduced: changes in map structure, and changes in the mapping of data vectors. A dissimilarity measure for two maps has been proposed, based on the expected value of distances between pairs of representative data points on both maps [7]. This approach can determine how much two maps differ, but it cannot pinpoint the differences. The similarity between two data sets in terms of a SOM can be shown using data hit histograms [8], that show the frequency with which data vectors are mapped to nodes. This can indicate changes in the data mapping, but it is difficult to interpret these simply by comparing the data hit histograms. If a vector is mapped into a dense area of the SOM, a small change in the data may cause it to be mapped to a different node. In a spare area, however, the same magnitude of change might not cause a different mapping. Another drawback is that it cannot identify the origin of the 'migrants' in the previous map.

3 Self-organizing Maps

A SOM is an artificial neural network that performs unsupervised competitive learning [9]. Importantly, SOMs allow the visualization and exploration of a high-dimensional data space by non-linearly projecting it onto a lower-dimensional manifold, most commonly a 2-D plane [10]. Artificial neurons are arranged on a low-dimensional grid. Each neuron has an n-dimensional prototype vector, m_i, also known as a weight or codebook vector, where n is input data dimensionality. Each neuron is connected to neighbouring neurons, determining the topology of the map. In the map space, neighbours are equidistant.

SOMs are trained by presenting a series of data vectors to the map and adjusting the prototype vectors accordingly. The prototype vectors are initialized to differing values, often randomly. The training vectors can be taken from the data set in random order, or cyclically. At each training step t, the BMU (Best Matching Unit) b_i for training data vector x_i, i.e. the prototype vector m_j closest to the training data vector x_i, is selected from the map according to Equation 1:

$$\forall j, \qquad \|x_i - m_{b_i}(t)\| \leq \|x_i - m_j(t)\|. \qquad (1)$$

The prototype vectors of b_i and its neighbours are then moved closer to x_i:

$$m_j(t+1) := m_j(t) + \alpha(t) h_{b_{ij}}(t)[x_i - m_j(t)], \qquad (2)$$

where $\alpha(t)$ is the learning rate and $h_{b_{ij}}(t)$ is the neighbourhood function (often Gaussian) centered on b_i. Both α and the radius of $h_{b_{ij}}$ decrease after each step. SOMs have been shown to cope with large, high-dimensional data sets.

The final orientation of the SOM is sensitive to the initial values of the prototype vectors and the sequence of training vectors [10]. Various training runs with the same data can produce rotated or inverted maps, since node indices are not related to initial prototype vector values.

3.1 Clustering of SOMs

Clustering is often used to simplify dealing with the complexities of real, large data sets. For example, it may be easier to devise marketing strategies based on groupings of customers sharing similar characteristics because the number of groupings/clusters can be small enough to make the task manageable.

Two kinds of clustering methods based on the SOM have been introduced: direct clustering and two-level clustering (hybrid). In direct clustering, each map unit is treated as a cluster, its members being the data vectors for which it is the BMU. This approach has been applied to market segmentation [11]. A disadvantage is that the map resolution must match the desired number of clusters, which must be determined in advance. In contrast, in two-level clustering the units of a trained SOM are treated as 'proto-clusters' serving as an abstraction of the data set [12]. Their weight vectors are clustered using a traditional clustering technique, such as k-means, to form the final clusters. Each data vector belongs to the same cluster as its BMU. Adding an extra layer simplifies the clustering task and reduces noise, but may yield higher distortion.

4 Visualizing Cluster Change Using SOMs

We now propose training and visualization techniques that allow cluster changes such as migration of individuals between clusters, the introduction of new clusters, and the disappearance of clusters to be detected and interpreted.

4.1 Training the Maps

We consider two data sets, the first containing data from period t and the second data from period $t+1$. For maps trained using different data sets to be comparable, their orientations must be the same. As seen in §3, this is sensitive to the initial values of the prototype vectors and the sequence of training vectors. To preserve the orientation of the map, some data points can be mapped onto fixed coordinates in map space [10]. However, this can distort the map, so that it does not follow the distribution of the data set. This approach thus cannot be used to detect changes in cluster structure. Consequently, we propose a joint method to preserve the orientations of the maps. First, a map trained with the first data set is used as the initial map for the second. The orientation of the second map will thus match that of the first. Secondly, batch, rather than sequential, training is used: all training vectors are presented to the map and the prototype vectors are updated 'simultaneously' using averages. There is thus no sequence of training vectors that can change map orientation during training. In summary:

1. Normalize both data sets using the same techniques and parameters
2. Initialize a SOM for the first data set
3. Train the SOM using the first data set
4. Initialize a SOM for the second data set using the trained first SOM
5. Train the second SOM using the second data set
6. Map data vectors from each data set to the trained maps
7. Cluster both maps using the k-means clustering algorithm.

Since k-means is sensitive to the initial cluster centroids and can get trapped in local minima, multiple runs of k-means are performed and the optimal result is chosen for each number of clusters. The optimal clustering result for different numbers of clusters is selected using the Davies-Bouldin index [13].

4.2 Visualizing Changes in the Maps

It is not possible to compare the prototype vectors of the maps based on unit location in map space, since a given unit might be represent a different part of the data space in each map. This can be caused by SOM sensitivity or a change of data distribution. Other techniques are needed for linking the maps.

Linking the Maps Using Colours. The units of the second map can be labelled in terms of the units in the first. The units of the first map are labelled using indices from 1 to n, as shown in the left map in Figure 1. For each prototype vector x in the second map, the BMU of x in the first map is found, and the unit labelled using the index of the BMU, as shown in the right map in Figure 1. The second map can thus be explained in terms of the first map.

It is difficult to interpret or detect changes from this visualization, but it can be used as a basis for others. Rather than using the index of the BMU in the first map, its *colour* can be used. For example, if the first map is coloured using a different colour for each cluster, as shown in the top-left map in Figure 2, the second map, as shown in the top-right map, can be coloured using this technique.

Visualizing Changes of Cluster Structure. Three map visualizations are produced, as shown in Figure 2. The first (top left) illustrates the clustering result

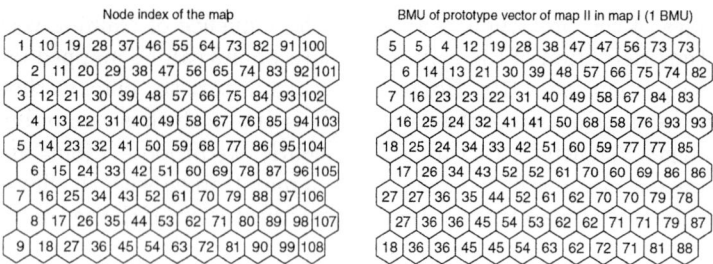

Fig. 1. The BMUs of the prototype vector of the second map

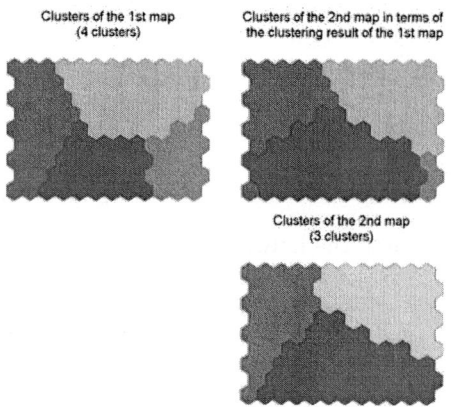

Fig. 2. The visualization of the clusters of maps

of the first map, acquired using two-level clustering. The second (top right) is a visualization of the second map in terms of the first clustering result, i.e. the second map coloured using the cluster borders of the first map in the data space. The last (bottom right) is a visualization of the independent clustering result of the second map.

This visualization can show the changes of data distribution in the clusters. If the cluster area becomes larger in the second map, it means that more data is assigned to that cluster in the second data set than in the first. For example, in Figure 2, the light green cluster shrank, the blue cluster grew, and the magenta cluster almost vanished in the second data set.

The independent clustering result of the second map can differ from the clustering result of the second map visualized using of the clustering result of the first. However, since both visualizations are based on the same map, the units are linked by position. These maps are not linked by colours, as the independent clustering result have a different number of clusters or different cluster colours. In Figure 2, it can be seen that the second map has only three clusters: the magenta cluster from the first data set has indeed vanished.

5 Experiments

Two types of data set were used in these experiments. Synthetic data sets were used to evaluate the ability of the proposed technique to visualize known cluster changes, such as migration of individuals between clusters, the introduction of new clusters, and the disappearance of clusters. Experiments on these data sets showed that the proposed approach can reveal changes of distribution, detect new clusters, and indicate the disappearance of clusters. Full descriptions of these experiments can be found in [14].

A data set from the World Development Indicators (WDI) database[1] from the World Bank was used to test the approach using real-world data. The WDI data is multi-variate time-series data with 574 indicators covering 205 countries, from 1960 to 2003. This experiment clustered countries based on selected indicators that reflect different aspects of development. Cluster changes from one period to another period were then visualized.

Inspection of the WDI data revealed many missing values. First, almost all indicators are not recorded for every year, since this would not be cost-effective. Secondly, the same indicator might be recorded at different years for different countries; therefore, comparing year-by-year is not possible. Moreover, some countries started recording the indicators later than others. In fact, some countries did not record some indicators at all.

An initial set of indicators that reflect different aspects of development were selected based on the work of Kaski and Kohonen [15] and the Millennium Development Goal of the United Nations[2]. This set of indicators was then filtered based on the availability of the indicators in terms of the missing values described earlier. The resultant set of indicators is shown in Table 1.

Table 1. Selected Indicators

Age dependency ratio	Mortality rate, infant	Birth rate, crude
Mortality rate, under-5	Daily newspapers	Physicians
Death rate, crude	Population ages 0-14	Illiteracy rate, adult female
Population growth	Illiteracy rate, adult total	Pupil-teacher ratio, primary
Immunization, measles	School enrollment, primary	Inflation, consumer prices
School enrollment, primary, female	Inflation, food prices	School enrollment, secondary
School enrollment, secondary, female	Labor force, children 10-14	Labor force, female
School enrollment, tertiary	Labor force, total	Televisions per 1,000 people
Life expectancy at birth		

Although SOMs can handle some missing values, too many can affect the map training, since it may disturb the ordering process [15]. Therefore, the several

---
[1] http://www.worldbank.org/data/onlinedatabases/onlinedatabases.html
[2] http://www.un.org/millenniumgoals/

strategies were used to handle missing values. Yearly values were grouped for 10-year periods, and latest available value available used. Indicators with missing values for more than 1/3 of the countries in the chosen periods were removed. Since some countries start recording late, or did not record some indicators at all, countries for which more than 1/3 of the values are missing in all periods were removed from the training data set. Further details of this preprocessing can be found in [14].

The 1980s and the 1990s were chosen for the experiments because they are the most complete and recent data sets. Country names were replaced with 3-digit ISO 3166 country codes[3] for visualization purposes. This research was implemented using SOM Toolbox[4] , MATLAB, and MS Access.

6 Results and Discussion

Figure 3 shows the mapping of 1980s data points onto the 1980s map. If countries are mapped nearby, they have similar characteristics, in this case similar development status. For example, the bottom-right corner consists of OECD (Organization for Economic Co-operation and Development) and developed countries, such as the United States of America, Australia, and Japan.

The visualization of the 1990s map in terms of the clustering result of the 1980s map (Figure 4, top right), shows that the blue cluster in the bottom-left corner of the 1980s map disappeared, while the magenta cluster grew.

The positions of these regions can be linked to the mapping in Figure 3. The magenta region consists of the OECD countries. Since the magenta region grew in the 1990s map, it can be said that more countries in the 1990s had similar development status to that of OECD countries in the 1980s.

The missing blue cluster contains four South American countries: Brazil, Argentina, Nicaragua, and Peru. Figure 5 shows the mapping of the countries from the 1990s data set onto the first map. In this map, the new locations of these countries can seen: all have moved towards OECD countries, except Nicaragua that has moved towards African nations. In the 1980s, these countries were suffering economic difficulties due to debt crisis—this period is known as the "lost decade" for many South American countries. However, South America was the world's second fastest-growing region between 1990 and 1997 [16]. This explains the missing cluster in the 1990s.

The proposed visualization methods have been tested with both synthetic and real-world data sets. The results show that they can reveal lost clusters, new clusters, and changes in the size of clusters. The methods can explain the new clustering result in terms of the previous clustering results. If there is a new cluster in the second data set, one can detect which cluster this new cluster came from in the first data set.

---

[3] http://www.niso.org/standards/resources/3166.html
[4] MATLAB 5 library freely available from http://www.cis.hut.fi/projects/somtoolbox/

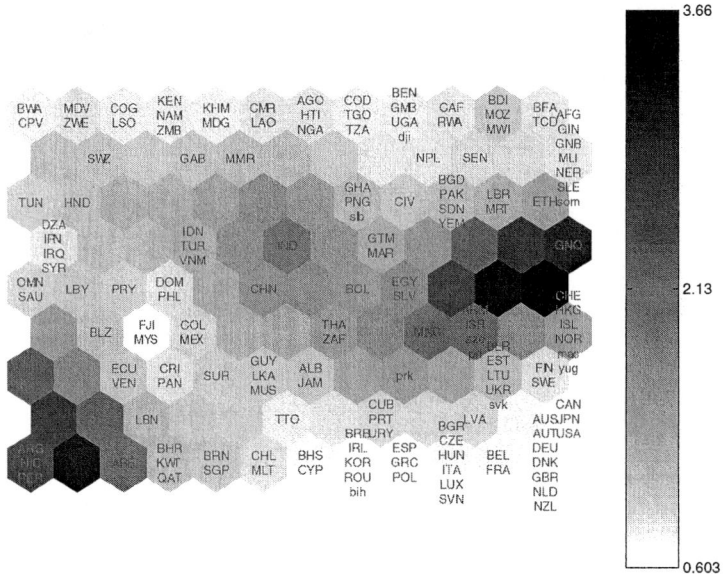

Fig. 3. The mapping of the 1980s data set onto the 1980s map

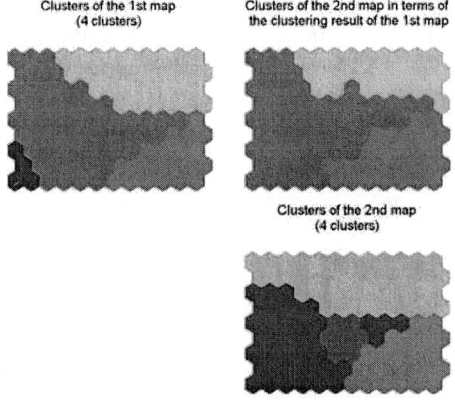

Fig. 4. The visualization of the clusters of maps of the "Decades-80-90" data sets

The visualization of the second map using the cluster borders of the first map can show changes in the data distribution in terms of the previous clustering. It can indicate if clusters disappear, but cannot show new clusters.

Comparing the independent clustering of the second map with the clustering of the first map is not easy, since they are not linked. However, the visualization of the second map in terms of the clustering of the first map can be used to relate the independent clustering of the second map with the clustering result of the first map. This comparison can reveal new or missing clusters.

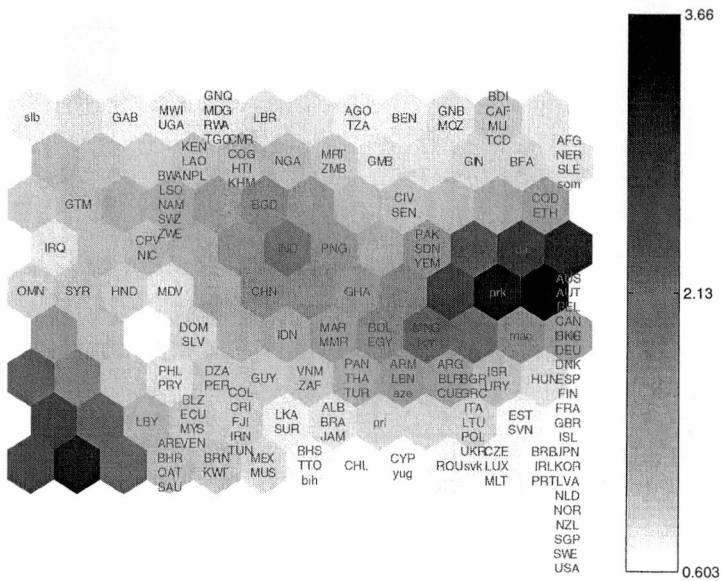

Fig. 5. The mapping of the 1990s data set onto the 1980s map

The efficiency of two-level clustering allows for interactive data exploration: one can experiment with any number of clusters, without having to recluster the whole data set. Only the SOM prototype vectors need to be clustered.

7 Conclusion

The proposed SOM-based visualization methods are capable of revealing various types of cluster changes in two related data sets from different time periods. They are capable of explaining the clustering result of the second data set in terms of the clustering result of the first data set, by using colour and positional linking. They can show structural changes such changes in distributions, new clusters, and missing clusters. Experiments using a real-world data set have shown that the methods are capable of indicating actual changes such as the the change in economic fortunes of South American countries between the 1980s and 1990s.

References

1. Fayyad, U.: Data mining grand challenges. In: Proceedings of the 8th Pacific-Asia Conference on Advances in Knowledge Discovery and Data Mining (PAKDD 2004). Number 3056 in Lecture Notes in Computer Science, Sydney, Australia, Springer-Verlag (2004) 2 (keynote speech).

2. Guralnik, V., Srivastava, J.: Event detection from time series data. In: Proceedings of the fifth ACM SIGKDD international conference on Knowledge discovery and data mining, ACM Press (1999) 33–42
3. Yamanishi, K., Takeuchi, J.: A unifying framework for detecting outliers and change points from non-stationary time series data. In: Proceedings of the eighth ACM SIGKDD international conference on Knowledge discovery and data mining, ACM Press (2002) 676–681
4. Ganti, V., Gehrke, J., Ramakrishnan, R.: A framework for measuring changes in data characteristics. In: Proceedings of the Eighteenth ACM SIGACT-SIGMOD-SIGART Symposium on Principles of Database Systems, May 31 - June 2, 1999, Philadelphia, Pennsylvania, ACM Press (1999) 126–137
5. Dong, G., Li, J.: Efficient mining of emerging patterns: discovering trends and differences. In: Proceedings of the fifth ACM SIGKDD international conference on Knowledge discovery and data mining, ACM Press (1999) 43–52
6. Liu, B., Hsu, W., Han, H.S., Xia, Y.: Mining changes for real-life applications. In Kambayashi, Y., Mohania, M.K., Tjoa, A.M., eds.: Data Warehousing and Knowledge Discovery, Second International Conference, DaWaK 2000, London, UK, September 4-6, 2000, Proceedings. Volume 1874 of Lecture Notes in Computer Science., Springer (2000) 337–346
7. Kaski, S., Lagus, K.: Comparing Self-Organizing Maps. In von der Malsburg, C., von Seelen, W., Vorbrüggen, J.C., Sendhoff, B., eds.: Proceedings of ICANN96, International Conference on Artificial Neural Networks, Bochum, Germany, July 16–19. Lecture Notes in Computer Science, vol. 1112. Springer, Berlin (1996) 809–814
8. Vesanto, J.: SOM-based data visualization methods. Intelligent Data Analysis **3** (1999) 111–126
9. Kohonen, T.: Self-organized formation of topologically correct feature maps. Biological Cybernetics **43** (1982) 59–69
10. Kohonen, T.: Self-Organizing Maps. Volume 30 of Springer Series in Information Sciences. Springer, Berlin, Heidelberg (2001) (Third Edition 2001).
11. Dolnicar, S.: The use of neural networks in marketing: market segmentation with self organising feature maps. In: Proceedings of WSOM'97, Workshop on Self-Organizing Maps, Espoo, Finland, June 4–6. Helsinki University of Technology, Neural Networks Research Centre, Espoo, Finland (1997) 38–43
12. Vesanto, J., Alhoniemi, E.: Clustering of the Self-Organizing Map. IEEE Transactions on Neural Networks **11** (2000) 586–600
13. Davies, D.L., Bouldin, D.W.: A cluster separation measure. IEEE Transactions on Pattern Analysis and Machine Intelligence **1** (1979) 224–227
14. Denny: Visualizations of cluster changes by comparing self-organizing maps. Master's minor thesis, School of Computer Science and Software Engineering, Monash University, 900 Dandenong Road Caulfield, Victoria 3145, Australia (2004)
15. Kaski, S., Kohonen, T.: Structures of welfare and poverty in the world discovered by the Self-Organizing Map. Report A24, Helsinki University of Technology, Faculty of Information Technology, Laboratory of Computer and Information Science, Espoo, Finland (1995)
16. Altbach, E.: Growth and regional integration in Latin America: will Japan miss the boat? Report 41, Japan Economic Institute (JEI), Washington, DC (1998)

An Incremental Data Stream Clustering Algorithm Based on Dense Units Detection

Jing Gao[1], Jianzhong Li[2], Zhaogong Zhang[2], and Pang-Ning Tan[1]

[1] Dept. of Computer Science & Engineering, Michigan State University,
East Lansing, MI 48824-1226 USA
{gaojing2, ptan}@cse.msu.edu
[2] Dept. of Computer Science & Technology, Harbin Institute of Technology,
Harbin, 150001 China
{lijz, zhangzhaogong}@mail.banner.com.cn

Abstract. The data stream model of computation is often used for analyzing huge volumes of continuously arriving data. In this paper, we present a novel algorithm called DUCstream for clustering data streams. Our work is motivated by the needs to develop a single-pass algorithm that is capable of detecting evolving clusters, and yet requires little memory and computation time. To that end, we propose an incremental clustering method based on dense units detection. Evolving clusters are identified on the basis of the dense units, which contain relatively large number of points. For efficiency reasons, a bitwise dense unit representation is introduced. Our experimental results demonstrate DUCstream's efficiency and efficacy.

1 Introduction

In recent years, data stream model is motivated by many applications that continuously generate huge amount of data at unprecedented rate [1]. In this paper, we will focus on the stream clustering problem, which is a central task of data stream mining.

Recently this problem has attracted much attention. O'Callaghan et. al. [2] study the k-median problem over data streams. Aggarwal et. al. [3] present a framework of clustering evolving data streams, which analyzes the clusters over different portions of the stream. However this framework can not give online response of queries of macro clusters. Nasrouni et. al. [7] design an immune system learning model to find evolving clusters in data streams. But this algorithm is not space and time efficient due to the use of AIS model.

In static data environment, many clustering algorithms have been designed [4, 5, 6]among which grid-based clustering is an efficient method. This approach partitions the data space into many units and perform clustering on these units [6]. Recently, Park et.al. [8] propose a statistical grid-based method which identifies evolving clusters as a group of adjacent dense units in data stream environments. But their work is focusing on partitioning dense units and maintaining their distributions.

In this paper, we propose an efficient data stream clustering algorithm DUCstream. We partition the data space into units and only keep those units which contain relatively large number of points. An incremental clustering algorithm is presented based on these dense units. The clustering results are represented by bits to reduce the memory requirements. Extensive experiments indicate that our framework can obtain high-quality clustering with little time and space.

2 Problem Statement

We begin by defining the stream clustering problem in a formal way.

Suppose S is a d-dimensional numerical space. For each dimension, we partition it into non-overlapping rectangular units. The density of a unit u is defined as the number of points that belong to it, i.e. $\text{den}(u) = |v_i|v_i \in u|$. The relative density of u is defined as follows: $\text{rel\_den}(u) = \text{den}(u)/|D|$, where $\text{den}(u)$ is the density and D is the data set we observe. If u's relative density is greater than the density threshold γ, then u is referred to as a dense unit. As defined in [6], a cluster is a maximal set of connected dense units in d-dimensions.

A data stream is a set of points from data space S that continuously arrives. We assume that data arrives in chunks $X_1, X_2, \ldots, X_n, \ldots$, at time stamps $t_1, t_2, \ldots, t_n, \ldots$. Each of these chunks fits in main memory. Suppose that each chunk contains m points, and the current time stamp is t. We use $\text{den}(u)$ to denote the overall density of u with respect to the t chunks that has been seen so far. The density of u with respect to the i-th chunk is denoted as $\text{den}^i(u)$. The relative density of a unit u is $\text{rel\_den}(u) = \text{den}(u)/(mt)$. If u's relative density is greater than the density threshold γ, then u is referred to as a dense unit at time t. At time t, the clustering result R is all the clusters found in the t chunks of data visited so far. Our goal is to compute the clustering results when the data stream continuously arrives, i.e. obtain $R_1, R_2, \ldots, R_n, \ldots$, where R_i represents the result of clustering $X_1, X_2, \ldots, X_i$.

3 Algorithm Description

3.1 Basic Idea

In brief, we will find the dense units and cluster these units. First, we consider what units should be maintained thus introduce the concept of local dense units.

Suppose that each chunk contains m points, and the current time stamp is t. If unit u begins to be maintained at time i, the local relative density of u is $\text{loc\_den}(u) = \text{den}(u)/(m(t-i+1))$, where $\text{den}(u)$ is the density of u. If u's local relative density is greater than the density threshold γ, then u is referred to as a local dense unit at time t. The following proposition holds on.

Proposition 1. *For any dense unit u at time t, it must be recorded as a local dense unit at time $i(1 \leq i \leq t)$.*

Proof. Suppose that a dense unit u is not recorded as a local dense unit at time $1, 2, \ldots, t$ and each chunk contains m points. We recall that the number of points that belong to u in the i-th chunk is $\text{den}^i(u)$. Then $\text{den}^i(u) < \gamma m$. Therefore at time t, $\text{den}(u) = \sum_{i=1}^{t} \text{den}^i(u) < \gamma m t$. so u is not a dense unit at current time, contrary to the hypothesis. The conclusion is accordingly established.

In other words, local dense units are candidate dense units, which may become dense in the future. Therefore we maintain all the local dense units and pick up dense units among them to do clustering. We call this process dense units detection. The following proposition analyzes the error of our algorithm.

Proposition 2. *Assume that a certain unit u's density gradually increases so that its density with respect to the i-th chunk is ipm where m is the number of points belonging to each chunk, p is a constant from 0 to 1 that indicates the amount of increase. At the time from $(1 + \sqrt{1 + 8\gamma/p})/2$ to γ/p, this unit can not be successfully detected as a dense unit.*

Proof. According to the definition of local dense units, we will not keep unit u as long as $ipm < \gamma m$, i.e., $i < \gamma/p$. However, when its density reaches γm, u becomes a dense unit at that time. Suppose at time k, u's density is equal to γm. Then $\sum_{i=1}^{k} ipm = \gamma m$, i.e., $\frac{k(k-1)}{2}p = \gamma$. It can be derived that $k = (1 + \sqrt{1 + 8\gamma/p})/2$. Therefore the time range when error occurs is as stated.

Another issue is how to get the right results with little time and memory. To lighten the computational and storage burden, we propose to represent the clustering results in bits. Suppose that the dense units are sorted by their density and each of them is assigned a unique id. The Clustering Bits (CB) of a cluster r is a $0-1$ bit string $a_n, \ldots, a_1$, where a_i is a bit and n is the number of dense units. $a_i = 1$ if and only if the i-th dense unit is in cluster r, otherwise $a_i = 0$. We can benefit from the use of Clustering Bits in both the time and space usage.

3.2 Stream Clustering Framework

Based on the above two points, we summarize our stream clustering algorithm in Figure (1). We refer to this algorithm as DUCstream (**D**ense **U**nits **C**lustering for data **stream**). The data structures used in the algorithm include: L, the local dense units table; Q_a, the added dense units id list; Q_d, the deleted dense units id list; R_i, the clustering result $\{c_1, \ldots, c_s\}$ at time stamp i.

The important components in this framework entail:

1. map_and_maintain(X_i, L): This procedure maps each data point in X_i into the corresponding unit. For one of these units u, if it is in L, update the corresponding item, otherwise if u is a local dense unit, insert it into L. After that, scan L once and decide Q_a and Q_d.
2. create_clusters(Q): We use a depth-first search algorithm to create clusters as described in [6]. They identify the clusters as the connected components of the graph whose vertices represent dense units and whose edges correspond to the common faces between two vertices.

```
DUCstream Algorithm:
Input: Data chunks $X_1, X_2, \ldots, X_n, \ldots$
Output: Clustering results $R_1, R_2, \ldots, R_n, \ldots$
Method:
1. Create a new empty table $L$;
2. $(L, Q_a, Q_d)$=map_and_maintain$(X_1, L)$;
3. $R_1$=create_clusters$(Q_a)$;
4. i=2;
5. Repeat until the end of the data stream
5.1 $(L, Q_a, Q_d)$=map_and_maintain$(X_i, L)$;
5.2 $R_i$=update_clusters$(R_{i-1}, Q_a, Q_d)$;
5.3 $i = i + 1$;
```

Fig. 1. SemiSOD algorithm framework

3. update_clusters(R_{i-1}, Q_a, Q_d): We get the clustering result R_i in an incremental manner stated as follows.

For each added dense unit u, one of following occurs: **Creation:** If u has no common face with any old dense units, a new cluster is created containing u; **Absorption:** There exits one old dense unit u' such that u has common face with u', then absorb u into the cluster u' is in; **Mergence:** There exist multiple old dense units $w_1, w_2, \ldots, w_k (k > 1)$ that have common faces with u, then merge the clusters these dense units belong to. Absorb u into the new cluster.

For each deleted dense unit u, suppose it is contained in cluster c, we can distinguish the following cases: **Removal:** If there are no other dense units in c, i.e. the cluster becomes empty after deleting u, we remove this cluster; **Reduction:** All other dense units in c are connected to each other, then simply delete u from c; **Split:** All other dense units in c are not connected to each other, this leads to the split of cluster c.

After processing all the units in Q_a, Q_d, we can obtain the new clustering result R_i.

4 Empirical Results

The data set is KDD'99 Intrusion Detection Data, which is partitioned into chunks each consisting of $1K$ points. We first examine the time complexity of DUCstream compared with the baseline methods STREAM [2] and CluStream [3]. To make the comparison fair, we make the number of clusters all five in these algorithms. Figure (2) shows that DUCstream is about four to six times faster than STREAM and CluStream. This is attributed to our use of dense units detection, Clustering Bits and good design of incremental update algorithm.

DUCstream maintains the local dense units and current clustering results in main memory. Since the clustering results, represented by Clustering Bits, cost very little space, we only keep track of the number of local dense units to monitor the memory usage. Figure (3) demonstrates that after a certain time, a steady

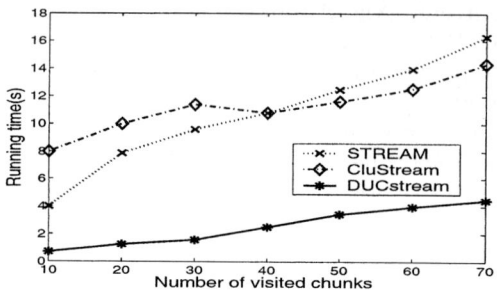

Fig. 2. Running time

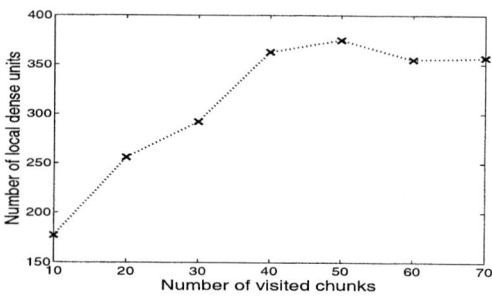

Fig. 3. Memory usage

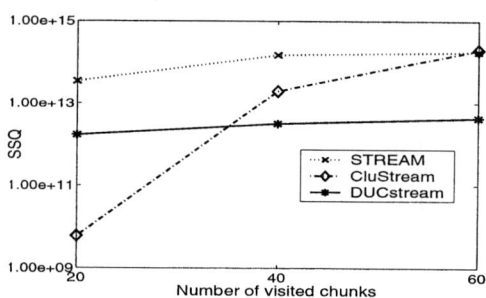

Fig. 4. Quality Comparison

state is reached as for the number of local dense units. In general, the algorithm only requires a negligible amount of memory even when the data stream size becomes sufficiently large.

We then compare DUCstream with STREAM and CluStream using the measurement SSQ, the sum of square distance. Figure (4) shows that the clustering quality of DUCstream is always better than that of STREAM because we capture the characteristics of clusters more precisely using the dense units compared with only maintaining k centers. For CluStream, it performs better when the horizon is small but the accuracy tends to be lower when the horizon becomes larger.

5 Conclusion

In this paper, we propose an efficient data stream clustering algorithm based on dense units detection. This is an incremental, one-pass density-based algorithm, which finds high-quality clusters with considerably little time and memory in the data stream environment. It discards noisy and obsolete units through dense units detection. The clustering result is updated using the changed dense units. We also introduce a bitwise clustering representation to update and store away the clustering results efficiently. Empirical results prove that this algorithm has good quality while cost surprisingly little time. The problem of finding arbitrary-shaped clusters is an interesting future work.

Acknowledgment

The work was partially supported by IRGP grant #71-4823 from the Michigan State University.

References

1. B. Babcock, S. Babu, M. Datar, R. Motwani and J. Widom. Models and issues in data stream systems. In *Proceedings of the 21st ACM Symposium on Principles of Database Systems*, pages 1–16, 2002.
2. L. O'Callaghan, A. Meyerson, R. Motwani, N. Mishra and S. Guha. Streaming-data algorithms for high-quality clustering. In *Proceedings of IEEE International Conference on Data Engineering*, pages 685–696, 2002.
3. C. Aggarwal, J. Han, J. Wang and P. S. Yu. A framework for clustering evolving data streams. In *Proceedings of the International Conference on Very Large Data Bases*, pages 81–92, 2003.
4. M. Ester, H.P. Kriegel, J. Sander and X. Xu. A density-based algorithm for discovering clusters in large spatial databases with noise. In *Proceedings of the ACM SIGKDD international conference on Knowledge discovery and data mining*, pages 226–231, 1996.
5. M. Ester, H.P. Kriegel, J. Sander, M. Wimmer and X. Xu. Incremental clustering for mining in a data warehousing environment. In *Proceedings of the International Conference on Very Large Data Bases*, pages 323–333, 1998.
6. R. Agrawal, J. Gehrke, D. Gunopulos, and P. Raghavan. Automatic subspace clustering for high dimensional data for data mining applications. In *Proceedings of the ACM International Conference on Management of Data*, pages 94–105, 1998.
7. O. Nasraoui, C. Cardona, C. Rojas and F. Gonzlez. TECNO-STREAMS: Tracking evolving clusters in noisy data streams with a scalable immune system learning model. In *Proceedings of the IEEE International Conference on Data Mining*, pages 235–242, 2003.
8. N.H. Park and W.S. Lee. Statistical grid-based clustering over data streams. *ACM SIGMOD Record*, 33(1):32–37, 2003.

Visual Interactive Evolutionary Algorithm for High Dimensional Data Clustering and Outlier Detection

Lydia Boudjeloud and François Poulet

ESIEA Recherche,
38, rue des docteurs Calmette et Guérin, 53000 Laval - France
{boudjeloud, poulet}@esiea-ouest.fr

Abstract. Usual visualization techniques for multidimensional data sets, such as parallel coordinates and scatter-plot matrices, do not scale well to high numbers of dimensions. A common approach to solve this problem is dimensionality selection. Existing dimensionality selection techniques usually select pertinent dimension subsets that are significant to the user without loose of information. We present concrete cooperation between automatic algorithms, interactive algorithms and visualization tools: the evolutionary algorithm is used to obtain optimal dimension subsets which represent the original data set without loosing information for unsupervised mode (clustering or outlier detection). The last effective cooperation is a visualization tool used to present the user interactive evolutionary algorithm results and let him actively participate in evolutionary algorithm searching with more efficiency resulting in a faster evolutionary algorithm convergence. We have implemented our approach and applied it to real data set to confirm this approach is effective for supporting the user in the exploration of high dimensional data sets.

1 Introduction

In most existing data mining tools, visualization is only used during two particular steps of the process: in one of the first steps to view the original data and in one of the last steps to view the final results. Some new methods called Visual data Mining have recently appeared [13], [15], trying to involve more significantly the user in the data mining process and using more intensively the visualization [1]. Usual visualization techniques for multidimensional data sets, such as parallel coordinates [7] or scatter-plot matrices [4] do not scale well to high dimensional data sets. For example, Figure 1 shows a subset of the Lung Cancer data set [8] with one hundred dimensions (among 12533 dimensions), the user cannot detect any pertinent information from the visualization. Even with low numbers of elements, high dimensionality is a serious challenge for current display techniques.

To overcome this problem, one promising approach is dimensionality selection [10]. The basic idea is to select some pertinent dimensions without loosing too much information and then to treat the data set in this subspace. Most of these methods focus on supervised classification and evaluate potential solutions in terms of predictive accuracy. Few works [9], deal with unsupervised classification where we do not have prior information to evaluate potential solution.

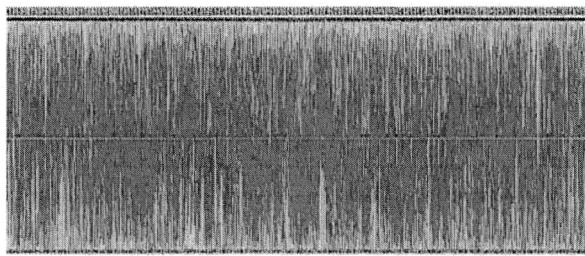

Fig. 1. One hundred dimensions of the lung cancer data set (12533 dimensions, 32 elements) with parallel coordinates

We present a semi-interactive algorithm we have developed, integrating automatic algorithm, interactive evolutionary algorithm and visualization. First evolutionary algorithm generates pertinent dimension subsets, according to the user choice (clustering or outlier detection). The data are displayed in these dimension subsets using parallel coordinates. The user can interactively choose the visualization that seems significant and selected dimension subsets are then in input of the next evolutionary algorithm generation and so on until having optimal visualization. The originality of our approach is to combine evolutionary algorithm to obtain optimal dimension subsets which represent the original data set without loosing information for unsupervised mode (clustering or outlier detection) with a visualization tool to present the user interactive evaluation and let him actively participate in evolutionary algorithm searching with resulting in a faster evolutionary algorithm convergence. We present some results obtained with several high dimensional data sets.

This paper is organized as follows. The next section describes some existing interactive evolutionary algorithms and a brief overview of dimensionality selection methods. We describe our evaluation functions for outlier detection and clustering and then we present some results obtained by our new interactive evolutionary algorithm in section 3 before the conclusion and future work.

2 Interactive Evolutionary Algorithm for Dimension Selection

Interactive Evolutionary Algorithm (IEA) can be defined [14] as an optimization method that adopts evolutionary algorithm (EA) based on subjective human evaluation. It is simply an EA technique whose fitness function is replaced by human. Evolutionary algorithms have attracted attention in the data mining and knowledge discovery processes. They are used for performing some pre-processing and post-processing steps of the knowledge discovery process and then extract high-level knowledge from data. They focus on dimension selection and pruning of a set of classifiers used as black box fitness function. If the user is involved, he contributes to a faster evolutionary algorithm convergence.

A wide number of approaches for dimension selection have been proposed [10]. Dimension selection algorithms can broadly be classified into three categories according to the search strategy used: exhaustive search, heuristic search and randomized search. Evolutionary algorithms [11] such as genetic algorithms [6] can also be used for dimension selection. We use genetic algorithm for dimension selection: the individual represents a small subset of dimensions. At first step, the initial population is ready; it is evaluated by a distance-based function for outlier detection and validity indexes for clustering and presented to the user for validation. The originality of our approach is to combine both user interactive validation and automatic validation (black box fitness function) for a fast algorithm convergence. The advantage is the proposed solutions are not biased by the user choice or automatic fitness function, but both are considered to generate next evolutionary algorithm generation.

2.1 Clustering Fitness Function

The goal of clustering is to partition a data set into subgroups such that objects in each particular group are similar and objects in different groups are dissimilar. With most of the algorithms, the user has first to choose the number of clusters. To determine number of clusters we usually use validity indexes [12] that are based on the minimization of the sum of squared distances within (SSW) the clusters and the maximization of the sum of squared distances between (SSB) the clusters. We use this technique first to find the optimal number of clusters. Then, we try to obtain (with K-means [5]) the optimal validity index values in the dimension subsets that are generated by evolutionary algorithm to optimize clustering in this dimension subset [3]. Each individual evaluation is carried out with the best index according to [12]. We use the values obtained to classify the genetic algorithm individuals. Some individuals are then visualized with the parallel coordinates and presented to the user to determine and select the most significant ones for clustering.

2.2 Outlier Detection Fitness Function

An outlier is a data subset, an observation or a point that is considerably dissimilar, distinct or inconsistent with the remainder of data. The main problem is to define this dissimilarity between objects characterizing an outlier. Typically, it is estimated by a function computing the distance between objects, the next task is to determine the

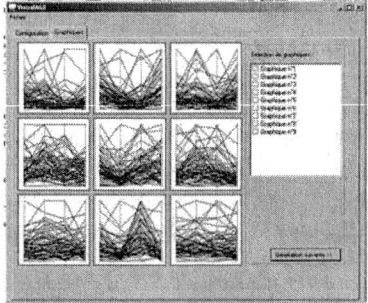 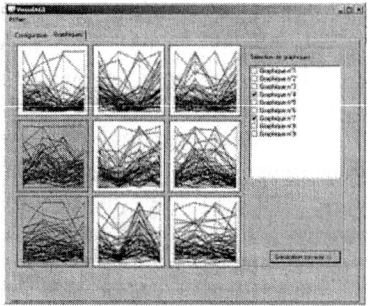

Fig. 2. Visualized generation and selected individuals

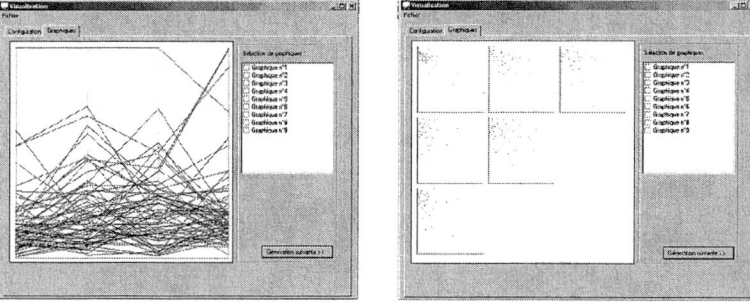

Fig. 3. Optimal solution visualization (parallel coordinates and scatter-plot matrices)

objects farthest from the mass. We choose an Euclidean distance-based outlier detection procedure as evaluation of each evolutionary individual. The procedure output is the outlier point and the distance between this point and the remainder of data in the subset of attributes (selected dimensions). We use the obtained distance to classify the genetic algorithm individuals [2]. The farthest element will be considered as an outlier element, the whole data will be visualized in dimension subsets and presented to the user for validation and to consider them for the next generation.

3 Some Results

After the user has chosen the problem (outlier detection or clustering) and data set, we have the view shown in fig.2 with 9 visualization screens of evolutionary individuals. Randomly 9 individuals are presented to the user who chooses the visualizations that seem significant. In our example (outlier detection), we want to find visualizations where we can see element different from the whole data set.

For example, visualization number 4 and 7 are selected because they contain element that has extreme values and can be outlier. The corresponding dimensions are in input of the next genetic algorithm generation. Once this step is performed we operate the standard genetic operators (crossover and mutation) to create new generations guided by user choice and so on, until we obtain optimal subset displayed using parallel coordinates and scatter-plot matrices in the final view (figure 3).

3.1 Algorithm Parameters

Our genetic algorithm starts with a population of 50 individuals (chromosomes), every individual is made of 4 genes (user-defined) because we want to allow visual interpretation. Once the whole population has been evaluated and sorted, we operate a crossover on two parents chosen randomly. Then, one of the children is muted with a probability of 0.1 and is substituted randomly for an individual in the second part of the population, under the median. We run our algorithm for 10000 generations and each 100 generations, we propose 9 randomly chosen individuals for user interactive evaluation. Our algorithm ends after a maximum number of iterations or after the user satisfaction.

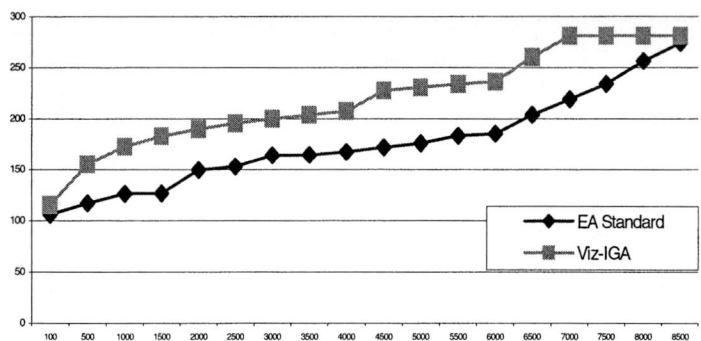

Fig. 4. Convergence of the standard EA vs. Viz-IGA

3.2 Evaluation of Convergence

We evaluate how human involvement speeds up the convergence of the EA search. Since our approach deals with subjective fitness value combined with black box fitness depending on the application task (clustering or outlier detection), we compare convergence of the evolutionary algorithm described in [2] and Viz-IGA. The role of human in Viz-IGA is to select the best candidates in the 4-D visualization, while in the GA [2] user only validates the final result. We obtain the same results in some minutes with Viz-IGA (about ten times faster than automatic GA as shown in Figure 4).

4 Conclusion and Future Work

Traditional visualization techniques for multidimensional data sets do not scale well to high dimensional data sets, the user cannot detect any pertinent information. To overcome the problem of high dimensionality, one promising approach is dimensionality selection. We have proposed to use user perception to overcome drawbacks of dimension selection process, according his choice of the unsupervised learning problem. We have implemented semi-interactive algorithm, integrating automatic algorithm, interactive evolutionary algorithm and visualization. First evolutionary algorithm generates pertinent dimension subsets without loosing too much information. Some dimension subsets are visualized using parallel coordinates and the user interactively chooses the visualization that seems significant. The selected dimensions are in input of the next evolutionary algorithm generation and so on until having optimal visualization according to the unsupervised problem.

The originality of our approach is to combine evolutionary algorithm to obtain optimal dimension subsets which represent the original data set without loosing information for unsupervised mode (clustering or outlier detection) with a visualization tool to propose the user interactive evaluation and let him actively participate in evolutionary algorithm searching with more efficiency resulting in a faster evolutionary algorithm convergence.

Of course, we have improved convergence of algorithm with the user interactive involvement, but we think that if we have other fitness function easier to compute than different black boxes fitness functions, we can optimize the algorithm.

References

1. Aggarwal C., Towards effective and interpretable data mining by visual interaction, in SIGKDD Eplorations 3(2), pp 11-22, accessed from www.acm.org/sigkdd/explorations/.
2. Boudjeloud L., Poulet F., A Genetic Approach for Outlier Detection in High-Dimensional Data-Sets, in Modeling, Computation and Optimization in Information Systems and Management Sciences, Le Thi H., Pham D.T. Eds, Hermes Sciences Publishing, 2004, 543-550.
3. Boudjeloud L., Poulet F., Attributes selection for high dimensional data clustering, to appear in proc. of International Symposium on Applied Stochastic Models and Data Analysis, ASMDA'05, May 2005, Brest, France.
4. Carr D. B., Littlefield R. J., Nicholson W. L. , Scatter-plot matrix techniques for large N, Journal of the American Statistical Association 82(398), pp 424-436, Littlefield.
5. Hartigan, J.A., Wong, M.A. (1979), A K-means clustering algorithm: Algorithm AS 136. Applied Statistics, 28, pp 126-130, 1979.
6. Holland J., Adaptation in Natural and Artificial Systems. University of Michigan Press, Ann Arbor, 1975.
7. Inselberg A., The Plane with Parallel Coordinates, Special Issue on computational Geometry, vol 1, pp69-97, 1985.
8. Jinyan L., Huiqing L., Kent ridge bio-medical data set repository, 2002, http://sdmc.-lit.org.sg/GEDatasets. accessed Dec.2004.
9. Kim Y., Nick Street W., Menczer F., Evolutionary Model Selection in Unsupervised Learning, Intelligent data analysis, 6, pp 531-556, IOS Press, 2002.
10. Liu H., Motoda H., Feature detection for knowledge discovery and data mining, Kluwer Academic Publishers, 1998.
11. Michalewicz Z., Genetic Algorithms + Data Structures = Evolution Programs, Springer-Verlag, New York, Third edition, 1996.
12. Milligan G., Cooper M. (1985), An examination of procedures for determining the number of classes in a data set, Psychometrika, vol.52, n°2, p 159-179, 1985.
13. Poulet F., Visualisation in data mining and knowledge discovery, in proc. HCP'99, 10th Mini Euro Conference "Human Centred Processes" ed. P. Lenca, pp 183-192, 1999.
14. Takagi H., Interactive Evolutionary Computation: Fusion of the Capacities of EC Optimization and Human Evaluation, Proceedings of the IEEE, Vol.89, No.9, pp.1275-1296, 2001.
15. Wong P., Visual data mining, in IEEE Computer Graphics and Application, 19(5), 20-21, 1999.

Approximated Clustering of Distributed High-Dimensional Data

Hans-Peter Kriegel, Peter Kunath, Martin Pfeifle, and Matthias Renz

University of Munich, Germany
{kriegel, kunath, pfeifle, renz}@dbs.ifi.lmu.de

Abstract. In many modern application ranges high-dimensional feature vectors are used to model complex real-world objects. Often these objects reside on different local sites. In this paper, we present a general approach for extracting knowledge out of distributed data sets without transmitting all data from the local clients to a server site. In order to keep the transmission cost low, we first determine suitable local feature vector approximations which are sent to the server. Thereby, we approximate each feature vector as precisely as possible with a specified number of bytes. In order to extract knowledge out of these approximations, we introduce a suitable distance function between the feature vector approximations. In a detailed experimental evaluation, we demonstrate the benefits of our new feature vector approximation technique for the important area of distributed clustering. Thereby, we show that the combination of standard clustering algorithms and our feature vector approximation technique outperform specialized approaches for distributed clustering when using high-dimensional feature vectors.

1 Introduction

One of the primary data mining tasks is clustering. Clustering aims at partitioning the data set into distinct groups, called clusters, while maximizing the intra-cluster similarity and minimizing the inter-cluster similarity [8]. Traditionally, the clustering algorithms require full access to the data which is going to be analyzed. All data has to be located at the site where it is processed. Nowadays, large amounts of heterogeneous, complex data reside on different, independently working computers which are connected to each other via local or wide area networks. Examples comprise distributed mobile networks, sensor networks or supermarket chains where check-out scanners, located at different stores, gather data unremittingly. Furthermore, international companies such as DaimlerChrysler have some data which are located in Europe and some data located in the US and Asia. Those companies have various reasons why the data cannot be transmitted to a central site, e.g. limited bandwidth or security aspects.

Many of these real-world distributed data sets consist of objects modeled by high-dimensional feature vectors. For instance, a starting point for applying clustering algorithms to distributed unstructured document collections is to create a vector space model, alternatively known as a bag-of-words model [13], where each document is represented by a high-dimensional feature vector. Other examples for high-dimensional feature vectors representing distributed complex objects can be found in the area of image retrieval [12], and molecular biology [4].

The requirement to extract knowledge from distributed data, without a prior unification of the data, created the rather new research area of Distributed Knowledge Discovery in Databases (DKDD). In this paper, we present a general approach which helps to extract knowledge out of high-dimensional feature vectors spread over several sites. To get specific, we demonstrate the benefits of our approach for distributed density-based clustering. Our approach tries to describe local feature vectors as accurately as possible with a certain number of granted bytes. These approximations are sent to a server site, where the global server clustering is carried out based on a suitable distance function measuring the similarity between the locally determined feature vector approximations.

The remainder of the paper is organized as follows: In Section 2, we present the related work in the area of distributed clustering. In Section 3, we explain how to form the local approximations which are sent to a central server site. Then, in Section 4, a meaningful similarity measure for the feature vector approximation is introduced. In Section 5, we demonstrate the suitability of our feature vector approximation technique and the corresponding distance function. In Section 6, we close this paper with a short summary and a few remarks on future work.

2 Related Work

Distributed Data Mining (DDM) is a dynamically growing area within the broader field of KDD. Generally, many algorithms for distributed data mining are based on algorithms which were originally developed for parallel data mining. In [10], some state-of-the-art research results related to DDM are summarized. Whereas there already exist algorithms for distributed classification and association rules, there is a lack of algorithms for distributed clustering.

In [5] the "collective hierarchical clustering algorithm" for vertically distributed data sets was proposed which applies single link clustering. In contrast to this approach, we concentrate in this paper on horizontally distributed data sets.

In [14] the authors presented a technique for centroid-based hierarchical clustering for high-dimensional, horizontally distributed data sets by merging clustering hierarchies generated locally. Unfortunately, this approach can only be applied for distance-based hierarchical distributed clustering approaches, whereas our aim is to introduce a generally applicable approach.

In [6, 7], density-based distributed clustering algorithms were presented which are based on the density-based partitioning clustering algorithm DBSCAN. The idea of these approaches is to determine suitable local objects representing several other local objects. Based on these representatives a global DBSCAN algorithm is carried out. These approaches are tailor-made for the density-based distributed clustering algorithm DBSCAN.

The goal of this paper is to introduce an approach which is generally applicable to DDM. To get specific, we demonstrate the benefits of our approach for distributed clustering algorithms. In contrast to the above specific distributed clustering approaches, our approach is not susceptible to an increasing number of local clients. It does only depend on the overall allowed transmission cost, i.e. on the number of bytes we are allowed to transmit from the local clients to a server. In order to keep these transmission cost low, we introduce in the following section a suitable client-side approximation technique for describing high-dimensional feature vectors.

3 Client-Side Approximation

The hybrid-approximation approach which we propose in this section is quite similar to the idea of the IQ-tree [2] which is an index structure especially suitable for managing high-dimensional feature vectors. First, we divide the data set into a set of partitions represented by minimum bounding rectangles (MBRs) of the points located in the corresponding region in the data space. This kind of data set approximation is further elaborated in Section 3.1. In Section 3.2, we describe each single feature vector by an approximation hierarchy where in each level of the hierarchy K more bits are used to describe the feature vector more accurately.

3.1 Data Set Approximation

The goal of this section is to find a rough description of the complete data set by means of some (flat) directory pages which conservatively approximate the complete data space. The problem of finding these MBRs is related to clustering. We are not mainly interested in the clusters themselves but rather in a partitioning of the data space into rectangular cuboids. Similar, to directory pages in an index structure, these cuboids should be formed as quadratic as possible for efficient query processing [3]. We can achieve such cuboids with only a little variation of the lengths of the edges by applying the k-means clustering algorithm [11]. Thereby the data set is approximated by k centroids, and each vector is assigned to its closest centroid. All feature vectors which are assigned to the same centroid form a cluster and are approximated by an MBR of all the vectors of this cluster. As desired, the form of these MBRs tend to be quadratic as the centroid of a cluster tends to be close to the middle of the MBR. Thus, the k-means clustering algorithm indirectly also minimizes the average length of the space diagonals of the k MBRs.

3.2 Feature Vector Approximation

After having partitioned the local data space into k clusters represented by MBRs, we express each feature vector v w.r.t. to the lower left corner of its corresponding minimum bounding rectangle $MBR_{Cluster}(v)$.

Definition 1 *Feature Vector*
Each feature v_i of a d-dimensional feature vector $v = (v_1,..., v_d)^t \in IR^d$ is represented by a sequence of bytes $<b_{i,1},..., b_{i,m}>$ where each byte consists of w bits. The feature value v_i is calculated by

$$v_i = \sum_{j=1}^{m} val(b_{i,j}) \text{, where } val(b_{i,j}) = b_{i,j} \cdot 2^{w(m-j)}$$

For clarity, we assume in this paper that each feature of a d-dimensional feature vector is represented by a byte string of length m. We will describe each feature vector by a conservative hierarchy of approximations where in each level we use some more bytes to approximate the feature vector more closely. By traversing the complete approximation hierarchy, we can reconstruct the correct feature vector.

The client first computes a byte ranking of all the bytes $b_{i,j}$ of v. Then the most meaningful bytes are transmitted to the server along with positional information of the

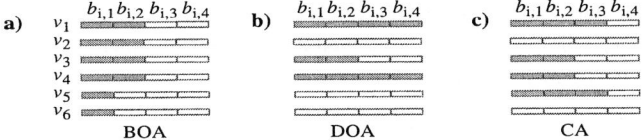

Fig. 1. Approximation techniques. ($L = 10$ Bytes)

bytes. By means of this additional positional information, the server can construct an accurate server side approximation of v.

Definition 2 *Ranking and Approximation Function*
Let W be the set of all byte sequences of length $m \cdot d$. Let $v = (v_1, ..., v_d)^t \in IR^d$ be a feature vector where each feature v_i is represented by a sequence of bytes $<b_{i,1},...,b_{i,m}>$. Then, we require a byte *ranking function* $f_{rank}: IR^d \to W$ and a feature vector *approximation function* $f_{app}: W \times \{0...m \cdot d\} \to [IR \times IR]^d$ to have the following properties:

- $f_{rank}(v) = <b_1, ..., b_{m \cdot d}>$ where $b_l = b_{\pi(i,j)}$, for a bijective ranking function $\pi_{rank}: \{1...d\} \times \{1...m\} \to \{1...m \cdot d\}$
- $f_{app}(f_{rank}(v), 0) = MBR_{Cluster}(v)$, $f_{app}(f_{rank}(v), L_1) \subseteq f_{app}(f_{rank}(v), L_2)$ iff $L_1 \geq L_2$, and $f_{app}(f_{rank}(v), m \cdot d)) = v$

After having received a certain number of L bytes the server can compute the approximation area $A = f_{app}(f_{rank}(v), L)$. In the following subsections, we present three approximation techniques of high-dimensional feature vectors, i.e. the *byte-oriented*, the *dimension-oriented*, and the *combined* approximation technique (cf. Figure 1). All three approaches fulfill rather obviously the properties stated in Definition 2. Nevertheless, they differ in the way they actually define the ranking and approximation functions. In the following, we assume that the cluster MBR of a feature vector v $MBR_{Cluster}(v) = [MBR\_l_1 \times MBR\_u_1] \times ... \times [MBR\_l_d \times MBR\_u_d]$ has already been transmitted to the server. Furthermore, we assume that v is defined according to Definition 1.

3.2.1 Byte-Oriented Approximation (BOA)

As the first bytes of each feature contain the most significant information, we rank the bytes $b_{i,j}$ by means of the bijective function $\pi: \{1...d\} \times \{1...m\} \to \{1...m \cdot d\}$ according to their j-positions, i.e. $\pi(i,j) < \pi(i',j')$ iff $(j < j')$ or $(j = j'$ and $i < i')$.

The server computes the approximation area $a = f_{app}(f_{rank}(v), L) = [l_1, u_1] \times ... \times [l_d, u_d]$ as follows:

$$l_i = MBR\_l_i + \sum_{j=1}^{m} \begin{cases} val(b_{i,j}), & \text{if } (\pi(i,j) \leq L) \\ 0, & \text{else} \end{cases}$$

$$u_i = min\left(MBR\_u_i, MBR\_l_i + \sum_{j=1}^{m} \begin{cases} val(b_{i,j}), & \text{if } (\pi(i,j) \leq L) \\ (2^w - 1) \cdot 2^{w(m-j)}, & \text{else} \end{cases}\right)$$

3.2.2 Dimension-Oriented Approximation (DOA)

In the above approach, we considered the first bytes of each dimension to build the approximation. In this approach, we select significant dimensions and then transmit all bytes of the selected features to the server. The dimension oriented approximation approach (cf. Figure 1b) selects $\lceil L/m \rceil$ dimensions i having the highest values v_i. Thus, we rank the bytes $b_{i,j}$ by means of the bijective function $\pi: \{1...d\} \times \{1...m\} \to \{1...m \cdot d\}$ as follows: $\pi(i,j) < \pi(i',j')$ iff $(v_i > v_i')$ or $(v_i = v_i'$ and $i < i')$ or $(v_i = v_i'$ and $i = i'$ and $j < j')$.

The big advantage of this technique is that it implies an upper bound for those dimensions which have not been selected for transmission. Thus, we can shrink the approximation area also for those dimensions for which we have not received any bytes. This shrinking is possible due to the ranking agreement between the clients and the server that the value of the dimensions for which we have not received any bytes is equal or smaller to the smallest value for which we have already received some bytes. Let $i' \in \{1,..,d\}$ now be the transmitted dimension with the smallest feature value of all transmitted dimensions.

Then, the server computes the approximation area $a = f_{app}(f_{rank}(v), L) = [l_1, u_1] \times ... \times [l_d, u_d]$ as follows:

$$l_i = MBR\_l_i + \sum_{j=1}^{m} \begin{cases} val(b_{i,j}) &, \text{if } (\pi(i,j) \leq L) \\ 0 &, \text{else} \end{cases}$$

$$u_i = min\left(MBR\_u_i, MBR\_l_i + \sum_{j=1}^{m} \begin{cases} val(b_{i,j}) &, \text{if } (\pi(i,j) \leq L) \\ val(b_{i',j}) &, \text{if } (\pi(i,j) > L) \wedge (\pi(i',j) \leq L) \\ (2^w - 1) \cdot 2^{w(m-j)} &, \text{else} \end{cases}\right)$$

3.2.3 Combined Approximation (CA)

This approach combines the two previous approaches. According to Definition 1, each byte $b_{i,j}$ of v can be assigned to a value $val(b_{i,j}) = b_{i,j} \cdot 2^{w(m-j)}$. Now, we can rank the set of bytes $\{b_{i,j}: i = 1,..,d; j = 1,..,m\}$ according to their value $val(b_{i,j})$, and transmit the L bytes having the highest ranking values. Thus the bijective function $\pi: \{1...d\} \times \{1...m\} \to \{1...m \cdot d\}$ is defined as follows: $\pi(i,j) < \pi(i',j')$ iff $(val(b_{i,j}) > val(b_{i',j'}))$ or $(val(b_{i,j}) = val(b_{i',j'})$ and $i < i')$ or $(val(b_{i,j}) = val(b_{i',j'})$ and $i = i'$ and $j < j')$.

Let now $b_{i',j'}$ be the byte with the L highest $val(b_{i',j'})$. Then, the server computes the approximation area $a = f_{app}(f_{rank}(v), L) = [l_1, u_1] \times ... \times [l_d, u_d]$ as follows:

$$l_i = MBR\_l_i + \sum_{j=1}^{m} \begin{cases} val(b_{i,j}) &, \text{if } (\pi(i,j) \leq L) \\ 0 &, \text{else} \end{cases}$$

$$u_i = min(MBR\_u_i, MBR\_l_i + offset), \text{ where}$$

$$offset = \sum_{j=1}^{m} \begin{cases} val(b_{i,j}) &, \text{if } (\pi(i,j) \leq L) \\ 0 &, \text{if } (\pi(i,j) > L) \wedge (j < j') \\ val(b_{i',j'}) &, \text{if } (\pi(i,j) > L) \wedge (j = j') \\ (2^w - 1) \cdot 2^{w(m-j)} &, \text{if } (\pi(i,j) > L) \wedge (j > j') \end{cases}$$

The example presented in Figure 2 demonstrates the conservative approximation areas for the three proposed approaches. The figure shows clearly that the combined

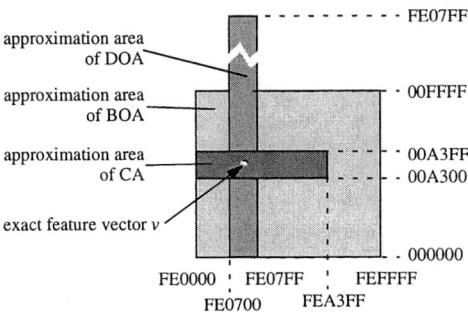

Fig. 2. Approximation areas (BOA, DOA, CA)

approach leads to a smaller approximation area than the byte-oriented and the dimension-oriented approach.

4   Approximated Clustering Based on Fuzzy Distance Functions

In this section, we will show how to compute the similarity between two feature vector approximations. The most straightforward approach is to use the box center to compute the distance between two box approximations. This center oriented box distance approximates the exact distance between the feature vectors rather accurate if the boxes are rather small and do not overlap.

On the other hand, imagine that we have two rather big boxes where the box centers are identical. The center oriented distance would assign a zero distance to the approximated feature vectors, although the exact distance between the feature vectors might be very high. Therefore, it is better to generally use the expectation value of the exact distances between the feature vectors rather than the distances between the box centers. This distance expectation value is based on the distance distribution function $P_d: O \times O \rightarrow (IR_0^+ \rightarrow [0..1])$, which assigns a probability value p to each possible distance τ (cf. Figure 3a). The value p indicates the probability that the exact distance between the feature vectors is smaller than τ. Figure 3b shows how we can compute P_d for two feature vectors based on two arbitrary conservative approximations $A = f_{app}(f_{rank}(v), L)$ and $A' = f_{app}(f_{rank}(v'), L')$. First, we measure those portions of the area A' which are overlapped by a sphere around $x \in A$ with radius τ. Summing up all these values for all $x \in A$ yields the probability $P_d(v,v')(\tau)$ that the distance $d(v, v')$ is smaller than τ. The following lemma describes formally how to compute P_d for two approximated feature vectors.

Lemma 1 *Distance Distribution Function.* Let $A = f_{app}(f_{rank}(v), L)$ and $A' = f_{app}(f_{rank}(v'), L') \in [IR \times IR]^d$ be two arbitrary conservative approximations of the feature vectors v, $v' \in IR^d$. Let $R(x, \tau)$ denote a sphere around the feature vector $x \in IR^d$ with radius $\tau \in IR$. Then the *distance distribution function* $P_d: IR^d \times IR^d \rightarrow (IR_0^+ \rightarrow [0..1])$ based on the approximations A and A' can be computed as follows.

$$P_d(v, v')(\tau) = \frac{\int_A |A' \cap R(x, \tau)| dx}{|A| \cdot |A'|}$$

As already mentioned clustering algorithms can only handle unique distance values. In order to put clustering methods into practice, we extract an aggregated value which we call distance expectation value. The distance expectation value $E_d: O \times O \to I\!R_0^+$

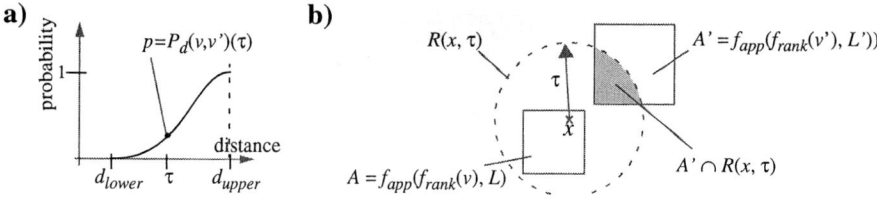

Fig. 3. Computation of the distance distribution function P_d. **a)** distance distribution function **b)** computation of the probability $P_d(v,v')(\tau)$

represents the similarity between two box approximations in the best possible way by one single value $E_d(v, v') = \int_{-\infty}^{\infty} ((P'_d(v, v')(x)/dx) \cdot x) dx$, where $P'_d(v, v')$ denotes the derivation of $P_d(v, v')$.

Practically, we can compute this distance expectation value between two approximated feature vectors by means of monte-carlo sampling. Thereby, we create randomly feature vectors located in the boxes and compute the average distance of these randomly created feature vectors to each other, Obviously, the higher the sample rate the more accurate is the computation of the distance expectation value. Note, that the center oriented distance can be regarded as a distance expectation value where only one sample pair, i.e. the box centers, is used.

5 Experiments

In this section, we evaluate the performance of our approach with a special emphasis on the overall transmission cost. The tests are based on an artificial dataset ART and two real world data sets PLANE and PDB which were distributed to two clients:

ART dataset. The artificial dataset ART consists of 1000 feature vectors, equally distributed in a 30-dimensional vector space.

PLANE dataset. The PLANE dataset consists of 1000 high-resolution 3D CAD objects provided by our industrial partner, an American airplane manufacturer. Each object is represented by a 42-dimensional feature vector which is derived from the cover sequence model as described in [9].

PDB dataset. This 3D protein structure dataset is derived from the Brookhaven Protein Data Bank (PDB). The 1000 objects are represented by 3D shape histograms [1] resulting in a 120-dimensional feature vector per object.

5.1 Quality of the Feature Vector Approximation Techniques

In a first experiment, we examined the quality of the three approximation techniques *BOA*, *DOA* and *CA* (cf. Section 3). For each feature vector, we transmitted once L bytes (measured in percent of all bytes of a feature vector) to the server which then constructs the approximations based on the transmitted data. Figure 4 depicts how the approxima-

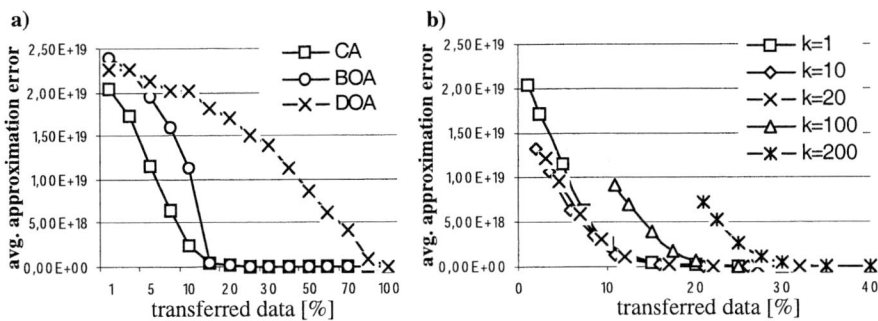

Fig. 4. Average approximation error (ART dataset) **a)** varying approximation techniques ($k = 1$) **b)** varying k parameter (CA approach)

tion error depends on the transmission cost. The error is measured by the average length of the diagonal of the approximation areas.

Figure 4a shows that the average approximation error rapidly decreases for the *BOA* approach as well as for the *CA* approach. For high values of L, the *DOA* approach performs worst. Only for very small values of L it outperforms the *BOA* approach. However, our *CA* approach yields to the best results, especially for low transmission cost.

Furthermore, we examined the *CA* approach for a varying parameter k used for the k-means based pre-clustering of the client sites. Figure 4b shows that if we initially transmit only the pre-clustering information of the feature vectors, i.e. the dataset approximations (cf. Section 3.1), the approximation quality increases slowly with increasing k. Obviously, an increasing k parameter yields higher transfer cost. In contrast to the dataset approximation approach, the quality increases more rapidly when we increase the amount of transmitted data of the feature vector approximations (cf. Section 3.2). Figure 4b shows that we achieve the best trade-off between accuracy and transfer overhead when we set $k = 10$, especially for low transfer cost.

5.2 Distance Measures

In this section, we investigate the accuracy of the two distance measures, $mid(A, A')$ and $exp(A, A')$. The distance function $mid(A, A')$ denotes the distance between the center points of the approximations A and A' and the distance function $exp(A, A')$ denotes the expected distance of the feature vectors approximated by A and A' (cf. Section 4). For the computation of the expected distance $exp(A, A')$, we used monte-carlo sampling with a sample rate s. For measuring the quality we summed up the quadratic distance error of the examined distance measures $exp(A, A')$ and $mid(A, A')$ with respect to the exact distance of the feature vectors. Figure 5 depicts the average quadratic distance error of all feature vector approximations.

In the first experiment, we observed the behavior of $exp(A, A')$ for a varying sample rate s. Figure 5a shows that the distance function $exp(A, A')$ reflects the exact distance between the feature vectors much more accurately than the distance function $mid(A, A')$, already for a sample rate $s > 2$. Figure 5b shows that the difference between the two

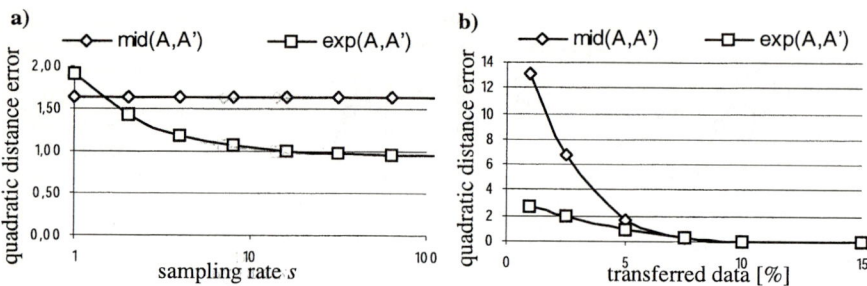

Fig. 5. Evaluation of the Distance Measures (ART dataset) **a)** varying sampling rates s (5% transferred) **b)** varying approximation accuracy ($s=50$)

distance measures $exp(A, A')$ and $mid(A, A')$ increases with decreasing transfer cost. Therefore it is especially important to use the $exp(A, A')$ distance measure when only small transfer cost are allowed.

5.3 Density-Based Clustering

In a last experiment, we compared a standard DBSCAN run based on the $exp(A, A')$ measure to the distributed clustering approach introduced in [7]. We measured the quality of the approximated clustering result by the quality criterion used in [7]. Figure 6 shows clearly that for a certain amount of transferred information our approach performs much better, i.e. our approach yields higher quality values than the approach presented in [7]. Note that the approach of [7] was especially designed for achieving high-quality distributed clusterings based on little transmitted information. We would like to point out that this experiment shows that our approximation technique for high-dimensional feature vectors can beneficially be used as basic operation for distributed data mining algorithms.

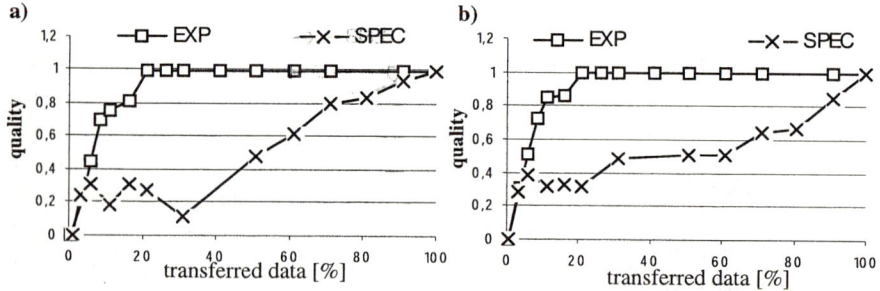

Fig. 6. Clustering quality ($s = 10, k = 10$) **a)** PLANE dataset **b)** PDB dataset

6 Conclusion

In this paper, we presented a novel technique for approximating high-dimensional distributed feature vectors. In order to generate suitable approximations, we enhanced the idea of state-of-the-art index structures for high-dimensional data which approximate each single feature vector by a certain number of granted bytes. Based on this technique we can limit the transmission cost considerably while only allowing a small decrease of the quality. We demonstrated the benefits of our technique for the important area of distributed clustering.

In our future work, we will show that also other distributed data mining algorithms benefit from our high-dimensional feature vector approximation technique.

References

1. Ankerst M., Kastenmüller G., Kriegel H.-P., Seidl T.: *3D Shape Histograms for Similarity Search and Classification in Spatial Databases*. Proc. 6th Int. Symposium on Large Spatial Databases (SSD'99), Hong Kong, China, in: Lecture Notes in Computer Science, Vol. 1651, Springer, 1999, pp. 207-226.
2. Berchtold S., Böhm C., Jagadish H. V., Kriegel H.-P., Sander J.: *Independent Quantization: An Index Compression Technique for High Dimensional Data Spaces*.ICDE'00, 2000.
3. Beckmann N., Kriegel H.-P., Schneider R., Seeger B.: *The R*-tree: An Efficient andRobust Access Method for Points and Rectangles*. Proc. ACM SIGMOD'90, 1990.
4. Golub, T., Slonim, D.K., Tamayo, P., Huard, C., Gaasenbeek, M., Mesirov, J., Coller, H., Loh, M. L., Downing, J., Caligiuri, M., Bloomeld, C., Lander, E.: *Molecular classification of cancer: Class discovery and class prediction by gene expression monitoring*. Science, 286, pp. 531-537.
5. Johnson E., Kargupta H.: *Hierarchical Clustering From Distributed, Heterogeneous Data*. In Zaki M. and Ho C., editors, Large-Scale Parallel KDD Systems. Lecture Notes in Computer Science, colum 1759, pp. 221-244. Springer-Verlag, 1999
6. Januzaj E., Kriegel H.-P., Pfeifle M.: *DBDC: Density Based Distributed Clustering*. Proc. 9th Int. Conf. on Extending Database Technology (EDBT 2004), Heraklion, Greece, 2004, pp. 88-105.
7. Januzaj E., Kriegel H.-P., Pfeifle M.: *Scalable Density-Based Distributed Clustering*. Proc. 7th European Conference on Principles and Practice of Knowledge Discovery in Databases (PKDD), Pisa, Italy, 2004.
8. Jain A. K., Murty M. N., Flynn P. J.: *Data Clustering: A Review*. ACM Computing Surveys, Vol. 31, No. 3, Sep. 1999, pp. 265-323.
9. Kriegel H.-P., Brecheisen S., Kröger P., Pfeifle M., Schubert M.: *Using Sets of Feature Vectors for Similarity Search on Voxelized CAD Objects*. Proc. ACM SIGMOD Int. Conf. on Management of Data (SIGMOD'03), San Diego, CA, 2003, pp.587-598.
10. Kargupta H., Chan P. (editors): *Advances in Distributed and Parallel Knowledge Discovery*. AAAI/MIT Press, 2000.
11. McQueen J.: *Some Methods for Classification and Analysis of Multivariate Observation*. Proc. 5th Berkeley Symp. on Math. Statist. and Prob., Vol. 1, 1965
12. Shawney H., Hafner J.: *Efficient Color Histogram Indexing*. Proc. Int. Conf. On Image Processing, 1994, pp. 66-70.
13. Salton G., McGill M. J.: *Introduction to Modern Retrieval*. McGraw-Hill Book Company, 1983.
14. Samatova N.F., Ostrouchov G., Geist A., Melechko A.V.: RACHET: *An Efficient Cover-Based Merging of Clustering Hierarchies from Distributed Datasets*. Distributed and Parallel Databases 11(2): pp. 157-180; Mar 2002.

Improvements of IncSpan: Incremental Mining of Sequential Patterns in Large Database

Son N. Nguyen, Xingzhi Sun, and Maria E. Orlowska

School of Information Technology and Electrical Engineering,
The University of Queensland, QLD 4072, Australia
{nnson, sun, maria}@itee.uq.edu.au

Abstract. In reality, sequence databases are updated incrementally. The changes on the database may invalidate some existing sequential patterns and introduce new ones. Instead of recomputing the database each time, the incremental mining algorithms target efficiently maintaining the sequential patterns in the dynamically changing database.

Recently, a new incremental mining algorithm, called **IncSpan** was proposed at the International Conference on Knowledge Discovery and Data Mining (KDD'04). However, we find that in general, IncSpan fails to mine the *complete* set of sequential patterns from an updated database. In this paper, we clarify this weakness by proving the incorrectness of the basic properties in the IncSpan algorithm. Also, we rectify the observed shortcomings by giving our solution.

Keywords: Sequential patterns, Incremental mining, Algorithm.

1 Introduction

Discovering sequential patterns from databases is of great importance in many application domains (e.g., fault detection in a network, web access pattern analysis, and plan failure identification, etc.). Since the research problem of discovering sequential patterns was first introduced by Argawal et al. in [2], many mining algorithms [3, 4, 5, 6] have been proposed for efficiently finding frequent sequential patterns from the sequence databases.

In reality, a database is often dynamic. With the evolution of databases, some existing sequential patterns would be invalid and some new sequential patterns might be introduced. Thus, maintaining sequential patterns (over a significantly long period) becomes essential for sequential pattern mining.

Generally, the change on a sequential database can be categorized as 1) deleting records, 2) inserting new records, and 3) appending new items on the existing records. In this paper, discussions on the change of database only refer to the last two categories (i.e., we do not consider deleting records). Thus, the corresponding research problem of pattern maintenance can be described as follows. Given a sequence database D and the set FS of sequential patterns in D, when D evolves to D' (with the updated part db known, i.e., inserting and appending part), how to *efficiently* find the set FS' of sequential patterns in D' ?

The naive approach to solve this problem is to apply a certain sequential mining algorithm A on D' to re-compute the sequential patterns. Obviously, it wastes computational resources because the previous mining result is not utilized in this subsequent problem. Another alternative, which will be discussed in this paper, is to incrementally update FS into FS'. The basic idea is to first investigate the new input data db while maximizing the use of a previous mining result FS. Various incremental update algorithms [7, 8, 9] have been designed to improve efficiency by reducing the number of scans on D'.

In general, an incremental update algorithm A' should satisfy the following two conditions.

1. Correctness: The mining result of A' should be identical to FS', where FS' is the set of frequent sequential patterns discovered by the traditional sequential mining algorithm A from the updated database D'.
2. Efficiency: Generally, for maintaining sequential patterns in a dynamically changing database, the time of applying A' is significantly less than that of re-computing D' by using A.

Recently, Cheng et al. [1] proposed a new incremental update algorithm, called IncSpan. IncSpan is developed on the basis of the sequential mining algorithm PrefixSpan. With the aims of improving performance, the redundant semi-frequent patterns (i.e., patterns that are *"almost frequent"* in D) are introduced and maintained as candidates of newly appearing sequential patterns in the updated database D'. In addition, some optimization techniques are applied in the pattern matching and projected database generation. The experimental results in [1] show that IncSpan significantly outperforms the non-incremental sequential mining algorithm [3] and the previously proposed incremental algorithm [7].

However, in this paper, we argue that in general, the algorithm IncSpan cannot find the complete set of frequent sequential patterns in the updated database D', i.e., it violates the correctness condition. Particularly, we give counter examples to prove that the foundations of IncSpan, and its three key properties presented in [1], are incorrect. The main purpose of this paper is to identify the weakness of IncSpan, and to propose a correct solution which can guarantee the completeness of the mining result.

The remainder of the paper is organised as follows. Section 2 first gives the formal definition of the incremental update problem and then describes the algorithm IncSpan. In Section 3, we prove that IncSpan fails to find the complete set of sequential patterns in the updated database. Based on IncSpan, our solution is proposed in Section 4. Finally, we conclude this paper in Section 5.

2 Summary of IncSpan Approach

For the completeness of this presentation and to establish our notation, we first formally define the problem of incremental sequential patterns mining.

2.1 Incremental Sequential Pattern Mining

Let $I = \{i_1, i_2, ..., i_k\}$ be a set of k distinct literals called items. A subset of I is called an itemset. A sequence $s = <t_1, t_2, ..., t_m>$ ($t_i \subseteq I$) is an ordered list. Without loss of generality, we assume that the items in each itemset are sorted in certain order such as alphabetic order (a, b, c ...). Formally, a sequence $\beta = <b_1, b_2, ..., b_n>$ is a subsequence of another sequence $\alpha = <a_1, a_2, ..., a_m>$, denoted $\beta \subseteq \alpha$, if and only if $\exists$ $i_1, i_2, ..., i_n$ such that $1 \leq i_1 < i_2 < ... < i_n \leq m$ and $b_1 \subseteq a_{i1}; b_2 \subseteq a_{i2}; ...; b_n \subseteq a_{in}$. In this case, we also call α is supersequence of β and α contains β.

A sequence database, $D = \{s_1, s_2, ..., s_n\}$, is a set of sequences. The support of a sequence α in D is the number of sequences in D which contain α, denoted as $sup_D(\alpha)$ = |{s | s $\in$ D and $\alpha \subseteq s$ }|. Given a support threshold, min_sup, a sequence is frequent if its support is no less than min_sup. The set of frequent sequential patterns, denoted as FS, includes all the frequent sequences in D.

The sequential pattern mining is to discover the complete set FS when the sequence database D and min_sup are given.

The incremental sequential pattern mining is formalized as follows. Given the original database D, db is the incremental part which is added to D. As the result, the new database is created, denoted as D' = D + db. There are two scenarios for this processing.

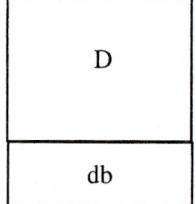
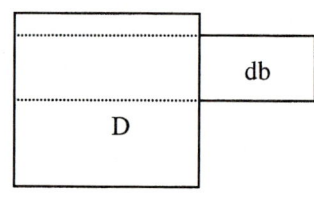

Fig. 2.1. Insert scenario **Fig. 2.2.** Append scenario

The Insert scenario (Figure 2.1) means that the new sequences are inserted into the original database. These old sequences are still unchanged, but the total number of the sequences in D' is increased. In the Append scenario (Figure 2.2), some old sequences are appended with new sequences. And the total number of the sequences in D' is unchanged. If we consider Insert scenario is a special case of Append scenario (i.e., db is appended with the empty sequences), we can combine two scenarios and formulate the problem as follows.

Given a sequence $s = < t_1, t_2, ..., t_m> \in$ D and another sequence $s_a = < t'_1, t'_2, ..., t'_n> \in$ db, if s concatenates with s_a in D', the new sequence s' is called an appended sequence, denoted as s'= s + s_a. Obviously, s' is always not empty. If we assume that s is empty, then we can treat the Insert scenario as the Append scenario. We denote LDB = {s' | s' $\in$ D' and s' is appended with items/itemsets}. We denote ODB = {s | s $\in$ D and s is appended with items/itemsets in D'}, denote NDB = {s | s $\in$ D and s is not appended with items/itemsets in D'}. The set of frequent sequences in D' is denoted as FS'. As a result, we have the following formula.

D' = D + db = (ODB + NDB) + db = (ODB + db) + NDB = LDB + NDB.

Example 2.1: A sample sequence database D and an appended part db are given to explain the notations defined above

Table 2.1. A sample sequence database D and the appended part

Seq ID	Original Part	Appended Part
0	(a)(h)	(c)
1	(eg)	(a)(bce)
2	(a)(b)(d)	(ck)(l)
3	(b)(df)(a)(b)	∅
4	(a)(d)	∅
5	(be)(d)	∅

Given min_sup = 3, we have:
ODB = {SeqID: 0-2 in D}; NDB = {SeqID:3-5 in D}; LDB = {SeqID: 0-2 in D'};
FS={<(a)>:4, <(b)>:3, <(d)>:4, <(b)(d)>:3};
FS'={<(a)>:5, <(b)>:4, <(d)>:4, <(b)(d)>:3, <(c)>:3, <(a)(b)>:3, <(a)(c)>:3};

Problem statement: Given D, D', min_sup, and FS, the incremental sequential pattern mining is to mine the set FS' in D' based on FS, rather than to recompute D' from scratch.

2.2 IncSpan Approach

In this part, we introduce IncSpan [1], the recently proposed approach for the incremental sequential pattern mining. In [1], the authors present the idea of buffering semi-frequent patterns as the following description, study its properties, and design solutions of how to mine and maintain FS incrementally based on PrefixSpan approach.

Buffering Semi-frequent Pattern: Given a factor $\mu \leq 1$, a sequence is semi-frequent if its support is less than min_sup but no less than $\mu *$ min_sup; a sequence is infrequent if its support is less than $\mu *$ min_sup. The set of semi-frequent sequential pattern includes all the semi-frequent sequences in D and D', denoted as SFS and SFS' respectively. For example, given $\mu = 0.6$, according to Example 2.1, we have SFS={<(e)>:2, <(a)(b)>:2, <(a)(d)>:2} and SFS'={<(e)>:2, <(a)(d)>:2, <(be)>:2}.

PrefixSpan Approach: PrefixSpan approach [3] is one of the most efficient methods for mining sequential patterns. It uses prefix sequence, suffix sequence and p-projected database concepts to discover the complete set of sequential patterns when the sequence database D and min_sup are given, more detail can refer to [3]. The prototype of PrefixSpan algorithm which is used in the following sections is **PrefixSpan(p, D|p, $\mu *$ min_sup, FS, SFS)**, where p is a sequence, D|p is p-projected database of D. This routine is called recursively to mine the complete FS, SFS in D|p.

It is assumed in [1] that the sequences in SFS are "almost frequent" in D, most of the frequent subsequences in the appended database (D') will either come from SFS or they are already frequent in original database D. According to [1], the SFS' and FS" in D' are derived from the following cases:

1. A pattern p which is frequent in D, is still frequent in D'
2. A pattern p which is semi-frequent in D, becomes frequent in D'
3. A pattern p which is semi-frequent in D, still semi-frequent in D'
4. Appended data db brings new frequent / semi-frequent items
5. A pattern p which is infrequent in D, becomes frequent in D'
6. A pattern p which is infrequent in D, becomes semi-frequent in D'

Case (1) - (3). There exsists information in FS and SFS, so one can update the support number and project D' to find all frequent / semi-frequent sequences which are generated from FS and SFS.

Case (4). Property 2.1: An item which does not appear in D and is brought by db has no information in FS and SFS. **Solution 2.1:** Scan LDB for single items. Then use the new frequent item as prefix to construct projected database and discover frequent sequences recursively by PrefixSpan approach.

Case (5). Property 2.2: An infrequent sequence p' in D becomes frequent in D', all of its subsequences must also be frequent in D'. Then at least one of its prefix subsequences, p, is in FS. **Solution 2.2:** Start from its prefix p in FS and construct p-projected database on D', use PrefixSpan approach, IncSpan will discover p'.

IncSpan provides the pruning technique based on the following theorem.

Theorem 2.1. *For a frequent pattern p, if its support in LDB $sup_{LDB}(p) < (1 - \mu) * min\_sup$, then there is no sequence p' having p as prefix changing from infrequent in D to frequent in D'.*

This theorem provides an effective bound to decide whether it is necessary to project the whole database D', which can reduce the number of projections.

Case (6). Property 2.3: An infrequent sequence p' in D becomes semi-frequent in D', all of its subsequences must also be semi-frequent in D'. Then at least one of its prefix subsequences, p, is in FS or SFS. **Solution 2.3:** Start from its frequent prefix p in FS or SFS, and construct p-projected database on D', use PrefixSpan approach, IncSpan will discover p'.

IncSpan Algorithm Outline: Given an original database D, an appended database D', a threshold min_sup, a buffer ratio μ, a set of frequent sequences FS and a set of semi-frequent sequences SFS, IncSpan wants to discover only the set of frequent sequences FS' in D' [1]. The basic idea of algorithm are described as follows.

Step 1: Scan LDB for single items, as show in case (4).

Step 2: Check every pattern in FS and SFS in LDB to adjust the support.

> Step 2.1: If a pattern becomes frequent, add it to FS'. Then check whether it meets the projection condition according to Theorem 1. If so, use it as prefix to project database D', as show in case (5).
>
> Step 2.2: If a pattern is semi-frequent, add it to SFS'.

Algorithm Outline: IncSpan(D', min_sup, μ, FS, SFS)
Input: An appended database D', min_sup, FS and SFS in D
Output: FS' in D'
Method:
1:   FS' = $\emptyset$; SFS' = $\emptyset$;
2:   Scan the LDB for new single items
3:          Add new frequent items into FS'
4:          Add new semi-frequent items into SFS'
5:   **For** each new item i in FS' **do**
6:          **PrefixSpan**(i, D'|i, μ * min_sup, FS', SFS')
7:   **For** every pattern p in FS or SFS **do**
8:          Check $\Delta sup(p) = sup_{db}(p)$
9:          **If** $sup_{D'}(p) = sup_D(p) + \Delta sup(p) \geq min\_sup$
10:                 **Insert**(FS', p)
11:                 **If** $supLDB(p) \geq (1 - \mu) * min\_sup$
12:                        **PrefixSpan**(p, D'|p, μ * min_sup, FS', SFS')
13:          **Else**
14:                 **Insert**(SFS', p)
15: **Return;**

3 Critical Observations

After the IncSpan approach is reviewed in Section 2, in this section, we show that in general, IncSpan provides incomplete results. Particularly, we prove that the solutions in IncSpan for Case 4-6 are incorrect by giving counter examples.

Claim 3.1 (Incorrectness of Solution 2.1 for Case (4)): Scanning LDB cannot find the complete set of new single frequent / semi-frequent items in D'.

Proof: Generally, scanning LDB can only discover the new single frequent / semi-frequent items in terms of LDB. Since the support is counted as number, for the single items that are infrequent ones in LDB and D but become frequent / semi-frequent in D', Solution 2.1 fails to discover them. The following example illustrates the incompleteness of frequent single items. The example for semi-frequent single items can be created by following the same idea.

Counter example 3.1: This example is generated from Example 2.1 with small change in appended part: item (f) is appended in SeqID 0-1. Remember that min_sup is 3 and μ is 0.6. If the IncSpan algorithm scans only LDB, the new frequent item (f) in D' cannot be discovered. As a result, IncSpan loses all new frequent sequences which have (f) as a prefix.

Claim 3.2 (Incorrectness of Property 2.2 for Case (5)): In IncSpan, if an infrequent sequence p' in D becomes frequent in D', it is possible that none of its prefix subsequence p is in FS.

Proof: A counter example is illustrated as follows.

Counter example 3.2: This example generates from Example 2.1 with a small change in original part (SeqID 3-5 are deleted). Given min_sup = 3; μ = 0.6

In this example, <(a)(c)> is infrequent in D but becomes frequent in D'. However, its prefix subsequences, <(a)>, is not in FS, in this case, FS is a empty set.

Claim 3.3 (Incorrectness of Property 2.3 for Case (6)): In IncSpan, if an infrequent sequence p' in D becomes semi-frequent in D', it is possible that none of its prefix subsequence p is in FS or SFS.

Proof: A counter example is illustrated as follows.

Counter example 3.3: This example generates from Example 2.1 with a small change in original part, and appended part. Given min_sup = 3; $\mu = 0.6$

In this example, <(be)> is infrequent in D but becomes semi-frequent in D'. However, its prefix subsequences, <(b)>, is not in SFS. In this case, SFS = {<(a)>:2, <(e)>:2}.

Table 3.1. A sample sequence database D and the new appended part

Seq ID	Original Part	Appended Part
0	(a)(h)	(c)(f)
1	(eg)	(a)(bce)(f)
2	(a)(b)(d)	(ck)(l)
3	(b)(df)(a)(b)	⊘
4	(a)(d)	⊘
5	(be)(d)	⊘

Table 3.2. A deleted sequence database D and the appended part

Seq ID	Original Part	Appended Part
0	(a)(h)	(c)
1	(eg)	(a)(bce)
2	(a)(b)(d)	(ck)(l)

Table 3.3. A deleted sequence database D and the new appended part

Seq ID	Original Part	Appended Part
0	(a)(h)	(c)
1	(eg)	(a)(bce)
2	(a)(be)(d)	(ck)(l)

Claim 3.4 (Extension of Theorem 2.1): In order to apply the pruning technique based on Theorem 2.1 for any pattern p in FS or SFS. Theorem 2.1 can be extended as follows. The difference between the following theorem and Theorem 2.1 is that we can apply for not only frequent pattern p in D, but also any other pattern p in D.

Theorem 3.1: For any pattern p in D, if its support in LDB $sup_{LDB}(p) < (1 - \mu) * min\_sup$, then there is no sequence p' having p as prefix changing from infrequent in D to frequent in D'.

Proof: p' was infrequent in D, so $sup_D(p') < \mu * min\_sup$ (i)
If $sup_{LDB}(p) < (1 - \mu) * min\_sup$ then
$sup_{LDB}(p') \leq sup_{LDB}(p) < (1 - \mu) * min\_sup$ because $p \subset p'$.
Since $sup_{LDB}(p') = sup_{ODB}(p') + sup_{db}(p')$ because LDB = ODB + db, then
$sup_{db}(p') \leq sup_{LDB}(p') < (1 - \mu) * min\_sup$ (ii)
Combining (i) and (ii), we have $sup_{D'}(p') = sup_D(p') + sup_{db}(p') < min\_sup$. So p' cannot be frequent in D'.

4 Proposed the Complete Solution

With all above observations in mind, we present the following algorithm for improvement of IncSpan, denoted as IncSpan+.

Given an original database D, an appended database D', a minimum support min_sup, a buffer ratio μ, a set of frequent sequences FS and a set of semi-frequent sequences SFS, we want to mine the set of frequent sequences FS', and the set of semi-frequent sequences SFS' in D'.

Algorithm Outline: IncSpan+(D', min_sup, μ, FS, SFS)
Input: An appended database D', min_sup, FS and SFS in D
Output: FS', SFS' in D'
Method:

1:     FS' = $\emptyset$; SFS' = $\emptyset$;
2:     Determine LDB; Total number of sequences in D', adjust the min_sup if it is changed due to the increasing of total number of sequences in D'.
3:     Scan the whole D' for new single items
4:         Add new frequent items into FS'
5:         Add new semi-frequent items into SFS'
6:     **For** each new item i in FS' **do**
7:         **PrefixSpan**(i, D'|i, μ * min_sup, FS', SFS')
8:     **For** each new item i in SFS' **do**
9:         **PrefixSpan**(i, D'|i, μ * min_sup, FS', SFS')
10:    **For** every pattern p in FS or SFS **do**
11:        Check $\Delta sup(p) = sup_{db}(p)$
12:        **If** $sup_{D'}(p) = sup_D(p) + \Delta sup(p) \geq min\_sup$
13:            **Insert**(FS', p)
14:            **If** $supLDB(p) \geq (1 - \mu) * min\_sup$
15:                **PrefixSpan**(p, D'|p, μ * min_sup, FS', SFS')
16:        **ElseIf** $sup_{D'}(p) \geq \mu * min\_sup$
17:            **Insert**(SFS', p)
18:            **PrefixSpan**(p, D'|p, μ * min_sup, FS', SFS')
19:    **Return;**

This algorithm follows the same spirit as IncSpan, using PrefixSpan approach to maintain both FS' and SFS'. However, new proposed algorithm ensures the correctness of mining result in the updated database, as is proven in Claim 4.1.

Claim 4.1 (Correctness of IncSpan+): The IncSpan+ outputs the complete set of frequent pattern FS' and the complete set of semi-frequent pattern SFS' in the updated database D'.

Proof: The complete FS' and SFS' sets come exactly from two sources:

Case (4.1): From new frequent / semi-frequent single items and all of frequent / semi-frequent supersequences which have these new single items as the prefix;

Case (4.2): From FS, SFS, and sequences that have the prefix in FS or SFS.

As can be seen from the outline of IncSpan+, line from 3-9 will discover all frequent sequences and semi-frequent sequences in D' corresponding to case (4.1). That means all frequent sequences / semi-frequent sequences in D', which have the first new frequent / semi-frequent item as the prefix, will be discovered. These new frequent / semi-frequent items are not included in FS and SFS.

Line 10-18 in IncSpan+ will discover all frequent / semi-frequent sequences which have their prefix in FS or SFS, and all of them correspond to case (4.2).

Claim 4.1 proves the correctness of our proposed IncSpan+. Compared with the original approach, IncSpan+ has the following improvements:

1. IncSpan+ can find the complete FS', which guarantees the correctness of the mining result.
2. IncSpan+ can find the complete SFS', which is helpful in incrementally maintaining the frequent patterns for further database updates.

5 Conclusion

This paper clarified the weakness of the recent work [1] in the context of the incremental mining sequential patterns. We proved that IncSpan, the incremental mining approach in [1], cannot find the complete set of sequential patterns in the updated database. The solution, IncSpan+ was proposed to rectify the observed shortcoming. IncSpan+ not only guarantees the correctness of the incremental mining result, but maintains the complete set of semi-frequent sequences for future updates.

References

1. Cheng H., Yan X., Han J.: IncSpan: Incremental mining of sequential patterns in large database. Proc. ACM KDD Conf. on Knowledge Discovery in Data, Washington (KDD'04), 2004
2. Agrawal R., Srikant R.: Mining sequential patterns. Proc. 11th IEEE Int. Conf. on Data Engineering (ICDE'95), 1995
3. Pei J., Han J., Mortazavi-Asl B., Wang J., Pinto H., Chen Q., Dayal U., Hsu M.: Mining sequential patterns by Pattern-Growth: The PrefixSpan approach. IEEE Transactions on Knowledge and Data Engineering, Vol.16, No. 10, 2004
4. Srikant R., Agrawal R.: Mining sequential patterns: Generalizations and performance improvements. Proc. 5th IEEE Int. Conf. on Extending Database Technology (IDBT'96)

5. Zaki M.: SPADE: An efficient algorithm for mining frequent sequences, Machine Learning, 40: (31-60), 2001
6. Ayres J., Gehrke J.E., Yiu T., Flannick J.: Sequential pattern mining using bitmaps. Proc. ACM KDD Conf. on Knowledge Discovery in Data (KDD'02), 2002
7. Parthasarathy S., Zaki M., Ogihara M., and Dwarkadas S.: Incremental and interactive sequence mining. Proc. 8th Int. Conf. on Information and Knowledge Management (CIKM'99), 1999
8. Masseglia F., Poncelet P., Teisseire M.: Incremental mining of sequential patterns in large database. Data & Knowledge Engineering, 46: (97-121), 2003
9. Zhang M., Kao B., Cheung D., Yip C.: Efficient algorithms for incremental update of frequent sequences. Proc. of Pacific-Asia Conf. on Knowledge Discovery and Data Mining (PAKDD'02), 2002

Efficient Sampling: Application to Image Data

Surong Wang, Manoranjan Dash, and Liang-Tien Chia

Center for Multimedia and Network Technology,
School of Computer Engineering,
Nanyang Technological University, Singapore 639798
{pg02759741, asmdash, asltchia}@ntu.edu.sg

Abstract. Sampling is an important preprocessing algorithm that is used to mine large data efficiently. Although a simple random sample often works fine for reasonable sample size, accuracy falls sharply with reduced sample size. In KDD'03 we proposed EASE that outputs a sample based on its 'closeness' to the original sample. Reported results show that EASE outperforms simple random sampling (SRS). In this paper we propose EASIER that extends EASE in two ways. 1) EASE is a halving algorithm, i.e., to achieve the required sample ratio it starts from a suitable initial large sample and iteratively halves. EASIER, on the other hand, does away with the repeated halving by directly obtaining the required sample ratio in one iteration. 2) EASE was shown to work on IBM QUEST dataset which is a categorical count dataset. EASIER, in addition, is shown to work on continuous data such as Color Structure Descriptor of images. Two mining tasks, classification and association rule mining, are used to validate the efficacy of EASIER samples *vis-a-vis* EASE and SRS samples.

Keywords: Sampling, frequency estimation, classification, association rule mining, image processing.

1 Introduction

As the size of stored data is increasing day-by-day thanks to cheaper storage devices and increasing number of information sources such as Internet, the need for scalability is intensifying. Often sampling is used to reduce the data size while remaining the underlying structure. Use of a simple random sample, however, may lead to unsatisfactory results. The problem is that such a sample may not adequately represent the entire data set due to random fluctuations in the sampling process. The difficulty is particularly apparent at small sample sizes.

In [1] we proposed EASE (Epsilon Approximation Sampling Enabled) that outputs a sample. It starts with a relatively large simple random sample of transactions and deterministically trim the sample to create a final subsample whose distance from the complete database is as small as possible. For reasons of computational efficiency, it defines the subsample as *close* to the original database if the high-level aggregates of the subsample normalized by the total number of data points are close to the normalized aggregates in the database. These normalized aggregates typically correspond to 1-itemset or 2-itemset supports in the association-rule setting or, in the setting of a contingency table, relative marginal or cell frequencies. The key innovation of EASE lies in the method by which the final subsample is obtained. Unlike FAST [2], which obtains the final subsample by trimming away outliers in a process of quasi-greedy descent, EASE uses an approximation method to obtain the final subsample by repeated halving. Unlike FAST, the EASE provides a guaranteed upper bound on the distance between the initial sample and final subsample. In addition, EASE can process transactions on the fly, i.e., a transaction is examined only once to determine whether it belongs to the final subsample. Moreover, the average time needed to process a transaction is proportional to the number of items in that transaction. We showed that EASE leads to much better estimation of frequencies than SRS. Experiments in the context of both association-rule mining and classical contingency-table analysis indicated that EASE outperforms both FAST and SRS.

However, EASE has some limitations.

- Due to its halving nature, EASE has certain granularity in sample ratio. In [1] an ad-hoc solution was proposed where the size of the initial sample is so chosen that by repeated halving of several rounds, one obtains the required size. In this paper we propose a modified algorithm (EASIER) that outputs the required sample size in just one iteration, thus saving both time and memory (detailed analysis for time and memory are given in later sections). By doing so, EASIER no longer guarantees an upper bound on the distance between the initial sample and final subsample. However, experimental results in various domains (including, transactional data, image data, and audio data) suggest that EASIER attains the accuracy of EASE or even better.
- EASE was built especially for categorical count data, e.g., transactional data. It was natural to extend it to continuous data. In this paper we show how EASIER is used for image data that has continuous attribute values (Color Structure Descriptor - CSD). The CSD values are discretized using a simple discretization technique, and EASIER is successfully applied.[1]

We validate and compare the output samples of EASE[2], SRS and EASIER by performing two important data mining tasks: classification and association rule mining. Support vector machine (SVM) is used as the classifier to image data due to its high classification accuracy and strong theoretical foundation [3]. SVM classifier results show that EASIER samples outperforms SRS samples in accuracy and EASE samples in time. EASIER achieves the same or even better accuracy than EASE. Similar results are obtained for association rule mining. We also report the association rule mining results for IBM QUEST data set [4] in order to compare and contrast with the earlier reported results [1].

---

[1] We have also applied to audio data successfully, but due to space constraint we will not discuss it any further.
[2] There exist a few variants of EASE. Owning to time constraint and the goal of generalization, we use EASE for comparison.

The rest of the paper is organized as follows. In Section 2, we briefly review the ε-approximation method and EASE algorithm. In Section 3 we introduce the new EASIER algorithm and analyze the performance vis-a-vis EASE. The application of image processing is discussed in Section 4. In Section 5 experimental results are presented. Conclusion and future work are given in Section 6.

Notation. Denote by D the database of interest, by S a simple random sample drawn without replacement from D, and by I the set of all items that appear in D. Let $N = |D|$, $n = |S|$ and $m = |I|$. Also denote by $\mathcal{I}(D)$ the collection of itemsets that appear in D; a set of items A is an element of $\mathcal{I}(D)$ if and only if the items in A appear jointly in at least one transaction $t \in D$. If A contains exactly $k(\geq 1)$ elements, then A is sometimes called a k-itemset. In particular, the 1-itemsets are simply the original items. The collection $\mathcal{I}(S)$ denotes the itemsets that appear in S; of course, $\mathcal{I}(S) \subseteq \mathcal{I}(D)$. For $k \geq 1$ we denote by $\mathcal{I}_k(D)$ and $\mathcal{I}_k(S)$ the collection of k-itemsets in D and S, respectively.

For an itemset $A \subseteq I$ and a transactions set T, let $n(A;T)$ be the number of transactions in T that contain A. The support of A in D and in S is given by $f(A; D) = n(A;D)/|D|$ and $f(A; S) = n(A;S)/|S|$, respectively. Given a threshold $s > 0$, an item is frequent in D (resp., in S) if its support in D (resp., in S) is no less than s. We denote by $L(D)$ and $L(S)$ the frequent itemsets in D and S, and $L_k(D)$ and $L_k(S)$ the collection of frequent k-itemsets in D and S, respectively. Specifically, denote by S^i the set of all transactions in S that contains item A_i, and by r_i and b_i the number of red and blue transactions in S^i respectively. Red means the transactions will be kept in final subsample and blue means the transactions will be deleted. Q is the penalty function of r_i and b_i. f_r denotes the ratio of red transactions, i.e., the sample ratio. Then the ratio of blue transactions is given by $f_b = 1 - f_r$.

2 Epsilon-Approximation Method and EASE Algorithm

In order to obtain a good representation of a huge database, ε-approximation method is used to find a small subset so that the supports of 1-itemset are close to those in the entire database. The sample S_0 of S is an ε-approximation if its discrepancy satisfies $Dist(S_0, S) \leq \varepsilon$. The discrepancy is computed as the distance of 1-itemset frequencies between any subset S_0 and the superset S:

$$Dist_\infty(S_0, S) = \max_{A \in I_1(S)} |f(A; S_0) - f(A; S)|. \tag{1}$$

Given an $\varepsilon > 0$, Epsilon Approximation Sampling Enabled (EASE) algorithm is proposed to efficiently obtain a sample set S_0 which is an ε-approximation of S. S is obtained from the entire dataset D by using SRS. A repeated halving method keeps about half of the transactions in each round. Each halving iteration of EASE works as follows:

1. In the beginning, uncolor all transactions.
2. Color each transaction in S as red or blue. Red means the transaction is selected in sample S_0 and blue means the transaction is rejected.

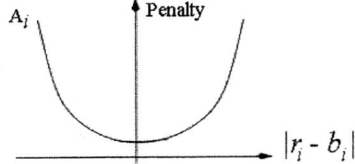

Fig. 1. The penalty function for the halving method: penalty as a function of $|r_i - b_i|$

3. The coloring decision is based on a penalty function Q_i for item A_i. Q_i is low when $r_i = b_i$ approximately, otherwise Q_i increases exponentially in $|r_i - b_i|$. The shape of Q_i is depicted in Figure 1. Q_i for each item A_i is as follows:

$$Q_i = Q_i^{(j)} = Q_{i,1}^{(j)} + Q_{i,2}^{(j)} \tag{2}$$

$$Q_{i,1}^{(j)} = (1+\delta_i)^{r_i}(1-\delta_i)^{b_i} \qquad Q_{i,2}^{(j)} = (1-\delta_i)^{r_i}(1+\delta_i)^{b_i} \tag{3}$$

where $Q_i^{(j)}$ means the penalty of ith item in jth transaction and δ_i controls the steepness the penalty plot. The initial values of $Q_{i,1}$ and $Q_{i,2}$ are both 1.

Suppose the $(j+1)$-th transaction is colored as red (or blue), then the corresponding penalty function $Q_i^{(j||r)}$ (or $Q_i^{(j||b)}$) is:

$$\begin{array}{ll} Q_{i,1}^{(j||r)} = (1+\delta_i)Q_{i,1}^{(j)} & Q_{i,2}^{(j||r)} = (1-\delta_i)Q_{i,2}^{(j)} \\ Q_{i,1}^{(j||b)} = (1-\delta_i)Q_{i,1}^{(j)} & Q_{i,2}^{(j||b)} = (1+\delta_i)Q_{i,2}^{(j)} \end{array} \tag{4}$$

The penalty function of current transaction is the summation of penalties for all items. If $Q^{(j||b)} = \sum_i Q_i^{(j||b)}$ is less than $Q^{(j||r)} = \sum_i Q_i^{(j||r)}$, the $(j+1)$-th transaction will be colored blue and deleted. Otherwise, it will be colored red and added to the sample. The initial value of δ_i is $\sqrt{1 - \exp\left(-\ln(2m)/n\right)}$ where m is the number of items in original dataset and n is the initial sample size. The details can be found in [1].

3 New and Modified EASE: EASIER

EASE is a good sampling algorithm that outperforms SRS, but it has some disadvantages. In this section we analyzed the problems of EASE and proposed the new algorithm EASIER to avoid these problems.

3.1 Without Halving

In EASE the halving process has certain granularity. It can only compute a subset approximately half the size of S. If a different sample ratio is wanted other than half the size, we have to run the halving procedure several times with a proper initial random sample set S of data set D. This will consume more time and memory due to multiple halving iterations. In order to directly obtain a sample set of any sample ratio in one pass, the halving round is modified to select red transactions with a probability which is proportional to the desired final sample size. This will remove the need to store several levels of penalties. If we want to obtain a sample set from S with sample ratio r_s directly, the ratio of red transactions is $f_r = r_s$ and the ratio of blue transactions is $f_b = 1 - r_s$. Then we have $r_i = f_r \cdot |S^i|$ and $b_i = f_b \cdot |S^i|$. So $\frac{r_i}{f_r} = \frac{b_i}{f_b} = |S^i|$. As $\frac{r_i}{f_r} + \frac{b_i}{f_b} = 2|S^i|$, we use $\frac{r_i}{2f_r} = \frac{b_i}{2f_b}$ ($\frac{r_i}{2f_r} + \frac{b_i}{2f_b} = |S^i|$). As the objective of halving is to minimize $|r_i - b_i|$, and $r_i + b_i = |S^i|$ for each item i, our new method will be modified to minimize $\left|\frac{r_i}{2f_r} - \frac{b_i}{2f_b}\right|$ instead of $|r_i - b_i|$. The modified penalty Q_i of j-th transaction is:

$$Q_i = Q_i^{(j)} = Q_{i,1}^{(j)} + Q_{i,2}^{(j)} \tag{5}$$

$$Q_{i,1}^{(j)} = (1+\delta_i)^{\frac{r_i}{2f_r}}(1-\delta_i)^{\frac{b_i}{2f_b}} \qquad Q_{i,2}^{(j)} = (1-\delta_i)^{\frac{r_i}{2f_r}}(1+\delta_i)^{\frac{b_i}{2f_b}} \tag{6}$$

Suppose the $(j+1)$-th transaction is colored as r (or b), the corresponding penalty function $Q_i^{(j||r)}$ (or $Q_i^{(j||b)}$) in Equation 4 is changed to:

$$\begin{aligned} Q_{i,1}^{(j||r)} &= (1+\delta_i)^{\frac{r_i+1}{2f_r}}(1-\delta_i)^{\frac{b_i}{2f_b}} = (1+\delta_i)^{\frac{1}{2f_r}}(1+\delta_i)^{\frac{r_i}{2f_r}}(1-\delta_i)^{\frac{b_i}{2f_b}} \\ &= (1+\delta_i)^{\frac{1}{2f_r}} Q_{i,1}^{(j)} \end{aligned} \tag{7}$$

$$\begin{aligned} Q_{i,2}^{(j||r)} &= (1-\delta_i)^{\frac{1}{2f_r}} Q_{i,2}^{(j)} \\ Q_{i,1}^{(j||b)} &= (1-\delta_i)^{\frac{1}{2f_b}} Q_{i,1}^{(j)} \qquad Q_{i,2}^{(j||b)} = (1+\delta_i)^{\frac{1}{2f_b}} Q_{i,2}^{(j)} \end{aligned} \tag{8}$$

The computation process of $Q_{i,1}^{(j||r)}$ is given in Equation 7. Other penalty functions are computed with a similar procedure and the results are shown in Equation 8. The overall penalty is calculated as described in section 2.

For $Q_i^{(final)}$, we cannot guarantee $Q_i^{(final)} \le 2m$. As δ_i is a very small value, $(1+\delta_i)^{\frac{1}{2f_r}}$ and $(1+\delta_i)^{\frac{1}{2f_b}}$ are both close to 1. So $Q_i^{(final)}$ is close to $2m$. According to [1] the value of $\left|\frac{r_i}{2f_r} - \frac{b_i}{2f_b}\right|$ is close to:

$$\frac{\ln(2m)}{\ln(1+\delta_i)} + \frac{|S^i|\ln(1/(1-\delta_i^2))}{\ln(1+\delta_i)} \tag{9}$$

Therefore, the same δ_i, as in section 2, is used in EASIER. In algorithm 1 the completed EASIER algorithm is given. The penalty for each item i of a transaction is calculated only once. So it does not need to store the penalty for each halving iteration, and thus results in a reduction of memory from $O(mh)$ for EASE to $O(m)$. The time for processing one transaction is bounded by $O(T_{max})$ for EASIER whereas EASE requires $O(hT_{max})$ where T_{max} denotes the maximal transaction length in T. Thus, unlike EASE, EASIER is independent of sample size.

Algorithm 1. EASIER Sampling

Input: D, n, m, f_r
Output: S_0, the transactions in red color

1: **for** each item i in D **do**
2: $\quad \delta_i = \sqrt{1 - \exp\left(-\frac{\ln(2m)}{n}\right)}$;
3: $\quad Q_{i,1} = 1; \quad Q_{i,2} = 1$;
4: **end for**
5: **for** each transaction j in S **do**
6: $\quad$ color transaction j red;
7: $\quad Q^{(r)} = 0; \quad Q^{(b)} = 0$;
8: $\quad$ **for** each item i contained in j **do**
9: $\quad\quad Q_{i,1}^{(r)} = (1 + \delta_i)^{\frac{1}{2f_r}} Q_{i,1}; \quad Q_{i,2}^{(r)} = (1 - \delta_i)^{\frac{1}{2f_r}} Q_{i,2}$;
10: $\quad\quad Q_{i,1}^{(b)} = (1 - \delta_i)^{\frac{1}{2f_b}} Q_{i,1}; \quad Q_{i,2}^{(b)} = (1 + \delta_i)^{\frac{1}{2f_b}} Q_{i,2}$;
11: $\quad\quad Q^{(r)} += Q_{i,1}^{(r)} + Q_{i,2}^{(r)}; \quad Q^{(b)} += Q_{i,1}^{(b)} + Q_{i,2}^{(b)}$;
12: $\quad$ **end for**
13: $\quad$ **if** $Q^{(r)} < Q^{(b)}$ **then**
14: $\quad\quad Q_{i,1} = Q_{i,1}^{(r)}; \quad Q_{i,2} = Q_{i,2}^{(r)}$;
15: $\quad$ **else**
16: $\quad\quad$ color transaction j blue;
17: $\quad\quad Q_{i,1} = Q_{i,1}^{(b)}; \quad Q_{i,2} = Q_{i,2}^{(b)}$;
18: $\quad$ **end if**
19: $\quad$ **if** transaction j is red **then**
20: $\quad\quad$ set $S_0 = S_0 + \{j\}$;
21: $\quad$ **end if**
22: **end for**

4 Image Processing Application

In [1] we showed the application of EASE over association rule mining and classical contingency table analysis, and results showed that EASE performs better than SRS. In this paper we include another important application, image processing, mainly to show that EASE and EASIER are applicable to continuous data.

With the availability of the Internet and the reduction in price of digital cameras, we are experiencing a high increase in the amount of multimedia information. The need to analyze and manage the multimedia data efficiently is essential. For applications like machine learning and classification of multimedia, the training data is important and the use of a good sampling set would influence the final results significantly. EASIER is a suitable sampling algorithm for such application. This fast and efficient process can select a representative sample set from a large database based on a set of visual features dynamically.

4.1 Color Structure Descriptor

The tremendous growth of multimedia content is driving the need for more efficient methods for storing, filtering and retrieving audiovisual data. MPEG-7 is

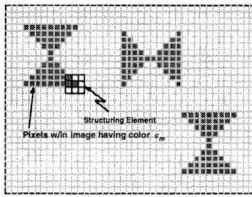

(a) Highly structured color plane (b) un-structured color plane

Fig. 2. Example of structured and un-structured color

a multimedia standard, which improves multimedia management by providing a rich set of standardized descriptors and description schemas for describing multimedia content. Based on one or several visual descriptors, users can efficiently browse the image database or retrieve similar images. Color Structure Descriptor (CSD) is one such descriptor in MPEG-7. It is defined to represent images by both the color distribution of the images (like a color histogram) and the local spatial structure of the color. CSD can distinguish the images which have similar histograms and different color spatial distribution. Figure 2 [5] illustrates this using a pair of images. Figure 2(a) is a highly structured color image and Figure 2(b) is a un-structured color image. As the number of foreground color pixels is the same, they cannot be distinguished with the traditional color histograms. But the CSDs of these two images are very different because the distributions of color are different. Compared with other color descriptors, CSD has the detailed color information and can achieve better retrieval results [6]. The format of CSD is identical to a color histogram. It is a 1-D array of eight bit-quantized values and can be 256-, 128-, 64- or 32-bin. In our experiments, 256-bins CSD is used.

4.2 Application of EASIER

We apply EASIER to find the representative samples from huge image databases according to CSD. Because EASIER is based on the calculation of the frequency of each item, the format of CSD is changed. Each CSD descriptor has 256 bins and each bin has an 8-bits numerical value. To handle non-binary data, the numerical value in a bin is converted into an indication on which bit position is set to one in a string of 256 binary bits (allocated for each bin) which has been initialized to zero. After mapping, there will be a total of $256 \times 256 = 65536$ (one-bit) items. The positions of each non-zero bit in this 65536 items list is subsequently converted into a vector and this vector is usually of a length of 100.

In order to reduce the number of items in the data set, the 8-bit bin value of the original CSD is re-quantized into a 4-bit quantized representation. The re-quantization is non-uniform and the suggested CSD amplitude quantization table in MPEG-7 standard [5] is used. This effectively reduces the number to $16 \times 256 = 4096$ items for each CSD descriptor. A smaller number of bits to represent the data will result in a loss of accuracy in the percentage of each color representation but this action will result in a significant reduction in the number of items for each CSD descriptor. Experimental results have shown that the retrieval accuracy of the quantized data is close to the original CSD data.

5 Experimental Results

In this section we compare the performance of EASIER, EASE and SRS in the context of classification and association rule mining. Both real-world data (CSDs of image database) and synthetic data (IBM QUEST data) are used for testing.

Primary metrics used for the evaluation are accuracy, sampling time, execution time and memory requirement. Sampling time is the time taken to obtain the final samples. Since association rule mining [4, 7] is focused on finding the frequent itemsets, i.e., the itemsets satisfying the minimum support, we have used a metric to measure the accuracy of our data reduction methods. In particular:

$$accuracy = 1 - \frac{|L(D) - L(S)| + |L(S) - L(D)|}{|L(D)| + |L(S)|} \qquad (10)$$

where, as before, $L(D)$ and $L(S)$ denote the frequent itemsets from the database D and the sample S, respectively. We used Apriori[3] [4] to compare the three algorithms in fair manner by computing the frequent item sets.

To verify the representativeness of the samples, a classification algorithm is used to test the classification performance of different training sample sets. For the choice of classification algorithm, SVM is a good candidate for the image classification applications [8]. The training set of SVM is selected by EASE, EASIER and SRS. Other remaining images are used for testing the SVM.

5.1 Image Processing (CSD)

We use COIL-100[4] as the image database. It has 7200 color images of 100 objects: for each object there are 72 images where each image is taken at pose intervals of 5 degrees. So the descriptor database includes 7200 CSD descriptors. All three algorithms, EASIER, EASE and SRS start from the whole dataset. As the halving process has a certain granularity and EASE cannot achieve the specific sample ratio, in each iteration we first ran EASE with a given number of halving times. Then we use the actual sample ratio from the EASE samples to generate EASIER samples. As EASIER is probabilistic, it does not guarantee the exact sample size. Hence, the actual EASIER sample size is used to generate SRS sample. Note that although EASE and EASIER sample sizes are not exactly the same, the difference is very little. For image classification, because EASE is based on halving, the final sample ratios of training sets are predetermined to be 0.5, 0.25, 0.125 and 0.0625.

---

[3] http://fuzzy.cs.uni-magdeburg.de/~borgelt/apriori.html
[4] http://www1.cs.columbia.edu/CAVE/research/softlib/coil-100.html

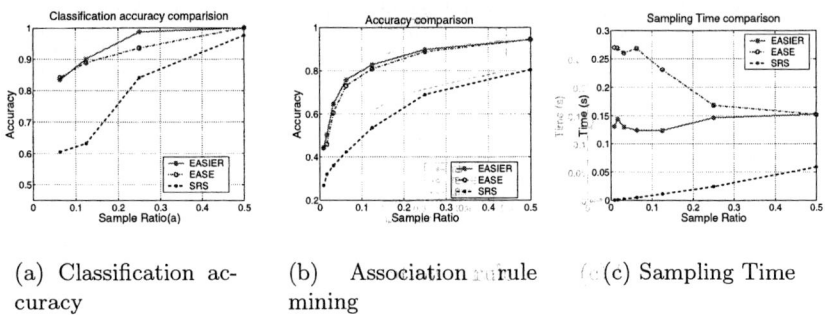

(a) Classification accuracy (b) Association rule mining (c) Sampling Time

Fig. 3. Performance of original CSD data extracted from COIL image set

For association rule mining, we applied halving one to seven times, hence the sample ratios are 0.5, 0.25, 0.125, 0.0625, 0.03125, 0.015625 and 0.0078125.

Since results of EASIER and EASE change with the input sequence of data, for each sample ratio we run EASIER and EASE for 50 times and in each iteration the input descriptors are shuffled randomly. SRS is also run 50 times over this shuffled data. The association rule mining results of EASIER, EASE and SRS are computed as an average over these 50 runs. The minimum support value for Apriori is set to 0.77% and we evaluated only the 1-item frequent set for CSD data. Otherwise there are too many frequent itemsets. However, for IBM transaction data, we use all frequent itemsets. The image classification results are average of 10 runs.

The results of original CSD data (65536 items) are shown in Figure 3. Figure 3(a) and Figure 3(b) shows the correct rate of image classification and the accuracy of association rule mining respectively. EASIER achieves similar accuracy as EASE which is better than SRS for both classification and association rule mining, especially when the sample ratio is very small. For a sample ratio 0.0625, EASIER achieves 83.5% classification rate, while EASE achieves 82.3% and SRS achieves only 59.7%. Figure 3(c) shows the sampling time. EASIER outperforms EASE for all sample ratios smaller than 0.5. It requires an almost fixed amount of time whereas EASE requires more time as the sample ratio falls.

The results of re-quantized CSD data (4096 items) are shown in Figure 4. Accuracy of EASIER is similar to EASE with less running time. For example, for a sample ratio of 0.000785, EASIER achieves 85.7% accuracy of association rule mining, while the accuracy of EASE is 85.4% and SRS achieves only 71.5%. The sampling time of EASIER does not change with the sample ratio. As the number of items is reduced, the running time of EASIER is reduced and closer to SRS.

Memory consumption comparison: In EASE, the memory required for storing the penalty function increases with the halving times. For example, when we applied seven halvings to CSD data, the required memory for storing the penalty of CSD items is about 9MB. But for EASIER, the memory is only about 3MB.

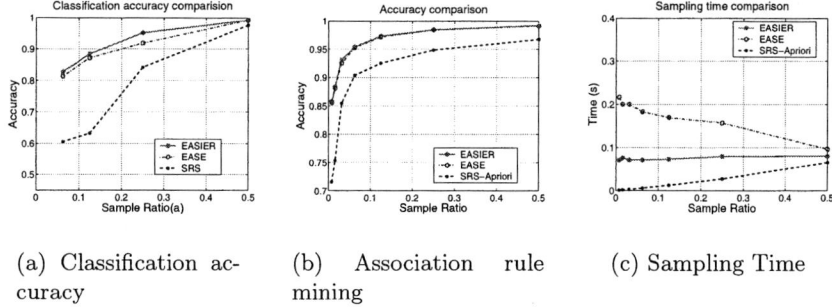

(a) Classification accuracy (b) Association rule mining (c) Sampling Time

Fig. 4. Performance of re-quantized CSD data extracted from COIL image set

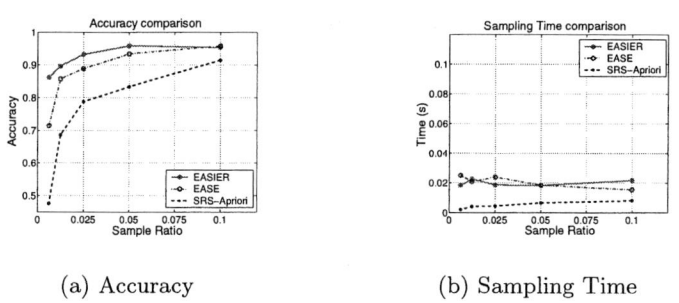

(a) Accuracy (b) Sampling Time

Fig. 5. Performance of IBM QUEST transaction data

5.2 IBM Transaction Data

In order to further compare the performance of the three algorithms, the IBM QUEST data [1] is used to test the performance of association rule mining. The dataset has total 98,040 transactions and the total number of items is 1000. The average length of these transactions is 10, and the average length of potentially frequent item sets is 4. The minimum support value is set to 0.77%. All three algorithms start from a 20% simple random sample S of the original database. One to five halvings are applied for EASE. Thus the final sample ratios is 0.1, 0.05, 0.025, 0.0125 and 0.00625 of the whole dataset. The three algorithms generate samples using the described setting. All three algorithms run 50 times for each sample and the results are the average over these 50 runs. For EASIER and EASE, in each iteration a different random sample is used as the initial sample.

Figure 5(a) shows that the accuracy of EASIER is better than EASE in small sample ratio. For ratio 0.00625, the accuracy of EASIER is about 86.2% while EASE has only 71.2% accuracy. The SRS gives the worst accuracy of 41.2%. The sampling time of the three methods are very similar as shown in Figure 5(b).

6 Conclusion

In this paper we proposed a new sampling algorithm, EASIER. The algorithm is similar to its predecessor EASE but it reduces the requirements for time and memory. In EASE, the sampling time and memory are increased when the sample ratio is reduced. However, in EASIER the running time is almost fixed and the memory is independent of the sample ratio. Another improvement is, due to its halving nature, in EASE we must change the size of initial sample to obtain some specific ratios. But using EASIER, any sample ratio can be obtained directly from the orginal set. We have evaluated the performance of EASIER using both real-world and synthetic data. Experiments show that EASIER is a good approximation algorithm which can obtain better sampling results with almost fixed time and even better accuracy than EASE.

In this paper we have applied EASIER to image applications with continuous features. As EASIER can flexibly generate representative samples of huge image database, it is used to select the training sets for an SVM classifier in image domain. The performance shows that an EASIER sample represents the original data much better than a simple random sample. EASIER is an online algorithm where the incoming transactions are processed once and a decision is taken regarding its participation in the final sample. This scheme is very conducive for stream data processing. The idea is to maintain a sample for the stream data dynamically. Just like reservoir sampling [9], each incoming transaction is processed in a given amount of time. But unlike reservoir sampling where this decision is made based solely on probability, EASIER makes informed decisions. The early idea is to maintain a ranking among the selected transactions in the reservoir sample. When a new streaming transaction arrives, its rank is determined by calculating the change in distance of the reservoir sample from the actual data. If its rank is higher than the lowest rank among the reservoir transactions, it is selected.

References

1. Brönnimann, H., Chen, B., Dash, M., Haas, P., Scheuermann, P.: Efficient data reduction with EASE. In: Proc. 9th Int. Conf. on KDD. (2003) 59–68
2. Chen, B., Haas, P., Scheuermann, P.: A new two-phase sampling based algorithm for discovering association rules. In: Proc. Int. Conf. on ACM SIGKDD. (2002)
3. Chapelle, O., Halffiner, P., Vapnik, V.N.: Support vector machine for histogram based image classification. IEEE Trans. on Neutral Network **10** (1999)
4. Agrawal, R., Srikant, R.: Fast algorithms for mining association rules. In: Proc. Int. Conf. on VLDB. (1994)
5. ISO/IEC15938-8/FDIS3: Information Technology - Multimedia Content Description Interface - Part 8:. (Extraction and use of MPEG-7 descriptions)
6. Ojala, T., Aittola, M., Matinmikko, E.: Empirical evaluation of mpeg-7 xm color descriptors in content-based retrieval of semantic image categories. In: Proc. 16th Int. Conf. on Pattern Recognition. (2002) 1021–1024

7. Han, J., Pei, J., Yin, Y.: Mining frequent patterns without candidate generation. In: Proc. Int. Conf. on ACM SIGMOD. (2000)
8. Jin, R., Yan, R., Hauptmann, A.: Image classification using a bigram model. In: AAAI Spring Symposium on Intelligent Multimedia Knowledge Management. (2003)
9. Vitter, J.: Random sampling with a reservoir. ACM Trans. Math. Software (1985)

Cluster-Based Rough Set Construction*

Qiang Li and Bo Zhang

Department of Computer Science and Technology,
Tsinghua University, Beijing 100084, China
lq@s1000e.cs.tsinghua.edu.cn

Abstract. In many data mining applications, cluster analysis is widely used and its results are expected to be interpretable, comprehensible, and usable. Rough set theory is one of the techniques to induce decision rules and manage inconsistent and incomplete information. This paper proposes a method to construct equivalence classes during the clustering process, isolate outlier points and finally deduce a rough set model from the clustering results. By the rough set model, attribute reduction and decision rule induction can be implemented efficiently and effectively. Experiments on real world data show that our method is useful and robust in handling data with noise.

1 Introduction

Cluster analysis is an important task in data mining. It is widely used in a lot of applications, including pattern recognition, data analysis, image processing, etc. By clustering, one can discover overall pattern distributions and interesting correlations among data attributes. Unlike classification, clustering does not rely on predefined classes and class-labeled training examples. Conventional clustering categorizes objects precisely into one of the clusters based on their attributes, including partitioning methods such as k-means and k-medoids, hierarchical methods such as agglomerative and divisive algorithms, and density-based, grid-based methods, etc[1]. However, based on the clustering results, it needs further generate descriptions for each class, induce conceptual interpretations or decision rules.

Rough set theory, proposed by Pawlak[2], has been received considerable attention in the field of data mining since it provides a tool to treat the roughness of concepts mathematically[3]. To combine the clustering with rough set theory, we proposes a new method to define the equivalent relation on objects according to the closeness of their attributes in the initial stage of a bottom-up clustering. So the objects are divided into cliques and outliers. A clique is a multielement set of objects with compact density distribution, viz. its variance is below some given

---

* Supported by National Natural Science Foundation of China (60135010, 60321002) and Chinese National Key Foundation Research & Development Plan (2004CB318108).

threshold. An outlier is an isolated object introduced by noise or incomplete information. According to rough set theory, the cliques form the elementary concepts of the given data set. Clustering results can be interpreted in the terms of lower approximation and upper approximation of rough set theory. Rules are inducted based on the elementary concepts to describe the clustering results. Finally, clusters based on different groups of attributes are integrated in the same rough set model, by means of a family of equivalent relations. Attribute reduction and clustering results comparison can be implemented by the mathematical tools of rough set.

The paper is organized as follows. Section 2 reviews some important clustering algorithms, and describes the agglomerative approach we choose to construct the rough set. Section 3 introduces the rough set theory and its current applications to clustering. Then in section 4 we develop the new method to define the equivalence classes in the initial stage of clustering and its applications to analysis the clustering results. Experiments of artificial and real life data sets are described in section 5. The conclusion is given in section 6.

2 Review of Clustering Algorithms

Assume that an object is represented by an N-dimensional feature vector. Each component of the vector is an attribute of the object. Let $V = \{v_1, \ldots, v_Q\}, v_i \in R^N$, be a set of objects. The *prototype* (or *centroid*) for each cluster C_k is represented by c_k. The vectors are standardized independently in each component to the N-cube $[0, 1]^N$. This permits each attribute of the objects have the same influence on the clustering.

K-means clustering and fuzzy c-means clustering are the most popular statistical clustering algorithms. The name K-means originates from the value K, i.e. the number K of clusters. It is given in advance and K centroids of the clusters are chosen randomly. Then each object v_i is assigned to the cluster whose centroid c_k is the nearest one among the K clusters to v_i. The new centroid vectors of the clusters are calculated as follows:

$$c_k = \frac{\sum_{j \in C_k} v_j}{\|C_k\|}, k = 1, \ldots, K. \qquad (1)$$

The above process continues until all centroids of the clusters are invariant during the iteration. K-means algorithms are very fast and are often used as the initialization of many other clustering algorithms. But its disadvantages are obvious: the number K is hard to be determined; the clusters formed may be very different for different initially given centroids and for different input orders of the objects; only unit spherical-shaped clusters can be found and discovering clusters with arbitrary shapes is difficult.

In fuzzy C-means clustering, each object has various contributions to all clusters. This is represented by the fuzzy membership values u_{jk} so that each object v_j belongs to multiple clusters $c_k, k = 1, \ldots, C$ in some degree. The feature vectors and membership values are chosen to minimize an objective function, the weighting squared error, during the iteration process:

$$J(U, c) = \sum_{j=1}^{Q} \sum_{k=1}^{C} u_{jk}^m \|v_j - c_k\|^2, \tag{2}$$

where U is a membership matrix, u_{jk} is its element, c is a set of cluster prototype vectors, $c = \{c_1, \ldots, c_C\}$, and the exponent m of u_{jk} is a severe constriction for cluster overlap, usually assigned a value among 1.5 to 2.5. The convergence of the membership matrix U does not assure the minimum of the objective function $J(U, c)$ defined in (2) due to the local minima and saddle points.

C.G. Looney [4] proposed a new agglomerative clustering algorithm to overcome many perplexing problems in clustering, such as (i) determine the optimal number K of clusters; (ii) prevent the selection of initial prototypes from affecting the clustering results; (iii) prevent the order of the cluster merging from affecting the clustering results; (iv) permit the clusters to form more natural shapes rather than only unit spheres. This approach can be easily generalized to fuzzy or non-fuzzy bottom-up clustering algorithms. Looney's agglomerative clustering algorithm is as follows:

Step 1: Standardize independently each component of the vectors to the N-cube $[0, 1]^N$.

Step 2: Determine a relatively large number K of prototypes uniformly and randomly via an empirically-derived formula: $K = \max\{6N + 12 \log_2(Q), Q\}$, avoiding the chosen prototype seeds affect the clustering.

Step 3: Thin the set of initial cluster prototypes out by successively finding the two closest prototypes and deleting one if their distance is less than the threshold $\tau = 1/(K)^{1/N}$.

Step 4: Assign vectors to a cluster by means of minimal distance to the prototype of the cluster.

Step 5: Compute cluster prototypes, by means of non-fuzzy formula (1) or the fuzzy C-means formula (2) or *modified weighted fuzzy expected value* formula in [4].

Step 6: Main loop goto step 4, until clusters converge.

Step 7: Eliminate empty clusters and any other clusters with p or fewer vectors. $p = 1$ works well on the first selection or p is selected by interaction. Reassigned the vectors of the eliminated clusters.

Step 8: Merge clusters that are too close together. The mean or *modified weighted fuzzy expected value* of the set of all distances $\{d(r) : r = 1, \ldots, K(K-1)/2\}$ between all pairs of centroids is computed and designated by D. A multiplier, $\beta = 0.5$, is suggested but the user can change it interactively. Each pair of clusters having $d(r) < \beta D$ is merged.

Step 9: Interact with the user to gradually increase the parameters β and p, goto step 7 to iterate the eliminating and merging clusters until the clustering result is acceptable.

We will present a new approach of constructing equivalence classes based on the above Looney agglomerative clustering algorithm and analyze the clustering results in section 4.

3 Rough Set Theory and Its Application to Clustering

Rough set theory has widely been used in data mining. This section briefly introduces the preliminary of rough set theory and current application in clustering.

Let U be a universe (a finite non-empty set of objects, $U = \{v_1, \ldots, v_{\|U\|}\}$) and $A = \{a_1, \ldots, a_{\|A\|}\}$ be a finite non-empty set of attributes(features). Each attribute $a \in A$ defines an information function $f_a : U \longrightarrow V_a$, where $V_a = \{f_a(v_i) \| v_i \in U\}$ called the domain of attribute a. The two-tuples (U, A) can be seen as an *information system* S (or an approximation space).If $A = C \cup D, C \cap D = \varnothing$, information system (U, A) called a decision table, where the elements of C are conditional attributes, the elements of D are decision attributes.

For any subset $R \subseteq A$, the equivalence relation $ind(R)$ can be defined as: $\{(v_i, v_j) | v_i \in U, v_j \in U, \forall a (a \in R \Rightarrow f_a(v_i) = f_a(v_j))\}$. The equivalence relation $ind(R)$ partitions the set U into disjoint subsets, denoted by $U/ind(R) = E_1, \ldots, E_m$, where E_i is an *equivalence class* of $ind(R)$. Each E_i is called *elementary set in* R because the elements in the same equivalence class are indistinguishable under R.

An arbitrary set $X \subseteq U$ can be represented by its lower and upper approximations using the elementary sets of R.The lower approximation $\underline{ind(R)}X$ is the union of all element sets which are subsets of X. The upper approximation $\overline{ind(R)}X$ is the union of all element sets which intersect the X non-empty. The pair $(\underline{ind(R)}X, \overline{ind(R)}X)$ is the rough set of X.

Recently some attempts have been made to introduce rough set into the clustering for managing imprecise concepts or induce cluster rules. Lingras and West[5] propose a variation of the K-means clustering algorithm based on the properties of rough sets. The proposed algorithm represents each cluster C_p as an interval set by using the lower bound $\underline{C_p}$ and upper bound $\overline{C_p}$. For each object v_t, let $d(v_t, c_i)$ be the distance to the centroid of cluster C_i. Let $d(v_t, c_i) = \min_{1 \le s \le K} d(v_t, c_s)$ and $T = \{j | d(v_t, c_i) - d(v_t, c_j) \le threshold$ and $i \ne j\}$.

1. If $T = \varnothing$, $v_t \in \underline{C_i}$.
2. Otherwise, $T \ne \varnothing$, $v_t \in \overline{C_i}$ and $v_t \in \overline{C_j}, \forall j \in T$. Furthermore, v_t is not part of any lower bound.

The above rough K-means algorithm can give unsupervised descriptions of imprecise cluster concepts. Its computation is simpler than that of the cut-subset method of fuzzy C-means clustering.

Huang et al.[6] propose a clustering method based the decision table. The clustering uses the selected subset of condition attributes and is evaluated by the decision attribute. The procedure starts from a small pre-specified number c of clusters. If all the variances of the decision attribute of the cluster are below the given threshold, the clustering results are accepted. Otherwise, increase the number c of clusters and the clustering algorithm continues. Then the attributes are discretized into a few intervals and the decision rules are inducted according to the rough data model(RDM). Actually, this can be done more effectively by supervised learning, which assigns different weight values to conditional attributes for computing the classification surface.

Hirano et al.[7] presente a new algorithm which directly uses the rough sets to implement clustering. Each object v_i forms an equivalence relation R_i by partitioning all objects into two equivalence classes: $\{\{v_j|d(v_i,v_j) \leq threshold\}, \{v_j| d(v_i,v_j) > threshold\}\}$, denoted by $[v_i]_{R_i}$ and $\overline{[v_i]_{R_i}}$ respectively, where $d(v_i, v_j)$ is the distance between two objects. Then modify each equivalence relation R_i by $R'_i = \{\{v_j \mid \|[v_i]_{R_i} \cap [v_j]_{R_j}\| + \|\overline{[v_i]_{R_i}} \cap \overline{[v_j]_{R_j}}\| \geq \rho\|U\|\}, \{v_j|others\}\}$, where ρ is another threshold. The intersection of all modified equivalence classes embodies the clustering results. Obviously this approach can not form more natural cluster rather than the sphere shape.

Expectation maximization (EM) clustering can only produce convex clusters and is sensitive to initial conditions. P. Mitra and S.K. Pal et al.[9] suggest initializing the parameters of EM algorithm by rough set. Each feature of an object is represented in terms of the membership to three fuzzy linguistic sets *low*, *medium* and *high*. Threshold the fuzzy membership values to obtain a high dimensional binary feature vector for each object. This forms an attribute-value table whose reducts and logic rules are generated by the discernibility function of rough set theory. The initial paremeters of EM are estimated as follows: the number of component Gaussian density functions (K) is the number of the distinct logic rules; the weights, means and diagonal covariance matrices of component Gaussians can be computed from the construction of fuzzy sets and derivation of logic rules.

4 Construction of the Rough Set Based on Clustering

In this section, we present an approach to construct the rough set during the bottom-up clustering process. It is obvious that the data objects can be divided into outlier points and the points distributed densely over some small spaces, called *cliques* here. Each cluster usually consists of some cliques and outliers, but the objects within the same clique may belong to different clusters due to the different given number K of clusters or some clusters including fewer objects may be deleted. Thus we can regard a clique as a part of a multielement equivalence class and a outlier point as a singleton equivalence class.

Definition 1. Let the variance of a multielement set E_v of objects be σ_v^2, if $\sigma_v^2 \leq \rho$, we called E_v a *clique*, where ρ is a threshold experimentally assigned. If E_v is a maximum clique, i.e. E_v is not a proper subset of other cliques, E_v forms a part of a multielement equivalence class.

Definition 2. A *clique expansion* is formed by merging two of existed cliques or clique expansions that are close enough each other. The detail criterion of generating clique expansions is shown below. The maximums of clique expansions are the multielement equivalence classes of objects.

Definition 3. An object v_i not belonging to any clique or clique expansion is called an *outlier* point. $\{v_i\}$ is a singleton equivalence class.

A direct attempt is that the distances of all pairs of objects are computed and formed the distance matrix $D = \{d_{ij}\}$, where d_{ij} is the distance of two objects v_i and v_j. The cliques and outlier points can be discovered by netting clustering on the distance matrix. But this method can't avoid the chaining effect of spanning tree. Thus we will try another way to alter the Looney agglomerative clustering algorithm to construct the maximal cliques and the outlier points in the clustering process. Let the $CliquesEx$ be the set of the intermediate results of singletons, cliques and clique expansions during the clustering. For a set S of objects, $\overline{S}$ and $\underline{S}$ are its upper approximation and lower approximation under the equivalence classes $CliquesEx$. The key issues in the algorithm implementation are described as follows:

Clique Initialization. For a cluster C_t as the intermediate result of clustering process, compute the variance $\sigma^2_{C_t}$ about its centroid. If the $\sigma^2_{C_t} \leqslant \rho$, the cluster C_t is accepted as a clique and added to $CliquesEx$.

Clique Expansion. When two clusters C_s, C_t are merged in the clustering process, compare all pair $<E_i, E_j> \subseteq (CliquesEx \times CliquesEx)$ for $E_i \subseteq \overline{C_s}$ and $E_j \subseteq \overline{C_t}$. Let c_S be the centroid of the set S in a distance space. If the sphere defined with radius $(\|E_i\|\sigma_{E_i} + \|E_j\|\sigma_{E_j})/(\|E_i\| + \|E_j\|)$ and center point $c_{E_i \cup E_j}$ contains a certain percentage of the objects in the two cliques respectively, we construct a clique expansion $E_k = E_i \cup E_j$. Add E_k to $CliquesEx$ and eliminate E_i and E_j from $CliquesEx$.

Clique Elimination. For any $E_i \in CliquesEx$, if there exist $E_j \in CliquesEx$ and $E_i \subset E_j$, eliminate E_i from $CliquesEx$.

Our algorithm CRS (Cluster-based Rough Set construction) is described as follows:

Step 1: Initialize the clustering, including step 1 to step 4 of Looney agglomerative clustering algorithm. Let the set $CliquesEx$ be empty.

Step 2: Execute the conventional clustering algorithms, such as K-means or fuzzy C-means, till the clusters converge.

Step 3(Clique Initialization): Compare the variance of each cluster with the threshold ρ, add the clusters satisfied the criteria to the set $CliquesEx$.

Step 4: Eliminate the clusters with p or fewer objects. Reassigned the vectors of the eliminated clusters to the nearest clusters.

Step 5: Merge each possible pair $<C_s, C_t>$ of the clusters according to the distances of the pair of the centroids. Then merge all pairs $<E_i, E_j>$ that $E_i, E_j \in CliquesEx$, $E_i \subseteq \overline{C_s}$, $E_j \subseteq \overline{C_t}$ and $<E_i, E_j>$ satisfies the condition of clique expansion. Delete E_i and E_j from the set $CliquesEx$ after they are merged and add $E_i \cup E_j$ to $CliquesEx$.

Step 6: Interact with the user to gradually increase the parameters β and p, goto step 4 to iterate the eliminating and merging until the clustering result is acceptable.

This algorithm only add the computational complexity on the step 5 than the bottom-up clustering algorithms. It is obvious that algorithm CRS has the similar efficiency with Looney's agglomerative clustering algorithm. The algorithm CRS is applicable to more natural cluster shapes, while several algorithms in section 3 are only limited to the case of mixed Gaussian models. If objects have decision attributes just like the Wisconsin breast cancer data, let M be the number of all distinct decision attribute values. The algorithm CRS is iterated when the current number of clusters is not less the M. We adopt the total decision errors as the clustering validity measure:

$$V(j) = \sum_{m=1}^{K_j} \text{DError}_m$$

where the K_j is the number of clusters in the j-th iteration of the step 6 of CRS and DError defined in next section. If there is not a decision attribute, the modified Xie-Beni (MXB) measure[4] can be used as the evaluation of validity.

5 Experimental Results

The first experiment is a simple two-dimensional data set used by Looney[4] to demonstrate the clique initialization and merge. We directly use the coordinate values not standardized to $[0,1]^2$. Let the threshold $\rho = 0.75$. After the step 1-3 of algorithm CRS, all the elements of the $CliquesEx$ are singletons. In the step 5 clusters and the elements of $CliquesEx$ are merged. Then we have the result: three clusters, and five multielement equivalence classes and one singleton equivalence classes (outlier point) in the $CliquesEx$, depicted in Figure 1.

The next data set is the Wisconsin breast cancer data[8] to diagnose breast masses based on the digital images of the sections of Fine Needle Aspiration(FNA) samples. Ten real-valued features are computed for each cell nucleus: radius, texture, perimeter, area, smoothness, compactness, concavity, concave points, symmetry and fractal dimension. The mean, standard error, and "worst" or largest (mean of the three largest values) of these features were computed for each image, resulting in 30 features. There are 569 instances in the data set, labeled for two classes: benign with 357 instances and malignant with 212 instances.

Given the initial number $K = 569$ of clusters, assign the prototypes of the clusters randomly and uniformly, then thin the set of clusters by deleting the close ones, we get 325 cluster prototypes. After the initial K-means clustering and delete all empty cluster, we get 33 clusters as the first stage of clustering result presented in Table 1, where the column 'count' is the total elements of each cluster, the column 'decision value' is the benign or malignant which the most

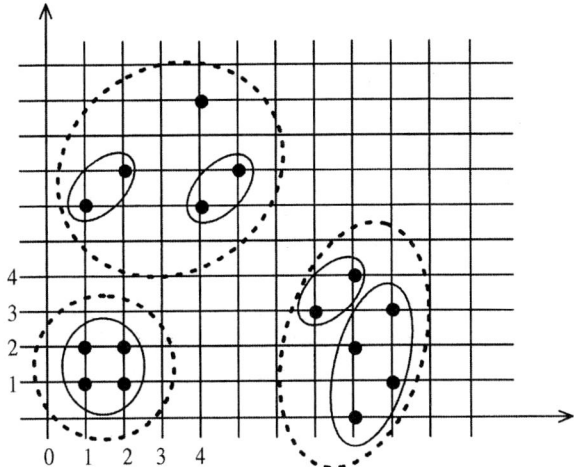

Fig. 1. Experiment 1: get three clusters and one singleton and five multielement equivalence classes

Table 1. Original Cliques and singletons

No.	Count	σ	decision value	DError	No.	Count	σ	decision value	DError
1	1	0	M	0	18	7	0.567	B	0
2	67	0.308	B	0.045	19	1	0	M	0
3	4	0.459	M	0	20	5	0.335	M	0
4	3	0.370	M	0	21	3	0.437	M	0
5	13	0.39	M	0	22	45	0.405	B	0.067
6	7	0.545	B	0	23	41	0.419	B	0
7	27	0.415	M	0.148	24	20	0.402	M	0
8	1	0	M	0	25	12	0.388	M	0
9	12	0.394	M	0	26	11	0.449	M	0
10	67	0.396	B	0.045	27	1	0	M	0
11	18	0.450	M	0	28	1	0	M	0
12	73	0.322	B	0.014	29	43	0.414	B	0
13	7	0.414	M	0	30	3	0.728	B	0
14	33	0.358	M	0.152	31	4	0.605	M	0
15	1	0	M	0	32	4	0.603	M	0
16	1	0	M	0	33	25	0.351	M	0
17	8	0.445	M	0					

elements of the cluster belong to, and the column 'DError' is the exception ratio of the decision value of the cluster. Since the maximum standard deviation of the clusters is 0.728, we choose the threshold $\rho = 0.530$, i.e. the square of 0.728. Now each cluster satisfies the criteria to form cliques, $CliquesEx$ is initialized by these 33 sets.

Then in the step 4-6 of algorithm CRS, eliminations with $p = 1, 3$ and 7, followed by merging clusters with $\beta = 0.2, 0.25, 0.3, 0.5$ and 0.66, respectively,

Table 2. The final clique expansions

No. of clique expansion	origins in the Table 1
1	2,12,17
2	10,29
3	5,14
4	7,24
5	9,33

brought K down to 16, 13, 10, 5 and 2, and total decision errors V down to 0.471, 0.413, 0.385, 0.302 and 0.226 respectively. The final cluster sizes are 316 and 253. And we get the clique expansions from the original $CliquesEx$ presented in Table 2. And there are seven singletons as the outlier points of the data set.

6 Conclusions

This paper proposes a new approach to construct rough sets during the bottom-up clustering process. With the rough set theory, we can identify the outliers and the concept granularities to interpret the constructions of clusters. Experiments on the artificial and real life data show the algorithm effective and efficient. It remains as a future work to investigate the attribute reduction and rule induction based on rough set to compare and improve the clustering according to different groups of attributes.

References

1. Han, Jiawei and Kamber, Micheline: Data mining: Concepts and Techniques. Morgan Kaufmann Publishers, San Francisco, (2001)
2. Pawlak, Z. Rough sets, theoretical aspects of reasoning about data. Kluwer Academic Publishers, Dordrecht, (1991)
3. Pawlak, Z. Rough set theory and its applications to data analysis, International Journal of Cybernetics and Systems, 29(1998), 661-688
4. Looney, Carl G.: Interactive clustering and merging with a new fuzzy expected value. Pattern Recognition, 35(2002) 2413–2423
5. Lingras, P.J. and West, C.: Interval Set Clustering of Web Users with Rough K-means. Journal of Intelligent Information System, 23(2004) 5–16
6. Huang, Jin-jie et al: A fast approach to building rough data model through G-K fuzzy clustering. Proceedings of the Second International Conference on Maching Learing and Cybernetics, IEEE Publisher (2003), 1559–1564
7. Hirano, Shoji et al: A rough set-based clustering method with modification of equivalence relations. Proceedings of PAKDD 2001, LNAI 2035, Springer Verlag, (2001), 507–512

8. Wolberg, W.H. and Mangasarian, O.L.: Multisurface method of pattern separation for medical diagnosis applied to breast cytology. Proceedings of the National Academy of Sciences of the USA, 87(1990) 9193-9196. See also: Wisconsin Breast Cancer Data, http://www.cs.wisc.edu/~olvi/uwmp/cancer.html
9. Mitra, P., Pal, S.K. and Siddiqi, M.A.: Non-convex clustering using expectation maximization algorithm with rough set initialization. Pattern Recognition Letters, 24(2003) 863-873

Learning Bayesian Networks Structures from Incomplete Data: An Efficient Approach Based on Extended Evolutionary Programming

Xiaolin Li[1], Xiangdong He[2], and Senmiao Yuan[1]

[1] College of Computer Science and Technology, Jilin University,
Changchun 130012, China
lixl@email.jlu.edu.cn, yuansenmiao@hotmail.com
[2] VAS of China Operations, Vanda Group,
Changchun 130012, China
hexd_163@163.com

Abstract. This paper describes a new data mining algorithm to learn Bayesian networks structures from incomplete data based on extended Evolutionary programming (EP) method and the Minimum Description Length (MDL) metric. This problem is characterized by a huge solution space with a highly multimodal landscape. The algorithm presents fitness function based on expectation, which converts incomplete data to complete data utilizing current best structure of evolutionary process. Aiming at preventing and overcoming premature convergence, the algorithm combines the restart strategy into EP. The experimental results illustrate that our algorithm can learn a good structure from incomplete data.

1 Introduction

The Bayesian belief network is a powerful knowledge representation and reasoning tool under conditions of uncertainty. Recently, learning the Bayesian network from a database has drawn noticeable attention of researchers in the field of artificial intelligence. To this end, researchers developed many algorithms to induct a Bayesian network from a given database [1], [2], [3], [4], [5], [6].

Very recently, researchers have begun to tackle the problem of learning the network from incomplete data. A major stumbling block in this research is that when in closed form expressions do not exist for the scoring metric used to evaluate network structures. This has led many researchers down the path of estimating the score using parametric approaches such as the expectation-maximization (EM) algorithm [7]. However, it has been noted [7] that the search landscape is large and multimodal, and deterministic search algorithms find local optima. An obvious choice to combat the problem is to use a stochastic search method.

This paper developed a new data mining algorithm to learn Bayesian networks structures from incomplete data based on extended Evolutionary Programming (EP) method and the Minimum Description Length (MDL) metric. The algorithm presents fitness function by using expectation, which converts incomplete data to complete data utilizing current best structure of evolutionary process. Another important characteristic of our algorithm is that, in order to preventing and overcoming premature convergence, we combine the restart technology [8] into EP. Furthermore, our algorithm, like some previous work, does not need to impose the restriction of having a complete variable ordering as input.

We'll begin by briefly introducing Bayesian network and MDL metric. Next we will introduce the restart-EP method. In section 4, we will describe the algorithm based on the restart-EP method and the MDL metric. In the end, we will conduct a series of experiments to demonstrate the performance of our algorithm and sum up the whole paper in section 5 and 6, respectively.

2 Bayesian Network and MDL Metric

2.1 Bayesian Network

A Bayesian network is a directed acyclic graph (DAG), nodes of which are labeled with variables and conditional probability tables of the node variable given its parents in the graph. The joint probability distribution (JPD) is then expressed by the formula:

$$p(x_1,...,x_n) = \sum_{i=1}^{n} p(x_i \mid \pi(x_i)) \quad (1)$$

where $\pi(x_i)$ is the configuration of X_i's parent node set $\Pi(X_i)$.

2.2 The MDL Metric

The MDL metric [9] is derived from information theory. With the composition of the description length for network structure and the description length for data, the MDL metric tries to balance between model accuracy and complexity. Using the metric, a better network would have a smaller score. Similar to other metrics, the MDL score for a Bayesian network, S, is decomposable and could be written as in equation 2. The MDL score of the network is simply the summation of the MDL score of $\Pi(X_i)$ of every node X_i in the network.

$$MDL(S) = \sum_{i} MDL(X_i, \Pi(X_i)) \quad (2)$$

According to the resolvability of the MDL metric, equation 2 can be written when we learn Bayesian networks from complete data as follow:

$$MDL(S) = N \sum_{i=1}^{N} \sum_{X_i, \Pi(X_i)} P(X_i, \Pi(X_i)) \log P(X_i, \Pi(X_i)) \quad (3)$$

$$- \sum_{i=1}^{N} \frac{\log N}{2} \| \Pi(X_i) \| (\| X_i \| -1)$$

3 Restart-EP

Although EP was first proposed as an evolutionary algorithm to artificial intelligence, it has been recently applied to many numerical and combinatorial optimization problems successfully.

One of EP's key features is its self-adaptation scheme. In EP, mutation is typically the only operator used to generate new offspring. The mutation is often implemented by adding a random number from a certain distribution to the parent. An important parameter of the Gaussian distribution is its standard deviation (or equivalently the variance). In the widely used self-adaptation scheme of EP, this parameter is evolved, rather than manually fixed, along with the objective variables.

Premature convergence is a serious issue in evolutionary algorithms since it might significantly degrade the overall performance. EP is easy to fall into local optimums. When a point enters the absorption domain of the certain local optimum, the factors and of many individuals diminish rapidly because of self-adaptation scheme.

We define a quantity which characterize the premature convergence. Suppose population $P = \{p_i = (x_i, \eta_i)\}_{i=1}^{m}$ have arranged by the fitness. p_1 denotes the most excellent individual.

$$mean = \frac{1}{k}\sum_{i=1}^{k}\max_{1 \leq j \leq n} \eta_i(j) \quad (4)$$

Where $k = [0.3 \times m]$, m is population size, $[\ ]$ denotes the integer function.

The main process of restart strategy is as follows. The population variety is monitored dynamicly in the evolutionary process. When the population variety decreases to a certain finitude, we consider that the trend of premature convergence appears. Then initialize afresh the population & comeback the population variety. So the evolution can progress effectively.

We combine the restart strategy into EP. When $mean$ is less than a positive number $threshold$ which is confirmed beforehand, we consider that the evolution has danger of premature convergence and initializes afresh the population. Based on previous analysis, we only initialize afresh the factors τ and τ'. Moreover, the individuals can get rid of the absorption domain of a local optimum and prevent premature convergence. We do not initialize afresh the objective vectors, which can withhold the evolutionary information better.

4 Learning Bayesian Network from Incomplete Data

The algorithm we propose is shown below.
1. Set to 0.
2. Create an initial population, Pop(t), of PS random DAGs. The initial population size is PS.
3. Convert incomplete data to complete data utilizing a DAG of the initial population randomly

4. Each DAG in the population Pop(t) is evaluated using the MDL metric.
5. While t is smaller than the maximum number of generations G
   a) Each DAG in Pop(t) produces one offspring by performing mutation operations. If the offspring has cycles, delete the set of edges that violate the DAG condition. If choices of set of edges exist, we randomly pick one choice.
   b) The DAG in Pop(t) and all new offspring are stored in the intermediate population Pop'(t). The size of Pop'(t) is 2*PS.
   c) Conduct a number of pair-wise competitions over all DAGs in Pop'(t). Let S_i be the DAG being conditioned upon, q opponents are selected randomly from Pop'(t) with equal probability. Let S_{ij}, $1 \leq j \leq q$, be the randomly selected opponent DAGs. The S_i gets one more score if $D_i(S_i) \leq D_i(S_{ij})$, $1 \leq j \leq q$. Thus, the maximum score of a DAG is q.
   d) Select PS DAGs with the highest scores from Pop'(t) and store them in the new population Pop(t+1).
   e) Compute *mean* of Pop(t+1). Initialize afresh the factors τ and τ' of every individual if *mean* < *threshold*.
   f) Increase t by 1
6. Return the DAG with lowest MDL metric found in any generation of a run as the result of the algorithm.

5 Experimental Results and Analyses

We have conducted a number of experiments to evaluate the performance of our algorithm. The learning algorithms take the data set only as input. The data set is derived from ALARM network (http://www.norsys.com/netlib/alarm.htm).

Firstly, we generate 5,000 cases from this structure and learn a Bayesian network from the data set ten times. Then we select the most perfect network structure as the final structure. We also compare our algorithm with a classical GA algorithm. The algorithms run without missing data. The MDL metric of the original network structures for the ALARM data sets of 5,000 cases is 81,219.74.

The population size PS is 30 and the maximum number of generations is 5,000. We employ our learning algorithm to solve the ALARM problem. The value of q is set to be 5. We also implemented a classical GA to learning the ALARM network. The one-point crossover and mutation operations of classical GA are used. The crossover probability p_c is 0.9 and the mutation probability p_m is 0.01. The MDL metric for our learning algorithm and the classical GA are delineated in Figure 1.

From Figure 1, we see that the value of the average of the MDL metric for restart-EP is 81362.1 and the value of the average of the MDL metric for the GA is 8,1789.4. We find our learning algorithm evolves good Bayesian network structures at an average generation of 4210.2. The GA obtains the solutions at an average generation of 4495.4. Thus, we can conclude that our learning algorithm finds better network structures at earlier generations than the GA does. Our algorithm can also prevent and overcome the premature convergence.

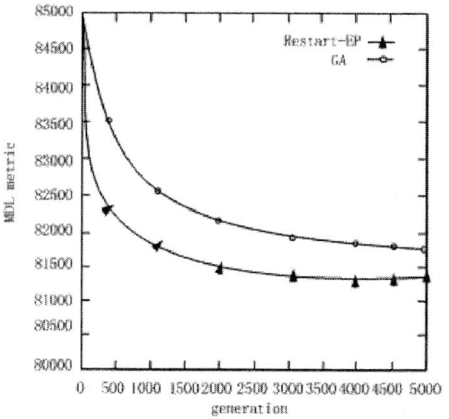

Fig. 1. The MDL metric for the ALARM network

Our algorithm generates 1000, 10000 cases from the original network for training and testing. The algorithm runs with 10%, 20%, 30%, and 40% missing data. The experiment runs ten times for each level of missing data. Using the best network from each run we calculate the log loss. The log loss is a commonly used metric appropriate for probabilistic learning algorithms. Figure 2 shows the comparison of log loss between our algorithm and reference [10].

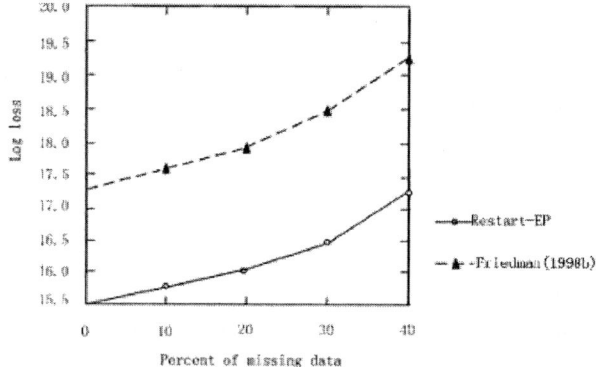

Fig. 2. The Comparison of log loss

As can be seen from figure 2, the algorithm finds better predictive networks at 10%, 20%, 30%, and 40% missing data than reference [10] does.

6 Conclusions

In this paper we describe a novel evolutionary algorithm for learning Bayesian networks from incomplete data. This problem is extremely difficult for deterministic algorithms and is characterized by a large, multi-dimensional, multi-modal search space. The experimental results show that our learning algorithm can learn a good structure from incomplete data.

References

1. J Suzuki. A construction of Bayesian networks from databases based on a MDL scheme, Proc of the 9th Confon Uncertainty in Artificial Intelligence. San Mateo, CA: Morgan Kaufmann, (1993) 266-273,
2. Y Xiang, S K M Wong. Learning conditional independence relations from a probabilistic model, Department of Computer Science, University of Regina, CA, (1994) Tech Rep: CS-94-03,.
3. D Heckerman. Learning Bayesian network: The combination of knowledge and statistic data, Machine Learning, Vol. 20, No. 2, (1995), 197-243,.
4. Cheng J, Greiner R, Kelly J. Learning Bayesian networks from data: An efficient algorithm based on information theory, Artificial Intelligence. Vol.137, No.1-2, (2002) 43-90.
5. Lam, W. and Bacchus, F., Learning Bayesian belief networks: An algorithm based on the MDL principle, Computational Intelligence, Vol 10, No.4, 1994.
6. P. Larranaga, M. Poza, Y. Yurramendi, R. Murga, and C. Kuijpers, Structure Learning of Bayesian Network by Genetic Algorithms: A Performance Analysis of Control Parameters, IEEE Trans. Pattern Analysis and Machine Intelligence, vol.
7. Friedman, N. (1998a). The Bayesian Structural EM Algorithm. Proceedings of the Fourteenth Conference on Uncertainty in Artificial Intelligence, Madison, WI, Morgan Kaufmann Publishers.
8. Eshelman L.J., The CHC adaptive search algorithm, In: Foundations of Genetic Algorithms. San Mateo: Kaufmann Publishers Inc, (1991) 265-283.
9. W. Lam and F. Bacchus. Learning Bayesian belief networks: an algorithm based on the MDL principle, Computational Intelligence, Vol.10, No.4, (1994) 269–293.
10. Friedman, N. (1998b). Learning Belief Networks in the Presence of Missing Values and Hidden Variables. Fourteenth International Conference on Machine Learning, Vanderbilt University, Morgan Kaufmann Publishers, (1997).

Dynamic Fuzzy Clustering for Recommender Systems

Sung-Hwan Min and Ingoo Han

Graduate School of Management, Korea Advanced Institute of Science and Technology,
207-43 Cheongrangri-dong, Dongdaemun-gu, Seoul 130-722, Korea
shmin@kgsm.kaist.ac.kr

Abstract. Collaborative filtering is the most successful recommendation technique. In this paper, we apply the concept of time to collaborative filtering algorithm. We propose dynamic fuzzy clustering algorithm and apply it to collaborative filtering algorithm for dynamic recommendations. We add a time dimension to the original input data of collaborative filtering for finding the fuzzy cluster at different timeframes. We propose the dynamic degree of membership and determine the neighborhood for a given user based on the dynamic fuzzy cluster. The results of the evaluation experiment show the proposed model's improvement in making recommendations.

1 Introduction

Recommender systems are used by e-commerce sites to suggest products and to provide consumers with information to help them decide which products to purchase. Due to an explosion of e-commerce, recommender systems are rapidly becoming a core tool for accelerating cross-selling and strengthening customer loyalty. Collaborative filtering (CF) is the most successful recommendation technique, which has been used in a number of different applications such as recommending movies, articles, products, Web pages [5]. CF is a general approach to personalized information filtering. CF systems work by collecting user feedback in the form of ratings for items in a given domain and exploit the similarities and differences among the profiles of several users in determining how to recommend an item [3, 2].

Finding neighbors in traditional CF is crucial for accurate recommendations because recommendations are based on the ratings of an active user's neighbors. But the current CF algorithms are not adaptive to these situations dynamically, which results in the false recommendations. Our research focuses on these situations.

This paper presents a new approach to collaborative filtering based on dynamic fuzzy cluster. We add a time dimension to the original input data of collaborative filtering for finding the fuzzy cluster at different timeframes. We propose the dynamic degree of membership and determine the neighborhood for a given user based on the dynamic fuzzy cluster. The proposed model is expected to find the active user's neighbors dynamically according to his or her changing pattern by using dynamic fuzzy cluster and improve recommendations.

2 Fuzzy Clustering

The most widely used fuzzy clustering algorithm is the fuzzy c-means (FCM) algorithm proposed by Bezdek [1,6]. FCM is a clustering method in which an object can be a member of different classes at the same time. FCM aims to determine cluster centers v_i (i=1, 2,...,c) and the fuzzy partition matrix U by minimizing the objective function J defined as follows:

$$J_m(U,V;X) = \sum_{j=1}^{n} \sum_{i=1}^{c} u_{ij}^m \|x_j - v_i\|^2 \tag{1}$$

where n is the number of individuals to be clustered, c is the number of clusters and u_{ij} is degree of membership of individual j in cluster i. The exponent m is used to control the fuzziness of membership of each datum. $\|x_j - v_i\|$ is the Euclidean norm between x_j and v_i. The FCM algorithm is as follows:

Step 1. Initialize u_{ij} by generating random numbers in the interval [0, 1] such that

$$\sum_{i=1}^{c} u_{ij} = 1 \; \forall j = 1,...,n \tag{2}$$

Step 2. Compute the fuzzy cluster centroid v_i for I = 1, 2, ..., c according to the following Eq. (3)

$$v_i = \frac{\sum_{j=1}^{n}(u_{ij})^m x_j}{\sum_{j=1}^{n}(u_{ij})^m} \tag{3}$$

Step 3. Update the degree of membership u_{ij} using

$$u_{ij} = \frac{1}{\sum_{k=1}^{c}(\frac{d_{ij}}{d_{kj}})^{2/m-1}} = \frac{1}{\sum_{k=1}^{c}(\frac{\|x_j - v_i\|}{\|x_j - v_k\|})^{2/m-1}} \tag{4}$$

Step 4. If the improvement in $J_m(U,V;X)$ is less than the given threshold ε, then stop. Otherwise go to step 2.

In this paper, the FCM algorithm is used in order to cluster users. We modify the FCM algorithm and apply it to CF for dynamic recommendations.

3 Dynamic Fuzzy Clustering for CF

This paper suggests a dynamic fuzzy cluster based collaborative filtering which is adaptive to users' changing patterns in order to improve recommendations. The procedure for the proposed model is as follows:

Step 1. Data Preparation: Add a time dimension to the original input data and reduce item dimension by using hierarchy information.
Step 2. User Clustering: Apply the FCM algorithm to produce p partitions of users. This is used as a cluster base for finding time-variant fuzzy cluster.
Step 3. Dynamic fuzzy cluster: Find fuzzy cluster at different timeframes for a given active user and compute the dynamic degree of membership.
Step 4. Neighbor Selection: Determine the neighborhood for a given user based on the dynamic fuzzy cluster.
Step 5. Recommendation: Predict the active users' rating unanswered based on neighborhood ratings.

Input data of a CF problem is usually a user-to-rating matrix. In order to detect the dynamic cluster change for an active user, we need to add a time dimension to the original input data. Table 1 shows item ratings for an active user at different timeframes. As shown in Table 1, each row is too sparse. To solve this problem, input data reduction (item dimension reduction) methods are needed. In this paper, we use a hierarchy of items whose leaf nodes represent items and non-leaf nodes represent a higher-level category to reduce the dimension of input data space.

Table 1. Time-to-Item Rating Matrix of an active user

Timeframe	Item1	Item2	Item3	Item4	...	Item n
T1	1					
T2		5	2			
T3				5		3
...					...	
Tn						

The rating for a category is defined as the average ratings for the items in that category as follows [7].

$$cr_{a,k} = \sum_{i \in category\ .k} \frac{1}{RN_k} r_{a,i} \qquad (5)$$

Table 2. Time-to-Category Rating Matrix

| Timeframe | Category 1 | | | | | Category n | |
	Item 1	Item 4	Item 7			Item 2	Item i
T1	1		3				
		2					
T2			1			5	
		1					
T3		5					1
		5				1	
...							

In above equation, $CR_{a,k}$ is the derived rating of category k of user A, RN_k is the number of rated items that belong to that category, and $r_{a,i}$ is the rating of item i of user A. These derived ratings of non-leaf level nodes are incorporated in computing the fuzzy cluster at different timeframes. Table 2 shows an example of category ratings.

Customers with similar interests are clustered by the FCM algorithms and this output is used as a base for detecting dynamic cluster change. In the FCM process, category ratings, calculated using item hierarchy information, are used as input data in order to extend the FCM into the dynamic FCM. Fuzzy cluster, defined as the degree of membership u_{ij}, is computed in this step. Crisp cluster of a given user is also determined. Crisp cluster of a given user j (CC_j) is defined as the cluster with the largest degree of membership for a given user as follows.

$$CC_a = k, \text{ if } u_{ka} = \max_i \{u_{ia}\} \text{ for user a} \tag{6}$$

Information on the crisp cluster of all users is used in neighbor selection step. Input data shown in Table 2 is used to find the fuzzy cluster at different timeframes for a given active user. One user may belong to the same fuzzy cluster at different timeframes, but another user may belong to the different fuzzy cluster at different timeframes. Different clusters at different timeframes means that the user may have a time-variant pattern. Sum of degree of membership (su_{ij}) is computed as follows.

$$su_{ij} = \sum_{t=1}^{T} w(t) \bullet u(t)_{ij} \tag{7}$$

where $u(t)_{ij}$ is the degree of membership of individual j in cluster i at timeframe t and w(t) is the weighting function which is used to weight $u(t)_{ij}$ differently according to timeframe. Dynamic degree of membership of individual j in cluster i (du_{ij}) is defined as follows.

$$du_{ij} = \frac{su_{ij}}{\sum_{i=1}^{c} su_{ij}} \tag{8}$$

We select neighbors among users in the cluster i where the dynamic degree of membership of active user (du_{ia}) is larger than zero. Number of neighbors selected in each cluster is proportional to du_{ij}. Once the neighborhood is selected, traditional collaborative filtering algorithm is used to generate recommendation from that.

4 Experimental Evaluation

We conducted experiments to evaluate the proposed model. For experiments we used the EachMovie database, provided by Compaq Systems Research Center (http://www.research.compaq.com/SRC/eachmovie). We assumed that the items are classified into a multi-level (hierarchical) category, and we used the category information to compute the similarity between an active user's ratings at different timeframes. In this experiment we used genre data as category information. We used

MAE(Mean Absolute Error) as our choice of evaluation metric to report prediction experiments because it is commonly used and easy to interpret..

First we selected 1200 users with more than 100 rated items. We divided the data set into a training set and a test portion. To compare the performance of the proposed dynamic fuzzy cluster based CF (DFCF) algorithm we used the traditional CF (TCF) algorithm as the benchmark model. The traditional CF recommendation employs the Pearson nearest neighbor algorithm. We also experimented using both fuzzy CF (FCF) algorithm and crisp cluster based CF (CCF) algorithm. In CCF algorithm, crisp cluster is determined by using Eq.(6) and the degree of membership in the crisp cluster is defined as 1 while the degree of membership in other clusters is zero.

Table 3. Performance Results (MAE)

No. of cluster	Model			
	FCF	CCF	DFCF	TCF
c=2	0.19919	0.19918	0.19918	0.19917
c=3	0.19924	0.19935	0.19915	
c=4	0.19926	0.19945	0.19908	
c=5	0.19926	0.19964	0.19883	
c=6	0.19927	0.19991	0.19859	
c=7	0.19926	0.20033	0.19856	
c=8	0.19927	0.20103	0.19864	
c=9	0.19927	0.20187	0.19861	
c=10	0.19927	0.20271	0.19873	
c=15	0.19943	0.20621	0.19899	
c=20	0.19985	0.22301	0.19918	
c=30	0.20021	0.24821	0.19927	

Table 4. Paired t-test

	p-value			
	DFCF	CCF	DFCF	TCF
DFCF		0.033**	0.029**	0.031**
CCF		.	0.335	0.463
FCF			.	0.428
TCF				.

** Significant at the .05 level

Table 3 presents the performance of the competing models according to the metric of MAE of recommendation. It can be observed that the proposed dynamic fuzzy cluster based CF algorithm outperforms the traditional CF algorithm. When the number of cluster is 7, the performance of the proposed model is best. When the number of clusters is 7, prediction quality of the FCF is worse than TCF but the difference is small while prediction quality of CCF is worst. It can also be observed from the table that as the number of clusters increase the quality tends to be inferior in case of CCF. In addition, a set of pairwise t-tests in Table 4 indicates that the differences were statistically significant. DFCF reflects better user preference than other models at the 5% significance level. These results show that the proposed DFCF algorithm is more accurate than the traditional CF algorithm.

5 Conclusion

Due to the explosion of e-commerce, recommender systems are rapidly becoming a core tool to accelerate cross-selling and strengthen customer loyalty. This study focused on improving the performance of recommender system by dynamic fuzzy clustering. We modified the FCM algorithm and applied it to CF for dynamic recommendations. We conducted an experiment to evaluate the proposed model on the Each-Movie data set and compared them with the traditional CF algorithm. The results show the proposed model's improvement in making recommendations.

In this paper, we applied the concept of time to CF algorithm and proposed a new model which is adaptive to users' changing patterns. In our future work, we intend to evaluate our model using other data set. We would also like to develop a model considering the change in cluster structure.

References

1. Bezdek, J.C., (1981). Pattern Recognition with Fuzzy Objective Function Algorithm. Plenum, New Your.
2. Breese, J.S., Heckerman, D., Kadie, C. (1998). Empirical Analysis of Predictive Algorighms for Collaborative Filtering. Proceedings of the 14th Conference on Uncertainty in Artificial Intelligence (UAI-98), pp. 43-52.
3. Herlocker, J.L., Konstan, J.A. and Riedl, J., (2000). Explaining collaborative filtering recommendations. Proceedings on the ACM 2000 Conference on Computer Supported Cooperative Work, (pp. 241–250). Philadelphia.
4. Sarwar,B.M., Konstan,J.A., Borchers,A., Herlocker,J.L., Miller,B.N., Riedl,J. (1998). Using filtering agents to improve prediction quality in the grouplens research collaborative filtering system. Proceedings of CSCW'98. Seattle, WA.
5. Schafer, J.B., Konstan, J.A. and Riedl, J. (2001). Electronic Commerce Recommender Applications. Data Mining and Knowledge Discovery 5(1/2), pp. 115-153.
6. Xie, X.L., Beni, G.A., (1991). Validity measure for fuzzy clustering. IEEE Trans. Pattern Anal. Machine Intell. 3(8), 841-846.
7. Yu, K.A., et al. (2000). Improving the performance of collaborative recommendation by using muli-level similarity computation. IASTED Interenational Conference on Artificial Intelligence and Soft Computing, July 2000.
8. http://www.research.compaq.com/SRC/eachmovie

Improving Mining Quality by Exploiting Data Dependency

Fang Chu, Yizhou Wang, Carlo Zaniolo, and D. Stott Parker

University of California, Los Angeles, CA 90095, USA
{fchu, wangyz, zaniolo, stott}@cs.ucla.edu

Abstract. The usefulness of the results produced by data mining methods can be critically impaired by several factors such as (1) low quality of data, including errors due to contamination, or incompleteness due to limited bandwidth for data acquisition, and (2) inadequacy of the data model for capturing complex probabilistic relationships in data. Fortunately, a wide spectrum of applications exhibit strong dependencies between data samples. For example, the readings of nearby sensors are generally correlated, and proteins interact with each other when performing crucial functions. Therefore, dependencies among data can be successfully exploited to remedy the problems mentioned above. In this paper, we propose a unified approach to improving mining quality using Markov networks as the data model to exploit local dependencies. Belief propagation is used to efficiently compute the marginal or maximum posterior probabilities, so as to clean the data, to infer missing values, or to improve the mining results from a model that ignores these dependencies. To illustrate the benefits and great generality of the technique, we present its application to three challenging problems: (i) cost-efficient sensor probing, (ii) enhancing protein function predictions, and (iii) sequence data denoising.

1 Introduction

The usefulness of knowledge models produced by data mining methods critically depends on two issues. (1) *Data quality*: Data mining tasks expect to have accurate and complete input data. But, the reality is that in many situations, data is contaminated, or is incomplete due to limited bandwidth for acquisition. (2) *Model adequacy*: Many data mining methods, for efficiency consideration or design limitation, use a model incapable of capturing rich relationships embedded in data. The mining results from an inadequate data model will generally need to be improved.

Fortunately, a wide spectrum of applications exhibit strong dependencies between data samples. For example, the readings of nearby sensors are correlated, and proteins interact with each other when performing crucial functions. Data dependency has not received sufficient attention in data mining research yet, but it can be exploited to remedy the problems mentioned above. We study this in several typical scenarios.

Low Data Quality Issue. Many data mining methods are not designed to deal with noise or missing values; they take the data "as is" and simply deliver the best results obtainable by mining such imperfect data. In order to get more useful mining results, contaminated data needs to be cleaned, and missing values need to be inferred.

Data Contamination. An example of data contamination is encountered in optical character recognition (OCR), a technique that translates pictures of characters into a machine readable encoding scheme. Current OCR algorithms often translate two adjacent letters " ff " into a "#" sign, or incur similar systematic errors.

In the OCR problem, the objective is not to ignore or discard noisy input, but to identify and correct the errors. This is doable because the errors are introduced according to certain patterns. The error patterns in OCR may be related to the shape of individual characters, the adjacency of characters, or illumination and positions. It is thus possible to correct a substantial number of errors with the aid of neighboring characters.

Data Incompleteness. A typical scenario where data is incomplete is found in sensor networks where probing has to be minimized due to power restrictions, and thus data is incomplete or only partially up-to-date. Many queries ask for the minimum/maximum values among all sensor readings. For that, we need a cost-efficient way to infer such extrema while probing the sensors as little as possible.

The problem here is related to filling in missing attributes in data cleansing [5]. The latter basically learns a predictive model using available data, then uses that model to predict the missing values. The model training there does not consider data correlation. In the sensor problem, however, we can leverage the neighborhood relationship, as sensor readings are correlated if the sensors are geographically close. Even knowledge of far-away sensors helps, because that knowledge can be propagated via sensors deployed in between. By exploiting sensor correlation, unprobed sensors can be accurately inferred, and thus data quality can be improved.

Inadequate Data Model Issue. Many well known mining tools are inadequate to model complex data relationships. For example, most classification algorithms, such as Naive Bayes and Decision Trees, approximate the posterior probability of hidden variables (usually class labels) by investigating on individual data features. These discriminative models fail to model the strong data dependencies or interactions.

Take protein function prediction as a concrete classification example. Proteins are known to interact with some others to perform functions, and these interactions connect genes to form a graph structure. If one choose to use Naive Bayes or Decision Trees predict unknown protein functions, he is basically confined to a tabular data model, and thus have lost rich information about interactions.

Markov networks, as a type of descriptive model, provide a convenient representation for structuring complex relationships, and thus a solution for handling probabilistic data dependency. In addition, efficient techniques are available to do inference on Markov networks, including the powerful *Belief Propagation* [15] algorithm. The power in modeling data dependency, together with the availability of efficient inference tools, makes Markov networks very useful data models. They have the potential to enhance mining results obtained from data whose data dependencies are underused.

Our Contributions. The primary contribution of this paper is that we propose a unified approach to improving mining quality by considering data dependency extensively in data mining. We adopt Markov networks as the data model, and use belief propagation for efficient inference. This paper may also contribute to data mining practice with our investigations on some real-life applications.

Outline. We describe Markov networks in the next section. Also discussed there are pairwise Markov networks, a special form of Markov network. Pairwise Markov networks not only model local dependency well, but also allow very efficient computation by belief propagation. We then address the above-mentioned examples in sections 3 and 4.[1] We conclude the paper with related work and discussion in Section 5.

2 Markov Networks

Markov networks have been successfully applied to many problems in different fields, such as artificial intelligence [10], image analysis [13] and turbo decoding [7]. They have the potential to become very useful tools of data mining.

2.1 Graphical Representation

The Markov network is naturally represented as an undirected graph $G = (V, E)$, where V is the vertex set having a one-to-one correspondence with the set of random variables $X = \{x_i\}$ to be modeled, and E is the undirected edge set, defining the neighborhood relationship among variables, indicating their local statistical dependencies. The local statistical dependencies suggest that the joint probability distribution on the whole graph can be factored into a product of local functions on cliques of the graph. A clique is a completely connected subgraphs (including singletons), denoted as X_C. This factorization is actually the most favorable property of Markov networks.

Let C be a set of vertex indices of a clique, and let $\mathcal{C}$ be the set of all such C. A *potential function* $\psi_{X_C}(x_C)$ is a function on the possible realization x_C of the clique X_C. Potential functions can be interpreted as "constraints" among vertices in a clique. They favor certain local configurations by assigning them a larger value.

The joint probability of a graph configuration $p(\{x\})$ can be factored into

$$P(\{x\}) = \frac{1}{Z} \prod_{C \in \mathcal{C}} \psi_{X_C}(x_C) \tag{1}$$

where Z is a normalizing constant: $Z = \sum_{\{x\}} \prod_{C \in \mathcal{C}} \psi_{X_C}(x_C)$

2.2 Pairwise Markov Networks

Computing joint probabilities on cliques reduces computational complexity, but still, the computation may be difficult when cliques are large. In a category of problems

---

[1] Due to space limit, we only discuss two applications here: cost-efficient sensor probing and enhancing protein function predictions. Please refer to our technical report [3] for another application in sequence data denoising.

where our interest involves only pairwise relationships among the variables, we can use use *pairwise Markov networks*. A pairwise Markov network defines potentials functions only on pairs of nodes that are connected by an edge.

In practical problems, we may observe some quantities of the underlying random variables $\{x_i\}$, denoted as $\{y_i\}$. The $\{y_i\}$ are often called evidence of the random variables. In the text denoising example discussed in Section 1, for example, the underlying segments of text are variables, while the segments in the noisy text we observe are evidence. These observed external evidence will be used to make inferences about values of the underlying variables. The statistical dependency between x_i and y_i is written as a joint compatibility function $\phi_i(x_i, y_i)$, which can be interpreted as "external potential" from the external field.

Another type of potential functions are defined between neighboring variables. The compatibility function $\psi_{ij}(x_i, x_j)$ which captures the "internal binding" between two neighboring nodes i and j. An example of pairwise Markov networks is illustrated in Figure 1(a), where the white circles denote the random variables, and the shaded circles denote the evidence. Figure 1(b) shows the potential functions $\phi()$ and $\psi()$.

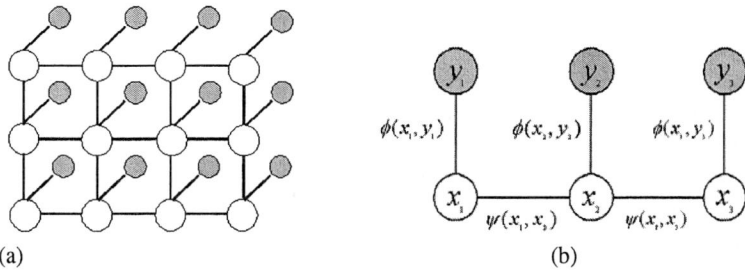

Fig. 1. Example of a Pairwise Markov Network. In (a), the white circles denote the random variables, and the shaded circles denote the external evidence. In (b), the potential functions $\phi()$ and $\psi()$ are showed

Using the pairwise potentials defined above and incorporating the external evidence, the overall joint probability of a graph configuration in Eq.(1) is approximated by

$$P(\{x\}, \{y\}) = \frac{1}{Z} \prod_{(i,j)} \psi_{ij}(x_i, x_j) \prod_i \phi_i(x_i, y_i) \qquad (2)$$

where Z is a normalization factor, and the product over (i, j) is over all compatible neighbors.

2.3 Solving Markov Networks

Solving a Markov network involves two phases:
- *The learning phase,* a phase that builds up the graph structure of the Markov network, and learns the two types of potential functions, $\phi()$'s and $\psi()$'s, from the training data.

– *The inference phase,* a phase that estimates the marginal posterior probabilities or the local maximum posterior probabilities for each random variable, such that the joint posterior probability is maximized.

In general learning is an application-dependent statistics collection process. It depends on specific applications to define the random variables, the neighborhood relationships and further the potential functions. We will look at the learning phase in detail with concrete applications in Sections 3-4.

The inference phase can be solved using a number of methods: simulated annealing [6], mean-field annealing [11], etc. These methods either take an unacceptably long time to converge, or make oversimplified assumptions such as total independence between variables. We choose to use the Belief Propagation method, which has a computation complexity proportional to the number of nodes in the network, assumes only local dependencies, and has proved to be effective on a broad range of Markov networks.

2.4 Inference by Belief Propagation

Belief propagation (BP) is a powerful inference tool on Markov networks. It was pioneered by Judea Pearl [10] in belief networks without loops. For Markov chains and Markov networks without loops, BP is an exact inference method. Even for loopy networks, BP has been successfully used in a wide range of applications[8]. We give a short description of BP in this subsection.

The BP algorithm iteratively propagates "messages" in the network. Messages are passed between neighboring nodes only, ensuring the local constraints, as shown in Figure 2. The message from node i to node j is denoted as $m_{ij}(x_j)$, which intuitively tells how likely node i thinks that node j is in state x_j. The message $m_{ij}(x_j)$ is a vector of the same dimensionality as x_j.

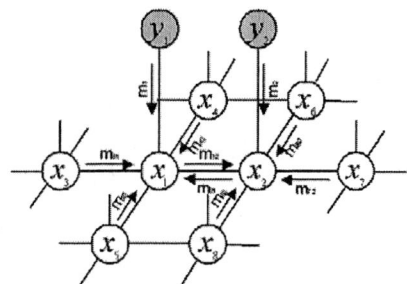

Fig. 2. Message passing in a Markov network

There are two types of message passing rules:

– *SUM-product rule*, that computes the marginal posterior probability.
– *MAX-product rule*, that computes the maximum a posterior probability.

For discrete variables, messages are updated using the SUM-product rule:

$$m_{ij}^{t+1}(x_j) = \sum_{x_i} \phi_i(x_i, y_i)\psi_{ij}(x_i, x_j) \prod_{k \in N(i), k \neq j} m_{ki}^t(x_i) \quad (3)$$

or the MAX-product rule,

$$m_{ij}^{t+1}(x_j) = \max_{x_i} \phi_i(x_i, y_i)\psi_{ij}(x_i, x_j) \prod_{k \in N(i), k \neq j} m_{ki}^t(x_i) \quad (4)$$

where $m_{ki}^t(x_i)$ is the message computed in the last iteration of BP, k runs over all neighbor nodes of i except node j.

BP is an iterative algorithm. When messages converge, the final belief $b(x_i)$ is computed. With the SUM-product rule, $b(x_i)$ approximates the marginal probability $p(x_i)$, defined to be proportional to the product of the local compatibility at node i ($\phi(x_i)$), and messages coming from all neighbors of node i:

$$b_i(x_i)_{SUM} = x_i \phi_i(x_i, y_i) \prod_{j \in N(i)} m_{ji}(x_i) \quad (5)$$

where $N(i)$ is the neighboring nodes of i.

If using the MAX-product rule, $b(x_i)$ approximates the maximum a posterior probability:

$$b_i(x_i)_{MAX} = \arg\max_{x_i} \phi_i(x_i, y_i) \prod_{j \in N(i)} m_{ji}(x_i) \quad (6)$$

3 Application I: Cost-Efficient Sensor Probing

In sensor networks, how to minimize communication is among the key research issues. The challenging problem is how to probe a small number of sensors, yet to effectively infer the unprobed sensors from the known. Cost-efficient sensor probing represents a category of problems where complete data is not available, but has to be compensated by inference.

Our approach here is to model a sensor network with a pairwise Markov network, and use BP to do inference. Each sensor is represented by a random variable in the Markov network. Sensor neighborhood relationships are determined by spatial positions. For example, one can specify a distance threshold so that sensors within the range are neighbors. Neighbors are connected by edges in the network.

In the rest of this section, we study a rainfall sensornet distributed over Washington and Oregon [9]. The sensor recordings were collected during 1949-1994. We use 167 sensor stations which have complete recordings during that period.

3.1 Problem Description and Data Representation

The sensor recordings were collected in past decades over two states along the Pacific Northwest. Since rain is a seasonal phenomena, we split the data by week and build a Markov network for each week.

We need to design the potential functions $\phi_i(x_i, y_i)$ and $\psi_{ij}(x_i, x_j)$ in Eq. (2) in order to use belief propagation. One can use Gaussian or its variants to compute the potential functions. But, in the sensornet we study, we find that the sensor readings are overwhelmed by zeroes, while non-zero values span a wide range. Clearly Gaussian is not a good choice for modeling this very skewed data. Neither are mixtures of gaussian, due to limited data. Instead, we prefer to use discrete sensor readings in the computation. The way we discretize data is given in section 3.3.

The $\phi()$ functions should tell how likely we observe a reading y_i for a given sensor x_i. It is natural to use the likelihood function:

$$\phi_i(x_i, y_i) = P(y_i|x_i) \tag{7}$$

The $\psi()$ functions specify the dependence of sensor x_j's reading on its neighbor x_i.

$$\psi_{ij}(x_i, x_j) = P(x_j|x_i) \tag{8}$$

3.2 Problem Formulation

A theoretical analysis of the problem will that the problem fits well into the maximum a posterior (MAP) estimation on a Markov chain solvable by belief propagation.

Objective: MAP

Let X to be the collection of all underlying sensor readings, Y the collection of all probed sensors. Using Bayes' rule, the joint posterior probability of X given Y is:

$$P(X|Y) = \frac{P(Y|X)P(X)}{P(Y)} \tag{9}$$

Since $P(Y)$ is a constant over all possible X, we can simplify this problem of maximizing the posterior probability to be maximizing the joint probability

$$P(X, Y) = P(Y|X)P(X) \tag{10}$$

Likelihood

In a Markov network, the likelihood of the readings Y depends only on those variables they are directly connected to:

$$P(Y|X) = \prod_{i=1}^{m} P(y_i|x_i) \tag{11}$$

where m is the number of probed sensors.

Prior

Priors shall be defined to capture the constraints between neighboring sensor readings. By exploiting the Markov property of the sensors, we define the prior to involve only the first order neighborhood. Thus, the prior of a sensor is proportional to the product of the compatibility between all neighboring sensors:

$$P(X) \propto \prod_{(i,j)} P(x_j|x_i) \tag{12}$$

Solvable by BP

By replacing Eqs.(11) and (12) into the objective Eq.(10), we have the joint probability to be maximized:

$$P(X,Y) = \frac{1}{Z} \prod_{(i,j)} P(x_j|x_i) \prod_{i=1}^{N} P(y_i|x_i) \qquad (13)$$

Looking back at the $\phi()$ and $\psi()$ functions we defined in Eqs.(7) and (8), we see that the objective function is of the form:

$$P(X,Y) = \frac{1}{Z} \prod_{(i,j)} \psi(x_i, x_j) \prod_{i=1}^{N} \phi(x_i, y_i) \qquad (14)$$

where Z is a normalizing constant.

This is exactly the form in Eq.(2), where the joint probability over the pairwise Markov network is factorized into products of localized potential functions. Therefore, it is clear that the problem can be solved by belief propagation.

3.3 Learning and Inference

The learning part is to find the $\phi()$ and $\psi()$ functions for each sensor, as defined in Eqs.(7) and (8). The learning is straight-forward. We discretize the sensor readings in the past 46 years, use the first 30 years for training and the rest 16 years for testing. In the discrete space, we simply count the frequency of each value a sensor could possibly take, which is the $\phi()$, and the conditional frequencies of sensor values given its neighbors, which is the $\psi()$.

We use a simple discretization with a fixed number of bins, 11 bins in our case, for each sensor. The first bin is dedicated to zeroes, which consistently counts for over 50% of the populations. The 11 bins are assigned in a way that give roughly balanced number of readings in each bin. This very simple discretization method has been shown to work well in the sensor experiments. More elaborated techniques can be used which may further boost the performance, such as histogram equalization that gives balanced bin population with adaptive bin numbers.

For inference, belief propagation does not guarantee to give the exact maximum a posterior distribution, as there are loops in the Markov network. However, loopy belief propagation still gives satisfactory results, as we will see shortly.

3.4 Experimental Results

We evaluate our approach using Top-K queries. A Top-K query asks for the K sensors with the highest values. It is not only a popular aggregation query that the sensor community is interested in, but also a good metric for probing strategies as the exact answer requires contacting all sensors.

We design a probing approach in which sensors are picked for probing based on their local maximum a posterior probability computed by belief propagation, as follows.

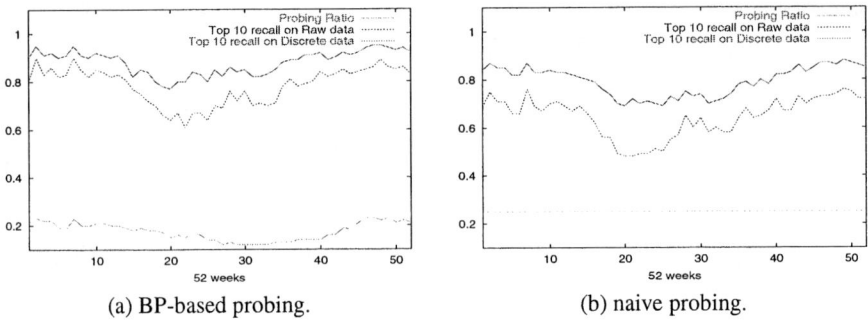

Fig. 3. Top-K recall rates vs. probing ratios. (a): results obtained by our BP-based probing; (b) by the naive probing. On average, BP-based approach probed 8% less, achieves 13.6% higher recall rate for raw values, and 7.7% higher recall rate for discrete values

BP-based Probing:

1. Initialization: Compute the expected readings of sensors using the training data. As the initialization, pick the top M to probe. (We set $M = 20$ in our experiments.)
2. Probe the selected sensors.
3. True values acquired in step 2 become external evidence in the Markov network. Propagate beliefs with all evidence acquired so far.
4. Again, pick the top sensors with the highest expectations for further probing, but this time use the updated distributions to compute expectations. When there are ties, pick them all.
5. Iterate steps 2-4, until beliefs in the network converge.
6. Pick the top K with the highest expectations according to BP MAP estimation.

As a comparative baseline, we have also conducted experiments using a naive probing strategy as follows:

Naive Probing:

1. Compute the expectations of sensors. Pick the top 25% sensors.
2. Probe those selected sensors.
3. Pick the top K.

Performance of the two approaches is shown in Figure 3 (a) and (b), respectively. On each diagram, the bottom curve shows the probing ratio, and the two curves on the top show the recall rates for raw values and discrete values, respectively. We use the standard formula to compute recall rate. Let S denotes the top-K sensor set returned, and T the true top-K set. then:

$$\text{Recall} = \frac{|S \cap T|}{|T|} \qquad (15)$$

Since the sensor readings are discretized in our experiments, we can compute S and T using raw values, or discrete values. Discrete recall demonstrates the effectiveness of BP, while raw recall may be of more interest for real application needs. As can be seen from Figure 3, raw recall is lower than discrete recall. This is due to error introduced in the discretization step. We expect raw recall to be improved when a more elaborated discretization technique is adopted.

It shows clearly in Figure 3 that BP-based approach outperforms the naive approach in terms of both recall rates, while requiring less probing. On average, the BP-based approach has a discrete recall of 88% and a raw recall of 78.2%, after probing only 17.5% sensors. The naive recall has a discrete recall of only 79.3%, a raw recall of only 64.6%, after probing 25% sensors.

The results shown in Figure 3 are obtained for $K = 10$. The relative performance remains the same for other values $K = 20, 30, 40$.

In our technical report [3], we give a closer look on how sensor beliefs change over iterations, and further discussions on how belief propagation works.

4 Application II: Enhancing Protein Function Predictions

Local data dependency can not only help infer missing values, as in the sensor example, but can also be exploited to enhance mining results. Many data mining methods, for efficiency consideration or design limitation, use a model incapable of capturing rich relationships embedded in data. Most discriminative models like Naive Bayes and SVM belong to this category. Predictions of these models can be improved, by exploiting local data dependency using Markov networks. The predictions are used as the likelihood proposal, and message passing between variables refines and reinforces the beliefs. Next we show how to improve protein function predictions in this way.

4.1 Problem Description

Proteins tend to localize in various parts of cells and interact with one another, in order to perform crucial functions. One task in the KDD Cup 2001 [2] is to predict protein functions. The training set contains 862 proteins with known functions, and the testing set includes 381 proteins. The interactions between proteins, including the testing genes, are given. Other information provided specifies a number of properties of individual proteins or genes that encodes the proteins. These include the chromosome on which the gene appears, phenotype of organisms with differences in this gene, etc.

Since information about individual proteins or genes are fixed features, it becomes crucial how to learn from interactions. According to the report of the cup organizers, most competitors organized data in relational tables, and employed algorithms that deal with tabular data. However, compared with tables, graphical models provide a much more natural representation for interacting genes. With a Markov network model, interactions can be modeled directly using edges, avoiding preparing a huge training table. Interacting genes can pass messages to each other, thus getting their beliefs refined together.

In the next of this section, we show a general way of enhancing a weak classifier by simply leveraging local dependency. The classifier we use is Naive Bayes, which is learned from the relational table. We build a Markov network, in which genes with interactions are connected as neighbors. The $\phi()$ function prediction comes from Naive Bayes, and the $\psi()$ are learned from gene interactions.

4.2 Learning Markov Network

We separate the learning of each function, as focusing on one function a time is easier. There are 13 function categories, hence we build 13 Markov networks. To prepare the initial beliefs for a network, we first learn a Naive Bayes classifier, which output a probability vector $b_0()$, indicating how likely a gene will perform the function in question.

Each gene i maps to a binary variable x_i in the Markov network. First we design the $\phi()$ potentials for $\{x_i\}$. One can set the Naive Bayes prediction $b_0()$ to be $\phi()$. But this way the Naive Bayes classifier is over trusted, make it harder to correct the misclassifications. Instead, we adopt a generalized logistic function, shown in Eq.(16), to blur the margin between the belief on two classes, yet still keeping the prediction decision. In the experiments, we set $a = 0.75$, $b = 0.125$, $\alpha = 6$, and $\beta = 0.5$.

$$f = \frac{a}{1 + e^{-\alpha(x-\beta)}} + b \tag{16}$$

The $\psi()$ potentials are learned from protein interactions. Interactions are measured by the correlation between the expression levels of the two encoding genes. At first we tried to related the functions of two genes in a simple way: a positive correlation indicates that with a fixed probability both or neither genes perform the function, while a negative correlation indicates that one and only one gene perform the function. This will leads to a simple fixed $\psi()$ function for all interacting genes. But, a close look at the interaction tells that 25% of the time this assumption is not true. In reality, sometimes two genes participating in the same function may be negatively correlated; a more influential phenomena is that genes may participate in several functions, hence the correlation is a combined observation involving multiple functions.

We decided to learn the distribution of correlation values for three groups of interactions, separately: (a)FF: a group for protein pairs that both perform the function, (b)FNF: a group for pairs that one and only one performs the function, and (c)NFNF: a group for protein pairs that neither performs the function. Thus, the potential function $\psi_{i,j}$ defines how likely to observe a correlation value given for genes x_i and x_j, under different cases where x_i and x_j each has the function or not. In our technical report, we plot the distributions of correlation values learned for two functions. The distribution histograms show that correlation distributions differ among the three groups, and are specific to functions as well.

4.3 Experiments

Naive Bayes does not perform well on this problem, because it does not model the gene interactions sufficiently, and thus cannot fully utilize the rich interaction information. Taking the average predictive accuracy of all classifiers, one per function, the overall accuracy of Naive Bayes is 88%. Belief propagation improves this to 90%.

To exemplify how misclassifications get corrected due to message passing, we show a subgraph of genes in Figure 4. The white circles represent genes(variables), and the shaded circles represent external evidence. Only training genes have corresponding external evidence. The 1's or 0's in the circles tell whether a gene has the function in question or not. For interested readers, we also put the gene ID below the circle. The subgraph contains four training genes and five testing genes. All these testing genes

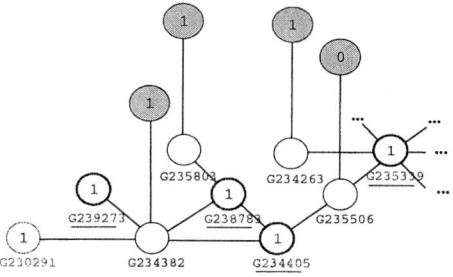

Fig. 4. A subgraph in which testing genes got correct class labels due to message passing

were misclassified by Naive Bayes. After receiving strong beliefs from their neighboring genes, four out of five testing genes were correctly classified. The other test gene 'G230291' was misclassified by both, but Naive Bayes predicted 0% for it to have the function (which is the truth), while belief propagation increased this belief to 25%.

We also evaluated our approach using the score function originally used in the 2001 KDD cup [2]. First we picked out all the functions we predicted for a gene. If more functions are predicted than the true number (which is actually the number of duplicates of that gene in the test table provided), we remove the ones with the smallest confidence. The final score is the ratio of correct predictions, including both positive and negative predictions. Our final score is 91.2%, close to the Cup winner's 93.6%. Although the winner scored reasonably high, they organized data in relational tables and didn't fully explore gene interactions. We expect that their method could perform better if integrated with our approach to exploit local dependencies between genes.

The Cup winner organized data in relational tables, which is not designed at all for complex relationships. To make up for this, they manually created new features, such as computing "neighbors" within k ($k > 1$) hops following neighbor links. Even so, these new features can only be treated the same as the other individual features. The rich relationship information in the original graph structure was lost. Graphical models, on the other hand, are natural models for complex relationships. Markov networks together with belief propagation provides a general and powerful modeling and inference tool on problems satisfying local constraints, such as protein function prediction.

5 Related Work and Discussions

Data dependency is present in a wide spectrum of applications. In this paper, we propose a unified approach that exploits data dependency to improve mining results, and we approach this goal from two directions: (1) improving quality of input data, such as by correcting contaminated data and by inferring missing values, and (2) improving mining results from a model that ignores data dependency.

Techniques for improving data quality proposed in the literature have addressed a wide range of problems caused by noise and missing data. For better information retrieval from text, data is usually filtered to remove noise defined by grammatical errors [12]. In data warehouses, there has been work on noisy class label and noisy attribute detection based on classification rules [16] [14], as well as learning from both labeled and unlabeled data by assigning pseudo-classes for the unlabeled data [1] using boosting ensembles. All this previous work has its own niche concerning data quality. Our work is more general in that it exploits local data constraints using Markov networks.

A pioneering work in sensor networks, the BBQ system [4] has studied the problem of cost-efficient probing. However, their method relies on a global multivariate Gaussian distribution. Global constraints are very strict assumptions, and are not appropriate in many practical scenarios.

The primary contribution of this paper is to propose a unified approach to improving mining quality by considering data dependency extensively in data mining. This paper may also contribute to data mining practice with our investigations on several real-life applications. By exploiting data dependency, clear improvements have been achieved in data quality and the usefulness of mining results.

Acknowledgement

We would like to thank Zhenyu Liu and Ka Cheung Sia for preparation of the sensor data and helpful discussions about the probing problem.

References

1. K. Bennett, A. Demiriz, and R. Maclin. Exploiting unlabeled data in ensemble methods. In *Proc. of the 8th ACM SIGKDD Intl. Conf. on Knowledge Discovery and Data Mining, pp. 289-296*, 2002.
2. J. Cheng, C. Hatzis, H. Hayashi, M.-A. Krogel, S. Morishita, D. Page, and J. Sese. Kdd cup 2001 report. In *SIGKDD Explorations, 3(2):47–64*, 2001.
3. Fang Chu, Yizhou Wang, Carlo Zaniolo, and D.Stott Parker. Improving mining quality by exploiting data dependency. Technical report, UCLA Computer Science, 2005.
4. A. Deshpande, C. Guestrin, S. Madden, J. Hellerstein, and W. Hong. Model-driven data acquisition in sensor networks. In *In Proc. of the 30th Int'l Conf. on Very Large Data Bases (VLDB 04)*, 2004.
5. I. Guyon, N. Natic, and V. Vapnik. Discovering informative patterns and data cleansing. In *AAAI/MIT Press, pp. 181-203*, 1996.
6. S. Kirkpatrick, C. Gelatt, and M. Vecchi. Optimization by simulated annealing. In *Science, vol. 220, no.4598*, 1983.
7. R. McEliece, D. MacKay, and J. Cheng. Turbo decoding as an instance of pearl's 'belief propagation' algorithm. In *IEEE J. on Selected Areas in Communication, 16(2), pp. 140-152*, 1998.
8. K. Murphy, Y. Weiss, and M. Jordan. Loopy belief propagation for approximate inference: an empiricial study. In *Proc. Uncertainty in AI*, 1999.
9. University of Washington. http://www.jisao.washington.edu/data_sets/widmann/.
10. J. Pearl. *Probabilistic reasoning in intelligent systems: networks of plausible inference*. Morgan Kaufmann publishers, 1988.
11. C. Peterson and J. Anderson. A mean-field theory learning algorithm for neural networks. In *Complex Systems, vol.1*, 1987.

12. G. Salton and M. McGill. *Introduction to modern information retrieval.* McGraw Hill, 1983.
13. R. Schultz and R. Stevenson. A bayesian approach to image expansion for improved definition. In *IEEE Trans. Image Processing, 3(3), pp. 233-242*, 1994.
14. Y. Yang, X. Wu, and X. Zhu. Dealing with predictive-but-unpredictable attributes in noisy data sources. In *In Proc. of the 8th European Conf. on Principles and Practice of Knowledge Discovery in Databases (PKDD 04)*, 2004.
15. J. Yedidia, W. Freeman, and Y. Weiss. Generalized belief propagation. In *In Advances in Neural Information Processing Systems (NIPS), Vol 13, pp. 689-695*, 2000.
16. X. Zhu, X. Wu, and Q. Chen. Eliminating class noise in large datasets. In *In Proc. of the 20th Int'l Conf. Machine Learning (ICML 03)*, 2003.

Feature Selection for High Dimensional Face Image Using Self-organizing Maps

Xiaoyang Tan[1,2], Songcan Chen[2,3], Zhi-Hua Zhou[1], and Fuyan Zhang[1]

[1] National Laboratory for Novel Software Technology,
Nanjing University, Nanjing 210093, China
[2] Department of Computer Science and Engineering,
Nanjing University of Aeronautics & Astronautics, Nanjing 210016, China
[3] Shanghai Key Laboratory of Intelligent Information Processing,
Fudan University, Shanghai 200433, China
{x.tan, s.chen}@nuaa.edu.cn
{zhouzh, fyzhang}@nju.edu.cn

Abstract. While feature selection is very difficult for high dimensional, unstructured data such as face image, it may be much easier to do if the data can be faithfully transformed into lower dimensional space. In this paper, a new method is proposed to transform the high dimensional face images into low-dimensional SOM topological space, and then identify important local features of face images for face recognition automatically using simple statistics computed from the class distribution of the face image data. The effectiveness of the proposed method are demonstrated by the experiments on AR face databases, which reveal that up to 80% local features can be pruned with only slightly loss of the classification accuracy.

1 Introduction

Face recognition has been an active research area of computer vision and pattern recognition for decades. Many classical face recognition methods have been proposed [1] to date and have obtained success. In most of the subspace-type face recognition method [1,2], each face is represented by a single vector formed by concatenating pixels of the face image in row scan way. Such a representation makes the dimensionality of the feature space very high. On the other hand, the number of available training samples is generally very limited. In some extreme case, only one image is available per person [3-7]. This makes it important to investigate feature selection to improve the generalization of the recognition system.

This, however, can be a very difficult task for some complex data such as face image due to the sparseness nature of the high dimensional feature space. To circumvent that problem, we therefore propose to transform the high dimensional, unstructured face image data to lower dimensional space first, and then select features in the latter space. The task of feature selection may be much easier due to the simplification of the feature space. In previous work, we have found that the SOM (self-organizing maps, [8]) topological space is suitable for face representation [6], and an SOM-based face representation model called "SOM-face" has been proposed.

In this paper, we extended the work [6] in two aspects: (1) a novel method of automatically selecting important local features from the face image for recognition is proposed, and (2) investigate the problem of how much of the local features can be reduced without losing useful information in class prediction. The paper proceeds as follows. After briefly reviewing the SOM-face model in section 2, we described the proposed method in section 3. The experiments are reported in section 4. Finally, conclusions are drawn in section 5.

2 The SOM-Face Model

The essence of SOM-face [6] is to express each face image as a function of local information presented in the image. This is achieved by dividing the image into k different local sub-blocks at first, each of which potentially preserves some structure information of the image. Then, a self-organizing map (SOM) neural network is trained using the obtained sub-blocks. The reconstruction of a face in the SOM topological space is called SOM-face (see Fig.1). Note that any face localized in the same way above can be projected onto the quantized lower dimensional space to obtain its compact but robust representation. The main advantage of such a representation is that, in the SOM-face, the information contained in the face is distributed in an orderly way and represented by several neurons instead of only one neuron corresponding to a weight vector, so the common features of different classes can be easily identified.

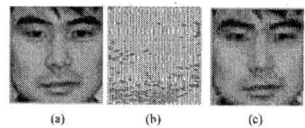

Fig. 1. Example of an original image, its projection and the reconstructed image a) Original face image. b) The distribution of image in the topological space. c) "SOM face" reconstructed

3 Feature Selection in the SOM Topological Space

SOM mapping makes it feasible for us to analyze the degrees of importance of different local areas based on simple statistics computed from the empirical distribution. Here, three criterions are proposed to measure the goodness of sub-blocks of each face, i.e., face frequency (FF), a χ^2 statistic (CHI) and neuron purity (NP).

- Face frequency criterion (FF)

Face frequency is derived from the concept of document frequency in automatic text analysis field [10]. Here it means the number of different faces a neuron attracts. Based on this simple statistics, two strategies can be applied to perform feature selection. The first strategy (FF-1) is based on the assumption that the rare neurons (i.e., the neurons with low FF value) may be either non-informative for recognition or not influential in global performance. The degree of importance of a neuron can be calculated as a *non-decreasing* function of its FF value.

On the other hand, the FF-value can be regarded as an indicator of the distribution of the sub-blocks of all the faces in SOM topological space. That is, big FF-values indicate much overlap among the distributions of different classes and hence inducing low discriminability, whereas small values indicate little overlap and hence high discriminability. In this sense, the degree of importance of a neuron can also be calculated as a *non-increasing* function of its FF value. This strategy is named FF-2.

- χ^2 statistic (CHI)

The χ^2 statistic tries to measure the dependence between class and term. Consider a two-way contingency table of a neuron t and a face c, where A is the number of times the t and c co-occur, B is the number of time the t occurs without c, C is the number of times the c occurs without t, and N is the total number of faces, then the χ^2 statistic between the t and c is defined to be:

$$\chi^2(t,c) = \frac{N \times (AD-CB)^2}{(A+C) \times (B+D) \times (A+B) \times (C+D)} \quad (1)$$

The χ^2 statistic is zero if t and c are independent. In this study, we computed the pairwise χ^2 statistic between each neuron and each training face, and then measured the final neuron-goodness score according to the maximal rule:

$$\chi_{max}^2(t,c) = \max_{i=1}^{C} \{\chi^2(t,c_i)\} \quad (2)$$

- Neuron purity (NP,[9])

Neuron purity is a measure to quantify the degree of separability of a neuron. Let the number of sub-blocks of class c attracted by neuron t be λ_c, then the probability of class c in neuron t is given as $p_{ct} = \frac{\lambda_{ct}}{\sum_{c \in C} \lambda_{ct}}$ and the degree of separability (or the purity) of a neuron t is defined to be:

$$NP_t = \sqrt{\frac{K_t}{K_t-1} \sum_{i=1}^{K_t} (P_{it} - 1/K)^2} \quad (3)$$

where K_t is number of classes attracted by neuron t.

4 Experiments

The AR face database [11] is used in the experiments, which contains over 4,000 color face images of 126 people's faces, among them, a subset of 400 images from 100 different subjects was used. Some sample images are shown in Fig.2.

Fig. 2. Sample images for one subject of the AR database [10]

Before the recognition process, each image was cropped and resized to 120x165 pixels and then converted to gray-level images, which were then processed by a histogram equalization algorithm. A sub-block size of 5x3 was used and only the neutral expressions images (Fig.2a) of the 100 individuals were used for training, while the other three were used for testing.

We ranked the neurons according to the goodness value obtained with different criteria, and removed those neurons whose goodness values are below some predefined threshold. Since different faces have different set of neurons, the resulting neurons after pruning are also different face by face and only the left neurons would be used for classification. The relationship between the top 1 recognition rate and the total remaining local features of the probe set are displayed in Fig.3.

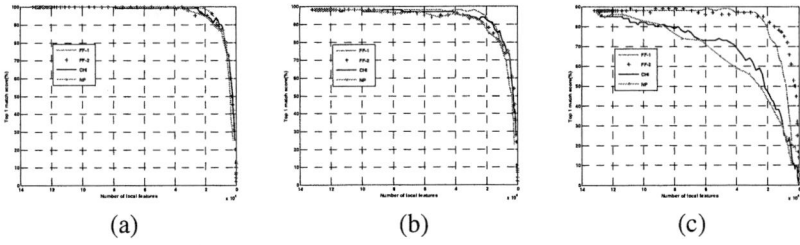

Fig. 3. Top 1 recognition rate vs. total remaining unique local feature count (a)-(c) are corresponding to the smile, anger and scream images (Figs.3b, 3c, and 3d)

It can be observed from Fig. 3 that on the first two probe sets, the four compared criterions (FF-1, FF-2, CHI and NP) have similar effect on the performance of the recognition system, and 80% or more of the sub-blocks can be safely eliminated with almost no loss in recognition accuracy. On more challenging probe sets (Fig.2d), two FF-type criterions perform much better than the other two (CHI and NP). This observation indicates that the criterions which make use of the class information do not necessarily lead to excellent performance.

5 Conclusions

In this paper, a novel feature selection method for high dimensional face images based on their SOM transformation is proposed. Experiments on AR database reveal that up to 80% sub-blocks of a face can be removed from the probe set without loss of the classification accuracy. This could be particularly useful when a compact representation of face is needed, such as in the application of smart card, where the storage capability is very limited.

Acknowledgement

This work was supported by the National Science Foundation of China under the Grant No. 60271017, the National Outstanding Youth Foundation of China under the Grant No. 60325207, and the Jiangsu Science Foundation Key Project under the Grant No. BK2004001.

References

1. Zhao W., Chellappa R., Phillips P. J., and Rosenfeld A.. Face Recognition: A Literature Survey. *ACM Computing Survey*, 2003,35(4): 399-458.
2. Turk M. and Pentland A. Eigenfaces for recognition. *Journal of Cognitive Neuroscience*, 1991, 3(1): 71-86.
3. Wu J. and Zhou Z.-H. Face recognition with one training image per person. *Pattern Recognition Letters*, 2002, 23(14): 1711-1719.
4. Chen S. C., Zhang, D. Q., and Zhou Z.-H. Enhanced $(PC)^2A$ for face recognition with one training image per person. *Patt. Recog. Lett.*, 2004, 25(10):1173-1181.
5. Chen S. C., Liu J., and Zhou Z.-H. Making FLDA applicable to face recognition with one sample per person. *Pattern Recognition*, 2004,37(7): 1553-1555
6. Tan X., Chen S.C., Zhou Z.-H., and Zhang F. Recognizing partially occluded, expression variant faces from single training image per person with SOM and soft kNN ensemble. IEEE Trans. Neural Networks, in press.
7. Martinez A.M. Recognizing imprecisely localized, partially occluded, and expression variant faces from a single sample per class. IEEE TPAMI, 2002, 25(6): 748-763
8. Kohonen T. *Self-Organizing Map*, 2nd edition, Berlin: Springer-Verlag, 1997.
9. Singh S., Singh M., and Markou M. Feature selection for face recognition based on data partitioning. In: *Proc. 15th ICPR*, Quebec, Canada, 2002, 680-683.
10. Yang Y. and Pedersen J. O. A comparative study on feature selection in text categorization. In: *Proc. 14th ICML*, Nashville, TN, 1997, 412-420.
11. Martínez A. M. and Benavente R. The AR face database. *CVC Technical Report*, no.24, June 1998.

Progressive Sampling for Association Rules Based on Sampling Error Estimation

Kun-Ta Chuang, Ming-Syan Chen, and Wen-Chieh Yang

Graduate Institute of Communication Engineering, National Taiwan University,
Intelligent Business Technology Inc., Taipei, Taiwan, ROC
doug@arbor.ee.ntu.edu.tw
mschen@cc.ee.ntu.edu.tw
slavayang@gmail.com

Abstract. We explore in this paper a progressive sampling algorithm, called *Sampling Error Estimation* (*SEE*), which aims to identify an appropriate sample size for mining association rules. *SEE* has two advantages over previous works in the literature. First, *SEE* is highly efficient because an appropriate sample size can be determined without the need of executing association rules. Second, the identified sample size of *SEE* is very accurate, meaning that association rules can be highly efficiently executed on a sample of this size to obtain a sufficiently accurate result. This is attributed to the merit of *SEE* for being able to significantly reduce the influence of randomness by examining several samples with the same size in one database scan. As validated by experiments on various real data and synthetic data, *SEE* can achieve very prominent improvement in efficiency and also the resulting accuracy over previous works.

1 Introduction

As the growth of information explodes nowadays, reducing the computational cost of data mining tasks has emerged as an important issue. Specifically, the computational cost reduction for association rule mining is elaborated upon by the research community [6]. Among research efforts to improve the efficiency of mining association rules, sampling is an important technique due to its capability of reducing the amount of analyzed data [2][5][7].

However, using sampling will inevitably result in the generation of incorrect association rules, which are not valid with respect to the entire database. In such situations, how to identify an appropriate sample size is key to the success of the sampling technique. Progressive sampling is the well-known approach to determine the appropriate sample size in the literature. Progressive sampling methods are based on the observation that when the sample size exceeds a size N_s, the model accuracy λ obtained by mining on a sample will no longer be prominently increased. The sample size N_s can therefore be suggested as the appropriate sample size. In general, the sample size N_s can be identified from the "*model accuracy curve*", which is in essence the model accuracy versus the sample size

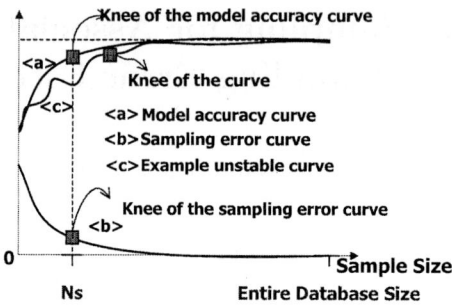

Fig. 1. The illustrated model accuracy curve and sampling error curve

[3]. Curve <a> in Figure 1 illustrates an example of the *model accuracy curve* versus the sample size. It can be observed that the model accuracy will stay in a plateau (the accuracy is no longer much improved) when the sample size exceeds N_s. Thus, the goal of progressive sampling algorithms is to efficiently estimate the *model accuracy curve* and then identify the point (N_s, λ_i) of this curve, where λ_i is the corresponding model accuracy for the sample size N_s.

However, previous progressive sampling algorithms for association rules mainly suffer from two problems. First, to measure the model accuracy of each sample size, previous progressive sampling algorithms have to resort to the execution of association rules either on samples [2] or on the entire database [3], which is, however, very costly. Second, for efficiency reasons, previous algorithms usually evaluate the model accuracy of a sample size by only executing association rules on a sample with this size, and the phenomenon of *randomness* [4] is thus not considered. *Randomness* refers to the phenomenon that mining on samples of the same size may obtain different results. In fact, *randomness* will affect the determination of the accuracy of obtained association rules for each sample size. Thus previous works will generate an unstable curve, like curve <c> in Figure 1, to estimate the *model accuracy curve*, and the resulted sample size may not be a proper choice.

To remedy these problems, we devise in this paper an innovative algorithm, referred to as *Sampling Error Estimation* (abbreviated as *SEE*), to identify the appropriate sample size without the need of executing association rule mining either on several samples or on the entire database, thus significantly improving the execution efficiency. The fundamental concept of algorithm *SEE* is to estimate the *model accuracy curve* by generating a curve of *sampling errors* versus the sample size. *Sampling errors* stem from the phenomenon that the proportion (also referred to as *support* in association rule research) of each item in the sample will deviate from its population proportion, and *sampling error* is indeed the reason for incurring incorrect association rules in the sample. In general, the smaller the *sampling error* of the sample, the higher accuracy can be obtained by mining on this sample. By calculating those *sampling errors* which will influence the obtained model accuracy, the shape of the *sampling error curve* can reflect the shape of the *model accuracy curve*. Curve <b> in Figure 1 illustrates the *sampling error curve*. Moreover, *SEE* can greatly reduce the influence of *randomness* and correctly measure *sampling errors* of each sample size. Thus the *sampling error curve* can be employed to better estimate the *model accuracy curve*. This is attributed to the merit that *SEE* can calculate *sampling errors* of each sample size from a number of samples of this size in one database scan. As validated by experiments on various real data and synthetic data, algorithm *SEE* can achieve very prominent improvement in efficiency and also the resulting accuracy over previous works.

2 Sampling Errors for Association Rules

2.1 Descriptions of Sampling Errors

Sampling errors, referring to the phenomenon that the *support* of each item in the sample will deviate from its *support* in the entire data [4], will result in incorrectly identifying whether an item is a *frequent* item. The inference can be made from the following discussion. Suppose that we have 10,000 transactional records and 1,000 records of them contain the item {bread}. Hence the *support* of {bread} is 10%. In general, the phenomenon of *sampling errors* can be observed from the distribution of the item support in samples. When we generate a lot of samples of the same size, the sampling distribution, i.e., the *support* of one specified item among these samples, will approximately follow a *normal* distribution with *mean* equal to the support of this item in the entire database. Figure 2(a) shows an example of the distribution of the support of the item {bread} in samples. The *support* of {bread} in samples of the same size will follow a *normal* distribution with *mean* 10% because {bread} occurs 10% in the entire database. Suppose that the *minimum support* is specified as *8%* and thus the item type of {bread} is *frequent* because its support in the entire database is larger than *8%*. As a result, the shadow region in Figure 2(a) can represent the probability of incorrectly identifying {bread} as a *non-frequent* item. Moreover, the larger the shadow region, the larger probability we will obtain the incorrect item type.

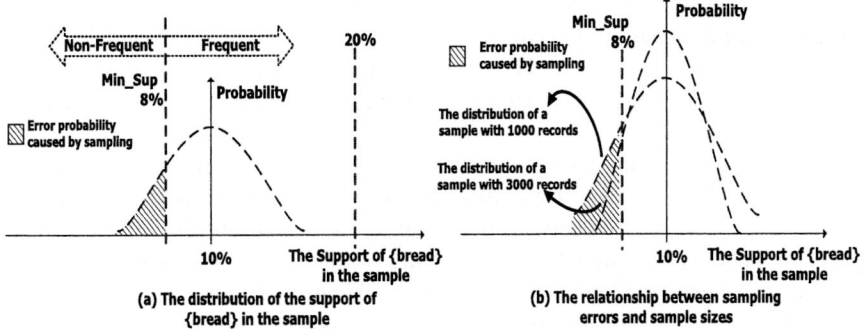

Fig. 2. The phenomenon of sampling errors

In practice, for one specified item, the probability of obtaining the incorrect item type relies on the sample size. More specifically, the variance of the item support in samples is inversely proportional to the square root of the sample size [4], and thus the influence of *sampling errors* will decrease as the sample size increases. As shown in Figure 2(b), the support of the item {bread} will have a smaller variance in a sample of 3,000 records than in a sample of 1,000 records. We can observe that the sample of 3,000 records will have a smaller error probability (the shadow region is smaller) of identifying {bread} as a *non-frequent* item. Thus for each item, the error probability of identifying its item type will decrease as the sample size increases.

In essence, mining association rules will generate a lot of itemsets. We know that as the sample size decreases, all itemsets will have the larger probability of incorrectly identifying their item types since *sampling errors* also increase as the sample size decreases. Therefore, if *sampling errors* cannot be significantly decreased when the sample size is larger than a sample size s_n, s_n can be suggested as the appropriate sample size for association rules.

In fact, such a size can be determined by generating a curve of *sampling errors* versus the sample size. In addition, the corresponding sample size at the convergence point of the curve will be suggested as the appropriate sample size. In the following, we present the method to measure *sampling errors* of each size.

2.2 Measurement of Sampling Errors

Since *sampling errors* stem from the difference between the proportion of each item in the sample and its population proportion, the *sampling error of item i_n* in the sample S, can be defined as:

Definition 1: (*Sampling error of item i_n in the sample S*)

$$SE(i_n, S) = |Sup(i_n, S) - Sup(i_n, D)|,$$

where $Sup(i_n, S)$ and $Sup(i_n, D)$ denote supports of item i_n in the sample S and in the entire database D, respectively.

Furthermore, for evaluating the accuracy of association rules, *sampling errors* of all itemsets are calculated because *sampling errors* of all itemsets will influence the result of association rules. Hence, *sampling errors* of the sample S can be defined as the root mean square sum of *sampling errors* of each occurred itemsets in the entire database.

In practice, using this measurement to evaluate the model accuracy of association rules will suffer from two problems. The first problem is that examining *sampling errors* of all itemsets is inefficient because the number of itemsets is huge. The second problem is that calculating *sampling errors* of several itemsets is indeed unnecessary. Consider the example shown in Figure 2(a). If the *minimum support* is specified as 20%, the error probability of identifying the item type of the item {bread} will be close to zero. Thus *sampling errors* of this item will not influence the accuracy of association rules.

To remedy the first problem, we employ the solution that only *sampling errors* of 1-itemsets will be calculated. Indeed, calculating *sampling errors* of 1-itemsets cannot completely response the model accuracy of association rules. However, it will be generally sufficient for the following reason. The property of association rules, i.e., *downward closure property* [6], shows that *sampling errors* of 1-itemsets will influence the accuracy of 2-itemsets, 3-itemset, etc. In other words, inaccurately identifying the item type of 1-itemsets will incur the error identification of item types of 2-itemsets, 3-itemsets, an so on. It can be expected that *sampling errors* of 1-itemsets will dominate the accuracy of *frequent* itemsets. Thus the model accuracy will not significantly increase when the sample size is larger than a size whose corresponding *sampling errors* of 1-itemsets will no longer significantly decrease. Therefore calculating *sampling errors* of 1-itemsets will be a good pilot to determine whether a sample size is sufficient for mining association rules.

Furthermore, to resolve the second problem, i.e., some itemsets will be irrelevant to the model accuracy if their supports are far away from the *minimum support*, we take into account the relationship between *sampling errors* and the *minimum support*. Suppose that p denotes the sample ratio, i.e. $\frac{|S|}{|D|}$, where $|S|$ and $|D|$ are sizes of S and D, respectively. Since sampling may cause changes of item supports, three distinct cases of the support change will be considered when the range of item supports (0~1) is divided into 3 intervals, i.e., <A> $[0, p \cdot Min\_Sup]$, <B> $[p \cdot Min\_Sup, Min\_Sup]$, and <C> $[Min\_Sup, 1]$. Case (1) consists of those items whose supports change from <C> to <B> after sampling. Those items are identified as *frequent* in D but *non-frequent* in S. Case (2) consists of those items whose supports change from <B> to <C> after sampling. Those items are identified as *non-frequent* in D but *frequent* in S. In addition, case (3) consists of items identified as *frequent* in both D and S.

Note that the model accuracy is usually calculated as the combination of *recall* and *precision* [1]. In addition, *F-score* is a widely-used measurement which combines *recall* and *precision*. *F-score* of the result obtained by mining on S, is defined as $\frac{(\beta^2+1) \cdot P \cdot R}{\beta^2 \cdot P + R}$, where P and R are *precision* and *recall*, respectively. β is a weighted value and it is usually set as 1 to fairly consider *precision* and *recall*. *Precision* $P(S)$ and *recall* $R(S)$ of *frequent* itemsets obtained in the sample S are defined as

$$P(S) = \frac{|FI_a(S)|}{|FI_a(S)|+|FI_b(S)|}; R(S) = \frac{|FI_a(S)|}{|FI_a(S)|+|FI_c(S)|},$$

where $FI_a(S)$ consists of itemsets which belong to case (3). $FI_b(S)$ consists of itemsets belonging to case (2) and $FI_c(S)$ consists of itemsets belonging to case (1). $|FI_a(S)|$, $|FI_b(S)|$ and $|FI_c(S)|$ are their corresponding sizes. The accuracy of the set of *frequent* itemsets obtained by mining on the sample S is thus formulated as $F(S) = \frac{2 \cdot P(S) \cdot R(S)}{P(S)+R(S)}$. In essence, we can observe that only those itemsets belonging to $FI_a(S)$, $FI_b(S)$ and $FI_c(S)$ will affect the accuracy of *frequent* itemsets whereas other itemsets will not. Consequently, the *association rules-related sampling errors* of the sample S can be defined:

Definition 2: (*Association rules-related sampling errors of the sample S*). Suppose that there are M 1-itemsets, $\{a_1, a_2, ..., a_M\}$, belonging to cases (1)/(2)/(3). *Sampling errors* of the sample S which will influence the accuracy of association rules, can be defined as:

$$A\_SE(S) = \sqrt{\left(\sum_{k=1}^{M} SE(a_k, S)^2\right)/M}.$$

Furthermore, as mentioned previously, only mining on a sample is inadequate to evaluate the correct mining accuracy of the corresponding sample size due to the phenomenon of *randomness*. We use L samples of the same size to measure *sampling errors* of this size. Definition 3 follows.

Definition 3: (*Association rules-related sampling errors of sample size $|S|$*). Suppose that $A\_SE(S_k)$ denotes *association rules-related sampling errors* of the kth sample of the sample size $|S|$, where $1 \leq k \leq L$. *Association rules-related sampling errors* of the sample size $|S|$ is defined as:

$$A\_SSE(|S|) = \left(\sum_{k=1}^{L} A\_SE(S_k)\right)/L.$$

Therefore, we calculate $A\_SSE(|S|)$ to measure *sampling errors* of each sample size which is given by a sampling schedule, and then the curve of $A\_SSE(|S|)$ versus the sample size can be used to estimate the curve of the model accuracy versus the sample size.

3 Algorithm SEE

3.1 Pseudocode of Algorithm SEE

Algorithm *SEE* will generate a curve of *sampling errors* versus the sample size immediately after one database scan, and then suggest the corresponding sample size at the convergence point of this curve as the appropriate sample size for association rules. To measure *sampling errors*, frequencies (or said *support count*) of each item in the entire database and in each sample will be required in *SEE*. To efficiently acquire such information, *SEE* is devised as a two phases algorithm: (1) the *database scan phase*, in which the database is scanned once and simultaneously the *frequency* of each item in the entire database and in each sample is stored. (2) The *convergence detection phase*, in which *sampling errors* of each sample size are calculated, and then the appropriate sample size is identified from the curve of *sampling errors* versus the sample size. The pseudo code of SEE is outlined below:

Algorithm SEE: SEE(D,L,$\mathbb{R}$,minSup)

$SEE[n][m]$: store sampling errors of the mth sample of size s_n.

//**The database scan phase**

01. while has next transaction t_d
02.   for every item i_k in t_d
03. $IList[i_k] \to freq_D + +;$
04.   for $n = 1$ to P
05.   for $m = 1$ to L
06.   if $(rv[n][m].\text{next} < \frac{s_n}{|D|})$
07.   for every item i_k in t_d
08. $IList[i_k] \to freq_S[n][m] + +;$

//**The convergence detection phase**

01. for $n = 1$ to P
02.   for $m = 1$ to L
03.   e=0;count_item=0;
04.   for each item i_k in $IList$ {
05.   if i_k belongs to case (1)/(2)/(3)
06.   e+=$\left(\frac{IList[i_k] \to freq_S[n][m]}{s_n} - \frac{IList[i_k] \to freq_D}{|D|} \right)^2;$
07.   count_item++;
08. $SEE[n][m] = \sqrt{\frac{e}{\text{count\_item}}};$
09. $A\_SSE(s_n) = \frac{\sum_{j=1}^{L} SEE[n][j]}{L};$
10.   if $(s_n, A\_SSE(s_n))$ is the convergence point
11.   report s_n as the appropriate sample size; program terminated;

3.2 Complexity Analysis of Algorithm SEE

Time Complexity: Suppose that $|I|$ is the number of distinct items in the entire database D. The time complexity of SEE is $O(|D| \times P \times L + P \times L \times |I|)$, which is linear with respect to the entire database size $|D|$.

Space Complexity: The space complexity is $O(L \times P \times |I|)$. The major space requirement is used to store the frequency of each item in each sample, i.e., $IList[i_k] \to freq_S[n][m]$.

4 Experimental Results

We assess the quality of algorithm SEE in Windows XP professional platform with 512Mb memory and 1.7G P4-CPU. For comparison, the corresponding results of algorithm RC-S [2] are also evaluated. In the literature, RC-S can be deemed as the most efficient method to date. The utilized association rule mining algorithm is $Eclat$ [6]. Furthermore, in all experiments, the user-specified parameter α of algorithm RC-S is set to one, which is the same as its default value addressed in [2]. All necessary codes are implemented by Java and complied by Sun jdk1.4.

4.1 Methods for Comparison

To demonstrate that progressive sampling can improve the performance of mining association rules, in all experiments, the execution time of each algorithm will consist of two costs: (1) the time of executing progressive sampling algorithm to determine the appropriate sample size; (2) the time of executing association rules on a sample with the identified size. Moreover, since the scale of *sampling errors* is different from the model accuracy and the self-similarity, all curves shown in experimental results are normalized to [0,1] scale.

In our experiments, the curve "*Normalized Model Accuracy*" denotes the curve of the normalized *model accuracy* of frequent itemsets versus the sample size. Note that the *model accuracy* of a sample size is calculated as the average *F-Scores* from 20 runs with this size. In addition, the curve "*RC-S*" denotes the normalized *self-similarity curve* which is generated by algorithm *RC-S*. Note that *sampling errors* will decrease as the sample size increases, and the model accuracy will increase as the sample size increases. Thus the curve "*Inv_NSEE*" shows *the inverse sampling errors*, i.e., 1-*normalized* $A\_SSE(s_n)$, of each sample size s_n. In addition, algorithms *RC-S* and *SEE* all aim to estimate curves "*Normalized Model Accuracy*", and thus we can estimate the effectiveness of *SEE* and *RC-S* by observing the difference between "*Inv_NSEE*"/"*RC-S*" and "*Normalized Model Accuracy*". Note that the quantitative analysis of this difference can be defined as the root mean square error, $\sqrt{\frac{1}{P}\sum_{i=1}^{P}[v(s_i) - \varphi(s_i)]^2}$, where P is the number of distinct sample sizes in the sampling schedule, and $v(s_i)$ denotes the normalized model accuracy of the sample size s_i and $\varphi(s_i)$ denotes the normalized score of the sample size s_i in the curve "*Inv_NSEE*"/"*RC-S*". The root mean square error will be shown as the value "*Curve Error*" in each experiment.

Furthermore, we use an arithmetic sampling schedule with $\mathbb{R}=\{0.05 \times |D|, 0.1 \times |D|, ..., 0.95 \times |D|\}$. In addition, since curves "*Normalized Model Accuracy*", "*RC-S*", and "*Inv_NSEE*" are all monotonically increasing, the appropriate sample size identified in each curve will be the smallest sample size whose corresponding normalized score exceeds 0.8, meaning that we have up to 20% improvement when the sample size is larger than the identified size.

4.2 Experiments on Real Data and Synthetic Data

Experiments on the Parameter Sensitivity. In this experiment, we observe the parameter sensitivity of algorithm *SEE*, and the large real data set, POS is utilized. First, Figure 3(a) shows the sensitivity analysis of the parameter L, which is the number of samples used to evaluate the corresponding *sampling errors* of a sample size. In Figure 3(a), the y-axis represents the score distance of each sample size s_i, which is defined as $|v(s_i) - \varphi(s_i)|$, where $v(s_i)$ denotes the

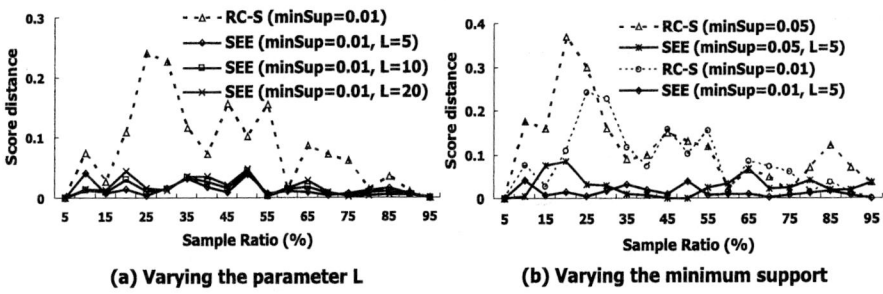

Fig. 3. The sensitivity analysis of different parameters

normalized model accuracy of the sample size s_i and $\varphi(s_i)$ denotes the normalized score of s_i in the curve "Inv_NSEE"/"RC-S". We can observe that curves SEE with different L are all with smaller score distances than the corresponding score distance of RC-S, meaning that SEE can more correctly estimate the variation of the *model accuracy curve*. Moreover, $L = 5$ is sufficient to obtain an accurate *sampling error curve* because the differences between $L = 5$, 10, and 20 are very small. Thus we can use acceptable memory to store frequencies of each item in 5 samples of the same size, showing the practicability of algorithm SEE. Furthermore, we observe the influence of the *minimum support* in Figure 3(b). Results of two different *minimum supports* are shown. We can observe that algorithm SEE has the smaller score distance than that of algorithm RC-S under different *minimum supports*. Moreover, changing the *minimum support* will not obviously influence the result of algorithm SEE, indicating that SEE is robust under different parameters of association rules.

Experiments on Synthetic Data. The observations on various synthetic data are shown in Figure 4. We generate four different synthetic data with different "the average transaction length", "the average length of maximal patterns" and "number of different items" (denoted as **T**, **I**, **N** in the name of the generated data, respectively). The number of transactions is set to 50,000,000 in all generated data to mimic the large database.

We observe that SEE can save a lot of time and obtain a sufficiently correct model result. On the other hand, RC-S may have a smaller execution time in some cases but the obtained model accuracy will be not so acceptable, showing the ability of SEE to balance the efficiency and the model accuracy.

Furthermore, the execution time of four synthetic data, which database sizes vary from 5×10^7 to 2×10^8, are shown in Figure 5. In this experiment, "the average transaction length" is set to 15, and "the average length of maximal patterns" is set to 3. In addition, "the number of different items" is set to 10^6. In this experiment, algorithms SEE and RC-S similarly suggest sample ratios 40%~45% as the appropriate sample ratios for association rules, which can achieve a 96% model accuracy when we execute association rules on a sample of the identified sample size. In practice, we observe that the execution time of

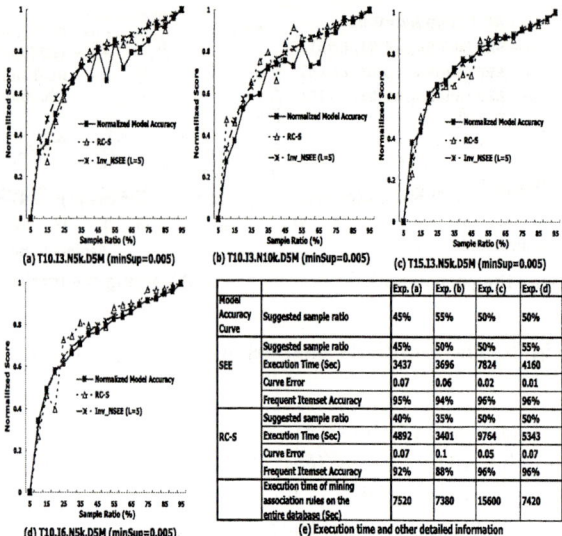

Fig. 4. Experiments on various synthetic data

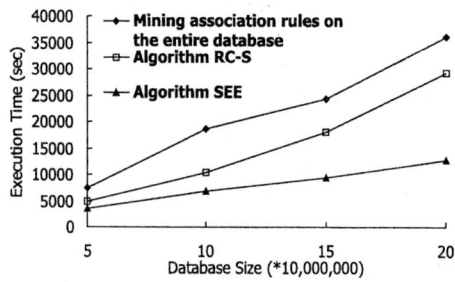

Fig. 5. The execution time on various database size

SEE is smaller than that of *RC-S* in all cases. When the database size increases, algorithm *RC-S* cannot effectively reduce the execution time because it will suffer from the need of considerable I/O operations (executing association rules on several samples). On the other hand, algorithm *SEE* only requires I/O operations of one database scan, showing high execution efficiency.

5 Conclusion

In this paper, we devise a progressive sampling algorithm, *SEE*, to identify an appropriate sample size for mining association rules with two advantages over previous works. First, *SEE* is highly efficient because the appropriate sample size can be identified without the need of executing association rules on samples and on the entire database. Second, the identified sample size will be a proper sample size since it is determined as the corresponding sample size at the convergence point of the *sampling error curve*, which can effectively estimate the *model accuracy curve*. As shown by experiments on various real data and synthetic data, the efficiency and the effectiveness of *SEE* significantly outperform previous works.

Acknowledgments

The work was supported in part by the National Science Council of Taiwan, R.O.C., under Contracts NSC93-2752-E-002-006-PAE.

References

1. R. Baeza-Yates and B. Ribeiro-Neto. *Modern Information Retrieval*. Addison-Wesley, 1999.
2. S. Parthasarathy. Efficient progressive sampling for association rules. In *Proc. of ICDM*, 2002.
3. F. Provost, D. Jensen, and T. Oates. Efficient progressive sampling. In *Proc. of SIGKDD*, 1999.
4. R. L. Scheaffer, W. Mendenhall, and R. L. Ott. *Elementary Survey Sampling*. Duxbury Press, 1995.
5. H. Toivonen. Sampling large databases for association rules. In *Proc. of VLDB*, 1996.
6. M. J. Zaki, S. Parthasarathy, M. O., and W. Li. New algorithms for fast discovery of association rules. In *Proc. of SIGKDD*, 1997.
7. M.J. Zaki, S. Parthasarathy, Wei Li, and M. Ogihara. Evaluation of sampling for data mining of association rules. In *Int. Workshop on Research Issues in Data Engineering*, 1997.

CLeVer: A Feature Subset Selection Technique for Multivariate Time Series*

Kiyoung Yang, Hyunjin Yoon, and Cyrus Shahabi

Computer Science Department, University of Southern California,
Los Angeles, CA 90089, U.S.A
{kiyoungy, hjy, shahabi}@usc.edu

Abstract. Feature subset selection (FSS) is one of the data pre-processing techniques to identify a subset of the original features from a given dataset before performing any data mining tasks. We propose a novel FSS method for Multivariate Time Series (MTS) based on Common Principal Components, termed *CLeVer*. It utilizes the properties of the principal components to retain the correlation information among original features while traditional FSS techniques, such as Recursive Feature Elimination (RFE), may lose it. In order to evaluate the effectiveness of our selected subset of features, classification is employed as the target data mining task. Our experiments show that *CLeVer* outperforms RFE and Fisher Criterion by up to a factor of two in terms of classification accuracy, while requiring up to 2 orders of magnitude less processing time.

1 Introduction

Feature subset selection (FSS) is one of the techniques to pre-process the data before we perform any data mining tasks, e.g., classification or clustering. FSS is to identify a subset of original features from a given dataset while removing irrelevant and/or redundant features [1]. The objectives of FSS are to improve the prediction performance of the predictors, to provide faster and more cost-effective predictors, and to provide a better understanding of the underlying process that generated the data [2].

The FSS methods choose a subset of the original features to be used for the subsequent processes. Hence, only the data generated from those features need to be collected ignoring all the other features. This makes FSS different

---

* This research has been funded in part by NSF grants EEC-9529152 (IMSC ERC), IIS-0238560 (PECASE) and IIS-0307908, and unrestricted cash gifts from Microsoft. Any opinions, findings, and conclusions or recommendations expressed in this material are those of the author(s) and do not necessarily reflect the views of the National Science Foundation. The authors would like to thank Dr. Carolee Winstein and Jarugool Tretiluxana for providing us the BCAR dataset and valuable feedbacks, and Thomas Navin Lal for providing us the BCI MPI dataset. The authors would also like to thank the anonymous reviewers for their valuable comments.

from feature *extraction*, where the correspondence information to the original features are in general not maintained, so all the original variables are required to be measured.

A time series is a series of observations, $x_i(t); [i = 1, \cdots, n; t = 1, \cdots, m]$, made sequentially through time where i indexes the measurements made at each time point t [3]. It is called a univariate time series when n is equal to 1, and a multivariate time series (MTS) when n is equal to, or greater than 2. An MTS item is naturally represented in an $m \times n$ matrix, where m is the number of observations and n is the number of *variables*, e.g., sensors. However, the state of the art FSS techniques, such as Recursive Feature elimination (RFE) [2], require each item to be represented in one row. Consequently, to utilize these techniques on MTS datasets, each MTS item needs to be first transformed into one row or column vector. However, since each of variables is considered separately during this *vectorization*, the correlation information among the features might be lost in the previous FSS method for MTS datasets [4].

In this paper, we propose a novel feature subset selection method for multivariate time series (MTS)[1] named ***CLeVer*** (descriptive **C**ommon principal component **L**oading based **V**ariable subset selection method). ***CLeVer*** utilizes the property of the principal components and common principal components to retain the correlation information among the variables.

2 Proposed Method

CLeVer is a novel variable subset selection method for multivariate time series (MTS) based on common principal component analysis (CPCA) [5,6]. Figure 1 illustrates the entire process of ***CLeVer***, which involves three phases: (1) principal components (PCs) computation per MTS item, (2) descriptive common principal components (DCPCs) computation per label[2] and their concatenation, and (3) variable subset selection using K-means clustering on DCPC loadings of variables. Each of these phases is described in the subsequent sections. Table 1 lists the notations used in the remainder of this paper, if not specified otherwise.

2.1 PC and DCPC Computations

The first and second phases (except the concatenation) of ***CLeVer*** are incorporated into Algorithm 1. It obtains both PCs and then DCPCs consecutively. The required input to Algorithm 1 is a set of MTS items with the same label.

Though there are n PCs for each item, only the first $p(< n)$ PCs, which are adequate for the purpose of representing each MTS item, are taken into consideration. Algorithm 1 takes in the threshold (δ) to determine p. That is,

---

[1] For multivariate time series, each *variable* is regarded as a feature [4]. Hence, the terms *feature* and *variable* are interchangeably used throughout this paper, when there is no ambiguity.
[2] The MTS datasets considered in our analysis are composed of labeled MTS items.

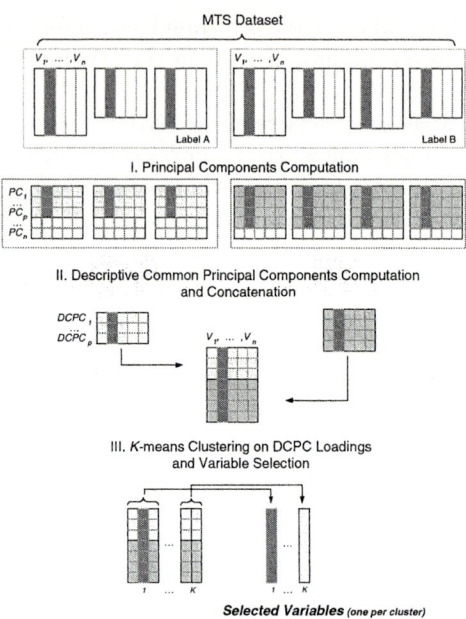

Fig. 1. The process of *CLeVer*

Table 1. Notations used in this paper

Symbol	Definition
N	number of MTS items in an MTS dataset
n	number of variables in an MTS item
K	number of clusters for K-means clustering
p	number of PCs for each MTS item to be used for computing DCPCs

for each input MTS item, p is determined to be the minimum value such that the ratio of the variances explained by its first p PCs to the total variance exceeds the provided threshold δ for the first time (Lines 3∼10). Since the MTS items can have different values for p, p is finally determined as their maximum value (Line 11).

All MTS items are now described by their first p principal components. Let them be denoted as L_i ($i = 1, \ldots, N$). Then, the DCPCs that *agree most closely* with all N sets of p PCs are successively defined by the eigenvectors of the matrix $H = \sum_{i=1}^{N} L_i^T L_i$ [5]. That is, $SVD(H) = SVD(\sum_{i=1}^{N} L_i^T L_i) = V \Lambda V^T$, where rows of V^T are eigenvectors of H and the first p of them define p $DCPCs$ for N MTS items. This computation of DCPC is captured by Lines 16∼17.

Algorithm 1. *ComputeDCPC*: PC and DCPC Computations

Require: MTS data groups with N items and δ {a predefined threshold}
1: $DCPC \leftarrow \emptyset$;
2: $H[0] \leftarrow \emptyset$;
3: **for** $i=1$ to N **do**
4:    $X \leftarrow$ the ith MTS item;
5:    $[U, S, U^T] \leftarrow$ SVD(correlation matrix of X);
6:    $loading[i] \leftarrow U^T$;
7:    $variance \leftarrow diag(S)$;
8:    $percentVar \leftarrow 100 \times (variance / \sum_{j=1}^{n} variance_j)$;
9:    $p_i \leftarrow$ number of the first p $percentVar$ elements whose cumulative sum $\geq \delta$;
10: **end for**
11: $p \leftarrow max(p_1, p_2, \ldots, p_N)$;
12: **for** $i=1$ to N **do**
13:    $L[i] \leftarrow$ the first p rows of $loading[i]$;
14:    $H[i] \leftarrow H[i-1] + (L[i]^T \times L[i])$;
15: **end for**
16: $[V, S, V^T] \leftarrow$ SVD(H);
17: $DCPC \leftarrow$ first p rows of V^T;

2.2 Variable Subset Selection

CLeVer utilizes a clustering method to group the *similar* variables together and finally to select the least redundant variables. First, the DCPCs per label are computed by Algorithm 1 and are concatenated column-wise if the MTS dataset has more than one label. Subsequently, K-means clustering is performed on the columns of the concatenated DCPC loadings, each column of which holds the one-to-one correspondence to the original variables. Then, the column vectors with the similar pattern of contributions to each of the DCPCs will be clustered together.

The next step is the *actual* variable selection, which decides the representatives of clusters. Once the clustering is done, one column vector closest to the centroid vector of each cluster is chosen as the representative of that cluster. The other columns within each cluster therefore can be eliminated. Finally, the corresponding original variable to the selected column is identified, which will form the selected subset of variables with the least redundant and possibly the most related information for the given K. For details, please refer to [6].

3 Performance Evaluation

We evaluate the effectiveness of ***CLeVer*** in terms of classification performance and processing time. We conducted several experiments on three real-world datasets: HumanGait, BCAR, and BCI MPI. After obtaining a subset of variables using ***CLeVer***, we performed classification using Support Vector Machine (SVM) with linear kernel and leave-one-out cross validation for BCAR and 10 fold stratified cross validation for the other two datasets. Subsequently, we com-

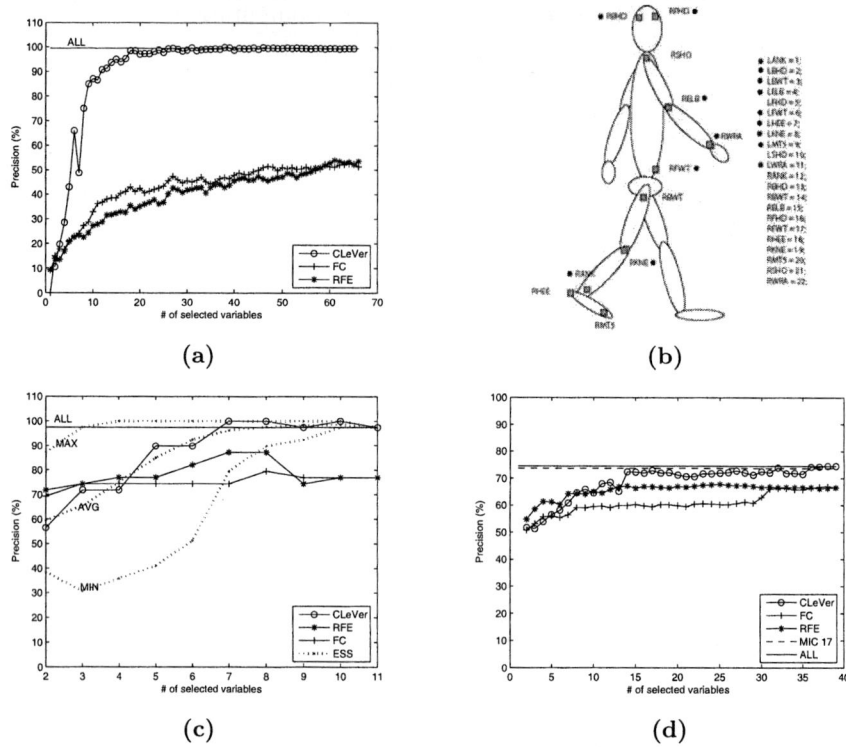

Fig. 2. (a) Classification Evaluation for HumanGait dataset (b) 22 markers for the HumanGait dataset. The markers with a filled circle represent 16 markers from which the 27 variables are selected by *CLeVer*, which yields the same performance accuracy as using all the 66 variables. Classification Evaluations for (c) BCAR dataset and (d) BCI MPI dataset

pared the performance of *CLeVer* with those of Recursive Feature Elimination (RFE) [2], Fisher Criterion (FC), Exhaustive Search Selection (ESS), and using all the available variables (ALL). The algorithm of *CLeVer* for the experiments is implemented in $Matlab^{TM}$. SVM classification is completed with *LIBSVM* [7].

Classification Performance. For the MTS dataset to be fed into SVM, each MTS item is vectorized using the upper triangle of its correlation matrix for *CLeVer*, Exhaustive Search Selection (ESS), and using all the variables (ALL). For RFE and FC, we vectorized each MTS item as in [4]. That is, each variable is represented as the autoregressive (AR) fit coefficients of order 3 using the forward backward linear prediction [8].

Figure 2(a) presents the generalization performances on the HumanGait dataset. The X axis is the number of selected subset of variables, i.e., the number of clusters K, and the Y axis is the classification accuracy. It shows that a subset of 27 variables selected by *CLeVer* out of 66 performs the same as the one using all the variables, which is 99.4% accuracy. The 27 variables selected by *CLeVer* are from only 16 markers (marked with a filled circle in Figure 2(b)) out of 22, which would mean that the values generated by the remaining 6 markers does not contribute much to the identification of the person. The performances by RFE and FC is much worse than the ones using *CLeVer*. Even when using all the variables, the classification accuracy is around 55%. Figure 2(c) illustrates *CLeVer* consistently outperforms RFE and FC when the number of selected features is more than 4. The 7 variables selected by *CLeVer* produce about 100% classification accuracy, which is even better than using all the 11 variables which is represented as a horizontal solid line. This implies that *CLeVer* never eliminates useful information in its variable selection process. Figure 2(d) represents the performance comparison using the BCI MPI dataset[3]. It depicts that when the number of selected variables is less than 10, RFE performs better than *CLeVer* and FC technique. When the number of selected variables is greater than 10, however, *CLeVer* performs far better than RFE. Using the 17 variables selected by *CLeVer*, the classification accuracy is 72.85%, which is very close to the performance of *MIC 17*, i,e., the known 17 motor imagery channels, whose accuracy is 73.65%.

Processing Time. The processing time for *CLeVer* includes the time to perform Algorithm 1 and the average time to perform the clustering and obtain the variable subsets while varying K from 2 to the number of all variables for each dataset. The processing time for RFE and FC includes the time to obtain 3 autoregressive fit coefficients and perform the feature subset selection. Overall, *CLeVer* takes up to 2 orders of magnitude less time than RFE, while performing better than RFE up to about a factor of two. The detailed results are provided in [6].

References

1. Liu, H., Yu, L., Dash, M., Motoda, H.: Active feature selection using classes. In: Pacific-Asia Conference on Knowledge Discovery and Data Mining. (2003)
2. Guyon, I., Elisseeff, A.: An introduction to variable and feature selection. Journal of Machine Learning Research **3** (2003) 1157–1182
3. Tucker, A., Swift, S., Liu, X.: Variable grouping in multivariate time series via correlation. IEEE Trans. on Systems, Man, and Cybernetics, Part B **31** (2001)
4. Lal, T.N., Schröder, M., Hinterberger, T., Weston, J., Bogdan, M., Birbaumer, N., Schölkopf, B.: Support vector channel selection in BCI. IEEE Trans. on Biomedical Engineering **51** (2004)
5. Krzanowski, W.: Between-groups comparison of principal components. Journal of the American Statistical Association **74** (1979)

---

[3] Unlike in [4] where they performed the feature subset selection per subject, the whole items from the 5 subjects were utilized in our experiments. Moreover, the regularization parameter C_s was estimated via 10 fold cross validation from the training datasets in [4], while we used the default value, which is 1.

6. Yang, K., Yoon, H., Shahabi, C.: Clever: a feature subset selection technique for multivariate time series. Technical report, University of Southern California (2005)
7. Chang, C.C., Lin, C.J.: Libsvm – a library for support vector machines. http://www.csie.ntu.edu.tw/~cjlin/libsvm/ (2004)
8. Moon, T.K., Stirling, W.C.: Mathematical Methods and Algorithms for Signal Processing. Prentice Hall (2000)

Covariance and PCA for Categorical Variables

Hirotaka Niitsuma and Takashi Okada

Department of Informatics, Kwansei Gakuin University,
2-1 Gakuen-cho, Sanda 669-1323, Japan,
niitsuma@mub.biglobe.ne.jp, okada-office@ksc.kwansei.ac.jp

Abstract. Covariances from categorical variables are defined using a regular simplex expression for categories. The method follows the variance definition by Gini, and it gives the covariance as a solution of simultaneous equations using the Newton method. The calculated results give reasonable values for test data. A method of principal component analysis (RS-PCA) is also proposed using regular simplex expressions, which allows easy interpretation of the principal components.

1 Introduction

There are large collections of categorical data in many applications, such as information retrieval, web browsing, telecommunications, and market basket analysis. While the dimensionality of such data sets can be large, the variables (or attributes) are seldom completely independent. Rather, it is natural to assume that the attributes are organized into topics, which may overlap, i.e., collections of variables whose occurrences are somehow correlated to each other.

One method to find such relationships is to select appropriate variables and to view the data using a method like Principle Components Analysis (PCA) [4]. This approach gives us a clear picture of the data using KL-plot of the PCA. However, the method is not settled for the data including categorical data. Multinomial PCA [2] is analogues to PCA for handling discrete or categorical data. However, multinomial PCA is a method based on the parametric model and it is difficult to construct a KL-plot for the estimated result. Multiple Correspondence Analysis (MCA) [3] is analogous to PCA and can handle discrete categorical data. MCA is also known as homogeneity analysis, dual scaling, or reciprocal averaging. The basic premise of the technique is that complicated multivariate data can be made more accessible by displaying their main regularities and patterns as plots ("KL-plot") . MCA is not based on a parametric model and can give a "KL-plot" for the estimated result. In order to represent the structure of the data, sometimes we need to ignore meaningless variables. However, MCA does not give covariances or correlation coefficients between a pair of categorical variables. It is difficult to obtain criteria for selecting appropriate categorical variables using MCA.

In this paper, we introduce the covariance between a pair of categorical variables using the regular simplex expression of categorical data. This can give a

2 Gini's Definition of Variance and Its Extension

Let us consider the contingency table shown in Table 1, which is known as Fisher's data [5] on the colors of the eyes and hair of the inhabitants of Caithness, Scotland. The table represents the joint population distribution of the categorical variable for eye color x_{eye} and the categorical variable for hair color x_{hair}:

$$x_{hair} \in \{\text{ fair red medium dark black}\}$$
$$x_{eye} \in \{\text{ blue light medium dark}\}. \tag{1}$$

Table 1. Fisher's data

		x_{hair}				
		fair	red	medium	dark	black
x_{eye}	blue	326	38	241	110	3
	light	688	116	584	188	4
	medium	343	84	909	412	26
	dark	98	48	403	681	85

Before defining the covariances among such categorical variables, $\sigma_{hair,eye}$, let us consider the variance of a categorical variable. Gini successfully defined the variance for categorical data [6].

$$\sigma_{ii} = \frac{1}{2N^2} \sum_{a=1}^{N} \sum_{b=1}^{N} (x_{ia} - x_{ib})^2 \tag{2}$$

where, σ_{ii} is the variance of the i-th variable, x_{ia} is the value of x_i for the a-th instance, and N is the number of instances. The distance of a categorical variable between instances is defined as $x_{ia} - x_{ib} = 0$ if their values are identical, and $= 1$ otherwise. A simple extension of this definition to the covariance σ_{ij} by replacing $(x_{ia} - x_{ib})^2$ to $(x_{ia} - x_{ib})(x_{ja} - x_{jb})$ does not give reasonable values for the covariance σ_{ij} [8]. In order to avoid this difficulty, we extended the definition based on scalar values, $x_{ia} - x_{ib}$, to a new definition using a vector expression [8]. The vector expression for a categorical variable with three categories $x_i \in \{r_1^i, r_2^i, r_3^i\}$ was defined by placing these three categories at the vertices of a regular triangle.

A regular simplex can be used for a variable with more than four categories. This is a straightforward extension of a regular triangle when the dimension of space is greater than two. For example, a regular simplex in the 3-dimensional space is a regular tetrahedron. Using a regular simplex, we can extend and generalize the definition of covariance to

Definition 1. *The covariance between a categorical variable $x_i \in \{r^i_1, r^i_2, ... r^i_{k_i}\}$ with k_i categories and a categorical variable $x_j \in \{r^j_1, r^j_2, ... r^j_{k_j}\}$ with k_j categories is defined as*

$$\sigma_{ij} = \max_{L^{ij}} (\frac{1}{2N^2} \sum_{a=1...N} \sum_{b=1...N} (\mathbf{v}^{k_i}(x_{ia}) - \mathbf{v}^{k_i}(x_{ib})) L^{ij} (\mathbf{v}^{k_j}(x_{ja}) - \mathbf{v}^{k_j}(x_{jb}))^t), \quad (3)$$

where $\mathbf{v}^n(r_k)$ is the position of the k-th vertex of a regular $(n-1)$-simplex [1]. r^i_k denotes the k-th element of the i-th categorical variable x_i. L^{ij} is a unitary matrix expressing the rotation between the regular simplexes for x_i and x_j.

Definition 1 includes a procedure to maximize the covariance. Using Lagrange multipliers, this procedure can be converted into a simpler problem of simultaneous equations, which can be solved using the Newton method. The following theorem enables this problem transformation.

Theorem 1. *The covariance between categorical variable x_i with k_i categories and categorical variable x_j with k_j categories is expressed by*

$$\sigma_{ij} = trace(A^{ij} L^{ij^t}), \quad (4)$$

where A^{ij} is $(k_i - 1) \times (k_j - 1)$ matrix :

$$A^{ij} = \frac{1}{2N^2} \sum_a \sum_b (\mathbf{v}^{k_i}(x_{ia}) - \mathbf{v}^{k_i}(x_{ib}))^t (\mathbf{v}^{k_j}(x_{ja}) - \mathbf{v}^{k_j}(x_{jb})). \quad (5)$$

L^{ij} is given by the solution of the following simultaneous equations.

$$A^{ij} L^{ij^t} = (A^{ij} L^{ij^t})^t$$
$$L^{ij} L^{ij^t} = \mathbf{E} \quad (6)$$

Proof. Here, we consider the case where $k_i = k_j$ for the sake of simplicity. Definition 1 gives a conditional maximization problem :

$$\sigma_{ij} = \max_{L^{ij}} \frac{1}{2N^2} \sum_a \sum_b (\mathbf{v}^{k_i}(x_{ia}) - \mathbf{v}^{k_i}(x_{ib})) L^{ij} (\mathbf{v}^{k_j}(x_{ja}) - \mathbf{v}^{k_j}(x_{jb}))^t$$
$$\text{subject to} \quad L^{ij} L^{ij^t} = \mathbf{E} \quad (7)$$

The introduction of Lagrange multipliers Λ for the constraint $L^{ij} L^{ij^t} = \mathbf{E}$ gives the Lagrangian function:

$$V = trace(A^{ij} L^{ij^t}) - trace(\Lambda^t L^{ij} L^{ij^t} - \mathbf{E}),$$

where Λ is $k_i \times k_i$ matrix. A stationary point of the Lagrangian function V is a solution of the simultaneous equations (6). □

Instead of maximizing (3) with constraint $L^{ij} L^{ij^t} = \mathbf{E}$, we can get the covariance by solving the equations (6), which can be solved easily using the Newton method.

Application of this method to Table 1 gives

$$\sigma_{hair,hair} = 0.36409, \sigma_{eye,hair} = 0.081253, \sigma_{eye,eye} = 0.34985 \quad (8)$$

We can derive a correlation coefficient using the covariance and variance values of categorical variables in the usual way. The correlation coefficients for x_{eye}, x_{hair} for Table 1 is 0.2277.

3 Principal Component Analysis

3.1 Principal Component Analysis of Categorical Data Using Regular Simplex (RS-PCA)

Let us consider categorical variables $x_1, x_2...x_J$. For the a-th instance, x_i takes value x_{ia}. Here, we represent x_{ia} by the vector of vertex coordinates $\mathbf{v}^{k_i}(x_{ia})$. Then, the values of all the categorical variables $x_1, x_2...x_J$ for the a-th instance can be represented by the concatenation of the vertex coordinate vectors of all the categorical variables:

$$\mathbf{x}(a) = (\mathbf{v}^{k_1}(x_{1a}), \mathbf{v}^{k_2}(x_{2a}), ..., \mathbf{v}^{k_J}(x_{Ja})). \quad (9)$$

Let us call this concatenated vector the *List of Regular Simplex Vertices* (LRSV). The covariance matrix of LRSV can be written as

$$\mathcal{A} = \frac{1}{N} \sum_{a=1}^{N} (\mathbf{x}(a) - \bar{\mathbf{x}})^t (\mathbf{x}(a) - \bar{\mathbf{x}}) = \begin{bmatrix} A^{11} & A^{12} & ... & A^{1J} \\ A^{21} & A^{22} & ... & A^{2J} \\ ... & ... & ... & ... \\ A^{J1} & A^{J2} & ... & A^{JJ} \end{bmatrix}. \quad (10)$$

where $\bar{\mathbf{x}} = \frac{1}{N} \sum_{a=1}^{N} \mathbf{x}(a)$ is an averege of the LRSV. The equation (10) shows the covariance matrix of LRSV. Since its eigenvalue decomposition can be regarded as a kind of Principal Component Analysis (PCA) on LRSV, we call it the *Principal Component Analysis using the Regular Simplex for categorical data* (RS-PCA).

When we need to interpret an eigenvector from RS-PCA, it is useful to express the eigenvector as a linear combination of the following vectors. The first basis set, d, shows vectors from one vertex to another vertex in the regular simplex. The other basis set, c, show vectors from the center of the regular simplex to one of the veritices.

$$\mathbf{d}^{k_j}(a \to b) = \mathbf{v}^{k_j}(b) - \mathbf{v}^{k_j}(a) \qquad a, b = 1, 2...k_j \qquad (11)$$

$$\mathbf{c}^{k_j}(a) = \mathbf{v}^{k_j}(a) - \frac{\sum_{b=1}^{k_j} \mathbf{v}^{k_j}(b)}{k_j} \qquad a = 1, 2...k_j \qquad (12)$$

Eigenvectors defined in this way change their basis set depending on its direction to the regular simplex, but this has the advantage of allowing us to grasp its meaning easily. For example, the first two principal component vectors from the data in Table 1 are expressed using the following linear combination.

$$\begin{aligned}
\mathbf{v}_1^{rs-pca} &= -0.63 \cdot \mathbf{d}^{eye}(medium \to light) - 0.09 \cdot \mathbf{c}^{eye}(blue) - 0.03 \cdot \mathbf{c}^{eye}(dark) \\
&\quad -0.76 \cdot \mathbf{d}^{hair}(medium \to fair) + 0.07 \cdot \mathbf{d}^{hair}(dark \to medium) \quad (13) \\
\mathbf{v}_2^{rs-pca} &= 0.64 \cdot \mathbf{d}^{eye}(dark \to light) - 0.13 \cdot \mathbf{d}^{eye}(medium \to light) \\
&\quad -0.68 \cdot \mathbf{d}^{hair}(dark \to medium) + 0.30 \cdot \mathbf{c}^{hair}(fair) \quad (14)
\end{aligned}$$

This expression shows that the axis is mostly characterized by the difference between $x^{eye} = light$ and $x^{eye} = medium$ values, and the difference between $x^{hair} = medium$ and $x^{hair} = fair$ values. The KL-plot using these components is shown in Figure 1 for Fisher's data. In this figure, the lower side is mainly occupied by data with values: $x^{eye} = medium$ or $x^{hair} = medium$. The upper side is mainly occupied by data with values $x^{eye} = light$ or $x^{hair} = fair$. Therefore, we can confirm that $(\mathbf{d}^{eye}(medium \to light) + \mathbf{d}^{hair}(medium \to fair))$ is the first principal component. In this way, we can easily interpret the data distribution on the KL-plot when we use the RS-PCA method.

Multiple Correspondence Analysis (MCA) [7] provides a similar PCA methodology to that of RS-PCA. It uses the representation of categorical values as an

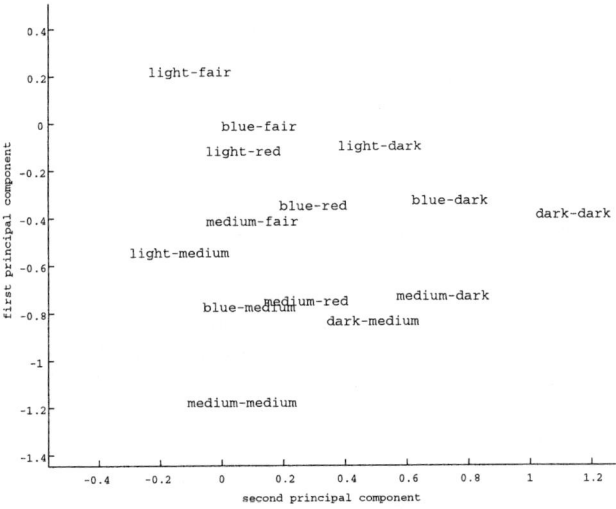

Fig. 1. KL-plot of Fisher's data calculated using RS-PCA. A point is expressed by a pair of eye and hair categories: $x^{eye} - x^{hair}$

indicator matrix (also known as a dummy matrix). MCA gives a similar KL-plot. However, the explanation of its principal components is difficult, because their basis vectors contain one redundant dimension compared to the regular simplex expression. Therefore, a conclusion from MCA can only be drawn after making a great effort to inspect the KL-plot of the data.

4 Conclusion

We studied the covariances between a pair of categorical variables based on Gini's definition of the variance for categorical data. The introduction of the regular simplex expression for categorical values enabled a reasonable definition of covariances, and an algorithm for computing the covariance was proposed. The regular simplex expression was also shown to be useful in the PCA analysis. We showed these merits through numerical experiments using Fisher's data.

The proposed RS-PCA method is mathematically similar to the MCA method, but it is much easier to interpret the KL-plot in RS-PCA than in MCA.

Acknowledgments. This research was partially supported by the Ministry of Education, Culture, Sport, Science and Technology, of Japan, with a Grant-in-Aid for Scientific Research on Priority Areas, 13131210 and a Grant-in-Aid for Scientific Research (A) 14208032.

References

1. F. Buekenhout and M. Parker. The number of nets of the regular convex polytopes in dimension ≤ 4. *Disc. Math.*, 186:69–94, 1998.
2. W. Buntine. Variational extensions to EM and multinomial PCA. In T. Elomaa, H. Mannila, and H. Toivonen, editors, *Machine Learning: ECML 2002. LNAI 2430*, pages 23–34. Springer-Verlag, 2002.
3. S.-E. Clausen. *Applied correspondence analysis: an introduction*. Thousand Oaks: Sage Publ, 1998.
4. K. Diamantaras and S. Kung. *Principal Component Neural Networks*. Wiley, New York, 1996.
5. R. A. Fisher. The precision of discriminant functions. *Annals of Eugenics (London)*, 10:422–429, 1940.
6. C. W. Gini. Variability and Mutability, contribution to the study of statistical distributions and relations. Studi Economico-Giuridici della R. Universita de Cagliari (1912). Reviewed in: R. J. Light and B. H. Margolin: An Analysis of Variance for Categorical Data. *J. American Statistical Association*, 66:534–544, 1971.
7. J. C. Gower and D. J. Hand. *Biplot*. Chapman and Hall, London, 1996.
8. T. Okada. A note on covariances for categorical data. In K. S. Leung, L. W. Chan, and H. Meng, editors, *Intelligent Data Engineering and Automated Learning - IDEAL 2000 LNCS 1983*, pages 150–157, 2000.

ADenTS: An Adaptive Density-Based Tree Structure for Approximating Aggregate Queries over Real Attributes*

Tianyi Wu, Jian Xu, Chen Wang, Wei Wang, and Baile Shi

Fudan University, China
{tywu, xujian, chenwang, weiwang1, bshi}@fudan.edu.cn

Abstract. In many fields and applications, it is critical for users to make decisions through OLAP queries. How to promote accuracy and efficiency while answering multiple aggregate queries, e.g. COUNT, SUM, AVG, MAX, MIN and MEDIAN? It has been the urgent problem in the fields of OLAP and data summarization recently. There have been a few solutions such as *MRA-Tree* and *GENHIST* for it. However, they could only answer a certain aggregate query which was defined in a particular data cube with some limited applications. In this paper, we develop a novel framework ***ADenTS***, i.e. **A**daptive **Den**sity-based **T**ree **S**tructure, to answer various types of aggregate queries within a single data cube. We represent the whole cube by building a coherent tree structure. Several techniques for approximation are also proposed. The experimental results show that our method outperforms others in effectiveness.

1 Introduction

Answering aggregate queries in data cubes approximately and efficiently is one of the most essential techniques for data warehousing and data mining. Consider a database of a charitable organization with 2 dimensions age and salary. Each data point in the age-salary space corresponds to a donator and the attribute value represents the amount of money the person donated. Moreover, consider for instance in a 3-dimensional geographical database, geologists may be particularly interested in a geological feature such as readings of snowfall or wind speed at different locations. Typically, aggregate functions include COUNT, SUM, AVERAGE, MAX, MIN, MEDIAN, etc.

Most developed algorithms could only adopt a certain defined aggregation in the course of cube modelling. But real applications often call for methods that can support different aggregations in a single cube. For example, for the database of charitable organization mentioned above, we may have aggregate queries as follows: *what is the total number of people whose age is between* 40 *and* 50, *whose*

---

* This research is supported in part by the Key Program of National Natural Science Foundation of China (No. 69933010 and 60303008), China National 863 High-Tech Projects (No. 2002AA4Z3430 and 2002AA231041).

salary is above $10K$, *and whose donation is between* $1K$ *and* $2K$ *and please give the average amount of money donated by people whose donation falls in the range* $5K - 10K$ *and* $15K - 20K$.

The method we proposed in this paper aims at constructing an effective tree structure by pre-computation of data cubes. Several techniques are presented for efficient aggregate query approximation. Generally speaking, our method has the following properties: (1) it is able to answer a variety of types of aggregate queries within a single data cube; (2) while most traditional algorithms in this category do not allow a user to specify a query range over the value dimension, our algorithm does not have such constraint; (3) empirical evidence indicates that our method leads to good accuracy in query processing.

The remainder of this paper is organized as follows. Section 2 describes related work in this area. Section 3 and 4 include the algorithms for tree construction and query approximation. Next we discuss in section 5 a heuristic approach for the tree construction. Experimental results are presented in Section 6.

2 Related Work

Existing algorithms for aggregate query approximation include quadtree structure [4], histogram techniques [2,3], kernel density estimator [2,5], clustering techniques [5], and wavelet decomposition [1,6].

Multi-Resolution aggregate tree (*MRA-tree*) [4] introduces a quadtree-like (or *R-tree* like) multi-dimensional index structure that answers aggregate queries like COUNT, SUM, MIN, MAX, AVG by selectively traversing the nodes of the tree in a top-down fashion. Nodes at different levels of the tree correspond to space partitioning in different granularities. An essential idea of the approach is that it provides 100% intervals of confidence on the value of the aggregate. Furthermore, the algorithm introduces a tree node traversal policy in order to reduce the uncertainty of the aggregate value as fast as possible. However, this algorithm sometimes cannot provide good aggregate estimation because the interval of confidence is too wide and it suffers from information redundancy.

Another common approach to approximate aggregate queries is using probability density function or kernel density estimators to represent a data set. For example, gaussian mixture model [5] is utilized for such representation. Given a fixed gaussian model number k, the algorithm generates k clusters represented by gaussian models based on the data set. A high compression ratio is achieved. Nevertheless, the types of queries it can answer are limited and the quality of the model is completely dependent upon the similarity between the actual probability distribution of the points and gaussian distribution.

The histogram technique *GENHIST* [2] is designed to approximate the density of multi-dimensional datasets, i.e. to answer COUNT queries. It creates buckets of variable size and allows them to overlap. In order to make the data points in a bucket uniformly distributed, the algorithm removes a proper number of points from the dense grids and therefore the density of the data point in space becomes smoother. The method is able to achieve high accuracy while storing comparatively small number of values.

3 Tree Structure for Aggregate Query Answering

To the best of our knowledge, our method, unlike all previous ones, addresses the problem of answering multiple types of aggregate queries in a single data cube by maintaining an effective index structure.

3.1 Basic Definitions and Notations

We denote a d-dimensional data set consisting of n data points to be $D(n,d)$. A data point $<P_0, P_1, \cdots, P_{d-1}>$ in $D(n,d)$ have the form of $<loc, value>$, where $<P_0, P_1, \cdots, P_{d-2}>$ is regarded as in the $(d-1)$-dimensional loc space R^{loc}. We also regard P_{d-1} as in the R^{value} domain, representing the value attribute associated with the point. Generally, users are interested in the value attribute and therefore aggregate queries are performed over this dimension.

The proposed structure is a modified binary tree (Figure 1). For each node, we maintain five fields: $Area$, $Count$, Max, Min, and $Distrb$. Given a tree node N, We have $Area(N)$ corresponds to a (hyper-)rectangular area and $Area(N) \subseteq R^{loc}$. We denote $PSet(N)$ as a set of data points associated with node N. $PSet(N)$ is a subset of the data points falling in Area(N). How $PSet$ is calculated will be discussed in the following section. Formally, we have the followings: (1) $PSet(N) \subseteq \{P | \forall P \in D(n,d) \land P_0 P_1 \cdots P_{d-2} \in Area(N)\}$; (2) $Count(N) = ||PSet(N)||$; (3) $Max(N) = max\{P_{d-1} | \forall P \in PSet(N)\}$, $Min(N) = min\{P_{d-1} | \forall P \in PSet(N)\}$; (4) $\forall N_1 \forall N_2 \Rightarrow PSet(N_1) \cap PSet(N_2) = \Phi$, $D(n,d) = \{P | \exists N \Rightarrow P \in PSet(N)\}$. For a node N, we also maintain $Distrb(N)$ in order to store the probability distribution function over the value dimension R^{value} of $PSet(N)$.

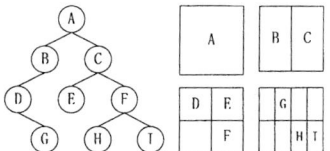

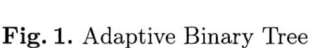

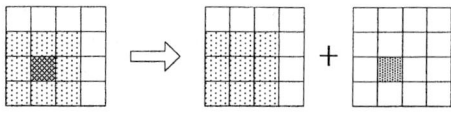

Fig. 1. Adaptive Binary Tree

Fig. 2. High Bump Removal

3.2 Algorithm

The main algorithm for building up the tree structure can be divided into two major steps. The first step is the adaptive construction of density binary tree, which updates the $Area$ field for every node. The second step is to iteratively update all tree nodes with $Count$, Min, Max, and $Distrb$.

Adaptive Construction of the Density Binary Tree. Two parameters h and p are specified initially. Here h represents the maximum height of the tree and p specifies a density threshold for the creation of a new node. The outline of the step is shown in figure 3.

Input: dataset $D(n,d)$, maximum tree height h, splitting density threshold p;
Output: density binary tree TR;
Method: ADTreeCons($D(n,d), h, p$)
1:   $depth \leftarrow$ current tree depth;
2:   IF $depth \geq h$ or $n < p$;
3:      RETURN Null;
4:   create a new node N and set $Area(N)$ to the space corresponding to $D(n,d)$;
5:   $k \leftarrow depth \bmod (d-1)$;
6:   divide $D(n,d)$ half through the k-th dimension into two subspaces of D_1 and D_2;
7:   $N.LeftChild \leftarrow$ ADTreeCons($D_1(n1,d), h, p$);
8:   $N.RightChild \leftarrow$ ADTreeCons($D_2(n2,d), h, p$);
9:   RETURN N;

Fig. 3. Algorithm for Adaptive Construction of the Density Binary Tree

---

Input: density binary tree TR, dataset $D(n,d)$, and maximum grid size S;
Output: updated tree TR;
Method: ADTreeUpdate($TR, D(n,d), S$)
1:   $level \leftarrow$ bottom of the tree;
2:   REPEAT
3:      FOR each N in TR at current level
4:         $p \leftarrow$ density in $Area(N)$;
5:         $q \leftarrow$ density in the neighborhood of $Area(N)$;
6:         IF $p > q$ //**high bump removal**
7:            $PSet(N) \leftarrow$ random set of $(p-q)$ points removed from $Area(N)$;
8:            Update $Count(N), Max(N), Min(N), Distrb(N)$ from $PSet(N)$;
9:      $level \leftarrow level - 1$;
10: UNTIL (grid size at $level > S$ OR $level = 0$);
11: update the $level$ by removing the remained points;
12: RETURN TR;

Fig. 4. Algorithm for Bottom-Up Updating the Density Binary Tree

Bottom-Up Updating. Nodes at a same level have two properties: (1) they have same shape (figure 1); (2) their density $Count$ is greater than the threshold. Thus for any node N, $Area(N)$ can be regarded as one of the populated grids picked up from a particular partitioning of the $(d-1)$-dimensional space R^{loc}.

The detailed algorithm is described in figure 4. The bottom-up process can be derived from *GENHIST* [2]. Given a populated grid N, we compare its density with their surrounding ones. If its density is higher than its neighbors, we randomly remove a certain number of data points from $Area(N)$ to make sure that the density of the remaining points in $Area(N)$ is equal to the average density of the surrounding grids. We call this process as "High Bump Removal" (figure 2). Intuitively, the dense areas become sparser and the density of the new data set comes to be smoother in R^{loc}. After we obtain *PSet* for the node N, *Count*, *Min*, *Max*, and *Distrb* can be easily acquired according to the previous definitions. The distribution function can be specified either by users or by implementing unbiased parameter estimation.

4 Aggregate Query Approximation

An aggregate query consists of three parts: (1) a query type identification specifying the type of aggregate query over the value dimension, (2) a $(d-1)$-dimensional user query region Q ($\subseteq R^{loc}$), and (3) a specified range of the *value* dimension denoted as T ($\subseteq R^{value}$). For example, for the charitable organization database, a query can be $\{\text{COUNT}, Q(40-50, 10K+), T(1K-2K)\}$.

4.1 COUNT

We traverse the tree from the root. Suppose that the i-th node visited is $N^{(i)}$. There are four kinds of relation between Q and $Area(N^{(i)})$ [4], as shown in figure 5. For case (d), we simply return 0 and stop traversing. Otherwise, based on the assumption that the data points in $PSet(N^{(i)})$ are uniformly distributed in the R^{loc} space, an estimated aggregate value $E^{(i)}_{count}$ corresponding to the node $N^{(i)}$ is calculated using the function $E^{(i)}_{count} = Count(N^{(i)}) \int_T p^{(i)}(\xi) d\xi \times \frac{||Area(N^{(i)}) \cap Q||}{||Area(N^{(i)})||}$, where $p^{(i)}(\xi)$ is the probability density distribution information derived from $Distrb(N^{(i)})$. If totally k nodes are visited, $E_{count} = \sum_{i=1}^{k} E^{(i)}_{count}$ is returned as the approximation of the COUNT query.

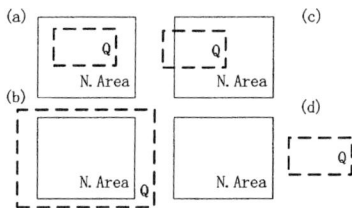

Fig. 5. Relation of Node

Fig. 6. Example

200	200	200	200
200	N_1 1000	N_2 600	200
200	200	200	200

4.2 SUM and AVERAGE

The SUM and AVERAGE queries differ from the COUNT queries in that different estimated aggregate values are calculated. As for SUM, we have $E_{sum}^{(i)} = \int_T Count(N^{(i)}) \xi p^{(i)}(\xi) d\xi \times \frac{||Area(N^{(i)}) \cap Q||}{||Area(N^{(i)})||}$ and $E_{sum} = \sum_{i=1}^{k} E_{sum}^{(i)}$. As for AVERAGE, we have $E_{average} = \frac{E_{sum}}{E_{count}}$.

4.3 MAX and MIN

Due to the continuous property of the value dimension, it is unnecessary to specify a range T for MAX or MIN queries. Without loss of generality, we discuss how MAX queries are answered.

Obviously we have $E_{max}^{(i)} = Max(N^{(i)})$ for case (b) and $E_{max}^{(i)} = -\infty$ for case (d). Otherwise, when no correct answer can be guaranteed, the problem of approximating $E_{max}^{(i)}$ can be stated as follows. Given $n = Count(N^{(i)})$ points and we randomly pick up $m = \frac{||Area(N^{(i)}) \cap Q||}{||Area(N^{(i)})||} \times n$ ($1 \leq m \leq n$), what is the expected maximum value among the m values? Assume that the n data points have value $V_1, \cdots, V_n$ and $V_i \leq V_{i+1}$ ($1 \leq i \leq n$), then a straightforward solution is $E_{max}^{(i)}(m) = \sum_{i=m}^{n} P(max = V_i) \times V_i = \frac{\binom{m-1}{m-1} \times V_m + \binom{m}{m-1} \times V_{m-1} + \cdots + \binom{n-1}{m-1} \times V_n}{\binom{n}{m}}$.
However, such computational cost is extremely high. Note that the $E_{max}^{(i)}(m)$ is a function that monotonically increases with respect to m, and $E_{max}^{(i)}(1) = E_{max}^{(i)}(max(V_1)) = E_{max}^{(i)}(V_1) = E^{(i)}\xi$, where $E^{(i)}\xi$ is the expectation of the function $Distrb(N^{(i)})$. Also note that $E_{max}^{(i)}(n) = Max(N^{(i)})$. Therefore, we come up with a linear estimation: $E_{max}^{(i)}(m) = E_{max}^{(i)}(1) + (m-1) \times \frac{E_{max}^{(i)}(n) - E_{max}^{(i)}(1)}{n} = \frac{1}{n} \times Max(N^{(i)}) + \frac{n-m+1}{n} \times E^{(i)}\xi$. This estimation takes probability distribution information stored in each node into consideration and does not require us to pre-compute $V_1, \cdots, V_n$. The final approximation for the MAX query is $E_{max} = max\{E_{max}^{(i)}, 1 \leq i \leq k\}$.

4.4 MEDIAN

Probability distribution information is very useful in the approximation of many other types of aggregate queries. Typically, we discuss how MEDIAN queries are approximated. For the i-th visited tree node $N^{(i)}$, we calculate the median of $PSet(N^{(i)})$ denoted as $E_{median}^{(i)}$ by solving the equation $\int_{\xi \leq x \wedge \xi \in T} p^{(i)}(\xi) d\xi = \int_{\xi > x \wedge \xi \in T} p^{(i)}(\xi) d\xi$. The final approximation of the MEDIAN query is $E_{median} = \frac{\sum_{i=1}^{k} Count(N^{(i)}) \times E_{median}^{(i)}}{\sum_{i=1}^{k} Count(N^{(i)})}$.

5 Heuristics for the Construction of Tree Structure

The goal of "high bump removal" process in algorithm *ADTreeUpdate* is to improve the uniformity of the density of the whole data set. Nevertheless, if two children nodes with a same parent are denser than their neighboring grids, there will be interference between the two corresponding adjacent high bumps. To minimize the problem, we take the advantage of the binary tree structure and update every pair of adjacent dense children simultaneously instead of updating them one by one. This process is called "double-bump removal", as shown in figure 7.

Input: density binary tree TR, dataset $D(n,d)$, and maximum grid size S;
Output: updated tree TR;
Method: ADTreeUpdate-Double(TR, $D(n,d)$, S)
1:    $level \leftarrow$ bottom of the tree;
2:    mark all nodes in TR as "single";
3:    REPEAT
4:       FOR each pair of nodes N_1 and N_2 in TR at current level
5:          $p_1 \leftarrow$ density in $Area(N_1)$;
6:          $p_2 \leftarrow$ density in $Area(N_2)$;
7:          $q \leftarrow$ density in the neighborhood of $Area(N_1) \cup Area(N_2)$;
8:          IF $p_1 > q$ AND $p_2 > q$ //**double-bump removal**
9:             $PSet(N_1) \leftarrow$ random set of $(p_1 - q)$ points removed from $Area(N)$;
10:           Update $Count(N_1)$, $Max(N_1)$, $Min(N_1)$, $Distrb(N_1)$ from $PSet(N_1)$;
11:           $PSet(N_2) \leftarrow$ random set of $(p_2 - q)$ points removed from $Area(N)$;
12:           Update $Count(N_2)$, $Max(N_2)$, $Min(N_2)$, $Distrb(N_2)$ from $PSet(N_2)$;
13:           mark N_1 and N_2 as "double";
14:       FOR each node N marked as "single" in TR at current level
15:          IF N is denser than its neighbors
16:            Remove the High Bump and update N
17:    $level \leftarrow level - 1$;
18:    UNTIL (grid size in level $> S$ OR $level = 0$);
19:    update the $level$ by removing the remained points;
20:    RETURN TR;

Fig. 7. Improved Algorithm for Updating of the Density Binary Tree

For instance, suppose that there is a data set whose density distribution is depicted in figure 6. Suppose that N_1 and N_2 are two nodes with the same parent. The original algorithm will result in $Count(N_1) = 1000 - \frac{200 \times 7 + 600}{8} = 750$ and $Count(N_2) = 600 - \frac{200 \times 7 + 750}{8} = 331$. After the heuristic approach, the result will be better with $Count(N_1) = 800$ and $Count(N_2) = 400$.

6 Experiments

6.1 Methodology

We applied ***ADenTS*** to real database obtained from the US Forest Cover Type and synthesized query workloads. The implementation of our algorithm is on an

IBM 1.5GHz CPU and 256MB of DDR main memory with Windows XP and Microsoft Visual C++ 6.0.

Data sets with 3 to 5 dimensions are generated. For each projected data set and each supported type of query, 1000 random queries are created with average selectivity 1% to form a query workload. Queries with data point selectivity of less than 0.1% are disregarded because small selectivity would seriously degenerates the effectiveness of all algorithms. The accuracy is measured by relative error. Note that the relative error of MIN, MAX queries should be calculated by

$$RelativeError = \frac{|CorrectAnswer - ApproximatedAnswer|}{|RangeofValueDimension|}.$$

6.2 Comparison with MRA-Tree and GENHIST

Due to the apparent difference of the type of queries three algorithms can support, we categorize the experimental results into three groups according to: (1) queries that can be only answered by *ADenTS*; (2) queries that can be answered by both *ADenTS* and *MRA-Tree*; and (3) queries that can be answered by *ADenTS* and *GENHIST*.

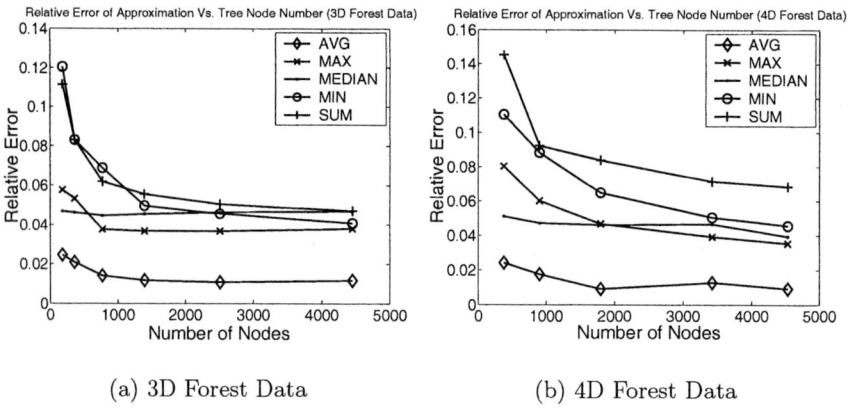

(a) 3D Forest Data   (b) 4D Forest Data

Fig. 8. Performance of ADenTS

In figure 8 and 9, we plot the relation of the tree size with query answering accuracy for the 3- and 4-dimeniosnal datasets and all supported queries with *ADenTS*, i.e. COUNT, SUM, AVG, MAX, MIN, MEDIAN. For case (2), it is stated that *MRA-Tree* does not support queries with a specified value range and that it does not support MEDIAN queries, either. Therefore, we generate four types of queries with no specified range over the value dimension so that they can be answered by *MRA-Tree*. In figure 11 we demonstrate that for COUNT, SUM, AVG, and MAX queries, our method can produce results better than *MRA-Tree* in 3 to 5 dimensional data sets. This is due to the fact that in *ADenTS* the

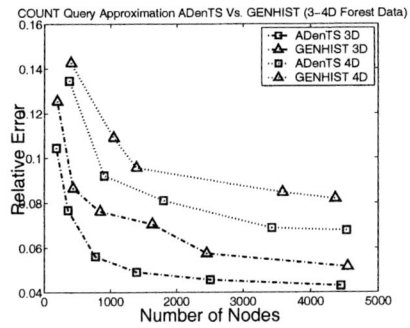
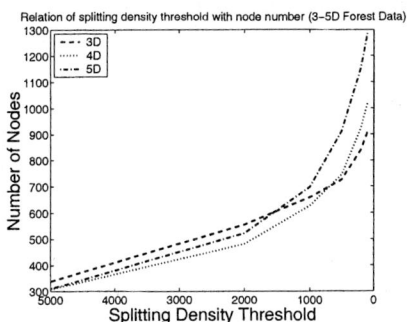

Fig. 9. Performance of ADenTS Vs. GENHIST

Fig. 10. Relation of Parameters

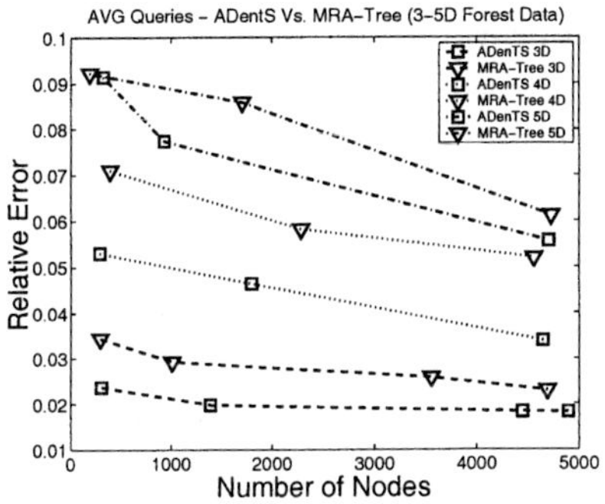

(a) AVG Query

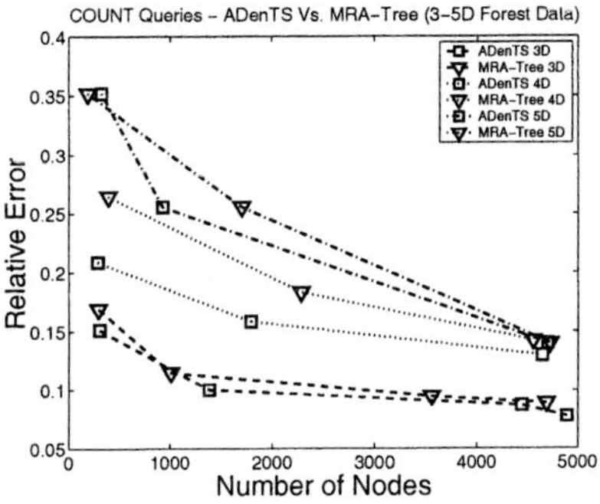

(b) COUNT Query

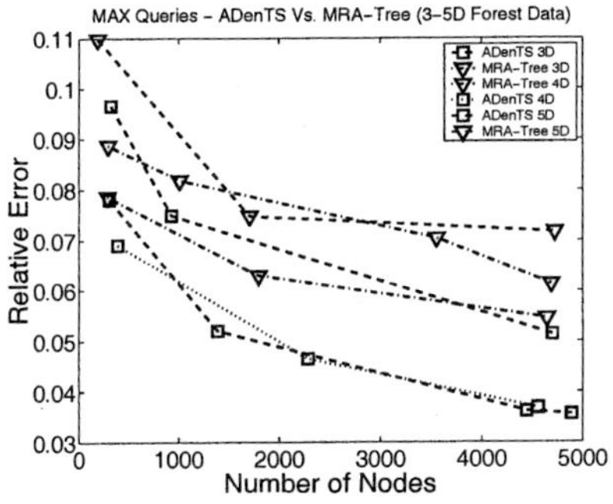

(c) MAX Query

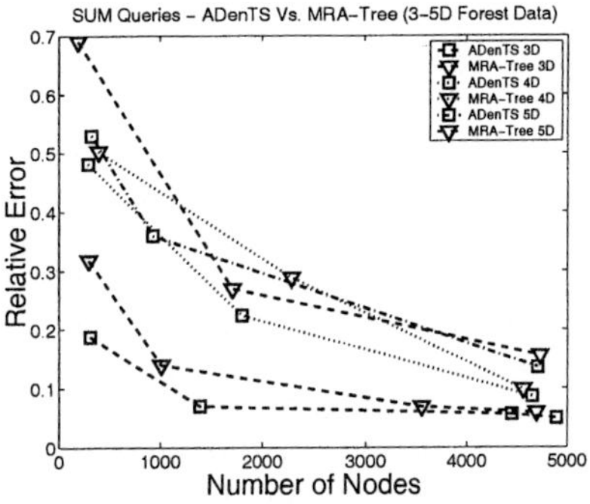

(d) SUM Query

Fig. 11. Performance of AdenTS Vs. MRA-Tree

uniform assumption can be better satisfied than in the simple grid partitioning in *MRA-Tree*. Furthermore, unlike *MRA-Tree*, the data points related to the nodes in **ADenTS** are disjoint and contains no information redundancy. In case (3), *GENHIST* can answer only COUNT queries. The way our method approximate COUNT queries is essentially similar to *GENHIST* with respect to the bump removing procedure. However, because **ADenTS** introduces the process of double-bump removal, it outperforms *GENHIST* slightly, as shown in figure 9. We also present the relation of the number of tree nodes with respect to the splitting density threshold ρ. The three curves drawn in figure 10 display that, as the splitting density threshold is decreased, more and more nodes with relatively less data points will be created.

7 Conclusion

In this paper, we presented **ADenTS**, an effective tree structure for aggregate query answering in multi-dimensional data cubes. The main target of our method is to support various types of aggregate queries without loss of accuracy. Briefly speaking, The tree is built through an adaptive density-based approach in top-down and bottom up fashion. Our method inherits the advantages of both the Multi-Resolution Aggregate tree (*MRA-Tree*) structure and the histogram technique *GENHIST*. In addition, it outperforms them by answering more kinds of queries within a single data structure and meanwhile achieving good accuracy, which can be well demonstrated in the experimental evaluation.

References

1. Chakrabarti, M. N. Garofalakis, R. Rastogi, and K. Shim. Approxmate Query Processing Using Wavelets. In *Proc. 26th Int. Conf. on Very Large Data Base (VLDB'00)*, Cairo, Egypt, Sep. 2000.
2. D. Gunopulos, G. Kollios, and V. J. Tsotras. Approximating Multi-dimensional Aggregate Range Queries over Real Attributes. In *Proc. ACM SIGMOD 19th Int. Conf. on Management of Data (SIGMOD'00)*, Dallas, USA, May 2000.
3. J. Lee, D. Kim, and C. Chung. Multi-dimensional Selectivity Estimation Using Compressed Histogram Information. In *Proc. ACM SIGMOD 18th Int. Conf. on Management of Data (SIGMOD'99)*, Philadelphia, USA, Jun. 1999.
4. I. Lazaridis and S. Mehrotra. Progressive Approximate Aggregate Queries with a Multi-Resolution Tree Structure. In *Proc. ACM SIGMOD 20th Int. Conf. on Management of Data (SIGMOD'01)*, Santa Barbara, USA, May 2001.
5. J. Shanmugasundaram, U. Fayyad, and P. S. Bradley. Compressed Data Cubes for OLAP Aggregate Query Approximation on Continuous Dimensions. In *Proc. ACM SIGKDD 6th Int. Conf. on Knowledge Discovery and Data Mining (KDD'99)*, San Diego, USA, Aug. 1999.
6. J. S. Vitter and M. Wang. Approximate Computation of Multi-dimensional Aggregates of Sparse Data Using Wavelets. In *Proc. ACM SIGMOD 18th Int. Conf. on Management of Data (SIGMOD'99)*, Philadelphia, USA, Jun. 1999.

Frequent Itemset Mining with Parallel RDBMS

Xuequn Shang[1] and Kai-Uwe Sattler[2]

[1] Department of Computer Science, University of Magdeburg,
P.O.BOX 4120, 39106 Magdeburg, Germany
shang@iti.cs.uni-magdeburg.de
[2] Department of Computer Science and Automation, Technical University of Ilmenau
kus@tu-ilmenau.de

Abstract. Data mining on large relational databases has gained popularity and its significance is well recognized. However, the performance of SQL based data mining is known to fall behind specialized implementation. We investigate approaches based on SQL for the problem of finding frequent patterns from a transaction table, including an algorithm that we recently proposed, called *Ppropad* (Parallel PROjection PAttern Discovery). *Ppropad* successively projects the transaction table into frequent itemsets to avoid making multiple passes over the large original transaction table and generating a huge sets of candidates. We have built a parallel database system with DB2 and made performance evaluation on it. We prove that data mining with SQL can achieve sufficient performance by the utilization of database tuning.

1 Introduction

Mining frequent pattern in transaction databases has been studied popularly in data mining research. Most of the algorithms used today typically employ sophisticated in-memory data structures, where the data is stored into and retrieved from flat files. The integration of data mining with database systems is an emergent trend in database research and development area. This is particularly driven by explosion of the data amount stored in databases such as Data Warehouses during recent years, and database systems provide powerful mechanisms for accessing, filtering, and indexing data, as well as SQL parallelization. In addition, SQL-aware data mining systems have ability to support ad-hoc mining, ie. allowing to mine arbitrary query results from multiple abstract layers of database systems or Data Warehouses.

However, from the performance perspective, data mining algorithms that are implemented with the help of SQL are usually considered inferior to algorithms that process data outside the database systems. On the other hand recently most major database systems have included capabilities to support parallelization. This fact motivated us to develop a new parallel SQL-based algorithm which avoids making multiple passes over the large original input table and complex joins between the tables, and to examine how efficiently SQL based frequent itemset mining can be parallelized and speed up using parallel database system.

The remainder of this paper is organized as follows. In section 2, we briefly discuss frequent pattern mining algorithms that employ SQL queries. *Ppropad* algorithm is explained in section 3. Section 4 presents several experiments that assess the performance of the algorithms based on synthetic datasets. We conclude the paper in section 5 and give a brief outlook on future work.

2 Frequent Pattern Mining with SQL

2.1 The Problem of Frequent Pattern Mining

The frequent Pattern mining problem can be formally defined as follows. Let $I = \{i_1, i_2, ..., i_m\}$ be a set of items, and DB be a transaction database, where each transaction T is a set of items and $T \subseteq I$. An unique identifer, called TID, is assigned with each transaction. A transaction T contains a pattern P, a set of items in I, if $P \subseteq T$. The support of a pattern P is the number of transactions containing P in DB. We say that P is a frequent pattern if P's support is not less than a predefined minimum support threshold ξ.

Most of the previous studies adopt an *Apriori*-like candidate set generation-and-test approach [2, 4, 5], which is based on an anti-monotone *Apriori* heuristic: if any length k pattern is not frequent in the database, its super-pattern of length $(k+1)$ can never be frequent. Recently, an FP-tree based frequent pattern mining method [3], called FP-growth, developed by Han et al. achieves high efficiency, compared with *Apriori*-like approach. Pramudiono et al. reported parallel execution of FP-$growth$ on shared nothing environment [6].

2.2 Frequent Pattern Mining Based on SQL

Before data can be mined with SQL, it has to be made available as relational tables. Transaction data, as the input, is transformed into the first normal form table T with two column attributes: transaction identifier (*tid*) and item identifier (*item*). The support counters of frequent items can be kept in a separate table F (*item, count*).

Almost all frequent pattern mining algorithms with SQL consist of a sequence of steps proceeding in a bottom-up manner. The result of the kth step is the set of frequent itemsets, denoted as F_k. The first step computes frequent 1-itemsets F_1. The candidate generation phase computes a set of potential frequent k-itemsets C_k from F_{k-1}. The support counting phase filters out those itemsets from C_k that appear more frequently in the given set of transactions than the minimum support and stores them in F_k.

3 Algorithm for *Ppropad*

We proposed a SQL-based algorithm, called PROjection PAttern Discovery, or *Propad* for short [7]. Like the FP-growth method it adopts the divide-and-conquer strategy and successively transforms the original transaction table into

Table 1. A transaction database DB and $\xi = 3$

TID	Items	Frequent Items
1	a, c, d, f, g, i, m, o	a, c, f, m, o
2	a, b, c, f, l, m, n	a, b, c, f, m
3	b, f, h, j, n	b, f
4	$b, c, k, o,$	b, c, o
5	a, c, e, f, l, m, o	a, c, f, m, o

a set of frequent item-related projected tables. Then we separately mine each one of the tables as soon as they are built. Let us give an example with five transactions in Table 1. The support threshold is set to 3.

Before the algorithm is given, let us define the projected transaction table:

Definition 1. *In order to avoid repetitiousness and to ensure each frequent item is projected to at most one projected table, we suppose items in alphabetical order. Let i be a frequent item. A i-related projected transaction table, is denoted as PT_i, that collects all frequent items (larger than i) in the transactions containing i and the support of these items satisfies the minimum support threshold.*

The mining process can be regarded as a process of frequent pattern growth, which is facilitated by projecting transaction tables in a top-down fashion. In our approach, we are trying to find all frequent patterns with the respect to one frequent item, which is the base item of the tested projected table. For each frequent item i we traverse the transaction table to find all frequent items that occur with i. All items that are locally frequent with i will participate in building the i projected table. To describe the process of projecting, let's first examine the example in Table 1 as follows.

- At the first level we simply gather the count of each item and items that satisfy the minimum support are inserted into the transformed transaction table TF that has the same schema as transaction table T. It means that only frequent 1-items are included in the table TF.
- At the second level, for each frequent 1-item i (except the last one) in the table TF we construct its respective projected transaction table PT_i. This is done by two phases. The first step finds all frequent items that co-occur with i and are larger than i from TF. The second step finds the local frequent items. Only those local frequent items are collected into the PT_i. Frequent 1-items are regarded as the prefixes, frequent 2-patterns are gained by simply combining the prefixes and their local frequent itemsets. For instance, we get the frequent 1-items $\{a, b, c, f, m, o\}$ and their respective projected transaction tables PT_a, PT_b, PT_c, PT_f, PT_m. For the table PT_a, its local frequent item are $\{c, f, m\}$ stored in F. The frequent 2-patterns are $\{\{a,c\}, \{a,f\}, \{a,m\}\}$.
- At the next level, to each frequent item j in the projected transaction table PT_i we recursively construct its projected transaction table PT_i_j and gain

its local frequent items. For example, c is the first frequent item in PT_a. We get the frequent 3-patterns are {{a,c,f}, {a,c,m}}. The similar procedure goes on until one projected transaction table is filtered. A filtered projected table is that each transaction in the table only maintains items that contribute to the further construction of descendants. We construct $TEMP$ and finally filter it out due to only one frequent item in it. The item set {a,c,f,m} is contained in the frequent 4-itemsets.

3.1 Parallel *Propad* Approach

The Parallel *Propad* Approach, *Ppropad* for short, we proposed consists of two main stages.

Stage one is the construction of the transformed transaction table TF that includes all frequent 1-items. In order to enumerate the frequent items efficiently, the transaction data is partitioned uniformly correspond to transaction *tid* among the available processors. In a partitioned database, this can be done automatically.

Stage two is the actually mining of the table by projecting. In the *Propad* approach, the projecting process is facilitated by depth first approach. Since the processing of the projection of one frequent itemsets is independent from those of others, it is natural to consider it as the execution unit for the parallel processing. We divide the frequent items of the table TF among the available nodes in a round-robin fashion. Each node is given an approximately equal number of items to read and analyze. As a result, the items is spilt in p equal size. Each node locally constructs the projected transaction tables associated with the items in hand until the the search for frequent patterns associated with the items terminates.

4 Performance Evaluation

In our experiment we built a parallel RDBMS: IBM DB2 UDB EEE version 8.1 on multiple nodes. We configure DB2 EEE to execute in a shared-nothing architecture that each node has exclusive access to its own disk and memory. Four nodes were employed in our experiments. Each node runs the Linux operation system on Intel Xeon 2.80Ghz.

4.1 Datasets

We use synthetic transaction data generation with program described in *Apriori* algorithm paper [1] for experiments. The nomenclature of these data sets is of the form TxxIyyDzzzK, Where xx denotes the average number of items present per transaction, yy denotes the average support of each item in the data set and zzzK denotes the total number of transactions in K (1000's). We report experimental results on four data sets, they are respectively T25I20D100K that are relatively dense, T10I4D100K that is very sparse. (Here we have chosen the dataset T10I4D100K, because for this dataset, the experiment runs for 10 passes and we want to see how these approaches perform when mining long pattern.) Transaction data is partitioned uniformly by hashing algorithm corresponds to transaction ID among processing nodes.

4.2 Performance Comparison

In this subsection, we describe our algorithm performance compared with K-Way join. Figure 1 (a) shows the execution time for T10I4D100 with the minimum support of 0.1% and 0.06% on each degree of parallelization. We can drive that *Propad* is faster than K-Way join as the minimum support threshold decreases. This is because for datasets with long patterns, joining k-copies of input table for support counting at higher passes is quite significant though the cardinality of the C_k decreases with the increase in the number of passes. The speedup ration is shown in Figure 1 (b). Figure 1 (c) shows the execution time and speedup ration for T25I20D100K with the minimum support of 0.2% and 0.1% on each degree of parallelization. The speedup ratio shown in Figure 1 (d) seems to decrease with 4 processing nodes. It might be caused by the communication overhead.

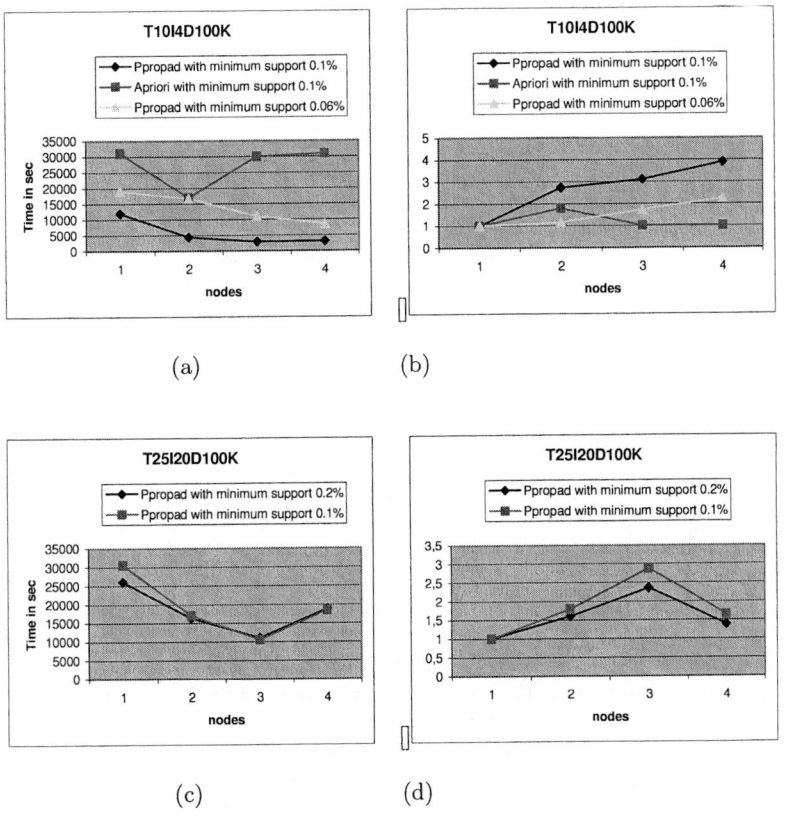

Fig. 1. Execution time (left) Speedup ration (right)

From the results we can see that the *Ppropad* approach has better parallelization than K-Way join. This is because for K-Way join approach with many large tables and a wide variety of tables and columns involved in joins, it can be difficult or impossible to choose the table's partitioning key such that all significant queries can be executed without heavy inter-partition communication. While, *Ppropad* approach avoids complex joins between tables.

5 Summary and Conclusion

In this paper, we implemented the parallelization of SQL based algorithm, *Ppropad*, to mine frequent itemsets from databases. Rather than *Apriori*-like method it adopts the divide-and-conquer strategy and projects the transaction table into a set of frequent item-related projected tables. Experimental study shows that the *Ppropad* algorithm can get better speedup ratio than K-Way join based on *Apriori*-like on all data sets, that means it is parallelized well.

There remain lots of further investigations. Since the round robin fashion of frequent items partition among the nodes, the load balancing is a problem when the extreme skew exists in data. We would like to examine how to absorb such skew. We also plan to check our parallel SQL based frequent pattern mining approach on more nodes. In addition, we'd like to investigate the effect of intra parallelism under SMP environment.

References

1. R. Agarwal and R. Shim. Developing tightly-coupled data mining application on a relational database system. In *Proc.of the 2nd Int. Conf. on Knowledge Discovery in Database and Data Mining*, Portland,Oregon, 1996.
2. R. Agarwal and R. Srikant. Fast algorithm for mining association rules. In *Proc. 1994 Int. Conf. Very Large Data Bases (VLDB'94)*, pages 487–499, Santiago, Chile, Sept. 1994.
3. J. Han, J. Pei, and Y. Yin. Mining frequent patterns without candidate generation. In *Proc. 2000 ACM-SIGMOD Int. Conf. Management of Data (SIGMOD'00)*, pages 1–12, Dallas, TIX, May 2000.
4. R. Meo, G. Psaila, , and S. Ceri. A new sql like operator for mining association rules. In *Proc. Of the 22nd Int. Conf. on Very Large Databases*, Bombay, India, Sept. 1996.
5. J.S. Park, M.S. Chen, and P.S. Yu. An effective hash-based algorithm for mining association rules. In *Proc. 1995 ACM-SIGMOD Int. Conf. Management of Data (SIGMOD'95)*, pages 175–186, San Jose, CA, May 1995.
6. I. Pramudiono and M. Kitsuregawa. Parallel fp-growth on pc cluster. In *Proc. Advances in Knowledge Discovery and Data Mining: 7th Pacific-Asia Conference, PAKDD 2003*, pages 467–473, Seoul, Korea, April 2003. Springer-Verlag.
7. X. Shang and K. Sattler. Depth-first frequent itemset mining in relational databases. In *Proc. ACM Symposium on Applied Computing SAC 2005*, New Mexico, USA, March 2005.

Using Consensus Susceptibility and Consistency Measures for Inconsistent Knowledge Management

Ngoc Thanh Nguyen and Michal Malowiecki

Institute of Information Science and Engineering,
Wroclaw University of Technology, Poland
{thanh, malowiecki}@pwr.wroc.pl

Abstract. Conflicts may appear during knowledge processing, where some knowledge pieces are different but they refer to the same subject. Consensus methods are useful in processing inconsistent knowledge. However, for almost conflict situations consensus may be determined but it is not always sensible. In this paper we investigate the aspect of reasonableness of consensus. For this aim we define two notions: consensus susceptibility and consistency measure. Owing to them one may get to know when it is worth to determine a consensus for a conflict. We show the dependencies between consistency and the consensus susceptibility for conflict situations. Some results of the analysis are presented.

1 Introduction

In many practical situations in order to solve a problem one often has to gather knowledge from different resources for realizing the task. Nowadays owing to modern computer technologies gathering knowledge is not a hard task at all, but there may be two features of this knowledge, which often make the decision making process difficult. The first feature is related to the big amount of knowledge which on one hand contains many useful elements, but on the other hand it often contains also a lot of useless elements. For these problems many methods for information filtering or ordering have been proposed. The second feature refers to the consistency of the gathered knowledge. Some elements of this knowledge may refer to the same subject, but they are not coherent. Hunter [3] describes the resources of inconsistent knowledge as situations in which one obtains "too much" information.

It is known that the consistency of knowledge bases is very important because inconsistent knowledge may provide inconsistent conclusions [6]. Inconsistent knowledge often leads to conflict uprising. For example in multiagent systems, where sources of knowledge are as various as methods for its acquisition, the inconsistency of knowledge often leads to conflicts among agents, for which the resolution is required. Consensus methods seem to be very useful in conflict solving [1],[7],[9]. Consensus methods are related to conflict models presented in [7],[13]. In this approach the body of conflict is a set of agents, the conflict subject consists of contentious issues and the conflict profile is a collection of tuples representing the participants' opinions referring to some issue. Information system tools seem to be good for representing conflicts [14]. By a conflict profile (representing a conflict situation) we understand a set of knowledge versions representing different opinions on some matter, which are generated by different resources such as agents functioning in sites of a distributed system or experts of some field including this matter.

In purpose to solve a conflict the management system should determine a proper version of knowledge for the common matter. This final knowledge version is called a consensus of the given conflict profile. The main subject of this paper consists of consideration of reasonableness of potential consensus. In other words, we consider two aspects of consensus reasonableness: the first is related to so called *consensus susceptibility* for conflict profiles and the second refers to measuring up the degree of *knowledge consistency*. In the first aspect we present some criteria which allow to state if for a given profile representing a conflict situation the chosen consensus is sensible or good enough. In the second aspect we present another approach for evaluating the possibility for effective conflict solving, which is based on measuring up the consistency degree representing the coherence level of the profile' elements. How to understand the consistency notion? People use it very intuitively to describe some divergences in various occurrences. Researchers often use this term but they do not define what it means. The definition has been provided by Helpern & Moses [2] and Nieger [5]: *Knowledge consistency is a property that a knowledge interpretation has with respect to a particular system.* The need for measures of knowledge consistency has been announced earlier in the aspect of solving conflicts in distributed environments [9].

The formal investigations on conflicts have been initiated by Pawlak [13]. In work [7] the author has proposed a conflict model which enables processing multi-value and multi-attribute conflicts. Conflicts have been also investigated by using non-classical logics such as paraconsistent logics [3] or four-valued logics [4].

In this paper we present in short the consensus problem (Section 2). Next we define the notion of consensus susceptibility (Section 3) and show several consistency functions (Section 4). In Section 5 the relationships between consensus susceptibility and the consistency degree of conflict profiles are presented. Some conclusions and directions for future works are included in Section 6.

2 Consensus Choice Problem

In this section we present a brief overview of consensus choice problem. The wide description of this problem may be found in [8]. We assume that the subject of interests is a finite universe U of objects. Let $\Pi(U)$ denote the set of subsets of U. By $\hat{\Pi}_k(U)$ we denote the set of k-element subsets (with repetitions) of set U for $k \in N$, and let $\hat{\Pi}(U) = \bigcup_{k>0} \hat{\Pi}_k(U)$. Each element of set $\hat{\Pi}(U)$ is called a *profile*. The structure of this universe is a distance function

$$\delta: U \times U \to \Re^+,$$

which satisfies the following conditions:

Nonnegative: $(\forall x,y \in U)[\delta(x,y) \geq 0]$,
Reflexive: $(\forall x,y \in U)[\delta(x,y) = 0 \text{ iff } x=y]$,
Symmetrical: $(\forall x,y \in U)[\delta(x,y) = \delta(y,x)]$.

Let us notice that the above conditions are only a part of metric conditions. Metric is a good measure of distance, but its conditions are too strong [8]. A space (U,δ) defined in this way does not need to be a metric space. Therefore we will call it a *distance space*.

Let $X, X_1, X_2 \in \hat{\prod}(U)$, $x \in U$, and

$$\delta(x,X) = \sum_{y \in X} \delta(x,y),$$

$$\delta^n(x,X) = \sum_{y \in X} [\delta(x,y)]^n \text{ for } n \in N,$$

$$\overline{\delta}(x, X) = \sum_{y \in X} [\delta(x, y) - \frac{1}{\text{card}(X)}\delta(x, X)]^2.$$

Definition 1. *By a consensus choice function in space* (U,δ) *we mean a function*

$$c: \hat{\prod}(U) \to \Pi(U).$$

For $X \in \hat{\prod}(U)$, the set $c(X)$ is called the *representation* of the profile X, where an element of $c(X)$ is called a *consensus* (or a *representative*) of the profile X. Let C_s denote the set of all consensus choice functions in a space $s = (U,\delta)$.

In the following definition symbol $\dot{\cup}$ denotes the sum operation on sets with repetitions, and symbol $*$ denotes the multiple occurrence of an element in a set with repetitions (that is a multiset).

Definition 2. *The consensus choice function* $c \in C_s$ *satisfies the postulate of:*

1. *Condorcet consistent (Cc), iff*
   $(c(X_1) \cap c(X_2) \neq \emptyset) \Rightarrow (c(X_1 \dot{\cup} X_2) = c(X_1) \cap c(X_2))$
2. *Faithful (Fa), iff*
   $c(\{x\}) = \{x\}$
3. *Unanimous (Un), iff*
   $c(\{n *x\}) = \{x\}$
4. *Reliability (Re) iff*
   $c(X) \neq \emptyset$
5. *Consistency (Co) iff*
   $(x \in c(X)) \Rightarrow (x \in c(X \dot{\cup} \{x\}))$
6. *Quasi-unanimous (Qu) iff*
   $(x \notin c(X)) \Rightarrow ((\exists n \in N)(x \in c(X \dot{\cup} \{n*x\})))$
7. *Proportion (Pr) iff*
   $(X_1 \subseteq X_2 \wedge x \in c(X_1) \wedge y \in c(X_2)) \Rightarrow (\delta(x,X_1) \leq \delta(y,X_2))$
8. *1-Optimality* (O_1) *iff*
   $(x \in c(X)) \Rightarrow (\delta(x,X) = \min_{y \in U} \delta(y,X))$

9. *2-Optimality* (O_2) *iff*
$(x \in c(X)) \Rightarrow (\delta^2(x,X) = \min_{y \in U} \delta^2(y,X))$.

Below we present some results of postulates' analysis [8].

Let S be the set of all spaces $s = (U, \delta)$, where the set U is finite and the distance function δ is nonnegative, reflexive and symmetric. Each space s uniquely determines the set C_s of all consensus functions in this space. Then S should not be the model for the following formulas, in which the names of postulates are treated as atomic predicates:

a) $(\forall c \in C_s)(C \Rightarrow C')$

where $C, C' \in \{Re(c), Co(c), Qu(c), Pr(c), O_1(c), O_2(c)\}$ and $C \neq C'$

b) $(\exists c \in C_s)(Re(c) \land Co(c) \land Qu(c) \land Pr(c) \land O_1(c) \land O_2(c))$

c) $(\exists c \in C_s)(O_1(c) \land O_2(c))$

But S should be the model for the following formulas:

d) $(\exists c \in C_s)(Re(c) \land Co(c) \land Qu(c) \land Pr(c) \land O_1(c) \land Cc(c))$

e) $(\exists c \in C_s)(Re(c) \land Co(c) \land Qu(c) \land O_2(c) \land Cc(c))$

f) $(\forall c \in C_s)((O_1(c) \land Re(c)) \Rightarrow Co(c))$

g) $(\forall c \in C_s)((O_1(c) \land Re(c)) \Leftrightarrow (Pr(c) \land Qu(c) \land Re(c)))$

h) $(\forall c \in C_s)((O_1(c) \land Re(c)) \Leftrightarrow (Pr(c) \land Qu(c) \land Re(c) \land Co(c)))$

i) $(\forall c \in C_s)((Re(c) \land Pr(c) \land Qu(c)) \Rightarrow (Fa(c) \land Un(c)))$

j) $(\forall c \in C_s)((O_2(c) \land Re(c)) \Rightarrow (Co(c) \land Qu(c)))$

k) $(\forall c \in C_s)((O_2(c) \land Re(c)) \Rightarrow (Fa(c) \land Un(c)))$.

The above results show that 2 consensus functions, i.e. O_1 and O_2, play very important role. They satisfy a large number of postulates, such as Condorcet consistent, Faithful, Unanimous, Reliability, Consistency, Quasi-unanimous and Proportion. Therefore, these 2 choice criteria are the main subject in the consideration of consensus susceptibility in this paper.

In this paper we do not deal with concrete structures of objects belonging to universe U. In consensus theory some object structures of have been investigated, for example, linear orders (rankings); n-trees; semillatices; partitions of sets etc.

3 Susceptibility to Consensus

In this section we present the definition, criteria and their analysis for profiles' susceptibility to consensus. It often happens that in a distance space for each profile one may always determine its consensus (for example using consensus function O_1 or O_2). However, one should give the answer for the following question: Is the consensus sensible and may it be accepted as the solution of the conflict? In other words, is the profile susceptible to consensus? It seems, to the best of the author's knowledge, that the notion "susceptibility" is a new one in the consensus theory and has not been investigated up to now.

Below we present an example illustrating the above problem.

Example 1. [10] Let space (U,δ) be defined as follows: $U=\{a,b\}$ where a and b are objects of some type, and distance function ∂ is given as

$$\delta(x,y) = \begin{cases} 0 & \text{for } x = y \\ 1 & \text{for } x \neq y \end{cases}$$

for $x,y \in U$. Let X be a profile, as a set with repetitions, where $X = \{50*a, 50*b\}$, that is each of objects a and b occurs 50 times in the profile. Assume that X represents the result of some voting, in which 100 voters take part, each of them gives one vote (for a or b). There are 50 votes for a and 50 votes for b. It is easy to note that for profile X the consensus should be equal to a or b. But it intuitively seems that none of them is a good consensus, because there is lack of a compromise in this conflict situation. Let us consider now another profile $X' = \{50*a, 51*b\}$. For this profile the only consensus should be b and it seems to be a good consensus, that means this profile is susceptible to consensus.

The above example shows that although consensus may always be chosen for a conflict profile, it does not have to be a good one. We define below the notion of profile' susceptibility to consensus.

For given distance space (U,δ), $X \in \hat{\prod}(U)$ and card$(X)=k$. We firstly define the following values:

$$\hat{\delta}^i(X) = \frac{\sum_{x,y \in X}(\delta(x,y))^i}{k(k+1)} \quad \text{for } i=1,2.$$

$$\hat{\delta}^i_x(X) = \frac{\sum_{y \in X}(\delta(x,y))^i}{k} \quad \text{for } i=1,2.$$

$$\hat{\delta}^i_{\min}(X) = \min_{x \in U} \hat{\delta}^i_x(X) \quad \text{for } i=1,2.$$

$$\hat{\delta}^i_{\max}(X) = \max_{x \in U} \hat{\delta}^i_x(X) \quad \text{for } i=1,2.$$

The interpretations of these values are the following:

- $\hat{\delta}^i(X)$ – This value serves to representing the average distance of all distances between elements of profile X. The sum of these distances is expressed by the numerator of the quotient. However, one can ask a question: Why in the denominator is not k^2, but $k(k+1)$? The answer is: In the numerator each distance $\delta(x,y)$, where $x \neq y$, occurs exactly twice, while each distance $\delta(x,y)$, where $x=y$, occurs exactly only one time. Because $\delta(x,y) = 0$ for $x = y$, then adding such distance does not change the value of numerator, however, in determining the average each distance should be taken into account twice. Thus the denominator should be $k(k+1)$, but not k^2.
- $\hat{\delta}^i_x(X)$ – This value represents the average distance of all distances between object x and the elements of profile X.

- $\hat{\delta}^i_{min}(X)$ – The minimal value of $\hat{\delta}^i_x(X)$ for $x \in U$.
- $\hat{\delta}^i_{max}(X)$ – The maximal value of $\hat{\delta}^i_x(X)$ for $x \in U$.

Definition 3. *Let* $X \in \hat{\prod}(U)$ *be a profile. We say that profile X is susceptible to consensus in relation to postulate O_i for i=1,2 (or O_i-susceptible to consensus) if and only if the following inequality takes place:*

$$\hat{\delta}^i(X) \geq \hat{\delta}^i_{min}(X).$$

The idea of the above definition relies on such intuition that because value $\hat{\delta}^i(X)$ represents the average distance in profile X, and $\hat{\delta}^i_{min}(X)$ represents the average distance from the consensus to the elements of the profile, then X is susceptible to consensus (i.e. is possible to determine a "good enough" consensus for X), if the second value is not greater then the first. Satisfying the above inequality means that the elements of profile X are "dense" enough for determining a good consensus. In other words, opinions represented by these elements are consistent enough for determining a good compromise.

Henceforth if $i = 1$ then we will not write the index i.

For the profile X from Example 1 the above defined values are calculated as follows:

$$\hat{\delta}(X) = \frac{2 \times 50 \times 50}{100 \times 101} = \frac{50}{101}$$

and

$$\hat{\delta}_{min}(X) = \frac{50}{100} = \frac{1}{2}.$$

Of course $\frac{50}{101} < \frac{1}{2}$, then profile X is not O_1-susceptible to consensus. For profile X' we have the following:

$$\hat{\delta}(X') = \frac{2 \times 50 \times 51}{101 \times 102},$$

and

$$\hat{\delta}_{min}(X') = \frac{50}{101}.$$

Thus $\hat{\delta}(X') = \hat{\delta}_{min}(X')$, it means that profile X' is O_1-susceptible to consensus. Similarly we can state that profile X is not O_2-susceptible to consensus, but profile X' is O_2-susceptible to consensus. Definition 3 is then consistent with the intuition.

Definition 4. *Let* $X \in \hat{\prod}(U)$ *be a profile. We say that X is i-regular for i = 1,2 if and only if for each pair of objects $x,y \in U$ the following equality takes place:*

$$\hat{\delta}^i_x(X) = \hat{\delta}^i_y(X).$$

Notice that profile X defined in Example 1 is i-regular for $i = 1,2$, while profile X' is not i-regular (or i-irregular) for $i = 1,2$.

Below we present some results of the analysis [10].

Theorem 1. *Each i-regular profile X, where $\mathrm{card}(X) > 1$, is not O_i-susceptible to consensus for $i = 1,2$.*

Theorem 1 implies that if a profile is regular, then it is not worth to determine the consensus.

Theorem 2. *Let $X, X' \in \hat{\prod}(U)$ be such profiles that $X' = X \dot{\cup} \{x\}$ for some $x \in X$ and X is i-regular, then profile X' is O_i-susceptible to consensus for $i = 1,2$.*

Theorem 2 shows that if profile X is i-regular, then adding to it any its element should cause that the new profile is O_i-susceptible to consensus. The practical sense of this theorem is the following: if in given conflict situation any of the opinions does not dominate the others but in the additional voting one of them dominates the rest, then it is possible to determine a sensible consensus.

Thus profile X can not be O_1-susceptible to consensus. In Example 1 for profile X (the number of voters is equal to 100) one can note that it is neither O_1-susceptible, nor O_2-susceptible to consensus.

Below we present the notion of susceptibility to consensus of a profile in the context of other profile.

Definition 5. *Profile X is susceptible to consensus in the context of profile Y if $X \subset Y$ and $\hat{\delta}_{\max}(X) \leq \hat{\delta}_{\min}(Y)$.*

The above definition serves in such situations when profile X is not susceptible to consensus but its context (profile Y) is more non-susceptible to consensus. In other words, the conflict encompassed by profile X is not meaningful in the relation to the conflict represented by profile Y. In this case the consensus determined for profile X could be acceptable.

4 Consistency Functions for Conflict Profiles

By symbol C we denote the consistency function of profiles. This function has the following signature:

$$C: \hat{\prod}(U) \rightarrow [0,1].$$

where $[0,1]$ is the closed interval of real numbers between 0 and 1.

The idea of this function is relied on measuring up the consistency degree of profile's elements. The requirements for consistency are expressed and analysed in work [12].

In this section we present five consistency functions [12], which are defined as follows:

Let $X = \{x_1, \ldots, x_M\}$ be a profile. We assume that $M > 1$ because if $M = 1$ then the profile X is a homogeneous one. We introduce the following parameters:

- The matrix of distances between the elements of profile X:

$$D^X = \left[\delta_{ij}^X\right] = \begin{bmatrix} \delta(x_1, x_1) & \cdots & \delta(x_1, x_M) \\ \vdots & \ddots & \vdots \\ \delta(x_M, x_1) & \cdots & \delta(x_M, x_M) \end{bmatrix}$$

- The vector of average distances between an element to the rest:

$$W^X = \left[w_i^X\right] = \left(\frac{1}{M-1}\sum_{j=1}^{M}\delta_{j1}^X, \frac{1}{M-1}\sum_{j=1}^{M}\delta_{j2}^X, \ldots, \frac{1}{M-1}\sum_{j=1}^{M}\delta_{jM}^X\right)$$

- Diameters of sets X and U:

$$Diam(X) = \max_{x,y \in X} \delta(x,y),$$

$$Diam(U) = \max_{x,y \in U} \delta(x,y) = 1;$$

- The maximal element of vector W^X:

$$Diam(W^X) = \max_{1 \le i \le M} w_i^X$$

representing this element of profile X, which generates the maximal sum of distances to other elements.

- The average distance in profile X:

$$\bar{\delta}(X) = \frac{1}{M(M-1)}\sum_{i=1}^{M}\sum_{j=1}^{M}d_{ij}^X = \frac{1}{M}\sum_{i=1}^{M}w_i^X$$

- The sum of distances between an element x of universe U and the elements of set X:

$$\delta(x, X) = \sum_{y \in X} \delta(x, y)$$

- The maximal sum of distances from an element from profile X to other its elements:

$$\delta_{max}(X) = \max_{x \in X}\left(\frac{\delta(x,X)}{M}\right)$$

These parameters are now applied for the defining the following consistency functions:

$$C_1(X) = 1 - Diam(X)$$
$$C_2(X) = 1 - Diam(W^X)$$
$$C_3(X) = 1 - \bar{\delta}(X)$$
$$C_4(X) = 1 - \delta_{min}(X)$$
$$C_5(X) = 1 - \delta_{max}(X)$$

Values of these functions reflect accordingly:

- $C_1(X)$ – the maximal distance between two elements of profile X.
- $C_2(X)$ – the maximal average distance between an element of profile X and other elements of this profile.
- $C_3(X)$ – the average distance between elements of X.
- $C_4(X)$ – the minimal average distance between an element of universe U and elements of X.
- $C_5(X)$ – the maximal average distance between an element of profile X and elements of this profile.

5 Consistency Versus Consensus Susceptibility

In this section we present some dependencies between profile consistency and consensus susceptibility. These dependencies show that in general the two notions are coherent. The following properties are true:

Proposition 1. *For each $j=1,...,5$ if $C_j(X) = 1$ then profile X is susceptible to consensus in relation to postulate O_i for $i=1,2$.*

This property shows the coherence between consistency measures and consensus susceptibility: In case when a profile has maximal consistency then it is also susceptible to consensus. However, if a profile has minimal consistency then not necessarily it should not be susceptible to consensus. The following properties show that the intuition is true only for functions $C_2,..., C_5$ and false for function C_1.

Proposition 2. *It is not true that if $C_1(X) = 0$ then profile X is not susceptible to consensus in relation to postulate O_i for $i=1,2$.*

Proposition 3. *For each $j=2,...,5$ if $C_j(X) = 0$ then profile X is not susceptible to consensus in relation to postulate O_i for $i=1,2$.*

The reason of the property included in Proposition 2 is that function C_1 takes into account only these elements of the profile, for which the distance is maximal. The behavior of other elements are not interested for this measure. Nevertheless function C_1 has many practical applications.

Another aspect of the relationship between consistency measures and consensus susceptibility is based the investigation of the behavior of consistency in situations when a profile is susceptible to consensus. It turned out that the behaviors of defined consistency functions are not similar.

Proposition 4. *If a profile X is susceptible to consensus in relation to postulate O_1 then*:

a) $C_3(X) > 0.5$, $C_4(X) > 0.5$, *and*
b) $C_2(X) > 0$, $C_5(X) > 0$.

Presented above properties show the strong relationship between consensus susceptibility in relation to postulate O_1 to consistency functions C_3 and C_4; weaker relationship to functions C_2 and C_5; and very weak relationship to functions C_1.

The proofs of the above propositions are given in the technical report [11].

6 Conclusions

In this paper we have presented a brief overview of consensus choice problem; the notions of consensus susceptibility and consistency measures for conflict profiles and some aspects of the relationships between them. We have shown that if a conflict profile is not susceptible to consensus then there is no need to determine it because the consistency value is too low. Owing to this in many practical situations one may avoid determining consensus since consensus choice algorithms are often complex. The future works should concern the deeper analysis of the dependencies between the two criteria for consensus reasonableness.

References

1. Barthelemy, J.P., Janowitz, M.F.: A Formal Theory of Consensus, SIAM J. Discrete Math. **4** (1991) 305-322.
2. Helpern, J.Y., Moses, Y.: Knowledge and Common Knowledge in Distributed Environment. Journal of the Association for Computing Machinery **37** (2001) 549-587.
3. Hunter, A.: Paraconsistent Logics. In: D. Gabbay, P. Smets (eds), Handbook of Defeasible Reasoning and Uncertain Information. Kluwer Academic Publishers (1998) 13-43.
4. Loyer, Y., Spyratos, N., Stamate, D.: Integration of Information in Four-Valued Logics under Non-Uniform Assumption. In: Proceedings of 30th IEEE International Symposium on Multiple-Valued Logic (2000) 180-193.
5. Nieger, G.: Simplifying the Design of Knowledge-based Algorithms Using Knowledge Consistency. Information & Computation **119** (1995) 283-293.
6. Ng, K.C., Abramson, B.: Uncertainty Management in Expert Systems. IEEE Expert Intelligent Systems and their Applications **27** (1990) 29-48.
7. Nguyen, N.T.: Consensus Choice Methods and their Application to Solving Conflicts in Distributed Systems. Wroclaw University of Technology Press (2002)
8. Nguyen, N.T.: Using Distance Functions to Solve Representation Choice Problems. Fundamenta Informaticae **48**(4) (2001) 295-314.
9. Nguyen, N.T.: Consensus System for Solving Conflicts in Distributed Systems. Journal of Information Sciences **147** (2002) 91-122
10. Nguyen, N.T.: Criteria for Consensus Susceptibility in Conflicts Resolving. In Inuiguchi M., Tsumoto S., Hirano S. (eds): Rough Set Theory and Granular Computing. Series Studies in Fuzziness and Soft Computing **125** (2003) 323-333 (Springer).
11. Nguyen, N.T., Malowiecki, M.: Deriving Consensus for Conflict Situations with Respect to its Susceptibility. Technical Report PRE 268. Wroclaw University of Technology (2004)
12. Nguyen, N.T., Malowiecki, M.: Consistency Measures for Conflict Profiles. LNCS Transactions on Rough Sets **1** (2004) 169-186.
13. Pawlak, Z.: An Inquiry into Anatomy of Conflicts. Journal of Information Sciences **108** (1998) 65-78.
14. Skowron, A., Deja, R., On Some Conflict Models and Conflict Resolution. Romanian Journal of Information Science and Technology **5** (2002) 69-82.

WLPMiner: Weighted Frequent Pattern Mining with Length-Decreasing Support Constraints

Unil Yun and John J. Leggett

Department of Computer Science, Texas A&M University,
College Station, TX77843, U.S.A
{yunei, leggett}@cs.tamu.edu

Abstract. Two main concerns exist for frequent pattern mining in the real world. First, each item has different importance so researchers have proposed weighted frequent pattern mining algorithms that reflect the importance of items. Second, patterns having only smaller items tend to be interesting if they have high support, while long patterns can still be interesting although their supports are relatively small. Weight and length decreasing support constraints are key factors, but no mining algorithms consider both the constraints. In this paper, we re-examine two basic but interesting constraints, a weight constraint and a length decreasing support constraint and propose weighted frequent pattern mining with length decreasing constraints. Our main approach is to push weight constraints and length decreasing support constraints into the pattern growth algorithm. For pruning techniques, we propose the notion of Weighted Smallest Valid Extension (WSVE) with applying length decreasing support constraints in weight-based mining. The WSVE property is applied to transaction and node pruning. WLPMiner generates more concise and important weighted frequent patterns with a length decreasing support constraint in large databases by applying the weighted smallest valid extension.

1 Introduction

To overcome problems of Apriori-based algorithms [1, 2], such as generation and test of all candidates and repeatedly scanning a large amount of the original database, pattern growth based approaches [10, 11, 12, 13, 14] were developed. FP-tree based methods mine the complete set of frequent patterns using a divide and conquer method to reduce the search space without generating all the candidates. An association mining algorithm generates frequent patterns and then makes association rules satisfying a minimum support. Two main limitations of the traditional approach exist when mining frequent patterns.

The first limitation is that all items are treated uniformly, but real items have different importance. For this reason, weighted frequent pattern mining algorithms [3, 4, 5, 6] have been suggested. The items are given different weights in the transaction database. These algorithms focus on satisfying the downward closure property because this property is usually broken when different weights are applied to the items according to their significance. The second limitation is that most of the previous mining algorithms use a constant support constraint irrespective of the length of the discovered patterns. The key observation here is that long patterns can be interesting although their support is low and short patterns can be interesting if they have high support. LPMiner [7] and Bamboo [8] have addressed these issues. As stated in [7, 8, 9], the downward closure property can not be used with length decreasing support constraints. For this reason, the smallest valid extension (SVE) property was introduced to prune the search space. The weight constraint and the length decreasing support constraint are key factors, but no mining algorithms consider both constraints.

Table 1. Transaction database TDB

TID	Set of items
100	a, c, d, f
200	a, b, c, d
300	b, c, f
400	b, c, d, f
500	b, c, d, g

In this paper, we re-examine two basic but interesting constraints, a weight constraint and a length decreasing support constraint and propose weighted frequent pattern mining with length decreasing support constraints. The simple way to use a length decreasing support constraint is to set min $_{l>0}$ f(l) and remove patterns which do not satisfy the length decreasing constraints. However, it takes a lot of time to generate frequent patterns. WLPMiner takes into account two features, a weight constraint and a length decreasing support constraint in real world instead of only resetting the minimum support. This allows WLPMiner to generate more meaningful patterns. For pruning techniques, we propose 1) the notion of Weighted Smallest Valid Extension (WSVE) to apply to both the length decreasing support constraints and weight constraints and 2) a weight range is used as a supplement to maintain the downward closure property.

Our Contributions: The main contributions of this paper are: 1) incorporation of two key features for real datasets, a weight constraint and a length decreasing support constraint 2) introduction of the concept of the weighted smallest valid extension property, 3) description of pruning techniques, transaction pruning and node pruning, using the weighted smallest valid extension property, 4) implementation of our algorithm, WLPMiner, and 5) execution of an extensive experimental study to compare the performance of our algorithm with BAMBOO [8] and WFIM [3].

The remainder of the paper is organized as follows. In section 2, we describe the problem definition and related work. In Section 3, we present WLPMiner (Weighted Frequent Pattern Mining with Length decreasing support constraints). Section 4 shows the extensive experimental study and results. Finally, conclusions are presented in section 5.

2 Problem Definition and Related Work

2.1 Problem Definition

Let $I = \{i_1, i_2..., i_n\}$ be a unique set of items. A transaction database, TDB, is a set of transactions in which each transaction, denoted as a tuple <tid, X>, contains a unique tid and a set of items. A pattern is called a k-pattern if it contains k items. A pattern $\{x_1, x_2,..., x_n\}$ is also represented as $x_1, x_2,..., x_n$. The support of a pattern is the number of transactions containing the pattern in the database.

2.1.1 Weighted Frequent Itemset Mining

In weighted frequent itemset mining, a weight of an item is a non-negative real number that shows the importance of the item. We can use the term, *weighted itemset* to represent a set of weighted items. A simple way to obtain a weighted itemset is to calculate the average value of the weights of the items in the itemset. The weight of each item is assigned to reflect the importance of each item in the transaction database. A weight is given to an item within a weight range, $W_{min} \leq W \leq W_{max}$. We want to give a balance between the two measures of weight and support. Therefore, we use a minimum weight constraint (min_weight) like a minimum support (min_sup) in order to prune items which have lower weights.

Table 2. Example of sets of items with different WRs

Item	a	b	c	d	f	g
Support	2	4	5	3	3	1
Weight ($0.9 \leq WR_1 \leq 1.1$)	1.1	1.0	0.8	1.0	0.7	0.9
Weight ($0.4 \leq WR_2 \leq 0.8$)	0.5	0.8	0.6	0.4	0.7	0.6
Weight ($0.2 \leq WR_3 \leq 0.7$)	0.6	0.3	0.5	0.7	0.5	0.2

An itemset is a weighted infrequent itemset if, following pruning, condition 1 or condition 2 below is satisfied. If an itemset does not satisfy both of these, it is called a weighted frequent itemset. In the pruning conditions, a maximum weight (MaxW) is defined as the value of the maximum weight of items in a transaction database and a minimum weight (MinW) is defined as the value of the minimum weight of a conditional pattern in a conditional database.

Pruning Condition 1: (support < min_sup && weight < min_weight)
The support of an itemset is less than a minimum support and the weight of an itemset is less than a minimum weight threshold.

Pruning Condition 2: (support * MaxW (MinW) < min_sup)
In a transaction database, the value of multiplying the support of an itemset with the maximum weight (MaxW) among items in the transaction database is less than a minimum support. In conditional databases, the value of multiplying the support of an itemset with the minimum weight (MinW) of a conditional pattern in the FP-trees is less than a minimum support.

Example 1: Table 1 shows the transaction database TDB. Table 2 shows example sets of items with different weights. The frequent list is <a:2, b:4, c:5, d:3, f:3, g:1>. The minimum support threshold (min_sup) is 2. The columns in Table 3 show the set of weighted frequent itemsets after pruning weighted infrequent itemsets using pruning condition1 and pruning condition 2 by applying different WRs.

2.1.2 Frequent Itemset Mining with a Length Decreasing Support Constraint

The length decreasing support constraint was suggested from the observation that long patterns can be interesting although their support is low. Meanwhile, short patterns become interesting if they have high support. Given a transaction database TDB and its length decreasing support constraint function f(x) that satisfies $1 \le f(x+1) \le f(x) \le |TDB|$ for any positive integer x, frequent itemset mining with length decreasing support constraint f(x) requires finding all frequent itemsets that satisfy support (Y) $\ge$ f (|Y|).

The itemsets found are called frequent itemsets with a length decreasing support constraint. However, the downward closure property can not be used to find a frequent itemset with a length decreasing support constraint because an infrequent itemset may become a frequent itemset in the next step using a length decreasing support constraint. In order to prune the search space, the Smallest Valid Extension (SVE) property was defined. Given an itemset P such that support (P) < f (|P|), then f^{-1}(support (P)) = min (l| f (l) $\le$ support (P)) is the minimum length that a super itemset of P must have before it can potentially satisfy the length deceasing support constraint. The SVE property can be used to prune the conditional FP trees.

Example 2: We use the transaction database TDB in Table 1 and assume that a length decreasing support constraint function f (x) is f (x) = 4 (for x $\le$ 1), f (x) = 3 (for 2 $\le$ x $\le$ 3), and f (x) = 2 (for x > 3). The frequent list is: frequent_list = <a:2, b:4, c:5, d:3, f:3, g:1>. By using a constant minimum support threshold, the frequent itemsets are {ac:2, bc:2, bd3, bf:2, cd:3, cf:3, df:2, bcd:2, bcf:2}. We can easily show that the final frequent itemset using length support constraint f(x) is {bc:4, bd:3, cd:3 and cf:3}.

Table 3. Weighted frequent itemsets with different WRs

TID	WFI list (0.9 ≤ WR$_1$ ≤ 1.1) min_weight = 0.9	WFI list (0.4 ≤ WR$_2$ ≤ 0.8) min_weight = 0.4	WFI list (0.2 ≤ WR$_3$ ≤ 0.6) min_weight = 0.2
100	a, c, d, f	c, d, f	c
200	a, b, c, d	b, c, d	b, c
300	b, c, f	b, c, f	b, c
400	b, c, d, f	b, c, d, f	b, c
500	b, c, d	b, c, d	b, c

2.2 Related Work

No association rule mining algorithms exist that consider both a weight constraint and a length decreasing support constraint, both of which are characteristics of real datasets. Additionally, most of the weighted frequent itemset mining algorithms [4, 5, 6] suggested so far have used Apriori-like approaches. Therefore, previous algorithms for weighted association rule mining generate many candidates and scans of the database. Recently, we have suggested **WFIM (Weighted Frequent Itemset Mining)** algorithm [3]. WFIM is the first algorithm to use the pattern growth approach in weighted itemset mining. WFIM can push weight constraints into the pattern growth algorithm while maintaining the downward closure property. A minimum weight and a weight range are defined and items are given different weights within the weight range. However, the WFIM did not take into account a length decreasing support constraint. In **WARM (Weighted Association Rule Mining)** [4], the problem of breaking the downward closure property is solved by using a weighted support and developing a weighted downward closure property. However, this algorithm is still based on the Apriori algorithm which uses a candidate generation and test mechanism. **WAR (Weighted Association rules)** [5] generates the frequent items without considering the weights and then does post-processing during the rule generation step. The WAR algorithm does not concern mining frequent itemsets, so this technique is a post-processing approach. **LPMiner** [7] is the first algorithm to find itemsets that satisfy a length decreasing support constraint. It defines and uses the Smallest Valid Extension (SVE) property. Assigning weights according to the importance of the items is one of the main considerations for real datasets. However, LPMiner dose not consider the importance of the items even thought it takes into account length decreasing support constraints. **BAMBOO** [8] pushed the length decreasing support constraint deeply into closed itemset mining in order to generate more concise itemsets. While BAMBOO outperforms LPMiner, it is only concerned with the length decreasing support constraint. BAMBOO also does not consider different weights of items within patterns.

In this paper, we propose an efficient weighted pattern growth algorithm with a length decreasing support constraint. We use a prefix tree structure to construct conditional databases. Our algorithm adopts an ascending weight order method and a bottom-up traversal strategy.

3 Weighted Pattern Mining with a Length Decreasing Support Constraint

In this section, we suggest an efficient weighted frequent pattern mining algorithm with a length decreasing support constraint, called WLPMiner. Our approach is to push a weight constraint and a length decreasing support constraint into the pattern growth algorithm. We introduce the weighted smallest valid extension (WSVE) property to prune the search space. In addition, we show that the effect of combining a weight constraint and a length decreasing support constraint generates fewer but important patterns.

3.1 Weighted Smallest Valid Extension (WSVE) Property and Pruning by the WSVE

Definition 3.1. Weighted Smallest Valid Extension (WSVE) Property
Given a conditional pattern P such that (support (P) * weight (P)) < f (|P|), then f^{-1} (support (P) * weight (P)) = min (l| f (l) ≤ (support (P) * weight (P))) is the minimum

length that a super pattern of P must have before it can potentially satisfy the length deceasing support constraint.

Lemma 3.1. As the value, x, of multiplying the support with the weight of pattern P decreases, the value of the inverse function, $f^{-1}(x)$, of a length decreasing support constraint function increases. Meanwhile, as the value of x increases, the value of the inverse function of a length decreasing support constraint function, $f^{-1}(x)$ decreases.

The WSVE property allows us to prune the search space. It considers not only a support measure but also a weight measure. From the WSVE property, if the pattern P is a weighted infrequent pattern, the length of any superset of the pattern P should have at least f^{-1}(support (P) * weight (P)). If not, the superset is also a weighted infrequent pattern, so it can be pruned.

Lemma 3.2 Given a conditional pattern, X, and a pattern in the conditional database, Y, weight (X) is greater than or equal to a weight of a pattern, Y within a transaction t in the conditional database.

In the weighted smallest valid extension property, weight (X) is used as a weight parameter of the inverse function of a length decreasing support constraint. WLPMiner uses an ascending weight ordered prefix tree and the tree is traversed using a bottom-up strategy. Therefore, the weight of a conditional pattern, X, is always greater than or equal to the weight of an item, Y of a transaction, t, within a conditional database.

Lemma 3.3. The following formula is always satisfied: f^{-1}(support (X) * weight (X)) $\leq f^{-1}$ (support (X) * weight (Y)) $\leq f^{-1}$(support (X+Y) * weight (Y)).

As stated in lemma 3.2, weight (X) is always greater than or equal to weight (Y) and weight (X+Y) is always greater than or equal to weight (Y). From lemma 3.1, we know that f^{-1} (weight (X)) $\leq f^{-1}$(weight (X+Y)) $\leq f^{-1}$(weight (Y)). Finally, f^{-1}(support (X) * weight (X)) is less than or equal to f^{-1}(support (X) * weight (Y)) and f^{-1}(support (X) * weight (Y)) is less than or equal to f^{-1}(support (X+Y) * weight (Y)). In lemma 3.3, we see that f^{-1} (support (X) * weight (X)) is the minimum length for a superset (X+Y) of a conditional pattern (X). Weighted frequent patterns with a length decreasing support constraint should satisfy the WSVE property, although other patterns may also satisfy the WSVE property.

Using the weighted smallest valid extension property, we suggest two pruning techniques, transaction pruning and node pruning. Transaction pruning is applied before constructing FP-tree, while node pruning is used after building FP-tree.

Definition 3.2. Transaction Pruning by Weighted Smallest Valid Extension (WSVE)
Given a length decreasing support constraint f(l), and a conditional database D` with regard to a conditional pattern X, a pattern Y ∈ D` can be pruned from D` if ((support (X) * weight (X)) < f (|X| + |Y|)).

The transaction pruning method is used to remove candidate transactions of a conditional database. It uses the weighted smallest valid extension property. Separate local FP-trees are built for all patterns that contain the conditional pattern. From the WSVE property, any superset (X+Y) of a conditional pattern (X) must have a length of at least f^{-1}(support (X) * weight (X)). We can remove any patterns (Y) with a length of less than $f^{-1}(\text{support}(X) * \text{weight}(X)) - |X|$. This formula can be rewritten as: $|X| + |Y| < f^{-1}(\text{support}(X) * \text{weight}(X))$, which is the same as (support (X) * weight (X)) $< f(|X| + |Y|)$.

Definition 3.3. Node Pruning by Weighted Smallest Valid Extension (WSVE)
Given a length decreasing support constraint f(l), a conditional pattern database D` with regard to a conditional pattern P, and the FP tree, T, built from D`, a node v in T can be pruned from T if $h(v) + |P| < f^{-1}(\text{support}(I(v)) * \text{weight}(I(v)))$.

The node pruning method reduces nodes of a conditional local FP-tree. Assume that I(v) is the item stored in this node and h(v) is the height of the longest path from the root to a leaf node in which path, the node v should be located. From the weighted smallest valid extension property, we can see that a node that contributes to a weighted frequent pattern, should adhere to the following formula: $h(v) + |P| \geq f^{-1}$ (support (I(v)) * weight (I(v))). Therefore, we can define node pruning by the weighted smallest valid extension in definition 3.1. This formula can be rewritten as: f (h(v) + |P|) ≤ (support (I(v)) * weight (I(v))). We can remove a node if $h(v) + |P| < f^{-1}$ (support (I(v)) * weight (I(v))). Assume that the transactions of the conditional database are sorted in decreasing transaction length and traverse each transaction in that order. Let t be a transaction and l (t) be its length. For practical considerations, we can use $l(t) + |P| < f^{-1}$ (support (I (v)) * weight (I (v))) instead of $h(v) + |P| < f^{-1}$ (support (I (v)) * weight (I (v))).

Definition 3.4. Weighted Frequent Pattern with Length decreasing support
A pattern is a weighted frequent pattern with length decreasing support constraints if all of the following pruning conditions are satisfied. If a pattern does not satisfy anyone of them, thepattern is called a weighted infrequent pattern with length decreasing support constraints.

Pruning condition 1: (support ≥ f (maxLength) ‖ weight ≥ min_weight)
Pruning condition 2: (support * MaxW (MinW) ≥ f (maxLength)
Pruning condition 3: Transaction pruning by the WSVE property
Pruning condition 4: Node pruning by the WSVE property

In a transaction database, the value of multiplying the support of a pattern with a maximum weight (MaxW) among items in the transaction database is less than f (|maxLength|). In conditional databases, the value of multiplying the support of a pattern with a minimum weight (MinW) of a conditional pattern in the conditional database is less than f (|maxLength|), a length decreasing minimum support.

In WLPMiner, an ascending weight order method and a bottom-up traversal strategy are used in mining weighted frequent patterns. WLPMiner defines weighted Smallest Valid Extension property and prunes transactions and nodes by the WSVE property. The performance of pruning conditions 1 and 2 may not be good since the minimum support for the longest pattern of the length decreasing support constraint must be used in order to keep downward closure property. However, performance can be improved by using these pruning conditions with the weighted smallest valid extension property and the weight range. The weighted smallest valid extension and the downward closure property are both used to prune the search space.

3.2 WLPMiner Algorithm

The WLPMiner algorithm uses a weight range and a minimum weight. Items are given different weights within the weight range. We now the weighted frequent pattern mining process and present the mining algorithm.

WLPMiner algorithm: Mining weighted frequent patterns with a length decreasing support constraint

---

Input: (1) A transaction database: TDB
(2) f(x): a length decreasing support constraint function
(3) weights of the items within weight range : w_i
(4) minimum weight threshold : min_weight

Output: (1) WFP: the complete set of Weighted Frequent Patterns that satisfy the length decreasing support constraint.
Begin

1. Let WFP be the set of weighted frequent patterns that satisfy the length decreasing support constraint. Initialize WFP ← 0;
2. Scan TDB once to find the global weighted frequent items satisfying the following definition: A pattern is a Weighted Frequent Pattern (WFP) if the following pruning conditions 1 and 2 are not satisfied.
Condition 2.1: (support < f(maxLength) && weight < min_weight)
The support of a pattern is less than a minimum support and the weight of a pattern is less than a minimum weight constraint.
Condition 2.2: (support * MaxW < f(maxLength))
In a transaction database, the value of multiplying the support of a pattern with a maximum weight (MaxW) of each item in the transaction database is less than a minimum support.
3. Sort items of WFP in weight ascending order. The sorted weighted frequent item list forms the weighted frequent list.
4. Scan the TDB again and build a global FP-tree using weight_order.
5. Call WLPMiner (FP-tree, 0, WFP)

Procedure WLPMiner (Tree, α, WFP)

1: for each a_i in the header of Tree do
2: set β = α U a_i;
3: get a set $I_β$ of items to be included in β conditional database, $CDB_β$;
4: for each item in $I_β$, compute its count in β conditional database;
5: for each b_j in $I_β$ do
6: if (sub (β b_j) < f(maxLength) && weight (β b_j) < min_weight) delete b_j from $I_β$;
7:   if (sub (β b_j) * MinW < f(maxLength)) delete b_j from $I_β$;
8: end for
9: $CDB_β$ ← transaction_pruning_by_WSVE (β, $CDB_β$);
10: $Tree_β$ ← FP_Tree_Construction ($I_β$, $CDB_β$)
11: $Tree_β$ ← node_pruning_by_WSVE (β, $Tree_β$);

12:   if Tree$_\beta \neq 0$ then
13:     call WLPMiner (Tree$_\beta$, β, WFP)
14:   end if
15: end for

In the WLPMiner algorithm, TDB is scanned once, weighted frequent items satisfying condition 2.1 and condition 2.2 are found and these items are sorted in weight ascending order. The WLPMiner algorithm then calls the recursive procedure WLPMiner (Tree, α, WFP). Lines 6 and 7 generate weighted frequent patterns with a length decreasing constraint. Line 9 conducts transaction pruning by the WSVE property. If a pattern in a conditional database satisfies the transaction pruning, it is inserted into a local FP-tree. After a local FP-tree is constructed in line 10, node pruning by the WSVE property is carried out in line 11. WLPMiner algorithm adopts the bottom-up divide and conquer paradigm to grow the current prefix. If the local FP-tree is not empty, the procedure WLPMiner (Tree$_\beta$, β, WFP) is called recursively in line 13.

Table 4. Characteristics of datasets

Data sets	Size	#Trans	#Items	A.(M.) t. l.
Connect	12.14M	67557	150	43 (43)
T10I4D100K	5.06M	100K	1000	10 (31)
T10I4Dx	10.12-50.6M	200K-1000K	1000	10 (31)

4 Experiments

In this section, we present our performance study over various datasets. WLPMiner is the first weighted frequent pattern mining algorithm that considers both weight constraints and length decreasing support constraints which are characteristics of real datasets. We report our experimental results on the performance of WLPMiner in comparison with recently developed algorithms such as BAMBOO and WFIM. Our results show that WLPMiner not only generates more concise and important result sets, but also has much better performance than recently developed mining algorithms through incorporating a length decreasing support constraint into weighted frequent pattern mining. Moreover, WLPminer has good scalability of the number of transactions. In our experiments, we compared WLPMiner with BAMBOO [8] which is a frequent pattern mining algorithm with a length decreasing support constraint. We also compared WLPMiner with WFIM [3] that is a weighted frequent pattern mining algorithm developed recently. We used one real dataset and one synthetic dataset that are popularly used in pervious experiments [3, 8, 11, 12, 13]. Table 4 shows the characteristic of two datasets used for performance evaluation. The real dataset used is the connect dataset available in the UCI machine learning repository [15]. The connect dataset is very dense and includes game state information. The synthetic dataset was generated from IBM dataset generator. We used T10I4D100k which is very sparse and contains 100,000 transactions. However, the synthetic datasets T10I4Dx contain 200k to 1000k transactions. To test scalability, T10I4Dx datasets have been popularly used in the previous performance evaluations [3, 8, 12, 13]. WLPMiner was written in C++. Experiments were performed on a sparcv9 processor operating at 1062 MHz, with 2048MB of memory. All experiments were performed on a Unix machine. In our experiments, a random generation function generates weights for each item. When running WLPMiner, the minimum support was determined as the cut off value for the maximum pattern length under the corresponding length decreasing support constraint.

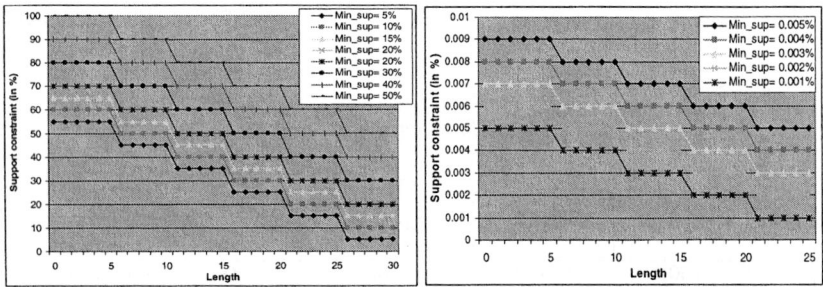

Fig. 1. Support constraint (connect)   **Fig. 2.** Support constraint (T10I4DnK)

Fig. 1 and Fig. 2 show various length decreasing support constraints used in the performance evaluation for different datasets. These length decreasing support constraints are the same as those used in BAMBOO [8]. LPMiner [7] and BAMBOO [8] are recently developed mining algorithms using length decreasing support constraints. Bamboo outperforms LPMiner in terms of runtime and the number of frequent patterns. Therefore, we compared WLPMiner with BAMBOO.

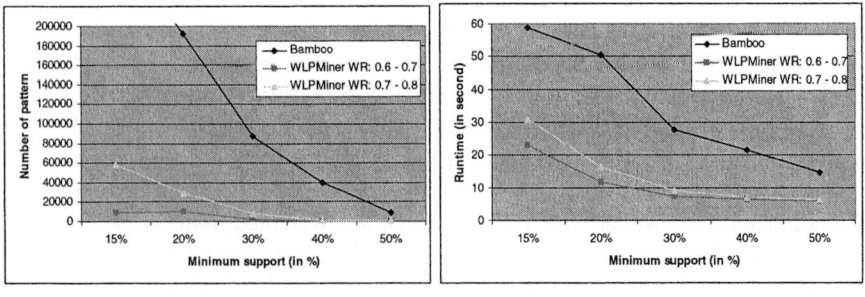

Fig. 3. Number of patterns (connect)   **Fig. 4.** Runtime (connect)

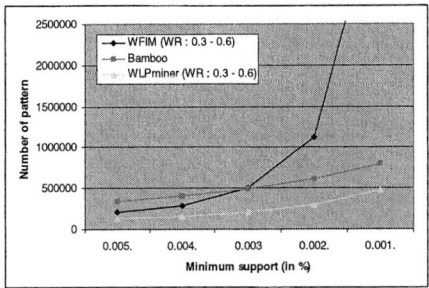 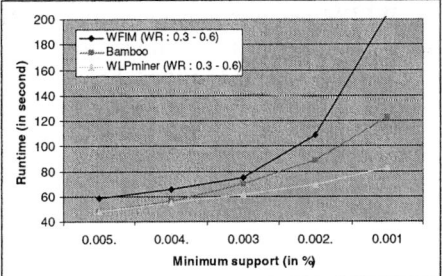

Fig. 5. Number of patterns (T10I4D100K)   **Fig. 6.** Runtime (T10I4D100K)

Fig. 3 to Fig. 6 show the comparison results for datasets connect and T10I4D100K. In Fig. 3 and Fig 5, we can see WLPMiner generates smaller weighted frequent patterns than that of BAMBOO. In Fig. 4 and Fig 6, the runtime for finding weighted frequent patterns is shown in the two datasets. From Fig. 4 and Fig. 6, we see that WLPMiner is much faster than BAMBOO. Although WLPMiner and BAMBOO algorithms use a length decreasing support constraint, WLPMiner outperforms BAMBOO because WLPMiner uses the Weighted Smallest Valid Extension property and incorporates weight constraints into length decreasing support constraints. WLPMiner and WFIM [3] are both weighted frequent pattern mining algorithms. We used several weight ranges to test the algorithms. Fig. 5 compares the number of weighted frequent patterns of WLPMiner with those of WFIM and BAMBOO. Fig. 6 shows the runtime of the algorithms under the same weight range. In Fig. 5 and Fig. 6, WLPMiner generates fewerr but important patterns. In addition, it can be several orders of magnitude faster than WFIM since the Weighted Smallest Valid Extension property for the length decreasing support constraint is effective and efficient in pruning the result set in both the connect and T10I4D100K dataset.

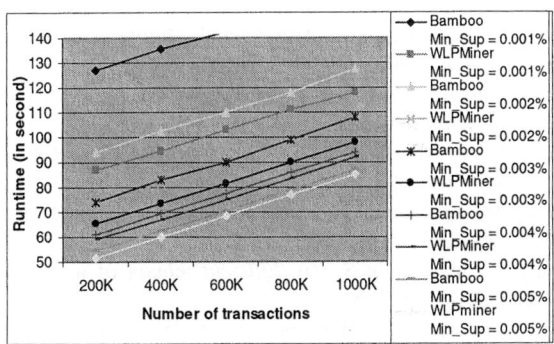

Fig. 7. Runtime (T10I4Dx) (WR: 0.3 - 0.6)

Scalability Test
To test the scalability with the number of transactions, T10I4DxK datasets are used. WLPMiner scales much better than previous mining algorithms. In this scalability

test, WLPMiner is compared with BAMBOO. BAMBOO shows linear scalability with the number of transactions from 200k to 1000k. However, WLPMiner is much more scalable than BAMBOO. From Fig 7, the difference between the two algorithms becomes clear. We first tested the scalability in terms of base size from 200K tuples to 1000K tuples and different minimum support of 0.001% to 0.005%. From Fig. 7, we can see that WLPMiner has much better scalability in terms of base size. The slope ratio for each different minimum support is almost similar.

5 Conclusion

We developed the WLPMiner algorithm that integrates a weight constraint measure with a length decreasing support constraint measure for mining frequent patterns. The key insights achieved in this paper are the high performance of the WSVE property and the use of a weight range in the weight constraint. We show that combining a weight constraint with a length decreasing support constraint improves performance in terms of the number of patterns and runtime. The extensive performance analysis shows that WLPMiner is efficient and scalable in weighted frequent pattern mining. In future work, the WSVE property will be used with different pruning techniques suggested in other algorithms using length decreasing support constraints.

Acknowledgement

We would like to thank Dr. George Karypis and Dr. Jianyong Wang for providing executable code for the BAMBOO algorithm.

References

[1] Rakesh Agrawal, Tomasz Imieliński, Arun Swami, *Mining association rules between sets of items in large databases*, ACM SIGMOD, May 1993.
[2] Rakesh Agrawal, Ramakrishnan Srikant, *Fast Algorithms for Mining Association Rules in Large Databases*, VLDB, September 1994.
[3] Unil Yun, John J. Leggett, *WFIM: Weighted Frequent Itemset Mining with a weight range and a minimum weight*, SDM`05, April 2005.
[4] Feng Tao, *Weighted Association Rule Mining using Weighted Support and Significant framework*. ACM SIGKDD, Aug 2003.
[5] Wei Wang, Jiong Yang, Philip S. Yu, *Efficient mining of weighted association rules (WAR)*, ACM SIGKDD, Aug 2000.
[6] C. H. Cai, Ada Wai-Chee Fu, C. H. Cheng, and W. W. Kwong, *Mining association rules with weighted items*. IDEAS'98, Cardiff, Wales, UK, July 1998.
[7] Masakazu Seno, George Karypis, *LPMiner: An Algorithm for Finding Frequent Itemsets Using Length-Decreasing Support Constraint*, ICDM`01, Nov. 2001.
[8] Jianyong Wang, George Karypis, *BAMBOO: Accelerating Closed Itemset Mining by Deeply Pushing the Length-Decreasing Support Constraint*, SDM`04, April 2004
[9] Masakazu Seno, George Karypis, *Finding Frequent Patterns Using Length Decreasing Support Constraints*, University of Minnesota, Minneapolis, MN 55455, Technical Report 03-003, 2003.

[10] Jiawei Han, Jian Pei, Yiwen Yin, *Mining frequent patterns without candidate generation*, ACM SIGMOD, May 2000.
[11] Jian Pei, Jiawei Han, *CLOSET: An Efficient Algorithm for Mining Frequent Closed Itemsets*, DMKD, May 2000.
[12] Jianyong Wang, Jiawei Han, Jian Pei, *CLOSET+: searching for the best strategies for mining frequent closed itemsets*, ACM SIGKDD, August 2003.
[13] Guimei Liu, Hongjun Lu, Yabo Xu, Jeffrey Xu Yu, *Ascending Frequency Ordered Prefix-tree: Efficient Mining of Frequent Patterns*. DASFAA, 2003.
[14] Junqiang Liu, Yunhe Pan, Ke Wang, Jiawei Han, *Mining frequent item sets by opportunistic projection*, ACM SIGKDD, July, 2002.
[15] http://www.ics.uci.edu/~mlearn/MLRepository.html.

A Framework for Incorporating Class Priors into Discriminative Classification

Rong Jin and Yi Liu

Department of Computer Science and Engineering, Michigan State University,
East Lansing, MI48824, U.S.A.
{rongjin, liu3}@cse.msu.edu

Abstract. Discriminative and generative methods provide two distinct approaches to machine learning classification. One advantage of generative approaches is that they naturally model the prior class distributions. In contrast, discriminative approaches directly model the conditional distribution of class given inputs, so the class priors are only implicitly obtained if the input density is known. In this paper, we propose a framework for incorporating class prior proportions into discriminative methods in order to improve their classification accuracy. The basic idea is to enforce that the distribution of class labels predicted on the test data by the discriminative model is consistent with the class priors. Therefore, the discriminative model has to not only fit the training data well but also predict class labels for the test data that are consistent with the class priors. Experiments on five different UCI datasets and one image database show that this framework is effective in improving the classification accuracy when the training data and the test data come from the same class proportions, even if the test data does not have exactly the same feature distribution as the training data.

1 Introduction

Machine learning approaches to classification usually fall either into the discriminative (or conditional modeling) category, or the generative category. Discriminative classification directly attempts to model $p(y\,|\,\vec{x})$ where $\vec{x}$ is the vector of input features and y is the class label. Generative approaches model the joint distribution, split into the class prior and the class conditional density: $p(y,\vec{x}) = p(y)p(\vec{x}\,|\,y)$. One difference between them is the conditional model usually only focuses on the relationship between the input features and the class label, while the generative model has to explain both how the inputs are generated and how the class label is associated with the input data. One usually finds that state-of-the-art conditional models perform better than generative models on classification problems. More detailed studies of the comparison of conditional models and generative models can be found in [8].

However, compared to discriminative approaches, generative approaches have the advantage in that they are able to explicitly make use of class priors for predicting class labels. Knowledge of the class priors or proportions can be very useful in several contexts. Assume that the learning algorithm is provided with labeled training data and unlabeled test data. First, if the training data does not have the same proportions of points in each class as the test data, incorporating knowledge of class priors can help create a classifier that predict classes with the correct probabilities on the test data. Second, sometimes the class conditional probabilities $p(\vec{x}\,|\,y)$ are somewhat different from the training data to the test data. This is a situation which most learning algorithms are not designed for, but it occurs in practice due to the fact that the labeling process can be biased. Incorporating knowledge of the class priors can make a learning algorithm more robust to inaccuracies in the class conditional probabilities obtained from training data. In the extreme case, the inputs of the training data and the test data may be completely random numbers and only the class priors are consistent through the whole dataset. The best strategy of predicting class labels for the test data would be to always predict the class label with the highest prior. Unfortunately, since discriminative models focus on learning the mapping from input data to class label, $p(y\,|\,\vec{x})$, without representing the input density $p(\vec{x})$, most methods can't directly model or take advantage of the marginal class distribution, $p(y)$.

In this paper, we propose a framework that is able to *explicitly* incorporate class prior information into discriminative learning. This framework is based on the assumption that the class priors give a reasonably accurate description for the class distribution of the test data. Therefore, the discriminative model learned from the training data should not only explain the class labels for the training data well but also predict the class labels for the test data in such a way that the distribution of the predicted class labels for the test dataset is also coherent with the class priors. Clearly, this framework differs from the traditional approach for discriminative learning where the objective is to make the class labels predicted by the model consistent with the assigned class labels on the training dataset. Furthermore, our framework is able to utilize both training data and testing data in the construction of discriminative models, while traditional approaches for discriminative learning only take advantage of training data.

This framework can be useful when the training data and the test data have the same class priors but do not have exactly the same feature distributions and therefore the model learned from the training data may not be appropriate for the test data. Differences between the training data and the test data can be caused by the fact that either the sampling for the training data is quite different from the sampling for the test data, or the amount of training data is too small to give a good representation for the whole dataset. The other interesting aspect of this framework is that it allows the discriminative model to use a mixture of training data and test data. Thus, this framework is able to deal with learning problems in which only a small number of training examples are available and the majority of instances are unlabelled. Unlike previous works on the combination of labeled data and unlabeled data, which mainly focus on the generative model, the framework provides room for the discriminative model to take advantage of unlabeled data.

The rest of this paper is arranged as follows: section 2 will discuss related work. The formal description of the framework is presented in Section 3. Section 4 describes the empirical study of this framework. Conclusions and future work are presented in Section 5.

2 Related Work

As already discussed in the introduction, the main idea of this paper is to incorporate class prior information into discriminative learning. Therefore, it combines some aspects of learning both discriminative and generative models. There have been several studies on the improvement of discriminative models using information from a generative model [6]. The motivation of that work is based on the observation that sometimes the generative model captures properties of the input distribution that are meaningful to the discrimination problem. Therefore, if we can influence the discriminative model with a generative model that is specific to the problem, for example choosing a discriminative function that is coherent with the generative density, we may be able to gain better performance than using a generic discriminative model. Approaches based on this idea, such as combining the support vector machine with fisher kernels derived from a generative model, have shown significant improvement in classification problems. Unlike that work, in our framework we don't change the kernel function; instead, we only consider the class prior information as an extra hint to be used by the discriminative model.

Since this framework is taking advantage of both the training data and the test data, it is strongly related to the work on learning from the mixture of labeled and unlabeled data [9]. Many of works on this problem assume some form of generative model, which is used to explain both the labeled data (i.e. the inputs and the label) and the unlabeled data (i.e. just the inputs). In cases where only a small amount of data are labeled, a model learned from this data can be quite skewed and the incorporation of unlabeled data can help avoid idiosyncrasies of the labeled data to some extent. Unlike this work, our framework focuses on incorporating unlabeled data into discriminative training. Other works on learning from the mixture of labeled and unlabeled data have focused on using unlabeled data for model regularization and model selection. One example is the transductive SVM [7], where the classification margin is influenced both by the labeled and unlabeled data. Unlike their work, in this framework, we refine the learned model by only examining the discrepancy between the class distribution of the predicted labels of unlabeled data and the 'true' class priors.

3 Incorporating Class Priors into Discriminative Training

The basic logic behind this framework can be simply understood as follows: consider the case when the test data has quite different patterns from the training data. This situation can happen if there is very little training data, or as in many real applications, if the training data and testing data come from different sources. Then, applying the discriminative model that is learned from the training data directly to label the test data will be problematic. If we have prior knowledge on the class distribution for the test data, we may be able to find out the fact that the test data are noisy by simply examining the difference between the class priors and the distribution of the class labels for the test data predicted by the discriminative model. If there is a significant discrepancy between these two distributions, we will suspect that the learned model may not be appropriate for the test data and needs to be adjusted. In order to refine the discriminative model, we need to do two things: first, we can adjust the probability of classes for the test data computed from the original discriminative model in such a way that the averaged distribution of the predicted class labels for the test data is shifted toward the class priors. Then, the test data with the adjusted class probabilities will be included in the training data and a discriminative model will be retrained over the 'enlarged training dataset'. The procedures of adjusting the class probabilities for test data using priors and retraining the discriminative model will be carried out iteratively until it reaches some local maximum.

3.1 Model Description

The essence of a discriminative model is the computation of the conditional probability for a class label y given the input vector $\vec{x}$, i.e. $p(y|\vec{x})$. The learning of a discriminative model can be formalized as the search for a model that maximizes the log-likelihood of the training data, i.e.

$$L = \sum_{i \in Train} \log p(y_i | \vec{x}_i, M) \tag{1}$$

where M stands for a discriminative model, $\vec{x}_i$ is the ith training data point and y_i stands for its class label.

In order to incorporate class prior information into a model, a discriminative model will not only have to explain the training data well but also to predict class labels for the test data in such a way that the distribution of predicted class labels for the test data is consistent with class priors. Therefore, we need an extra term in Equation (1) that can account for the discrepancy between the two distributions. In the following sections, we will discuss three different approaches. To this end, for every instance in the test dataset, an unknown distribution over class labels is introduced. This represents the estimated distribution over classes, which will incorporate both the prior class constraints and the model predictions. Moreover, we will see that it considerably simplifies the computation in optimization. Let r_k be this estimated class distribution for the kth data point in the test set, and value $r_{k,y}$ be the probability for the kth test data point to be in class y. To enforce the consistency between class priors of training data and test data, we impose the following constraint on the estimated class probability $r_{k,y}$, i.e.,

$$\forall y \quad \frac{1}{N_{test}} \sum_{k \in Test} r_{k,y} = p_y \tag{2}$$

Now, the next step is to connect the estimated class probability $r_{k,y}$ to our objective function. Of course, we want the distribution of class labels predicted by model M, i.e., $p(y|\vec{x}_k, M)$, to be consistent with the estimated class distribution $r_{k,y}$. Therefore, the objective in (1) can be modified as:

$$L' = \sum_{i \in Train} \log p(y_i | \vec{x}_i, M) + \sum_{k \in Test} \sum_y r_{k,y} \log \frac{p(y|\vec{x}_k, M)}{r_{k,y}} \tag{3}$$

In above, the KL divergence is introduced as the measurement of consistency between the estimated class distribution $r_{k,y}$ and the distribution of the predicted class labels $p(y|\vec{x}_k, M)$. By maximizing (3) under constraints (2), we ensure that the discriminative model is consistent with the estimated class distribution $r_{k,y}$, which indirectly forces consistency between the class priors of test data and of training data.

3.2 Finding the Optimal Solution

As indicated in Equation (3), the objective function contains two different sets of parameters, namely the model parameters M and the estimated class distribution $r_{k,y}$. Therefore, we can optimize the objective function in (3) by alternatively freezing one set of parameters. More specifically, we will first optimize the objective function in (3) using only the discriminative model parameters, and then search for the estimated class distributions that optimize (3) under the constraints in (2). It is not difficult to see that the strategy used in the optimization exactly corresponds to the intuition stated at the beginning of this section.

In the first step of optimization, the $r_{k,y}$ are held fixed (as target distributions for the test data) so the constraint in (2) is not relevant. Thus, the discriminative model can be trained with almost no modification except that both the training data and the test data are fed into the learning module. Of course, any discriminative classifier that accepts distributions as targets can be used here.

For the second step of optimization, we need to find the set of estimated class distributions that maximizes the objective function in (3) subject to the constraints in (2). Since parameters for the discriminative model are frozen, the objective function in (3) is simplified as:

$$L'' = \sum_{i \in Test} \sum_{y} r_{k,y} \log \frac{p(y|\vec{x}_k, M)}{r_{k,y}} \qquad (3')$$

The problem of maximizing (3') under the constraints in (2) is exactly the same problem as solved in maximum entropy (ME) models [1]. The original version of maximum entropy model is to find a set of probabilities that not only maximize the entropy function and but also satisfy a set of linear constraints. This can be extended to the case when the objective function is not an entropy function but a KL divergence between the distribution to be optimized and a set of given probabilities, i.e. a *minimum relative entropy* (MRE) problem, which is exactly our problem.

4 Experiments

In this experiment, we examined the effectiveness of our model in terms of using class priors to improve classification accuracy. More specifically, we would like to address two scenarios of application for this framework:

1) *A scenario of a small number of labeled examples.* In this case, we will expose the system to a small number of labeled examples together with a large number of unlabeled examples. Under the assumption that a reliable estimation of class priors

are available, we can examine how well this model is able to improve the model by taking into count the large number of unlabeled data.
2) *A scenario of heterogeneous training and testing examples.* In this case, we assume that the training data are somehow different from testing data in some respects. Therefore, a classification model learned from training data is not appropriate for the testing data. By inspecting the discrepancy between the class priors and the class distribution of the predicted labels of unlabeled data, we expect to adjust our model to the characteristics of the testing data.

Table 1. UCI datasets used in our experiments

Data Set	Number of Feature	Number of Class	Number of Instance
Ecoli	7	5	327
Wine	13	3	178
Pendigit	16	10	2000
Iris	14	3	154
Glass	10	5	204

4.1 Experiment Design

The discriminative model used for the experiment is the conditional exponential model [1], in which conditional probability $p(y|\vec{x})$ is defined as $p(y|\vec{x}) = \exp(\vec{\lambda}_y \cdot \vec{x})/Z(\vec{x})$, where $\vec{\lambda}_y$ is the weight vector for class y and $Z(\vec{x})$ is the normalization factor. A conjugate gradient [10] is used to find the appropriate weight vectors.

To illustrate the effectiveness of our framework on the two different scenarios mentioned before, we tested the algorithm against two different groups of datasets. For the first scenario, we use five UCI datasets as the testbed. We use a small portion of each UCI dataset as training examples and leave majority of the dataset as testing examples. The detailed information about the five datasets is listed in Table 1. For the second scenario, we tested our algorithm on both the synthesized data that are generated from the above five UCI datasets and real image data. To simulate the difficult circumstance in which test data and training data have different feature distributions, for every feature, we uniformly randomly generate a weight factor ranging from 1 to 1.5 and multiple it with the corresponding feature of the testing data. By this 'corruption' procedure, the weights of the exponential model learned from the training data will not be appropriate for the test data because the scale of the test are changed. By testing our algorithm against the synthesized datasets, we are able to see how effectively our framework is able to adjust the model parameters $\vec{\lambda}_y$ according to the large number of unlabeled data.

The other dataset that we used for the second scenario is the image dataset. We use the images downloaded from the image directory of Google as the training examples and images from Corel image database as testing examples [3]. Six image categories are used in the experiment, i.e., category 'food', 'building', 'bird', 'sky', 'fish' and 'fruit & vegetable'. Each category contains 100 training images and 100 testing images. The training images are acquired by querying the image directory of Google with the name of categories as the query words. The top ranked 100 images from Google are used as the training examples. The testing images are collected by randomly sampling 100 images out of the corresponding categories from Corel database. Apparently, images downloaded from Google image database will be considerably different from images from Corel database. The extended color co-occurrence matrix [5] is used for image representation, which have shown its effectiveness in image classification. For each image, totally 500 image features are extracted. More detailed discussion about image classification can be found in [2, 4, 11].

Table 2. Classification Errors for UCI datasets when 25% data are used for training

Data Set	No Prior	Empirical Estimate	Intermediate Estimate	Optimal Estimate
Ecoli	16.1%	20.6%	16.7%	16.0%
Wine	15.1%	15.0%	9.1%	8.0%
Pendigit	8.8%	12.4%	8.7%	8.0%
Iris	5.6%	16.0%	4.5%	3.7%
Glass	9.8%	14.2%	3.9%	2.7%

Table 3. Classification Errors for UCI datasets when 10% data are used for training

Data Set	No Prior	Empirical Estimate	Intermediate Estimate	Optimal Estimate
Ecoli	32.4%	26.9%	21.2%	21.8%
Wine	20.8%	26.0%	15.0%	15.1%
Pendigit	11.8%	17.9%	11.8%	11.6%
Iris	7.5%	23.6%	5.3%	4.4%
Glass	5.7%	27.8%	2.2%	2.6%

The key component in this framework is the knowledge of class priors. To examine the impact of class prior accuracy on classification performance, we introduce three different ways of estimating class priors:

1) 'Empirical Estimate': Estimate the class priors only based on the training data. Since we use small portion of the data as training, this estimate of class prior can be quite inaccurate for the test data.

2) 'Optimal Estimate': Estimate the class priors based on the test data. Of course, this estimate gives the exact class distribution for the test data and is not realistic. However, the performance under this estimate gives a sense of an upper bound performance of our framework.
3) 'Intermediate Estimate': Estimate the class priors based on all the data including the test data and training data. Definitely, this estimate will be better than the first case and worse than the second case.

4.2 Scenario 1: A Small Number of Labeled Examples

In this experiment, we set the training size to be 10% and 25% of the total dataset, respectively. The averaged classification errors based on cross validation for the proposed algorithm are listed in Table 2 and 3. We also included the classification results when no class priors information is used.

First, by comparing the performance listed in Table 2 to what listed in Table 3, it is clear that, by decreasing the amount of training examples from 25% to 10%, all learning methods on most UCI datasets suffers degradation in performance except the 'Glass' dataset. Second, comparing the proposed framework using different estimators of class priors, it is clear that the new framework with optimal estimator appears to have the best performance while the intermediate estimator gives the second best performance. The new framework with these two estimators of the class priors appears to substantially outperform the baseline model, i.e., the simple discriminative model without using class priors. This fact indicates that our algorithm is effective in improving the performance of discriminative classifier with reliable estimates of class priors. Third, the proposed algorithm with empirical estimates appears to perform significantly worse than the original discriminative model without using class priors. Since the empirical estimator bases its estimates on the empirical class distribution of training data and only small portion of training examples are available in the study, the empirical estimates usually gives poor estimation of class priors, which results in poor performance of the proposed algorithm. Based on this fact, we can see that, it is very important to our algorithm to have accurate estimates of class priors.

4.3 Scenario 2: Heterogeneous Training Data and Testing Data

In this subsection, we will test our algorithm against the case when the testing data have different feature distributions from the training data. This is a practically relevant scenario, which is rarely studied in machine learning. First, we will test the propose algorithm on the synthesized datasets, which are generated from the five UCI datasets. The 'corruption' procedure has already been described in section 4.1. The results for the proposed algorithm with three different estimators of class priors together with the discriminative model without class priors are listed in Table 4 and 5.

First, by comparing the results in Table 4 and 5 to the results in Table 2 and 3, it is clear that, by multiplying the testing data with a random weight factor, the performance of the discriminative model without using class priors suffers from a severe degradation. On the contrary, the proposed algorithm appears to suffer from much smaller degradation for all three different estimators. This fact indicates that, the proposed algorithm is robust to the 'corruption' on the features. Second, it is clear that, the proposed framework with all three different estimates is significantly better than the discriminative model without class priors. Unlike the results presented in previous scenario, where the empirical estimates gives rise to poor performance due to its inaccurate estimation on class priors, for this scenario, even the empirical estimate is able to result in a better performance than the original discriminative model for most datasets. The fact that the proposed algorithm is able to outperform the discriminative model without class priors indicates that our algorithm is effective in dealing with the situation when the testing data are quite different from the training data.

In addition to testing our algorithm against the synthesized dataset, we also examine our algorithm over the problem of image classification. The details of image databsets have already been described in the section 4.1. The downloaded 600 images are used as training examples and the 600 images from Corel database are used as testing instances. The classification error for the discriminative model without class priors is 66.9% and 60.9% for the proposed model assuming that the class priors equal to 1/6. Again, with accurate knowledge on class priors, we are able to decrease the classification error significantly. Notice that, the classification errors in this experiment is quite high, over 60%, while some image classification works show extremely good performance over Corel datasets. We believe the main reason for that is because the images from Google do not resemble the images from Corel in many cases. Table 6 shows two images of category 'bird' from both Google and Corel database. Apparently, the training examples are very different from testing images, either from the viewpoint of color or from the viewpoint of texture. We think that is the reason why this task is so hard and causes so large testing errors.

5 Conclusions

In this paper, we proposed a framework that is able to incorporate class prior information into training a discriminative model. This algorithm can also be thought as a machine learning algorithm which allows a discriminative model to use a mixture of labeled and unlabeled data. In the empirical study over five different UCI datasets and Corel image database, this algorithm is able to improve the performance of the conditional exponential model significantly when the number of training examples is small and when the test data are heterogeneous from the training data. Therefore, we conclude that the new algorithm is able to help the performance even with inaccurate estimation for class priors and the improvement depends on the accuracy of the estimation. Usually large improvements were found when accurate class priors were incorporated into training but these improvements vanished when the class priors had substantial inaccuracies. Thus, more research work is needed in order to study how to improve the classification accuracy in case of inaccurate class priors.

Acknowledgement

We thank Dr. Zoubin Ghahramani and Dr. Huan Liu for their discussion and suggestion on this work.

References

1. Berger, A., S.D. Pietra, and V.D. Pietra, *A Maximum Entropy Approach to Natural Language Processing.* Computational Linguistics, 1996. **22**(1).
2. Chapelle, O., P. Haffner, and V. Vapnik, *Support Vector Machine for Histogram Based Image Classification.* IEEE Transaction on Neutral Network, 1999. **9**.
3. Corporation, C., *Corel Clipart & Photos.* 1999.
4. Goh, K.-S., E. Chang, and K.-T. Cheng. *Svm Binary Classifier Ensembles for Image Classification.* in Proceedings of the tenth international conference on Information and knowledge management. 2001.
5. Huang, J., et al. *Image Indexing Using Color Correlograms.* in Proccedings of IEEE Computer Vision and Pattern Recognition Conference. 1997.
6. Jaakkola, T. and D. Haussler. *Exploiting Generative Models in Discriminative Classifiers.* in Advance in Neutral Information Processing System 11. 1998.
7. Joachims, T. *Transductive Inference for Text Classification Using Support Vector Machines.* in Proceedings of The Sixteenth International Conference on Machine Learning (ICML 99). 1999.
8. Ng, A. and M. Jordan. *On Discriminative Vs. Generative Classifiers: A Comparison of Logistic Regression and Naive Bayes.* in Advances in Neural Information Processing Systems 14. 2002.
9. Seeger, M. *Learning with Labeled and Unlabeled Data.* Technical report Edinburgh University, 2001
10. Shewchuk, J. *An Introduction to the Conjugate Gradient Method without the Agonizing Pain.* Techinical Report School of Computer Science, Carnegie Mellon Unversity, 1994
11. Teytaud, O. and D. Sarrut. *Kernel Based Image Classification.* in Proceedings of International Conference on Artificial Neural Networks. 2001.

Increasing Classification Accuracy by Combining Adaptive Sampling and Convex Pseudo-Data

Chia Huey Ooi and Madhu Chetty

Gippsland School of Computing and Information Technology,
Monash University, Churchill, VIC 3842, Australia
{chia.huey.ooi, madhu.chetty}@infotech.monash.edu.au

Abstract. The availability of microarray data has enabled several studies on the application of aggregated classifiers for molecular classification. We present a combination of classifier aggregating and adaptive sampling techniques capable of increasing prediction accuracy of tumor samples for multiclass datasets. Our aggregated classifier method is capable of improving the classification accuracy of predictor sets obtained from our maximal-antiredundancy-based feature selection technique. On the Global Cancer Map (GCM) dataset, an improvement over the highest accuracy reported has been achieved by the joint application of our feature selection technique and the modified aggregated classifier method.

Keywords: Arcing, microarray, tumor classification, boosting.

1 Introduction

Perturbed training sets, not unknown in the area of traditional machine learning, have been used to increase classification accuracy by combining or *aggregating* several classifiers. There are several types of classifier aggregation methods, such as bootstrapping, convex pseudo-data, boosting and arcing. Whilst the first two are simply different versions of pure bagging (bootstrap aggregating), the last two belong to the category of adaptive sampling techniques. Although various flavors of bootstrapping, convex pseudo-data and boosting have been applied independently for molecular classification [1–4], none have satisfactorily addressed the problems presented by truly multiclass datasets (>5 classes), namely, poor classification accuracy and large predictor sets. Nor has an appropriate classifier-independent (filter-based) feature selection technique been applied prior to employing either aggregating or adaptive sampling methods.

This paper demonstrates how a method combining pseudo-convex data and arcing, with the aid of a maximal-antiredundancy-based feature selection technique, can improve the molecular classification accuracy of multiclass datasets. Moreover, the improvement is achieved using a comparatively small predictor set size.

2 Perturbation Methods

The original training set T, upon which feature selection and perturbation methods are to be implemented, consists of N genes and M samples. Prior to training set perturbation and/or classifier training, feature selection is employed to form a subset of genes, called the *predictor set*, S from the total of N genes. The criterion for choosing the members of S is given in section 2.3.

2.1 Existing Methods

Convex Pseudo-Data (CPD). Breiman [5] suggested the use of the convex pseudo-data method to form B perturbed training sets. Here, each of the perturbed training set is created as follows:

1. Pick two samples **x** and **x'** randomly from the original training set T.
2. Select a random number v from the interval [0, d], where $0 \le d \le 1$.
3. Generate a new *pseudo-sample* **x"** by linearly combining **x** and **x'**.

$$\mathbf{x}'' = (1-v)\mathbf{x} + v\mathbf{x}' \quad (1)$$

4. Repeat step 1 until step 3 given above M times.

B is traditionally assigned values within the range [10, 200].

Arcing. Breiman [6] has also established a technique for adaptive sampling, arc-x4, where B classifiers are produced as follows:

1. Initialize the sampling probability, p_j for training sample j to $1/M$ for $j = 1, 2, \ldots, M$.
2. Sample T with replacement based on the current sampling probabilities, p_j for each sample in order to form a perturbed training set of the same size, T''.
3. Train a base classifier employing features from the predictor set used, and information from perturbed training set T''.
4. Use classification results from step 3 to update the sampling probabilities, p_j using the following equation.

$$p_j = \frac{1+m_j}{\sum_{j=1}^{M}(1+m_j)} \quad (2)$$

where m_j denotes the number of times sample j has been misclassified by the classifier.
5. Repeat step 2 until step 4 given above B times.

Again, the range for B in case of arcing is the same as that of CPD. Modifications have also been made to ensure that the class composition of T'' remains the same as that of T. After B classifiers have been created this way, the test samples are predicted by unweighted voting of those B classifiers.

2.2 Proposed Methods

For the work reported in this paper, the following modifications to the existing perturbation methods are implemented to investigate their impact on classification accuracy and optimal predictor set size.

Fixed Mixing Parameter CPD (FMCPD). In existing CPD, the value of v is picked randomly within the range $[0, d]$. We propose a modified version of CPD where the value of v is fixed to the upper limit of the interval, d in step 2 of the existing CPD method. We call this adaptation the fixed mixing parameter CPD (FMCPD).

Combining CPD or FMCPD with Arcing. Our second proposed modification involves combining the CPD (or FMCPD) and arcing techniques in order to take advantage of both techniques. B classifiers are formed as follows:

1. Same as step 1 in *Arcing*.
2. Sample T with replacement based on the current sampling probabilities, p_j for each sample in order to form an intermediate perturbed training set of the same size, T'. Then, from T', use either CPD or FMCPD to construct a final perturbed training set of the same size, T''.
3. Same as step 3 in *Arcing*.
4. Same as step 4 in *Arcing*.
5. Repeat step 2 until step 4 given above B times.

Finally, the test samples are predicted by unweighted voting of the B classifiers trained in the procedures above.

2.3 Pre-aggregating Feature Selection

Predictor sets used to train the classifiers are obtained through a maximal-antiredundancy-based feature selection technique proposed in [7]. In this technique, the antiredundancy-based predictor set score is defined to measure the goodness of predictor set S.

$$W_{AC,S} = (V_S)^\alpha \cdot (U_S)^{1-\alpha} \qquad (3)$$

V_S denotes the measure of relevance for S and is computed by averaging the BSS/WSS (between-class sum of squares/within-class sum of squares) ratios of the members of S. The BSS/WSS measure, first used in [1] for multiclass tumor classification, is a modification of the F-ratio statistics for one-way ANOVA (Analysis of Variance). U_S represents the measure of antiredundancy for S.

$$U_S = \frac{1}{|S|^2} \sum_{i,j \in S} 1 - |R(i,j)| \qquad (4)$$

with $R(i, j)$ being the Pearson product moment correlation coefficient between members i and j of S. As can be seen from equation (3), α and $(1-\alpha)$ represent the importance placed upon relevance and antiredundancy respectively in determining the membership of S.

This predictor set scoring approach is employed to find the optimal predictor set of sizes $P = 2, \ldots, 150$.

3 Results

Existing and proposed methods were tested on a 14-class oligonucleotide microarray (Affymetrix) dataset of 14 different primary tumor types ("Global Cancer Map", or GCM) [8]. GCM contains $N = 16063$ features (genes), and is considered to be the largest publicly available microarray dataset in terms of the total number of features and classes. 198 tumor samples of the GCM dataset were split into a training set of 144 samples and a test set of 54 samples.

A total of 9 optimal predictor sets are considered in the experiment, each predictor set being obtained based on the α values of 0.1, 0.2, ..., 0.9 respectively. In this paper we use the split of training and test sets similar to the split used in the original studies on the dataset [8, 9]. Except for the runs employing pure arcing technique, B is set to 25 throughout the experiments. A SVM-based multi-classifier, DAGSVM, is used as the base classifier. This classifier uses substantially less training time compared to either the standard algorithm or Max Wins, and has been shown to produce comparable accuracy to both of these algorithms [10].

3.1 Arcing

First, we applied the arcing technique on the GCM dataset in order to investigate whether arcing by itself, without either CPD or FMCPD, is able to produce any significant improvement over classification results of the original single classifier (SC). For this purpose, experiments were conducted using pure arcing technique with $B = 25$ and $B = 50$. For the particular dataset employed, identical sets of results are obtained from pure arcing runs for both values of B and are shown in Tables 1 and 2.

Table 1. Optimal classification accuracy ($P \leq 150$) from pure arcing method (SC denotes single classifier)

B \ α	0.1	0.2	0.3	0.4	0.5	0.6	0.7	0.8	0.9
25 or 50	75.9	77.8	79.6	83.3	81.5	77.8	79.6	74.1	61.1
SC	75.9	75.9	77.8	83.3	79.6	75.9	75.9	72.2	61.1

Table 2. Corresponding optimal predictor set size ($P \leq 150$) from pure arcing method

B \ α	0.1	0.2	0.3	0.4	0.5	0.6	0.7	0.8	0.9
25 or 50	46	135	75	85	118	144	125	145	150
SC	51	92	75	85	109	144	125	145	150

Arcing produced only a slight improvement of classification accuracy for 6 out of the 9 predictor sets with respect to the single classifier. In terms of predictor size, in fact, there is no improvement with arcing except for a small decrease in the case of $\alpha = 0.1$.

3.2 CPD-Arcing

The CPD methods may or may not incorporate arcing for forming perturbed training sets. The methods employing exclusively CPD are labeled as **CNU***d* (Convex pseudo-data, **N**o arcing, **U**nfixed *v*), while the methods incorporating arcing are tagged as **CRU***d* (**C**onvex pseudo-data, a**R**cing, **U**nfixed *v*).

The full set of results is presented here for only CRU*d* methods. The values of *d* implemented in the experiments are 0.5, 0.625, 0.75 and 1.0. The results for CPD methods demonstrate that at *d*=1.0 the CPD-Arcing method produces the classification accuracy that are better than those of the single classifier except for the predictor set formed using α=0.4 (Tables 3 and 4).

Table 3. Optimal classification accuracy ($P \leq 150$) from CRU*d* methods

d \ α	0.1	0.2	0.3	0.4	0.5	0.6	0.7	0.8	0.9
0.5	77.8	81.5	83.3	81.5	83.3	79.6	75.9	75.9	68.5
0.625	77.8	81.5	81.5	81.5	83.3	79.6	79.6	74.1	63.0
0.75	77.8	81.5	83.3	83.3	83.3	79.6	79.6	75.9	64.8
1.0	79.6	81.5	83.3	81.5	85.2	79.6	81.5	79.6	64.8
SC	75.9	75.9	77.8	83.3	79.6	75.9	75.9	72.2	61.1

Table 4. Corresponding optimal predictor set size ($P \leq 150$) from CRU*d* methods

d \ α	0.1	0.2	0.3	0.4	0.5	0.6	0.7	0.8	0.9
0.5	73	102	75	102	108	139	125	143	150
0.625	68	104	85	97	103	138	150	138	133
0.75	73	105	80	85	128	122	145	142	130
1.0	76	113	75	106	130	123	139	146	145
SC	51	92	75	85	109	144	125	145	150

In terms of the average of accuracies across different α values, CRU1.0 (i.e. $d = 1.0$) gives the best overall results.

3.3 FMCPD-Arcing

Like CPD, FMCPD methods may or may not incorporate arcing to construct perturbed training sets. The methods employing exclusively FMCPD are labeled as **CNF***v* (**C**onvex pseudo-data, **N**o arcing, **F**ixed *v*), while the methods incorporating arcing are tagged as **CRF***v* (**C**onvex pseudo-data, a**R**cing, **F**ixed *v*).

The full set of results is shown only for the CRF*v* methods. In the experiments, several values of *v* were used: 0.5, 0.625, and 0.75. The results from the FMCPD-Arcing experiments show that at *v*=0.5 and 0.625, the classification accuracy are improved (Table 5) with a minimal increase in predictor set size (Table 6) as compared to the accuracy achieved by single classifier (without any classifier aggregation) for the cases of α=0.3, 0.4, 0.5. Whereas for α=0.6, the increase in accuracy is accompanied by a slight *decrease* in corresponding predictor set size.

Table 5. Optimal classification accuracy ($P \leq 150$) from CRFv methods

v \ α	0.1	0.2	0.3	0.4	0.5	0.6	0.7	0.8	0.9
0.5	79.6	81.5	**79.6**	**85.2**	85.2	81.5	79.6	79.6	64.8
0.625	79.6	81.5	**79.6**	**85.2**	87	81.5	79.6	79.6	70.4
0.75	72.2	66.7	70.4	77.8	77.8	74.1	77.8	75.9	70.4
SC	75.9	75.9	77.8	83.3	79.6	75.9	75.9	72.2	61.1

Table 6. Corresponding optimal predictor set size ($P \leq 150$) from CRFv methods

v \ α	0.1	0.2	0.3	0.4	0.5	0.6	0.7	0.8	0.9
0.5	120	102	78	98	128	126	127	138	140
0.625	108	109	78	105	126	134	111	150	80
0.75	81	79	75	141	105	84	104	149	90
SC	51	92	75	85	109	144	125	145	150

By averaging the test accuracies across different α values, CRF0.625 (i.e. $v = 0.625$) gives the best overall results. The improvements attained by the FMCPD-Arcing methods over the single classifier, either in terms of accuracy increase or predictor set size, are more impressive than those achieved by the CPD-Arcing methods.

4 Discussion

The results from the previous section show that while arcing improves the classification accuracy of only 6 out of the 9 predictor sets with respect to the single classifier, the FMCPD-Arcing methods increase the accuracy for all 9 predictor sets. Moreover, the increase in accuracy produced through arcing is of smaller magnitude than the improvement through the FMCPD-Arcing method. We therefore conclude that arcing by itself, without either CPD or FMCPD, does *not* bring about any significant improvement over the results from the original single classifier. It is the combination of FMCPD and arcing that is crucial for the vast improvement in classification accuracy.

A further inspection of the results from the previous section leads us to the following two questions:

1. How much of the improvement in accuracy is the effect of fixing?
2. How much additional improvement is due to combining arcing with the CPD or FMCPD technique?

To answer the first question, a comparison between the FMCPD and the original CPD methods, with or without arcing is required and is presently given in Section 4.1. For the second question, a double contemplation of FMCPD-Arcing vs. FMCPD and CPD-Arcing vs. CPD is necessary and is presented subsequently in Section 4.2.

4.1 The Effects of Fixing the Mixing Parameters in CPD

Two sets of comparison must be made to study the effect of fixing the mixing parameter v on classification accuracy. One is a comparison between CRFv and CRUd methods, while the other is a comparison between CNFv and CNUd methods. The value of v which gave the best overall results in accuracy for the FMCPD-Arcing methods, 0.625; and the value of d which produced the best average of accuracies for the CPD-Arcing methods, 1.0 are used in the comparisons.

Table 7. Optimal classification accuracy ($P \leq 150$) from the CRF0.625, CNF0.625, CRU1.0 and CNU1.0 methods using predictor sets formed based on different values of α

Method \ α	0.1	0.2	0.3	0.4	0.5	0.6	0.7	0.8	0.9
CRF0.625	79.6	81.5	79.6	85.2	87.0	81.5	79.6	79.6	70.4
CNF0.625	79.6	79.6	77.8	85.2	85.2	81.5	79.6	81.5	66.7
CRU1.0	79.6	81.5	83.3	81.5	85.2	79.6	81.5	79.6	64.8
CNU1.0	77.8	83.3	81.5	85.2	85.2	81.5	79.6	77.8	70.4

Table 8. Corresponding optimal predictor set size ($P \leq 150$) from the CRF0.625, CNF0.625, CRU1.0 and CNU1.0 methods using predictor sets formed based on different values of α

Method \ α	0.1	0.2	0.3	0.4	0.5	0.6	0.7	0.8	0.9
CRF0.625	108	109	78	105	126	134	111	150	80
CNF0.625	118	135	110	144	130	138	130	137	132
CRU1.0	76	113	75	106	130	123	139	146	145
CNU1.0	38	102	91	138	103	138	112	141	150

These 4 methods (Tables 7 and 8) will be used to compare and contrast
- the results from CRF0.625 with those from CRU1.0 to examine the effects of fixing v in methods incorporating the arcing technique; and
- the results from CNF0.625 with those from CNU1.0 to investigate the effects of fixing v in methods with no arcing.

The comparison is carried out as follows. In comparing method A to method B (A vs. B), if the accuracy from method A is greater than that from method B, then method A is said to be better than method B, and vice versa. If, however, both methods give equal accuracy rate, then the method providing a smaller predictor set is considered to be superior.

For methods integrating the arcing technique, fixing v does not create any considerable effects, improving the accuracy in only those predictor sets which are based on α values of 0.2, 0.4, 0.5, 0.6 and 0.9. On the other hand, for methods with no arcing, fixing v clearly deteriorates the classification accuracy for the majority of predictor sets, with only those based on α values of 0.1 and 0.8 showing improved accuracy, while the 6 other predictor sets produce worse accuracy (Table 9, rows 1 and 2).

Table 9. Comparison of different methods using predictor sets formed based on different values of α. '+' means the first method performs better than the second method in the comparison first method vs. second method; '−' indicates that the first method performs worse; '=' denotes equal performance from both methods

α Comparison	0.1	0.2	0.3	0.4	0.5	0.6	0.7	0.8	0.9
CRF0.625 vs. CRU1.0	−	+	−	+	+	+	−	−	+
CNF0.625 vs. CNU1.0	+	−	−	−	−	=	−	+	−
CRF0.625 vs. CNF0.625	+	+	+	+	+	+	+	−	+
CRU1.0 vs. CNU1.0	+	−	+	−	−	−	+	+	−
CRF0.625 vs. CNU1.0	+	−	−	+	+	+	+	+	+

In short, from these comparisons it may be construed that fixing the mixing parameter v does not contribute to the improvement in accuracy for methods with arcing, and in fact produces adverse effects for methods without arcing.

4.2 The Effects of Incorporating Arcing in CPD and FMCPD

As in case of the study on the effects of fixing v, two sets of comparison are necessary in order to investigate the effects of incorporating the arcing technique on classification accuracy. The first is a comparison between CRFv and CNFv methods, while the second is a comparison between CRUd and CNUd methods. Again, the value of v which gave the best overall results in accuracy for the FMCPD-Arcing methods, 0.625; and the value of d which produced the best average of accuracies for the CPD-Arcing methods, 1.0 are used in the comparisons. The same 4 methods (Tables 7 and 8) used in the previous comparisons will be employed, in a different combination, to compare and contrast

- the results from CRF0.625 with those from CNF0.625 to study the effects of incorporating the classifier arcing technique into FMCPD methods; and
- the results from CRU1.0 with those from CNU1.0 to analyze the effects of incorporating the classifier arcing technique into original CPD methods.

The first set of comparison indicates that FMCPD benefits greatly from the integration of arcing. CRF0.625 outperforms CNF0.625 in all but one (α=0.8) of the predictor sets of varying α values. The same, however, cannot be said for the second set of comparison. Combining arcing with original CPD yields mixed results, with CRU1.0

giving a better accuracy in 4 predictor sets (α=0.1, 0.3, 0.7 and 0.8), and performing worse than CNU1.0 in the remaining 5 predictor sets (Table 9, rows 3 and 4).

Incorporating arcing produces significant improvement in the classification accuracy only when v is fixed. In other words, either one of the modifications **by itself** (i.e. incorporating arcing or fixing v) is **not** useful in improving classification performance. To prove this, another comparison is made between CRF0.625 and CNU1.0 to show the benefits of simultaneously combining arcing and fixing v (Table 9, bottom-most row). The former is a combination of arcing and FMCPD at v=0.625, while the latter, a pure CPD method at d=1.0. Recall that the values of v and d for each of the method respectively are values that have been empirically shown to yield the best overall performance in classification accuracy across different values of α from 0.1 to 0.9.

CRF0.625 performs better than CNU1.0 in 7 of the predictor sets, either by producing better accuracy or similar accuracy but with a smaller predictor set in each case. In fact, CRF0.625 is able to achieve a test set classification accuracy of the GCM dataset that is higher than previously published [7–9, 11], 87.0%, with corresponding predictor set size of 126 genes.

4.3 The Effects of Antiredundancy Factor (1–α)

With CRF0.625, although the best accuracy of 87.0% is obtained at α=0.5, the predictor set size required is an unwieldy 126 genes (Figure 1). At α=0.4, an accuracy of 85.2%, also higher than any previously reported, is achieved using only a considerably smaller 105-gene predictor set. In a previous study [7], we found an optimal point in the trade-off between accuracy (83.3%) and predictor set size (85 genes) at α=0.4. Although the results are not so clear-cut in case of the CRF0.625 method, from the plot in Figure 1 we can still say that the trend towards high accuracy and correspondingly small predictor set points to the same α value of 0.4.

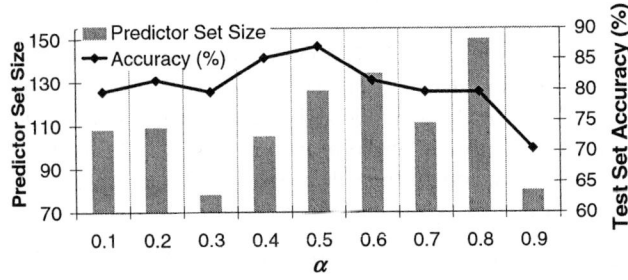

Fig. 1. Plot of best accuracy and the corresponding predictor set size vs. α for CRF0.625

For a microarray dataset whose molecular classification was proclaimed to reach the maximum accuracy barrier of 77.8% using all of its 16000 genes in the predictor set [8, 9], an improvement of more than 7% in accuracy, employing only 105 of those 16063 genes, is significant indeed. This has been done with the aid of an appropriate feature selection technique, and an astute modification of adaptive sampling and classifier aggregation methods.

5 Conclusion

A precise combination of a modified CPD and arcing methods implemented together with the maximal-antiredundancy-based feature selection technique has improved the classification accuracy of a highly complex and multiclass microarray dataset. The increase in classification accuracy has been achieved along with a reasonably small predictor set. The underlying factors (fixing v; the simultaneous incorporation of arcing) behind the improved performance of the hybrid method of FMCPD-Arcing have also been elucidated. Although the results presented are derived from one dataset, this dataset is so far considered to be the most 'difficult' microarray dataset (largest in term of the number of classes and among the largest in term of total number of features). Therefore the performance of our proposed methods is most likely reproducible in other multiclass datasets of less complexity.

References

1. Dudoit, S., Fridlyand, J., Speed, T.: Comparison of discrimination methods for the classification of tumors using gene expression data. JASA 97 (2002) 77–87
2. Dettling, M., Bühlmann, P.: Boosting for tumor classification with gene expression data. Bioinformatics 19 (2003) 1061–1069
3. Dettling, M.: BagBoosting for Tumor Classification with Gene Expression Data. Bioinformatics Advance Access. Oct 5, 2004. Bioinformatics, doi:10.1093/bioinformatics/bth447
4. Tan, A.C., Gilbert, D.: Ensemble machine learning on gene expression data for cancer classification. Applied Bioinformatics 2 (3 Suppl) (2003) S75-83
5. Breiman, L.: Using convex pseudo-data to increase prediction accuracy. Technical Report 513, Statistics Department, U.C. Berkeley, March 1998
6. Breiman, L.: Arcing classifiers. Annals of Statistics 26 (1998) 801–824
7. Ooi, C.H., Chetty, M., Gondal, I.: The role of feature redundancy in tumor classification. International Conference on Bioinformatics and its Applications (ICBA'04) (2004). In press
8. Ramaswamy, S., Tamayo, P., Rifkin, R., Mukherjee, S., Yeang, C.H., Angelo, M., Ladd, C., Reich, M., Latulippe, E., Mesirov, J.P., Poggio, T., Gerald, W., Loda, M., Lander, E.S., Golub, T.R.: Multi-class cancer diagnosis using tumor gene expression signatures. Proc. Natl. Acad. Sci. 98 (2001) 15149–15154
9. Rifkin, R., Mukherjee, S., Tamayo, P., Ramaswamy, S., Yeang, C.H., Angelo, M., Reich, M., Poggio, T., Lander, E.S., Golub, T.R., Mesirov, J.P.: An Analytical Method for Multiclass Molecular Cancer Classification. SIAM Review 45 No. 4 (2003) 706–723
10. Platt, J.C., Cristianini, N., Shawe-Taylor, J.: Large Margin DAGs for Multiclass Classification. Advances in Neural Information Processing Systems 12 (2000) 547–553
11. Linder, R., Dew, D., Sudhoff, H., Theegarten D., Remberger, K., Poppl, S.J., Wagner, M.: The "Subsequent Artificial Neural Network" (SANN) Approach Might Bring More Classificatory Power To ANN-based DNA Microarray Analyses. Bioinformatics Advance Access. July 29, 2004. Bioinformatics, doi:10.1093/bioinformatics/bth441

Kernels over Relational Algebra Structures

Adam Woźnica, Alexandros Kalousis, and Melanie Hilario

University of Geneva, Computer Science Department,
Rue General Dufour 24, 1211 Geneva 4, Switzerland
{woznica, kalousis, hilario}@cui.unige.ch

Abstract. In this paper we present a novel and general framework based on concepts of relational algebra for kernel-based learning over relational schema. We exploit the notion of foreign keys to define a new attribute that we call instance-set and we use this type of attribute to define a tree like structured representation of the learning instances. We define kernel functions over relational schemata which are instances of $\Re$-Convolution kernels and use them as a basis for a relational instance-based learning algorithm. These kernels can be considered as being defined over typed and unordered trees where elementary kernels are used to compute the graded similarity between nodes. We investigate their formal properties and evaluate the performance of the relational instance-based algorithm on a number of relational datasets.

1 Introduction

Learning from structured data has recently attracted a great deal of attention within the machine learning community ([1]). The reason for this is that it is in general hard to represent most of real world data as a flat table. Recently it has been also realized that one strength of the kernel-based learning paradigm is its ability to support input spaces whose representation is more general than attribute-value ([2, 3, 4]). The latter is mainly due to the fact that the proper definition of a kernel function enables the structured data to be embedded in some linear feature space without the explicit computation of the feature map. The main advantage of this approach is that any propositional algorithm which is based on inner products can be applied on the structured data.

In this paper we bring kernel methods and learning from structured data together. First we propose a novel database oriented approach and define our algorithms and operations over relational schema where learning examples come in the form of interconnected relational tables. There exists a single main relation each tuple of which gives rise to a relational instance that spans through the relations of the relational schema. Second we define a family of kernel functions over relational schemata which are generated in a "syntax-driven" manner in the sense that the input description specifies the kernel's operation. We show that the resulting kernels can be considered as kernels defined on typed, unordered trees and analyze their formal properties. In this paper we concentrate on the instance-based learning paradigm and we exploit these kernels to define a relational distance, however since the kernels are valid in terms of their mathematical properties any kernel-based algorithm could be used. Finally we report the results of an instance-based learner on a number of standard relational benchmark datasets.

2   Description of the Relational Instance

Consider a general relational schema that consists of a set of relations $\mathcal{R} = \{R\}$. Each tuple, $R_{i\_}$, of a relation R represents a relationship between a set of values $\{R_{ij}\}$ of the set of attributes $\{R_{\_j}\}$ related via R. The *domain*, $D(R_{\_j})$, of attribute $R_{\_j}$ is the set of values that the attribute assumes in relation R. An attribute $R_{\_j}$ is called a *potential key* of relation R if it assumes a unique value for each instance of the relation. An attribute $X_{\_i}$ of relation X is a *foreign key* if it references a potential key $R_{\_j}$ of relation R and takes values in the domain $D(R_{\_j})$ in which case we will also call the $R_{\_j}$ a *referenced key*. The association between $R_{\_j}$ and $X_{\_i}$ models one-to-many relations, i.e. one element of R can be associated with a set of elements of X. A *link* is a quadruple of the form $l(R, R_{\_k}, X, X_{\_l})$ where either $X_{\_l}$ is a foreign key of X referencing a potential key $R_{\_k}$ of R or vice versa. We will call the set of attributes of a relation R that are not keys (i.e. referenced keys, foreign keys or attributes defined as keys but not referenced) *standard attributes* and denote it with $\{S_{\_j}\}$. The notion of links is critical for our relational learner since it will provide the basis for the new type of attributes, i.e. the instance-set type that lies in the core of our relational representation.

For a given referenced key $R_{\_k}$ of relation R we denote by $L(R, R_{\_k})$ the set of links $l(R, R_{\_k}, X, X_{\_f_k})$ in which $R_{\_k}$ is referenced as a foreign key by $X_{\_f_k}$ of X. We will call the multiset of X relations, denoted as $L(R, R_{\_k})\{1\}$, the *directly dependent relation* of R for $R_{\_k}$. By $L(R, \_) = \cup_k L(R, R_{\_k})$ we note the list of all links in which one of the potential keys of R is referenced as a foreign key by an attribute of another relation. Similarly for a given foreign key $R_{\_f_k}$ of R, $L^{-1}(R, R_{\_f_k})$ will return the link $l(R, R_{\_f_k}, X, X_{\_k})$ where $X_{\_k}$ is the potential key of X referenced by the foreign key $R_{\_f_k}$. We will call relation X the *directly referenced relation* of R for $R_{\_f_k}$ and denoted it as $L^{-1}(R, R_{\_f_k})\{1\}$. If R has more than one foreign keys then by $L^{-1}(R, \_) = \cup_{f_k} L^{-1}(R, R_{\_f_k})$ we denote the set of all links of R defined by the foreign keys of R, and by $L^{-1}(R, \_)\{1\}$ the corresponding multiset of relations to which these foreign keys refer.

To define a classification problem one of the relations in $\mathcal{R}$ should be defined as the *main relation*, M. Then one of the attributes of this relation should be defined as the class attribute, $M_{\_c}$, i.e. the attribute that defines the classification problem. Each instance, $M_{i\_}$, of the M relation will give rise to one *relational instance*, $M_{i\_}^+$, i.e. an instance that spans the different relations in $\mathcal{R}$. To get the complete description of $M_{i\_}^+$ one will have to traverse possibly the whole relational schema according to the associations defined in the schema. More precisely given instance $M_{i\_}$ we create a relational instance $M_{i\_}^+$ that will have the same set of standard attributes $\{S_{\_j}\}$ and the same values for these attributes as $M_{i\_}$ has. Furthermore each link $l(M, M_{\_k}, R, R_{\_f_k}) \in L(M, \_) \cup L^{-1}(M, \_)$ adds in $M_{i\_}^+$ one attribute of type *instance-set*. The value of an attribute of type instance-set is defined based on the link l and it will be the set of instances (actually also relational instances) with which $M_{i\_}$ is associated in relation R when we follow the link l (these are retrieved directly by a simple SQL query). By recursive application of this procedure at each level we obtain the complete description of the relational instance $M_{i\_}^+$.

We should note here that the relational instance $M_{i\_}^+$ can be seen as a tree like structure whose root contains $M_{i\_}$. Each node at the second level of the tree is an instance from some relations $R \in \mathcal{R}$ related via a link $l(M, M_{\_k}, R, R_{\_f_k})$ with instance $M_{i\_}$. In the same way nodes at the d level of the tree are also instances from a given relation. Each of these instances is related with one of the instances found in nodes of the $d-1$ level. In other words M_i^+ is a tree where each node is one tuple from one of the relations that are part of the description of the relational instance and the connections between the nodes are determined by the foreign key associations defined within the relational schema. This means that the resulting tree is *typed* (i.e. each node is of a given type determined by one of the relations within the relational schema) and *unordered* (i.e. the order of instances appearing as children of a given node is not important). To limit the size of the resulting tree and to make the computation of our algorithms less expensive we sometimes prune the tree to a specific depth d.

Traversing the relational schema in order to retrieve the complete description of a given relational instance can easily produce self replicating loops. In order to avoid that kind of situation we will have to keep track of all the instances of the different relations that appear in a given path of the recursion; the moment an instance appears a second time in the given recursion path the recursion terminates.

Having an adequate way to handle attributes of type instance-set is the heart of the problem that should be tackled by any relational learning algorithm that could exploit the relational structure that we have sketched thus far.

3 Kernels

A kernel is a symmetric function $k : X \times X \to \mathbb{R}$, where X is any set, such that for all $x, y \in X$, $k(x, y) = <\phi(x), \phi(y)>$ where ϕ is a mapping from X to a feature space Φ embedded with an inner product, actually a *pre-Hilbert* space.

We should note here that the definition of kernels does not require that the input space X be a vector space –it can be any kind of set which we can embed in the feature space Φ via the kernel. This property allows us to define kernels on any kind of structures that will embed these structures in a linear space. The attractiveness of kernels lies in the fact that one does not need to explicitly compute the mappings $\phi(x)$ in order to compute the inner products in the feature space.

Examples of kernels defined on vector spaces are the polynomial kernel $k_{P_{p,a}}(x,y) = \frac{(<x,y>+a)^p}{\sqrt{(<x,x>+a)^p}\sqrt{(<y,y>+a)^p}}$ where $a \in \mathbb{R}, p \in \mathbb{N}^+$ and the Gaussian RBF kernel $k_{G_\gamma}(x,y) = e^{-\gamma \|x-y\|^2}$ where $\gamma \in \mathbb{R}$. This two kernels are the ones we are going to use in our experiments.

3.1 Kernels on Relational Instances

In order to define a kernel on the relational instances we will distinguish two parts, R_{is}, R_{iset}, in each relational instance $R_{i\_}$ found in a relation R. R_{is} denote the vector of standard attributes $\{S_{\_j}\}$ of R, let $D_s = |\{S_{\_j}\}|$; R_{iset} denotes the vector of attributes that are of type instance-set and for a relation R are given by $L(R, \_) \cup L^{-1}(R, \_)$, let $D_{set} = |L(R, \_) \cup L^{-1}(R, \_)|$.

Let $R_{i\_} = (R_{is}, R_{iset}) \in X = X_{\{S_{\_j}\}} \times X_{set}$ where $X_{set} = X_{set_1} \times X_{set_2} \times ... \times X_{set_{D_{set}}}$ and $R_{is} \in X_{\{S_{\_j}\}}, R_{iset} \in X_{set}$. Given this formalism we defined two relational kernels: *direct sum kernel* ($k_\Sigma(.,.)$) and the kernel which is derived by direct application of the $\Re$-*Convolution* kernel ([2]) on the set X ($k_\Re(.,.)$). Since these kernels are defined over multi-relational instances they are computed following the same recursion path as the retrieval of a multi-relational instance.

The *direct sum kernel* is obtained by exploiting the fact that the direct sum of kernels is itself a kernel, [5], which would give the following kernel on the set X (if $|\{S_{\cdot j}\}| \neq 0$) $k_\Sigma(R_{i\_}, R_{j\_}) = k_s(R_{is}, R_{js}) + \sum_{l=1}^{D_{set}} K_{set}(R_{iset_l}, R_{jset_l})$ where $k_s(.,.)$ can be any type of elementary kernel defined on the set $\{S_{\cdot j}\}$ of the standard attributes of R and $K_{set}(.,.)$ is a kernel between sets which will be defined in Section 3.2. If $|\{S_{\cdot j}\}| = 0$ then the kernel defined over standard attributes vanishes and we obtain $k_\Sigma(R_{i\_}, R_{j\_}) = \sum_{l=1}^{D_{set}} K_{set}(R_{iset_l}, R_{jset_l})$. It is obvious that the value of $K_\Sigma(.,.)$ is affected by the number of attributes that are of type instance-set since it contains a sum of kernels defined on these attributes. In order to factor out that effect among different relations we use a normalized version of k_Σ (if $|\{S_{\cdot j}\}| \neq 0$) defined as:

$$K_\Sigma(R_{i\_}, R_{j\_}) = \frac{k_\Sigma(R_{i\_}, R_{j\_})}{1 + D_{set}} \quad (1)$$

If $|\{S_{\cdot j}\}| = 0$ we have $K_\Sigma(R_{i\_}, R_{j\_}) = \frac{k_\Sigma(R_{i\_}, R_{j\_})}{D_{set}}$.

An alternative kernel is derived by the direct application of the $\Re$-*Convolution* kernel as described in [2]. The main idea in the $\Re$-Convolution kernel is that composite objects consist of simpler parts that are connected via a relation $\Re$. Kernels on the composite objects can be computed by combining kernels defined on their constituent parts. Let $x \in X$ be a composite object and $\boldsymbol{x} = x_1, ..., x_D \in X_1 \times ... \times X_D$ its constituent parts. Then we can represent the relation $\boldsymbol{x}$ *are the parts of* x by the relation $\Re$ on the set $X_1 \times X_2 \times ... \times X_D \times X$ where $\Re(\boldsymbol{x}, x)$ is true iff $\boldsymbol{x}$ are the parts of x. Let $\Re^{-1}(x) = \{\boldsymbol{x} : \Re(\boldsymbol{x}, x)\}$, a composite object can have more than one decomposing possibilities. Then the $\Re$-Convolution kernel is defined as:

$$k_\Re(x, y) = \sum_{\boldsymbol{x} \in \Re^{-1}(x), \boldsymbol{y} \in \Re^{-1}(y)} \prod_{d=1}^{D} K_d(x_d, y_d) \quad (2)$$

Since we defined only one way to decompose a relational instance $R_{i\_}$ the sum in the equation 2 vanishes and we obtain the product of kernels defined over attributes of type instance-set and the kernels defined on standard attributes (only if standard attributes are present). In case $|\{S_{\cdot j}\}| \neq 0$ the resulting $\Re$-Convolution kernel is defined as $k_\Re(R_{i\_}, R_{j\_}) = k_s(R_{is}, R_{js}) \prod_{l=1}^{D_{set}} K_{set}(R_{iset_l}, R_{jset_l})$ otherwise we obtain: $k_\Re(R_{i\_}, R_{j\_}) = \prod_{l=1}^{D_{set}} K_{set}(R_{iset_l}, R_{jset_l})$. Again it is obvious that the value of $k_\Re(.,.)$ is affected by the number of attributes that are of type instance-set so we opted for the following normalization version of the kernel:

$$K_\Re(R_{i\_}, R_{j\_}) = \frac{k_\Re(R_{i\_}, R_{j\_})}{\sqrt{k_\Re(R_{i\_}, R_{i\_}) k_\Re(R_{j\_}, R_{j\_})}} \quad (3)$$

These two kernels, $K_\Re(.,.), K_\Sigma(.,.)$, are the ones with which we are going to experiment and on which we are going to base our distance computations. Having a kernel it is straightforward to compute the distance in the feature space Φ in which the kernel computes the inner product as $d(\phi(x), \phi(y)) = \sqrt{K(x,x) - 2K(x,y) + K(y,y)}$.

We already mentioned in the Section 2 that a relational instance can be considered as a tree-like structure where each node is one tuple from one of the relations and connections between nodes are determined by the foreign key associations. This makes the relational kernel a kernel over trees. The input trees are *typed* which results in the definition of a graded similarity. The similarity of nodes of different type, i.e. nodes coming from different relations, is zero. The similarity of nodes of the same type is determined on the basis of the attributes found on the relation associated with the given type. At the same time input trees are *unordered* which means that the order of comparison of the descendants is not important, the only constraint being that only descendants of the same type can be compared. In other words the subtree comparison is meaningful only between subtrees that are rooted on nodes that come from the same relation.

3.2 Kernels on Sets

To complete the definition of the kernel on the relational structure we define here a kernel over sets of instances by exploiting the $\Re$-Convolution kernel from equation 2 (we put $\Re$ be $\boldsymbol{x} \in \Re^{-1}(x) \Leftrightarrow \boldsymbol{x} \in x$). Consequently we obtain the *cross product kernel* $K_{set}(X,Y) = \sum_{\boldsymbol{x} \in X, \boldsymbol{y} \in Y} K_{\Sigma|\Re}(\boldsymbol{x}, \boldsymbol{y})$ where $K_{\Sigma|\Re}(.,.)$ is either $K_{\Sigma}(.,.)$ or $K_{\Re}(.,.)$. The computation of the final kernel is based on recursive alternating applications of $K_{\Sigma}(.,.)$ or $K_{\Re}(.,.)$ and $K_{set}(.,.)$.

The procedure of computing the kernel on the variables of type instance-set is sensible to cardinality variations; sets with larger cardinality will dominate the solution. This leads us to the issue of normalization of the cross product kernel, so that we obtain:

$$K_{norm}(X,Y) = \frac{K_{set}(X,Y)}{f_{norm}(X) f_{norm}(Y)} \quad (4)$$

where $f_{norm}(x)$ is a normalization function which is nonnegative and takes non-zero values. Different choices of $f_{norm}(x)$ give rise to different normalization methods, [6]. By putting $f_{norm}(X) = |X|$ we obtain the *Averaging* normalization method ($k_{\Sigma_A}(.,.)$). Defining $f_{norm}(X) = \sqrt{k_{set}(X,X)}$ we get the *Normalization in the feature space* ($k_{\Sigma_{FS}}(.,.)$). The obtained functions are valid kernels since the explicit representation of the feature space can be constructed, Section 3.3.

3.3 Feature Space Induced by Relational Kernels

In order to get a new insight into the behavior of the relational kernels defined so far we will specify the feature space associated with them. We start with the definition of the feature space induced by the cross product kernel defined in Section 3.2. Lets assume $\Phi_{\Sigma|\Re}$ (i.e. Φ_{Σ} or $\Phi_{\Re}$) is an embedding function into a feature space $F_{\Sigma|\Re}$ (F_{Σ} or $F_{\Re}$) for the kernel $K_{\Sigma|\Re}$ (K_{Σ} or $K_{\Re}$) on the right hand of the definition of the cross product kernel so that $K_{\Sigma|\Re}(\boldsymbol{x}, \boldsymbol{y}) = <\Phi_{\Sigma|\Re}(\boldsymbol{x}), \Phi_{\Sigma|\Re}(\boldsymbol{y})>$. It is easy to show that the feature space induced by this kernel is given by $\Phi_{set}(X) = \sum_{x \in X} \Phi_{\Sigma|\Re}(x) \in F_{\Sigma|\Re}$, [7]. Similarly the feature space induced by kernel from equation 4 where $f_{norm}(X) = |X|$ is given by $\Phi_{set}(X) = \frac{\sum_{x \in X} \Phi_{\Sigma|\Re}(x)}{|X|}$. It is clear that this normalization method amounts to computing the inner product, in the feature space induced by the elementary kernels, between the two centroids of the corresponding sets. In case $f_{norm}(X) = \sqrt{K_{set}(X,X)}$ the feature space is given by $\Phi_{set}(X) = \frac{\sum_{x \in X} \Phi_{\Sigma | \Re}(x)}{\|\sum_{x \in X} \Phi_{\Sigma | \Re}(x)\|}$. So this normalization method computes the cosine of the angle between the two normalized resultants of the vectors of the two sets.

Now we define the feature space associated with the direct sum (K_Σ) and the $\Re$-Convolution ($K_\Re$) kernels. Lets assume that Φ_Σ ($\Phi_\Re$) is an embedding function into a feature space F_Σ ($F_\Re$) for kernel K_Σ ($K_\Re$). Let also $\phi_s, \phi_{set_1}, \ldots, \phi_{set_{|D_{set}|}}$ be embedding functions into feature spaces $F_s, F_{set_1}, \ldots, F_{set_{|D_{set}|}}$ of the kernels $k_s, k_{set_1}, \ldots, k_{set_{|D_{set}|}}$ which constitute the K_Σ and $K_\Re$ kernels, respectively. It is easy to show that $F_\Sigma = F_s \oplus F_{set_1} \oplus \cdots \oplus F_{set_{|D_{set}|}}$ and $F_\Re = F_s \otimes F_{set_1} \otimes \cdots \otimes F_{set_{|D_{set}|}}$ where $\oplus$ denotes the direct sum and $\otimes$ denotes the tensor product of vector spaces. In other words the $F_\Re$ is constructed by computing all the possible products of all the dimensions of its constituent spaces, where each product becomes a new dimension of $F_\Re$. In contrast the F_Σ is constructed by a simple concatenation of the dimensions of its constituent spaces. It is obvious that $dim(F_\Sigma) = dim(F_s) + \sum_{i=1}^{|D_{set}|} dim(F_{set_i})$ and $dim(F_\Re) = dim(F_s) \prod_{i=1}^{|D_{set}|} dim(F_{set_i})$. In order to get an explicit feature space representation induced by the relational kernel one has to recursively combine the feature spaces induced by the kernel on sets and the direct sum or the $\Re$-Convolution kernels.

An important result is that instance based learning in the feature space induced by the $\Re$-Convolution kernel, $F_\Re$, should be more difficult than in this induced by the direct sum kernel, F_Σ. This is because the dimensionality of $F_\Re$ is much higher than F_Σ (this holds if the elementary kernels induce a feature space of finite dimensionality, otherwise they are both of infinite dimension). On the other hand the $\Re$-Convolution kernel is more expressive since it accounts for feature interactions.

3.4 Time Complexity

Here we analyze the time complexity of the relational kernel defined above. Let $TrI = \{TrI_1, TrI_2, \ldots, TrI_n\}$ be a set of tree representations of the relational instances in a given relational schema. Let also TrR be a tree representation (with the depth d) of the relational schema at the "relation" level where each node is a relation and the connections between the nodes are again determined by the foreign key associations. In case there are loops in the relational schema the depth is limited to an arbitrary value so that a valid tree is constructed. It is worth noting that depths of each tree in TrI are at most d. Having defined TrI and TrR let BF_I be the maximal out-degree of all nodes in all trees in the set TrI while BF_R be the maximal out-degree of all nodes in the TrR. It is easy to show the computation of the relational kernel between two tree representation of relational instances is proportional to $O((BF_I{}^2)^{d-1}) = O(BF_I{}^{2(d-1)})$ (here we assume that the root of a tree is at level 1). The overall time complexity is proportional to $O(BF_R{}^{d-1}BF_I{}^{2(d-1)})$. This complexity is dominated by BF_I since $BF_R << BF_I$. This is the pessimistic estimate of the time complexity and more accurate would be acquired if the average branching factors were used.

An interesting analysis arises from the comparison of the computational complexity of the cross product kernel with that of the inner product computed directly in the feature space. It is easy to show that the former complexity is proportional to $O(|A||B|(N+p))$ whereas the latter is proportional to $O(2\binom{p+N-1}{p})(|A|+|B|))$ where A and B are two finite sets, the elementary kernel is polynomial elementary kernel with the exponent p (without the bias towards lower order monomial) and input space is $\mathbb{R}^N$. This means that for sets with high cardinality and for low values of N and p it is better to explicitly map the instances to the feature space and compute the inner product there.

4 Experiments

We will compare the selected kernel-based distance measures on a number of relational problems: musk - version 1, diterpenes and mutagenesis. In the diterpene dataset [8] the goal is to identify the type of diterpenoid compound skeletons given their ^{13}C-NMR-Spectrum. The musk dataset was described in [9]; here the goal is to predict the strength of synthetic musk molecules. We worked with version 1 of the dataset. The Mutagenesis dataset was introduced in [10]. The application task is the prediction of mutagenicity of a set of 230 aromatic and heteroaromatic nitro-compounds. We worked with the "regression friendly" version of the dataset. We defined two different versions of the learning problem. In *version 1* the examined compounds (in the *main* relation) consist of atoms (in the *atom* relation) which constitute bonds (in the *bound* relation). The recursion depth was limited to four. In *version 2* the compounds consist of bonds while bonds consists of atoms and the recursion level was limited to three. In both versions the recursion depth was limited because of the time complexity of the algorithm. All the results are given in table 1.

For diterpenes and musk datasets the computation of relational kernel can be simply reduced to computing kernels on sets of vectors requiring thus no recursion. In these cases the $K_\Sigma(.,.)$ and $K_\Re(.,.)$ relational kernels are equivalent (up to a normalization term) so we report results only for the former. In the mutagenicity problem it will be possible to move beyond a single level comparison of the instances and have many levels of recursion. We report results for different set normalization schemes; the subscript $_A$ will denote averaging and the subscript $_{FS}$ feature space normalization. Here we give results for $p = \{2, 3\}, a = 1$ (normalized polynomial kernel $k_{P_{p,a}}(.,.)$) and for $\gamma = \{0.01, 0.001\}$ (Gaussian RBF $k_{G_\gamma}(.,.)$). In the experiments we want to explore the effect of different elementary kernels, the effect of different kernel set normalizations as well as the relative performance of the $K_\Sigma(.,.)$ and $K_\Re(.,.)$ kernels. We will experiment with a single nearest neighbor.

We estimate accuracy using stratified ten-fold cross-validation and control for the statistical significance of observed differences using McNemar's test (sig. level=0.05). We also establish a ranking schema of the different kernel-based distance measures, based on their relative performance as determined by the results of the significance tests, as follows: in a given dataset if kernel-based distance measure a is significantly better than b then a is credited with one point and b with zero points; if there is no significant difference then both are credited with half point.

Table 1. Accuracy and rank results on the benchmark datasets

Elementary kernel	DITERPENES	MUSK (VERSION 1)	MUTAGENESIS (VERSION 1)	MUTAGENESIS (VERSION 2)
Relational kernel	K_{Σ_A}		$K_{\Sigma_{FS}}$	
$k_{P_{p=2,a=1}}$	91.22 (5.5)	85.87 (3.5)	84.04 (3.5)	81.91 (3.5)
$k_{P_{p=3,a=1}}$	91.75 (6)	88.04 (3.5)	83.51 (3.5)	84.57 (3.5)
$k_{G_{\gamma=0.01}}$	86.69 (2.5)	83.69 (3.5)	82.45 (3.5)	83.51 (3.5)
$k_{G_{\gamma=0.001}}$	83.30 (0.5)	81.52 (3.5)	81.91 (3.5)	84.04 (3.5)
Relational kernel	$K_{\Sigma_{FS}}$		$K_{\Re_{FS}}$	
$k_{P_{p=2,a=1}}$	90.82 (4.5)	85.87 (3.5)	82.98 (3.5)	85.11 (3.5)
$k_{P_{p=3,a=1}}$	91.68 (6)	88.04 (3.5)	82.45 (3.5)	84.04 (3.5)
$k_{G_{\gamma=0.01}}$	86.76 (2.5)	83.69 (3.5)	81.91 (3.5)	84.57 (3.5)
$k_{G_{\gamma=0.001}}$	83.03 (0.5)	81.52 (3.5)	79.79 (3.5)	84.57 (3.5)
Default Accuracy	29.81	51.09	66.49	

5 Results

To compare the different elementary kernels we fix a dataset and average the ranks of k_P and k_G, ignoring their parameter settings. There is an advantage of the polynomial over the Gaussian RBF elementary kernel for diterpenes dataset - the average rank of polynomial kernels is 5.5 (for Gaussian RBF 1.5). For both formulations of mutagenesis and musk 1 the average rank of polynomial kernels is 3.5 (3.5).

The different normalization methods for kernels over sets also do not appear to have an influence on the final results. For diterpenes *Averaging* had an average rank of 3.625 over the different elementary kernels and *Feature space normalization* an average rank of 3.375. For musk 1 the corresponding figures were 3.5 and 3.5. One explanation for this might be that the two denominators in the explicit feature space representations of the normalized relational kernels from Section 3.3 are correlated, which makes sense since sets of higher cardinality will have probably a higher $\|\sum_{x \in X} \Phi_{\Sigma|\Re}(x)\|$, at least for the datasets we examined.

The final dimension of comparison is the relative performance of $K_{\Sigma}(.,.)$ and $K_{\Re}(.,.)$. Here again it did not have a big influence on the final results: for both formulations of the mutagenesis problem $K_{\Sigma}(.,.)$ and $K_{\Re}(.,.)$ had an average rank of 3.5. This is rather surprising since as we have seen before instance-based learning in the space induced by the R-Convolution kernel should be harder than in the space induced by the direct sum kernel. However the R-Convolution kernel is more expressive than the direct sum kernel because it accounts for feature interactions. The trade-off between hardness of learning in space of higher dimensionality and the higher expressiveness might explain similar performance of the $\Re$-Convolution and the direct sum kernel.

To situate the performance of our relational learner to other relational learning systems we give the best results reported in the literature on the same benchmark datasets. All the results denote the accuracy and all have been estimated with ten fold cross-validation. The best result for the musk 1 dataset is 92.40 % (IAPR algorihtm) and it was reported in [9]. In comparison our best kernel gave 88.04 % of accuracy. For the diterpenes dataset the best accuracy was achieved using the *DeS* algorithm which comes from [3]. The authors got 97.10 % whereas our best kernel gave 91.75 % of accuracy. For the mutagenesis dataset we obtained 85.11 % of accuracy while the best result from the literature was 90.4 % and was taken from [11].From the results reported above we can see that our kernel-based learner compares favorably with the results achieved by special-purpose algorithms applied to structured data.

6 Related Work

One of the first systems exploiting the concepts of relational algebra and foreign keys was *MIDOS*, [12]. However [12] is focused on the KDD subgroup discovery task.

The most relevant kernel in our context is the $\Re$-Convolution kernel which was mentioned in Section 3. To our best knowledge our kernel is the first time the original $\Re$-Convolution kernel, [2], was applied to the type of relational structures we considered here.

Gärtner et al. in [3] proposed a framework that allows the application of kernel methods to different types of structured data e.g. sets, trees, graphs, lists. The representation formalism used was that of typed λ-calculus. The representation framework allows for the modeling of arbitrary complex objects which however is not at all a trivial task. Under this framework the authors explicitly defined kernels on sets and multisets. In [3] elementary (atomic) kernels are defined for each attribute separately, while our kernel assumes elementary kernels defined on the level of relations thus treating relations as indivisible objects. Besides of this in [3] a kernel over tuples of objects is always defined as a direct sum of its constituent parts whereas in our framework one is able to use either the direct sum kernel or the $\Re$-Convolution, which have different representational powers.

The kernels described in [4] and in [13] can be considered as specialized $\Re$- Convolution kernels where instances are considered to be labeled ordered directed trees. The idea of these kernels is based on the notion of a number of common subtrees in a tree i.e. the kernel function is the inner product in the space which describes the number of occurrences of all possible subtrees. The main difference between [4] and [13] is that the former is applicable only to trees where no node shares its label with any of its siblings. [13] overcomes this limitation by defining the substructures of a tree as a tree such that there is a descendants order preserving mapping from vertices in the substructure to vertices in the tree. There are two main differences between our kernel and kernels defined in [13]. First the trees considered in [13] are labeled trees, i.e. each node is characterized by a discrete label so two nodes are either the same or different, there is no graded similarity. In our case however nodes are not labeled but typed which results in the definition of a graded similarity. Second the trees in [13] are ordered whereas in our case there is no order restriction, the only restriction imposed is that comparison is performed only between subtrees rooted at nodes of the same type, i.e. same relation, and only descendants of the same type can be compared.

7 Discussion and Future Work

In this paper we proposed a kernel based relational instance based learner which, contrary to most of the previous relational approaches that rely on different forms of typed logic, builds on notions from relational algebra. Thus we cover what we see as an important gap in the current work on multirelational learning bringing it closer to the database community. We concentrated here on the instance-based learning paradigm however our kernel can be plugged to any kernel-based classification algorithm.

Our kernel functions can be considered as instances of the $\Re$-Convolution kernel in the sense that we define a kernel on a composite object by means of kernel on the parts of objects. On the other hand our kernels could be also seen as being defined over typed and unordered trees. Since in other areas of computational biology many problems can be described using similar structures we believe that our kernel could also useful there.

Central to the whole approach was the definition of appropriate kernels on the new type of attributes i.e. the instance-set type. We believe that there is still a lot to be gained in classification performance if more refined kernels are used for this type of attributes. We have followed a rather simple approach where the kernel between two sets was simply the sum of all the pairwise kernels defined over all the pairs of elements of the two sets. A more elaborate approach would take into account only the kernels computed over specific pairs of elements based on some mapping relation of one set to the other defined on the feature space.

References

1. Dzeroski, S., Lavrac, N.: Relational Data Mining. Springer-Verlag New York, Inc. (2001)
2. Haussler, D.: Convolution kernels on discrete structures. Technical report, UC Santa Cruz (1999)
3. Gärtner, T., Lloyd, J., Flach, P.: Kernels and distances for structured data. Machine Learning (2004)
4. Collins, M., Duffy, N.: Convolution kernels for natural language. In Dietterich, T.G., Becker, S., Ghahramani, Z., eds.: Advances in Neural Information Processing Systems 14, Cambridge, MA, MIT Press (2002)
5. Schölkopf, B., Smola, A.J.: Learning with Kernels: Support Vector Machines, Regularization, Optimization, and Beyond. MIT Press, Cambridge, MA (2002)
6. Gärtner, T., Flach, P., Kowalczyk, A., Smola, A.: Multi-instance kernels. In Sammut, C., ed.: ICML02, Morgan Kaufmann (2002)
7. Shawe-Taylor, J., Cristianini, N.: Kernel Methods for Pattern Analysis. Cambridge University Press (2004)
8. Dzeroski, S., Schulze-Kremer, S., Heidtke, K.R., Siems, K., Wettschereck, D.: Applying ILP to diterpene structure elucidation from ^{13}c NMR spectra. In: Inductive Logic Programming Workshop. (1996) 41–54
9. Dietterich, T.G., Lathrop, R.H., Lozano-Perez, T.: Solving the multiple instance problem with axis-parallel rectangles. Artificial Intelligence **89** (1997) 31–71
10. Srinivasan, A., Muggleton, S., King, R., Sternberg, M.: Mutagenesis: ILP experiments in a non-determinate biological domain. In Wrobel, S., ed.: Proceedings of the 4th International Workshop on Inductive Logic Programming. Volume 237. (1994) 217–232

11. Mahé, P., Ueda, N., Akutsu, T., Perret, J.L., Vert, J.P.: Extensions of marginalized graph kernels. In: ICML 2004. (2004)
12. Wrobel, S.: An algorithm for multi-relational discovery of subgroups. In: PKDD '97: Proceedings of the First European Symposium on Principles of Data Mining and Knowledge Discovery, Springer-Verlag (1997) 78–87
13. Kashima, H., Koyanagi, T.: Kernels for semi-structured data. In: ICML 2002. (2002)

Adaptive Nonlinear Auto-associative Modeling Through Manifold Learning

Junping Zhang[1,2] and Stan Z. Li[3]

[1] Intelligent Information Processing Laboratory,
Department of Computer Science and Engineering,
Fudan University, Shanghai 200433, China
jpzhang@fudan.edu.cn
[2] The Key Laboratory of Complex Systems and Intelligence Science,
Chinese Academy of Sciences
[3] National Laboratory of Pattern Recognition & Center for Biometrics and
Security Research Institute of Automation, CAS, Beijing, 100080, China
szli@nlpr.ia.ac.cn

Abstract. We propose adaptive nonlinear auto-associative modeling (ANAM) based on Locally Linear Embedding algorithm (LLE) for learning intrinsic principal features of each concept separately and recognition thereby. Unlike traditional supervised manifold learning algorithm, the proposed ANAM algorithm has several advantages: 1) it implicitly embodies discriminant information because the suboptimal parameters of ANAM are determined based on error rate of the validation set. 2) it avoids the curse of dimensionality without loss accuracy because recognition is completed in the original space. Experiments on character and digit databases show that the advantages of the proposed ANAM algorithm.

1 Introduction

Much manifold learning literature has been published for discovering intrinsic information embedded in the high-dimensional space[1]. Two major algorithms (LLE and ISOMAP)[2, 3] are devoted to discover some intrinsic regularity underlying in the high-dimensional data. Also some algorithms based on manifold learning are proposed for supervised learning. However, most supervised manifold learning algorithms assume that data can be projected into the same subspace and recognition without considering the properties of concepts [4]. The disadvantage of the supervised manifold learning approach is that the separability of data would be impaired because data from different classes would be overlapped in the low-dimensional subspace [5]. Based on our observation, we assume that data are projected different subspaces are more suitable than being projected a common subspace if data contain remarkably distinct concepts, for example, character and digit.

Assuming that data manifold of each concept is generated by some intrinsic principal features, we propose adaptive nonlinear auto associative modeling (ANAM) for learning intrinsic features and recognition (Section 2). First, the low-dimensional subspace of each class are attained with LLE algorithm. Second, based on the error rates of validation set, the parameters of each ANAM are adaptively obtained by computing minimum error rate of the validation set. Consequently, a ANAM-classifier is developed without LLE algorithm. The proposed ANAM will not lead to the overlapped data in the low-dimensional subspace and loss corresponding accuracy. Therefore, it partially overcomes the curse of dimensionality. Experiments (Section 3) on several character and digit databases show the advantages of the proposed ANAM algorithm.

2 Adaptive Nonlinear Auto-associative Modeling

To establish the mapping and inverse mapping relationship of ANAM between the observed data and the corresponding low-dimensional one, locally linear embedding (LLE) algorithm [2] is first used to form the corresponding low-dimensional one Y ($Y \subset \mathbb{R}^d$) of the training set X ($X \subset \mathbb{R}^N, N \gg d$) in the paper. Then the data set (X, Y) is used for modelling the subsequently ANAM.

The main principle of LLE algorithm is to preserve local neighborhood relation of data in both the embedded Euclidean space and the intrinsic one. Each sample in the observation space is a linearly weighted average of samples under neighbor constrain. Thus, we obtain the corresponding low-dimensional one Y of the original data X in the embedding space. And the completed set (X, Y) is used for the subsequent model of ANAMs.

While the mapping idea of unknown sample in the LLE framework can't obtain the optimal mapping solution based on our experiments, in addition, it is used to avoid to calculate the parameters of inverse mapping matrices and mapping matrices of ANAM simultaneously.

To construct ANAMs, the forward mapping and inverse mapping matrices need to be estimated. In the proposed algorithm, we utilize mis-classified rate on the validation set to adjust model parameter to obtain the suboptimal model.

First, validation set $V' \in \mathbb{R}^N$ are mapped into the corresponding low-dimensional one $V'_d \in \mathbb{R}^d$ with LLE mapping idea for avoiding the simultaneous computation of the mapping and inverse mapping matrices. After V'_d is obtained, the reconstruction procedure is then formulated with inverse mapping matrices of ANAM. On the basis of weierstrass approximation theorem, the inverse mapping formula in the i-th ANAM would be achieved with nonlinear polynomial function as follows:

$$v'_x(i) = \sum_{j=1}^{n_i} \beta_j(i) k_{\text{rec}}(y_j(i), v'_y(i)) \quad y_j(i) \in Y(i), v'_y(i) \in \mathbb{R}^d, v'_x(i) \in \mathbb{R}^N \quad (1)$$

where $v'_x(i)$ is a reconstructed sample through the i-th ANAM, n_i is the number of samples used to construct the i-th ANAM, $\mathcal{B}(i) = \{\beta_j(i)\}$ is the $N \times n_i$ weighted inverse mapping matrix or reconstruction matrix of the i-th ANAM, and $v'_y(i)$ is the low-dimensional validation sample based on LLE algorithm. Without loss of generality, let the reconstruction kernel function be Gaussian kernel as:

$$k_{\text{rec}}(y_j(i), v'_y(i)) = \exp(- \parallel y_j(i) - v'_y(i) \parallel /2\sigma^2_{\text{rec}}(i)) \quad (2)$$

For computational simplicity, parameters $\sigma^2_{rec}(i)$ of all concepts are set to the same value in the proposed ANAM algorithm. Once the validation sample is auto-associated through different ANAMs, the similarity measure can be used for recognition as follows:

$$C(v_x) = \arg\max_i(\exp(-\|v_x - v'_x(i)\|)) \quad i = 1,\cdots,L \quad (3)$$

Where L denotes the number of concepts. The geometrical explanation on Formula (3) is that sample is re-projected to the original space with the ANAM of same concept is closer to the original sample than these reconstructed samples through the ANAMs of different concepts.

Considering the geometrical property of Gaussian kernel, it is not difficult to see that the suboptimal parameter $\sigma^2_{optrec}(i)$ can be adaptively obtained through searching some value which is related to the minimum recognition error rate of validation set.

After the parameters of inverse mapping matrices are obtained, the mapping function of validation set can be formulated as:

$$v^*_y(i) = \sum_{j=1}^{n_i} \alpha_i k_{map}(x_j(i), v_x(i)), x_i \in X; v_x(i) \in \mathbb{R}^N; v^*_y(i) \in \mathbb{R}^d \quad (4)$$

Where $\mathcal{A} = \alpha_i$ is a $d \times n_i$ weighted mapping matrix, $k_{map}(x_j(i), v_x(i))$ denotes the similarity metric of data $v_x(i)$ with sample $x_j(i)$ as follows:

$$k_{map}(x_j(i), v_x(i)) = \exp(-\|x_j(i) - v_x(i)\|/2\sigma^2_{map}(i)). \quad (5)$$

And the reconstruction matrix is the same as Eq.(1). The only difference is that $v'_x(i)$, $v'_y(i)$ is alternative with $v^*_x(i)$, $v^*_y(i)$ in Eq. (1) and Eq.(3), and the suboptimal parameters $\sigma^2_{optmap}(i)$ of mapping matrices is adaptively computed based on the error rate of validation set with fixed reconstruction parameters $\sigma^2_{optrec}(i)$.

It is worthy noting that given the completed data set (X, Y), the computation of the weighted mapping matrix $\mathcal{A}(i)$ and the weighted inverse mapping matrix $\mathcal{B}(i)$ are calculated as follows:

$$\mathcal{A}(i) = Y(i) \cdot (k_{map}(x_j(i), x_k(i)))^{-1}, j, k = 1,\ldots,n_i \quad i = 1,\ldots,L \quad (6)$$
$$\mathcal{B}(i) = X(i) \cdot (k_{rec}(y_j(i), y_k(i)))^{-1}, i, j = 1,\ldots,n_i \quad i = 1,\ldots,L \quad (7)$$

Until then, test sample is projected into low-dimensional space and reconstructed the corresponding set in the original space with ANAM, and recognition is completed based on Eq. (3). Different from ANN Bourlard proposed[6], the proposed ANAN generalizes the model into high-dimensional nonlinear data and avoids the problem of convergence neural network often suffers.

3 Experiments

Experiments are carried out on four databases to evaluate the recognition ability of the proposed ANAM approach. Two sets are UCI character database[7] and OCR (optical character recognition) database[8], and the other two sets are OPTDigits databases

Table 1. Experimental Databases and Data Partitions

	UCI	OCR	OPTDigits	PENDigits
The number of Samples	20,000	16,280	5,620	10,992
Original Dimensions	16	26	64	16
ClassNum	26	26	10	10
Training Set	300*26	250*26	300*10	600*10
Validation Set	50*26	50*26	823	1,494
Test Set	10,900	8,480	1,797	3,498

Table 2. The Average Error Rates and Standard Deviations of several algorithms for ANAM, K-nearest Neighbor (K-NN, K=3)

CLASSIFIER	UCI %	OCR %	OPTDIGITS %	PENDIGITS %
ANAM	6.99 ± 0.34 (10DIM)	10.79 ± 0.56 (10DIM)	1.28 (10 DIM)	4.26 (10 DIM)
K-NN	10.10	10.5	2.00 (K=1)	2.26 (K=1)
MLP	20.7	23.8	—	—

and PENDigits database from UCI repository[9, 10]. The details on the mentioned four databases and data partitions are illustrated in Table 1. It is noticeable that each dimension in three datasets except OPTDigits Database was linearly scaled to [0,1] in this experiments. The former two databases are randomly partitioned three disjointing sets, that is, training set, validation set and test set. And the final results are the average of 10 repetitions. Meanwhile, the training set and validation set of the latter two databases are randomly partitioned disjointing sets, and test set has been separated in the original databases.

In our experiments, training sets from different concepts are used for building the different low-dimensional subspaces with LLE algorithm separately, validation set is used for searching the suboptimal parameters of ANAMs based on the error rate, and test set is used for evaluating the generalization performance of the proposed ANAM algorithm.

Moreover, several additional parameters need to be predefined. Without loss of generality, the neighbor parameter K is set to 50 for all the four databases. The ranges of mapping parameter σ^2_{map} and reconstruction parameter σ^2_{rec} are both set to $[10^{-5}, 10^{10}]$, and the size of each step is $10^{0.5}$ so the optimal parameter can be adaptively searched.

The experimental results on these databases are reported as in Table 2. For comparing the recognition performance between the proposed NAMs and other known state-of-the-art algorithms, experimental results from [8] are cited.

By analyzing the above results, it is not difficult to see that the proposed ANAM algorithm is comparable with other algorithm. For UCI letter and OPTDigits databases, the error rates of the proposed algorithm is lowest when comparing with other algorithms. For instances, in UCI character database, the error rates of NAMs is about 69.20% of the K-NN, 33.76% of the MLP. Furthermore, our proposed NAMs for the four databases using fewer features (10 dimensions) to model intrinsical feature spaces.

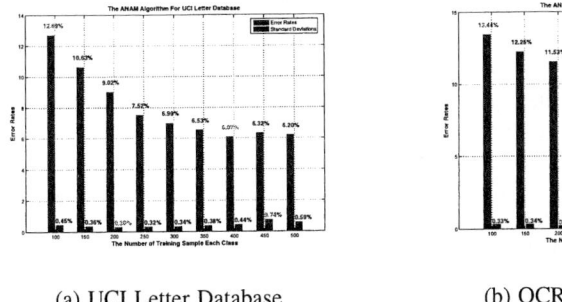

(a) UCI Letter Database (b) OCR Letter Database

Fig. 1. The Influence of Training Samples

Actually, we also investigate the influence of training sample with the fixed number of validation set for the error rates of test set (The experimental results also are the average of 10 runs). For example, the results in UCI and OCR database are illustrated as Figure 1. It can be seen that the error rate gradually decreases as the number of training sample increases. For example, when the number of training sample is equal to 400, the error rate and deviation are 6.07% ± 0.49% in the UCI Letter Database.

4 Discussions/Conclusions

In this paper, we propose ANAM for modeling different concepts and classifying samples which belong to remarkably distinct concepts. Unlike the other supervised manifold learning approaches, the advantages of the proposed ANAM is that it overcomes the curse of dimensionality without loss accuracy. And the discriminant information is implicity embodied because the parameters are determined based on the error rate of validation set. In the further study, we will consider how to combine cluster algorithm with ANAM for improving recognition rate.

References

1. J. Zhang, Stan. Z. Li, and J. Wang,"Manifold Learning and Applications in Recognition," in *Intelligent Multimedia Processing with Soft Computing*. Yap Peng Tan, Kim Hui Yap, Lipo Wang (Ed.), Springer-Verlag, Heidelberg, 2004.
2. S. T. Roweis, and K. S. Lawrance,"Nonlinear Dimensionality reduction by locally linear embedding," *Science*, 2000, 290, pp: 2323-2326.
3. J. B. Tenenbaum, D. Silva, V.& Langford, "A global geometric framework for nonlinear dimensionality reduction," *Science*, 2000, 290, pp: 2319-2323.
4. J. Zhang, H. Shen, and Z. H. Zhou,"Unified Locally Linear Embedding and Linear Discriminant Analysis Algorithm (ULLELDA) for Face Recognition," Advances in Biometric Personal Authentication. Stan Z. Li, Jianhuang Lai, Tieniu Tan, Guocan Feng, Yunhong Wang (Ed.), LNCS 3338, Springer-Verlag, 2004, pp: 209-307.

5. P. M. Baggenstoss,"Class-Specific Classifier: Avoiding the Curse of Dimensionality," *IEEE A& E Sysmtes Magazine*, Vol 19, No. 1, 2004, pp: 37-52
6. H. Bourlard, Y. Kamp, "Auto-Association by Multilayer Perceptrons and Singular Value Decomposition," *Biological Cybernetics*, 59, 1988, pp: 291–294.
7. P. W. Frey, D. J. Slate,"letter recognition using holland-style adaptive classifiers," *Machine Learning*, 6, 1991, pp: 161–182.
8. S. Kumar, J. Ghosh, M. Crawford,"A Bayesian Pairwise Classifier for Character Recognition,"*Cognitive and Neural Models for Word Recognition and Document Processing*, Nabeel Mursheed (Ed), World Scientific Press, 2000.
9. F. Alimoglu, E. Alpaydin,"Methods of Combining Multiple Classifiers Based on Different Representations for Pen-based Handwriting Recognition," *Proceedings of the Fifth Turkish Artificial Intelligence and Artificial Neural Networks Symposium* (TAINN 96), June 1996, Istanbul, Turkey. http://www.cmpe.boun.edu.tr/ alimoglu/tainn96.ps.gz
10. C. L. Blake and C. J. Merz. UCI repository of machine learning databases, 1998. http://www.ics.uci.edu/ mlearn/MLRepository.html.

Maximizing Tree Diversity by Building Complete-Random Decision Trees

Fei Tony Liu[1], Kai Ming Ting[1], and Wei Fan[2]

[1] School of Computing and Information Technology,
Monash University Churchill, Victoria 3842 Australia
{Tony.Liu, KaiMing.Ting}@infotech.monash.edu.au
[2] IBM T.J. Waston Research, Hawthorne, NY 10532
weifan@us.ibm.com

Abstract. One of the ways to lower generalization error of decision tree ensemble is to maximize tree diversity. Building complete-random trees forgoes strength obtained from a test selection criterion. However, it achieves higher tree diversity. We provide a taxonomy of different randomization methods and find that complete-random test selection produces diverse trees and other randomization methods such as bootstrap sampling may impair tree growth and limit tree diversity. The well accepted practice in constructing decision trees is to apply bootstrap sampling and voting. To challenge this practice, we explore eight variants of complete-random trees using three parameters: ensemble methods, tree height restriction and sample randomization. Surprisingly, the most accurate variant is very simple and performs comparably to *Bagging* and *Random Forests*. It achieves good results by maximizing tree diversity and is called ***Max-diverse Ensemble***.

1 Introduction

Random tree ensembles introduce different random elements to construct diversified decision trees. For classification problems, results from these trees are combined by an ensemble method to produce the final prediction. *Random Decision Trees* [8] is one that is constructed without conventional test selection criteria, which questions the utility of these heuristics that are widely employed in many decision tree learning algorithms. *The underlying argument is that they are effective to compute accurate single trees but there is no guarantee on the final accuracy of a tree ensemble.*

As it stands, there is no creditable report known to us that extensively analyses and compares complete-random trees with other decision tree ensembles. This paper aims to explore complete-random trees and compare them with *Bagging* [3] and *Random Forests* [5] which are widely accepted and use techniques such as randomized feature selection, bootstrap sampling and voting. The fundamental objective of randomization in tree construction is to create diversity. After all, there is no point in combining a forest of identical trees. Section 2 of this paper discusses how increasing tree diversity can lower the generalization error of tree ensembles. Since there are many randomization methods, a systematic framework to characterize each method is necessary to guide the research in this area. A taxonomy is provided in section 3 and the focus of our study is *sample randomization* and *complete-random test selection*. Section 4 describes the random tree learning process, section 5 provides the experimental settings and results, and follows by conclusions in the last section.

2 Tree Diversity

In Breiman's analysis [5] on strength and correlation, he gives an upper bound on the generalization error $PE^* \leq \overline{\rho}\frac{(1-s^2)}{s^2}$, where s is the strength of the set of trees and $\overline{\rho}$ is the mean correlation. The implication of this upper bound is that no ensemble can do better than the boundary given its strength and correlation. Generally, this upper bound is applicable to classifier based ensemble, including complete-random trees the subject of this paper. Lowering PE^* can be achieved by either minimizing $\overline{\rho}$ or increasing s. Building complete-random trees forgoes strength obtained from a test selection criterion. However, it helps to achieve higher tree diversity.

3 Different Categories of Randomization

The taxonomy of tree randomizations is summarized as follows:

1. **Randomization before model induction**
   (a) Sample randomization
       e.g. *Bootstrap sampling* [3]
   (b) Feature randomization
       e.g. *Randomized Trees* [1] and *Random Subspace* [9]
   (c) Data perturbation
       e.g. *Output Smearing* and *Output Flipping* [4]
2. **Randomization during model induction**
   (a) Partial-random test selection
       e.g. *Tree Randomization* [6] and *Random Forests* [5]
   (b) Complete-random test selection
       e.g. *Random Decision Trees* [8]

This paper focuses on sub-categories (1a) *sample randomization* and (2b) *complete-random test selection* to investigate whether complete-random test selection produces good results.

4 Random Tree Ensemble

For the experiment, our implementation is based on *C4.5* release 8 [10] with modifications to cater for bootstrap sampling, multiple trees, complete-random test selection, tree height restriction, random split point selection for continuous features and ensemble methods. The *tree height restriction* is originated from [8]. Let k be the total number of features, setting tree height to $\frac{k}{2}$ is called half height tree. Alternatively, unrestricted tree growth is called full height tree. Consider a rule or a branch in a tree, when selecting i features from k features, there are $C_i^k = \frac{k!}{i! \cdot (k-i)!}$ unique feature combinations. To use only a single value of i, $i = \frac{k}{2}$ produces the largest number of combinations. Fan et. al. [8] uses this argument as the basis to choose the tree height limit of $\frac{k}{2}$, but allowing any value of i is more desirable as it gives the maximum choice or diversity. Thus, the total number of possible unique combinations to include any value of i is $T(i) = \sum_i C_i^k$. Since $T(k) > T(\frac{k}{2}) > C_{\frac{k}{2}}^k$, setting tree height to k produces maximum diversity. For *continuous feature split point selection*, random split point is determined by randomly selecting two different sample values and assigning it as the mid point between the two. This increases the possible split points from $l-1$ to $\sum_{i=0}^{l-1} i$, l is the number of distinct feature values. Hence, it increases diversity. *Missing values* for probability averaging are handled by: 1. growing missing value branches; 2. classifying them with reduced weight $w = w_p \frac{n_{missing}}{n_{total}}$, where w_p is the classification weight from the parent node, $n_{missing}$ is the number of missing value samples and n_{total} is node size. This avoids disruption of the usual weight disseminating routine in handling missing values.

At classification phase, posterior probability estimation or class label is generated using these counts. To predict a class given a test case z, the predicted class c_p is obtained by:

1. Probability averaging, $c_p = arg\max_c(\frac{1}{N} \sum_{i=1}^{N} (w \frac{n_{h_i,c}}{n_{h_i}}))$
2. Voting, $c_p = arg\max_c(\frac{1}{N} \sum_{i=1}^{N} I(\frac{n_{h_i,c}}{n_{h_i}}))$.

where N is the number of trees, $I()$ is an indicator function. Relevant to z, $n_{h_i,c}$ is the count of class c for tree h_i and n_{h_i} is the leaf size for h_i. Probability averaging are reported to cause overfitting [7]. However, it is worth noting that none of the empirical evaluations are conducted in the context of complete-random trees.

5 Experiments

There are three parameters in the experiments. The followings are the abbreviations used in the experiments:

Ensemble Methods	Tree height restriction	Sample randomization
Probability averaging	**F**ull height	**O**riginal training samples
Voting	**H**alf height	**B**ootstrap training samples

In total, there are eight possible variants from these three parameters. Each variant is represented by three letters, for example "VFO" refers to a random trees ensemble with parameters **V**oting, **F**ull height tree and **O**riginal training samples.

Table 1. The average error results are listed with asterisk(s)* indicating best error rate(s) among different methods

Data set	size	PFO	PFB	PHO	PHB	VFO	VFB	VHO	VHB	Bagging	Random Forests
abalone	4177	29.8	29.9	31.7	31.7	29.9	29.7	31.7	31.7	*29.1	*29.1
anneal	898	*0.9	1.7	2.6	2.9	1.0	1.8	3.7	3.9	3.2	14.8
audiology	226	19.5	*18.5	21.7	19.0	22.6	22.1	26.2	23.8	20.8	37.5
autos	690	24.3	22.9	24.4	22.4	26.7	22.0	25.3	22.9	*16.2	20.5
balance	205	13.8	13.6	13.9	13.8	13.6	*13.4	18.2	14.6	15.5	15.5
breast-w	625	*2.4	2.7	2.7	2.7	3.0	3.2	2.9	3.2	3.4	3.1
breast-y	699	25.5	25.5	*23.4	25.5	24.8	27.6	25.1	27.0	26.6	26.9
chess	286	*1.6	1.9	2.5	2.8	2.7	4.1	4.9	5.1	4.8	4.9
cleveland	20000	*41.3	41.9	43.9	44.3	42.6	43.6	43.6	44.2	43.6	42.2
coding	3196	*16.3	*16.3	*16.3	*16.3	18.7	23.9	19.2	23.8	33.5	27.5
credit-a	303	*12.3	12.6	12.9	13.5	13.3	15.1	14.1	14.2	13.2	13.6
credit-g	3186	25.3	25.7	27.1	27.3	27.3	29.2	29.0	29.3	*24.7	26.9
DNA	131	28.8	28.9	28.8	28.8	16.1	14.5	16.1	14.4	*7.1	12.9
echo	1066	32.9	34.3	*32.1	32.8	33.7	36.7	34.5	35.1	35.2	35.9
flare	1000	18.5	19.0	17.5	17.4	18.2	18.3	17.5	*17.1	17.5	17.6
glass	214	26.2	26.2	34.6	33.1	28.0	30.0	37.9	36.0	35.2	*21.9
hayes-roth	160	44.4	40.0	48.1	47.5	53.8	46.9	58.1	47.5	17.5	*17.5
hepatitis	155	15.3	15.7	16.0	15.7	17.3	*15.0	16.0	15.7	24.7	16.3
horse-colic	368	18.8	16.9	19.1	17.2	19.9	17.2	20.4	17.4	*14.2	16.3
hypothyroid	3163	2.2	2.4	4.7	4.7	2.3	2.4	4.7	4.7	*0.9	1.3
ionosphere	351	9.4	9.7	9.4	9.7	11.7	15.9	11.7	15.3	6.8	*5.7
iris	150	4.7	6.0	7.3	7.3	*4.0	5.3	11.3	11.3	6.7	6.0
led24	3200	*28.4	28.8	*28.4	29.0	36.8	37.4	36.5	36.5	28.5	29.8
liver	345	30.7	31.8	38.3	37.1	29.5	31.6	38.5	37.7	*28.1	29.2
lymph	148	15.5	15.4	15.5	*14.7	18.2	15.4	18.9	15.4	21.6	17.4
nursery	12960	*2.0	2.3	5.4	5.5	2.2	1.9	7.1	5.2	6.4	4.7
pima	768	24.7	23.6	28.6	28.1	24.3	25.4	29.4	29.5	24.7	*23.4
primary	339	56.1	55.2	*53.1	*53.1	54.6	55.8	55.5	55.2	55.5	*53.1
segment	2310	2.9	3.1	5.1	4.9	3.3	3.7	6.6	6.4	2.4	*2.3
sick	3163	7.6	7.8	9.3	9.3	8.1	8.0	9.3	9.3	2.1	3.7
solar	323	30.0	29.7	27.2	27.5	29.6	30.6	29.3	28.8	27.2	*24.7
sonar	208	*13.4	16.8	*13.4	16.8	23.5	26.9	23.5	26.9	20.1	19.6
soybean	683	6.0	5.7	5.7	5.6	5.6	6.0	6.0	5.7	6.2	*5.4
threeOf9	512	*0.2	0.8	11.3	11.1	8.2	2.7	13.1	12.9	3.3	2.2
tic-tac-toe	958	*9.4	10.4	24.6	23.8	18.8	27.5	27.5	26.2	29.4	26.4
vehicle	846	27.1	29.2	27.4	29.3	27.9	27.1	29.3	28.4	24.9	*25.2
vote	435	5.3	5.3	5.3	5.3	5.1	6.2	4.8	6.0	*4.6	4.8
waveform	5000	*14.1	14.2	14.8	14.9	14.3	14.6	14.2	14.1	16.3	14.7
wine	178	2.3	1.7	2.3	1.7	2.3	2.8	2.3	2.3	5.6	*1.1
zoo	101	*2.0	4.0	*2.0	3.0	3.0	3.0	3.0	3.9	6.9	7.9
Mean		17.3	17.5	19.0	18.9	18.7	19.1	20.7	20.2	17.4	17.8

Table 2. Summary of pairwise comparison *(wins, losses, draws)* reading from top to left. The number of significant wins and losses is bold faced, based on a sign test of 95% confidence level

	Max-diverse Ensemble PFO	PFB	PHO	PHB	VFO	VFB	VHO	VHB	Bagging
Random Forests	21,19,0	20,19,1	17,23,0	17,23,0	14,**26**,0	11,**29**,0	11,**29**,0	11,**29**,0	19,20,1
Bagging	23,17,0	22,18,0	20,20,0	18,22,0	18,22,0	15,25,0	12,**28**,0	13,**27**,0	
VHB	**30**,8,2	**29**,6,5	25,10,5	**29**,6,5	**29**,10,1	22,15,3	14,20,6		
VHO	**32**,6,2	**33**,6,1	**30**,4,6	**30**,14,5	**27**,8,5	21,15,4			
VFB	**28**,10,2	**27**,11,2	19,21,0	19,19,2	**26**,13,1				
VFO	**28**,11,1	23,15,2	21,18,1	16,21,3					
PHB	25,10,5	23,9,8	15,16,9						
PHO	25,6,9	25,11,4							
PFB	25,11,4								

In this experiment, the main aim are to investigate : 1. the main contributing factors among the eight possible variants; 2. if complete-random trees overfits. All variants will be compared with the benchmarking *Bagging* and *Random Forests*. The results are assessed by a sign test using 95% confidence level to determine whether the wins are statistically significant. Forty data sets are selected from the UCI repository [2]. Their data sizes range from one hundred to twenty thousand. This experiment uses ten thousand trees for each ensemble to see if any variants overfit. *Tenfold cross-validation* is conducted for each data set and the average error rate is reported.

5.1 Results

The average error is shown in table 1 and a pairwise comparison summary is presented in table 2. Comparing variants with the benchmark classifiers, we summarize the results as follows:

– PFO, PFB and PHO perform comparable to *Bagging*.
– PFO and PFB perform comparable to *Random Forests*.
– PFO has the most wins against the two benchmark classifiers having 23 and 21 out of 40.
– PFO has thirteen data sets with the best error rates as marked with asterisks in table 1 *Random Forests* has elevens and *Bagging* has eights.

For each of the three parameters, we summarize the results as follows:

– all probability averaging variants are significantly better than their voting counterparts according to the *sign test*.
– full height tree performs better than half height tree.
– bootstrap sampling impairs accuracy as suggested earlier on in section 3.

The results above suggest that the most accurate variant PFO is comparable to the benchmark classifiers and the probability averaging is the main contributing factor to complete-random decision trees. We call PFO "***Max-diverse Ensemble***". *Max-diverse Ensemble* performs better against all variants and has the lowest mean-error rate 17.3% as shown in table 1. Regarding overfitting, none of the data sets suffers from overfitting in general. It dispels the concern of using probability averaging with complete-random trees causes overfitting.

6 Conclusions

In this paper, we first discuss that maximizing tree diversity is a way to lower generalization error. Then, we provide a taxonomy on tree randomization as a systematic framework to characterize existing tree randomization methods. We find that complete-random test selection produces diverse trees. Finally, we thoroughly investigate the complete-random decision trees by exploring eight possible variants. The most accurate variant *Max-diverse Ensemble* has the maximum diversity according to our analysis and uses only simple probability averaging without any feature selection criterion or other random elements. For future work, it would be valuable to determine situations where *Max-diverse Ensemble* would perform better than other methods and vice versa.

References

1. Yali Amit and Donald Geman. Shape quantization and recognition with randomized trees. *Neural Computation*, 9(7):1545–1588, 1997.
2. C.L. Blake and C.J. Merz. Uci repository of machine learning databases, 1998.
3. Leo Breiman. Bagging predictors. *Machine Learning*, 24(2):123–140, 1996.
4. Leo Breiman. Randomizing outputs to increase prediction accuracy. *Machine Learning*, 40(3):229–242, 2000.
5. Leo Breiman. Random forests. *Machine Learning*, 45(1):5–32, 2001.
6. Thomas G. Dietterich. An experimental comparison of three methods for constructing ensembles of decision trees: Bagging, boosting, and randomization. *Machine Learning*, 40(2):139–157, 2000.
7. Pedro Domingos. Bayesian averaging of classifiers and the overfitting problem. In *Proc. 17th International Conf. on Machine Learning*, pages 223–230. Morgan Kaufmann, San Francisco, CA, 2000.
8. Wei Fan, Haixun Wang, Philip S. Yu, and Sheng Ma. Is random model better? on its accuracy and efficiency. *Third IEEE International Conference on Data Mining*, pages 51–58, 2003.
9. Tin Kam Ho. The random subspace method for constructing decision forests. *IEEE Transactions on Pattern Analysis and Machine Intelligence*, 20(8):832–844, 1998.
10. J. R. Quinlan. *C4.5 : programs for machine learning*. Morgan Kaufmann Publishers, San Mateo, Calif., 1993.

SETRED: Self-training with Editing

Ming Li and Zhi-Hua Zhou

National Laboratory for Novel Software Technology,
Nanjing University, Nanjing 210093, China
{lim, zhouzh}@lamda.nju.edu.cn

Abstract. Self-training is a semi-supervised learning algorithm in which a learner keeps on labeling unlabeled examples and retraining itself on an enlarged labeled training set. Since the self-training process may erroneously label some unlabeled examples, sometimes the learned hypothesis does not perform well. In this paper, a new algorithm named SETRED is proposed, which utilizes a specific data editing method to identify and remove the mislabeled examples from the self-labeled data. In detail, in each iteration of the self-training process, the local *cut edge weight* statistic is used to help estimate whether a newly labeled example is reliable or not, and only the reliable self-labeled examples are used to enlarge the labeled training set. Experiments show that the introduction of data editing is beneficial, and the learned hypotheses of SETRED outperform those learned by the standard self-training algorithm.

1 Introduction

In many practical machine learning applications, obtaining a fully labeled data set is usually difficult. The requirement of lots of human expertise makes the labeling process fairly expensive. A more feasible way is to label just a small part of data set, leaving a huge amount of examples in data set unlabeled. The learner itself should find a way to exploit the merit of unlabeled data.

Semi-supervised learning is to learn a hypothesis by combining information in both labeled and unlabeled data. *Self-training* [10] is a well-known semi-supervised algorithm. In self-training process, a base learner is firstly trained on labeled set. Then, iteratively, it attempts to choose to label several examples that it is most confident of in the unlabeled set. After that it enlarges its labeled training set with these self-labeled examples. Since the labeled set is usually insufficient for learning, misclassifying a certain amount of unlabeled data is unavoidable. Thus, the enlarged labeled set for the learner to learn in the next iteration could contain much noise. Once those noisy examples are added into the learner's training set, there is no chance for the self-trained learner to reconsider the validity of those incorrect labels, and the mislabeled examples will keep on affecting the learner in the following iterations. If the distribution the learner has caught is badly distorted by those mislabeled examples, the generalization ability degrades as the self-training process goes on. Therefore, it is obvious that identifying and removing the mislabeled examples in each iteration might help improve the generalization ability of the learned hypothesis.

In this paper, a new self-training style algorithm named SETRED (SElf-TRaining with EDiting) is proposed. SETRED introduces a data editing technique to the self-training process to filter out the noise in the self-labeled examples. Specifically, after labeling some examples chosen from the unlabeled set, SETRED actively identifies the possibly mislabeled examples with the help of some local information in a neighborhood graph, and keeps those mislabeled examples from being added to the learner's training set, hence a less noisy training set is obtained. Actually, SETRED could be considered as a semi-supervised algorithm that utilizes an active-learning-like technique to improve its performance. Experiments on ten UCI data sets show that SETRED is more robust than the standard self-training algorithm, and the generalization ability of its learned hypotheses outperform those learned by standard self-training.

The rest of the paper is organized as follows. Section 2 briefly reviews some works on learning from labeled and unlabeled data. Section 3 presents SETRED. Section 4 reports the experiment result. Finally, Section 5 concludes and issues some future work.

2 Learning from Labeled and Unlabeled Data

An effective way to utilize unlabeled data in assistance of supervised learning is known as *semi-supervised learning* [13], where an initial hypothesis is learned from the labeled set and then refined through information derived from the unlabeled set.

In some methods only one base learner is used, which uses the unlabeled examples iteratively according its own knowledge. Such methods include using Estimation-Maximization approach to estimate posterior parameters of a generative model, such as Naive Bayes, by assigning each unlabeled example a soft label, i.e. a probability for each class [11]; using the unlabeled data to search for a better structure of Bayesian Network [2]; using a transductive inference for support vector machines on a special test set [4]. The self-training algorithm [10] is of this kind, where in each iteration the learner converts the most confidently predicted unlabeled example of each class into a labeled training example.

In some other methods, the unlabeled data is utilized with more learners. A representative is the co-training paradigm proposed by Blum and Mitchell [1]. In co-training, two base learners are trained within the multi-view framework, i.e. two sets of independent attributes, each of which is sufficient for classification. One base learner iteratively labels several examples which it is most confident of from its point of view, and feeds them to the other learner. The co-training paradigm has already been successfully applied to many areas such as nature language processing [12].

Besides semi-supervised learning, there is another effective way to use the unlabeled data, that is, *active learning*. Different from semi-supervised learning choosing confident examples to label by itself, active learning actively chooses most problematic examples from the unlabeled set and asks a teacher for the labels. Two major techniques in active learning are uncertainty-base sampling [5] and committee-base sampling [14].

Since semi-supervised learning and active learning utilize unlabeled data in different way, their merits can be combined through some specifically designs. McCallum and Nigam [6] combined semi-supervised EM with committee-based sampling in text classification. Muslea et al.[9] employed co-testing [8] to choose unlabeled examples to query, and used co-EM [10] to boost the accuracy of the hypotheses. Zhou et al. [16] combined co-training with co-testing in content-based image retrieval.

3 SETRED

Let L and U denote the labeled and unlabeled set drawn from the same distribution $D(\mathbf{X}, Y)$, respectively, where $\mathbf{X}$ is a p-dimensional feature vector while Y is a class label. In standard self-training process, a learner keeps on choosing to label a small set of its most confident examples, say L', from U and retraining itself on $L \cup L'$. Self-training requires neither estimation and maximization of some posterior probability nor a sufficient and conditional independent attributes, so it is much easier to use than semi-supervised EM and standard co-training. However, due to the small size of L, the generalization ability of the initial hypothesis may be poor. Consequently, L' may contain much noise because the learner may incorrectly assign labels to some unlabeled examples, and the generalization ability of the final hypothesis will be hurt by the accumulation of such noise in each iteration of the training process. Therefore, it is obvious that if the mislabeled examples in L' could be identified in the self-training process, especially in the early iterations, the learned hypothesis is expected to perform better.

Data editing is a technique which attempts to improve the quality of the training set through identifying and eliminating the training examples wrongly generated in the human labeling process. Some useful data editing methods have been studied in [3][15]. In those works, another learner is used to improve the quality of the training set before the wanted learner are trained. In a recent work, Muhlenbach et al. [7] proposed a method based on a statistical method called *cut edge weight* statistic [17] to identify mislabeled examples in the training set. Here this data editing method is employed to identify the examples possibly mislabeled by the learner in the self-training process.

In detail, SETRED initiates the self-training process by firstly learning a hypothesis form the labeled set. In each self-training iteration, the base learner detects unlabeled examples on which it makes most confident prediction and labels those examples according to the prediction. Then, for each possible label y_j (where j ranges from 1 to the number of possible labels), k_j examples are selected and added to L' according to the prediction confidence of their labels, keeping the class distribution in L' similar to that in L. For instance, if there are 4 positive and 16 negative examples in L, then L' contains 1 positive and 4 negative examples.

After L' is formed, the identification of mislabeled examples are performed on the learner's potential training set $L \cup L'$. Firstly, SETRED constructs a neighborhood graph [17] that expresses certain local information from all the examples in $L \cup L'$. A neighborhood graph is a graph in p-dimensional feature space where a distance metric could be defined. Each example in the graph is a vertex and there exists an edge between two vertices a and b if the distance between a and b satisfies Eq. 1. An edge connecting two vertices that have different labels is called *cut edge*.

$$\text{Dist}(a, b) \leq \max(\text{Dist}(a, c), \text{Dist}(b, c)) \quad (1)$$

Then SETRED identifies the mislabeled examples based on their neighborhood in the graph. The neighborhood of an example is a set of examples it connected to with edges in graph. Intuitively, most examples possess the same label in a neighborhood. So if an example locates in a neighborhood with too many cut edges, this example should be considered *problematic*. Thus, cut edge plays an important role for identifying mislabeled examples. To explore the information of cut edges, SETRED associates every $(x_i, \hat{y}_i)$ in L' with a local cut edge weight statistic J_i defined in Eq. 2.

$$J_i = \sum_{x_j \in N_i} w_{ij} I_{ij} \quad (2)$$

where N_i is the neighborhood of x_i, w_{ij} is the weight on the edge between x_i and x_j and I_{ij} are i.i.d random variables according to the Bernouilli law of parameter $P(y \neq \hat{y}_i)$.

By definition, J_i describes the relationship between the adjacency of an x_i and other vertices in its neighborhood and the fact that they have the same label [17]. Similarly, a null hypothesis H_0 that can be tested with J_i is defined as every examples in $L \cup L'$ is independently labeled according to the marginal distribution $D(Y)$. H_0 specifies a case that the label y_i is assigned to each example x_i without considering any information from x_i, i.e. for any example (x_i, y_i), the probability of examples in its neighborhood possessing labels other than y_i is expected to be no more than $1 - P(y = \hat{y}_i)$ under H_0. Hence, a good example will be *incompatible* with H_0. To test H_0 with J_i, the distribution of J_i under H_0 is need. The distribution of J_i can be approximated to a normal distribution with mean μ_i and variance σ_i^2 estimated by Eq. 3 and Eq. 4, if the size of neighborhood is big and the weights are not too unbalanced, otherwise a simulation must be proceed [7].

$$\mu_{i|H_0} = (1 - P(y = \hat{y}_i)) \sum_{x_j \in N_i} w_{ij} \quad (3)$$

$$\sigma_{i|H_0}^2 = P(y = \hat{y}_i)(1 - P(y = \hat{y}_i)) \sum_{x_j \in N_i} w_{ij}^2 \quad (4)$$

Therefore, if the observation value of J_i that associates with an example $(x_i, \hat{y}_i)$ in L' locates in the left rejection region, then there are significantly less

Table 1. Pseudo-code describing the SETRED algorithm

Algorithm: SETRED
Input: the labeled set L, the unlabeled set U,
the left rejection threshold θ, the maximum number of iterations M
Output: the learned hypothesis h
Progress:
    Create a pool U' by randomly selecting examples from U
    $h \leftarrow Learn(L)$
    Repeat for M iterations:
        $L' \leftarrow \phi$
        for each possible label y_j **do**
            h chooses k_j most confident examples from U'
            Add the chosen examples to L' after giving them the label y_j
        Build a neighborhood graph G with $L \cup L'$
        for each $x_i \in L'$ **do**
            Compute the observation value o_i of J_i
            Find the neighborhood N_i of x_i in G
            Compute the distribution function of J_i under H_0
            if o_i locates in the left rejection region specified by θ
                $L' \leftarrow L' - \{(x_i, \hat{y}_i)\}$
        $h \leftarrow Learn(L \cup L')$
        Replenish U' by randomly selecting examples from U
    End of Repeat

---

cut edges than expected under H_0, hence it is a *good* example. In contrast, if the observation value locates in places other than the left rejection region, then lots of examples in the neighborhood disagree with its label, hence it could be regarded as a *mislabeled* example. The left rejection region is specified by a pre-set parameter θ.

Once the possibly mislabeled examples in L' are identified, SETRED simply discards those examples, keeping the good ones intact. Consequently, a filtered set L'' is obtained. Note that SETRED does not try to relabel the identified mislabeled examples in order to avoid introducing new noise to the data set. Finally, SETRED finishes the current iteration by relearning a hypothesis on $L \cup L''$. SETRED stops self-training process after the pre-set maximum times of iteration M is reached. The pseudo-code of SETRED is shown in Table 1.

Note that a pool of unlabeled examples smaller than U is used in the algorithm. Blum and Mitchell [1] suggested to choose examples from a smaller pool in stead of the whole unlabeled set. For convenience, we adopt this strategy directly without verification. Furthermore, it is worth noticing that the SETRED could be regarded as a type of active semi-supervised learning algorithm that actively identify the bad examples from the self-labeled set. Absence of the teacher, SETRED just discards the problematic data after identification instead of asking the teacher for labels as in the standard active learning scenario.

4 Experiments

In order to test the performance of SETRED, ten UCI data sets are used. Information on these data sets are tabulated in Table 2.

Table 2. Data set summary

Data set	Size	Attribute	Class	Class distribution
australian	690	15	2	44.5%/55.5%
breast-w	699	9	2	65.5%/34.5%
colic	368	22	2	63.0%/37.0%
diabetes	768	9	2	65.1%/34.9%
german	1000	20	2	70.0%/30.0%
heart-statlog	227	13	2	55.5%/44.5%
hepatitis	155	19	2	20.6%/79.4%
ionosphere	351	34	2	35.9%/64.1%
vehicle	846	18	4	25.1%/25.7%/25.7%/23.5%
wine	178	13	3	33.1%/39.9%/27.0%

For each data set, 25% data are kept aside to evaluate the performance of learned hypothesis, while the remaining 75% data are partitioned into labeled set and unlabeled set under the unlabel rate 90%, i.e. just 10% (of the 75%) data are used as labeled examples while the remaining 90% (of the 75%) data are used as unlabeled examples. Note that the class distributions in these splits are similar to that in the original data set.

Since SETRED exploits local information to identify mislabeled examples, the learner that utilizes local information is expected to benefit a lot from this method. Therefore in the experiments, the Nearest Neighbor classifier is used as the base learner. Unlike those probabilistic model such as Naive Bayes, whose confidence for an example belonging to a certain class can be measured by the output probability in prediction, the Nearest Neighbor classifier has no explicitly measured confidence for an example. Here for a Nearest Neighbor classifier, the most confidently predicted unlabeled example with label y_j is defined as the unlabeled example which is the nearest to labeled examples with label y_j while far away from those with labels other than y_j. The pre-set parameter θ that specifies the left rejection region of the distribution of J_i is fixed on 0.1, the same as that in [7]. The self-training process stops when either there are no unlabeled examples available or 40 iterations have been done.

For comparison, the standard self-training, namely Self-training, is run on the same labeled/unlabeled/test splits as those used for evaluating SETRED. Same as SETRED, the maximum iteration is also 40. Moreover, two base lines, denoted by NN-L and NN-A respectively, are used for comparison. One is a Nearest Neighbor trained only from the labeled set L, and the other is the one that trained from $L \cup U$ provided the true label of all the examples in U. Note that

Table 3. Average error rate on the experimental data sets (50 runs)

Data set	NN-A	NN-L	SETRED	Self-training	SETRED-imprv.	Self-imprv.
australian	.185	.188	.167	.170	11.3%	9.4%
breast-w	.046	.046	.038	.038	17.9%	16.9%
colic	.194	.237	.191	.209	19.3%	11.8%
diabetes	.298	.330	.320	.335	3.1%	-1.6%
german	.185	.339	.349	.357	-2.8%	-5.2%
heart-statlog	.237	.248	.209	.226	15.8%	8.7%
hepatitis	.161	.186	.208	.157	-11.9%	15.7%
ionosphere	.143	.228	.197	.254	13.6%	-11.4%
vehicle	.298	.412	.399	.413	2.9%	-0.3%
wine	.048	.090	.066	.079	26.6%	12.8%

NN-L is the initial state of both SETRED and Self-training before they utilize any information from the unlabeled examples. NN-A is the ideal state of SETRED and Self-training since every examples chosen in self-training process are given the correct label and all the examples available in the unlabeled set are used.

Experiments are carried out on each data set for 50 runs. In each run, all the four learners are trained and evaluated on the randomly partitioned labeled/unlabeled/test splits. In Table 3, the first four columns are the average error rates of NN-A, NN-L, SETRED and Self-training respectively over 50 runs on each data set. The last two columns denoted by "SETRED-imprv." and "Self-imprv." respectively show the performance improvements of SETRED and Self-training over NN-L, which is computed by the error rate of learned hypothesis of SETRED and Self-training over the error rate of the learned hypothesis of NN-L.

Table 3 shows that SETRED benefits much from the unlabeled examples since the performance improvements are evident in 8 data sets, except that it goes worse on *german* and *hepatitis*. The two-tailed paired *t*-test under the significant level of 95% shows that all the improvement of performance are significant. Note that on 4 data sets SETRED performs even better than NN-A which is able to access all the information of the unlabeled examples. In contrast, although the performance of the learned hypothesis of Self-training improves on 6 data sets, only on five the improvements are significant, including *australian, breast-w, colic, heart-statlog* and *hepatitis*. Furthermore, Table 3 also shows that SETRED outperforms Self-training on 9 data sets, among which significance is evident in 6 data sets under a two-tailed pair-wise *t*-test with the significance level of 95%. This evidence supports our claim that SETRED is robust to noise in the self-labled examples hence achieves better performance than Self-training.

Interestingly, Self-training does benefit from the unlabeled exmaples on *hepatitis*, while the performance of SETRED degrades. One possible explanation is that SETRED suffers imbalance of the data set. In *hepatitis* data set, there are only 32 positive examples out of 155 examples in all, which is only 20.6% of the total. Recall the method we used for identifying mislabeled examples, one can only be regarded as a good example only if there exists a significantly large num-

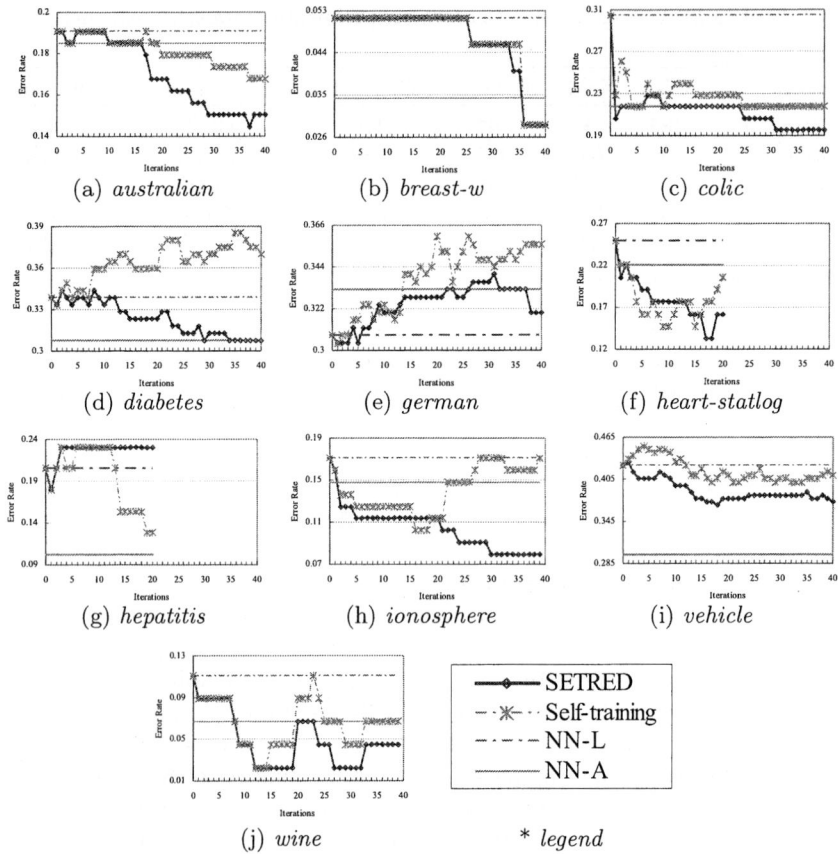

Fig. 1. Median performance on experimental data

ber of examples having the same label in its neighborhood. Since the data set is unbalanced, a correctly labeled positive examples could be easily mis-identified as mislabeled examples and rejected to be added to the labeled set for further training, due to the lack of neighbors possessing the same label. The percentage of the negative examples in the labeled set increases as the self-training process goes on, hence less chance for a correctly labeled positive examples available for further training. The more the distribution of the training set is distorted, the easier for the learner to be misled. Consequently, the performance degrades. Similarly, the error rate of hypothesis learned via SETRED climbs up to 0.349 from the initial error rate of 0.339 on *german*, in which the negative examples are only 30%. The imbalance of this data set might also account for the performance degradation of SETRED.

For further evaluation, one run out of 50 with a median performance on each data set is investigated carefully. Fig. 1 gives plot of error rate versus number of iterations of the median performance on each data set respectively. Note that SETRED and Self-training stop before the maximum number of iterations is reached on several data set such as *heart-statlog* and *hepatitis*, due to no more unlabeled examples available for further training. In most cases except for *german* and *hepatitis*, SETRED outperforms Self-training, and the error rates of the learned hypothesis by SETRED usually go lower than or converge to the error rates of NN-A. These are consistent with the average performance of the 50 runs on the experimental data sets.

In Fig. 1(g), the error rate curve of SETRED climes up to a high level after a few iterations and remains unchanged. By contrast, the curve of Self-training drops when many unlabeled examples have been self-labeled and used for further training. This supports the explanation above for the SETRED's failure that SETRED suffers the imbalance data. Once the correctly labeled positive examples are rejected by SETRED, the misclassified positive examples in the test set, which are probably be correctly classified after more positive examples are learned, will remain being misclassified in the following iterations.

In summary, the experiments show that SETRED can benefit from the unlabeled examples. SETRED is robust to the noise introduced in self-labeling process and its learned hypothesis outperforms that learned via standard self-training.

5 Conclusion

In this paper, a novel self-training style algorithm named SETRED, which incorporates data editing technique to learn actively from the self-labeled examples, is proposed. In detail, SETRED firstly learns from labeled examples and then iteratively chooses to label a few unlabeled examples, on which the learner is most confident in prediction, and adds those self-labeled examples to its labeled set for further training. In each iteration, SETRED does not completely accept all the self-labeled examples that might be highly noisy. Instead, it actively identifies the possibly mislabeled examples from those self-labeled examples by testing a predefined null hypothesis with the local cut edge weight statistic associated with each self-labeled example. If the test indicates a left rejection, the example is regarded as a good example, otherwise it is a possible mislabeled example which should be kept from adding to the learner's training set. The experiment results on 10 UCI data sets show that SETRED is able to benefit from the information provided by unlabeled examples, and it is robust to the noise introduced in the self-labeling process hence the generalization ability of its learned hypothesis is better than that learned via standard self-training, which is easily affected a lot by those noise. Since SETRED is sensitive to imbalance data, exploring a way to solve this problem will be done in future. Since SETRED uses a Nearest Neighbor as base learner, extending this idea to other base leaners will also be future work.

It is noteworthy that the SETRED algorithm is not only a robust self-training style algorithm but shedding a light on a possible way to handle the noise introduced in the learning process by incorporating an active-learning-like technique to refine the self-labeled examples in semi-supervised learning scenario, hence obtaining better performance of the learned hypothesis. Different from others, the work is done when the teacher to assign labels to the problematic examples is absent. In the future work, theoretical verification of this method will be done, which might help to understand the functionality of this method. Moreover, extending this method to classic semi-supervised learning algorithms, such as co-training [1], or searching for more suitable active learning methods for those algorithms to improve their performance will also be the future work.

Acknowledgement

This work was supported by the National Science Foundation of China under the Grant No. 60473046, the Jiangsu Science Foundation Key Project under the Grant No. BK2004001, the Foundation for the Author of National Excellent Doctoral Dissertation of China under the Grant No. 200343, and the Graduate Innovation Program of Jiangsu Province.

References

1. Blum, A., Mitchell, T.: Combining labeled and unlabeled data with co-training. In: Proceedings of the 11th Annual Conference on Computational Learning Theory, New York, NY (1998) 92–100
2. Cohen, I., Cozman, F.G., Sebe, N., Cirelo, M.C., Huang, T.S.: Semisupervised learning of classifier: theory, algorithms, and their application to human-computer interaction. IEEE Transactions on Pattern Analysis and Machine Intelligence **26** (2004) 1553–1567
3. Jiang, Y., Zhou, Z.-H.: Editing training data for kNN classifiers with neural network ensemble. In: Yin, F., Wang, J., Guo, C. (eds.): Lecture Notes in Computer Science, Vol.3173. Springer, Berlin (2004) 356–361
4. Joachims, T.: Transductive inference for text classification using support vector machines. In: Proceedings of the 16th International Conference on Machine Learning, San Francisco, CA (1999) 200–209
5. Lewis, D., Gale, W.: A sequential algorithm for training text classifiers. In: Proceedings of the 17th ACM International Conference on Research and Development in Information Retrieval, Dublin, Ireland (1994) 3–12
6. McCallum, A., Nigam, K.: Employing EM in pool-based active learning for text classification. In: Proceedings of the 15th International Conference on Machine Learning, Madison, WI (1998) 359–367
7. Muhlenbach, F., Lallich, S., Zighed, D.A.: Identifying and handling mislabelled instances. Journal of Intelligent Information Systems **39** (2004) 89–109
8. Muslea, I., Minton, S., Knoblock, C.A.: Selective sampling with redundant views. In: Proceeding of the 17th International Conference on Machine Learning, Stanford, CA (2000) 621–626
9. Muslea, I., Minton, S., Knoblock, C.A.: Active + semi-supervised learning = robust multi-view learning. In: Proceeding of the 19th International Conference on Machine Learning, Sydney, Australia (2002) 435–442
10. Nigam, K., Ghani, R.: Analyzing the effectiveness and applicabilbity of co-training. In: Proceedings of the 9th International Conference on Information and Knowledge Management, Washington, DC (2000) 86–93

11. Nigam, K., McCallum, A., Thrun, S., Mitchell, T.: Text classification from labeled and unlabeled documents using EM. Machine Learning **39** (2000) 103–134
12. Sarkar, A.: Applying co-training methods to statistical parsing. In: Proceedings of the 2nd Annual Meeting of the North American Chapter of the Association of Computational Linguistics. Pittsburgh, PA (2001) 95–102
13. Seeger, M.: Learning with labeled and unlabeled data. Technical Report, University of Edinburgh, Edinburgh, UK (2001)
14. Seuong, H., Opper, M., Sompolinski, H.: Query by committee. In: Proceedings of the 5th ACM Workshop on Computational Learning Theory, Pittsburgh, PA (1992) 287–294
15. Wilson, D.R.: Asymptotic properties of nearest neighbor rules using edited data. IEEE Transactions on Systems, Man and Cybernetics, **2** (1972) 408–421
16. Zhou, Z.-H., Chen, K.-J., Jiang, Y.: Exploiting unlabeled data in content-based image retrieval. In: Boulicaut, J.-F., Esposito, F., Giannotti, F., Pedreschi, D. (eds.): Lecture Notes in Artificial Intelligence, Vol.3201. Springer, Berlin (2004) 525–536
17. Zighed, D.A., Lallich, S., Muhlenbach, F.: Separability index in supervised learning. In: Elomaa, T., Mannila, H., Toivonen, H. (eds.) Lecture Notes in Artificial Intelligence, Vol.2431. Springer, Berlin (2002) 475–487

Adjusting Mixture Weights of Gaussian Mixture Model via Regularized Probabilistic Latent Semantic Analysis

Luo Si[1] and Rong Jin[2]

[1] School of Computer Science, Carnegie Mellon University,
5000 Forbes Ave, Pittsburgh PA, U.S.A
lsi@cs.cmu.edu
[2] Department of Computer Science and Engineering, Michigan State University,
East Lansing, MI, U.S.A
rongjin@cse.msu.edu

Abstract. Mixture models, such as Gaussian Mixture Model, have been widely used in many applications for modeling data. Gaussian mixture model (GMM) assumes that data points are generated from a set of Gaussian models with the same set of mixture weights. A natural extension of GMM is the probabilistic latent semantic analysis (PLSA) model, which assigns different mixture weights for each data point. Thus, PLSA is more flexible than the GMM method. However, as a tradeoff, PLSA usually suffers from the overfitting problem. In this paper, we propose a regularized probabilistic latent semantic analysis model (RPLSA), which can properly adjust the amount of model flexibility so that not only the training data can be fit well but also the model is robust to avoid the overfitting problem. We conduct empirical study for the application of speaker identification to show the effectiveness of the new model. The experiment results on the NIST speaker recognition dataset indicate that the RPLSA model outperforms both the GMM and PLSA models substantially. The principle of RPLSA of appropriately adjusting model flexibility can be naturally extended to other applications and other types of mixture models.

1 Introduction

Mixture models, such as Gaussian Mixture Model, have been widely used throughout the applications of data mining and machine learning. For example, Gaussian Mixture model (GMM) has been applied for time series classification [8], image texture detection [7] and speaker identification [9]. In these tasks, the GMM model assumes that data points from a specific object or class (e.g., a speaker in speaker identification) are generated from a pool of Gaussian models with fixed mixture weights; it estimates mixture models from the training data using a maximum likelihood method; it predicts test data with the classes that generate the test data with the largest probabilities.

One general problem of modeling data with GMM is that GMM uses the same set of mixture weights for all the data points of a particular class, which limits the power of the mixture model in fitting the training data accurately. In contrast, a probabilistic latent semantic analysis (PLSA) [5][6] model allows each data point to choose its own mixture weights. Apparently, PLSA model is more flexible than GMM model in that a different set of mixture weights is introduced for each data point. However, as a tradeoff, PLSA has a substantially larger parameter space than the GMM model; the excessive freedom of assigning data point dependant mixture weights invites the PLSA model to the potential overfitting problem given the limited amount of training data.

In this paper, we propose a *regularized probabilistic latent semantic analysis* (RPLSA) model that addresses the overfitting problem in PLSA by regularizing the mixture weights. In particular, a regularization term is introduced in RPLSA, which punishes the objective function in RPLSA when different data points of the same class choose mixture weights that are far away from each other. It is an intermediate model between GMM and PLSA: different mixture weights are allowed for data points; but similar mixture weights are favored for different data points in the same class.

Empirical study for the application of speaker identification was conducted to show the effectiveness of the new RPLSA model. The NIST 1999 speaker recognition evaluation dataset with 539 speakers were used and the experiment results indicate that the RPLSA model achieves better results than both the GMM and PLSA. Furthermore, careful analysis shows that the advantage of RPLSA comes from the power of properly adjusting model flexibility.

2 Previous Research of Mixture Model

In this section, we only survey the most related research of mixture model.

2.1 Gaussian Mixture Model

GMM is one of the most widely used mixture modeling techniques [4][7][8][9]. It is a simple model and is reasonably accurate when data are generated from a set of Gaussian distributions. Let $X^i = \{x_t, 1 \le t \le T^i\}$ denote the feature vectors for data points from the ith class (e.g., a particular speaker). They are modeled by a total number of J Gaussians as follows:

$$P(X^i | \theta^i_{GMM}) = \prod_{t=1}^{T^i} \sum_{j=1}^{J} P(z_j) P_{z_j}(x_t | u_j, \Sigma_j) \quad (1)$$

where θ^i_{GMM} includes all the model parameters, i.e., $\{ P(z_j), u_j, \Sigma_j, 1 \le j \le J \}$. $P_{z_j}(x_t | u_j, \Sigma_j)$ is the Gaussian distribution for the j-th class, with a mean vector u_j and a covariance matrix Σ_j as:

$$P_{z_j}(x_t | u_j, \Sigma_j) = \frac{1}{(2\pi)^{D/2} |\Sigma_j|^{1/2}} \exp\{-\frac{1}{2}(x_t - u_j)^T \Sigma_j^{-1}(x_t - u_j)\} \quad (2)$$

where D is the dimension of the feature vector x_t. Usually Σ_j is set to be a diagonal matrix as $diag\{\sigma^2_{jd} : 1 \le d \le D\}$ in order to reduce the size of the parameter space [4].

It can be seen from Equation (1) that the data points of a specific class are generated from multiple Gaussian models with an identical set of mixture weights (i.e., $P(z_j)$). This constraint may not be valid in the data modeling process. For example, in speaker identification, mixture weights for a vowel can be significantly different from the mixture weights for a consonant. Therefore, it is important to incorporate data point dependent mixture weights into the framework of mixture models.

2.2 Probabilistic Latent Semantic Analysis

Unlike the Gaussian Mixture Model, the probabilistic latent semantic analysis model (PLSA) allows for data point specific mixture weights. Formally, the likelihood of training data for the ith class is written as:

$$P(X^i | \theta^i_{PLSA}) = \prod_{t=1}^{T^i} \sum_{j=1}^{J} P(z_j | d_t) P_{z_j}(x_t | u_j, \Sigma_j) \tag{3}$$

where θ^i_{PLSA} includes $\{u_j, \Sigma_j, 1 \le j \le J; P(z_j | d_t), 1 \le j \le J, 1 \le t \le T^i\}$. Note that a dummy variable d_t is introduced for every data point, and therefore the mixture weights $P(z_j | d_t)$ are data point dependent. The PLSA model was originally proposed for the probabilistic semantic indexing (PLSI) technique of information retrieval [5][6]. Both PLSI and PLSA allow data point specific mixture weights, but the PLSI model is based on multinomial distributions to model documents while the PLSA model is used here for modeling continuous data with Gaussian distributions. Note that the PLSA model shares the same idea with the tied-mixture model technique [1], which assumes that speech data is generated from a common pool of Gaussian models and each data point can choose its own mixture weights independently.

Because the mixture weights are data point dependent, PLSA is capable to fit training data better than GMM. However, a potential problem with PLSA is that it has a significantly larger parameter space than GMM, thus is prone to overfitting training data. To alleviate this problem, a maximum posterior (MAP) smoothing technique can be used for estimating PLSA. In particular, priors are introduced for parameters in the Gaussian models, and the parameters are estimated by maximizing the posterior of training data:

$$\log P(\theta^i_{PLSA} | X^i) \propto \sum_{t=1}^{T^i} \log(\sum_{j=1}^{J} P(z_j | d_t) P_{z_j}(x_t | u_j, \Sigma_j))$$
$$+ A \sum_{j=1}^{J} \log P(u_j | u_0, \Sigma_0) + B \sum_{j=1}^{J} \sum_{d=1}^{D} \log P(\sigma^2_{jd} | a_{0d}, \beta_{0d}) \tag{4}$$

The first item on the right hand side is the likelihood of training data. The next two items are the conjugate priors for the means and variances in the Gaussian models. A and B are two constants that adjust the weights of priors. $P(u_j | u_0, \Sigma_0)$ is a Gaussian distribution with mean u_0 and variance Σ_0 as a diagonal matrix $diag\{\sigma^2_{0d}\}$; $P(\sigma^2_{jd} | a_{0d}, \beta_{0d})$ is an inverse gamma distribution with parameters a_{0d}, β_{0d} as:

$$P(\sigma_{jd}^2 \mid a_{0d}, \beta_{0d}) \propto (\sigma_{jd}^2)^{-(a_{0d}+1)} e^{\frac{-\beta_{0d}}{\sigma_{jd}^2}} \qquad (5)$$

Although maximum posterior smoothing can alleviate the overfitting problem in some extent, the PLSA model still suffers from the excessive freedom of assigning totally independent data point specific mixture weights. To further address this problem, a novel method of regularizing mixture weights is proposed in this paper.

2.3 Latent Dirichlet Allocation

Latent Dirichlet Allocation (LDA) [2] is a generative model for collections of discrete data such as text. In LDA, each item (document) of a class (text collection) is modeled as a finite mixture over a set of topics (mixture models). LDA shares a common feature with the new research in this paper in that both of them choose moderate amount of model flexibility. LDA assumes that the mixture weights of items in a class are generated from a common Dirichlet distribution so that the weights for different data points in the same class are coupled instead of being chosen independently.

However, LDA model requires sophiscated variational methods to calculate the model parameters both in training and testing phrases, which is time consuming and thus limits the application of LDA in practical work. Furthermore, LDA model does not work well when each item contains a very small number of data points (like documents contain small number of words by average, or in speaker identification each item of a speaker utterance is a single vector of acoustic features in multi-dimensional space). Specifically consider the extreme case when each item only contains a single data point. LDA models a class X^i with single data point items as:

$$P(X^i \mid \theta_{LDA}) = \prod_{t=1}^{T^i} \int \left(P(u \mid \alpha)(\sum_{j=1}^{J} P(z_j \mid u) P_{z_j}(x_t \mid u_j, \Sigma_j)) \right) du \qquad (6)$$

Where $P(u \mid \alpha)$ is the Dirichlet distribution that generates the mixture weights for all data points. By switching the order of integration and summation and integrating out the parameter u, Equation (6) becomes:

$$P(X^i \mid \theta_{LDA}) = \prod_{t=1}^{T^i} \sum_{j=1}^{J} \frac{\alpha_j}{\sum \alpha_{j'}} P_{z_j}(x_t \mid u_j, \Sigma_j) \qquad (7)$$

This is essentially a GMM model if we set $\alpha_j / \sum_{j'} a_{j'}$ as the mixture weight $P(z_j)$ in the GMM model.

3 Regularized Probabilistic Latent Semantic Analysis Model

From the above research, we find that both GMM and PLSA are two extreme cases of the mixture model family: GMM uses the same set of mixture weights for all data points of the same class, thus lacking flexibility; PLSA model allows each data point to choose its own mixture weights and therefore is prone to overfitting training data.

A better idea is to develop an algorithm that can properly adjust the amount of model flexibility so that not only the training data can be fit well but also the model is robust to overfitting problems. This is the motivation of the regularized probabilistic latent semantic analysis model (RPLSA).

3.1 Model Description

Similar to the PLSA model, RPLSA allows each data point to choose its own mixture weights. At the meantime, it requires mixture weights from different data points to be similar in order to avoid overfitting. This is realized by assuming that there is a common set of mixture weights and mixture weights for different training data points should be close to the common set of mixture weights, formally as:

$$\log P(\theta^i_{RPLSA} | X^i) \propto \sum_{t=1}^{T^i} \log(\sum_{j=1}^{J} P(z_j | d_t) P_{z_j}(x_t | u_j, \Sigma_j)) + A \sum_{j=1}^{J} \log P(u_j | u_0, \Sigma_0)$$
$$+ B \sum_{j=1}^{J} \sum_{d=1}^{D} \log P(\sigma^2_{jd} | \alpha_{0d}, \beta_{0d}) - C \sum_{t=1}^{T^i} \sum_{j=1}^{J} P_c(z_j) \log \frac{P(z_j | d_t)}{P_c(z_j)} \quad (8)$$

Compared to the PLSA model in Equation (4), the above equation introduces a new regularization term, i.e., $C \sum_{t=1}^{T^i} \sum_{j=1}^{J} P_c(z_j) \log \frac{P(z_j | d_t)}{P_c(z_j)}$, into the objective function. It is a weighted sum of the Kullback-Leibler (KL) divergence between the common mixture weights (i.e., $P_c(z_j)$) and the mixture weights that are specific to each data point (i.e., $P(z_j | d_t)$). C is the regularization constant that controls the amount of model flexibility.

The role of the regularization term is to enforce mixture weights for different data points to be close to each other. In general, the closer the data-dependent mixture weights are to the common set of mixture weights, the smaller the KL divergence will be. Thus, by adjusting the constant C, we are able to adjust the flexibility of the RPLSA model: A small C will lead to a large freedom in assigning different mixture weights to different data points, thus exhibiting a behavior similar to the PLSA model; A large C will strongly enforce different data points to choose similar mixture weights, thus close to the behavior of the GMM method. Therefore, the RPLSA model connects the spectrum of mixture models between GMM and PLSA.

3.2 Parameter Estimation

The Expectation-Maximization (EM) algorithm [1] is used to estimate the model parameters of the RPLSA model. In the E step, the posterior probability of which mixture model each data point belongs to is calculated as follows:

$$P'(z_j | d_t) = \frac{P(z_j | d_t) P(x_t | u_j, \Sigma_j = diag\{\sigma^2_{jd}\}) + C P_c(z_j)}{\sum_{j'} P(z_{j'} | d_t) P(x_t | u_{j'}, \Sigma_{j'} = diag\{\sigma^2_{jd}\}) + C} \quad (9)$$

In the M step, the $P^{new}(z_j|d_t)$, u_j^{new} and Σ^{new} parameters are updated using Equations (10), (11) and (12) separately.

$$P^{new}(z_j|d_t) = P'(z_j|d_t) \tag{10}$$

$$u_{jd}^{new} = \frac{\sum_{t=1}^{T^i} P'(z_j|d_t)\sigma_{jd}^{-2} x_{td} + A\sigma_{0d}^{-2} u_{0d}}{\sum_{t'=1}^{T^i} P'(z_j|d_{t'})\sigma_{jd}^{-2} + A\sigma_{0d}^{-2}} \tag{11}$$

$$\left[\sigma_{jd}^{new}\right]^2 = \frac{\sum_{t=1}^{T^i} P'(z_j|d_t)(x_{td}-u_{jd})^2 + 2B\beta_{0d}}{\sum_{t'=1}^{T^i} P'(z_j|d_{t'}) + 2B(\alpha_{0d}+1)} \tag{12}$$

where u_{jd} and σ_{jd} are the dth element of the mean and variance respectively for the jth mixture, and x_{td} is the dth element of the feature vector x_t.

Finally, the common set of mixture weights is updated as follows:

$$P_c^{New}(z_j) \propto \exp\{1/T^i * \sum_{t=1}^{T^i} \log(P'(z_j|d_t))\} \tag{13}$$

which is essentially the geometric mean of the corresponding mixture weights that are attached to each data point. Note that choice of adaptively adjusting the common set of mixture weights in Equation (13) is different from the method that simply selecting a prior distribution of the mixture weights and estimating the model with maximum posterior smoothing. It can be imagined that the same set of prior of mixture weights (e.g., the Dirichlet prior distribution with uniform parameter values of the mixture weights) does not fit data with different characteristics. The adaptive estimation of the common set of mixture weights in RPLSA is a more reasonable choice.

The parameter estimation procedure for PLSA is a simplified version of that for RPLSA. In the expectation step, the posterior probability is calculated by a similar formula as Equation (9) without the factor of the regularization item. In the maximization step, the new parameters $P^{new}(z_j|d_t)$, u_j^{new} and Σ^{new} of PLSA are updated in a similar way as the Equations (10), (11) and (12).

3.3 Identification

The RPLSA model is different from the GMM model in that some parameters $P(z_j|d_t)$ need to be estimated for the test data in the identification phase. A plug-in EM procedure is used to accomplish this. Specifically, the EM algorithm described in Section 3.2 is rerun to estimate $P(z_j|d_t)$ for each test data point while all the other parameters are fixed. With the estimated new mixture weights, we can identify the test item (e.g., a vector of acoustic features) for a particular class (e.g., a speaker in the training set) whose model has the largest generation probabilities of test item X^{test} as:

$$ID\_Rst(X^{test}) = \arg\max_i P(X^{test} | \theta^i_{RPLSA}) \qquad (14)$$

The identification process of PLSA is almost the same as the procedure of RPLSA, which is not described due to space limit.

4 Experimental Results

This section shows empirical study that demonstrates the advantage of the new regularized probabilistic latent semantic model (RPLSA). Specifically, three models of GMM, PLSA and RPLSA are compared for the application of speaker identification.

4.1 Experiment Methodology

The experiments were conducted on the NIST 1999 speaker recognition evaluation dataset[1]. There are a total of 309 female speakers and 230 male speakers. The speech signal was pre-emphasized using a coefficient of 0.95. Each frame was windowed with a Hamming window and set to 30ms long with 50% frame overlap. 10 mel frequency cepstral coefficients were extracted from each speech frame. Both the training data and the test data come from the same channel. The training data consists of speech data of 7.5 seconds for each training speaker.

We present experiment results to address two issues: 1) Will the proposed RPLSA be more effective than the GMM and the PLSA models? 2) What is the power of the RPLSA model? What is the behavior of the RPLSA model with different amount of model flexibility by choosing different values for the regularization parameter C?

4.2 Experiment Results of Different Algorithms

The first set of experiments was conducted to study the effectiveness of the three mixture models. The numbers of mixture models were chosen by cross-validation for the three models. Specifically, 30 mixtures for GMM model, 50 for both PLSA and RPLSA. The smoothing prior parameters of PLSA and RPLSA were set as follows: u_0 is the mean value of the training data; Σ_0 is identity matrix; a_{0d} is 1 and β_{0d} is twice the variance of the dth value of the training data. The smoothing constants in Equations (4) and (8) were set as: A is $|T_i|/(10*J)$ and B is $|T_i|/J$ (where $|\bullet|$ indicates the number of items within a class). The regularization constant C of RPLSA was set to be 20 by cross-validation.

To compare the algorithms in a wide range we tried various lengths of test data. The results are shown in Table1. Clearly, both PLSA and RPLSA are more effective than the GMM in all cases. This can be attributed to the fact that both PLSA and RPLSA relax the constraint on mixture weights imposed by GMM. Furthermore, the RPLSA model outperforms the PLSA model. This is consistent with the motivation of

---

[1] http://www.nist.gov/speech/tests/spk/

the RPLSA model as it automatically adjusts the model flexibility for better recognition accuracy.

Table 1. Speaker identification errors for the Gaussian mixture model (GMM), the probabilistic latent semantic analysis model (PLSA) and the regularized probabilistic latent semantic analysis model (RPLSA)

Test Data Length	GMM	PLSA	RPLSA
2 Sec	37.8%	33.9%	**31.2%**
3 Sec	31.5%	24.7%	**21.8%**
5 Sec	27.3%	22.5%	**20.1%**

Table 2. Speaker identification errors for the smoothed Gaussian mixture model (GMM), the probabilistic latent semantic analysis model (PLSA) with uniform Dirichlet prior ($\alpha = 100$) and the RPLSA model

Test Data Length	GMM (Smoothed)	PLSA (Dirichlet Prior)	RPLSA
2 Sec	36.1%	33.2%	**31.2%**
3 Sec	30.2%	24.3%	**21.8%**
5 Sec	26.0%	22.3%	**20.1%**

To further confirm the hypothesis that RPLSA model has advantage than both the GMM and PLSA methods, two more sets of experiments were conducted. The first set of extended experiments was to train a GMM model with smoothed Gaussian model parameters like that used for PLSA (Two smoothed items of Gaussian model parameters like that of Equation (4) were introduced into the GMM objective function with A and B roughly tuned to be five times smaller than that of the RPLSA setting). The second set of extended experiments was to regularize the mixture weights in

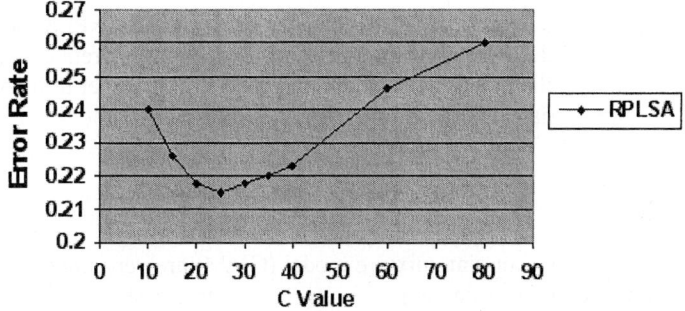

Fig. 1. Behavior of RPLSA Model with Different Values of Regularization Const

PLSA using a Dirichlet prior as described in Section 3.2. It is different from the regularization scheme of Equation (9) in that a Dirichlet prior uses a fixed set of common mixture weights (uniform) that is unable to adapt to the training data. The modified PLSA is trained with a new likelihood function of Equation (4) with an additional item of a Dirichlet prior with the parameter values of 100 (roughly tuned).

It can be seen from Table 2 that the new versions of GMM and PLSA give very small improvement of the original algorithms. The behavior of GMM model can be explained as that GMM has a much smaller parameter space than PLSA and RPLSA, smoothing does not give too much help. The results of the PLSA model with uniform Dirichlet prior indicates that the simple method of smoothing the mixture weights with a single prior does not successfully solve the overfitting problem.

4.3 Study the Behavior of the RPLSA Method

The new proposed RPLSA is an intermediate model between GMM and PLSA: different mixture weights are allowed for each data point; but similar mixture weights for different data points are encouraged. The RPLSA is the bridge to connect a spectrum of mixture models with two extreme cases of GMM and RPLSA models. Therefore, it is very interesting to investigate the behavior of the RPLSA method with different amount of model flexibly and its relationship with the GMM and RPLSA models.

Specifically, different values of parameter C in the RPLSA model of Equation (8) were investigated. 3 seconds' test data was used in this set of experiments. The detailed results are shown in Figure 1.

According to previous analysis in Section 3.1, we know that a smaller C value gives more freedom to the data points in choosing their own mixture weights, which leads to a behavior closer to that of the PLSA model. This is consistent with the observation from Figure 1. When C is as small as 10, RPLSA acquires a similar recognition accuracy with that of PLSA. On the other hand, a larger value for C makes RPLSA behave more like GMM. As indicated in Figure 1, a larger C leads to worse recognition accuracy.

For the middle part of the curve, with C ranging from 15 to 40, RPLSA acquires the best recognition accuracy; this suggests that RPLSA with reasonable amount of model flexibility reaches a better trade-off between enough model flexibility and model robustness.

The experiments in this section show the behavior of the new RPLSA model with different amount of model flexibility. It is consistent with our theoretical analysis that RPLSA has advantage than the GMM model and the RPLSA model in its better ability to adjust the appropriate amount of model flexibility.

5 Conclusion

Mixture models such as Gaussian mixture model (GMM) are very important tools for data mining and machine learning applications. However, classic mixture models like GMM have limitations in their modeling abilities as all data points of an object are required to be generated from a pool of mixtures with the same set of mixture weights. Previous research such as the probabilistic latent semantic analysis (PLSA) model has been proposed to release this constraint. PLSA allows totally independent data point specific mixture weights. But the excessive model flexibility makes PLSA tend to suffer from the overfitting problem.

This paper proposes a new regularized PLSA (RPLSA) model: On one hand, it is similar to the original PLSA model in that a different set of mixture weights is used for different data points; on the other hand, it is similar to GMM in that mixture weights for different data points are required to be similar to each other. In particular, the new model has the ability in adjusting the model flexibility of the mixture weights through the regularization term. Experiment results for speaker identification application have shown that the new RPLSA model outperforms both the GMM and the PLSA models substantially. Choosing the appropriate amount of modeling flexibility is a general problem for all mixture modeling techniques. The new research in this paper can be naturally incorporated with other types of mixture models than the GMM model and be applied for other applications.

Acknowledgements

We thank Alex Waibel and Qin Jin for their helpful discussion of this work.

References

1. Bellegarda J. R., Nahamoo, D.: Tied mixture continuous parameter modeling for speech recognition, IEEE Trans. Acoustic., Speech, Signal Processing, vol. 38, (1990).
2. Blei, D., Ng, A., Jordan., M.: Latent Dirichlet allocation. Journal of Machine Learning Research. (2003) 993-1022.
3. Dempster, A. P., Laird N. M., Rubin D. B.: Maximum likelihood from incomplete data via the EM algorithm. Journal of the Royal Statistical Society, B39. (1977) 1-38.
4. Furui, S.: An overview of speaker recognition technology. Automatic speech and speaker Recognition. Edited by Lee, C., Soong, F., Paliwal, K. Kluwer Academic Press. (1996).
5. Hofmann, T.: Probabilistic latent semantic analysis. In Proceedings of the 15th Conference on Uncertainty in AI (UAI). (1999).
6. Hofmann, T.: Probabilistic Latent Semantic Indexing. In Proceedings of the 22nd International Conference on Research and Development in Information Retrieval (SIGIR) (1999).
7. Permuter, H., Francos J., Jermyn I. H.: Gaussian mixture models of texture and colour for image database retrieval. in Proc . ICASSP, vol. 1. (2003) 25-88.
8. Povinelli R. J., Johnson M. T., Lindgren A. C., Ye J. J.: Time Series Classification Using Gaussian Mixture Models of Reconstructed Phase Spaces. IEEE Transactions on Knowledge and Data Engineering. Vol. 16. No6. (2004).
9. Reynolds, D. A.: Speaker identification and verification using Gaussian mixture speaker models. Speech Communication (17) (1998) 91-108.

Training Support Vector Machines Using Greedy Stagewise Algorithm

Liefeng Bo, Ling Wang, and Licheng Jiao

Institute of Intelligent Information Processing,
Xidian University, Xi'an 710071, China
{blf0218, wliiip}@163.com

Abstract. Hard margin support vector machines (HM-SVMs) have a risk of getting overfitting in the presence of the noise. Soft margin SVMs deal with this problem by the introduction of the capacity control term and obtain the state of the art performance. However, this disposal leads to a relatively high computational cost. In this paper, an alternative method, greedy stagewise algorithm, named GS-SVMs is presented to deal with the overfitting of HM-SVMs without the introduction of capacity control term. The most attractive property of GS-SVMs is that its computational complexity scales quadratically with the size of training samples in the worst case. Extensive empirical comparisons confirm the feasibility and validity GS-SVMs.

1 Introduction

Hard margin support vector machines have a risk of getting overfitting in the presence of the noise [1]. To deal with this problem, soft margin SVMs [2] introduce the capacity control parameter that allows a little training error to obtain the large margin. This is a highly effective mechanism for avoiding overfitting, which leads to good generalization performance. Though very successful, we can identify some shortages of soft margin SVMs:

① The training procedure of soft margin SVMs amounts to solving a constrained quadratic programming. Although the training problem is, in principle, solvable, in practice it is intractable by the classical optimization techniques, e.g. interior point method because their computational complexity usually scales cubically with the size of training samples.

② Capacity control parameter depends on the task at hand; hence there is no foolproof method for determining it before training. Usually, we have to resort to a cross validation procedure, which is wasteful in computation [3].

In the past few years, a lot of fast iterative algorithms were presented for tackling the problem ①. Probably, the most famous method among them is sequential minimization optimization algorithm (SMO), which is proposed by Platt [4] and further improved by Keerthi [5]. Some other examples include SVM^{light} [6], SimpleSVM [7],

SVMTorch [8], and so on. These algorithms proved to be effective and boosted the development of SVMs.

In this paper, an alternative method, greedy stagewise algorithm, named GS-SVMs is presented to deal with the overfitting of HM-SVMs. Instead of employing the capacity control term, GS-SVMs attempts to control the capacity of hypothesis space by algorithm itself. In summary, the proposed algorithms possess the following two attractive properties:

① The computational complexity of GS-SVMs is $O(nl)$, where l and n are the size of training samples and support vectors, respectively. Even in the worst situation that all the training samples are the support vectors, the computational complexity of GS-SVMs is only $O(l^2)$.

② No extra capacity control parameter is required.

2 Greedy Stagewise Algorithm for SVMs

The Wolfe dual of hard margin SVMs

$$\min\left(\frac{1}{2}\sum_{i,j=1}^{l}\alpha_i\alpha_j y_i y_j K(\mathbf{x}_i,\mathbf{x}_j) - \sum_{i=1}^{l}\alpha_i\right) \tag{1}$$

$$s.t. \quad 0 \le \alpha_i \quad i=1,\cdots,l,$$

can be regarded as a loss function induced by reproducing kernel Hilbert space norm. This allows us to approximate it using greedy algorithm. Due to the room limitation, the detailed interpretation is ignored and the interested reader can refer to [9]. Though HM-SVM is, in principle, solvable by the classical optimization technique, in practice it suffers from two serious problems: (1) their computational complexity usually scales cubically with the size of training samples; (2) there often is a risk of getting overfitting due to no capacity control term. Here, we will deal with the two problems by greedy stagewise algorithm, which attempts to approximate (1) quickly while avoids the overfitting. Greedy stagewise algorithm [10] can be described as the following.

For $m=1,2,\cdots l$,

$$(w_m,\beta_m) = \arg\min_{w,\beta}\left(\sum_{i=1}^{l}L\left(y_i, f_{m-1}(\mathbf{x}_i) + wK(\mathbf{x}_i,\mathbf{x}_\beta)\right)\right) \tag{2}$$

$$s.t. \quad \beta \ne \beta_j \quad j=1,2,\cdots m-1$$

and then

$$f_m = f_{m-1} + w_m K(\mathbf{x},\mathbf{x}_{\beta_m}). \tag{3}$$

where $L(\Box)$ denotes loss function, $f_0 \equiv 0$ and the constraint terms guarantee that each basis function is used once at most.

For SVMs, w takes the form αy_β, $\alpha \geq 0$. Using the loss function (1) we have

$$(\alpha_m, \beta_m) = \arg\min_{\alpha,\beta} \left(\frac{1}{2}\sum_{i=1}^{m-1}\sum_{j=1}^{m-1} \alpha_{\beta_i}\alpha_{\beta_j} y_{\beta_i} y_{\beta_j} K(\mathbf{x}_{\beta_i}, \mathbf{x}_{\beta_j}) - \sum_{i=1}^{m-1} \alpha_{\beta_i} + \alpha y_\beta \sum_{i=1}^{m-1} \alpha_{\beta_i} y_{\beta_i} K(\mathbf{x}_{\beta_i}, \mathbf{x}_\beta) + \frac{1}{2}\alpha^2 K(\mathbf{x}_\beta, \mathbf{x}_\beta) - \alpha \right)$$

s.t. $\alpha \geq 0$
$\beta \neq \beta_j \quad j = 1, 2, \cdots m-1$ (4)

Note that the first two terms of (4) can be ignored. Define the gradient vector

$$\mathbf{g}_\beta^m = \begin{cases} -1 & \text{if } m = 0 \\ y_\beta \sum_{j=1}^{m} \alpha_{\beta_j} y_{\beta_j} K(\mathbf{x}_{\beta_j}, \mathbf{x}_\beta) - 1 & \text{if } m \geq 1 \end{cases}$$ (5)

We can reformulate (4) as

$$(\alpha_m, \beta_m) = \arg\min_{\alpha,\beta} \left(\frac{1}{2}\alpha^2 K(\mathbf{x}_\beta, \mathbf{x}_\beta) + \alpha \mathbf{g}_\beta^{m-1} \right)$$

s.t. $\alpha \geq 0$
$\beta \neq \beta_j \quad j = 1, 2, \cdots m-1$ (6)

(6) can be solved in two steps. In the first step, we fix β and compute the minimal value h_β^{m-1} of (6) with respect to α. In the second step, we compute β_m by minimizing h_β^{m-1} with respect to β, and then compute α_m in terms of β_m. Fixing β, we have the subproblem

$$\min_{\alpha} \left(\frac{1}{2}\alpha^2 K(\mathbf{x}_\beta, \mathbf{x}_\beta) + \alpha \mathbf{g}_\beta^{m-1} \right)$$

s.t. $\alpha \geq 0$ (7)

Since (7) is a single variable quadratic programming, we can give its analytical solution, i.e.

$$\alpha_\beta = \begin{cases} -\mathbf{g}_\beta^{m-1}/K(\mathbf{x}_\beta, \mathbf{x}_\beta), & \text{if } -\mathbf{g}_\beta^{m-1}/K(\mathbf{x}_\beta, \mathbf{x}_\beta) > 0 \\ 0, & \text{if } -\mathbf{g}_\beta^{m-1}/K(\mathbf{x}_\beta, \mathbf{x}_\beta) \leq 0 \end{cases}$$ (8)

According to the positive definite property of kernel function, we have $K(\mathbf{x}_\beta, \mathbf{x}_\beta) > 0$. Thus (8) can be further simplified as

$$\alpha_\beta = \begin{cases} -\mathbf{g}_\beta^{m-1}/K(\mathbf{x}_\beta, \mathbf{x}_\beta), & \text{if } \mathbf{g}_\beta^{m-1} < 0 \\ 0, & \text{if } \mathbf{g}_\beta^{m-1} \geq 0 \end{cases}$$ (9)

Combining (7) and (9), we get

$$h_\beta^{m-1} = \min_{\alpha \geq 0}\left(\frac{1}{2}\alpha^2 K(\mathbf{x}_\beta,\mathbf{x}_\beta) + \alpha g_\beta^{m-1}\right) = \begin{cases} -(g_\beta^{m-1})^2/2K(\mathbf{x}_\beta,\mathbf{x}_\beta), & if\ g_\beta^{m-1} < 0 \\ 0, & if\ g_\beta^{m-1} \geq 0 \end{cases}. \quad (10)$$

In GS-SVMs, each basis function corresponds to a specified training sample and vice versa. Hence, if the basis function $K(\mathbf{x},\mathbf{x}_\beta)$ does not appear in f_m, we say its corresponding training sample $\mathbf{x}_\beta$ unused. From (10), we can derive that if the gradients of all the unused training samples are larger than zero, the loss function (1) will stop decreasing. Hence we will terminate the algorithm if the above condition is satisfied.

Considering (9) and (10), we can obtain the parameter pairs (α_m, β_m) by the following equations

$$\beta_m = \arg\min_{\beta \in Q}(h_\beta^{m-1}). \quad (11)$$

$$\alpha_m = -g_{\beta_m}^{m-1}/K(\mathbf{x}_{\beta_m},\mathbf{x}_{\beta_m}). \quad (12)$$

Thus the greedy stagewise algorithm for SVMs (GS-SVMs) can be described as

---
Algorithm 1: GS-SVMs
1. Set $f_0(\mathbf{x}) = 0$, $\alpha = \mathbf{0}$, $\mathbf{g}^0 = -\mathbf{1}$, $\mathbf{h}^0 = -\mathbf{1}$, $Q = \{1, 2, \cdots l\}$, $P = \varnothing$;
2. For $m = 1$ to l, do:
3.   If $\min_{\beta \in Q}(g_\beta^{m-1}) \geq 0$, stop;
4. $\beta_m = \arg\min_{\beta \in Q}(h_\beta^{m-1})$, $\alpha_m = -g_{\beta_m}^{m-1}/K(\mathbf{x}_{\beta_m},\mathbf{x}_{\beta_m})$;
5. $P = P \cup \{\beta_m\}$, $Q = Q - \{\beta_m\}$;
6. $g_\beta^m = g_\beta^{m-1} + \beta_m y_{\beta_m} y_\beta K(\mathbf{x}_\beta,\mathbf{x}_{\beta_m})$, $\beta \in Q$;
7.   Update h_β^{m-1}, $\beta \in Q$ according to (4.9);
8. $f_m(\mathbf{x}) = f_{m-1}(\mathbf{x}) + \alpha_m y_{\beta_m} K(\mathbf{x},\mathbf{x}_{\beta_m})$;
9. End For
10. End Algorithm

Fig. 1. Pseudo code of GS-SVMs

Updating g_β^k, $\beta \in Q$ is an operation of cost $O(l)$ and successive n update incurs a computational cost of $O(nl)$, where n is the size of support vector. Besides that, the memory requirement of GS-SVMs is only $O(l)$.

3 Empirical Comparison

In all the experiments, the kernel matrix is constructed by Gaussian kernel $K(\mathbf{x}_i, \mathbf{x}_j) = \exp(-\theta \|\mathbf{x}_i - \mathbf{x}_j\|^2)$. Following [4], we compare the number of kernel evaluations of GS-SVMs and SMO, which is an effective measure of the algorithm's speed. For the sake of fair comparison, we use the same data sets and kernel parameter as in [4]. Note that the number of kernel evaluations of SMO in Table 1. denotes the average number under the different capacity control parameters.

Table 1. Number of kernel evaluations of GS-SVMs and SMO. Each unit corresponds to 10^6 kernel evaluations. SMO-1 and SMO-2 correspond to SMO-Modification 1 and SMO-Modification 2 in [5]

Problems	Size	θ	Dim	SMO-1	SMO-2	GS-SVMs
Adult-1	1605	0.05	123	29.518	17.375	0.845
Adult-4	4781	0.05	123	344.977	231.349	6.791
Adult-7	16100	0.05	123	856.212	698.864	73.014
Web-1	2477	0.05	300	11.543	11.187	0.439
Web-4	7366	0.05	300	79.415	79.008	3.224
Web-7	24692	0.05	300	691.419	703.495	31.981

From Table 1, we can see that GS-SVMs obtain the speedup range from 10 to 30 on the different data sets. In order to validate the performance of GS-SVMs, we compare it with hard magin and Soft margin SVMs on the fifteen benchmark data sets that are from UCI machine learning repository [11]. One-against-one method is used to extend binary classifiers to multi-class classifiers.

On each data set, ten-fold cross validation is run. The average accuracy of ten-fold cross validation is reported in Table 2. For each training-test pair, ten-fold cross validation is performed on training set for tuning free parameters. The detailed experiment setup is the following:

(a) For soft margin SVMs, Kernel width and capacity control parameter are chosen from intervals $\log 2(\theta) = [-8,-7,\cdots,7,8]$ and $\log 2(C) = [-1,0,1,\cdots 8,9,10]$. This range is enough for our problems. The number of trainings on each training-test pair needed by this method is $10 \times 17 \times 12 = 2040$.

(b) For GS-SVMs and HM-SVMs, Kernel width is chosen from interval $\log 2(\theta) = [-8,-7,\cdots,7,8]$. The number of trainings on each training-test pair needed by this method is $10 \times 17 = 170$.

The two-tailed t-tests also indicate that GS-SVMs are significantly better than SVMs on Glass and worse than SVMs on Liver. As for the remaining data sets, GS-SVMs and SVMs obtain the similar performance. Hence we have the conclusion that GS-SVMs are significantly better in speed than SMO and comparable in performance with SMO.

Tabel 2. Accuracy of GS-SVMs, HM-SVMs and SVMs

Problems	Size	Dim	Class	GS-SVMs	HM-SVMs	SVMs
Australian	690	15	2	**84.93**	78.55	84.49
German	1000	20	2	74.20	69.30	**75.40**
Glass	214	9	6	71.54	68.66	66.81
Heart	270	13	2	**83.70**	76.67	83.23
Ionosphere	351	34	2	94.00	94.00	**94.02**
Iris	150	4	3	95.33	92.00	**96.00**
Liver	345	6	2	66.03	61.69	**71.29**
Page	5473	10	4	96.45	96.50	**96.93**
Diabetes	768	8	2	**77.21**	70.55	77.08
Segment	2310	18	7	**97.32**	96.84	97.01
Splice	3175	60	3	**96.72**	96.31	96.25
Vowel	528	10	11	98.29	**99.05**	**99.05**
WDBC	569	30	2	**97.72**	96.49	97.54
Wine	178	13	3	**98.89**	96.64	**98.89**
Zoo	101	10	7	**97.09**	96.09	96.09
Mean	/	/	/	88.63	85.96	**88.67**

4 Conclusion

This paper proposes a greedy stagewise algorithm, named GS-SVMs to deal with the overfitting of HM-SVMs. Empirical comparisons confirm the feasibility and validity of GS-SVMs.

References

1. Boser, B., Guyon, I. and Vapnik, V.: A training algorithm for optimal margin classifiers, In D. Haussler, Proceedings of the 5th Annual ACM Workshop on Computational Learning Theory, ACM Press, (1992) 144-152
2. Cotes, C. and Vapnik, V.: Support vector networks, Machine Learning 20 (1995) 273-279
3. Tipping, M.: parse Bayesian learning and the relevance vector machine, Journal of Machine Learning Research 1 (2001) 211-244
4. Platt, J.: Fast training of support vector machines using sequential minimal optimization In B. Schölkopf, C. J. C. Burges, and A. J. Smola, editors, Advances in Kernel Methods --- Support Vector Learning, Cambridge, MA: MIT Press, (1999) 185-208
5. Keerthi, S.S., Shevade, S.K., Bhattacharyya, C. and Murthy, K.R.K.: Improvements to Platt's SMO algorithm for SVM classifier design. Neural Computation13 (2001) 637-649
6. Joachims, T.: Making large-scale SVM learning practical, Advances in Kernel Methods- Support Vector learning, Cambridge, MA: MIT Press, (1999) 169-184
7. Vishwanathan, S.V.N., Smola, A.J. and Murty. M.N.: SimpleSVM. In Proceedings of the Twentieth International Conference on Machine Learning, 2003
8. Collobert, R. and Bengio,S.: SVMTorch: Support Vector Machines for Large-Scale Regression Problems Journal of Machine Learning Research,1 (2001) 143-160

9. Girosi, F.: An equivalence between sparse approximation and support vector machines, Neural Computation. 10 (1998) 1455-1480
10. Friedman, J.H.: Greedy Function Approximation: A gradient boosting machine. Annals of Statistics, 29 (2001) 1189-1232
11. Blake, C.L. and Merz, C.J.: UCI repository of machine learning databases (1998).

Cl-GBI: A Novel Approach for Extracting Typical Patterns from Graph-Structured Data

Phu Chien Nguyen, Kouzou Ohara, Hiroshi Motoda, and Takashi Washio

The Institute of Scientific and Industrial Research, Osaka University,
8-1 Mihogaoka, Ibaraki, Osaka, 567-0047, Japan
{chien, ohara, motoda, washio}@ar.sanken.osaka-u.ac.jp

Abstract. Graph-Based Induction (GBI) is a machine learning technique developed for the purpose of extracting typical patterns from graph-structured data by stepwise pair expansion (pair-wise chunking). GBI is very efficient because of its greedy search strategy, however, it suffers from the problem of overlapping subgraphs. As a result, some of typical patterns cannot be discovered by GBI though a beam search has been incorporated in an improved version of GBI called Beam-wise GBI (B-GBI). In this paper, improvement is made on the search capability by using a new search strategy, where frequent pairs are never chunked but used as pseudo nodes in the subsequent steps, thus allowing extraction of overlapping subgraphs. This new algorithm, called Cl-GBI (Chunkingless GBI), was tested against two datasets, the promoter dataset from UCI repository and the hepatitis dataset provided by Chiba University, and shown successful in extracting more typical patterns than B-GBI.

1 Introduction

In recent years, discovering frequent patterns of graph-structured data, i.e., frequent subgraph mining or simply graph mining, has attracted much research interest because of its broad application areas such as bioinformatics [2, 12], cheminformatics [8, 10, 15], etc. Moreover, since these patterns can be used as input to other data mining tasks (e.g., clustering and classification [6]), the graph mining algorithms play an important role in further expanding the use of data mining techniques to graph-based datasets.

AGM [8] and a number of other methods (AcGM [9], FSG [10], gSpan [15], FFSM [7], etc.) have been developed for the purpose of enumerating all frequent subgraphs of a graph database. However, the computation time increases exponentially with input graph size and minimum support. This is because the kernel of frequent subgraph mining is subgraph isomorphism, which is known to be NP-complete [5].

On the other hand, existing heuristic algorithms, which are not guaranteed to find the complete set of frequent subgraphs, such as SUBDUE [4] and GBI (Graph-Based Induction) [16], tend to find an extremely small number of patterns. Both the two methods use greedy search to avoid high complexity of the subgraph isomorphism problem. GBI extracts typical patterns from graph-structured data by recursively chunking two adjoining nodes. Later an improved version called B-GBI (Beam-wise Graph-Based Induction) [12] adopting the beam search was proposed to increase the search space, thus extracting more discriminative patterns while keeping the computational complexity within a tolerant level.

Since the search in GBI is greedy and no backtracking is made, which patterns are extracted by GBI depends on which pairs are selected for chunking. There can be many patterns which are not extracted by GBI. B-GBI can help alleviate this problem, but cannot solve it completely because the chunking process is still involved.

In this paper we propose a novel algorithm for extracting typical patterns from graph-structured data, which does not employ the pair-wise chunking strategy. Instead, the most frequent pairs are regarded as new nodes and given new node labels in the subsequent steps but none of them is chunked. In other words, they are used as pseudo nodes, thus allowing extraction of overlapping subgraphs. This algorithm, now called Chunkingless Graph-Based Induction (or Cl-GBI for short), was evaluated on two datasets, the promoter dataset from UCI repository [1] and the hepatitis dataset provided by Chiba University, and shown successful in extracting more typical substructures compared to the B-GBI algorithm.

2 Graph-Based Induction Revisited

2.1 Principle of GBI

GBI employs the idea of extracting typical patterns by stepwise pair expansion as shown in Fig. 1. In the original GBI, an assumption is made that typical patterns represent some concepts/substructures and "typicality" is characterized by the pattern's frequency or the value of some evaluation function of its frequency. We can use statistical indices as an evaluation function, such as frequency itself, Information Gain [13], Gain Ratio [14] and Gini Index [3], all of which are based on frequency. In Fig. 1 the shaded pattern consisting of nodes 1, 2, and 3 is thought typical because it occurs three times in the graph. GBI first finds the 1→3 pairs based on its frequency, chunks them into a new node 10, then in the next iteration finds the 2→10 pairs, chunks them into a new node 11. The resulting node represents the shaded pattern.

It is possible to extract typical patterns of various sizes by repeating the above three steps. Note that the search is greedy and no backtracking is made. This means that in enumerating pairs no pattern which has been chunked into one node is restored to the original pattern. Because of this, all the "typical patterns" that exist in the input graph are not necessarily extracted. The problem of extracting all the isomorphic subgraphs is known to be NP-complete. Thus, GBI aims at extracting only meaningful typical patterns of a certain size. Its objective is not finding all the typical patterns nor finding all the frequent patterns.

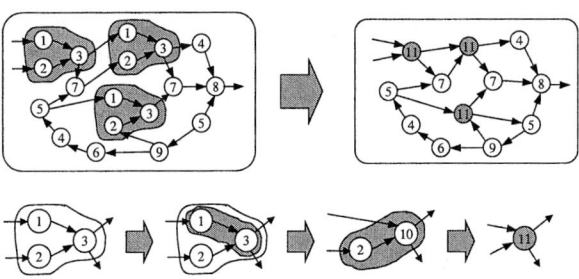

Fig. 1. Principle of GBI

As described earlier, GBI can use any criterion that is based on the frequency of paired nodes. However, for finding a pattern that is of interest any of its subpatterns must be of interest because of the nature of repeated chunking. In Fig. 1 the pattern 1→3 must be typical for the pattern 2→10 to be typical. Said differently, unless pattern 1→3 is chunked, there is no way of finding the pattern 2→10. Frequency measure satisfies this monotonicity. However, if the criterion chosen does not satisfy this monotonicity, repeated chunking may not find good patterns even though the best pair based on the criterion is selected at each iteration. To resolve this issue GBI was improved to use two criteria, one for frequency measure for chunking and the other for finding discriminative patterns after chunking. The latter criterion does not necessarily hold monotonicity property. Any function that is discriminative can be used, such as Information Gain [13], Gain Ratio [14] and Gini Index [3], and some others.

2.2 Beam-Wise Graph-Based Induction (B-GBI)

Since the search in GBI is greedy and no backtracking is made, which patterns (subgraphs) are extracted by GBI depends on which pair is selected for chunking. There can be many patterns which are not extracted by GBI. In Fig. 2, if the pair B–C is selected for chunking beforehand, there is no way to extract the substructure A–B–D even if it is a typical pattern.

A beam search is incorporated into GBI in B-GBI [12] within the framework of greedy search in order to relax this problem, increase the search space, and

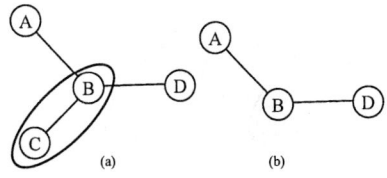

Fig. 2. Missing patterns due to chunking order

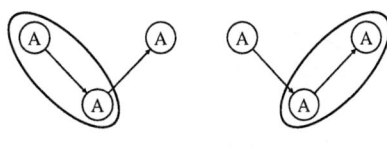

Fig. 3. Two different pairs representing identical patterns

extract more discriminative patterns while still keeping the computational complexity within a tolerant level. A certain number of pairs ranked from the top are selected to be chunked individually. To prevent each branch from growing exponentially, the total number of pairs to be chunked (the beam width) is fixed at every time of chunking. Thus, at any iteration step, there is always a fixed number of chunking performed in parallel.

Another improvement made in conjunction with B-GBI is canonical labeling. GBI assigns a new label to each newly chunked pair. Because it recursively chunks pairs, it happens that the new pairs that have different labels happen to be the same pattern. A simple example is shown in Fig. 3. They represent the same pattern but the way they are constructed is different. To identify if two pairs represent the same pattern, each pair is represented by canonical label [5] and they are regarded identical only when the label is the same.

3 Problem Caused by Chunking in B-GBI

As described in Section 2.2, B-GBI increases the search space by running GBI in parallel. As a result, B-GBI can help alleviate the problem of overlapping subgraphs, but cannot solve it completely because the chunking process is still involved. It happens that some of the overlapping patterns are not discovered by B-GBI. For example, suppose in Fig. 2 the pair B–C is most frequent, followed by the pair A–B. When $b = 1$, there is no way that the pattern A–B–D is discovered because the pair B–C is chunked first, but by setting $b = 2$, the pair A–B can be chunked in the second beam and if the substructure A–B–D is frequent enough, there is a chance that the pair (A–B)–D is chunked at next iteration. However, setting b very large is prohibitive from the computational point of view.

Any subgraph that B-GBI can find is along the way in the chunking process. Thus, it happens that a pattern found in one input graph is unable to be found in the other input graph even if it does exist in the graph. An example is shown in Fig. 4, where even if the pair A – B is selected for chunking and the substructure D – A – B – C exists in the input graphs, we may not find that substructure because an unexpected pair A – B is chunked (see Fig. 4(b)). This causes a serious problem in counting the frequency of a pattern.

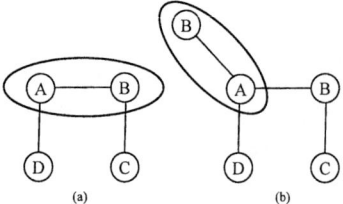

Fig. 4. A pattern is found in one input graph but not in the other

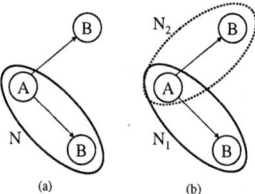

Fig. 5. An example of frequency counting

Complete graph mining algorithms such as AGM [8], AcGM [9], FSG [10], gSpan[15], FFSM [7], etc. do not face the problem of overlapping subgraphs since they can find all frequent patterns in a graph database. However, these methods are designed to find existence or non-existence of a certain pattern in one transaction and not to count how many times a certain pattern appear in one transaction. They also cannot give information on the positions of each pattern in any transaction of the graph database which is required by domain experts.

Heuristic algorithms for graph mining such as SUBDUE [4], GBI [16] and GREW [11], on the other hand, are designed for the purpose of enumerating typical patterns in a single large graph. Specially, B-GBI [12] can find (not all) typical patterns in either a single large graph or a graph database. However, all of them are not designed to detect the positions of patterns in any graph transaction. In Section 4, we will introduce a novel algorithm that can overcome the problem of overlapping subgraphs imposed on both GBI and B-GBI. The proposed algorithm, called Cl-GBI (Chunkingless Graph-Based Induction), employs a "chunkingless chunking" strategy, where frequent pairs are never chunked but used as pseudo nodes in the subsequent steps, thus allowing extraction of overlapping subgraphs. It can also give the positions of patterns present in each graph transaction as well as be applied to find frequent patterns in either a single large graph or a graph database.

4 Chunkingless Graph-Based Induction (Cl-GBI)

4.1 Approach

The basic ideas of Cl-GBI are described as follows. Those pairs that connect two adjoining nodes in the graphs are counted and a certain fixed number of pairs (the beam width) ranked from the top are selected. In B-GBI, each of the selected pairs is registered as one node and this node is assigned a new label. Then, the graphs in the respective state are rewritten by replacing all the occurrences of the selected pair with a node with the newly assigned label (pair-wise chunking).

In Cl-GBI, we also register the selected pairs as new nodes and assign new labels to them. But those pairs are never chunked and the graphs are not "compressed" nor copied into respective states as in B-GBI. In the presence of the pseudo nodes (i.e., newly assigned-label nodes), we count the frequencies of pairs consisting of at least one new pseudo node. The other is either one of pseudo nodes including those already created in the previous steps or an original one. In other words, the other is one of the existing nodes. Among the remaining pairs (after selecting the most frequent pairs) and the new pairs which have just been counted, we select the most frequent pairs, with the number equal to the beam width specified in advance, again and so on.

These steps are repeated in a predetermined number of times, each of which is referred to as a level. Those pairs that satisfy a typicality criterion (e.g., pairs whose information gain exceeds a given threshold) among all the extracted pairs are the output of the algorithm.

A frequency threshold is used to reduce the number of pairs being considered to be typical patterns. Another possible method to reduce the number of pairs is to eliminate those pairs whose typicality measure is low even if their frequency count is above the frequency threshold. The two parameters, beam width and number of levels, control the search space. Frequency threshold is another important parameter.

As in B-GBI, the Cl-GBI approach can handle both directed and undirected graphs as well as both general and induced subgraphs. It can also extract typical patterns in either a single large graph or a graph database.

4.2 Algorithm of Cl-GBI

Given a graph database, two natural numbers b (beam width) and N_e (number of levels), and a frequency threshold θ, the new "chunkingless chunking" strategy repeats the following three steps.

Step 1. Extract all the pairs consisting of two connected nodes in the graphs, register their positions using node id (identifier) sets, and count their frequencies. From the 2nd level on, extract all the pairs consisting of two connected nodes with at least one node being a new pseudo node.

Step 2. Select the b most frequent pairs from among the pairs extracted at Step 1 (from the 2nd level on, from among the unselected pairs in the previous levels and the newly extracted pairs). Each of the b selected pairs is registered as a new node. If either or both nodes of the selected pair are not original nodes but pseudo nodes, they are restored to the original patterns before registration.

Step 3. Assign a new label to each pair selected at Step 2 but do not rewrite the graphs. Go back to Step 1.

These steps are repeated N_e times (N_e levels). All the pairs extracted at Step 1 in all the levels (i.e. level 1 to level N_e), including those that are not used as pseudo nodes, are ranked based on a typicality criterion using a discriminative function such as Information Gain, Gain Ratio or Gini Index. It is worth noting that those pairs that have frequency count below a frequency threshold θ are eliminated, which means that there are three parameters b, N_e, θ to control the search in Cl-GBI.

The output of Cl-GBI algorithm is a set of ranked typical patterns, each of which comes together with the positions of every occurrence of the pattern in each transaction of the graph database (given by the node id sets) as well as the number of occurrences.

4.3 Implementation Issues of Cl-GBI

The first issue concerns with frequency counting. To count the number of occurrences of a pattern in a graph transaction, the canonical labeling employed in [12] is adopted. However, only canonical labeling cannot solve the problem completely as shown in Fig. 5. Suppose that the pair A → B is registered as a

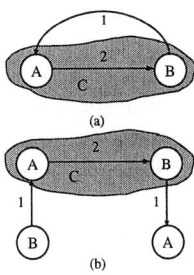 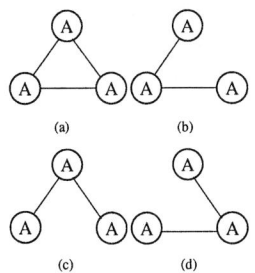

Fig. 6. Relations between a pseudo node and its embedded nodes

Fig. 7. Three occurrences but counted once

pseudo node N in the graph shown in Fig. 5(a). How many times the pair N → B should be counted here? If only the canonical label is considered, the answer is 2 because there are two pseudo nodes N_1 and N_2 as shown in Fig. 5(b), and both $N_1 \to B$ and $N_2 \to B$ are counted separately. However, the pair N → B should be counted once. Our solution is to incorporate the canonical label with the node id set. If both the canonical label and the node id set are identical for two subgraphs, we regard that they are the same and count once.

The second issue regards the relations between a pseudo node and those nodes which are embedded inside. Think of the pseudo node C and two embedded nodes A, B in Fig. 6(a). What are the relations between C and A or C and B? In the case of enumerating frequent induced subgraphs, there is not any relation between C and A nor C and B. This is because a pair in this case must consist of two nodes and all links between them. However, in the case of extracting frequent general subgraphs, there is still a link between C and A as well as a link between C and B. To differentiate between the graphs shown in Fig. 6(a) and Fig. 6(b), a flag indicating whether it is a self-loop or not is required.

In summary, a pair consists of 6 elements: labels of two nodes and label of the link between them, information of which two nodes inside the pair the link is connected to, and a self-loop flag. In the case of enumerating frequent induced subgraphs, all links between two nodes should be considered.

4.4 Unsolved Problem of Cl-GBI

We found that there is still a problem in frequency counting that the use of both the canonical label and the node id set cannot solve. Think of the graph in Fig. 7(a). The three subgraphs A–A–A illustrated in Figs. 7 (b), (c), and (d) share the same canonical label and the same node id set. Our current Cl-GBI cannot distinguish between these three. However, this problem arises only when extracting general subgraphs. It causes no problem in the case of enumerating frequent induced subgraphs.

5 Experiments

To assess the performance of the Cl-GBI approach, we conducted some experiments on both synthetic and real-world graph-structred datasets. The proposed Cl-GBI algorithm was implemented in C^{++}. Since the current implementation is very naive, we did not evaluate the computation time. It should be noted that all graphs/subgraphs reported here are connected ones.

In the first experiment, we verified that Cl-GBI is capable of finding all frequent patterns in a single graph that other graph mining algorithms cannot. An example of such a single graph is shown in Fig. 8(a). Suppose that the problem here is to find frequent induced subgraphs that occur at least 3 times in the graph. Fig. 8(c) shows an example of frequent induced subgraph which has the support of 3.

The current algorithms that are designed for extracting frequent patterns in a single graph such as GBI [16], B-GBI [12], SUBDUE [4], or GREW [11], etc. cannot discover the pattern shown in Fig. 8(c) because three occurrences of this pattern are not disjoint, but overlapping. Meanwhile, the complete graph mining algorithms like AcGM [9], FSG [10], gSpan [15], FFSM [7], etc., in case that they are adapted to find frequent patterns in a single graph, also cannot find that pattern because of the monotonic nature. Since the pattern shown in Fig. 8(b) occurs only once in the graph and thus cannot be extracted, the pattern shown in Fig. 8(c) which is one of its super-graph is also unable to be found. The proposed Cl-GBI algorithm, on the other hand, can find all 36 frequent induced subgraphs, including the one shown in Fig. 8(c), with $b = 3, N_e = 5$.

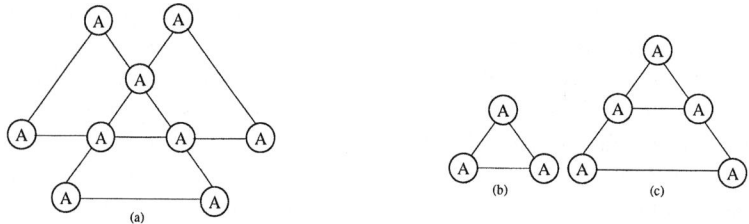

Fig. 8. An example of finding frequent patterns in a single graph

In the second experiment, we evaluated the performance of Cl-GBI on the promoter dataset from UCI repository [1] and the hepatitis dataset provided by Chiba University. Since only the number of frequent induced subgraphs discovered by Cl-GBI is evaluated in this experiment, it is not required to use the whole datasets. Therefore, we used only positive instances of the two datasets, i.e., "promoter sequences" in the case of promoter dataset and "patients who have response to interferon therapy" in the case of hepatitis dataset. These promoter and hepatitis datasets were converted to undirected and directed graph database, respectively. The former contains 53 undirected graphs having the

Table 1. Number of frequent induced subgraphs obtained from the promoter dataset

Algorithm	Number of discovered patterns	Parameters
Cl-GBI	1888	$b=5, N_e = 5$
Cl-GBI	4638	$b=6, N_e = 20$
B-GBI	580	$b = 5$
AcGM	4638	N/A

same size of 57, while the latter has 56 directed graphs having the average size of 75.4. It should be noted that the promoter dataset is usually converted to directed graphs [12] since information on the order of nucleotides is important. We compared the number of frequent induced subgraphs discovered by Cl-GBI with B-GBI [12] and AcGM [9] given the frequency threshold of 50%. B-GBI is an improved version of GBI, while AcGM can extract all frequent induced subgraphs.

Table 1 shows some experimental results obtained from the promoter dataset. It is shown that Cl-GBI can find more frequent patterns than B-GBI given the same beam width. Also, as is easily predicted, this algorithm can find all the frequent patterns by setting b and N_e large enough. One of the nice aspects of B-GBI is that the size of the input graph keeps reducing progressively as the chunking proceeds, and thus the number of pairs to be considered also progressively decreases accordingly. In the case of Cl-GBI, the number of pairs to be considered keeps increasing because the number of pseudo nodes keeps increasing as the search proceeds. Thus, it is important to select appropriate values for b and N_e.

For the directed graph database representing the hepatitis dataset, Cl-GBI extracts 4439 frequent patterns with $b = 5$, $N_e = 10$ and B-GBI finds 870 frequent patterns with $b = 5$. Meanwhile, since the current version of AcGM [9] has not been implemented to handle directed graphs, we cannot use it to find frequent patterns in this graph dataset. However, even if we consider the graphs in this dataset as undirected graphs, AcGM cannot give the results due to the large graph size and the large number of links.

In the third experiment, we modified Cl-GBI to simulate B-GBI in the following way. The frequency threshold θ is used for selecting b (the beam width) most frequent pairs only and those pairs that have frequency count below θ are now not eliminated. The promoter dataset (both promoter and non-promoter sequences) were converted to directed graphs as in [12] and we obtained 106 directed graphs having the same size of 57 which were classified equally into 2 classes: positive class and negative class.

We set $\theta = 50\%$ and $b=5$. Suppose that a pattern is called typical if its information gain [13] is greater than or equal to 0.01, i.e., the typicality measure is information gain and the typicality threshold is set as 0.01. Cl-GBI finds 3269 typical patterns in 5 levels, including 8 patterns having information gain greater than 0.19. Meanwhile, the number of typical patterns discovered by B-GBI is 3045 in 18 levels, however, and only 4 patterns among them have information gain greater than 0.19. Within 10 levels, Cl-GBI finds 13407 typical patterns which include 35 patterns having information gain greater than 0.19. This justifies the fact that Cl-GBI can find more typical patterns than B-GBI.

6 Conclusion

A novel algorithm, Chunkingless Graph-Based Induction (Cl-GBI), was introduced for the purpose of discovering typical patterns in either a single large graph or a graph database. The proposed method employs a "chunkingless chunking" strategy which helps overcome the problem of overlapping subgraphs. Also, Cl-GBI can give the correct number of occurrences of a pattern as well as their positions in each transaction of the graph database. Experiments conducted on both synthetic and real-world graph-structured data confirm its effectiveness. For future work we plan to employ some heuristics to speed up the Cl-GBI algorithm in order to extract larger typical subgraphs and apply the method to some application domains.

References

1. Blake, C.L., Keogh, E., and Merz, C.J. 1998. UCI Repository of Machine Learning Database, http://www.ics.uci.edu/~mlearn/MLRepository.html.
2. Borgelt, C. and Berthold, M.R. 2002. Mining Molecular Fragments: Finding Relevant Substructures of Molecules. In: *Proc. ICDM 2002*, pp.51–58.
3. Breiman, L., Friedman, J.H., Olshen, R.A., and Stone, C.J. 1984. *Classification and Regression Trees*, Wadsworth & Brooks/Cole Advanced Books & Software.
4. Cook, D. J. and Holder, L. B. 1994. Substructure Discovery Using Minimum Description Length and Background Knowledge, *Artificial Intelligence Research*, Vol. 1, pp. 231–255.
5. Fortin, S. 1996. The Graph Isomorphism Problem, *Technical Report TR96-20*, Department of Computer Science, University of Alberta, Edmonton, Canada.
6. Gaemsakul, W., Matsuda, T., Yoshida, T., Motoda, M., and Washio, T. 2003. Classifier Construction by Graph-Based Induction for Graph-Structured Data, In: *Proc. PAKDD 2003*, pp. 52–62.
7. Huan, J., Wang, W., and Prins, J., 2003. Efficient Mining of Frequent Subgraphs in the Presence of Isomorphism, In: *Proc. ICDM 2003*, pp. 549–552.
8. Inokuchi, A., Washio, T., and Motoda, H. 2003. Complete Mining of Frequent Patterns from Graphs: Mining Graph Data, *Machine Learning*, Vol. 50, No. 3, pp. 321–354.
9. Inokuchi, A., Washio, T., Nishimura, K., and Motoda, H. 2002. A Fast Algorithm for Mining Frequent Connected Subgraphs, *IBM Research Report RT0448*, Tokyo Research Laboratory, IBM Japan.
10. Kuramochi, M. and Karypis, G. 2004. An Efficient Algorithm for Discovering Frequent Subgraphs, *IEEE Trans. Knowledge and Data Engineering*, Vol. 16, No. 9, pp. 1038-1051.
11. Kuramochi, M. and Karypis, G. 2004. GREW–A Scalable Frequent Subgraph Discovery Algorithm, In: *Proc. ICDM 2004*, pp. 439–442.

12. Matsuda, T., Motoda, H., Yoshida, T., and Washio, T. 2002. Mining Patterns from Structured Data by Beam-wise Graph-Based Induction, In: *Proc. DS 2002*, pp. 422–429.
13. Quinlan, J.R. 1986. Induction of Decision Trees, *Machine Learning*, Vol. 1, pp. 81–106.
14. Quinlan, J.R. 1993. *C4.5: Programs for Machine Learning*, Morgan Kaufmann Publishers.
15. Yan, X. and Han, J. 2002. gSpan: Graph-Based Structure Pattern Mining, In: *Proc. ICDM 2002*, pp. 721–724.
16. Yoshida, K. and Motoda, M. 1995. CLIP: Concept Learning from Inference Patterns, *Artificial Intelligence*, Vol. 75, No. 1, pp. 63–92.

Improved Bayesian Spam Filtering Based on Co-weighted Multi-area Information

Raju Shrestha and Yaping Lin

Department of Computer and Communication, Hunan University,
Changsha 410082, P.R. China
raju_shrestha@hotmail.com
yplin@hunu.edu.cn

Abstract. Bayesian spam filters, in general, compute probability estimations for tokens either without considering the email areas of occurrences except the body or treating the same token occurred in different areas as different tokens. However, in reality the same token occurring in different areas are inter-related and the relation too could play role in the classification. In this paper we incorporated this novel idea, co-relating multi-area information by co-weighting them and obtaining more effective combined integrated probability estimations for tokens. The new approach is compared with individual area-wise estimations and traditional separate estimations in all areas, and the experimental results with three public corpora showed significant improvement, stability, robustness and consistency in the spam filtering with the proposed estimation.

1 Introduction

Spam, also known as junk, is one of the greatest challenges to the email world these days. Spam not only wastes the time, but also wastes bandwidth, server space and some contents like pornographic contents are even harmful to under-aged recipients. Many anti-spam filtering techniques are being already proposed and in use to fight with ever growing spams. However, the new non-stop clever tricks of spammers necessitates further improvement in the filtering approaches. Among many machine learning based spam filters, Bayesian filters are the most popular and widely used because of efficient training, quick classification, easy extensibility, adaptive learning and fewer false positives.

Sahami [1] first employed the Naive Bayes algorithm to classify messages as spam or legitimate. In the series of papers, Androutsopoulos [2, 3, 4] extended Naive Bayes filter by investigating the effect of different features and training-set sizes on the filter's performance. Paul Graham in [5, 6] defined various tokenization rules, treating tokens in different parts of emails separately, and computed token probabilities and combined spam probability based on Bayes rule, but in a different way. Gary Robinson, in [7] suggested enhancements to Paul's approach by proposing Bayesian approach of handling rare words and in [8] further recommended to use Fisher's inverse chi-square function for combining probability estimations. This paper presents a new and novel approach which, on the top of all evolutions and enhancements in Bayesian approach, takes into account of the relation between the same tokens occurred in different areas of the email.

The rest of the paper is organized as follows. In Sect. 2, we review popular evolutions and variants of statistical spam filtering algorithms. Sect. 3 presents the main idea of this paper. Sect. 4 describes the experiments and analysis. Finally, Sect. 5 presents the conclusion of the paper and possible future work.

2   Statistical Bayesian Filtering Algorithms

2.1   Naive Bayes (NB) Algorithm

Naive Bayes algorithm is the simplified version of Bayes theorem with the assumption of feature independence. It computes the probability of a $Class \in$ {Spam, Legitimate} given an $Email$ as:

$$P(Class|Email) = P(Class) \prod_i P(t_i|Class) \quad (1)$$

Where $P(Class)$ is the prior probability for the $Class$. $P(t_i|Class)$ is the conditional probability of the token t_i given the $Class$, which is calculated as in [9] using the following formula.

$$\frac{1 + no(t, Class)}{\sum_i no(t_i, Class) + |V|} \quad (2)$$

Where $no(t, Class)$ is the number of occurrences of the token t in the $Class$, $\sum_i no(t_i, Class)$ is the total number of occurrences of all the tokens in the $Class$ and $|V|$ is the size of the vocabulary. The filter classifies an email as Spam or Legitimate according to whether the $P(\text{Spam}|Email)$ is greater than $P(\text{Legitimate}|Email)$ or not. Many implementations have shown that the algorithm is fairly robust and powerful in filtering spams and outperforms many other knowledge base filters [4, 10].

2.2   Paul Graham's (PB) Algorithm

Paul Graham [5] calculated the probability estimates for a given email being spam and legitimate, given a token appears in that using the formulas:

$$p(t) = P(\text{Spam}|t) = \frac{\frac{tbad}{nbad}}{\frac{tbad}{nbad} + \frac{2*tgood}{ngood}}, \quad P(\text{Legitimate}|t) = 1 - P(\text{Spam}|t) \quad (3)$$

Where $tbad$ and $tgood$ are the number of times the token t occurred in all the spam and legitimate emails respectively, and $nbad$ and $ngood$ are the number of spam and legitimate emails respectively. $tgood$ is multiplied by 2 to bias towards legitimate emails. The combined probability for spam is obtained using Bayesian approach using the formula:

$$P(\text{Spam}|Email) = \frac{\prod_i P(\text{Spam}|t_i)}{\prod_i P(\text{Spam}|t_i) + \prod_i P(\text{Legitimate}|t_i)} \quad (4)$$

Paul used only 15 most interesting tokens in computing the combined probability for spam. A test email is classified as spam if the combined probability is greater than a defined threshold value of 0.9.

2.3 Gary Robinson's (GR) Algorithm

Gary in [7] pointed out several drawbacks with Paul's algorithm and suggested several improvements:

- Gary proposed consistent and smooth way of dealing rare words by using the Bayesian approach to compute the token probability guesstimate, termed as degree of belief :

$$f(t) = \frac{s * x + n * p(t)}{s + n} \quad (5)$$

Where $p(t)$ is the Paul's probability estimation (3) for the token t, s is the strength to be given to background information, x is the assumed probability for an unknown token and n is the number of emails containing the token t.

- In [8], Gary further suggested to use Fisher's inverse chi-square function to compute combined probabilities using the formulas:

$$H = C^{-1}(-2\ln \prod_i f(t), 2n), \text{ and } S = C^{-1}(-2\ln \prod_i (1 - f(t)), 2n) \quad (6)$$

$C^{-1}()$ is the Fisher's inverse chi-square function used with $2n$ degrees of freedom. The combined indicator of spamminess or hamminess for the email as a whole is then obtained using the formula:

$$I = \frac{1 + H - S}{2} \quad (7)$$

The email is classified as spam if the indicator value I is above some threshold value otherwise as legitimate.

3 Bayesian Spam Filtering Based on Co-weighted Multi-area Information

In this section, we present our new approach of co-weighted multi-area information along with preprocessing and feature extraction techniques.

3.1 Preprocessing

Due to the prevalence of headers, html and binary attachments in modern emails, pre-processing is required on email messages to allow effective feature extraction. We use following preprocessing steps:

- The whole email structure is divided into 4 areas: (1.) Normal header comprising of 'From', 'Reply to', 'Return-path', 'To', 'Cc', (2.) Subject, (3.) Body, and (4.) Html tags comprising of <FONT>, <A> and <IMG> tags.
- All other headers and html tags (except those mentioned above) are ignored and hence removed from the email text.
- All binary attachments, if any, are ignored and so removed.

3.2 Feature Extraction or Tokenization

Our approach considers tokens as the sole features for the spam filtering. The remaining text after preprocessing is tokenized using the following tokenizer rules:

- All terms constituting alphanumeric characters, dash(-), underscore(_), apostrophe('), Exclamation(!), asterisk(*) and currency signs(like $,£etc.) are considered valid tokens and tokens are case-sensitive.
- IP addresses, domain names, money values (numbers separated by comma and/or with currency symbols) are considered valid tokens. Pure numbers are ignored.
- For domain name, it is broken into sub-terms (like www.hnu.net is broken into www.hnu.net, www.hnu, hnu.net, www, hnu and net) and the sub-terms are also considered valid tokens.
- Spammer's one of newest tricks of non-HTML text, interspersed with HTML tags like "You can b<!-x->uy val<!-abc->ium <b>here!<b>" is handled and obtains the text as "You can buy valium here " and tokenize it normally.

3.3 Main Idea and Algorithm Description

The main idea in our approach lies in the fact that the same token occurred in different areas of an email are inter-related. So treating the token occurred in one area separately from that occurring in other areas like in all previous algorithms described in Sect.2 wouldn't reflect the realistic estimation. In this paper, we relate the individual area-wise token probability estimations by co-weighting and obtain the combined integrated estimate for the token. The estimation steps are described in details below.

Let $ns(t, a)$ and $nl(t, a)$ be the number of occurrences of token t in the area a of spam and legitimate emails respectively, Ns and Nl be the number of spam and legitimate emails respectively. Then the probability estimation for spam given the token t and the area a is computed as:

$$p(t, a) = \frac{\frac{ns(t,a)}{Ns}}{\frac{ns(t,a)}{Ns} + \frac{nl(t,a)}{Nl}} \qquad (8)$$

This estimation corresponds to PG's probability estimation (3), but without bias factor. Next, the GR's degree of belief estimation $f(t, a)$ is computed using (5), replacing $p(t)$ with $p(t, a)$ and using $n = ns(t, a) + nl(t, a)$. s and x

are, like in GR's algorithm, the belief factor and the assumed probability for an unknown token whose values are determined while tuning the filter for optimal performance. Then the combined probability estimation for the token is calculated by co-weighting the individual estimations corresponding to different areas:

$$f(t) = \sum_i w(t, a_i) * f(t, a_i) \qquad (9)$$

Where $w(t, a_i)$ is the weight factor for the token t in the area a_i. The weight factor for the token t corresponding to area a is computed by the ratio of the number of occurrences of the token in that area and total number of occurrences of the token in all areas in all spam emails:

$$w(t, a) = \frac{ns(t, a)}{\sum_i ns(t, a_i)}, \quad \sum_i w(t, a_i) = 1. \qquad (10)$$

Since the combined estimate $f(t)$ co-relates the area-wise estimations according to token occurrences, it represents better and more realistic estimation. Moreover, since fixed number of interesting tokens as suggested in PG's algorithm is unrealistic and unreasonable, we consider all tokens whose probability values are above and below certain offset value $PROB\_OFFSET$ from the neutral 0.5 as interesting. If it gives less than predefined $MIN\_INTTOKENS$ of tokens, the range is extended to that number. Now values for those interesting tokens are used to obtain the final indicator I of spamminess and hamminess using (7), whereby H and S are calculated by Fisher's inverse chi-square functions (6). Finally the email is classified as spam if I is greater than certain threshold value, $SPAM\_THRESHOLD$, otherwise classified as legitimate.

4 Experiments and Analysis

First we will introduce the corpora collection and performance measures used for performance comparisons and then discuss the experiments and analysis.

4.1 Corpora Collection

This paper used three publicly available corpora and from each corpus, training and test datasets are prepared by randomly picking two-thirds of the total corpus data as training dataset and the rest one-third as test dataset. The corpora are:

1. *Ling Spam* corpus which was made available by Ion Androutsopoulos [2] and has been used in a considerable number of publications. It is composed of 481 spams, and 2,412 legitimate emails.
2. *Spam Assassin* corpus used to optimize the open source SpamAssassin filter. It contains 1,897 spam and 4,150 legitimate emails.
3. *Annexia/Xpert* corpus, synthesis of 10,025 spam emails from Annexia spam archives and 22,813 legitimate emails from X-Free project's Xpert mailing list. We randomly picked 7,500 spam and 7,500 legitimate emails from this corpus.

4.2 Performance Measures

Let N_S and N_L be the total number of spam and legitimate email messages to be classified by the filter respectively, and $N_{X \to Y}$ the number of messages belonging to class X that the filter classified as belonging to class Y ($X, Y \in$ {Spam(S), Legitimate(L)}). Then the seven performance measures are calculated as shown below in four categories:

1. *Weighted Accuracy (WAcc) and Weighted Error (WErr):*

$$WAcc = \frac{N_{S \to S} + \lambda N_{L \to L}}{N_S + \lambda N_L}, \quad WErr = \frac{N_{S \to L} + \lambda N_{L \to S}}{N_S + \lambda N_L} \quad (11)$$

The measures are calculated in our experiments with two reasonable values of λ: 9 and 99 [2].

2. *Total Cost Ratio (TCR):*

$$TCR = \frac{WErr^b}{WErr} = \frac{N_S}{\lambda N_{L \to S} + N_{S \to L}} \quad (12)$$

The weighted accuracy and error rates of the baseline are calculated as:

$$WAcc^b = \frac{\lambda N_L}{\lambda N_L + N_S}, \quad WErr^b = \frac{N_S}{\lambda N_L + N_S} \quad (13)$$

3. *Spam Recall (SR) and Spam Precision (SP):*

$$SR = \frac{N_{S \to S}}{N_S}, \quad SP = \frac{N_{S \to S}}{N_{S \to S} + N_{L \to S}} \quad (14)$$

4. *False Positive Rate (FPR) and False Negative Rate (FNR):*

$$FPR = \frac{N_{L \to S}}{N_L}, \quad FNR = \frac{N_{S \to L}}{N_S} \quad (15)$$

4.3 Experiments and Analysis

We have performed experiments on the filter application we developed in Java. All experiments are carried out five times for all three datasets by randomly picking training and test datasets as described above in Sect. 4.1 and the average results are reported. Our experiments consist of two parts:

1. First, thorough and exhaustive tests on the filter are performed with all three corpora datasets in order to determine optimal values of the filter parameters. The performance varies widely on varying the parameter combinations. Because of the different type and nature of contents, different datasets behaves differently in some parameter combinations. So we searched for parameter combination that gives compromised result with reduced false positive, high accuracy, and stable and consistent result for all three corpora dataset

and we found the values 0.9, 19, 0.42, 0.6 and 0.35 for $SPAM\_THRESHOLD$, $MIN\_INTTOKENS$, $PROB\_OFFSET$, s and x respectively. As false positives are much worse than false negatives, $SPAM\_THRESHOLD$ of 0.9 reasonably biases towards legitimates. The $MIN\_INTTOKENS = 19$ and $PROB\_OFFSET = 0.42$ combination provides the effective number of interesting tokens for optimal performances. The assumed probability value for unknown token $x = 0.35$ also slightly biases towards legitimates with 0.6 belief factor. we use the parameter combination in the second experiment.

2. In the second part of the experiment, the filter is tested with four individual email areas: normal headers only, subject only, body only and html tags only, then with all areas but treating same tokens occurred in different areas as different tokens and finally with our new approach of co-weighted multi-area information. All those tests are carried out with the same parameter combination we obtained in the first part of the experiment above and performance

Table 1. Test results (SR, SP, FPR and FNR) with individual area-wise and co-weighted multi-area based estimations

Areas	Measures	Ling Spam	Spam Assassin	Annexia/Xpert
Normal Headers Only	SR	0.00000	0.62180	0.76680
	SP	NA	0.98500	1.00000
	FPR	0.00000	0.00430	0.00000
	FNR	1.00000	0.37820	0.23320
Subject Only	SR	0.43750	0.40510	0.60040
	SP	1.00000	0.99220	1.00000
	FPR	0.00000	0.00140	0.00000
	FNR	0.56250	0.59490	0.39960
Body Only	SR	0.81880	0.84340	0.92880
	SP	1.00000	0.98890	0.99610
	FPR	0.00000	0.00430	0.00360
	FNR	0.18120	0.15660	0.07120
Html Tags Only	SR	0.00000	0.42090	0.22280
	SP	NA	0.97440	0.97720
	FPR	0.00000	0.00510	0.00520
	FNR	1.00000	0.57910	0.77720
All but Separate Areas	SR	0.83120	0.88290	0.96920
	SP	1.00000	0.99470	1.00000
	FPR	0.00000	0.00220	0.00000
	FNR	0.16880	0.11710	0.03080
Co-weighted Multi-areas	SR	0.83750	0.88610	0.97240
	SP	1.00000	0.99640	1.00000
	FPR	0.00000	0.00140	0.00000
	FNR	0.16250	0.11390	0.02760

Table 2. Weighted accuracy rates, error rates and total cost ratios with individual area-wise and co-weighted multi-area based estimations

Areas	Measures	$\lambda = 9$			$\lambda = 99$		
		Ling Spam	Spam Assassin	Annexia Xpert	Ling Spam	Spam Assassin	Annexia Xpert
	$WAcc^b$	0.97837	0.95168	0.90000	0.99799	0.99541	0.99000
	$WErr^b$	0.02163	0.04832	0.10000	0.00201	0.00459	0.01000
Normal Headers Only	$WAcc$ $WErr$ TCR	0.97837 0.02163 1.00000	0.97760 0.02240 2.15700	0.97668 0.02332 4.28816	0.99799 0.00201 1.00000	0.99394 0.00606 0.75870	0.99767 0.00233 4.28816
Subject Only	$WAcc$ $WErr$ TCR	0.98783 0.01217 1.77778	0.96988 0.03012 1.60406	0.96004 0.03996 2.50250	0.99887 0.00113 1.77778	0.99583 0.00417 1.10105	0.99600 0.00400 2.50250
Body Only	$WAcc$ $WErr$ TCR	0.99608 0.00392 5.51724	0.98830 0.01170 4.13072	0.98964 0.01036 9.65251	0.99964 0.00036 5.51724	0.99496 0.00504 0.91198	0.99572 0.00428 2.33863
Html Tags Only	$WAcc$ $WErr$ TCR	0.97837 0.02163 1.00000	0.96720 0.03280 1.47319	0.91760 0.08240 1.21359	0.99799 0.00201 1.00000	0.99230 0.00770 0.59679	0.98708 0.01292 0.77399
All but Separate Areas	$WAcc$ $WErr$ TCR	0.99635 0.00365 5.92593	0.99228 0.00772 6.25743	0.99692 0.00308 32.46753	0.99966 0.00034 5.92593	0.99730 0.00270 1.70350	0.99969 0.00031 32.46753
Co-weighted Multi-areas	$WAcc$ $WErr$ TCR	0.99648 0.00352 6.15385	0.99312 0.00688 7.02222	0.99724 0.00276 36.23188	0.99967 0.00033 6.15385	0.99804 0.00196 2.34074	0.99972 0.00028 36.23188

measures are reported for two values of λ: 9 and 99. Test results are given in the Tables 1 and 2. Performance measures independent of λ (SR, SP, FPR and FNR) are given in Table 1 and λ dependent measures ($WAcc^b$, $WErr^b$, $WAcc$, $WErr$ and TCR) are given separately in Table 2 for both values of λ. The comparative results based on weighted accuracy and total cost ratio are shown graphically in Figs. 1 and 2.

On analyzing the experiments and results, we observed the followings:

- *With individual area-wise estimations:*
  * Filtering based on individual area-wise estimations with datasets having no data in that area, like headers and tags in Ling Spam, is meaningless.
  * Even with Spam Assassin and Annexia/Xpert datasets, the filter classifies all those emails without html tags as legitimates resulting much higher false negatives. The penalty for these errors are high with $\lambda = 99$

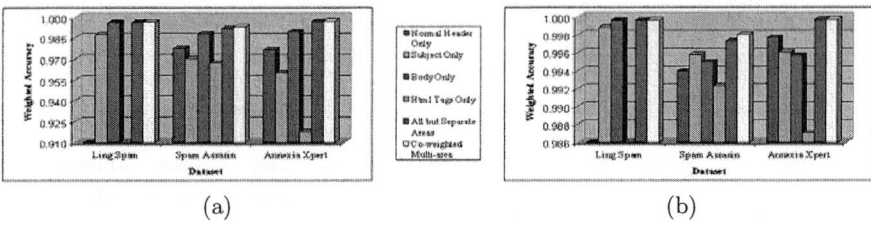

Fig. 1. Classification accuracies with individual area-wise and co-weighted multi-area based estimations, for (a) $\lambda = 9$ and (b) $\lambda = 99$

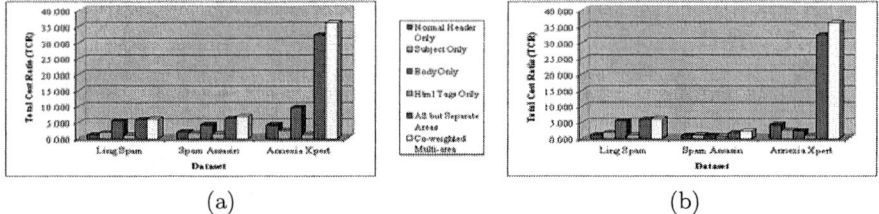

Fig. 2. Total cost ratios with individual area-wise and co-weighted multi-area based estimations, for (a) $\lambda = 9$ and (b) $\lambda = 99$

resulting $TCR < 1$ for both datasets indicating html only test ineffective. However with $\lambda = 9$, TCR values are slightly greater than 1.

* For Ling Spam, body only test resulted better performance than subject only test with higher accuracies, SR, and TCR, lower FNR, zero FPR and 100% SP. The presence of less spammy tokens in subjects of spam emails leads to higher number of false negatives with subject only test.
* For Spam Assassin, normal headers, body, and html tags only tests resulted higher false positives because of heavy presence of spammy tokens in those areas especially in hard hams. This causes heavy penalties in their accuracies with $\lambda = 99$, while even with higher FNR and lower SR, subject only test resulted higher accuracy and TCR because of fewer false positives. With lesser penalty with $\lambda = 9$, the accuracies and TCRs for body only is higher because of relatively fewer false negatives.
* For Annexia/Xpert, normal headers and subject only tests resulted zero false positives, however with high false negatives because of relatively low presence of spammy tokens in spam emails. Html only test resulted high false positives because of presence of spammy tokens in html tags of legitimate emails, and at the same time very high false negatives because of miss-classifying spam emails without html tags as legitimates. With lower value of $\lambda = 9$, better accuracy and TCR values are resulted with body only test, then with normal headers only, subject only and html tags only in the descending order. But with $\lambda = 99$, the order goes: normal headers only, subject only, body only and html tags only tests.

- *With all areas but treating same tokens in different areas as different tokens*, the test resulted better performance than individual area-wise estimation for both values of λ, in terms of all performance measures in all three datasets due to the combined positive effect of all areas.
- *With our new approach*, the performance is further improved with even lesser false positives as well as lesser false negatives in all three datasets, resulting better performance values of $WAcc$, SR, SP, FPR, FNR and TCR for both values of λ. TCR value is almost the same with Ling Spam and Annexia/Xpert datasets for both values of λ, however the value is three times higher for $\lambda = 9$ than for $\lambda = 99$. This is because of higher penalty on false positives with higher λ value.

Thus the experiments showed the proposed approach of incorporating the co-relation between tokens in different areas results significant improvement in the performance of the filter, and at the same time exhibit more stable, robust and consistent performances with all three corpora.

5 Conclusion

In this paper we present the new approach to statistical Bayesian filter based on co-weighted multi-area information. This new algorithm co-relates the area-wise token probability estimations using weight coefficients, which are computed according to the number of occurrences of the token in those areas. Experimental results showed significant improvement in the performance of spam filtering than using individual area-wise as well as using separate estimations for all areas. Moreover, the performances are much more stable and consistent with all three datasets.

Future developments may include integrating our approach with phrase-based and/or other lexical analyzers and with rich feature extraction methods which can be expected to achieve even better performance.

References

1. Sahami, M., Dumais, S., Heckerman, D., Horvitz, E.: A bayesian approach to filtering junk e-mail. In: Learning for Text Categorization, Wisconsin (1998)
2. Androutsopoulos, I., Koutsias, J., Chandrinos, K.V., Paliouras, G., Spyropoulos, C.D.: An evaluation of naive bayesian anti-spam filtering. In: Proceedings of 11th European Conference on Machine Learning (ECML 2000), Barcelona (2000) 9–17
3. Androutsopoulos, I., Koutsias, J., Chandrinos, K.V., Paliouras, G., Spyropoulos, C.D.: An experimental comparison of naive bayesian and keyword-based anti-spam filtering with personal e-mail messages. In: Proceedings of 23rd ACM International Conference on Research and Development in Information Retrieval, Athens, Greece, ACM Press, New York, US (2000) 160–167
4. Androutsopoulos, I., Koutsias, J., Chandrinos, K.V., Paliouras, G., Spyropoulos, C.D.: Learning to filter spam e-mail: A comparison of a naive bayesian and a memory-based approach. In: Proceedings of 4th European Conference on Principles and Practice of Knowledge Discovery in Databases, Lyon, France (2000) 1–13

5. Graham, P.: A plan for spam. (2002) http://www.paulgraham.com/spam.html.
6. Graham, P.: Better bayesian fitering. In: Proceedings of the First Annual Spam Conference, MIT (2003) http://www.paulgraham.com/better.html.
7. Robinson, G.: Spam detection (2003) http://radio.weblogs.com/0101454/stories/2002/09/16/spamDetection.html.
8. Robinson, G.: A statistical approach to the spam problem. Linux Journal, Issue-107 (2003) http://www.linuxjournal.com/article.php?sid=6467.
9. Mitchell, T.M.: Machine learning. McGraw-Hill (1997) 177–184
10. Provost, J.: Naive-bayes vs. rule-learning in classification of email. Technical report, Department of Computer Science, The University of Texas at Austin (1999)

An Efficient Framework for Mining Flexible Constraints

Arnaud Soulet and Bruno Crémilleux

GREYC, CNRS - UMR 6072, Université de Caen,
Campus Côte de Nacre, F-14032 Caen Cédex France
Forename.Surname@info.unicaen.fr

Abstract. Constraint-based mining is an active field of research which is a key point to get interactive and successful KDD processes. Nevertheless, usual solvers are limited to particular kinds of constraints because they rely on properties to prune the search space which are incompatible together. In this paper, we provide a general framework dedicated to a large set of constraints described by SQL-like and syntactic primitives. This set of constraints covers the usual classes and introduces new tough and flexible constraints. We define a pruning operator which prunes the search space by automatically taking into account the characteristics of the constraint at hand. Finally, we propose an algorithm which efficiently makes use of this framework. Experimental results highlight that usual and new complex constraints can be mined in large datasets.

1 Introduction

Mining patterns under various kinds of constraints is a key point to get interactive and successful KDD processes. There is a large collection of constraints which are useful for the user and the latter needs an independent tool to tackle various and flexible queries. The outstandingly useful constraint of frequency is often used in practice. Furthermore, we have efficient algorithms to extract patterns satisfying it. Nevertheless, in the context of constraint-based mining, supplementary constraints like interestingness measures or syntactic constraints have to be added to achieve relevant and desired patterns. The number of elementary constraints and their combinations is too important to build a particular solver dedicated to each constraint. These observations are sound motivations to design and implement a general solver which is able to mine, in a flexible way, patterns checking various and meaningful constraints.

Let us recall that constraint-based mining remains a challenge due to the huge size of the search space which has to be explored (it exponentially increases according to the number of features of the data). Classical algorithms are based on pruning properties in order to reduce the search space. But, unfortunately, these properties are deeply linked to the constraints and many constraints (e.g., average [13], variance [10], growth rate [6]) have been studied individually [1]. Section 2.2 overviews the main classes of constraints and recalls that efficient algorithms are devoted to some of these classes. Several approaches [16, 2] are based on condensed representations of frequent patterns which are easier to mine. We will see that our work uses this approach but without limitation to the classical classes of constraints. The paradigm of inductive databases [8] proposes to handle constraints by reusing existing algorithms (for instance, a complex constraint is decomposed into several constraints having suitable properties like monotonicity). However, the decomposition and the optimization of constraints remain non-trivial tasks [5]. To the best of our knowledge, there is no existing framework presenting at the same time flexibility and effective computation.

In this paper, we present a new general framework to soundly and completely mine patterns under a constraint specified by the user as a simple parameter. We think that this framework brings three meaningful contributions. First, it allows a large set of constraints: the constraints are described by combinations of SQL-like aggregate primitives and syntactic primitives (see Section 3.1). This formalism deals with the most usual constraints (e.g., monotonous, anti-monotonous and convertible ones) and allows to define more original new constraints (e.g., the area constraint which is on the core of our running example). Furthermore, this formalism also enables to combine constraints with boolean operators. Second, thanks to an automatic process to compute lower and upper bounds of a constraint on an interval and a general *pruning operator*, the constraint is pushed in the extraction step. Finally, we provide an algorithm called MUSIC (Mining with a User-Specified Constraint) which allows the practical use of this framework. MUSIC guarantees an efficient pruning to offer short run-time answers and facilitate the iterative process of KDD. We developed a prototype to implement this algorithm.

This paper is organized in the following way. Section 2 introduces the basic notations and related work. A running example (i.e., the area constraint) shows the common difficulties of constraint-based mining and the key ideas of our framework. Section 3 depicts the set of constraints that we address, details the theoretical framework and defines the pruning operator. Section 4 indicates how to use it by providing the MUSIC algorithm. Finally, Section 5 presents experimental results showing the efficiency of MUSIC on various constraints.

2 Context and Motivations

2.1 Notation

Let us first introduce the basic notations. A transactional dataset $\mathcal{D}$ is a triplet $(\mathcal{A}, \mathcal{O}, R)$ where $\mathcal{A}$ is a set of attributes, $\mathcal{O}$ is a set of objects and $R \subseteq \mathcal{A} \times \mathcal{O}$ is a binary relation between the attributes and the objects. $(a, o) \in R$ expresses that the object o has the attribute a (see for instance Table 1 where $A, \ldots, F$ denote the attributes and $o_1, \ldots, o_6$ denote the objects). A pattern X is a subset of attributes.

The aim of constrained patterns mining is to extract all patterns present in $\mathcal{D}$ and checking a predicate q. The minimal frequency constraint is likely the most usual one (the frequency of a pattern X is the number of objects in $\mathcal{D}$ that contain X, i.e. $count(X) \geq \gamma$ where γ is a threshold). Many algorithms since [1] efficiently mine this constraint by using closure or free (or key) patterns [3, 14].

Table 1. Example of a transactional data set

$\mathcal{D}$

Objects	Attributes
o_1	A B E F
o_2	A E
o_3	A B C D
o_4	A B C D E
o_5	D E
o_6	C F

2.2 Related Work

Many works have been done with various complex constraints (e.g., average [13], variance [10], growth rate [6]) providing particular approaches. More generally, we can distinguish several classes of constraints. A well-known class of constraints is based on *monotonicity*. A constraint q is anti-monotone (resp. monotone) according to the specialization of the attributes if whenever $X \subseteq Y$ then $q(Y) \Rightarrow q(X)$ (resp. $q(X) \Rightarrow q(Y)$). For instance, the minimal frequency constraint is anti-monotonous. In this case, the search space can be efficiently pruned by a general level-wise algorithm [12]. Another class is the *convertible* constraints. Such a constraint uses an ordering relation on the attributes in order to obtain properties of monotonicity on the prefixes [15] (typically, the minimal average constraint q_8 is convertible, see Section 3.1 for its exact definition). Let us note that Wang et al. introduce in [18] a method dedicated to the aggregate constraints (e.g. the minimal frequency constraint or the average q_8).

There are specific algorithms devoted to these different classes of constraints. Unfortunately, the combination of constraints may require again a particular algorithm. For example, a conjunction of two convertible constraints may lead to a no convertible constraint, and a particular algorithm has to been developed. So, several approaches attempt to overcome these difficulties. The *inductive databases* framework [8] proposes to decompose complex constraints into several constraints having good properties like monotonicity. This approach needs to apply non-trivial reformulations and optimizations of constraints [5]. Introduced in [10], the concept of *witness* provides properties to simultaneously prune patterns under different kinds of constraints. Nevertheless, this approach does not propose a method to automatically obtain witnesses. Thus, instead of building a particular algorithm to mine patterns, a particular algorithm to find witnesses is needed. By exploiting equivalence classes (i.e., a set of patterns having the same outcome with respect to the constraint), condensed representations [4] enable powerful pruning criteria during the extraction which greatly improve the efficiency of algorithms [3, 14]. But only few works exploit the equivalence classes with monotonous and anti-monotonous classes [2] or other constraints [9, 16]. Our work follows this approach but it addresses a much more general set of constraints (see Section 3.1).

2.3 Problem Statement and Key Idea of Our Framework

Let us come back on the example given by Table 1. Assume that we are interested in all subsets of $\mathcal{A}$ having an *area* greater than 4 i.e. $count(X) \times length(X) \geq 4$ (where *length* is the cardinality of the pattern). Recently, [7] have dedicated efficient approaches to only mine the closed patterns that check this constraint. The constraint of area is difficult because it is neither monotone ($area(A) \geq 4$ but $area(ABF) < 4$), nor anti-monotone ($area(B) < 4$ but $area(AB) \geq 4$), nor convertible (no ordering relation exists). None decomposition of this constraint benefits from the properties of these classes of constraints. Thus, the practical approach is to mine all patterns with their own frequency and then to post-process them by checking the constraint on each pattern. Unfortunately, this method fails with large datasets due to a too much number of candidate patterns.

From this running example, we now indicate how to take into account the characteristics of the constraint to present our pruning strategy of the search space. The main idea is based on the definition of lower and upper bounds of the constraint on an interval, the latter after allows the pruning. We can notice that if $X \subseteq Z \subseteq Y$, the area of the pattern Z can be bounded by $count(Y) \times length(X) \leq count(Z) \times length(Z) \leq count(X) \times length(Y)$. We note that if $count(Y) \times length(X) \geq 4$, the area of the pattern Z is larger than 4 and Z checks the constraint. In this example, with $X = AB$ and $Y = ABCD$, the area of $count(ABCD) \times length(AB)$ is equal to 4 and the patterns AB, ABC, ABD, $ABCD$ have an area larger than 4. Thus, it is not necessary to check the constraint for these four patterns. Similarly, when $count(X) \times length(Y)$ is strictly smaller than 4, the area of the pattern Z ($X \subseteq Z \subseteq Y$) is inevitably smaller than 4. In these two cases, the interval $[X, Y]$ can be pruned for this constraint. Also, the patterns AB, ABC, ABD and $ABCD$, which are included between AB and $ABCD$, satisfy the constraint. Instead of outputting these four patterns, it is more judicious to only output the corresponding interval $[AB, ABCD]$. This one can be seen as a condensed representation of the patterns with respect to the constraint. This idea - mining a *representation* of the constrained patterns - is generalized in the next section to a large set of constraints.

3 Pruning the Search Space by Pruning an Interval

3.1 The Set of Constraints

Our work deals with the set of constraints $\mathcal{Q}$ recursively defined by Table 2. Examples of constraints of $\mathcal{Q}$ are given at the end of this section. We claim that $\mathcal{Q}$ defines a very large set of constraints.

$\mathcal{L}_\mathcal{A}$ (resp. $\mathcal{L}_\mathcal{O}$) denotes the language associated with the attributes $\mathcal{A}$ (resp. the objects $\mathcal{O}$) i.e. the powerset $2^\mathcal{A}$ (resp. the powerset $2^\mathcal{O}$). The set of constraints $\mathcal{Q}$ is based on three spaces: the booleans $\mathfrak{B}$ (i.e., $true$ or $false$), the positive reals $\Re^+$ and the patterns of $\mathcal{L} = \mathcal{L}_\mathcal{A} \cup \mathcal{L}_\mathcal{O}$. In addition to the classical operators of these domains, the function $count$ denotes the frequency of a pattern, and $length$ its cardinality. Given a function $val : \mathcal{A} \cup \mathcal{O} \to \Re^+$, we extend it to a pattern X and note $X.val$ the set $\{val(a)|a \in X\}$. This kind of function is used with

Table 2. Set of constraints $\mathcal{Q}$

Constraint $q \in \mathcal{Q}$	Operator(s)	Operand(s)
$q_1 \theta q_2$	$\theta \in \{\wedge, \vee\}$	$(q_1, q_2) \in \mathcal{Q}^2$
θq_1	$\theta \in \{\neg\}$	$q_1 \in \mathcal{Q}$
$e_1 \theta e_2$	$\theta \in \{<, \leq, =, \neq, \geq, >\}$	$(e_1, e_2) \in \mathcal{E}^2$
$s_1 \theta s_2$	$\theta \in \{\subset, \subseteq, =, \neq, \supseteq, \supset\}$	$(s_1, s_2) \in \mathcal{S}^2$
constant $b \in \mathfrak{B}$	-	-
Aggregate expression $e \in \mathcal{E}$	**Operator(s)**	**Operand(s)**
$e_1 \theta e_2$	$\theta \in \{+, -, \times, /\}$	$(e_1, e_2) \in \mathcal{E}^2$
$\theta(s)$	$\theta \in \{count, length\}$	$s \in \mathcal{S}$
$\theta(s.val)$	$\theta \in \{sum, max, min\}$	$s \in \mathcal{S}$
constant $r \in \Re^+$	-	-
Syntactic expression $s \in \mathcal{S}$	**Operator(s)**	**Operand(s)**
$s_1 \theta s_2$	$\theta \in \{\cup, \cap, \setminus\}$	$(s_1, s_2) \in \mathcal{S}^2$
$\theta(s_1)$	$\theta \in \{f, g\}$	$s_1 \in \mathcal{S}$
variable $X \in \mathcal{L}_\mathcal{A}$	-	-
constant $l \in \mathcal{L} = \mathcal{L}_\mathcal{A} \cup \mathcal{L}_\mathcal{O}$	-	-

the usual SQL-like primitives *sum*, *min* and *max*. For instance, $sum(X.val)$ is the sum of *val* of each attribute of X. Finally, f is the intensive function i.e. $f(O) = \{a \in \mathcal{A} | \forall o \in O, (a, o) \in \mathcal{R}\}$, and g is the extensive function i.e. $g(A) = \{o \in \mathcal{O} | \forall a \in A, (a, o) \in \mathcal{R}\}$. We give now some examples of constraints belonging to $\mathcal{Q}$ and highlighting the generality of our framework.

$q_1 \equiv count(X) \geq \gamma \times |\mathcal{D}|$
$q_2 \equiv count(X) \times length(X) \geq 2500$
$q_3 \equiv X \subseteq A$
$q_4 \equiv X \supseteq A$
$q_5 \equiv length(X) \geq 10$
$q_6 \equiv sum(X.val) \geq 500$
$q_7 \equiv max(X.val) < 50$
$q_8 \equiv sum(X.val)/length(X) \geq 50$
$q_9 \equiv min(X.val) \geq 30 \wedge max(X.val) \leq 90$
$q_{10} \equiv max(X.val) - min(X.val) \leq 2 \times length(X)$
$q_{11} \equiv length(X \setminus A) > length(X \cap A)$
$q_{12} \equiv q_2 \wedge q_3$
$q_{13} \equiv q_6 \vee \neg q_5$
$q_{14} \equiv q_5 \wedge q_7$

Starting from a constraint of $\mathcal{Q}$, the following sections explain how to get sufficient conditions to prune the search space.

3.2 Bounding a Constraint on an Interval

This section indicates how to automatically compute lower and upper bounds of a constraint of $\mathcal{Q}$ on an interval without enumerating each pattern included in the interval. These bounds will be used by the pruning operator (see Section 3.3).

Let X and Y be two patterns. The interval between these patterns (denoted $[X, Y]$) corresponds to the set $\{Z \in \mathcal{L}_\mathcal{A} | X \subseteq Z \subseteq Y\}$. In our running example dealing with the area constraint, Section 2.3 has shown that $count(Y) \times length(X)$ and $count(X) \times length(Y)$ are respectively a lower bound and an upper bound of the constraint for the patterns included in the interval $[X, Y]$. At a higher level, one can also notice that $\forall Z \in [X, Y]$, we have $(count(Y) \times length(X) \geq 4) \leq q(Z) \leq (count(X) \times length(Y) \geq 4)$ with respect

Table 3. The definitions of $\lfloor . \rfloor$ and $\lceil . \rceil$

$e \in \mathcal{E}_i$	Operator(s)	$\lfloor e \rfloor \langle X, Y \rangle$	$\lceil e \rceil \langle X, Y \rangle$
$e_1 \theta e_2$	$\theta \in \{\wedge, \vee, +, \times, \cup, \cap\}$	$\lfloor e_1 \rfloor \langle X, Y \rangle \theta \lfloor e_2 \rfloor \langle X, Y \rangle$	$\lceil e_1 \rceil \langle X, Y \rangle \theta \lceil e_2 \rceil \langle X, Y \rangle$
$e_1 \theta e_2$	$\theta \in \{>, \geq, \supset, \supseteq, -, /, \setminus\}$	$\lfloor e_1 \rfloor \langle X, Y \rangle \theta \lceil e_2 \rceil \langle X, Y \rangle$	$\lceil e_1 \rceil \langle X, Y \rangle \theta \lfloor e_2 \rfloor \langle X, Y \rangle$
θe_1	$\theta \in \{\neg, count, f, g\}$	$\theta \lceil e_1 \rceil \langle X, Y \rangle$	$\theta \lfloor e_1 \rfloor \langle X, Y \rangle$
$\theta(e_1.val)$	$\theta \in \{min\}$	$\theta(\lceil e_1 \rceil \langle X, Y \rangle .val)$	$\theta(\lfloor e_1 \rfloor \langle X, Y \rangle .val)$
$\theta(e_1)$	$\theta \in \{length\}$	$\theta \lfloor e_1 \rfloor \langle X, Y \rangle$	$\theta \lceil e_1 \rceil \langle X, Y \rangle$
$\theta(e_1.val)$	$\theta \in \{sum, max\}$	$\theta(\lfloor e_1 \rfloor \langle X, Y \rangle .val)$	$\theta(\lceil e_1 \rceil \langle X, Y \rangle .val)$
$c \in E_i$	-	c	c
$X \in \mathcal{L}_\mathcal{A}$	-	X	Y

to $false < true$. Thus, the area constraint is bounded on the interval. Those bounds only depend on the patterns X and Y and their definitions are the same for any interval $[X, Y]$.

Let us generalize this approach for any constraint q of $\mathcal{Q}$. For that, we define two operators denoted $\lfloor . \rfloor$ and $\lceil . \rceil$ (see Table 3). Starting from q and $[X, Y]$, the recursive application of these operators leads to compute one boolean with $\lfloor . \rfloor$ (noted $\lfloor q \rfloor \langle X, Y \rangle$) and one boolean with $\lceil . \rceil$ (noted $\lceil q \rceil \langle X, Y \rangle$). Property 1 will show that $\lfloor q \rfloor \langle X, Y \rangle$ (resp. $\lceil q \rceil \langle X, Y \rangle$) is a lower bound (resp. an upper bound) of the interval $[X, Y]$ for q. In other words, these operators enable to automatically compute lower and upper bounds of $[X, Y]$ for q. This result stems from the properties of increasing and decreasing functions. In Table 3, the general notation E_i designates one space among $\mathfrak{B}$, $\mathfrak{R}^+$ or $\mathcal{L}$ and $\mathcal{E}_i$ the associated expressions (for instance, the set of constraints $\mathcal{Q}$ for the booleans $\mathfrak{B}$). Several operators given in Table 2 must be split into several operators of Table 3. For instance, the equality $e_1 = e_2$ is decomposed to $(e_1 \leq e_2) \wedge (e_1 \geq e_2)$. In Table 3, the functions are grouped by monotonous properties according to their variables. For instance, the operators $-$, $/$ and $\setminus$ are increasing functions according to the first variable and decreasing functions according to the second variable.

Let us illustrate $\lfloor . \rfloor$ and $\lceil . \rceil$ on the area constraint: $\lfloor count(X) \times length(X) \geq 4 \rfloor \langle X, Y \rangle = \lfloor count(X) \times length(X) \rfloor \langle X, Y \rangle \geq \lceil 4 \rceil \langle X, Y \rangle = \lfloor count(X) \rfloor \langle X, Y \rangle \times \lfloor length(X) \rfloor \langle X, Y \rangle \geq 4 = count(\lceil X \rceil \langle X, Y \rangle) \times length(\lfloor X \rfloor \langle X, Y \rangle) \geq 4 = count(Y) \times length(X) \geq 4$. Symmetrically, $\lceil count(X) \times length(X) \geq 4 \rceil \langle X, Y \rangle$ is equal to $count(X) \times length(Y) \geq 4$.

Property 1 shows that $\lfloor q \rfloor$ and $\lceil q \rceil$ are bounds of the constraint q.

Property 1 (bounds of an interval). *Let q be a constraint, $\lfloor q \rfloor$ and $\lceil q \rceil$ are respectively a lower bound and an upper bound of q i.e. given an interval $[X, Y]$ and a pattern Z included in it, we have $\lfloor q \rfloor \langle X, Y \rangle \leq q(Z) \leq \lceil q \rceil \langle X, Y \rangle$.*

This property justifies that $\lfloor . \rfloor$ and $\lceil . \rceil$ are respectively named the lower and upper bounding operators (due to space limitation the proof is not given here, see [17]). Contrary to most frameworks, these top-level operators allow us to directly use constraints containing conjunctions or disjunctions of other constraints. Besides, they compute quite accurate bounds. In the particular case of monotonous constraints, these bounds are even *exact*. They have other meaningful properties (linearity or duality) which are not developed here (details in [17]).

3.3 Pruning Operator

In this section, we define an operator, starting from a constraint q, which provides a condition to safely prune or not an interval. As for the area constraint, there are two different strategies to prune an interval $[X, Y]$ by using the bounds of q. If a lower bound of q on $[X, Y]$ is equal to $true$ (i.e., the lower bound checks q), all the patterns included in $[X, Y]$ check q because they are all greater than $true$. We say that we *positively* prune the patterns of $[X, Y]$. Conversely, we can *negatively* prune $[X, Y]$ when an upper bound is $false$ because all the patterns of $[X, Y]$ do not check q. Note that witnesses [10] already exploit these two kinds of pruning. We define now the pruning condition for q and the pruning operator.

Definition 1 (pruning operator). *Let q be a constraint, the pruning condition for q, denoted by $[q]$, is equal to $\lfloor q \rfloor \vee \neg \lceil q \rceil$. $[.]$ is called the pruning operator.*

This definition is linked to the two ways of pruning: $\lfloor q \rfloor$ is associated with the positive pruning, and $\neg \lceil q \rceil$ to the negative one. For instance, the pruning condition for the area constraint on an interval $[X, Y]$ is $(count(Y) \times length(X) \geq 4) \wedge (count(X) \times length(Y) < 4)$. This conjunction corresponds to the two cases allowing us to prune intervals (see Section 2.3).

The following key theorem will be used extensively.

Theorem 1. *Let q be a constraint and $[X, Y]$ an interval, if $[q]\langle X, Y \rangle$ is true, then all the patterns included in $[X, Y]$ have the same value for q.*

Due to space limitation, the proof is not proposed here (see [17]).

Whenever the pruning condition is $true$ on an interval $[X, Y]$, we know the value of any pattern of $[X, Y]$ by checking only one pattern. Thereby, $[X, Y]$ can be pruned without having to check the constraint on whole patterns of $[X, Y]$. Section 4 details how to prune the search space with the pruning condition.

Let us note that the converse of Theorem 1 is false because $\lfloor . \rfloor$ (resp. $\lceil . \rceil$) does not give the greatest lower (resp. the least upper) bound. However, in practice, the pruning operator often provides powerful pruning (see Section 5).

4 MUSIC: A Constraint-Based Mining Algorithm

MUSIC (Mining with a User-Specified Constraint) is a level-wise algorithm which takes advantage of the pruning operator to efficiently mine constrained patterns and get a representation of these patterns. It takes one constraint q belonging to $\mathcal{Q}$ as input and one additional anti-monotonous constraint q_{AM} to benefit from the usual pruning of level-wise algorithm [1] (line 4). The completeness with respect to q is ensured by sticking $true$ for q_{AM}. MUSIC returns in output all the intervals containing the patterns checking $q \wedge q_{AM}$. MUSIC is based on tree key steps: the creating of the generators similar to one used in [1] (line 14), the evaluation of candidates by scanning the dataset in order to compute the extension of patterns (line 3), and finally the testing candidates (lines 7-12). Its main originality is that the "tested candidates" are different from generators and they are intervals instead of patterns.

MUSIC (a constraint $q \in \mathcal{Q}$, an anti-monotonous constraint q_{AM}, a dataset $\mathcal{D}$)

1. $Cand_1 := \{a \in \mathcal{A} | q_{AM}(a) = true\}$; $\mathcal{R}es := \emptyset$; $\mathcal{A}Cand_1 := \emptyset$; $k := 1$
2. while $Cand_k \cup \mathcal{A}Cand_k \neq \emptyset$ do
3.   for each $X \in Cand_k$ do $X.extension := \{o \in \mathcal{O} | X \subseteq f(o)\}$
4. $\mathcal{G}en_k := \{X \in Cand_k | q_{AM}(X) = true \text{ and } X \text{ is } free\}$
5. $\mathcal{A}Cand_{k+1} := \emptyset$
6.   for each $X \in \mathcal{G}en_k \cup \mathcal{A}Cand_k$ do
7.   if $[q]\langle X, f(X.extension)\rangle = true$ then
8.   if $q(X) = true$ then $\mathcal{R}es := \mathcal{R}es \cup \{[X, f(X.extension)]\}$
9.   else do
10.   if $q(X) = true$ then $\mathcal{R}es := \mathcal{R}es \cup \{[X, X]\}$
11. $\mathcal{A}Cand_{k+1} := \mathcal{A}Cand_{k+1} \cup \{X \cup \{a\} | a \in h(X) \setminus X\}$
12.   od
13.   od
14. $Cand_{k+1} :=$ APRIORI-GEN$(\mathcal{G}en_k)$
15. $k := k + 1$
16. od
17. return $\mathcal{R}es$

MUSIC uses intervals already proposed in [11], where the left bound is a generator (i.e., a free pattern or an additionnal candidate) and the right one, its closure (i.e., a closed pattern). The closed patterns are exactly the fixed points of the closure operator $h = f \circ g$. An important property on the extension stems from the closure: $g(X) = g(h(X))$. Moreover, as the closure operator is extensive, any pattern is a subset of its closure and the interval $[X, h(X)]$ has always a sense. The pruning condition is pushed into the core of the mining by applying it on the intervals defined above. Such an approach enables a powerful pruning criterion during the extraction thanks to the use of an anti-monotonous constraint based on the *freeness* [3] (line 4). If an interval $[X, h(X)]$ satisfies the pruning condition, all the patterns in $[X, h(X)]$ are definitively pruned. Otherwise, some patterns of $[X, h(X)]$ are added as *additional* candidates to repeat the same process on shorter intervals (line 11). The computation of the extension for each pattern (line 3) is sufficient to deduce the values of all the primitives given by Table 2 even if they depend on the dataset like g or *count*. In particular, the frequency of a pattern X is $length(g(X))$ and the closure of X is computed with $f(g(X))$. Note that the additionnal candidates have a low cost because they belong to $[X, h(X)]$ and their extension is equal to $g(h(X))$.

Due to the lack of place, we provide here only an intuitive proof of the correction of MUSIC (see [17] for a formal proof). All the patterns are comprise between a free pattern and its closure. As MUSIC covers all the free patterns, all the intervals $[X, h(X)]$ are checked by the pruning condition. There are two cases. First, if the pruning condition is true, all the patterns included in $[X, h(X)]$ are checked. Otherwise, the patterns included in $[X, h(X)]$ are enumerated level by level until that the interval between it and its closure can be pruned (that always arises because $[q]\langle X, X\rangle = true$). Thus, the whole search space is covered.

5 Experimental Results

The aim of our experiments is to measure the run-time benefit brought by our framework on various constraints (i.e., constraints $q_1, \ldots, q_{14}$ defined in Section 3.1). All the tests were performed on a 700 GHz Pentium III processor with Linux and 1Go of RAM memory. The used dataset is the version of mushroom coming from the FIMI repository[1]. The constraints using numeric values were applied on attribute values (noted val) randomly generated within the range [1,100]. We compare our algorithm with an APRIORI-like approach (i.e., mining all patterns according to q_{AM} and using a filter to select patterns checking q).

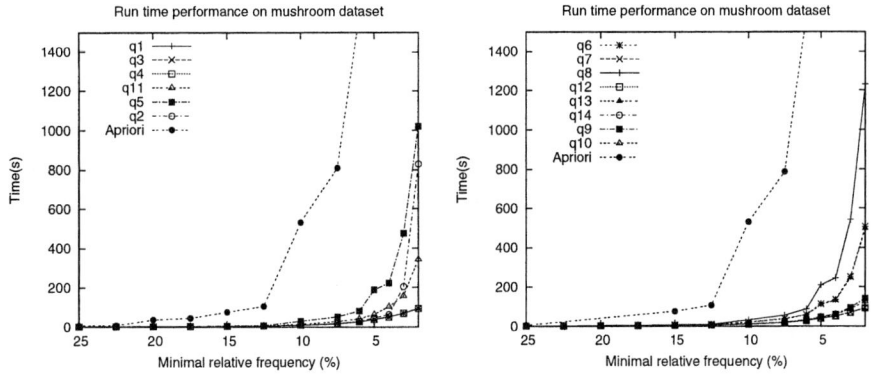

Fig. 1. Mining patterns with various constraints on mushroom dataset

Figure 1 plots the comparison between MUSIC and APRIORI run-times according to $q_1, \ldots, q_{14}$. With the constraint q_1, MUSIC has no additional candidates and its behavior is similar than [3]. That shows the abilities of MUSIC towards usual constraints. The best performances are achieved by the constraints q_3, q_4, q_7, q_9 and q_{10} because the number of additional candidates in these cases is very low. On the other hand, the three worst performances (i.e., q_2, q_5 and q_8) are obtained with the constraints including $length$. It is interesting to observe that the run-time performances are independent of the complexity of the constraint (i.e., the number of combinations). For instance, a very complex constraint such as q_{10} is quickly mined and a conjunction of constraint such as $q_2 \wedge q_3$ has better results than q_2 alone. This fact can be explained with the improvement of the selectivity of the constraint. Additional results are given in [17]. The good

---

[1] http://fimi.cs.helsinki.fi/data/

behavior of MUSIC with complex constraints allows the user to ask a broad set of queries and to mine constrained patterns which were intractable until now.

6 Conclusion

In this paper, we have proposed a new and general framework to efficiently mine patterns under constraints based on SQL-like and syntactic primitives. This framework deals with boolean combinations of the usual constraints and allows to define new complex constraints. The efficiency of the approach relies on the pruning of the search space on intervals which are took into account by a general pruning operator. Starting from this approach, MUSIC algorithm mines soundly and completely patterns under a primitive-based constraint given by the user as a simple parameter. The experimental results show that MUSIC clearly outperforms APRIORI with all constraints. New tough constraints can be mined in large datasets. We think that our algebraisation is an important step towards the integration of the constraint-based mining in database systems.

Further work addresses optimization of specific primitives like *length*. About the generality of our framework, we would like also to know if other primitives used to define constraints should be useful to achieve successful KDD processes from real world data set. We think that our ongoing work on geographical datasets is a good way to test new expressive queries specified by a geographer expert and the usefulness of the primitives.

Acknowledgements. This work has been partially founded by the ACI "BINGO".

References

[1] R. Agrawal and R. Srikant. Fast algorithms for mining association rules. In *Proc. 20th Int. Conf. Very Large Data Bases, VLDB*, pages 432–444, 1994.
[2] F. Bonchi and C. Lucchese. On closed constrained frequent pattern mining. In *proceedings of ICDM'04*, pages 35–42, 2004.
[3] J. F. Boulicaut, A. Bykowski, and C. Rigotti. Free-sets: a condensed representation of boolean data for the approximation of frequency queries. *Data Mining and Knowledge Discovery journal*, 7(1):5–22, 2003.
[4] T. Calders and B. Goethals. Minimal k-free representations of frequent sets. In *proceedings of PKDD'03*, pages 71–82. Springer, 2003.
[5] L. De Raedt, M. Jäger, S. D. Lee, and H. Mannila. A theory of inductive query answering. In *proceedings of ICDM'02*, pages 123–130, Maebashi, Japan, 2002.
[6] G. Dong and J. Li. Efficient mining of emerging patterns: Discovering trends and differences. In *Knowledge Discovery and Data Mining*, pages 43–52, 1999.
[7] K. Gade, J. Wang, and G. Karypis. Efficient closed pattern mining in the presence of tough block constraints. In *proceedings of ACM SIGKDD*, pages 138–147, 2004.
[8] T. Imielinski and H. Mannila. A database perspective on knowledge discovery. In *Communication of the ACM*, pages 58–64, 1996.
[9] B. Jeudy and F. Rioult. Database transposition for constrained (closed) pattern mining. In *proceedings of KDID'04*, pages 89–107, 2004.

[10] D. Kiefer, J. Gehrke, C. Bucila, and W. White. How to quickly find a witness. In *proceedings of ACM SIGMOD/PODS 2003 Conference*, pages 272–283, 2003.
[11] M. Kryszkiewicz. Inferring knowledge from frequent patterns. In *proceedings of Soft-Ware 2002*, Lecture Notes in Computer Science, pages 247–262, 2002.
[12] H. Mannila and H. Toivonen. Levelwise search and borders of theories in knowledge discovery. *Data Mining and Knowledge Discovery*, 1(3):241–258, 1997.
[13] R. T. Ng, L. V. S. Lakshmanan, and J. Han. Exploratory mining and pruning optimizations of constrained associations rules. In *proceedings of SIGMOD*, 1998.
[14] N. Pasquier, Y. Bastide, R. Taouil, and L. Lakhal. Discovering frequent closed itemsets for association rules. *Lecture Notes in Computer Science*, 1999.
[15] J. Pei, J. Han, and L. V. S. Lakshmanan. Mining frequent item sets with convertible constraints. In *proceedings of ICDE*, pages 433–442, 2001.
[16] A. Soulet, B. Crémilleux, and F. Rioult. Condensed representation of emerging patterns. In *proceedings of PAKDD'04*, pages 127–132, 2004.
[17] A. Soulet and B. Crémilleux. A general framework designed for constraint-based mining. Technical report, Université de Caen, Caen, France, 2004.
[18] K. Wang, Y. Jiang, J. X. Yu, G. Dong, and J. Han. Pushing aggregate constraints by divide-and-approximate. In *proceedings of ICDE*, pages 291–302, 2003.

Support Oriented Discovery of Generalized Disjunction-Free Representation of Frequent Patterns with Negation

Marzena Kryszkiewicz[1] and Katarzyna Cichoń[1, 2]

[1] Institute of Computer Science, Warsaw University of Technology,
Nowowiejska 15/19, 00-665 Warsaw, Poland
mkr@ii.pw.edu.pl

[2] Institute of Electrical Apparatus, Technical University of Lodz,
Stefanowskiego 18/22, 90-924 Lodz, Poland
cichon@p.lodz.pl

Abstract. The discovery of frequent patterns has attracted a lot of attention in the data mining community. While an extensive research has been carried out for discovering positive patterns, little has been offered for discovering patterns with negation. An amount of frequent patterns with negation is usually huge and exceeds the number of frequent positive patterns by orders of magnitude. The problem can be significantly alleviated by applying the generalized disjunction-free literal sets representation, which is a concise lossless representation of all frequent patterns, both with and without negation. In this paper, we offer new efficient algorithm *GDFLR-SO-Apriori* for discovering this representation and evaluate it against the *GDFLR-Apriori* algorithm.

1 Introduction

Discovering of frequent patterns in large databases is an important data mining problem. The problem was introduced in [1] for a sales transaction database. Frequent patterns were defined there as sets of items that are purchased together frequently. Frequent patterns are commonly used for building association rules. For example, an association rule may state that 80% of customers who buy fish also buy white wine. This rule is derivable from the fact that fish occurs in 5% of sales transactions and set {fish, white wine} occurs in 4% of transactions. Patterns and association rules can be generalized by admitting negation. A sample rule with negation could state that 75% of customers who buy coke also buy chips and neither beer nor milk. Admitting negation usually results in abundance of mined patterns, which makes analysis of the discovered knowledge infeasible. It is thus preferable to discover and store a possibly small fraction of patterns from which one can derive all other significant patterns when required. This problem was addressed in [3-4], where a generalized disjunction-free literal sets representation (GDFLR) was offered as a lossless representation of all frequent patterns, both with and without negation. GDFLR is by orders of magnitude more concise than all frequent patterns [3]. To the best of our knowledge, no other lossless representation of frequent patterns with negation was proposed. In this paper, we offer new algorithm *GDFLR-SO-Apriori* for discovering the GDFLR representation and evaluate it against the *GDFLR-Apriori* algorithm proposed in [3].

The layout of the paper is as follows: Section 2 recalls basic notions of frequent positive patterns and patterns with negation, as well as methods of inferring frequencies (or supports) of patterns from frequencies of other patterns. Section 3 recalls the GDFLR representation and the *GDFLR-Apriori* algorithm. Our main theoretical contribution is presented in Section 4, where we first propose an ordering for groups of patterns with negation, then examine the properties of patterns following this ordering, and finally use these properties to construct the *GDFLR-SO-Apriori* algorithm. The performance of the *GDFLR-SO-Apriori* and *GDFLR-Apriori* algorithms is evaluated in Section 5. Section 6 concludes the obtained results.

2 Basic Notions

2.1 Frequent Patterns

Let us analyze sample transactional database D presented in Table 1, which we will use throughout the paper. Each row in this database reports items that were purchased by a customer during a single visit to a supermarket.

Table 1. Sample database D

Id	Transaction
T_1	{abce}
T_2	{abcef}
T_3	{abch}
T_4	{abe}
T_5	{acfh}
T_6	{bef}
T_7	{h}
T_8	{af}

As follows from Table 1, items a and b were purchased together in four transactions. The number of transactions in which set of items $\{x_1, ..., x_n\}$ occurs is called its *support* and is denoted by $sup(\{x_1, ..., x_n\})$. A set of items is called a *frequent pattern* if its support exceeds a user-specified threshold (*minSup*). Otherwise, it is called *infrequent*. Clearly, the support of a pattern never exceeds the supports of its subsets. Thus, subsets of a frequent pattern are also frequent, and supersets of an infrequent pattern are infrequent.

2.2 Positive Pattern, Pattern with Negation, Variations of a Pattern

Aside from searching for statistically significant sets of items, one may be interested in identifying frequent cases when purchase of some items excludes purchase of other items. A pattern consisting of items $x_1, ..., x_m$ and negations of items $x_{m+1}, ..., x_n$ will be denoted by $\{x_1, ..., x_m, -x_{m+1}, ..., -x_n\}$. The *support of pattern* $\{x_1, ..., x_m, -x_{m+1}, ..., -x_n\}$ is defined as the number of transactions in which all items in set $\{x_1, ..., x_m\}$ occur and no item in set $\{x_{m+1}, ..., x_n\}$ occurs. E.g., $\{a(-b)(-e)\}$ is supported by two transactions in D. A pattern X is called *positive*, if it does not contain any negated item. Otherwise, X is called a *pattern with negation*. A pattern obtained from X by negating any number of items in X is called a *variation of X*. E.g., {ab} has four distinct variations (including itself): {ab}, {a(-b)}, {(-a)b}, {(-a)(-b)}.

2.3 Calculating Supports of Patterns with Negation

One can note that for a pattern X and an item x, the number of transactions in which X occurs is the sum of the number of transactions in which X occurs with x and the number of transactions in which X occurs without x. Hence $sup(X \cup \{(-x)\}) = sup(X) - sup(X \cup \{x\})$ [6]. Multiple usage of this property enables calculation of the supports of patterns with any number of negated items from the supports of positive patterns [7]:

$$sup(\{x_1,\ldots,x_m\} \cup \{-x_{m+1},\ldots,-x_n\}) = \Sigma_{Z \subseteq \{x_{m+1},\ldots,x_n\}} (-1)^{|Z|} \times sup(\{x_1,\ldots,x_m\} \cup Z) \quad \text{(Eq. 1)}$$

Nevertheless, the knowledge of the supports of only frequent positive patterns may be insufficient to derive the supports of all frequent patterns with negation [6-7].

2.4 Reasoning About Positive Patterns with Generalized Disjunctive Rules

A *generalized disjunctive rule* based on a positive pattern $X = \{x_1, \ldots, x_n\}$ is defined as an expression of the form $x_1 \ldots x_m \rightarrow x_{m+1} \vee \ldots \vee x_n$, where $\{x_1, \ldots, x_m\} \cap \{x_{m+1}, \ldots, x_n\} = \emptyset$ and $\{x_{m+1}, \ldots, x_n\} \neq \emptyset$. Please note that one can build $2^{|X|} - 1$ generalized disjunctive rules from pattern X. We will say that a transaction *supports* rule $r: x_1 \ldots x_m \rightarrow x_{m+1} \vee \ldots \vee x_n$ if it contains all items in $\{x_1, \ldots, x_m\}$ and at least one item in $\{x_{m+1}, \ldots, x_n\}$. We will say that a transaction *violates* rule r if it contains all items in $\{x_1, \ldots, x_m\}$ and no item in $\{x_{m+1}, \ldots, x_n\}$. The number of transactions violating rule r will be called its *error* and will be denoted by $err(r)$. It was shown in [3] that $err(r)$ is determinable from the supports of subsets of $\{x_1, \ldots, x_m, x_{m+1}, \ldots, x_n\}$:

$$err(x_1 \ldots x_m \rightarrow x_{m+1} \vee \ldots \vee x_n) = \Sigma_{Z \subseteq \{x_{m+1},\ldots,x_n\}} (-1)^{|Z|} \times sup(\{x_1,\ldots,x_m\} \cup Z) \quad \text{(Eq. 2)}$$

The following equation follows immediately from Eq. 1 and Eq. 2:

$$err(x_1 \ldots x_m \rightarrow x_{m+1} \vee \ldots \vee x_n) = sup(\{x_1,\ldots,x_m\} \cup \{-x_{m+1},\ldots,-x_n\}) \quad \text{(Eq. 3)}$$

Hence, the error of a generalized disjunctive rule based on a positive pattern X equals the support of X's particular (exactly one) variation with negation.

Rule $x_1 \ldots x_m \rightarrow x_{m+1} \vee \ldots \vee x_n$ is an implication ($x_1 \ldots x_m \Rightarrow x_{m+1} \vee \ldots \vee x_n$) if $err(x_1 \ldots x_m \rightarrow x_{m+1} \vee \ldots \vee x_n) = 0$. Clearly, if $x_1 \ldots x_m \rightarrow x_{m+1} \vee \ldots \vee x_n$ is an implication, then $x_1 \ldots x_m z \rightarrow x_{m+1} \vee \ldots \vee x_n$ and $x_1 \ldots x_m \rightarrow x_{m+1} \vee \ldots \vee x_n \vee z$, which are based on a superset of $\{x_1, \ldots, x_n\}$, are also implications.

The knowledge of such implications can be used for calculating the supports of patterns on which they are based. For example, $ac \Rightarrow b \vee f$ implies that the number of transactions in which $\{ac\}$ occurs equals the number of transactions in which $\{ac\}$ occurs with b plus the number of transactions in which $\{ac\}$ occurs with f minus the number of transactions in which $\{ac\}$ occurs both with b and f. Hence, $sup(\{abcf\}) = sup(\{abc\}) + sup(\{acf\}) - sup(\{ac\})$, which means that the support of pattern $\{abcf\}$ is determinable from the supports of its proper subsets. In general, if there is an implication based on a positive pattern, then the support of this pattern is derivable from the supports of its proper subsets [5]. Each such pattern is called a *generalized disjunctive set*. Otherwise, it is called a *generalized disjunction-free set*.

3 Representing Frequent Patterns with Negation

3.1 Generalized Disjunction-Free Literal Representation GDFLR

A *generalized disjunction-free literal representation* (GDFLR) was introduced in [3] as a concise representation of all frequent patterns, both with and without negation.

GDFLR consists of the following components:

- the main component (*Main*) containing each positive pattern (stored with its support) that has at least one frequent variation and is neither generalized disjunctive nor has support equal 0;
- the infrequent border (*IBd*$^-$) containing each positive pattern all variations of which are infrequent and all proper subsets of which belong to *Main*;
- the generalized disjunctive border (*DBd*$^-$) containing each positive pattern (stored with its support and/or implication) that is either generalized disjunctive or has support equal 0, has at least one frequent variation, and all its proper subsets belong to *Main*.

GDFLR is a lossless representation of all frequent patterns. A formal presentation of this model and its properties can be found in [3]. In particular, it has been proved there that each element in GDFLR has all its proper subsets in the main component. Another important property of GDFLR is that its elements are guaranteed to contain no more than $\lfloor \log_2(|D| - minSup) \rfloor + 1$ items [3].

3.2 Sample Usage of the GDFLR Representation

Fig. 1 depicts the GDFLR representation discovered in D for *minSup* = 1. We will illustrate how to use this representation to evaluate unknown patterns. Let us consider pattern $\{a(-c)(-e)f\}$. We note that $\{acef\}$, which is a positive variation of the evaluated pattern, has subset $\{cef\}$ in the infrequent border. This means that all supersets of $\{cef\}$ and all their variations, including $\{acef\}$ and $\{a(-c)(-e)f\}$, are

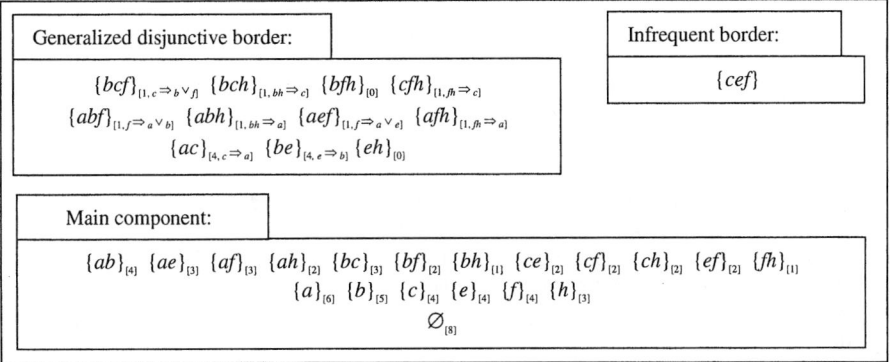

Fig. 1. The *GDFLR* representation of frequent patterns with negation (*minSup* = 1)

infrequent. Now, we will consider pattern $\{bef(-h)\}$. The positive variation $\{befh\}$ of $\{bef(-h)\}$ does not have any subset in the infrequent border, so $\{bef(-h)\}$ has a chance to be frequent. Since, $sup(\{bef(-h)\}) = sup(\{bef\}) - sup(\{befh\})$, we need to determine the supports of two positive patterns $\{bef\}$ and $\{befh\}$. $\{bef\}$ has subset $\{be\}$ in the generalized disjunctive border, the implication of which is $e \Rightarrow b$. Hence, $ef \Rightarrow b$ is an implication for $\{bef\}$. Thus, $sup(bef) = sup(ef) = 2$ (please, see the main component for pattern $\{ef\}$). Pattern $\{befh\}$ also has a subset, e.g. $\{eh\}$, in the generalized disjunctive border. Since $sup(\{eh\}) = 0$, then $sup(\{befh\})$ equals 0 too. Summarizing, $sup(\{bef(-h)\}) = 2 - 0 = 2$, and thus $\{bef(-h)\}$ is a frequent pattern.

3.3 Discovering the GDFLR Representation with the *GDFLR-Apriori* Algorithm

In this section, we recall the *GDFLR-Apriori* algorithm [3], which discovers GDFLR. Creation of candidate elements and calculation of their supports in *GDFLR-Apriori* are assumed to be carried out as in the *Apriori*-like algorithms [2] that discover all frequent positive patterns. *GDFLR-Apriori*, however, differs from such algorithms by introducing additional tests classifying candidates to the *Main*, IBd^- or DBd^- components, respectively. In the algorithm, we apply the following notation:

Notation for *GDFLR-Apriori*
• X_k – candidate k item positive patterns;
• $X.sup$ – the support field of pattern X;

Algorithm *GDFLR-Apriori*(support threshold *minSup*);
$Main = \{\}$; $DBd^- = \{\}$; $IBd^- = \{\emptyset\}$;   // initialize GDFLR
if $\|D\| > minSup$ **then begin**
$\emptyset.sup = \|D\|$; move $\emptyset$ from IBd^- to $Main_0$; $X_1 = \{1$ item patterns$\}$;
for ($k = 1$; $X_k \neq \emptyset$; k++) **do begin**
calculate the supports of patterns in X_k within one scan of database D;
forall candidates $X \in X_k$ **do begin**
/* calculate the errors of all generalized disjunctive rules based on X (by Eq. 2) */
$Errs = Errors\text{-}of\text{-}rules(X, Main)$;
if $max(\{X.sup\} \cup Errs) \leq minSup$ **then**   // all variations of X are infrequent (by Eq. 3)
add X to IBd^-_k
elseif $min(\{X.sup\} \cup Errs) = 0$ **then**   // there is a generalized disjunctive variation of X
add X to DBd^-_k
else add X to $Main_k$ **endif**
endfor;
/* create new $(k+1)$-candidates by merging k item patterns in *Main* */
$X_{k+1} = \{X \subseteq I \| \exists Y, Z \in Main_k \, (\|Y \cap Z\| = k-1 \wedge X = Y \cup Z)\}$;
/* remain only those candidates that have all k item subsets in *Main* */
$X_{k+1} = X_{k+1} \setminus \{X \in X_{k+1} \| \exists Y \subseteq X \, (\|Y\| = k \wedge Y \notin Main_k)\}$
endfor
endif;
return $<\cup_k Main_k, \cup_k DBd^-_k, \cup_k IBd^-_k>$;

After initializing the GDFLR components, the *GDFLR-Apriori* algorithm checks whether the number of transactions in database D is greater than *minSup*. If so, then ∅ is frequent. Hence, ∅, which is a generalized disjunction-free set, is inserted into *Main*$_0$. Next, all items are stored as candidates in X_1. Now, the following steps are performed level-wise for all *k* item candidates, for $k \geq 1$:

- Supports of all candidates in X_k are determined during a pass over the database.
- For each candidate *X* in X_k, the errors *Errs* of all generalized disjunctive rules based on *X* are calculated from the supports of *X*'s subsets in accordance with Eq. 2. Since $\{X.sup\} \cup Errs$ equals the set of the supports of all variations of *X* (by Eq. 3), the condition $max(\{X.sup\} \cup Errs) \leq minSup$ checks if all variations of *X* are infrequent. If so, *X* is found an element of the infrequent border *IBd*$^-$. Otherwise, at least one variation of *X* is frequent. Then, the condition $min(\{X.sup\} \cup Errs) = 0$ checks if *X* generalized disjunctive or its support equals 0. If so, *X* is found an element of the border *DBd*$^-$. Otherwise, it is found an element of *Main*.
- After all candidates in X_k were classified to respective components of GDFLR, candidates X_{k+1} longer by one item are created. Since all proper subsets of each element in GDFLR must belong to the *Main* component, the creation of *k*+1 item candidates is restricted to merging of pairs of *k* item patterns in *Main*. In addition, the newly created candidates that have missing *k* item subsets in *Main* are found not valid GDFLR elements, and thus are discarded from X_{k+1}.

The algorithm ends when there are no more candidates to evaluate.

Please note that the most critical operation in the *GDFLR-Apriori* algorithm is the calculation of errors of a given candidate pattern *X*. As follows from Eq. 2, the error of rule $x_1 \ldots x_m \to x_{m+1} \vee \ldots \vee x_n$ built from *X*, which has *n* items in consequent, requires the knowledge of the supports of *X* and its $2^n - 1$ proper subsets. Taking into account that one can built $\binom{|X|}{n}$ distinct rules from *X* that have *n* items in their consequents, the calculation of the errors of all rules based on *X* that may have 1 to |X| items in their consequents requires $\Sigma_{n=1..|X|} \binom{|X|}{n} (2^n - 1)$ accesses to proper subsets of *X*.

4 New Approach to Computing the GDFLR Representation

Our goal is to speed up the discovery of GDFLR by efficient re-use of the information of the supports of subsets when calculating the errors of rules built from a candidate pattern. Since, the calculation of the errors of rules built from pattern *X* is equivalent to the determination of the supports of *X*'s variations with negation, we will focus only on the latter task. First we will propose an ordering of *X*'s variations. Based on this ordering, we will propose a new method of calculating the support of each variation from the supports of two patterns. Eventually, we will offer new *GDFLR-SO-Apriori* algorithm, which will apply this method for fast discovery of GDFLR.

4.1 Enumerating, Clustering and Calculating Supports of Pattern Variations

In this paper, we define the following ordering of the variations of pattern X:

Let $0 \leq n < 2^{|X|}$. n^{th} *variation of pattern* X ($V_n(X)$) is defined as this variation of X that differs from X on all and only bit positions with value 1 in the binary representation of n. For variation $V_n(X)$, n is called its *(absolute) ordering number*.

Let $0 \leq i < |X|$. i^{th} *cluster* ($C_i(X)$) for pattern X is defined as the set of all variations of X such that i is the leftmost bit position with value 1 in the binary representation of their ordering numbers. Please note that X, which is 0th variation of X, does not belong to any cluster $C_i(X)$, $0 \leq i < |X|$, since the binary representation of its ordering number does not contain any bit position with value 1.

Let $X = \{abc\}$. $V_5(X) = V_{2^2+2^0}(X) = V_{(101)_2}(X) = \{(-a)b(-c)\}$; that is, 5th variation of X differs from X on positions 2 and 0. Table 2 enumerates all variations of X. The variations of X that are different from X can be split to $|X| = 3$ clusters: $C_0(X) = \{V_{(001)_2}(X)\}$; $C_1(X) = \{V_{(010)_2}(X), V_{(011)_2}(X)\}$; $C_2(X) = \{V_{(100)_2}(X), V_{(101)_2}(X), V_{(110)_2}(X), V_{(111)_2}(X)\}$. Note that the ordering numbers of variations in cluster $C_i(X)$, $i \in \{0, ..., |X|-1\}$, can be expressed as $2^i + j$, where $j \in \{0, ..., 2^i - 1\}$ (see Table 2).

Let $0 \leq i < |X|$ and $j \in \{0, ..., 2^i - 1\}$. j^{th} *variation of pattern* X *in cluster* $C_i(X)$ is defined as $V_{2^i+j}(X)$. For variation $V_{2^i+j}(X)$, j is called its *ordering number in cluster* $C_i(X)$ (or *relative ordering number*).

Table 2. Absolute and relative ordering of variations of pattern $X = \{abc\}$

variation V of pattern X	ordering number n of variation V	$|X|$ bit binary representation of n	cluster $C_i(X)$ including variation V	j - ordering number of variation V in $C_i(X)$	binary representation of j	absolute versus rel. ordering of variation V
$\{(\ a)(\ b)(\ c)\}$	0	–	–	–	–	–
$\{(\ a)(\ b)(-c)\}$	1	$(001)_2$	$C_0(X)$	0	$(000)_2$	$1 = 2^0 + 0$
$\{(\ a)(-b)(\ c)\}$	2	$(010)_2$	$C_1(X)$	0	$(000)_2$	$2 = 2^1 + 0$
$\{(\ a)(-b)(-c)\}$	3	$(011)_2$	$C_1(X)$	1	$(001)_2$	$3 = 2^1 + 1$
$\{(-a)(\ b)(\ c)\}$	4	$(100)_2$	$C_2(X)$	0	$(000)_2$	$4 = 2^2 + 0$
$\{(-a)(\ b)(-c)\}$	5	$(101)_2$	$C_2(X)$	1	$(001)_2$	$5 = 2^2 + 1$
$\{(-a)(-b)(\ c)\}$	6	$(110)_2$	$C_2(X)$	2	$(010)_2$	$6 = 2^2 + 2$
$\{(-a)(-b)(-c)\}$	7	$(111)_2$	$C_2(X)$	3	$(011)_2$	$7 = 2^2 + 3$

Corollary 1. Let X be a pattern. The set of all variations of X consists of X and all variations in the clusters $C_i(X)$, where $i \in \{0, ..., |X| - 1\}$:

$$V(X) = \{X\} \cup \bigcup_{i=0..|X|-1} C_i(X) = \{X\} \cup \bigcup_{i=0..|X|-1,\, j=0..2^i-1} \{V_{2^i+j}(X)\}.$$

Note that two distinct variations $V_j(X)$ and $V_{2^i+j}(X)$, $j \in \{0, ..., 2^i-1\}$, of a non-empty pattern X differ only on position i; namely, the item on i^{th} position in $V_{2^i+j}(X)$ is negation of the item on i^{th} position in $V_j(X)$. In addition, $V_j(X)$ and $V_{2^i+j}(X)$ do not differ from X on positions greater than i. These observations imply Theorem 1.

Theorem 1. Let X be a non-empty pattern, $i \in \{0, ..., |X| - 1\}$ and $j \in \{0, ..., 2^i-1\}$. Then: $sup(V_{2^i+j}(X)) = sup(V_j(X \setminus \{X[i]\})) - sup(V_j(X))$.

Table 3. Calculation of supports of consecutive variations of $X = \{abc\}$

i – X's cluster no.	$X \setminus \{X[i]\}$	j – rel. ordering number of X's variation in $C_i(X)$	support calculation for j^{th} variation of X in cluster $C_i(X)$ (that is, for variation $V_{2^i+j}(X)$)
0	$\{ab\}$	0	$sup(V_{2^0+0}(X)) = sup(V_0(X \setminus X[0])) - sup(V_0(X))$ /* $sup(\{ab(-c)\}) = sup(\{ab\}) - sup(\{abc\})$ */
1	$\{ac\}$	0	$sup(V_{2^1+0}(X)) = sup(V_0(X \setminus X[1])) - sup(V_0(X))$ /* $sup(\{a(-b)c\}) = sup(\{ac\}) - sup(\{abc\})$ */
		1	$sup(V_{2^1+1}(X)) = sup(V_1(X \setminus X[1])) - sup(V_1(X))$ /* $sup(\{a(-b)(-c)\}) = sup(\{a(-c)\}) - sup(\{ab(-c)\})$ */
2	$\{bc\}$	0	$sup(V_{2^2+0}(X)) = sup(V_0(X \setminus X[2])) - sup(V_0(X))$ /* $sup(\{(-a)bc\}) = sup(\{bc\}) - sup(\{abc\})$ */
		1	$sup(V_{2^2+1}(X)) = sup(V_1(X \setminus X[2])) - sup(V_1(X))$ /* $sup(\{(-a)b(-c)\}) = sup(\{b(-c)\}) - sup(\{ab(-c)\})$ */
		2	$sup(V_{2^2+2}(X)) = sup(V_2(X \setminus X[2])) - sup(V_2(X))$ /* $sup(\{(-a)(-b)c\}) = sup(\{(-b)c\}) - sup(\{a(-b)c\})$ */
		3	$sup(V_{2^2+3}(X)) = sup(V_3(X \setminus X[2])) - sup(V_3(X))$ /* $sup(\{(-a)(-b)(-c)\}) = sup(\{(-b)(-c)\}) - sup(\{a(-b)(-c)\})$ */

Corollary 2. Let X be a non-empty pattern and $i \in \{0, ..., |X| - 1\}$. The support of each variation in $C_i(X)$ is determinable from the support of a variation of $X \setminus \{X[i]\}$ and the support of either X or a variation of X belonging to a cluster $C_l(X)$, where $l < i$.

Table 3 illustrates a systematic way of calculating the supports of consecutive variations of a pattern X based on Theorem 1 and Corollary 2. Please note that the knowledge of the support of X and the supports of all variations of all proper $|X|-1$ item subsets of X suffices to calculate the supports of all variations of X in this way.

4.2 Algorithm *GDFLR-SO-Apriori*

In this section, we offer new *GDFLR-SO-Apriori* algorithm for discovering GDFLR. It differs from *GDFLR-Apriori* only in that it determines and uses the supports of variations instead of the errors of rules built from candidate patterns. The differences between *GDFLR-SO-Apriori* and *GDFLR-Apriori* are highlighted in the code below.

Additional notation for *GDFLR-SO-Apriori*

- $X.Sup$ – table storing supports of all variations of pattern X; note: $|X.Sup| = 2^{|X|}$.
  Example: Let $X = \{ab\}$, then:
  $X.Sup[0] = X.Sup[(00)_2] = sup(\{(\ a)(\ b)\})$; $X.Sup[1] = X.Sup[(01)_2] = sup(\{(\ a)(-b)\})$;
  $X.Sup[2] = X.Sup[(10)_2] = sup(\{(-a)(\ b)\})$; $X.Sup[3] = X.Sup[(11)_2] = sup(\{(-a)(-b)\})$.

Algorithm *GDFLR-SO-Apriori*(support threshold *minSup*);

$Main = \{\}$; $DBd^- = \{\}$; $IBd^- = \{\varnothing\}$; // initialize GDFLR
if $|D| > minSup$ **then begin**
  $\varnothing.Sup[0] = |D|$; move $\varnothing$ from IBd^- to $Main_0$; $X_1 = \{1$ item patterns$\}$;
  for $(k = 1; X_k \neq \varnothing; k++)$ **do begin**
    calculate supports of patterns in X_k within one scan of D
    forall candidates $X \in X_k$ **do**
      Calculate-supports-of-variations(X, $Main_{k-1}$);
      if $max(\{X.Sup[i]| i = 0..2^k-1\}) \leq minSup$ **then** add X to IBd^-_k
      elseif $min(\{X.Sup[i]| i = 0..2^k-1\}) = 0$ **then** add X to DBd^-_k
      else add X to $Main_k$ **endif**
    endfor;
    /* create new $(k+1)$-candidates by merging k item patterns in *Main* */
    $X_{k+1} = \{X \subseteq I| \exists Y, Z \in Main_k\ (|Y \cap Z| = k-1 \wedge X = Y \cup Z)\}$;
    /* remain only those candidates that have all k item subsets in *Main* */
    $X_{k+1} = X_{k+1} \setminus \{X \in X_{k+1}| \exists Y \subseteq X\ (|Y| = k \wedge Y \notin Main_k\}$
  endfor
endif;
return $<\cup_k Main_k, \cup_k DBd^-_k, \cup_k IBd^-_k>$;

In particular, the *Errors-of-rules* function was replaced by the *Calculate-supports-of-variations* procedure. *Calculate-supports-of-variations* determines the supports of variations of candidate pattern X in two loops. The external loop iterates over clusters of variations of X, the internal loop iterates over variations within a current cluster. The supports of variations are determined in accordance with Theorem 1. As follows from the code, the *Calculate-supports-of-variations* procedure requires only $|X|$ accesses to the proper subsets of a given candidate pattern X instead of $\Sigma_{n=1..|X|} \binom{|X|}{n} (2^n - 1)$ accesses, which would be carried out by the equivalent *Errors-of-rules* function in the *GDFLR-Apriori* algorithm.

procedure *Calculate-supports-of-variations*(k-pattern X, $Main_{k-1}$);

/* assert 1: $X.Sup[0]$ stores support of pattern X */
/* assert 2: all $k-1$ item subsets of X are in $Main_{k-1}$ and*/
/*   supports of all their variations are known */
for $(i = 0; i < k; i++)$ **do begin**   // focus on cluster $C_i(X)$
  $Y = X \setminus \{X[i]\}$; find Y in $Main_{k-1}$;   // $Y \subset X$ is accessed once per cluster
  for $(j = 0; j < 2^i; j++)$ **do**
    $X.Sup[2^i + j] = Y.Sup[j] - X.Sup[j]$;   // calculate support of jth variation in cluster $C_i(X)$
  endfor;
return;

5 Experimental Results

The *GDFLR-SO-Apriori* and *GDFLR-Apriori* algorithms were implemented in C++. The experiments were carried out on the benchmark *mushroom* and *connect-4* data sets. *mushroom* contains 8124 transactions; each of which consists of 23 items; the total number of distinct items in this data set is 119. *connect-4* contains 67557 transactions of length 43 items; the total number of distinct items is 129. The runtime results for both algorithms are presented graphically in Fig. 2 in logarithmic scale.

We observe that *GDFLR-SO-Apriori* performs faster than *GDFLR-Apriori* on these data sets, especially for low support threshold values. In particular, in the case of *mushroom*, *GDFLR-SO-Apriori* performs faster than *GDFLR-Apriori* by two orders of magnitude for *minSup* = 10%, while in the case of *connect-4*, *GDFLR-SO-Apriori* performs faster than *GDFLR-Apriori* by an order of magnitude for *minSup* = 40%.

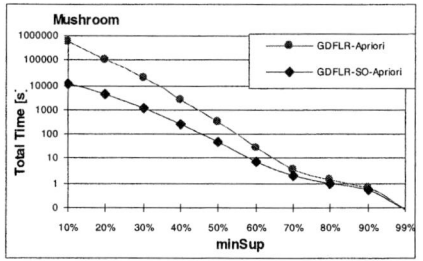
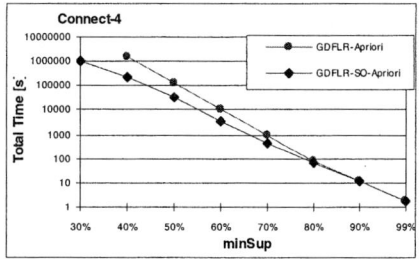

Fig. 2. Duration of *GDFLR-SO-Apriori* and *GDFLR-Apriori* (log. scale)

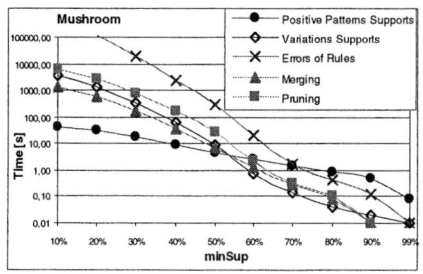
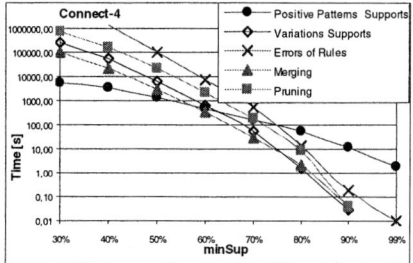

Fig. 3. Duration of particular phases of *GDFLR-SO-Apriori* and *GDFLR-Apriori* (log. scale)

Fig. 3 shows the duration of particular phases of the algorithms in logarithmic scale. The common phases in both algorithms are: calculating supports of positive patterns (PPS), merging (M) and pruning (P). In addition, the *GDFLR-SO-Apriori* algorithm performs the phase of calculating supports of variations (VS), while *GDFLR-Apriori* carries out the analogous phase that calculates errors of rules (E). As follows from Fig. 3, for low support threshold values (less than 50% for *mushroom*, and less than 60% for *connect-4*, respectively), phase (E) is most time consuming; phase (P) is longer than phase (VS), and (VS) is longer than phase (M); phase (PPS) is least time consuming for both algorithms. Concluding, for low threshold values, the performance of *GDFLR-Apriori* depends mainly on the performance of phase (E), which is longer than phase (P), while the performance of *GDFLR-SO-Apriori* depends mainly on the performance of phase (P), which is longer than analogous to (E) phase

(VS). For high threshold values, the runtime of phase (PPS) strongly dominates the runtimes of all other phases. Hence, for high threshold values, *GDFLR-SO-Apriori* is faster than *GDFLR-Apriori* in lower degree than in the case of low threshold values.

6 Conclusions

We have offered new *GDFLR-SO-Apriori* algorithm for discovering the GDFLR representation of all frequent patterns. The experiments prove that *GDFLR-SO-Apriori* is faster than the *GDFLR-Apriori* algorithm by up to two orders of magnitude for low support threshold values. The speed-up was obtained by replacing time-consuming operation of calculating the errors of rules built from candidate patterns with efficient operation of calculating the supports of variations of candidate patterns.

References

[1] Agrawal, R., Imielinski, T., Swami, A.: Mining Associations Rules between Sets of Items in Large Databases. In: Proc. of the ACM SIGMOD, Washington, USA (1993) 207–216
[2] Agrawal, R., Mannila, H., Srikant, R., Toivonen, H., Verkamo, A.I.: Fast Discovery of Association Rules. In: Advances in KDD. AAAI, Menlo Park, California (1996) 307-328
[3] Kryszkiewicz, M.: Generalized Disjunction-Free Representation of Frequent Patterns with Negation. JETAI, Taylor & Francis Group, UK (2005) 63-82
[4] Kryszkiewicz, M.: Reasoning about Frequent Patterns with Negation. Encyclopedia of Data Warehousing and Mining, Information Science Publishing, Idea Group, Inc. (in print)
[5] Kryszkiewicz, M., Gajek, M.: Concise Representation of Frequent Patterns based on Generalized Disjunction-Free Generators. In: Proc. of PAKDD'02 (2002) 159–171
[6] Mannila, H., Toivonen, H.: Multiple Uses of Frequent Sets and Condensed Representations. In: Proc. of KDD'96, Portland, USA (1996) 189-194
[7] Toivonen, H.: Discovery of Frequent Patterns in Large Data Collections. Ph.D. Thesis, Report A-1996-5, University of Helsinki (1996)

Feature Selection Algorithm for Data with Both Nominal and Continuous Features

Wenyin Tang and Kezhi Mao

School of Electrical & Electronic Engineering,
Nanyang Technological University,
Nanyang Avenue, Singapore, 639798

Abstract. Wrapper and filter are two commonly used feature selection schemes. Because of its computational efficiency, the filter method is often the first choice when dealing with large dataset. However, most of filter methods reported in the literature are developed for continuous feature selection. In this paper, we proposed a filter method for mixed data with both continuous and nominal features. The new algorithm includes a novel criterion for mixed feature evaluation, and a novel search algorithm for mixed feature subset generation. The proposed method is tested using a few benchmark real-world problems.

1 Introduction

In many real world problems such as in medical and business, the data obtained is mixed, containing both continuous and nominal features. When employing feature selection algorithm for such mixed data, the common practice is to regard nominal features as numeric, ignoring the difference between them. Another common practice is to preprocess the mixed features into single type, e.g. correlation based feature selection (CFS) [6]. A well-known filter method Relief algorithm [4] deals with this problem by using Hamming distance for nominal features while Euclidean distance distance is used for continuous features. [1] employed a generalized Mahalanobis distance for mixed feature evaluation. This method linearly combines the contributions from continuous and nominal feature subsets, ignoring the association between features of different types and is unsuitable in some applications.

In this paper, we proposed an error probability based measure for mixed feature evaluation. For a mixed feature subset, the entire feature space is first divided into a set of homogeneous subspaces based on nominal features. The merit of the mixed feature subset is then measured based on sample distributions in the homogeneous subspaces spanned by continuous features. The strength of this method is that it avoids transformation of feature types, and takes the association between both types of features into consideration.

Besides feature evaluation criterion, the search algorithm also needs to be carefully designed for mixed data. Here, we proposed a mixed forward selection (MFS) search algorithm. MFS applies SFS as an embedded selecting scheme.

The basic idea of MFS to select features from one type of features, while fixing the other type of features as starting set of SFS. MFS selects both types of features in parallel to complete one iteration. By selectively shrinking fixed sets iteratively, MFS makes an in-depth search into the mixed feature space.

This paper is organized as follows. The proposed mixed feature evaluation method are introduced in Section 2. Section 3 introduces a new search procedure for mixed feature space, Mixed Forward Selection (MFS). Section 4 tests proposed method using a few benchmark real-world mixed datasets. Some concluding remarks are given in Section 5.

2 Proposed Evaluation Method for Mixed Data

2.1 The Basic Idea

Consider mixed feature subset, which is mixture of continuous features $\mathbf{X}$ with p features and nominal features $\mathbf{Z}$ with q features $[\mathbf{X}, \mathbf{Z}] = [x_1, \cdots, x_p, z_1, \cdots, z_q]$. The joint error probability for mixed features, denoted as $P_e(\mathbf{X}, \mathbf{Z})$, is given as Eq.(1).

$$P_e(\mathbf{X}, \mathbf{Z}) = \sum_{\mathbf{Z}} \int_{\mathbf{X}} \left[1 - \max_j P(y_j | \mathbf{X}, \mathbf{Z}) \right] p(\mathbf{X}, \mathbf{Z}) d\mathbf{X} \tag{1}$$

Rearranging Eq.(1) yields:

$$P_e(\mathbf{X}, \mathbf{Z}) = 1 - \sum_{\mathbf{Z}} \int_{\mathbf{X}} \max_j p(y_j, \mathbf{X}, \mathbf{Z}) d\mathbf{X} \tag{2}$$

$$= \sum_{\mathbf{Z}} P(\mathbf{Z}) - \sum_{\mathbf{Z}} \int_{\mathbf{X}} \max_j p(y_j, \mathbf{X} | \mathbf{Z}) P(\mathbf{Z}) d\mathbf{X}$$

$$= \sum_{\mathbf{Z}} P(\mathbf{Z}) \int_{\mathbf{X}} \left[1 - \max_j P(y_j | \mathbf{X}, \mathbf{Z}) \right] p(\mathbf{X}) d\mathbf{X}$$

$$= \sum_{\mathbf{Z}} P(\mathbf{Z}) P_e(\mathbf{X} | \mathbf{Z})$$

Eq.(2) shows that the error probability of entire mixed feature set is the weighted sum of conditional error probability of continuous features given nominal features. For ease of representation, we first define a multi-nominal variable z to replace nominal feature subset $\mathbf{Z}$. The multi-nominal variable z contains N possible distinct values with frequencies n_i, $i = 1, 2, \cdots, N$ and each distinct value represents a distinct combination of nominal feature subset $\mathbf{Z}$. Thus, Eq.(2) can be rewritten as:

$$P_e(\mathbf{X}, \mathbf{Z}) = \sum_{i=1}^{N} P(\mathbf{z} = i) P_e(\mathbf{X} | \mathbf{z} = i) \tag{3}$$

Eq.(3) now gives a very simple expression for the mixed feature evaluation criterion. To obtain $P_e(\mathbf{X}, \mathbf{Z})$, we can first decompose the mixed feature space into a set of homogeneous feature subspaces based on **z**, and the conditional error probability of each subspace, $P_e(\mathbf{X}|\mathbf{z}=i)$ is measured based on continuous features **X**.

2.2 Error Probability Estimation

Error Probability Estimation Based on k Nearest Neighbor (KNN).
Consider a sample set $\mathcal{D}_i$ corresponding to $\mathbf{z}=i$ has n_i training pairs $[\mathbf{x}(j), c(j)]$, $j=1,2,\cdots,n_i$. Error probability of set $\mathcal{D}_i$ estimated by KNN is given as:

$$P_e^{KNN}(\mathbf{X}|\mathbf{z}=i) = \frac{1}{n_i}\sum_{j=1}^{n_i} P_e(\mathbf{x}(j)) \tag{4}$$

where $P_e(\mathbf{x}(j))$, the error probability of sample $\mathbf{x}(j)$, is estimated using the proportion of neighbor(s) having inconsistent class label with $\mathbf{x}(j)$ in the k nearest neighbors' vicinity:

$$P_e(\mathbf{x}(j)) = 1 - P(c(j)|\mathbf{x}(j)) = 1 - \frac{k_{c(j)}}{k} \tag{5}$$

where $k_{c(j)}$ is number of neighbor(s) having the same class labels with $\mathbf{x}(j)$ in its k nearest neighbors' vicinity.

Error Probability Estimation Based on Mahalanobis Distance. Mahalanobis distance, denoted as d, can be constructed as global characterizations for the overlap of random samples drawn from two different distributions [2].

Considering sample set $\mathcal{D}_i$ corresponding to $\mathbf{z}=i$, error probability of $\mathcal{D}_i$ in continuous feature space **X** estimated by Mahalanobis distance, denoted as $P_e^{Maha}(\mathbf{X}|\mathbf{z}=i)$, is normalized into $[0,1]$ as given in Eq.(6).

$$P_e^{Maha}(\mathbf{X}|\mathbf{z}=i) = e^{-\alpha \cdot d^{(i)}} \tag{6}$$

where $d^{(i)}$ is Mahalanobis distance between two classes based on sample set $\mathcal{D}_i$ in continuous feature space **X**; α is a parameter. If error probability is plotted against Mahalanobis distance, the curve will have a bigger curvature with a larger α. We choose a default value of α as 0.25, which produces a satisfactory performance in the experimental study.

3 MFS for Mixed Feature Selection

In this section, we propose a new search procedure, Mixed Forward Selection (MFS). MFS applies SFS as an embedded selecting scheme. In order to deal with the scaling problem, MFS always selects features from a single typed feature subset, while fixing the other types of features as the initial feature subset of the forward selection procedure. Some notations need to be declared here. At the first step of SFS, the non-empty initial feature subset, called *fixed set*, is denoted as

$\mathbf{X}_{fixed}$ and $\mathbf{Z}_{fixed}$ for continuous and nominal feature set respectively, while the corresponding ranking results are denoted as $\mathbf{Z}_{ranked}$ and $\mathbf{X}_{ranked}$ respectively.

MFS selects nominal and continuous feature separately but simultaneously. Although MFS selects two types of features separately, selections of both types of features are linked via fixed sets, which are updated based on the ranking results of SFSs in the current iteration. By selectively shrinking fixed sets iteratively, MFS makes an in-depth search into the mixed feature space. MFS algorithm is summarized as follow.

1. Initialize: fixed sets $\mathbf{X}_{fixed} = \mathbf{X}$, $\mathbf{Z}_{fixed} = \mathbf{Z}$; searching depthes $l_p = l_q = 1$ for continuous and nominal features respectively.
2. Rank each type of features using SFS in parallel.

SFS ranks $\mathbf{X}$ with nonempty beginning set $\mathbf{Z}_{fixed}$	SFS ranks $\mathbf{Z}$ with nonempty beginning set $\mathbf{X}_{fixed}$
(1) Input: $\mathbf{X}, P_e$	(1) Input: $\mathbf{Z}, P_e$
(2) Initialize: $\mathbf{S} = \mathbf{Z}_{fixed}$	(2) Initialize: $\mathbf{S} = \mathbf{X}_{fixed}$
(3) For $i = 1$ to p	(3) For $i = 1$ to q
$\quad \mathbf{x}_i = \arg\min_j P_e(\mathbf{S} \cup \mathbf{x}_j)$	$\quad \mathbf{z}_i = \arg\min_j P_e(\mathbf{S} \cup \mathbf{z}_j)$
$\quad \mathbf{S} = \mathbf{S} \cup \mathbf{x}_i$	$\quad \mathbf{S} = \mathbf{S} \cup \mathbf{z}_i$
$\quad \mathbf{X} = \mathbf{X} - \mathbf{x}_i$	$\quad \mathbf{Z} = \mathbf{Z} - \mathbf{z}_i$
End	End
(4) Store: $\mathbf{X}^{(l_p)}_{ranked} = \mathbf{S}$	(4) Store: $\mathbf{Z}^{(l_q)}_{ranked} = \mathbf{S}$

3. Update fixed feature sets

Update $\mathbf{X}_{fixed}$	Update $\mathbf{Z}_{fixed}$
If $l_p < p$	If $l_q < q$
$\quad \mathbf{X}_{fixed} = \mathbf{X}^{(l_p)}_{ranked}\left[1 \ldots p - l_p\right]$	$\quad \mathbf{Z}_{fixed} = \mathbf{Z}^{(l_q)}_{ranked}\left[1 \ldots q - l_q\right]$
$\quad l_p = l_p + 1$	$\quad l_q = l_q + 1$
End	End

Repeat 2.

Unlike SFS which produce only one step-optimum feature subset, MFS outputs a number of step-optimum feature subsets at each step. To select the best feature subset, classification results are employed. Note that, error estimation method (cross-validation in our experiments) can only be based on the training set. It means that the training part of the whole dataset is further divided into training set and testing set to evaluate the classification accuracy of the induction algorithm. Although this wrapper-like scheme induces extra computations compared with the filter method, the computations involved are far fewer than that of that of the wrapper method because of the limited number of candidate feature subsets in evaluation.

The stopping criterion in MFS algorithm is either a predefined feature subset size or cross-validation error rate. Details will be discussed in the experimental study.

4 Experimental Study

The performances of proposed method: MFS with criteria for mixed feature evaluation are evaluated on two benchmark real-world mixed datasets: *crx* and *bridges* (multi-class problem) [9].

We use cross-validation as the accuracy estimation method in this experimental study for the ease of comparison. The rule of thumb is to keep at least 30 samples in each fold (see [11], chapter 5). Hence, 5-fold cross-validation are employed in our experiment. The average error rate over 10 trials are reported in graphes. For the sample classification, naive Bayes (nB) is used as the classifier, where the probabilities for nominal features are estimated using counts, and those for continuous features are estimated using Parzen window density estimation [3].

4.1 Australian Credit Screening Dataset

Australian credit screening Dataset (*crx*) is downloaded from UCI Machine Learning Repository [9]. The task is to determine whether a credit card should be given to an applicant.

To assess the performance of our methods for *crx* and compare our methods with Relief algorithm [10] and Generalized Mahalanobis distance based forward selection algorithm [1], again, 5-fold cross validation is used to evaluate the classification error rates. The experimental results presented in Fig.1 show that MFS based methods achieve better results than the other two methods mentioned.

4.2 Pittsburgh Bridges Dataset

In order to further assess the performance of our method for the multi-class problem, we choose Pittsburgh bridges dataset (*bridges*) from UCI Machine Learning Repository [9].

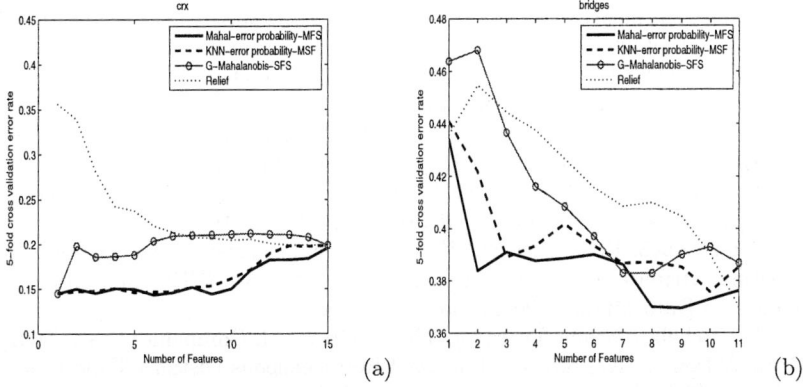

Fig. 1. 5-fold cross validation error rate for mixed datasets: (a) *crx*, (b) *bridge*

The experimental results of our methods and other methods including Relief algorithm [10] and Generalization Mahalanobis distance [1] based forward selection algorithm are shown in Fig.1. This graph shows that the average CV error rates of our methods decrease faster than other two methods. Moreover, the lowest the average CV error rates of our methods are 37.0% and 37.6% when the selected feature size are 8 and 11 features respectively, while the lowest average error rate of Generalization Mahalanobis distance based forward selection algorithm is 38.3% when 7 features are selected. This experiment shows good performance of MFS for mixed feature selection.

5 Conclusion

In this paper we have presented a filter method for mixed feature selection. The performances of our method were tested on a few benchmark real world mixed datasets. The experimental results showed that the our method was more suitable for mixed feature selection problems than some other well-known filter methods including Relief algorithm and Generalized Mahalanobis distance based sequential forward selection.

References

1. A. Bar-Hen and J. J. Daudin, Generalization of Mahalanobis Distance in Mixed Case, Journal of Multivariate Analysis, (1995), 53, pp 332–342.
2. P. A. Devijver and J. Kittler: Pattern Recognition, A Statistical Approach, Prentice-Hall International, INC.,London, (1982).
3. R. O. Duda, P. E. Hart and D. G. Stork: Pattern Classification, A Wiley-Interscience Publication, (2001).
4. I. Kononenko: Estimating Attributes: Analysis and Extensions of RELIEF, ECML, (1994), pp.171–182.
5. M. Robnik, I. Kononenko:Theoretical and Empirical Analysis of ReliefF and RReliefF, Machine Learning, (2003), 53, pp. 23–26.
6. M. A. Hall: Correlation-based Feature Selection for Machine Learning, A Dissertation submit to Department of Computer Science, University of Waikato, Hamilton, NewZealand. (1999).
7. M. A. Hall: Correlation-based Feature Selection for Discrete and Numeric Class Machine Learning, ICML, (2000), pp. 359–366.
8. L. C. Molina, L. Belanche, and A. Nebot: Feature Selection Algorithms: A Survey and Experimental Evaluation, ICDM, (2002), pp.306–313.
9. P. M. Murphy, D. W. Aha: UCI repository of machine learning databases, http://www.ics.uci.edu/~mlearn/MLRepository.html, (1995).
10. R. Florez-Lopez: Reviewing RELIEF and its extensions: a new approach for estimating attributes considering high-correlated features. IEEE International Conference on Data Mining, (2002), pp. 605 – 608.
11. T. M. Mitchell: Machine Learning, The McGraw-Hill Companies, Inc., (2001).
12. D. R. Wilson, T. R. Martinez: Improved Heterogeneous Distance Functions. Journal of Artificial Intelligence Research, (1997), 6, pp. 1–34.

A Two-Phase Algorithm for Fast Discovery of High Utility Itemsets

Ying Liu, Wei-keng Liao, and Alok Choudhary

Electrical and Computer Engineering Department,
Northwestern University, Evanston, IL, USA 60208
{yingliu, wkliao, choudhar}@ece.northwestern.edu

Abstract. Traditional association rules mining cannot meet the demands arising from some real applications. By considering the different values of individual items as utilities, utility mining focuses on identifying the itemsets with high utilities. In this paper, we present a Two-Phase algorithm to efficiently prune down the number of candidates and precisely obtain the complete set of high utility itemsets. It performs very efficiently in terms of speed and memory cost both on synthetic and real databases, even on large databases that are difficult for existing algorithms to handle.

1 Introduction

Traditional Association rules mining (ARM) [1] model treat all the items in the database equally by only considering if an item is present in a transaction or not. Frequent itemsets identified by ARM may only contribute a small portion of the overall profit, whereas non-frequent itemsets may contribute a large portion of the profit. In reality, a retail business may be interested in identifying its most valuable customers (customers who contribute a major fraction of the profits to the company). These are the customers, who may buy full priced items, high margin items, or gourmet items, which may be absent from a large number of transactions because most customers do not buy these items. In a traditional frequency oriented ARM, these transactions representing highly profitable customers may be left out. Utility mining is likely to be useful in a wide range of practical applications.

Recently, a *utility mining* model was defined [2]. Utility is a measure of how "useful" an itemset is. The goal of utility mining is to identify high utility itemsets that drive a large portion of the total utility. Traditional ARM problem is a special case of utility mining, where the utility of each item is always 1 and the sales quantity is either 0 or 1.

There is no efficient strategy to find all the high utility itemsets due to the nonexistence of "downward closure property" (anti-monotone property) in the *utility mining* model. A heuristics [2] is used to predict whether an itemset should be added to the candidate set. We refer this algorithm as MEU (Mining using Expected Utility) for the rest of this paper. However, the prediction usually overestimates, especially at the beginning stages, where the number of candidates approaches the number of all the combinations of items. Such requirements can easily overwhelm the available

Table 1. A transaction database

(a) Transaction table. Each row is a transaction. The columns represent the number of items in a particular transaction. TID is the transaction identification number

TID \ ITEM	A	B	C	D	E
T_1	0	0	18	0	1
T_2	0	6	0	1	1
T_3	2	0	1	0	1
T_4	1	0	0	1	1
T_5	0	0	4	0	2
T_6	1	1	0	0	0
T_7	0	10	0	1	1
T_8	3	0	25	3	1
T_9	1	1	0	0	0
T_{10}	0	6	2	0	2

(b) The utility table. The right column displays the profit of each item per unit in dollars

ITEM	PROFIT ($) (per unit)
A	3
B	10
C	1
D	6
E	5

(c) Transaction utility (TU) of the transaction database

TID	TU	TID	TU
T_1	23	T_6	13
T_2	71	T_7	111
T_3	12	T_8	57
T_4	14	T_9	13
T_5	14	T_{10}	72

memory space and computation power of most of the machines. In addition, MEU may miss some high utility itements when the variance of the itemset supports is large.

The challenge of utility mining is in restricting the size of the candidate set and simplifying the computation for calculating the utility. In order to tackle this challenge, we propose a Two-Phase algorithm to efficiently mine high utility itemsets. A performance study has been conducted on real and synthetic data, obtaining significant improvement in terms of speed and accuracy over the best existing algorithm [2]. Our algorithm easily handles very large databases that existing algorithms cannot handle.

The rest of this paper is organized as follows. Section 2 overviews the related work. In Section 3, we propose the Two-Phase algorithm. Section 4 presents our experimental results and we summarize our work in section 5.

2 Related Work

Researches that assign different weights to items have been proposed in [3, 4, 5, 6]. These weighted ARM models are special cases of utility mining.

A concept, *itemset share*, is proposed in [7]. It can be regarded as a utility because it reflects the impact of the sales quantities of items on the cost or profit of an itemset. Several heuristics have been proposed.

A utility mining algorithm is proposed in [8], where the concept of "useful" is defined as an itemset that supports a specific objective that people want to achieve. The definition of utility and the goal of his algorithm are different from those in our work.

3 Two-Phase Algorithm

We start with the definition of a set of terms that leads to the formal definition of utility mining problem. The same terms are given in [2].

- $I = \{i_1, i_2, ..., i_m\}$ is a set of items.
- $D = \{T_1, T_2, ..., T_n\}$ be a transaction database where each transaction $T_i \in D$ is a subset of I.
- $o(i_p, T_q)$, *local transaction utility value*, represents the quantity of item i_p in transaction T_q. For example, $o(A, T_8) = 3$, in Table 1(a).
- $s(i_p)$, *external utility*, is the value associated with item i_p in the Utility Table. This value reflects the importance of an item, which is independent of transactions. For example, in Table 1(b), the external utility of item A, $s(A)$, is 3.
- $u(i_p, T_q)$, *utility*, the quantitative measure of utility for item i_p in transaction T_q, is defined as $o(i_p, T_q) \times s(i_p)$. For example, $u(A, T_8) = 3 \times 3 = 9$, in Table 1
- $u(X, T_q)$, *utility of an itemset X in transaction* T_q, is defined as $\sum_{i_p \in X} u(i_p, T_q)$, where $X = \{i_1, i_2, ..., i_k\}$ is a k-itemset, $X \subseteq T_q$ and $1 \le k \le m$.
- $u(X)$, *utility of an itemset X*, is defined as $\sum_{T_q \in D \wedge X \subseteq T_q} u(X, T_q)$. (3.1)

Utility mining is to find all the itemsets whose utility values are beyond a user specified threshold. An itemset X is a *high utility itemset* if $u(X) \ge \varepsilon$, where $X \subseteq I$ and ε is the minimum utility threshold, otherwise, it is a *low utility itemset*. For example, in Table 1, $u(\{A, D, E\}) = u(\{A, D, E\}, T_4) + u(\{A, D, E\}, T_8) = 14 + 32 = 46$. If $\varepsilon = 120$, $\{A, D, E\}$ is a low utility itemset.

To address the drawbacks in MEU, we propose a novel Two-Phase algorithm. In Phase I, we define *transaction-weighted utilization* and propose a model — *transaction-weighted utilization mining*. This model maintains a *Transaction-weighted Downward Closure Property*. Thus, only the combinations of high transaction-weighted utilization itemsets are added into the candidate set at each level during the level-wise search. Phase I may overestimate some low utility itemsets, but it never underestimates any itemsets. In phase II, only one extra database scan is performed to filter the overestimated itemsets.

3.1 Phase I

Definition 1. (Transaction Utility) The *transaction utility of transaction* T_q, denoted as $tu(T_q)$, is the sum of the utilities of all the items in T_q: $tu(T_q) = \sum_{i_p \in T_q} u(i_p, T_q)$. Table 1 (c) gives the transaction utility for each transaction in Table 1.

Definition 2. (Transaction-weighted Utilization) The *transaction-weighted utilization of an itemset X*, denoted as $twu(X)$, is the sum of the transaction utilities of all the transactions containing X: $twu(X) = \sum_{X \subseteq T_q \in D} tu(T_q)$. (3.2)

For the example in Table 1, $twu(AD) = tu(T_4) + tu(T_8) = 14 + 57 = 71$.

Definition 3. (High Transaction-weighted Utilization Itemset) For a given itemset X, X is a *high transaction-weighted utilization itemset* if $twu(X) \ge \varepsilon'$, where ε' is the user specified threshold.

Theorem 1. (**Transaction-weighted Downward Closure Property**) Let I^k be a k-itemset and I^{k-1} be a $(k-1)$-itemset such that $I^{k-1} \subset I^k$. If I^k is a high transaction-weighted utilization itemset, I^{k-1} is a high transaction-weighted utilization itemset.

Proof: Let T_{I^k} be the collection of the transactions containing I^k and $T_{I^{k-1}}$ be the collection containing I^{k-1}. Since $I^{k-1} \subset I^k$, $T_{I^{k-1}}$ is a superset of T_{I^k}. According to Definition 2, $twu(I^{k-1}) = \sum_{I^{k-1} \subseteq T_q \in D} tu(T_q) \geq \sum_{I^k \subseteq T_p \in D} tu(T_p) = twu(I^k) \geq \varepsilon'$ ❑

The *Transaction-weighted Downward Closure Property* indicates that any superset of a low transaction-weighted utilization itemset is low in transaction-weighted utilization. That is, only the combinations of high transaction-weighted utilization $(k-1)$-itemsets could be added into the candidate set C_k at each level.

Theorem 2. Let *HTWU* be the collection of all high transaction-weighted utilization itemsets in a transaction database D, and *HU* be the collection of high utility itemsets in D. If $\varepsilon' = \varepsilon$, then $HU \subseteq HTWU$.

Proof: $\forall X \in HU$, if X is a high utility itemset, then

$$\varepsilon' = \varepsilon \leq u(X) = \sum_{X \subseteq T_q} u(X, T_q) = \sum_{X \subseteq T_q} \sum_{i_p \in X} u(i_p, T_q) \leq \sum_{X \subseteq T_q} \sum_{i_p \in T_q} u(i_p, T_q) = \sum_{X \subseteq T_q} tu(T_q) = twu(X)$$

Thus, X is a high transaction-weighted utilization itemset and $X \in HTWU$. ❑

According to Theorem 2, we can utilize the *Transaction-weighted Downward Closure Property* in our *transaction-weighted utilization mining* in Phase I by assuming $\varepsilon' = \varepsilon$ and prune those overestimated itemsets in Phase II.

Figure 1 shows the search space of Phase I. The level-wise search stops at the third level, one level less than MEU. (For larger databases, the savings should be more evident.) *Transaction-weighted utilization mining* model outperforms MEU in several aspects:

1) **Less candidates** — When ε' is large, the search space can be significantly reduced at the second level and higher levels. As shown in Figure 1, four out of 10 itemsets are pruned because they all contain item A. However, in MEU, the prediction hardly prunes any itemset at the beginning stages.
2) **Accuracy** — Based on Theorem 2, if we let $\varepsilon' = \varepsilon$, the complete set of high utility itemsets is a subset of the high transaction-weighted utilization itemsets discovered by our *transaction-weighted utilization mining* model. However, MEU may miss some high utility itemsets when the variation of itemset supports is large.
3) **Arithmetic complexity** — One of the kernel operations in the Two-Phase algorithm is the calculation for each itemset's transaction-weighted utilization as in equation 3.2, which only incurs add operations rather than a number of multiplications in MEU. Thus, the overall computation is much less complex.

3.2 Phase II

In Phase II, one database scan is performed to filter the high utility itemsets from high transaction-weighted utilization itemsets identified in Phase I. The number of high

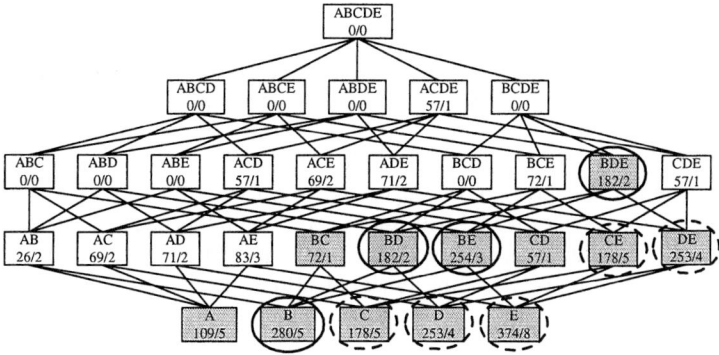

Fig. 1. Itemsets lattice related to the example in Table 1. ε' = 120. Itemsets in circles (solid and dashed) are the high transaction-weighted utilization itemsets in *transaction-weighted utilization mining model*. Gray-shaded boxes denote the search space. Itemsets in solid circles are high utility itemsets. Numbers in each box are transaction-weighted utilization / number of occurrence

transaction-weighted utilization itemsets is small when ε' is high. Hence, the time saved in Phase I may compensate for the cost incurred by the extra scan in Phase II.

In Figure 1, the high utility itemsets ({B}, {B, D}, {B, E} and {B, D, E}) are covered by the high transaction-weighted utilization itemsets. One database scan is performed in Phase II to prune 5 of the 9 itemsets since they are not high utility itemsets.

4 Experimental Evaluation and Performance Study

We run all our experiments on a 700-MHz Xeon 8-way shared memory parallel machine with a 4 Gbytes memory.

4.1 Synthetic Data from IBM Quest Data Generator

We use a synthetic database [9], T20.I6.D1000K. However, the IBM Quest data generator only generates the quantity of 0 or 1 for each item in a transaction. We randomly generate the quantity of each item in each transaction, ranging from 1 to 5. Utility tables are also synthetically created by assigning a utility value to each item randomly, ranging from 0.01 to 10.00. Observed from real world databases that most items are in the low profit range, we generate the utility values using a log normal distribution.

In Table 2, the number of candidate itemsets generated by Phase I at the first database scan decreases dramatically as the threshold goes up. However, the number of candidates generated by MEU is always 499500. We don't provide the exact numbers for MEU because it actually takes an inordinate amount of time (longer than 10 hours) to complete the second scan. Observed from Table 2, the *Transaction-weighted Downward Closure Property* in transaction-weighted utilization mining model can help prune candidates very effectively.

4.2 Real-World Market Data

We also evaluated the Two-Phase algorithm using a real world data from a major grocery chain store. There are 1,112,949 transactions and 46,086 items in the database. Each transaction consists of the products and the sales volume of each product purchased by a customer at a time point. The utility table describes the profit of each item.

We evaluate the scalability of our algorithm by varying the threshold. As shown in Table 3, it is fast and scales well. MEU doesn't work with this dataset because the number of 2-itemset candidates is so large (over 2 billion) that it overwhelms the memory available to us. Actually, very few machines can afford such a huge memory cost.

Table 2. The number of candidate itemsets generated by Phase I of Two-Phase algorithm vs. MEU

Databases Threshold		T20.I6.D1000K	
		Phase I	MEU
0.5%	1st scan	315615	499500
	2nd scan	18653	-
1%	1st scan	203841	499500
	2nd scan	183	-
1.5%	1st scan	135460	499500
	2nd scan	8	-
2%	1st scan	84666	499500
	2nd scan	1	-

Table 3. Scalability in execution time with varying minimum utility threshold of the real world database

Threshold	Running time (seconds)	# Candidates	# High transaction-weighted utilization (Phase I)	# High utility (Phase II)
1%	25.76	11936	9	2
0.75%	33.3	23229	26	3
0.5%	53.09	69425	80	5
0.25%	170.49	627506	457	17
0.1%	1074.94	7332326	3292	80

5 Conclusions

This paper proposed a Two-Phase algorithm that discovers high utility itemsets highly efficiently. The transaction-weighted utilization mining we proposed not only restricts the search space, but also covers all the high utility itemsets. Only one extra database scan is needed to filter the overestimated itemsets. Our algorithm requires fewer database scans, less memory space and less computational cost compared to the best existing utility mining algorithm. It can easily handle very large databases for which other existing algorithms are infeasible.

Acknowledgements

This work was supported in part by NSF grants CCF-0444405, CNS-0406341, CCR-0325207, DOE grant DE-FC02-01ER25485 and Intel Corp.

References

1. Agrawal and R. Srikant: Fast algorithms for mining association rules. 20th VLDB (1994)
2. Hong Yao, Howard J. Hamilton, and Cory J. Butz: A Foundational Approach to Mining Itemset Utilities from Databases. SDM (2004)
3. C.H. Cai, Ada W.C. Fu, C.H. Cheng, and W.W. Kwong: Mining Association Rules with Weighted Items. IDEAS (1998)

4. W. Wang, J. Yang, and P. Yu: Efficient Mining of Weighted Association Rules (WAR). 6th KDD (2000)
5. Feng Tao, Fionn Murtagh, and Mohsen Farid: Weighted Association Rule Mining using Weighted Support and Significance Framework. 9th KDD (2003)
6. S. Lu, H. Hu, and F. Li: Mining weighted association rules. Intelligent Data Analysis, 5(3) (2001), 211-225
7. B. Barber and H.J.Hamilton: Extracting share frequent itemsets with infrequent subsets. Data Mining and Knowledge Discovery, 7(2) (2003), 153-185
8. Raymond Chan, Qiang Yang, Yi-Dong Shen: Mining high utility Itemsets. ICDM (2003)
9. IBM data generator, http://www.almaden.ibm.com/software/quest/Resources/index.shtml

On Multiple Query Optimization in Data Mining*

Marek Wojciechowski and Maciej Zakrzewicz

Poznan University of Technology,
Institute of Computing Science,
ul. Piotrowo 3a, 60-965 Poznan, Poland
{marek, mzakrz}@cs.put.poznan.pl

Abstract. Traditional multiple query optimization methods focus on identifying common subexpressions in sets of relational queries and on constructing their global execution plans. In this paper we consider the problem of optimizing sets of data mining queries submitted to a Knowledge Discovery Management System. We describe the problem of data mining query scheduling and we introduce a new algorithm called *CCAgglomerative* to schedule data mining queries for frequent itemset discovery.

1 Introduction

Multiple Query Optimization (MQO) [10] is a database research area which focuses on optimizing a set of queries together by executing their common subexpressions once in order to save execution time. The main tasks in MQO are common subexpression identification and global execution plan construction. When common subexpressions have been identified, they can be executed just once and materialized for all the queries, instead of being executed once for each query. A specific type of a query is a Data Mining Query (DMQ) [7], describing a data mining task. It defines constraints on the data to be mined and constraints on the patterns to be discovered. DMQs are submitted for execution to a Knowledge Discovery Management System KDDMS [7], which is a DBMS extended with data mining functions. Traditional KDDMSs execute DMQs serially and do not try to share any common subexpressions.

DMQs are often processed in batches of 10-100 queries. Such queries may show many similarities about data or pattern constraints. If they are executed serially, it is likely that many I/O operations are wasted because the same database blocks may be required by multiple DMQs. If I/O steps of different DMQs were integrated and performed once, then we would be able to decrease the overall execution cost of the whole batch. Traditional MQO methods are not applicable to DMQs. DMQs perform huge database scans, which cannot and should not be materialized. Moreover, DMQs usually have high memory requirements that make it difficult to dynamically materialize intermediate results. One of the methods we proposed to process batches of DMQs is Apriori Common Counting (ACC), focused on frequent itemset discovery

---

* This work was partially supported by the grant no. 4T11C01923 from the State Committee for Scientific Research (KBN), Poland.

queries [1]. ACC is based on Apriori algorithm [2], it integrates the phases of support counting for candidate itemsets – candidate hash trees for multiple DMQs are loaded into memory together and then the database is scanned once. Basic ACC [11] assumes that all DMQs fit in memory, which is not the common case, at least for initial Apriori iterations. If the memory can hold only a subset of all DMQs, then it is necessary to schedule the DMQs into subsets, called phases [12]. The way such scheduling is done determines the overall cost of batched DMQs execution. To solve the scheduling problem, in [12] we proposed an "initial" heuristic algorithm, called *CCRecursive*.

2 Related Work

To the best of our knowledge, apart from the ACC method discussed in this paper, the only other multiple query processing scheme for frequent pattern discovery is Mine Merge, presented in one of our previous papers [13]. In contrast to ACC, Mine Merge is independent of a particular frequent itemset mining algorithm. However, it was proven very sensitive to data distribution and less predictable than ACC. A MQO technique based on similar ideas as ACC has been proposed in the context of inductive logic programming, where similar queries were combined into query packs [4].

Somewhat related to the problem of multiple data mining query optimization is re-using results of previous queries to answer a new query, which can be interpreted as optimizing processing of a sequence of queries independently submitted to the system. Methods falling into that category are: incremental mining [5], caching intermediate query results [9], and reusing materialized results of previous queries provided that syntactic differences between the queries satisfy certain conditions [3] [8].

3 Preliminaries and Problem Statement

Data mining query. A *data mining query* is a tuple $DMQ = (R, a, \Sigma, \Phi, \beta)$, where R is a relation, a is an attribute of R, Σ is a condition involving the attributes of R, Φ is a condition involving discovered patterns, β is the min. support threshold. The result of the DMQ is a set of patterns discovered in $\pi_a \sigma_\Sigma$, satisfying Φ, and having support $\geq \beta$.

Problem statement. Given a set of data mining queries $DMQ = \{dmq_1, dmq_2, ..., dmq_n\}$, where $dmq_i = (R, a, \Sigma_i, \Phi_i, \beta_i)$, Σ_i has the form "$(l^i_{1min} < a < l^i_{1max}) \vee (l^i_{2min} < a < l^i_{2max}) \vee .. \vee (l^i_{kmin} < a < l^i_{kmax})$", $l^i_* \in dom(a)$ and there exist at least two data mining queries $dmq_i = (R, a, \Sigma_i, \Phi_i, \beta_i)$ and $dmq_j = (R, a, \Sigma_j, \Phi_j, \beta_j)$ such that $\sigma_{\Sigma_i} R \cap \sigma_{\Sigma_j} R \neq \emptyset$. The problem of *multiple query optimization* of DMQ consists in generating such an algorithm to execute DMQ which has the lowest I/O cost.

Data sharing graph. Let $S = \{s_1, s_2, ..., s_k\}$ be a set of *distinct data selection formulas* for DMQ, i.e., a set of selection formulas on the attribute a of the relation R such that for each i,j we have $\sigma_{s_i} R \cap \sigma_{s_j} R = \emptyset$, and for each i there exist integers a, b, ..., m, such that $\sigma_{\Sigma_i} R = \sigma_{s_a} R \cup \sigma_{s_b} R \cup ... \cup \sigma_{s_m} R$. We refer to the graph $DSG = (V,E)$ as to a *data sharing graph* for the set of data mining queries DMQ if and only if $V = DMQ \cup S$, $E = \{(dmq_i, s_j) \mid dmq_j \in DMQ, s_j \in S, \sigma_{\Sigma_i} R \cap \sigma_{s_j} R \neq \emptyset\}$.

Example. Consider the following example of a data sharing graph. Given a database relation $\mathcal{R}_1 = (attr_1, attr_2)$ and three data mining queries: $dmq_1 = (\mathcal{R}_1, "attr_2", "5 < attr_1 < 20", \emptyset, 3)$, $dmq_2 = (R_1, "attr_2", "10 < attr_1 < 30", \emptyset, 5)$, $dmq_3 = (\mathcal{R}_1, "attr_2", "15 < attr_1 < 40", \emptyset, 4)$. The set of distinct data selection formulas is: $S = \{s_1 = "5 < attr_1 < 10", s_2 = "10 < attr_1 < 15", s_3 = "15 < attr_1 < 20", s_4 = "20 < attr_1 < 30", s_5 = "30 < attr_1 < 40"\}$. The data sharing graph for $\{dmq_1, dmq_2, dmq_3\}$ is shown in Fig. 1. Ovals represent DMQs and boxes represent distinct selection formulas.

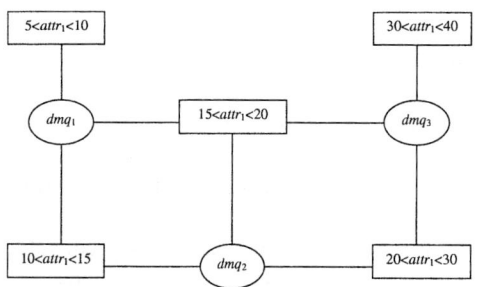

Fig. 1. Sample data sharing graph for a set of data mining queries

Apriori Common Counting (Fig. 2). ACC executes a set of data mining queries by integrating their I/O operations. First, for each data mining query we build a separate hash tree for 1-candidates. Next, for each distinct data selection formula we scan its corresponding database partition and we count candidates for all the data mining queries that contain the formula. Such a step is performed for 2-candidates, 3-candidates, etc. Notice that if a given distinct data selection formula is shared by many data mining queries, then its corresponding database partition is read only once.

```
for (i=1; i<=n; i++)          /* n = number of data mining queries */
    C_1^i = {all 1-itemsets from σ_{s1∪s2∪...∪sk}R, ∀s_j∈ S: (dmq_i,s_j)∈ E}   /* generate 1-candidates */
for (k=1; C_k^1 ∪ C_k^2 ∪..∪ C_k^n ≠ ∅; k++) do begin
    for each s_j∈ S do begin
        CC= ∪C_k^i: (dmq_i,s_j)∈ E;         /* select the candidates to count now */
        if CC≠ ∅ then count(CC, σ_{sj}R); end
    for (i=1; i<=n; i++) do begin
        F_k^i = {C ∈ C_k^i | C.count ≥ minsup^i};      /* identify frequent itemsets */
        C_{k+1}^i = generate_candidates(F_k^i); end
end
for (i=1; i<=n; i++) do  Answer^i = ∪_k F_k^i;      /* generate responses */
```

Fig. 2. Apriori Common Counting

4 Data Mining Query Scheduling

The basic ACC algorithm assumes that memory is unlimited and therefore the candidate hash trees for all DMQs can completely fit in memory. If, however, the memory is limited, ACC execution must be partitioned into multiple *phases*, so that in each phase only a subset of DMQs is processed. In such a case, the key question to answer is: which data mining queries from the set should be executed together in one phase and which data mining queries can be executed in different phases? We will refer to the task of data mining queries partitioning as to *data mining query scheduling*.

There are several issues to be addressed when scheduling data mining queries. First of all, it is obvious that the number of data mining queries which can be included in the same phase is restricted by the actual memory size. Memory requirements of individual data mining queries are determined by sizes of their candidate hash trees, which in turn are dependent on underlying data characteristics and on candidate sizes. Since the sizes of candidate hash trees change between Apriori iterations, the scheduling should be performed at the beginning of every iteration, not only before data mining query set execution starts.

Another observation concerns the nature of ACC. Scheduling of DMQs should be based on inter-query similarities. Queries which operate on separate database partitions should be performed in separate phases, while queries which operate on significantly overlapping database partitions could benefit from being executed in the same phase. To measure the level of "overlapping" we can use cost estimation features of existing cost-based query optimizers.

A scheduling algorithm requires that sizes of candidate hash trees are known in advance. They can be estimated in two ways. We can find an upper bound for the number of candidates knowing the number of frequent itemsets from the previous Apriori iteration. Unfortunately, typical upper bounds are far from actual sizes of the candidate hash trees. Another approach is to first generate all the candidate hash trees, measure their sizes, save them to disk, schedule the data mining queries, and then load the required trees from disk. This method introduces the cost of materialization.

5 CCAgglomerative Scheduling Algorithm

The *CCAgglomerative* algorithm first transforms the data sharing graph into a *gain graph*, which contains (1) vertices being the original data mining queries and (2) two-vertex edges whose weights describe gains that can be reached by executing the connected queries in the same phase. Due to the restricted size of this paper we skip the algorithm of gain graph generation. A sample gain graph for the earlier discussed set of data mining queries is shown in Fig. 3. For example, putting the data mining queries dmq_1 and dmq_2 in the same phase will allow us to save 9000 I/O cost units.

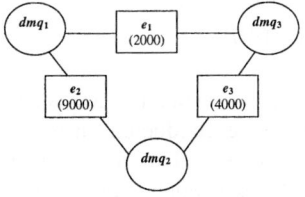

Fig. 3. Sample gain graph

An initial schedule is created by putting each data mining query into a separate phase. Next, the algorithm processes the edges sorted with respect to the decreasing weights. For each edge, the algorithm tries to combine phases containing the connected data mining queries into one phase. If the total size of all the data mining queries in such phase does not exceed the memory size, the original phases are replaced with the new one. Otherwise the algorithm simply ignores the edge and continues. The *CCAgglomerative* algorithm is shown on Fig. 4.

```
CCAgglomerative(G=(V,E), E contains 2-node edges only):
begin
    Phases ← ∅
    for each v in V do Phases ← Phases ∪ {{v}}
    sort E = {e₁ , e₂ ,..., eₖ} in desc. order with respect to eᵢ.gain, ignore edges with zero gains
    for each eᵢ = (v₁, v₂) in E do begin
        phase₁ ← p ∈ Phases such that v₁ ∈ p
        phase₂ ← p ∈ Phases such that v₂ ∈ p
        if treesize(phase₁ ∪ phase₂) ≤ MEMSIZE then
            Phases ← Phases – {phase₁}
            Phases ← Phases – {phase₂}
            Phases ← Phases ∪ {phase₁ ∪ phase₂}
        end if
    end
    return Phases
end
```

Fig. 4. *CCAgglomerative* Algorithm

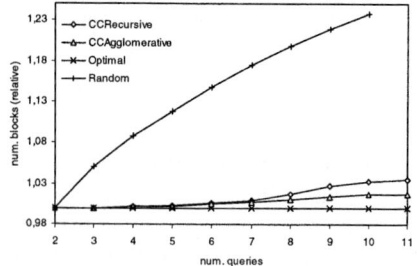
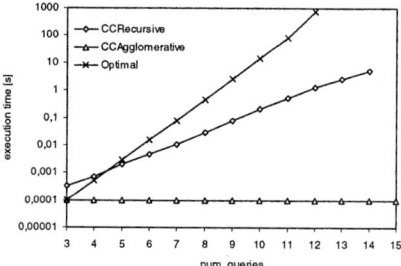

Fig. 5. Accuracy of data mining query scheduling algorithms

Fig. 6. Execution time of data mining query scheduling algorithms

6 Experimental Evaluation

We performed several experiments using the MSWeb dataset from the UCI KDD Archive [6]. The experiments were conducted on a PC AMD Duron 1.2 GHz with 256 MB of RAM. The datasets resided in flat files on a local disk. Memory was intentionally restricted to 10kB-50kB. Each experiment was repeated 100 times.

Fig.5 shows disk I/O costs of schedules generated by the optimal scheduling algorithm, by the *CCAgglomerative* algorithm, by the *CCRecursive* algorithm, and by a random algorithm (which randomly builds phases from queries). *CCAgglomerative* has outperformed the other heuristic approach and achieved a very good accuracy. For example, for the set of 10 data mining queries, the *CCAgglomerative* algorithm misses the optimal solution by only 1.5%. Fig. 6 presents execution times for the optimal scheduling algorithm, *CCRecursive*, and *CCAgglomerative* (the execution time for *CCAgglomerative* includes the time required to build the gain graph). Notice that the optimal algorithm needed ca. 1000s to schedule 12 data mining queries, *CCRecursive* showed exponential execution time, while *CCAgglomerative* (polynomial wrt. the number of queries) still needed just about 0.0001s even for 15 queries.

7 Conclusions

The paper addressed the problem of optimizing sets of multiple data mining queries. We showed that in order to apply Apriori Common Counting in a restricted memory system, it is required to schedule data mining queries into separate phases. The way such scheduling is performed influences the overall cost of executing the set of data mining queries. We presented the new heuristic scheduling algorithm, called *CCAgglomerative* which significantly outperforms the other existing approach, *CCRecursive*, yet it provides a very good accuracy.

References

1. Agrawal R., Imielinski T., Swami A: Mining Association Rules Between Sets of Items in Large Databases. Proc. of the 1993 ACM SIGMOD Conf. on Management of Data, 1993.
2. Agrawal R., Srikant R.: Fast Algorithms for Mining Association Rules. Proc. of the 20th Int'l Conf. on Very Large Data Bases (1994)
3. Baralis E., Psaila G.: Incremental Refinement of Mining Queries. Proceedings of the 1st DaWaK Conference (1999)
4. H. Blockeel, L. Dehaspe, B. Demoen, G. Janssens, J. Ramon, H. Vandecasteele: Improving the Efficiency of Inductive Logic Programming Through the Use of Query Packs, Journal of Artificial Intelligence Research, Vol. 16 (2002)
5. Cheung D.W., Han J., Ng V., Wong C.Y.: Maintenance of Discovered Association Rules in Large Databases: An Incremental Updating Technique. Proc. of the 12th ICDE (1996)
6. Hettich S., Bay S. D.: The UCI KDD Archive [http://kdd.ics.uci.edu]. Irvine, CA: (1999)
7. Imielinski T., Mannila H.: A Database Perspective on Knowledge Discovery. Communications of the ACM, Vol. 39, No. 11 (1996)
8. Morzy T., Wojciechowski M., Zakrzewicz M.: Materialized Data Mining Views. Proceedings of the 4th PKDD Conference (2000)
9. Nag B., Deshpande P.M., DeWitt D.J.: Using a Knowledge Cache for Interactive Discovery of Association Rules. Proc. of the 5th KDD Conference (1999)
10. Sellis T.: Multiple query optimization. ACM Transactions on Database Systems, Vol. 13, No. 1 (1988)
11. Wojciechowski M., Zakrzewicz M.: Evaluation of Common Counting Method for Concurrent Data Mining Queries. Proc. of the 7th ADBIS Conference (2003)
12. Wojciechowski M., Zakrzewicz M.: Data Mining Query Scheduling for Apriori Common Counting. Proc. of the 6th Int'l Baltic Conf. on Databases and Information Systems (2004)
13. Wojciechowski M., Zakrzewicz M.: Evaluation of the Mine Merge Method for Data Mining Query Processing. Proc. of the 8th ADBIS Conference (2004)

USAID: Unifying Signature-Based and Anomaly-Based Intrusion Detection

Zhuowei Li[1,2], Amitabha Das[1], and Jianying Zhou[2]

[1] School of Computer Engineering, Nanyang Technological University,
50 Nanyang Avenue, Singapore 639798
[2] Institute for Infocomm Research,
21 Heng Mui Keng Terrace, Singapore 119613
zhwei.li@pmail.ntu.edu.sg, asadas@ntu.edu.sg, jyzhou@i2r.a-star.edu.sg

Abstract. Most intrusion detection techniques suffer from either an inability to detect unknown intrusions, or unacceptably high false alarm rates. However, there lacks a general basis to analyze and find solutions to these problems. In this paper, we propose such a theoretical basis for intrusion detection, which makes it possible to systematically express and analyze the detection performance metrics such as the detection rate and false alarm rate in a quantified manner. Most importantly, the insights gained from the basis lead to the proposal for a new intrusion detection technique – USAID. USAID attempts to exploit the advantages of both techniques, and overcome their respective shortcomings. The experimental results show that USAID can achieve uniform level of efficiency to detect both known (99.78%) and new intrusions (98.18%), with a significantly reduced false alarm rate (1.45%). Most significantly, the performance of USAID is superior to all the participants in KDD'99 if the anomalies detected by USAID can be categorized correctly.

1 Introduction

In general, there exist two approaches for intrusion detection: signature-based (a.k.a. misuse detection) and anomaly-based intrusion detection. In principle, signature-based intrusion detection (SID) works reliably on well-known intrusions, but it is *incapable of detecting new intrusions*. In addition, it is also limited by several practical problems: *signature updating bottleneck, intrusion variation detection* (Rubin et al.[11]), and *too many false alarms* (Julisch[5]).

Consequently, anomaly-based intrusion detection (AID) has become the research focus as it is a useful alternative to SID, being capable of detecting novel intrusions. Besides that it can not identify intrusions, AID suffers from *higher false alarm rate*, the *difficulty in determining whether the anomalies are caused by intrusions* (Li et al.[8]), *concept drifting problem*, and *mimicry attacks* (Wagner et al.[12]). Most importantly, the *computational cost of intrusion detection must be reduced considerably* before it can be usefully employed in practical systems. This is more so as the size of the audit trails keep on growing.

In this paper, we try to find solutions to these problems. One possible solution may lie in the fact that only partial knowledge is used in each approach of intrusion detection. For example, SID only uses the knowledge about well-known intrusions (e.g., Snort[10]). For AID, it is the knowledge about the normal behaviors of the computing resources (e.g., LERAD[9], MADAM ID[7]). In our research, we try to use all the available knowledge about the behaviors (*intrusive and normal*) in the historical data to detect intrusions.

Our main contributions in this paper are two-fold. First, we propose a theoretical basis for intrusion detection. The basis then allows us to systematically analyze the detection performance to identify root causes for the problems in SID and AID. Secondly, the insights gained from the analysis naturally leads us to a new technique, named as USAID, which **Unifies SID and AID**.

The remaining parts are organized as follows. the nomenclature is introduced in section 2. Then, the existing intrusion detection approaches are analyzed systematically in section 3. Section 4 describes USAID. Experimental results showing the effectiveness of USAID and related work are presented in section 5 and section 6 respectively. Lastly, we conclude the paper and layout the future work.

2 Theoretical Basis for Intrusion Detection

In our theoretical basis for intrusion detection, we build behavior models like SID and AID, which use a *feature vector* $FV = \{F_1, F_2, \ldots, F_m\}$, where F_i is a feature in the feature set. In general, a feature F_i can be categorized into *nominal*, *discrete* or *continuous* one. For example, 'name' is nominal, 'TCP port number' is discrete and 'SYN rate within 2 seconds' is continuous. A feature vector can contain any number of nominal, discrete, and/or continuous features. Besides that, we assume that there are *training audit trails*, which are constituted with labeled **normal** audit trails and **intrusive** audit trails.

2.1 Definitions

For any feature F, there is a meaningful domain $Dom(F)$ called the feature space. Any value occurring in the audit trails is a *feature value* v_F. With respect to the training audit tails, a feature value v_F will be labeled as *normal*, *suspicious* or *anomalous*. Specifically, if it only occurs in the normal audit trails, it is *normal*. If it only occurs in the anomalous audit trails, for example, in the intrusion signatures, it is *anomalous*. Otherwise, it is labeled as '*suspicious*'. We will refer to 'normal', 'suspicious', or 'anomalous' as the **NSA label** of the feature value v_F, denoted as $L(v_F) \in \{'N', 'S', 'A'\}$ using the first letter of the label.

Feature Ranges. For any discrete/continuous feature F, the interval between v_F^1 and v_F^2 is defined as *a feature range* $R_F = [v_F^1, v_F^2]$. For the sake of uniformity, each nominal feature value is also referred to as a feature range. Thus, $R_F \subseteq Dom(F)$. The concept of **NSA labels** can be extended to the feature range:

$$L(R_F) = \text{`}N\text{'} \Leftrightarrow \forall v_F(v_F \in R_F \wedge L(v_F) = \text{`}N\text{'})$$
$$L(R_F) = \text{`}A\text{'} \Leftrightarrow \forall v_F(v_F \in R_F \wedge L(v_F) = \text{`}A\text{'})$$
$$L(R_F) = \text{`}S\text{'} \Leftrightarrow \exists v_F^1 \exists v_F^2 (v_F^1 \in R_F \wedge v_F^2 \in R_F \wedge L(v_F^1) = \text{`}N\text{'} \wedge L(v_F^2) = \text{`}A\text{'})$$

Where, all the feature values occur in the training audit trails.

Next, for the feature F, we can collect a series of feature ranges from $Dom(F)$: $\{R_F^1, R_F^2, \ldots, R_F^m\}$, such that $R_F^i \neq R_F^j$ and $L(R_F^j) \neq L(R_F^{j+1})$ ($1 \leq i, j \leq m$, and $i \neq j$). Furthermore, using the following rules, we can partition the feature space $Dom(F)$ into three feature subspaces: $N(F)$, $S(F)$ and $A(F)$.

$$N(F) = \{R_F^j|\ 1 \leq j \leq m,\ L(R_F^j) = \text{`}N\text{'}\}$$
$$S(F) = \{R_F^j|\ 1 \leq j \leq m,\ L(R_F^j) = \text{`}S\text{'}\}$$
$$A(F) = \{R_F^j|\ 1 \leq j \leq m,\ L(R_F^j) = \text{`}A\text{'}\}$$

We also define $\Omega(F) = N(F) \cup S(F) \cup A(F)$ so that $\Omega(F)$ is the collection of all feature ranges found in the audit trails.

Definition 1 (compound feature). *A compound feature F_{12} is an ordered pair $\{F_1, F_2\}$, and $\Omega(F_{12})$ is a subset of the cartesian product of $\Omega(F_1)$ and $\Omega(F_2)$, such that each element in $\Omega(F_{12})$ actually represents at least one element in the audit trails. For the sake of expression, $F_{12} = F_1 \times F_2$.*

Intuitively, similar to its component features, a feature range of F_{12} has an NSA label with respect to its representive feature instance(s), and the compound feature space can also be partitioned into three feature subspaces, i.e., $\Omega(F_{12}) = N(F_{12}) \cup S(F_{12}) \cup A(F_{12})$. Note that the suspicious compound feature ranges can potentially shrink with respect to component feature ranges as the combinations of two 'suspicious' feature ranges may be 'normal' or 'anomalous'.

In summary, the compound feature built from two atomic features shows similar behaviours as any of its component atomic features. Therefore, we can treat the compound feature as an atomic one to build higher order compound features. Using this recursive procedure, the feature vector FV for intrusion detection can be converted into an equivalent n-order compound feature $F_{1\ldots n}$. In USAID, each compound feature range of $F_{1\ldots n}$ is defined as a **behavior signature** in the behavior models for intrusion detection.

2.2 An Illustrative Example

Let us assume that $FV = \{F_1, F_2, F_3\}$. The example instances of the feature vector are listed below in Table 1. The feature ranges for every feature are listed as follows. For saving space, we will often use 'N' and 'I' respectively to denote the statuses 'normal' and 'intrusion'.

Table 1. The instances of the feature vector

INDEX	F_1	F_2	F_3	STATUS
1	TCP	1	0.01	normal
2	ICMP	2	0.04	normal
3	NETBIOS	6	0.10	intrusion$_3$
4	TCP	4	0.08	intrusion$_1$
5	UDP	5	0.06	intrusion$_2$
6	UDP	8	0.14	intrusion$_4$
7	NETBIOS	6	0.10	normal
8	UDP	7	0.02	normal
9	UDP	8	0.14	intrusion$_2$

- For F_1,
  $R_{F_1}^1 = TCP, N\&I_1 \Rightarrow \text{`}S\text{'}$
  $R_{F_1}^2 = ICMP, N \Rightarrow \text{`}N\text{'}$
  $R_{F_1}^3 = UDP, N\&I_2\&I_4 \Rightarrow \text{`}S\text{'}$
  $R_{F_1}^4 = NETBIOS, N\&I_3 \Rightarrow \text{`}S\text{'}$
- For F_2,
  $R_{F_2}^1 = [1,2], N \Rightarrow \text{`}N\text{'}$;
  $R_{F_2}^2 = [3,5], I_1\&I_2 \Rightarrow \text{`}A\text{'}$;
  $R_{F_2}^3 = [6,6], N\&I_3 \Rightarrow \text{`}S\text{'}$;
  $R_{F_2}^4 = [7,7], N \Rightarrow \text{`}N\text{'}$;
  $R_{F_2}^5 = [8,8], I_2\&I_4 \Rightarrow \text{`}A\text{'}$;
- For F_3,
  $R_{F_3}^1 = [0.01, 0.05], N \Rightarrow \text{`}N\text{'}$;
  $R_{F_3}^2 = (0.05, 0.09], I_1\&I_2 \Rightarrow \text{`}A\text{'}$;
  $R_{F_3}^3 = (0.09, 0.14], N\&I_3\&I_4 \Rightarrow \text{`}S\text{'}$;

- $F_{23} = F_2 \times F_3$
  $R_{F_{23}}^1 = R_{F_2}^1 \times R_{F_3}^1, N \Rightarrow \text{`}N\text{'}$;
  $R_{F_{23}}^2 = R_{F_2}^2 \times R_{F_3}^2, I_1\&I_2 \Rightarrow \text{`}A\text{'}$;
  $R_{F_{23}}^3 = R_{F_2}^3 \times R_{F_3}^3, N\&I_3 \Rightarrow \text{`}S\text{'}$;
  $R_{F_{23}}^4 = R_{F_2}^4 \times R_{F_3}^1, N \Rightarrow \text{`}N\text{'}$;
  $R_{F_{23}}^5 = R_{F_2}^5 \times R_{F_3}^3, I_2\&I_4 \Rightarrow \text{`}A\text{'}$;
- $F_{123} = F_1 \times F_{23}$
  $R_{F_{123}}^1 = R_{F_1}^1 \times R_{F_{23}}^1, N \Rightarrow \text{`}N\text{'}$;
  $R_{F_{123}}^2 = R_{F_1}^2 \times R_{F_{23}}^1, N \Rightarrow \text{`}N\text{'}$;
  $R_{F_{123}}^3 = R_{F_1}^1 \times R_{F_{23}}^2, I_1 \Rightarrow \text{`}A\text{'}$;
  $R_{F_{123}}^4 = R_{F_1}^3 \times R_{F_{23}}^2, I_2 \Rightarrow \text{`}A\text{'}$;
  $R_{F_{123}}^5 = R_{F_1}^4 \times R_{F_{23}}^3, N\&I_3 \Rightarrow \text{`}S\text{'}$;
  $R_{F_{123}}^6 = R_{F_1}^3 \times R_{F_{23}}^4, N \Rightarrow \text{`}N\text{'}$;
  $R_{F_{123}}^7 = R_{F_1}^3 \times R_{F_{23}}^5, I_2\&I_4 \Rightarrow \text{`}A\text{'}$;

The feature subspaces are: $N(F_{123}) = \{R_{F_{123}}^1, R_{F_{123}}^2, R_{F_{123}}^6\}$, $S(F_{123}) = \{R_{F_{123}}^5\}$, and $A(F_{123}) = \{R_{F_{123}}^3, R_{F_{123}}^4, R_{F_{123}}^7\}$.

3 Theoretical Analysis

In our analysis, we will focus on a compound feature F since the feature vector for intrusion detection can be compounded into a higher-order compound feature. In the **ideal** scenario, which assumes hypothetical complete knowledge, the three feature subspaces are determined without any misclassification: $N_i(F)$, $A_i(F)$ and $S_i(F)$. Similarly, in the **real** scenario where we possess only partial knowledge of them, three feature subspaces are $N_r(F)$, $A_r(F)$ and $S_r(F)$.

Due to the quality of the training audit trails (i.e., incompleteness and incorrect labels), there are two defects in the behavior model: (1) *model inaccuracy* and (2) *model incompleteness*. We use the following subsets of feature ranges to quantify the inaccuracy: $NA_1(F)$, $NS_1(F)$, $SN_1(F)$, $SA_1(F)$, $AN_1(F)$ and $AS_1(F)$. Every subset name is defined as: the first letter is the real feature subspace, the second letter represents the ideal one, and the subscript '1' indicates that it is caused by 'model inaccuracy'. As the behavior model is incomplete, there are some unknown feature ranges: $N_f(F)$, $A_f(F)$ and $S_f(F)$. The influence of incompleteness in the behavior model is quantified as: $NS_2(F)$ and $AS_2(F)$, where the subscript '2' indcates that is it caused by 'model inaccuracy'. Then,

$$N_r(F) - NS(F) + N_f(F) - NA(F) + AN(F) = N_a(F)$$
$$A_r(F) - AS(F) + A_f(F) - AN(F) + NA(F) = A_a(F)$$
$$S_r(F) + NS(F) + AS(F) + S_f(F) = S_a(F)$$

where, '+' and '-' denote set union and difference operations respectively.

3.1 Performance Analysis

In this subsection, we quantify and analyze the detection performance based on the detection results and the principles laid out so far. The detection performance is mainly represented by the detection rate and false alarm rate in the detection phase. In our discussion, we will assume that a fraction α of the feature ranges labeled as 'suspicious' will be detected as 'anomalous'.

Signature-Based Intrusion Detection. In SID, only $A_r(F)$ is known (a.k.a. the intrusion signature base). The behaviors which do not match $A_r(F)$ are regarded as 'normal' behaviors. Therefore, its detection performance is,

$$DR = \frac{|A_r(F)| - |AS_1(F) + AS_2(F)| * (1-\alpha) - |AN_1(F)|}{|A_i(F)| + |AS_1(F) + AS_2(F)| * \alpha - |NA_1(F)| + |S_f(F)| * \alpha}$$
$$FAR = \frac{|AS_1(F) + AS_2(F)| * (1-\alpha) + |AN_1(F)|}{|A_r(F)|}$$

Where, $|\ldots|$ represents the size of a set of feature ranges.

Considering that $A_a(F)$ includes all the intrusions and their variations, the incapability to detect new intrusions as well as intrusion variations and the signature updating problem are due to the limited size and quality of $A_r(F)$. $AS_1(F)$, $AN_1(F)$ and $AS_2(F)$ lead to too many false alarms in practice.

Anomaly-Based Intrusion Detection. In AID, $N_r(F_i)$ is known beforehand in the normal run of a process, and the behaviors that violate $N_r(F_i)$ are regarded as 'anomalous', $S_r(F_i) = \Phi$. Therefore, its detection performance is,

$$DR = \frac{|A_i(F)| + |S_f(F)| * \alpha}{|A_i(F)| + |S_f(F)| * \alpha + |NS_1(F) + NS_2(F)| * \alpha + |NA_1(F)|}$$
$$FAR = \frac{|N_f(F)| + |S_f(F)| * (1-\alpha)}{|A_i(F)| + |N_f(F)| + |S_f(F)|}$$

Obviously, the higher false alarm rate is largely rooted in $N_f(F)$ and $S_f(F)$. Mimicry attacks try to utilized $NS_1(F)$ and $NS_2(F)$. Concept drifting problem is to enlarge $NS_1(F)$, $NS_2(F)$ and $NA_1(F)$. Conversely, anomaly context identification tries to shrink the above sets as well as $S_f(F)$.

Conclusively, the quality issue in the behavior models leads to the problems in SID and AID. In our research, we try to utilized all available knowledege instead of the partial knowledge used in SID (i.e., $A_r(F)$) and AID (i.e., $N_r(F)$).

Unifying Signature-Based and Anomaly-Based Intrusion Detection. In USAID, all three *real* feature subspaces are known in advance: $N_r(F)$, $S_r(F)$

and $A_r(F)$. Except the known feature ranges in all three feature subspaces, other feature ranges are detected as 'suspicious' in USAID. As before, we can deduce the detection performance of USAID as follows.

$$DR = 1 - \frac{|NS_1(F) + NS_2(F)| * \alpha + |NA_1(F)| + |SA_1(F)| * (1-\alpha)}{|A_i(F)| + |S_i(F)| * \alpha}$$

$$FAR = \frac{|N_f(F)| + |S_f(F)| + AS_1(F) + AS_2(F)| * (1-\alpha) + |SN_1(F)| * \alpha + |AN_1(F)|}{|N_f(F)| + |A_r(F)| + |A_f(F)| + |S_f(F)| + |S_r(F)| * \alpha}$$

It is clear that USAID will achieve similar detection rate as AID. Like AID, if the behavior model is accurate and complete, $DR = 100\%$ and $FAR = 0$. On the other hand, other than detecting anomalies, USAID can identify the intrusions in $A_r(F)$ as in SID. In summary, even though it is still limited by the quality issue of the behavior model, USAID provides advantages of both SID and AID.

4 A Novel Intrusion Detection Technique: USAID

In this section, we apply USAID for intrusion detection. The architecture of USAID consists of three modules as indicated in Figure 1. The first module extracts the three feature subspaces of every feature. Module 2 is to construct the signature base. The last module incorporates the detection mechanism.

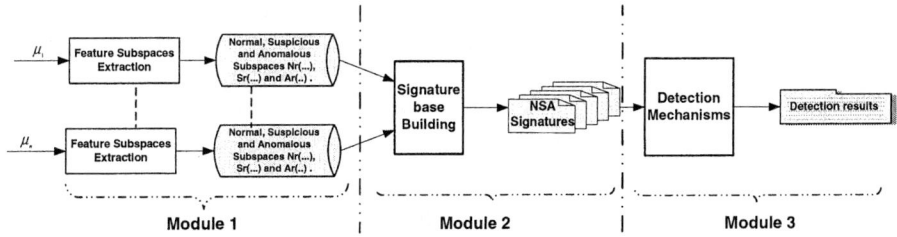

Fig. 1. A general framework for USAID

4.1 Feature Subspaces Extraction

Step 1: Feature Value Collection. In the labeled training audit trails, the feature values are collected for every feature. The statuses (*normal and/or intrusions*) of every feature value are also collected into its status list. Based on its status list, every feature value is assigned an NSA label. Note that this step is applied to nominal, discrete and continuous features, but the following two steps are only applicable to discrete and continuous features.

Step 2: Feature Value Clustering. The objective of this step is to form initial feature ranges for every feature by clustering the neighboring feature values. For a discrete feature, two feature values x_1 and x_2 are *neighboring* if

$|x_1 - x_2| = 1$[1]. For a continuous feature, two feature values x_1 and x_2 are *neighboring* if $|x_1 - x_2| \le \delta$. If several neighboring feature values have the same NSA label, they will be combined to form an *initial feature range*. As a special case, if, for a feature value, its neighbors have different NSA labels from itself, it forms an initial feature range itself. Every initial feature range thus formed inherits the NSA label of the feature values falling within it.

Step 3: Feature Range Generalization. Under most scenarios, the initial feature ranges will not cover all of the feature space. Any outside feature subspace is named as an *uncovered subspace*. Comparing to the neighboring definition of feature values, two feature ranges are *neighboring* if there is no other feature range(s) between them. Then, an uncovered subspace between two neighboring feature ranges is processed as follows: if the two feature ranges have the same NSA label, a new feature range will be formed to cover the two feature ranges as well as the uncovered subspace; otherwise, the uncovered space is divided equally and allocated to these two defined feature ranges. The NSA labels of the initial feature ranges will be inherited by the newly extended or combined feature ranges. Ultimately, all the *known* feature space of every feature will be covered by well-defined feature ranges.

4.2 Building Behavior Signatures

Initially, the NSA signature base (i.e., the behavior models) is empty. The following procedures are performed resursively on each feature instance. First, we construct a signature by replacing the feature values in the instance with respective feature ranges. Then, the signature is inserted into the NSA signature base with the status of the feature instance. Finally, based on the accumulated status list, every signature is assigned an NSA label.

4.3 Detection Mechanisms Via Signatures

We first extract a feature instance from the test audit trails at a time, and the feature value of every feature is replaced by its corresponding feature range to construct a temporary signature. We then try to search for the temporary signature in the NSA signature base, and if found, the current instance is assigned the same status list as that of the stored signature. Otherwise, the detection result is 'anomaly'.

5 Experiments on USAID

We have chosen a typical dataset from KDD CUP 1999 contest, in which every record is an instance of a specific feature vector (Table 2). The dataset meets the requirements of USAID: *labeled audit trails* and *a feature vector*. The sizes

---

[1] If the distance is larger than 1, there is an uncovered space $|x_1 - x_2 - 1|$.

Table 2. Features in the connection records

Types (41)	Features
nominal (9)	protocol type, service, flag, land, logged_in, root shell, su_attempted, is_hot_login, is_guest_login
discrete (15)	duration, src_bytes, dst_bytes, wrong_fragments, urgent, hot, num_failed_logins, num_compromised, num_root, num_file_creations, num_shells, num_access_files, num_outbound_cmds, count, srv_count
continuous (17)	serror_rate, srv_serror_rate, rerror_rate, srv_rerror_rate, same_srv_rate, diff_srv_rate, srv_diff_host_rate, dst_host_count, dst_host_srv_count, dst_host_same_srv_rate, dst_host_diff_srv_rate, dst_host_same_src_port_rate, dst_host_srv_diff_host_rate, dst_host_serror_rate, dst_host_srv_serror_rate, dst_host_rerror_rate, dst_host_srv_rerror_rate

of the dataset are as follows: *training dataset: 4898431 instances, test dataset: 311029 instances*. For a detailed description of the datasets, please refer to [1]. In addition, as the precision for continuous features in our experimental datasets is 0.01, we set the neighboring threshold $\delta = 0.01$.

5.1 Experimental Results

First, let us talk about the dataset quality. In the training dataset, there are several illegal records (e.g. instance 4817100). In the test dataset, we found several instances (whose indices are 136489 and 136497) with illegal combination between 'TCP' protocol type and 'ICMP' service. Therefore, they are discarded in our experiments. Moreover, there is an intrusion 'spy' that is not documented properly (in instances 1381226, 1381227).

Feature Ranges of Every Feature. In our experiments, the average numbers of normal, suspicious and anomalous feature ranges of all features in the feature vector are $\overline{N} : \overline{S} : \overline{A} = 23 : 18 : 12$. Given the relatively large number of 'suspicious' feature ranges, it is clear that only one feature from our selected feature vector is not enough for intrusion detection.

The NSA Signature Base. In it, the numbers of the normal, suspicious and anomalous signatures are $N : S : A = 60371 : 58 : 2779$. Even though some suspicious signatures still exist, the detection capability has been improved much in comparison to any single feature. The existence of suspicious signatures indicates that the features in our experiments are not enough to detect all known intrusions. Note that the high ratio of N and A, namely $N : A = 60371 : 2779 \approx 21.7$ is quite significant since it indicates that the detection speed of SID will be faster. In addition, the total of possible signatures due to feature ranges of all 41 features are 8.38×10^{33}. In contrast, our signature base is compact enough. We further sense that searching future intrusion signatures via negative selection algorithm [4] from 8.38×10^{33} signatures is a mission impossible.

Signature Variations of a Behavior. At the same time, we observed that most intrusions cause more than one signature. For example, for portsweep, the number of signatures is 941; ipsweep, 72; satan, 389. The observation indicates that the intrusion variations do exist to a significant extent. In the NSA signature base, some signatures are shared by several intrusions, such as portsweep and neptune. Even though the number of shared signatures is small (Figure 2.(A)), this

	Normal	probe	DOS	U2R	R2L
normal	---	47	16	8	0
probe	47	33	10	2	0
DOS	16	10	0	0	0
U2R	8	2	0	0	0
R2L	0	0	0	0	0

(A) Statistics of Shared Signatures.

	0	1	2	3	4	A
0	57101	423	53	0	0	3015
1	2	2341	7	0	4	1818
2	5143	14	215142	0	0	11167
3	8	0	0	0	0	69
4	7817	2	0	0	0	6939

(B) Detection Results of USAID.

Fig. 2. Statistics of Shared Signatures and Detection Results

phenomenon shows that, under some scenarios (e.g., inadequate no. of features), it is difficult to identify some intrusions correctly. If the response strategies for the intrusions with overlapped signatures are much different, then generating responses to such intrusions may lead to disastrous results.

Shared Signatures among Behaviors. Also in Figure 2.(A), we enumerate the numbers of signatures shared between different intrusion categories. The normal category will share many signatures with other intrusion categories. This is the main source of false alarms or false negatives in intrusion detection. In this table, the signatures of R2L intrusions are not shared with other intrusion categories, whereas only 'probe' intrusions have shared signatures with each other category. One possible reason for this phenomenon lies in the proportions of signatures in every category, that is, $normal : probe : DOS : U2R : R2L = 60432 : 1661 : 1255 : 65 : 42$. In addition, the major principles of 'probe' intrusions are similar to each other [6], that's why the shared 'probe' signatures are significant in Figure 2.(A). The strategies behind probe and DOS are similar, but it is different from the one behind U2R and R2L. Therefore, probe and DOS can be classified in one class, and U2R and R2L in another class [6]. The differences between these two classes explain why there are few signatures shared by them.

Detection Results from the Test Dataset. We eliminate two new intrusions, namely, snmpgetattack and mailbomb, from the detection results. This is because the information in the feature vector is not enough to detect these two intrusions.

Detection Performance. Quantitatively, the false alarm rate is 1.45%, the detection rate for known intrusions 99.78%, the detection rate for most new intrusions 98.18%. We also evaluate the USAID performance in comparison with the participants of KDD'99 Classifier Learning Contest[1], in which every entry will be assigned a detection cost, and an average cost per entry is calculated for comparison. The lower the average cost per entry is, the higher rank the classifier.

The detection results of USAID are summarized in Figure 2.(B), in which the first 5 columns constitute the confusion matrix, and the last column includes the numbers of detected anomalies. Its horizontal dimension is the predicted class of every test example, and the vertical dimension is its actual class. In the performance comparisons, the detected anomalies will be processed in two ways. First, these anomalies are classified correctly to their actual intrusion categories. For example, the number of correctly predicated entries of 'probe' is 2341+1818=4159. In such case, the performance of USAID is scored 0.1355, which is much better than the 1st rank of KDD'99, 0.2331. Secondly, under the worst scenario, these anomalies are classified incorrectly into the intrusion categories with highest cost. For instance, as cost(R2L,probe)=4 is highest in row 'R2L', the anomalies detected from actual 'R2L' will be detected as category 'normal'. The performance in the worst scenario is scored 0.3283, which is ranked 19th among all the participants. Note that almost half of the R2L intrusions, which are detected poorly in KDD'99, are detected as anomalies in USAID. In summary, USAID is expected to achieve better performance than all the participants of KDD'99 if the detected anomalies are categorized correctly.

6 Related Work

In USAID, two intrusion detection approaches are unified and their respective problems can be solved partially. The research work in [2] also shows the effectiveness of this combination, in which an algorithm (*similar to the negative selection algorithm in[4]*) is proposed to generate the artificial anomalies. An intrusion detection system is then built on the synthetic datasets. Actually, it only relies on the partial knowledge as well, and it lacks flexibility to fine-tune the model online. Other obvious advantages of USAID over [2] are the mechanisms for intrusion identification and anomaly context identification

Since the output of an intrusion detection technique can be considered to be a compound feature in our general feature vector, USAID is a multiple classifier ensembler [3]. In this aspect, USAID is similar to the research work in [3], in which one classifier is used to detect known intrusions, and another classifier tries to classify the new intrusions. However, [3] depends on the assumption that the first classifier can detect known intrusions accurately. That's not true since there are significant intrusion variations as shown in our experiments.

7 Conclusions and Future Work

In this paper, we proposed a theoretical basis for intrusion detection, in which we unified signature-based and anomaly-based intrusion detection and systematically analyzed the hard problems faced by the researchers on intrusion detection. Our experimental results have also shown that the detection performance of USAID are encouraging. Specifically, most *new and known* intrusions are detected in USAID, and the false alarm rate is 1.451%. In our future work, we will continue research on our theoretical basis for intrusion detection.

References

1. C. Elkan. Results of the kdd'99 classifer learning contest. http://www.cs.ucsd.edu/users/elkan/clresults.html, 1999.
2. W. Fan, M. Miller, S. Stolfo, W. Lee, and P. Chan. Using artificial anomalies to detect unknown and known network intrusions. In *Proceedings of First IEEE International Conference on Data Mining (ICDM'01)*, pages 123–130, 2001.
3. W. Fan and S. Stolfo. Ensemble-based adaptive intrusion detection. In *Proceedings of SIAM International Conference on Data Mining 2002 (SDM2002)*, 2002.
4. S. Hofmeyr and S. Forrest. Architecture for an artificial immune system. *Evolutionary Computation*, 8(4):443–473, 2000.
5. K. Julisch. Clustering intrusion detection alarms to support root cause analysis. *ACM Transaction on Information and System Security*, 6(4):443–471, 2003.
6. K. Kendall. A database of computer attacks for the evaluation of intrusion detection systems. Master thesis, Massachusetts Institute of Technology, June 1999.
7. W. Lee and S. Stolfo. A framework for contructing features and models for intrusion detection systems. *ACM Transactions on Information and System Security*, 3(4):227–261, Nov. 2000.
8. Z. Li and A. Das. Visualizing and identifying intrusion context from system calls trace. In *Proceedings of 20th Annual Computer Security Applications Conference*. IEEE Computer Society, Dec. 2004.
9. M. Mahoney and P. Chan. Learning Nonstationary Models of Normal Network Traffic for Detecting Novel Attacks. In *SIGKDD 2002*, July 23-26 2002.
10. M. Roesch. Snort - lightweight intrusion detection for networks. In *Proceedings of USENIX LISA*, 1999.
11. S. Rubin, S. Jha, and B. Miller. Automatic generation and analysis of nids attacks. In *Proceedings of 20th Annual Computer Security Applications Conference*, Tucson, AZ, USA, Dec. 2004. IEEE Computer Society.
12. D. Wagner and P. Soto. Mimicry attacks on host-based intrusion detection systems. In *Proceedings of the 9th ACM conference on Computer and communications security*, pages 255–264, 2002.

Mining Mobile Group Patterns: A Trajectory-Based Approach

San-Yih Hwang[1], Ying-Han Liu[1], Jeng-Kuen Chiu[1], and Ee-Peng Lim[2]

[1] Department of Information Management National Sun Yat-Sen University,
Kaohsiung, Taiwan 80424
[2] School of Computer Engineering Nanyang Technological University,
Singapore 639798, Singapore

Abstract. In this paper, we present a group pattern mining approach to derive the grouping information of mobile device users based on a trajectory model. Group patterns of users are determined by distance threshold and minimum time duration. A trajectory model of user movement is adopted to save storage space and to cope with untracked or disconnected location data. To discover group patterns, we propose ATGP algorithm and TVG-growth that are derived from the Apriori and VG-growth algorithms respectively.

1 Introduction

Behavior research on sociology show that peer pressure and group conformity can affect the buying behaviors of individuals [1]. With a good knowledge of groups a customer belongs to, one can derive common buying interests among customers, and develop group-specific pricing models or marketing strategies for personalized services. There are many ways one can determine the groups a person belongs to, for example, by the set of product items s/he purchased, his/her occupation or income, and/or the places s/he visited. As implied by the loads of research in spatial-temporal databases [7], the information about users' locations over time can play a crucial role in determining the groups. As mobile phones and other similar devices become widely used, users' locations of errors usually less than 1km can be gathered by mobile communication operators using the existing communication infrastructure. With more accurate positioning technologies, the errors can be reduced even further.

In this research, we are interested in discovering groups of users such that users in the same group are geographically close to one another for significant amounts of time. Finding such grouping information of mobile users, based on the spatio-temporal distances among them, is known as "Group patterns mining", originally proposed by Wang et al. [9, 10]. Previous research represents the movement data of an object as a synchronous time series of locations. This representation, however, has the following three pitfalls:

1. To maintain accurate location tracking, the frequency of sampling users' locations must be high. As a result, the movement database can be become huge.
2. Moving objects may be disconnected from time to time voluntarily or involunarily. It is therefore not realistic to assume that the location information is present for each time point.

3. Lastly, it is almost impossible to have perfectly synchronized sampling of users' locations in reality due to clock differences of base stations conducting the sampling and the locations of moving objects.

To deal with the first and the third problems, a trajectory-based model to represent object movements can be adopted instead [4, 5, 6, 8]. A trajectory is a function that maps time to locations. To represent object movement, a trajectory can be decomposed into a set of linear functions, one for each disjoint time interval. The derivative of each linear function yields the direction and speed in the associated time interval. Various approaches have been proposed to accurately induce the trajectory of an object from its location update data while saving storage space using dead-reckoning policies [11] or regression techniques [2]. In this paper, we use trajectories for modeling moving objects and develop efficient algorithms for discovering mobile group patterns from trajectory data. Furthermore, we address the second problem by allowing the trajectory of each object not to cover the entire location tracking period.

The rest of the paper is organized as follows. In Section 2, we formally define the mobile group pattern discovery problem in the context of using trajectories to represent moving objects. In Section 3, we describe the algorithms for discovering mobile group patterns. Finally, we conclude in Section 4.

2 Problem Definition

A trajectory is a set of piecewise linear functions, each of which maps from a disjoint time interval to an n-dimensional space. That is, one can perceive a piece of a trajectory as a set of n linear functions of the time variable t, one for each dimension, and the trajectory may change speed and direction at finitely many time instants. Each linear piece can be represented as a conjunction of linear constraints using the time variable and coordinate variables. A trajectory is a disjunction of all its linear pieces. For example, a trajectory of the user moving on a 2-D space may consist of 3 linear pieces as shown below:

$[(x = 2t - 40) \wedge (y = -t + 23) \wedge (0 \leq t < 21)]$
$\vee [(x = 2) \wedge (y = -t + 23) \wedge (21 \leq t < 22)]$
$\vee [(x = 0.5t - 9) \wedge (y = 1) \wedge (22 \leq t \leq 30)]$

An *object movement database D* consists of a set of trajectories, one for each object. That is, $D = Y_{i=1}^{M} T_i$, where M is the number of moving objects. Each linear piece in a trajectory T_i is a set of 4-tuples: (*reference_point, velocity, start_time, end_time*), denoting the location function $f(t) = velocity \times t + reference\_point$ during time interval [*start_time, end_time*). Table 1 shows the trajectories of three example objects in a 2-dimensional space. Note that the trajectory of each object may become untraceable at some time points, resulting in a sequence of non-continuous linear pieces. As shown in Table 1, moving object o_1 is disconnected in time [5, 6) and [9, 10), object o_2 is disconnected in time [5, 6), and object o_3 is untraceable during time interval [8, 10).

Table 1. An example object movement database

	reference_point	velocity	start_time	end_time
	(1,1)	(3,1)	0	3
o_1	(7,-11)	(1,5)	3	5
	(10,-3)	(4,3)	6	9
	(2,2)	(2,1)	0	3
o_2	(2,-13)	(2,6)	3	5
	(-4,5)	(3,2)	6	10
o_3	(2,4)	(3,1)	0	3
	(17,-5)	(-2,4)	3	5
	(12,35)	(-1,-4)	5	8

Definition 1. Given a group of objects G and a maximum distance threshold *max_dis*, we say objects in G are *geographically close* at a time point t if every pair of objects in G are no farther than *max_dis* apart, and *geographically far* at t if there exists one pair of objects in G whose distance is larger than *max_dis*. We also say objects in G is *geographically decided* if they are either close or far, and *geographically undecided* otherwise.

Definition 2. Given a group of objects G and a minimal time duration threshold *min_dur*, a time interval $[t,t+k]$ is called a close interval of G if

1. objects in G are geographically close at any time point in $[t, t+k]$,
2. objects in G are not geographically close at time $t-\varepsilon$, where ε is an arbitrarily small positive number,
3. objects in G are not geographically close at time $t+k+\varepsilon$, where ε is an arbitrarily small positive number, and
4. $k \geq min\_dur$.

A far interval can be similarly defined.

A group of objects G, *max_dis*, and *min_dur* are said to form to a group pattern, denoted $P=(G, max\_dis, min\_dur)$ [9]. Given an object movement database, a group pattern may have multiple close intervals and multiple far intervals, within which the geographical property associated with G can be decided. For the time points not covered by any close or far intervals, aggregated as the undecided intervals, the geographical property associated with G is not clear. We quantify the significance of a group pattern by the proportion of the total length of close intervals and *estimated* close subintervals of the undecided intervals.

Definition 3. Let P be a group pattern with n close intervals $c_1, c_2, ..., c_n$, m far intervals $f_1, f_2, ..., f_m$, and k undecided intervals $u_1, u_2, ...u_k$. The weight of P is defined as:

$$\text{weight}(P) = \frac{L_{close} + L_{undecided} \cdot \frac{L_{close}}{L_{close} + L_{far}}}{L_{close} + L_{far} + L_{undecided}} = \frac{L_{close}}{L_{close} + L_{far}}, \quad (1)$$

where $L_{close} = \sum_{i=1}^{n} c_i$, $L_{far} = \sum_{i=1}^{m} f_i$, and $L_{undecided} = \sum_{i=1}^{k} u_i$.

The weight represents the proportion of the time when users of P (are expected to) stay close. Thus, the larger is the weight, the more significant is the group pattern.

Definition 4. Given a threshold *min_wei*, a group pattern $P = <G, \text{max\_dis}, \text{min\_dur}>$, is valid if the weight of P exceeds the threshold *min_wei*.

The problem is how to identify all valid group patterns given a trajectory-based object movement database and the thresholds *max_dis*, *min_dur*, and *min_wei*.

3 The Algorithms

The geographical property of a mobile group (i.e., close or far) is determined by the distances between all pairs of objects at any time point. Here, the distance function of o_1 and o_2 for each corresponding linear piece (i.e., with the same time interval) can be easily computed. For example, suppose the location functions of objects o_1 and o_2 between time 0 and 3 are as follows:

Location of object o_1 at time t: $(1 + 3t, 1 + t)$
Location of object o_2 at time t: $(2 + 2t, 2 + t)$

The Euclidean distance between o_1 and o_2 when $0 \leq t < 3$ is $\sqrt{(1-t)^2 + (1)^2}$

A complete distance function between o_1 and o_2 whose location data listed in Table 1 is the following:

$$\text{distance}_{o_1,o_2}(t) = \sqrt{(1-t)^2 + 1^2}, 0 \leq t < 3$$

$$\sqrt{(-5+t)^2 + (-2+t)^2}, 3 \leq t < 5$$

undecided, $5 \leq t < 6$

$$\sqrt{(-14-t)^2 + (8-t)^2}, 6 \leq t < 9$$

undecided, $9 \leq t \leq 10$

Given a distance function $dist(t)$ of two objects o_1 and o_2 within an interval I, we would like to identify the subintervals I' in I such that $dist(t) \leq \text{max\_dis}$, $t \in I'$. This can be done by computing the roots of the equation $dist(t) = \text{max\_dis}$. In case of Euclidean distance, there will be two roots, denoted t_a and t_b, where $t_a \leq t_b$. When both t_a and t_b are real numbers, we have $dist(t) \leq \text{max\_dis}, \forall t_a \leq t \leq t_b$. Obviously, within the time interval $I \cap [t_a, t_b]$, the distance between o_1 and o_2 is no more than *max_dis*, and at any time in $I - [t_a, t_b]$, the distance between o_1 and o_2 is greater than *max_dis*. The weight of a mobile group can be subsequently decided by looking at each pair of objects in the group, and the function for computing the weight of a group c_k is named Group-Weight(c_k). Group-Weight(c_k) starts with synchronizing the linear pieces of the trajectory in each trajectory of c_k such that each trajectory has the same set of time segments, followed by computing the close and far intervals in each time segment. The pseudo-code of Group-Weight(c_k) is omitted due to space limitation.

Given two group patterns, $P = <G, max\_dis, min\_dur>$ and $P' = <G', max\_dis, min\_dur>$, P' is called a sub-group pattern of P if $G' \subseteq G$. The Apriori property within the context of mobile group pattern mining states that any sub-group pattern of a valid group patterns must also be valid. In [9], both AGP and VG-growth algorithms are based on this Apriori property. To re-use both algorithms, we must ensure that Apriori property still holds for trajectory-based movement data.

Theorem 1. [Apriori property for group patterns] Given a database D and thresholds *max_dis*, *min_dur*, and *min_wei*, if a group pattern is valid, all of its subgroup patterns will also be valid.

Based on the AGP algorithm [9], we develop an algorithm called Apriori Trajectory-based Group Pattern Mining, abbreviated ATGP, whose pseudo-code is shown in Fig. 1. In the algorithm, we use C_k to denote the set of candidate k-groups, and G_k to denote the set of valid k-groups. From G_1, the set of all distinct objects, the algorithm first computes C_2, the pair set of objects in G_1. This algorithm performs join operation to generate candidate k groups C_k from G_{k-1} (C_k=Generate_Candidate_Groups(G_{k-1})), and the generated candidates are verified by computing their weights.

Input: D, *max_dis*, *min_dur*, and *min_wei*
Output: all valid groups G
01    $G=\emptyset$; G_1=all distinct objects;
02   **for** ($k = 2$; $G_{k-1} \neq \emptyset$; k++)
03   C_k = **Generate_Candidate_Groups**(G_{k-1}); $G_k=\emptyset$;
04   **for each** candidate k-group $c_k \in C_k$
05   c_k.weight = **Group-Weight**(c_k, *max_dis*, *min_dur*);
06   **if** (c_k.weight >= *min_wei* } $G_k=G_k \cup c_k$;
07       $G = G$ Y G_k;
08   **return** G;

Fig. 1. Algorithm ATGP

In [WL03], Wang et al. proposed a data structure VG-graph whose edges represent all valid 2-groups and an algorithm VG-growth that traverses VG-graph to identify all valid groups. We adapt VG-graph to work for trajectory-based object movement database and call the resultant data structure TVG-graph, and the traversal algorithm TVG-growth. Similar to VG-graph, the edges of TVG-graph are determined by the set of valid 2-groups. However, since each object may have some untraceable periods, every edge in TVG-graph is associated with a set of close intervals as well as another set of far intervals. The group mining procedure of TVG-growth remains the same as that of VG-growth, however, the set of close and far intervals associated with each edge has to be updated as the recursive mining procedure proceeds [3].

4 Conclusion

In this paper, we reported a novel approach that discovers moving object group patterns from a database comprising trajectories of moving objects. Furthermore, our research allows non-continuous trajectories which model the disconnected behavior of moving objects.

In this work, the location of an object at a time point was assumed to be either accurately determined or completely unknown. In some applications, location data of an object may incur different degrees of uncertainties over time. Our future work includes mining mobile group patterns by considering the inherent uncertainty of location data.

References

1. D.R. Forsyth. *Group Dynamics*. Wadsworth, Belmont, CA, 1999.
2. V. Guralnik, J. Srivastava. "Event Detection from Time Series Data." Proceedings of ACM International Conference on Knowledge Discovery and Data Mining (KDD2000), 2000.
3. Y.-H. Liu "Mining Mobile Group Patterns: A Trajectory-based Approach," *master thesis, Department of Information Management, National Sun Yat-sen U.*, available at http://etd.lib.nsysu.edu.tw/ETD-db/ETD-search/view_etd?URN=etd-0730104-102312.
4. H. Mokhtar, J. Su, and O.H. Ibarra, "On moving object queries", *Proceedings of the ACM Symposium on Principles of Database Systems (PODS)*, 2002.
5. H.K. Park, J.H. Son, M.-H. Kim, "An Efficient Spatiotemporal Indexing Method for Moving Objects in Mobile Communication Environments." *Proc. Of Int'l. Conf. on Mobile Data Management (MDM*2003), 2003.
6. S. Saltenis, C.S. Jensen, S.T. Leutenegger, M.A. Lopez, "Indexing the positions of continuously moving objects," *Proc. of 2000 ACM SIGMOD Conference*, 2000.
7. S. Shekhar and S. Chawla, "Introduction to Spatial Data Mining," Chapter 7, *Spatial Databases: A Tour*, Prentice Hall, New Jersey, 2003.
8. M. Vazirgiannis and O. Wolfson, "A spatiotemporal model and language for moving objects on road networks", *Proc. of Symposium on Spatial and Temporal Databases* (SSTD), 2001.
9. Y. Wang, E.-P. Lim, and S.-Y. Hwang, "On Mining Group Patterns of Mobile Users." *Proc. Of the 14th International Conference on Database and Expert Systems Application (DEXA 2003)*, 2003.
10. Y. Wang, E.-P. Lim, and S.-Y. Hwang, "Effective Group Pattern Mining Using Data Summarization," 9*th International Conference on Database Systems for Advanced Application* (DASFAA2004), 2004.
11. O. Wolfson, A. P. Sistla, S. Chamberlain, Y. Yesha, "Updating and querying databases that track mobile units," *Distributed and Parallel Databases*, 1999.

Can We Apply Projection Based Frequent Pattern Mining Paradigm to Spatial Co-location Mining?

Yan Huang, Liqin Zhang, and Ping Yu

Department of Computer Science and Engineering,
University of North Texas,
P.O. Box 311366, Denton, Texas 76203
{huangyan, lzhang, py0003}@unt.edu

Abstract. A co-location pattern is a set of spatial features whose objects are frequently located in spatial proximity. Spatial co-location patterns resemble frequent patterns in many aspects. Since its introduction, the paradigm of mining frequent patterns has undergone a shift from a generate-and-test based frequent pattern mining to a projection based frequent pattern mining. However for spatial datasets, the lack of a transaction concept, which is critical in frequent pattern definition and its mining algorithms, makes the similar shift of paradigm in spatial co-location mining very difficult. We investigate a projection based co-location mining paradigm. In particular, we propose a projection based co-location mining framework and an algorithm called **FP-CM**, for **FP-growth Based Co-location Miner**. This algorithm only requires a small constant number of database scans. It out-performs the generate-and-test algorithm by an order of magnitude as shown by our preliminary experiment results.

1 Introduction

We focus on a recent spatial data mining problem: finding spatial features that tend to be located in spatial proximity. This problem is also referred to as *spatial co-lcoation patterns mining* [7, 4, 2, 10, 9]. Let $\mathcal{F} = \{f_1, f_2, \ldots, f_l\}$ be a set of spatial features. consider a number of l spatial datasets $\{SD_1, SD_2, \ldots, SD_l\}$, such that $SD_i, i \in [1, l]$ contains all and only the objects that have the spatial feature f_i, Let $\mathcal{R}$ be a given spatial neighbor relation (e.g. distance less than 1.5 miles). A set of spatial features $X \subseteq \mathcal{F}$ is a co-location if its value $im(X)$ of an interesting measure, is above a threshold min_im. The problem of finding the complete set of co-location patterns is called the co-location mining problem. Mining *spatial co-location patterns* is an important spatial data mining task with broad applications.

Spatial co-location patterns resemble frequent patterns [5], a more general problem of mining association rules [1] in many aspects. Since its introduction, the problem of mining frequent patterns from large databases, has been subject of numerous studies. The paradigm of frequent pattern mining algorithms has undergone a fundamental shift from generate-and-test approaches [1] to projection based approaches [5]. Projection based approaches have major advantages over generate-and-test approaches and avoids multiple database scans by compressing transactional data into compact structures. However, the lack of pre-materialized transactions becomes a major obstacle in adopting projection based algorithms in spatial co-location pattern mining. A natural question to ask is: can we push the same paradigm shift for mining spatial co-location patterns?

Many algorithms for co-location mining proposed in literature [7, 4, 10, 9, 3] employ an generate-and-test co-location mining paradigm, which utilizes the anti-monotone property of interestingness measures. In a clustering-based map overlay approach[4, 3], every spatial feature is treated as a map layer and point-data in each layer are clustered into regions. In a reference feature based approach [7], transactions are created according to different algorithms, then a level wise algorithm is applied . Under this model, a frequent pattern based algorithm can be applied straightforwardly due to the fact that the interestingness measure is defined based on the generated transactions. In distance based approaches [9, 10], the number of instances for each spatial feature set is used to define the interestingness measure. In an event centric model [10], a participation index was defined as the interestingness measure. The participation index of a pattern is defined as the minimal participation ratio of the objects of each feature in the pattern.

The contribution of this work is to study how to use a projection based paradigm for event based spatial co-location pattern mining (CM). We proposed a projection based framework for CM, which can incorporate any fast frequent pattern mining algorithm. In particular, we developed an FP-growth based algorithm for spatial co-location mining(FP-CM) based on the proposed framework. We provide preliminary experiment results to show that the FP-CM is an order of magnitude faster than the generate-and-test algorithm *Co-location Miner*.

Paper Outline: Section 2 recalls important concepts of co-location and frequent pattern mining. Section 3 proposes our projection based FP-CM framework and a FP-growth based co-location mining algorithm. We present the preliminary experimental results in section 4 and summarize our work and present future work in section 5.

2 Background

We review basic concepts of co-location patterns, a traditional generate-and-test co-location mining algorithm, and a projection based frequent pattern mining algorithm in this section.

In an event centric model [10], a participation index was defined as the interestingness measure. For a set of spatial features $X \subseteq \mathcal{F}$, a set of objects $\{o_1, o_2, \ldots, o_k\}$ is an *instance* of X iff $(\forall i, i \in [1, k], o_i \in SD_i)$ and $(\forall i \forall j, 0 < i < j \leq k, (o_i, o_j) \in \mathcal{R})$. The *participation ratio* $pr(f, X)$ of a feature f in a pattern X is defined as:

$$pr(f, X) = \frac{\text{number of objects of f that participate in any instance of } X}{\text{total number of objects of } f}$$

The *participation index* of a pattern X is defined as: $pi(X) = \min_{\forall f \in X}\{pr(f, X)\}$. Because of the downward closure property of the participation index [10], a generate-and-test mining paradigm was employed by previous algorithms, e.g. *Co-location Miner*. This approach generates the candidate size $k + 1$ co-location set based on the size k co-location set. The candidate size $k + 1$ co-location set includes all and only those size $k + 1$ spatial feature set whose size k subsets are all co-locations. Then it uses spatial joins on the instances of size k co-locations to generate the instances of the size $k + 1$ candidates and calculate the participation indexes for them. It prunes false candidates before starting the next iteration.

Projection based frequent pattern mining utilizes a highly condensed prefix-tree structure to compress frequent patterns and employs a pattern fragment growth method for mining the complete set of frequent patterns from the prefix-tree. Due to the reduced number of database scans, this algorithm is very fast compared with traditional generate-and-test algorithms [5]. (We refer readers to [5] for the details of the FP-growth algorithm). However, a FP-growth based algorithm can not be used directly in spatial co-location mining due to the lack of transactions in spatial datasets. Transactionizing spatial datasets and establishing the relationship between *support* and *participation index* to develop a complete and correct projection based co-location mining algorithm is non-trivial.

3 A Projection Based Co-location Mining Framework

Our proposed framework is shown in Figure 3. A transactional database TD_i is created for each spatial feature f_i. Any fast maximal frequent pattern mining algorithm may be applied to each transactional database TD_i to find maximal frequent patterns $MFP_s = \cup_{i=1...K} MFP_i$ using a support threshold $min\_sup = min\_pi$. The mined maximal frequent patterns MFP_i are combined by a pattern combining component to generate a superset of all the co-location patterns. Finally, a pattern filtering component filters out the false candidate co-locations.

Based on the projection based framework, we develop an algorithm called **FP-CM**, for **FP-growth Based Co-location Miner**. This algorithm consists of four components: transactionization, maximal frequent pattern mining, combining patterns, and pattern filtering.

1. *Transactionization (step 2)*
   For each spatial feature f, we create a transactional database TD_f as follows. For each object o of f, a transaction containing all other spatial features whose object(s) is(are) within neighbor $\mathcal{R}$ of o is created.
2. *Maximal Frequent Pattern Mining (step 3)*
   For each transactional database TD_f, we find all maximal frequent patterns

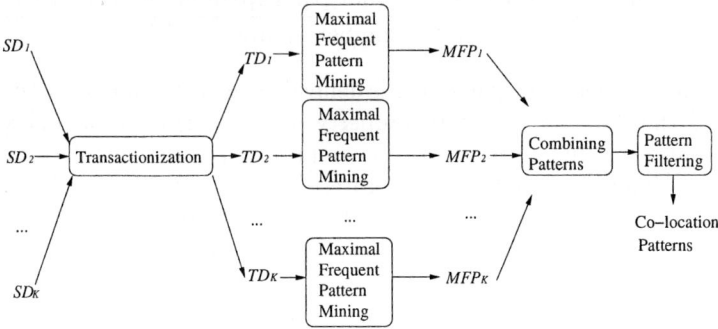

Fig. 1. Projection Based Co-location Pattern Mining Framework

---

Algorithm 1. FP-CM

1: **for** $i = 1$ to K **do**
2:   $TD[i] \leftarrow transactionize(SD[1], SD[2], \ldots, SD[K])$;
3:   $MFP[i] \leftarrow FP-growth(TD[i], min\_pi)$;
4: **end for**
5: $i \leftarrow 1$;
6: $C[1] \leftarrow \{1, 2, \ldots, K\}$;
7: **while** $C[i] \neq \emptyset$ **do**
8:   $C[i+1] \leftarrow apriori\_gen(C[i])$ /*refer to [1]*/;
9:   $C[i+1] \leftarrow prune(C[i+1], MFP[1], MFP[2], \ldots, MFP[K])$;
10:   $i \leftarrow i + 1$;
11:   $C \leftarrow C \cup C[i]$;
12: **end while**
13: $P \leftarrow multi-way-spatial-join-prune(C, min\_pi)$;
14: **return** P;

---

based on the FP-growth frequent pattern mining algorithm using a support threshold $min\_sup_f = min\_pi$ in this step.

3. *Combining Patterns (step 5-12)*

   The basic structure of the combining pattern step is the level-wise structure of CM [10]. However, it does not require expensive spatial joins to calculate participation indexes. Instead, it consults the MFPs to prune the majority of the false candidate patterns. This step will produce a superset of the true co-location patterns to feed to the next pattern filtering step.

   The prune step (step 9) works as follows. For each candidate pattern C, $\forall f \in C$, if MFP_f does not contain a superset of $(C - f)$, then C is pruned. This will not falsely delete any true patterns since $pr(f, X) \geq min\_pi$ implies $(C - f)$ is frequent and should have a superset in MFP_f.

4. *Pattern Filtering (step 13)*

   Once we reduce the total number of candidate co-location patterns from

$2^{\#features}$ to a small superset of the true co-location patterns, we can use hash-based spatial join techniques [6] and multi-way spatial joins [8] to filter the patterns. We hash spatial datasets into buckets using a grid [6] and then use a multi-way spatial join which is based on a backtracking search heuristic [8] to find all the maximal cliques. We keep the list of all the candidate co-location patterns from the previous step and register the cliques to their corresponding candidate co-location patterns. Finally we calculate the actual participation indexes for each candidate co-location pattern and return the set of all co-location patterns found.

The FP-CM algorithm requires a small number of database scans. One database scan is required to transactionize the spatial data, then FP-growth based maximal pattern mining requires two or a few database scans depending on the average size of the FP-trees. Combining patterns involves only spatial features and the maximal frequent pattern sets and usually is a memory based step. Finally, the pattern filtering step using gridding and multi-way spatial joins requires two more database scans. So the total number of database scans of FP-CM algorithm is bounded by a small constant.

4 Experiment Results

We implemented both the co-location miner (CM) and FP-growth based co-location miner (FP-CM) using C++ and all the experiments are carried out on a Pentium IV 2.4GHz machine with 1GB memory, running the Debian linux operating system. Our experiments are extensive and the results are consistent. Limited by space, we only report representative results for various parameters. Our dataset generator is similar to [1].

We use a notation like $|P|50.PS5.|F|100.|I|24k.min\_pi0.4$ to denote an experiment with 50 pre-generated patterns whose average size is 5 and the number of features participating in a pattern is 100, 24k spatial objects, and minimum participation index threshold is 0.4. Since the time for computing size 2 co-locations are the same (one database scan) for both algorithms, we only report the time for calculating size 3 or more co-location patterns.

1. *Effect of thresholds:*
   As Figure 2 (a) shows, FP-CM is much faster than CM for all the threshold range in [0.5,0.2]. The advantage of FP-CM over CM increases when the participation threshold decreases due to the increased number of candidate patterns and associated spatial joins CM has to perform. FP-CM is an order of magnitude faster than CM when the participation index threshold is low.
2. *Effect of total number of Objects:*
   We compare the scalability of the two algorithms when the total number of objects increase from 5k to 50k. As shown in Figure 2 (b), FP-CM is 5 to 40 times faster than CM and the running time of the FP-CM algorithms remains almost the same while the running time of the CM increases dramatically.

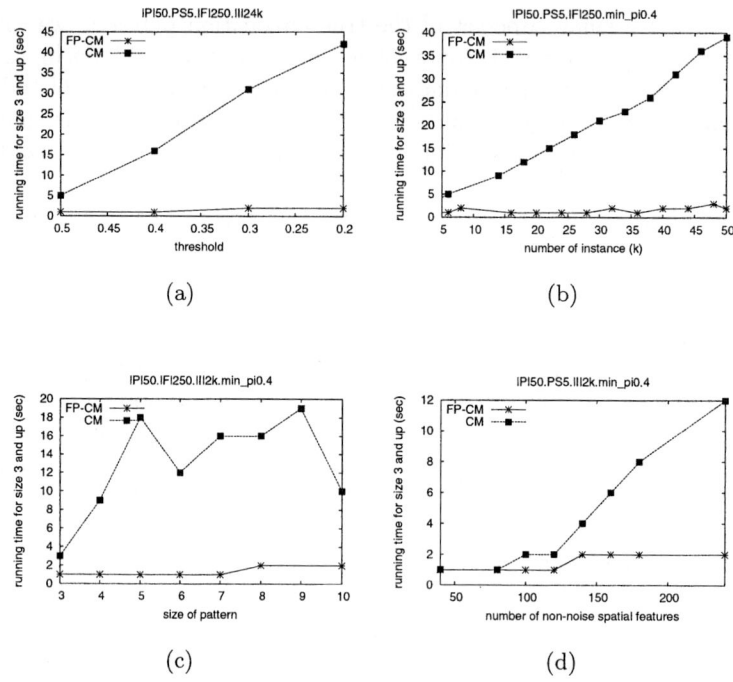

Fig. 2. Performance Comparison of **CM** and **FP-CM**

3. *Effect of Average Maximal Pattern Size:*
   Figure 2 (c) shows the result when the size of the pattern ranges from 3 to 10. FP-CM is 3 to 18 times faster than CM. The running time of FP-CM is stable as the size of the patterns increases while the the running time of CM highly correlates with the total number of co-locations found.
4. *Effect of Number of Patterns:*
   We range the number of non-noise spatial features from 50 to 250 as shown in Figure 2 (d). FP-CM is up-to 12 times faster than CM when the number of non-noise spatial features increases.

5 Conclusion and Future Work

In this paper we proposed a projection based framework for mining spatial co-locations, which is flexible in incorporating any fast maximal frequent pattern mining algorithm developed in literature to help spatial co-location mining. In particular, we developed a complete and correct FP-tree based algorithm for spatial co-location mining. It combines the salient features of FP-tree based maximal frequent pattern mining [5] and fast multi-way spatial joins [8] to reduce the total number of database scans into a small constant. Our experiment results showed that the FP-CM is an order of magnitude faster than a generate-and-test algorithm *Co-location Miner*.

The proposed projection based co-location mining framework could be treated as a new data-driver partitioning of spatial datasets according to the objects of each spatial features. Compared with traditional spatial partition approaches [11], this approach does not have the problem of combinatorial explosion of temporary candidate patterns needed to be maintained by the algorithm before all the partitions are processed as acknowledged by the authors in [11]. In future work, comparing various partition based co-location mining algorithms would be an interesting and imperative research direction.

References

1. R. Agarwal and R. Srikant. Fast Algorithms for Mining Association Rules. In *Proc. of the 20th Int'l Conference on Very Large Data Bases*, 1994.
2. N.A.C. Cressie. *Statistics for Spatial Data*. Wiley and Sons, 1991.
3. V. Estivill-Castro and I. Lee. Data Mining Techniques for Autonomous Exploration of Large Volumes of Geo-referenced Crime Data. In *Proc. of the 6th International Conference on Geocomputation*, 2001.
4. V. Estivill-Castro and A. Murray. Discovering Associations in Spatial Data - An Efficient Medoid Based Approach. In *Proc. of the Second Pacific-Asia Conference on Knowledge Discovery and Data Mining*, 1998.
5. Jiawei Han, Jian Pei, and Yiwen Yin. Mining frequent patterns without candidate generation. In *ACM SIGMOD Intl. Conference on Management of Data*, 2000.
6. D. J. DeWitt J. M. Patel. Partition Based Spatial-Merge Join. In *Proc. of the ACM SIGMOD Conference on Management of Data*, June 1996.
7. K. Koperski and J. Han. Discovery of Spatial Association Rules in Geographic Information Databases. In *Proc. of the 4th International Symposium on Spatial Databases*, 1995.
8. Nikos Mamoulis and Dimitris Papadias. Multiway spatial joins. *ACM Trans. Database Syst.*, 26(4):424–475, 2001.
9. Y. Morimoto. Mining Frequent Neighboring Class Sets in Spatial Databases. In *Proc. ACM SIGKDD International Conference on Knowledge Discovery and Data Mining*, 2001.
10. S. Shekhar and Y. Huang. Co-location Rules Mining: A Summary of Results. In *Proc. 7th Intl. Symposium on Spatio-temporal Databases*, 2001.
11. Xin Zhang, Nikos Mamoulis, David W. Cheung, and Yutao Shou. Fast Mining of Spatial Collocations. In *Proc. of the ACM SIGKDD International Conference on Knowledge Discovery and Data Mining*, 2004.

PatZip: Pattern-Preserved Spatial Data Compression

Yu Qian, Kang Zhang, and D. T. Huynh

Department of Computer Science,
The University of Texas at Dallas, Richardson, TX 75083-0688, USA
{yxq012100, kzhang, huynh}@utdallas.edu

Abstract. This paper presents a compression method, PatZip, to improve the efficiency of spatial pattern mining methods. PatZip can avoid overcompression and stop automatically before pattern is destroyed. Compared with existing compression methods, PatZip is deterministic and its result is reproducible, and original data can be easily recovered. The compression process is data-driven and parameter-free, and requires only $O(nlogn)$ time for n data points.

1 Introduction

Spatial data mining presents new challenges due to the large size and the complex structure of spatial data. A common approach to such challenges is to perform some form of compression on the initial databases and then process the compressed data [7]. General requirements for a quality compression include:

- **Minimal information loss.** Information needed by the mining methods should be preserved in the compressed data while the size of the data is significantly reduced. To minimize the information loss, there should be a mapping between the compressed data and the original data so that the original data can be recovered when necessary.
- **Efficient.** For data mining purposes, a compression becomes worthwhile only if it can make the mining process more efficient. In most cases, compression methods cannot be slower than mining methods. Otherwise, the efficiency of the whole process is decreased.
- **Data-driven.** In many cases there is no prior knowledge about the given data. A general compression method should not make assumptions on data features but let the data speak for themselves.

This paper addresses spatial data compression problem from the above perspectives and presents a compression method called PatZip that meets all above requirements. PatZip extends the idea of GraphZip [14] that merges closest data iteratively based on a nearest neighbor graph by adding an automatic termination scheme. In PatZip, the compression process stops when the distortion caused by the merging grows out of proportion. While most existing compression methods require user-specified parameters on the size of compressed data, PatZip can detect when compression should stop so that overcompression or undercompression caused by wrong inputs can be avoided. The termination detection is based on a comparison between predicted distortion and actual distortion as the compression continues. Since the predicted distortion is computed by assuming that merging happens only inside patterns, if the actual distortion is greater than the predicted value, some patterns have been wrongly merged together and compression stops before such a merging happens. This paper studies and discovers the properties of PatZip with comprehensive experimental studies. The discovered properties imply that PatZip can be used to compress source data for pattern mining. PatZip has been applied to assist FAÇADE [13], a clustering method, to discover spatial patterns effectively.

The rest of this paper is organized as follows. Section 2 reviews related work. The process of PatZip is described in Section 3. Section 4 studies the properties of PatZip through experiments. Section 5 concludes the paper.

2 Related Work

There are two general approaches to spatial data compression: summary construction and sampling [4]. The representative summary construction methods include micro-clustering and vector quantization. K-means [11] and Birch [16] are two representative micro-clustering methods while LBG algorithm [10] and PNN algorithm [5] are two classic vector quantization methods. Sampling approaches include random sampling (also called uniform sampling) and biased sampling. There have been some improved versions [1, 6, 9, 12] of these methods proposed in recent years.

2.1 Sampling

Random sampling may be the most popular size reduction method used in data mining. Its idea is straightforward: given an integer k and a data set with n data points, choose k data points randomly as the substitution for the original n points. Random sampling has advantages on efficiency and generality. Its disadvantages, however, are apparent: the inaccuracy introduced by sampling variance, the missing of small clusters, and the inability of reproducing or repeat the execution process. Random sampling does not provide a mapping between the compressed data and the original data, so the original data are usually recovered with a nearest neighbor classification, which is computational expensive when the sample is big.

Palmer and Faloutsos [12] propose a density-biased sampling method to solve the small-cluster-missing problem caused by random/uniform sampling. The heart of their method is the use of group size to bias the sample. Kollios *et al.* [9] point out that density-biased sampling is more accurate for cluster detection than uniform sampling, especially when noise exists. Their approach can detect either clusters or outliers efficiently through setting different density thresholds, which, however, is not a guided process when lacking knowledge about the given data.

2.2 Micro-clustering

Theoretically speaking, all efficient data clustering methods can be used as micro-clustering to reduce data size. The two most widely used methods are: *k*-means [11] and Birch [16]. The disadvantages of k-means include: low efficiency when k is big, results relying on the initial point selection, and an input, i.e., k, from the user about the size of the compressed data, which is usually difficult to set. Bradley et al. [1] propose an approach to perform k-means clustering more effectively by finding good initial points for it. Bradley's approach, however, introduces more parameters and is still not efficient when k is big.

BIRCH [16] is an efficient clustering algorithm for large databases, which is sensitive to data input order. Several concepts like radius, diameter, and centroid are used in Birch to describe the distance properties of a cluster, which leads to inaccurate compression for non-sphere regions. GraphZip [14] is a recently published graph-based micro-clustering method. PatZip and GraphZip use a similar way generating representative points of the original data. PatZip, however, can detect the termination point of the compression automatically while GraphZip fixes the size of compressed data in the algorithm. To our knowledge, PatZip is the first compression algorithm that does not require user input on size of compressed data.

2.3 Vector Quantization

The idea of vector quantization is to identify a set of possible vectors which are representative of the information/data to be encoded. The set of the representative vectors is called codebook. Aiming at minimizing the distortion between the codebook and the original data, the codebook generation method decides the quality of compression.

The most popular code book generation method is the LBG algorithm [10]. Its idea on representative point generation is exactly the same as that of k-means while their usages are different. While k-means is for unsupervised learning, LBG is supervised. Grouping in LBG will be used to encode future data and the original grouping will remain unchanged. This is also the main difference between micro-clustering and vector quantization methods. PNN (Pairwise Nearest Neighbor) algorithm [5] is another well-known codebook generation method for vector quantization. In each step of PNN, two closest vectors are merged and the process is repeated until the desired size of the codebook is reached. To minimize the average distortion, PNN requires $O(n^3)$ time for n data points to complete the generation of the codebook [6].

PatZip is distinguished from the aforementioned approaches for several desirable properties. First of all, PatZip is data-driven and completely automatic. Both summary construction and sampling approaches require users to provide the size of summary/sample. As a result, one has to guess how many points can represent the patterns, which not only increases burden on users but also causes overcompression or undercompression. Summary construction methods usually contain additional parameters for the models they use, which are also hard to set without prior knowledge. Secondly, PatZip is fast. Its requires only $O(nlogn)$ time for n data points. Thirdly, PatZip is deterministic. Its result is reproducible and original data can be recovered from compressed data.

3 PatZip

PatZip is a recursive process and contains multiple running cycles. It represents the original data set with a *1-neareset neighbor graph* and takes the graph as the input of the first running cycle. In each running cycle, new data points will be generated by substituting the points produced in the previous running cycle. Each running cycle is called a *compression stage* and the new points generated are called *representative points* of the substituted points. The iterative process ends when the size of the data drops to *1* or a termination condition is satisfied. Section 3.1 will introduce how the representative points are generated through adapting GraphZip [14] and Section 3.2 will explain the automatic termination of compression, which is the major contribution of this paper.

3.1 Compression

Before presenting the process of PatZip, let us first introduce the concept of *k*-nearest neighbor graph briefly. Generally, each vertex of a *k*-nearest neighborhood graph represents a data item. For each pair of data items, if either of them is among the *k*-most similar data items of the other, there exists an edge between the two corresponding vertices. In a spatial database, the data items are the points in a metric/dimensional space and the similarity of two data points is usually measured by the Euclidean distance between them.

Algorithm *PatZip* (Data Set *D*)
begin
  Construct *1*-nearest neighbor graph *G* for *D*;
  Create an empty data set *D'*;
  For each connected-component *C* of *G*:
    Generate a point *p* that is located at the center of *C*;
    Add *p* to *D'* and update the mapping file;
  if $|D'|=1$ **return** *D'*;
  else *PatZip(D')*;
end

Fig. 3.1. The *PatZip* algorithm without termination condition

As shown in Fig. 3.1, PatZip is an iterative process, which accepts the output of the previous running cycle as the input and replaces each connected component with its centroid point. When the number of points of the original data set, *n*, is decreased to *1*, i.e., the final representative point, the iteration stops. The point *p* in Fig. 3.1 is called the *representative point* of *C*, and the element points of *C* are called *member points* of *p*.

The time complexity of PatZip can be decided in a similar way to that of GraphZip [14]. It depends on the time to construct the *1*-nearest neighbor graph and find the connected components. Constructing a *1*-nearest neighbor graph can be considered as the problem of constructing all-nearest-neighbors, i.e., given a set *S* of *n* points in $\Re^d$, we want to compute for each point *p* of *S* another point of *S* that is closest to *p*. There have been many literatures on efficiently finding all-nearest neighbors. The first *O(nlogn)* time algorithm for the all-nearest-neighbors problem for an arbitrarily fixed dimension *d* was given by Clarkson [3], using randomization. Vaidya [15] solves the problem deterministically, in *O(nlogn)* time. Vaidya's algorithm can be implemented in the algebraic computation tree model and is, therefore, optimal. To find the connected components in an undirected graph with *n* vertices and *m* edges requires

$O(n+m)$ when using a DFS or BFS tree. Since $m \leq kn$ in a k-nearest neighbor graph while $k=1$ in PatZip, i.e., $m \leq n$, the time complexity of the first iteration of PatZip is $O(nlogn)+O(n+n)=O(nlogn)$. Now let us analyze how many iterations PatZip needs to run before reaching the final representative point.

Theorem 3.1. PatZip requires $O(log(n))$ iterations to compress the size of a given data set from n to 1.

Proof. Assume that after the first iteration, there are X_1 connected components in the 1-nearest graph of the given data set. According to the definition of k-nearest neighbor graph, every node has at least k edges connected. Since $k \geq 1$, there is no isolated node in a k-nearest neighbor graph. In other words, each connected-component in the 1-nearest neighbor graph has at least two nodes, i.e., $X_1 \leq n/2$. Generally, let X_i denote the number of connected components after i iterations and $X_0=n$, we have $X_i \leq X_{i-1}/2$. Solving the formula $X_t=1$, i.e., $n(1/2)^t=1$ based on the recursive expression for variable t leads to $t \leq log(n)$.

Theorem 3.1 indicates that the total time complexity of PatZip is the sum of all of the $log(n)$ iteration steps:

$$O(nlogn+(n/2)log(n/2)+...+O(1))$$
$$<O(nlogn+(n/2)log(n)+...+log(n))$$
$$=O((n+n/2+...+1)log(n))$$
$$=O(2nlogn)=O(nlogn).$$

The algorithm depicted in Fig. 3.1 can be easily extended to accept a user input to specify the termination condition when necessary. The condition could be a lower bound of the size of the compressed data or a maximum allowable compression ratio. By judging if the compressed data set satisfies the user-specified constraint, PatZip can terminate under user's control. The better way, however, is to let PatZip terminate automatically through estimating the compression distortion, as described in the following section.

3.2 Termination Condition

The process of PatZip is data-driven. The original data points will be recursively merged until the size of the data reaches 1, i.e., the final representative point. In most pattern mining applications, however, it is desirable to stop before reaching the final representative point so that spatial patterns can be preserved. For example, spatial clustering methods should be applied to a set of compressed data that still contains the cluster information. This section will show how PatZip can detect the termination condition of a compression before losing spatial patterns.

The basic idea of detecting the termination condition is to judge the distortion caused by compression. The distortion definition used in this paper is the popular square sum used by many previous approaches [5, 6, 16]: let G denote a set of n data points: $X_1, X_2,..., X_n$, and their corresponding representative points $C(X_1), C(X_2), ..., C(X_n)$, where $C(X_i)$ is a function that returns the representative point of X_i, and $dist(X_i, C(X_i))$ is the Euclidean distance between X_i and $C(X_i)$. Distortion is defined as the radius r of G:

$$r = (\sum_{i=1}^{n} dist\,(X_i, C(X_i))^2 / n)^{1/2} \qquad (1)$$

We compute a distortion value at every compression stage and check the value increment. If the value increases out of proportion, then we think one or more spatial patterns have been destroyed, compression should stop. To justify this method, let us see first why the distortion value would increase out of proportion when spatial patterns are destroyed in continuous compressions. According to the definition, distortion is the average distance from member points to their representative point. Since inter-cluster distances are significantly larger than intra-cluster distances, a representative point that represents the data points from different clusters would have a much bigger distortion than a point that represents the data points within the same cluster. The more data points from different clusters are mixed up, the faster the distortion value increases. The increase will not be gradual but sharp when patterns are destroyed. At the beginning of the compression, the nearest neighbor of a data point is in the same cluster and merging happens only inside the cluster. As compression continues, two representative points belonging to different clusters may be merged. There are two possibilities in this case. First, the whole cluster has been merged into one point A which will be merged with point B belonging to another cluster if the compression continues. Second, due to various shapes of clusters, as the compression continues, a point representing part of cluster A may be nearer to a point representing part of cluster B than to other representative points of cluster A. In either case, points A and B have accumulated some member points from previous compression stages. Merging A and B would cause the distortion value increase substantially so that this "illegal" merging can be detected and avoided. This rule holds for clusters of different shapes, densities, and sizes, with the only assumption that inter-cluster distances must be bigger than intra-cluster distances. That is, when two sets of points are merged together and distortion value does not increase sharply, the two sets must belong to the same cluster, or neither of them represents a significant number of data points, implying a merge between outliers and thus not affecting the true patterns. In summary, a compression stage whose distortion values increase disproportionally is considered a termination stage before which compression should stop.

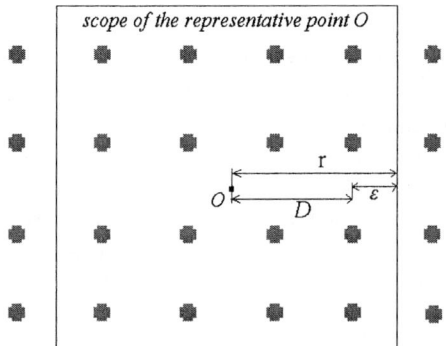

Fig. 3.2. The approximation of distortion

Then let us quantify what is a "disproportional increase" when distortion value increases as compression continues. A naive idea is to estimate the distortion increase with a fixed threshold. When the increasing ratio of distortion exceeds the threshold, compression terminates. Setting a reasonable threshold, however, is difficult. By observing the relation between the increase of the distortion and the decrease of the data size, we devise a more accurate data-driven solution without using any threshold: predicting the distortion value of each compression stage with the reducing number of data points at the stage. The predicted value is computed by assuming that merges happen only inside clusters so that when a merge cross different clusters occurs, the distortion would be greater than the predicted value. That is, when the actual distortion is bigger than the predicted value, compression stops. According to its definition, distortion is the average distance between member points and representative point, which can be approximated as the radius of the representative point's scope[1]. Because the area of the whole spatial data set is fixed, when the number of representative points decreases, the average scope of each representative point increases, which causes the radius, i.e., the distortion, increases. The approximation process is presented formally as follows. Given a d-dimensional data set and two sequential compression stages A_{i-1}, and A_i, let S_{i-1} denote the size of the data set at stage A_{i-1}, S_i the size at stage A_i, R_{i-1} the average area of the representative point at stage A_{i-1}, and R_i the average area at stage A_i, we have $R_{i-1}S_{i-1}=R_iS_i$ because the total area is fixed. Then we have

$$S_{i-1}/S_i = R_i/R_{i-1} = r_i^d/r_{i-1}^d \qquad (2)$$

where r_{i-1} and r_i are the radii of the average areas of the representative points at stages A_{i-1}, and A_i, respectively. Let D_{i-1} denote the actual distortion at stage A_{i-1}, and P_i the predicted distortion at stage A_i, $r_{i-1}=D_{i-1}$ in a continuous space. For the discrete space of the spatial data set, however, we need to consider the interval between the data points, i.e.:

$$r_{i-1}=D_{i-1}+\varepsilon, \; r_i=P_i+\varepsilon \qquad (3)$$

where ε is the half of the average interval between two neighbor points of the original data set, as shown in Fig. 3.2. Based on Formulas (2) and (3), we obtain the approximate definition of predicted distortion as follows.

Definition 3.1. (Predicted Distortion)
The *predicted distortion* at compression stage A_i is:

$$P_i = (D_{i-1} + \varepsilon)(S_{i-1}/S_i)^{1/d} - \varepsilon \qquad (4)$$

Similarly, we can approximate ε with D_1 because D_1 is the distortion of the first compression stage and ε is the radius of the area of this stage. Finally, we have:

$$P_i = (D_{i-1} + D_1)(S_{i-1}/S_i)^{1/d} - D_1 \quad \forall i>1 \qquad (5)$$

Definition 3.2. (Termination Stage, Termination Condition)
A compression stage A_t is the *termination stage* of compression when the following termination condition satisfies: $\forall i<t, D_i \leq P_i$ and $D_t > P_t$.

Definition 3.2 can be easily justified as follows. Because P_i is computed as the average distortion with uniformly distributed data points, at the beginning of compression, say, at stage A_i, the value of D_i is mainly affected by intra-cluster distance and smaller than the average, i.e., P_i. When patterns are mixed up at stage A_t, D_t will be mainly affected by inter-cluster distances, which are typically much larger than the average, i.e., P_t.

---

[1] Scope of a representative point covers the area occupied by all the member points of the representative point.

In summary, compression will terminate before the stage whose actual distortion is bigger than the predicted value. The self-terminated version of PatZip is described in Fig. 3.3. The effect of the approximation will be evaluated in Section 4.

4 Properties of PatZip

This section will investigate the properties of PatZip. Section 4.1 will demonstrate the compression results while Section 4.2 will evaluate the termination scheme.

Algorithm *PatZip* (Data Set *D*)
begin
Construct *1*-nearest neighbor graph *G* for *D*;
Create an empty data set *D'*;
For each connected-component *C* of *G*:
   Generate a point *p* that is located at the center of *C*;
   Compute distortion of *p* using *C* and mapping file;
Add *p* to *D'*
Add distortion of *p* to *ad*, i.e., the actual distortion;
Compute the predicted distortion *pd* with Formula (5);
if *(pd<ad)* or *|D|=1* **return** *D*;
else
update the mapping file;
PatZip(D');
end

Fig. 3.3. The self-terminated *PatZip* algorithm

4.1 Experimental Results

Three different testing data sets have been collected for our experiments and visualized in Fig. 4.1 (a). DS1 is a benchmark data set of Birch [16], containing

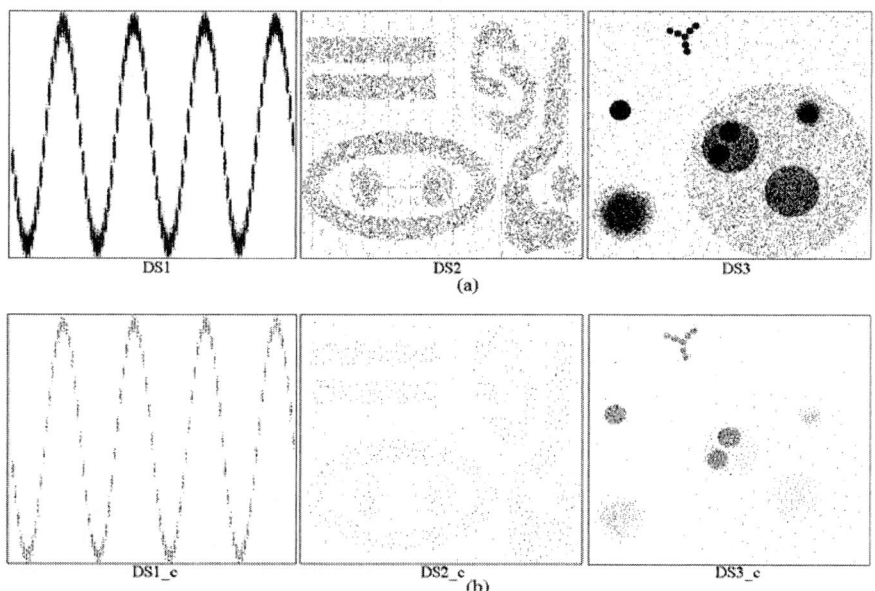

Fig. 4.1. (a) The 6 testing data sets; **(b)** The 6 compressed results after applying PatZip (compression ratio: 50)

100,000 points. DS2 is a benchmark used by CHAMELEON [8], which contains clusters of different shapes and 10,000 data points. DS3 is our synthetic data set that simulates the data set used by DataBubble [2]. DS3 contain clusters of different densities and 1 million data points.

The compressed results for the 3 data sets after applying PatZip are shown in Fig. 4.1 (b). To make the visual examination easier, the results shown in Fig. 4.1(b) should not contain too few points, so we stop PatZip at compression ratio of about 50 before reaching the termination stage. Fig. 4.1(b) shows that the shape and density of the patterns are preserved while their sizes are significantly reduced, and no natural clusters are mixed. We believe that this result is general enough to be applicable to different spatial pattern mining algorithms for improving efficiency.

4.2 Accurate Termination

This section will evaluate the termination condition described in Section 3.2. Fig. 4.2 compares the predicted distortion and the actual distortion for DS1~3 at different compression stages. It clearly shows the termination stage when the actual distortion exceeds the predicted distortion. Table 4.1 lists the sizes of the remaining data when compression stops and the actual number of clusters in the corresponding data sets. We expect each compressed data set to have a size close to but not smaller than the actual number of clusters, because the closer they are, the higher efficiency will be achieved for later pattern mining methods. If the approximation is correct, compression will stop at the stage before patterns are mixed up. As mentioned in Section 3.2, two conditions can cause compression stop: first, a whole cluster has been merged into only one point. If the compression continues, this cluster has to be merged with others according to the process of PatZip. For example, the compressed version of DS1 has 117 data points while its actual number of clusters is 100. Most of these clusters have been each represented by only one point. If the compression continues, the 117 points will be merged into 26 points, i.e., 74 clusters will be merged into others. This would cause a sharp increase in distortion and compression will stop. The second possibility is: even if every cluster has more than 1 representative point, if compression continues, the representative points of different clusters could still be merged together due to various shapes of the clusters. For example, the compressed version of DS2 has 55 points while it has only 9 clusters. Each cluster may have more than 1 representative point but the compression cannot proceed. This is because for non-spherical clusters, points of different clusters could be nearest to each other. PatZip should not merge such points and should stop. Generally, the closeness between the actual number of clusters and the size of the remaining data depends on the discrepancy of cluster sizes and shapes of the original data. For a data set containing clusters of similar sizes and spherical shapes, the distortion increment at the termination point would be more remarkable, compared with those with clusters of various shapes and sizes. In such

Table 4.1. Comparison between actual number of clusters and size of compressed data

Data Set	Size after compression	Actual number of clusters
DS1	117	100
DS2	55	9
DS3	37	15

cases, the size of the remaining data, i.e., the approximated number of clusters, would be closer to the actual number of clusters.

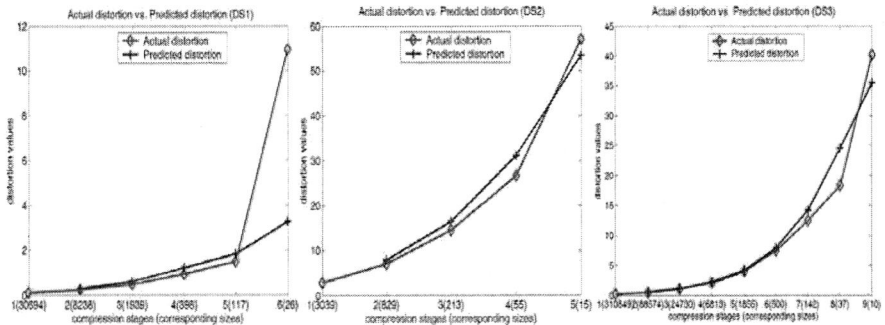

Fig. 4.2. The comparison between the predicted distortion and actual distortion

5 Conclusion

This paper presents a simple but effective data compression method, called PatZip, to produce a compact representation of the data to scale up the spatial pattern mining process. PatZip has a series of desirable properties: a deterministic result with high cohesiveness, a fast compression speed, and a data-driven process without requiring any prior knowledge about the data. The compression can terminate automatically before breaking the spatial patterns. These properties and the corresponding experimental studies make us believe that PatZip should appeal to a large collection of pattern discovery methods. In the future we will investigate the impact of different data on prediction of termination stages and how to optimize PatZip for more pattern mining algorithms.

References

1. Bradley, P. S., Fayyad, U., and Reina, C., Scaling clustering algorithms to large databases. In *Proc. of 4th International Conf. on Knowledge Discovery and Data Mining (KDD'98)*, 1998, pp. 9-15.
2. Breunig, M. M., Kriegel, H., Kroger, P., and Sander, J., Data bubbles: quality preserving performance boosting for hierarchical clustering. In *Proc. of Int'l Conf. on Management of Data (SIGMOD'01)*, 2001, pp. 79-90.
3. Clarkson, K. L., Fast algorithms for the all nearest neighbors problem. In *Proc. of 24th Annual Symposium on Foundations of Computer Science*, 1983, pp. 226-232.
4. DuMouchel, W., Volinsky, C., Johnson, T., Cortez, C., and Pregibon, D., Squashing Flat Files Flatter. In *Proc. of 5th Int'l Conf. on Knowledge Discovery and Data Mining (KDD'99)*, AAAI press, 1999, pp. 6-15.
5. Equitz, W. H., A new vector quantization clustering algorithm. *IEEE Trans. on Acoustics, Speech, and Signal Processing*, vol. 37, no. 10, 1989, pp. 1568-1575.

6. Franti, P., Kaukoranta, T., Shen, D., and Chang, K., Fast and memory efficient implementation of the exact PNN. *IEEE Trans. on Image Processing*, vol. 9, no. 5, 2000, pp. 773-777.
7. Han, J., Kamber, M., and Tung, A. K. H., Spatial clustering methods in data mining: a survey. H. Miller and J. Han (eds.), *Geographic Data Mining and Knowledge Discovery*, Taylor and Francis, 2001.
8. Karypis, G., Han, E., and Kumar, V., CHAMELEON: a hierarchical clustering algorithm using dynamic modeling. *IEEE Computer*, vol.32, 1999, pp. 68-75.
9. Kollios, G., Gunopulos, D., Koudas, N. and Berchtold, S., Efficient biased sampling for approximate clustering and outlier detection in large data sets. *IEEE Trans. on Knowledge and Data Eng.*, vol. 15, no. 5, 2003, pp. 1-18.
10. Linde, Y., Buzo, A., and Gray, R. M., An algorithm for vector quantizer design, *IEEE Trans. on Commun.*, vol. 28, 1980, pp. 84-95.
11. MacQueen, J., Some methods for classification and analysis of multivariate observations. In *Proc. of the Fifth Berkeley Symposium on Mathematical Statistics and Probability*, 1967, pp. 281-297.
12. Palmer, C. R. and Faloutsos, C., Density biased sampling: an improved method for data mining and clustering. In *Proc. of Int'l Conf. of Management of Data (SIGMOD'00)*, 2000, pp. 82-92.
13. Qian, Y., Zhang, G., and Zhang, K., FAÇADE: a fast and effective approach to discovery of dense clusters in noisy spatial data (demo abstract). In *Proc. of Int'l Conf on Management of Data (SIGMOD' 04)*, 2004, pp. 921-922.
14. Qian, Y. and Zhang, K., GraphZip: A fast and automatic compression method for spatial data clustering. In *Proc. of 19th Annual ACM Symposium on Applied Computing (SAC' 04)*, 2004, pp. 571-575.
15. Vaidya, P. M., An *O(nlogn)* algorithm for the all-nearest-neighbors problem. *Discrete & Computational Geometry* 4, 1989, pp. 101-115.
16. Zhang, T., Ramakrishnan, R., and Linvy, M., BIRCH: an efficient data clustering method for very large databases. In *Proc. of Int'l Conf. on Management of Data (SIGMOD' 96)*, 1996, pp.103-114.

A Likelihood Ratio Distance Measure for the Similarity Between the Fourier Transform of Time Series

G.J. Janacek, A.J. Bagnall, and M. Powell

School of Computing Sciences, University of East Anglia, Norwich, England

Abstract. Fast Fourier Transforms (FFTs) have been a popular transformation and compression technique in time series data mining since first being proposed for use in this context in [1]. The Euclidean distance between coefficients has been the most commonly used distance metric with FFTs. However, on many problems it is not the best measure of similarity available. In this paper we describe an alternative distance measure based on the likelihood ratio statistic to test the hypothesis of difference between series. We compare the new distance measure to Euclidean distance on five types of data with varying levels of compression. We show that the likelihood ratio measure is better at discriminating between series from different models and grouping series from the same model[1].

1 Introduction

The growth in size and number of longitudinal databases has lead to an increase in interest in time series data mining (TSDM) [6]. Two fundamental issues in any TSDM task are how to measure the similarity between time series and how to represent the data compactly without discarding important information. A common approach to the compression problem is to transform the data series so that the majority of the variation in the series can be captured in a small number of terms. In this paper we concentrate on fast fourier transforms (FFTs), probably the most popular transformation used in time series data mining (for example, see [1, 3, 7, 9, 10]). The basic method is to take the FFT of each series, retain a fixed number of coefficients, then measure similarity between series as the Euclidean distance between the retained parameters. The objectives of this paper are to introduce an alternative distance measure based on the likelihood ratio statistic for testing the significance of differences between series and to demonstrate that this measure produces better results on types of data for which an FFT approach should be appropriate. We maintain that, if the problem is complex enough to merit the use FFTs, then the likelihood ratio statistic will tend to give better results than Euclidean distance. Informally, this is because the likelihood ratio is better able to detect consistent differences between

---

[1] This work is supported by an EPSRC CASE award with Masterfoods Europe.

small coefficients that may be undetected with Euclidean distance because of fluctuations in the larger fourier terms.

2 FFT in Time Series Data Mining

For a real valued time series y, defined over discrete intervals $y(t), t = 1, \cdots, n$, the fourier transform represents y as a linear combination of sinusodal functions. After transformation a series is commonly compressed by retaining only the first of the FFT coefficients [1]. The Euclidean distance between the first f_c coefficients of series x and y with coefficients (p, q) and (r, s) is then

$$d_E(x_f, y_f) = 2 \cdot \sum_{i=1}^{f_c} (p_i - r_i)^2 + (q_i - s_i)^2 \qquad (1)$$

For complex mining problems where the use of FFT is justified, Euclidean distance can give too much preference to small variations in the largest coefficients, masking more complex differences in the wider spectrum. To overcome this problem we define a new distance measure, based on a test statistic for a hypothesis test of whether two series are significantly different, derived from the periodogram of a series. If series y has fourier coefficients (p_i, q_i) then the periodogram of y is the sequence $a_i = p_i^2 + q_i^2$. If the data is stationary, each a_i can be thought of as an observation of an independent random variable A_i with exponential density

$$g(a) = \frac{1}{2\alpha_i} \exp\left(-\frac{a}{2\alpha_i}\right) \qquad i = 2, 3, \cdots, n-1.$$

Since we have independence, the likelihood of our series is

$$L(a) = \prod_{i=1}^{n-1} \frac{1}{2\alpha_i} \exp\left(-\frac{a_i}{2\alpha_i}\right).$$

We can use the likelihood function to determine the similarity of two series by constructing a likelihood ratio hypothesis test. Assume for simplicity that the two series are the same length and have periodograms a_i and b_i. Assuming only the first f_c coefficients are retained, the likelihood ratio statistic can be simplified to

$$d_L(x_f, y_f) = 4 \sum_{i=1}^{f_c} \{2 \log(a_i + b_i) - \log a_i - \log b_i\} \qquad (2)$$

The major benefit of basing the distance on the likelihood ratio statistic is that it asymptotically follows a known distribution, and so could be used to not only measure the distance between series, but also test whether that distance is significant. d_L can also be better at discriminating between series for problems where an FFT approach would seem to be appropriate: it is less influenced by small fluctuations in the larger coefficients, and better at detecting consistent variation in the smaller coefficients. A more complete description of the distance measure and the experimentation is provided in [4].

3 Experimentation

Experiments with simulated data are designed to test two things over a class of model, $\mathcal{M}$. Firstly, we test whether the likelihood ratio distance metric is better at discriminating between data from models in $\mathcal{M}$. Secondly, we test whether any detected difference in discrimination effects the clustering and classification accuracy. To measure how well each distance metric discriminates, for each pair of models we measure the percentage difference, D, in the average distance of series from the same cluster to the average distance between series from different clusters. 100 random model pairs are generated to create 100 within and between distance estimates for d_E and d_L for different coefficient sizes. To estimate how well each measure would classify series, we form a correctness function C which is 1 if the distances between series from the same model are less than all the differences from series of different models, and 0 otherwise.

3.1 Sinusoidal and AR(1) Data

We demonstrate the benefits of using d_L on data from stationary order one auto-regressive models (AR(1) models), which take the form

$$x(t) = \phi x(t-1) + \varepsilon \qquad (3)$$

where ε is a random variable with a standard normal distribution and $\phi \in (-1, 1)$. AR(1) models have been used extensively in TSDM research [2,5] and are a good basis for testing how well a distance metric measures similarity based on change. For each run, two random AR(1) models X and Y were selected with $\phi \in (-1, 1)$. Figure 1 shows boxplots for the average (over 100 observations) within and between distance when the first 4 coefficients are retained. The left hand figure shows that, when using d_E, although the median for within distance is lower than the median between distance, there is a large amount of overlap between the distributions. In contrast, when using d_L, the largest difference in the between distance is lower than the median of the within distances, and there is clearly much greater discrimination than with Euclidean distance. The results presented in Table 1 show that d_L provides a better means of discriminating between series than d_E. The difference is significant at all levels. It is also interesting to note that the discriminatory power of d_E actually decreases with the number of coefficients retained. This is caused by the fact it gives too great a weight to small fluctuations in the larger parameters, and the more coefficients retained the greater the chance of this resulting in an incorrect grouping. We also consider a class of model where an autoregressive structure discriminates the models, but where a common cyclical trend may cause a distance metric to be unable to detect differences. Let

$$x(t) = \sum_{j=1}^{r} a_j (\sin(b_j + c_j \cdot t)) + \varepsilon, \qquad (4)$$

where ε is a random variable with a standard normal distribution. The parameters a, b and c control the amplitude, offset and frequency of the curves

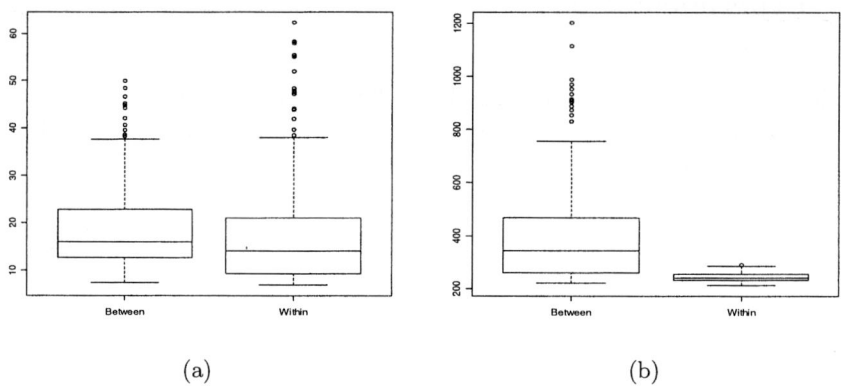

Fig. 1. Distribution of the average difference between different and within the same models for (a) Euclidean distance and (b) Likelihood ratio distance

Table 1. Percentage difference between within and between cluster series, D, and number of correct classifications, C, for AR(1) and combination series of length 1024. Low values of D indicate better discrimination

Coefficients	AR(1) Data				AR(1)+sinusoidal Data			
	D		C		D		C	
	d_E	d_L	d_E	d_L	d_E	d_L	d_E	d_L
2	93.43%	66.99%	1302	3090	96.20%	82.55%	1223	2152
4	93.34%	66.40%	873	3925	96.19%	82.27%	936	3925
16	93.32%	65.74%	298	5622	95.88%	81.68%	539	5622
64	93.86%	67.95%	202	6661	96.57%	83.98%	347	4079
128	94.87%	72.99%	180	6858	97.54%	87.63%	280	4262
256	96.72%	81.42%	161	6727	98.64%	92.37%	275	3894
512	99.07%	81.07%	281	7131	99.54%	91.61%	325	5221

respectively. Suppose $y(t)$ is an AR process as defined in Equation 3. Let the class of Sinusoidal and AR(1) models be $z(t) = x(t) + y(t)$. Our objective is to detect whether the distance metrics can detect the difference in autocorellation structure even when series have the same sine wave series. Hence for a particular experiment involving two models we randomly generate a single sine model of the form given in Equation 4 and combine it with two separate AR(1) models. Table 1 shows the the average observed value of statistics C and D for both d_E and d_L. Although the extra cyclical trend of the sine data reduces the power of both distance metrics, d_L is still better at detecting the difference between series. The following clustering experiments also show the benefit of using d_L. k models are randomly selected and l series are generated from each model. Clusters are found using k-means and partitioning around the medoid (PAM) (both restarted 100 times at random initial data points) and the accuracy is measured

Table 2. Clustering accuracy averaged over 100 runs for two clusters of combination series

	PAA		d_E		d_L	
Coefficients	k Means	PAM	k Means	PAM	k Means	PAM
512	59.40%	56.90%	59.75%	56.20%	68.00%	72.05%
256	60.15%	57.05%	59.55%	57.55%	67.30%	69.55%
128	61.25%	57.10%	61.55%	57.25%	73.80%	78.25%

against the known true clustering. The process is repeated 100 times to estimate the average clustering accuracy for compression ratios (i.e. different numbers of retained coefficients). Table 2 shows the clustering results for Piecewise Aggregate Approximation (PAA) (with Euclidean distance), d_E and d_L. Given that the difference between series is in the autocorellation structure, PAA is not a suitable transformation and hence the clusters found should simply reflect the differences in the common sinusoidal component. The results shown in Table 2 demonstrate that although there are differences in performance of the clustering algorithms, the clusters formed with d_L are consistently more like the true clusters than those formed using d_E.

3.2 ECG Data

ECG data has commonly been used in TSDM [2, 5, 6] and has the characteristic that it has a underlying cyclical trend that is not the true cause in the differences between series from different clusters. To show how d_L can produce better clusters than d_E we cluster an ECG data set first used in [5]. The data consists of 70 series of ECG measurements of patients with malignant ventricular arrhythmia (V), normal (N) or superventricular arrhythmia (S). Table 3 shows the clustering accuracy results with both d_L and d_E distance measures. In each cell of Table 3 the first number is the k-means accuracy and the second that achieved with PAM. The results are comparable to those reported in [5] for PAM with FFT and Euclidean distance. Although there is variation in performance of the clustering algorithms, in all but two cases (were the performance was equal) using d_L rather than d_E gave a higher accuracy. From the dendrograms (not shown because of space restrictions, see [4])the level three clusters formed with nearest neighbour linkage demonstrate the superiority of the d_L distance measure. The clusters formed by d_E are (N,V,N,V,S,S,S), (V,V,S) and (V,V,S,N,C,V,N). The clusters formed by d_L, (V,V,S,V,V,S,V), (N,N,N,N,S) and (S,N,S,S) are much closer to the correct classification.

3.3 Motor Current Data

The simulated motor current data set used in [8] consists of 420 series of length 1500. We repeatedly randomly selected two pairs of series from different classes and measured the statistics D and C described in Section 3. The average D value (percentage difference of within class and between class distance) for d_L

Table 3. Clustering accuracy on ECG Data for three different experiments with k-means and PAM

Retained	Three clusters		Two Clusters, N and V		Two Clusters, N and S	
Coefficents	d_E	d_L	d_E	d_L	d_E	d_L
256	39%\44%	53%\61%	58%\69%	69%\69%	60%\60%	68%\88%
128	39%\44%	47%\64%	60%\69%	71%\69%	58%\60%	65%\88%
64	46%\49%	49%\63%	63%\58%	71%\71%	58%\60%	53%\63%
32	46%\41%	43%\44%	65%\54%	64%\71%	55%\63%	63%\65%

was 82.83%, whereas with d_E the average within distance was actually higher than the between difference (D=104.43%, averaged over 5000 repetitions). The number correct, C, was also higher for d_L (C=850) than with d_E (C= 507). The superiority of d_L is also evident when accuracy is measured with a 1-nearest neighbour classifier on the first 16 coefficients. d_E gave a classification accuracy of 14%, whereas d_L achieved only 9.8%. These accuracies are better than those reported for a time delay neural network in [8].

4 Conclusion

In this paper we have described an alternative distance metric for use with FFTs in time series data mining. The metric, d_L, is based on the likelihood ratio for testing the null hypothesis that the series are from the same process. It has the desirable property of asymptotically following a known distribution and of not being overwhelmed by small variations in the larger coefficients. We have shown this is true for simulated and real world data.

References

1. R. Agrawal, C. Faloutsos and A. N. Swami. Efficient similarity search in sequence databases. In *Proceedings of the 4th FODO*, 1993.
2. A. J. Bagnall and G. J. Janacek. Clustering time series from arma models with clipped data. In *Proceedings of 10th ACM KDD*, 2004.
3. C. Faloutsos, M. Ranganathan and Y. Manolopoulos. Fast subsequence matching in time-series databases. In *Proceedings of ACM SIGMOD Conference*, 1994.
4. G. J. Janacek, A. J. Bagnall and M. Powell. A likelihood ratio distance measure for the similarity between the fourier transform of time series. CMP-C05-01, UEA, 2005.
5. K. Kalpakis, D. Gada, and V. Puttagunta. Distance measures for effective clustering of ARIMA time-series. In *Proceedings of the ICDM*, 2001.
6. E. Keogh and S. Kasetty. On the need for time series data mining benchmarks: A survey and empirical demonstration. In the *Proceedings of 8th ACM KDD*, 2002
7. F. Morchen. Time series feature extraction for data mining using DWT and DFT. Technical Report 3, Philipps-University, Marburg, 2003.

8. R. Povinelli, M. Johnson and J. Ye. Time series classification using Gaussian mixture models of reconstructed phase spaces. *IEEE T. KDE*, 16(6), 2004.
9. M. Vlachos, C. Meet, and Z. Vagena. Identifying similarities, periodicities and bursts for online search queries. In *ACM SIGMOD ICMD*, 2004.
10. Y. Wu, D. Agrawal, and A. El Abbadi. A comparison of DFT and DWT based similarity search in time-series databases. In *9th ACM CIKM*, 2000.

The TIMERS II Algorithm for the Discovery of Causality

Howard J. Hamilton and Kamran Karimi

Department of Computer Science,
University of Regina,
Regina, Saskatchewan,
Canada S4S 0A2
`{hamilton, karimi}@cs.uregina.ca`

Abstract. We present the Temporal Investigation Method for Enregistered Record Sequences II (TIMERS II), which can be used to classify the relationship between a decision attribute and a number of condition attributes as instantaneous, causal, or acausal. In this paper we consider it possible to refer to both previous and next values of attributes in temporal rules, and thus enhance the definition of acausality. We also present a new algorithm for distinguishing between causality and acausality.

1 Introduction

In this paper we present the Temporal Investigation Method for Enregistered Record Sequences II (TIMERS II), which can be used to classify the relationship between a decision attribute and a number of condition attributes as instantaneous, causal, or acausal.

Instantaneous rules are normal decision rules. An example rule is: if {(Outlook$_t$ = sunny) AND (Temperature$_t$ > 20)} then (Play$_t$ = yes), where t indicates the time step of observing the attribute's value. For causality and acausality, the results are temporal decision rules. For the causal case, the decision attribute's value is causally determined by the condition attributes, whose values all appear in the past relative to the decision attribute. An example is: If {(outlook$_{t-1}$ = sunny) then (outlook$_t$ = sunny). The index t-1 indicates that the attribute's value is seen in the previous time step.

For an acausal relationship, values at time steps bigger than t are used in the process of predicting the decision attribute at time t. In TIMERS II it is also possible for some condition attributes to have happened in the past. An example acausal rule is: if {(outlook$_{t-1}$ = overcast) AND (outlook$_{t+1}$ = rainy) then (outlook$_t$ = rainy). In an acausal relation, the decision attribute's value is not caused by the condition attributes, but just happens to be seen together over time. In this case there may have been hidden common causes that affected all the attributes in the same rule. The same method can be used for linear spatial data, where "back" and "forward" can be used to indicate the relative position of an attribute's observation.

The formal definitions of instantaneous and causal sets of rules are given in [2]. In TIMERS II a set of rules is acasual if the current value of the decision attribute relies on the future value of at least one condition attribute [3].

The rest of this paper is organised as follows. Section 2 introduces the TIMERS II algorithm. Section 3 presents a number of experimental results obtained from TIMERS II. Section 4 concludes the paper.

2 The TIMERS II Algorithm

We consider there to be an *order of conceptual simplicity* among the three types of the relations, with instantaneous being the simplest type of relationship, followed by acausality, and then causality being the most complex. Hence, instantaneous $<_{simplicity}$ acausal $<_{simplicity}$ causal. The intuition behind this ordering is that as we move from instantaneous to acausal and then to causal, more claims are being made about the relationship. As a principle, we try to explain a relationship with the simplest possible type. As we will see in Section 3, this ordering is used to choose a winning relations type when the results of the three tests are close.

Temporalisation was introduced in [2]. TIMERS II introduces the *sliding position* temporalisation as explained in [3]. The temporalisation operator *Temporalise(w, pos, D, d)* takes as input a window size w, the position of the decision attribute within the window *pos*, the input records D, and the decision attribute d, and outputs temporalised records. The TIMERS II algorithm is shown in Figure 1.

Input: A sequence of sequentially ordered data records D, minimum and maximum temporalisation window sizes α and β, where $0 < \alpha \le \beta$, a minimum accuracy threshold ac_{th}, a decision attribute d, and a confidence level cl. The attribute d can be set to any of the observable attributes in the system, or the algorithm can be tried on all attributes in turn. *Preference* determines whether the user prefers higher accuracy or a simpler method.
Output: A set of accuracy values and a verdict as to the nature of the relationship among the decision attribute and the condition attributes. It could be spontaneous, causal, or acausal.
RuleGenerator() is a function that receives input records, generates decision trees, rules, or any other representation for predicting the decision attribute, and returns the training or predictive accuracy, as well as the size of the generated rules.

TIMERS II(D, α, β, Ac_{th}, d, cl, $preference$) {
   ac_i = RuleGenerator(D, d); // instantaneous accuracy; window size = 1
   for ($win = \alpha$ to β)
      for ($pos = 1$ to win)
         ($ac_{w,pos}$, $ruleSize_{w,pos}$) = RuleGenerator(*Temporalise*(win, pos, D, d), d)
   ac_c = max($ac_{\alpha,\alpha}$, ..., $ac_{\beta,\beta}$) // best causal result
   ac_a = max($ac_{\alpha,pos1}$, ..., $ac_{\beta,pos2}$), $\forall ac_{x,pos}$, $1 \le pos < x$ // best acausal result
   if (max(ac_i, ac_c, ac_a) < ac_{th}) **then** stop.   // Maybe there is not enough related information?
   Verdict = "for attribute " + d + ", "
   Relation = RelationType(cl, (ac_i, $ruleSize_i$), (ac_a, $ruleSize_a$), (ac_c, $ruleSize_c$), $preference$)
   Case relation **of**
      INSTANTANEOUS: verdict += "the relation is instantaneous"
      ACAUSAL: verdict += "the relation is acausal" // an element from the future is present
      CAUSAL: verdict += "the relation is causal" // all condition attributes are from the past
   end case
   return verdict.
}

Fig. 1. TIMERS II algorithm for discovering the nature of a relationship

TIMERS II has been implemented in an application programme called TimeSleuth [1]. TIMERS II first performs the instantaneous test. Since it may not be obvious which window size is most appropriate for a particular dataset, TIMERS II tries a range of window sizes. The resulting temporalised data are fed to a rule generator which comes up with decision rules, and returns the accuracy and also the complexity of the rules. These measures are used to decide on a relation type. For an analysis of the time and space requirements of TIMERS II, see [3].

We use the accuracy and complexity of the rules obtained from each method to choose the best relation type that applies to the data. Normally the method with the highest accuracy value would be selected. However, it may happen that the accuracy values are close to each other. In such cases we choose the simpler relationship because the gains from choosing another relationship type may not be worth the extra complexity. Users can employ their discretion in making this decision. However, TIMERS II includes a statistical method. The RelationType() routine uses accuracy intervals to make a judgment about the type of the relationship. Using the confidence level provided by the user in the cl parameter, and assuming normal distribution, it constructs a confidence interval for the accuracy [4]. Then it checks to see if the corresponding intervals overlap. If they do, the method with the simpler type of relationship will be chosen provided it has simpler rules. The intuition is that even if the simpler method has resulted in less accuracy, it could have *potentially* produced better or the same results. After selecting a winner between the first two methods, the winning relation type is tested against the third relation type using similar comparison of intervals, to determine the final winner.

As an example, suppose with a confidence level of 90%, we have: the instantaneous accuracy ac_i = 32.5%, $interval_{aci}$= [31%, 34%], the acausal accuracy ac_a = 35%, $interval_{aca}$ = [33%, 37%], and the causal accuracy ac_c = 37%, $interval_{acc}$ = [35%, 39%]. For simplicity of the example we assume all methods have the same size of rules. Because the confidence intervals of the instantaneous method and the acausal methods intersect, instantaneous is chosen because it is considered simpler. Since the intervals of the instantaneous and causal methods do not overlap, the causal method is chosen as the final verdict because of its higher accuracy value. This example also shows the special case when the every two intervals are overlapping. In this case, starting with the first two or the last two methods give different results. In the first case, as shown above, we choose the method with the highest accuracy. But when starting from right to left (higher accuracy value to lower values) we choose the simplest method. We leave the decision about which direction to follow to the user. In the TimeSleuth programme the user can choose between "Prefer simpler method" (right to left) and "Prefer higher accuracy" (left to right) options. See [3] for more details.

Here is how the algorithm to choose a method works. To determine which method/relation type to choose, we sort the accuracy values in either ascending order (preferring higher accuracy), or descending order (preferring simpler method). This different ordering simplifies the algorithm, because we do not need to worry about the direction after this point. Starting with the two methods with the lowest (or highest) accuracy values, we test to see if there is an overlap among their confidence intervals. If so, then we choose the simpler method. The choice of the simpler method depends on both the conceptual complexity of the relation as defined above, and also the size of the rules that are needed to express the relationship. In our method the more space needed for the rules, the more complex that relationship. We use the number of conjuncts in the rules to measure their size, as in the Minimum Description Length (MDL) principle. We make the decision as to which method to choose the following way: If a conceptually simpler method overlaps with a conceptually more complex method, but at the same time requires more space to represent the rules, then priority is given to the more complex rule. In other words, for a simpler method to over-ride a more complex method, not only should there be an overlap between their accuracy intervals, but the simpler method should result in fewer or shorter rules. While our assumed order of complexity is subjective, including the size of rules adds an objective element to the complexity measure. If there is no overlap in the accuracy intervals, we choose the method with the better accuracy value. A winner is thus selected among the first two methods. This winning relation type is then compared with the third method to determine the final method. Figure 2 shows how the best method is selected.

Input: A confidence level cl, three accuracy values corresponding to the instantaneous, acausal, and causal methods: ac_i, ac_a, ac_c, and their corresponding size of rules: $ruleSize_i$, $ruleSize_a$, $ruleSize_c$, a preference p for higher accuracy vs. a simpler method.
Output: A verdict as to the best relationship type.
//info[].method contains one of INSTANTANEOUS, CAUSAL, or ACSUAL. info[]. Accuracy is the best
//accuracy value. info[].interval contains the interval of the accuracy value, computed using a confidence value

Function RelationType(cl, (ac_i, $ruleSize_i$), (ac_a, $ruleSize_a$), (ac_c, $ruleSize_c$), p) {
  // initialise the info[] structure
  forEach (method = INSTANTANEOUS, ACAUSAL, CAUSAL)
    info[method] = (method , accuracy$_{method}$, ruleSize$_{method}$, Interval$_{method}$ =
ComputeAccuracyInterval(accuracy$_{method}$))
  // if preference is given to higher accuracy, then start the search from lower accuracy values
  if (p == HIGHER_ACCURACY)
    sort_Ascending(info[]); // sort in ascending order of accuracy.
  else // SIMPLER_METHOD
    sort_Descending(info[])
  winner = 1
  for (count = 2 to 3)
    if (overlap(info[winner].interval, info[count].interval)) {// if overlap, then choose the simpler method
      if (info[count].method <$_{simplicity}$ info[winner].method **and**
        info[count].ruleSize ≤ info[winner].ruleSize) **then**
        winner = count
    }
    else { // if no overlap, choose the method with higher accuracy
      if (info[count].accuracy > info[winner].accuracy) **then**
        winner = count
    }
  return info[winner].method //one of INSTANTANEOUS, ACAUSAL, or CAUSAL
}

Fig. 2. Selecting the best relationship

If needed, this algorithm can also select the best window size based on a number of accuracy values obtained in either the acausal or causal case. The order of simplicity is then determined by the window size, with bigger window sizes considered less simple. In the TIMERS II algorithm in Figure 1, we use the window size that gives the maximum accuracy.

3 Experimental Results

We report on experiments using two temporal datasets. The first one is generated by an artificial life programme called URAL, and involves an artificial robot moving left, right, up and down on an 8 × 8 board. The goal is for us to discover the effects of moving the robot. The position is expressed by a x and y pair. We used 2500 records for training, and 500 for testing the rules (to compute the predictive accuracy). This data comes from a controlled environment with no exceptions.

Each record in the robot dataset contains x and y position values at any given time and the direction of movement at that time. We set the decision attribute to be the current value of x, and the other three attributes are set as the condition attributes. There is no relationship between the current value of x on one hand, and the current values of y, direction of the movement, or the presence of food on the other hand. So we predict that an instantaneous test (window size of 1) will give poor results. From our understanding of the domain we know that the current value of x depends on the previous value of x, and the previous direction of movement. We expect the method to classify the relationship as a causal one. The acausal hypothesis says that you can tell where you were before if you know where you are now, and which direction you are will be going next. This hypothesis is clearly wrong, as we could have ended at the current position from a different number of previous positions. Hence we do not expect to get good results with our acausality test. The results are shown in Table 1. Even though an acausal method may have been used, the output rules may not have any references to attributes that appear after the decision attribute. In this case the rules are considered to be causal, seen in under "Actual rules."

Considering the result with a window size of 2, we declare the relation to be causal. With any position bigger than 1, the previous record which contains the relevant information for accurate prediction of current x value, is included in the temporalised data. The method discovers the correct temporal relation between the current value of x and the previous x and movement direction, with results having 100% accuracy with sliding positions of 2 or more. In other words, even with an acausal test, the rules are all causal, because they only contain attributes from the previous time step.

The second series of experiments concerns a real-world dataset from weather observations in Louisiana [5], and hence interpreting the dependencies and relationships is harder. It includes 343 training records, each containing the air temperature, the soil temperature, humidity, wind speed and direction, and solar radiation, gathered hourly. 38 other records were used for testing the rules and estimating predictive accuracy. We set the soil temperature to be the decision attribute. The results are shown in Table 1.

The relationship is not instantaneous, as observed by relatively poor results with a window size of 1 (instantaneous test). The accuracy goes up after temporalisation, implying that there is a temporal relationship at work. This relation is not causal, and the current value of the soil temperature just happens to change relative to its past values. Since the accuracy values in causal and acausal tests are not much different, TimeSleuth declares the relationship between the soil temperature and other attributes to be acausal.

Table 1. TIMERS II's accuracy result with the robot and weather data

Win	Position	Type of test	Robot Data			Weather Data		
			Training Accuracy	Predictive Accuracy	Actual Rules	Training Accuracy	Predictive Accuracy	Actual Rules
1	1	Instant	19.7%	20.4%	Instant	27.7%	23.7%	Instant
2	1	Acausal	56.2	55.7%	Acausal	75.1%	59.5%	Acausal
2	2	Causal	100%	100%	Causal	82.7%	67.6%	Causal
3	1	Acausal	57.6%	55.6%	Acausal	85.3%	75.0%	Acausal
3	2	Acausal	100%	100%	Causal	82.4%	72.7%	Acausal
3	3	Causal	100%	100%	Causal	86.8%	77.8%	Causal
4	1	Acausal	58.4%	58.1%	Acausal	85.3%	74.3%	Acausal
4	2	Acausal	100%	100%	Causal	85.9%	74.3%	Acausal
4	3	Acausal	100%	100%	Causal	83.2%	74.3%	Acausal
4	4	Causal	100%	100%	Causal	84.4%	71.4%	Causal

4 Concluding Remarks

TIMERS II provides a method to discover and distinguish between instantaneous, causal, and acausal relationships between a decision attribute and a set of condition attributes. Our method is based on the passage of time between causes and effects. We generalised the ability to refer to attribute values from other time steps so that a rule can refer to condition attribute' values that appear before and after the decision attribute. This ability results in an enhancement to the definition of an acausal relationship. We also provided an algorithmic method of distinguished between instantaneous, causal, and acausal relations.

One can apply the same temporal considerations to associations, so the values of a number of attributes from different time steps can be associated together. However, in an association we do not have a distinguished decision attribute, observed at a reference time (the current time). So defining the future and the past may not be straightforward.

TimeSleuth is available from http://www.cs.uregina.ca/~karimi/downloads.html.

References

1. Karimi, K., and Hamilton, H.J. TimeSleuth: A Tool for Discovering Causal and Temporal Rules, *The 14th IEEE International Conference on Tools with Artificial Intelligence (ICTAI 2002)*, Washington DC, November, 2002, pp. 375-380.

2. Karimi, K., and Hamilton, H.J., Distinguishing Causal and Acausal Temporal Relations, *The Seventh Pacific-Asia Conference on Knowledge Discovery and Data Mining (PAKDD'2003)*, Seoul, South Korea, April/May 2003, pp. 234-240.
3. Karimi, K., and Hamilton, H.J., An Extension to the TIMERS Method, *Technical Report CS-2005-02*, Department of Computer Science, University of Regina, Canada, March 2005.
4. Witten, I.A., and Frank, E., *Data Mining: Practical Machine Learning Tools and Techniques with Java Implementations*, Morgan Kaufmann, 2000.
5. http://typhoon.bae.lsu.edu/datatabl/current/sugcurrh.html. Contents change with time.

A Recent-Biased Dimension Reduction Technique for Time Series Data

Yanchang Zhao, Chengqi Zhang, and Shichao Zhang

Faculty of Information Technology, University of Technology, Sydney, Australia
{yczhao, chengqi, zhangsc}@it.uts.edu.au

Abstract. There are many techniques developed for tackling time series and most of them consider every part of a sequence equally. In many applications, however, recent data can often be much more interesting and significant than old data. This paper defines new recent-biased measures for distance and energy, and proposes a recent-biased technique based on DWT for time series in which more recent data are considered more significant. With such a recent-biased technique, the dimension of time series can be reduced while effectively preserving the recent-biased energy. Our experiments have demonstrated the effectiveness of the proposed approach for handling time series.

1 Introduction

Analyzing time series is a challenging topic in the field of data modelling and mining. In many applications, such as stock market, one concerns more about the recent data than what happened long ago. Besides the global trend, the recent data are very important to judge the similarity between time series and more significant to predict and make decisions than the detail of old data. For example, for a stockbroker, the long-term (say, six years) trend of stock price and the detailed variances in the last month of a stock are important, but the variance in a certain month four years ago is of little significance. In such kind of scenarios, a mechanism which favors the recent is called for. Nevertheless, most of the techniques for time series give equal significance to all data in the series. In this paper, we design a recent-biased technique to tackle the above problem. With our method, recent data are given more significance and kept with finer resolution, while old data with coarser resolution. Weights for DWT coefficients are derived from a decaying function, and then the coefficients with the largest k weights are chosen as the representation of the time series. Our technique is different from SWAT [1] in that the largest k coefficients are kept with our method while only a single coefficient is maintained at each level with SWAT. Our technique is also different from the traditional method of keeping the largest k coefficients and the RAM-DS algorithm [8], because with our technique the largest k coefficients are obtained from the weights only and has nothing to do with specific time series. However, the subsets of largest coefficients are different for different time series with both the tradition method and RAM-DS. The same subset of coefficients for all time series is used in [6], but the subset of coefficients are obtained from all time series without weighting.

2 Related Work

In this section, related work on similarity measures, dimension reduction and recent-biased techniques for time series data will be introduced. Euclidian Distance and other L_p-norms are popular to measure the distance between time series. Another measure, DTW (Dynamic Time Warpping), is designed to handle time series with some time shifts.

Because time series is usually of very high dimension, the dimension has to be reduced to improve the efficiency of computation. Popular techniques include PCA, DFT (Discrete Fourier Transform), DWT (Discrete Wavelet Transform) [6], Landmark [7], major minima and maxima [4], PIP [3], PAA [5], etc. With DWT, a time series can be represented by a rough sketch by keeping only the first coefficients. Some researchers propose to use the largest coefficients to preserve the optimal amount of energy, or to choose the same subset of the coefficients for all time series for the ease of computing similarity [6].

As to recent-biased techniques, Bulut et al designed a structure named SWAT [1] to process queries over data streams that are biased towards the more recent values. A time weighting function is defined in [8] so that the old data values are with lower weights, and then more resources can be utilized to explore more recent data with finer granularities. Cohen et al uses decay functions to maintain time-decaying stream aggregates [2].

3 A Recent-Biased Technique for Dimension Reduction

Our idea comes from the observation that recent data are usually more important than ancient data. Considering the time series shown in Figure 1, which pair of S_1S_2 and S_1S_3 is more similar? In many applications, we care more about recent data than ancient data, then S_1 and S_3 is more similar than S_1 and S_2, since the difference between S_1 and S_3 happened long ago. However, the L_p-norm distance between S_1 and S_2 is 1 , while that between S_1 and S_3 is also 1. So the two pairs are of the same similarity according to L_p-norm distance. If using $(\frac{1}{8}, \frac{2}{8}, \frac{3}{8}, \frac{4}{8}, \frac{5}{8}, \frac{6}{8}, \frac{7}{8}, 1)$, as the weights, the distance between S_1 and S_2 is 0.75, while that between S_1 and S_3 is 0.25. Therefore, S_1 is more similar to S_3 than to S_2 when a biased distance measure is used.

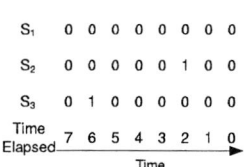

Fig. 1. An Example of Time Series

Since old data become less significant as time goes by, a bias can be given to recent data. A simple idea is to give larger weights to more recent data. Actually, decaying functions are used widely for processing time series data.

For example, exponential decaying functions are used for time series to find recent frequent itemsets adaptively, and to explore temporal and support count granularities in data streams [8]. In addition to exponential functions, polynomial and ployexponential decaying functions are also considered to maintain time-decaying stream aggregates [2]. Similarly, we define a recent-biased function as follows to help keep more recent data at finer scales. Recent-biased function $\mathbf{B} = b(t), t \geq 0$ is a monotonously decreasing function with $b(0) = 1$ and $b(+\infty) = 0$, where t is the time elapsed till now.

The most common used decay function is exponential function, $b_t = d^t$, where t is the time elapsed, and d is the decay factor, $0 < d < 1$, $\alpha > 0$. Linear decay can be get with linear function $b_t = \frac{n-t}{n}$, where $0 \leq t \leq n$ and n is the length of time series. Based on the above recent-biased function, the recent-biased L_p-norm distance with bias on recent is defined in the following.

Definition 1 (Recent-Biased Distance & Energy). *The recent-biased distance between time series $\mathbf{S}$ and $\mathbf{S}'$ is defined as*

$$Dist(\mathbf{S}, \mathbf{S}') = \frac{\|(\mathbf{S} - \mathbf{S}') \bullet \mathbf{B}\|}{\|\mathbf{B}\|} \quad (1)$$

where '$\bullet$' stands for the operator of inner product, '$\|\cdot\|$' denotes L_p-norm, and $\mathbf{B}$ is a recent-biased vector. If L_2-norm is used, the recent-biased energy of $\mathbf{S}$ is defined as

$$E(\mathbf{S}) = \frac{\|\mathbf{S} \bullet \mathbf{B}\|^2}{\|\mathbf{B}\|^2} \quad (2)$$

3.1 Recent-Biased Dimension Reduction

In this paper, Haar wavelet transform is used because it is very simple and widely used and is of linear time complexity. For Discrete Wavelet Transform, there are two different ways for choosing coefficients, the first k or the largest k coefficients. If the first k coefficients are selected, the global trend and variation can be preserved. With the largest coefficients kept, the parts of large energy are preserved and the energy is better kept, which is better for compressing a single signal. Nevertheless, when dealing with multiple time series, more storage space is required to keep the positions of coefficients and the distance computations is more complex. To keep more detail for recent data and preserve the recent-biased energy, a recent-biased technique based on DWT is designed in the following. Instead of keeping the largest or the first coefficients, the largest recent-biased coefficients are kept, which are computed from the weights only.

The DWT coefficients for a time series with 8 values are shown in Figure 1. We can make the average to be zero by normalizing the time series, so the average C_0 is zero and is not considered here. Suppose that the time series are normalized that the average is zero. The original time series is $\mathbf{S}=(-C_{11}-C_{21}-C_{31}, -C_{11}-C_{21}+C_{31}, -C_{11}+$

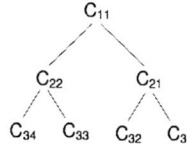

Fig. 2. Coefficients of DWT

$C_{21} - C_{32}, -C_{11} + C_{21} + C_{32}\ C_{11} - C_{22} - C_{33}$,
$C_{11} - C_{22} + C_{33}, C_{11} + C_{22} - C_{34}, C_{11} + C_{22} + C_{34})$
from recent data to old data. Assume that $\mathbf{B} = b(t), t \geq 0$ is the bias function, so the recent-biased energy of $\mathbf{S}$ is

$$\begin{aligned}
E(\mathbf{S}) &= \|\mathbf{B} \bullet \mathbf{S}\|^2 / \|\mathbf{B}\|^2 \\
&= \tfrac{1}{\|\mathbf{B}\|^2}((-C_{11} - C_{21} - C_{31})^2 b_0^2 + (-C_{11} - C_{21} + C_{31})^2 b_1^2 \\
&\quad + (-C_{11} + C_{21} - C_{32})^2 b_2^2 + (-C_{11} + C_{21} + C_{32})^2 b_3^2 \\
&\quad + (C_{11} - C_{22} - C_{33})^2 b_4^2 + (C_{11} - C_{22} + C_{33})^2 b_5^2 \\
&\quad + (C_{11} + C_{22} - C_{34})^2 b_6^2 + (C_{11} + C_{22} + C_{34})^2 b_7^2) \\
&= \tfrac{1}{\|\mathbf{B}\|^2}(C_{11}^2 \sum_{i=0}^{7} b_i^2 + C_{21}^2 \sum_{i=0}^{3} b_i^2 + C_{22}^2 \sum_{i=4}^{7} b_i^2 \\
&\quad + C_{31}^2 \sum_{i=0}^{1} b_i^2 + C_{32}^2 \sum_{i=2}^{3} b_i^2 + C_{33}^2 \sum_{i=4}^{5} b_i^2 + C_{34}^2 \sum_{i=6}^{7} b_i^2 \\
&\quad + 2C_{11}C_{21}(b_0^2 + b_1^2 - b_2^2 - b_3^2) + 2C_{11}C_{22}(-b_4^2 - b_5^2 + b_6^2 + b_7^2) \\
&\quad + 2C_{21}C_{31}(b_0^2 - b_1^2) + 2C_{21}C_{32}(-b_2^2 + b_3^2) \\
&\quad + 2C_{22}C_{33}(b_4^2 - b_5^2) + 2C_{22}C_{34}(-b_6^2 + b_7^2))
\end{aligned} \tag{3}$$

In order to preserve the energy as large as possible, the coefficients with largest weights will be kept, and other coefficients are set to zero. It is difficult to tell which C_{ij} is of the greatest importance from Formula (3). To make it easy, we only consider C_{ij}^2 while ignoring those $C_{i_1 j_1} C_{i_2 j_2}$, where $i_1 \neq i_2$ or $j_1 \neq j_2$. Then, the question becomes to choose C_i with the largest coefficients in $C_{11}^2 \sum_{k=0}^{7} b_k^2 + C_{21}^2 \sum_{k=0}^{3} b_k^2 + C_{22}^2 \sum_{k=4}^{7} b_k^2 + C_{31}^2 \sum_{k=0}^{1} b_k^2 + C_{32}^2 \sum_{k=2}^{3} b_k^2 + C_{33}^2 \sum_{k=4}^{5} b_k^2 + C_{34}^2 \sum_{k=6}^{7} b_k^2$. The weight of C_{ij} is

$$w_{ij} = \sum_{k=(j-1)\cdot 2^{L-i+1}}^{j\cdot 2^{L-i+1}-1} b_k^2,\ 1 \leq i \leq L, 1 \leq j \leq 2^i, L = \lceil \log_2 n \rceil \tag{4}$$

where n is the length of time series. Since b_i is a recent-biased function, it is monotonously decreasing with the increase of i. Therefore, C_{11} is of the largest coefficient, $\sum_{k=0}^{7} b_k^2$, and it is the first one to choose. The second one is C_{21}, whose weight is $\sum_{k=0}^{3} b_k^2$. The third one will be C_{22} (or C_{31}) if $\sum_{k=4}^{7} b_k^2$ is greater (or less) than $\sum_{k=0}^{1} b_k^2$. Different bias functions will probably lead to different sets of the largest k coefficients. Given the specific bias function, the weight of DWT coefficients can be calculated immediately with Formula (4) and ready for use for all time series. With our method, only the weighted function is used to decide which coefficients to keep, so the same subset of coefficients are chosen for all time series, which is different from the traditional method of keeping different subsets of the largest coefficients for different time series.

The Euclidean distance between the extracted k coefficients is used as the similarity between time series and the recent-biased Euclidean distance and energy are calculated with the following formulae.

$$Dist(\mathbf{S}, \mathbf{S}') \approx \frac{(\sum_{ij} w_{ij}(C_{ij} - C'_{ij})^2)^{\frac{1}{2}}}{(\sum_{ij} w_{ij})^{\frac{1}{2}}} \tag{5}$$

$$E(\mathbf{S}) \approx \frac{\sum_{ij} w_{ij} C_{ij}^2}{\sum_{ij} w_{ij}} \tag{6}$$

where C_{ij} is in the first k coefficients. If the bias function is set to $b_i = 1, i \geq 0$, that is, there is no bias, then the weights for those coefficients in Figure 1 are 8, 4, 4, 2, 2, 2, 2 from top to bottom. So those coefficients at higher level will be chosen first, which is the same as traditional DWT with the first coefficients.

3.2 Complexity Analysis

Assume the length of time series is n, and there are m time series. The time complexity for computing weights from bias function are $O(n)$ and the time complexity for getting the largest k weights is $O(kn)$. The time complexity of DWT for a time series is $O(n)$. Therefore, the total time complexity for processing m time series is $O(n + kn + mn)$, i.e., $O((m + k + 1)n)$. As to space complexity, the space requirement for keeping the positions of the largest k coefficients is $O(k)$, and there are k coefficients for each time series, so the total space complexity is $O(k + mk)$, i.e., $O((m + 1)k)$.

4 Experimental Results

Effectiveness of our algorithm for capturing recent details is shown in Figure 3. The original time series (see Figure 3a) is "leleccum" from Matlab, and the first 4096 values are kept. Linear bias function $b(t) = \frac{n-t}{n}$ is used, where n=4096. The reconstructed times series after keeping the recent-biased largest k coefficients are shown in Figure 3b-h. These figures show clearly that the more recent data are preserved with more details while the older data kept with a coarser scale.

The Nasdaq indices from "Yahoo! Finance" (http://finance.yahoo.com/) is used to test the accuracy of our technique in experiments. The close prices of indices from Jun 1988 to Oct 2004 are chosen and each time series is composed of 4096 points. To evaluate the effectiveness of our technique, we design a criterion to measure the precision of approximation after dimension reduction. Assume that $\mathbf{S}$ and $\mathbf{S'}$ are respectively the original and reconstructed time series. The error of approximation between $\mathbf{S'}$ and $\mathbf{S}$ is defined as $Err(\mathbf{S'}, \mathbf{S}) = \frac{E(\mathbf{S'}-\mathbf{S})}{E(\mathbf{S})}$, where $E(\mathbf{S})$ is the recent-biased energy of $\mathbf{S}$ defined in Formula (2). The experimental result for accuracy is shown in Figure 4. The horizontal axis stands for k, the number of coefficients kept, and the vertical axis stands for the error rate. The solid line denotes the error rate of recent-biased DWT, while the dotted denotes that of traditional DWT with the first coefficients. It is clear that the accuracy gets improved as more coefficients are kept. Exponential bias functions are used, and the decay factor is $d = 1 - \frac{1}{1+10^\alpha}$. From Figure 4a to 4d, the decay function becomes less biased on recent with the increase of α. When the bias is large, higher accuracy can be achieved with our method than with traditional DWT with the first coefficients. When the bias is tiny (see Figure 4d), our method becomes nearly the same as traditional DWT with the first coefficients.

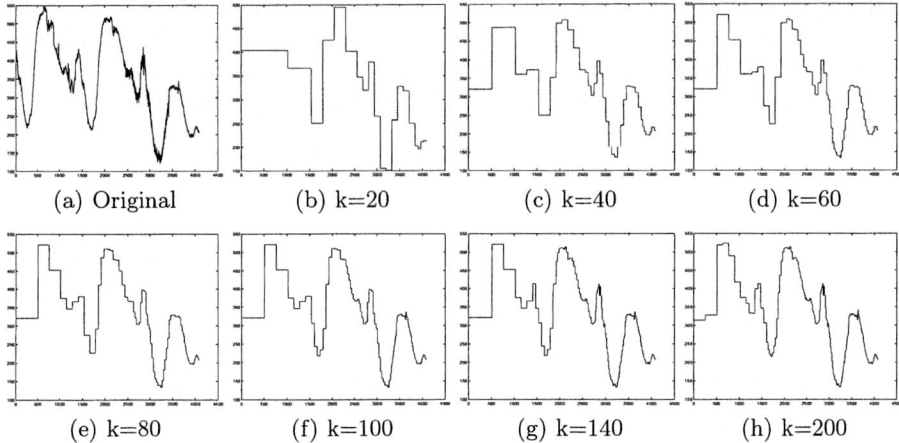

Fig. 3. Effectiveness. The original time series is shown in (a), and the reconstructed times series by keeping the recent-biased largest k coefficients are shown in (b)-(h)

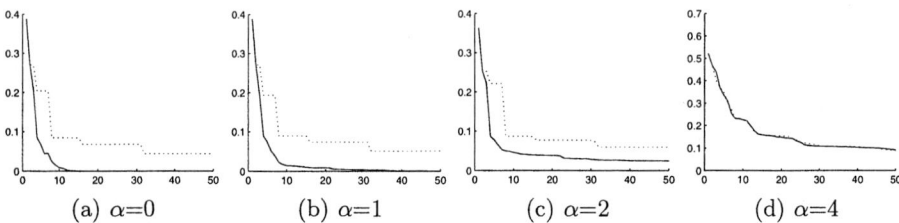

Fig. 4. Accuracy. The horizontal axis stands for the number of coefficients, and the vertical axis stands for the accuracy. The solid line denotes the error rate of recent-biased DWT, while the dotted denotes that of DWT with the first coefficients

5 Conclusions

We have designed a recent-biased technique for time series, which gives greater weights to more recent data and also preserves more details of recent data. Our experiment shows that the recent-biased technique is very efficient and effective to handle time series. Our future work includes combining our recent-biased idea with DFT, PIP, PAA and other dimension reduction techniques for time series data, and extending dynamic time warpping to a recent-biased measure.

References

1. A. Bulut and A. K. Singh: SWAT: Hierarchical Stream Summarization in Large Networks. Proc. of the 19th Int. Conf. on Data Engineering, Bangalore, India, 2003.
2. E. Cohen and M. Strauss: Maintaining time-decaying stream aggregates. Proc. of the 22nd ACM Symposium on Principles of Database Systems, 2003.

3. T. Fu, F. Chung, V. Ng and R. Luk: Pattern Discovery from Stock Time Series Using Self-Organizing Maps. Workshop Notes of KDD'01 . San Francisco, CA, USA, 2001.
4. E. Fink, K. B. Pratt, and H. S. Gandhi: Indexing of Time Series by Major Minima and Maxima. Proc. of the IEEE Int. Conf. on Systems, Man, and Cybernetics, 2003.
5. E. Keogh, K. Chakrabarti, et al: Dimensionality Reduction for Fast Similarity Search in Large Time Series Databases. Knowledge and Information Systems 3(3), 2000.
6. F. Mörchen: Time series feature extraction for data mining using DWT and DFT. In Technical Report No. 33, Philipps-University Marburg, 2003.
7. C.-S. Perng, H. Wang, et al: Landmarks: A New Model for Similarity-Based Pattern Querying in Time Series Databases. Proc. of the 16th ICDE, San Diego, CA, 2000.
8. W-G. Teng, M-S. Chen, and P.S. Yu: Using Wavelet-based Resource-aware Mining to Explore Temporal and Support Count Granularities in Data Streams. Proc. of the SIAM Int. Conf. on Data Mining, April 2004.

Graph Partition Model for Robust Temporal Data Segmentation

Jinhui Yuan*, Bo Zhang, and Fuzong Lin

State Key Laboratory of Intelligent Technology and System,
Department of Computer Science and Technology, Tsinghua University,
Beijing 100084, P.R.China
yuan-jh03@mails.tsinghua.edu.cn
{dcszb, linfz}@mail.tsinghua.edu.cn

Abstract. This paper proposes a novel temporal data segmentation approach based on a graph partition model. To find the optimal segmentation, which maintains maximal connectivity within the same segment while keeping minimum association between different ones, we adopt the min-max cut as an objective function. For temporal data, a linear time algorithm is designed by importing the temporal constraints. With multi-pair comparison strategy, the proposed method is more robust than the existing pair-wise comparison ones. The experiments on TRECVID benchmarking platform demonstrate the effectiveness of our approach.

1 Introduction

Temporal data mining is a rapidly evolving area of research. Similar to other application fields, data segmentation is the fundamental task to temporal data mining. Only after the segmentation, can the subsequent analysis of temporal data, such as clustering, classification and association rules mining, be carried out. Conventionally, temporal data are modelled as time series. To the day, numerous approaches for segmenting time series have been proposed [1]. However, with pair-wise comparison strategy, most of them are sensitive to noises. To overcome the drawbacks, this paper proposes a novel temporal data segmentation method based on graph partition model. In the method, multi-pair comparison is performed. Meanwhile, a linear time algorithm is designed by importing the temporal constraints. The experiments show that the proposed approach is more robust than existing ones but less efficient.

The rest of this paper is organized as follows. In the next section, the graph partition model with temporal constraints is introduced. In Section 3, we propose a temporal segmentation algorithm based on the above model. In Section 4, the proposed algorithm is evaluated on the TRECVID benchmarking platform and compared to existing approaches. Finally, the conclusion is made in Section 5.

---

* Supported by National Natural Science Foundation of China (60135010, 60321002) and Chinese National Key Foundation Research & Development Plan (2004CB318108).

2 Graph Partition Model

Clustering based on graph partition model has attracted a great interest recently [2]. Although so far there has been no good algorithms of polynomial time for the general graph partition problem, linear time algorithm exists for the one with the temporal constraints.

2.1 Computing Segmentation with Graph Cuts

Given an undirected, weighted graph $G = G(V, E)$ with a set of nodes V, a set of edges E. Assume $|V| = N$, namely, there are N nodes in graph G. Let w_{ij} denote the weight of edge $e(i,j) \in E$, i.e., the similarity between nodes i and j. The larger the w_{ij}, the more similar between nodes i and j. To introduce the graph partition model more clearly, we first present some graph terminologies. A graph with 14 nodes and the related terminologies are depicted in Figure 1.

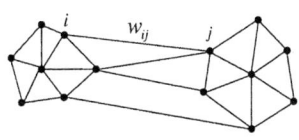
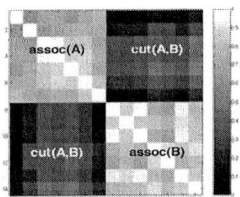

Fig. 1. *Left*: A graph with 14 nodes. *Right*: The visualization of similarity matrix of the left graph. w_{ij} is defined as the reciprocal of Euclidean distance of the nodes i and j. The stronger the connectivity between i and j, the brighter the entry (i, j) is

Definition 1. The similarity matrix $\mathbf{W}$ is a $N \times N$ symmetric matrix, in which entry w_{ij} represents the similarity of nodes i and j.

Definition 2. The cut which divides graph G into subgraphs A and B is defined as: $cut(A, B) = \sum_{i \in A, j \in B} w_{ij}$.

Definition 3. The association of subgraph A is defined as: $assoc(A) = \sum_{i,j \in A} w_{ij}$.

Given a data set, a graph can be constructed by treating each sample as a node and linking an edge between each pair of the nodes. By defining the weight of the edge as the similarity of the samples, clustering can be formulated as a graph partition problem. In this way, the graph partition model helps to define more appropriate criteria. Initially, minimum cut is proposed to be a partition objective function. To avoid skewed cut, other objectives such as ratio cut, normalized cut and min-max cut are proposed successively [2]. From the point of view of clustering, min-max cut, defined by Equation 1, which tries to minimize the association between the two subgraphs while maximize the association within each subgraph, gives the best criterion:

$$Mcut(A,B) = \frac{cut(A,B)}{assoc(A)} + \frac{cut(A,B)}{assoc(B)}. \quad (1)$$

Unfortunately, this problem is NP-complete because of its combinatoric nature. An approximate optimal solution can be yielded by spectral graph theory[2]. However, it still is not able to deal with a huge amount of data because of the large calculation during the process of matrix spectral decomposition.

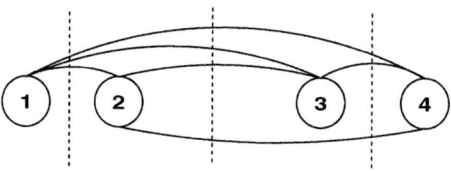

Fig. 2. A full graph with 4 nodes. The broken line indicates the positions of the feasible cuts. With temporal constraints, the size of feasible set is 3, otherwise, 10

2.2 Cuts with Temporal Constraints

When applied to temporal data, clustering based on graph partition model must satisfy some temporal constraints. For example, in temporal data segmentation, the method should guarantee the temporal continuity of each cluster. More pricisely, once two unadjacent samples belong to a cluster, any one between them must be grouped into the same cluster. By importing this temporal constraints, a feasible cut can only occur at one of the $N-1$ possible positions between any two adjacent samples. Thus, the size of feasible set is reduced from exponential to $N-1$, as illustrated in Figure 2. To get the optimal solution, we just need to compute the $N-1$ possible values, and then select the minimum one through a linear search. Formally, we define $score(i)$ as the objective function of the i-th feasible cut:

$$score(i) = Mcut(\{1, 2, \cdots, i\}, \{i+1, i+2, \cdots, N\}). \quad (2)$$

Then the cut with minimal score is the optimal solution.

3 Temporal Segmentation Algorithm

In Section 2, we only introduce the problem of partitioning a graph into two subgraphs. To segment data into more than two segments, we can partition the data recursively. In summary, the segmentation algorithm of the temporal data consists of the following steps:

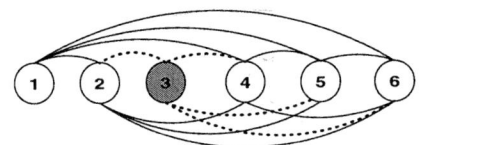
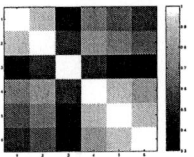

Fig. 3. *Left*: The full graph of a temporal sequence, in which the third sample is affected by noises. *Right*: The similarity matrix of the right graph

Step 1. Given a data set, construct a weighted graph $G = G(V,E)$. Treat each sample as a node and link each other by an edge.
Step 2. Compute w_{ij}, the weight of each edge, to obtain similarity matrix **W**.
Step 3. Calculate scores of the $N-1$ feasible cuts according to Equation 2. The cut with minimum score is the optimal partition.
Step 4. Recursively partition the segmented sequences if necessary.

Similar to the spectral clustering algorithms, the weight w_{ij} is usually defined as [2]:

$$w_{ij} = sim(i,j) \times \begin{cases} e^{-\frac{\|i-j\|_2^2}{\sigma^2}} & \text{if } |i-j| < \frac{r}{2} \\ 0 & \text{otherwise} \end{cases} \quad (3)$$

where r is the size of the active weight matrix, outside of which the entries contribute little to objective function. Meanwhile, it is not necessary to involve all the entries while computing each *score*. Different from Equation 2, we redefine $score(i) = Mcut(\{i-\frac{r}{2}, \cdots, i\}, \{i+1, \cdots, i+\frac{r}{2}\})$, and thus the computation is constrained in a matrix of size $r \times r$. With a temporal sequence of length N, considering the overlap of two successive active matrices, the overall time complexity is $O(Nr)$. Compared with spectral clustering methods, it is much more efficient.

Another prominent advantage of the approach is the robustness, since it makes decisions via multi-pair comparison strategy. As shown in Figure 3, the third sample is affected by noises and varies greatly from the second one. The methods based on pairwise comparison may consider it the boundary of two different segments and thus cause an over-segmentation. While with the proposed approach, the strong connectivity among the samples before and after the third one makes it unlikely to separate the sequence to two parts. Noises disturb the segmentation little.

4 Experiments

In this section, the proposed approach is implemented to perform video temporal segmentation and evaluated on the TRECVID benchmarking platform [3]. The 2003 test collections for shot boundary detection task is used for training, and the 2004 collection for testing. F_1 measure is adopted to rank the performance of

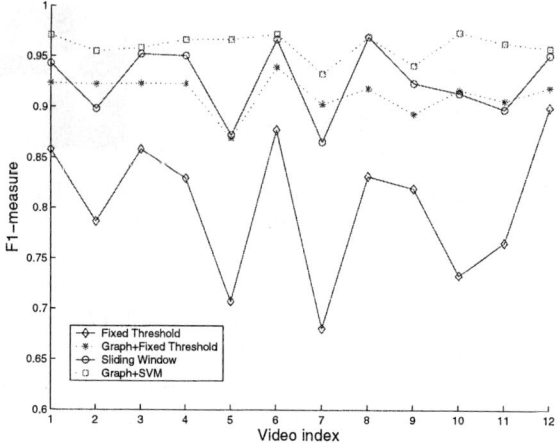

Fig. 4. Performance comparison of the four different approaches over 12 videos

the different algorithms [6]. Two graph-based methods and two non-graph ones are implemented for comparison, and they are:

1. **Fixed Threshold**: Directly compare content variation with an appropriate threshold ϵ_2. If the variation is below ϵ_1, a shot boundary is declared.
2. **Sliding Window**: Employ a window of width $2w+1$ to slide along the temporal axis, and takes λ times of the average variation within the window as ϵ_1. The other procedure is the same as the 1-st method [5].
3. **Graph+Fixed Threshold**: Directly compare the *score* with an appropriate threshold ϵ_2. If the score is below ϵ, a shot boundary is declared.
4. **Graph+SVM**: Firstly, the 3-rd approach with a low enough ϵ_2 is employed to select the boundary candidates. Then extract the score features around the candidates and train a support vector machine to distinguish boundaries and non-boundaries(abbreviated for the page limitation, see details in [7]).

Totally 10928, including 3048 positive and 7880 negative, 11-dimension samples are used to train SVM of a Gauss kernel function. The training process spends about 481 seconds and 1308 support vectors left. After the training on 2003 data set, the best parameter settings for each method are as follows: $\epsilon_1=0.85$ for Fixed Threshold, $w=4$ and $\lambda=3.0$ for Sliding Window, $r=10$ for both the graph based methods, $\epsilon_2=0.18$ for Graph+Fixed Threshold, for the SVM method,the standard deviation $std=1.0$. Their performance on the 2004 data set is depicted in Figure 4 and the elapsed time is in Table 1. Of the 12 videos, the fifth and the seventh video have more abrupt illumination changes and more great object or camera movement. As Figure 4 shows, non-graph based methods suffer evident drops of performance over these difficult videos. Both the graph-based methods perform excellently and robustly. In Graph+Fixed Threshold, even employing a single global threshold ϵ_2, the performance is comparable

Table 1. Elapsed time of the four algorithms (Unit: second). They are all evaluated on an Intel Pentium 1.51GHz machine with 256M memory. The 12 testing videos consist of total 583,623 frames

Fixed Threshold	Sliding Window	Graph+FH	Graph+SVM
25.6	26.0	74.3	100.3

to that of Sliding Window. Graph+SVM method performs best on all the videos but spends about three times of time more than non-graph based ones.

5 Conclusions and Discussions

In this paper, we propose a novel robust temporal data segmentation algorithm based on graph partition model. The method segments temporal sequence via multi-pair comparison strategy, and therefore is more robust to various noises than the existing pairwise comparison ones. By importing temporal constraints, linear time algorithm is designed to seek the optimal graph partition. The experiments on the TRECVID benchmarking platform show that the proposed approach is more robust. However, this paper has only focused on detecting abrupt transitions. To deal with gradual transitions, incorporating multi-scale strategy may be a promising direction.

References

1. E. Keogh, et al, Segmenting time series: a survey and novel approach, Data Mining in Time Series Databases, World Scientific Publishing Company, 2003
2. C. Ding, et al, A min-max cut algorithm for graph partitioning and data clustering. Proc. IEEE Int'l Conf. Data Mining, pp.107-114, 2001
3. A. F. Smeaton, et al, TRECVID: Evaluating the effectiveness of information retrieval tasks on digital video. In Proceedings of the ACM MM'04,pp. 652-655,2004
4. M.R. Naphade,et al, A high-performance shot boundary detection algorithm using multiple cues, International Conference on Image Processing, vol.3, 1998
5. Boon-Lock Yeo and Bede Liu, Rapid scene analysis on compressed video, IEEE Transactions on Circuits and Systems for Video Technology, vol.5, pp.533-544, 1995
6. C. J. van Rijsbergen. Information Retireval. Butterworths, London, 1979
7. Jinhui Yuan, et al, A robust temporal data segmentation method base on graph partition model, Technical Report, Department of Computer Science and Technology, Tsinghua University, 2004

Accurate Symbolization of Time Series

Xinqiang Zuo and Xiaoming Jin

School of Software,
Tsinghua University, Beijing, China
zuoxq04@mails.tsinghua.edu.cn
xmjin@tsinghua.edu.cn

Abstract. Symbolization is a useful method for mining time series. As our experimental results demonstrated, the previous methods are not accurate enough due to their limitations in handling a prevalent kind of time series in which similar movements are often with different lengths. This paper considers the accuracy issue of symbolization of time series. We propose a novel approach that emphasizes the meaning of each movement in the time series, regardless of the length or shift of it. To make the proposed approach more practicable, we also provide a semiautomatic method for setting the parameters. The nature of the problem and the performance of our approach had been analyzed on both real data and synthetic data. Experimental results justified the superiority of our approach over the previous one and gave some useful empirical conclusions.

1 Introduction

Recently, there has been a renewed interest in managing and mining time series [1, 2, 3]. Symbolization of time series is a very useful tool in this field, and had been extensively studied for various data mining tasks [2, 3, 4, 5, 6, 7]. Firstly, it is very important in analyzing time series on the high level. Secondly, it provides a tie between the task of mining time series and most techniques that were originally designed for handling symbolic sequence.

Briefly, symbolization can be viewed as a process that classifies each individual subseries into a typical subseries. Obviously, its performance can be evaluated by its capability in accurately retrieving and representing the subjective notion on the movements in time series. A simple solution is to choose the typical subseries manually based on the domain expert's analysis and explanation, and then use this domain knowledge in symbolizing time series. However, manual extraction of the typical subseries is usually a very difficult, or even impossible task in many applications. Another well known method is to automatically discover the typical subseries by clustering all the subseries in a sliding window with fixed size [2]. Such method can symbolize time series in the unsupervised way. However, an implicit assumption that all the similar movements are with the same length is essential for the development of this kind of methods. Intuitively, this is not the case for most real-world applications where there are many movements in time series that are similar but with different lengths. Therefore, as we demonstrated, the methods that considering sliding windows with fixed size cannot symbolize time series accurately. Moveover, selection of the window size is a very difficult task, especially when little about it can be known apriori. Our experiments show that when the window size is not correctly set, the performance will reduce to the results of random selection.

In this paper, we propose a novel approach to solve the problem above. The key idea of our approach is to adjust the window size dynamically according to the data. This strategy enables us to focus on each individual subseries with relatively "optimal" length rather than a fixed one, whereupon similarity measurement that allows an elastic shifting of the time axis can be applied. The performance of our approach was evaluated on both synthetic and real data. The experimental results justified the superiority of our approach.

2 Problem Descriptions

A time series $T = T(1), T(2), ..., T(N)$ is a sequence of real values in which each value corresponds to a time point. $T(n)$ stands for the value at n-th sampling time. The value might be of various dimensions. $|T| = N$ denotes the length of T. The subseries from time s to time e is defined as $T[s, e] = T(s), T(s+1), ..., T(e)$.

Symbolization of time series can be formally described as: Given a time series T, symbolization is to convert T into a symbolic sequence $S = S(1), S(2), ..., S(M)$, where each $S(i)(1 \leq i \leq M)$ coming from a predefined alphabet Σ represents the movement (or content) of one subseries.

Since we focus on the accuracy of symbolization of time series, we apply the standard F_1 model as the measurement of it. Given the resulting symbolic sequence of symbolization S_g and the "true" symbolic sequence corresponding to the time series S_t, the F_1 measure combines $recall(r)$ and $precision(p)$ with an equal weight in the following form: $F_1(r, p) = 2rp/(r+p)$ where r is defined to be the ratio of the "true" symbols in S_g by the total number of symbols in S_t, i.e. $r = \text{LCS}(S_g, S_t)/|S_t|$, and p is the ratio of the "true" symbols in S_g by the total number of symbols in S_g, i.e. $p = \text{LCS}(S_g, S_t)/|S_g|$. LCS is the longest common subsequence of the two sequences defined as follows: Given two sequences A and B, and a parameter ε,

$$\text{LCS}_\varepsilon(A, B) = \begin{cases} 0 & A = \phi \vee B = \phi \\ 1 + \text{LCS}_\varepsilon(\text{H}(A), \text{H}(B)) & (A(|A|) - B(|B|)) < \varepsilon \\ \max\{\text{LCS}_\varepsilon(\text{H}(A), B), \text{LCS}_\varepsilon(A, \text{H}(B))\} & \text{otherwise} \end{cases}$$

3 Symbolization Approach

Our approach is to expand a subseries tentatively, and determine whether the current subseries should be classified into a category, or a new category contains it should be formed, according to the distances between it and each category. In

```
1.  S = ϕ, s = 1, e = W_min  //S is the symbolic sequence.
    //s and e is the start and end of the tentative subseries.
2.  while(e ≤ |T|) //If T has not been processed completely.
3.      T_c=the normalization result of (T[s,e])
4.      mindist=d_avg(T_c,the nearest category C)
5.      if(mindist ≤ d_c)
6.          classify T[s,e] into C
7.          S=S+Σ_i //Σ_i is the symbol corresponding to C.
8.      else
9.          if(mindist ≥ d_c or |e − s + 1| ≥ W_max)
10.             generate a new category C_n that contains T[s,e]
11.             S=S+Σ_n //Σ_n is the symbol appointed to C_n.
12.         else
13.             e = e+ΔL //the tentative subseries is expanded ΔL steps.
14.     if(classifying or generating subprocess has been operated)
15.         s = e  e = s + W_min − 1 //re-initialize the tentative subseries.
```

Fig. 1. Symbolization algorithm

order to accommodate the subseries that are similar but out of phase, we adopt dynamic time warping (DTW), which allows an elastic shifting of the time axis, as the measurement method. Given two subseries A and B, formally, $D(A, B)$ for A and B is defined as follows:

$$D(A, B) = d(A(|A|), B(|B|)) + \min\{D(H(A), H(B)), D(H(A), B), D(A, H(B))\}$$

where $d(A(|A|), B(|B|))$ is the L_1 distance between two points $A(|A|)$ and $B(|B|)$ (i.e. $d(A(|A|), B(|B|)) = |A(|A|) - B(|B|)|$). The parameters used in our approach are d_c, d_g, W_{min}, W_{max}, ΔL and Σ. d_c and d_g are the thresholds classifying subseries into the category and generating a new category respectively. W_{min} and W_{max} are used to restrict the length of subseries, and ΔL is the sliding step of the subseries at each time. Σ is the alphabet containing symbols used to represent the subseries. The overall approach is illustrated in detail as follows and formally in Fig 1.

From the beginning of the time series, we initialize the tentative subseries $T[s, e]$ with the length W_{min}. Then our approach finds the nearest category C of $T[s, e]$ such that $d_{avg}(T[s, e], C) = \frac{1}{|C|} \sum_{p \in C} D(T[s, e], p)$ is minimum ($|C|$ is the number of the subseries belong to C, and p is a subseries). If $d_{avg}(T[s, e], C) \leq d_c$ then $T[s, e]$ will be classified into C, and the symbol representing C will be appended to the symbolic sequence S which begins with empty one. When $d_{avg}(T[s, e], C) \geq d_n$ or $|T[s, e]| \geq W_{max}$, a new category C_n contains $T[s, e]$ will be formed, then a symbol from Σ will be appointed to C_n, and similarly, this symbol will be appended to S. If any of the previous two conditions is satisfied, $T[s, e]$ will be re-initialized. Otherwise, $T[s, e]$ will be expanded ΔL steps, i.e. $T[s, e]$ is expanded to $T[s, e+\Delta L]$. When the time series is processed completely, the whole symbolic sequence S has been generated.

The normalization step shown in symbolization algorithm removes the impacts of baseline and scaling factor by going beyond the absolute value and emphasizing the trend or shape. There are many normalization methods which have been proposed, e.g. normalize time series T to $T'(i) = (T(i)-\min(T))/(\max(T)-\min(T))$ where $\min(T)$ and $\max(T)$ are the minimum and maximum of T.

To make the proposed approach more practicable, we propose a semiautomatic method to set the parameters by sketching: Manually appointing similar subseries $T[s1, s2], T[s3, s4]$ and dissimilar subseries $T[d1, d2], T[d3, d4]$ in the time series, then d_c and d_n can be estimated as follows: $d_c = \mathrm{D}(T[s1, s2], T[s3, s4]) \cdot ce1$, $d_n = \mathrm{D}(T[d1, d2], T[d3, d4]) \cdot ce2$ where $ce1$ and $ce2$ are the coefficients that revise d_c and d_n respectively to remove the influence of randomization of selections, e.g. $ce1$ and $ce2$ could take values within the ranges [1,1.2] and [0.8,1]. W_{min} and W_{max} can be taken by estimating the range of the length of subseries in time series. Consider ΔL, greater ΔL will result in improvement in time performance, whereas decrease in accuracy of the algorithm. Generally it can be set, e.g. 1 to 5, based on the length of the time series and the accuracy required.

Since the symbolization method is an unsupervised process, the meaning of each typical subseries cannot be known apriori. Then Σ is selected to be a group of simple identifiers without meaning, e.g. "a"..."z", before symbolization. And then a post-processing procedure can convert the resulting symbolic sequence by rewriting each symbol to a more meaningful one provided by domain expert's explanations on the resulting typical subseries.

4 Experimental Evaluation

In this section we first compare the performance of our approach with the fixed-window method. The synthetic data set was used so that we could control the "true" symbolic sequence. The data set was generated as follows: It totally consisted of 100 time series. Each time series was connected by 100 subseries. Each subseries was randomly selected from the three basic shapes including Sine, Cosine and Straight line and then extended or contracted in the time axis and y-axis respectively. Finally Gaussian noise was added to the whole time series.

In this experiment, hierarchical clustering was used as the clustering method in the fixed-window method. And for an exhaustive comparison, each time series was symbolized for 2125 times by the fixed-window method with the various parameters including window width from 6 to 30, clustering number from 2 to 6 and start position from 1 to $w - 1$ where w is the window width and one time by our approach. So $100 \times (2125 + 1) = 212600$ times symbolizations were done by us. The parameters in our approach were set using the method described in Sect. 3 as follows: $d_c = 1.45; d_n = 5.5; W_{min} = 6; W_{max} = 30; \Delta L = 1$. F_1 measure proposed in Sect. 2 was used to evaluate the accuracy of the two methods.

The results of the symbolizations were modified in order to utilize F_1 measure. The "true" symbolic sequence S_t could be obtained during the process generating the synthetic time series. S_g was generated as follows: For the typical subseries

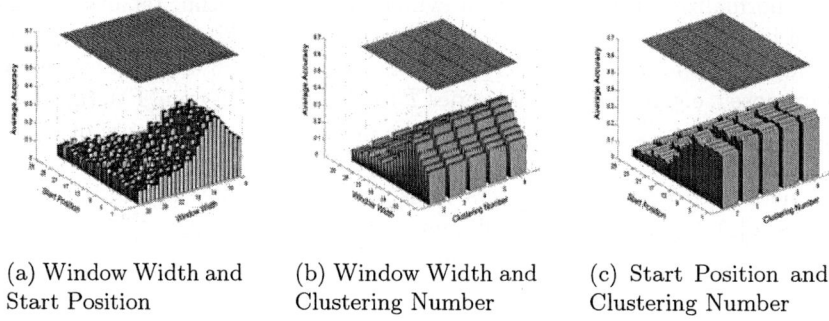

(a) Window Width and Start Position

(b) Window Width and Clustering Number

(c) Start Position and Clustering Number

Fig. 2. The comparison from three aspects, the histograms display the results of the fixed-window method, and the plane above is the results of our approach. Each histogram presents results at two dimensions, maximized in another dimension

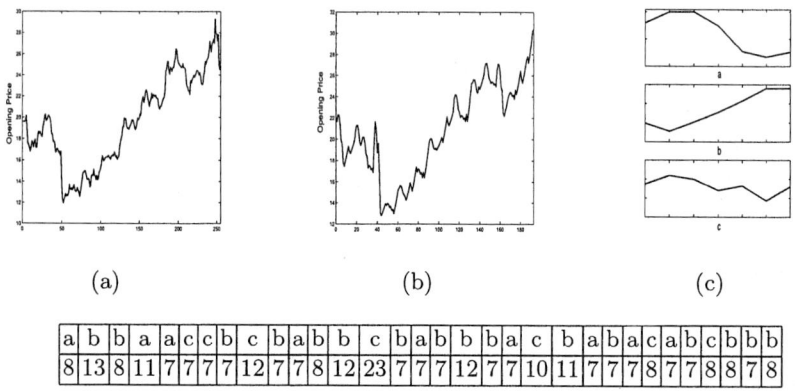

(a) (b) (c)

a	b	b	a	a	c	c	b	c	b	a	b	b	c	b	a	b	b	b	a	c	b	a	b	a	c	a	b	c	b	b	b
8	13	8	11	7	7	7	7	12	7	7	8	12	23	7	7	7	12	7	7	10	11	7	7	7	8	7	7	8	8	7	8

Fig. 3. Real data (a), the reconstructed time series(b), the typical subseries(c), and the table shows the resulting symbolic sequence, together with the lengths of the subseries in the original time series corresponding to each symbol

$T[s, e]$ of each category C_i generated, we found the nearest shape B_j from the the three basic shapes such that $D(T[s, e], B_j)$ in minimum. If $D(T[s, e], B_j) \leq d_c$, all the appearances of the symbol representing C_i in symbolic sequence S were replaced by the symbol of B_j, otherwise replaced by '$' which was not one of the symbols representing the three basic shapes.

Then we give the results of comparison between two methods. The averages of the results of 100 time series described in Fig. 2 show that our approach can yield substantially greater accuracy than the fixed-window method. There were none of the results whose accuracies were greater than our approach. We also made the comparison with the best results of the fixed-window method. We got the result that the inferiorities of our approach comparing with the best results of the method existed only at only 33 in 100 time series. Furthermore, selection of the appropriate window size to obtain the best result is a very difficult task, e.g. in this experiment. We do not display the figure to save space.

Finally, in order to gain an intuition about the effectiveness of our approach in real-world applications, we used stock price data series, which corresponded to the opening prices of the stocks from Nov. 25th in 2002 to Nov. 24th in 2003, as an instance to explain the effectiveness. Each time series consisted of 252 real numbers. Due to space limitation, we only show one group of results in Fig. 3.

The table in Fig. 3 lists the resulting symbolic sequence and the length of subseries corresponding to each symbol. Replacing each subseries partitioned with the typical subseries of the category which contained it, we could obtain a reconstructed time series shown in Fig. 3(b). The accuracy of our method can be empirically justified through the visual analysis of the similarity between the original series in Fig. 3(a) and the reconstructed version in Fig. 3(b).

5 Conclusions

In this paper, we consider the accuracy issue of symbolization of time series. And then a novel approach is proposed, together with a semiautomatic method for setting the parameters used. The experimental results justified the superiority of our approach over the fixed-window method. Finally, stock price data was used to illustrate the effectiveness of our approach in real-world applications. Our approach can be expanded in the following directions: multidimensional space, other similarity measurement, and incremental approach for streaming data. Though our approach decides the boundaries of subseries only locally, it is simple comparing with the difficulty in the global criterion, moveover, it can obtain the acceptable results demonstrated in our experiments. In the future we plan to explore a more effective method by considering the global criterion.

Acknowledgement

The work was supported by the NSFC 60403021 and the 973 Program 2004CB-719400. We thank the anonymous reviewers and E. Keogh for their helpful comments.

References

1. D. Berndt and J. Clifford. Using dynamic time warping to find patterns in time series. pages 229-248. In Proc. of KDD Workshop, 1994.
2. G. Das, K. Lin, H. Mannila, G. Renganathan, and P. Smyth. Rule discovery from time series. pages 16-22. In Proc. of KDD 98.
3. Y. zhu and D. Shasha. Fast approaches to simple problems in financial time series streams. Workshop on management and processing of datastreams, 2003.
4. X. Jin, Y. Lu, and C. Shi. Distribution discovery: local analysis of temporal rules. pages 469-480. In Proc. of PAKDD 2002.

5. Z. Yao, L. Gao, and X. S. Wang. Using triangle inequality to efficiently process continuous queries on high-dimensional streaming time series. SSDBM 2003.
6. J. Lin, E. Keogh, S. Lonardi, and B. Chiu. A symbolic representation of time series, with implications for streaming algorithms. DMKD 2003.
7. E. Keogh, J. Lin, and W.Truppel. Clustering of time series subsequences is meaningless: Implications for previous and future research. pages 115–122. ICDM 2003.

A Novel Bit Level Time Series Representation with Implication of Similarity Search and Clustering

Chotirat Ratanamahatana[1], Eamonn Keogh[1], Anthony J. Bagnall[2], and Stefano Lonardi[1]

[1] Dept. of Computer Science & Engineering, Univ. of California,
Riverside, CA 92521 USA
{ratana, eamonn, stelo}@cs.ucr.edu
[2] School of Computing Sciences, University of East Anglia,
Norwich, UK
ajb@cmp.uea.ac.uk

Abstract. Because time series are a ubiquitous and increasingly prevalent type of data, there has been much research effort devoted to time series data mining recently. As with all data mining problems, the key to effective and scalable algorithms is choosing the right representation of the data. Many high level representations of time series have been proposed for data mining. In this work, we introduce a new technique based on a bit level approximation of the data. The representation has several important advantages over existing techniques. One unique advantage is that it allows raw data to be directly compared to the reduced representation, while still guaranteeing lower bounds to Euclidean distance. This fact can be exploited to produce faster exact algorithms for similarly search. In addition, we demonstrate that our new representation allows time series clustering to scale to much larger datasets.

1 Introduction

Time series are a ubiquitous and increasingly prevalent type of data. Because of this fact, there has been much research effort devoted to time series data mining in the last decade [1],[2],[3],[4]. As with all data mining problems, the key to effective and scalable algorithms is choosing a suitable representation of the data. Many high level representations of time series have been proposed for data mining. In this work, we introduce a novel technique based on a bit level approximation of the data. As we will show, our clipped representation has several important advantages over existing techniques. The proposed approach is not only a new representation; it is a new type of representation. For data adaptive, non-data adaptive, and model-based approaches, the user has a choice (implicit or explicit) of the compression ratio. This allows the user to fine tune the parameters to achieve the ideal compression/ fidelity tradeoff for their particular application.

In contrast, with the clipped representation, the data *itself* dictates the compression ratio; the user has no choice to make. This may be seen as somewhat of a disadvantage (although removing parameters from a data mining task is often a good thing

[5]). However, this lack of flexibility is counterbalanced by another unique property of the clipped representation. For all other dimensionality reduction approaches, we must transform the query into the same representation as the dimensionality reduced database, i.e. having a loss of fidelity for the candidate matches stored in the index and a loss of fidelity for the query. This in turn produces weak lower bounds, and thus weak pruning power. In contrast, the clipped representation is unique in that the original raw query can be compared directly to the clipped candidate sequences, thus producing tighter lower bounds, greater pruning power and faster query by content.

2 The Clipped Representation

Our proposed representation works by replacing each real valued data point with a single bit. gives the visual intuition.

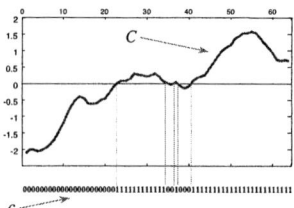

Fig. 1. A time series, C, of length 64, is converted to the clipped representation, c, by observing each element of C; if its value is strictly above zero, the corresponding bit is set to 1, and to 0 otherwise

More formally, we can define c, the clipped representation of C as:

$$c(i) = \begin{cases} 1 & \text{if } C(i) > \mu \\ 0 & \text{otherwise} \end{cases} \quad (1)$$

where μ is the mean of C. Since the importance of normalizing the data before attempting any clustering, classification or indexing [3] is well-established, we can simply assume $\mu = 0$, without loss of generality for the rest of this work. Note that this representation has been considered before in the statistical community [6], but its utility for data mining, namely, the ability to lower bound distance functions, is first documented here.

2.1 Lower Bounding Euclidean Distance

Suppose we have 2 time series, a query $Q = Q_1, Q_2, \ldots, Q_i, \ldots, Q_n$, and a candidate match $C = C_1, C_2, \ldots, C_j, \ldots, C_n$. The Euclidean Distance can simply be used to compare the two time series. However, if we have a clipped time series c, and a raw time series Q, we can also lower bound the squared Euclidean distance between C and Q, using equation 2) below. Due to space limitations, the proof of this *LB_clipped* is omitted and can be found in [7]. However, gives its visual intuition.

$$LB\_clipped(Q,c) \equiv \sum_{i=1}^{n} \begin{cases} Q_i^2 & \text{if } (Q_i > 0 \text{ and } c_i = 0) \text{ or} (Q_i \le 0 \text{ and } c_i = 1) \\ 0 & \text{Otherwise} \end{cases} \quad (2)$$

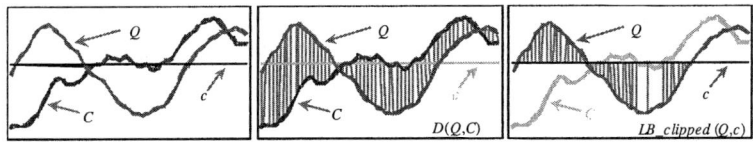

Fig. 2. The distance returned by both LB_clipped(Q, c) and $D(Q,C)$ is the sum of squared lengths of the gray hatch lines. Because every hatch line for $LB\_clipped(Q,c)$ is matched with corresponding line in $D(Q,c)$ which is at least as long, we must have $LB\_clipped(Q,c) \le D(Q,c)$

2.2 Run Length Encoding

Consider the clipped sequence c, which we have been using as a running example. Its value is **0000000000000000000000111111111111001000111111111111111111111111**. Note that we could write this as 22#**0**, 11#**1**, 2#**0**, 1#**1**, 3#**0**, 24#**1**, which we can interpret as 22 zeros followed by 12 ones, etc. The shorter format allows us to fit more data in main memory. In fact, we can be even terser; because we always toggle from zero to one or vice versa, so we only need to record the parity of the first bit, giving us 22#**0**, **11,2,1,3,24**. This classic lossless compression technique is known as Run Length Encoding (RLE). To make the representation even shorter, we can represent the parity bits of 0 and 1 with two special characters, e.g. "@" and "!," respectively; our run length encoding now can be represented as @**22,11,2,1,3,24**. We can use this to further reduce the clipped representation of the data. Note that while the example above illustrates the idea with ASCII characters, we actually do RLE at the bit level.

2.3 Numerosity Reduction

Even though the run length-encoding scheme itself gives an impressive compression ratio, we can improve it by numerosity reduction on sliding windows. This step is motivated by observing that while applying a sliding window on the streaming data, time series in consecutive sliding windows are very often identical in the clipped representation, except for the first and the last values that are omitted and added, respectively. If the time series in each sliding window has this property, we can exploit this fact and just record the maximum amount of time this property has consecutively been observed, along with a special character, $, that represents this reduction. Consider the run length encoding from our example in the previous section and let the encoding of the next five sliding windows be:

@22,11,2,1,3,24@*21*,11,2,1,3,*25*@*20*,11,2,1,3,*26*@*19*,11,2,1,3,*27*@*18*,22,2,1,3,27,1@*17*,22,2,1,3,2 7,*2*.

We can readily see that the first four windows are very similar and can be reduced to one since the only values differ from each other are the first and the last (italicized for clarity). However, the 5th window cannot be combined with the previous one since the last bit has changed from 1 to 0, but it can be combined with its next window. As a result, the final encoding with numerosity reduction becomes @22,11,2,1,3,24$3@18,22,2,1,3,27,1$1.

As before, although we demonstrate the idea with ASCII text, we actually encode everything at the bit level. With the Power Demand dataset of size 10,000 data points, numerosity reduction together with Huffman coding yields a huge compression ratio of 1057:1. Note that while the factor of 32 to 1 achieved by clipping is lossy, the remaining factor of approximately 33 to 1 is lossless with respect to the clipped data.

3 Empirical Evaluations

In this section, we will provide an extensive empirical comparison among the raw and various representations of compressed data in two major data mining tasks, time series indexing and clustering. Twelve datasets were used in our indexing experiments, and two were used for clustering experiments (only subsets of results are shown here due to space limitations). We also tested on a wide range of both real and synthetic datasets. The datasets range from 66 Kilobytes to 2 Gigabytes in size (see [7] for complete details).

3.1 Experimental Methodology

For indexing, we will demonstrate the superiority of our clipped representation in terms of number of disk accesses. We compare our proposed method with the classic Piecewise Aggregate Approximation (PAA) and Discrete Fourier Transform (DFT), all preserving similar compression ratio. We then demonstrate that clipped series can produce clusters similar to those obtained with the raw data when clustering a very large real world database introduced in section 3.3. We show that clipping performs favorably when compared to clustering with unclipped data since clustering can be done faster and with much less memory requirement.

For similarity search, we performed all experiments over a range of query lengths. Since we want to include PAA in our experiments, the query length is somewhat limited. We therefore consider query lengths of 256 and 512 data points. We tested our approach on a variety of twelve datasets with various properties within the data, obtained from the UCR Time Series Data Mining Archive [8]. The sizes of the datasets range from 6,875 data points to 198,400 data points. Leaving-one-out cross validation is used; on each run, we randomly pick a query from a database, create a run-length encoding with numerosity reduction for the rest of the data, and determine the resultant compression ratio. We then create PAA and DFT on the same data and with the same compression ratio (or with smaller compression ratio, in favor of PAA and DFT) then measure the number of random disk accesses for the nearest neighbor queries of all methods. To determine the number of dimensionality reduction (m) in PAA and DFT in these cases, we assume that each value in PAA and DFT can be represented by only *two* bytes (instead of 4 or 8 bytes) to demonstrate that our results are still competitive among all the approaches. In addition, to avoid any possibility of implementation bias, the number of I/O disk accesses of each method is measured instead of recording the actual running time. This is done by first computing the lower bound distances using LB_clipped and Euclidean distance, between a query and all the sequences in the dataset. Then to retrieve the nearest neighbor, each sequence is visited in the order according to the lower bound values. We count the number of times the real disk accesses must be made. These numbers also indicates the tightness of the lower bounds for each representation. The results are averaged over 100 separate runs for each dataset. For simplicity, we only report results for one-nearest neighbor queries.

3.2 Indexing Results

As noted above, the amount of compression is dictated by the data itself. For the twelve datasets considered the compression ratios range between 60.2:1 to 1,089.5:1. We compare different representations in terms of I/O random disk accesses during the process of the 1-nearest neighbor retrieval of a query time series. In particular, in each run, we reduce the dimensionality of the data from n to m using Clipped, PAA, and DFT representations, and build their indices on the reduced spaces based on their lower bounds between each subsection (sliding window) of the time series and the query. To allow a visual comparison, we normalize each experiment on each dataset by the worst performing algorithm; the raw numbers are available in [7]. Fig. 3 shows the number of disk accesses with lower bounding the Euclidean distance, using the three dimensionality-reduction techniques over the range of query lengths of 256 and 512 data points. In general, the results show that the clipped representation greatly outperforms or at least is comparable to the other approaches, expressing the superiority in its tightness of the lower bounds. Again, we would like to emphasize that our results here are obtained by conservatively assuming only *two*-byte requirement to represent each number in PAA and DFT. If we assume 4 or 8 bytes or without the parameter m adjusted, the results will be much improved.

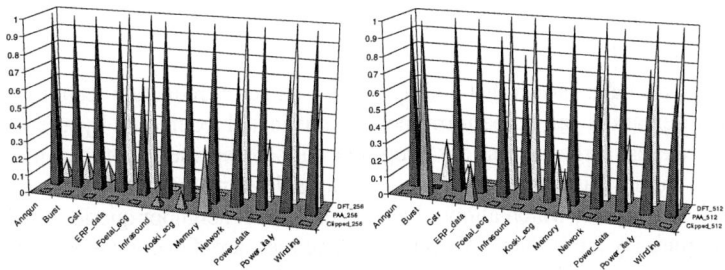

Fig. 3. Number of disk accesses with lower bounding of Euclidean distance, normalized by the worst performing approach, using the 3 representations for query lengths of 256 and 512 points

3.3 General Compression-Based Clustering

We examine a class of problems where a DFT approach should produce good results, and show that clipping is better than the most commonly used DFT approach described in [2].

To demonstrate how clipping can help with a real world large dataset, we cluster optical recording data from a bee's olfactory system [9]. The data consists of 980 images, each image containing of 688x520 measurements. If we consider each position in the image as a time series, the data consists of 357,760 time series of length 980. Preliminary analysis has shown that clustering the series based on similarity in time produces results that have a sensible physiological interpretation [9]. We cluster with k-means (with k set to 16) restarted 50 times from random initial centroids, and take as the best clustering the one with the lowest within-cluster variation.

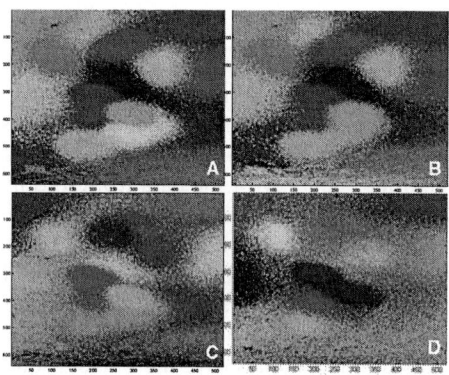

Fig. 4. A) 16 clusters produced using all 2GB of raw data. B) Clusters formed using the clipped data with 32:1 compression ratio. The spatial cluster co-occurrences between this plot and A) shows its effectiveness in the clipped data reduction technique. C) Clusters formed using PAA with 59 coefficients, giving 20:1 compression ratio. D) Clusters formed using first 17 DFT coefficients, giving 29.7:1 compression ratio

4 Conclusions

In this paper, we have shown that a simple dimensionality reduction technique, i.e. the clipped representation, can outperform more sophisticated techniques by a few orders of magnitude. We have shown that our proposed clipped representation can improve the compression ratio by a wide margin, while being able to maintain or increase the tightness of its lower bound, which allows even faster nearest neighbor queries, especially in ones that require Dynamic Time Warping distance measure. Other than producing faster exact algorithms for similarity search, we have also demonstrated that our clipped representation approach can support clustering and scale to much larger datasets.

Acknowledgements. This research was partly funded by the NSF under grant IIS-0237918.

References

1. Aach, J. & Church, G. Aligning gene expression time series with time warping algorithms. Bioinformatics(17) (2001) 495-508.
2. Berndt, D., Clifford, J. Using dynamic time warping to find patterns in time series. AAAI-94 Workshop on Knowledge Discovery in Databases. (1994) 229-248.
3. Keogh E., Kasetty, S. On the Need for Time Series Data Mining Benchmarks: A Survey and Empirical Demonstration. In the 8th ACM SIGKDD (2002) 102-111.
4. Yi B K., Faloutsos, C. Fast time sequence indexing for arbitrary Lp norms. VLDB (2000) 385-94.
5. Keogh, E., Lonardi, S., Ratanamahatana, CA. Towards Parameter-Free Data Mining. In proceedings of SIGKDD (2004).
6. Kedem B., Slud, E. On Goodness of Fit of Time Series Models: An Application of Higher Order Crossings, Biometrika, Vol 68, (1981) 551-556
7. Ratanamahatana, C.A., Keogh, E., Bagnall, A.J., Lonardi, S. A Novel Bit Level Time Series Representation with Implication of Similarity Search and Clustering. (2004) [http://www.cs.ucr.edu/downloads/techrpt/TR_clippedpaper.pdf].
8. Keogh E., Folias, T. The UCR time Series Data Mining archive. (2002) [http://www.cs.ucr.edu/~eamonn/TSDMA].
9. Galan, R.F., Sachse, S., Galizia, C.G., Herz, A.V.M. "Odor-driven attractor dynamics in the antennal lobe allow for simple and rapid olfactory pattern classification." Neural Computation (2004)
10. Bagnall, A. J., Janacek, G. Clustering time series from ARMA models with clipped data, SIGKDD (2004).

Finding Temporal Features of Event-Oriented Patterns

Xingzhi Sun, Maria E. Orlowska, and Xue Li

School of Information Technology and Electrical Engineering,
The University of Queensland, QLD, 4072, Australia
{sun, maria, xueli}@itee.uq.edu.au

Abstract. A major task of traditional temporal event sequence mining is to predict the occurrences of a special type of event (called target event) in a long temporal sequence. Our previous work has defined a new type of pattern, called event-oriented pattern, which can potentially predict the target event within a certain period of time. However, in the event-oriented pattern discovery, because the size of interval for prediction is pre-defined, the mining results could be inaccurate and carry misleading information. In this paper, we introduce a new concept, called temporal feature, to rectify this shortcoming. Generally, for any event-oriented pattern discovered under the pre-given size of interval, the temporal feature is the minimal size of interval that makes the pattern interesting. Thus, by further investigating the temporal features of discovered event-oriented patterns, we can refine the knowledge for the target event prediction.

1 Introduction

An important task of temporal event sequence mining is to predict the occurrences of a special type of event, called target event, in a long sequence. Substantial work [1–4] has been done for target event prediction in a long temporal event sequence. Particularly, in our previous work [3], we have defined a new type of pattern, called event oriented pattern, to address this research problem. In general, the discovered event-oriented pattern P is represented as the prediction rule $r = \left\{ P \stackrel{T}{\Longrightarrow} e \right\}$, where e is target event type and T is the temporal constraint indicating the time period for target event prediction. The prediction rule r can be interpreted as: if pattern P occurs, the target event is likely to occur within a T-sized interval.

The temporal constraint T plays an important role in the target event prediction because it is closely related to the sensibility of the prediction rule. So, in our event-oriented pattern discovery problem, T is pre-defined by domain experts to ensure that all prediction rules discovered under the parameter T are sensible in the given application domain. However, in the event-oriented pattern discovery, an important observation (as will be explained later) is that if a rule $P \stackrel{T_1}{\Longrightarrow} e$ is interesting in term of the interestingness measures, for any $T_2 > T_1$, the rule $P \stackrel{T2}{\Longrightarrow} e$ is interesting as well. Thus, pre-defining T may cause the side-effect on the accuracy of prediction.

Let us consider the following example in the application of telecommunication network fault analysis. Assume that an event-oriented pattern P can lead to the trouble report (TR) within 12 hours. However, in the process of pattern discovery, domain experts set the temporal constraint T as 24 hours (which is believed sensible to the application domain). According to the observation, P is discovered to be interesting and a corresponding rule r is in the form of $P \stackrel{24hrs}{\Longrightarrow} e$. Compared with the fact $P \stackrel{12hrs}{\Longrightarrow} e$, the rule r is lack of accuracy in term of the size of interval for prediction. Such inaccurate prediction rules sometimes carry misleading information. For example, if domain experts take 24 hours as the interval in which P will lead to TRs, they may underestimate the emergency that TRs will occur and therefore, fail to take corresponding precautions in time.

In this paper, we introduce a novel concept, called temporal feature, to address the above problem. Given an event-oriented pattern P, the temporal feature of P, denoted as $T^{(P)}$, is defined as the minimal size of interval that makes P interesting in terms of the interestingness measures. Based on this new concept, we can solve the problem in two steps. First, as done in our previous work, a set of event-oriented patterns are discovered under a given size of interval T, where T is believed as the maximal sensible size of interval for prediction in the application domain. Then, we further analyze the mining result, i.e., for each interesting pattern P (discovered under the parameter T), we try to find its temporal feature $T^{(P)}$. In this case, the corresponding prediction rule of P is refined from $P \stackrel{T}{\Longrightarrow} e$ to $P \stackrel{T^{(P)}}{\Longrightarrow} e$, which carries more accurate temporal information for the target event predication.

The rest of this paper is organized as follows. Section 2 gives the background of mining event-oriented patterns. In Section 3, the problem of discovering temporal features is formulated and further discussed. In Section 4, we propose approaches to find temporal features of event-oriented patterns. Section 5 shows the experiment results. Finally, we conclude this paper in Section 6.

2 Background

In this section, we review some key concepts of discovering event-oriented patterns in a long temporal sequence.

Let us consider a finite set E of *event types*. An *event* is a pair (a, t) where $a \in E$ and t is a *timestamp*. We define one special event type e as the *target event type*. Any event of type e is called a *target event*. A *temporal event sequence*, or *sequence* in short, is a list of events totally ordered by their timestamps.

Let a *window* w be $[t_s, t_e)$, where t_s and t_e are the start time and the end time of w respectively. The *window size* of w is defined as $t_e - t_s$. A *sequence fragment* $f(S, w)$ is the part of sequence S determined by w.

A *pattern* P is defined as either a set or a list of event types. A sequence fragment $f(S,w)$ is called to *contain* a pattern P if there exist events in $f(S,w)$ that can match P. Given a pattern P and a sequence S, the approach in [1] can identify all the occurrences, called *minimal occurrences (OWs)*, of P in S.

A *prediction rule* is in the form of $r = \left\{ P \stackrel{T}{\Longrightarrow} e \right\}$, where e is the target event type, P is a *event-oriented pattern*, and T is the size of interval. Generally, a prediction rule can be interpreted as: the occurrence of P is likely to lead to the target event in a T-sized interval.

To evaluate the significance of prediction rules, we use *support* and *confidence*. Before giving the definition of support and confidence, we first formally define the dataset prior to target events.

Given a sequence S, for each target event (e, t_i) in S, we establish a T-sized window $w_i = [t_i - T, t_i]$ and get the sequence fragment $f_i = f(S, w_i)$. The set of these sequence fragments $D = \{f_1, f_2, \ldots, f_m\}$ is called the *local dataset* of target event type e.

Definition 2.1 (Support). Given a sequence S and a prediction rule $r = \left\{ P \stackrel{T}{\Longrightarrow} e \right\}$, the *support* of r is defined as $Supp(r) = \frac{|\{f_i \in D | f_i \text{ contains } P\}|}{|D|}$, where D is the local dataset of e.

Definition 2.2 (Confidence). Given a sequence S and a prediction rule $r = \left\{ P \stackrel{T}{\Longrightarrow} e \right\}$, the *confidence* of r is defined as $Conf(r) = \frac{\sum_i \# \text{ MO in } f_i}{\# \text{ MO in } S}$, where f_i is the sequence fragment in the local dataset D.

The formal definition of finding event-oriented pattern is presented as follows: given a sequence S, a target event type e, a window size T, two thresholds s_0 and c_0, to find any event-oriented pattern in the form of $r = \left\{ P \stackrel{T}{\Longrightarrow} e \right\}$ such that $Supp(r) \geq s_0$ and $Conf(r) \geq c_0$.

3 Temporal Features

In this section, we first give the formal definition of the temporal feature. Then, we prove that the temporal feature exists for any interesting event-oriented pattern. Finally, we formulate the problem of finding temporal features.

Definition 3.1 (Temporal feature). Given an interesting event-oriented pattern P (discovered from the sequence S under a window size T and threshold s_0 and c_0), the *temporal feature* of P, denoted as $T^{(P)}$, is the minimal window size that makes P interesting, i.e., 1) both $Supp(P \stackrel{T^{(P)}}{\Longrightarrow} e) \geq s_0$ and $Conf(P \stackrel{T^{(P)}}{\Longrightarrow} e) \geq c_0$ hold, and 2) there does not exist $T^{(P)'} < T^{(P)}$, satisfying $Supp(P \stackrel{T^{(P)'}}{\Longrightarrow} e) \geq s_0$ and $Conf(P \stackrel{T^{(P)'}}{\Longrightarrow} e) \geq c_0$.

To prove the existence of temporal feature, we first give the following property of support and confidence.

Claim 3.1 (Monotony of support and confidence with window size T). Given a pattern P, for any two window sizes T_1 and T_2 with $T_1 \geq T_2$, we have $Supp(P \stackrel{T_1}{\Longrightarrow} e) \geq Supp(P \stackrel{T_2}{\Longrightarrow} e)$ and $Conf(P \stackrel{T_1}{\Longrightarrow} e) \geq Conf(P \stackrel{T_2}{\Longrightarrow} e)$.

Proof. The proof is straightforward according to Definitions 2.1 and 2.2. □

Claim 3.2 (Existence of temporal feature). For any interesting pattern P (discovered from the sequence S under a window size T and threshold s_0 and c_0), the temporal feature $T^{(P)}$ exists.

Proof. First, for a given pattern P, we define the *temporal feature w.r.t support* as the minimal window size that makes the pattern P interesting in terms of support (i.e., the support of P is no less than the given support threshold). Similarly, the *temporal feature w.r.t. confidence* is defined as the minimal window size that makes P interesting in terms of confidence. Thanks to Claim 3.1, for any interesting pattern P discovered under the window size T, there exist the temporal feature w.r.t support and the temporal feature w.r.t confidence, denoted as $T_{Supp}^{(P)}$ and $T_{Conf}^{(P)}$ respectively. According to Definition 3.1, the temporal feature of P is $Max\{T_{Supp}^{(P)}, T_{Conf}^{(P)}\}$. □

Problem Definition. The problem of finding temporal features can be formulated as: given a set PS of interesting pattern event-oriented patterns (discovered from the sequence S under a window size T and threshold s_0 and c_0), for any pattern $P \in PS$, to refine the corresponding prediction rule $P \stackrel{T}{\Longrightarrow} e$ to a new rule $P \stackrel{T^{(P)}}{\Longrightarrow} e$, where $T^{(P)}$ is the temporal feature of P.

4 Finding Temporal Features

In this section, we discuss how to find the temporal feature of a single interesting pattern P. According to the proof of Claim 3.2, our task is naturally decomposed into two sub-tasks, i.e., finding temporal feature w.r.t. support and finding temporal feature w.r.t. confidence.

Finding Temporal Feature w.r.t. Support. Suppose that the total number of target events is m. For each target event $v_i = (e, t_i)$ where i is from 1 to m, we first find a window size T_i such that T_i is the minimal window size that makes the sequence fragment in $[t_i - T_i, t_i)$ contain P. We call T_i the *initial window size* of the target event v_i in terms of the pattern P. In this case, for all target events from v_1 to v_m, we can create a vector of initial window size (in terms of P), denoted as $VS^{(P)}$. Then, we could sort $VS^{(P)}$ on the ascending order, and by definition, the temporal feature w.r.t. support $T_{Supp}^{(P)}$ should be the $([s_0 * m] + 1)$-th element of $VS^{(P)}$, where s_0 is the support threshold.

Finding Temporal Feature w.r.t. Confidence. Suppose the total number of occurrences of a pattern P in S is N. Let us consider one target event $v_i = (e, t_i)$. Suppose that P occurs k times in the interval $[t_i - T, t_i)$, where is T the pre-given window size for discovering P. We can find k special window sizes $(T_1, T_2...T_k)$ such that for each T_j $(j = 1 \cdots k)$, 1) P occurs j times in the sequence fragment in $[t_i - T_j, t_i)$ and 2) T_j is the minimal window size that holding condition 1). For all target events, we can create a vector $VC^{(P)}$ which consists of all such special window sizes in terms of pattern P. After sorting $VC^{(P)}$ on the ascending order and $T_{Conf}^{(P)}$ should be the $([c_0 * N] + 1)$-th element of $VC^{(P)}$, where c_0 is the confidence threshold.

Improvement. Suppose that the temporal feature w.r.t. support of P, $T_{Supp}^{(P)}$, has been found. According to Definition 3.1, the temporal feature $T^{(P)}$ should be a value in the interval $[T_{Supp}^{(P)}, T]$, where T is the pre-given window size. Based on this observation, we do the improvement as follows. Let N_{Supp} be the number of occurrences of P in the $T_{Supp}^{(P)}$-sized windows before target events. During the reverse scan for creating the vector $VC^{(P)}$, for any special window with size $T_i <= T_{Supp}^{(P)}$, we increase N_{Supp} by 1 but do not record T_i in the $VC^{(P)}$. After the scan, if N_{Supp} is no less than $[c_0 * N] + 1$, there is no need to compute $T_{Conf}^{(P)}$ as we already find the temporal feature $T^{(P)} = T_{Supp}^{(P)}$. Otherwise, let $N_\triangle$ be $[c_0 * N] + 1 - N_{Supp}$. Then the $N_\triangle$-th smallest element in $VC^{(P)}$ is $T_{Conf}^{(P)}$ (note that it is also the temporal feature $T^{(P)}$).

5 Empirical Results

In this section, we the show experiment results of the application of telecommunication network fault analysis. The telecommunication event database contains 120,312 events, covering 190 event types. The population of target events is 2,317. In the experiment, all patterns to be further investigated are discovered under the conditions $T = 12hrs$, $s_0 = 5\%$, and $c_0 = 20\%$. The experiment results show that the temporal features of those event-oriented patterns are ranged from 40% to 97% as large as T. Table 1 gives some samples of our mining results, illustrating that how the temporal features improve the accuracy of prediction rules.

Let us take the first pattern $\{A1, E2\}$ as an example. The initial prediction rule $\{A1, E2\} \stackrel{12hrs}{\Longrightarrow} e$ with support 24.46% and confidence 40.04%. However, with the thresholds $s_0 = 5\%$ and $c_0 = 20\%$, a window size of 4.83 hrs can already makes $\{A1, E2\}$ interesting. That is, the rule $\{A1, E2\} \stackrel{4.83hrs}{\Longrightarrow} e$ is believed to be significant enough for predicting the target events. Compared with the initial predicting rule, the size of prediction interval is only 40.25% as large as T. So,

Table 1. Examples of temporal features

P	T (hrs)	$Supp(P \overset{T}{\Rightarrow} e)$	$Conf(P \overset{T}{\Rightarrow} e)$	T_P (hrs)	$\frac{T_P}{T}$
$\{A1, E2\}$	12	24.46%	40.04%	4.83	40.25%
$\{E1, F1\}$	12	7.32%	25.56%	6.32	52.67%
$\{A2, F1, F2\}$	12	5.03%	22.17%	8.15	67.92%

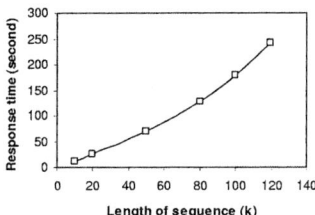

Fig. 1. Performance evaluation

the new prediction rule is more accurate and sensible in term of the size of interval for prediction. Another observation is that for some prediction rules whose interestingness measures are close to the thresholds (e.g., the prediction rule of pattern $\{A2, F1, F2\}$), finding temporal features could also considerably reduce the size of prediction interval, with marginal decrease on the values of interestingness measures.

For the efficiency study, we report the response time for finding the temporal features from sequences with different length. Figure 1 shows such a result under the condition $T = 12hrs$, $s_0 = 5\%$, and $c_0 = 20\%$. We could see our approach is not linearly scalable in terms of the length of sequence.

6 Conclusions

As an extension of our previous work, this paper aims to improve the accuracy of the prediction rule in terms of the temporal constraint. We defined a new concept called temporal feature, and formulated the research problem of finding temporal features. Approaches was proposed to solve the identified problem. We also reported the experiment results of a real dataset.

The significance of this work is summarized as two points. First, rather than taking the pre-given window size for granted, we improve the accuracy of the prediction rules by minimizing the size of the prediction interval. It is also interesting to extend such a concept to other type of knowledge for perdition. A question could be: does there exist the minimal window size to make the knowledge interesting? Secondly, the identified problem can be utilized in the post-analysis of event-oriented mining results. For a discovered pattern which domain experts are specially interested in, we can borrow the idea of temporal feature to find the minimal window sizes that make this pattern interesting for different thresholds of interestingness measures. Such analysis could provide detailed information of the relevance between the pattern and the target event.

References

1. Mannila, H., Toivonen, H., Verkamo, A.I.: Discovery of frequent episodes in event sequences. Data Mining and Knowledge Discovery **1** (1997) 259–289
2. Weiss, G.M., Hirsh, H.: Learning to predict rare events in event sequences. In: Proc. 4th KDD. (1998) 359–363
3. Sun, X., Orlowska, M.E., Zhou, X.: Finding event-oriented patterns in long temporal sequences. In: Proc. 7th PAKDD. (2003) 15–26
4. Sun, X., Orlowska, M.E., Li, X.: Finding negative event-oriented patterns in long temporal sequences. In: Proc. 8th PAKDD. (2004) 212–221

An Anomaly Detection Method for Spacecraft Using Relevance Vector Learning

Ryohei Fujimaki[1], Takehisa Yairi[2], and Kazuo Machida[2]

[1] The Univ. of Tokyo, Aero. and Astronautics
[2] The Univ. of Tokyo, RCAST

Abstract. This paper proposes a novel anomaly detection system for spacecrafts based on data mining techniques. It constructs a nonlinear probabilistic model w.r.t. behavior of a spacecraft by applying the *relevance vector regression* and *autoregression* to massive telemetry data, and then monitors the on-line telemetry data using the model and detects anomalies. A major advantage over conventional anomaly detection methods is that this approach requires little *a priori* knowledge on the system.

1 Introduction

Anomaly detection is a key issue in the development of recent advanced spacecraft. The space environment is very harsh for spacecraft due to a variety of factors such as direct radiation, great temperature difference, and so on. In addition, the space is so distant from the earth that it is extremely difficult to directly inspect and repair a damaged component. Therefore, early detection of anomalous symptoms is important to avoid disastrous situations such as loss of control. Although several anomaly detection/diagnosis methods using modern reasoning techniques have been developed, they have difficulties in acquiring accurate and complete models and knowledge of the spacecraft systems and in monitoring the system behavior exhaustively and efficiently.

In this paper, we propose a new anomaly detection method for spacecraft based on data mining technique, *autoregressive model* and the *relevance vector regression*, and constructs a predictive model for each time series in the telemetry data. Then, it monitors online telemetry data and detect anomalies by checking the probability density of the observation.

2 Conventional Approaches to Anomaly Detection for Spacecraft

Limit-checking checks whether the value is within the pre-defined upper and lower limits. Though the limit-checking can be applied to any types of spacecraft, it lacks flexibility and expressiveness and suffers from false alarm problem.

In the model-based fault detection and diagnosis method, system models are utilized to simulate the spacecraft behavior and examine the validity of the actual telemetry data. This approach would provide an ideal performance if an accurate and complete model and infinite computational power were available. In practice, however, both of them are not available.

Expert systems also have been developed for this purpose. The knowledge is generally represented in the form of "if-then" production rules. Though the expert systems are powerful and flexible, it has a difficulty in preparing a set of accurate and complete knowledge on the spacecraft.

In summary, the above methods have a common problem that they are too dependent on the knowledge of human experts.

A reasonable approach to this problem is the application of data mining and machine learning techniques to the telemetry data. Actually, some researchers have developed anomaly detection methods for spacecraft using regression tree learning[7], temporal pattern clustering[8], association rule mining[9].

3 Proposed Anomaly Detection System

3.1 Autoregressive Model

Autoregressive (AR) model is the most basic data mining technique for time-series data[3][10][11]. For the purpose of applying AR model to anomaly detection problems for spacecraft, we define AR model as

$$\tau^j_{ARk} = \boldsymbol{\Theta}^j \mathbf{x}_k \tag{1}$$

where $j = \{1, \cdots, s\}$ represents the jth series of telemetry data, $\boldsymbol{\Theta}^j = (\Theta^j_0, \Theta^j_{1,k-1}, \cdots, \Theta^j_{1,k-p}, \cdots, \Theta^j_{s,k-1}, \cdots, \Theta^j_{s,k-p})$ is a row vector of AR coefficients, $\mathbf{x}_k = (1, \tau^1_{k-1}, \cdots, \tau^1_{k-p}, \cdots, \tau^s_{k-1}, \cdots, \tau^s_{k-p})$ is the data vector of all p previous time series, and the notation AR represent that the target value τ is based on AR model. This modified AR model implies that the value of a target time-series depends not only on the past values of itself but also on those of other series. The capability of modeling the relationships among some series is a great advantage. Fig.1 is the concept of this model. We made use of the framework of the relevance vector learning to extend this model to nonlinear and probabilistic.

3.2 Relevance Vector Regression

The relevance vector regression (RVR) originally proposed by Tipping[5] is a state of the art kernel-based nonlinear regression learning method[1][2][4][6].

We write a sample of N training pairs as $\{\mathbf{x}_n, t_n\}_{n=1}^N$ for the jth telemetry series, corresponding $\mathbf{x}_k$ and τ^j_k. Hereinafter, we deal with the jth series of telemetry data and omit the notation j for simplification.

The RVR model assumes that the targets are samples from a distribution model with additive independent zero-mean Gaussian noise, with variance σ^2,

$$\mathbf{t} = \mathbf{y} + \boldsymbol{\epsilon} = \boldsymbol{\Phi}\mathbf{w} + \boldsymbol{\epsilon} \tag{2}$$

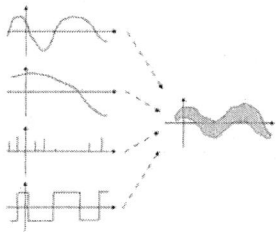

Fig. 1. Concept of our AR model

where $\mathbf{t} = (t_1, \ldots, t_N)$, $\mathbf{y}$, $\boldsymbol{\Phi} = (\phi_1, \ldots, \phi_M)$, $\mathbf{w} = (w_0, \ldots, w_M)^T$, $\phi_m(\mathbf{x}) = K(\mathbf{x}, \mathbf{x}_m)$ and $\boldsymbol{\epsilon} = (\epsilon_1, \ldots, \epsilon_n)^T$ represent the target vector, the vector of the predicted value, the $N \times M$ design matrix, the weights, the kernel function, and an error vector, respectively.

To achieve the sparsity, M independent hyperparameters, $\boldsymbol{\alpha} = (\alpha_1, \ldots, \alpha_M)^T$ over $\mathbf{w}$ are indroduced. Then, after maximizing logarithm likelihood w.r.t. $\boldsymbol{\alpha}$ and σ^2, we obtain a conditional distribution model for a new datum $\mathbf{x}_*$ as

$$p(t_*|\mathbf{t}, \boldsymbol{\alpha}_{MP}, \sigma^2_{MP}) = \mathcal{N}(t_*|y_*, \sigma^2_*) \qquad (3)$$

where y_* is the predict value of the new target t_* and σ^2_* is the variance of the prediction. See [1][4][6] for more details of the sparse Bayesian learning.

3.3 Anomaly Detection System

The anomaly detection system based on the proposed method operates as follows,

1. (Learning) Learn the relevance vector autoregressive model using a set of validated normal telemetry data.
2. (Prediction) Compute the next probable range of the target series.
3. (Monitoring) Obtain and check the (pseudo-)telemetry data.
4. (Alarming) Give an alarm if the data is out of the predicted range.
5. Repeat steps 2-4

The system is supposed to give many false alarms if we directly apply the redefined AR model Eq.(1) and the RVR Eq.(3). The reasons are,

1. The AR model completely cannot be modeled as the complicate spacecraft system,

$$t_*(\mathbf{x}_*) = t_{AR*}(\mathbf{x}_*) = t_{TRUE*}(\mathbf{x}_*) + \epsilon_{AR} = y_*(\mathbf{x}_*) + \epsilon_*(\mathbf{x}_*) \quad p(\epsilon_*) = \mathcal{N}(0, \sigma^2_*) \qquad (4)$$

where ϵ_{AR} is the modeling error of the AR model.
2. The RVR adopts the *relevance* vector for the prototypes of the model formed as Eq.(2)[1].

---

[1] This is a great advantage of the RVR for the execution speed of anomaly detection.

As the result, some training data can be mistaken as anomaly though all training data are normal because these data have relatively so large ϵ_{AR} that the system regards them as the data which is far from the prototype data.

We evaluated the variance of the difference between another data set $\{\mathbf{x}_i, t_i\}_{i=1}^{N_2}$ and corresponding prediction values, $y_*(\mathbf{x}_i)$.

$$\hat{\sigma}_{AR}^2 = \frac{1}{N_2}\sum_i^{N_2}(t_i - y_*(\mathbf{x}_i))^2, \quad \sum_i^{N_2}\epsilon(\mathbf{x}_i) \longrightarrow 0 \qquad (5)$$

If the AR model completely describes the system behavior, (5) must be zero in theory. Therefore, we extended (3) as

$$p(t_*|\mathbf{t}, \boldsymbol{\alpha}_{MP}, \sigma_{MP}^2) = \mathcal{N}(y_*, \sigma_*^2 + \hat{\sigma}_{AR}^2). \qquad (6)$$

4   Experiment and Discussion

We performed an experiment with telemetry data obtained from an orbital rendezvous simulation. This telemetry consists of 27 time-series variables in total, 13 of which are from *position and attitude control subsystem*, and the rest are from *propulsion subsystem*. In more detail, the former group consists of 12 numerical observation time-series variables regarding the position and attitude of the vehicle and one command sequence. The latter group consists of 14 discrete-valued time-series variables, each of which indicates the command input to each of the 14 thruster engines.

In this experiment, we assumed a scenario where the power of fourth thruster engine used for the pitch control falls to zero at time 250 [*sec*]. With this scenario, we performed following two scenarios.

Comparison of Proposed Method with normal RV Autoregressive Model First, we have compared the proposed method with the *normal* relevance vector autoregressive model.

Fig.2 , Fig.3 show the results of anomaly detection in the series which represents the pitch angle. The solid line in the upper figure shows the pitch angle and the dotted line shows the predicted range, and the solid line in the lower figure shows the probability density of the observation. The system gives the alarm when the probability became lower than the computed limit[2]. We can see that the normal model gives many false alarms. On the other hand, the proposed method correctly gives alarms after a little time delay. The proposed method also succeeded in detecting anomalies in the series representing the pitch rate as shown in Fig.4.

---

[2] In this experiment, we adopted the probability density value of $\sqrt{\sigma_*^2 + \sigma_{AR}^2}$ as limit.

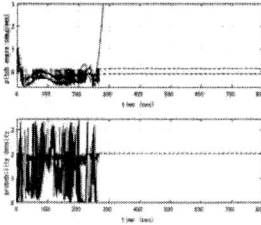
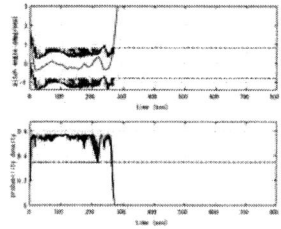

Fig. 2. Result of anomaly detection by normal RV AR model. The upper graph shows the confidence of target series in the telemetry data

Fig. 3. Result of anomaly detection by the proposed method. The lower graph shows the probability density

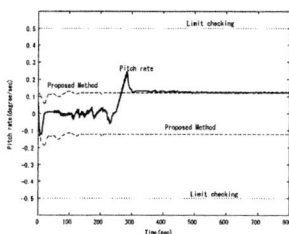

Fig. 4. Comparison of the proposed method with conventional limit-checking. The solid line represents the pitch rate and the dotted lines represent the upper and lower bounds given by each method

Comparison with Conventional Limit-Checking. We compared the proposed method with the conventional limit-checking. We set the limit on the standard deviation for the proposed method and on the maximum absolute value in the normal phase for the limit-checking. Fig.4 shows the result. The conventional limit-checking fails to detect slight anomalies like this case. On the other hand, the proposed method is capable of detecting this anomaly because it can dynamically estimate proper range of the target series.

We only showed the results with respect to the series which included the anomalies due to limitations of space, though we have ran the experiments against all series.

5 Conclusion

This paper proposed a new anomaly detection method based on the *relevance vector regression* and *autoregressive model*.

First, we extended the traditional AR model (3.1) and adopted the relevance vector frameworks for learning this model. In addition, we extended this RV AR model for the purpose of removing false alarm (3.3).

Compared with the conventional anomaly detection method, this method has great advantages. First, the proposed method requires little a priori knowledge on the spacecraft system. Therefore, it can be applied to various kinds of spacecraft.

We performed an experiment with telemetry data obtained from an orbital rendezvous simulation and confirmed the efficiency of the propos4ed method.

Acknowledgements

The authors would like to thank Japan Aerospace Exploration Agency (JAXA) for providing the simulation telemetry data.

References

1. A.C. Faul and M.E. Tipping. Analysis of Sparse Bayesian Learning, In T.G. Dietterich, S. Becker and Z. Ghahramanim editors, Advances in Neural Information Processing Systems 14, pages 383-389, MIT Press, 2002.
2. C.M. Bishop and M.E. Tipping, Variational Relevance Vector Machines, In C. Boutilier and M. Goldszmidt, editors, Proceedings of the 16th Conference on Uncertainty in Artificial Intelligence, page 46-53, Morgan Kaufmann, 2000.
3. K.-R. Muller, A.J. Smola, G.Ratsch, B. Scholkopf, J. Kohlmorgen, and V. Vapnik, Predicting time series with support vector machines. In W. Gerstner, A. Germond, M. Hasler, and J.-D. Nicoud, editors, Artificail Neural Networks, ICANN'97, page 999-1004, Berlin, 1997, Springer Lecture Notes in Computer Science, Vol. 1327
4. M.E. Tipping, Sparse Bayesian Learning and the Relevance Vector Machinem Journal of Machine Learning Researchm 1:211-244, 2001.
5. M.E. Tipping The Relevance Vector Machine. In S.A. Solla, T.K. Leen, and K.-R. Muller, editors, Advances in Neural Information Processing Systems 12, pages 652-658. MIT Press 2000.
6. M.E. Tipping, and A. C. Faul (2003). Fast marginal likelihood maximisation for sparse Bayesian models. In C. M. Bishop and B. J. Frey (Eds.), Proceedings of the Ninth International Workshop on Artificial Intelligence and Statistics, Key West, FL, Jan 3-6
7. Minoru Nakatsugawa, Takehisa Yairi, Naoki Ishihama, Koichi Hori and Shinichi Nakasuka "Supporting Anomaly Detection from Satellite Telemetry Data by Regression Trees" The 24th International Symposium on Space Technology and Science (ISTS), 2004
8. Takehisa Yairi, Shiro Ogasawara, Koichi Hori, Shinichi Nakasuka, and Naoki Ishihama "Summarization of Spacecrafts Telemetry Data By Extracting Significant Temporal Patterns" The Eighth Pacific-Asia Conference on Knowledge Discovery and Data Mining (PAKDD2004), pp.240-244, 2004
9. Takehisa Yairi, Yoshikiyo Kato and Koichi Hori, "Fault Detection by Mining Association Rules from House-keeping Data", Proc. of International Symposium on Artificial Intelligence, Robotics and Automation in Space (i-SAIRAS 2001)
10. U.U. Muller, A. Schick and W. Wefelmeyer(2004) , Efficient prediction for linear and nonlinear autoregressive models, 2004, (Submitted paper)
11. W.D. Penny and S.J. Roberts, Bayesian Methods for Autoregressive Models, IEEE Workshop on Neural Networks for Signal Processing, Sydney Australia, December 2000.

Cyclic Pattern Kernels Revisited

Tamás Horváth

Fraunhofer Institute for Autonomous Intelligent Systems,
Schloß Birlinghoven, D-53754 Sankt Augustin, Germany
tamas.horvath@ais.fraunhofer.de

Abstract. The cyclic pattern kernel (CPK) is a powerful graph kernel based on patterns formed by simple cycles of labeled graphs. In a recent work, we proposed a method for computing CPK which is restricted to graphs containing polynomial number of simple cycles. In this work, we present two approaches relaxing this limitation. We first show that for graphs of bounded treewidth, CPK can be computed in time polynomial in the number of cyclic patterns, which in turn can be exponentially smaller than that of simple cycles. We then propose an alternative CPK based on the set of relevant cycles which is known to be enumerable with polynomial delay and its cardinality is typically only cubic in the number of vertices. Empirical results on the NCI-HIV dataset indicate that there is no significant difference in predictive performance between CPK based on simple cycles and that based on relevant cycles.

1 Introduction

Recently, there is an increasing interest in supervised concept learning problems, where instances are *labeled undirected graphs*. This problem is motivated by various practical problems e.g. in computational chemistry. Besides rule-based learning algorithms, *kernel methods* (see, e.g., [17]), in particular, *support vector machines* [20], have become a popular approach for this task. The crucial step of graph kernel approaches is to design some *effectively* computable *positive definite function* $k : \mathcal{G} \times \mathcal{G} \to \mathbb{R}$, called *kernel*, where $\mathcal{G}$ is the set of labeled graphs representing the instances of the underlying learning problem. Many recent graph kernels rely on embedding $\mathcal{G}$ into a feature space defined by the set of *frequent subgraphs* occurring in $\mathcal{G}$ (see, e.g., [5]).

Graph kernels based on frequent patterns involve, however, the problem of finding an optimal trade-off between predictive power and runtime, as both these conflicting requirements depend on the choice of the frequency threshold. Therefore, as an alternative to graph kernels based on frequent subgraphs, a new graph kernel, called the *cyclic pattern kernel* (CPK), has been proposed recently in [10]. CPK is based on embedding the graphs into a Boolean feature space made up of *cyclic patterns* independent of their frequencies, where cyclic patterns are strings formed by the labels of *simple cycles*. On the NCI-HIV[1] biochemical domain, CPK turned out to outperform frequent subgraph-based approaches in accuracy measured by the area under the ROC curve [4].

---
[1] http://cactus.nci.nih.gov/

In [10] we show that CPK cannot be computed efficiently (neither by a closed form nor by any other algorithm). For a restricted graph class, we propose an algorithm based on explicitly embedding the graphs into the feature space. Although the features are only the *patterns* of simple cycles, they are computed by enumerating *all* simple cycles of the graphs. The reason is that, in contrast to simple cycles [15], cyclic patterns cannot be enumerated in output-polynomial time [10], i.e. in time polynomial in the combined size of the input and output. Since graphs may have exponentially many simple cycles, the method in [10] is restricted to graphs with a polynomial number of simple cycles. This limitation is rather severe because even graphs with small number of cyclic patterns may contain exponentially many simple cycles. Furthermore, to decide whether a graph meets the above requirement, one has to count its simple cycles which is #P-complete [19].

In this paper, we present two approaches relaxing the above limitation. We first show that cyclic patterns for graphs of *bounded treewidth* can be enumerated with polynomial delay. Hence, CPK can be computed in time polynomial in the number of cyclic patterns for this class of graphs. Treewidth [16] is a measure of tree-likeness of graphs. The class of bounded treewidth graphs includes many practically relevant graph classes (see, e.g., [3] for an overview). To show our result, we use the positive result [1] on the regular-language-constrained simple path problem for bounded treewidth graphs.

We then propose an alternative CPK based on the set of *relevant cycles* [14]. A cycle is relevant if it belongs to a *minimum basis* of the graph's *cycle space*. Although in worst case, the number of relevant cycles of a graph can be exponential in the number n of its vertices, relevant cycles have important advantages over simple cycles; (i) They can be counted in time *polynomial* in n and are enumerable with polynomial delay [21], and (ii) their number is typically only cubic in n [9]. We present empirical results which indicate that there is *no* significant difference in predictive performance between CPK defined by simple cycles and that defined by relevant cycles. Hence, utilizing the above nice properties of relevant cycles, a more robust CPK can be obtained.

The paper is organized as follows. In Section 2, we recall some basic notions and the definition of CPK. In Section 3, we show that for graphs of bounded treewidth, CPK can be computed in time polynomial in the number of cyclic patterns. In Section 4, we define CPK based on relevant cycles and evaluate it empirically. Finally, in Section 5, we conclude and list some problems for future work. Due to space limitations, proofs are only sketched or even omitted in this version.

2 Graphs, Kernels, and the Cyclic Pattern Kernel

In this section we recall some necessary notions. For further details, the reader is referred to [6, 10, 17].

Graphs. For a set S, $[S]^k$ denotes the family of k-subsets of S, i.e., $[S]^k = \{S' \subseteq S : |S'| = k\}$. A *labeled undirected graph* is a quadruple $G = (V, E, \Sigma, \lambda)$, where V is a finite set of *vertices*, $E \subseteq [V]^2$ is a set of *edges*, Σ is a finite set of *labels*, and $\lambda : V \cup E \to \Sigma$ is a function assigning a label to each element of $V \cup E$. $|V|$ and $|E|$ are denoted by n and m, respectively. Two vertices of G are *adjacent* if they are connected by an edge. The *degree* of a vertex $v \in V$ is the number of vertices adjacent to v. Unless otherwise stated, in this paper by graphs we always mean *labeled undirected* graphs. A *graph database* $\mathcal{G}$ is a set of disjoint graphs, and $|\mathcal{G}|$ denotes the number of graphs in $\mathcal{G}$.

Let $G = (V, E, \Sigma, \lambda)$ be a graph. A graph $G' = (V', E', \Sigma, \lambda')$ is a *subgraph* of G, if $V' \subseteq V$, $E' \subseteq E$, and $\lambda'(x) = \lambda(x)$ for every $x \in V' \cup E'$. A sequence $w = \{v_0, v_1\}, \{v_1, v_2\}, \ldots, \{v_{k-1}, v_k\}$ of edges of G forms a *simple path* if the v_i's are all distinct. G is *connected* if there is a (simple) path between any pair of its vertices. A *connected component* of G is a maximal subgraph of G that is connected. A vertex $v \in V$ is an *articulation* (also called *cut*) vertex, if its removal increases the number of connected components of G. G is *biconnected* if it contains no articulation vertex. A *biconnected component* (or *block*) of G is a maximal subgraph that is biconnected.

Let G be a graph. A subgraph C of G forms a *cycle* if each of its vertices has even degree. If, furthermore, C is connected and each of its vertices has degree 2 then C is a *simple cycle* of G. We denote by $\mathcal{S}(G)$ the set of simple cycles of G. Two simple cycles C and C' of G are considered to be the same iff C or its reverse is a cyclic permutation of C'. We note that the number of simple cycles can grow *faster* than 2^n.

It holds that the biconnected components of a graph G are pairwise edge disjoint and form thus a partition on the set of G's edges. This partition, in turn, corresponds to the following equivalence relation on the set of edges: two edges are equivalent iff they belong to a common simple cycle. This property of biconnected components implies that an edge of a graph belongs to a simple cycle iff its biconnected component contains more than one edge. Edges not belonging to simple cycles are called *bridges*. The subgraph of a graph G formed by its bridges is denoted by $\mathcal{B}(G)$. Clearly, each bridge of a graph is a singleton biconnected component, and $\mathcal{B}(G)$ is a forest.

The graphs $G_1 = (V_1, E_1, \Sigma, \lambda_1)$ and $G_2 = (V_2, E_2, \Sigma, \lambda_2)$ are *isomorphic* if there is a bijection $\varphi : V_1 \to V_2$ such that (i) for every $u, v \in V_1$, $\{u, v\} \in E_1$ iff $\{\varphi(u), \varphi(v)\} \in E_2$, (ii) $\lambda_1(u) = \lambda_2(\varphi(u))$ for every $u \in V_1$, and (iii) $\lambda_1(\{u, v\}) = \lambda_2(\{\varphi(u), \varphi(v)\})$ for every $\{u, v\} \in E_1$.

Kernel Methods. Kernel methods (see, e.g., [17]) are a theoretically well-founded class of statistical learning algorithms that have received considerable attention recently also in the data mining community. Algorithms in this broad class (e.g., support vector machines, Gaussian processes, etc.) have proved to be powerful tools in various real-world data mining applications. Since kernel methods are not restricted to the attribute-value representation used by most data mining algorithms, many of these applications involve datasets given in some non-vectorial representation formalism such as graphs (see, e.g., [8,17]). In general, kernel methods are composed of two components:

(i) A *domain specific* function Φ *embedding* the underlying instance space X into a high (possibly infinite) dimensional inner product *feature space* $\mathcal{F}$ (usually a Hilbert-space) and
(ii) a *domain independent* algorithm aimed at discovering *patterns* (e.g., classification, clustering, etc.) in the embedded data, where patterns are restricted to linear functions defined in terms of inner products between the points of the embedded input data.

One of the attractive computational properties of kernel methods is that in many cases, patterns can be computed in time independent of the dimension of $\mathcal{F}$. In such cases, the inner product of the feature vectors can be calculated by a *kernel* without explicitly performing or even knowing the embedding function, where a kernel is a function of the form $\kappa : X \times X \to \mathbb{R}$ satisfying $\kappa(x,y) = \langle \Phi(x), \Phi(y) \rangle$ for every $x, y \in X$.

To simplify the description, we define a function that will be used many times in what follows. Let U be a set and $\kappa_\cap : 2^U \times 2^U \to \mathbb{N}$ be the function defined by

$$\kappa_\cap : (S_1, S_2) \mapsto |S_1 \cap S_2| \tag{1}$$

for every $S_1, S_2 \subseteq U$. The proof of the following proposition follows directly from the definitions.

Proposition 1. $\kappa_\cap$ *is a kernel.*

The function defined in (1) is called the *intersection kernel*[2].

The Cyclic Pattern Kernel. Finally, we recall the definition of the cyclic pattern kernel (CPK) introduced in [10]. Let Σ, Γ be alphabets, and π be a mapping from the set of simple cycles and trees labeled by Σ to Γ^* such that (i) π maps two graphs to the same string iff they are isomorphic and (ii) π can be computed in polynomial time. We note that such Γ and π always exist and can easily be constructed (see, e.g., [10,22]). Using π, the set of *cyclic* and *tree patterns* of G is defined by

$$P_\mathcal{S}(G) = \{\pi(C) : C \in \mathcal{S}(G)\} \tag{2}$$
$$P_\mathcal{T}(G) = \{\pi(T) : T \text{ is a maximal tree of } \mathcal{B}(G)\}, \tag{3}$$

respectively. The *cyclic pattern kernel* for a graph database $\mathcal{G}$ is then defined by

$$\kappa_\mathcal{S}(G_1, G_2) = \kappa_\cap(P_\mathcal{S}(G_1), P_\mathcal{S}(G_2)) + \kappa_\cap(P_\mathcal{T}(G_1), P_\mathcal{T}(G_2)) \tag{4}$$

for every $G_1, G_2 \in \mathcal{G}$. Since $P_\mathcal{S}(G)$ and $P_\mathcal{T}(G)$ are disjoint for every G, $\kappa_\mathcal{S}$ is a kernel by Proposition 1.

Unfortunately, unless P = NP, $\kappa_\mathcal{S}$ cannot be computed in polynomial time [10]. In fact, one can show that computing $\kappa_\mathcal{S}$ is at least as hard as counting simple cycles of length k in a graph. This problem is #W[1]-complete [7], and is therefore unlikely to be fixed-parameter tractable.

Because of the high complexity, CPK is computed in [10] by (i) explicitly performing the embedding into the feature space for every graph, and then by (ii) calculating the inner product of the obtained feature vectors. To perform the embedding for a graph G, $P_\mathcal{S}(G)$ is computed by enumerating all elements of $\mathcal{S}(G)$. The reason is that while $\mathcal{S}(G)$ can be enumerated with linear delay [15], $P_\mathcal{S}(G)$ cannot be enumerated in output-polynomial time (unless P = NP) [10]. Thus, the algorithm computing CPK in [10] is polynomial just in $|\mathcal{S}(G)|$ rather than in $|P_\mathcal{S}(G)|$. Since $|\mathcal{S}(G)|$ can be exponential in n, the method in [10] is restricted to graphs with polynomial number of simple cycles. To decide whether the graphs in the database satisfy this condition, one has to count their simple cycles which is #P-complete in general [19].

---

[2] We note that intersection kernels are often defined in a more general way (see, e.g., [17]).

3 Graphs of Bounded Treewidth

Restricting CPK to graphs of polynomial number of simple cycles is rather severe; graphs containing exponentially many simple cycles may have polynomially or even constant many cyclic patterns. Below we give an example of such graphs.

Example 1. Let $G = (V, E, \{a\}, \lambda)$ with $V = \{u_1, \ldots, u_n, v_1, \ldots, v_n, w_1, \ldots, w_n\}$ and

$$E = \bigcup_{i=1,\ldots,n} \{\{v_i, u_i\}, \{u_i, v_{(i \bmod n)+1}\}, \{v_i, w_i\}, \{w_i, v_{(i \bmod n)+1}\}\} \ .$$

G contains $2^n + n$ simple cycles, which in turn form only two different cyclic patterns.

This, as well as other examples from real-world datasets motivate us to deal with the problem of listing cyclic patterns of a graph without enumerating the possibly exponentially large set of all its simple cycles. More precisely, we consider the problem whether cyclic patterns can be enumerated with polynomial delay. The following proposition states that, in contrast to simple cycles, this problem is most likely intractable.

Proposition 2 ([10]). *Unless P = NP, cyclic patterns cannot be enumerated in output-polynomial time.*

The proof of the above proposition is based on a polynomial-time reduction from the NP-complete Hamiltonian cycle problem. This and many other NP-hard computational problems become, however, polynomially solvable when restricted to graphs of *bounded treewidth* (see, e.g., [2] for an overview). Treewidth [16] is a measure of tree-likeness of graphs. More precisely, a *tree decomposition* of a graph $G = (V, E, \Sigma, \lambda)$ is a tree $T = (V', E', 2^V, \mu)$ such that (i) $\bigcup_{v \in V'} \mu(v) = V$, (ii) for every $e \in E$ there is a vertex $v \in V'$ satisfying $e \subseteq \mu(v)$, and (iii) for every $u, v, w \in V'$ it holds that $\mu(u) \cap \mu(v) \subseteq \mu(w)$ whenever w is a vertex on the simple path between u and v. The *width* of T is $\max_{v \in V'} |\mu(v)| - 1$, and the *treewidth of G* is the width of a tree decomposition of G with the smallest width. Clearly, the treewidth of a tree is 1, and the treewidth of a simple cycle of length at least 3 is 2.

Treewidth proved to be a useful tool in the design of graph algorithms. It has wide algorithmic applications because many problems that are hard on arbitrary graphs become easy for graphs of bounded treewidth. The class of bounded treewidth graphs includes many practically relevant graph classes (see, e.g., [3] for an overview).

We note that even graphs with small treewidth may have exponentially many simple cycles. For instance, one can easily see that the treewidth of the graph in Example 1 is 2 for every $n > 1$.

Using the positive result in [1] on the regular-language-constrained simple path problem for graphs of bounded treewidth, in this section we show that CPK for graphs of *bounded treewidth* can be computed in time polynomial in the number of cyclic patterns. To prove this result, by (2) and (4) it is sufficient to show that cyclic patterns can be enumerated with polynomial delay for graphs of bounded treewidth. We start with the definition of the regular-language-constrained simple path problem ([1, 13]).

Algorithm 1. ENUMERATING CYCLIC PATTERNS

Require: bounded treewidth graph $G = (V, E, \Sigma, \lambda)$ with $\lambda : E \to \Sigma$ and integer $N > 0$
Ensure: set P of cyclic patterns of G such that $|P| = \min\{N, |P_S(G)|\}$ and for every $p, q \in P$
it holds that p is neither a cyclic permutation of q nor that of q^{-1}

1: $k := 0, P := \emptyset$
2: **while** $k < N$ and $|E| > 2$ **do**
3:    let $M_k = (S_k, \Sigma, \delta_k, s, F_k)$ be a DFA such that

$$L(M_k) = \begin{cases} \emptyset & \text{if } k = 0 \\ \{p \in \Sigma^* : \exists q \in P \text{ s.t. } p \text{ is a cyclic permutation of } q \text{ or } q^{-1}\} & \text{otherwise} \end{cases}$$

4:    let $e = \{u, v\}$ be some arbitrary edge of G
5:    let $M'_k = (S'_k, \Sigma, \delta'_k, s, \{f\})$ be the NFA such that
    - $S'_k = S_k \cup \{f\}$ for some new state $f \notin S_k$,
    - $\delta'_k = \delta_k \cup \{(a, x, f) : \exists b \in S_k, x \in \Sigma \text{ s.t. } \delta_k(a, x) = b \text{ and } \delta_k(b, \lambda(e)) \in S_k \setminus F_k\}$
6:    $p := \text{REG\_SIP}(G \setminus \{e\}, M'_k, u, v)$
7:    **if** p is the empty path **then** remove e from G
8:    **else**
9:       $P := P \cup \{w \cdot \lambda(e)\}$, where w is the string corresponding to p
10:      $k := k + 1$
11:   **endif**
12: **endwhile**
13: **return** P

Given (i) an edge-labeled graph $G = (V, E, \Sigma, \lambda)$, i.e., λ is a function mapping E to Σ, (ii) a source and a target vertex $s, t \in V$, respectively, and (iii) a regular language[3] $L \subseteq \Sigma^*$, *find* a simple path $p = e_1, e_2, \ldots, e_n$ from s to t such that $\lambda(e_1) \cdot \ldots \cdot \lambda(e_n) \in L$, or *print* 'NO' if such a path does not exist. While this problem is NP-complete in general, for graphs of bounded treewidth the following positive result holds.

Theorem 1 ([1]). *The regular-language-constrained simple path problem can be solved in polynomial time for graphs of bounded treewidth.*

Using this result, we can state the following theorem.

Theorem 2. *Let $G_i = (V_i, E_i, \Sigma, \lambda_i)$ be bounded treewidth graphs for $i = 1, 2$. Then $\kappa_S(G_1, G_2)$ can be computed in time polynomial in*

$$\max\{|V_1|, |V_2|, |P_S(G_1)|, |P_S(G_2)|\} \ .$$

Proof sketch. By (4), it is sufficient to show that cyclic patterns can be enumerated with polynomial delay because $P_T(G_i)$ ($i = 1, 2$) is bounded by $|V_i|$ and can be computed efficiently [18].

In order to apply the result provided by Theorem 1, we first note that each graph G with treewidth tw can be transformed into an edge labeled graph G' with treewidth tw such that there is a bijection between the sets of cyclic patterns of G and G'.[4] Thus, we may assume wlog that G_1 and G_2 are edge labeled graphs of bounded treewidth.

---

[3] For basic notions of formal languages and finite automata, the reader is referred, e.g., to [12].
[4] We note that we do not need to know cyclic patterns in order to compute CPK.

The algorithm enumerating N cyclic patterns of a graph of bounded treewidth is given in Algorithm 1. In each iteration of the loop (step 2), we compute a deterministic finite automaton (DFA) $M_k = (S_k, \Sigma, \delta_k, s, F_k)$ that accepts the patterns computed so far in P, as well as each cyclic permutation of q and q^{-1} for every $q \in P$, where q^{-1} denotes the reverse of q. S_k, Σ, δ_k, s, and F_k denote the set of states, the input alphabet, the transition function, the initial state, and the set of final states of M_k, respectively. Both the size of M_k and the time required to construct M_k is bounded by $O(nk)$. In step 4 we select an arbitrary edge $e = \{u, v\}$ of G, and then, in step 5, construct a nondeterministic finite automaton (NFA) M'_k recognizing the language $L_k = \{w \in \Sigma^* : w \cdot \lambda(e) \in \Sigma^* \setminus L(M_k)\}$. In step 6, we call the subroutine given in [1] that decides in polynomial time whether or not there is a simple path between u and v such that the string defined by this path belongs to L_k. If such a simple path does not exist, we remove e from G, as in this case there is no simple cycle containing e that defines a new cyclic pattern. Otherwise, we add the new cyclic pattern to P (step 9), and repeat the loop. If N cyclic patterns have been found or G contains at most two edges, we stop the algorithm and return P containing $\min\{N, |P_S(G)|\}$ cyclic patterns. □

4 CPK Based on Relevant Cycles

Consider again the graph G in Example 1, but now with $\Sigma = \{a, 0, 1\}$ and with λ labeling the u_i's by 0, the w_i's by 1, and each other vertex and edge by a. One can see that G has exponentially many cyclic patterns. In this section we empirically investigate whether another, possibly smaller set of cyclic structures can also be applied without significant loss of predictive performance. In particular, we consider cyclic patterns based on the *relevant* cycles [14] of a graph.

In order to recall the definition of relevant cycles, we start with some basic notions from algebraic graph theory. Let $G = (V, E, \Sigma, \lambda)$ be a graph and consider the set of cycles of G. Since cycles are subgraphs of G (such that every vertex has even degree), each cycle $C = (V', E', \Sigma, \lambda')$ of G can be represented by the incidence vector $\vec{C}$ of its edges. That is, the components of $\vec{C}$ are indexed by E, and for every $e \in E$, $\vec{C}_e = 1$ if $e \in E'$ and it is 0 otherwise. It holds that the set of vectors corresponding to the cycles of G forms a vector space, called the *cycle (vector) space*, over the field GF(2).[5] Thus, vector addition in the cycle space corresponds to the symmetric difference of the sets of edges of the cycles represented by the vectors. The dimension of the cycle space of G is its *cyclomatic number* $\nu(G) = m - n + c(G)$, where $c(G)$ denotes the number of connected components of G. To represent the cycle space of G, one can consider one of its bases, which has minimum length. The length of a basis is the sum of the number of edges of the cycles represented by the vectors belonging to the basis. Since the minimum basis of a graph's cycle space is not unique, the cyclic structure of a graph is described by the union of all minimum bases of its cycle space [14]. This canonical set of cycles is called the set of *relevant cycles*. In [21], it is shown that *relevant cycles*

---

[5] GF(2) is the binary field with the elements 0 and 1. Addition is defined by $0 \oplus 0 = 0, 0 \oplus 1 = 1$, and $1 \oplus 1 = 0$. Multiplication is given by $0 \otimes 0 = 0, 0 \otimes 1 = 0$, and $1 \otimes 1 = 1$.

can be enumerated with polynomial delay and counted in time polynomial in the order of G. Although the set of relevant cycles of a graph can be exponential in the number of its vertices in worst case, its cardinality is typically only cubic in n [9].

To measure the predictive performance of CPK based on relevant cycles, in our experiments we used monotone increasing subsets of simple cycles that can be generated by relevant cycles. More precisely, for a graph G and integer $k \geq 1$, let $\mathcal{R}_k(G)$ denote the set

$$\mathcal{R}_k(G) = \begin{cases} \text{the set of relevant cycles of } G & \text{if } k = 1 \\ \{C \oplus C' \in \mathcal{S}(G) : C \in \mathcal{R}_{k-1}(G) \text{ and } C' \in \mathcal{R}_1(G)\} & \text{otherwise} . \end{cases}$$

Since $\mathcal{R}_1(G) \subseteq \mathcal{S}(G)$, it holds that $\mathcal{R}_1(G) \subseteq \mathcal{R}_2(G) \subseteq \ldots \subseteq \mathcal{R}_{\nu(G)}(G) = \mathcal{S}(G)$. We note that the set of relevant cycles of a graph is the union of the relevant cycles of its biconnected components. Since biconnected components of a graph are enumerable with linear delay [18], and relevant cycles of a biconnected graph are enumerable with polynomial delay [21], $\mathcal{R}_k(G)$ can be computed in time polynomial in $|\mathcal{R}_k(G)|$ for any arbitrary graph G and $k > 0$.

For a graph database $\mathcal{G}$ and integer $k \geq 1$, the CPK based on $\mathcal{R}_k(G)$, denoted $\kappa_{\mathcal{R}_k}$, is then defined by

$$\kappa_{\mathcal{R}_k}(G_1, G_2) = \kappa_\cap(P_{\mathcal{R}_k}(G_1), P_{\mathcal{R}_k}(G_2)) + \kappa_\cap(P_\mathcal{T}(G_1), P_\mathcal{T}(G_2))$$

for every $G_1, G_2 \in \mathcal{G}$, where $P_{\mathcal{R}_k}(G) = \{\pi(C) : C \in \mathcal{R}_k(G)\}$ and $P_\mathcal{T}(G)$ is defined by (3) for every $G \in \mathcal{G}$. The remarks above along with Proposition 1 imply that $\kappa_{\mathcal{R}_k}$ is a kernel that can be computed in time polynomial in $\max\{n_1, |\mathcal{R}_k(G_1)|, n_2, |\mathcal{R}_k(G_2)|\}$, where n_1 and n_2 denote the number of vertices of G_1 and G_2, respectively.

4.1 Empirical Evaluation

To evaluate the predictive performance of CPK based on relevant cycles, we used the same NCI–HIV dataset and evaluation method as in [10]. We briefly describe both the dataset and the method, and refer the reader to [10] for further details.

Each compound in the NCI-HIV dataset is described by its molecular graph and by its activity against HIV, which is one of the categories *confirmed inactive* (CI), *moderately active* (CM), and *active* (CA). The NCI-HIV dataset contains 42689 molecules, 423 of which are active, 1081 are moderately active, and 41185 are inactive. The total number of vertices and edges in this dataset is 1951154 and 2036712, respectively. Table 1 shows the total number of cycles of different types in this graph database, as well as the total number of patterns defined by them.

Table 1. Number of cycles and cyclic patterns of different type in the NCI-HIV domain

$\mathcal{R}_1$	$P_{\mathcal{R}_1}$	$\mathcal{R}_2$	$P_{\mathcal{R}_2}$	$\mathcal{R}_3$	$P_{\mathcal{R}_3}$	$\mathcal{S}$	$P_\mathcal{S}$
132559	998	181367	2274	205829	3713	376125	6204

Table 2. AUC for different tasks and costs. $\bullet_X$ (resp. $\circ_X$) denotes a significant win at a 5% (resp. 10%) level wrt. X

cost	$\kappa_{\mathcal{R}_1}$	$\kappa_{\mathcal{R}_2}$	$\kappa_{\mathcal{R}_3}$	κ_S
		CA/CM problem		
1.0	0.801($\pm$ 0.045)	0.815($\pm$ 0.031) $\bullet_{\kappa_{\mathcal{R}_1}}$	0.814($\pm$ 0.032)	0.813($\pm$ 0.033)
2.5	0.821($\pm$ 0.046)	0.830($\pm$ 0.041) $\bullet_{\kappa_{\mathcal{R}_1}}$	0.829($\pm$ 0.039)	0.827($\pm$ 0.042)
		CACM/CI problem		
1.0	0.754($\pm$ 0.022)	0.771($\pm$ 0.023) $\bullet_{\kappa_{\mathcal{R}_1}}$	0.778($\pm$ 0.018) $\bullet_{\kappa_{\mathcal{R}_1}}$	0.778($\pm$ 0.019) $\bullet_{\kappa_{\mathcal{R}_1}}$
35.0	0.795($\pm$ 0.028)	0.800($\pm$ 0.027)	0.804($\pm$ 0.024) $\bullet_{\kappa_{\mathcal{R}_1}}$	0.805($\pm$ 0.025) $\bullet_{\kappa_{\mathcal{R}_1}}$ $\bullet_{\kappa_{\mathcal{R}_2}}$
		CA/CMCI problem		
1.0	0.892($\pm$ 0.032)	0.907($\pm$ 0.025) $\bullet_{\kappa_{\mathcal{R}_1}}$	0.908($\pm$ 0.027) $\bullet_{\kappa_{\mathcal{R}_1}}$	0.908($\pm$ 0.027) $\bullet_{\kappa_{\mathcal{R}_1}}$
100.0	0.929($\pm$ 0.026)	0.929($\pm$ 0.032) $\circ_{\kappa_S}$	0.926($\pm$ 0.029) $\bullet_{\kappa_S}$	0.922($\pm$ 0.030)
		CA/CI problem		
1.0	0.911($\pm$ 0.034)	0.925($\pm$ 0.029) $\bullet_{\kappa_{\mathcal{R}_1}}$	0.925($\pm$ 0.027) $\bullet_{\kappa_{\mathcal{R}_1}}$	0.926($\pm$ 0.028) $\bullet_{\kappa_{\mathcal{R}_1}}$
100.0	0.937($\pm$ 0.024)	0.939($\pm$ 0.019)	0.936($\pm$ 0.018)	0.934($\pm$ 0.020)

In order to evaluate the predictive performance of CPK based on relevant cycles, we compared $\kappa_{\mathcal{R}_1}, \kappa_{\mathcal{R}_2}, \kappa_{\mathcal{R}_3}$, and κ_S with each other on the following classification problems: distinguish CA from CM (CA/CM), CA and CM from CI (CACM/CI), CA from CM and CI (CA/CMCI), and CA from CI (CA/CI). We used a modified version of the SVM-light [11] support vector machine with the same misclassification cost parameters as used in [5, 10]. For each problem and for each misclassification cost, we performed a 5-fold cross-validation and measured the predictive performance using the mean of the areas under the ROC curve (AUC) [4]. We used 5% and 10% significance levels in the comparisons (see [10] for the details). The results are given in Table 2. They indicate that $\kappa_{\mathcal{R}_2}$ can be used in most of the cases without a significant loss of predictive performance wrt. κ_S, and that $\kappa_{\mathcal{R}_3}$ was never outperformed significantly by κ_S. Hence, the alternative definition of CPK allows one to apply it to graph databases containing graphs even with exponentially many simple cycles.

5 Conclusion and Future Work

In this paper, we have presented two approaches relaxing the complexity limitation of computing CPK based on the set of simple cycles. We have shown that for graphs of bounded treewidth, CPK can be computed in time polynomial in the number of cyclic patterns. We then proposed an alternative CPK based on the smaller set of relevant cycles and compared its predictive performance with that based on simple cycles. Empirical results on the NCI-HIV dataset indicate that there is no significant difference between the two CPK's. Since the number of relevant cycles of a graph is typically only cubic in its order, the CPK proposed in this work can be applied even to graphs with exponentially many simple cycles. In addition, relevant cycles can be counted efficiently in contrast to simple cycles. This allows one to decide in polynomial time whether the graphs in a database contain polynomial number of relevant cycles. Hence, utilizing the above nice properties of relevant cycles, we have presented a more robust CPK.

For future work, we are going to perform experiments with other real-world graph databases. Furthermore, since $\mathcal{R}_k(G)$ is usually significantly larger than $P_{\mathcal{R}_k}(G)$ (see, e.g., Table 1), we are going to investigate, whether $\kappa_{\mathcal{R}_k}(G_1, G_2)$ can be computed in time polynomial in $|P_{\mathcal{R}_k}(G_1)|$ and $|P_{\mathcal{R}_k}(G_2)|$ for every G_1 and G_2.

Acknowledgments

The author thanks Thomas Gärtner and Stefan Wrobel for interesting discussions of this work. This work was supported in part by the DFG project (WR 40/2-1) *Hybride Methoden und Systemarchitekturen für heterogene Informationsräume*.

References

1. C. Barrett, R. Jacob, and M. Marathe. Formal-language-constrained path problems. *SIAM Journal on Computing*, 30(3):809–837, 2000.
2. H. L. Bodlaender. A tourist guide through treewidth. *Acta Cybernetica*, 11(1-2):1–22, 1993.
3. H. L. Bodlaender. A partial k-arboretum of graphs with bounded treewidth. *Theoretical Computer Science*, 209(1–2):1–45, 1998.
4. A. Bradley. The use of the area under the ROC curve in the evaluation of machine learning algorithms. *Pattern Recognition*, 30(7):1145–1159, 1997.
5. M. Deshpande, M. Kuramochi, and G. Karypis. Frequent sub-structure based approaches for classifying chemical compounds. In *Proc. of the 3rd IEEE Int. Conf. on Data Mining*, pages 35–42. IEEE Computer Society, 2003.
6. R. Diestel. *Graph theory*. Springer-Verlag, New York, 2nd edition, 2000.
7. Flum and Grohe. The parameterized complexity of counting problems. *SIAM Journal on Computing*, 33(4):892–922, 2004.
8. T. Gärtner. A survey of kernels for structured data. *SIGKDD Explorations*, 5(1):49–58, 2003.
9. P. M. Gleiss and P. F. Stadler. Relevant cycles in biopolymers and random graph. In *Proc. of the 4th Slovene Int. Conf. in Graph Theory*, 1999.
10. T. Horváth, T. Gärtner, and S. Wrobel. Cyclic pattern kernels for predictive graph mining. In *Proc. of the 10th ACM SIGKDD Int. Conf. on Knowledge Discovery and Data Mining*, pages 158–167, 2004.
11. T. Joachims. Making large–scale SVM learning practical. In B. Schölkopf, C. J. C. Burges, and A. J. Smola, editors, *Advances in Kernel Methods — Support Vector Learning*, pages 169–184, Cambridge, MA, 1999. MIT Press.
12. H. Lewis and C. Papadimitriou. *Elements of the Theory of Computation*. Prentice Hall, New York, 2nd edition, 1997.
13. A. O. Mendelzon and P. T. Wood. Finding regular simple paths in graph databases. *SIAM Journal on Computing*, 24(6):1235–1258, 1995.
14. M. Plotkin. Mathematical basis of ring-finding algorithms at CIDS. *J. Chem. Doc.*, 11:60–63, 1971.
15. R. C. Read and R. E. Tarjan. Bounds on backtrack algorithms for listing cycles, paths, and spanning trees. *Networks*, 5(3):237–252, 1975.

16. N. Robertson and P. D. Seymour. Graph minors. II. Algorithmic Aspects of Tree-Width. *J. Algorithms*, 7(3):309–322, 1986.
17. J. Shawe-Taylor and N. Cristianini. *Kernel Methods for Pattern Analysis*. Cambridge University Press, 2004.
18. R. Tarjan. Depth-first search and linear graph algorithms. *SIAM Journal on Computing*, 1(2):146–160, 1972.
19. L. G. Valiant. The complexity of enumeration and reliability problems. *SIAM Journal on Computing*, 8(3):410–421, 1979.
20. V. Vapnik. *Statistical Learning Theory*. John Wiley, 1998.
21. P. Vismara. Union of all the minimum cycle bases of a graph. *The Electronic Journal of Combinatorics*, 4(1):73–87, 1997.
22. M. Zaki. Efficiently mining frequent trees in a forest. In *Proc. of the 8th ACM SIGKDD Int. Conf. on Knowledge Discovery and Data Mining*, pages 71–80. ACM Press, 2002.

Subspace Clustering of Text Documents with Feature Weighting K-Means Algorithm

Liping Jing[1,2], Michael K. Ng[1], Jun Xu[2], and Joshua Zhexue Huang[2]

[1] Department of Mathematics,
The University of Hong Kong, HongKong, China
`mng@maths.hku.hk`
[2] E-Business Technology Institute,
The University of Hong Kong, Hong Kong, China
`{lpjing, fxu, jhuang}@eti.hku.hk`

Abstract. This paper presents a new method to solve the problem of clustering large and complex text data. The method is based on a new subspace clustering algorithm that automatically calculates the feature weights in the k-means clustering process. In clustering sparse text data the feature weights are used to discover clusters from subspaces of the document vector space and identify key words that represent the semantics of the clusters. We present a modification of the published algorithm to solve the sparsity problem that occurs in text clustering. Experimental results on real-world text data have shown that the new method outperformed the *Standard KMeans* and *Bisection-KMeans* algorithms, while still maintaining efficiency of the k-means clustering process.

Keywords: Subspace Clustering, Text Mining, High Dimensional Data, Feature Weighting, Cluster Interpretation.

1 Introduction

Clustering text documents into different category groups is an important step in indexing, retrieval, management and mining of abundant text data on the Web or in corporate information systems. Among others, the challenging problems of text clustering are big volume, high dimensionality and complex semantics. In this paper we are interested in solutions to the first two problems while use of ontology provides promising solutions to the third problem [1].

In text document clustering, a document is often transferred to a vector $< t_1, t_2, \cdots, t_n >$. A set of documents are represented as a matrix where each row indicates a document and each column represents a term or word in the vocabulary of the document set. In this model, clustering algorithms such as the *Standard KMeans* [2] and its varieties [3, 4], as well as the hierarchical clustering methods [5, 6], are used to cluster text data. In many real applications, the vocabulary and the number of documents are very large, which results in a very large matrix. On the other hand, the clusters in a document set are categorized by different subsets of terms, which makes the matrix sparse. The sparsity is dependent of the differences of semantics of the clusters in the document set.

To effectively cluster large and sparse text data requires the clustering algorithms to be efficient, scalable and able to discover clusters from subspaces of the *vector* space model (VSM). Scalable subspace clustering methods are made good candidates for text clustering [7], while other clustering algorithms often fail to produce satisfactory clustering results.

In this paper, we present a study of using the feature weighting k-means algorithm, denoted as *FW-KMeans*, to cluster text data [8,9]. *FW-KMeans* is a subspace clustering algorithm that identifies clusters from subspaces by automatically assigning large weights to the variables that form the subspaces in which the clusters are formed. The new algorithm is based on the extensions to the standard k-means algorithm so it is efficient and scalable to large data set. We propose a modification to the original *FW-KMeans* to handle highly sparse text data where many words do not appear in documents of certain categories. This situation makes the original *FW-KMeans* unsolvable because the weights for these terms turn to infinite. By introducing a constant σ to the distance function, the problem is solved, the convergence of the algorithm is guaranteed, and its efficiency is preserved.

We propose a method to calculate σ from the distribution of the data set because σ can affect the significance of the feature weights of the modified *FW-KMeans*. We have used different data sets from the *20-Newsgroups* to test the clustering performance of the new algorithm, and compared our results with those of the *Standard KMeans* and *Bisection KMeans* algorithms. The experimental results from different data sets have shown that our new algorithm outperformed the others. Beyond the clustering performance, the other advantage of the new algorithm is able to identify a subset of key words in each cluster. We present analysis of these key words and show how they can be used to present the semantics of clusters which can help understand the discovered clusters.

A similar work was reported in [10], which used a different feature weighting k-means algorithm in [8] to cluster text documents, while the sparsity problem was not discussed. The concept vector approach [11] is similar to the subset of features identified with weights in our proposed method. However, the concept vector for each cluster was obtained by associating with a word cluster that was separately generated. This process potentially affects the running time and complexity. Besides, the clustering algorithm used is the *Standard Spherical KMeans*.

The rest of the paper is organized as follows. Section 2 discusses subspace clustering with the feature weighting k-means algorithm and presents a modification to handle the sparsity problem in text clustering. Section 3 defines clustering evaluation methods that are used in experiments. The comparison studies and feature analysis are presented in Section 4. Finally, we draw some conclusions and point out future work in Section 5.

2 Subspace Clustering with Feature Weighting k-Means

In the *VSM*, a set of documents are represented as a set of vectors $\mathbf{X} = \{\mathcal{X}_1, \mathcal{X}_2, \ldots, \mathcal{X}_n\}$. Each vector $\mathcal{X}_j$ is characterized by a set of m terms or words, $(t_1, t_2, \ldots, t_m)$. Here, the terms can be considered as the features of the vector space and m as the number of dimensions representing the total number of terms in the vocabulary. Assume that several categories exist in $\mathbf{X}$, each category of documents is characterized by a subset of terms in the vocabulary that corresponds to a subset of features in the vector space. In this sense, we say that a cluster of documents is situated in a subspace of the vector space.

To discover clusters of documents from different subspaces, it is important that the clustering algorithm has the capability of subspace clustering. The feature weighting k-means algorithm that we have recently developed [9] and also reported by others [8] is able to cluster data in a subspace by automatically weighting features in the k-means clustering process. Using the k-means clustering process, the new algorithm clusters n objects into k clusters by minimizing the following objective function:

$$F(W, Z, \Lambda) = \sum_{l=1}^{k} \sum_{j=1}^{n} \sum_{i=1}^{m} w_{l,j} \lambda_{l,i}^{\beta} d(z_{l,i}, x_{j,i}) \quad (1)$$

where $d(z_{l,i}, x_{j,i})$ is a dissimilarity measure between object X_j and cluster center Z_l in feature i; $w_{l,j} = 1$ indicates that object j is assigned to cluster l and otherwise $w_{l,j} = 0$; $\lambda_{l,i}$ is the weight to feature i in cluster l; and $\beta > 1$ is a given parameter.

The unknowns W and Z are solved in the same way as the *Standard KMeans* algorithm. Each feature weight λ is solved by:

$$\lambda_{l,i} = \frac{1}{\sum_{t=1}^{m} \left[\frac{\sum_{j=1}^{n} \tilde{w}_{l,j} d(\tilde{z}_{l,i}, x_{j,i})}{\sum_{j=1}^{n} \tilde{w}_{l,j} d(\tilde{z}_{l,t}, x_{j,t})} \right]^{1/(\beta-1)}} \quad (2)$$

where $\tilde{w}_{l,j}$ and $\tilde{z}_{l,i}$ are the known values obtained from the previous iterative steps. (refer to [8, 9] for details of the clustering algorithm.)

There are totally $m \times k$ weights produced by the algorithm. In each cluster m weights are assigned to m features and the weight of a feature is inversely proportional to the dispersion of values of that feature. The larger the dispersion, the smaller the weight. This indicates that the values of a good feature in a cluster are very close to the value of the cluster center in that feature. In text clustering, this implies that a good term or word appears in the majority of the documents of a cluster with similar frequency. Therefore, a large weight identifies a key term in a cluster.

However, two special situations cause zero dispersion, which makes λ infinite. One is that a word does not occur in any document in that cluster and the other is that the word appears in each document with the same frequency. Table 1 shows examples of the two cases where the term t_4 appears in each document of the first cluster two times and the term t_3 does not appear in any document in the first cluster. To calculate the weights for these two terms, their weights λ become infinite so the objective function (1) cannot be minimized properly.

Table 1. An example of a data set in VSM: feature-object; the entry value is feature frequency

		t_0	t_1	t_2	t_3	t_4
C_0	x_0	1	2	3	0	2
	x_1	2	3	1	0	2
C_1	x_2	0	0	1	3	2
	x_3	0	0	2	1	3

To solve this problem, we modify the objective function by introducing a constant σ to the dissimilarity measure as below:

$$F_1(W, Z, \Lambda) = \sum_{l=1}^{k} \sum_{j=1}^{n} \sum_{i=1}^{m} w_{l,j} \lambda_{l,i}^{\beta} [d(z_{l,i}, x_{j,i}) + \sigma] \quad (3)$$

Fixing $\tilde{W}$ and $\tilde{Z}$ and using the Lagrange multiplier technique to minimize F_1 with respect to Λ, we obtain

$$\lambda_{l,i} = \frac{1}{\sum_{t=1}^{m} \left[\frac{\sum_{j=1}^{n} \tilde{w}_{l,j}[d(\tilde{z}_{l,i},x_{j,i})+\sigma]}{\sum_{j=1}^{n} \tilde{w}_{l,j}[d(\tilde{z}_{l,t},x_{j,t})+\sigma]} \right]^{1/(\beta-1)}} \quad (4)$$

We can easily verify that $\sum_{i=1}^{m} \lambda_{l,i} = 1$ and $1 \leq l \leq k$.

With the introduction of σ, the dispersion of a feature in a cluster can never be zero so all $\lambda_{l,i}$ can be calculated in (4). The features with zero dispersion will have the maximal weight in the cluster, while the weights of other features will be smaller, depending on the value of the dispersion. For example in Table 1, the features t_3 and t_4 will have the largest weight in the first cluster. To identify the cluster, term t_4 is apparently more important than term t_3. The two different terms can be easily separated in post-processing. When extracting important features to represent different clusters, we remove t_3 type features but retain t_4 type features.

The value of the parameter β has been discussed in [9] and [8]. Here, we discuss how to choose σ because it will affect the values of weights. From (4), we can see that if σ is too larger than $d(\tilde{z}_{l,i}, x_{j,i})$, the weights will be dominated by σ and $\lambda_{l,i}$ will approach to $\frac{1}{m}$. This will make the clustering process back to the standard k-means. If σ is too small, then the gap of the weights between the zero dispersion features and other important features will be big, therefore, undermining the importance of other features.

To balance we calculate σ based on the average dispersion of the entire data set for all features as follows:

$$\sigma = \frac{\sum_{j=1}^{\hat{n}} \sum_{i=1}^{m} d(x_{j,i}, o_i)}{\hat{n} \cdot m} \quad (5)$$

where o_i is the mean feature value of the entire data set. In practice we use a sample instead of the entire data set to calculate σ. (5% sample is used according to the sampling theory [12].) Experimental results have shown that this selection of σ is reasonable to produce satisfactory clustering results and identify important features of clusters.

From the above description, we can summarize that this subspace clustering method has the following two major advantages:

1. It is efficient and scalable to cluster large and sparse text data in subspaces.
2. From the weights, the subset of key words in each cluster can be identified, which helps the interpretation of the clustering results.

3 Clustering Evaluation

In this work we use four different external cluster validation methods to evaluate the clustering performance of our approach in clustering real world text data and compare our results with the results of other clustering methods. They are *accuracy*, *entropy*, *F1 score (FScore)* [6], and normalized mutual information *(NMI)* [13] which are defined as follows.

Given a data set with k classes C_h, we use a clustering algorithm to cluster it into k clusters S_l, where $1 \leq l, h \leq k$. Let n_h, n_l be the numbers of documents in class C_h and in cluster S_l respectively, $n_{h,l}$ be the number of documents appearing in both class C_h and cluster S_l, n be the total number of documents in the data set, and k is the number of clusters equal to the number of classes. Table 2 shows the four evaluation functions used in this paper:

Table 2. Evaluation functions

Accuracy	$\frac{\sum_{l=1}^{k} n_{ll}}{n}$
Entropy	$\sum_{l=1}^{k} \frac{n_l}{n} \left(-\frac{1}{\log k} \sum_{h=1}^{k} \frac{n_{h,l}}{n_l} \cdot \log \frac{n_{h,l}}{n_l} \right)$
NMI	$\frac{\sum_{h,l} n_{h,l} \log \left(\frac{n \cdot n_{h,l}}{n_h n_l} \right)}{\sqrt{(\sum_h n_h \log \frac{n_h}{n})(\sum_l n_l \log \frac{n_l}{n})}}$
FScore	$\sum_{h=1}^{k} \frac{n_h}{n} \cdot \max_{1 \leq l \leq k} \left\{ \frac{2 \cdot n_{hl}/n_h \cdot n_{hl}/n_l}{n_{hl}/n_h + n_{hl}/n_l} \right\}$

4 Experiment Results and Discussion

4.1 Text Datasets

To demonstrate the effectiveness of the *FW-KMeans* on different structured text data, we built 6 datasets from the *20-Newsgroups* collection[1] with different characteristics in sparsity, dimensionality and class distribution.

---

[1] http://kdd.ics.uci.edu/databases/20newsgroups/20newsgroups.html.

Table 3 lists the 6 datasets. The source column gives the class categories of each dataset and n_d indicates the number of documents in each class. Data sets A2 and A4 contain categories with very different topics while datasets B2 and B4 consist of categories in similar topics. Sparsity of the former datasets is bigger than that of the later datasets because there are more overlapping words in the later datasets to describe the similar topics. Datasets A4-U and B4-U contain unbalanced classes.

Table 3. Summary of text datasets

DataSet	Source	n_d	DataSet	Source	n_d
A2	alt.atheism	100	B2	talk.politics.mideast	100
	comp.graphics	100		talk.politics.misc	100
A4	comp.graphics	100	B4	comp.graphics	100
	rec.sport.baseball	100		comp.os.ms-windows	100
	sci.space	100		rec.autos	100
	talk.politics.mideast	100		sci.electronics	100
A4-U	comp.graphics	120	B4-U	comp.graphics	120
	rec.sport.baseball	100		comp.os.ms-windows	100
	sci.space	59		rec.autos	59
	talk.politics.mideast	20		sci.electronics	20

The raw data were preprocessed using the *Bow* toolkit [14]. The preprocessing steps include removing the headers, the stop words, and the words that occur in less than three documents or greater than the average number of documents in each class, as well as stemming the left words with the Porter stemming function. The standard $tf \cdot idf$ term weighting was used to represent the document vector.

4.2 Cluster Analysis

We used three k-means type algorithms, *FW-KMeans*, *Bisection-KMeans* and *Standard KMeans* to cluster the 6 datasets. Table 4 shows the clustering results evaluated in the 4 evaluation measures defined in Section 3. Since the k-means type algorithms are known to be sensitive to the choice of an initial partition, for sparse and high-dimensional text data, randomly selecting initial cluster centers usually does not lead to a good clustering. In these experiments, we first randomly sampled 5% of documents from a data set and used the farthest k points between two classes in the sample data as the initial center for each cluster [15]. Experimental results have shown that this initialization strategy performed well.

Table 4 gives the comparisons of three clustering algorithms on the 6 text datasets. The 4 figures in each cell represent the values of Accuracy, Entropy, Fscore and NMI respectively. We can see that *FW-KMeans* performed the best in most cases. For the balanced datasets, *Standard KMeans* was worst. The *Bisection-KMeans* performed slightly better than the *FW-KMeans* on A2 and A4 which are less overlap because the classes in them are separate with each other, therefore sharing the small set of similar words. For the datasets B2 and

Table 4. Comparisons of *FW-KMeans* with *Standard KMeans* and *Bisection-KMeans*.(Bold-face shows the best performance in the three methods)

	A2	B2	A4	B4	A4-U	B4-U
FW-KMeans	0.96	0.905	0.8975	**0.8621**	**0.9591**	**0.9197**
	0.2057	**0.4014**	0.2509	**0.3574**	**0.1513**	**0.2314**
	0.9599	0.9043	0.9003	**0.8631**	**0.9591**	**0.9205**
	0.7961	**0.6050**	0.7554	**0.6467**	**0.8480**	**0.7385**
Bisection-KMeans	**0.965**	0.88	**0.9375**	0.7017	0.8954	0.6087
	0.2146	0.5294	**0.1919**	0.6195	0.2830	0.5357
	0.9650	0.8800	**0.9376**	0.7049	0.8961	0.6586
	0.7857	0.4706	**0.8083**	0.3822	0.7126	0.3793
Standard KMeans	0.895	0.735	0.6	0.5689	0.95	0.8729
	0.4028	0.7121	0.6375	0.7492	0.1721	0.3459
	0.8938	0.7150	0.6146	0.5564	0.9498	0.8707
	0.6070	0.3246	0.4180	0.2721	0.8292	0.6346

B4, the *FW-KMeans* performed much better because of its capability of subspace clustering by feature weighting.

For the unbalanced A4-U and B4-U, both *FW-KMeans* and *Standard KMeans* performed reasonably well while the performance of *Bisection-KMeans* clearly deteriorated. This was because the *Bisection-KMeans* needs to choose a branch to split at each step, and usually, the largest cluster is chosen. This resulted in artificial division of some inherent large classes in the early stage so the mistake could not be corrected in the later stage. This can be shown by the following two confusion matrices from dataset B4-U. The large classes C_0 and C_1 were divided into separate clusters by the *Bisection-KMeans*. However, the *FW-KMeans* algorithm recovered them accurately.

	S_0	S_1	S_2	S_3
C_0	21	2	70	27
C_1	30	0	11	59
C_2	1	53	1	4
C_3	0	10	4	6

Confusion Matrix produced by
Bisection-KMeans on B4

	S_0	S_1	S_2	S_3
C_0	109	9	0	2
C_1	3	95	1	1
C_2	0	3	54	2
C_3	2	1	0	17

Confusion Matrix produced by
FW-KMeans on B4

4.3 Feature Analysis

Equation (4) in Section 2 shows that $\lambda_{l,i}$ is inversely related to the ratio of the dispersion along feature i to the total dispersion of all features in cluster l. The more compact (smaller dispersion) the cluster is along feature i, the bigger the weight of feature i. This implies that the term or word of feature i appears evenly in all documents of the cluster. Therefore, this word is an important identifier for this cluster.

We can use the weights produced by the *FW-KMeans* algorithm to identify the important terms or words in each cluster. However, because of the special case of zero dispersion in certain features, the largest weights may identify some words which do not occur in the documents of the cluster. In this case we ignored these words. In fact, in sparse text data, many of such words can be identified.

For example, after preprocessing we got 1322 features for dataset B4. In the 4 clusters generated by the *FW-KMeans* algorithm, 381 words do not appear in the first cluster and 363, 318 and 301 features do not appear in the documents of other 3 clusters respectively. The percentage is a little less than 30%. Although the weights for these features are large, they do not represent the semantics of the cluster.

Word weights are divided into five intervals:

	weight intervals	word num
0~1:	(0,1e-08]	8
1~2:	(1e-08,1e-07]	280
2~3:	(1e-07,1e-06]	433
3~4:	(1e-06,1e-05]	188
4~5:	(1e-05,1)	32

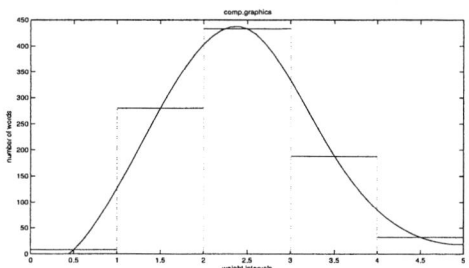

Fig. 1. Word distribution according to word weights

After removing the words that do not appear in the corresponding clusters, we divided the rest words into groups according to the intervals of the weights. The left table of Figure 1 shows the number of words in each interval and the right figure plots the distribution of words in category *comp.graphics* of dataset B4. Most words have relatively small weights (over 75%). Given a weight threshold we identified 220 words that we considered important. This is less than 17% of the total words. These are the words which contributed most to the semantics of the cluster so we can use these words to interpret the cluster.

We plotted the weights of these words in different clusters as shown in Figure 2. The horizontal axis is the index of the 220 words and the vertical lines indicate the values of the weights. It is clear that each cluster has its own subset of key words because the lines do not have big overlapping in different clusters. Category groups *comp.graphics* and *comp.os.ms-windows* have some overlapping because the two topics are close to each other. So do the topics *rec.autos* and *sci.electronics*. However, we can still distinguish them easily.

On the right side of Figure 2, ten words are listed for each cluster. These words have larger weights and are noun. They are strongly related to the topic of each cluster. They were identified based on the weights and the word functions in sentences. In fact, they can also be manually identified interactively from the left side graph. It is clear that each set of words is essentially correlated to only one of the four topics: *comp.graphics*, *comp.os.ms-windows*, *rec.autos* and *sci.electronics*. However, some high-weight words can be related to more than one topic if the topics are close to each other. For example, word 'request' has higher weight in two clusters but the topics of the two clusters are closely related (Graphics and Windows). We remark that the words identified by the weights and function analysis can improve the interpretability of the clustering results.

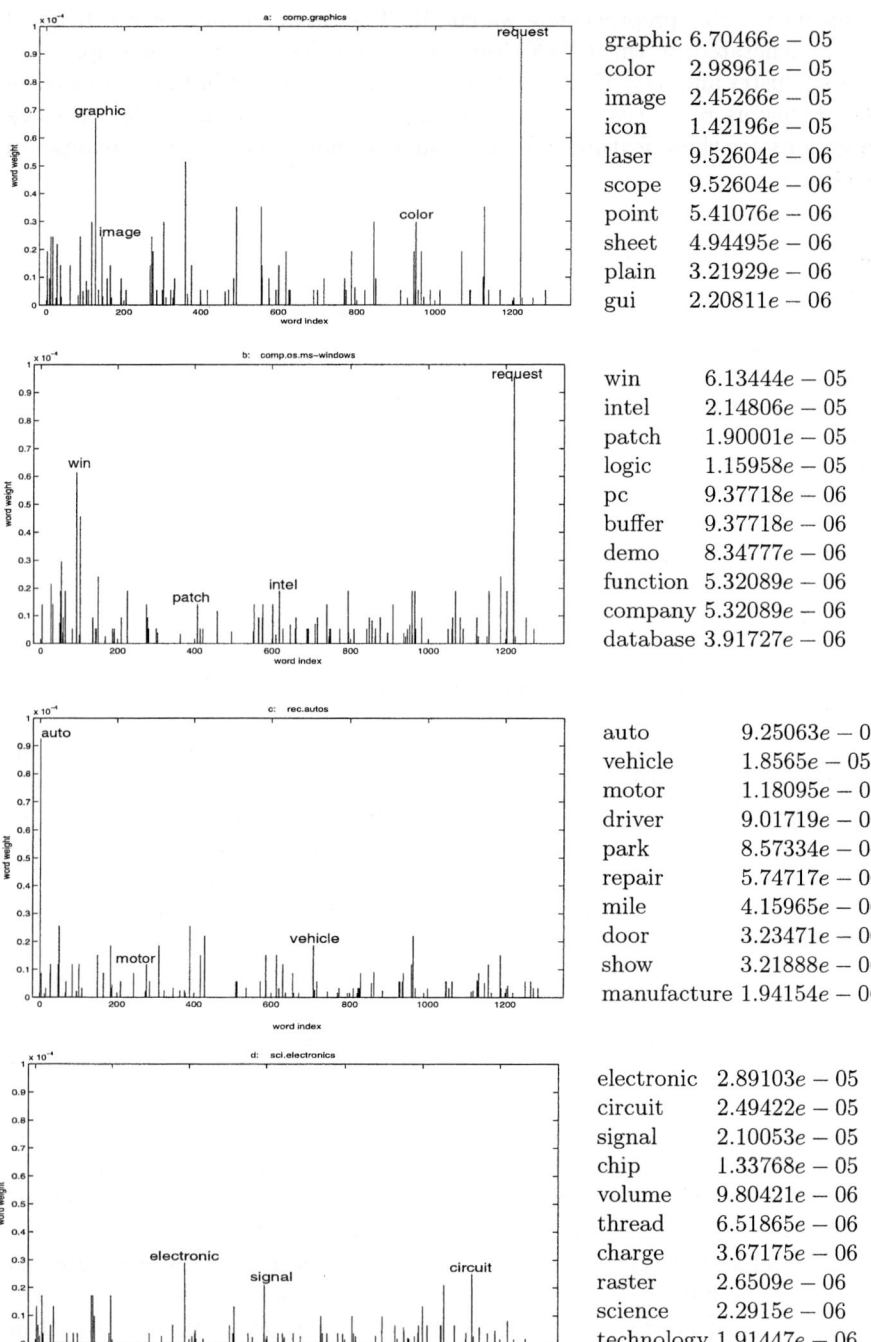

Fig. 2. The four concept vectors corresponding to a clustering of the B_4 dataset into 4 clusters. For each concept vector, the top ten words with the corresponding weights are shown on the right

We further studied whether the important words extracted is reasonable by scoring each feature with the *foil gain* function [16] and retaining only the top-scoring features for every cluster. Then, we compared them with the features identified by high weights. The result is shown in Table 6. The first line shows the precision of each cluster after performing the *FW-KMeans* algorithm. Here, C is the set of the important words extracted by *FW-KMeans*. A is the first $|C|$ words with higher foil-gain scores obtained by the original class label. B is the first $|C|$ words with higher foil-gain scores obtained by the cluster. The entry terms show the percentage of $A \cap B$ and $A \cap C$. From the table 5 we can see that most of the words with higher weights also have higher foil-gain scores. This further verifies that the words identified by high weights are important.

Table 5. Compare the foil-gain scores and weights of the important words for each cluster in the $B4$ dataset

	C_0	C_1	C_2	C_3
Precision (%)	75.78	83.02	97.59	95.12
$A \cap B$ (%)	90.45	92.96	94.64	89.94
$A \cap C$ (%)	60.91	61.3	63.69	62.26

5 Conclusions and Future Work

In this paper we have discussed the method to use the *FW-KMeans* algorithm to cluster text data in high dimensionality and sparsity. We have presented the modification to the original *FW-KMeans* to solve the sparsity problem that occurs in text data where different sets of words appear in different clusters. The capability of subspace clustering of the *FW-KMeans* algorithm has a clear advantage in clustering such text data. The experiment results have shown that the subspace clustering method was superior to the standard k-means and the *Bisection-KMeans* that cluster data on the entire space.

In the next step we will conduct a scalability benchmark test of our method on very large text data. We also plan to integrate ontology as background knowledge to enhance our method in text clustering and mining. The ontology will sever several purposes in the clustering process, including data preprocessing, selection of initial cluster centers, determination of the number of clusters k, and interpretation of clustering results.

References

1. A. Hotho, A. Maedche, and S. Staab, "Ontology-based text document clustering," *In Proceedings of the IJCAT 2001 Workshop on Text Learning: Beyond Supervision*, 2001.

2. J. MacQueen, "Some methods for classification and analysis of multivariate observations," *In proceedings of 5th berkeley symposium on mathematical statistics and probability*, pp. 281–297, 1967.
3. S. Dhillon, J. Fan, and Y. Guan, "Efficient clustering of very large document collections," *Data mining for scientific and engineering applications*, pp. 357–381, 2001.
4. M. Steinbach, G. Karypis, and V. Kumar, "A comparison of document clustering techniques," *Proc. Text ming workshop, KDD*, 2000.
5. O. Duda, E. Hart, and G. Stork, "Pattern classification," *John Wiley & Sons, 2nd edition*, 2000.
6. Y. Zhao and G. Karypis, "Comparison of agglomerative and partitional document clustering algorithms," *Technical report ♯02-014, University of Minnesota*, 2002.
7. L. Parsons, E. Haque, and H. Liu, "Subspace clustering for high dimensional data: a review," *SIGKDD Explorations*, vol. 6, no. 1, pp. 90–105, 2004.
8. H. Frigui and O. Nasraoui, "Unsupervised lerning of prototypes and attribute weights," *Pattern recognition*, vol. 37, no. 3, pp. 567–581, 2004.
9. Y. Chan, K. Ching, K. Ng, and Z. Huang, "An optimization algorithm for clustering using weighted dissimilarity measures," *Pattern recognition*, vol. 37, no. 5, pp. 943–952, 2004.
10. H. Frigui and O. Nasraoui, "Simultaneous clustering and dynamic keyword weighting for text documents," *Survey of text mining, Michael Berry, Ed, Springer*, pp. 45–70, 2004.
11. S. Dhillon and S. Modha, "Concept decompositions for large sparse text data using clustering," *Machine learning*, vol. 42, no. 1, pp. 143–175, 2001.
12. P. Hague and P. Harris, "Sampling and statistics," *Kogan Page*, 1993.
13. Z. Shi and J. Ghosh, "A comparative study of generative models for document clustering," *In SDW workshop on clustering high dimensional data and its applications*, May, 2003.
14. A. McCallum, "Bow: A toolkit for statistical language modeling, text retrieval, classification and clustering," 1996. [Online]. Available: http://www.cs.cmu.edu/mccallum/bow
15. I. Katsavounidis, C. Kuo, and Z. Zhang, "A new initialization technique for generalized lioyd iteration," *IEEE signal proceeding, Letters 1(10)*, pp. 144–146, 1994.
16. M. Mitchell, "Machine learning," *McGRAW-HILL international editions*, 1997.

Using Term Clustering and Supervised Term Affinity Construction to Boost Text Classification

Chong Wang and Wenyuan Wang

Department of Automation, Tsinghua University, Beijing, 100084, P.R.China
wangchong99@mails.tsinghua.edu.cn
wwy-dau@tsinghua.edu.cn

Abstract. The similarity measure is a crucial step in many machine learning problems. The traditional cosine similarity suffers from its inability to represent the semantic relationship of terms. This paper explores the kernel-based similarity measure by using term clustering. An affinity matrix of terms is constructed via the co-occurrence of the terms in both unsupervised and supervised ways. Normalized cut is employed to do the clustering to cut off the noisy edges. Diffusion kernel is adopted to measure the kernel-like similarity of the terms in the same cluster. Experiments demonstrate our methods can give satisfactory results, even when the training set is small.

1 Introduction

The performance of many machine learning algorithms depends on a good metric that reflects the relationship between the data in the input space. In the classical Vector Space Model (VSM), each text document is represented as vector of terms. These vectors define an input space where each distinct term represents an axis of the space. The cosine similarity defined in this space can give an effective approximation of similarities between text documents. However, cosine similarity fails to discover the semantic relationship of terms such as synonymy and polysemy.

In [1], Ferrer et al. used co-occurrence analysis where a semantic relation is assumed between terms whose occurrence patterns in the documents of corpus are correlated. In this paper, we extend the similar idea as [1], and try to get a more semantic kernel K, so that $x^T K y$ can better than $x^T y$ through the analysis of the term co-occurrences. Considering that when the training set is small, we introduce a supervised method to adjust the weight between the terms to improve the construction of term affinity matrix, called Supervised Affinity Construction. However, the affinity matrix will still have noises. We apply clustering to the term affinity matrix to decrease the noise and get semantic "small worlds" [1], which may discover semantic relationships. The final similarity matrix of terms is gained through diffusion kernels [3]. Experiments results show that the proposed method is better than the cosine similarity with either Nearest Neighbor classifier or SVM classifier, and both the clustering and supervised affinity construction contribute to the improvement.

2 Term Affinity Matrix

Humans write articles using terms, so terms can be regarded as the basic element of text documents.

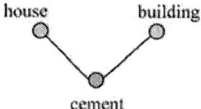

Fig. 1. An example to explain that the co-occurrence will be useful

In the traditional VSM model, the "house" and "building" will have no relationship with each other, and based on the cosine similarity, the inner product of "house" and "building" will lead to zero, even though they really have the similar meaning. In figure 1, this relationship can be represented by a linked graph and how to measure this relationship is introduced in section 3.

2.1 Unsupervised Term Affinity Matrix Construction via Co-occurrence

Given a collection of text documents D, $d_j \in D$, $1 \leq j \leq n$, $|D| = n$ and a word vocabulary T, $t_i \in T$, $1 \leq i \leq m$, $|T| = m$, a text document can be represented as $\boldsymbol{d}_j = [w_{ij}]_{1 \leq i \leq m}^T$ based on a Boolean model [5], where $w_{ij} = 1\ or\ 0$ means whether the term t_i occurs in the text document d_j or not. The term-document matrix is defined as $B = [\boldsymbol{d}_j]_{1 \leq j \leq n}$. As we can see, a text document $\boldsymbol{d}_j$ will contribute to a part of the term affinity matrix G as $G_j = \boldsymbol{d}_j \boldsymbol{d}_j^T$. So G is gained by formulation 1

$$G = \sum_{j=1}^{n} G_j = \sum_{j=1}^{n} \boldsymbol{d}_j \boldsymbol{d}_j^T = BB^T \qquad (1)$$

Apparently, G is a symmetric matrix.

2.2 Supervised Term Affinity Matrix Construction

Until now, the term affinity matrix construction process is still an unsupervised technique. When the training set is small, the statistical characteristic of co-occurrence information may be lost. In order to incorporate the supervised information, we have treated the document in the same class as a "huge" document. So, there will be only $|C|$ "huge" text document in our problem ($|C|$ is the number of classes). Another view is that the class label can be treated as "cement" in Figure 1, and it connects the two terms together. In order to add more discriminative information, we modify the weight of each pair in the "huge" documents the like follows. The assumption here is that the fewer classes the term pair (t_i, t_j) occurs together in, the more weight it will get in the affinity matrix. Based on G obtained by formulation 1, its element g_{ij} can be modified as follows

$$g_{ij} \leftarrow g_{ij} + \alpha(1 - k_{ij}/|C|)I_{ij} \quad (2)$$

where k_{ij} is the number of classes in which the term pair (t_i, t_j) occurs together, and I_{ij} is the indicating function to represent whether k_{ij} is above zero or not. This leads to our supervised affinity matrix construction. In this formulation, α controls the amounts to which the supervised information should be incorporated. Typically, $\alpha = 0$ will lead to the original unsupervised method.

3 Clustering and Diffusion Kernels

3.1 Clustering as a Denoising Process

Since some term pairs happen to be in the same document without any relationship, a clustering step is needed to eliminate noises. Those terms highly connected will probably fall into the same cluster, and the noisy links will be cut off. It is easy to transform the term affinity matrix into a graph (V, E) by taking the each term as a node and connecting each pair of terms by an edge. The weight on that edge should reflect the likelihood that two terms belong to one cluster. Based on the term affinity matrix G in section 2, we can define the graph edge weight connecting two nodes i and j as $(W = [w_{ij}]_{i \times j})$

$$y = \begin{cases} 1 & \text{if } i = j \\ \frac{g_{ij}}{g_m + \beta} & \text{otherwise} \end{cases} \quad (3)$$

where $g_m = \max(g_{ij})_{i \neq j}$, and $\beta > 0$.

It is difficult to apply k-means clustering on W. So, the normalized cut (Ncut) algorithm [6] proposed for image segmentation by solving an eigenvalue problem is adopted. A recursive two-way partition will partition the graph into more pieces or clusters. After the clustering, some elements of W will become zero to indicate that there is no link among the corresponding clusters.

3.2 Term Similarity Measure Through Diffusion Kernels

Diffusion kernel is a method to generate valid kernel on graphs [3]. Suppose the K is the kernel matrix, the inner product of two documents becomes

$$k(\boldsymbol{x}, \boldsymbol{y}) = \boldsymbol{x}^T K \boldsymbol{y} = (L\boldsymbol{x})^T L\boldsymbol{y} \quad (4)$$

where $K = L^T L$. The diffusion kernel can be generated like follows($H = [h_{ij}]_{i \times j}$)

$$h_{ij} = \begin{cases} w_{ij} & \text{if } i \neq j \\ -\sum_{k, k \neq i} w_{ik} & \text{otherwise} \end{cases} \quad (5)$$

H is called the Laplacian matrix. First diagonalize $H = P^T D P$, which is always possible because H is symmetric. Then compute

$$K = L^T L = e^{\gamma H} = P^T e^{\gamma D} P \qquad (6)$$

where γ is a positive decay factor. $L = e^{\gamma D/2} P$ gives us the diffusion kernel representation. Another advantage of clustering is that the computation of formulation 7 is reduced. Thus, let $\boldsymbol{y} = L\boldsymbol{x}$, the similarity of two documents in the feature space becomes

$$\cos \tilde{\theta}_{ij} = \frac{<\boldsymbol{y}_i, \boldsymbol{y}_j>}{\parallel \boldsymbol{y}_i \parallel \cdot \parallel \boldsymbol{y}_j \parallel} \qquad (7)$$

4 Experiments

4.1 Experiment Settings

Two well-known data sets **Reuters-21578**[1] (Reuters) and **20 Newsgroups**[2] are used in our experiments. For Reuters-21578, we only chose the most frequent 25 topics, about 9,000 documents and for 20 Newsgroups, we chose the five classes about computer, about 5,000 documents as our experiment data set. Stop words and punctuation were removed from the documents and the Porter stemmer was applied to the terms. The terms in the documents were weighted according to the widely used *tfidf* scheme. Two classifiers, nearest neighbor and support vector machine, are used in our experiments. We use F1 measure [4] to evaluate the results.

4.2 Experiment Results

We present results pertaining to two experiments. In the first experiments, we compare the term-clustering based (unsupervised) methods against traditional inner product under classifier Nearest Neighbor and SVM. In the second experiment, we compare supervised term affinity method against unsupervised term affinity and traditional methods when the size of training data varies.

In the first experiment, we randomly selected 70% of text document for the training and the rest for testing. The number of the term clusters varies from 1 to more than 2000. Figure 2 and 3 show the results under Nearest Neighbor and SVM, respectively and the axis of the number of clusters is represented as a logarithmic scale. We use TIP and TCB to refer to *T*raditional *I*nner *P*roduct method and our *T*erm *C*lustering *B*ased method, respectively. (Note: TIP has no relationship with the number of clusters, just for the ease of comparison). For SVM, the baseline is the result using the linear kernel. This experiment was conducted on Retuers-21578 data set. From the figure 2, we see that our TCB outperforms the TIP most of the time under classifiers: Nearest Neighbor

---

[1] http://www.daviddlewis.com/resources/testcollections
[2] http://www-2.cs.cmu.edu/ TextLearning/datasts.html

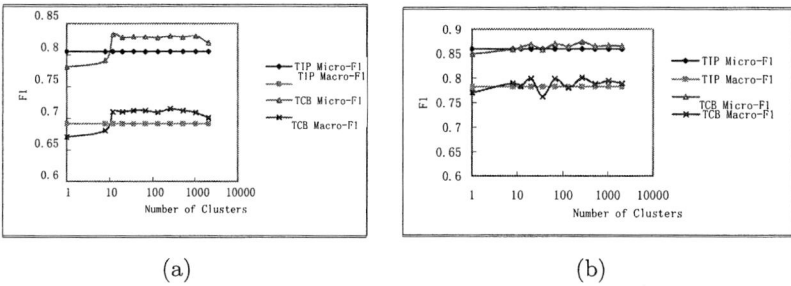

Fig. 2. Comparison of Micro-F1 and Macro-F1 between TIP and TCB on Retuers-21578 data set. (a)Nearest Neighbor. (b)Support Vector Machine, linear kernel in TIP

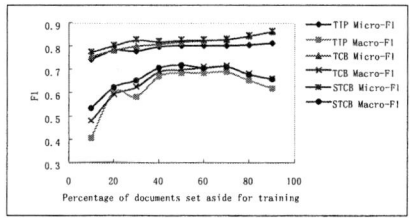

Fig. 3. Effect of size of training data set on Micro-F1 and Macro-F1. Classifier: Nearest Neighbor, Retuers-21578 data set

and SVM, and the best improvement is above 4%. We find that clustering is an importance step for our approach; when the number of cluster is too small, the performance is worse than the TIP because of the noisy edges in the graph. Typically, when the number of clusters is only 1 (no clustering), the performance has been reduced about 2%~3%. The experiments result of Nearest Neighbor is not sensitive to the number of clusters in a large range, since the most of the noisy edges in the graph are cut off at the beginning. TCB Macro-F1 of SVM is a little sensitive. One of the reasons may be that the choice of parameters of SVM in our experiment is not very appropriate.

In the second experiment, we studied the effectiveness of the supervised term affinity construction. We use STCB to represent *S*upervised *T*erm affinity construction combined *C*lustering *B*ased method. The number of term clusters and the parameter α in formulation 2 were both well tuned in our experiment. We selected the Nearest Neighbor as the classifier. This experiment was conducted on both Retuers-21578 and 20 Newsgroups data sets. Figure 3 shows the results on Retuers-21578 and Figure 4 on 20 Newsgroup (we present the results of 20 Newsgroup separately because the curves of Micro-F1 and Macro-F1 overlap too much).

From the figure 3 and 4, we see that when the training data set is small, STCB performs better than TCB and TCB performs better than TIP, and the best improvement is about 5%. The supervised information does help giving pos-

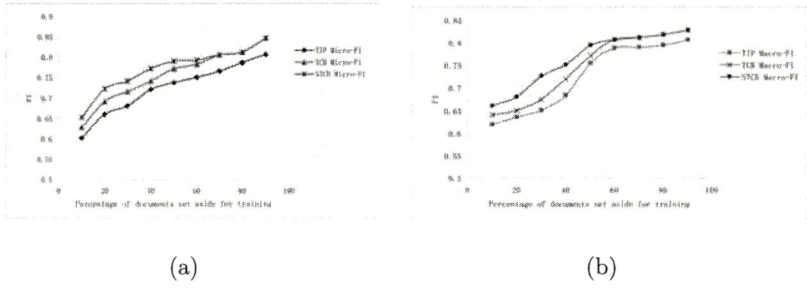

Fig. 4. Effect of size of training data set on (a)Macro-F1, (b)Macro-F1. Classifier: Nearest Neighbor, 20 Newsgroup data set

itive results in this situation. When the training data set is large (>60%~70%), STCB performs equally to TCB, both better than TIP. Actually, the parameter α in formulation 2 is zero now, so STCB has become TCB. The supervised information does little help when the training data is enough. In this case, the training data can already give reasonable statistical characteristic of co-occurrence information.

5 Conclusions

In this paper, we introduce a term clustering based kernel to enhance the performance of text classification. The term affinity matrix is constructed via the co-occurrence of the terms in both unsupervised and supervised ways. Normalized cut is used to cluster the terms by cutting off the noisy edges. The final kernel is generated through the exponential matrix operation of diffusion kernel. Experiment result shows our proposed method can explore the semantic relationship between terms effectively.

Acknowledgements

We would like to express our special appreciation to Ni Lao, and Lu Wang for their insightful suggestions.

References

1. Ferrer, R. and Sole, R.V.: The small world of human language. In Proceedings of the Royal Society o f London Series B- Biological Sciences, pages 2261-2265, 2001
2. Kandola, J., Taylor, J.S, Cristianini, N. and Davis.: Learning Semantic Similarity. In Proceedings of Neural Information Processing Systems. 2002
3. Kondor, R.I. and Lafferty, J.: Diffusion kernels on graphs and other discrete structures. In Proceedings of International Conferecne on Machine Learning (ICML 2002), 2002

4. Rijsbergen, C.J.: Information Retrieval. Butterworths, 2nd Edition, 1979
5. Salton, G. and Michael, J.: Introduction to Modern Information Retrieval. McGraw-Hill, 1983
6. Shi, J. and Malik, J.: Normalized Cuts and Image Segmentation. IEEE Transactions on Pattern Analysis and Machine Intelligence, 22(8), 888-905, August 2000

Technology Trends Analysis from the Internet Resources

Shin-ichi Kobayashi[1], Yasuyuki Shirai[2], Kazuo Hiyane[2], Fumihiro Kumeno[2], Hiroshi Inujima[1], and Noriyoshi Yamauchi[1]

[1] Graduate School of Information, Production and Systems,
Waseda University, Kitakyushu, Fukuoka 808-0135, Japan
kobayash@suou.waseda.jp
{inujima-hiroshi, yamauchi-noriyoshi}@waseda.jp
[2] Mitsubishi Research Institute, Inc., Tokyo, 100-8141, Japan
{shirai, hiya, kumeno}@mri.co.jp

Abstract. In business or technology planning, it is strongly required to grasp the overall technology trends and predict what will happen in the near future. In this paper, we propose the method where we can detect and analyze the technology trends from the Internet resources.

1 Introduction

Many attempts have been made to organize and present the relationships between technical concepts for advanced use of the Internet content and academic papers [1, 2, 3, 4, 5, 6, 7, 8]. Based on these researches, we need further enhancement of the methods to grasp not only the current relationships between concepts but also the near future trends of the specified technology field. In this paper, we propose some metrics as leading indicators of changes over time, by which we can analyze the current situation and the future trends. The architecture of our system is shown in Figure 1.

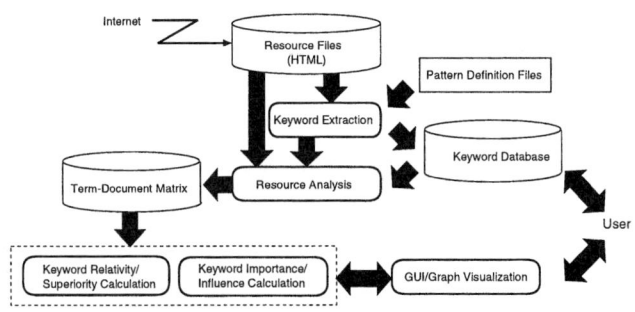

Fig. 1. The Architecture of the System

2 Technical Trend Maps and Trend Indicators

2.1 Keyword Extraction

This module extracts keywords from resources using pattern matching, including regular expressions, and stores the keywords together with their phases in a keyword database. The patterns depend on the resource type and are defined by the user. This module extracts keywords for nine different phases: organization, technical term, system, protocol, business service, person, hardware, social phenomenon/trend, and business term. By considering indicators that are discussed later, in terms of individual phases, it is possible to recognize the target field from a global perspective (for example, organizationally driven, technologically driven, standards-centered, etc.).

2.2 Resource Analysis

This module determines whether or not the keywords in the dictionary appear in each resource and generates a term-document matrix. If we consider n documents with m keywords, the size of the matrix is $m \times n$. Each element f_{ij} is 1 if keyword i is included in document j and 0 if not. In addition, $\mathbf{t}_i = (f_{i1}, \cdots, f_{ij}, \cdots, f_{in})$ is the term vector relating to keyword i, and $\mathbf{d}_j = (f_{1j}, \cdots, f_{ij}, \cdots, f_{mj})$ is the document vector relating to document j.

2.3 Keyword Relativity and Superiority

The relativity and superiority are defined between two keywords in a group of specified resources.

Relativity. The relativity $R(i,j)$ of keywords i and j is defined by the similarity of the group of documents in which they appear. $R(i,j)$ is defined by the well-known metrics, that is, the cosine of the term vectors $\mathbf{t}_i$ and $\mathbf{t}_j$, as shown below.

$$R(i,j) = \frac{\mathbf{t}_i \cdot \mathbf{t}_j}{\|\mathbf{t}_i\| \cdot \|\mathbf{t}_j\|}$$

Superiority. The objective of superiority is to extract a hierarchical relationship between two keywords in the group of defined resources. In order to define superiority, the complementary concept i^c for keyword i is defined. The complementary concept i^c is a hypothetical concept that bears no similarity (no relationship) to i, that is, $\mathbf{t}_{i^c} = (1 - f_{i1}, \cdots, 1 - f_{ij}, \cdots, 1 - f_{in})$. For two keywords i and j, assume i is a superior concept to j, then in accordance with the relativity between i and the complementary concept of j would be higher than the relativity between j and the complementary concept of i. Accordingly, we define the superiority $P(i,j)$ of i with respect to j as follows:

$$P(i,j) = \arccos(R(i,j^c)) - \arccos(R(i^c,j))$$

where $-\frac{\pi}{2} \leq P(i,j) \leq \frac{\pi}{2}$. If $P(i,j) < 0$, i is called to be superior to j.

2.4 Keyword Importance and Influence

The importance and influence indicators show the status or positioning of keywords within a specified resource.

Importance. Importance is an indicator of relative importance within a specified resource and uses the well-known indicators *tf* and *idf*. The importance indicator $W_R(i)$ of keyword i in a resource group R (with n documents) is defined as follows:

$$W_R(i) = tf_R(i) \cdot \left(\log \frac{n}{df_R(i)} + 1\right)$$

where $tf_R(i)$ is a total number of appearances of i in R and $df_R(i)$ is a number of documents in R that include i.

Influence. Suppose a keyword group A has been defined for a resource group R. The influence $I_A(i)$ of keyword $i \in A$ with respect to A indicates the degree of influence of i for other keywords in A. The evaluation considers the degree to which the appearance of i in the documents reduces the entropy with respect to the other keywords in A, namely $(A \setminus \{i\})$.

First, the entropy relating to appearance of a keyword in a certain keyword group S is indicated by $H(S)$ as follows:

$$H(S) = -p(S) \log p(S) - p(\bar{S}) \log p(\bar{S})$$

where $p(S)$ ($p(\bar{S})$) is a probability of any of the keywords in S being (not being) included in the document.

In addition, $H(S \mid i)$ is the entropy relating to the appearance of S under the condition that the appearance of keyword i is known. $H(S \mid i)$ is defined as follows:

$$H(S \mid i) = - \sum_{x=S,\bar{S}} \sum_{y=i,\bar{i}} p(x,y) \log p(x \mid y)$$

where $p(x, y)$ is a joint probability of x and y, and $p(x \mid y)$ is a conditional probability of x based on y.

In this situation, the influence $I_A(i)$ of $i \in A$ with respect to a keyword group A is defined as follows:

$$I_A(i) = \frac{H(A \setminus \{i\}) - H(A \setminus \{i\} \mid i)}{H(A \setminus \{i\})}$$

$I_A(i)$ indicates the proportion of reduction of the entropy for $A \setminus \{i\}$ as a result of knowing i. The greater $I_A(i)$ is, the greater the influence of i on the appearance of the other keywords of A. We consider that a highly influential keyword could induce structural changes in the target field after a certain time. In other words, influence is one of the leading indicators of time change. The nature of the change cannot be determined uniformly, but typically, there is an increase or decrease in the importance of the highly influential keyword itself or in a different keyword that has a high degree of relativity or superiority with the highly influential keyword.

2.5 Trend Analysis as Time Series

By continuously calculating the indicators discussed in section 2.3 and 2.4 for the specified set of keywords that characterize the field in a time series, it is possible to obtain data on past trends and future developments in that field. We consider the changes on the average values for importance, influence, relativity and superiority of keywords in the target field or phase as indicators for time series analysis, as well as the changes of importance and influence for the specified keyword.

3 Experiments and Evaluation

This section describes experiments using actual Internet resources, ITmedia[1] news articles (2001 - 2003), targeting the categories of broadband and mobile communications. To identify time series trends in fields and phases, significant or representative keywords in that field are extracted according to their importance values. We consider that the trends of that field can be revealed by analyzing the trends of that group of keywords.

Analysis of Overall Trends in the Broadband Field. We have selected 19 keywords according to their importance values. Figure 2 shows the overall trends of indicators in time-series from period 1 (2001/07 - 2001/12) to period 13 (2003/07 - 2003/12).

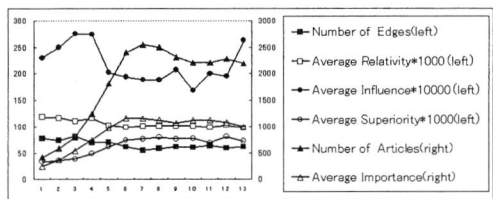

Fig. 2. Overall Trends of the Broadband Field

The average influence showed high peaks in periods 3 and 4. Many keywords with a high degree of influence appeared during these periods. As if in response, the average importance increased through period 6 and subsequently stabilized. Based on the meanings for the indicators shown in section 2.3, 2.4 and 2.5, this can be interpreted as the high influence of this trend, which first appeared in periods 3 and 4 and became a leading indicator of the broadband "boom" that followed. In actuality, the media began reporting widely on the potential for

---

[1] One of the most famous IT news site in Japan (http://www.itmedia.co.jp).

new platforms and services at that time, such as WiFi hot spots, IPv6, MPEG-4 video transmission, etc. Moreover, the fact that the average superiority gradually increased indicates that some kind of hierarchic structure had developed between the important keywords.

Detailed Analysis of the Broadband Field. Figure 3 shows the trends in importance and influence for each keyword. Noteworthy in the broadband field is the conspicuous growth in the importance of wireless LAN. Wireless LAN was followed in importance by DSL, mobile phones, IP telephoning, all of which showed steady increases in importance, which matched the situation at the time. The rapid growth of wireless LAN can be predicted from the influence value. Both the absolute value and the influence of IEEE 802.11i and WiFi hot spots increased rapidly up through period 4, following which the importance of wireless LAN increased.

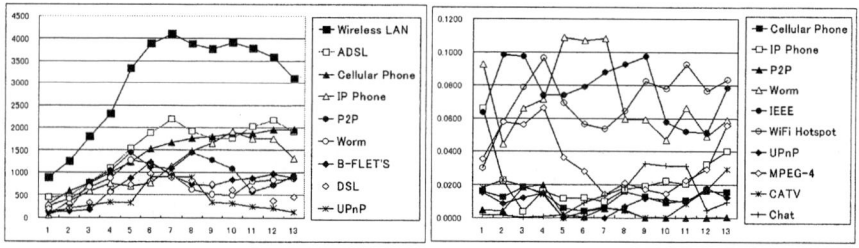

Fig. 3. Trends of Importance/Influence of the Broadband Field

As for the the relationships of the keywords to one another during the specified period, we found that wireless LAN has the highest relativity value and a high superiority value with respect to IEEE and WiFi hot spots. In this sense as well, the hypothesis noted in the definition of influence is supported.

Comparison with Human Experts' Estimation. In order to reinforce the performance of this method, we surveyed the estimations of 10 researchers who are particularly interested in and have minimum 10 years' working experience in these fields so that we can compare the results of this method with human experts' opinions.

First evaluation is about the importance and the influence. Keywords can be relatively characterized in terms of the importance and the influence. For example, if a keyword has high importance and influence, then it can be classified as an important and noteworthy concept. We evaluated this method by comparing the system's and human experts' answers about the classification of each keyword. Table 1 shows the results for this comparison. Although the value of concordance rate are not extremely high in themselves, taking into consideration that the answers by human experts hardly agreed with each other, we conclude that our system makes good performance and has ability to judge the status of each technology concept properly. Second evaluation is for the superiority. We surveyed human experts' opinions about the structural configuration of the organizations, at the time of 2001 and 2003. While the results of the system say that higher hierarchization with respect to organizations achieved at the end of 2003 rather than at 2001 in the broadband field, eight of ten experts agreed with this trend.

Table 1. The concordance rate of the system and human experts

	Case 1	Case 2
Broadband	53%	61%
Mobile	50%	57%

Case 1: Concordance rate of the method's quadrant and human experts' quadrants regarding all keywords. Case 2: Concordance rate of the method's quadrant and human experts' quadrants regarding major keywords (over 60% of human experts' answers agreed)

4 Summary

In this paper, we proposed the methods to characterize technology concepts and technology trends based on the Internet news resources, introducing some new metrics to grasp time series trends for technology fields or phases.

Some experimental results including comparison with human experts proved the validity of this method and the use of indicators for technology trend analysis.

References

1. Aizawa, A. and Kageura, K.: Calculating Associations between Technical Terms Based on Co-ocurrences in Keyword Lists of Academic Papers. *Trans. IEICE* Vol.J83-D-I, No.11, 2000 (in Japanese)
2. Watabe, I. and Kawaoka, T.: The Degree of Association between Concepts using the Chain of Concepts. *Proc. of IEEE International Conference on Systems, Man & Cybernetics*, 2001
3. Boyack, K.W. and Borner, K.: Indicator Assisted Evaluation and Funding of Research. *JASIST*, Vol.54, Issue.5, 2003
4. Hassan, E.: Simultaneous Mapping of Interactions between scientific and technological Knowledge Bases. *JASIST*, Vol.54, Issue.5, 2003
5. Maedche,A., Pekar,V. and Staab,S.: Ontology Learning Part One - On Discovering Taxonomic Relations from the Web. In Zhong,N., Liu,J. and Yao,Y.Y., editor, *Web Intelligence*, Springer Verlag, 2002
6. Morris, S. A., Yen, G., Wu, Z., and Asnake, B.: Timeline visualization of research fronts. *JASIST*, Vol.55, Issue.5, 2003
7. Chen,C, and Kuljis, J.: The Rising Landscape:A Visual Exploration of Superstring Revolution in Physics. *JASIST*, Vol.54, Issue.5, 2003
8. Kontostathis, A., Galitsky, L. M., Pottenger, W. M., Roy, S., Phelps, D.: A Survey of Emerging Trend Detection in Textual Data Mining. Berry, M. W., editor,*Survey of Text Mining*, Springer Verlag, 2004

Dynamic Mining Hierarchical Topic from Web News Stream Data Using Divisive-Agglomerative Clustering Method[1]

Jian-Wei Liu, Shou-Jian Yu, and Jia-Jin Le

College of Information Science & Technology
Donghua University
liujw@mail.dhu.edu.cn

Abstract. Given the popularity of Web news services, we focus our attention on mining hierarchical topic from Web news stream data. To address this problem, we present a Divisive-Agglomerative clustering method to find hierarchical topic from Web news stream. The novelty of the proposed algorithm is the ability to identify meaningful news topics while reducing the amount of computations by maintaining cluster structure incrementally. Our streaming news clustering algorithm also works by leveraging off the nearest neighbors of the incoming streaming news datasets and has ability of identifying the different shapes and different densities of clusters. Experimental results demonstrate that the proposed clustering algorithm produces high-quality topic discovery.

1 Introduction

On most Web pages, vast amounts of useful knowledge are embedded into text. Given such large sizes of text datasets, mining tools, which organize the text datasets into structured knowledge, would enhance efficient document access. Given that the Web has become a vehicle for the distribution of information, many news organizations are providing newswire services through the Internet. Given this popularity of the Web news services, we have focused our attention on mining topic from news streams.

In this paper, we propose a mining algorithm that supports the identification of meaningful topics from news stream data. We introduce the algorithm that works in a single-pass manner, where the documents are processed sequentially, one at a time. The algorithm explore dynamically divisive and agglomerative characteristic in whole document clustering process.

Conventional divisive clustering algorithm such as K-means and K-medoid algorithms is much faster than the hierarchical technique but not as accurate, while conventional agglomerative clustering algorithm is completely accurate but is very CPU intensive when the data set has a large number of data points. Our method which

---

[1] This work was sponsored by the National Science Foundation of China (NSFC) under the grant No.60273085 and the state High-tech Research and Development project (863) under the grant No.2002AA4Z3430.

combines the strengths of each method attempts to remedy this problem by first performing the dynamic divisive clustering and then dynamic agglomerative clustering as the new incoming documents arrive, and both clustering phases equip a stopping criteria based on Hoeffding bound[1], thereby achieving a reasonable run time without sacrificing accurate acquirement. The system tests existing clusters by descending order of diameters, looking for a possible binary split. If no cluster deserves division, then the system searches for possible aggregation of clusters.

The main problem, then, with both of these methods is that their inability of identifying the different shapes and different densities of clusters. Our streaming news clustering algorithm is further able to work by leveraging off the nearest neighbors of the incoming streaming news datasets and has ability of identifying the different shapes and different densities of clusters.

Main features of our algorithm include the use of similarity as distance measure, a splitting criteria supported by the Hoefding bound, a stopping criteria based on the divisive coefficient and an agglomerative phase which decreases the number of unneeded clusters.

The rest of this paper is organized as follows. In Section 2, we introduce our problem definition. In Section 3 we represent our proposed method, In Section 4 we provide experimental results. In Section 5 we summarize our works.

2 Problem Definition

The following section is terminology and definitions. These are necessary to concretely define the problem at hand, and to explain our proposed solution.

Definition 1. $Similarity(d_i, d_j)$ And $Dissimilarity(d_i, d_j)$: To measure closeness between two documents, we use the Cosine metric, which measures the similarity of two vectors according to the angle between them. The cosine of the angles between two n-dimensional document vectors (d_i and d_j) is defined by

$$similarity(d_i, d_j) = \frac{\sum_{i=1}^{n} x_i \cdot x_j}{\|x_i\|_2 \cdot \|x_j\|_2} \tag{1}$$

the dissimilarity between two documents is represented as

$$Dissimilarity(d_i, d_j) = 1 - \frac{\sum_{i=1}^{n} x_i \cdot x_j}{\|x_i\|_2 \cdot \|x_j\|_2} \tag{2}$$

(4)

Definition 2. Similar: If $similarity(d_i, d_j) > \varepsilon$, then a document d_i is referred to as similar to a document d_j.

Definition 3. ζ-Neighborhood $N_\zeta(d_i)$: ζ-neighborhood for a document d_i is defined as a set of documents $\{x: similarity(x, d_i) \geq \zeta\}$.

3 Proposed Mining Algorithms

3.1 Streaming News Model

According to different application requirements, the scope of news streams can be categorized into two different alternatives. The landmark window identifies certain starting landmark in the stream and incorporates sequences since that point until the present. The sliding window is featured with a fixed length time-span of news the sequences. In this paper, we use logical-based landmark windows [2], also known as count-based or sequence-based landmark windows. The length of window is defined in terms of the number of news sequences.

3.2 Dynamic Divisive-Agglomerative Document Clustering

Over all different clustering techniques of hierarchical models known in literature, Agnes [3] proceeds by a series of fusions, and Diana [3] is a hierarchical clustering technique that constructs the hierarchy from the top (large cluster) to bottom (several clusters).

Agglomerative Nesting (Agnes). Agnes proceeds by a series of fusions. Initially, all objects are apart–each object forms a small cluster by itself. At the first step, two closest or minimally dissimilar objects are merged to form a cluster. Then the algorithm finds a pair of objects with minimal dissimilarity. If there are several pairs with minimal dissimilarity, the algorithm picks a pair of objects at random.

Divisive Analysis (Diana). Diana [3] is a hierarchical clustering technique, but its main difference with the agglomerative method (Agnes) is that it constructs the hierarchy in the inverse order. Initially, there is one large cluster consisting of all n objects. At each subsequent step, the largest available cluster is split into two clusters until finally all clusters, comprise of single objects. Thus, the hierarchy is built in n-1 steps.

One problem that usually arises with these sorts of models is the definition of a minimum number of observations that are necessary to assure convergence. Techniques based on the Hoeffding bound [1] can be applied to solve this problem, and have in fact be successfully used in online decision trees [8] [9].

Divisive-Agglomerative clustering algorithm use a statistical result known as Hoeffding bounds or additive Chernoff bounds. After n independent observations of a real-valued random variable r with range R, the Hoeffding bound ensures that, with confidence $1-\delta$, the true mean of r is at least $\bar{r} - \varepsilon$, where $\bar{r}$ is the observed mean of the samples and

$$\varepsilon = \sqrt{\frac{R^2 \ln(1/\delta)}{2n}} \qquad (3)$$

This is true irrespective of the probability distribution that generated the observations.

Figure 1 shows our proposed dynamic divisive-agglomerative clustering algorithm. We use count-based landmark window model supposed in 3.1. Initially, we assume that only news documents $\{d_1, d_2, \cdots, d_{w_{t_i}}\}$ in now window is available. Thus, this news documents itself forms a singleton cluster. Adding a new document to existing cluster structure proceeds in three phases: neighborhood search, dynamic divisive phase and dynamic agglomerative phase.

Input: A set of streaming news document, w_{t_i} is the length of the landmark window.

1. Get next news documents $\{d_1, d_2, \hbar, d_{w_{t_i}}\}$ in now window

2. Neighborhood search: Given a new incoming document $\{d_1, d_2, \hbar, d_{w_{t_i}}\}$, obtain $\{N_\varsigma(d_1), N_\varsigma(d_2), \hbar, N_\varsigma(d_{w_{t_i}})\}$ by performing a neighborhood search, and find the cluster C_N which can host a new document $d_i \in \{d_1, d_2, \hbar, d_{w_{t_i}}\}$, that mean to identify $C_N \supset N_\varsigma(d_i)$. Then add d_i to the cluster C_K

3. Choose next cluster $C_k \psi$ in descendant order of diameters or sum of pairwise distances within each cluster (initially, next news document $\{d_1, d_2, \hbar, d_{w_{t_i}}\}$ in now window is cluster which have a largest diameter)

4. Call Check-slitting () in cluster C_K

5. If we find a split point, goto 11 with a new cluster

6. If still exists a cluster $C_i \psi$ not yet tested for splitting goto 4

7. Choose next cluster $C_K \psi$ in ascendant order of diameters or or sum of pairwise distances within each cluster

8. Call Check-aggregate () in cluster C_K

9. If we found an aggregation then goto 11 with new cluster

10. If still exists a cluster $C_i \psi$ not yet tested for aggregation goto 8

11. If not end of data goto 1

Fig. 1. Divisive-Agglomerative Clustering Algorithm

4 Experiment Results

In this section, we list some experimental results for the proposed Web document clustering algorithm.

For the experiments we present here we selected 3000 web pages in four broad categories: business and finance, electronic communication and networking, sport, and manufacturing. These pages were downloaded, labeled, and archived. The word lists from all documents were filtered with a stop-list and "stemmed" using Porter's suffix-stripping algorithm [4] as implemented by [5].

The quality of a generated cluster hierarchy was determined by two metrics, precision and recall.

In our experiment, we find the best clustering result when the ζ value, which measure neighborhood for a document d_i, varies between 0.2 and 0.25, i.e., the algorithm guessed the exact number of clusters. If the value of ζ was too small, then the algorithm found a few large-size clusters. In contrast, many small-size clusters were identified if the value ζ is too large.

For the purpose of comparison, we decided to use spherical K-means clustering algorithm [6] and DC-tree clustering algorithm [7]. We compare the results of the various experiments across algorithms and represent in bellow Table1.

Table 1. Precision and recall comprison of our proposed algorithm, spherical K-means clustering algorithm and DC-tree clustering algorithm

Cluster Topic	Precision (our algorithm)	Recall (our algorithm)	Precision (K-means)	Recall (K-means)	Precision (DC-tree)	Recall (DC-tree)
congestion control	91.4	90.2	81.4	81.2	81.7	80.9
intrusion detection	89.4	88.1	79.5	80.7	68.7	71.4
Internet game	84.8	86.4	78.3	81.2	69.4	67.8
Credit fraud	92.3	89.6	76.8	79.9	72.3	75.4
Job opportunity	88.1	91.4	79.5	78.9	75.4	78.6
soccer	87.8	90.4	79.0	82.3	74.6	76.1
film	87.9	86.6	78.8	78.9	79.4	80.2
TV	84.9	88.7	80.3	79.6	69.8	73.4

Since the sizes of clusters can be of arbitrary numbers, clustering algorithms must be able to identify the clusters with wide variance in size. To test the ability of identifying clusters with different densities, we organized datasets where each dataset consists of document clusters with diverse densities. Then we perform clustering algorithm on this datasets. Table 2 shows the average values of precision and recall for all topics. As shown in Table 2, when the density of each cluster is not uniform, the accuracy of the modified K-means clustering algorithm degraded. In contrast, the accuracy of our algorithm remains similar. Then we can conclude that the proposed algorithm outperforms the modified K-means algorithm in terms of precision and recall. This is because the proposed algorithm measures similarity between a cluster and a neighborhood of a document while spherical K-means [6] and DC-tree [7] clustering measures similarity between a cluster and a document.

Table 2. Precision and recall of DC-tree clustering algorithm

Cluster Topic	Precision	Recall
Our algorithm	86.8	88.5
K-means	65.7	61.4
DC-tree	67.4	64.8

5 Conclusion

This work presents a divisive-agglomerative clustering algorithm that works online. The divisive-agglomerative clustering system hierarchically builds clusters of news document topic, using a single scan over the data. The experimental results suggest that the system exhibit dynamic behavior, adapting to changes in news document stream data. It also has ability of identifying the different shapes and different densities of clusters.

References

1. Hoeffding, W.: Probability Inequalities for Sums of Bounded Random Variables. Journal of the American Statistical Association (1963) 13-30.
2. Yunyue Zhu, Dennis Shasha: StatStream: Statistical Monitoring of Thousands of Data Streams in Real Time. In VLDB International Conference, pages 358–369, Hong Kong, China, August 2002.
3. Kaufman, L. and Rousseeuw, P.J.: Finding Groups in Data: An Introduction to Cluster Analysis. Wiley, New York. (1990).
4. M. F. Porter.: An Algorithm for Suffix Stripping. Program. (1980)14(3):130-137.
5. W. B. Frakes.: Stemming Algorithms. In W. B. Frakes and R. Baeza-Yates, editors, Informtion Retrieval Data Structures and Algorithms. (1992)131-160.
6. Inderjit S., Dhillon Dharmendra, S. Modha : Concept Decompositions for Large Sparse Text Data Using Clustering. Volume 42. (2001) 143 - 175
7. W.C. Wong, A. Fu, Incremental Document Clustering for Web Page Classification. Chapter in the book titled ``Enabling Society with Information Technology", Springer-Verlag. (2002) 101-110.
8. Domingos. P., Hulten, G.: Mining High-Speed Data Streams. In: Proceedings of the Sixth International Conference on Knowledge Discovery and Data Mining, Boston, MA, ACM Press (2000) 71-80.
9. Gama, J., Medas, P., Rodrigues, P.: Concept Drift in Decision-Tree Learning for Data Streams. In: Proceedings of the Fourth European Symposium on Intelligent Technologies and their implementation on Smart Adaptive Systems, Aachen, Germany, Verlag Mainz (2004) 218-225.

Collecting Topic-Related Web Pages for Link Structure Analysis by Using a Potential Hub and Authority First Approach

Leuo-Hong Wang[1] and Tong-Wen Lee[2]

[1] Evolutionary Computation Laboratory, Aletheia University, Taiwan
wanglh@email.au.edu.tw
[2] Evolutionary Computation Laboratory, Aletheia University, Taiwan
fa925722@email.au.edu.tw

Abstract. Constructing a base set consisting of topic-related web pages is a preliminary step for those web mining algorithms which use the link structure analysis technique based on HITS. However, except checking the anchor text of links and the content of pages, there has been few of research addressing other possibilities to improve topic relevance while collecting the base set. In this paper, we propose a potential hub and authority first (PHA-first) approach utilizing the concept of hub and authority to filter web pages. We investigate the satisfaction of dozens of users about the pages recommended by our method and HITS on different topics. The results indicate that our method is superior to HITS in most cases. In addition, we also evaluate the recall and precision measures of our method. The results show that our method is with relative high precision and low recall for all topics.

Keywords: Web mining, link structure analysis, hub and authority.

1 Introduction

PageRank [1], HITS [2] and SALSA [3] are three of the most popular link-structure based algorithms for web mining. No matter which method is used, a graph consisting of web pages and their hyperlinks will be constructed for further processing. Consequently, different graph analysis algorithms or traversing strategies are applied in order to measure the importance of each web page. Borodin and colleagues [4] further differentiated between these algorithms according to the scope of their derived target graph. In their opinion, PageRank is *query independent* because it operates on the whole web. On the other hand, both HITS and SALSA are *query dependent* since a base set consisting of web pages relevant to a given topic must be constructed from a query on the topic. Using filtering strategies is needed to construct such a base set since the amounts of irrelevant pages examined during construction are enormous. Whether the filtering strategies are good enough for excluding topic irrelevant pages as far as possible will significantly influence the recommendation results. However, eliminating navigational links is the only effort to filter irrelevant pages for most of the HITS based algorithms. In this paper, we will address the quality issue of the base set. The definition of *quality* used here is how *relevance* of the recommendation results to the given topic. We will measure the relevance by two different ways. The first is to calculate the precision and recall rates suggested by Vaughan [5]. The second is to investigate the quality of recommendation results via a user satisfaction survey.

In order to filter irrelevant web pages as far as possible, we propose a novel approach called potential hubs and authorities first (PHA-first) approach to construct a base set for further hub and authority analysis. Our PHA-first approach examines the outgoing and incoming links of the root set as usual. Instead of including all pages pointing to or pointed by the root set, we apply several inclusion rules, including co-citation, recommendation and relay to both the outgoing and incoming link sets.

The remainder of this paper is organized as follows. Section 2 introduces our PHA-first approach and the derived algorithm. Section 3 is our experimental results and discussion. And then, we draw a conclusion and describe the future work in the last section.

2 The Potential Hubs and Authorities First Approach

The hypertext-induced topic selection (HITS) algorithm proposed by Kleinberg is a very popular and effective algorithm to rank documents. However, the method still suffers from the problem of *topic drift* [6]. To reduce the impact of topic drift, various schemes have been proposed [6][7][8]. Although, these methods improved the performance of the original HITS, the underlying methods of base set construction were nearly the same.

According to [3], Lempel and Moran claimed that the quality of base set is an elementary requirement for any HITS-base algorithms. That motivated us to explore the possibility to improve the recommendation results by controlling the "quality" of the base set. The concept is actually not unusual. Chau et.al., [9] have proposed a priority first approach to collect topic relevant web pages, although the method was not HITS-based. Inspired by Chau's research, we try to develop a priority first approach based on the framework of HITS algorithm.

Reconsidering the definition of hub and authority, a good hub is a web page *recommending* many good quality pages and a good authority is a web page *cited* by many good quality pages. If we have constructed a root set of good quality web pages, then we can iteratively apply some inclusion rules which choose the pages in accordance with recommendation or co-citation to construct the base set. Based on the motivation, we come up with a novel base set construction method. Instead of filtering the candidate pages randomly[2][3] or by examining anchor text [7][9], our method applies three inclusion rules further utilizing the definition of hub and authority. Provided that a root set with good quality, such as a set consisting of the top dozens query results of a content-based search engine on a keyword, our inclusion rules will select *potential* hubs or authorities

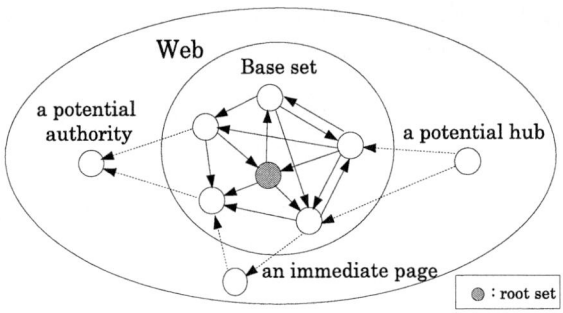

Fig. 1. Potential hubs, potential authorities and immediate nodes (relays)

into the base set as far as possible. Therefore, our method is named as potential hubs and authorities first (PHA-first) approach.

Assume $\mathcal{B}$ denotes the current base set, $B(p_i)$ as the set of all pages pointing to p_i and $F(p_i)$ as the set pages pointed by p_i. The three rules in our PHA-first approach are:

1. *The co-citation rule.* As shown in left side of Figure 1, every page p_a not belonging to $\mathcal{B}$, which is pointed by at least two pages in $\mathcal{B}$, will be added to $\mathcal{B}$.
2. *The recommendation rule.* As shown in right side of Figure 1, every page p_h not belonging to $\mathcal{B}$, which points to at least two pages in $\mathcal{B}$, will also be added to $\mathcal{B}$.
3. *The relay rule.* Two preceding rules aim at including pages with the *stronger* relevance to $\mathcal{B}$. However, as shown in the lower part of the Figure 1, some pages are immediate nodes between two pages in $\mathcal{B}$. We call these pages as *relays*. More formally, a relay is a page p_r not belonging to $\mathcal{B}$, but pointing to only one page in $\mathcal{B}$ and pointed by one page in $\mathcal{B}$ at the same time. Such pages will be added to $\mathcal{B}$.

Based on these three rules, our PHA-first approach works as shown in Figure 2. In Figure 2, p_h, p_a, p_{relay} are potential hubs, authorities and relays relative to the current base set S_q respectively. After adding these pages and deleting all irrelevant pages, the current iteration ends. Depending on how many links away from the root set we want to explore, we can decide the number of iterations. The default value for iterations is 3, which is larger than the setting of the CLEVER [7] algorithm by 1.

The rest of details on dealing with pages are listed as follows:

1. *Choice of the root set.* Because of iterative nature, our PHA-first approach can start from a small root set. In our opinion, we can easily control the quality of the root set if the number of pages in the root set is small enough.
2. *Deletion of irrelevant pages.* The main purpose of deleting irrelevant pages is to eliminate the navigational links. In addition, we count the number of

```
Input: a query string q; two integers t,w; R_q, the top t results of content-
based search engines on q
Output: a set of web pages S_q extended from R_q

PHA-first(q,t,w,R_q)
S_q ← R_q
Let B(p_i) denotes the set of all pages pointing to p_i
Let F(p_i) denotes the set of all pages p_i points to
while the number of iterations is less than w do
    Add p_h to S_q, where p_h ∈ B(p_i) ∩ B(p_j); ∀p_h ∉ S_q; p_i, p_j ∈ S_q, i ≠ j.
    Add p_a to S_q, where p_a ∈ F(p_i) ∩ F(p_j); ∀p_a ∉ S_q; p_i, p_j ∈ S_q, i ≠ j.
    Add p_relay to S_q, where p_relay = {p_r | p_r ∈ F(p_i) ∧ p_r ∈ B(p_j), ∀p_r ∉
    S_q, p_i, p_j ∈ S_q, i ≠ j}.
    Delete all irrelevant pages from S_q.
    The iteration count is added by one.
end while
Return S_q
```

Fig. 2. The PHA-first (potential hub and authority first) approach

out-going links for each page. Once the number is larger than a threshold value, we drop the page because it is very likely to be a portal.

After the base set $\mathcal{B}$ is constructed, the hub and authority values of each page can be calculated. The calculation is identical to the original HIT algorithm.

3 Experimental Results and Discussion

We have implemented two versions of PHA-first based algorithms. In the first version, we used 10 pages generated from three search engines (Yahoo!, Lycos and Teoma) as the root set. We call this version as PHA_{sr} ("sr" stands for a small root set). In the other version, the root set consisted of 200 pages found by searching from Yahoo!. We call this version as PHA_{200}. In addition, we have implemented the HITS algorithm for comparison.

We used 12 keywords in our experiments. The keywords are listed in Table 1. Instead of comparing the performance by using the traditional recall and precision rates, we use a web-version recall and precision rates proposed in [5]. Table 1 shows the recall and precision rates of three different algorithms.

In Table 1, the average number of web pages collected by PHA_{sr} is only 96. Comparing with the number of pages collected by HITS, it is about only 4% on average. Even so, both the recall as well as precision rates of PHA_{sr} are significantly superior to the ones of HITS. The higher recall rates mean that PHA_{sr} do not miss any important pages, which are capable of being found by HITS, even the initial root set is very small. More importantly, the higher precision rates mean that the linkage structure inside the base set still accurately reflects the relative importance of topic relevant pages.

Table 1. The recall and precision rates of three different algorithms

Keywords	PHA_{sr}			PHA_{200}			HITS		
	nodes	recall	precision	nodes	recall	precision	nodes	recall	precision
cheese	129	15%	0.1697	682	11%	0.2969	1757	11%	-0.1515
basketball	145	11%	0.2363	856	17%	0.9898	1613	0%	-0.1030
motorcycle	64	10%	0.2232	924	5.5%	0.5576	2252	26%	-0.2485
Harrypotter	110	19%	0.4421	657	13%	0.3333	1359	0%	-0.4362
Movies	53	21%	0.2481	808	0%	0.5454	3685	7%	0.1151
Search engine	220	13%	0.8112	3367	19%	0.7752	3765	19%	0.7879
Coffee	66	11%	0.4788	618	11%	0.5091	1869	5%	-0.2424
Blue	14	19%	0.2606	605	0%	0.8181	4249	0%	0.1393
Weather	77	8%	0.2242	1549	3%	0.4909	3005	15%	0.2969
Travel	92	4%	0.2848	1060	1%	0.3515	3410	9%	-0.0061
Cell phone	63	14%	-0.1090	751	0%	-0.2484	1853	0%	-0.8424
Notebook	113	0%	0.5151	829	0%	0.1878	1843	3%	-0.1151
Average	96	12.08%	0.32	1059	6.71%	0.47	2555	7.92%	-0.07

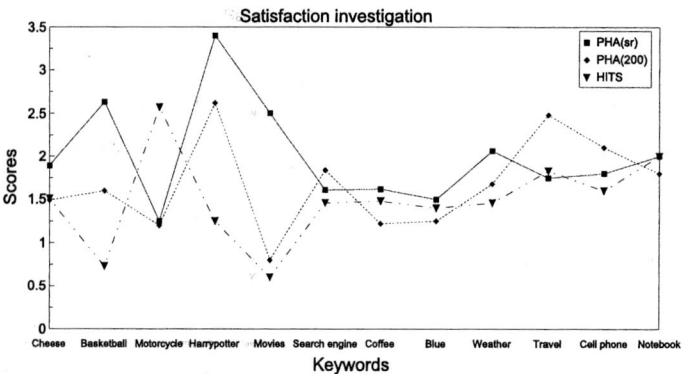

Fig. 3. Satisfaction investigation about PHA and HITS

Meanwhile, comparing with HITS, the number of web pages collected by PHA_{200} is about 40%. However, PHA_{200} got unusual high precision and terrible low recall. That is, the recommendation ranks match the expectation of respondents very well. But the recommendation pages are not good enough, at least not matching the best results collected from the commercial search engines. If we refer the moderate satisfaction (to be proposed later) of PHA_{200}, it is clear that our PHA-first approach is more suitable for a small root set. Since the satisfaction scores of PHA_{sr} is better in most cases.

It is surprised that almost all of the precision rates of HITS are negative. That implies the ranks recommended by HITS algorithm are very opposite to the expectation of respondents. In other words, it is possible to improve the results generated by HITS algorithm.

We also investigate satisfaction with three algorithms. There are a dozen of respondents to grade the top 5 hubs and authorities recommended by these three algorithms on a score from 0 to 4 (the highest score). Figure 3 shows the satisfaction scores of these three algorithms on those 12 keywords.

First of all, the average scores of PHA_{sr}, PHA_{200} and HITS are 2.0, 1.67 and 1.49 respectively. PHA_{sr} obtains better scores than the other two, and hence is more able to satisfy the requirement of respondents on average. Actually, in Figure 3, 8 of 12 scores are the highest for PHA_{sr}. Moreover, 3 to 4 of these 8 cases have statistically significant better scores.

4 Conclusion and the Future Work

In this paper, we propose a novel approach to construct a good quality base set for HITS based link structure analysis. We were motivated by the fact that the quality of base set will influence the final results of hub and authority analysis. Three inclusion rules to ensure the quality of base set have been presented. The rules are co-citation, recommendation, and relay. According to the experimental results, our method performs well on the measure of precision. In addition, the satisfaction investigation is also promising. However, comparing with commercial search engines, the diversity of pages recommended by our method is insufficient. We will try to increase the iterations performed by our method and to see whether the recall rate will be improved.

References

1. Brin, S., Page, L.: The anatomy of a large-scale hypertextual Web search engine. Computer Networks and ISDN Systems. **30** (1998) 107–117
2. Kleinberg, J.: Authoritative sources in a hyperlinked environment. Journal of the ACM. **46** (1999) 604–632
3. Lempel, R., Moran, S.: SALSA: the stochastic approach for link-structure analysis. ACM Trans. Inf. Syst. **19** (2001) 131–160
4. Borodin, A., Roberts, G., Rosenthal, J., Tsaparas, P.: Finding authorities and hubs from link structures on the World Wide Web. Proceedings of the 10th International World Wide Web Conference. (2001) 415–429
5. Vaughan, L.: New measurements for search engine evaluation proposed and tested. Inf. Process. Manage. **40**. Pergamon Press, Inc. (2004) 677–691
6. Bharat, K., Henzinger, M. R.: Improved algorithms for topic distillation in a hyperlinked environment. In: Proceedings of SIGIR-98, 21st ACM International Conference on Research and Development in Information Retrieval. (1998) 104–111
7. Chakrabarti, S., Dom, B., Indyk, H.: Enhanced hypertext categorization using hyperlinks. In: Proceedings of SIGMOD-98. ACM Press, New York, US (1998) 307–318
8. Wu, K.-J., Chen, M.-C., Sun, Y.: Automatic topics discovery from hyperlinked documents. Inf. Process. Manage. **40**. Pergamon Press, Inc. (2004) 239–255
9. Chau, M., Chen, H.: Comparison of Three Vertical Search Spiders. IEEE Computer. **36**. IEEE Computer Society Press. (2003) 56–62

A Top-Down Algorithm for Mining Web Access Patterns from Web Logs

Guo Jian-Kui[1], Ruan Bei-jun[1], Cheng Zun-ping[1], Su Fang-zhong[2], Wang Ya-qin[1], Deng Xu-bin[1], Shang Ning[1], and Zhu Yang-Yong[1]

[1] Department of Computer Science, Fudan University,
Shanghai 200433, China
gjk@fudan.edu.cn
[2] Department of Computer Science, FuZhou University,
Fuzhou 350002, China

Abstract. This paper proposes a new algorithm, called TAM-WAP(the shorthand for **T**op-down **A**lgorithm for **M**ining **W**eb **A**ccess **P**atterns), to mine interesting WAP from Web logs. TAM-WAP searches the P-tree database in the top-down manner to mine WAP. By selectively building intermediate data according to the features of current area to be mined, it can avoid stubbornly building intermediate data for each step of mining process. The experiments for both real data and artificial data show that our algorithm outperforms conventional methods.

Keywords: Sequential Mining, Web Mining, Web Usage Mining, Web Access Pattern, P-tree Database.

1 Introduction

Web usage mining is the application of data mining techniques to discover usage patterns from Web usage data, aiming at better understanding Web-based applications[1] (e.g., site structure simplification) and better serving their needs. Web usage data is the secondary data derived from the Web surfers' interactions with the Web, such as Web logs collected at different locations, business transactions, registration data and cookies.

WAP, first introduced in [2], is a sequence of events. By viewing each event (item) as a singleton set, we can regard WAP as a special instance of the conventional sequential pattern, originally introduced by Agrawal and Srikant in [3], whose elements are item-sets.

Therefore, the conventional algorithms, such as GSP[4] and PrefixSpan[5], can be used to mine WAP from Web logs. Unfortunately, GSP will suffer strongly from multiple database-scans when the length of sequential pattern grows long and PrefixSpan will generate many projected databases or false projected databases.

In order to solve those problems, a new algorithm WAP-mine[2] was specially designed. It first scans the database twice, transforming disk-resident database into a data tree, called WAP-tree, and then mines it in a bottom-up manner by recursively building the intermediate conditional bases and trees. The experimental results reported in [2] show that WAP-mine is in general an order of magnitude faster than GSP. However, the bottom-up tree traversal strategy obliges this algorithm to build intermediate data at each step of mining process. As support threshold goes low (the WAP-tree becomes large) such efforts will have negative impact on the performance.

In this paper, we develop a new algorithm, called TAM-WAP, for mining WAP. Instead of stubbornly building intermediate data for each step of mining process, TAM-WAP selectively builds intermediate data according to the features of current area by a prediction method. The outline of the paper is as follows. In Section 2, we propose our top-down mining algorithm TAM-WAP. Section 3 reports our experimental results. Section 4 is the conclusion.

2 TAM-WAP: Mining WAP from Web Logs

2.1 Definitions

We suppose that the necessary data preparation tasks have been applied to the original Web logs (the reader is referred to [7,8] for the details of such tasks) and the ultimate result of these tasks is a set of access sequences, called (Web) access database.

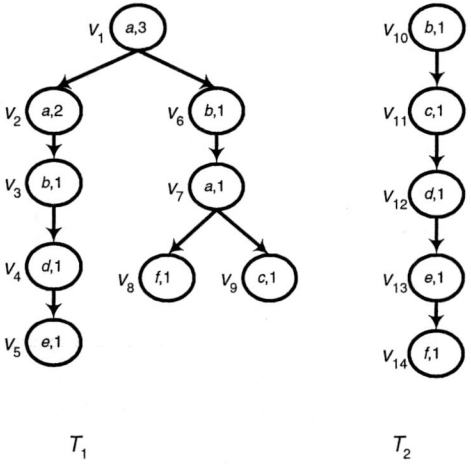

Fig. 1. a P-tree database

Definition 1 (P-tree and P-tree database): A pattern tree (P-tree) is defined to be a directed rooted tree $T= <V(T), E(T)>$, where $V(T)$ is a node-set, where each node is associated with an event and its count, written as $v.event$ and $v.count$ respectively. $E(T) \subset V(T) \times V(T)$ is a set of edges. Furthermore, T must be supposed to the following constraint: For any node $v \in V(T)$, $v.count \geq \sum_{c_i \in children(v)} c_i.count$. A set of P-tree is called a P-tree database. The node set consisting of all roots of P-trees is called the root set of D, denoted by $R(D)$. Figure 1 depicts a small P-tree database composed of two P-trees T_1 and T_2. The label beside each node is the node identifier and the letter and number (separated by comma) inside each node are the *event* and *count* respectively.

Let $p=<v_1,v_2,...,v_m>$ be a path in a P-tree database D. we use $seq(p)$ to denote the access sequence $s=<e_1,e_2,...,e_m>$ ($e_i=v_i.event$, $1\leq i \leq m$). If p is a rooted path in D, we also denote s as $seq(v_m|D)$.

Definition 2 (Counting node, projected tree and database): Let α be an access sequence and D a P-tree database. Let v be a node in $V(D)$. If $\alpha \subseteq seq(v|D)$ and there is no ancestor v' of v such that $\alpha \subseteq seq(v'|D)$, then (1) v is called a counting node of α, or α-node for short, (2) $v.count$ is called the count of α recorded in v, (3) a sub-tree rooted at a child of v is called an α-projected tree of D. All α-projected trees constitute a P-tree database, called α-projected database of D, denoted as $D|_\alpha$.

A projected database is exactly a part of the original P-tree database. This is the key feature exploited by our mining algorithm for reducing the cost of building intermediate data.

Defintion 3 (ξ-pattern and frequent event of P-tree database): Let α be an access sequence and D a P-tree database. N_α denotes the set of all α-nodes in D. The support of α in D is defined to be $support_D(\alpha)= \sum_{v_i \in N_\alpha} v_i.count$. Given a positive integer ξ as support threshold, if $support_D(\alpha) \geq \xi$, then α is called a ξ-pattern or simply access pattern of D. Let e be an event. If $<e>$ is an access pattern of D, then e is called a frequent event of D.

2.2 Compact Database and Prediction Method

Definition 4 (Compact database): Let D be a P-tree database. If (1) there exits at least one pair of nodes x and y ($x \neq y$) satisfying the following conditions:

(a) $x.event = y.event$, and
(b) they have the same parent or $x, y \in R(D)$,

or (2) there exists at least one node whose event is not frequent, then we say D is compressible. If not, we say D is incompressible or compact.

The appropriate condition for building compact database is that the time saved by mining compact database must be greater than the time wasted in building the compact database. Based on various experiments and analysis, we found a practical method defined as the following subroutine.

```
Subroutine1: CheckCompact(TFN, TN, NTV, pcr)
1   if (TFN/TN ≥ 0.75) then
2         return true
3   else begin
4         P=TFN×7×pcr;
5         if ((NTV-TN-P) ≤ 0) then
6               return true
7         else
8               return false
9   end
```

If CheckCompact returns true, that means it is not worthwhile to build the compact database. Otherwise, it is better to build a compact one.

2.3 TAM-WAP Algorithm

Now, we give TAM-WAP algorithm for mining all access patterns in P-tree database (To discover all access patterns of a P-tree database, call this algorithm with $\alpha = <>$).

```
Algorithm 1: TAM-WAP (Hα, ξ, α, occ_list, pcr)
Input:   Hα, the header table of current projected data-
         base, the support threshold ξ, an access-pattern
         prefix α, an event occurrence list occ_list.
Output:  all ξ-pattern having prefix α.
Method:
1    for each element a_k in Hα do begin
2      if (a_k.support ≥ ξ ) then
3        begin
4          output β=α ∪ a_k.event and ak.support;
5          if ((CheckCompact(TFN,TN,NTV, pcr)==false) then
6          begin
7            build the compact database Dβ;
8            call TAM-WAP(Hβ , ξ ,α ,occ_list , r);
9          end
10         else
12           call TAM-WAP(Hβ , ξ ,α ,occ_list , pcr);
13       end
14   end
```

3 Experiments

3.1 Experimental Setup

We used VC++6.0 to implement these algorithms. All experiments were conducted on an 800MHZ Celeron PC machine with 512MB of main memory, running Microsoft Windows 2000 professional. All datasets were stored in flat text files of NTFS on hard disk.

3.2 Performance

The result is shown in Figure 2. We can see the performance of TAM-WAP is better than WAP-mine.

3.3 Scale-Up Experiments

Figure 3 shows that when the support threshold is fixed as a percentage TAM-WAP has better scalability than WAP.

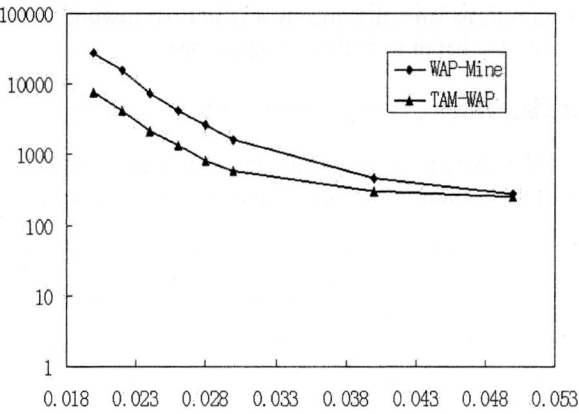

Fig. 2. TAM-WAP >WAP-Mine: Execution Time

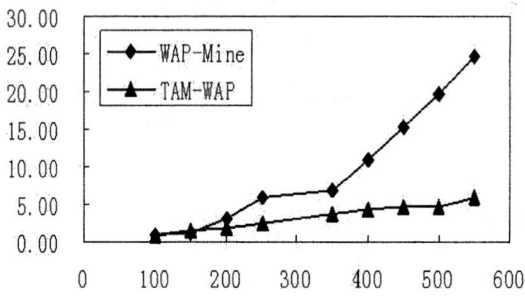

Fig. 3. Scale-up with support thresholds as percentages: 5%

4 Conclusion

In this paper, we propose a new mining algorithm, which visits the data tree in a top-down manner and selectively builds intermediate data according to the features of current area.

In addition, the results of our study are useful for association rule mining area. To solve the same problem for the well-known algorithm FP-growth[10], Two top-down algorithms have been proposed in [11,12]. They discover frequent patterns without building intermediate data and the experiments were conducted only on a few artificial datasets. Hence, our study implies there exists potential improvements for these top-down algorithms.

References

1. Jaideep Srivastava, Robert Cooley, Mukund Deshpande, Pang-Ning Tan. Web Usage Mining: Discovery and Applications of Usage Patterns from Web Data. SIGKDD Explorations 1(2), 2000, pp. 12-23.

2. J. Pei, J.W. Han, B. Mortazavi-asl and H. Zhu. Mining Access Patterns Efficiently from Web Logs. In Proceedings of Pacific Asia Conference on Knowledge Discovery and Data Mining PAKDD 2000, Kyoto, Japan, April 18-20, 2000, LNCS 1805, pp. 396-407.
3. R. Agrawal and R. Srikant. Mining sequential patterns. In Proc. 1995 Int. Conf. Data Engineering (ICDE'95), Taipei, Taiwan,Mar, 1995, pp. 3-14.
4. R.Srikant and R. Agrawal. Mining sequential patterns: Generalizations and performance improvements. In Proc. 5th Int. Conf. Extending Database Technology (EDBT'96), Avignon, France, Mar. 1996, pp. 3-17.
5. Jian Pei, Jiawei Han, Behzad Mortazavi-Asl, Helen Pinto, Qiming Chen, Umeshwar Dayal, Meichun Hsu. PrefixSpan: Mining Sequential Patterns by Prefix-Projected Growth. In Proceedings of the 17th International Conference on Data Engineering, April 2-6, 2001, Heidelberg, Germany, pp. 215-224.
6. Yi Lu, C.I. Ezeife: Position Coded Pre-order Linked WAP-tree for Web Log Sequential Pattern Mining. In Proceedings of the 7th Pacific-Asia Conference, PAKDD 2003, Seoul, Korea, April 30 - May 2, 2003, pp. 337 – 349.
7. S.K. Madria, S.S. Bhowmick, W.K. Ng, and E.P. Lim. Research Issues in Web Data Mining. In Proceedings of the 1st International Conference on Data Warehousing and Knowledge Discovery (DAWAK99), Florence, Italy, August 30- September 3 1999, LNCS 1676, pp. 303-312.
8. Robert Cooley, Bamshad Mobasher, Jaideep Srivastava. Data Preparation for Mining World Wide Web Browsing Patterns. Knowledge and Information Systems 1(1), 1999, pp.5-32.
9. Zijian Zheng, Ron Kohavi, Llew Mason. Real world performance of association rule algorithms. KDD 2001,pp. 401-406.
10. J. Han, J. Pei, and Y. Yin. Mining frequent patterns without candidate generation. In SIGMOD'2000, Dallas, Tx, May 2000, pp.1-12.
11. Yabo Xu, Jeffrey Xu Yu, Guimei Liu, Hongjun Lu: From Path Tree To Frequent Patterns: A Framework for Mining Frequent Patterns. ICDM 2002, pp. 514-521.
12. Ke Wang, Liu Tang, Jiawei Han, Junqiang Liu. Top Down FP-Growth for Association Rule Mining. PAKDD 2002: pp.334-340.

Kernel Principal Component Analysis for Content Based Image Retrieval

Guang-Ho Cha

Department of Multimedia Science, Sookmyung Women's University,
53-12 Chongpa-dong 2-ga, Yongsan-gu, Seoul 140-742, South Korea
ghcha@sookmyung.ac.kr

Abstract. Kernel principal component analysis (PCA) has recently been proposed as a nonlinear extension of PCA. The basic idea is to first map the input space into a *feature space* via a nonlinear map and then compute the principal components in that feature space. This paper illustrates the potential of kernel PCA for dimensionality reduction and feature extraction in content-based image retrieval. By the use of Gaussian kernels, the principal components were computed in the feature space of an image data set and they are used as new dimensions to approximate images. Extensive experimental results show that kernel PCA performs better than linear PCA in content-based image retrievals.

1 Introduction

Content-based image retrieval (CBIR) supports image searches based on visual features such as color, texture, and shape. In a CBIR system, these features are extracted and stored as feature vectors. During the retrieval process, the feature vector of the query image is computed and matched against those in the database. The returned images should be similar to the query image. This *similarity* (or *nearest neighbor*) indexing/retrieval problem can be solved efficiently when the feature vectors have low or medium dimensionalities (e.g., less than 10) by the use of existing indexing methods such as the R*-tree [1] and the HG-tree [2]. So far, however, there has been no efficient solution to this problem when the feature vectors have high dimensionalities, say over 100 [3]. So the issue is to overcome the *curse of dimensionality*. Motivated by this phenomenon, the approach to reduce the dimensinoality of image feature vectors has been attempted by the use of some dimensionality reduction techniques such as *principal component analysis* [4, 7].

Principal component analysis (PCA) [5] has been widely used for re-expressing multidimensional data. It allows researchers to reorient the data so that the first few dimensions account for as much of the available information as possible. If there is substantial redundancy present in the data set, then it may be possible to account for most of the information in the original data set with a relatively small number of dimensions. In other words, PCA finds out for the original data set the new structure given by the linear combination of the original variables. However, one cannot assert that *linear* PCA will always detect all structure in a given data set. By the use of suitable *nonlinear* features, one can extract more information. In this paper, we investigate the potential of a nonlinear form of PCA for dimensionality reduction and feature extraction in content-based image retrieval.

2 Kernel PCA

PCA is an orthogonal basis transformation. The new basis is found by diagonalizing the covariance matrix C of a centered data set $\{x_i \in R^N \mid i = 1, ..., m\}$, defined by

$$C = \frac{1}{m}\sum_{j=1}^{m} x_j x_j^t, \quad \sum_{j=1}^{m} x_j = 0$$

The coordinates in the Eigenvector basis are called *principal components*. The principal components are given by the linear combination of the original variables. The size of an Eigenvalue λ corresponding to an Eigenvector v of C equals the amount of variance in the direction of v. Furthermore, the directions of the first n Eigenvectors corresponding to the biggest n Eigenvalues cover as much variance as possible by n orthogonal directions. In many applications, they contain the most interesting information.

Clearly, one cannot assert that linear PCA will always detect all structure in a given data set. Moreover, it can be very sensitive to "wild" data ("outliers"). By the use of suitable *nonlinear* features, one can extract more information. Kernel PCA is very well suited to extract interesting nonlinear structures in the data [13].

The purpose of our work is therefore to consider the potential of kernel PCA for dimensionality reduction and feature extraction in CBIR. Kernel PCA first maps data into some feature space F via a (usually nonlinear) function Φ and then performs linear PCA on the mapped data. As the feature space F might be very high dimensional, kernel PCA employs Mercer kernels instead of carrying out the mapping Φ explicitly. A Mercer kernel is a function $k(x, y)$ which for all data sets $\{x_i\}$ gives rise to a positive matrix $K_{ij} = k(x_i, y_j)$ [11]. Using function k instead of a dot product in input space corresponds to mapping the data with some Φ to a feature space F, i.e., $k(x, y) = (\Phi(x) \cdot \Phi(y))$.

To perform PCA in feature space F, we need to find Eigenvalues $\lambda > 0$ and Eigenvectors $v \in F-\{0\}$ satisfying $\lambda v = C' v$ with the covariance matrix C' in F, defined as

$$C' = \frac{1}{m}\sum_{j=1}^{m} \Phi(x_j)\Phi(x_j)^t, \quad \sum_{j=1}^{m} \Phi(x_j) = 0$$

Substituting C' into the Eigenvector equation, we note that all solutions v must lie in the span of Φ-images of the sample data. This implies that we can consider the equivalent equation

$$\lambda(\Phi(x_j) \cdot v) = (\Phi(x_j) \cdot C'v) \quad \text{for all } j = 1, ..., m \quad (1)$$

and that there exist coefficients $\alpha_1, ..., \alpha_m$ such that

$$v = \sum_{j=1}^{m} \alpha_j \Phi(x_j). \quad (2)$$

Substituting C' and (2) into (1), and defining $m \times m$ Gram matrix $K_{ij} = k(\Phi(x_i), \Phi(y_j)) = k(x_i, y_j)$, we arrive at a problem which is cast in terms of dot product. Solve

$$m\lambda\alpha = K\alpha$$

for nonzero Eigenvalues λ and Eigenvectors $\alpha = (\alpha_1, ..., \alpha_m)^t$ subject to normalization condition $\lambda^k(\alpha^k \cdot \alpha^k) = 1$. To extract nonlinear principal components for the Φ-image of a test point x, we compute the projection onto the k-th component by

$$v^k \cdot \Phi(x) = \sum_{j=1}^{m} \alpha_j^k k(x, x_j) = \sum_{j=1}^{m} \alpha_j^k \Phi(x)\Phi(x_j) \cdot$$

3 Performance Measures

In this section, we consider the measures to assess the performance of kernel PCA. In our work, we use the Gaussian kernel $k(x, x') = \exp(-(\|x-x'\|^2/2\sigma^2))$ because it is widely used in content-based image retrieval and pattern recognition [9, 14]. In tradional (document) information retrieval, performance is often measured by using precision and recall [10, 12]. Recall measures the ability of the system to retrieve useful items, while precision measures its ability to reject useless items. For a given query, let T the total number of relevant items available, R_r the number of relevant items retrieved, and T_r the total number of retrieved items. Then precision is defined as R_r / T_r, and recall as R_r / T.

Precision and recall can also be applied to image retrieval. In IBM QBIC that performs similarity retrieval as opposed to exact match, *normalized* precision and recall have been suggested [6]. These reflect the positions in which the set of relevant items appear in the retrieval sequence (ordered by some similarity measure). If there are T relevant images in the database, then for an ideal retrieval, all T relevant items occur in the first T retrievals (in any order). Faloutsos et al [6] define this as IAVRR, the ideal AVRR (average rank of all relevant, retrieved images). It is the maximum when all relevant images are retrieved on the top: IAVRR = $(0 + 1 + ... + (T-1)) / T$ (where the first position is the 0-th). The ratio of AVRR to IAVRR gives a measure of the effectiveness of the retrieval. In an ideal case of retrieval, this ratio would be 1.

For example, if the relevant images for the query image A are defined as:

A46, A18, A101, A52, A35, A102

so that $T = 6$, and a CBIR system returns, in order:

A102, A109, A50, **A18**, A74, **A46**, **A52**, A57, A17, **A35**, A63, A16, A58, **A101**

then relevant items appear at 0, 3, 5, 6, 9 and 13. The AVRR for this is therefore $(0 + 3 + 5 + 6 + 9 + 13)/6 = 6$. The IAVRR would be $(0 + 1 + 2 + 3 + 4 + 5)/6 = 2.5$. Thus AVRR / IAVRR is 2.4.

If the order of retrieval matters, Kendall's tau can be used to provide measures of association between two sets of data [8]. Kendall's tau can be viewed as a coefficient of disorder. For example, consider the following two rankings, where both have selected the same four images, but have placed them in a different order:

$$\begin{matrix} 1 & 2 & 3 & 4 \\ 2 & 1 & 4 & 3 \end{matrix}$$

Tau is calculated as

(no. of pairs in order − no. of pairs out of order) / (total no. of possible pairs)

For this example, 2 in the bottom row is followed by 1, 4, and 3, 2-1 is out of order, scoring -1, and 2-4, 2-3 are in order, scoring +1 each. Similarly, 1 is followed by 4 and 3. Both are in order, scoring +1 each. Finally, 4 is followed by 3, scoring -1. The number of in-order pairs is four, and out-of-order pairs is two, therefore the total is +2, divided by the maximum number of in-order pairs, $N(N-1)/2$, which here is 6, since $N = 4$. The value of tau is therefore 2/6, or 0.3333. This gives a measure of the "disarray" or difference in ranking, between the two. It ranges from -1, which represents complete disagreement, through 0, to +1, complete agreement.

For each experiment in our work, we report the average of the above measures over 100 k nearest neighbor (k-NN) queries. For k-NN queries, precision and recall are the same because $T = T_r$, i.e., the total number of relevant items and the total number of retrieved items are the same. There we compute only the precision measure as a representative. The ratio of AVRR to IAVRR gives a measure that how much the results are close to the top. Kendall's tau provides a measure of the order for the k-NN search results.

4 Experimental Results

To demonstrate the effectiveness of kernel PCA, we performed an extensive experimental evaluation for kernel PCA and compared it to linear PCA. For our experiments we used 13,724 256-color images of U.S. stamps and photos. To obtain feature vectors for experiments, we used four MPEG-7 visual features: (1) color structure descriptor (256 dimensions), (2) homogeneous texture descriptor (30 dimensions), (3) edge histogram descriptor (80 dimensions), and (4) region-based shape descriptor (35 dimensions). These descriptors are general descriptors that can be used in CBIR.

We applied kernel PCA and linear PCA to those four data sets consisting of MPEG-7 visual descriptors, respectively, in order to reduce their dimensionality. We posed k nearest neighbor queries to 3 kinds of data sets, i.e., (1) the original data set, (2) the data set whose dimensionality is reduced by kernel PCA, and (3) the data set whose dimensionality is reduced by linear PCA. In all experiments, the numbers of nearest neighbors to find were 20, 40, 60, 80 and 100 and we averaged their results. 100 random k-NN queries were processed and the results were averaged.

Tables 1 − 4 show that the experimental results for four MPEG-7 visual features. The first column of each table represents the dimensionality of the original data and the dimensionalities of the transformed data after the dimensionality reduction. In addition, the percentages of the variance in each original variable we retain are also provided in the first column of each table. The performance of kernel PCA is better than that of linear PCA with respect to all three performance parameters, i.e., precision, the ratio of AVRR to IAVRR, and Kendall's tau. In terms of precision and AVRR/IAVRR, kernel PCA is 10% − 20% better than linear PCA. With respect to Kendall's tau, kernel PCA is better than linear PCA more than 50%. These experimental results indicate that kernel PCA can be successfully employed as a generalized nonlinear extension of linear PCA.

Table 1. Color structure experiments

original dim = 256	Linear PCA			Kernel PCA		
	Precision	AVRR/IAVRR	Tau	Precision	AVRR/IAVRR	Tau
95% (dim = 164)	0.62	3.45	0.42	0.73	3.05	0.70
90% (dim = 113)	0.59	3.65	0.37	0.67	3.17	0.62
85% (dim = 77)	0.53	3.70	0.38	0.64	3.25	0.63

Table 2. Homogeneous texture experiments

original dim = 30	Linear PCA			Kernel PCA		
	Precision	AVRR/IAVRR	Tau	Precision	AVRR/IAVRR	Tau
95% (dim = 14)	0.60	3.50	0.41	0.71	3.15	0.69
90% (dim = 10)	0.56	3.75	0.37	0.66	3.30	0.64
85% (dim = 9)	0.51	3.80	0.36	0.62	3.45	0.61

Table 3. Edge histogram experiments

original dim = 80	Linear PCA			Kernel PCA		
	Precision	AVRR/IAVRR	Tau	Precision	AVRR/IAVRR	Tau
95% (dim = 30)	0.59	3.48	0.45	0.69	3.27	0.61
90% (dim = 25)	0.54	3.45	0.42	0.64	3.27	0.60
85% (dim = 22)	0.50	3.68	0.39	0.61	3.35	0.58

Table 4. Region-based shape experiments

original dim = 35	Linear PCA			Kernel PCA		
	Precision	AVRR/IAVRR	Tau	Precision	AVRR/IAVRR	Tau
95% (dim = 15)	0.57	3.63	0.42	0.68	3.35	0.62
90% (dim = 11)	0.51	3.73	0.40	0.62	3.47	0.60
85% (dim = 10)	0.50	3.85	0.37	0.61	3.45	0.59

5 Conclusion

In this paper, we described the potential of kernel PCA for dimensionality reduction and feature extraction in content-based image retrieval. Through the use of Gaussian kernel, a kernel PCA was able to work effectively within the feature space of image data sets, thereby producing a good performance. Compared with linear PCA, kernel PCA showed better performance with respect to the retrieval quality as well as the retrieval precision in content-based image retrieval. Therefore, we can conclude that kernel PCA can be successfully employed as a generalized nonlinear extension of linear PCA.

Acknowledgments

This work was supported by Korea Research Foundation Grant (KRF-2003-041-D20438).

References

1. N. Beckmann, H.-P. Kriegel, R. Schneider, and B. Seeger, The R*-tree: An efficient and robust access method for points and rectangles, *Proc. of ACM SIGMOD Conf.* (1990) 322-331
2. G.-H. Cha and C.-W. Chung, A New Indexing Scheme for Content-Based Image Retrieval, *Multimedia Tools and Applications*, Vol. 6 (1998) 263-288
3. G.-H. Cha, X. Zhu, D. Petkovic, and C.-W. Chung, An Efficient Indexing Method for Nearest Neighbor Searches in High-Dimensional Image Databases, *IEEE Trans. on Multimedia* (2002) 76-87
4. K. Chakrabarti and S. Mehrotra, Local Dimensionality Reduction: A New Approach to Indexing High Dimensional Spaces, *Proc. of the Int'l Conf. on VLDB* (2000) 89-100
5. K.I. Diamantaras and S.Y. Kung, *Principal Component Neural Networks*, Wiley, New York, (1996)
6. C. Faloutsos et al., Efficient and Effective Querying by Image Content, *Journal of Intelligent Information Systems*, Vol. 3 (1994) 231-262
7. K.V.R. Kanth, D. Agrawal, and A. Singh, A., Dimensionality Reduction for Similarity Searching in Dynamic Databases, *Proc. of the ACM SIGMOD Conf.* (1998) 166-176
8. J. Payne, L. Hepplewhite, and T J stoneham, Texture, Hhuman perception and Information Retrieval Measures, *Proc. of ACM SIGIR 2000 Workshop* (2000)
9. J. Peng and D.R. Heisterkamp, Kernel indexing for Relevance Feedback Image Retrieval, *Proc. of IEEE Int'l Conf. on Image Processing* (2003) 733-736
10. C.J.V Rijsbergen, *Information Retrieval*, Butterworths, London (1979)
11. S. Saitoh, *Theory of Reproducing Kernels and its Applications*, Longman Scientific & Technical, Harlow, England (1988)
12. G. Salton and M.J. McGill, *Introduction to Modern Information Retrieval*, McGraw-Hill, New York (1983)
13. B. Schölkopf, A. Smola, and K. Müller, Nonlinear Component Analysis as a Kernel Eigenvalue Problem, *Neural Computation*, Vol. 10 (1998) 1299-1319
14. B. Schölkopf et al., Comparing Support Vector Machines with Gaussian Kernels to Radial Basis Function Classifiers, *IEEE Trans. on Signal Processing*, Vol. 45 (1997) 2758-2765

Mining Frequent Trees with Node-Inclusion Constraints

Atsuyoshi Nakamura and Mineichi Kudo

Graduate School of Information Science and Technology, Hokkaido University,
Kita 14, Nishi 9, Kita-ku, Sapporo, 060-0814, Japan
{atsu, mine}@main.ist.hokudai.ac.jp

Abstract. In this paper, we propose an efficient algorithm enumerating all frequent subtrees containing all special nodes that are guaranteed to be included in all trees belonging to a given data. Our algorithm is a modification of TreeMiner algorithm [10] so as to efficiently generate only candidate subtrees satisfying our constraints. We report mining results obtained by applying our algorithm to the problem of finding frequent structures containing the name and reputation of given restaurants in Web pages collected by a search engine.

1 Introduction

Frequent structure mining is one of the most popular way of data mining because of understandability of its analyzed results. The idea of finding frequent patterns is simple and easy, but for massive databases efficient algorithms to do this task are necessary, and it is not trivial to develop such algorithms. After Agrawal and Srikant developed efficient algorithm *Apriori*, various efficient algorithms have been developed for frequent itemsets [1], subsequences [2], subtrees [3, 10], subgraphs [6] and so on.

People sometimes have a certain point of view from which they want to analyze data. In such cases, they want to find frequent structures that satisfy certain constraints. This can be done by selecting structures satisfying the constraints after enumerating all frequent ones, but it is not efficient when a lot of more frequent other structures exist. In order to efficiently find the frequent structures satisfying constraints without enumerating unnecessary frequent ones, some algorithms have been also developed for itemsets [9] and subsequences [5].

In this paper, we consider a kind of constrained problem for frequent subtrees. Data we mine is a set of *labeled rooted ordered* trees, each of which contains just d special nodes labeled a different label belonging to a set of d special labels. Our mining problem is to find all frequent *embedded* subtrees [10] containing all d special nodes.

This research is motivated by *wrapper* induction [8, 4]. A *wrapper* is a program that extracts information necessary for some purpose from Web pages. Most wrappers extract necessary information by pattern matching around the information. Wrapper induction, automatic wrapper construction from training data, can be done by finding common patterns around the information in the data. Most Web pages are HTML documents of which tag structures can be represented by DOM-trees, so finding frequent subtrees of DOM-trees containing all nodes with necessary information can be used as a kind of wrapper induction methods, though some additional information like contents of the information and distance between nodes are necessary to construct high-precision wrappers [7].

In this paper, we propose an efficient algorithm enumerating all frequent subtrees containing all special nodes. Our algorithm is an extension of *TreeMiner* algorithm proposed by Zaki [10]. We modified TreeMiner algorithm so as to efficiently generate only candidate subtrees satisfying our constraints. We also report mining results obtained by using our algorithm for the problem of finding frequent structures containing the name and its reputation of given Ramen (lamian, Chinese noodles in Soup) shops in Web pages collected by a search engine.

2 Problem Statement

2.1 Notions and Notations

In this paper, all trees we deal with are labeled ordered trees defined as follows. A *rooted* tree $T = (N, B)$ is a connected acyclic graph with a set N of vertices and a set B of directed edges $(u, v) \in N \times N$ that represent *parent-child* relation, which satisfies the condition that every vertex but just one vertex (*root*) has just one parent vertex. For a tree, a vertex and an edge are called a *node* and a *branch*, respectively. An *ordered* tree $T = (N, B, \preceq)$ is a rooted tree (N, B) with partial order $\preceq$ on N representing a sibling relation, where the order is defined just for all the pairs of children having the same parent. Let L be the set of labels. A *labeled* ordered tree $T = (N, B, \preceq, l)$ is an ordered tree $(N, B, \preceq)$ of which nodes are labeled by the label function $l : N \to L$.

An *id* of a node of a tree $T = (N, B, \preceq, l)$ is its position in a preorder traversal of the tree, where the passing order of children of the same parent follows order $\preceq$. Note that, in our notation, any node with subscript i represents the node of id i.

The parent-child relation, which is defined by set B of branches, induces another partial order $\leq$, an *ancestor-descendant* relation, by extending the relation so as to satisfy reflexivity and transitivity. If nodes t_i and t_j are not comparable in order $\leq$, t_j is said to be a *left-collateral* of t_i when $i > j$ and a *right-collateral* of t_i when $i < j$.

A *scope* of a node t is defined as $[l, r]$ using the minimum id l and the maximum id r among the ids of t's descendant nodes. Note that l is always t's id.

A *string encoding* of a tree T is a sequence of labels in $L \cup \{-1\}$ generated by starting with null string, appending the label of the node at its first visit, and appending -1 when we backtrack from a child in a depth-first traversal. For example, tree S_2 in Fig. 1 is encoded as a b 1 -1 -1 2 -1.

We abuse notation and use the same symbol to represent both a tree and its string encoding.

Size of a tree T is the number of nodes in T. A *size-k prefix* of a tree is the subtree composed of the nodes of which id is less than k.

An *embedded* subtree considered by [10] is defined as follows.

Definition 1. *Let $T = (N_T, B_T, \preceq_T, l_T)$ and $S = (N_S, B_S, \preceq_S, l_S)$ be labeled ordered trees. Assume that N_T and N_S are represented as $\{t_0, t_1, ..., t_{n-1}\}$ and $\{s_0, s_1, ..., s_{m-1}\}$, respectively. If there is an one-to-one mapping $i : j \mapsto i_j$ from $\{0, 1, ..., m-1\}$ to $\{0, 1, ..., n-1\}$ satisfying the following conditions, S is called an* embedded subtree *of T and i is called an* embedding mapping.

1. *(Label preserving)*
$$l_S(s_j) = l_T(t_{i_j}) \text{ for } j \in \{0, ..., m-1\}.$$
2. *(Ancestor-descendant relation preserving)*
$$(s_j, s_k) \in B \Rightarrow t_{i_j} \preceq t_{i_k} \text{ for } j, k \in \{0, 1, ..., m-1\}.$$
3. *(Sibling relation preserving)*
$$j \leq k \Leftrightarrow i_j \leq i_k \text{ for } j, k \in \{0, 1, ..., m-1\}.$$

Let D denote a set of labeled ordered trees. For given D and a *minimum support* $0 < \sigma \leq 1$, a subtree S is *frequent* if the rate of trees in D that have S as an embedded subtree is at least σ.

Zaki [10] considered the problem of enumerating all frequent subtrees and developed an efficient algorithm for this problem.

2.2 Trees with Special Nodes

In this paper, we assume that there are d different special labels which are not members of L, and that every tree in D has just d special nodes labeled different special labels.

The notions of size and size-k prefix defined above are extended as follows. *Size* of a tree is the number of *non*-special nodes. For example, the size of tree T in Fig. 1 is 4. A *size-k prefix* of a tree is the subtree composed of k *non*-special

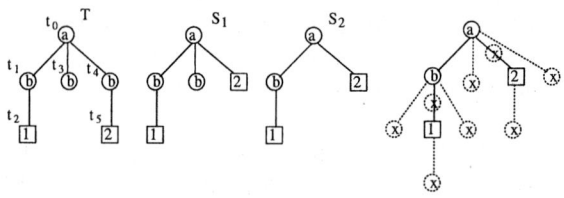

Fig. 1. Left 3 trees: Example of trees with special nodes, which are denoted by squares. The size of tree T is 4, and trees S_1 and S_2 are its size-3 and size-2 prefixes, respectively. Rightmost tree: Prefix equivalence class for tree S_2

nodes with k smallest ids and all special nodes. For example, trees S_1 and S_2 in Fig. 1 are the size-3 and size-2 prefixes of tree T, respectively.

We consider the following problem.

Problem 1 (Constrained tree mining problem). For given D and a minimum support σ, enumerate all frequent subtrees that have all special nodes.

3 Candidate Subtree Generation

Since all embedded subtrees of a frequent subtree are also frequent, enumerating frequent subtrees in size-increasing order reduces candidates, and as a result it is efficient. Zaki's efficient method [10] for enumerating frequent subtrees in size-increasing order is based on a notion of *prefix equivalence class*. We first describe his method in the next subsection, then propose modified version of the method for the constrained problem.

3.1 Method for the Problem Without Constraints

A *prefix equivalence class* for a size-$(k-1)$ tree P in a size-k tree set G, that is denoted by $[P]_G$, is the set of size-k trees in G of which size-$(k-1)$ prefix is P. We let $[G]_{\text{pre}}$ denote the set of all prefix equivalence classes. Every tree $T \in [P]_G$ is distinguished from other trees in $[P]_G$ by its last node, the node with the maximum id. The last node of T can be uniquely represented in $[P]_G$ by a pair (x, i) of its label x and id i of its parent node in P, because the node must be the last child of its parent node in order to preserve tree's prefix.

Since two size-k subtrees of a size-$(k+1)$ frequent subtree T created by removing[1] one of the last two nodes of T are also frequent, and the size-$(k-1)$ prefixes of the two nodes coincides with each other, all size-$(k+1)$ frequent subtrees can be enumerated by joining two size-k frequent subtrees that belong to the same prefix equivalence class. The relation between the last two nodes of size-$(k+1)$ tree is parent-child relation or not, thus join operator must generate trees of the both relations if possible.

When P_1 and P_2 are different size-$(k-1)$ trees, trees generated by join operator from $[P_1]_G$ and $[P_2]_G$ are trivially different because their size-$(k-1)$ prefixes are preserved. Thus, by applying join operator to every pair of frequent size-k trees in each prefix equivalence class, we can enumerate all size-$(k+1)$ candidates of frequent trees without duplication.

3.2 Method for the Constrained Problem

Our method for the constrained tree mining problem is based on the same idea as Zaki's method mentioned above. Only part we have to consider is how to deal with special nodes.

---

[1] Note that, when the removed node has child nodes, the new parent of those child nodes is the parent of the removed node.

A *prefix equivalence class* $[P]_G$ in a size-k tree set G for a size-$(k-1)$ tree P, and the set $[G]_{\text{pre}}$ of all prefix equivalence classes in G are defined similarly, though the definition of a size-$(k-1)$ prefix is different. The prefix equivalence class for tree S_2 in Fig. 1 is the rightmost tree shown in the same figure, which indicates that there are 8 positions for the last node of a tree in the class. The difference from the case without constraints is that the position of the last node of a tree in each class cannot be uniquely determined by its parent node. For example, there are three positions for the last node with its parent node labeled a in the prefix equivalence class for tree S_2.

To overcome this problem, we specify the position of the last node by its inserted position in the string encoding of its prefix. For example, in the above case, when the last node is between the nodes labeled a and 2, the tree is encoded as "a b 1 -1 -1 x 2 -1 -1" and the prefix of its last node is encoded as "a b 1 -1 -1 2 -1". In this case, the position of the last node is specified as $(4,6)$, which means that the label x is inserted right after the 4th character and -1 is inserted right after the 6th character, where a string begins with the 0th character. Therefore, three positions for the last node with its parent node labeled a is specified as $(4,4), (4,6)$ and $(6,6)$.

Let a tree be represented by a pair $(P, (x, (i_1, j_1)))$ of its prefix P and last node $(x, (i_1, j_1))$, where x is its label and (i_1, j_1) is its inserted position in the string encoding of P. We define join operators $\otimes_{\text{in}}$ and $\otimes_{\text{out}}$ on two trees in the same prefix equivalence class as follows.

Definition 2. *Let $(P, (x, (i_1, j_1)))$ and $(P, (y, (i_2, j_2)))$ be two trees in the same prefix equivalence class.*

$$(P, (x, (i_1, j_1))) \otimes_{in} (P, (y, (i_2, j_2))) \stackrel{def}{=} ((P, (x, (i_1, j_1))), (y, (i_2+1, j_2+1)))$$

$$(P, (x, (i_1, j_1))) \otimes_{out} (P, (y, (i_2, j_2))) \stackrel{def}{=} ((P, (x, (i_1, j_1))), (y, (i_2+2, j_2+2)))$$

Proposition 1. *Let T be a size-$(k+1)$ tree ($k \geq 2$) and let $(P, (x, (i_1, j_1)))$ and $(P, (y, (i_2, j_2)))$ be the trees generated by removing the second last and the last non-special nodes, respectively. Then just one of the following cases holds.*

Case I $(P, (x, (i_1, j_1))) \otimes_{in} (P, (y, (i_2, j_2))) = T$ *and* $i_1 \leq i_2, j_2 \leq j_1$
Case II $(P, (x, (i_1, j_1))) \otimes_{out} (P, (y, (i_2, j_2))) = T$ *and* $j_1 \leq i_2$

Proof. Omitted due to space limitation. □

Examples of size-4 trees of case I and case II are shown in Fig. 2.

The next proposition says that all size-$(k+1)$ candidate subtrees can be enumerated by generating $T_1 \otimes_{\text{in}} T_2$ for all $(T_1, T_2) \in \mathcal{P}_{\text{in}}$ and $T_1 \otimes_{\text{out}} T_2$ for all $(T_1, T_2) \in \mathcal{P}_{\text{out}}$ without duplication, where each of $\mathcal{P}_{\text{in}}$ and $\mathcal{P}_{\text{out}}$ consists of a pair of size-k frequent subtrees belonging to the same prefix equivalence class.

Proposition 2. *Let F_k denote the set of all size-k frequent subtrees for $k \geq 2$. Let*

$$\mathcal{P}_{in} = \{((P, (x, (i_1, j_1))), (Q, (y, (i_2, j_2)))) \in F_k \times F_k : P = Q, i_1 \leq i_2, j_2 \leq j_1\},$$
$$\mathcal{P}_{out} = \{((P, (x, (i_1, j_1))), (Q, (y, (i_2, j_2)))) \in F_k \times F_k : P = Q, j_1 \leq i_2\}.$$

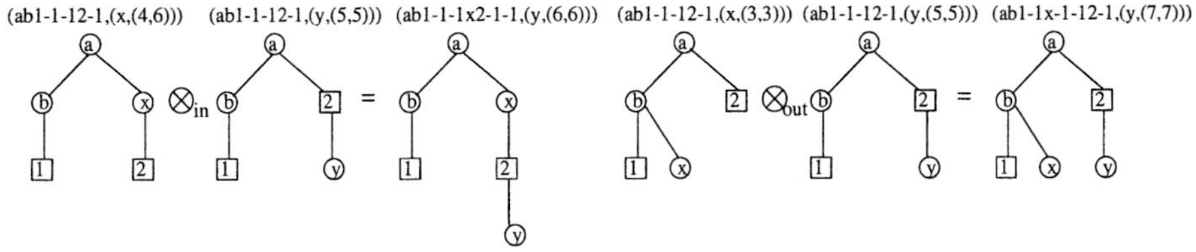

(ab1-1-12-1,(x,(4,6))) (ab1-1-12-1,(y,(5,5))) (ab1-1-1x2-1-1,(y,(6,6))) (ab1-1-12-1,(x,(3,3))) (ab1-1-12-1,(y,(5,5))) (ab1-1x-1-12-1,(y,(7,7)))

Fig. 2. Operation examples for two join operators

Define

$$C_{k+1} = \bigcup_{(T_1,T_2)\in \mathcal{P}_{in}} \{T_1 \otimes_{in} T_2\} \cup \bigcup_{(T_1,T_2)\in \mathcal{P}_{out}} \{T_1 \otimes_{out} T_2\}$$

Then, $F_{k+1} \subseteq C_{k+1}$ and $|C_{k+1}| = |\mathcal{P}_{in}| + |\mathcal{P}_{out}|$.

Proof. Omitted due to space limitation. □

4 ConstrainedTreeMiner Algorithm

4.1 Algorithm

ConstrainedTreeMiner algorithm in Fig. 3 enumerates all frequent embedded subtrees containing all special nodes efficiently. Its algorithmic structure is the same as TreeMiner algorithm [10], but its data structure and join operations are different as mentioned in the previous section.

The basic algorithmic structure is as follows. First, by executing procedure Enumerate-F_2 described in the next subsection, the algorithm creates the set F_2 of size-2 frequent subtrees and divides it into the set $[F_2]$pre of its prefix equivalence classes while creating the scope-list[2] $\mathcal{L}(S)$ of $S \in F_2$. For each $[P] \in [F_2]$pre, all larger frequent trees having prefix P can be created from $[P]$ and $\{\mathcal{L}(S) : S \in [P]\}$ by recursively applying Enumerate-Frequent-Subtrees procedure. In Enumerate-Frequent-Subtrees procedure, one size larger candidate subtree is created by joining two subtrees S_1 and S_2 with the same prefix P using operators $\otimes_{in}$ and $\otimes_{out}$, and frequency counting for the candidate is done by joining elements in $\mathcal{L}(S_1)$ and $\mathcal{L}(S_2)$ similarly as the original TreeMiner algorithm does.

Here, we assume that the least common ancestor of all special nodes in any tree in D is a non-special node for the sake of simplicity. Note that slight modification is necessary to deal with the case that the least common ancestor is a special node.

---

[2] An element (t, m, s) of scope-list $\mathcal{L}(S)$ of size-k tree S represents one S-embeddable position in some tree T belonging to D, where t is the id of T, m is a sequence of ids of the *non*-special nodes in T in which the size-$(k-1)$ prefix of S can be embedded, and s is the scope of the last *non*-special node in the S-embeddable position.

ConstrainedTreeMiner(D,minsup)
begin
  Enumerate-F_2(D,minsup)
  for all $[P] \in [F_2]$pre do
    Enumerate-Frequent-Subtrees($[P], \{\mathcal{L}(S) : S \in [P]\}$,minsup)
  enddo
end

Enumerate-Frequent-Subtrees($[P], \{\mathcal{L}(S) : S \in [P]\}$,minsup)
begin
  for all $(P,(x,(i_1,j_1))) \in [P]$ do
    $S_1 = (P,(x,(i_1,j_1)))$
    $[S_1] = \emptyset$
    for all $(P,(y,(i_2,j_2))) \in [P]$ do
      $S_2 = (P,(y,(i_2,j_2)))$
      if $i_1 \leq i_2 \leq j_2 \leq j_1$ then
        Create $\mathcal{L}(S_1 \otimes_{in} S_2)$
        if $S_1 \otimes_{in} S_2$ is frequent then $[S_1] \leftarrow [S_1] \cup \{(S_1,(y,(i_2+1,i_2+1)))\}$
      if $j_1 \leq i_2$ then
        Create $\mathcal{L}(S_1 \otimes_{out} S_2)$
        if $S_1 \otimes_{out} S_2$ is frequent then $[S_1] \leftarrow [S_1] \cup \{(S_1,(y,(i_2+2,i_2+2)))\}$
    enddo
    Enumerate-Frequent-Subtrees($[S_1], \{\mathcal{L}(S) : S \in [S_1]\}$,minsup)
  enddo
end

Fig. 3. ConstrainedTreeMiner algorithm

4.2 Procedure Enumerate-F_2

Procedure Enumerate-F_2, which creates the set F_2 of size-2 frequent subtrees and scope-lists for its elements, is the following process for each tree T in D. (See Fig. 4 for examples of each step.) After the execution of this process, $[F_2]$pre is constructed by selecting all prefix equivalence classes for frequent size-2 trees. Note that F_1 is also constructed by selecting all prefixes for which the prefix equivalence classes have at least 'minsup' different trees.

Step 0 Obtain the paths p_i from the root node to each special node labeled i in T. (In many cases, the paths are given.)

Step 1 Create a subtree U composed of all the paths $p_1, p_2, ..., p_d$. Let a one-to-one mapping $i : j \mapsto i_j$ denote the embedding mapping from the set $\{0, 1, .., m-1\}$ of node ids of U to the set of node ids of T. Let u_l, the node in U with id l, denote the least common ancestor of all special nodes.

Step 2 Create an embedded subtree S of U composed of the root node and all special nodes.

Step 3 Attach *insertable position* (s_u, e_u) for S to each node u of U. *Insertable positions* are calculated as follows. Set v to 0 initially. Starting from the root node, do a depth-first traversal. For each node u, set s_u to the value of v at

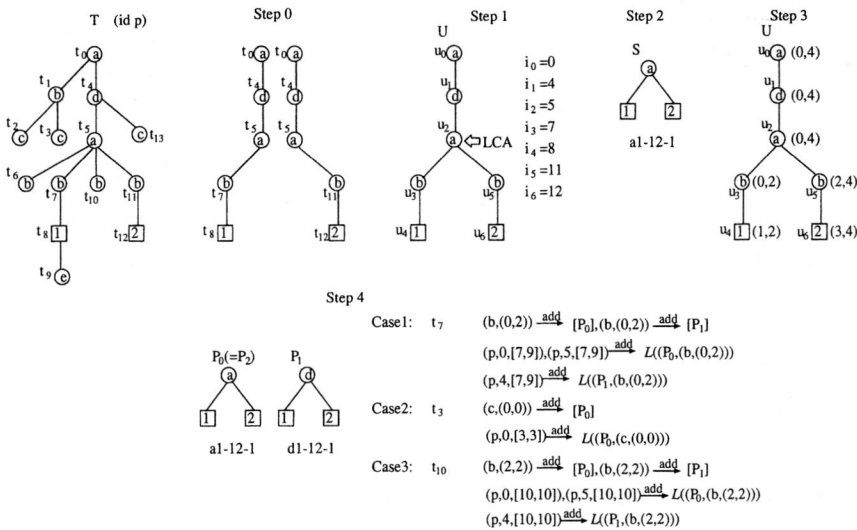

Fig. 4. Example of each steps of the procedure enumerating all size-2 frequent subtrees and creating its scope-list

the first visit, and set e_u to the value of v at the last visit. When visiting a special node at the first and last times, add 1 to v before setting s_u and e_u.

Step 4 For all node t_h in T, do the followings. Let t_{i_k} be the least ancestor of t_h among all nodes in $\{t_{i_0}, t_{i_1}, ..., t_{i_{m-1}}\}$. Let $C = \{j : i_j < h, u_j \text{ is a child of } u_k\}$. Let $l' = \min\{l, k\}$. Let P_j denote a tree that is created from S by replacing the root node label with u_j for $j \in \{0, 1, ..., l\}$. Let b denote the tree id of T, and let x and s denote the label and the scope of node t_h, respectively.

Case 1 $i_k = h$

Add $(x, (s_{u_k}, e_{u_k}))$ to $[P_j]$ and add (b, i_j, s) to $\mathcal{L}((P_j, (x, (s_{u_k}, e_{u_k}))))$ for all $j \in \{0, 1, ..., l'\}$ but $j = k$.

Case 2 $i_k \neq h$ and $C = \emptyset$

Add $(x, (s_{u_k}, s_{u_k}))$ to $[P_j]$ and add (b, i_j, s) to $\mathcal{L}((P_j, (x, (s_{u_k}, s_{u_k}))))$ for all $j \in \{0, 1, ..., l'\}$.

Case 3 $i_k \neq h$ and $C \neq \emptyset$

Let $j^* = \max C$. Add $(x, (e_{u_{j^*}}, e_{u_{j^*}}))$ to $[P_j]$ and add (b, i_j, s) to $\mathcal{L}((P_j, (x, (e_{u_{j^*}}, e_{u_{j^*}}))))$ for all $j \in \{0, 1, ..., l'\}$.

5 Application to DOM-Tree Analysis

We conducted an experiment of extracting common structures from DOM-trees of HTML documents. The HTML documents we used in our experiment are Web pages containing the information about a given Ramen (lamian, Chinese noodles in Soup) Shop. Among the most popular 100 Ramen-shops introduced

#	Pattern	σ
30	body #text(1) -1 #text(2) -1	1.0
27	body #text(1) -1 #text(2) -1 a -1 br -1	0.9
25	body #text(1) -1 #text(2) -1 br -1 br -1 br -1 br -1 br -1	0.8
24	body #text(1) -1 #text(2) -1 br -1 br -1 br -1 a -1 br -1	0.8
24	body #text(1) -1 br -1 #text(2) -1 br -1 a -1	0.8
24	body #text(1) -1 br -1 #text(2) -1 br -1 br -1	0.8
24	body #text(1) -1 br -1 #text(2) -1 a -1 br -1	0.8
24	body br -1 #text(1) -1 br -1 #text(2) -1 br -1	0.8
24	body #text(1) -1 #text(2) -1 hr -1 br -1	0.8
24	body #text(1) -1 #text(2) -1 a -1 a -1	0.8
24	body #text(1) -1 #text(2) -1 br -1 hr -1	0.8
20	table tr #text(1) -1 -1 tr td #text(2) -1 -1 -1	1.0
18	table tr td #text(1) -1 -1 -1 tr td #text(2) -1 -1 -1	0.9,0.8
18	table td #text(1) -1 -1 td -1 #text(2) -1	0.9
18	table #text(1) -1 td -1 td #text(2) -1 -1	0.9
17	table tr td #text(1) -1 -1 -1 td -1 td #text(2) -1 -1	0.8
16	table img -1 #text(1) -1 tr td #text(2) -1 -1 -1	0.8
19	td #text(1) -1 br -1 #text(2) -1	1.0,0.9,0.8
19	tr td #text(1) -1 -1 td #text(2) -1 -1	1.0,0.9
17	tr td #text(1) -1 -1 td -1 td #text(2) -1 -1	0.8
18	tr td #text(1) -1 -1 td(2) -1	1.0,0.9,0.8
16	html head meta -1 title #text(1) -1 -1 -1 body #text(2) -1 -1	1.0
15	html head meta -1 title #text(1) -1 -1 -1 body img -1 #text(2) -1 -1	0.9
15	html head meta -1 title #text(1) -1 -1 -1 body #text(2) -1 a -1 -1	0.9
14	html head meta -1 title #text(1) -1 -1 -1 body img -1 tr td -1 -1 #text(2) -1 img -1 -1	0.8
14	html head meta -1 title #text(1) -1 -1 -1 body strong -1 img -1 #text(2) -1 img -1 -1	0.8
14	html head meta -1 title #text(1) -1 -1 -1 body table tr td #text(2) -1 -1 -1 -1	0.8
14	html head meta -1 title #text(1) -1 -1 -1 body table td img -1 -1 #text(2) -1 -1 -1	0.8
14	html head meta -1 title #text(1) -1 -1 -1 body td -1 td -1 #text(2) -1 img -1 -1	0.8
13	html head meta -1 title #text(1) -1 -1 -1 body img -1 tr td -1 -1 #text(2) -1 tr td -1 -1 a -1 -1	0.8
13	html head meta -1 title #text(1) -1 -1 -1 body strong -1 img -1 #text(2) -1 tr td -1 -1 a -1 -1	0.8
13	html head meta -1 title #text(1) -1 -1 -1 body td -1 td -1 #text(2) -1 tr td -1 -1 a -1 -1	0.8
13	html head meta -1 title #text(1) -1 -1 -1 body img -1 img -1 #text(2) -1 tr td -1 -1 -1	0.8
13	html head meta -1 title #text(1) -1 -1 -1 body img -1 strong -1 #text(2) -1 tr td -1 -1 -1	0.8
13	html head meta -1 title #text(1) -1 -1 -1 body td -1 strong -1 #text(2) -1 tr td -1 -1 -1	0.8
13	html head meta -1 title #text(1) -1 -1 -1 body td -1 strong -1 #text(2) -1 img -1 -1	0.8
13	html head meta -1 title #text(1) -1 -1 -1 body img -1 strong -1 #text(2) -1 img -1 -1	0.8
13	html head meta -1 title #text(1) -1 -1 -1 body img -1 img -1 #text(2) -1 img -1 -1	0.8
13	html head meta -1 title #text(1) -1 -1 -1 body img -1 #text(2) -1 br -1 -1	0.8
13	html head meta -1 title #text(1) -1 -1 -1 body tr img -1 -1 #text(2) -1 -1	0.8
13	table tr td #text(1) -1 -1 -1 tr td(2) -1 -1	1.0,0.9,0.8
11	tbody tr td #text(1) -1 -1 -1 tr td(2) -1 -1	1.0
10	tbody tr #text(1) -1 -1 tr td(2) -1 -1 tr -1 tr td -1 -1	0.9,0.8
10	tbody td -1 td #text(1) -1 -1 tr td(2) -1 -1 td -1 td -1 td -1	0.9
10	tbody td -1 td #text(1) -1 -1 tr td(2) -1 -1 tr -1 tr td -1 -1	0.9,0.8
10	tbody tr td #text(1) -1 -1 tr td(2) -1 -1 td -1 td -1 td -1	0.9
9	tbody td -1 td #text(1) -1 -1 tr td(2) -1 -1 td -1 td -1 td -1 td -1	0.8
9	tbody tr td #text(1) -1 -1 tr td(2) -1 -1 td -1 td -1 tr td -1 -1	0.8
9	tbody tr td #text(1) -1 -1 tr td(2) -1 -1 td -1 td -1 td -1 td -1	0.8
9	tbody td -1 td #text(1) -1 -1 tr td(2) -1 -1 td -1 tr td -1 -1	0.8
9	tbody br -1 td #text(1) -1 -1 tr td(2) -1 -1 tr -1 tr td -1 -1	0.8
9	tbody br -1 td #text(1) -1 -1 tr td(2) -1 -1 td -1 td -1 td -1	0.8
9	tbody td br -1 -1 #text(1) -1 -1 td(2) -1	0.8
10	p #text(1) -1 #text(2) -1	1.0
9	p #text(1) -1 #text(2) -1 br -1	0.9
9	p #text(1) -1 br -1 #text(2) -1	0.9
8	p br -1 br -1 #text(1) -1 #text(2) -1 br -1	0.8
8	p #text(1) -1 br -1 #text(2) -1 br -1	0.8

Fig. 5. Maximal subtrees enumerated for $\sigma = 1.0, 0.9, 0.8$ and their frequencies

in a popular local town information magazine[3], we selected the most popular 10 Ramen shops with more than 15 Web pages retrieved by keyword search[4] using a shop name and its telephone number. Totally 301 pages were retrieved and we used 189 pages of them that contain the reputation about a target shop. For each DOM-tree of these 189 pages, we set a *name node* to the text node containing a target shop name and set a *reputation node* to the least common ancestor node[5]

---

[3] Hokkaido Walker 2002 NO.3.
[4] Google(www.google.co.jp) was used in the search.
[5] If the subtree rooted by the least common ancestor node contains reputations of other restaurants, the most informative one text node is selected as a representative instead of the least common ancestor node.

of the text nodes whose text contains the target reputation. We extracted one tree for one name node, so more than one trees might be extracted from one page. A tree in our experimental data is the subtree of which root is the least common ancestor (LCA) of two special nodes, a name node and a reputation node. The reason why we did not use the whole tree is that we did not want to extract trivial frequent structures like "html title -1 body ⋯ -1" that are common to all HTML documents. For each HTML-tag pair of a root node and a reputation node that appears at least 10 times, we applied our algorithm to enumerating all frequent subtrees containing all special nodes.

All *maximal* frequent subtrees for minimum support $\sigma = 1.0, 0.9, 0.8$ are shown in Fig. 5. Note that a *maximal* frequent subtree is a subtree such that its any super-tree is *not* frequent. Nodes corresponding to the tags with suffixes "(1)" and "(2)" are a name node and a reputation node, respectively. When the LCA tag of special nodes is 'table', 'tr' or 'tbody', trivial patterns were enumerated except 'table img -1 #text(1) -1 tr td #text(2) -1 -1 -1', a pattern that matches DOM-trees of pages using a small image at the beginning of each item. When the LCA tag of special nodes is 'body', 'td' or 'p', most pages contain 'br' tag between a shop name and its reputation. When the LCA tag of special nodes is 'html', the pages contain information of only one shop, and from an enumerated pattern 'html head meta -1 title #text(1) -1 -1 -1 body table td img -1 -1 #text(2) -1 -1 -1', we know that most reputations are placed in a table with a image.

6 Concluding Remarks

Structures found in our experiment appear to be too general to specify the place of necessary information. To construct a wrapper, other features of HTML documents like contents of the information and distance between nodes should be additionally used. We have developed such combined method using simpler patterns (path sequences) and text classification [7], which is able to extract necessary information from arbitrary Web pages retrieved by a search engine, while most conventional wrappers can do only from the pages in the same site as the training pages. Patterns extracted by ConstrainedTreeMiner possibly improve performance of the combined method in some cases.

References

1. R. Agrawal and R. Srikant. First algorithms for mining association rules. In *Proc. 20th Int'l Conf. on VLDB*, pages 487–499, 1994.
2. R. Agrawal and R. Srikant. Mining sequential patterns. In *Proc. 11th Int'l Conf. on Data Eng.*, pages 3–14, 1995.
3. T. Asai, K. Abe, S. Kawasoe, H. Arimura, H. Sakamoto, and S. Arikawa. Efficient substructure discovery from large semi-structured data. In *Proc. 2nd SIAM Int'l Conf. on Data Mining*, pages 158–174, 2002.

4. W. W. Cohen, M. Hurst, and L. S. Jensen. A flexible learning system for wrapping tables and lists in html documents. In *Proc. 11th Int'l World Wide Web Conf.*, pages 232–241, 2002.
5. M. Garofalakis, R. Rastogi, and K. Shim. Mining sequential patterns with regular expression constraints. *IEEE Transactions on Knowledge and Data Engineering*, 14(3):530–552, 2002.
6. A. Inokuchi, T. Washio, and H. Motoda. An apriori-based algorithm for mining frequent substructures from graph data. In *Proc. PKDD 2000*, pages 13–23, 2000.
7. H. Hasegawa, M. Kudo and A. Nakamura. Reputation Extraction Using Both Structural and Content Information. Technical Report TCS-TR-A-05-2, 2005, http://www-alg.ist.hokudai.ac.jp/tra.html.
8. N. Kushmerick. Wrapper induction:efficiency and expressiveness. *Artificial Intelligence*, (118):15–68, 2000.
9. R. Srikant, Q. Vu, and R. Agrawal. Mining association rules with item constraints. In *Proc. 3rd Int'l Conf. on Knowledge Discovery and Data Mining*, pages 67–73, 1997.
10. M. J. Zaki. Efficiently mining frequent trees in a forest. In *Proc. SIGKDD'02*, pages 71–80, 2002.

Author Index

Allen, Robert B.   173
Antonie, Maria-Luiza   240
Araki, Kenji   32
Ashrafi, Mafruz Zaman   125

Bac, Le Hoai   226
Bagnall, Anthony J.   737, 771
Bo, Liefeng   632
Bonchi, Francesco   114
Boudjeloud, Lydia   426

Cao, Tru H.   290
Cha, Guang-Ho   844
Chan, Tony F.   388
Chawla, Sanjay   155
Chen, Jianer   43
Chen, Ming-Syan   323, 505
Chen, Songcan   500
Chen, Zun-Ping   838
Chetty, Madhu   578
Chi, Nguyen Cam   21
Chia, Liang-Tien   452
Chiu, Jeng-Kuen   713
Choudhary, Alok   689
Chu, Fang   486
Chu, Jian   61
Chu, Yi-Hong   323
Chuang, Kun-Ta   323, 505
Church, W. Bret   155
Cichoń, Katarzyna   672
Coenen, Frans   216
Crémilleux, Bruno   661

Das, Amitabha   702
Dash, Manoranjan   107, 452
De Veaux, Richard   186
Deng, Zhi-Hong   150
Deng, Xu-bin   838

Ding, Chris   388
Dittenbach, Michael   399
Dung, Tran A.   290

Echizen-ya, Hiroshi   32

Fan, Wei   605
Feng, Boqin   280
Fujimaki, Ryohei   785

Gao, Jing   420
Gasmi, Gh.   81
Ge, Weiping   256
Guan, Jihong   361
Gunopulos, Dimitrios   333
Guo, Jian-Kui   838

Hamilton, Howard J.   744
Han, Ingoo   480
Harao, Masateru   50
He, Xiangdong   474
Hilario, Melanie   588
Hirata, Kouichi   50
Hiyane, Kazuo   820
Hoa, Nguyen Sinh   312
Hoáng, Thu   186
Horváth, Tamás   791
Hu, Xiaohua   173
Huang, Joshua   361
Huang, Joshua Zhexue   802
Huang, Shiying   71
Huang, Yan   719
Huh, Unna   15
Huo, Hua   280
Huynh, D. T.   726
Hwang, San-Yih   713

Inujima, Hiroshi   820

Janacek, Gareth J.   737
Janssens, Gerrit K.   234
Ji, Cong-Rui   150
Jia, Tao   61
Jiao, Licheng   632
Jin, Rong   568, 622
Jin, Xiaoming   764
Jing, Liping   802

Kalousis, Alexandros   588
Kao, Ben   143
Karimi, Kamran   744
Keogh, Eamonn   333, 771
Kim, Dae-Won   180
Kim, Sang-Wook   162, 203
Kim, Woo-Cheol   162
Kobayashi, Shin-ichi   820
Koh, Yun Sing   97
Kolippakkam, Deepak   107
Kriegel, Hans-Peter   432
Kryszkiewicz, Marzena   672
Kudo, Mineichi   850
Kumeno, Fumihiro   820
Kunath, Peter   432

Lavrač, Nada   2
Le, Cuong Anh   262
Le, Jia-Jin   333, 826
Lee, Doheon   180
Lee, Tong-Wen   832
Leggett, John J.   555
Leng, Paul   216
Li, Jianzhong   420
Li, Ming   611
Li, Qiang   464
Li, Stan Z.   599
Li, Xiaolin   474
Li, Xiaorong   256
Li, Xue   778
Li, Zhuowei   702
Liao, Wei-keng   689
Lim, Ee-Peng   713
Lim, Jong-Tae   180
Limère, Arthur   234
Lin, Fuzong   758

Lin, Jessica   333
Lin, Xuemin   343
Lin, Yaping   650
Liu, Fei Tony   605
Liu, Jian-We   333
Liu, Jian-Wei   826
Liu, Jun   372
Liu, Yi   568
Liu, Ying   689
Liu, Ying-Han   713
Liu, Zheng   343
Lonardi, Stefano   771
Loo, K. K.   143
Lu, Hongjun   343
Lucchese, Claudio   114
Lukov, Lior   155

Machida, Kazuo   785
Malowiecki, Michal   545
Mao, Kezhi   683
Matsumoto, Shotaro   301
Mayo, Michael   192
Meengen, Asavin   249
Min, Sung-Hwan   480
Minh, Hoang Ngoc   21
Mitchell, Tom   1
Mornouchi, Yoshio   32
Motoda, Hiroshi   639

Nakamura, Atsuyoshi   850
Nam-Huynh, Van   262
Nanni, Mirco   378
Ng, Michael K.   802
Nguifo, E. Mephu   81
Nguyen, Cao D.   290
Nguyen, Ngoc Thanh   545
Nguyen, Phu Chien   639
Nguyen, Son N.   442
Niitsuma, Hirotaka   523
Ning, Shang   838

Ohara, Kouzou   639
Okada, Takashi   523
Okumura, Manabu   269, 301
Ooi, Chia Huey   578
Orlowska, Maria E.   442, 778

Park, Sanghyun 203, 162
Parker, D.Stott 486
Pfeifle, Martin 432
Pölzlbauer, Georg 399
Poulet, François 426
Powell, Michael 737

Qian, Yu 726

Rak, Rafal 240
Ramamohanarao, Kotagiri 372
Ratanamahatana, Chotirat 771
Rauber, Andreas 399
Renz, Matthias 432
Rezgui, Jihen 91
Rountree, Nathan 97
Ruan, Bei-jun 838

Sattler, Kai-Uwe 539
Schaeffer, Satu Elisa 354
Shahabi, Cyrus 516
Shang, Xuequn 539
Shekhar, Shashi 136
Shi, Baile 256, 529
Shima, Yoshikazu 50
Shimazu, Akira 262
Shirai, Yasuyuki 820
Shrestha, Raju 650
Si, Luo 622
Slimani, Y. 81, 91
Smith, Kate 125
Son, Nguyen Hung 312
Song, Min 173
Song, Il-Yeol 173
Sörensen, Kenneth 234
Soulet, Arnaud 661
Squire, David McG. 410
Squire, Denny McG. 410
Stach, Wojciech 240
Su, Fang-Zhong 838
Su, HongYe 61
Sun, Xingzhi 442, 778

Takahashi, Kazuko 269
Takamura, Hiroya 269, 301

Tan, Pang-Ning 420
Tan, Xiaoyang 500
Tang, Shi-Wei 150
Tang, Wenyin 683
Taniar, David 125
Thammano, Arit 249
Ting, Kai Ming 605
Tong, Ivy 143
Trang, Tran 21
Tuan, Nguyen Anh 226
Tung, Thai Quang 180

Vanhoof, Koen 234
Vlachos, Michai 333

Wang, Chen 529
Wang, Chong 813
Wang, Leuo-Hong 832
Wang, Ling 632
Wang, Surong 452
Wang, Wei 256, 343, 529
Wang, Wenyuan 813
Wang, Ya-qin 838
Wang, Yizhou 486
Washio, Takashi 639
Webb, Geoffrey I. 71
Wojciechowski, Marek 696
Won, Jung-Im 162, 203
Woźnica, Adam 588
Wu, Tao 38
Wu, Tianyi 529

Xu, Jian 529
Xu, Jun 802

Yahia, S. Ben 91, 81
Yairi, Takehisa 785
Yamauchi, Noriyoshi 820
Yang, Kiyoung 516
Yang, Wen-Chieh 505
Yoo, Jin Soung 136
Yoon, Hyunjin 516
Yoon, Jee-Hee 162, 203
Yu, Jeffrey Xu 343
Yu, Ping 719
Yu, Shou-Jian 333, 826

Yuan, Jinhui 758
Yuan, Senmiao 474
Yun, Unil 555

Zaïane, Osmar R. 240
Zakrzewicz, Maciej 696
Zaniolo, Carlo 486
Zhang, Bo 464, 758
Zhang, Chengqi 751
Zhang, Fuyan 500
Zhang, Junping 599
Zhang, Kang 726
Zhang, Ling 38
Zhang, Liqin 719
Zhang, Lu 216
Zhang, Ming 150
Zhang, Nan 43
Zhang, Pusheng 136
Zhang, Shichao 751
Zhang, Yanping 38
Zhang, Ying 61
Zhang, Zhaogong 420
Zhao, Wei 43
Zhao, Yue 361
Zhao, Yanchang 751
Zhou, Jianying 702
Zhou, Shuigeng 361
Zhou, Zhi-Hua 500, 611
Zhu, Yang-Yong 838
Zuo, Xinqiang 764

Lecture Notes in Artificial Intelligence (LNAI)

Vol. 3518: T.B. Ho, D. Cheung, H. Liu (Eds.), Advances in Knowledge Discovery and Data Mining. XXI, 864 pages. 2005.

Vol. 3508: P. Bresciani, P. Giorgini, B. Henderson-Sellers, G. Low, M. Winikoff (Eds.), Agent-Oriented Information Systems II. X, 227 pages. 2005.

Vol. 3501: B. Kégl, G. Lapalme (Eds.), Advances in Artificial Intelligence. XV, 458 pages. 2005.

Vol. 3492: P. Blache, E. Stabler, J. Busquets, R. Moot (Eds.), Logical Aspects of Computational Linguistics. X, 363 pages. 2005.

Vol. 3488: M.-S. Hacid, N.V. Murray, Z.W. Raś, S. Tsumoto (Eds.), Foundations of Intelligent Systems. XIII, 700 pages. 2005.

Vol. 3452: F. Baader, A. Voronkov (Eds.), Logic for Programming, Artificial Intelligence, and Reasoning. XI, 562 pages. 2005.

Vol. 3419: B. Faltings, A. Petcu, F. Fages, F. Rossi (Eds.), Constraint Satisfaction and Constraint Logic Programming. X, 217 pages. 2005.

Vol. 3416: M. Böhlen, J. Gamper, W. Polasek, M.A. Wimmer (Eds.), E-Government: Towards Electronic Democracy. XIII, 311 pages. 2005.

Vol. 3415: P. Davidsson, B. Logan, K. Takadama (Eds.), Multi-Agent and Multi-Agent-Based Simulation. X, 265 pages. 2005.

Vol. 3403: B. Ganter, R. Godin (Eds.), Formal Concept Analysis. XI, 419 pages. 2005.

Vol. 3398: D.-K. Baik (Ed.), Systems Modeling and Simulation: Theory and Applications. XIV, 733 pages. 2005.

Vol. 3397: T.G. Kim (Ed.), Artificial Intelligence and Simulation. XV, 711 pages. 2005.

Vol. 3396: R.M. van Eijk, M.-P. Huget, F. Dignum (Eds.), Agent Communication. X, 261 pages. 2005.

Vol. 3394: D. Kudenko, D. Kazakov, E. Alonso (Eds.), Adaptive Agents and Multi-Agent Systems II. VIII, 313 pages. 2005.

Vol. 3392: D. Seipel, M. Hanus, U. Geske, O. Bartenstein (Eds.), Applications of Declarative Programming and Knowledge Management. X, 309 pages. 2005.

Vol. 3374: D. Weyns, H.V.D. Parunak, F. Michel (Eds.), Environments for Multi-Agent Systems. X, 279 pages. 2005.

Vol. 3371: M.W. Barley, N. Kasabov (Eds.), Intelligent Agents and Multi-Agent Systems. X, 329 pages. 2005.

Vol. 3369: V.R. Benjamins, P. Casanovas, J. Breuker, A. Gangemi (Eds.), Law and the Semantic Web. XII, 249 pages. 2005.

Vol. 3366: I. Rahwan, P. Moraitis, C. Reed (Eds.), Argumentation in Multi-Agent Systems. XII, 263 pages. 2005.

Vol. 3359: G. Grieser, Y. Tanaka (Eds.), Intuitive Human Interfaces for Organizing and Accessing Intellectual Assets. XIV, 257 pages. 2005.

Vol. 3346: R.H. Bordini, M. Dastani, J. Dix, A.E.F. Seghrouchni (Eds.), Programming Multi-Agent Systems. XIV, 249 pages. 2005.

Vol. 3345: Y. Cai (Ed.), Ambient Intelligence for Scientific Discovery. XII, 311 pages. 2005.

Vol. 3343: C. Freksa, M. Knauff, B. Krieg-Brückner, B. Nebel, T. Barkowsky (Eds.), Spatial Cognition IV. XIII, 519 pages. 2005.

Vol. 3339: G.I. Webb, X. Yu (Eds.), AI 2004: Advances in Artificial Intelligence. XXII, 1272 pages. 2004.

Vol. 3336: D. Karagiannis, U. Reimer (Eds.), Practical Aspects of Knowledge Management. X, 523 pages. 2004.

Vol. 3327: Y. Shi, W. Xu, Z. Chen (Eds.), Data Mining and Knowledge Management. XIII, 263 pages. 2005.

Vol. 3315: C. Lemaître, C.A. Reyes, J.A. González (Eds.), Advances in Artificial Intelligence – IBERAMIA 2004. XX, 987 pages. 2004.

Vol. 3303: J.A. López, E. Benfenati, W. Dubitzky (Eds.), Knowledge Exploration in Life Science Informatics. X, 249 pages. 2004.

Vol. 3301: G. Kern-Isberner, W. Rödder, F. Kulmann (Eds.), Conditionals, Information, and Inference. XII, 219 pages. 2005.

Vol. 3276: D. Nardi, M. Riedmiller, C. Sammut, J. Santos-Victor (Eds.), RoboCup 2004: Robot Soccer World Cup VIII. XVIII, 678 pages. 2005.

Vol. 3275: P. Perner (Ed.), Advances in Data Mining. VIII, 173 pages. 2004.

Vol. 3265: R.E. Frederking, K.B. Taylor (Eds.), Machine Translation: From Real Users to Research. XI, 392 pages. 2004.

Vol. 3264: G. Paliouras, Y. Sakakibara (Eds.), Grammatical Inference: Algorithms and Applications. XI, 291 pages. 2004.

Vol. 3259: J. Dix, J. Leite (Eds.), Computational Logic in Multi-Agent Systems. XII, 251 pages. 2004.

Vol. 3257: E. Motta, N.R. Shadbolt, A. Stutt, N. Gibbins (Eds.), Engineering Knowledge in the Age of the Semantic Web. XVII, 517 pages. 2004.

Vol. 3249: B. Buchberger, J.A. Campbell (Eds.), Artificial Intelligence and Symbolic Computation. X, 285 pages. 2004.

Vol. 3248: K.-Y. Su, J. Tsujii, J.-H. Lee, O.Y. Kwong (Eds.), Natural Language Processing – IJCNLP 2004. XVIII, 817 pages. 2005.

Vol. 3245: E. Suzuki, S. Arikawa (Eds.), Discovery Science. XIV, 430 pages. 2004.

Vol. 3244: S. Ben-David, J. Case, A. Maruoka (Eds.), Algorithmic Learning Theory. XIV, 505 pages. 2004.

Vol. 3238: S. Biundo, T. Frühwirth, G. Palm (Eds.), KI 2004: Advances in Artificial Intelligence. XI, 467 pages. 2004.

Vol. 3230: J.L. Vicedo, P. Martínez-Barco, R. Muñoz, M. Saiz Noeda (Eds.), Advances in Natural Language Processing. XII, 488 pages. 2004.

Vol. 3229: J.J. Alferes, J. Leite (Eds.), Logics in Artificial Intelligence. XIV, 744 pages. 2004.

Vol. 3228: M.G. Hinchey, J.L. Rash, W.F. Truszkowski, C.A. Rouff (Eds.), Formal Approaches to Agent-Based Systems. VIII, 290 pages. 2004.

Vol. 3215: M.G.. Negoita, R.J. Howlett, L.C. Jain (Eds.), Knowledge-Based Intelligent Information and Engineering Systems, Part III. LVII, 906 pages. 2004.

Vol. 3214: M.G.. Negoita, R.J. Howlett, L.C. Jain (Eds.), Knowledge-Based Intelligent Information and Engineering Systems, Part II. LVIII, 1302 pages. 2004.

Vol. 3213: M.G.. Negoita, R.J. Howlett, L.C. Jain (Eds.), Knowledge-Based Intelligent Information and Engineering Systems, Part I. LVIII, 1280 pages. 2004.

Vol. 3209: B. Berendt, A. Hotho, D. Mladenic, M. van Someren, M. Spiliopoulou, G. Stumme (Eds.), Web Mining: From Web to Semantic Web. IX, 201 pages. 2004.

Vol. 3206: P. Sojka, I. Kopecek, K. Pala (Eds.), Text, Speech and Dialogue. XIII, 667 pages. 2004.

Vol. 3202: J.-F. Boulicaut, F. Esposito, F. Giannotti, D. Pedreschi (Eds.), Knowledge Discovery in Databases: PKDD 2004. XIX, 560 pages. 2004.

Vol. 3201: J.-F. Boulicaut, F. Esposito, F. Giannotti, D. Pedreschi (Eds.), Machine Learning: ECML 2004. XVIII, 580 pages. 2004.

Vol. 3194: R. Camacho, R. King, A. Srinivasan (Eds.), Inductive Logic Programming. XI, 361 pages. 2004.

Vol. 3192: C. Bussler, D. Fensel (Eds.), Artificial Intelligence: Methodology, Systems, and Applications. XIII, 522 pages. 2004.

Vol. 3191: M. Klusch, S. Ossowski, V. Kashyap, R. Unland (Eds.), Cooperative Information Agents VIII. XI, 303 pages. 2004.

Vol. 3187: G. Lindemann, J. Denzinger, I.J. Timm, R. Unland (Eds.), Multiagent System Technologies. XIII, 341 pages. 2004.

Vol. 3176: O. Bousquet, U. von Luxburg, G. Rätsch (Eds.), Advanced Lectures on Machine Learning. IX, 241 pages. 2004.

Vol. 3171: A.L.C. Bazzan, S. Labidi (Eds.), Advances in Artificial Intelligence – SBIA 2004. XVII, 548 pages. 2004.

Vol. 3159: U. Visser, Intelligent Information Integration for the Semantic Web. XIV, 150 pages. 2004.

Vol. 3157: C. Zhang, H. W. Guesgen, W.K. Yeap (Eds.), PRICAI 2004: Trends in Artificial Intelligence. XX, 1023 pages. 2004.

Vol. 3155: P. Funk, P.A. González Calero (Eds.), Advances in Case-Based Reasoning. XIII, 822 pages. 2004.

Vol. 3139: F. Iida, R. Pfeifer, L. Steels, Y. Kuniyoshi (Eds.), Embodied Artificial Intelligence. IX, 331 pages. 2004.

Vol. 3131: V. Torra, Y. Narukawa (Eds.), Modeling Decisions for Artificial Intelligence. XI, 327 pages. 2004.

Vol. 3127: K.E. Wolff, H.D. Pfeiffer, H.S. Delugach (Eds.), Conceptual Structures at Work. XI, 403 pages. 2004.

Vol. 3123: A. Belz, R. Evans, P. Piwek (Eds.), Natural Language Generation. X, 219 pages. 2004.

Vol. 3120: J. Shawe-Taylor, Y. Singer (Eds.), Learning Theory. X, 648 pages. 2004.

Vol. 3097: D. Basin, M. Rusinowitch (Eds.), Automated Reasoning. XII, 493 pages. 2004.

Vol. 3071: A. Omicini, P. Petta, J. Pitt (Eds.), Engineering Societies in the Agents World. XIII, 409 pages. 2004.

Vol. 3070: L. Rutkowski, J. Siekmann, R. Tadeusiewicz, L.A. Zadeh (Eds.), Artificial Intelligence and Soft Computing - ICAISC 2004. XXV, 1208 pages. 2004.

Vol. 3068: E. André, L. Dybkjær, W. Minker, P. Heisterkamp (Eds.), Affective Dialogue Systems. XII, 324 pages. 2004.

Vol. 3067: M. Dastani, J. Dix, A. El Fallah-Seghrouchni (Eds.), Programming Multi-Agent Systems. X, 221 pages. 2004.

Vol. 3066: S. Tsumoto, R. Słowiński, J. Komorowski, J.W. Grzymała-Busse (Eds.), Rough Sets and Current Trends in Computing. XX, 853 pages. 2004.

Vol. 3065: A. Lomuscio, D. Nute (Eds.), Deontic Logic in Computer Science. X, 275 pages. 2004.

Vol. 3060: A.Y. Tawfik, S.D. Goodwin (Eds.), Advances in Artificial Intelligence. XIII, 582 pages. 2004.

Vol. 3056: H. Dai, R. Srikant, C. Zhang (Eds.), Advances in Knowledge Discovery and Data Mining. XIX, 713 pages. 2004.

Vol. 3055: H. Christiansen, M.-S. Hacid, T. Andreasen, H.L. Larsen (Eds.), Flexible Query Answering Systems. X, 500 pages. 2004.

Vol. 3048: P. Faratin, D.C. Parkes, J.A. Rodríguez-Aguilar, W.E. Walsh (Eds.), Agent-Mediated Electronic Commerce V. XI, 155 pages. 2004.

Vol. 3040: R. Conejo, M. Urretavizcaya, J.-L. Pérez-de-la-Cruz (Eds.), Current Topics in Artificial Intelligence. XIV, 689 pages. 2004.

Vol. 3035: M.A. Wimmer (Ed.), Knowledge Management in Electronic Government. XII, 326 pages. 2004.

Vol. 3034: J. Favela, E. Menasalvas, E. Chávez (Eds.), Advances in Web Intelligence. XIII, 227 pages. 2004.

Vol. 3030: P. Giorgini, B. Henderson-Sellers, M. Winikoff (Eds.), Agent-Oriented Information Systems. XIV, 207 pages. 2004.

Vol. 3029: B. Orchard, C. Yang, M. Ali (Eds.), Innovations in Applied Artificial Intelligence. XXI, 1272 pages. 2004.

Vol. 3025: G.A. Vouros, T. Panayiotopoulos (Eds.), Methods and Applications of Artificial Intelligence. XV, 546 pages. 2004.

Vol. 3020: D. Polani, B. Browning, A. Bonarini, K. Yoshida (Eds.), RoboCup 2003: Robot Soccer World Cup VII. XVI, 767 pages. 2004.

Vol. 3012: K. Kurumatani, S.-H. Chen, A. Ohuchi (Eds.), Multi-Agents for Mass User Support. X, 217 pages. 2004.